SELECTED SEMICONDUCTOR RESEARCH

SELECTED SEMICONDUCTOR RESEARCH

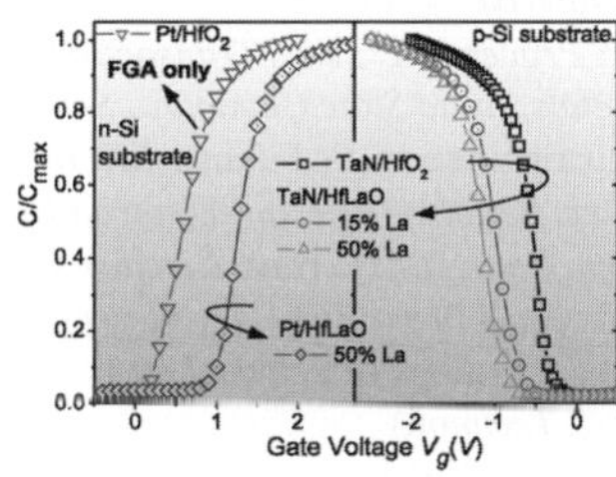

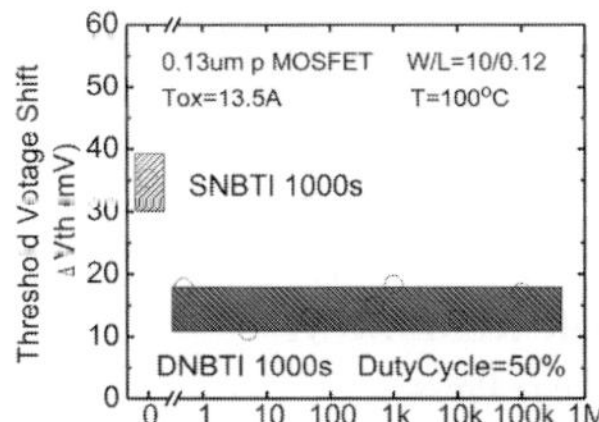

Ming-Fu Li

Professor of Microelectronics,
State Key Lab ASIC and System,
Fudan University, Shanghai, China

Former Professor of Electrical and Computer Engineering,
National University of Singapore, Singapore

Imperial College Press

Published by

Imperial College Press
57 Shelton Street
Covent Garden
London WC2H 9HE

Distributed by

World Scientific Publishing Co. Pte. Ltd.

5 Toh Tuck Link, Singapore 596224

USA office: 27 Warren Street, Suite 401-402, Hackensack, NJ 07601

UK office: 57 Shelton Street, Covent Garden, London WC2H 9HE

British Library Cataloguing-in-Publication Data
A catalogue record for this book is available from the British Library.

We are grateful to the following publishers for their permission to reproduce the articles found in this volume:

American Institute of Physics (*J. Appl. Phys.; Appl. Phys. Lett.*)
American Physical Society (*Phys. Rev. B*)
The Electrochemical Society (*ECS Proceedings; ECS Transaction*)
Elsevier B. V. (*Semicond. Semimetals; Solid State Commun.; Solid-State Electron.*)
Institute of Electrical and Electronic Engineers (*IEEE Trans. Electron Devices; IEEE Electron Device Lett.; IEEE Trans. Device Mater. Reliab.; IEEE Int. Electron Devices Meet., Tech. Dig.; Symp. VLSI Technol., Dig. Tech. Pap.; IEEE Trans. Circuits Syst.; Proceedings of ICES '99; IEEE Int. Symp. Circuits and Systems Proc.; IEEE Int. Reliab. Phys. Symp. Proc.*)
The Institution of Engineering and Technology (*Electron. Lett.*)
IOP Publishing Ltd. (*Semicond. Sci. Technol.*)
Japan Society of Applied Physics (*Jpn. J. Appl. Phys.; Japan 12th Workshop on Gate Stack Technology and Physics; Solid State Device and Materials*)
Polish Academy of Sciences, Institute of Physics (*Int. Conf. Phys. Semicond., 19th*)
Springer (*Analog Integrated Circuits and Signal Processing*)

SELECTED SEMICONDUCTOR RESEARCH

ISBN-13 978-1-84816-406-2
ISBN-10 1-84816-406-8

Printed by FuIsland Offset Printing (S) Pte Ltd. Singapore

Preface

This book reflects selected research achievements in semiconductors of mine from the period of 1982 to 2008. My career in research has been quite unusual. I graduated from the Physics Department, Fudan University, Shanghai, in 1960. However, I only published my first scientific paper in an international journal in 1982. It was most unfortunate that for a very long period of time, opportunities to study science were few and far between.

On the other hand, I was fortunate to meet some very kind teachers and friends whose invaluable advice and timely help, greatly influenced my scientific career. Late Professor Xide Xie, former president of Fudan University, introduced me to the fruitful field of semiconductors and gave me constant support during my 30 years of work in this field. Late Professor Kun Huang, former Director of Institute of Semiconductors, Chinese Academy of Sciences, gave me very strong support and encouragement. I was strongly influenced by his advice and research style. Professor Chih-Tang Sah of University of Illinois, and now R.C.Pittman Eminent Professor of University of Florida, introduced me to the area of deep defects in semiconductors, and gave me invaluable advice in various stages, deeply affecting the course of my research. Professor Peter Yu of University of California at Berkeley, invited me to join the research team for deep defects in III-V semiconductors at University of California at Berkeley and Lawrence Berkeley Lab, and also gave me invaluable help in my life. Finally I have had pleasant collaboration with Professor Dim-Lee Kwong, former Professor at University of Texas at Austin, and now Director of Institute of Microelectronics in Singapore. We have different backgrounds complementary to each other, and therefore very efficient collaboration for more than 6 years. I learned much from Dim-Lee about Si technology during this period.

I was also fortunate to have many highly talented students in my group. I not only supervised their research, but also learned from these young students. As I compiled this volume, many names and events that have left an indelible impression on me came to mind. All deserve to be mentioned in this book but I am constrained by the number of papers I can include. It is therefore inevitable that some excellent collaborators and students may be left out. I apologize for this and would like to assure them their important contributions are nevertheless deeply appreciated.

I wish to thank the National University of Singapore, Fudan University, Graduate School of Chinese Academy of Sciences, Institute of Semiconductors of Chinese Academy of Sciences, University of Science and Technology of China, and Institute of Microelectronics, Singapore where I have worked and obtained much support from them. I want to thank many funding agencies in Singapore and China, particularly the National Science and Technology Board and later A*STAR in Singapore, and National Science Foundation of China.

I would also like to thank Imperial College Press for giving me the opportunity to publish this collection of around 80 papers selected from more than 390 scientific publications of mine. Mr. V.K. Sanjeed, has been very cooperative and helpful in editing this book. This volume is a summary of my previous research achievements, which I hope to build on in time to come. I am expecting more breakthroughs in semiconductor science through collaborations with my colleagues in my home town Shanghai and my home university in Fudan.

Finally I am deeply indebted to my wife Xing-Zhen Qian and my son Hao-Hua Li. They have strongly supported and been very understanding of my work over the years.

Ming-Fu Li
in Shanghai, December 2009

Contents

Chapter 3. Analog Integrated Circuit Design   161

Chapter 4. CMOS Device Reliability 203

Chapter 5. CMOS Technology   309

Contents

Introduction

This chapter briefly introduces some background and history related to my selected papers in each chapter, published at different stages of my research career.

Chapter 1. Defects in Semiconductors

In 1980, on the recommendation of late Professor Lin-Zhao Qian of University of Science and Technology of China (USTC), I went to the Solid State Electronics Lab at University of Illinois, Urbana as a visiting scholar under the guidance of Professor Chih-Tang Sah, and started my research in deep defects in semiconductors. Professor Sah is a pioneer in CMOS device physics. He is also a pioneer in the junction transient measurements for semiconductor deep defects. I learned the transient measurements in Professor Sahs Lab and produced my first research papers, published in international journals (papers 1.1, 1.2). Although I published only a few papers in his lab, the experience was an eye-opener. I learned how to conduct scientific research in a top semiconductor lab (5 former graduate students and postdoctor researchers in Sahs lab finally became members of the National Academy of Engineering in the US) in a top US university. It influenced the rest of my scientific career. In the University of Illinois, I will never forget the honor of being a guest of the two-time Nobel Laureate Professor John Bardeen in his home. Professor Bardeen was awarded the Nobel Prize twice for his invention of transistors and his BCS theory of superconductivity. After I left Illinois in 1981, Professor Sah continuously gave me kind advice and encouragement, which deeply affected my research career.

After one and a half years stay in University of Illinois, I took a one month trip in the US on my own expense to visit many leading universities and semiconductor industries in the US. On the east coast, I visited the semiconductor Labs in MIT, Harvard, Cornell, Princeton Universities, the RCA Sarnoff Lab in Princeton, Bell Lab in Merry Hill, and IBM T.J. Watson Research Center in Yorktown Height. In the mid-east, I visited University of Wisconsin in Madison, and Argon National Lab in Illinois. On the west coast, I visited Berkeley and Stanford Universities, as well as Intel and Hewlett Packard. I took the Grey Hound buses all the way and lived in Chinese scholars homes to reduce my trip expenses because I earned very little income. This way, I got to know the rough picture of semiconductor research in the US.

In September 1981, I returned to China and taught semiconductor physics in the graduate school, Chinese Academy of Sciences (GSCAS) in Beijing. At that time the experimental research conditions in my lab was still very limited. The experimental work in paper 1.4 was done at the Institute of Physics, Chinese Academy of Sciences, in collaboration with my former students Jian-Xin Chen, Yu-Shu Yao and Guang Bai. At the same time, I started some research in theoretical studies of deep defects in semiconductors, in collaboration with two intelligent researchers, my friends and colleagues Shan-Yuan Ren and the late De-Qian Mao, two theorists in USTC. We published a series of papers in Chinese scientific journals as well as international journals (papers 1.3, 1.5), and other papers not included in this volume. During that period, I also had the opportunity to learn the beautiful Group theory by myself and its application to semiconductor physics.

At the same time, I taught all the graduate students from the Institute of Semiconductors, Chinese Academy of Sciences (ISCAS) and got to know the late Professor Kun Huang, former director of ISCAS, the father of semiconductor education and research in China, and a world renowned theoretical solid state physicist. I highly respect Professor Huangs pioneering works on the elegant non-radiative multi-phonon transition theory, polariton theory, and his classic book *Dynamical Theory of Crystal Lattices, Oxford 1954*, co-authored by Nobel Laureate Professor Max Born. I admired and was strongly influenced by Professor Huangs research style and spirit. He also gave me very strong support and encouragement and invited me to join ISCAS as an adjunct professor. I led a joint team of GSCAS and ISCAS in the research of experimental studies of deep defects and the theoretical studies in semiconductor band structures. Professor Huangs encouragement very much improved my self-confidence and drove me to dedicate my life to semiconductor research. I also immensely enjoyed working with my colleagues at ISCAS, two theorists Z.Q. Gu, B.S. Wang, and two experimentalists J. Zhou, J.L. Gao and produced some high quality papers (paper 1.10 and papers 2.2–2.4).

In 1984, the Nobel Laureate Professor C.N. Yang proposed to organize a Si workshop in Chinese Academy of Sciences. I was assigned to be the organizer of the workshop. I got to know Professor Peter Yu of UC Berkeley, a renowned expert in optical properties of semiconductors. He was the lecturer on semiconductor physics in that workshop. We became very close friends after that, and Peter gave me some invaluable help in my life. In 1986, Peter invited me to visit Berkeley to join a III-V semiconductor defect research team. The team also included two materials science researchers, Professor Eicke Weber and Professor Eugene Haller, and a theorist, Dr. V. Wallukewicz of Lawrence Berkeley Lab. At that time a very hot topic — DX center in compound semiconductors — first discovered by D.V. Lang at Bell Lab,[1] attracted many scientists worldwide. We studied the DX center at Berkeley and published a series of papers (papers 1.7, 1.8, 1.11). Particularly, for the first time, we used the diamond anvil cell technique to successfully measure the electrical and optical properties of the deep defects in GaAs under high pressure. We found that the optical ionization energy (0.4eV) is much larger than the electrical activation energy (0.1eV) of the pressure induced defect, very similar to the property of the DX center in AlGaAs compound semiconductors, and our experiments supported that the defect has a large lattice relaxation, proposed by Lang *et al.* The result we got was one of the experimental grounds which led to J. Chadis proposal of the negative U property of the DX center — a breakthrough in understanding of the defect property in semiconductors.[2,3] At UC Berkeley, I was also very fortunate to have the opportunity to learn optical characterization under the tutelage of Peter, who is a world renowned expert in Raman scattering.

In 1988, there was an international conference in Warsaw, Poland in the field of high pressure studies in semiconductors. The Warsaw conference invited a speaker from China to present works from the country. Recommended by Professor Kun Huang, I delivered the presentation at the Warsaw Conference to introduce research conducted in Peking University, led by

---

[1]D.V. Lang, K.A. Logan and M. Jaros, Phys.Review B, **19**, 1015(1979)
[2]Chadi and Chang, Phys. Rev. Lett. **61**, 873(1988)
[3]Chadi and Chang, Phys. Review B, **39**, 10366(1989)

Professor Guo-Gan Qin, the works in the ISCAS, led by Dr. X.S. Zhao, and the works by myself and my collaborators at UC Berkeley and at GSCAS (paper 1.9).

I joined the National University of Singapore (NUS) in 1991. I continued the research in DX center problem with my Ph. D student A.Y. Du and Masters student Y.Y. Luo at NUS and my collaboration with Berkeley professors. In this period, we published the papers 1.13 and 1.14. Co-authors Professor S.J. Chua and Professor T.C. Chong were my colleagues at NUS.

Our works in deep centers were summarized in two invited papers. Paper 1.12 was an invited presentation at the 5th International Conference on High Pressure in Semiconductor Physics in Kyoto. The second paper 1.15 was an invited review published in *Semiconductors and Semimetals*, edited by Dr. T. Suski of Institute of high pressure in semiconductors, Warsaw, and Professor W. Paul of Harvard University.

Chapter 2. Semiconductor Band Structures

I was always interested in semiconductor theory because I firmly believe that I should deepen my understanding of my experimental research. When I was at Berkeley, I got to know that the university had a world leading group in pseudo-potential calculation of semiconductor band structures, led by Professor Malvin Cohen and Professor Steven Louie. I approached Steven who generously gave me the self-consistent pseudo-potential calculation program source codes developed by his group and allowed me to participate in his group meetings and research. I therefore got some first-hand experience in the self-consistent pseudo-potential calculations, pioneered by Cohen and Louie. Steven suggested to me to modify the program by including the spin-orbit interaction to study the topic of linear k term of the zinc-blende semiconductor valence band at the Γ point, an interesting topic first raised by G. Dresselhause in 1955 from the consideration of the symmetry property of the crystal.[4] Our work on this topic led to paper 2.1, and was continued by Steven's Ph.D student M.P. Surh in another paper not included in this book.[5]

---

[4] G. Dresselhause, Phys. Rev. **100**, 580 (1955)
[5] M.P. Surh, M.F. Li and S. Louie, Phys. Rev. B, **43**, 4286 (1991)

When I returned to China, I led a joint team of GSCAS and ISCAS to continue the self-consistent pseudo-potential band structure calculation, with my ISGAS colleagues Z.Q. Gu and B.S. Wang, two very smart and dedicated theorists. This was the first self-consistent pseudo-potential calculation group in China. The collaboration led to papers 2.2 to 2.4. In paper 2.2, Z.G did the main calculations and I used group theoretical analysis to write the paper in a very compact form. Paper 2.3 was an experimental work mainly done by myself. In paper 2.4, Z.G and B.S and our Ph. D student J.Q. Wang spent a lot of time and effort to develop the very complicated computer program of quasi-particle pseudo-potential band structure calculation. Steven Louie's group at Berkeley pioneered the quasi-particle pseudo-potential calculation of semiconductor band structures in 1985.[6,7] This was a breakthrough in band structure calculations since it was based on a rigorous theoretical background of many electrons interaction on one hand, and gave the energy gap results in very good agreement with experiments on the other. The previous self-consistent pseudo-potential calculations based on local density approximation always underestimated the energy gap. Our paper 2.4 was the first produced in China on quasi-particle pseudo-potential calculation of semiconductor band structures with the program developed by ourselves, and with good agreement with the experiments.

When I joined NUS in Singapore, I continued this research area of semiconductor band structure calculations, in collaboration with my colleagues Professors T.C. Chong and Y.P. Feng and my students Teo Kie Leong and Wei-Jun Fang. The work in papers 2.5 and 2.6 was done by my student Wei-Jun Fan, in collaboration with T.C. Chong. T.C was familiar with optical laser problems and I was familiar with the band structure calculations. We co-supervised Wei-Jun to perform the quantum-well laser threshold calculation. At that time, GaN III-V compound alloy was a very hot topic for blue optoelectronic device applications. In our paper 2.6, I was very impressed by the result that in GaN/GaAlN quantum well, the heavy hole, light hole and S.O splitting hole were mixed. I understood that this valence band mixing effect has deep physical ground and should be universal, due to the symmetry lowering in the quantum well, from the group theoretical point of view. This led me to supervise another Ph. D student Y.T. Hou later in Si CMOS device quantum simulation of p-MOS transistors (papers 6.1, 6.2). The work in papers 2.7 and 2.8 was done by a Master of Engineering student Yee-Chia Yeo,

[6]M.S. Hybertsen and S.G. Louie, Phys. Rev. Lett., **55**, 1418 (1985)
[7]M.S. Hybertsen and S.G. Louie, Phys. Rev. B, **32**, 7005 (1985)

co-supervised by T.C and myself. I first supervised Yee-Chia's undergraduate final year project on the topic of ordering and disordering theory in alloy semiconductors. I quickly recognized Yee-Chia's extraordinary talent. His final year thesis led to a Physical Review paper. His contribution in papers 2.7 and 2.8 was to extend the work by Wei-Jun on cubic structure GaN to wurtzite structure GaN semiconductors. He then extended the method to discuss the strain effect and alloying effect and produced three papers in IEEE J.Quantum Electronics. For Yee-Chia's outstanding performance, I and T.C recommended him to the Electrical & Computer Engineering (ECE) Department as a NUS senior tutor in order to further his Ph.D studies at UC Berkeley in the CMOS device field, under the guidance of Professor Chenming Hu. I was happy to observe Yee Chia's career advance by leaps and bounds. Not surprisingly, he is now a leading researcher in CMOS devices.

Chapter 3. Analog Integrated Circuit Design

Although Analog IC design is not a major research area of mine, my first successful research was in this field. In 1975, during the later period of China's cultural revolution, I had the opportunity to work in a semiconductor integrated circuit factory at University of Science and Technology of China in Hefei. I proposed a novel way to fabricate integrated circuit operational amplifiers. The method was considerably different from the prevailing one used in the West, yet suitable for the technology in China at that point in time. My colleague Jin-Song Chen and I, together with other collaborators in the factory successfully designed and fabricated the integrated circuit with very good performance. The ICs were finally used in the Chinese satellites and I was awarded with a research achievement prize from the Chinese Academy of Sciences. This was my first success in semiconductor research which helped develop my strong interest in Analog IC design. Analog design is a beautiful art-form in spite of its lack of deep Physics. I enjoy working in this area but I never consider it a priority in my research.

The Analog IC design papers I wrote in my early research days, are all in Chinese and therefore cannot be included in this book. Chapter 3 includes selected papers I wrote when I was with NUS, ECE Department, VLSI Lab. The co-authors Professors Yong Chin Lim, Yong Lian and Ai-Qun Liu were my colleagues and close collaborators at NUS. X.Chen, Yajun Ha, X.W. Zhang, Uday Dasgupta, Aimin Xu and Zhenying Luo were my Master of Engineering students.

Chapter 4. CMOS Transistors (I) (Reliability)

I moved to the CMOS device research area in 1996, when the Singapore-based semiconductor manufacturing company Chartered Semiconductors (CSM) launched a CSM-NUS collaboration program to attract more NUS professors and students to be involved in CMOS research to support CSM's technology development. At that time NUS only had good semiconductor measurement equipment so I started my CMOS research in reliability characterization, with my first Ph.D student Binbin Jie in CMOS, and later Hao Guan, and some Masters students, and my colleagues Professors Byung Jin Cho and W.K. Chim (papers 4.1–4.5). After I worked in the CMOS area, I realized that it was a very interesting and fruitful field. I also realized the importance of CMOS technology development research in NUS. The problem was that we only had a very old and obsolete clean room at that time. In 1999, there was a Temasek Professorship program launched by Singapore National Science and Technology Board (NSTB) to improve scientific and engineering research in Singapore. I was the microelectronics division head at that time at ECE–NUS. I proposed a state-of-the-art CMOS device technology laboratory, using the Temasek program funding. Although many were opposed to it, the proposal was strongly supported by the Department Head Professor Daniel Chan, and finally approved by NSTB. After much searching, we invited Professor Dim-Lee Kwong, a young but very active expert in Si technology at University of Texas at Austin, to be a part-time Temasek Professor guiding technical research in the Lab. We also assigned a young NUS Professor Byung Jin Cho to be the Lab manager as he had a strong industry background. The lab, named Silicon Nano Device Lab (SNDL), was developed so rapidly and successfully that after several years, it was recognized as one of the world's best research Labs in the CMOS device area. This would not have been possible without the enthusiastic efforts of all its team members, and very strong support from Singapore government agencies like NSTB and later A*STAR. I learned a lot about CMOS technology in that period at SNDL. I very much enjoyed the work with Dim-Lee, my colleagues and students at SNDL. I had the most active and fruitful research output in that period. I worked in CMOS technology, device reliability and device quantum simulation areas simultaneously, and gave up my research in Analog IC design since I was too busy. Some very significant research was produced by my group in that time.

In CMOS device reliability, my Ph.D student Gan Chen discovered in 2001 that the negative bias temperature instability (NBTI) degradation for the p-MOSFET could be recovered when the stress gate voltage was released. I understood that this is a very important discovery and it should affect the dynamic stress degradation in real logic circuits since in the logic circuit, the stress applied to the p-MOSFET is on and off periodically. We designed a series of experiments and dynamic stress circuit to simulate the p-MOSFET worked in the logic circuit. Finally we reported for the first time in 2002 the recovery phenomena in NBTI degradation in modern CMOS devices, and the dynamic NBTI degradation life time is several times longer than the static NBTI degradation and is frequency independent (paper 4.7, 4.8). After four months in 2003, more than 5 institutes reported similar recovery phenomena in the *International Reliability Physics Symposium (IRPS)* 2003 in Dallas. This NBTI recovery phenomena later became the key to understanding BTI degradation and BTI measurement, attracting worldwide interest for many years even until now. The mechanism of NBTI recovery is still a matter of debate. Another very talented Ph.D student Chen Shen and a Master student Tian Yang continued to work on BTI and we published a series of important papers on BTI (papers 4.11–4.16).

At the end of 2006, I moved from NUS SNDL to the Microelectronics Department, Fudan University. I proposed to Fudan to establish a new research lab in electronic device reliability. The proposal got very strong support from Fudan University and a Fudan Professor Daming Huang expressed his interest in working with me on the new project. Daming had very good background on III-V semiconductor physics, and got his Ph.D from University of Illinois. Due to his effort, the new lab quickly bought all test equipment suggested by me and trained two Fudan students Wen-Jun Liu and Zhi-Ying Liu. Our first task in Fudan reliability lab was to develop a new measurement technique to measure the interface trap degradation under stress in MOS transistors without measurement delay in order to avoid the recovery during the measurement delay, a topic which is key to understanding the NBTI degradation mechanism but yet to be solved. We first tried to measure the interface trap induced bulk current using a fast method developed in SNDL NUS by my student Chen Shen. After some tries by Wen Jun, I realized that this method had some fundamental problems and a change in tactics was required. I suggested extending the charge pumping voltage range to the stress voltage, therefore bringing the charge pumping measurement near to on-the-fly stress condition. This idea led to papers 4.17 and 4.19.

Our results in CMOS device reliability were summarized in several invited presentations and papers in *Electrochemical Society (ECS) Meetings* in Paris in 2003 (paper 4.9), Chicago in 2007 and San Francisco in 2009 (not included), *IWDTF* in Tokyo (paper 4.10), and *IEEE Trans. Device and Materials Reliability (TDMR)* (paper 4.18).

Chapter 5. CMOS Transistors (II) (Technologies)

In the CMOS technology area, my first Ph.D student Hongyu Yu had gotten his Bachelors degree in Tsinghua University China, his Master degree in University of Toronto, and transferred to my group at NUS to continue his Ph.D study in 2001, co-supervised by Professor Dim-Lee Kwong. I was experienced in Physics but had not much research experience in CMOS technology before Hongyu came to my group. Fortunately Hongyu was a brilliant student, very innovative and self-motivated. We had broad discussions and learned together about CMOS technology, and produced some very important work on high-k dielectrics (papers 5.1–5.4, 5.8). My second Ph.D student who worked on CMOS technology was Xinpeng Wang. He also obtained his Bachelor degree in Tsinghua University and was Hongyu's classmate. Xinpeng was a very hardworking student and produced the most important CMOS technology work in my group. In 2003, a student Xiongfei Yu in my collaborator Professor Chun-Xiang Zhu's group at SNDL worked on metal Ta incorporation in HfO_2 for high-k gate dielectric. They found that HfTaO crystallization temperature is much higher than either HfO_2 or Ta_2O_5 high-k dielectric. The channel mobility and device reliability are also much improved. His work was very significant and published in *Symposium VLSI Technology* 2004. However the weakness of HfTaO gate dielectric is the large gate tunneling current. I noticed that the large gate tunneling current for HfTaO is probably due to the low electron tunneling barrier of Si to Ta_2O_5. I therefore suggested Xinpeng to replace Ta with La in metal incorporation in HfO_2 as his Ph.D research topic in high-k gate dielectric, since La_2O_3 has a large electron and hole tunneling barrier, large dielectric constant, different structure from HfO_2, and also has high crystallization temperature. The high-k gate dielectric La_2O_3 was first reported by my friend Professor Albert Chin in Taiwan Chiao Tung University,[8] but still suffered from moisture absorption in processing. After several months hard work by Xinpeng, we found that HfTaO not only had a higher crystallization temperature, better reliability and lower gate tunneling

---

[8] A. Chin. *et al*, Symp VLSI Tech, 2000, p.16

current as expected, but a very interesting and unexpected phenomenon. The flat-band voltage V_{FB} of the metal-HfLaO-Si capacitor could be modulated by changing La concentration in HfLaO, after high temperature annealing. Correspondingly, the threshold voltage V_{th} of n-MOS transistor decreased when the La concentration in HfLaO gate dielectric was increased. This was an extremely important finding because too high V_{th} is one of the most difficult obstacles in high-k/metal gate stack MOS transistors due to the Fermi-pinning effect in the gate first CMOS technology after 1000°C annealing. We first submitted this result to *Symp VLSI Tech* 2005, but the paper was rejected. We then submitted the paper with some additional information to *International Electron Device Meeting (IEDM)* 2005, however the paper was also rejected. I then planed to report this result in an invited talk at the 208th *ECS meeting* in Los Angeles in 2005. The paper was accepted and published (paper 5.11) but I was unable to present the work at the meeting due to a sudden dizzy spell. We only reported our result at the *International Semiconductor Device Research Symposium (ISDRS)* 2005 in Washington DC after *IEDM2005*.[9] Nevertheless, we still published our important results for the first time in the highly prestigious journal *Electron Device Letters* in January 2006 (paper 5.13). Since we assumed that the n-V_{FB} shift is due to Fermi pinning release of n-metal/HfLaO, it should also expect p-V_{FB} shift for p-metal/HfLaO. I therefore encouraged Xinpeng to work in that direction. In *Symp VLSI Tech* 2006, we reported for the first time the p-V_{FB} shift of p-metal/HfLaO in p-MOSFETs (paper 5.14). At the same symposium, Sematech and IBM also reported their work on n-V_{FB} shift effect in n-MOSFETs by La capping layer incorporation in HfLaO. La incorporation in HfO_2 has since become the main stream technology in developing gate first technology in high-k/metal-gate stacks to modulate the transistor V_{th}. We later discovered, with my Masters student J.D. Chen, that not only La, but many lanthanide elements (Er, Yb, Tb, Dy) have a similar effect. When incorporating these elements into HfO_2, the V_{FB} of the MOS system can be modulated (paper 5.15).

Another significant work in my group in CMOS technology area is the Schottky (metal) source/drain transistors. Dr. Shi-Yang Zhu from Fudan University, a visiting research fellow in my group in SNDL, has produced some very significant research on Schottky source/drain transistors. Y. Nishi first proposed the idea of replacing doped source/drain with metal in 1966 when he submitted a Japanese patent on this idea, which was later issued a patent

---

[9]X.P. Wang *et al*, ISDRS, 2005, Washington DC, p.242

in 1970[10] Lepselter and Sze at Bell Lab in 1963[11] published the first paper using Schottky-barrier contacts for the source and drain of a MOSFET. This idea has attracted much interest in recent years due to its inherent advantages of shallow s/d junction and low s/d resistance. However the main obstacle in Schottky s/d technology is the high Schottky electron barrier and the low on-current I_{ON} for n-MOSFETs. We noticed that Yb metal had the lowest work function and Yb/Si silicide was possibly a good candidate for Schottky n-MOSFET s/d material. Shi Yang successfully used Yb/Si silicide for the first time in Schottky s/d, combined with high-k/metal gate technology developed in SNDL, for n-MOSFETs and achieved a very low electron barrier and pretty high on-current and recorded I_{ON}/I_{OFF} ratio(papers 5.6, 5.7, 5.9). He also produced the first Schottky s/d transistor on Ge channel (paper 5.10).

The co-author Professor Chun-Xiang Zhu was my colleague and close collaborator at SNDL in ECE-NUS. Professor Albert Chin was a visiting Professor at SNDL and a Professor at National Chiao Tung University in Taiwan. We had a very pleasant and effective collaboration at SNDL during that period.

Our works in Si CMOS technology were summarized in three invited and keynote papers: the 207th *ECS meeting* (2005) in Quebec (paper 5.9), the 208th *ECS meeting* (2005) in Los Angeles (paper 5.11), and the 12th Workshop on *Gate Stack Technology and Physics*, (2007) in Mishima (paper 5.17).

Chapter 6. CMOS Transistors (III) (Quantum Simulations)

The quantum effects in nano-CMOS devices become more important when scaling down the device size because of the quantum confinement effect and quantum tunneling effect. Due to my semiconductor physics background, I am very interested in quantum simulation in nano-CMOS area. Fortunately I have some very talented students who can conduct the quantum simulation work with great success. My first student in quantum simulation was Yong-Tian Hou who is very smart and has a very good Physics background. He got his Bachelor and Master degrees in Peking University. He worked in a factory in Shen Zhen for several years before coming to my group at SNDL. His work mainly focused

---

[10]Y. Nishi, Insulated gate field effect transistor and its manufacturing method, Patent 587 527, 1970.
[11]Lepselter and Sze, Proc. IEEE, 50, 1462, 1968

on hole quantization and hole gate tunneling. The valence band mixing effect in hole quantization was not familiar to device physicists. Many device physicists simply used one valence band effective mass Schrodinger equation to simulate the hole quantization effect. I suggested to Yong Tian to read W.J. Fan's hole quantization work in GaAlN quantum well (paper 2.8) and develop six-band effective mass Schrodinger equations to count the band mixing effect in Si in p-MOSFETs. To my pleasant surprise, he developed the equations and programs much quicker than expected. Based on that, he developed a series of very significant works on Si hole quantization and hole tunneling (papers 6.1–6.4, 6.9) with very good agreement with experimental data.

My second student in quantum simulation was Tony Low, a local student at NUS. As a sophomore, he approached me and expressed his interest in doing research under my supervision. I gave him an old topic on gate quantum tunneling taking reflection effect into account to test his ability. I was pleasantly surprised with his superior understanding of physics and mathematics and his independent research ability. Tony became my undergraduate final year, Master and Ph.D student and has done a series of very significant and innovative works. His undergraduate final year project was to develop the program of Yong Tian's Si valence band six-band effective mass Schrodinger equation by self-consistent calculation coupled with Poisson equation. This work led to a publication in *IEEE Trans Electron Devices* (paper 6.6) — a very rare achievement for an NUS undergraduate student. Later he did some very important work in quantum simulation of ultra-thin body double-gate (UTB-DG) transistors (papers 6.7, 6.8, 6.10–6.14). Probably the most significant work from my group in nano device quantum simulation is that we first showed the importance of conduction valley competition effect in UTB-DG transistors. The semiconductor multi-valley conduction band may change the order of different valley energies due to different valley effective masses and therefore different quantization energies. In paper 6.8 in 2003, using effective mass approximation, we showed for the first time that in <100> Ge surface orientation, the lowest conduction valley is no longer L valley as in the bulk, but is Δ valley when the Si body thickness is beyond 5nm, due to the valley competition effect. As a result, the <110>, not <100> surface orientation exhibits highest on-current. Although this conclusion was debated in 2003, it has since been elaborated and confirmed by IBM group in an IEDM 2004 paper.[12] In 2006, we turned to III-V compound semiconductor

---

[12]Laux, Int. Electron Device Meeting (IEDM),Tech. Digest, 2004, p.135

InSb UTB-DG transistor quantum simulation. Due to the very high electron mobility, InSb attracted worldwide interest including the semiconductor wafer manufacturing giant Intel. I noticed that although bulk InSb had the lowest conduction valley at Γ with a very small effective mass, the valley was parabolic only in a very narrow region in k space. For an ultra-thin body, the electron wave function should have been spread in a pretty wide region; therefore the effective mass approximation used in Ge may not be appropriate for the discussion of valley competition in InSb I therefore asked my postdoctor Zhen-Gan Zhu to develop a pseudo-potential atomistic calculation program to simulate the UTB InSb band structure. We obtained the correct result that Γ valley remains the lowest conduction valley – different from the result previously obtained from the effective mass approximation. We also found that the on-current of InSb UTB-DG transistor has not much of an advantage over its Si counterpart, probably due to the low density of states in the conduction band. This conclusion became a consensus for III-V UTB transistors among theorists[13] but still remains a point of contention amongst experimentalists.

The co-authors Professors Ganesh Samudra and Yee-Chia Yeo were my colleagues and collaborators at SNDL.

---

[13] Cantley *et al*, IEDM Tech. Digest, 2007, p.113

Chapter 1

Defects in Semiconductors

306

IEEE TRANSACTIONS ON ELECTRON DEVICES, VOL. ED-29, NO. 2, FEBRUARY 1982

A New Method for the Determination of Dopant and Trap Concentration Profiles in Semiconductors

MING-FU LI AND CHIH-TANG SAH, FELLOW, IEEE

Abstract—A new method is described and experimentally verified using a computer-controlled data acquisition system. It is capable of measuring sharply peaked and rapidly varying impurity-concentration profiles in semiconductors, including situations where the trap density exceeds the dopant density, with resolution limited by the Debye length. It combines two well-known techniques: the constant-capacitance junction voltage transient and the quasi-static junction capacitance measurements as a function of dc bias voltage. We propose the acronym CCQS. New implementation methods of these two techniques are developed which offer better reliability, capability, and convenience than those reported previously. The quasi-static low-frequency capacitance is used to obtain experimentally the edge region correction which was estimated only theoretically in the past. The method is verified using gold-diffused p$^+$/n and ion-implanted n$^+$/p silicon diodes.

NOMENCLATURE

Subscript M refers to unknown diode to be measured and called the "measured diode" in the text. Subscript R refers to the reference diode.

A_M, A_R	Junction area of the diodes.
C_M, C_R	Capacitance of the junctions.
c_n, c_p	Electron and hole capture rates at the trapping centers.
e_n, e_p	Electron and hole emission rates at the trapping centers.
E_C	Bottom edge of the conduction band.
E_F	Fermi energy.
E_T	Energy level of the trapping centers.
N_{AA}, N_{DD}	Acceptor or donor dopant impurity concentration in the measured diode.
N_M	Defined in Table I.
N_R	Dopant impurity concentration in the reference diode.
N_{TT}	Trap concentration in the measured diode.
N_{TTI}	Image of the trap concentration in the measured diode as defined in (14).
V_M, V_R	Bias voltages applied to the measured and reference diodes.
V_{MI}	Voltage across the measured diode at $t = 0^+$ immediately after the trap filling pulse is applied.
ΔV_M	Change of the voltage across the measured diode after the filling pulse due to the emission of electrons trapped at the defect-impurity centers.
W_M, W_R	Space-charge layer thicknesses of measured and reference diodes.
W''	Difference between W and Y_T, $W'' = W - Y_T$.
Y_T	Edge layer thickness of the measured diode, $Y_T = W - W''$.
$\varepsilon_M, \epsilon_R$	Dielectric constants of the semiconductor of the measured and reference diodes.

I. INTRODUCTION

JUNCTION capacitance techniques have widely been used to determine both majority dopant impurity and trap concentration profiles in semiconductors [1]-[9]. Goto, Yamagisawa, Wada, and Takanashi [7] and Pals [8] made an important advancement by using constant capacitance voltage transient (CCVT) instead of the previous constant voltage capacitance transient (CVCT) method [1], [5], [6] to determine trap concentration profiles. Johnson *et al.* [9] combine the CCVT with a double-correlation technique proposed by Lefevre and Schulz [6]. The CCVT is especially advantageous over CVCT when the trap concentration N_{TT} is very high and rapidly varying with position. In this paper, we combine the CCVT method with a quasi-static C-V measurement to obtain three independent C-V relations. The quasi-static C-V is used to obtain the edge-region correction experimentally, which was only calculated theoretically in the past and has never been quantitatively verified by experiment. As will be pointed out in subsequent sections, the experimental and theoretical edge-region correction methods complement each other and when in combination, will improve the reliability and versatility of the transient C-V impurity profile measurement techniques. A real-time computer-controlled measuring system and a set of working equations in compact form are described in this paper to implement this method. The method is valid for a wide range of conditions, specifically for the case of very high trap concentration and abrupt changes of the spatial distributions of either or both of the trap centers and majority dopant impurities with space resolution limited by Debye length.

In Section II, we will illustrate the principle of the method. In Section III, experimental conditions and results will be discussed. The detailed electronic and circuit techniques

Manuscript received June 25, 1981; revised June 25, 1981. This work was supported in part by the Air Force Office of Scientific Research under Grant AFOSR-78-3714 and by the Rome Air Development Center under Contract F19628-77-0138.

M-F. Li was with the Solid State Electronics Laboratory, University of Illinois, Urbana. He is now with the Chinese University of Science and Technology, Peking, China.

C-T. Sah is with the Solid State Electronics Laboratory, University of Illinois, Urbana, IL 61801.

0018-9383/82/0200-0306$00.75 © 1982 IEEE

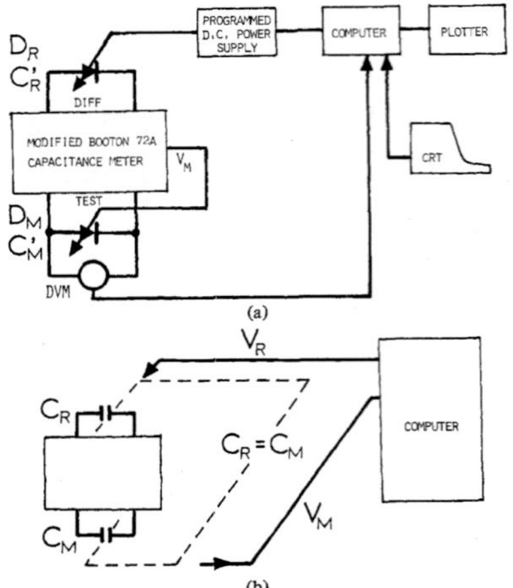

Fig. 1. Real-time computer-controlled constant-capacitance voltage transient experiment for impurity concentration profile measurements. (a) Block diagram. (b) Equivalent circuit.

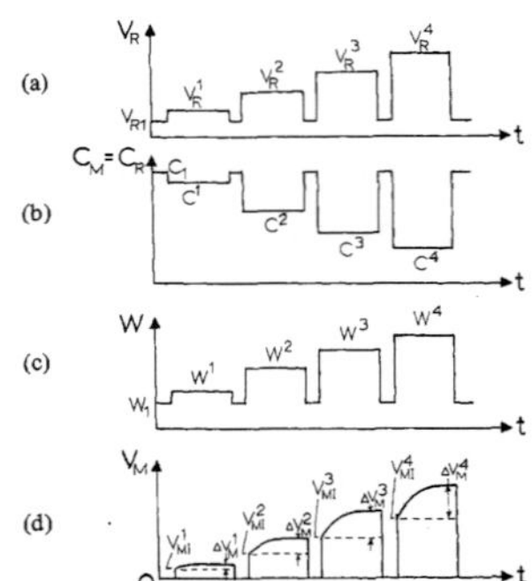

Fig. 2. Timing diagram of the voltage across the reference diode V_R, the capacitance of the reference and measured diodes $C_R = C_M$, the thickness of the space-charge layer W, and the voltage across the measured diode V_M.

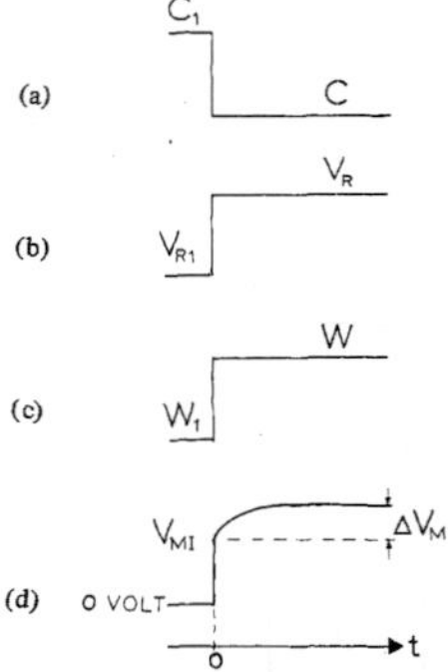

Fig. 3. Transient waveforms of the capacitance C; the voltage across the reference diode, V_R; the space-charge layer thickness, W; and the voltage across the measured diode, V_{MI} and ΔV_M.

developed in this work will be illustrated in another paper. All of the symbols used are listed in the Nomenclature.

II. Basic Principles

A real-time computer-controlled measuring system was set up to implement this method. A block diagram is shown in Fig. 1(a). A modified Boonton 72 capacitance meter is used as a "capacitance amplifier." A reference diode D_R is connected to the differential side of the capacitance meter input. Its bias voltage V_R is controlled by a computer through a programmable power supply (HP6106A). The measured diode D_M is connected to the TEST side of the capacitance meter input. D_M is treated as a voltage–capacitance converter. The analog output V_M of the capacitance meter is used as the bias voltage applying to the measured diode. A closed loop of the capacitance amplifier and D_M is formed, and they function like a "capacitance follower" to follow the "input capacitance" D_R. The method of using a feedback loop for carrier concentration measurement can be traced back to Miller [10]. A digital voltmeter (DVM) reads V_M and feeds the reading to a computer. The equivalent circuit of this setup is shown in Fig. 1(b).

When a computer data-acquisition program is running, a series of voltage pulses with gradually increased amplitude is applied to D_R. The timing diagram of V_R and corresponding timing diagrams of the diode capacitances C_M and $C_R (C_M = C_R)$, the space-charge layer thickness W, and the bias voltage of measured diode V_M are shown in Fig. 2(a)-(d).

A. Constant-Capacitance Analysis

Let us consider one of the transients of Fig. 2, which is shown in Fig. 3, in more detail. The reference diode has no trap centers. So, when bias voltage V_R changes suddenly from V_{R1} to V_R at $t = 0$, the junction capacitance C_R changes from C_1 to C, correspondingly. Since the control loop keeps $C_M = C_R$ at all times, C_M changes in exactly the same way as C_R. The width of the transition region of the measured diode W_M is given by

$$W_M = \frac{\epsilon_M A_M}{C_M} \qquad (1)$$

and it changes with time abruptly from W_1 to W, as shown in Fig. 3(c). The bias voltage of the measured diode also changes suddenly from zero at $t = 0^-$ to V_{M1} at $t = 0^+$, and then gradually increases to $V_{MI} + \Delta V_M$ at $t = \infty$. This increase is due to the thermal emission of trapped electrons from the recombination centers to conduction bands (or trapped holes

308   IEEE TRANSACTIONS ON ELECTRON DEVICES, VOL. ED-29, NO. 2, FEBRUARY 1982

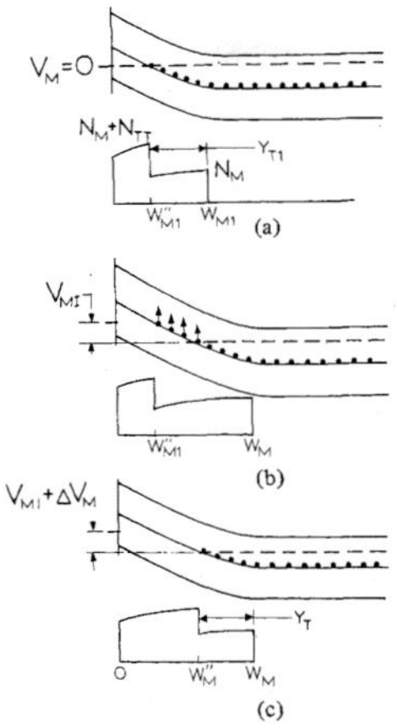

Fig. 4. Energy-band diagram and the space-charge distribution of a p^+/n junction on the n-side with electron trapping centers. (a) $t < 0$, steady-state zero bias with space-charge layer thickness W_{MI}. (b) $t = 0^+$, space-charge layer thickness changes from W_{MI} to W_M suddenly. (c) $t \gg e_n^{-1}$, quasi-equilibrium is reached.

TABLE I

Substrate Type	$q\phi$	Characteristics of Trapping Centers		N_M
N	$E_F - E_T$	Electron Trap	Acceptor	$N_{DD} - N_{TT}$
		$e_n \gg e_p$	Donor	N_{DD}
P	$E_T - E_F$	Hole Trap	Acceptor	N_{AA}
		$e_p \gg e_n$	Donor	$N_{AA} - N_{TT}$

to the valence bands) within the junction space-charge layer. To be definite, we consider a p^+/n abrupt junction and a majority-carrier (electron) traps in the upper half of the energy gap whose electron emission rate is much greater than hole

$$e_n \gg e_p. \tag{2}$$

The corresponding energy-band diagrams of Fig. 3 are shown in Fig. 4(a)–(c) where Y_T is the edge layer thickness [5], [12]. The meaning of $N_M(W)$ is defined in Table I.

When we integrate the Poisson equation according to the charge distribution illustrated in Fig. 4, we get [5, eq. (2)]

$$\frac{\epsilon}{q}(V + V_D) = \int_0^W x[N_M(x) + N_{TT}(x) - n_T(x, t)]\, dx \tag{3}$$

where V is the reverse applied voltage, V_D is the diffusion potential, $n_T(x, t)$ is the trapped electron concentration.

Applying (3) to the case of Fig. 2 to Fig. 4 for the measured diode, we have at $t = 0^+$

$$\frac{\epsilon_M}{q}(V_{MI} + V_{DM}) = \int_0^{W_M} x[N_M(x) + N_{TT}(x)]\, dx$$
$$- \int_{W_{MI}''}^{W_M} xN_{TT}(x)\, dx \tag{4}$$

and at $t \gg e_n^{-1}$

$$\frac{\epsilon_M}{q}(V_{MI} + \Delta V_M + V_{DM}) = \int_0^{W_M} x[N_M(x) + N_{TT}(x)]\, dx$$
$$- \int_{W_M''}^{W_M} xN_{TT}(x)\, dx \tag{5}$$

where subscript M refers to the parameter of the measured diode. Subtracting (4) from (5), we have

$$\Delta V_M = \frac{q}{\epsilon_M} \int_{W_{MI}''}^{W_M''} xN_{TT}(x)\, dx. \tag{6}$$

From (4) we also have

$$\frac{\partial V_{MI}}{\partial W_M} = \frac{q}{\epsilon_M} W_M N_M(W_M). \tag{7}$$

From (6) we have

$$\frac{\partial \Delta V_M}{\partial W_M} = \frac{q}{\epsilon_M} W_M'' N_{TT}(W_M'') \frac{\partial W_M''}{\partial W_M}\bigg|_{W_M}. \tag{8}$$

Applying (3) to the reference diode for the case of Fig. 2 to Fig. 4 and remembering that there are no trap levels in the reference diode, we have

$$\frac{\epsilon_R}{q}(V_R + V_{DR}) = \int_0^{W_R} xN_R(x)\, dx \tag{9}$$

where subscript R refers to the parameters of the reference diode, and $N_R(x)$ is the dopant impurity concentration in the reference diode. From (9) we have

$$\frac{\partial V_R}{\partial W_R} = \frac{q}{\epsilon_R} W_R N_R(W_R). \tag{10}$$

Since we always have

$$C_M = C_R$$

this gives

$$\frac{W_R}{A_R \epsilon_R} = \frac{W_M}{A_M \epsilon_M} \tag{11}$$

since $C = A\epsilon/W$, where A is the area of the junction. When we combine (7), (8), (10), and (11) we have

$$N_M(W_M) = N_R(W_R) \frac{\partial V_{MI}}{\partial V_R}\bigg|_{W_R} \tag{12}$$

where

$$\zeta = \left(\frac{A_R^2}{A_M^2}\right)\left(\frac{\epsilon_R}{\epsilon_M}\right) \tag{13}$$

and

$$N_{TT}(W_M'') = N_{TTI}(W_M)\left(\frac{W_M}{W_M''}\right)\left(\frac{\partial W_M''}{\partial W_M}\right)_{W_M}^{-1} \tag{14}$$

where

$$N_{TTI}(W_M) = N_R(W_R)\frac{\partial \Delta V_M}{\partial V_R}. \tag{15}$$

Equations (12)–(15) are a set of very compact relations between concentration profiles N_{TT}, N_M of the measured diode, and N_R of the reference diode. No approximation is made except the staircase electron distribution model of Fig. 4 based on depletion-layer approximation, so they are valid for a wide range of applications, such as 1) the "strongly compensated case" occuring in fast switching diodes where the dopant impurity concentration, N_M, is not much greater than the recombination center concentration: $N_M \simeq N_{TT}$; and 2) ion-implanted devices which may have abrupt changes in both N_{TT} and N_M with position.

In practice, we use a reference diode with uniform doping, $N_R(W_R) \equiv N_R$, whose junction area and substrate material are the same as those of the measured diode. The very slight temperature dependence of ϵ is neglected [11]. Then (11)–(15) are simplified to give

$$W_M = W_R = W \tag{11a}$$

$$N_M(W) = N_R\frac{\partial V_{MI}}{\partial V_R} \tag{12a}$$

$$N_{TT}(W'') = N_{TTI}(W)\left(\frac{W}{W''}\right)\left(\frac{\partial W''}{\partial W}\right)_W^{-1} \tag{14a}$$

$$N_{TTI}(W) = N_R\frac{\partial \Delta V_M}{\partial V_R}. \tag{15a}$$

However, if the reference diode has a position-dependent dopant impurity concentration, (11)–(15) must be used which can be readily solved using the real-time computer that obtained the data.

When experiments are done as shown in Fig. 2, we obtain the relationships between V_{MI}, ΔV_M, and V_R such as the data discussed later (Fig. 8). From (12a) and (15a), we can then obtain $N_M(W_2)$ and $N_{TTI}(W_2)$ as a function of W (Fig. 12, discussed later). The relation between the trap concentration before and after edge region correction is given by (14a). $N_{TTI}(W)$ is a directly measurable quantity from experiments before the correction. $N_{TT}(W'')$ is the true spatial distribution of the trap centers which is obtained from $N_{TTI}(W)$ using the edge region correction given by (14a). We can give a physical meaning to $N_{TTI}(W)$. It is the image of $N_{TT}(W'')$ with the amplitude-magnification factor (W''/W) $(\partial W''/\partial W)$ and the width-magnification factor $(\partial W/\partial W'')$, and the distance between the object and the image is the length of the edge region.

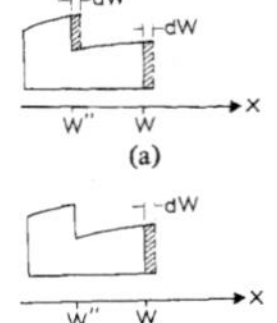

Fig. 5. Space-charge variations in the two capacitance measurement methods. (a) Quasi-static capacitance measurement. Bias voltage is decreased so slowly that electrons in the trapping center within the edge region is in quasi-equilibrium with the electrons in the conduction band at all times. (b) High-frequency capacitance measurement. Bias voltage is changed very fast so that electrons in the trapping centers do not have time to change.

From (14) and (14a), we have the following conclusion. The edge region correction is only important for two cases: 1) the thickness of the edge region is larger than about half of the total space-charge layer thickness and 2) $N_M(W)$ varies rapidly with position so that $\partial Y_T/\partial W$ is substantially different from zero. Equations similar to (14) have been obtained in [2] and [4] where a small-signal C-V measurement is used for trap concentration calculation. However, neither of these papers treated the important case of position variation of the dopant and majority-carrier concentrations. They assumed constant electron concentration (n_0 in Schulz' paper) and constant donor concentration (N_D in Kimerling's paper) which would not cover the practical situations.

B. Edge-Region Correction Analysis

In order to obtain N_{TT} from N_{TTI}, we have to know the relationship between W'' and W. This edge-region correction is very important. In the past, only a theoretical correction was developed [5], [6], [9]. In this paper, an experimental method is described to obtain W'' from W. In principle, there are three unknown functions of W to be determined. These are $N_M(W)$, $N_{TT}(W)$, and $Y_T(W)$. We have already obtained two independent relationships from experiments illustrated in Figs. 1, 2, and 3; they are (12a) and (15a). Thus only one more independent experimental measurement is needed.

The third experimental relationship we shall use is the differential or low-frequency capacitance [12], also known as the quasi-static capacitance. This dc differential capacitance is obtained as follows. First, we apply a reverse-bias voltage to the diode and wait long enough so that all the traps in the space-charge layer from $x = 0$ to $x = W''$ in Fig. 4 are empty of electrons. Then, we reduce the bias voltage at a sufficiently slow rate so that the traps can capture electrons even for those traps in the free carrier tail region of the edge region [17]. In other words, the bias reduction is sufficiently slow so that a quasi-equilibrium is established between the electrons trapped and the majority carriers even in the vicinity of the edge-region boundary W''.

The bias-voltage reduction can be divided into two parts shown in Fig. 5(a)

IEEE TRANSACTIONS ON ELECTRON DEVICES, VOL. ED-29, NO. 2, FEBRUARY 1982

310

$$dV_{\text{Bias}} = dV_M + d\Delta V_M$$

$$d\Delta V_M = (q/\epsilon)\, W'' N_{TT}(W'')\, dW''$$

$$dV_M = (q/\epsilon)\, W N_M(W)\, dW.$$

On the other hand, change of the charges stored in the capacitance during this procedure is

$$dQ = Aq\,[N_{TT}(W'')\, dW'' + N_M(W)\, dW].$$

Combine the above equations, we then have the quasi-static or dc- differential capacitance measured by this quasi-equilibrium procedure

$$C_{\text{LF}} = dQ/dV_{\text{Bias}}$$

$$= \frac{\epsilon A}{W} \left[1 + \frac{N_{TT}(W'')(dW''/dW)[(W''/W) - 1]}{N_M(W) + (dW''/dW)\, N_{TT}(W'')} \right]^{-1} \tag{16}$$

In [12], Sah and Reddi used the notation C_{DC}. We prefer C_{LF} here since C_{DC} sometimes refers to Q/V rather than dQ/dV.

At high frequencies such that $\omega \gg e_n$, the junction capacitance is simply given by

$$C_{\text{HF}} = \frac{\epsilon A}{W}. \tag{17}$$

To simplify the following analysis, we define a function $K(W)$ as the fractional difference between the low-frequency and high-frequency capacitances given by

$$K(W) = 1 - [C_{\text{HF}}(W)/C_{\text{LF}}(W)] \tag{18}$$

which is always less than unity since $C_{\text{HF}}(W) < C_{\text{LF}}(W)$. Then, from (14), (16), (17), and (18), we obtain the layer thickness ratio

$$\frac{W''}{W} = \frac{1 - K(W)}{1 + K(W)[N_M(W)/N_{TTI}(W)]}. \tag{19}$$

Since $N_M(W)$, $N_{TTI}(W)$, and $K(W)$ are experimentally measured quantities, thus W/W'' can be computed from the experimental data using (19).

C. Limitation of the Method

In the Appendix, a sensitivity analysis of (18) and (19) is given. It is shown there that if

$$\frac{W'' N_M(W)}{W N_{TTI}(W)} \gg 1$$

then W'' cannot be determined from W very precisely by the quasi-static experiment. In this case, the edge-region correction can be obtained using the integral equation [5]

$$\phi = \frac{q}{\epsilon} \int_{W''}^{W} (x - W'')\, N_M(x)\, dx \tag{20}$$

where ϕ is defined in Table I and in our case it is

$$\phi = [E_F - E_T]/q \tag{20a}$$

where E_F is Fermi energy. A constant Fermi energy throughout semiconductor is assumed which may not be a good ap-

proximation under certain conditions since the junction is at a nonequilibrium condition. Another serious problem is that (20) is a rather complicated integral equation of W'' if N_M is not constant. In some cases, reliable results could not be obtained as will be illustrated in Section III-B. Combining (19) and (20) as complement to each other will provide a method with improved reliability and versatility for profile measurement.

Another limitation of the present method is on the spatial resolution limited by Debye length L_D. If there is an abrupt change of N_M within a Debye length

$$L_D = \sqrt{\epsilon k T/q^2 N_M}$$

then it is important to consider the difference between majority-carrier density and impurity density [13]. The quantity N_M obtained by C–V measurement is the majority-carrier density and it is no longer equal to the impurity density in this case. For such a fine-grain analysis, the staircase model of Fig. 4, which has a sharp boundary of charge distribution at $x = W$, will no longer be a good approximation [14]–[16].

III. EXPERIMENTAL RESULTS AND DISCUSSIONS

The theoretical results of the new method just described are tested experimentally. In order to bring out the unique features of the new methods (high trap concentration, sharply peaked trap concentration, and experimental edge-region correction), experimental–theoretical correlations are made on two silicon diodes. The first diode is gold diffused with very high gold concentration. The second diode is ion implanted to give a sharply peaked trap-concentration profile which exceeds the dopant concentration. These are described below in two subsections.

A. Gold-Diffused Silicon p^+/n Diode Experiments

In this experiment, the gold-diffused experimental diode 40D3, originally used by Sah and Reddi in their 1964 paper [12], is selected. This is a silicon p^+/n diode which is gold-diffused at a high gold concentration. Its fabrication procedure is as follows. Boron is first diffused into the n-type silicon substrate of 0.65- to 0.67-$\Omega \cdot$ cm resistivity. The average dopant or phosphorus concentration is about 9×10^{15} cm^{-3} as estimated from the resistivity measurement. Boron was predeposited at 1000°C for 7 min and then diffused into the silicon substrate at 1020°C for 20 min. A very abrupt p^+/n junction was obtained under these conditions. Gold was then diffused in from the back surface at 1000°C for 16 min. The diffused p^+/n junction has a diameter of 15 mils or a junction area of $A = 1.14 \times 10^{13}$ cm^2.

A series of capacitance measurements were performed using a minicomputer- (HP-1000) controlled data-acquisition system. The software used is shown in the flow chart diagram in Fig. 6 which includes the data-acquisition program, the data-manipulation program, and the plotting program. The VSCTS (Voltage-Stimulated Capacitance Transient Spectra)-DLTS [18] of the data is obtained by calling the TSCAM program. The result which gave a single peak at 210 K when the sampling

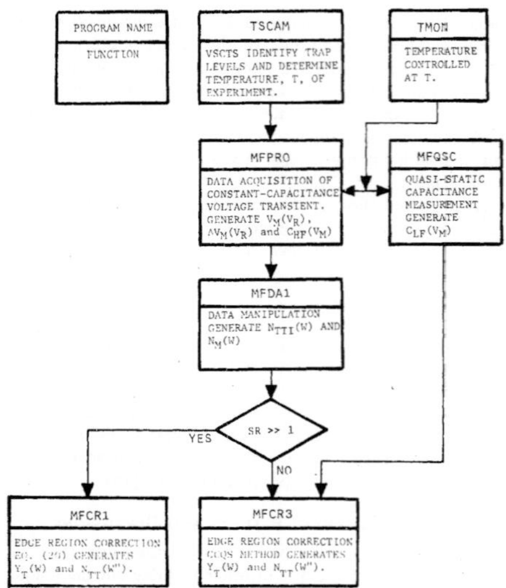

Fig. 6. Flow chart of the real-time computer-controlled impurity concentration profile measurement setup.

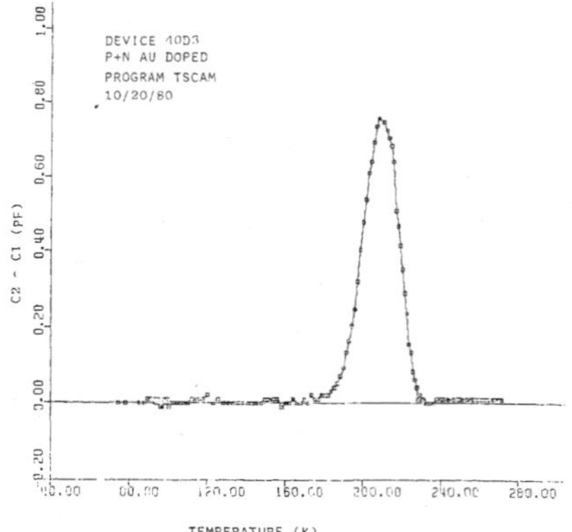

Fig. 7. Voltage-stimulated capacitance transient spectroscopy (VSCTS) of a gold-diffused silicon p$^+$/n diode (Unit 40D3). $t_1 \simeq 10$ ms and $t_2 = 0.6$ s.

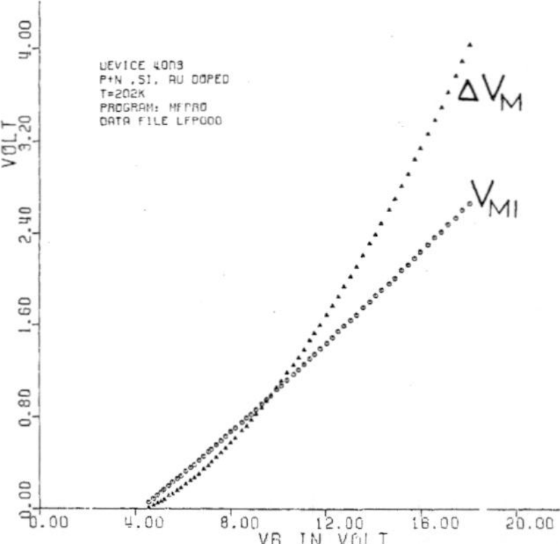

Fig. 8. V_{MI} and ΔV_M versus V_R curves obtained during the constant-capacitance voltage transient experiment on the gold-doped p$^+$/n diode.

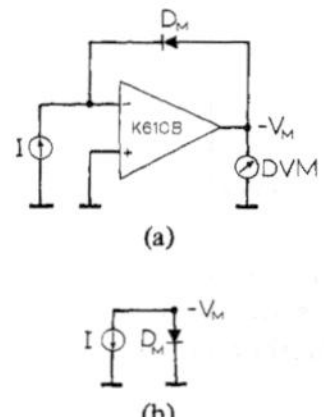

Fig. 9. Circuit diagram of a new quasi-static capacitance measurement method. (a) Block circuit. (b) Equivalent circuit diagram.

times of $t_1 = 2$ s and $t_2 = 7$ s were used is shown in Fig. 7. This peak comes from thermal emission of the electrons trapped at the gold acceptor level located at $E_C - 0.55$ eV in the n-type silicon base region of the diode. From this spectrum, a diode temperature of 202 K was selected to run the constant-capacitance voltage transient experiments which was performed by calling the MFPRO program. The V_{MI} and ΔV_M versus V_R data obtained in this experiment are shown in Fig. 8. High-frequency capacitance versus voltage data, C_{HF}-V, were also obtained in this experiment and shown later in Fig. 11. For a given V_R, the corresponding C_R ($= C_M$) is known since we use a precalibrated reference diode and the reverse-bias voltage V_M of the measured diode equals $V_{MI} + \Delta V_M$. Quasi-static capacitance versus bias voltage data, C_{LF}-V_M, were obtained by running the MFQSC program. The principle of the quasi-static capacitance versus V_M measurement is shown in Fig. 9. A high-input-impedance dc amplifier (Keithly model 610B electrometer set in the fast mode and 10^{12}-Ω range) was used to isolate the DVM from the measured diode. When constant current source I is applied to D_M, the bias voltage V_M of D_M increases or decreases according to

$$C_{LF}(dV_M/dt) = I.$$

Let $I = I_{in} > 0$ first to increase V_M to a high reverse bias.

312

IEEE TRANSACTIONS ON ELECTRON DEVICES, VOL. ED-29, NO. 2, FEBRUARY 1982

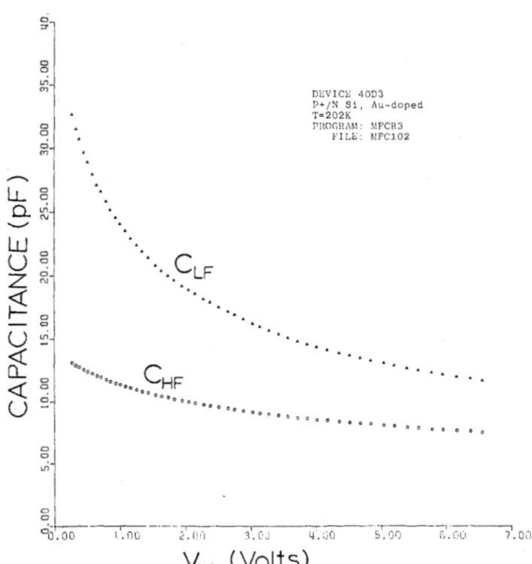

Fig. 10. High-frequency (C_{HF}) and quasi-static (C_{LF}) capacitance-voltage curves of the gold-diffused silicon p$^+$/n diode measured at 202 K.

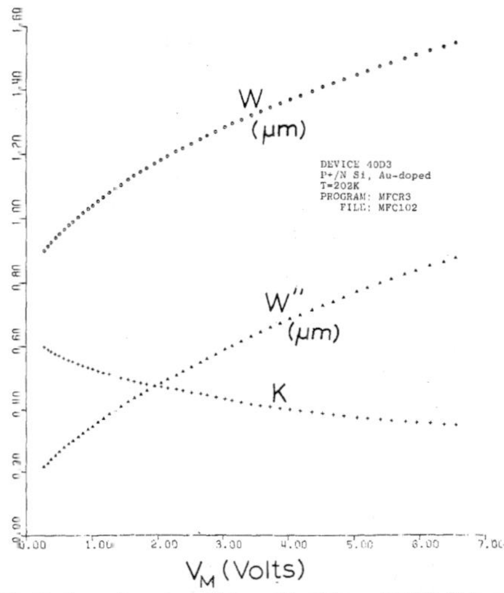

Fig. 11. Space-charge layer thickness W and the partial SCL thickness W'', and the fractional difference of high–low frequency capacitances K, as a function of the bias voltage V_M of the gold-diffused silicon p$^+$/n diode.

Then, I is reduced to a very small and negative value, $I = -I_{DE}$ in order to decrease V_M very slowly. In our previous work, it was shown that the capture rate of electrons by the trap is $e_n + c_n N$ where N is the electron density in the conduction band. In the neighborhood of the point W'' in the edge region (see Fig. 4), which is known as the "free-carrier tails," the capture rate approaches $2e_n$ since at that point $C_n N = e_n$ [19]. In order to satisfy the quasi-equilibrium condition in this tail region, $I_{DE} = 1$ pA was used. The corresponding reverse-bias voltage changed so slowly that within one decay time constant e_n^{-1}, the variation of the thickness of the space-charge layer was less than one tenth of the edge-layer thickness. The resulting C_{LF} versus V_M data obtained by running the MFWSC program are shown in Fig. 10. Different I_{DE} for different dV/dt were tried also and identical results of C_{LF} versus V data were obtained to confirm the quasi-equilibirum condition.

Program MFDA1 is a data-manipulation program which generates $N_M(W)$ and $N_{TTI}(W)$ using (12a) and (15a). Program MFCR1 is the edge-correction program which computes W'' from W using (19) and generates $N_{TT}(W'')$ from (14a). The results are plotted in Figs. 11 and 12. In Fig. 12, curve A is $N_M(W) = N_{DD}(W) - N_{TT}(W)$. It gives (1.6 to 2.0) $\times$ 10^{15} cm^{-3} which is in good agreement with the value of 1.7×10^{15} cm^{-3} obtained in [12]. Curve C is $N_{TT}(W'')$ computed from (14a). It gives (6.6 to 7.3) $\times$ 10^{15} cm^{-3} which compares favorably with the value of 7.3×10^{15} cm^{-3} obtained in [12]. The average donor concentration, $\bar{N}_{DD} = \bar{N}_M + \bar{N}_{TT}$, is about 9×10^{15} cm^{-3} in this experiment which is also consistent with the value obtained from substrate resistivity measurements.

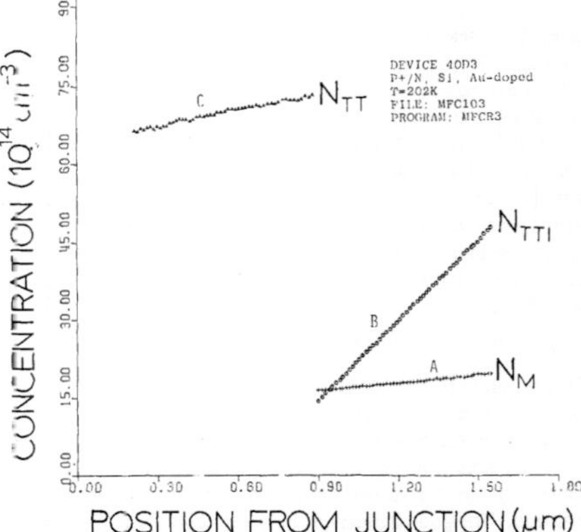

Fig. 12. N_{TTI}, N_{TT}, and N_M concentration profiles of the gold-diffused silicon diode obtained by the CCQS method.

In Fig. 11, curve A is W versus V_M which is obtained from $W = \epsilon A/C$. Curve B is W'' versus V_M which is obtained from (19). Curve C is K versus V_M. It is interesting to note that although K and N_{TTI} are strongly dependent on V_M, their contributions to (19) give almost a voltage-independent edge-

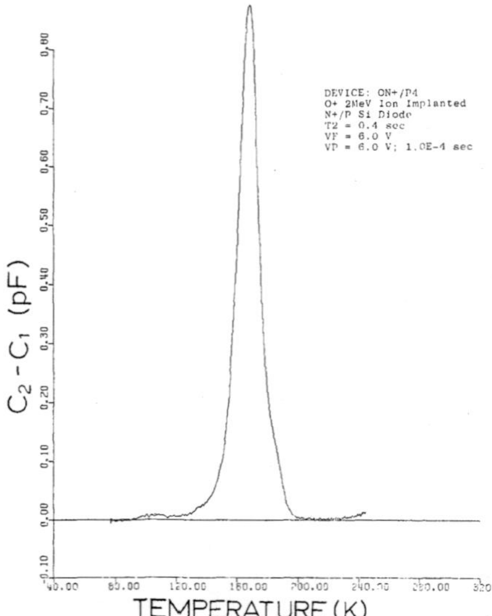

Fig. 13. Voltage-stimulated capacitance transient spectroscopy (VSCTS) of a diffused silicon n$^+$/p diode subsequently implanted by 2-MeV oxygen ion and then annealed at 400°C for 30 s (diode ON + P4).

layer thickness of $Y_T = W - W''$ which varies from 0.66 to 0.69 μm. This is expected since N_M is almost constant.

In order to compare experimental Y_T with estimated theoretical value from (20), we use $N_C = 2.8 \times 10^{19}$ (202/300)$^{3/2}$ = 1.6×10^{19} cm^{-3} [20] as the effective density of state at the conduction band edge of silicon at 202 K. Then, $(E_C - E_F)/q = -(kT/q) \log_e (\bar{N}_M/N_C) = 0.07$ V. From previous measurements, we have $(E_C - E_T)/q = 0.55$ V for the gold acceptor level in silicon [21]. Thus if we use $\bar{N}_M = 1.6 \times 10^{15}$ cm^{-3} as average concentration of $N_M(W)$ between $W = 0.2$ and 1.6 μm, (20) gives

$$Y_T = W - W'' = \sqrt{2e(E_F - E_T)/q^2 \bar{N}_M} = 0.63 \ \mu\text{m}.$$

This estimate is consistent with the value of 0.66 to 0.69 μm just obtained in our experiment that shows the edge-layer thickness is essentially a constant as we would expect since both the gold concentration and the phosphorus donor concentration in the substrate are essentially independent of position. The good agreement between theory and experiment is a strong confirmation of the validity of the theory and the experimental procedure just presented.

B. Ion-Implanted Silicon n$^+$/p Diode

The second diode we used to test our new method is an n$^+$/p-diffused silicon diode which is subsequently implanted by 2-MeV oxygen ion with a dose of 10^{12} ion/cm^2 to create a sharply peaked defect-trapping center profile. The diode

ON + P4, was fabricated by Dr. H. S. Fu while he was at the Solid State Electronics Laboratory at the University of Illinois. The oxygen ion implantation was made in 1971 at the AFCRL accelerator (courtesy of Dr. E. Davies). After ion implant, the diode was annealed at 400°C for 30 s to simulate the die bond condition.

The experimental results are shown in Figs. 13-15. Fig. 13 gives the VSCTS spectrum which shows a strong peak at 169 K at capacitance sampling times of $t_1 \simeq 10$ ms and $t_2 = 0.4$ s. This peak corresponds to a hole trapping level in the p-type silicon substrate.

Fig. 15(b) gives the final concentration profiles. Fig. 15(b) shows a decrease of $N_M(W)$ towards the physical p/n junction. This is confirmed by direct measurement of N_M near the junction in other ion-implanted devices in our laboratory. Curve N_{TT} is the hole trap concentration profile produced by the oxygen-ion implantation. Curve N_{TTI} is the image of N_{TT} and was obtained from experiments before applying the edge-region correction. The peaked trap profile N_{TT} is expected and in qualitative agreement with the range of the oxygen ion and the profile of the defect centers produced by the ion. More detailed and quantitative analyses are being undertaken currently to correlate the defect profiles with the implanted ion profiles.

It should be pointed out that no previous C-V method can measure concentraton profiles N_{TT} illustrated in this example. The integral equation (2) cannot be solved in this case using

314   IEEE TRANSACTIONS ON ELECTRON DEVICES, VOL. ED-29, NO. 2, FEBRUARY 1982

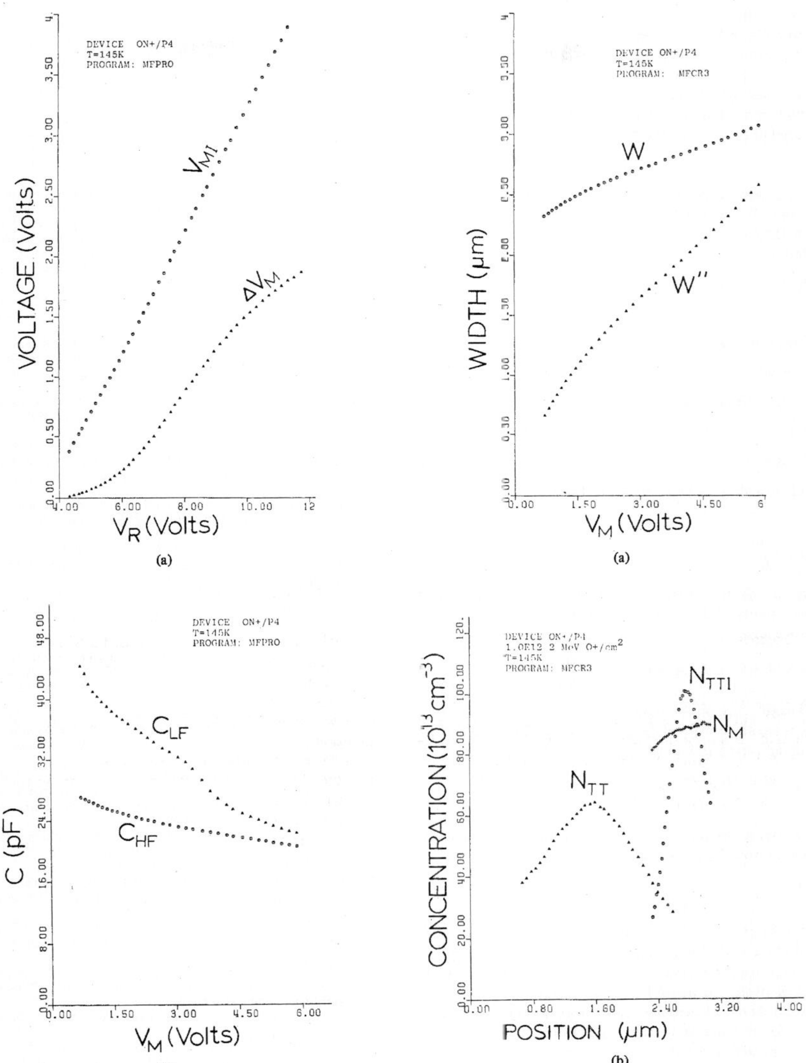

Fig. 14. Ion-implanted diode (ON + P4) measured at 145 K.   (a) V_{MI} and ΔV_M versus V_R curves.   (b) C_{HF} and C_{LF} versus V_M curves.

Fig. 15. Ion-implanted diode measured at 145 K.   (a) W and W'' versus V_M curves.   (b) N_{TTI}, N_{TT}, and N_M profiles.

previous methods. In order to evaluate W'' from (20), we need to know $N_M(W)$ in the range from $W = 0.5\text{-}3.3\,\mu m$. However, $N_M(W)$ cannot be determined by previous $C\text{-}V$ methods within $W = 0.5\text{-}2.2\,\mu m$ since it corresponds to a substantial forward-bias condition. In addition, $N_M(W)$ changes so rapidly and arbitrarily near the p/n junction that the extrapolation used in previous methods is unreliable.

APPENDIX

In this Appendix, we will provide a sensitivity analysis of (18) and (19) which can be used to assess the range within which they can be used to give accurate results. From (19), we define

$$R(W) = N_M(W)/N_{TTI}(W)$$

and

$$S(W) = W''/W = (1 - K)/(1 - KR)$$

or we have

$$K = (1 - S)/(1 + SR).$$

Since $0 < S < 1$ and $R > 0$, so that K will always satisfy

$$0 < K < 1. \tag{A1}$$

Let us now define the relative precision of a measurement of L as

$$\theta_L = \frac{\Delta L}{L}$$

where ΔL is the average or root-mean-square error of the measurements of L. It is easy to verify that up to the first order we have

$$\theta_S = (1 + SR)\,(\theta_{C\,\mathrm{HF}} + \theta_{C\,\mathrm{LF}}) - \frac{KR}{1 + KR}\,\theta_R. \tag{A2}$$

The second term on the right-hand side of (A2) is always small if θ_R is small. For small $\theta_{C\,\mathrm{HF}}$ and $\theta_{C\,\mathrm{LF}}$, then, the first term on the right-hand side of (A2) will be small only when

$$SR = \frac{W''}{W}\,\frac{N_M(W)}{N_{TTI}(W)} \gg 1. \tag{A3}$$

This is the necessary condition for accurate measurements of $N_{TT}(W'')$ from $N_{TTI}(W)$ data using (18), (19), and (14).

ACKNOWLEDGMENT

The authors would like to thank Dr. P. Chan for invaluable consultation on the implementation of the computer-controlled data acquisition hardware and software and Prof. G. G. Qin and C. K. Wang for helpful discussions and other members of the Solid State Electronics Laboratory at the University of Illinois for friendly assistance. Ms. Donna Stowe typed the manuscript with great skill and patience for which the authors are very grateful.

REFERENCES

[1] G. H. Glover, "Determination of deep levels in semiconductors from $C\text{-}V$ measurement," *IEEE Trans. Electron Devices*, vol. ED-19, pp. 138–143, 1972.

[2] M. Schulz, "Deep trap levels of ion-implanted germanium in Silicon measured by Schottky contact techniques," *Appl. Phys. Lett.*, vol. 23, pp. 31–33, 1973.

[3] C. H. Henry, H. Kukimoto, G. L. Miller, and F. R. Merritt, "Photo capacitance studies of the oxygen donor in GaP. Part II. Capture cross sections," *Phys. Rev.*, vol. B7, pp. 2486–2498, 1973.

[4] L. C. Kimerling, "Influence of deep traps on the measurement of free carrier distributions in semiconductors by junction capacitance techniques," *J. Appl. Phys.*, vol. 45, pp. 1839–1845, 1974.

[5] C. T. Sah and A. Neugroschel, "Concentration profiles of recombination centers in semiconductor junctions evaluated from capacitance transients," *IEEE Trans. Electron Devices*, vol. ED-13, pp. 1069–1074, 1976.

[6] H. Lefevre and M. Schulz, "Double correlation techniques (DDLTS) for the analysis of deep level profiles in semiconductors," *Appl. Phys.*, vol. 12, pp. 45–53, 1977.

[7] G. Goto, S. Yanagisawa, O. Wada, and H. Takanashi, "Determination of deep-level energy and density profiles in inhomogeneous semiconductors," *Appl. Phys. Lett.*, vol. 23, pp. 150–151, 1973.

[8] J. A. Pals, "Properties of Au, Pt, Pd and Rh levels in silicon measured with a constant capacitance technique," *Solid-State Electron.*, vol. 17, pp. 1139–1145, 1974.

[9] N. M. Johnson, D. J. Batelink, R. B. Gold, and J. F. Gibbons, "Constant capacitance DLTS measurement of defect density profiles in semiconductors," *J. Appl. Phys.*, vol. 50, pp. 4828–4833, 1979.

[10] G. L. Miller, "A feedback method for investigating carrier distributions in semiconductors," *IEEE Trans. Electron Devices*, vol. ED-19, pp. 1103–1108, 1972.

[11] M. Cardona, W. Paul, and H. Brooks, "Dielectric constant of Germanium and Silicon as a function of volume," *J. Phys. Chem. Solids*, vol. 8, pp. 204–206, 1959.

[12] C. T. Sah and V.G.K. Reddi, "Frequency dependence of the reverse biased capacitance of gold-doped silicon p^+/n step junction," *IEEE Trans. Electron Devices*, vol. ED-11, pp. 345–349, 1964.

[13] D. P. Kenneday, P. C. Murley, and W. Kleinfelder, "On the measurement of impurity atom distribution in silicon by the differential capacitance technique," *IBM J. Res. Devel.*, vol. 12, pp. 399–409, 1968.

[14] M. Nishida, "Depletion approximation analysis of the differential capacitance-voltage characteristics of an MOS structure with nonuniformly doped semiconductors," *IEEE Trans. Electron Devices*, vol. ED-26, pp. 1081–1085, 1979.

[15] M. Ruden, G. Spadini, H. Maes, W. Vandervorst, and R. Van Overstraeten, "Interpretation of $C\text{-}V$ measurements for determining the doping profile in semiconductors," *Solid-State Electron.*, vol. 23, pp. 65–71, 1980.

[16] W. C. Johnson and P. T. Panausis, "The influence of Debye length on the $C\text{-}V$ measurement of doping profiles," *IEEE Trans. Electron Devices*, vol. ED-18, pp. 965–973, 1971.

[17] H. G. Grimmeiss, L. A. Ledebo, and E. Meijer, "Capture from free carrier tails in the depletion region of junction barriers," *Appl. Phys. Lett.*, vol. 36, pp. 307–308, 1980.

[18] G. L. Miller, D. V. Lang, and L. C. Kimerling, "Capacitance transient spectroscopy," *Annu. Rev. Mat. Sci.*, vol. 7, pp. 377–447, 1977.

[19] C. T. Sah, "The equivalent circuit model in solid state electronics –Part I: The single energy level defect centers," *Proc. IEEE*, vol. 55, pp. 654–671, 1967.

[20] S. M. Sze, *Physics of Semiconductor Devices*. New York: Wiley, 1969.

[21] C. T. Sah, L. Forbes, L. L. Rosier, A. F. Tasch, Jr., and A. B. Tole, "Thermal emission rates of carrier at gold centers in silicon," *Appl. Phys. Lett.*, vol. 15, pp. 145–148, Sept. 1969.

Reprinted paper with permission from M.F. Li and C.T. Sah, Solid State Electronics,
Vol.25, pp.95–99 (1982). Copyright © 1982, Elsevier.

Solid-State Electronics Vol. 25, No. 2, pp. 95–99, 1982
Printed in Great Britain.

0038–1101/82/020095–05\$03.00/0
© 1982 Pergamon Press Ltd.

NEW TECHNIQUES OF CAPACITANCE–VOLTAGE MEASUREMENTS OF SEMICONDUCTOR JUNCTIONS†

MING-FU LI‡ and CHIH-TANG SAH‡

Department of Electrical Engineering, University of Illinois, Urbana, IL 61801, U.S.A.

(*Received* 11 *December* 1980; *in revised form* 2 *July* 1981)

Abstract—Two novel measurement techniques for C–V characteristics of a semiconductor junction are described which provide substantial improvement over previous methods. One is a new method to implement the constant-capacitance voltage transient experiment using the idea of a capacitance operational amplifier. The other is an improved method of quasi-static capacitance–voltage measurement using an integration scheme instead of the usual differentiation scheme, with substantial noise reduction and improvement of sensitivity.

1. A CAPACITANCE AMPLIFIER FEEDBACK LOOP FOR CONSTANT-CAPACITANCE VOLTAGE TRANSIENT (CCVT) MEASUREMENT

Recently, we reported a new method of computer-controlled trap concentration profile measurement of a semiconductor junction which overcame limitations of previous methods using experimentally measured correction for the edge region of a junction space charge layer[1]. This method employed two independent measurement techniques, both with improved and new methods. One of these two techniques is based on the CCVT idea first employed by Goto *et al.*[2] and Pals[3] with a new implementation to be described in this section. The other is based on the quasi-static method with its new implementation to be described in the next section.

In this section, we will describe the new method used to implement the CCVT measurement by a simple modification of a commerical 1 MHz capacitance meter (Boonton model 72A) which substantially improves the system accuracy.

Figure 1(a) is the schematic diagram of the method. Here, CB consists of a capacitance bridge, a one-MHz tuned amplifier with a gain of 10^2 to 10^3 and a phase-sensitive detector. The output d.c. voltage, V'_0, of CB is proportional to the difference of the differential input capacitances, C_R and C_M,

$$V'_0 = K_1(C_R - C_M).$$

DCVA is a d.c. voltage amplifier with a d.c. gain of about 10^5 within the original Boonton 72 capacitance meter. The circuits of CB and DCVA can be easily implemented by simple modifications of a Boonton 72 capacitance

meter. A complete description of the circuit is illustrated in the Appendix.

VCC in Fig. 1(a) is a diode whose capacitance is to be measured. It is treated as a voltage-capacitance converter. The combination of CB, DCVA and VCC is equivalent to a "Differential Input Capacitance Amplifier" to be denoted by DICA which is shown in Fig. 1(b). The small-signal transfer function of DICA can be expressed by

$$C_0 = K_2(C_R - C_M) \tag{1}$$

where C_0 is the output while C_R and C_M are the differential input of DICA and K_2 is the gain of the capacitance amplifier. The advantage of introducing the DICA idea here is that we can then use the existing theory of operational amplifiers to analyze the DICA performance. This analysis is given below.

Figure 2 is a "capacitance follower" configuration of the DICA. D_R is a reference diode which has little or no deep-level traps and whose bias voltage is varied by a computer-controlled programmable power supply. This reference diode is another V–C converter (varactor) and is treated as a capacitance input signal source, C_R.

The capacitance follower will track C_0 with C_R at all times to satisfy the constant capacitance condition during the transients of the CCVT measurements. The method of using a feed back loop for carrier concentration measurement can be traced back to Miller[4].

In order to realize a stable, precise and fast enough capacitance follower shown in Fig. 2, the following three conditions have to be satisfied.

(a) K_2 should approach infinity at d.c. This requirement is evident from eqn (1) since we have

$$C_M = C_R/[1 + K_2^{-1}]$$

when $C_M = C_0$. In our case, we have $K_2 > 10^7$ so that the error between C_M and C_R is less than 10^{-7}. This is much better than what is obtainable in an unmodified Boonton capacitance meter which is about 10^{-3}.

†This work was supported partially by the Rome Air Development Center (Contracts F19628-77-0138 and 19628-81-K-0016) and the National Science Foundation grant ECS-80-21114.

‡The authors are with the Solid State Electronics Laboratory of the University of Illinois, Urbana, IL 61801, U.S.A. LMF is on leave from the Graduate School of the University of Science and Technology of China, Peking and is supported by the Chinese Academy of Science.

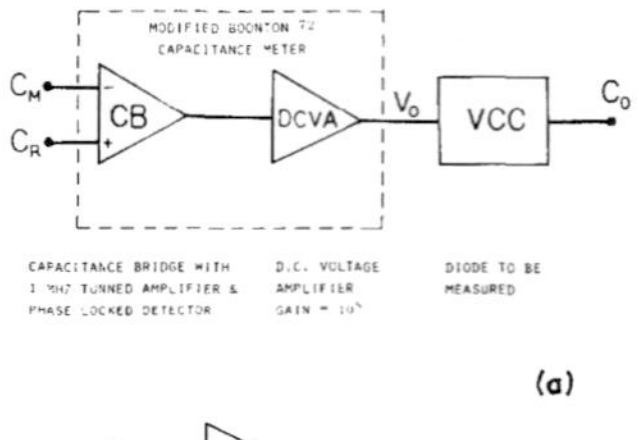

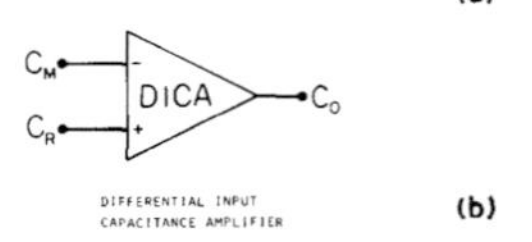

Fig. 1. The block diagram of the differential input capacitance amplifier. (a) CB = Capacitance Bridge with one-MHz tuned amplifier and phase locked detector. DCVA = d.c. Voltage Amplifier with 10^5 gain. VCC = Diode to be measured whose capacitance is $C_0 = C_0(V)$. (b) The differential input capacitance amplifier = DICA.

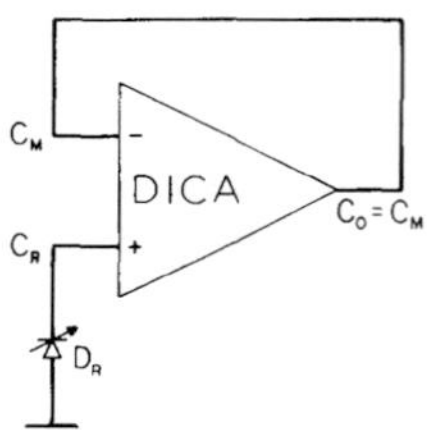

Fig. 2. DICA in the capacitance follower configuration controlled by a reference diode as the capacitance input signal source.

(b) K_2 is a function of frequency ω. It has to satisfy the Nyquist Criterion of stability and be fast enough for step transients[5]. This is accomplished by modifying the phase compensation networks in the Boonton 72 capacitance meter.

(c) The large signal transfer function,

$$C_0 = C_0(C_R, C_M) \qquad (2)$$

should be a monotonic function of C_M for any values of C_R, otherwise, the loop can cause latching to occur[6]. This latching problem can be explained using the dashed lines of the capacitance transfer characteristics, C_0 vs C_M, shown in Fig. 3. The input C_M and output C_0 must satisfy two relationships simultaneously. One is the transfer function given by eqn (2) and the dashed curves in Fig. 3. The other is $C_0 = C_M$ for the capacitance follower. When the transfer function is not a monotonic function of C_M, such as the dashed lines labeled TRAN' in Fig. 3(a), then we will have three intersection points, N, L and U as indicated in Fig. 3(a). Point N is the normal operating point which corresponds to $C_0 = C_R$.

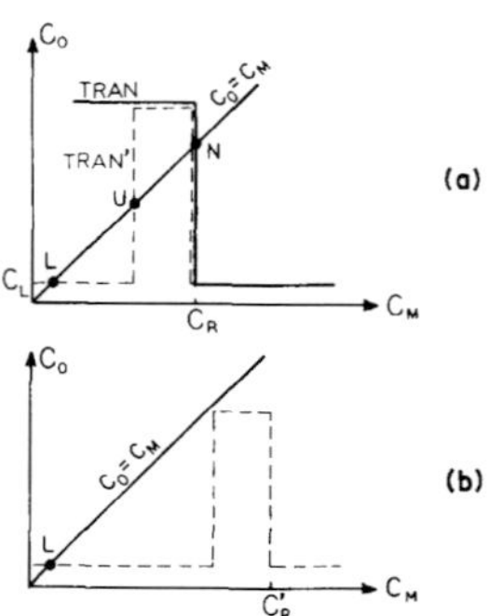

Fig. 3. Capacitance amplifier transfer characteristics used to explain the latch phenomena in a capacitance follower when there is an output inversion at low values of measured capacitance. C_0 = output capacitance and C_M = input or measured capacitance. TRAN' is the transfer characteristic before modification and TRAN is that after modification of the capacitance meter.

Point L is a latch state. It gives an output capacitance value, C_0, which is equal to the low saturation value of the system, C_L, and it is independent of the true capacitance C_R. The capacitance follower function of the system is destroyed if the system is latched at the state L. Whether the system will reside at the N or L state is determined by the initial condition. For example, when there is an accidental overflow of input signal, denoted by C_R, in Fig. 3(b), the system will move to the saturation point labeled L. After the overflow input signal is removed and the input returned to the normal value C_R, the output C_0 will still be latched at C_L and will not return to the state N. When the Boonton 72 capacitance meter is modified according to CB in Fig. 1, its transfer function is the dashed lines shown in Fig. 3. The inversion, as well as the saturation, are due to the saturation–inversion properties of the phase-locked detector used in the Boonton 72. We overcame this bistable problem by connecting two high speed diodes (in parallel and in opposite directions) in parallel with the feedback resistance of the tuned amplifier in the Boonton capacitance meter. These diodes clamp the tuned amplifier output to the ± 0.8 V range which prevents any inversion of the detector output. This modification results in a transfer function shown by the solid lines in Fig. 3(a), which is labeled TRAN.

We have already successfully implemented the computer-controlled CCVT experiments by the aforementioned modification which eliminated the latch problem. Results of C–V measurements and impurity profile measurements are reported elsewhere[1].

2. QUASI-STATIC C–V MEASUREMENT USING A CAPACITANCE INTEGRATOR

A new variation of the quasi-static capacitance measurement technique of semiconductor diodes is illustrated in Fig. 4. An electrometer (Keithley 610B) is

Capacitance-voltage measurements on junctions   97

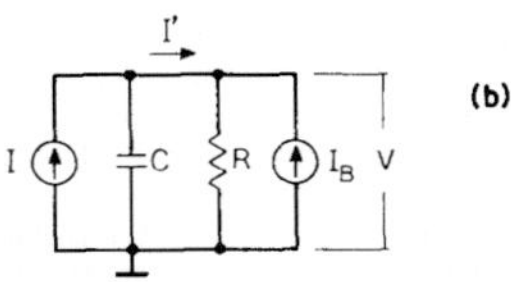

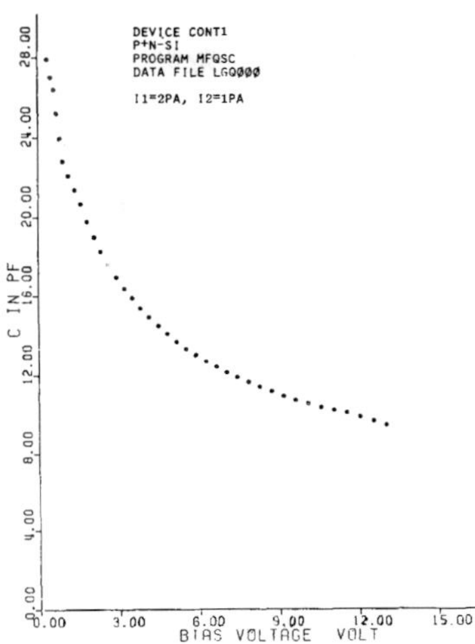

Fig. 5. Capacitance-voltage characteristics of a silicon $p+/n$ junction measured by the quasi-static capacitance method. The two input currents used were 1 pA and 2 pA.

Fig. 4. A new circuit for quasi-static capacitance-voltage measurements. (a) The circuit diagram and (b) the equivalent circuit showing the leakage currents. The K610B is at $10^{12}\,\Omega$, fast mode, guarded input.

used as a high gain and high input impedance d.c. amplifier. The Keithley 610B is set to the $10^{12}\,\Omega$ range and in fast mode with guard[7]. The semiconductor diode being measured is reverse biased by connecting it across the feedback path of the amplifier. A picoampere constant current source (Keithley 261) is used which can provide as low as 0.01 pA current. Point S at the input of the electrometer shown in Fig. 4 is the virtual ground since the K610B amplifier has voltage gain up to 10^5. The equivalent circuit of Fig. 4(a) is shown in Fig. 4(b) where $C(V)$ is the junction capacitance of the diode, $R(V)$ is the leakage resistance of the diode plus the measurement system, and I_B is the electrometer amplifier input current plus the internal current source of the ohmic range of the electrometer[7].

The capacitance to be measured is charged (or discharged) at two different currents so that at current I_1 we have

$$C(V_1)\frac{dV_1}{dt} = I_1 - I'(V_1) \qquad (3)$$

where $I'(V_1)$ is the sum of the leakage current through $R(V_1)$ of the diode and I_B. At I_2, we have

$$C(V_2)\frac{dV_2}{dt} = I_2 - I'(V_2). \qquad (4)$$

Subtracting (4) from (3), and setting $V_1 = V_2 = V$, we then obtain

$$C(V) = \frac{I_1 - I_2}{\dfrac{dV_1}{dt} - \dfrac{dV_2}{dt}} \qquad (5)$$

where both dV_1/dt and dV_2/dt are evaluated at the voltage V. The important result given by eqn (5) is that

the leakage current is cancelled so that any errors due to the leakage currents are eliminated.

Figure 5 shows the experimental C–V curve obtained by this method using a computer-controlled data acquisition and analysis system. The quasi-static C–V result agrees with the high-frequency C–V result very well in this diode which has no recombination and trapping centers in its junction space charge layer. It should be noted that in comparing the quasi-static and high-frequency (one MHz) results, the stray capacitances of the high-frequency and quasi-static measuring systems must be corrected carefully.

In comparison with the existing quasi-static method originally proposed by Kuhn[8], our method has the following unique characteristics.

(a) Our method employs a low-pass integration circuit while Kuhn's method employed a high-pass differential circuit. Thus, in our method, the noise at the very low currents is substantially reduced over Kuhn's method because the noise is largely averaged out by the integration circuit. In addition, the stability of the closed loop system is more easily established in an integration mode than in a differential mode since the feedback network in an integration mode is actually a phase lead network with phase lead within 90°[9], while in a differential mode it is just the opposite. The noise reduction is verified experimentally using a diode at $I = 1$ pA.

(b) In order to satisfy the quasi-static condition of measurement in semiconductor junctions, dV/dt must be made very small and the corresponding displacement current of the capacitance would also be low and become comparable with the leakage current of the diode and the

MING-FU LI and CHIH-TANG SAH

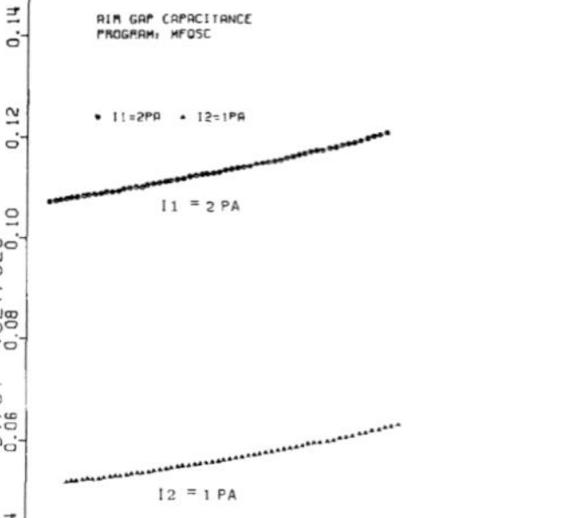

Fig. 6. The dV/dt vs voltage characteristics of a low loss airgap capacitance measured by the quasi-static capacitance method with input currents of 1 pA and 2 pA. The C–V characteristics are then obtained from eqn (5).

measurement system. If the leakage currents are not corrected, the capacitance results could be in large error. Our method is designed to eliminate the leakage current experimentally as indicated by eqn (5) where we took the data at two current levels. Figure 6 gives the data of dV_1/dt and dV_2/dt as a function of the bias voltage of a low-leakage and voltage-independent air-gap capacitor. The two sloped lines are in good agreement with eqns (3) and (4) if a constant leakage resistance of 5.4×10^{13} ohms is assumed. This example demonstrates the importance of leakage current on the measured capacitance value and the necessity to use a leakage current cancellation scheme such as that proposed here.

(c) In our scheme, a precise low current source is employed which is much easier to realize than a precise and ultra-slow voltage ramp required in the conventional quasi-static C–V implementations. In addition, the point S in our circuit in Fig. 4 is the virtual ground, thus, an extremely high output impedance of the current source is not needed.

Acknowledgement—Li Ming-Fu would like to extend his appreciation to Ms. Donna Stowe for her skilful and patient typing of the manuscript, and to other members of the Solid State Electronics Laboratory at the University of Illinois for their friendly assistance.

REFERENCES

1. Li Ming-fu and Sah Chih-Tang, A new method for the determination of dopant and trap concentration profiles in semiconductors. *IEEE Trans. Electron Dev.* **ED-29** (1982).

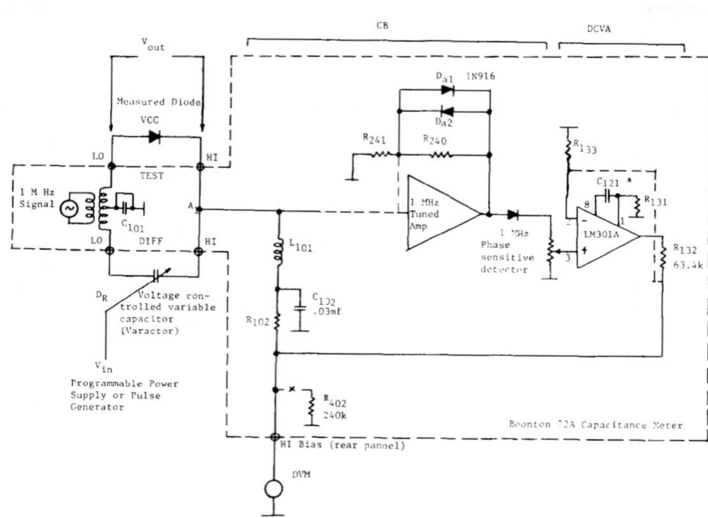

Fig. 7. Circuit diagram of modified Boonton 72A meter as a capacitance follower. The C121 is adjusted for fastest response without oscillation.

2. G. Goto, S. Yanagisawa, O. Wada and H. Taknashi, *Appl. Phys. Lett.* **23**, 150 (1973).
3. J. A. Pals, *Solid-St. Electron.* **17**, 1139 (1974).
4. G. L. Miller, *IEEE Trans. Electron Dev.* **ED-19**, 1103 (1972).
5. S. M. Shinners, *Modern Control System Theory and Application.* Addison-Wesley, Reading, Massachusetts (1972).
6. Li Ming-Fu and Chen Jin-Song, *J. China University Sci. Tech.* **8**, 119 (1978).
7. *Instruction Manual for Model 610B, 610BR Multi-range Electrometers.* Keithley Instruments, Inc., Cleveland, Ohio (1967).
8. M. Kuhn, *Solid-St. Electron.* **13**, 873 (1970).
9. J. G. Graeme, G. E. Tobey and L. P. Huelsman, *Operational Amplifiers: Design and Applications.* McGraw-Hill, New York (1971).
10. *Instruction Manual for Model 72A Capacitance Meter.* Boonton Electronics Corporation, Parsippany, NJ 07054, U.S.A. (1973).

APPENDIX

In this appendix we give a detailed description of modifying a Boonton 72A capacitance meter to a capacitance follower. Figure 7 is part of the circuit diagram of the Boonton 72A. Designations of components used are those given in the Boonton instrument manual [10]. Modifications made are as follows:

(1) Add two back-to-back diodes, D_1 and D_2, across the feedback resistor R_{240} to clamp the output of the 1 MHz tuned amplifier.

(2) Disconnect R_{132} from pin 2 of Op Amp LM301A, as shown by the dashed line in Fig. 7. The Op Amp is changed from its original closed loop configuration, with a gain of about 10, to an open loop configuration, with a gain of about 10^5.

(3) Connect R_{132} to R_{102} to complete the feedback loop. Disconnect R_{402} from the "HI Bias" terminal at the real panel of the Boonton 72A. Output of LM301A is then the power supply applied to measured diode VCC.

(4) Change C_{102} from 0.1 μF to 0.03 μF. Change C_{121} from 30 pF to 10^3 to 10^4 pF. $R_{132} + R_{102}$, C_{121} and C_{102} are the phase compensation components in our circuit.

(5) Short circuit C_{101}. Detach the bottom aluminum plate from the Boonton 72A chasis and keep the Boonton at least one foot

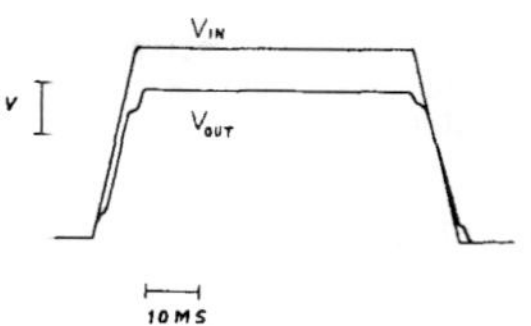

Fig. 8. Transient response of capacitance follower. Voltage waveform recorded from a HP-1200B oscilloscope.

away from the other metal chasis to avoid any additional coupling between the input and output.

Figure 8 is the waveform of the transient characteristics of the feedback system. We use a pulse generator HP8010A to generate a positive square voltage V_{in} as shown in Fig. 8 which is applied to the DIFF input in Fig. 7. The voltage response, V_{out}, across the measured diode, VCC, is also shown in Fig. 8 which was measured across the TEST terminal in Fig. 7 through a d.c. isolation network. The corresponding capacitance transient in this case is from 35 pF to 20 pF. The rise and fall times of the system are about 10 ms when the Boonton 72A C-meter is set to the 3 pF range. When the input, V_{in}, has a rise time faster than that shown in Fig. 8, large overshot and ringing appear in V_{out}. This oscillation can be suppressed by adjusting C_{121} so that the feedback loop has the fatest risetime without oscillation. For our experimental system, a C_{121} of 3000 pF was used.

Since the modified Boonton 72A C-meter is no longer operated in the capacitance meter mode, but instead it is operated as a capacitance amplifier, we can always use the most sensitive range to increase the loop gain and reduce noise, regardless of the value of the measured or unknown capacitance. However, it is easier to lock the system into the "capacitance follower" mode when the Boonton C-meter is set to the 30 pF range or less sensitive ranges. Thus, to begin a measurement, the Boonton is first set to the 30 pF range. Once the capacitance follower function becomes effective, the sensitivity is increased to the 3 pF range to give the higher system gain.

Solid State Communications, Vol. 48, No. 9, pp. 789–793, 1983.
Printed in Great Britain.

0038–1098/83 $3.00 + .00
Pergamon Press Ltd.

GLOBAL PREDICTIONS OF T_2 SYMMETRIC DEEP LEVEL WAVEFUNCTIONS IN SEMICONDUCTORS

Ming-fu Li

Graduate School, University of Science and Technology of China, Beijing, China

and

De-qiang Mao and Shang-yuan Ren

Department of Physics, University of Science and Technology of China, Hefei, China

(*Received* 10 *June* 1983 *by F. Bassani*)

Numerical results of T_2 symmetric SP^3 bonded deep level wavefunctions
due to short range defect potentials in Si, Ge, GaAs and InP are presented.
The general features of defect wavefunctions are insensitive to either the
band structure of the host or defect energy level. The total occupation
probability of wavefunction located on 0 . 1 . 2 shells around the defect
center is about 60–85%. This part of wavefunction may be expressed in
several simple symmetric combinations of SP^3 hybrid orbitals. The rest
part of the wavefunction extends diversely over a wide range of space.

IN [1], SUCCESSFUL CALCULATIONS of A_1
symmetric SP^3 bonded deep level wavefunctions for substitutional impurities in Si, GaP and GaAs were made.
The calculations are based on Koster–Slater Green's
function equation [2] and central cell defect potential
approximation [3]. Very good agreements are obtained
between the calculated results and the ESR experiments
for Si in [4, 5] and for GaAs and GaP in [6], revealing
the theoretical model to be correctly abstracting the
most important physics in the deep level problems. In
[7], the method was extended to T_2 symmetric deep
level wavefunctions in Si. In this paper, we describe the
results of T_2 symmetric deep level wavefunctions calculated for Ge, GaP, GaAs and InP. The results for Si are
also listed for comparison.

The energy E and wavefunction of deep levels in
semiconductors are determined by [8]

$$\det |1 - \hat{G}^0(E)\hat{V}| = 0, \tag{1}$$

$$\psi = \hat{G}^0(E)\hat{V}\psi \tag{2}$$

with normalization condition

$$-\left\langle \psi \left| \hat{V} \frac{\partial \hat{G}^0}{\partial E} \hat{V} \right| \psi \right\rangle = 1, \tag{3}$$

here $G^0(E)$ is the Green's function of the host crystal, V
is the defect potential. A nearest neighbour SP^3S^* ten
states model of Vogl *et al.* [9] is used for the host
Hamiltonian. The defect wavefunction is expanded in

$$|\psi_m^l\rangle = \sum_{n,k} |l, k, n, m\rangle\langle l, k, n|\psi\rangle, \tag{4}$$

here $|l, k, n, m\rangle$ are symmetric combinations of SP^3

hybrid orbitals. l stands for the irreducible representation of the defect level and is T_2 in this paper. k
indexes the kth shell around the point defect. n indexes
the nth representation and m is the partner index of the
irreducible representation.

When the central cell defect potential approximation is used [3], i.e.,

$$\hat{V} = \sum_l |l, 0, 1\rangle V_l\langle l, 0, 1|,$$

equations (1)–(3) reduce to [1]

$$\langle l, 0, 1|\hat{G}^0(E)|l, 0, 1\rangle = V_l^{-1}, \tag{5}$$

$$\langle l, k, n|\psi\rangle = \langle l, k, n|\hat{G}^0|l, 0, 1\rangle\langle l, 0, 1|\psi\rangle/$$
$$/\langle l, 0, 1|\hat{G}^0|l, 0, 1\rangle, \tag{6}$$

$$|\langle l, 0, 1|\psi\rangle|^2 = dE/dV_l. \tag{7}$$

Equation (5) shows that the defect potential is only
described by one parameter for definite symmetric type
wavefunction and is one–one correspondence to the
energy level E. Table 1 lists the group theoretical analysis of T_d symmetric irreducible representations of defect
wavefunctions in equation (4). The main results of wavefunctions in equations (6) and (7) for Si, Ge, GaP, GaAs
and InP are shown in Figs. 1 and 2.

The general features of the defect wavefunctions are
insensitive to the band structure of the host, and the
energy level position in the gap. The most important
part (part I) of the wavefunction is ascribed by one of all
irreducible representations with the following form

$$|T_2, 1, 1, 1\rangle = \frac{1}{\sqrt{4}}(|1, 5\rangle_1 + |2, 6\rangle_1 - |3, 7\rangle_1 - |4, 8\rangle_1),$$

Table 1. T_d symmetric irreducible representations of defect wavefunctions in equation (4) for 0–10th shells in diamond and zinc-blende structures

Shell type	Shell number k	Number of lattice sites	T_d symmetric representations of defect wavefunctions in equation (4)	Number of A_1 representation	Number of T_2 representation
A	0	1	$A_1 + T_2$	1	1
B	1, 7_2	4	$A_1 + T_2$	2	3
			$A_1 + E + T_2 + T_2 + T_1$		
C	4	6	$2 \times (A_1 + E + T_2 + T_2 + T_1)$	2	4
D	2, 3, 5, 6_1,	12	$2 \times (A_1 + E + T_2 + T_2 + T_1)$	3	7
	6_2, 7_1, 8		$A_1 + A_2 + E + E + T_2 + T_2 + T_2 + T_1 + T_1 + T_1$		
E	9, 10	24	$4 \times (A_1 + A_2 + E + E + T_2 + T_2 + T_2 + T_1 + T_1 + T_1)$	4	12

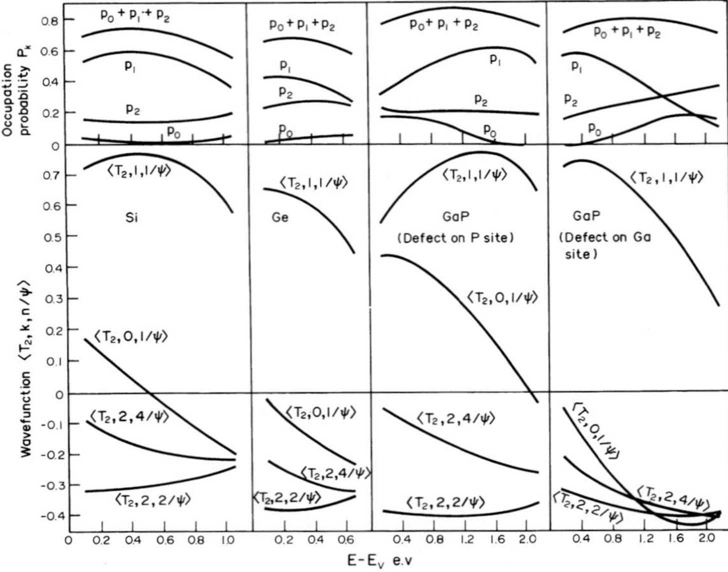

Fig. 1. Most important parts of wavefunctions $\langle T_2, k, n|\psi \rangle$ and occupation probabilities P_k of three nearest neighbour shells vs energy for Si, Ge and GaP. The vacancy level is the energy of $\langle T_2, 0, 1|\psi \rangle = 0$.

$$|T_2, 1, 1, 2\rangle = \frac{1}{\sqrt{4}}(|1, 5\rangle_1 - |1, 6\rangle_1 + |3, 7\rangle_1 - |4, 8\rangle_1),$$

$$|T_2, 1, 1, 3\rangle = \frac{1}{\sqrt{4}}(|1, 5\rangle_1 - |2, 6\rangle_1 - |3, 7\rangle_1 + |4, 8\rangle_1), \tag{8}$$

$|j, \mu\rangle_k$ in left side of equation (8) consist of the host SP^3 hybrid orbitals for four nearest neighbour sites $j = 1, 2, 3, 4$ in $k = 1$ shell, $\mu = 5, 6, 7, 8$ index four directions of the hybrid orbitals. Equation (8) is the T_2 symmetric combination of four hybrid orbitals located on four nearest neighbour sites and point toward the defect center. Equation (8) is in fact the defect molecule model wavefunction of a vacancy suggested by Coulson and Kearsly [10]. The occupation probability $|\langle T_2, 1, 1|\psi\rangle|^2$ is as large as 20–60% for most energy range in most of the host crystals. The second major part (part II) of the wavefunction consists of $|T_2, 2, n, m\rangle$ for $n = 2$ and 4, and may be expressed by

$$|T_2, 2, 2, 1\rangle = \frac{1}{\sqrt{8}}(|1, 4\rangle_2 + |2, 3\rangle_2 - |3, 2\rangle_2 - |4, 1\rangle_2$$

$$- |11, 2\rangle_2 - |10, 1\rangle_2 + |9, 3\rangle_2 + |12, 4\rangle_2),$$

$$|T_2, 2, 2, 2\rangle = \frac{1}{\sqrt{8}}(|5, 2\rangle_2 - |6, 3\rangle_2 - |7, 1\rangle_2 + |8, 4\rangle_2$$

$$+ |3, 2\rangle_2 - |2, 3\rangle_2 + |1, 4\rangle_2 - |4, 1\rangle_2),$$

$$|T_2, 2, 2, 3\rangle = \frac{1}{\sqrt{8}}(|9, 3\rangle_2 - |10, 1\rangle_2 + |11, 2\rangle_2$$

$$- |12, 4\rangle_2 - |7, 1\rangle_2 + |6, 3\rangle_2 + |5, 2\rangle_2 - |8, 4\rangle_2); \tag{9}$$

$$|T_2, 2, 4, 1\rangle = \frac{1}{\sqrt{4}}(|5, 2\rangle_2 + |7, 1\rangle_2 - |8, 4\rangle_2$$

$$- |6, 3\rangle_2),$$

$$|T_2, 1, 4, 2\rangle = \frac{1}{\sqrt{4}}(|9, 3\rangle_2 - |11, 2\rangle_2 - |12, 4\rangle_2$$

$$+ |10, 1\rangle_2),$$

$$|T_2, 2, 4, 3\rangle = \frac{1}{\sqrt{4}}(|1, 4\rangle_2 - |3, 2\rangle_2 + |4, 1\rangle_2$$

$$- |2, 3\rangle_2). \tag{10}$$

They are symmetric combinations of the host SP^3 hybrid orbitals located on twelve sites $j = 1$–12 of $k = 2$ shell. $\mu = 1, 2, 3, 4$ index four orbital directions in opposite to

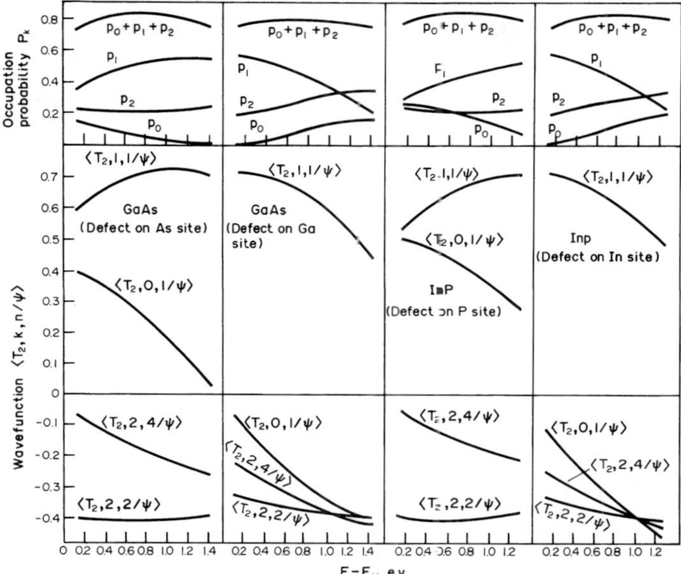

Fig. 2. The same key as Fig. 1 but for GaAs and InP.

$\mu = 5, 6, 7, 8$. There is one hybrid orbtial for one site and point toward the sites of $k = 1$ shell. They are dangling in character. The occupation probability of this part of wavefunction $|\langle T_2, 2, 2|\psi \rangle|^2 + |\langle T_2, 2, 4|\psi \rangle|^2$ is comparatively larger for III–V compound and Ge than for Si. For III–V compound, when the defect center is located on the cation site and the energy is close to the conduction band minimum, the occupation probability of part II wavefunction may even exceed that of part I. The third major part (part III) of the wavefunction is

$$|T_2, 0, 1, 1\rangle = \frac{1}{\sqrt{4}}(|0, 1\rangle_0 + |0, 2\rangle_0 - |0, 3\rangle_0 - |0, 4\rangle_0),$$

$$|T_2, 0, 1, 2\rangle = \frac{1}{\sqrt{4}}(|0, 1\rangle_0 - |0, 2\rangle_0 + |0, 3\rangle_0 - |0, 4\rangle_0),$$

$$|T_2, 0, 1, 3\rangle = \frac{1}{\sqrt{4}}(|0, 1\rangle_0 - |0, 2\rangle_0 - |0, 3\rangle_0 + |0, 4\rangle_0). \tag{11}$$

Equation (11) is the T_2 symmetric combinations of four SP^3 hybrid orbitals of the impurity located on the defect center and point toward the four nearest neighbour lattice sites. Equations (11) and (8) depict a picture of four bonding or antibonding hybrid orbital pairs between the on site and four nearest neighbour sites. When the defect energy is higher than the vacancy level, it is antibonding case, since there are opposite signs between $\langle T_2, 0, 1|\psi \rangle$ and $\langle T_2, 1, 1|\psi \rangle$; while the defect level is lower than the vacancy level, it is bonding case. In [7], we have mentioned for the defect wavefunctions in Si, the four hybrid orbitals of equation (8) are quasi-dangling since $\langle T_2, 0, 1|\psi \rangle$ are much smaller than $\langle T_2, 1, 1|\psi \rangle$. This is not always valid for zinc-blende structures. As shown in Figs. 1 and 2, $\langle T_2, 0, 1|\psi \rangle$ may be competitive with $\langle T_2, 1, 1|\psi \rangle$ when the defect center is located on the cation (or anion) site and the energy is close to the conduction band minimum (or valance band maxima).

Figure 3 is a typical diagram of wavefunction probabilities vs different shells. Here, the occupation probability P_k for the kth shell is defined by

$$P_k = \sum_n |\langle T_2, k, n|\psi \rangle|^2. \tag{12}$$

For the most cases, there is a strong peak at P_1 (curves 1, 2 of Fig. 3). The occupation probability for the nearest

Table 2. T_2 *symmetric vacancy levels in various semiconductors. Energy is measured in the unit of* eV *with zero energy for valence band maximum* E_v

	Ours	[6, 11]	[12]	[13]	[14]
Si	0.51	0.68			
Ge	0.06	0.11			
GaP(V_p)	2.06		1.54		
GaP(V_{Ga})	0.04		0.24	0.15	
GaAs(V_{As})	1.47	1.47	0.95	†	†
GaAs(V_{Ga})	− 0.08*	0.02	0.05	0.16	0.04
InP(V_p)	†		1.50		
InP(V_{In})	*		0.54		

* Resonance state in valence band.

† Resonance state in conduction band.

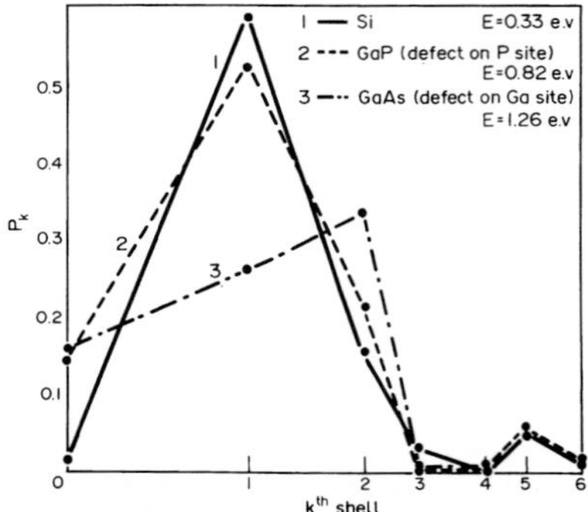

Fig. 3. The occupation probability P_k vs kth shell. For most cases, a strong peak is located at $k = 1$ shell. When the defect center is located on the cation site and the defect energy is close to the conduction band, the peak transfer to $k = 2$ shell (curve 3).

three shells $P_0 + P_1 + P_2$ is around 60–85%. The rest part of the wavefunction extends diversly over a wide range of distance in comparing with the lattice constant.

The vacancy level is determined by [11]

$$\langle T_2, 0, 1|\hat{G}^0(E)|T_2, 0, 1\rangle = 0, \tag{13}$$

Table 2 is the results for various materials. The results obtained by other authors are also listed. In contrary to the case of wavefunctions, the vacancy level determined by equation (13) is very sensitive to the detailed band structure of the host since the resulting Green's function is the sum of a lot of highly compensated contributions from different bands. The diverse results obtained from different authors are probably due to the tiny differences between the host band structures the

different authors used. Nevertheless, there are common characteristics between the results of various authors. For III—V compounds, for anion site vacancies the defect levels are near the conduction band while for cation site vacancies, the level are near the valance band.

In summary, we give for the first time the global predictions of T_2 symmetric defect level wavefunction behaviour in various energy range for various semiconductors. Contrary to A_1 symmetric states, unambiguous ESR experimental data are not available for direct comparison with calculated T_2 symmetric deep level wavefunctions. However, the results predicted are comparable and consistent with more elaborate self-consistent theoretical calculations of vacancies [8, 15] and H, Zn and Al [8] in Si, which may be attributed to special cases of our predictions. The bonding picture of LCAO defect wavefunctions depicted by equations (8)—(10) is in fact an extension of early work of Coulson and Kearsly [10] of vacancies but established on a more rigorous and quantitative way.

Acknowledgement — The authors gratefully acknowledge Miss Y.Q. Han for her assistance in preparing the manuscript.

REFERENCES

1. Shang-yuan Ren, Wei-min, Hu, O.F. Sankey & J.D. Dow, *Phys. Rev.* **B26**, 951 (1982); Shang-yuan Ren, (in press).
2. G.F. Koster & J.C. Slater, *Phys. Rev.* **96**, 1208 (1954).
3. H.P. Hjalmarson, P. Vogel, D.J. Wolford & J.D. Dow, *Phys. Rev. Lett.* **44**, 810 (1980).
4. G.W. Ludwig, *Phys. Rev.* **137**, A1520 (1965).
5. H.G. Grimmeis, E. Janzen, H. Ennen, O. Schirmer, J. Schneider, R. Worner, C. Holm, E. Sirtl & P. Wagner, *Phys. Rev.* **B24**, 4571 (1981).
6. J. Schneider & U. Kaufmann, *Defect and Radiation Effects in Semiconductors, 1980*, Inst. Phys. Conf. Ser. No. 59 p. 55 (1981).
7. Ming-fu Li, Shang-Yuan Ren & De-quiang Mao, (to be published in *Acta Physica Sinica*).
8. J. Bernholc, N.O. Lipari & S.T. Pantelides, *Phys. Rev.* **B21**, 3545 (1980); J. Bernholc, N.O. Lipari, S.T. Pantelides & M. Scheffler, *Phys. Rev.* **B26**, 5706 (1982).
9. P. Vogl, H.P. Hjalmarson & J.D. Dow, (to be published in *J. Phys. Chem. Solids*).
10. C.A. Coulson & M.J. Kearsly, *Proc. R. Soc.* **A241**, 433 (1957).
11. J. Bernholc, S.T. Pantelides, *Phys. Rev.* **B18**, 1780 (1978).
12. D.N. Talwar & C.S. Ting, *Phys. Rev.* **B25**, 2660 (1982).
13. M. Jaros & S. Brand, *Phys. Rev.* **B14**, 4497 (1976).
14. Jianbai Xia, *Chinese Journal of Semiconductors*, **4**, 1 (1983).
15. G.A. Baraff & M. Schluter, *Phys. Rev.* **B19**, 4965 (1979).

Reprinted with permission from M.F. Li, J.X. Chen, Y.S. Yao and G. Bai, J. Appl. Phys. Vol.58, pp.2599–2602 (1985). Copyright 1985, American Institute of Physics.

Au acceptor levels in Si under pressure

Ming-fu Li
Graduate School, University of Science and Technology of China, Beijing, China

Jian-xin Chen
Department of Electronics, Beijing Polytechnic University, Beijing, China

Yu-shu Yao
Institute of Physics, Academia Sinica, Beijing, China

Guang Bai
Graduate School, University of Science and Technology of China, Beijing, China

(Received 21 November 1984; accepted for publication 20 March 1985)

The hydrostatic pressure coefficient of Au acceptor levels E_T in Si was measured by transient capacitance method. Under the pressure range of 0–8 kbar, the pressure coefficient $\partial(E_c - E_T)/\partial P = -1.9$ meV/kbar. The electron capture cross section of Au acceptor centers is independent of pressure within experimental accuracy. For defect levels with defect potential of T_d symmetry, the uniaxial stress coefficient $\partial(\bar{E}_c - \bar{E}_T)/\partial F$ is isotropic and equal to one-third of corresponding hydrostatic pressure coefficient. By comparing the present result of hydrostatic pressure coefficient with the uniaxial stress coefficient reported by X. C. Yau, G. G. Qin, S. R. Zeng, and M. H. Yuan [Acta Phys. Sin. **33**, 377 (1984)], one concludes that the defect potential is far from T_d symmetry. Therefore, the Au acceptor levels are unlikely to have been originated by simple gold substitutional or interstitial configuration in Si.

I. INTRODUCTION

The deep levels in gold-doped silicon have been widely studied,[1-9] while the microstructure of gold-related deep levels is still unknown, with only some speculations giving rise to controversy.[8,10-13]

In this paper, we give a criterion to identify the microstructural symmetry properties of deep centers by combining hydrostatic pressure and uniaxial stress coefficient measurements. A careful hydrostatic pressure experiment for Au acceptor levels in Si was taken. The unaxial stress coefficient is quoted from Ref. 14. By comparing these two experimental results, a decisive conclusion about the symmetry property of Au acceptor centers in Si is obtained.

II. HYDROSTATIC PRESSURE EXPERIMENT

A. Principle

The emission rate e_n of electrons for Au acceptor levels in Si satisfies the following well-known relation:

$$e_n = g\sigma_n v_n N_c \exp[-(E_c - E_T)/kT]. \tag{1}$$

Here g is the degenerate factor, σ_n is the electron capture cross section, v_n is the average thermal velocity of electrons, N_c is the effective density of states at the bottom of the conduction band E_c, and E_T is the energy of Au acceptor levels. By neglecting the temperature dependence of effective mass, we have, from Eq. (1),[15,16]

$$\frac{\partial(E_c - E_T)}{\partial P} = -kT\left(\frac{\partial}{\partial P}(\ln e_n) - \frac{\partial}{\partial P}(\ln \sigma_n)\right). \tag{2}$$

By measuring the pressure dependence of e_n and σ_n, we obtain the hydrostatic pressure coefficient $\partial(E_c - E_T)/\partial P$ by Eq. (2).

B. Sample preparation

Silicon slices with $N_d = 2 \times 10^{16}$ phosphorus/cm^3 were used. The back sides of the wafers were heavily doped with phosphorus for good ohmic contact. Abrupt p^+n junctions were fabricated with $\phi = 600\,\mu$m. The gold was then evaporated on the back of the wafers and diffused at 940 °C for 20 min. Typical Au acceptor peak was detected by DLTS measurement and the Au concentration was estimated about 1×10 cm^{-3}.

C. High-pressure cell

The sample was placed in a high-pressure cell presurized with silicon oil. The temperature stability is less than ± 0.1 K. By Eq. (1),

$$\frac{de_n}{e_n} = \frac{(E_c - E_T)}{kT}\left(\frac{dT}{T}\right), \tag{3}$$

for $E_c - E_T = 0.54$ eV, $T = 249.5$ K and $dT = 0.1$ K, $de_n/e_n \approx 1\%$. Therefore, the temperature stabilization is vitally important for precise e_n measurement.

D. The measurements

The emission rate of electrons e_n is determined by measuring the capacitance transient signal at constant temperature.[17] Pulses were applied to the sample by a HP811 6A pulse generator. The capacitance was measured by a Boonton-72 meter. The transient signal was recorded by a HP3456A DVM and transmitted to a HP87 computer. The experimental procedure was completely controlled by the computer with high reproducibility and reliability. 20 repetitive transient signals were taken and averaged within a half minute. The data were then manipulated and plotted by the computer.

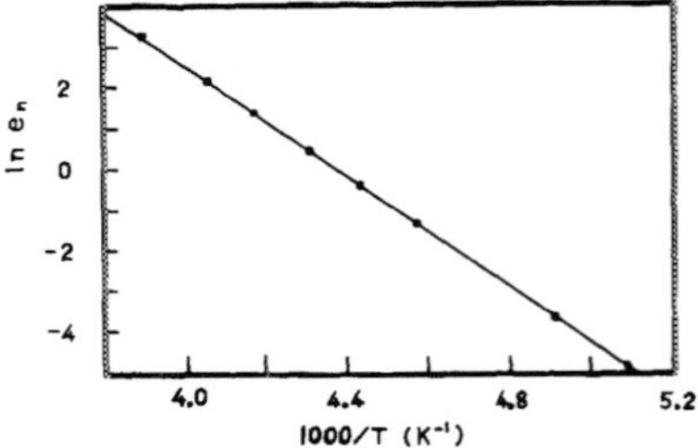

FIG. 1. Relation of $\ln e_n$ vs $1/T$ for Au acceptor levels in Si. By fitting $\ln (e_n/T^2) \sim 1/T$ to a straight line, the corresponding $E_c - E_T = 0.540$ eV. By fitting $\ln e_n \sim 1/T$ to a straight line, the corresponding $E_c - E_T = 0.578$ eV.

III. EXPERIMENTAL RESULTS

Figure 1 is the relation between e_n and temperature T under zero pressure for the Au acceptor levels in Si. The experimental points of $\ln(e_n/T^2) \sim 1/T$ least-square fitting to a straight line by the computer gives the energy level of $E_c - E_T = 0.540$ eV, which is consistent with the currently accepted value for Au acceptors in Si.[4,5,8,9] Figure 2 shows the raw data points of capacitance transient $\ln \Delta C$ versus time at $T = 249.5$ K under different hydrostatic pressure. Straight lines were fitted for various pressure and e_n were obtained from the slopes of the straight lines.[18] The relation between $\ln e_n$ versus pressure is plotted in Fig. 3. By least-square fit to a straight line for these points, we obtain $\partial /\partial p \ln e_n = 0.088$ kbar^{-1}. Figure 4 is the relation of capacitance transient amplitude $\Delta C_T(0)$ versus the filling pulse width τ to determine the electron capture cross section σ_n. The experimental points deviate from a straight line since the carrier tail region effect is essential for low bias voltage measurement.[19] The data in Fig. 4 and thus σ_n is pressure independent within experimental accuracy. By substituting the value of $\partial /\partial p \ln e_n$ and neglecting the second term in Eq. (2), we obtain

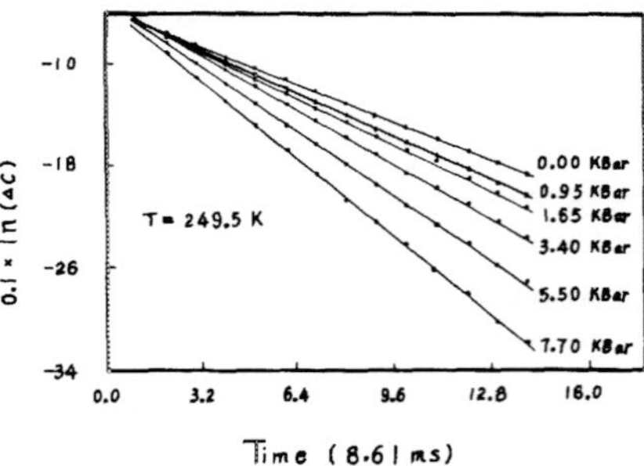

FIG. 2. Capacitance transient $\ln \Delta C$ vs time at $T = 249.5$ K under different hydrostatic pressure. The experimental points are the raw data taken from HP3456A DVM with 8.61 ms time interval between two consecutive DVM readings.

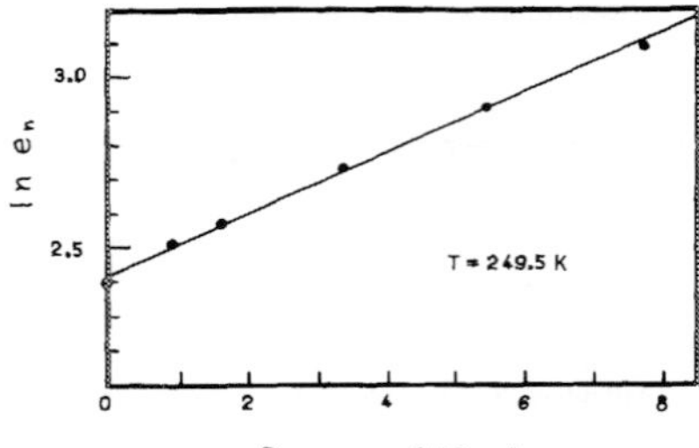

FIG. 3. The relation of electron emission rate e_n vs hydrostatic pressure p for Au acceptor levels in Si.

$$\frac{\partial (E_c - E_T)}{\partial P} \simeq -1.9 \text{ meV/kbar}. \tag{4}$$

IV. DISCUSSION

In the Appendix, a criterion is derived for testing the symmetric property of the defect potential. In cubic semiconductors, for defect levels with defect potential of T_d symmetry, the uniaxial stress coefficient $\partial (\bar{E}_c - \bar{E}_T)/\partial F$ [or $\partial (\bar{E}_T - \bar{E}_v)/\partial F$] is isotropic and equal to one-third of the corresponding hydrostatic pressure coefficient:

$$\frac{\partial (\bar{E}_c - \bar{E}_T)}{\partial F} = \left(\frac{1}{3}\right)\frac{\partial (E_c - E_T)}{\partial P}. \tag{5}$$

Here $\bar{E}_c$ and $\bar{E}_T$ are average energies of E_c and E_T under splitting, as defined by Eq. (A1). Therefore, it is significant to compare the hydrostatic pressure coefficient of this work with the uniaxial stress coefficient reported by Yao et al.[14]

The stress coefficient obtained by Yao et al. is of the magnitude of $\partial (E_{c1} - \bar{E}_T)/\partial F$, here E_{c1} is the minimum energy after splitting of the bottom of the conduction band under stress. By Herring and Vogl's deformation potential theory,[20] for uniaxial stress of (100) orientation,

$$\frac{\partial (\bar{E}_c - \bar{E}_T)}{\partial F_{100}} = \frac{\partial (E_{c1} - \bar{E}_T)}{\partial F_{100}} + \frac{1}{3}\Xi_u(S_{11} - S_{12}), \tag{6}$$

where Ξ_u is the shearing deformation potential constant,

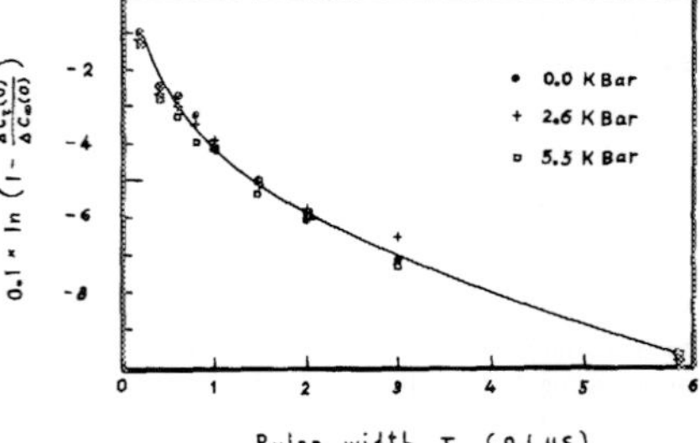

FIG. 4. The capacitance transient amplitude $\Delta C_r(0)$ vs filling pulse width for Au acceptor levels in Si under different hydrostatic pressure.

TABLE I. Comparison of uniaxial stress coefficients and hydrostatic pressure coefficient of Au acceptor levels in Si.

Ξ_u (eV)	$\dfrac{\partial(E_{c1}-\bar{E}_t)}{\partial F_{100}}$ (meV/kbar)	$\dfrac{\partial(\bar{E}_c-\bar{E}_T)}{\partial F_{100}}$ (meV/kbar)	$\dfrac{\partial(\bar{E}_c-\bar{E}_T)}{\partial F_{111}}$ (meV/kbar)	$\dfrac{\partial(E_c-E_T)}{\partial P}$ (meV/kbar)
9.2	-5.8	-2.8	-0.3	-1.9
11.4	-5.3	-1.6		

[a] Reference 14.

and S_{11} and S_{12} are compliance coefficients. For uniaxial stress of (111) orientation, E_c does not split, and thus

$$\frac{\partial(\bar{E}_c-\bar{E}_T)}{\partial F_{111}}=\frac{\partial(E_{c1}-\bar{E}_T)}{\partial F_{111}}. \tag{7}$$

We take $S_{11}=0.76\times10^{-12}$ cm^2 dyn and $S_{12}=-0.214\times10^{-12}$ cm^2/dyn.[21] The value of Ξ_u is quite diverse from different experiments. We use two typical values of 9.2 or 11.4 eV, respectively, according to Ref. 14. The final results of uniaxial stress coefficient, measured in Ref. 14 and reevaluated by Eqs. (6) and (7), are listed in Table I for comparison to our hydrostatic pressure coefficient. By Table I, the results are far from satisfying Eq. (5). Thus, the defect potential of Au acceptor levels in Si is far from T_d symmetry.

For a long time, experiments on the diffusion of gold into Si have led to the belief that electrically active gold sits on substitutional sites.[10] Suspicion of this assumption has been stimulated by recent theoretical and experimental results. Van Vechten et al., by thermodynamic analysis, proposed that Au acceptor level is a complex of interstitial Au and a vacancy rather than a simple substitutional impurity.[12] Lang et al.[8] argued that Au deep levels in Si are associated with some sort of complex structure. Their argument was based on a very wide variety of experimental analyses, although part of their assertions were refuted by later experiments of Ledebo et al.[26] and Morante.[27]

From the present work, a decisive conclusion is obtained that the defect potential of Au acceptor levels in Si is far from T_d symmetry. This gives a direct support to exclude the single substitutional or interstitial impurity of Au acceptor levels in Si.

ACKNOWLEDGMENTS

We would like to thank Professor Shou-an Hu for permission to use the high-pressure facility in his lab, S. Y. Ren and D. Q. Mao for theoretical discussion, and X. S. Wu and Y. Q. Han in our lab for technical assistance. M. F. Li would also like to extend his appreciation to Professor C. T. Sah for kindly providing him the original data of Ref. 4 for calibrating the system.

APPENDIX A

In this appendix, we derive the criterion of Eq. (5) for testing the symmetry property of defect potential in cubic semiconductors. The energy levels of defective semiconductors with common interests, such as the top of valence bands, the bottom of conduction bands and the bound states in the gap induced by defects are, in general, degenerate. We introduced the average of the splitting energies under stress $E_s^{(i)}$, $i=1,2,...n$, of originally degenerate level E_s:

$$\bar{E}_s=\sum_{i=1}^{n}E_s^{(i)}/n. \tag{A1}$$

In linear approximation, the average shift under stress tensor ϵ may be written as

$$\Delta\bar{E}_s=\frac{\partial\bar{E}_s}{\partial\epsilon_{xx}}\epsilon_{xx}+\frac{\partial\bar{E}_s}{\partial\epsilon_{yy}}\epsilon_{yy}+\frac{\partial\bar{E}_s}{\partial\epsilon_{zz}}\epsilon_{zz}$$
$$+\frac{\partial\bar{E}_s}{\partial\epsilon_{xy}}\epsilon_{xy}+\frac{\partial\bar{E}_s}{\partial\epsilon_{xz}}\epsilon_{xz}+\frac{\partial\bar{E}_s}{\partial\epsilon_{yz}}\epsilon_{yz}. \tag{A2}$$

For Hamiltonian with T_α group symmetry,[28] it is trivial to verify

$$\frac{\partial\bar{E}_s}{\partial\epsilon_{xx}}=\frac{\partial\bar{E}_s}{\partial\epsilon_{yy}}=\frac{\partial\bar{E}_s}{\partial\epsilon_{zz}},$$
$$\frac{\partial\bar{E}_s}{\partial\epsilon_{xy}}=\frac{\partial\bar{E}_s}{\partial\epsilon_{xz}}=\frac{\partial\bar{E}_s}{\partial\epsilon_{yz}}=0. \tag{A3}$$

We should emphasize that the key is to introduce the average of splitting energies. Equations (A3) are not valid for separate energy state of a degenerate level. A separate energy state of a degenerate level is not invariant under symmetry operation of group of Hamiltonian, while the configuration of splitting energies as a whole is invariant.

From Eq. (A3), Eq. (A2) is reduced to

$$\Delta\bar{E}_s=\frac{\partial\bar{E}_s}{\partial\epsilon_{xx}}\,\mathrm{Tr}\,\epsilon. \tag{A4}$$

Tr is the trace of tensor ϵ and is invariant under rotation. For uniaxial stress F, $\mathrm{Tr}\,\epsilon=F$ and for hydrostatic pressure P, $\mathrm{Tr}\,\epsilon=3P$. Thus, from Eq. (A4) we have

$$\frac{\partial\bar{E}_s}{\partial F}=\frac{1}{3}\frac{\partial E_s}{\partial P}. \tag{A5}$$

That is to say, the uniaxial stress coefficient of the average energy $\bar{E}_s$ is isotropic and equal to 1/3 of hydrostatic pressure coefficient of the same level.

For defect states in cubic semiconductors, let $\hat{H}=\hat{H}_0+\hat{V}$, where $\hat{H}_0$ is the Hamiltonian of the host and $\hat{V}$ is the defect potential. For defect potential with T_α group symmetry, we have, from Eq. (A5)

$$\frac{\partial(\bar{E}_c-\bar{E}_T)}{\partial F}=\left(\frac{1}{3}\right)\frac{\partial(E_c-E_T)}{\partial P},$$
$$\frac{\partial(\bar{E}_T-\bar{E}_v)}{\partial F}=\left(\frac{1}{3}\right)\frac{\partial(E_T-E_v)}{\partial P}. \tag{A6}$$

Applying Eq. (A6) to the case of Au accpetor levels in Si, the isotropic property of $\partial(\bar{E}_c-\bar{E}_T)/\partial F$ is hard to test reliably since the splitting of the bottom of the conduction band E_c under stress is large and strongly anisotropic. Further, the existing data for E_c splitting under stress are quite different for different experimental methods. Thanks are due to the (111) orientation of uniaxial stress with zero splitting of $\bar{E}_c$. By comparing the hydrostatic pressure coefficient $\partial(E_c-E_T)/\partial P$ and the uniaxial coefficient of (111) orientation $\partial(\bar{E}_c-\bar{E}_T)/\partial F_{111}$, whether the proportionality is 3 : 1 will give an experimentally reliable criterion for testing the T_α symmetry property of the defect potential.

[1]G. Bemsi, Phys. Rev. **111**, 1515 (1958).

[2]J. M. Fairfield and B. V. Gokhale, Solid State Electron. **8**, 685 (1968).

[3]R. R. Senechal and J. Basinski, J. Appl. Phys. **39**, 3723 (1968).

[4]C. T. Sah, L. Forbes, L. I. Rosier, A. F. Tasch, Jr., and A. B. Tole, Appl. Phys. Lett. **15**, 145 (1969).

[5]O. Engstrom and H. G. Grimmeiss, J. Appl. Phys. **46**, 831 (1975).

[6]J. Barbolla, M. Puguet, J. C. Brabant, and M. Brousseau, Phys. Status Solidi A **36**, 495 (1976).

[7]S. D. Brotherton and J. Bicknel, J. Appl. Phys. **49**, 667 (1978).

[8]D. V. Lang, H. G. Grimmeis, E. Meijer, and M. Jaros, Phys. Rev. B **22**, 3917 (1980).

[9]R. H. Wu and A. R. Peaker, Solid State Electron. **25**, 643 (1982).

[10]W. Wilcox and T. J. Lachapelle, J. Appl. Phys. **35**, 240 (1964).

[11]D. L. Kendall and D. B. Devries, in *Semiconductor Silicon*, edited by R. R. Haberecht and L. Kern (The Electrochemical Society, New York, 1969), p. 358.

[12]J. A. Van Vechten and C. D. Thurmond, Phys. Rev. B **14**, 3539 (1976).

[13]H. I. Ralph, J. Appl. Phys. **49**, 672 (1978).

[14]X. C. Yau, G. G. Qin, S. R. Zeng, and M. H. Yuan, Acta Phys. Sin. **33**, 377 (1984).

[15]A. Zylbersztejn, R. W. Wallis, and J. M. Benson, Appl. Phys. Lett. **32**, 764 (1981).

[16]W. Jantsch, K. Wunstel, O. Kumagai, and P. Vogl, Phys. Rev. B **25**, 5515 (1982).

[17]C. T. Sah, L. Forbes, L. L. Rosier, and A. F. Tasch, Jr., Solid State Electron. **13**, 759 (1970).

[18]The emission rate obtained in this way is actually the sum of electron and hole emission rate $e_n + e_p$. For $T = 249.5$ K, $e_n/e_p = 25$.[4] Therefore, we introduce 4% error if we neglect the influence of e_p.

[19]A. Zylbersztejn, Appl. Phys. Lett. **33**, 200 (1978).

[20]C. Herring and E. Vogl, Phys. Rev. **101**, 944 (1956).

[21]H. B. Huntington, Solid State Phys. **7**, 274 (1958).

[22]I. Balslev, Phys. Rev. **143**, 636 (1966).

[23]V. J. Tekippe, H. R. Chandrasekhar, P. Fisher, and A. K. Ramdas, Phys. Rev. B **6**, 2348 (1972).

[24]G. D. Watkins and F. S. Ham, Phys. Rev. B **1**, 4097 (1970).

[25]K. J. Schmidt-Tiedemann, *Proceedings of the International Conference on the Physics of Semiconductors*, Exeter (Institute of Physics, London, 1962), p. 191.

[26]L. A. Ledebo and Z. G. Wang, Appl. Phys. Lett. **42**, 680 (1983).

[27]J. R. Morante, J. E. Carceller, A. Herms, P. Cartujco, and J. Barbolla, Appl. Phys. Lett. **41**, 456 (1982).

[28]J. C. Slater, *Quantum Theory of Molecules and Solids*, Vol. 1, Appendix 12 (McGraw-Hill, New York, 1963); V. Heine, *Group Theory in Quantum Mechanics* (Pergamon, London, 1960).

PHYSICAL REVIEW B   VOLUME 32, NUMBER 10   15 NOVEMBER 1985

Binding energies of electrons by nitrogen pairs in GaP

Ming-fu Li
Graduate School, University of Science and Technology of China, Beijing, China

De-qiang Mao and Shang-yuan Ren
*Department of Physics, University of Science and Technology of China,
Hefei, China*
(Received 13 February 1985; revised manuscript received 3 June 1985)

Theoretical calculations of binding energies of electrons by nitrogen pairs in GaP are reported. The calculations are based on the Koster-Slater Green's-function equation and the central-cell defect-potential approximation of Hjalmarson *et al.* [Phys. Rev. Lett. **44**, 810 (1980)]. The defect-potential parameters V_s and V_p are adjusted to fit the experimental binding energies of electrons by single N impurity and seven $(NN)_i$ ($i = 1, 2, \ldots, 7$) pairs. The results are in general agreement for the first time with experiments for either range or ordering of binding energies, and thus strongly support the Hopfield-Thomas-Lynch model for isoelectronic traps. Besides, excited electronic states of $(NN)_1$, $(NN)_3$, $(NN)_5$, and $(NN)_6$ are reported. The energy value of the $(NN)_1$ excited state supports the speculation of Cohen *et al.*

The low-temperature fluorescence spectra of bound excitons by nitrogen pairs in GaP is a well-known experiment hitherto not well explained since the pioneer work of Thomas and Hopfield 19 years ago.[1] Cohen and co-workers[2,3] have reported detailed experimental results on luminescence excitation spectra of excitons bound to nitrogen pairs in GaP. A series of levels whose energies agree well with effective-mass calculation of the nS levels of an acceptor were detected. This supports the Hopfield-Thomas-Lynch (HTL) model[4] of bound excitons by isoelectronic traps. The problem which remains unsolved is the theoretical calculation of binding energies E_i of electrons for different $(NN)_i$ pairs, $i = 1, 2, 3, \ldots$, in order of increasing separation of two N atoms. Some theoretical attempts with different views[5-7] have been made for this problem but none of them succeeded quantitatively. To our knowledge,

Faulkner[5] was the first to make such an effort. He failed in getting binding energies with the correct magnitude or ordering. Recently, Brand and Jaros[7] made a pseuodopotential Green's-function calculation. Their results on a gross energy scale have obtained the correct range of binding energies but they found an incorrect ordering for different $(NN)_i$ pairs. In this Brief Report, we report our theoretical calculations for the same problem.

Our method is based on the Koster-Slater Green's-function equation and the central-cell defect-potential approximation.[8] The method has been used for successful calculations of point defect wave functions,[9,10] and subsequently extended to defect pairs in GaP[11] and divacancies in Si.[12] The defect potential $\hat{V}$ induced by an $(NN)_i$ pair in GaP may be written as

$$\hat{V} = V_s(|d,s\rangle\langle d,s| + |d',s\rangle\langle d',s|) + V_p(|d,p_x\rangle\langle d,p_x| + |d,p_y\rangle\langle d,p_y| + |d,p_z\rangle\langle d,p_z| + |d',p_x\rangle$$
$$\times\langle d',p_x| + |d',p_y\rangle\langle d',p_y| + |d',p_z\rangle\langle d',p_z|) . \tag{1}$$

Here, d and d' denote two sites of an $(NN)_i$ pair, respectively. $|d,s\rangle$ denotes the s atomic orbital located on the d site, etc. The Koster-Slater equation for determining the energy E_i of the ith nitrogen pair in GaP is the 8×8 determinant equation:

$$\det|\hat{G}^0(E)\hat{V} - \hat{I}| = 0 . \tag{2}$$

Here $\hat{G}^0(E)$ is the Green's function of the host crystals. A sp^3s^* ten states linear combination of atomic orbitals (LCAO) model, originally suggested by Vogl, Hjalmarson, and Dow,[13] is used and extended to next-nearest-neighbor approximation for the host band calculation. The special points method of Chadi and Cohen[14] is used for the Green's-function calculation. Ten special points in **k** space are selected and extended to 240 points by 24 T_d symmetry operations. At first sight, 240 special points are not enough since the binding energies concerned are very small. In fact, by theoretical analysis of our previous work,[15] the Green's-function matrix elements are nondivergent when the energy approaches the edge of the band gap. Therefore,

240 special points for the Green's-function calculation are expected to give reasonable results in gross feature for binding-energy calculations of $(NN)_i$ pairs.

We found that the binding energy of an $(NN)_1$ pair is sensitive to V_p in Eq. (1), while the binding energies of $(NN)_2$ to $(NN)_7$ pairs are insensitive to V_p, mainly determined by V_s. Thus, V_s is fitted by a compromise of binding energies of $(NN)_2$ to $(NN)_7$ pairs as well as a single nitrogen impurity level. V_p is actually adjusted to push the binding energy of $(NN)_1$ pair down to the experimental value $E_1 = 120$ meV. The values obtained are

$$V_s = -4.78 \text{ eV} , \quad V_p = -5.54 \text{ eV} . \tag{3}$$

For single nitrogen impurity in GaP, Eq. (2) is reduced to

$$\langle d,s|\hat{G}^0(E)|d,s\rangle = 1/V_s , \tag{4}$$

$$\langle d,T_2|\hat{G}^0(E)|d,T_2\rangle = 1/V_p . \tag{5}$$

Substituting Eq. (3) into Eq. (4) gives the binding energy of

TABLE I. The binding energies of the electrons for the nitrogen pairs in GaP.

| | Binding energy of electron[a] (meV) | | | | Coordinate of one nitrogen atom of pair with other at origin[c] |
| | Ground state | | Excited state | | |
	Experiment[b]	Our calculation	Experiment[b]	Our calculation	
$(NN)_1$	120	120	35	33	$(1\,1\,0)$
$(NN)_2$	113	53			$(2\,0\,0)$
$(NN)_3$	41	12		8	$(2\,1\,1)$
					$(2\,\bar{1}\,\bar{1})$
$(NN)_4$	18	25			$(2\,2\,0)$
$(NN)_5$	11	18		2	$(3\,1\,0)$
					$(2\,2\,2)$
$(NN)_6$	8	12		7	$(\bar{2}\,\bar{2}\,\bar{2})$
					$(3\,2\,1)$
$(NN)_7$	5	10(Doublet)			$(\bar{3}\,\bar{2}\,\bar{1})$

[a]The binding energy of electron is defined by $E_i = E_B - E_I$; here E_B is the energy of the bound exciton which is determined by the optical spectra.[3] E_I is the ionization energy of the hole in the $1S$ bound exciton state and is determined experimentally from the acceptor like exciton levels series limit.[3]
[b]Reference 3.
[c]In units of $a/2$. a is the lattice constant.

an electron to a single nitrogen impurity in GaP to be 10 meV, which is in good agreement with previous experimental data.[3,5,16] Since no T_2 symmetric single nitrogen level in the band gap was found experimentally, the V_p should not be strong enough to push the T_2 state down to band gap from a conduction resonant state. By actual calculation of Eq. (4), this leads to the following criterion,

$$|V_p| < 10.8 \text{ eV} \, , \tag{6}$$

which is satisfied by Eq. (3) in our calculations.

Table I lists our results for all seven $(NN)_i$ pairs which were reported unambiguously in previous experimental works.[1,3,16] Figure 1 gives the comparison of energy levels with data,[1,3] for our calculations and for calculations made by Brand and Jaros[7] and by Faulkner[5] respectively. Significant improvement has been achieved by our calculations over all of the previous theoretical work in the following aspects.

(1) The ordering of our electron binding energies agrees with the currently accepted experimental assignment. The only exception is E_3 of the $(NN)_3$ pairs.

(2) The electron binding energies of $(NN)_1$ to $(NN)_7$ range from 10–120 meV, in general agreement with experimental values of 5–120 meV. Five calculated values of seven $(NN)_i$ pairs, i.e., $(NN)_1$, $(NN)_4$, $(NN)_5$, $(NN)_6$, and $(NN)_7$, agree well with experimental values in one-to-one correspondence.

(3) For the $(NN)_1$ pair, it is reported for the first time to have an excited state E_1' with a binding energy of 33 meV. This seems to partly confirm the speculation of Cohen and Sturge.[3] From their luminescense excitation spectra, they speculated that there is an excited doublet electron state of the $(NN)_1$ exciton with binding energy of 35 meV. This is in general agreement with our theoretical prediction. Nevertheless, our theoretical prediction is a singlet rather than a doublet excited state. We should mention that Cohen *et al.* argued that if there is a bound excited state of the electron, it seems to rule out models of the isoelectronic trap which assumes that the primary particle moves in a

short-range potential of atomic dimensions. Our work reveals that their argument is not correct. We would also like to mention, as illustrated in Table I and Fig. 1, that there are also excited states of $(NN)_3$, $(NN)_5$, and $(NN)_6$, which were not reported in the previous experiments.

Our results are not satisfied in the following aspects.

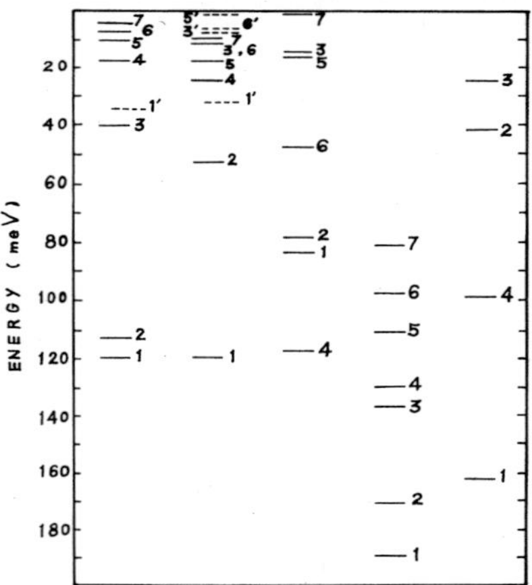

FIG. 1. The binding energies of electrons for the nitrogen pairs in GaP. The number 1 at the right side of an energy level denotes the level referred to the $(NN)_1$ pair in Table I. The number 1' with the dashed line level denotes an excited state of the $(NN)_1$ pair.

(1) There is an inversion in ordering between $(NN)_3$ and $(NN)_4$ and $(NN)_5$.

(2) There is a large discrepancy between calculated and experimental values for a $(NN)_2$ level.

(3) The experimental levels for widely separate pairs lie above those of the isolated defect, while our calculation always puts the pair level below the single N level.

The possible ways to improve these discrepancies between theoretical calculations and experiments within the scope of one electron theory are the following.

(1) As we pointed out in Refs. 10 and 17, the deep level is sensitive to the detailed band structure of the host; a more precise host model is very much expected to improve the calculations.

(2) For widely separate pairs, more special points are desirable but will greatly increase the computation time.

(3) Another source of defect potential other than Eq. (1), such as the $1/r^3$ strain potential,[6] is possible. But we feel that it is only of minor importance and we can add the strain potential as a perturbation to modify the electron binding energies.

In summary, we have made significant progress in understanding the nitrogen pair problem in GaP. Our calculations of binding energies of electrons by nitrogen pairs are for the first time in general agreement with experiments in either range or ordering, and thus strongly support the validity of HTL model for isoelectron traps. We are satisfied to obtain these results by such simple calculations. We also expose new excited states of $(NN)_i$ pairs for stimulating further experimental investigation of this interesting problem.

We are grateful to W. M. Hu for kindly providing us with the next-nearest-neighbor approximation linear combination of atomic orbitals Hamiltonian for GaP, and to Dr. C. Zhang for helpful discussions. Y. Q. Han's assistance in preparing the manuscript is also acknowledged. This work is supported by the Science Fund of the Chinese Academy of Sciences.

[1]D. G. Thomas and J. J. Hopfield, Phys. Rev. **150**, 680 (1966).

[2]E. Cohen, M. D. Sturge, N. O. Lipari, M. Altarelli, and A. Baldereshi, Phys. Rev. Lett. **35**, 1591 (1975).

[3]E. Cohen and M. D. Sturge, Phys. Rev. B **15**, 1039 (1977).

[4]J. J. Hopfield, D. G. Thomas, and R. T. Lynch, Phys. Rev. Lett. **17**, 312 (1966).

[5]R. A. Faulkner, Phys. Rev. **175**, 991 (1968).

[6]J. W. Allen, J. Phys. C **1**, L1136 (1968).

[7]S. Brand and M. Jaros, J. Phys. C **12**, 2789 (1979).

[8]H. P. Hjalmarson, P. Vogl, D. J. Wolford, and J. D. Dow, Phys. Rev. Lett. **44**, 810 (1980).

[9]S. Y. Ren, W. M. Hu, O. F. Sankey, and J. D. Dow, Phys. Rev. B **26**, 951 (1982); S. Y. Ren, Sci. Sin. A, 92 (1984).

[10]M. F. Li, S. Y. Ren, and D. Q. Mao, Acta Phys. Sin. **32**, 1263 (1983); M. F. Li, D. Q. Mao, and S. Y. Ren, Solid State Commun. **48**, 789 (1983).

[11]D. Q. Mao, M. F. Li, and S. Y. Ren, Acta Phys. Sin. **33**, 897 (1984).

[12]S. Y. Ren, D. Q. Mao, and M. F. Li, Acta Phys. Sin. **34**, 455 (1985).

[13]P. Vogl, H. P. Hjalmarson, and J. D. Dow, J. Phys. Chem. Solids **44**, 365 (1983).

[14]D. J. Chadi and M. L. Cohen, Phys. Rev. B **8**, 5747 (1973).

[15]M. F. Li, S. Y. Ren, and D. Q. Mao, Acta Phys. Sin. **33**, 738 (1984); J. Phys. C **17**, 3415 (1984).

[16]Zhao Xue-shu, Li Guo-hua, Han He-xiang, Wang Zhao-ping, Tang Ru-ming, and Hu Jing-zhu, Chin. Phys. Lett. **1**, 15 (1984).

[17]M. F. Li, S. Y. Ren, and D. Q. Mao, Acta Phys. Sin. **34**, 547 (1985).

 Solid State Communications, Vol.61, No.1, pp.13–15, 1987.   0038–1098/87 $3.00 + .00
Printed in Great Britain.   Pergamon Journals Ltd.

A NEW PROPOSED METHOD FOR DETERMINING INNER OR OUTER CROSSING LATTICE RELAXATION
OF DX CENTERS IN $Al_xGa_{1-x}As$ BASED ON PRESSURE EFFECTS

Ming-fu Li

Graduate School, University of Science and Technology of China, Beijing, China and
Center for Advanced Materials, Lawrence Berkeley Laboratory, Berkeley, CA 94720 USA

P.Y. Yu
Department of Physics, University of California, Berkeley and
Center for Advanced Materials, Lawrence Berkeley Laboratory, Berkeley, CA 94720, USA

(Received 26 June 1986 by H. Kamimura)

It is proposed that hydrostatic pressure measurements can be used to
settle the recent controversies surrounding the model of the DX
center in $Al_xGa_{1-x}As$ alloy. The method can in general
distinguish the two possible cases of lattice relaxations in defect
configuration coordinate diagrams: one with large lattice relaxation
which we labelled as "outer crossing," and the other with small
lattice relaxation labelled as "inner crossing".

The DX center in the $Al_xGa_{1-x}As$ alloys
system has attracted much interest in recent
years. Lang et al.[1] suggested in their
original work that the DX center is a complex
involving a donor atom and an anion vacancy.
The main feature of this model is the large
lattice relaxation (LLR) of the defect center
in order to explain many of the unusual
properties of the DX centers[1]. Recently,
Mizuta et al.,[2] based on their pressure
experiment and Hjalmarson and Drumond,[3]
based on theoretical analysis, both suggested
that the DX center is probably only a simple
substitutional donor, and not a complex.
Contrary to Lang et al., Hjalmarson and
Drumond suggested that a deep donor with small
lattice relaxation can explain most of the
characteristics associated with the DX
centers, such as persistent photoconductivity
(PPC). In this communication, we propose a
definitive way to resolve this controversy by
determining whether a defect has a LLR or
SLR. Our method is based on measuring the
pressure coefficients of two energies which
characterize the defect: the thermal barrier
energy E_B and the thermal ionization Energy
E_T. Since the pressure coefficients of
these energies are not yet available for the
DX center, we have illustrated our method by
applying it to the B traps in GaAs[4]. In
this case we concluded that the B hole traps
in GaAs have SLR associated with them.

Our analysis of the problem of lattice
relaxation of a defect starts with the
multiphonon theory of Huang and Rhys.[5] The
electron capture cross section of deep levels
at high temperature is approximated by:[5,6]

$$\sigma(T) = \sigma_\infty\, e^{-E_B/kT} \qquad (1)$$

$$E_B = (E_T - E_S)^2/4E_S \qquad (2)$$

where σ_∞ is the pre-exponential factor,
E_B is the thermal barrier energy, E_T is
the thermal ionization energy from deep level

to the conduction band, and $E_B = S\hbar\omega$ is
the lattice relaxation energy, S is the
Huang-Rhys factor and $\hbar\omega$ is the phonon
energy. It was pointed out by Hjalmarson and
Drumond that there are two possible solutions
of E_S for a given pair of values of E_T and
E_B:

$$E_S^{\pm} = E_T + 2E_B \pm 2\ (E_B^2 + E_TE_B)^{1/2} \qquad (3)$$

Fig. 1 shows the configuration coordinate (CC)
diagrams for the two possible cases. Fig.
1(a) depicts the case of SLR for the DX
center. It corresponds to the solution
E_S^- in Eq. (3) or in terms of the
dimensionless quantity $\epsilon_S = E_S/E_T$, it
corresponds to $\epsilon_S < 1$. In this case,
the minima of curves U_C and U_T are located
at the same side of the crossing of the two
curves. We labelled this case as "inner
crossing".[1,7] Fig. 1(b) depicts the case of
LLR suggested by Lang et al. It corresponds
to the solution E_S^+ in Eq. (3) or $\epsilon_S
> 1$. In this case, the minima of curves

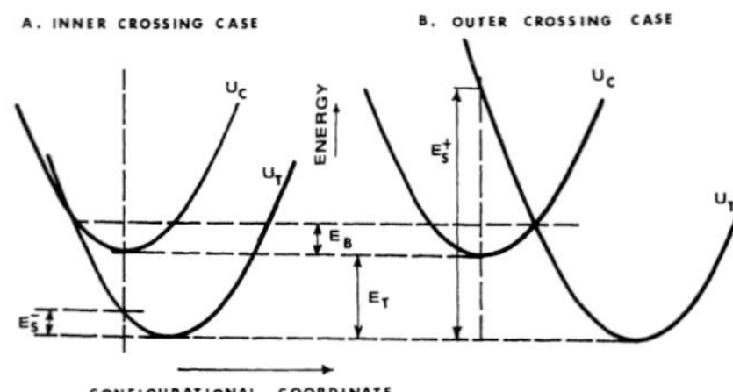

Fig. 1   Configuration coordinate diagram for
defect center in GaAs or
$Al_xGa_{1-x}As$. (a) "Inner crossing"
case, (b) "Outer crossing" case.

U_C and U_T are located at opposite sides of the crossing of the two curves. We labelled this case as "outer crossing".[1,7] From Eq. (2), we can derive the following equation for the hydrostatic pressure (p) coefficient of E_B:

$$\frac{dE_B}{dp} = \frac{(\varepsilon_S^{-1} - 1)}{2} \frac{dE_T}{dp} - \frac{(\varepsilon_S^{-2} - 1)}{4} \frac{dE_S}{dp} \qquad (4)$$

The first term on the RHS of Eq. (4) describes the contribution to dE_B/dp due to change of E_T under pressure. From Figure 1, this corresponds to a vertical shift between U_C and U_T. Evidently, the changes in E_B induced by the same shift in E_T have opposite signs for the two cases. This is essentially the method we propose to distinguish between the two cases described by Figs. 1(a) and (b). The second term on the RHS of Eq. (4) describes dE_B/dp due to variation of the lattice relaxation under pressure. This corresponds to a horizontal shift between the U_C and U_T curves in Fig. 1. Again, ΔE_B induced by the same shift ΔE_S have opposite signs for the outer crossing and inner crossing cases. The above conclusions about the relative signs of the contributions to dE_B/dp by the two terms dE_T/dp and dE_S/dp in Eq. (4) can also be obtained simply by noting that $(\varepsilon_S^{-1} - 1)$ and $(\varepsilon_S^{-2}-1)$ both change sign depending on whether ε_S is larger or smaller than 1. As shown by Barnes and Samara[4], dE_S/dp can be expressed as:[4]

$$\frac{dE_S}{dp} = -2E_S \frac{d \ln \omega}{dp} \qquad (5)$$

where $d \ln \omega/dp$ is the Gruneissen parameter γ of the phonon. For GaAs, the values of γ for different phonon modes can be found in Ref. 8. There are, however, no available data for the value of γ in $Al_xGa_{1-x}As$ or AlAs. For simplicity, we neglect the effect of a small fraction of Al, and use the value of $\gamma = 1.2$ in GaAs for the value γ in $Al_xGa_{1-x}As$ and obtain:

$$\frac{d \ln \omega}{dp} = 1.60 \times 10^{-3} \text{ kbar.}$$

Substituting this value into Eq. (5), we find:

$$\frac{dE_S}{dp} = -3.2 \times 10^{-3} E_S \text{ eV/kbar.}$$

Thus, dE_S/dp is uniquely determined by the value of E_S and is always negative. This can be understood from Fig. 1 since pressure always decreases the distance between the minima of U_C and U_T in the CC diagram.

For the DX center in $Al_xGa_{1-x}As$, with $x \sim 0.35$, we use the following values: $E_B = 0.20$ eV and $E_T = 0.15$ eV from Hjalmarson and Drumond. Substituting these values into Eqs. (3) and (4) we obtain these results for the two cases:

i) "Inner crossing" case:
$E_S^{\ddagger} = 0.021$ eV, $\varepsilon_S = 0.14$,
$dE_S/dp = -0.067$ meV/kbar,
and

$$\frac{dE_B}{dp} = +3.1 \left(\frac{dE_T}{dp} + 0.27 \text{ meV/kbar}\right) \qquad (6)$$

ii) "Outer crossing case":
$E_S^{\ddagger} = 1.08$ eV, $\varepsilon_S = 7.19$,
$$\frac{dE_S}{dp} = -3.5 \text{ meV/kbar}$$

and

$$\frac{dE_B}{dp} = -0.43 \left(\frac{dE_T}{dp} + 1.97 \text{ meV/kbar}\right) \qquad (7)$$

Eqs. (6) and (7) are plotted in Fig. 2(a). Now, dE_B/dp and dE_T/dp are experimentally measurable quantities. By measuring dE_B/dp and dE_T/dp and plotting the experimental point in Fig. 2, it should be possible to determine uniquely whether the deep level has a large or small lattice relaxation, except for the very coincident case that the experimental point lies at the intersection of the two lines. Lifschitz et al.,[9] from their Hall measurements, asserted that the energy difference between the DX level and the L minimum of the conduction band E_L is constant at specific range of X and pressure.

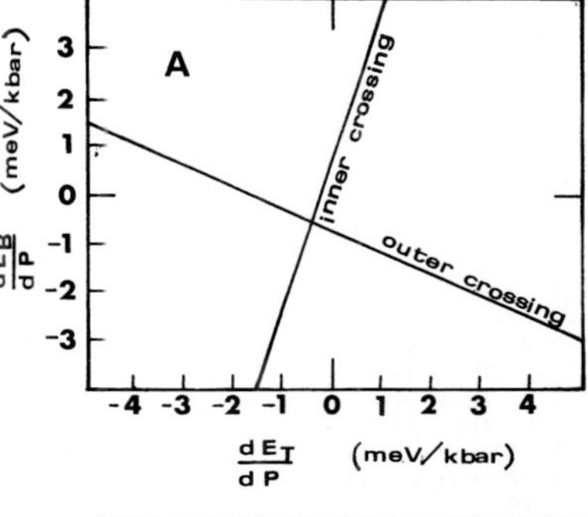

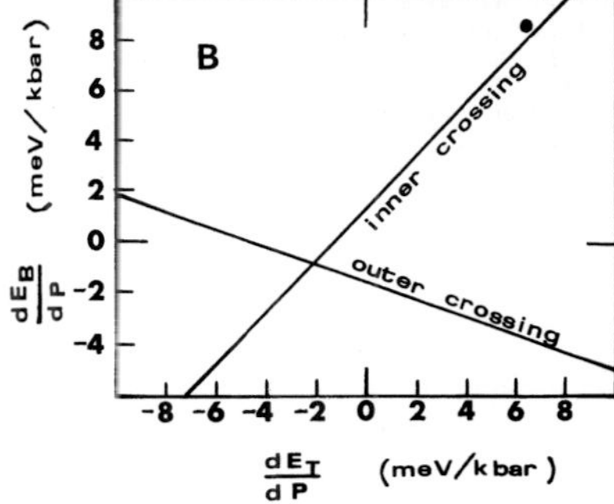

Fig. 2   Calculated dE_B/dp vs dE_T/dp curves for both "inner crossing" and "outer crossing" cases: (a) for the DX centers in $Al_xGa_{1-x}As$ with X = 0.35 and (b) for the B traps in GaAs. The solid circle in (b) is the experimental result of Ref. 4.

If their assertion is correct, we can roughly estimate dE_T/dp by:

$$\frac{dE_T}{dp} \doteq \frac{dE_\Gamma}{dp} - \frac{dE_L}{dp} \doteq (11.5-5.5) \text{ meV/kbar}$$

$$= 6 \text{ meV/kbar.}^9$$

Substituting into Eq. (6) and (7), we obtain $dE_B/dp = +19$ meV/kbar for the SLR case or -3meV/kbar for the LLR case. Thus, PPC should be strongly enhanced in the SLR case and suppressed in the LLR case. In the absence of direct experimental data for the DX center in $Al_xGa_{1-x}As$, we will apply our method to the B traps in GaAs as an illustration of its validity. In this case, the relevant pressure coefficients have been determined by Barnes and Samara.[4] According to Barnes and Samara, $E_B = 0.248$ eV, $E_T = 0.715$ eV for the B traps. Substituting into Eqs. (3) and (4) we obtain for the "inner crossing" case, $E_{\bar{S}} = 0.234$ eV, and

$$\frac{dE_B}{dp} = 1.03 \left(\frac{dE_T}{dp} + 1.52 \text{ meV/kbar}\right) \qquad \cdot(8)$$

and for the "outer crossing" case, $E_{\bar{S}} = 2.188$ eV, and

$$\frac{dE_B}{dp} = -0.34\left(\frac{dE_T}{dp} + 4.46 \text{ meV/kbar}\right) \qquad (9)$$

Equations (8) and (9) are plotted in Fig. 2(b). In this case, Barnes and Samara determined that $d(U_{T,min} - E_V)/dp = 4.9$ meV/kbar, where $U_{T.min}$ is the energy minimum of the U_T curve and E_V is the valence band edge. From Ref. 10 we obtain the pressure coefficient of the band gap in GaAs: $d(U_{C,min}-E_V)/dp = 11.4$ meV/kbar. Combining these two results we obtain $dE_T/dp = 6.5$ meV/kbar. Substitute it into Eq. (8) and (10), we find $dE_B/dp = +8.3$ meV/kbar for the "inner crossing" case and $dE_B/dp = -3.8$ meV/kbar for the "outer crossing" case. The experimental values of dE_B/dp determined by Barnes and Samara is +8.6 meV/kbar. We thus conclude that the "inner crossing" case is valid for the B traps in GaAs.

We are now in the process of determining the corresponding pressure coefficients for the DX center.

Acknowledgement--One of us (MFL) is deeply grateful to Professor Kun Huang for his stimulating discussion on multiphonon theory. He is also indebted to Dr. R.H. Wu and Dr. W.K. Ge for helpful discussions on DX centers. The work at the University of Science and Technology of China was supported by the science fund of Academia Sinica. The work at the Center for Advanced Materials, Lawrence Berkeley Laboratory was supported by the Director, Office of Basic Energy Sciences, Materials Science Division of the U.S. Department of Energy under Contract No. DE-AC03-76SF00098.

References

1. D.V. Lang and R.A. Logan, Phys. Rev. Lett. **39**, 635 (1977).
    D.V. Lang, R.A. Logan and M. Jaros, Phys. Rev. B**19**, 1015 (1979).

2. M. Mizuta, M. Tachikawa, H. Kukimoto and S. Minomura, Jpn. J. Appl. Phys. **24**, L143 (1985).

3. H.P. Hjalmarson and T.J. Drumond, Appl. Phys. Lett. **48**, 657 (1986).

4. C.E. Barnes and G.A. Samara, Appl. Phys. Lett. **43**, 677 (1983).

5. K. Huang and R. Rhys, Proc. R. Soc. **204**, 406 (1950).
    K. Huang, Progress in Physics, **1**, 31 (1981).

6. C.H. Henry and D.V. Lang, Phys. Rev. B**15**, 989 (1977).

7. W.K. Ge and R.H. Wu, Chinese Journal of Semiconductors, **7**, 254 (1986).

8. Landolt-Bornstein, _Numerical Data and Functional Relationships in Science and Technology_, Vol. III/17a (Springer-Verlag, 1982).

9. N. Lifshitz, A. Jayaraman, R.A. Logan and H.C. Card, Phys. Rev. B**21**, 670 (1980).

10. G. Feinleib, S. Groves, W. Paul and R. Zallen, Phys. Rev. **131**, 2070 (1963).

Lattice relaxation of pressure-induced deep centers in GaAs:Si

M. F. Li[a]

Department of Physics, University of California, Berkeley and Center for Advanced Materials, Lawrence Berkeley Laboratory, Berkeley, California 94720

P. Y. Yu

Department of Physics, University of California, Berkeley and Materials and Molecular Research Division, Lawrence Berkeley Laboratory, Berkeley, California 94720

E. R. Weber

Department of Materials Science and Mineral Engineering, University of California, Berkeley, California 94720 and Center for Advanced Materials, Lawrence Berkeley Laboratory, Berkeley, California 94720

W. Hansen

Center for Advanced Materials, Lawrence Berkeley Laboratory, Berkeley, California 94720

(Received 11 March 1987; accepted for publication 10 June 1987)

Deep centers induced by hydrostatic pressure in GaAs:Si have been studied by deep level transient spectroscopy and constant temperature capacitance transient techniques. The capture behavior of these centers has been studied in detail and found to be consistent with the multiphonon emission theory. The pressure coefficients of the ionization energy and the barrier height are consistent with the large lattice relaxation model proposed by D. V. Lang and R. A. Logan [Phys. Rev. Lett. **39**, 635 (1977)].

Recently Mizuta *et al.*[1] found that when GaAs containing shallow donors is subjected to hydrostatic pressure in excess of 20 kbar a deep center similar in properties to the *DX* center in AlGaAs alloys appeared. These results stimulated much discussion concerning the nature of this pressure-induced deep center (to be abbreviated as PIDC here) and of the related *DX* center. In particular the question of whether large lattice relaxation proposed by Lang and Logan[2] is necessary to explain the persistent photoconductivity (PPC) of the *DX* center in AlGaAs alloys and of the PIDC in GaAs has not been resolved. Several authors, such as Hjalmarson and Drummond[3] and Henning and Ansems,[4] have proposed alternate electronic mechanisms for PPC. The difference between the large lattice relaxation model and the model proposed by Hjalmarson and Drummond[3] is shown schematically in the inset of Fig. 1. Recently Li and Yu[5] proposed that hydrostatic pressure measurements can distinguish between these two models. In this letter we present measurements of the thermal activation energy and capture barrier height of the PIDC in Si-doped GaAs as a function of pressure. Our results for these PIDC are consistent with the large lattice relaxation model of Lang and Logan[2] but not with the model of Hjalmarson and Drummond.[3]

Our experiments were performed on Si-doped bulk GaAs crystals with $N_D - N_A = 2 \times 10^{17}$ cm^{-3}. Schottky diodes were fabricated by first evaporating an Au-Ge alloy on one side of the wafer and annealing at 450 °C for 1 min to form an ohmic contact. This was followed by evaporating Al on the other side of the wafer which was then cut into small 200×200 (μm)^2 chips. The cut sides of the samples were etched to reduce the reverse-biased leakage current before loading into a diamond anvil high-pressure cell. Details of

this cell and the technique for introducing wires into the cell have been described elsewhere.[6] Powder of calcium sulfate was used as the pressure transmitting medium. The pressure inside the cell was determined by measuring the fluorescence of ruby chips placed adjacent to the sample. The accuracy in the pressure measurement is better than 1 kbar. Deep level transient spectroscopy (DLTS) and constant temperature capacitance transient measurements were made using a Boonton model 72B capacitance meter and a dual-channel boxcar integrator. The temperature of the sample was measured with a calibrated Si diode in thermal contact with one of the diamond anvils. To minimize the temperature differ-

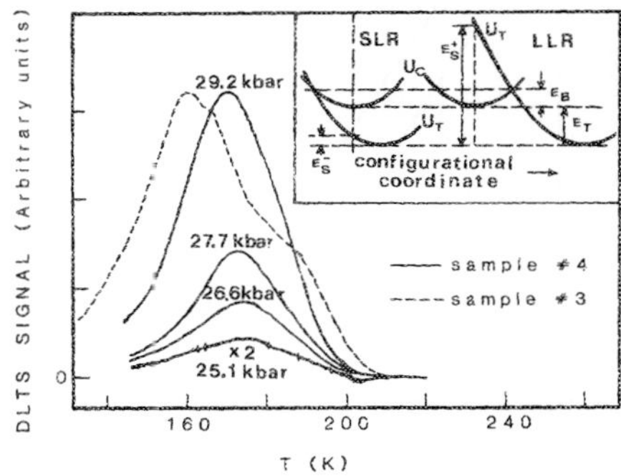

FIG. 1. DLTS spectra in two Si-doped GaAs under pressure. These spectra were obtained with window times of $t_1 = 1$ and $t_2 = 0.5$ ms while the width of the filling pulses was 10 ms. The inset shows the configuration coordinate diagrams for a defect center exhibiting small lattice relaxation (SLR) or large lattice relaxation (LLR).

[a] On leave from the Graduate School, University of Science and Technology of China, Beijing, People's Republic of China.

ence between the sample and the sensor, temperature scans in DLTS were made extremely slowly.

Figure 1 shows some typical DLTS spectra of two samples under pressure. Pulse widths (between 10 ms and 1 s) long enough for saturated transient amplitudes were used to avoid any possible distortion of the DLTS spectra due to temperature dependence of the capture rates. Our results are qualitatively similar to those reported by Mizuta et al.[1] Quantitatively there are significant differences. (1) The shift in our DLTS peaks with pressure is much smaller than that reported by Mizuta et al. (2) The density of our PIDC also increases at a slower rate with pressure. Mizuta et al.[1] found that at 30 kbar the amplitude of the capacitance transient, ΔC, becomes comparable to the junction capacitance C. To avoid complications due to pressure dependence ΔC and the PIDC concentration, we have limited the pressure we applied to the sample to below 29 kbar where $\Delta C / C$ is still less than 2%. We note that in sample No. 3 the DLTS spectra indicate the existence of several peaks as have also been observed by Mizuta et al. From the DLTS spectra the emission rates (e_n) of the PIDC were obtained as a function of temperature and pressure. Plots of e_n versus temperature at 29 kbar for both samples 3 and 4 are shown in Fig. 2. The capture rate (τ_e^{-1}) was measured by a standard majority-carrier pulse method[7] at constant temperatures corresponding to the DLTS peaks. In this case the transient signal amplitudes were recorded as a function of pulse widths. The capture time constant τ_e was determined by the half-signal point method of Lang.[8] The temperature dependence of the capture rates at 29 kbar in the two samples is also shown in Fig. 2. In sample 3 the DLTS was found to shift *en bloc* with

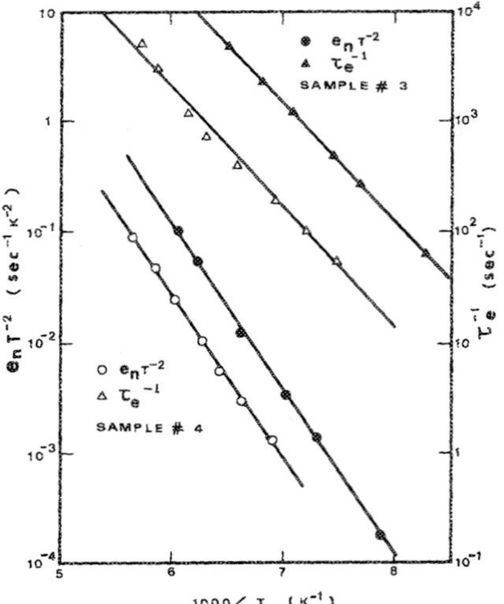

FIG. 2. Plots of the electron emission and capture rates vs temperature for the PIDC in two samples of GaAs under 29 kbar of pressure.

pressure so we assumed all the levels contributing to the DLTS peaks had the same pressure dependence. In spite of the multiplet nature of the DLTS peaks in sample 3 the slopes of the curves in Fig. 2 for samples 3 and 4 are quite similar.

We have interpreted the temperature dependence of the emission and capture rates of the PIDC in GaAs:Si with the multiphonon emission theory (MET).[9,10] The capture and emission rates are related to the capture cross section σ_n by

$$\tau_e^{-1} = \sigma_n \langle v \rangle n \tag{1}$$

and

$$e_n = N_c \langle v \rangle \sigma_n \exp(-E_T/KT) , \tag{2}$$

respectively. In Eqs. (1) and (2) $\langle v \rangle$ is the electron thermal velocity, n is the carrier concentration, N_c is the effective conduction-band density of states, E_T is the thermal ionization energy of the PIDC, and K is the Boltzmann constant. In the high-temperature limit of the MET the capture cross section depends on temperature as[9-11]

$$\sigma_{n\infty} = \sigma_{n\infty} \exp(-E_B/KT) , \tag{3}$$

where $\sigma_{n\infty}$ is the capture cross section at infinite temperature and E_B is the capture barrier height. Since our sample was lightly doped and the occupation of the trap was low, the effect of the Fermi level on the capture barrier height was negligible.[12] The quantity $N_c \langle v \rangle$ depends quadratically on temperature.[7] By fitting the curves in Fig. 2 with the above equations we obtained the values $E_T + E_B = 0.30 \pm 0.01$ eV and $E_B = 0.22 \pm 0.01$ eV for the PIDC in GaAs:Si at $P = 29$ kbar. The values for samples 3 and 4 are identical within experimental uncertainties. We note that our value for $E_T + E_B$ is in good agreement with the result of Mizuta et al.[1] but smaller than the corresponding value for the DX center in GaAlAs:Si by about 0.13 eV.[8]

We have also determined the pressure coefficients of the energies $E_T + E_B$ and E_B. Following previous work[11,13] we neglected the pressure dependence of the pre-exponential factors $\sigma_{n\infty}$, N_c, and $\langle v \rangle$ in Eqs. (1)–(3). From the pressure dependence of the DLTS peaks and of the capture rate, we obtained the pressure-induced shifts $\Delta(E_T + E_B)$ and ΔE_B as shown in Fig. 3. By a least-squares fit of these data points to a straight line we obtained $dE_B/dP = -2.1 \pm 0.4$ meV/ kbar and $d(E_T + E_B)/dP = -1.3 \pm 0.4$ meV/kbar. From these pressure coefficients we deduced that $dE_T/dP = 0.8$ meV/kbar.

Recently Li and Yu[5] proposed a method for determining whether the large lattice relaxation (LLR) model of Lang and Logan[2] or the small lattice (SLR) model of Hjalmarson and Drummond[3] applied to a deep center. According to this method the pressure coefficient dE_B/dP given by

$$\frac{dE_B}{dP} = \left(\frac{\epsilon_S^{-1} - 1}{2}\right)\frac{dE_T}{dP} - \left(\frac{\epsilon_S^{-2} - 1}{4}\right)\frac{dE_S}{dP} , \tag{4}$$

(where $\epsilon_S = E_S/E_T$ and E_S is the lattice relaxation energy as shown in Fig. 1) should be quite different for the two models. The reason is because in the LLR $E_S > E_T$ so that $\epsilon_S > 1$ while in the SLR model $\epsilon_S < 1$. In calculating dE_S/dP Li and Yu used the equation

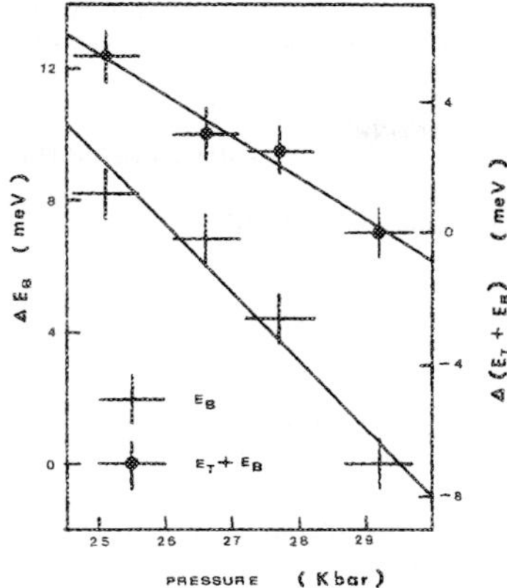

FIG. 3. Pressure-induced shifts of E_B and $E_T + E_B$ for the PIDC in GaAs vs pressure. The vertical and horizontal bars around the experimental points represent the estimated experimental errors. The straight lines drawn through the experimental points represent least-squares fits to the data points.

$$\frac{dE_S}{dP} = -2E_S \frac{d \ln \omega}{dP}, \qquad (5)$$

suggested by Barnes and Samara.[11] In Eq. (5) $\hbar\omega$ is the energy of the phonon in the MET and is usually assumed to be an optical phonon. However, for the DX centers in AlGaAs alloys Lang[8] has argued that the zone-edge transverse acoustic (TA) phonon is involved in the multiphonon emission. In GaAs the sign of $d \ln \omega/dP$ for the optical phonons is opposite to that for the TA phonon. In view of this uncertainty we do not use Eq. (5) to calculate dE_S/dP. Instead we note that in the same MET

$$\sigma_{n\infty} = \left(\frac{A}{|E_T - E_s|}\right)\left(\frac{4\pi E_B}{KT}\right)^{1/2}. \qquad (6)$$

Barnes and Samara have found that $\sigma_{n\infty}$ is independent of pressure for the B traps in GaAs and we have found this to be true also for the PIDC in GaAs. Using the result and Eq. (6) we obtain the following expression for the pressure coefficient of dE_S/dP:

$$\frac{dE_S}{dP} = \frac{dE_T}{dP} - \frac{E_T - E_S}{2E_B} \frac{dE_B}{dP}. \qquad (7)$$

Substituting the values of $E_B = 0.22$ eV, $E_T = 0.08$ eV, $dE_T/dP = 0.8$ meV/kbar, and $dE_B/dP = -2.1$ meV/kbar we obtained into the above equations, we get the following results for the PIDC in GaAs at 29 kbar: (a) In the LLR model, $dE_S/dP = -3.7$ meV/kbar and $dE_B/dP = -1.3$ meV/kbar. (b) In the SLR model, $dE_S/dP = 1.116$ meV/kbar and $dE_B/dP = -83$ meV/kbar. Clearly the experimental value of $dE_B/dP = -2.1$ meV/kbar is more consistent with the LLR model than with the SLR model. Thus we conclude that our results on the PIDC in GaAs are com-

pletely consistent with the MET in which the defect has a large lattice relaxation similar to that proposed by Lang and Logan[2] for the DX centers in the AlGaAs alloys. In addition substituting our values for E_T and E_B into the LLR model we obtained the value of 1 eV for the lattice relaxation energy E_S.

Finally our value of the pressure coefficient $d(E_T + E_B)/dP$ is more than a factor of 2 smaller than the value reported by Mizuta et al.[1] The reason for this difference is not clear. However, in both cases the value of dE_T/dP deduced using our value of dE_B/dP for the PIDC in GaAs under pressure is much smaller than the value one would deduce based on the proposal that the PIDC in GaAs is formed from the L conduction minima only.[14] If this proposal were true, dE_T/dP would be given by $dE_T/dP - dE_L/dP = 6$ meV/kbar. This is also in contrast to the results of pressure measurements on the DX center in GaAlAs alloys where the pressure coefficient of the ionization energy of the deep donor was found to be dominated by the L minima.[15,16] Thus qualitatively the PIDC in GaAs behaves like the DX center in GaAlAs alloys but quantitatively there are still significant differences between the two kinds of deep donors which need to be resolved by better theoretical understanding of these centers.

We wish to thank Professor K. Huang for valuable discussions on the MET, Dr. David Erskine for advice in using the diamond anvil cell, Professor Y. R. Shen for the loan of the boxcar integrator, Yihe Huang, J. Beeman, W. Tseng, C. Musgrave, M. Boenke, H. Lee, and Y. Lo for technical assistances. This work is supported by the Director, Office of Basic Energy Sciences, Materials Science Division of the U. S. Department of Energy under Contract No. DE-AC03-76SF00098.

M. Mizuta, M. Tachikawa, H. Kukimoto, and S. Minomura, Jpn. J. Appl. Phys. **24**, L143 (1985).

D. V. Lang and R. A. Logan, Phys. Rev. Lett. **39**, 635 (1977); D. V. Lang, R. A. Logan, and M. Jaros, Phys. Rev. B **19**, 1015 (1979).

H. P. Hjalmarson and T. J. Drummond, Appl. Phys. Lett. **48**, 657 (1986).

J. C. M. Henning and J. P. M. Ansems, Mater. Sci. Forum **10-12**, 429 (1986).

M. F. Li and P. Y. Yu, Solid State Commun. **61**, 13 (1987).

D. Erskine, P. Y. Yu, and G. Martinez, Rev. Sci. Instrum. **58**, 406 (1987).

G. L. Miller, D. V. Lang, and L. C. Kimerling, Annu. Rev. Mater. Sci. **7**, 377 (1977).

D. V. Lang, in *Deep Centers in Semiconductors*, edited by S. T. Pantelides (Gordon and Breach, New York, 1985), p. 489.

K. Huang and R. Rhys, Proc. R. Soc. **204**, 406 (1950); K. Huang, Prog. Phys. **1**, 31 (1981).

C. H. Henry and D. V. Lang, Phys. Rev. B **15**, 989 (1977).

C. E. Barnes and G. A. Samara, Appl. Phys. Lett. **43**, 677 (1983).

N. S. Caswell, P. M. Mooney, S. L. Wright, and P. M. Solomon, Appl. Phys. Lett. **43**, 1093 (1986).

R. H. Wallis, A. Z. Zylbersztejn, and J. M. Besson, Appl. Phys. Lett. **38**, 698 (1981).

M. Tachikawa, M. Mizuta, H. Kukimoto, and S. Minomura, Jpn. J. Appl. Phys. **24**, L821 (1985).

A. Saxena, Appl. Phys. Lett. **36**, 79 (1980).

N. Lifshitz, A. Jayaraman, R. A. Logan, and H. C. Card, Phys. Rev. B **21**, 670 (1980).

PHYSICAL REVIEW B   VOLUME 36, NUMBER 8   15 SEPTEMBER 1987-I

Photocapacitance study of pressure-induced deep donors in GaAs:Si

M. F. Li*
*Department of Physics, University of California, Berkeley, California 94720
and Center for Advanced Materials, Lawrence Berkeley Laboratory, Berkeley, California 94720*

Peter Y. Yu
*Department of Physics, University of California, Berkeley, California 94720
and Materials and Chemical Sciences Division, Lawrence Berkeley Laboratory, Berkeley, California 94720*

Eicke R. Weber
*Department of Materials Science and Mineral Engineering and Center for Advanced Materials,
Lawrence Berkeley Laboratory, Berkeley, California 94720*

W. Hansen
Center for Advanced Materials, Lawrence Berkeley Laboratory, Berkeley, California 94720
(Received 20 May 1987)

Photocapacitance transient measurements in GaAs:Si under pressures of 33 and 38 kbar are reported for the first time. The optical ionization energy of pressure-induced deep donors in GaAs
was determined to be 1.44 ± 0.04 eV. The low-temperature capture times of photoexcited carriers
were also measured and the results indicate that persistent photoconductivity would occur in
GaAs under pressures in excess of 30 kbar. These results show that qualitatively the pressure-
induced deep donors in GaAs are very similar to the DX centers in $Ga_{1-x}Al_xAs$ alloys in terms
of their optical properties.

The DX centers in $Ga_{1-x}Al_xAs$ alloys have recently
received much attention because of their influence on the
performance of modulation-doped field-effect transistors
and also because of their metastability. One significant
development has been the discovery by Mizuta, Tachikawa, Kikumot, and Minomura[1] of deep donors in GaAs
doped with Si or Sn under pressure with many properties
very similar to the DX centers in $Ga_{1-x}Al_xAs$.[2,3] Based
on their deep-level transient spectroscopy (DLTS) measurements Mizuta et al.[1] concluded that these pressure-
induced deep donors (PIDD's) in GaAs are identical to
the DX centers in $Ga_{1-x}Al_xAs$. To substantiate this conclusion it is necessary to compare all the known properties
of the DX centers with those of the PIDD's in GaAs. The
DX centers in GaAlAs alloys have many characteristic
properties. One of these characteristics is the large
difference between the optical ionization energy (E_n) and
thermal ionization energy (E_t). Another characteristic,
which is very important from the point of view of device
performance, is that the DX centers produce persistent
photoconductivity (PPC). Mizuta et al.[1] were unable to
determine the optical ionization threshold of the PIDD's
in GaAs under pressure because their high-pressure cell
has no optical access. Although Tachikawa et al.[4] demonstrated PPC due to the PIDD's in GaAs by illuminating
their sample with a light-emitting diode inside the high-
pressure cell, no quantitative measurements were reported. We have used a diamond-anvil cell to study the
PIDD's in GaAs:Si. Because of the transparency of the
diamond anvils we were able to use photocapacitance
transient techniques to study PPC quantitatively and to
determine E_n for the PIDD's in GaAs. We showed that
our optical results for the PIDD's in GaAs:Si are very

similar to those of the DX centers in $Ga_{1-x}Al_xAs$ alloys.

The samples used in our photocapacitance measurements were Schottky diodes fabricated from either GaAs
doped with 2×10^{17} cm^{-3} of Si or $Ga_{0.65}Al_{0.35}As$ doped
with 5×10^{16} cm^{-3} of Te. Ohmic contacts to the samples
were made by evaporating Au-Ge alloy on one side of a
wafer followed by annealing at 450 °C for one minute.
Schottky barriers were formed by evaporating Al onto the
other side. Chips of typically 200×200 μm^2 in size were
cut from the wafer. The cut sides of the chips were etched
to reduce the reverse-bias leakage current. The chips
were then mounted into a diamond-anvil cell with a soft
powder ($CaSO_4$) as the pressure-transmitting medium.
The details of loading the sample with leads into the
diamond-anvil cell have been determined by Erskine, Yu,
and Martinez.[5] A schematical diagram showing the sample and the wires inside the cell is shown in the inset of
Fig. 1. The diode was always placed with the Al side facing the incoming light. The cell was pressurized with a
hydraulic press at room temperature. The pressure was
determined by the standard ruby-fluorescence technique.[5]

Two optical measurements have been performed to
study the properties of the PIDD's in GaAs. In the first
experiment the dependence of the electron photoionization
cross section (σ_n^0) of deep centers on incident photon energy was measured. In the second experiment the thermal
capture rates of optically excited carriers were determined
at low temperatures. From these capture rates the decay
times of free carriers in PPB can be calculated.

In the first experiment light from a tungsten halogen
lamp is focused into a monochromator with a spectral
width of 7 nm. The radiation from the monochromator is
directed into the diamond-anvil cell and scattered by the

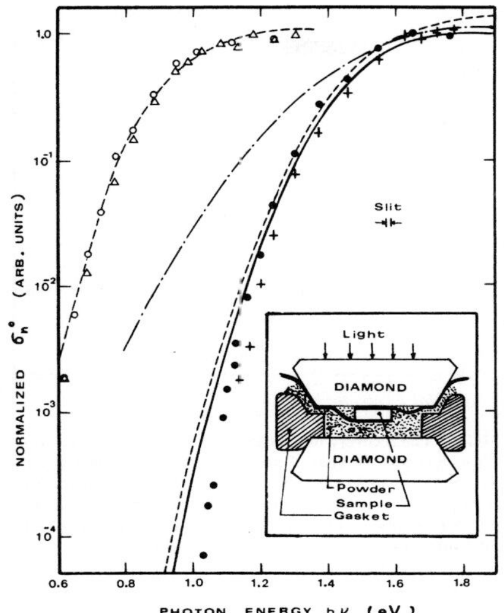

FIG. 1. The normalized photoionization cross-section spectra of the DX center in $Ga_{1-x}Al_xAs$ (open circles and open triangles) and of the PIDD's in GaAs under pressures of 33 kbar (solid circles) and 38 kbar (crosses). The open triangles are the data of Lang *et al.* reproduced from Ref. 3, while the open circles represent data obtained with the sample inside the diamond-anvil cell. The curves represent theoretical fits to the experimental points using Eq. (3) discussed in the text. The inset shows schematically the sample inside the diamond-anvil cell.

powder surrounding the sample. Since the diode is covered by metal electrodes on both the top and bottom, only the scattered light can enter the sample from the sides. As a result it is not possible to determine exactly the amount of light absorbed by the sample; however, it is still possible to measure the *relative* σ_n^0, provided we assume that the light scattering efficiency is constant over

the range of photon energies in this study. Except for this assumption we have corrected for the dispersion in the diamond absorption and in the lamp emission. As a test of the reliability of our measurement we have used the same setup to measure the dispersion of the σ_n^0 of $Ga_{1-x}Al_x$-As:Te at atmospheric pressure and compared our results with those reported by Lang, Logan, and Jaros.[3]

We have used the method of Chantre, Vincent, and Bois[6] to measure σ_n^0. The sample, whether $Ga_{1-x}Al_xAs$ at atmospheric pressure or GaAs under pressure, was first maintained at zero bias at room temperature in order to fill the traps and then cooled to 77 K in the dark. A reverse bias of 3 V was then applied to the diode. As the thermal emission rates of the DX center and the PIDD's in GaAs at 77 K were both negligible, the traps in the depletion layer remained occupied. Next the sample was illuminated with radiation from the monochromator and the rate of change in the diode capacitance was measured. Since the change in capacitance was proportional to the change in the electron concentration in the deep centers (n_T) within the depletion layer, we obtained in this way dn_T/dt. If Φ is the incident photon flux density, and σ_n^0 and σ_p^0 are, respectively, the electron and hole photoionization cross sections for the deep center, then it has been shown that dn_T/dt is given by[6]

$$dn_T/dt = \Phi(hv)[\sigma_p^0(hv)(N_T - n_T) - \sigma_n^0(hv)n_T] , \quad (1)$$

where N_T is the concentration of deep centers and hv is the photon energy. At the point where the light was first turned on, most of the deep centers in the depletion layer were occupied and so $n_T = N_T$ and Eq. (1) reduced to

$$dn_T/dt = \sigma_n^0(hv)\Phi(hv)N_T . \quad (2)$$

Using Eq. (2) and the measured dn_T/dt as a function of the incident photon energy hv we have deduced the normalized $\sigma_n^0(hv)$ spectra shown in Fig. 1. The open circles are results obtained from our $Ga_{1-x}Al_xAs$:Te sample located inside our diamond-anvil cell but with no pressure applied. The open triangles are the data of Lang *et al.*[3] on the DX centers in $Ga_{0.63}Al_{0.37}As$:Te. The excellent agreement between the two sets of results justifies our method of measuring the σ_n^0 of samples inside the diamond-anvil cell. The solid circles and crosses in Fig. 1 represent the σ_n^0 of PIDD's in GaAs:Si measured with the same procedure under pressures of 33 and 38 kbar, respectively.

To analyze the experimental results in Fig. 1 we have used the following expression obtained by Jaros[3] using the strong coupling model of Huang and Rhys[7]

$$\sigma_n^0(hv) \sim \frac{1}{hv}\int_0^\infty dE\,(E)^{1/2}\left[\frac{(1\pm e^{-2E/E_p})E^{1/2}}{|E_n|+E} + \frac{(1\mp e^{-2E/E_p})E_F^{1/2}}{|E_n|-E-(E_g+E_p)/2}\right]^2 \exp\left[\frac{hv-(|E_n|+E)^2}{-U}\right]U^{-1/2} . \quad (3)$$

In Eq. (3) E_p is the average optical (Penn) gap,[8] E_g is the band gap, E_F is the Fermi energy of the valance electrons, E_n is the optical ionization energy of the deep level measured from conduction band, and U is a function defined by

$$U = 2S(\hbar\omega)^2/\tanh(\hbar\omega/2k_BT) . \quad (4)$$

In Eq. (4) k_B is Boltzmann's constant, T is the temperature, and S is the Huang-Rhys factor defined by $E_S = S\hbar\omega$, where E_S is the lattice relaxation energy and $\hbar\omega$ is the phonon energy. The thermal ionization energy E_T of the deep center is related to its optical ionization energy by $E_n = E_T + E_S$. The choice of the $\pm$ and $\mp$ signs in Eq. (3) depends on the nodal character of the electron wave function. The upper and lower signs correspond to deep centers with valence-band-like and conduction-band-like wave functions, respectively.[9] The curves in Fig. 1 represent fits to the experimental points using Eq. (3). In fitting the data points for GaAs we have used the same values of $E_F = 11.5$ eV and $E_p = 5.2$ eV as Jaros.[9] For the band gap of GaAs at a pressure of 33–38 kbar we have used the approximate value of $E_g = 1.8$ eV.[10] It turns out that the shape of σ_n^0 is not very sensitive to the above energies anyway. The dispersion in σ_n^0 is mainly determined by E_n and $\hbar\omega$ at a given T. Lang[11] has argued that the DX center is coupled predominately to the transverse-acoustic (TA) phonon. We have considered both the longitudinal-optical (LO) phonon and the zone-edge TA phonon of GaAs in fitting the results for the PIDD's in GaAs. The phonon energies of GaAs under pressures of 33–38 kbar were 37 and 8 meV for the LO and TA phonons, respectively.[10,12] The remaining unknown parameters in Eq. (3) are E_S and E_n. Using the value of $E_T = 0.08$ eV obtained from DLTS measurements performed on the same samples[13] and the relation $E_n = E_S + E_T$ we reduce the adjustable parameters in fitting the data points for GaAs to E_n only. The resultant curves obtained by assuming that TA phonons are involved are shown as solid and broken lines in Fig. 1. The difference between the two curves is that the deep-center wave function was assumed to be conduction-band-like for the solid curve and valence-band-like for the broken curve. The values of E_n obtained are 1.48 and 1.4 eV, respectively. Since both curves fit the experimental results equally well we conclude that $E_n = 1.44 \pm 0.04$ eV. We note that the corresponding value for the DX center in $Ga_{1-x}Al_xAs$:Si obtained by Lang and Logan was 1.25 eV.[11] On the other hand, we could not obtain any reasonable fit to our results by assuming that the PIDD's couple to LO phonons. As an example, the dot-dashed line in Fig. 1 shows a plot of Eq. (3) using the LO-phonon energy and the value of $E_n = 1.40$ eV.

In the second experiment we have measured the thermal capture times of photoexcited free carriers by the PIDD's in GaAs:Si at a pressure of 38 kbar. The capture times are related to free-carrier lifetimes in PPC. The method we used to measure the capture times is very similar to those employed by Zhou, Ploog, and Gmelin.[14] Initially the sample inside the high-pressure cell was kept around liquid-nitrogen temperature under a reverse bias of 3 V. The sample was then illuminated with strong light to photoexcite electrons from the deep centers to the conduction band while the capacitance was monitored. This was carried out until the capacitance change became saturated. At this point we assumed that most of the deep centers in the depletion layer have been emptied. The light was then turned off and the carriers were allowed to be captured thermally by a sequence of filling pulses of zero bias and of durations $t_1, t_2, \ldots, t_n$ sec as shown schematically in the inset of Fig. 2. In between these filling pulses reverse-bias voltages of relatively short durations (~ 1 sec) were applied to measure the diode capacitance. Since, at low temperatures, the thermal emission rate of the PIDD's in GaAs under pressure was rather small[13] the change in capacitance during the negative-bias

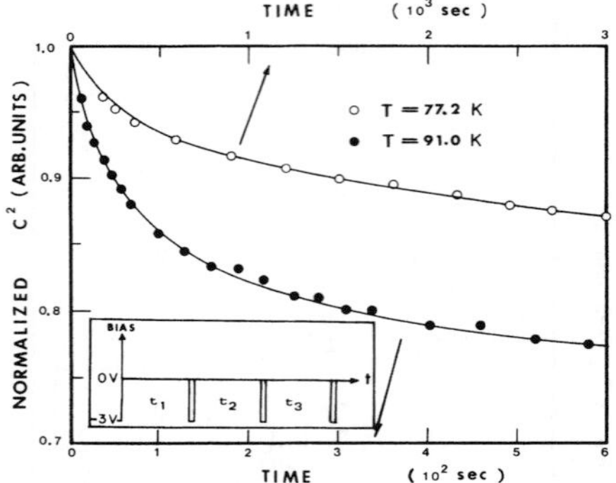

FIG. 2. The recovery in the capacitance after photoexcitation of GaAs:Si under 38 kbar of pressure at two different temperatures. Note the different time scales for the two temperatures. The solid curves have been drawn through the data points by hand. The inset shows the bias voltage sequence used in obtaining the time dependence of the capacitance recovery.

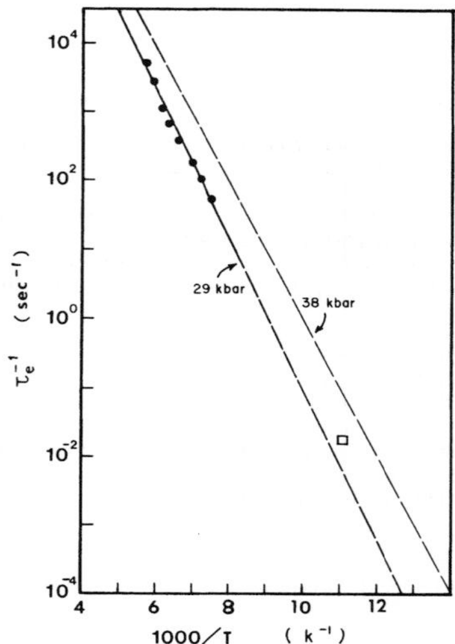

FIG. 3. The capture rate of carriers by the PIDD's in GaAs:Si plotted against $1/T$ for two different pressures. The solid circles are data points at 29 kbar reproduced from Ref. 13, while the open square is the result of photocapacitance measurement at 38 kbar. The broken lines represent results extrapolated from the 29-kbar experimental data.

pulses was negligible. The square of the capacitance (C^2) measured after the nth filling pulse was plotted as a function of the total time of the filling pulses ($t_1 + t_2 + \cdots + t_n$) in Fig. 2 for two different temperatures. Note the difference in the time scales for the two temperatures. From our DLTS measurements[13] and those reported by Mizuta et $al.$[1] we know that at pressures above 30 kbar trapping of free carriers in GaAs:Si is entirely dominated by the PIDD's. Thus the observed time dependence of C^2 is proportional to the time dependence of the bulk carrier concentration, and the trapping time of the free carriers can be determined from the plots in Fig. 2. Since the decay curves in Fig. 2 are not single exponentials we have used Lang's half-signal point method[11] to estimate the capture time of free carriers in GaAs in the presence of the PIDD's to be about 1 min at 91 K and over 1 h at 77 K. To compare these results with our DLTS measurements at higher temperatures,[13] we plotted the capture rate (τ_e^{-1}) vs $1/T$ in Fig. 3 for the PIDD's in GaAs at two different pressures. The solid circles represent the experimental data obtained by DLTS in Ref. 13. The broken lines are extrapolations of those experimental points to either lower temperatures or to higher pressures using the pressure coefficients reported in Ref. 13. The open square represents the result obtained in Fig. 2. Thus we see that the capture rates we obtained at low temperatures are quite consistent with the higher-temperature DLTS results.

In conclusion, we have performed photocapacitance measurements of the PIDD's in GaAs:Si under pressures of over 30 kbar. From these measurements we determined the photoionization thresholds and thermal capture times of carriers by the PIDD's in GaAs:Si. We found that the photocapacitance results of the PIDD's in GaAs under pressure were very similar to the DX centers in $Ga_{1-x}Al_xAs$ alloys. However, there are significant quantitative differences between the two centers. We determined the optical ionization energy of PIDD's in GaAs:Si to be 1.44 eV which is almost 0.2 eV higher than the corresponding value for the DX center in GaAlAs:Si. On the other hand, Mizuta et $al.$[1] have found that the DLTS activation energy of the PIDD's in GaAs was 0.2 eV lower than that of the DX center. Although the PIDD's in GaAs:Si under pressure is very similar in all respects to the DX center in $Ga_{1-x}Al_xAs$ alloys, these rather large differences in their energies should not be neglected.

We wish to acknowledge Professor S. G. Louis and Professor K. Huang for valuable discussions, Dr. E. Bauser for providing the GaAlAs:Te sample used in our experiment, and J Beeman and W. Shan for technical assistance. This research is supported by the Director, Office of Basic Energy Sciences, Materials Science Division of the U.S. Department of Energy under Contract No. DE-AC-03-76SF00098.

*Permanent address: Graduate School, University of Science and Technology of China, Beijing, People's Republic of China.

[1]M. Mizuta, M. Tachikawa, H. Kikumot, and S. Minomura, Jpn. J. Appl. Phys. **24**, L143 (1985).

[2]D. V. Lang and R. A. Logan, Phys. Rev. Lett. **39**, 635 (1977).

[3]D. V. Lang, R. A. Logan, and M. Jaros, Phys. Rev. B **19**, 1015 (1979).

[4]M. Tachikawa, T. Fujisawa, H. Kukimoto, A. Shibata, G. Oomi, and S. Minomura, Jpn. J. Appl. Phys. **24**, L893 (1985).

[5]D. Erskine, P. Y. Yu, and G. Martinez, Rev. Sci. Instrum. **58**, 406 (1987).

[6]A. Chantre, G. Vincent, and D. Bais, Phys. Rev. B **23**, 5335 (1981).

[7]K. Huang and R. Rhys, Proc. R. Soc. London, Ser. A **204**, 406 (1950).

[8]D. R. Penn, Phys. Rev. **128**, 2093 (1962).

[9]M. Jaros, Phys. Rev. B **16**, 3694 (1977).

[10]P. Y. Yu and B. Welber, Solid State Commun. **25**, 209 (1978).

[11]D. V. Lang, in $Deep$ $Centers$ in $Semiconductors,$ edited by S. T. Pantelides (Gordon and Breach, New York, 1985), p. 489.

[12]R. Tommer, E. Anastassakis, and M. Cardona, in $Light$ $Scattering$ in $Solids,$ edited by M. Balkanski, R. C. C. Leite, and S. P. S. Porto (Flammarion, Paris, 1976), p. 396.

[13]M. F. Li, P. Y. Yu, E. R. Weber, and W. Hansen, Appl. Phys. Lett. **51**, 349 (1987).

[14]B. L. Zhou, K. Ploog, and E. Gmelin, Appl. Phys. A **28**, 233 (1982).

Semicond. Sci. Technol. **4** (1989) 225–227. Printed in the UK

Some selected topics in high-pressure semiconductor research in China†

G G Qin†§ and M F Li‖
‡ Centre of Condensed Matter and Radiation Physics, CCAST (World Laboratory) and Department of Physics, Peking University, Beijing, China§
‖ Graduate School, Academia Sinica, Beijing, China

Received 30 September 1988, accepted for publication 8 December 1988

Abstract. We briefly summarise some aspects of high-pressure semi conductor research in China including: defect studies by capacitance transient under uniaxial stress and hydrostatic pressure; photoluminescence and other optical measurements; x-ray diffraction for phase transition studies.

The conduction (valence) bands of a many-valley semiconductor are split under uniaxial stress, hence the formula for the emission of electrons (holes) from a deep energy level to one conduction (valence) band is modified to the case of split conduction (valence) bands. Yao *et al* [1] did such a modification. According to the modified formulas and uniaxial stress deep level transient spectroscopy (USDLTS) measurements, the energy level position of the deep centre under uniaxial stress could be determined, irrespective of the splitting of the deep level (e.g., A centre) in Si [2] or the shift of the deep level without splitting (e.g., Au acceptor in Si). Yao *et al* [2] pointed out that in order to explain the pattern of split energy levels of the A centre obtained, in addition to the M^* defined in [3] a new parameter must be introduced, $L^* = (dE/d\varepsilon)_\perp$ where L^* is the change in the energy per unit strain along the direction perpendicular to the line connecting the two Si atoms of the Si–Si bond of the A centre and within the plane where the Si–Si bond lies. Selecting M^* and L^* for the best fit with experimental data gives $M^* = -7.4$ eV/unit strain and $L^* = 3.6$ eV/unit strain. Mou *et al* [4] measured the electron capture cross sections of A centre and Au acceptor in Si as functions of uniaxial stresses and found that the capture cross sections of the A centre have much stronger stress effects than those of the Au acceptor. Qin *et al* showed that by changing the bias and pulse conditions applied to the diode under test, pure neutrally and negatively charged A centres could be studied with USDLTS separately [5,6]. Keeping the sample temperature constant at a relatively higher value (125–148 K) with stress on the A centre will preferentially orient into the configuration with lower energy, and on maintaining this condition for a sufficiently long time, a Boltzman distribution between the

non-equivalent orientations can be obtained. Then the sample was quenched to 77 K to freeze out the orientations of the A centres and the stress-induced energy differences between non-equivalent orientations of neutral and negative A centres were obtained with the USDLTS separately.

Wang *et al* [7] detected three Pd-related levels $E_c - 0.18$ eV, $E_c - 0.22$ eV and $E_v + 0.33$ eV in Pd doped Si. From the experimental results they concluded that these levels belong to different Pd centres and that the $E_c - 0.18$ eV centre has C_{2v} symmetry, while $E_c - 0.22$ eV and $E_v + 0.33$ eV centres most probably have T_d symmetry. They suggested that the microstructure models for $E_c - 0.18$ eV and $E_c - 0.22$ eV centres for Pd atom with closed 4d shell occupy an off-centre position and an on-centre one, to a vacancy, respectively. Yao *et al* [8] studied the Cu level $E_c - 0.18$ in Si with USDLTS and found that there are two overlapping Cu-related deep levels with very similar activation energies (0.18 eV). The splitting of the level in USDLTS is due to the removing of an accidental degeneracy under uniaxial stress.

Li *et al* [9,10] have developed a novel method for determining the value of the shear deformation potential constant Ξ_u at the bottom of the conduction band for Si. The method is based on the [100] stress (F) dependence of the electron emission rate e_n from the deep level E_T to the split conduction band minima E_c. In the low stress limit when the splitting of the conduction band minima is small compared with kT, the deep level sees the average of the conduction band minima, $\partial \ln e_n/\partial F$ approaches

$$\frac{\partial}{\partial F} \frac{(\bar{E}_T - \bar{E}_c)}{KT}$$

and the conduction band splitting effect can be treated as a perturbation. In the large stress limit when the conduction band splitting is large compared with kT, the deep level only sees the lowest conduction band

---

† Based on a paper presented at The High Pressure in Semiconductor Physics conference organised by The High Pressure Research Centre of the Polish Academy of Sciences, Warsaw on 20–21 August 1988.
§ Address for correspondence.

G G Qin and M F Li

minima E'_c due to the Boltzmann distribution. $\partial \ln e_n/\partial F$ approaches

$$\frac{\partial}{\partial F}\frac{(\bar{E}_T - E'_c)}{KT}.$$

By subtraction of one value from the other, the valley splitting and thus the value of Ξ_u can be determined. For S° and S^+ levels in Si as reference, they obtained $\Xi_u = 10.7$–$11.3\,\text{eV}$ at $148.9\,\text{K}$ and 11.0–$11.6\,\text{eV}$ at $223.6\,\text{K}$ respectively. Li $et\ al$ [11] have further developed the method to study the symmetry properties of deep centres. The method is appropriate for the case of small energy splitting of the defect under stress. The key is to measure the stress coefficients of the average of the splitting energies for different crystal orientations. At low stress limit, $\partial \ln e_n/\partial F$ can be well fitted to linear and quadratic functions of stress F. The linear term corresponds to an average energy shift $(\bar{E}_c - \bar{E}_T)$ and the quadratic term is due to splitting of E_c. Using the value of Ξ_u obtained above and subtracting the quadratic term contribution, a very good linear relation for $\partial \ln e_n/\partial F$ versus F is obtained and the pressure coefficient $\partial(\bar{E}_c - \bar{E}_T)/\partial F$ can be deduced precisely.

The hydrostatic pressure behaviour of the emission coefficients of the deep levels e_n and e for the Au acceptor level E_t in Si has been investigated by Li $et\ al$ [12, 13]. The resulting pressure coefficient $\partial(E_c - E_T)/\partial p$ is $-1.9\,\text{meV}\,\text{kbar}^{-1}$. They have compared this result with the uniaxial stress coefficients of the same centre measured by Yau $et\ al$ [1] and concluded that the Au acceptor is not a simple gold substitutional or interstitial configuration in Si.

Recently Li has collaborated with the University of California at Berkeley to investigate the pressure-induced deep donors (PIDD) in GaAs and DX centres in AlGaAs [14–17]. They have successfully implemented capacitance transient techniques such as DLTS and photocapacitance under high pressure inside a diamond anvil cell. The photoionisation energy of PIDD in GaAs:Si obtained is $1.4\,\text{eV}$ and the thermal ionisation energy is $0.1\,\text{eV}$. The pressure behaviour of the DX centre in AlGaAs and the PIDD in GaAs both support the large lattice relaxation model of deep centres [18].

Zhao $et\ al$ [19] have investigated the photoluminescence (PL) of N and NN pairs in GaP samples under hydrostatic pressure at $77\,\text{K}$. When the pressure is lower than $33\,\text{kbar}$, the single N traps dominate the luminescence while above $33\,\text{kbar}$ the free exciton becomes important. Non-linear pressure shifts for the binding energies have been observed for the N and NN pairs bound excitons. Yang $et\ al$ [20,21] has successfully explained the experimental results by a Koster–Slater Green function calculation when the pressure dependence of the effective masses of the GaP conduction band is taken into account. Zhao $et\ al$ [22] have also investigated the PL of N traps in GaAs. When the pressure exceeds $25\,\text{kbar}$ the PL peak due to an N bound exciton was observed with pressure coefficient $2.8\,\text{meV}$

kbar^{-1} which is consistent with the theoretical calculation of Ren $et\ al$ for substitutional deep defects in GaAs [23].

Zhao $et\ al$ [24] and Wang $et\ al$ [25] have investigated the pressure effects of photoluminescence of AlGaAs–GaAs and GaAs–InGaAs quantum wells. The pressure behaviour of bound excitons, the band-edge shift and band-offset problems have been discussed.

Hu [26] has studied the phase transition of CdTe by high-pressure x-ray diffraction with a diamond anvil cell up to $392\,\text{kbar}$ at room temperature. It was found that the transformation sequence of CdTe is from the zincblende phase to the NaCl phase at $33.1\,\text{kbar}$, to β-Sn phase at $103\,\text{kbar}$ and to a new phase at $121.8\,\text{kbar}$. The new phase appears to have a simple orthorhombic structure (SGPmm2) with two atoms in the unit cell.

Zhu $et\ al$ [27] have developed some new techniques and methods for high-pressure experiments including a two-stage non-magnetic high-pressure equipment, a new method for producing a modulated magnetic field in a high-pressure chamber, photomodulated spectroscopy under high pressure, etc. These methods have been used for studying various physics problems in semimagnetic semiconductors [28, 29].

Acknowledgment

We acknowledge the support by the Science fund of Academia Sinica.

References

[1] Yao X C, Qin G G, Zeng S R and Yuan M H 1984 *Acta Phys. Sin.* **33** 377; 1986 *Chin. J. Phys.* **6** 279
[2] Yao X C, Mao J H and Qin G G 1987 *Phys. Rev.* B **35** 5734
[3] Watkins G D and Corbett J W 1961 *Phys. Rev.* **121** 1001
[4] Mou J X, Yao X C and Qin G G 1986 *Mater. Sci. Forum* **10–12** 481
[5] Qin G G, Yao X C and Mou J X 1985 *Solid State Commun.* **56** 201
[6] Yao X C, Mou J X and Qin G G 1986 *Chin. J. Semicond.* **7** 1139
[7] Wang L, Yao X C, Zhou J and Qin G G *Phys. Rev.* B to be published
[8] Yao X C, Wang L, Chen K M and Qin G G *Chin J. Semicond.* to be published
[9] Li M F 1985 *Acta Phys. Sin.* **34** 1549
[10] Li M F, Chen J X, Zhao X S and Li Y J 1986 *Mater. Sci. Forum* **10–12** 469
[11] Li M F, Li Y J, Zhao X S and Chen J X 1986 *18th Int. Conf. Physics of Semiconductors* (Singapore: World Scientific)
[12] Li M F, Chen J X, Yao Y S and Bai G 1985 *J. Appl. Phys.* **58** 2589

[13] Ren S Y, Mao D Q and Li M F 1986 *Chin. Phys. Lett.* **3** 313

[14] Li M F and Yu P Y 1987 *Solid State Commun.* **61** 13

[15] Li M F, Yu P Y, Weber E R and Hansen W L 1987 *Appl. Phys. Lett.* **51** 349

[16] Li M F, Yu P Y, Weber E R and Hansen W L 1987 *Phys. Rev. B* **36** 4531

[17] Li M F, Shan W, Yu P Y, Hansen W L, Weber E R and Bauser E 1988 *MRS Symp. Proc.* **104**

[18] Lang D V and Logan R A 1977 *Phys. Rev. Lett.* **39** 635

[19] Zhao X S, Li G H, Han H X, Wang Z P, Tang R M and Hu J Z 1984 *Chin. Phys. Lett.* **1** 15

[20] Yang G L 1985 *Chin. Phys. Lett* **2** 197

[21] Wang B S 1986 *Chin. Phys. Lett.* **3** 277

[22] Zhao X S, Li G H, Hang H X, Wang Z P, Tang R M and Che R Z 1984 *Acta Phys. Sin.* **33** 588

[23] Ren S Y, Dow J D and Wolford J 1982 *Phys. Rev. B* **25** 7661

[24] Zhao X S, Li G H, Hang H X, Wang Z P, Chen Z G, Sun D Z and Kong M Y 1986 *Chin. J. Semicond.* **7** 453

[25] Wang L J, Hou H Q, Zhou J M, Tang R M, Lu Z D, Wang Y Y and Huang Y to be published

[26] Hu J Z 1986 *Solid State Commun.* **63** 471

[27] Zhu H R, Shan W, Li Q G, Ju G L and Shen X C 1986 *Shanghai Institute of Technical Physics Annual Report* p 57

[28] Wei S, Shen X C, Zhao M G and Zhu H R *Acta Phys. Sin.* **35** 1290

[29] Jiang S, Shen X C, Li Q G, Zhu H R and Ju G L 1987–8 *Shanghai Institute of Technical Physics Annual Report* pp 124, 133

PHYSICAL REVIEW B   VOLUME 40, NUMBER 2   15 JULY 1989-I

Negative-U property of the DX center in $Al_xGa_{1-x}As$:Si

M. F. Li
Center of Theoretical Physics, Chinese Center of Advanced Science and Technology (World Laboratory), Beijing, China
and Graduate School, University of Science and Technology of China, Academia Sinica, P. O. Box 3908, Beijing, China

Y. B. Jia
Graduate School, University of Science and Technology of China, Academia Sinica, P. O. Box 3908, Beijing, China

P. Y. Yu
Department of Physics, University of California Berkeley, Berkeley, California 94720
and Materials and Chemical Sciences Division, Lawrence Berkeley Laboratory, Berkeley, California 94720

J. Zhou and J. L. Gao
Institute of Semiconductors, Academia Sinica, Beijing, China
(Received 30 January 1989)

We present two major points in this paper: (1) Statistics derived from the negative-U property of the DX center are not consistent with existing Hall experiments of Si-doped $Al_xGa_{1-x}As$. (2) The discrepancy between the negative-U model and Hall experiments can be improved if there exist two different donors SD and DX with comparable concentrations N_{SD} and N_{DX}. N_{SD}/N_{DX} increases with increasing N_{Si}. DX is a negative-U center binding two electrons and SD is a shallow donor. SD centers provide electrons and the electrons are captured by DX centers, and therefore the Fermi energy is no longer pinned to the DX energy level. Many experiments are reinterpreted in this view.

(I) Statistics derived from the negative-U property of the DX center are not consistent with existing Hall experiments.

The DX center has been one of the primary interests in the last decade for the electronic properties of III-V ternary semiconductors.[1,2] Recently, there has been significant progress since the negative-U argument of the DX center was proposed independently by Chadi and Chang,[3] and by Khachaturyan, Weber, and Kaminska.[4] The two-electron negative-U state, which is associated with a large lattice relaxation, explains a series of experiments including the absence of an ESR signal from DX centers.[4] Chadi and Chang's *ab initio* self-consistent calculations for substitutional donors in GaAs give a more quantitative description of the DX center. For a negatively charged center denoted by DX^-, a metastable resonant state with large lattice relaxation was found which is consistent with Theis, Mooney, and Wright's experiment.[5] A pressure of 18 kbar caused the DX^- center to be stable in the energy gap with large optical- and small thermal-ionization energies, a result that is consistent with recent pressure experiments.[6−8]

Although the negative-U model is very attractive in many aspects, conclusions derived from the current negative-U model are not consistent with existing Hall experiments, at least for Si in $Al_xGa_{1-x}As$. In the following, we focus our attention on the case of Si in $Al_xGa_{1-x}As$. According to the negative-U model,[3,4] Si occupying a Ga site has two possible bound electronic states: a shallow state E_d with no lattice relaxation or a very localized state E_{DX} which binds two electrons with large lattice relaxation and negative U. Following Baraff, Kane, and Schluter,[9] the energy difference between the positively charged Si center denoted by d^+ and the negatively charged Si center occupied by two electrons and denoted by DX^- is $2E_{DX}$. We have derived from grand-canonical-ensemble theory[10] the result that the probability of Si to be the DX^- center binding two electrons is

$$f_{DX^-} = \frac{1}{1+\exp[-2(E_F - E'_{DX})/kT]} , \tag{1}$$

when

$$E_d - E'_{DX} \gg kT . \tag{1a}$$

Here E_F is the Fermi energy of electrons. E'_{DX} is the free energy related to E_{DX} by

$$E'_{DX} = E_{DX} - S_{DX}T , \tag{2}$$

$$S_{DX} = (k \ln g_{DX})/2 . \tag{2a}$$

g_{DX} is the degeneracy factor of the DX^- state. According to Chadi and Chang's model, $g_{DX} = 4$. S_{DX} is the increase of entropy averaged to one electron. In deriving Eq. (1), we have neglected the probability f_{d^0} of Si being the d^0 state (the d^0 state is one electron with energy E_d with no lattice relaxation)

$$f_{d^0} = (\exp\{-[(E'_d - E'_{DX}) + (E_F - E'_{DX})]/kT\})f_{DX^-},$$

which is much smaller than f_{DX^-} when Eq. (1a) holds.

We temporarily suppose that there are no acceptors, no other shallow donors, and no other source to supply electrons. From the electrical neutrality condition, we have

$$f_{DX^-} = \frac{1}{2}\left[1 - \frac{n_0}{N_{Si}}\right] . \tag{3}$$

Here n_0 is the free-carrier concentration in the conduction band and N_{Si} is the doping concentration of Si. By substituting Eq. (3) into Eq. (1), we obtain

$$E_F = E'_{DX} - \frac{kT}{2}\ln\frac{1+(n_0/N_{Si})}{1-(n_0/N_{Si})} . \tag{3a}$$

Therefore, the following inequality is always valid:

$$f_{DX^-} \le \tfrac{1}{2} \tag{4}$$

or

$$E_F \le E'_{DX} . \tag{4a}$$

When increasing doping concentration N_{Si}, n_0 increases, E_F rises and is finally pinned to E'_{DX}. From Eq. (3), E_F is pinned to E'_{DX} when $n_0 \ll N_{Si}$ and the corresponding carrier concentration n_{0p} is determined by

$$n_{0p} = \left(\frac{1}{g_{DX}}\right)^{1/2} N_c \exp[-(E_c - E_{DX})/kT] . \tag{5}$$

Here N_c is the effective density of states at the bottom of the conduction band E_c. For the case of a direct band gap with Γ valley predominant, the preexponential factor $(1/g_{DX})^{1/2}N_c$ in Eq. (5) can be evaluated by taking $g_{DX} = 4$ and $m_\Gamma = (0.067 + 0.083x)m_0$.[11] Here m_0 is the free-electron mass and m_Γ is the effective mass of electrons at the Γ conduction valley. For instance, for $x \sim 0.3$, $(1/g_{DX})^{1/2}N_c \sim 3.6 \times 10^{17}$ cm^{-3} at room temperature. The large amount of temperature-dependent Hall experimental data reported by various laboratories does not satisfy Eq. (5) in the following three aspects:[12-15]

(1) For moderate doping concentration, $E_c - E_{DX}$ is independent of N_{Si} when x is fixed. Equation (5) predicts a saturated n_{0p} value independent of N_{Si}. The Hall electron concentration n_H determined by Hall experiments is always higher than n_{0p} evaluated by Eq. (5) and increases with increased N_{Si} with no saturation.

(2) For high-doping concentration, $E_c - E_{DX}$ increases with increased N_{Si}. The value of $E_c - E_{DX}$ can be determined from the temperature-dependent Hall data and n_0 can be estimated by Eq. (5). The carrier concentration determined by Hall measurement n_H is systematically higher than n_{0p}. For samples of $x \sim 0.3$, room-temperature n_H can reach as high as[13,14] 10^{18} cm^{-3}, the corresponding E_F is above the bottom of the conduction band E_c, and Eq. (4a) is no longer valid.

(3) The ratio n_H/n_{0p} varies from sample to sample, indicating a general trend to increase with increasing doping concentration but with some random exceptions.

For illustration, we show in Fig. 1 some typical experimental data of n_H vs T (Ref. 12) and compare with n_{0p} evaluated by Eq. (5). The aforementioned discrepancy between theory and experiment is evident.

The following effects have been considered to improve the theory. When the combination of the Γ, X, and L valleys' contributions to N_c is considered,[12] N_c is increased and temperature dependent. In high-doping concentration, the effective mass of electrons in the conduction band

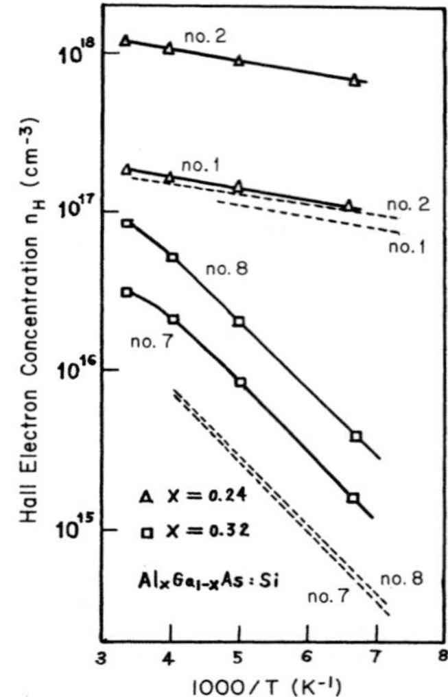

FIG. 1. Selected data of temperature-dependent Hall electron concentration n_H in Al$_x$Ga$_{1-x}$As, as quoted from Ref. 12. Samples No. 1: $x = 0.24$, $N_{Si} = 2.5 \times 10^{17}$ cm^{-3}. Sample No. 2: $x = 0.24$, $N_{Si} = 4.0 \times 10^{18}$ cm^{-3}. Sample No. 7: $x = 0.32$, $N_{Si} = 4.5 \times 10^{16}$ cm^{-3}. Sample No. 8: $x = 0.32$, $N_{Si} = 2.5 \times 10^{17}$ cm^{-3}. The dashed lines are evaluated from Eq. (5) with $g_{DX} = 4$ and $m_\Gamma = (0.067 + 0.083x)m_0$ (Ref. 11).

is slightly increased,[16] and thus N_c is increased. All these effects have been estimated to be too small to explain the experiments. The ratio of true electron concentration and Hall concentration n_T/n_H is in the range of 1–3,[12] which enlarges the discrepancy between theory and experiment. Another possible cause is the temperature dependence of $E_c - E'_{DX}$.[14] If $E_c - E_{DX}$ has a negative linear temperature coefficient, or a negative-entropy term in Eq. (2a), this will increase the preexponential factor in Eq. (5). The negative entropy means that when the defect is occupied by two electrons, the number of different microscopic states decreases. Such a picture is out of Chadi and Chang's model but is not impossible. However, this effect is independent of the doping concentration. In summary, all these considerations can at most to some extent improve the systematic discrepancy between theory and experiment but they are not able to explain all three aspects satisfactorily.

(II) The possible existence of shallow donors.

We suppose that there exist two different kinds of donors. The first kind of donor, denoted by DX with concentration N_{DX}, is the negative-U center which binds two electrons with energy level E_{DX} as suggested in Refs. 3 and 4. Here, we follow Lang, Logan, and Jaros to reserve the name of DX to the center itself[1] rather than use it exclusively to refer to the broken-bond configuration as did Chadi and Chang.[3] The second kind of donor, denoted by SD with concentration N_{SD}, is a shallow donor. The following equation is satisfied:

$$N_{SD} + N_{DX} + N_A = N_{Si} . \tag{6}$$

Here N_A is the concentration of compensating acceptors due to the amphoteric nature of Si in $Al_xGa_{1-x}As$. The previous deep-level transient spectroscopy (DLTS) and Hall experiments have led to extensive belief in early days that

$$N_{SD} \ll N_{DX} , \tag{7}$$

when Eq. (1a) holds. However, if DX binds two electrons, inequality (7) is not necessary and can be replaced by

$$N_{SD} - N_A < N_{DX} . \tag{8}$$

We like to emphasize that there is no existing experiment to our knowledge which excludes the possible existence of the SD center with concentration N_{SD} comparable to N_{DX} if DX binds two electrons. Many experiments should be reexplained. For instance, the electrons supplied by SD are first activated to conduction band, then captured by DX centers. Therefore, the carrier density n_0 is much lower than the electron density trapped in DX centers, as measured by DLTS or Hall experiments. The lack of an ESR signal[4] and the increase of mobility of conduction electrons after persistent photoconductivity effect[4,12] do not exclude the existence of SD centers. As long as Eq. (8) holds, the electrons supplied by SD are trapped by DX centers and, therefore, no ESR signal can be detected from either SD or DX centers. The far-infrared experiments of Theis et al.[17] should be carefully reinterpreted. They reported that the $1s$-$2p$ transition of shallow Si donors in $Al_xGa_{1-x}As$ was observed only after material is exposed to visible or near-visible light. The absorption showed strong correlation with $N_{Si} - N_A$, thus they concluded that the photoionized DX centers act as shallow donors. However, their conclusion was based on the assumption of Eq. (7). If Eq. (7) is not valid and N_{SD} is comparable to N_{DX}, their experiment should be carefully reexplained.

When both N_{SD} and N_A are comparable to N_{DX}, from electrical neutrality condition we have

$$f_{DX^-} = \frac{N_{DX^-}}{N_{DX}} = \frac{1}{2}\left[1 + \frac{N_{SD}}{N_{DX}} - \frac{n_0}{N_{DX}} - \frac{N_A}{N_{DX}}\right] , \tag{9}$$

$$N_{d^+} = N_{DX} - N_{DX^-} . \tag{9a}$$

Here N_{DX^-} is the concentration of negatively charged DX centers. N_{d^+} is the concentration of the positively charged DX center. When we substitute Eq. (9) into Eq. (1), we have

$$E_F = E'_{DX} + \frac{kT}{2}\ln\frac{1 + (N_{SD}/N_{DX} - N_A/N_{DX} - n_0/N_{DX})}{1 - (N_{SD}/N_{DX} - N_A/N_{DX} - n_0/N_{DX})} . \tag{10}$$

E_F may exceed E'_{DX} if N_{SD} is comparable to N_{DX}. The corresponding carrier concentration n_0 is

$$n_0 = \left[\frac{N_{DX} + N_{SD} - N_A - n_0}{N_{DX} - N_{SD} + N_A + n_0}\right]^{1/2} N_c \left[\frac{1}{g_{DX}}\right]^{1/2} \times \exp[-(E_c - E_{DX})/kT] . \tag{11}$$

When $n_0 \gg N_{DX} - N_{SD}$, Eq. (11) reduces to

$$n_0 = \left[\frac{N_{DX} + N_{SD} - N_A}{N_{DX} - N_{SD} + N_A}\right]^{1/2} N_c \left[\frac{1}{g_{DX}}\right]^{1/2} \times \exp[-(E_c - E_{DX})/kT] . \tag{11a}$$

Comparing Eq. (11a) with Eq. (5) clearly shows that the role of N_A is to decrease the preexponential factor of n_0. Therefore, consideration of N_A enlarges the discrepancy between theory and Hall experiments. On the contrary, if N_{SD}/N_{DX} increases with increasing N_{Si}, the existence of N_{SD} improves the consistency overall between theory and Hall experiments because the extra preexponential factor in Eqs. (11) or (11a) may be much larger than unity and increases with increasing N_{Si}.

The origin of SD is not yet clear. A possible candidate may be a complex Si_{sub}-Si_{int} pair or something else. Si_{sub}-Si_{int} is a reasonable candidate to explain the tendency of increasing N_{SD}/N_{DX} when N_{Si} is increased. Formation of silicon complexes have been reported by Maguire, Murray, and Newman for infrared measurement in GaAs.[18] An alternate possible explanation is that SD and DX are bistable states of one center and there is a very large barrier between the SD state and the DX state. The defect never overcomes the barrier at room temperature so there is no transfer from SD to DX or vice versa. Finally, it is interesting to note that Henning and Ansems[19] have recently argued from their optical experiments that the Si donor in $Al_xGa_{1-x}As$ has double faces. One face has large lattice relaxation and the other face has small lattice relaxation. Although the starting points and the methods are very different between our work and that of Henning and Asems, the two conclusions might be related. More detailed investigations and experiments are in progress in our laboratory.

We would like to acknowledge Dr. D. J. Chadi, Professor E. R. Weber, and Dr. J. C. M. Henning for kindly sending us copies of their work before publication. We are grateful to Professor K. Huang for his encouragement of this work and stimulating discussion. We also thank Dr. D. S. Chiang, Dr. Z. G. Wang, Dr. G. H. Li, Dr. B. F. Zhu, and Dr. H. Tang for valuable discussions. This work is supported by Natural Science Foundation of China. P. Y. Yu is supported by the Director, Office of Basic Energy Sciences, Materials Science Division of the U.S. Department of Energy under Contract No. DE-AC03-76SF-00098.

[1] D. V. Lang, R. A. Logan, and M. Jaros, Phys. Rev. B **19**, 1015 (1979).

[2] D. V. Lang, in *Deep Centers in Semiconductors,* edited by S. T. Pantelides (Gordon and Breach, New York, 1985), p. 489.

[3] D. J. Chadi and K. J. Chang, Phys. Rev. Lett. **61**, 873 (1988); and (unpublished).

[4] K. Khachaturyan, E. R. Weber, and M. Kaminska, in *Defects in Semiconductors,* edited by G. Ferenczi, Materials Science Forum Series, Vol. 38–41 (Trans. Tech. Publications, Aedermannsdorf, Switzerland, 1989), p. 1067.

[5] T. N. Theis, P. M. Mooney, and S. L. Wright, Phys. Rev. Lett. **60**, 361 (1988).

[6] M. Mizuta, M. Tachikawa, H. Kukimoto, and S. Minomura, Jpn. J. Appl. Phys. **24**, L143 (1985).

[7] M. F. Li, P. Y. Yu, E. R. Weber, and W. Hansen, Appl. Phys. Lett. **51**, 349 (1987); Phys. Rev. B **36**, 4531 (1987).

[8] D. K. Maude, J. C. Portal, L. Dmoswski, T. Foster, L. Eaves, M. Nathan, M. Heiblum, J. J. Haris, and R. B. Beall, Phys. Rev. Lett. **59**, 815 (1987).

[9] G. A. Baraff, E. O. Kane, and M. Schluter, Phys. Rev. B **21**, 5662 (1980).

[10] L. D. Landau and E. M. Lifshitz, *Statistical Physics* (Pergamon, London, 1958), p. 106. During revision of this paper we found that T. N. Theis *et al.* reached similar statistics of Eq. (1) in Ref. 4, p. 1073.

[11] *Landolt-Börnstein: Numerical Data and Functional Relationships in Science and Technology,* edited by O. Madelung (Springer-Verlag, Berlin, 1986), Vol. III, Part 2a, p. 137.

[12] N. Chand, T. Henderson, J. Klem, W. T. Messelink, R. Fisher, Y. C. Chang, and H. Morkoc, Phys. Rev. B **30**, 4481 (1984).

[13] H. Kunzel, K. Ploog, K. Wunstel, and B. L. Zou, J. Electron. Mater. **13**, 281 (1984); H. Kunzel, A. Fischer, J. Knecht, and K. Ploog, Appl. Phys. A **32**, 69 (1983).

[14] E. F. Schubert and K. Ploog, Phys. Rev. B **30**, 7021 (1984).

[15] T. Ishibashi, Seigo Tarucha, and H. Okamoto, Jpn. J. Appl. Phys. **21**, L476, (1982); T. Ishikawa, J. Saito, S. Sasa, and S. Hiyamizu, *ibid.* **21**, L675 (1982).

[16] M. Heiblum, M. V. Fischetti, W. P. Dumke, D. J. Frank, I. M. Anderson, C. M. Knoedler, and L. Orsterling, Phys. Rev. Lett. **58**, 816 (1987).

[17] T. N. Theis, T. F. Kuech, L. F. Palmateer, and P. M. Mooney, in *Gallium Arsenide and Related Compounds 1984,* Institute of Physics Conf. Ser. Vol. 74 (IOP, Bristol, 1984), p. 241.

[18] J. Maguire, R. Murray, and J. C. Newman, Appl. Phys. Lett. **50**, 516 (1987).

[19] J. C. M. Henning and J. P. M. Ansems, in Ref. 4, p. 1085.

Reprinted with permission from M.F. Li, P.Y. Yu and E R. Weber, Appl. Phys. Lett. Vol.59, pp.1197–1199 (1991). Copyright 1991, American Institute of Physics.

Simulation of effects of uniaxial stress on the deep level transient spectroscopy spectra of the *DX* center in AlGaAs alloys

Ming-fu Li
Department of Physics and Department of Materials Science and Mineral Engineering, University of California, Berkeley, California 94720

Peter Y. Yu
Department of Physics, University of California, Berkeley, California 94720 and Materials Sciences Division, Lawrence Berkeley Laboratory, Berkeley, California 94720

E. R.Weber
Department of Materials Science and Mineral Engineering, University of California, Berkeley, California 94720 and Materials Sciences Division, Lawrence Berkeley Laboratory, Berkeley, California 94720

(Received 1 April 1991; accepted for publication 1 July 1991)

Recently the effect of uniaxial stress on the deep level transient spectroscopy (DLTS) of the *DX* center in AlGaAs alloys have been reported by two separate groups. In both experiments no splitting of the DLTS peak was observed. We have analyzed the experimental results in terms of a large lattice relaxation model in which the *DX* center can have either a positive or a negative Coulomb energy U. We found that if the symmetry of the *DX* center depended on its charge state then its DLTS peak was not split by uniaxial stress in contrast to other defects with large lattice relaxation (such as the A center in Si).

Lattice distortion (LD) is an important property of many deep centers (DC) in semiconductors. Recently there is much interest in the *DX* center found in alloys of GaAlAs.[1] According to some microscopic models, such as those proposed by Morgan [2] and by Chadi and Chang [3] (the latter will be referred to as the CC model) the *DX* center is formed by a substitutional donor. Both models assume that when the donor is ionized the defect suffers no LD. After capturing one (the Morgan model) or two electrons (the CC model) the donor atom on Ga site or one of the nearest neighboring Ga atoms (in case of donors on As site) is displaced along one of the bond directions. While the CC model has been supported by many experimental results, it is by no means universally accepted.[4–7]

Uniaxial stress has been shown to be a very powerful technique for determining the symmetry of LD in the A center in Si.[8–10] To understand how uniaxial stress can help to determine the symmetry of LD in a DC, we represent the LD of the DC by configuration coordinate diagrams as shown in Fig. 1. Assume that the DC can be displaced along any one of the six equivalent (100) directions. Figure 1 (a) shows the diagram that is representative of DC such as the A center in Si under zero stress. The arrows represent emission and capture processes of DC displaced along two equivalent directions. Under a [100] stress [Fig. 1 (b)] the centers with displacement along the (100) and ($\bar{1}$00) directions [to be referred to as the (100) centers] will have emission barrier height (E_e) different from the (010) and (001) centers. This results in a splitting of the center's DLTS peaks. In general the relationship between the LD of a DC and its patterns of splitting under uniaxial stresses along the crystallographic axes can be predicted by the theory of Kaplyanski.[11]

Recently two groups have reported measurements of the effect of uniaxial stress on the DLTS spectra of the *DX*

center in AlGaAs alloys doped with either Te (Ref. 12) or Si (Ref. 13). Unlike the case of the A center in Si, *no splitting of the DLTS spectra* was observed in both experiments for uniaxial stress as large as 16 kbar. In this letter we will point out an important difference between the LD

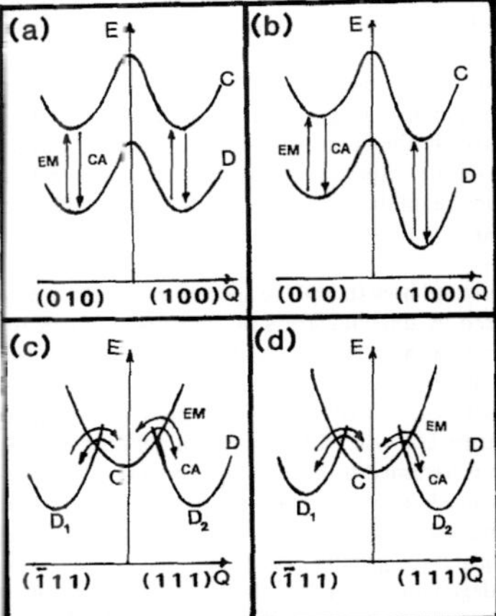

FIG. 1. Configuration coordinate diagram of two kinds of DC exhibiting large LD with [(b) and (d)] and without uniaxial stress [(a) and (c)]. (a) and (b) represent DC where the ground state (D) and the excited state (C) have the same amount of LD. In this case the LD is along the six (100) directions and the uniaxial stress in (b) is applied along the (100) direction. (c) and (d) represent the *DX* center where only D shows LD along the eight (111) directions while C is not relaxed. The stress in (d) is applied along the (111) direction. The arrows labeled as EM and CA represent respectively, emission and capture processes.

of the A center and that of the DX center which accounted for their different behaviors under uniaxial stress. We have simulated the effect of uniaxial stress on the DLTS spectra of the DX center based on large LD models and showed that the calculated spectra exhibit no stress-induced splitting in agreement with the experiments.

Figure 1(c) shows the configuration coordinate diagram for emission and capture processes in the DX center under zero stress. The main difference between Figs. 1(a) and 1(c) is that the symmetry of the A center is independent of the DC charge state.[10] During capture and emission processes the LD of the oxygen atom in the A center does not change significantly. The degeneracies between A centers with different directions of LD are splitted by uniaxial stress. Since there is a large barrier height between centers with different LD, the number of centers with different directions of LD does not change during DLTS experiments. This picture is valid for the A center in Si where the time constant for LD to change direction is $\approx 10^8$ s while the electron capture and emission time constants are $< 10^{-3}$ s at 90 K.[9] Thus during the DLTS experiments LD of the A center is frozen and defects with LD along different directions are not in thermal equilibrium. The situation is different for the DX centers. The symmetry of the DX center depends on its charge state. As shown in Fig. 1(c), when the center is positively charged (curve C), the defect is not relaxed and has symmetry T_d. When the DX center is occupied by one or two electrons, its symmetry is lowered. Figure 1(d) shows that uniaxial stress splits only the degeneracies of the neutral (Morgan model) or negatively charged (CC model) states. During the capture process, either the defect atom or one of its nearest neighbors undergoes LD. From DLTS experiments, one can infer that this LD in the DX center can occur in times of the order of the filling pulse durations. Similarly the reverse LD must occur during emission within the DLTS time window. Thus, it is possible for LD of the DX centers to change from a high energy direction to a lower energy direction through the intermediate undistorted T_d state during the DLTS experiment. Under this assumption, there is thermal equilibrium between defects with different LD and their populations are determined by Boltzmann statistics.

To understand how the DLTS spectra of the DX center will be affected by uniaxial stress in the above picture, we have performed numerical simulations appropriate for the DX centers in GaAlAs alloys doped with Te. We have considered both the positive-U ($+ U$) model of Morgan[1] and the negative-U ($- U$) CC model.[2] In both models the ground state is assumed to have a C_{3v} symmetry. The theory of Kaplyanskii is used to calculate the splitting pattern of the ground state. Under [111] and [110] stresses, the four-fold degenerate ground states are splitted into two states with degeneracies A_1 and A_2. We will denote these two sets of inequivalent DX centers as $DX1$ and $DX2$ and their corresponding energy minima as D_1 and D_2, respectively [see Fig. 1(d)]. These energies represent the total energy of the defect and therefore are not one electron energies. Since very little is known about the deformation

potentials of the DX centers we will consider three possible cases assuming that $D_2 < D_1$. For [111] stress, we will consider both when $DX1$ is the singlet (case 1) and when $DX1$ is the triplet (case 2) state. For [110] stress the ground state is always split into two doublets (case 3).

Experimentally it has been found that the DLTS spectra associated with DX centers often exhibit multiple peaks. For simplicity we will neglect this multiple peak structure. Furthermore, we have found that each DLTS peak can be approximated by a Gaussian function of the form:[12]

$$A_i \exp\{ - 0.5[(T_{pi} - T)/W]^2\} \qquad (1)$$

where T is the sample temperature, A_i and T_{pi} are, respectively, the population and peak positions of the level DXi ($i = 1$ or 2) and W is the half width of the DLTS peak. We have chosen W to be 6.5 K and the average value of T_{p1} and T_{p2} (equal to the zero stress DLTS peak position) to be around 140 K. These values roughly correspond to the experimental values for DX centers in $Ga_{1-x}Al_xAs$:Te with $x = 0.38$ and a DLTS time window of 21 ms/42 ms.[12] The ratio of A_1/A_2 is determined by the

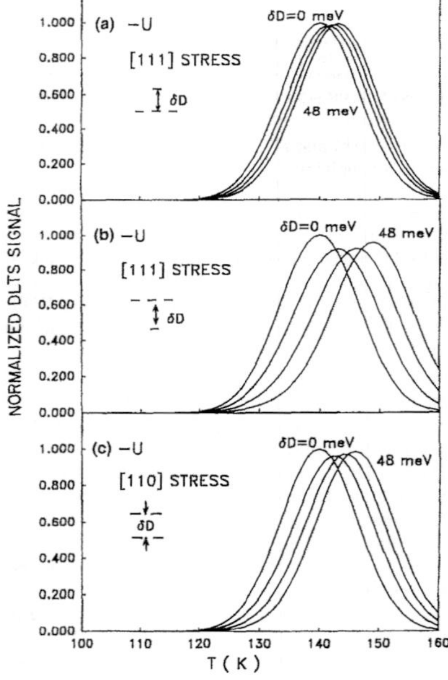

FIG. 2. Simulated DLTS spectra of the DX center under [111] and [110] uniaxial stresses for the $- U$ model of CC. In all three cases δD for the four curves increase in the sequence 0, 24, 36, and 48 meV towards the right. The sign of the deformation potential for the ground state is different in (a) and (b).

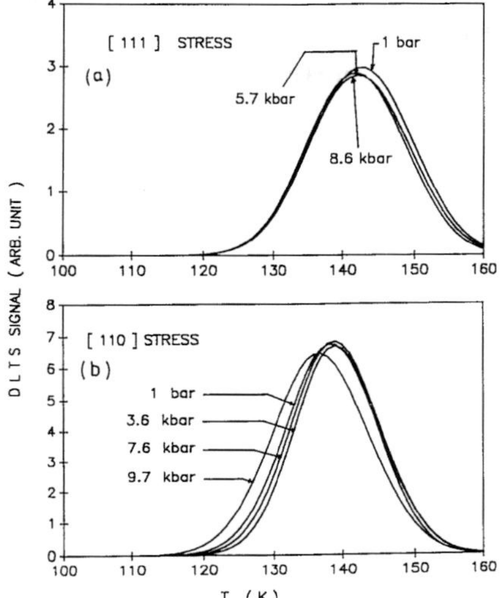

FIG. 3. Experimental uniaxial stress dependent DLTS peak of the *DX* center in AlGaAs:Te under (a) [111] and (b) [110] stresses obtained by deconvolving the experimental spectra in Ref. 12 into the sum of two Gaussian peaks. Only the higher temperature peak is reproduced here.

degeneracies of $DX1$ and $DX2$ and their relative occupation probability. By applying the grand canonical Gibbs distribution function, we obtain the probability of an LD along the direction i as:[14]

$$P_i = \exp(\Omega + \mu N - D_i)/K_b T \tag{2}$$

where Ω is a thermodynamical potential determined by normalization, μ is the chemical potential, N is the number of electrons captured on the DX center ($N = 1$ for $+ U$ case and $N = 2$ for $- U$ case) and K_b is the Boltzmann's constant. Thus, the relative occupation probability is given by:

$$P_1/P_2 = \exp[- (D_1 - D_2)/K_b T]. \tag{3}$$

The effect of uniaxial stress on the DX center is simulated by adjusting the value of $D_2 - D_1 = \delta D$. In general δD is proportional to δE_e where the constant of proportionality is dependent on the model for the effect of stress on the configuration coordinate diagram. For simplicity we will neglect the effect of uniaxial stress on the capture barrier heights of DX centers. We assume further that $\delta E_e = - (l/\tau)\delta D$ where $\tau = 1$ ($+ U$) or 2 ($- U$). The stress-induced splitting of the peak temperatures $\delta T_p = T_{p2} - T_{p1}$ is related to δE_e by the following approximate expression:

$$\delta T_p = \delta E_e (T_p/E_e). \tag{4}$$

Assuming the values of $E_e = 280$ meV and $T_p = 140$ K at zero stress, 1 meV of δE_e gives rise to 0.5 K of δT_p.

The calculated DLTS spectra for the three cases listed above are shown in Figs. 2(a)–2(c) for values of δD varying between 0 and 48 meV. Only the results for the $- U$ case are given because the results for the $+ U$ cases are similar. For all three uniaxial stress directions there are no resolvable splitting of the DLTS peak for both negative and positive U cases, in contrast to the A center in Si. There are no detectable broadenings of the DLTS peaks.

There are two peaks in the experimental DLTS spectra[12] so we have deconvoluted the spectra into the sum of two Gaussians for comparison with the calculated spectra. Only the higher temperature peak is replotted in Fig. 3 for several values of [111] and [110] stresses. Comparison of these spectra with Fig. 2 shows that the absence of any stress-induced splitting in the experimental DLTS spectra is consistent with the $- U$ model of CC. The calculated spectra show only shift of the DLTS spectra towards higher temperature while the experimental peak shows shifts to both higher and lower temperatures. The shifts in the experimental DLTS peak contain contribution by the hydrostatic component of the applied stress.[12] After this hydrostatic component induced shift is subtracted, the sign and values of δD can be determined by fitting the experimental spectra with the theoretical curves.

We thank Professor E. E. Haller for his interest in this work and for helpful discussions. P.Y.Y. was supported in part by the Miller Institute for Basic Research in Science. This research is supported by the Director, Office of Energy Research, Office of Basic Energy Sciences, Materials Science Division of the U.S. Department of Energy under contract BE-AC03-76SF00098.

[1] See D.V. Lang, in *Deep Centers in Semiconductors*, edited by S.T. Pantelides (Gordon and Breach, New York, 1986) p. 489; P. M. Mooney, J. Appl. Phys. **67**, RI (1990).

[2] T. N. Morgan, Mater. Sci. Forum **38–41** 1079 (1989).

[3] D. J. Chadi and K. J. Chang, Phys. Rev. Lett. **61**, 873 (1988); Phys. Rev. B **39**, 10366 (1989).

[4] M. Zazoui, S. L. Feng, and J. C. Bourgoin, Phys. Rev. B **41**, 8485 (1990).

[5] M. Mizuta and T. Kitano, Appl. Phys. Lett. **52**, 126 (1988).

[6] E. Yamaguchi, K. Shiraishi, and T. Ohno, in *Proceedings of the 20th International Conference on the Physics of Semiconductors*, edited by E. M. Anastassakis and J. D. Joannopoulos (World Scientific, Singapore, 1990) Vol. 1, p. 501.

[7] K. Khachaturyan, E. R. Weber, and E. G. Colas (unpublished).

[8] G. D. Watkins and J. W. Corbett, Phys. Rev. **121**, 1001 (1961).

[9] J. M. Meese, J. W. Farmer, and C. D. Lamp, Phys. Rev. Lett. **51**, 1286 (1983).

[10] G. G. Qin, X. C. Yau, and X. J. Mou, Solid State Commun. **56**, 201 (1985).

[11] A. A. Kaplyanskii, Opt. Spektrosk. **10**, 165 (1961); Opt. Spectrosc. (USSR) **10**, 33 (1961).

[12] M. F. Li, P. Y. Yu, E. Bauser, W. L. Hansen, and E. E. Haller (unpublished); Bull. Am. Phys. Soc. **36**, 503 (1991).

[13] Z. Wang, K. Chung, T. Miller, F. Williamson, and M. I. Nathan, Appl. Phys. Lett. **58**, 2366 (1991).

[14] L. D. Landau and E. M. Lifshitz, *Statistical Physics* (Pergamon Press, Oxford, 1958) p. 106, Eq. (35.2).

Proc. 5th Int. Conf. High Pressure in Semiconductor Physics, Kyoto, 1992
Jpn. J. Appl. Phys. Vol. 32 (1993) Suppl. 32-1, pp. 200–205

Probing the DX Center in GaAs and Related Alloys by Capacitance Transient Measurements under Stress

Ming-Fu Li, Center for Optoelectronics, Department of Electrical Engineering,National University of Singapore,Singapore 0511.

Peter Y. Yu, Department of Physics, University of California, Berkeley and Materials Science Division, Lawrence Berkeley Laboratory, Berkeley, CA 94720, USA.

This paper will review some recent results in the study of the DX center in GaAs and AlGaAs using capacitance transient techniques under hydrostatic and uniaxial stress. These measurements have established the DX center as the deep state of substitutional donors in GaAs or AlGaAs which is lowered into the band gap by alloying or pressure.When compared with the predictions of various models of the DX center, the stress results are shown to be consistent with the DX centers having a two electrons negatively charged ground state (negative U).Models of the DX center as effective mass state of donor associated with the L conduction band valleys are found to be inconsistent with stress experiments.

I. Introduction :

The DX center was first identified by Lang and Logan in $Al_xGa_{1-x}As$ with x>0.22 in 1977 [1]. During the last decade its properties have been studied extensively by both basic and applied researchers [2-4]. Lang et al. introduced the name DX center because they believed that it was a complex defect involving a donor atom D and an unknown constituent X.Since this center occurred only in alloys, it was believed that X was introduced during alloying. Some key characteristics of the DX center which distinguished it from other deep centers are: (1) a very large difference between its thermal ionization energy E_t (typically around 0.1 eV) and its optical ionization energy E_{op} (about 1 eV); and (2) a very small capture cross section which gives rise to persistent photoconductivity(PPC) at 77K. The electron capture cross section of large variety of deep levels in GaAs and GaP has been extensively studied by Henry and Lang[5],as summarized in Fig.1.The experimental data can be fitted by Multiphonon Emission Theory (MPE)[5,6] with an activation energy E_c at high temperature when kT is larger than the phonon energy. On the other hand,there is a low limit of the capture cross section of $10^{-21}cm^2$ at low temperatures due to optical capture. However, the capture cross section of the DX center can be several order of magnitude lower than 10^{-21} cm^2 which means that there is no optical capture for DX center. Lang has explained

these properties of the DX center in terms of large lattice relaxation and zone-edge TA phonon of 10 meV interaction with the DX center[2]. Since Lang et al's original paper in 1977, great progress has been made towards developing a microscopic model of the structure of DX center. In this paper,we will focus on some recent transient capacitance experiments which have helped to elucidate the nature of the DX center and in which high stress, both hydrostatic and uniaxial,has played an important role.

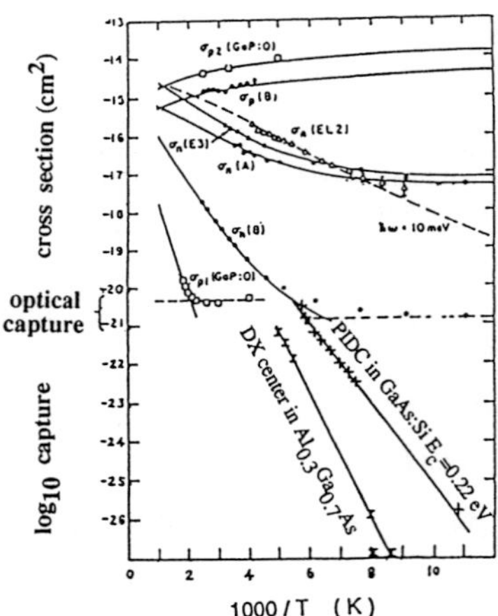

Fig.1. Capture behaviours of PIDC (+) in GaAS:Si under 29 kbar [11,12] and DX center (I) in $Al_{0.3}Ga_{0.7}As$[18]. They are different from catpure behaviours of most deep defects in GaAs and GaP [2,5] as illustrated in the context.

Jpn. J. Appl. Phys. Vol. 32 (1993) Suppl. 32–1

M.-F. Li and P. Y. Yu   201

II. Establishment of the DX center as due to substitutional donor :

Lifshitz et al. were the first to notice the existence of a relationship between pressure and Al alloying on the DX center[7]. They noted that 1 kbar of pressure has approximately the same effect on the conduction band of GaAs as increasing the Al concentration by 1 %. Since Lang et al. have found that the DX center appeared in $Al_xGa_{1-x}As$ when x>0.22, it was natural to ask whether the DX center will appear in GaAs under a pressure of more than 20 kbar ? This interesting question was answered by Mizuta et al. in 1985[8]. By applying quasi-hydrostatic pressure on n-type GaAs, Mizuta et al. found that a peak appeared in the deep level transient spectroscopy (DLTS) spectra at pressures above 24 kbar as shown in Fig.2. They identified this peak as the same DX center peak found by Lang et al. in $Al_xGa_{1-x}As$ when x>0.22. In a subsequent paper, Tachikawa et al. found evidence of PPC in GaAs:Si under pressure at 77 K [9], as depicted in the inset of Fig.2. Although these results strongly suggest that the new center found in GaAs under pressure is similar to the DX center in AlGaAs, more quantitative experimental results are needed to strengthen this conclusion. In particular, it is necessary to determine E_t and E_{op} of this pressure-induced

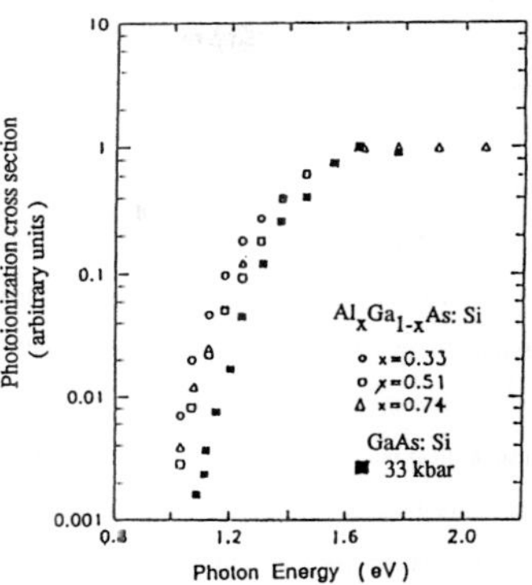

Fig.3 Photoicnization cross section spectra of the PIDC (■) in GaAs:Si under pressure of 33 kbar[12] at 77 K and of DX center in $Al_xGa_{1-x}As$ (o □ △) at 84 K.[13]

deep center (PIDC). Photoionization experiments are difficult to perform with the Bridgman anvil device used by Mizuta et al. so we have used a diamond anvil high pressure cell (DAC) instead to study this PIDC in GaAs. In order to perform transient capacitance measurements inside the DAC, we have used the method similar to those described by Erskine et al. [10] to introduce wires into the DAC. Our key experimental results for the PIDC in GaAs are summarized in Fig.1 and 3[11,12]. Fig.3 shows that the photcionization spectra of the PIDC in GaAs:Si is very similar to the DX center in AlGaAs:Si[13]. By fitting the experimental data with the theory of Lang, Logan and Jaros[1] we determined the E_{op} of the PIDC in GaAS:Si to be 1.4 eV. In fig.1, our experimental results show that the capture behavior of PIDC also has all the characteristics of the DX center with a capture activation energy E_c= 0.22 eV, giving rise to PPC at 77 K. Combined with the emission activation energy E_e of 0.30 eV deduced from the DLTS spectra[11], we found Et to be 0.08 eV, which is an order of magnitude smaller than E_{op}. We have thus conclusively shown that the PIDC found by Mizuta et al. in GaAs has all the important attributes of the DX center in AlGaAs. However, we noticed that there is quantitative difference for the DX center in GaAs and in AlGaAs. The origin of the quantitative difference between the DX centers in AlGaAs:Si and in GaAs:Si has recently been explained by

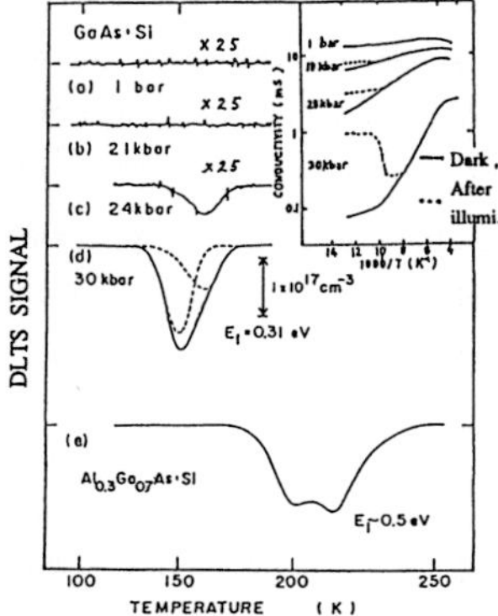

Fig.2 DLTS spectra taken for GaAs:Si under different pressure(a)-(d) and for $Al_{0.3}Ga_{0.7}As$:Si(e). Rate Window : 66 S⁻¹ with t_2/t_1 =2 [8]. The inset shows persistent photoconductivity effect of GaAs:Si under different pressures [9].

202   Jpn. J. Appl. Phys. Vol. 32 (1993) Suppl. 32–1   M.-F. Lɪ and P. Y. Yu

Mooney et. al[14] to be due to local environmental effect on the property of Si doped DX center.The more convincing experiment is again the pressure experiments by Calleja et al.[15] and Baba et al.[16]. The experiment by Calleja et al. is shown in Fig.4[15]. $Al_xGa_{1-x}As$ samples with x= 0, 0.04, 0.08 respectively were used and DLTS measurements were carried out under pressure. In fig.4 ,the lower temperature peak with emission activation energy 0.34 eV corresponds to DX center without Al atom as near neighbour and the experimental result is consistent with previous data of GaAs:Si[11,17].The higher temperature peak with emission activation energy 0.44 eV corresponds to DX center with Al atoms as near neighbours and the experimental results is also consistent with previous result of AlGaAs:Si:[18]. The 0.1 eV emission activation energy difference between DX centers with or without Al atoms as neighbours has been further confirmed by self-consistent theoretical calculation [19].

In addition to DX centers resulting from Group IV donors such as Si,the pressure dependence of DX centers due to Group VI donors such as Te has also been studied[20].The experiments were performed on $Al_xGa_{1-x}As$ epilayers,with x= 0.15, 0.25, 0.35 and doped with $5 \times 10^{16} cm^{-3}$ Te. The DLTS peak emerged at 16 kbar, 7 kbar and 1 bar for the x=0.15, 0.25 and 0.35 samples respectively. These experiments together with the results in GaAs:Si show that at ambient pressure the DX center in GaAs is a resonant state above the conduction band.As a result of the change in the conduction band structure caused by either alloying or pressure, the DX center emerges from the conduction band into the band gap and becomes the stable ground state of the donor.

III.Models of the DX center tested by stress measurements :

Once it became clear that the DX center was a simple substitutional donor in GaAs which exhibited a shallow-to-deep transformation as a result of change in the band structure induced by pressure or alloying, many models have been proposed to explain its properties.Both uniaxial and hydrostatic stress measurements are powerful techniques for testing these models since stress can modify the sample properties without changing the chemical properties.We will consider specifically tests of two aspects of the predictions by existing models.

a) The negative U property of the DX center :

Chadi and Chang [21] and Katchaturyan et al.[22] have both suggested that the ground state of the DX center has a negative Coulomb energy U (or -U). This means that the ground state will contain two electrons and should be negatively charged and diamagnetic.While this -U model is consistent with the majority of existing experiments on the DX center, including the absence of ESR signal[22],some of the experimental results have also been shown to be consistent with positive U (or +U) models.In fact the -U model is in conflict with Katchaturyan et al.'s magnetic susceptibility experiment which suggests that the DX center is paramagnetic[23].Fujisawa et al.[24] have tested the -U property of the DX center by applying pressure to GaAs samples co-doped with two donors of different binding energies:Ge and Si.The Si doping concentrations varies up to 2.6×10^{17} cm^{-3} while the Ge concentration is fixed at 1×10^{17} cm^{-3}.At 22 kbar,Ge in GaAs will be converted into DX centers while Si remains as shallow donors.In fact, the codoped samples show only one DLTS peak due to Ge.

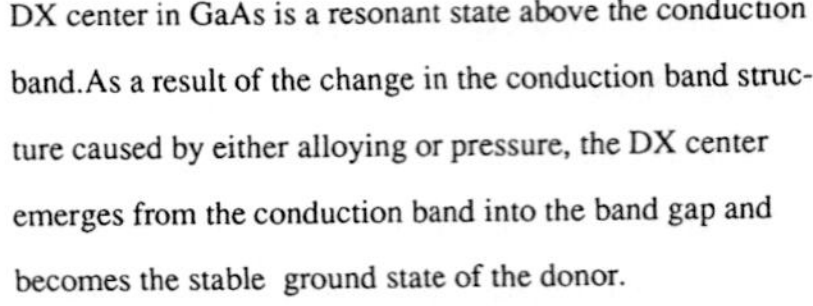

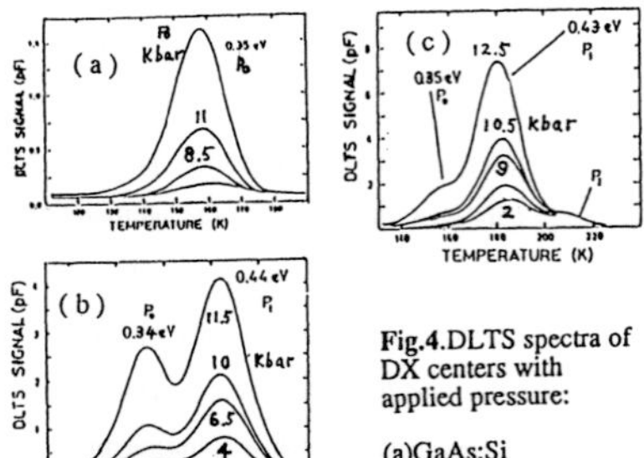

Fig.4.DLTS spectra of DX centers with applied pressure:

(a)GaAs:Si
(b)AlGaAs(X=0.04),
(c)AlGaAs(X=0.08).

Jpn. J. Appl. Phys. Vol. 32 (1993) Suppl. 32–1

M.-F. Lɪ and P. Y. Yu   203

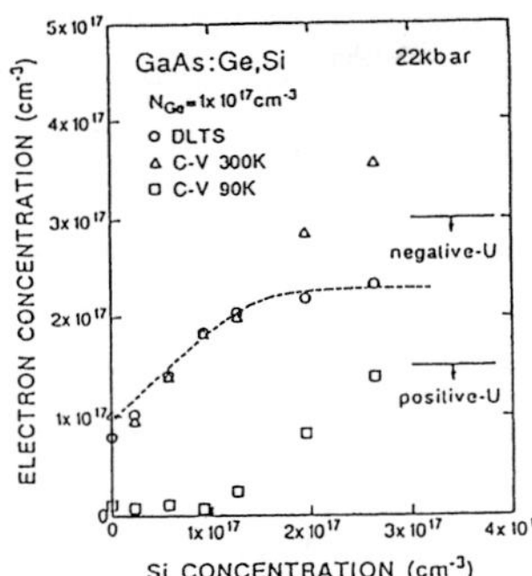

Fig.5 Electron concentration at Ge DX centers(o) and free carrier concentration(△ □) in Ge and Si codoped GaAs at 22 kbar as a function of Si donor concentration [24].

From the DLTS spectra, Fujisawa et al.found that the concentration of electrons trapped at the Ge DX centers increases with Si concentration and saturated at a value of 2.3×10^{17} cm^{-3} when the Si concentration is above 1.3×10^{17}cm^{-3} as shown in Fig.5. Fujisawa et al. estimated that the Ge donor concentration was at most 1.5×10^{17} cm^{-3} with a compensating acceptor concentration of 0.5×10^{17} cm^{-3}. Thus the saturated trapped-electron concentration of 2.3×10^{17} cm^{-3} cannot be explained by a +U model but can be explained by a -U model with a Ge DX center concentration of 1.15×10^{17}cm^{-3}.

b) Small or Large Lattice Relaxation in the DX center and the stress dependence of the DX energy state

The hydrostatic[20] and uniaxial[25] stresses dependence of Te doped DX center energy level has been studied systematically by DLTS method.For hydrostatic experiments,the experiments were performed on $Al_xGa_{1-x}As$ epilayers with x=0.15, 0.25 and 0.35 and doped with 5×10^{16}cm^{-3}, grown by liquid-phase epitaxy (LPE) on n$^+$ GaAs substrate.For uniaxial stress experiments,the LPE grown epitaxy $Al_{0.38}Ga_{0.62}As$ layers were grown on either (110) or (111) surface of n$^+$ GaAs rods with Te doping concentration 3×10^{17}cm^{-3}. Uniaxial stress up to 10 kbar were applied along the [111],[100] and [110] crystallographic direction respectively.Some typical data obtained by DLTS experiments are showned in Fig.7-8 and can be summarized as follows :a).In all cases,the hydrostatic pressure

coefficients of E_e, E_c and the thermal ionization energy E_t were found to change sign when the band gap of $Al_xGa_{1-x}As$ changes from direct to indirect.Furthermore, the DX center energy level appears to follow roughly the L conduction minima only when the band gap is direct. When the band gap is indirect the DX center energy level switches to follow the X and L minima. b).While the Te DX center is rather sensitive to hydrostatic pressure and often its DLTS peak splits into 2 or 3 peaks under hydrostatic pressure as low as 3 kbar, its DLTS peak exhibits no detectable splitting or broadening under uniaxial stress for stress as high as 10 kbar. We have compared these results with existing models of the DX center. Fenning et al[26],and Bourgoin et al[27] attempted to attribute the DX center to an effective mass donor with small lattice relaxation and wave functions constructed from the L conduction band minima only.Bourgoin et al. have futher suggested that E_c is given simply by the seperation between the L minima and the conduction band minima at gamma or X. Our stress experiments have shown that the DX center energy level does not simply follow L minima. Furthermore,in our uniaxial stress experiment,stress along the [111] direction lifts the degeneracy of four L minima and lowers one of them enough to become the lowest conduction minimum at stress beyond 6 kbar.

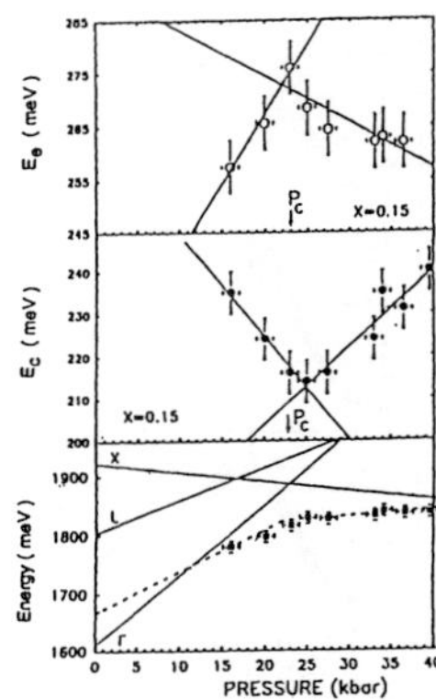

Fig.6 Pressure dependence of emission energy E_e, capture energy E_c and defect energy level E_{DX} relative to the top of the valence band for $Al_{0.15}Ga_{0.85}As$:Te.[20]

204   Jpn. J. Appl. Phys. Vol. 32 (1993) Suppl. 32–1

M.-F. LI and P. Y. YU

According to Bourgoin et al 's model, the capture barrier of DX center should vanish and the capture rate should greatly increases, which can be checked by using short pulse width DLTS experiment. We did not observe any sign of disappearence of the capture barrier in our experiment. Thus the stress experiments do not appear to support DX models based on deep center associated with a L valley effective mass state.

Recently Yamaguchi et al[28],based on their first principle pseudopotential calculation,suggest that the DX center is a simple substitutional donor with a deep A_1 ground state induced by the central-cell defect potential with the neighbouring bond relaxation correlated to the ionic radius of the substitution donor.Similar models have been suggested earlier by tight binding calculation[29,30].In Yamaguchi et al.'s model the lattice relaxation has T_d symmetry and it predicts no level splitting under uniaxial stress .Unlike the L valley effective mass model,this model does not predict that the DX center energy level follows the L conduction minima.These predictions of the Yamaguchi et al.'s model are in agreement with our stress experiments. The weak point of Yamaguchi's model is difficult in explaining the capture behaviour of the DX center. Yamaguchi also equated the capture barrier of the DX center to the seperation between the lowest conduction minimum and the L or X valley similar to the model of Bourgoin et al.. Therefore Yamaguchi et al.'s model also disagree with our uniaxial stress experiment with regard to this point.

In contrast to the small lattice relaxation model of DX center, Chadi and Chang (c.c)[21] proposed that when a Te donor became a DX center it captures two electrons (i.e. -U) and one of its nearest neighbor Ga atoms is displaced towards an interstitial site.At first glance, one may expect its DLTS peak to split under uniaxial stress since the local symmetry of the Te donor is C_{3v} rather than T_d in the c.c model. However,in the c.c model the symmetry-breaking is charge dependent. The positively charged Te donor (d+) has Td symmetry and should show no level splitting under uniaxial stress. The displacement of Ga atom near the Te donor occurs only during the filling pulse of DLTS experiment after the center has captured two electrons. During the emission phase of a DLTS experiment, the Ga atoms will return to their original unrelaxed state after releasing the two electrons back to the conduction band.Thus,it is possible for the DX centers to reorient themselves during the DLTS experiment by changing from a high energy displacement direction to a low energy displacement direction through the intermediate undistorted T_d state.Therefore, during the DLTS experiment,there is thermal equilibrium between DX centers with different lattice relaxation directions.Based on this assumption, we have been able to simulate the effect of uniaxial stress on the DLTS spectra of the DX center[31] within the c.c model and show that no resolvable splitting is produced in agreement with experiment.

Finally we discuss some experimental results of Te DX center which have been overlooked previously. As early as in Lang et al's original paper of DX center[1], those authors have noticed that some of Te doped AlGaAs samples showed two DLTS peaks. In our uniaxial stress experiments as shown in Fig.7, the two DLTS peaks do not show any broadening or further splitting under uniaxial stress. However suprisingly, in our hydrostatic pressure experiments, we found that the DLTS peak of Te DX center always split from one peak into 2 or 3 peaks under hydrostatic pressure,as shown in Fig.8. Combined with

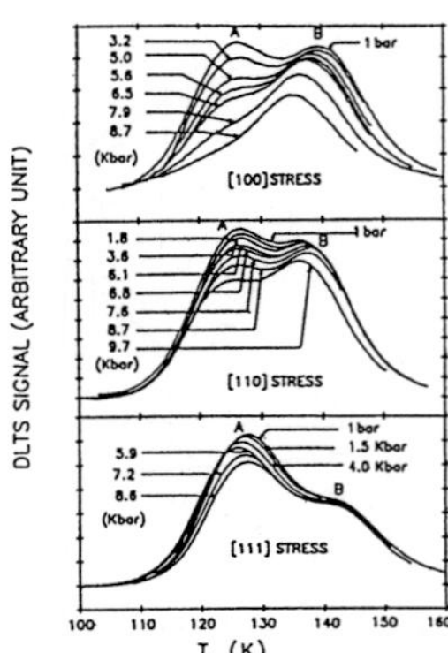

Fig.7 DLTS spectra of DX center in $Al_{0.38}Ga_{0.62}As$:Te under uniaxial stress along three directions,with DLTS window t_1=21 ms,t_2=42 ms,and filling pulse = 8 ms.[25]

Jpn. J. Appl. Phys. Vol. 32 (1993) Suppl. 32–1   M.-F. Li and P. Y. Yu   205

hydrostatic pressure and uniaxial stress experiments, we can rule out the possibility that the multi DLTS peaks are due to internal shearing stress involved in AlGaAs epitaxial layer . We can also rule out the possibility that the multi DLTS peaks correspond to emission of DX center electrons to different conduction valleys. If so,the DLTS peak should not split under hydrostatic pressure and will split or broaden under uniaxial stress due to the splitting of L and X conduction valleys. The only possibility is that there are more than one kind of DX center in Te doped AlGaAs with comparable con – centrations and very closed defect energy. One possibility is that different DX center corresponds to different number of Al atoms located at near Te impurity sites. Recently, Chadi has proposed another possibility by self-consistent pseudo-potential calculation that substitutional donors in AlGaAs may have two distinct negatively charged DX like deep donor states with closed energy levels[32]. The first state has a broken-bond atomic configuration while the second arises from a symmetric breathing mode atomic relaxation around the impurity. Further investigation of this multi-peak problem for Te DX center is apparently significant for exploring the nature of the DX center and related phenomena .

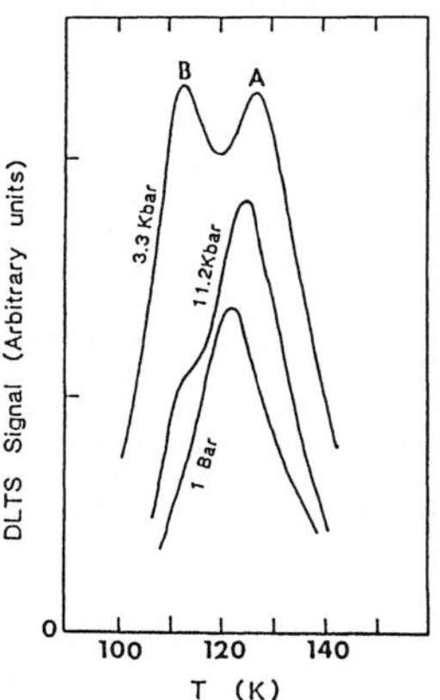

Fig.8 DLTS spectra of $Al_{0.35}Ga_{0.65}As$ under different pressure. One DLTS peak at 1 bar splits into 2 peaks under pressure. DLTS Window t_1=1 S, t_2 = 2 S.

We thank W.Shan ,E.R.Weber,E.Bauser,W.L.Hansen and E.E.Haller for their collaboration in this work. We thank Dr. D.J.Chadi for sending us a preprint. One of the authors (MFL) acknowledges the support of National University of Singapore to write this review. The experimental part of this work was supported by the Director, Office of Energy Research,Office of Basic Energy Sciences, Materials Sciences Division of the U.S. Department of Energy under Contract No.DE-AC03-76SF00098.

References

1.D.V.Lang,R.A.Logan and M.Jaros,Phys.Rev.B **19**,1015(1979)
2.D.V.Lang, in **Deep Centers in Semiconductors**, ed. by S.T.Pantelides(Gordon and Breach,New York,1985) p.489.
3.P.M.Mooney, J.Appl.Phys. **67**,R1 (1990).
4.J.C.Bourgoin(ed),**Physics of DX Centers in GaAS Alloys** (Vaduz:Sci. Tech. Publications,1990).
5.C.H.Henry and D.V.Lang,Phys.Rev.B **15**,989 (1977).
6.K.Huang and R.Rhys,Proc.R.Soc.204,406(1950).
7.N.Lifshitz,A.Jayaraman and R.A.Logan,Phys.Rev.**B21**, 670(1980).
8.M.Mizuta,M.Tachikawa,H.Kukimoto and S.Minomura, J.J.Appl.Phys. 24,L143 (1985).
9.M.Tachikawa,T.Fujisawa,H.Kukimoto,A.Shibata,G.Oomi and S.Minomura,J.J.Appl.Phys.**24**,L.893(1985).
10.D.Erskine,P.Y.Yu and G.Martinez,Rev.Sci.Instrum.58, 406(1987).
11.M.F.Li,P.Y.Yu,E.R.Weber and W.L.Hansen,Appl.Phys.Lett. **51**, 349 (1987).
12.M.F.Li,P.Y.Yu,E.R.Weber and W.L.Hansen,Phys.Rev. **B36**,4531 (1987).
13.R.Legros,P.M.Mooney and S.L.Wright, Phys.Rev.**B35**, 7505(1987).
14.P.M.Mooney,T.N.Theis and S.L.Wright,Appl.Phys.Lett. **53**,2546 (1988).
15.E.Calleja,F.Garcia,A.Gomez,E.Munoz,P.M.Mooney,T.N. Morgan and S.L.Wright,Appl.Phys.Lett.**56**,934(1990).
16.T.Baba,M.Mitzuta,T.Fujisawa,J.Yoshino and H.Kukimoto, J.J.Appl.Phys.**28**,L891(1989).
17.T.N.Theis,P.M.Mooney and S.L.Wright,Phys.Rev.Lett.**60**, 361(1988).
18.B.L.Zhou K.Ploog,E.Gemlin,X.O.Zheng and M.Schultz,Appl.Physics.**A28**,223(1982).
19.S.B.Zhang,Phys.Rev.**B44**,3417(1991).
20.W.Shan,P.Y.Yu,M.F.Li,W.L.Hansen and E.Bauser, Phys.Rev.**E40**,7831(1989).
21.D.J.Chadi and K.J.Chang,Phys.Rev.Lett.**61**,61,873(1988); Phys.Rev.**E39**,10366(1989).
22.K.A.Khachaturyan,E.R.Weber and M.Kaminska,in Defects in Semiconductors 15, ed. G.Ferenczi(Trans. Tech. Switzerland,1989)p.1067.
23.K.A.Khachaturyan,D.D.Awschalom,J.R.Rosen and E.R.Weber Phys.Rev. Lett.**63**,1311 (1989).
24.T.Fujisawa,J.Yoshino and H.Kukimoto,20th Int.Conf. on the Phys.Semiconductors,ed. E.M.Anastassakis,J.D. Joannapolos,World Scientific,1990, Vol.1,p.509.
25.M.F.Li,P.Y.Yu,E.Bauser,W.L.Hansen and E.E.Haller, Semicond. Sci. Technol. 6825(1991).
26.J.C.M.Henning and J.P.M.Ansems,Semicond.Sci.Technol. 2,1(1987).
27.J.C.Bourgoin,Solid State Phenomena,10,253,1989.
28.E.Yamaguchi,K.Sheraishi and T.Ohno,J.Phys.Soc.Japan,**60**, 3093 (1990).
29.H.P.Hjarmarson and T.J.Drumond,Appl.Phys.Lett.**48** 656(1986).
30.S.Y.Ren,J.D.Dow and J.Shen,Phys.Rev.B **38**,10677(1988).
31.M.F.Li,P.Y.Yu and E.R Weber,Appl.Phys.Lett.**59**,1197 (1991).
32.D.J.Chadi private communication,unpublished.

PHYSICAL REVIEW B   VOLUME 50, NUMBER 11   15 SEPTEMBER 1994-I

Two-electron state and negative-U property of sulfur DX centers in GaAs$_{1-x}$P$_x$

M. F. Li and Y. Y. Luo

Center for Optoelectronics, Department of Electrical Engineering, National University of Singapore, Singapore 0511

P. Y. Yu

*Department of Physics, University of California, and Materials Science Division, Lawrence Berkeley Laboratory,
Berkeley, California 94720*

E. R. Weber and H. Fujioka

*Department of Materials Science and Mineral Engineering, University of California, Berkeley, California 94720
and Materials Science Division, Lawrence Berkeley Laboratory, Berkeley, California 94720*

A. Y. Du and S. J. Chua

Center for Optoelectronics, Department of Electrical Engineering, National University of Singapore, Singapore 0511

Y. T. Lim

Hewlett-Packard Singapore, Singapore 0410
(Received 19 May 1994)

A new way to study the two-electron state in DX centers at atmospheric pressure is reported. It is based on the idea of codoping a GaAs$_{0.6}$P$_{0.4}$ sample with a uniform background of Te shallow donors (as source of free carriers) and a Gaussian distribution of sulfur DX centers by ion implantation. Using both capacitance-voltage profiling and deep-level transient spectroscopy measurements, we demonstrate that the ground state of the sulfur DX center traps two electrons and therefore has negative U.

The deep donor known as the DX center is one of the most important defects in III-V and II-VI compound semiconductors.[1,2] One interesting property of the DX center pointed out independently by Chadi and Chang,[3] and by Khachaturyan, Weber, and Kaminska[4] is that it traps two electrons in its ground state. Such systems are said to have a negative U (or $-U$) (Ref. 5) where U denotes the on-site Coulomb repulsion between the two electrons localized on the same impurity. Several experiments have claimed to demonstrate that DX centers can trap two electrons but most of the evidence is indirect and not definitive. The notable exception is experiments based on the idea of codoping a sample with two donors (one shallow and one deep) as first suggested by Khachaturyan, Weber, and Kaminska.[4] At least two such experiments have been reported, one by Fujisawa, Yoshino, and Kukimoto[6] and the other by Baj, Dmowski, and Stupinski.[7] In both cases hydrostatic pressure was used to bring the resonant deep DX centers from above the conduction band in GaAs samples into the band gap. In this paper we report an implementation of the codoping idea *at ambient pressure*. We choose Ga$_{1-x}$As$_x$P because Te and S have been shown to form shallow donor and DX centers, respectively, in this material at atmospheric pressure. The Te concentration was kept constant while a spatial variation of the S concentration is produced by ion implantation. Using C-V profiling and deep-level transient spectroscopy (DLTS), both in the dark and after light illumination, we demonstrate that the ground state of the sulfur DX centers traps two electrons and therefore has $-U$.

Hall effect,[8] electron spin resonance,[9] and DLTS (Ref.

10) measurements in GaAs$_{1-x}$P$_x$ alloys (with $x=0.2$–0.45) have shown that substitutional Te is a shallow donor while S forms a DX center ground state with optical ionization energy $E_{op}=1.5$ eV and thermal ionization energy $E_t=0.2$ eV. Furthermore, this ground state S exhibits persistent photoconductivity (PPC) at 77 K.[8] We therefore choose GaAs$_{0.6}$P$_{0.4}$ as our sample. We start with a wafer which is doped uniformly with Te and vary the S concentration spatially inside the wafer by ion implantation. A major advantage of our approach is that the S concentration can be varied with only one sample. Furthermore, this variation can be predicted theoretically from the known ranges of S ions in GaAs and GaP.[11] If the sample contains N_{Te} donors and N_A compensating acceptors before implantations, then the free-carrier concentration n will be independent of depth (x) into the sample as shown by the horizontal (broken) line $n=N_{Te}-N_A$ in Fig. 1(a). We assume that the dotted curve (labeled N_S) in the same figure represents the profile of implanted S DX centers. This profile can be calculated and also measured by using the PPC property of DX centers. When the sample is illuminated at 77 K by light of energy above E_{op} the DX centers are completely ionized. The resultant carrier concentration [labeled as n_{op} in Fig. 1(a)] will be given by $N_{Te}-N_A+N_S$. If n is measured in the dark instead (when the DX centers are filled with electrons) the resultant n_{dark} will depend on whether the DX centers trap one or two electrons. If the DX center traps two electrons, $n_{dark}=N_{Te}-N_A-N_S$ will have the shape shown in Fig. 1(a). Note that this curve should be a *mirror image* of the curve n_{op} with respect to

0163-1829/94/50(11)/7996(4)/$06.00   <u>50</u>   7996   ©1994 The American Physical Society

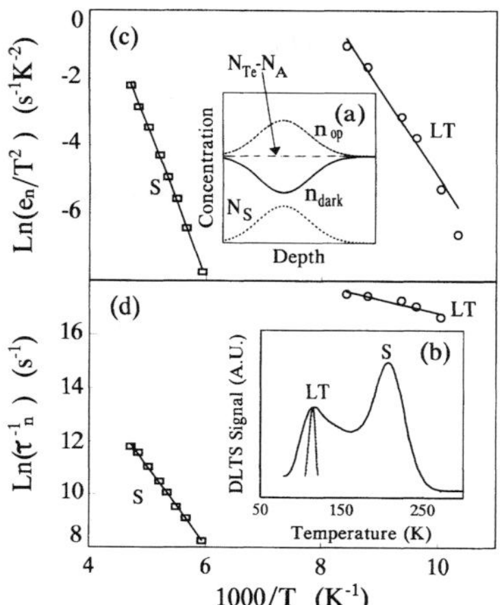

FIG. 1. (a) N_S represents the implanted S ion spatial profile. $N_{Te} - N_A$ is the free-carrier profile in the unimplanted sample; n_{dark} is the predicted free-carrier profile if each S DX center captures two electrons and n_{op} is the predicted profile after all the S DX centers have been emptied of electrons via PPC. (b) DLTS spectra (rate window $= 0.6$ ms) of $Ga_{1-x}As_xP$:Te sample after S implantation and RTA at 890 °C for 1 min. The dashed curve showing a peak at LT is a computer simulation of the LT peak in the DLTS spectrum. (c) and (d) Arrhenius plots of the emission rates (e_n) and capture rate $(\tau_n)^{-1}$ for both the S DX center and LT traps observed in (b).

the horizontal line $N_{Te} - N_A$. On the other hand, if the DX centers trap only one electron then $n_{dark} = N_{Te} - N_A$ will be a horizontal line as in the unimplanted sample. Other defects produced by ion implantation can complicate this scheme. However, we found that most of these defects can either be removed by annealing or be accounted for by methods to be described later.

Our experiment started with a high-quality epitaxial $GaAs_{0.6}P_{0.4}$ layer grown by vapor-phase epitaxy on a GaAs substrate (containing 10^{18} cm^{-3} or Te) and doped uniformly with Te to achieve a free-carrier concentration of $N_{Te} - N_A = 6.5 \times 10^{16}$ cm^{-3}. The same growth procedure is used for the production of light-emitting diodes at Hewlett-Packard Singapore. No deep level was detected in the wafer by DLTS. Sulfur ions were implanted into the wafer at 260 keV of energy with a dose density of 1×10^{12} cm^{-2}. According to the projected range table in Ref. 11, these ions form a Gaussian distribution centered at a depth of 2000 Å with a peak concentration of 5×10^{16} cm^{-3} and half-peak width of 800 Å. After implantation, the wafer was capped by Si_3N_4 and annealed [via rapid thermal annealing (RTA)] at 890 °C for 1 min to activate the S atoms and to remove as much as possible any defects introduced by the implantation. After removing the Si_3N_4 layer and cleaning the sample surface

by etching, Schottky diodes were fabricated by evaporating Au onto the $Ga_{1-x}As_xP$ epilayer and making Ohmic contact to the substrate with Au-Ge-Ni. Unfortunately etching removed part of the surface layer containing the S ions while not all the implantation-induced defects were annealed out. Some residual defects were detected in the DLTS spectra as shown in Fig. 1(b). The peak S is attributed to sulfur DX centers since its properties all agree well with those of S DX centers in $Ga_{1-x}As_xP$ reported previously.[10] A lower-temperature peak (labeled LT) is identified with residual implantation-induced defects. This peak is unusually broad suggesting that it may arise from several different traps. We have measured their capture and emission rates [labeled as (τ_n^{-1}) and e_n, respectively, in Figs. 1(c) and 1(d)] as a function of temperature. Note the rather nonlinear dependence of the emission rate at low temperatures. From these data we have determined their average emission and capture barrier heights and thermal ionization energy to be 0.226, 0.044, and 0.18 eV, respectively. When extrapolated to 77 K their capture time constant is much faster than 10^{-4} s while their emission time is estimated to be much longer than 10^3 s. Thus during the time duration (1 min) for the computer-controlled C-V measurements at 77 K these LT traps can only capture electrons but not emit them. On the other hand, the DX centers cannot capture nor emit electrons at 77 K. Thus we can determine separately the concentrations of the S DX centers and the LT traps by using their different properties.

The free-carrier concentration n is measured as a function of depth x from the C-V curves using these standard formulas:

$$n(x) = \left[\frac{2}{q\epsilon A^2} \right] \left[\frac{d}{dV} \left(\frac{1}{c^2} \right) \right]^{-1} \tag{1a}$$

and

$$x = \epsilon A / C , \tag{1b}$$

where q is the electronic charge, ϵ $(= 12.2)$ is the dielectric constant of $GaAs_{0.6}P_{0.4}$, A and C are, respectively, the area and capacitance of the Schottky diode. The Debye screening length in our sample is estimated to be around 90 Å. Since this is much smaller than the half-peak width of the S implantation profile, diffusion current can be neglected and Eq. (1) should be satisfied in the bulk of the sample.[12] The C-V curves in Fig. 2 were measured with a 1-MHz ac test signal under the following conditions (the carrier concentration for curve i will be denoted by n_i).

Curve 1: Diode cooled slowly in the dark to 77 K under zero bias and then C-V measured in the dark. Under this condition all electron traps are expected to be filled.

Curve 2: Diode cooled to 77 K under strong light (photon energy > 1.5 eV) illumination and large reverse bias in order to completely ionize both the DX centers and the LT traps. The light was then turned off and C-V measured in the dark by *decreasing* the bias voltage to zero. The carrier concentration measured (n_2) contains contributions from all the traps including LT since all traps are expected to be emptied (provided we neglect a small edge

7998   **BRIEF REPORTS**   <u>50</u>

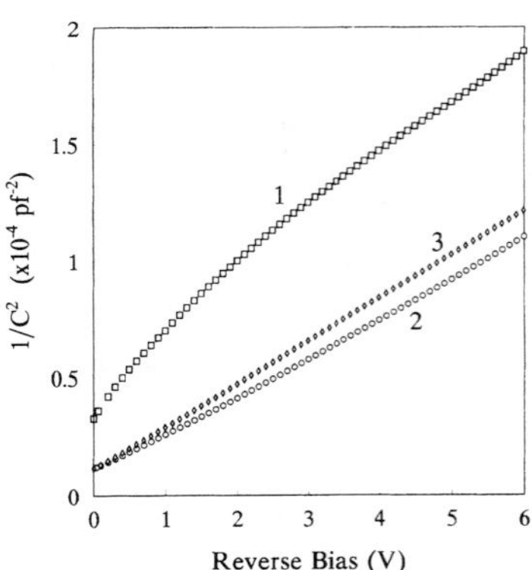

FIG. 2. Capacitance vs voltage curves of the $Ga_{1-x}As_xP$:(Te,S) sample under three different conditions: curve 1, in the dark; curve 2, after strong light illumination and depletion of all traps; and curve 3, under conditions such that only the LT traps are filled.

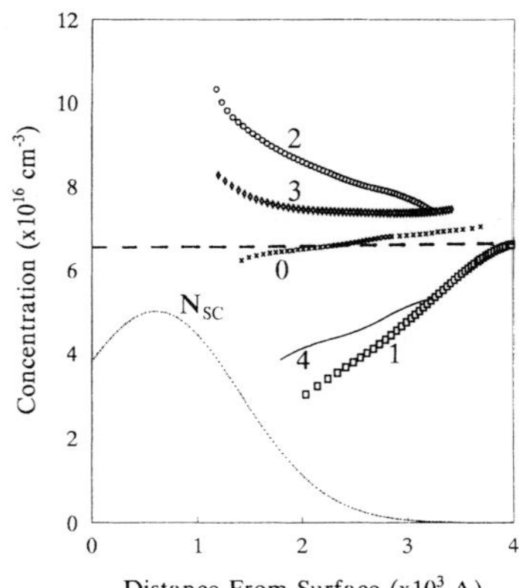

FIG. 3. The curves labeled 1–3 represent the free-carrier spatial profiles in $Ga_{1-x}As_xP$:(Te,S) calculated from the C-V curves in Fig. 2 using Eq. (1). Curve 0 represents the free-carrier profile measured in the unimplanted sample. Curve 4 is calculated by adding the difference between curves 2 and 3 to curve 1. Curve labeled N_{SC} is the S implantation profile calculated from the projection range table in Ref. 11 with the peak position shifted to 600 Å to agree better with the experimental curves because of etching away of a surface layer before fabrication of the Schottky diode. The broken horizontal line represents a constant carrier concentration equal to 6.5×10^{16} cm^{-3}.

effect due to the difference between the conduction depletion width and trap depletion width.[13]

Curve 3: After measuring curve 2 by decreasing the bias voltage to zero, a reverse bias voltage was *increased in the dark* from zero to a large value to obtain curve 3. Only the LT traps are filled while the *DX* levels should be empty because of PPC.

Not shown in Fig. 2 was a curve to be labeled *Curve 0*, which represents the carrier distribution in an unimplanted sample and is therefore equal to the background carrier concentration $n_0 = N_{Te} - N_A$. n_0 is almost independent of x at a constant value of 6.5×10^{16} cm^{-3} as expected. In the rest of our discussions we will assume that n_0 is independent of x for simplicity.

To understand these curves we assume that after implantation and annealing S ions are incorporated into our sample as: S *DX* and LT centers (concentrations equal to N_S and N_{LT}, respectively) as indicated by DLTS and possibly some shallow (hole trap) acceptor levels A' (concentration N_A,) which cannot be detected by DLTS in our n-type sample. The presence of other unusual centers cannot be completely ruled out but is highly unlikely. The photoluminescence spectra of the implanted and annealed sample at room temperature is identical to that of the unimplanted sample suggesting that no new recombination centers have been introduced.

We note that curve 1 is not constant but shows a spatial profile qualitatively consistent with $n_0 - N_S$. The curve labeled as N_{SC} in Fig. 3 is a plot of the calculated S distribution based on Ref. 11. Due to etching of the sample surface after annealing, the peak of this distribution is no longer at 2000 Å as predicted by theory. We have displaced the N_{SC} curve in Fig. 3 to peak at $x = 600$ Å to agree better with the experimental curves in that figure. The actual shape of N_S may also deviate slightly from the

calculated one because of diffusion of sulfur atoms during RTA. This effect is minimized by the short duration of RTA. Taking these factors into consideration, we see that *qualitatively curves 1 and 2 in Fig. 3 are consistent with the predictions [Fig. 1(a)] based on the S DX centers capturing two electrons.* However, there are deviations between the experimental results and the simple theoretical predictions of Fig. 1(a). Specifically curves 1 and 2 are not exact mirror images with respect to the horizontal line defined by $n_0 = 6.5 \times 10^{16}$ cm^{-3}.

To account for this difference we have to consider the possible effects due to the shallow acceptors A' and the LT trap levels.

First we will consider the effect of the LT traps. These can trap electrons as shown by their DLTS peak. To estimate their concentration we note that they were empty during the measurement of curve 2. However, during curve 3 they were filled because of their fast capture but slow emission rates. Hence N_{LT} is equal to $n_2 - n_3$. The contribution of N_{LT} to curve 1 can now be removed to produce a new curve (curve 4 in Fig. 3) by defining $n_4 = n_1 + N_{LT} = n_1 + (n_2 - n_3)$. Comparing curve 4 with curve 2 shows that they are almost mirror images of each other with respect to the horizontal line $n_0 = 6.5 \times 10^{16}$ cm^{-3} as predicted by the two-electron model for the S *DX* centers. This also suggests that the LT levels can capture electrons (such as during curves 1 and 3) but not contribute free carriers to curve 2. We therefore propose that the LT levels are associated with electron-trap acceptors whose energy levels lie at 0.18 eV below the conduction band as deduced from DLTS.

The shallow acceptors (hole traps) A' will compensate the shallow donors introduced by Te. Hence the background donor concentration after implantation is no longer $N_{Te} - N_A$ but becomes $(N_{Te} - N_A) - N_{A'}$. Under the conditions for curve 2 we expect that both the S DX and the LT centers are completely ionized. Thus n_2 in curve 2 will be equal to $(N_{Te} - N_A) + N_S - N_{A'}$. The difference between curves 2 and 0 $(n_2 - n_0)$ should be equal to $(N_S - N_{A'})$ and its maximum value from Fig. 3 is approximately 4×10^{16} cm^{-3}. Since the maximum S concentration is only 5×10^{16} cm^{-3} we conclude that the concentration of A' is at most 10^{16} cm^{-3} and is therefore negligible when compared to the two deep centers.

Finally, we like to point out that this two-electron state of the S DX center is also its ground state and therefore the S DX center has $-U$.[5] For a donor to be a $-U$ center, the energy of its two-electron state (E_2) has to be lower than the energy of its one-electron state (E_1). A donor with a $+U$ may be able to trap a second electron (an analog of the hydrogen ion H$^-$) and form a two-electron *excited state* but in this case $E_2 > E_1$. So far experiments on the S DX center in Ga$_{1-x}$As$_x$P have shown that its ground state has $E_t = 0.2$ eV and exhibits PPC. We have verified the existence of this ground state in our S implanted sample by DLTS. Now the question is whether this ground state is also the two-electron state

responsible for the dip in curve 1 of Fig. 3. If the S DX center has $+U$ then $E_2 > E_1$. Then this two-electron state should be *shallower* than the DX level but yet remains occupied at 77 K since this is the temperature of measurement for curve 1. Our DLTS spectra show that the LT levels (with $E_t = 0.19$ eV) are the only electron traps satisfying these conditions. However, *their concentration $N_{LT} = n_2 - n_3$ is too small to account for the decrease in carrier concentrations in curve 1 by a factor of 3.* In addition if the LT levels were indeed associated with the S DX centers they would have been reported also by Craven and Finn.[10] Hence we can rule out the possibility that the LT traps are the two-electron levels. This leaves us with no other candidates for the observed two-electron level in our implanted sample other than the S DX ground state.

In conclusion, when our results are combined with other existing experiments on S DX centers in Ga$_{1-x}$As$_x$P, we find the evidence for the ground state of the S DX center in GaAs$_{0.6}$P$_{0.4}$ to have a negative U to be very strong and convincing.

The authors would like to thank Professor Nathan Cheung, Dr. M. G. Craford, and Dr. A. G. Elliot for helpful discussions. The NUS group is supported by National University of Singapore Research Grant No. RP 3920621. The Berkeley group is supported by the Director, Office of Basic Energy Sciences, Materials Science Division of the U.S. Department of Energy under Contract No. DE-AC0376SF00098.

[1] D. V. Lang, in *Deep Centers in Semiconductors*, edited by S. T. Pantelides (Gordon and Breach, New York, 1986), p. 489.

[2] P. M. Mooney, J. Appl. Phys. **67**, R1 (1990); K. J. Malloy and K. Khachaturyan, in *Semiconductors and Semimetals*, edited by E. R. Weber (Academic, New York, 1993), Vol. 38, p. 235.

[3] D. J. Chadi and K. J. Chang, Phys. Rev. Lett. **61**, 873 (1988); Phys. Rev. B **39**, 10 366 (1989).

[4] K. A. Khachaturyan, E. R. Weber, and M. Kaminska, in *Defects in Semiconductors 15*, edited by G. Ferenczi (Trans. Tech., Swizerland, 1989), p. 1067.

[5] G. D. Watkins, in *Festkorperproblem*, edited by P. Gross (Vieweg, Braunschweig, 1984), Vol. XXIX, p. 163; G. A. Baraff, E. O. Kane, and M. Schluter, Phys. Rev. B **21**, 5662 (1980).

[6] T. Fujisawa, J. Yoshino, and H. Kukimoto, in *Proceedings of the 20th International Conference on the Physics of Semiconductors*, edited by E. M. Anastassakis and J. D. Joannapolos

(World Scientific, Singapore, 1990), p. 509; Jpn. J. Appl. Phys. **29**, L388 (1990).

[7] M. Baj, I. H. Dmowski, and T. Stupinski, Phys. Rev. Lett. **71**, 3529 (1993).

[8] M. G. Craford, G. E. Stillman, N. Holonyak, Jr., and J. A. Rossi, J. Electron. Mater. **20**, 3 (1991); Phys. Rev. **168**, 867 (1968).

[9] K. A. Khachaturyan, E. R. Weber, M. G. Craford, and G. E. Stillman, J. Electron. Mater. **20**, 59 (1991).

[10] R. A. Craven and D. Finn, J. Appl. Phys. **50**, 6334 (1979).

[11] J. F. Gibons, W. S. Johnson, and S. W. Mylroie, *Projected Range Statistics*, 2nd ed. (Dowden, Hutchingson and Rose, Stroudsburg PA, 1975).

[12] W. C. Johnson and P. T. Panousis, IEEE Trans. Electron Devices **ED-18**, 965 (1971).

[13] M. F. Li and C. T. Sah, IEEE Trans. Electron Devices **ED-29**, 306 (1982).

Reprinted with permission from A.Y. Du, M.F. Li, T.C. Chong and S.J. Chua, Appl. Phys. Lett.,
Vol.66, pp.1391–1393 (1995). Copyright 1995, American Institute of Physics.

Observation of carrier concentration saturation effect in n-type Al$_x$Ga$_{1-x}$As

A. Y. Du, M. F. Li,[a] T. C. Chong, and S. J. Chua
*Center for Optoelectronics, Department of Electrical Engineering, National University of Singapore,
0511 Singapore*

(Received 1 November 1994; accepted for publication 22 December 1994)

In a series of Al$_{0.3}$Ga$_{0.7}$As epitaxial layers with Si doping concentrations varied from 1×10^{17} to
1.5×10^{18} cm^{-3}, carrier concentration saturation effect was observed by Hall measurements. When
Si doping concentration was increased, the carrier concentration tended to saturate. This is due to the
negative U property of the donor DX center. The Fermi energy tends to be pinned at the free energy
level of the DX center. This carrier concentration saturation effect should not be limit to only n-
Al$_x$Ga$_{1-x}$As semiconductors. It is a general effect in n-type compound semiconductors when donor
impurities induce negative U DX levels, and will have a great influence in designing optoelectronic
and fast speed microelectronic devices. © *1995 American Institute of Physics.*

The Al$_x$Ga$_{1-x}$As/GaAs material system has received
much attention due to its superior properties in high speed
digital, high frequency microwave, and optoelectronic device
applications.[1–3] In this letter, we report a general property of
carrier concentration saturation effect in n-type
Al$_x$Ga$_{1-x}$As, due to the negative U property of the DX
center,[4,5] where U denotes the on-site Coulomb repulsion
between two electrons localized on the same center. This
carrier concentration saturation effect was predicted several
years ago by one of the authors,[6] and was confirmed in this
work by careful preparation of a series of molecular beam
epitaxy (MBE) grown samples. This effect will have a great
influence in designing microelectronic and optoelectronic devices.

During the last decade of worldwide research efforts,[7–9]
it has already been generally accepted that for most of donor
impurities in Al$_x$Ga$_{1-x}$As and other III–V or even II–VI
compound semiconductors, each donor center introduces not
only one shallow donor level which was well understood as
early as in the 1950's by the effective mass theory,[10] but also
introduces a deep DX level[11] simultaneously. The deep DX
level is located either in the conduction band or in the energy
gap, depending on the host conduction band structure. For
instance, in the case of Si doped Al$_x$Ga$_{1-x}$As, when $x>0.2$ at
ambient pressure[8] or at pressure higher than 22 kbar for
$x=0$,[12–14] the DX level appears in the energy gap. Otherwise,
the DX level is in the conduction band as a resonance state.[15]
Chadi and Chang,[4] and Khachaturyan, Weber, and
Kaminska,[5] independently proposed that the DX deep state is
a negative U center[16] with two electrons occupied on the
ground state with the following reaction

$$2d^0 \rightarrow d^+ + \text{DX}^- \tag{1}$$

where d^0 and d^+ represent neutral and ionized substitution
donors and DX$^-$ is the negatively charged DX center. This
negative U property of the DX center has recently been directly confirmed by four independent codoping experiments
for different samples in different laboratories.[17–20] On the
other hand, as discussed in Ref. 6, if there is only one kind of
donor with a negative U deep level, the electron concentra-

tion in the conduction band will saturate when doping concentration increases to a certain level. More specifically, consider Si donors in Al$_{0.3}$Ga$_{0.7}$As material. The probability of
Si to be the DX$^-$ center binding two electrons is given by:[6]

$$f_{\text{DX}} = \{1 + \exp[-2(E_F - E'_{\text{DX}})/kT]\}^{-1} \tag{2}$$

$$E'_{\text{DX}} = E_{\text{DX}} - (kT)(\ln g_{\text{DX}})/2. \tag{2a}$$

Here, E'_{DX} is the free energy and E_{DX} is the energy of the DX
state. E_F is the Fermi energy of electrons. $g_{\text{DX}}=4$ is the
degeneracy factor of the DX$^-$ state, according to the Chadi
and Chang model.[4] If there are no other kinds of donors, it is
obvious from Eq. (1) that the f_{DX} is always less than or at
most equal to 1/2, or from Eq. (2), we have

$$E_F \leqslant E'_{\text{DX}}. \tag{3}$$

With increasing doping concentration N_{Si}, the carrier concentration increases. E_F rises and is finally pinned at E'_{DX}.
The carrier concentration n_0 is determined by:[6]

$$n_0 = \left(\frac{N_{\text{DX}} - N_A - n_0}{N_{\text{DX}} + N_A + n_0}\right)^{1/2} N_C \left(\frac{1}{g_{\text{DX}}}\right)^{1/2} \exp[-(E_C - E_{\text{DX}})/kT]$$

$$\leqslant n_{0S} = \left(\frac{N_{\text{DX}} - N_A}{N_{\text{DX}} + N_A}\right)^{1/2} N_C \left(\frac{1}{g_{\text{DX}}}\right)^{1/2}$$

$$\times \exp[-(E_C - E_{\text{DX}})/kT]. \tag{4}$$

N_{DX} is the concentration of donors, N_A is the acceptor concentration. n_{0S} is the saturation carrier concentration, corresponding to the case of the Fermi level pinning. N_c is the
effective density of states at the bottom of the conduction
band, and E_c is the bottom of the conduction band.

To verify the above theory, samples must be carefully
prepared to avoid the electron conducting channels in the
AlGaAs/GaAs interface.[21,22] Very careful sample structures
have been designed and computer simulated. Six samples of
Al$_{0.3}$Ga$_{0.7}$As with varying doping concentration were grown
by MBE on {100} oriented Cr-doped semi-insulating GaAs
substrates in a Riber MBE 32P system. Two types of sample
structures were used. The type I structure (samples Nos. 2
and 6) consisted of a 0.4 μm thick, undoped GaAs buffer
layer, a 0.05 μm thick p-type Al$_{0.3}$Ga$_{0.7}$As buffer layer with
Be doping concentrations 3×10^{17} cm^{-3}, and finally a 2.7

---

[a]Electronic mail: elelimf@nusvm.bitnet

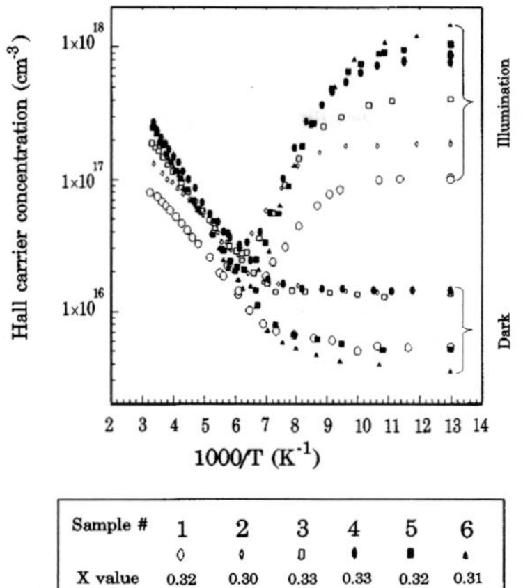

Sample #	1	2	3	4	5	6
	◊	◊	◻	◆	■	▲
X value	0.32	0.30	0.33	0.33	0.32	0.31

FIG. 1. Temperature dependence of the Hall electron concentration n_H for six Si-doped $Al_{0.3}Ga_{0.7}As$ samples with different doping concentrations. The N_H values at 77 K after illumination measured the doping concentration (N_{DX}-N_A).

μm thick n-type $Al_{0.3}Ga_{0.7}As$ active layer with Si doping. Type II structure (samples Nos. 1, 3, 4, and 5) consisted of a 0.2 μm thick, undoped GaAs buffer layer, a 0.5 μm thick undoped $Al_{0.3}Ga_{0.7}As$ buffer layer, and finally a 3.0 μm thick n-type $Al_{0.3}Ga_{0.7}As$ active layer with Si doping. The undoped GaAs buffer layer is to provide an atomically smooth surface for the growth of the AlGaAs layer, and to stabilize the substrate temperature for AlGaAs growth. The p-doped or undoped AlGaAs buffer layer is to prevent the formation of two-dimensional electron gas in the GaAs layer.[21] The GaAs buffer layers and the $Al_{0.3}Ga_{0.7}As$ layers were grown at a substrate temperature of 610 °C. The temperature was measured by an optical pyrometer. The Ga and Al effusion cells were controlled to keep the flux constant at a beam equivalent pressure of 4.0×10^{-7} and 7.7×10^{-8} Torr, respectively. The Si cell temperature T_{Si} was varied between 980 and 1140 °C to control the Si flux for different samples. The alloy composition of $Al_xGa_{1-x}As$ layer was determined by x-ray diffraction rocking curves using a Phillips DCD-3 X-ray Double Crystal Diffractometer. The resolution of this equipment is 2 arcsec and the accuracy of the x value of Al in $Al_xGa_{1-x}As$ is ±0.01 for $x=0.3$. The measured x values are within 0.30–0.33. After the sample growth, Ohmic contacts were formed by alloying indium dots at 400 °C in forming gas (12% H_2 in N_2) for 4 min. The samples were characterized by Hall effect measurements using the Van der Pauw technique in a magnetic field of 7.8 kG and in the temperature range of 77–300 K, under the dark condition after strong light illumination.

The Hall electron n_H is plotted in Fig. 1 for six different samples. The n_H is determined by

$$n_H = \frac{-1}{eR_H},$$

where R_H is the Hall coefficient and e is the electron charge. During the Hall measurement, the sample was kept in the dark and the temperature was decreased very slowly to assure thermal equilibrium being reached between the conduction band electrons and electrons captured by DX centers when the temperature was above 160 K. When the temperature was below 140 K, the electrons in the conduction band and the DX states were not in thermal equilibrium due to the very low capture and emission rates of the DX centers.[7] After the dark Hall measurements with the temperature being decreased from room temperature to 77 K, the sample was measured under strong white light illumination from a halogen tungsten light source and with the sample temperature increased very slowly from 77 to 180 K. Each Hall datum was taken after light illumination was turned off. The electrons persisted in the conduction band after turning off the light, due to the persistent photoconductivity effect.[7]

In Fig. 1, the electron concentrations measured after light illumination at 77 K represent the Si doping concentrations N_{DX}-N_A for different samples. For the six samples, Si doping concentration ranges from 1.04×10^{17} to 1.46×10^{18} cm^{-3}, with a change in doping level by a factor of $14.6/1.04=14$, and are in correct order as compared with the Si cell temperatures during the MBE growth. The electron concentrations at 300–160 K clearly show the saturation effect as predicted by Eq. (4). For samples No. 1 and No. 2, the doping levels are not high enough for carrier concentration saturation at 300 K. However, the carrier concentrations tend to saturate at lower temperature when n_0 is decreased. At $T=160$ K, the carrier concentration for the six samples changed by only a factor of 2 from 1.5×10^{16} to 3×10^{16} cm^{-3}. This is probably due to experimental error in the x value during sample preparation, or in the change of compensation ratio N_A/N_{DX} in Eq. (4). The carrier concentration at a temperature lower than 140 K shows no saturation. This is because the electrons between the conduction band and the DX levels are not in thermal equilibrium. In this case, the carrier concentration in the conduction band depends on the temperature decreasing rate, and cannot be accounted by Eq. (4).

The early Hall measurements of Si doped AlGaAs by Chand et al. did not show the apparent carrier concentration saturation effect.[6,23] We have estimated and found that the 0.2 μm thick buffer layer of undoped $Al_xGa_{1-x}As$ in Chand et al.'s samples were not thick enough to avoid the 2D electron modulation at the GaAs/AlGaAs interface, which may introduce large error in Hall carrier concentration measurement.[21,22]

Finally, we can estimate from Eq. (4), the absolute value of the saturation carrier concentration. For $Al_{0.3}Ga_{0.7}As$, the multiconduction valleys effect was considered.[23] The following data were used: $E(\Gamma)=1.80$ eV, $E(X)=1.91$ eV, $E(L)=1.92$ eV, $m(\Gamma)=0.093$ m_0, $m(X)=0.807m_0$, $m(L)=0.591m_0$. With these values of valley energies and effective masses,[24] N_c, the effective density of states of the conduction band was estimated. By substituting into Eq. (4), we obtain an E_{DX} energy level of 0.04–0.03 eV below the conduction band E_c which can fit all the experimental carrier concentration data. This value is slightly smaller than the value of 0.10±0.05 eV obtained by the deep level transient spectroscopy method.[7]

In conclusion, due to the negative U characteristic of the DX center, the Fermi level of n-AlGaAs will never exceed the free-energy level of the DX center. When the doping level is increased, the carrier concentration will saturate at a value determined by Eq. (4). This carrier saturation effect has been confirmed in our Hall measurements and will have great influence in designing optoelectronic and microelectronic devices. This carrier concentration saturation effect should not be limited to only n-AlGaAs. It is a general effect in n-type compound semiconductors when donor impurities induce negative U DX levels.

[1] P. M. Solomon and H. Morkoç, IEEE Trans. Electron Devices **ED-31**, 1015 (1984).

[2] M. E. Kim, A. K. Oki, G. M. Gorman, D. K. Vmemoto, and J. B. Camoa, IEEE Trans. Microwave Theory Tech. **MTT-37**, 1286 (1989).

[3] S. Adachi, J. Appl. Phys. **58**, R1 (1985).

[4] D. J. Chadi and K. J. Chang, Phys. Rev. Lett. **61**, 873 (1988); Phys. Rev. B **39**, 10366 (1989).

[5] K. A. Khachaturyan, E. R. Weber, and M. Kaminska, in *Defects in Semiconductors*, edited by G. Ferenczi (Trans. Tech., Switzerland, 1989), Vol. 15, p. 1067.

[6] M. F. Li, Y. B. Jia, P. Y. Yu, J. Zhou, and J. L. Gao, Phys. Rev. B **40**, 1430 (1989).

[7] D. V. Lang, in *Deep Centers in Semiconductors*, edited by S. T. Pantelides (Gordon and Breach, New York, 1985), p. 489.

[8] P. M. Mooney, J. Appl. Phys. **67**, R1 (1990).

[9] K. J. Molloy and K. Khachaturyan, in *Semiconductors and Semimetals*, edited by E. R. Weber (Academic, New York, 1993), Vol. 38, p. 235.

[10] W. Kohn, *Solid State Physics*, edited by F. Seitz, H. Ehrenreich, and D. Turnbull (Academic, New York, 1957), vol. 5, p. 257.

[11] T. N. Theis, T. F. Kuech, L. F. Palmateer, and P. M. Mooney, Inst. of Phys. Conf. Ser. **74**, 241 (1984).

[12] M. Mizuta, M. Tachikawa, H. Kukimoto, and S. Minomura, Jpn. J. Appl. Phys. **24**, L143 (1985).

[13] M. F. Li, P. Y. Yu, E. R. Weber, and W. Hansen, Appl. Phys. Lett. **51**, 349 (1987); Phys. Rev. B **36**, 4531 (1987).

[14] D. K. Moude, J. C. Portal, L. Dmowski, T. Foster, L. Eaves, M. Natan, M. Heiblum, J. J. Harris, and R. B. Beall, Phys. Rev. Lett. **59**, 815 (1987).

[15] T. N. Theis, P. M. Mooney, and S. L. Wright, Phys. Rev. Lett. **60**, 361 (1988).

[16] G. D. Watkins, in *Festkorperprobleme*, edited by P. Gross (Vieweg, Braunschweig, 1984), Vol. XXIX, p. 163.

[17] T. Fukisawa, J. Yoshino, and H. Kukimoto, Jpn. J. Appl. Phys. **29**, L388 (1990).

[18] M. Baj, I. H. Dmowski, and T. Stupinski, Phys. Rev. Lett. **71**, 3529 (1993).

[19] M. F. Li, Y. Y. Luo, P. Y. Yu, E. R. Weber, H. Fujioka, A. Y. Du, S. J. Chua, and Y. T. Lim, Phys. Rev. B **50**, 7996 (1994).

[20] D. K. Maude, U. Willke, M. L. Fille, J. C. Potal, and P. Gibart, Mater. Sci. Forum **117–118**, 441 (1993).

[21] J. E. Dmochowski, L. Dobaczewski, J. M. Langer, and W. Jantsch, Phys. Rev. B **40**, 9671 (1989).

[22] J. E. Dmochowski, Phys. Rev. **42**, 9709 (1990).

[23] N. Chand, T. Henderson, J. Klem, W. T. Messelink, R. Fisher, Y. C. Chang, and H. Morkoç, Phys. Rev. B **30**, 4481 (1984).

[24] *Landolt-Börnstein, Numerical Data and Functional Relationships in Science and Technology, Group III: Crystal and Solid State Physics* (Springer, New York, 1982), Vol. 22.

CHAPTER 5.2

High-Pressure Study of DX Centers Using Capacitance Techniques

Ming-fu Li

DEPARTMENT OF ELECTRICAL ENGINEERING
NATIONAL UNIVERSITY OF SINGAPORE
SINGAPORE

Peter Y. Yu

DEPARTMENT OF PHYSICS
UNIVERSITY OF CALIFORNIA
BERKELEY, CA
AND
MATERIALS SCIENCES DIVISION
LAWRENCE BERKELEY NATIONAL LABORATORY
BERKELEY, CA

I. Introduction

Pressure can change the band structure of a semiconductor without changing its symmetry or composition. Thus, pressure is a powerful tech-

458					MING-FU LI AND PETER Y. YU

nique for studying the influence of electronic band structures on the properties of defects in semiconductors. This is particularly true for zincblende-type semiconductors because the pressure dependence of the various high symmetry critical points in their lowest conduction band are known to obey the so-called Paul's Empirical Rule [1]. For example, the pressure coefficients of the energy of the conduction band at the Brillouin zone center (or Γ point) in this family of semiconductors are all *positive* and of the order of $\sim$100 meV/GPa in magnitude. On the other hand, the pressure coefficients of the conduction band at the zone edge in the [100] direction (or X point) are all *negative* and $\sim$10 meV/GPa in magnitude. As a result, if the X valleys are higher in energy than the Γ valley, then the conduction bandwidth decreases with pressure. In many semiconductors the location of the conduction-band minimum in the Brillouin zone can be made to change from the Γ point to the X point under sufficiently high pressure. This phenomenon is known as the Γ–X *crossover*. For example, this crossover occurs around 4 GPa in GaAs [2].

Defect centers in semiconductors are usually classified as shallow (or hydrogenic) and deep. A defect energy level whose wave function can be constructed out of the nearest band extremum is considered shallow. On the other hand, the wave function of a highly localized center can be expressed only as a linear combination of wave functions from a large region of the Brillouin zone. In some cases many bands may be involved. Such centers are said to be deep. The properties of shallow and deep centers are quite different, their pressure dependence being one of them. So far, this property has proven to be one of the most reliable ways to distinguish between shallow and deep defects. By definition, the wave function of a shallow center is constructed from the wave functions of its nearest band extremum. Hence, the pressure dependence of a shallow center should be identical to that of its nearest band extremum. On the other hand, the pressure dependence of a deep center can be quite different from that of its nearest band extremum. In addition to being a method for distinguishing shallow and deep defects, pressure can change the properties of a defect by changing the host band structure. In some cases pressure can convert a shallow defect into a deep one or vice versa. Thus, high pressure can play an important role in the study of defects, especially in deep centers whose nature is often poorly understood.

With the development of the diamond anvil cell (DAC), it is possible both to achieve high pressures and to obtain optical access to the sample [3]. Many shallow impurities and a few deep centers have been studied with the DAC using optical techniques. Many defects, especially deep centers, are important in semiconductor technology because of their electrical characteristics. Thus, it is desirable to develop methods for carrying out electrical

5.2   Study of DX Centers Using Capacitance Techniques   **459**

measurements on deep levels inside the DAC. Furthermore, such methods will enable electro-optical measurements to be performed inside the DAC. It should be noted that high-pressure electrical measurements have been performed using either the Bridgman cell or the large piston-cylinder type of high-pressure cell (see references in [3] for further details). The disadvantages of these cells are that the former provides no optical access so pressure calibration cannot be performed easily as with the ruby fluorescence technique, whereas the latter kind of cell is limited to pressures less than 2 GPa. In this article we will concentrate on the DAC because of its versatility and ability to reach the highest pressure. We shall discuss the techniques for performing transient capacitance and photocapacitance measurements inside the DAC developed at the University of California at Berkeley. The usefulness of these techniques will be illustrated by results obtained on DX centers found in GaAs and its alloys.

The organization of this article is as follows. In Section II we concentrate on the technique of introducing wires into the DAC for capacitance and photocapacitance measurements. In Section III we present a short introduction to various capacitance transient techniques. The experimental results on the DX centers obtained with these techniques are presented in Section IV where they are discussed in light of recent models.

II.   Techniques for Electrical Measurements on Samples Inside the DAC

1.   Introducing Wires into the DAC

The design of our DAC high-pressure cells has been described extensively in the literature [3] and so will not be repeated here. The major difficulty in performing electrical measurements on samples inside the DAC lies in making electrical contacts to the sample with wires [3]. High pressure is achieved in a DAC by pressing two diamond anvils onto a metal gasket as shown in Fig. 1. The sample and the pressure medium are confined in a hole drilled in the gasket. To make electrical contact to the sample inside the pressure medium, wires have to pass through the contact region between one of the diamond anvils and the gasket. Thus, the wires have to be insulated from the metallic gasket. Because of the small size of the sample to be contacted by the wires, the diameter of the wire is typically 20–50 μm. Such thin wires are necessarily quite fragile. Unless they are protected in some way, they are easily cut by the diamond anvil.

Various methods have been devised to solve these two problems. Insulating gaskets have been used to avoid the problem of shorting the wires through the gasket [4]. However, most insulating materials are hard rather

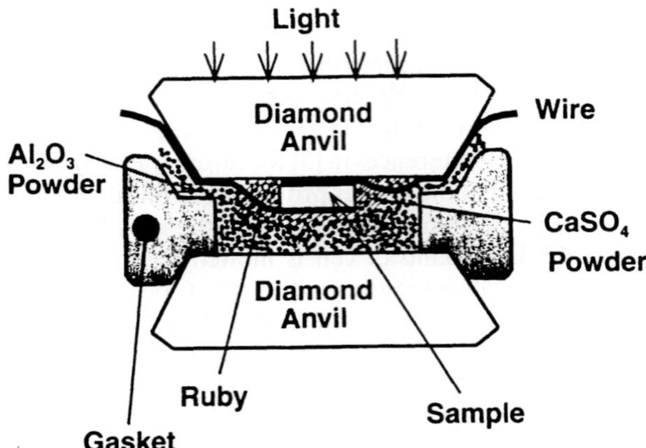

FIG. 1. Schematic diagram of the diamond anvils and sample inside the hole drilled in the metal gasket. In particular note the details of how the wire is insulated from the gasket by Al_2O_3 powders while the sample is surrounded by a softer powder ($CaSO_4$).

than malleable like a metal. Since they do not "flow" under stress, they cannot confine the pressure medium by forming a tight seal with the diamond anvils. Another approach is to insulate the gasket from the wires by coating it with an insulator. Typically, a thin layer of Al_2O_3 is deposited on a metal gasket by plasma spraying or sputtering [5]. The disadvantage of this technique is that most laboratories do not have plasma spraying or sputtering facilities. The gaskets have to be indented and then sent away to be coated with Al_2O_3. As an alternative to spraying and sputtering, we have developed the technique of applying Al_2O_3 in the form of a fine powder (the same powder used in polishing crystals, with grain sizes about 1 μm or less). The powder is compacted into a layer about 30 μm thick on the indented gasket using the diamond anvils themselves as the press. The shortcoming of this method is that the Al_2O_3 powder tends to spill over into the hole in the gasket, and so Al_2O_3 powder becomes the pressure medium. Since Al_2O_3 is a very hard material, it is not a good medium for a homogeneous pressure. Erskine *et al.* [6] alleviated this problem by adding a softer powder, such as $CaSO_4$, as the pressure medium inside the hole. A schematic diagram of the gasket, sample, and wire after installation inside the cell is shown in Fig. 1. Other groups have independently developed this technique [7, 8] or variations of this method, such as using a mixture of epoxy and Al_2O_3 powder to insulate the metal gasket [9]. The latter method has the advantage that the seal between the gasket and the diamond

5.2   STUDY OF DX CENTERS USING CAPACITANCE TECHNIQUES   **461**

remains tight at even low pressure. This allows a liquid pressure medium to be used for achieving greater pressure homogeneity.

The pressure homogeneity inside a DAC prepared in this way has been studied by Erskin *et al.* [6] using either the width of the ruby fluorescence line or the sharpness of the superconducting transition in superconductors such as Pb. Typical pressure inhomogeneity in cells prepared this way is less than 10%. Using this technique we have carried out capacitance measurements on semiconductor diodes inside the DAC. Since the resistance of these diodes under reverse bias is usually very high, the pressure medium must be a good insulator. In this respect, powders such as $CaSO_4$ have an advantage over liquid pressure media. However, diodes can also be destroyed by large pressure gradients. So far we have succeeded in making capacitance measurements on diodes at pressures up to about 4 GPa. Assuming that the pressure gradient inside the gasket hole is about 10%, the nonhydrostatic component of the pressure can be as high as 0.4 GPa. This stress is sufficient to damage some materials such as GaAs. In addition to standard capacitance measurements, we have made transient measurements, such as deep-level transient spectroscopy (DLTS), on samples inside the DAC.

2.   PERFORMING CAPACITANCE MEASUREMENTS INSIDE THE DAC

Some special consideration are relevant to the application of capacitance techniques to samples inside the DAC. Given the sample configuration in Fig. 1, it is unavoidable that there will be a rather large background stray capacitance. Thus, one advantage of transient capacitance experiments is that one will be measuring a small change in the diode capacitance. Since the background stray capacitance is unchanged during the experiment, it can be easily subtracted out. In DLTS measurements the sample temperature has to be monitored. Since it is difficult to put the thermometer inside the cell next to the sample, we have attached a calibrated Si diode thermometer to the diamond anvil nearest to the sample. To minimize the temperature difference between the sample and the diode sensor, the cell temperature is changed very slowly (usually at the rate of about 2 K/min or less). To achieve a short equilibration time between the sample and thermometer, the thermal mass of the cell should be minimal. We have achieved this by separating the lever system used to apply the pressure from the cylinders holding the anvils. Pressure is applied to the cell via a hydraulic press. Once the desired pressure has been reached, the pressure is maintained by a locking ring on the cell. The linewidth of DLTS spectra is inherently

462   MING-FU LI AND PETER Y. YU

rather broad and therefore the larger pressure inhomogeneity associated with a solid pressure medium is not a major disadvantage.

We mentioned earlier that one advantage of the DAC is that it will be possible to perform electro-optical measurements on samples inside the DAC. One such measurement we have carried out is photocapacitance. The problem with performing photocapacitance experiments inside the DAC results from the metal electrodes on the diodes. The metal overlayer prevents light from reaching the depletion layer of the sample directly. We have solved this problem by scattering the light in the pressure medium so that light can enter the diode from its sides. Unfortunately, it is not possible to determine exactly the amount of light absorbed by the sample.

III. Introduction to Capacitance Transient Techniques

Capacitance transient techniques are among the most important methods for characterizing deep defects in semiconductors [10–15]. Although these techniques have been extensively studied and described in the literature, we shall give a short introduction to them here for the benefit of readers not familiar with them. This also allows us to discuss the advantages and the problems of such measurements on samples inside DAC.

Let us consider a p^+n junction (or an n-Schottky barrier junction) with a reverse bias voltage V_b. In order to simplify the discussion we shall assume that the shallow donors are distributed uniformly throughout the n-type layer. The region $0 < x < W$ is completely depleted of free electrons (i.e., the electron concentration is zero). This assumption is known as the *depletion approximation*. The uniform space charge due to the ionized donors gives rise to a electric field which is linearly dependent on distance x as a result of the Gauss law. Alternatively, the electric potential V varies quadratically with x such that $V = V_b$ at $x = 0$ and $V = 0$ at $x = W$. In the region $x > W$, the potential is assumed to be identically zero. The free electron concentration n_0 is determined by the donor doping concentration N_d. The small signal junction capacitance C can be written as

$$C = \frac{A\epsilon}{W},\qquad(1)$$

where A is the area of the junction and ϵ is the permittivity of the semiconductor. The depletion layer thickness W and hence the capacitance C are both dependent on the bias voltage V_b. Let us assume that there are deep centers in the depletion region with concentration $N_t \ll N_d$. Suppose the

5.2 STUDY OF DX CENTERS USING CAPACITANCE TECHNIQUES **463**

bias voltage V_b is fixed but the occupancy of the deep centers is changed. As an illustration, let us assume that electrons are thermally excited from the deep centers into the conduction band. This changes the density of space charges (ρ) in the depletion layer. As a result, the width of the depletion layer will increase by ΔW, which leads to a change ΔC in the junction capacitance. From Eq. (1) the two quantities are related by

$$\Delta C/C = -\Delta W/W. \tag{2}$$

On the other hand, because the bias voltage is kept constant and

$$V_b \sim \rho W^2, \tag{3}$$

we obtain the relation

$$-2\Delta W/W = \Delta \rho/\rho. \tag{4}$$

Under our assumption that $N_d \gg N_t$, ρ is determined mainly by N_d while $\Delta \rho$ is equal to $(-e)\Delta n_t$, where Δn_t is the change in the concentration of electrons trapped on the deep levels. Substituting these results into Eq. (4) and then combining the resultant equation with Eq. (2), we obtain the simple expression

$$\Delta C/C = -\Delta n_t/2N_d. \tag{5}$$

In deriving Eq. (5) we have assumed that the concentration of acceptors in the p^+ region is much higher than N_d so that the change in the depletion width W caused by the change Δn_t occurs entirely in the n-doped region. We shall now consider three ways to apply Eq. (5) to study deep centers.

1. CAPACITANCE TRANSIENTS AT CONSTANT TEMPERATURE

The most obvious way to apply Eq. (5) is to keep the sample temperature constant and superimpose an applied voltage pulse on the bias voltage to induce a transient Δn_t in the deep center population [10]. In this way the rates of emission from and capture of carriers into the deep center can be determined from the transient in the capacitance. Such an experiment can be carried out in two phases. In phase I a *forward* bias (filling) pulse with amplitude V_p and pulse width τ is added to the *reverse* bias voltage V_b so that the total bias is given by

$$V_1 = V_b - V_p. \tag{6}$$

The change in bias from V_b to V_l results in the depletion width W also changing from W_b to W_l. As V_b is larger than V_l, so is W_b larger than W_l. Hence, the region between W_l and W_b is no longer depleted of carriers, and electrons will be *captured* from the conduction band into the deep level E_t with a capture rate C_n. By solving a simple rate equation we can easily show that the increase in the deep level population Δn_t is given by

$$\Delta n_t = N_t(1 - e^{-C_n \tau}), \tag{7}$$

assuming that $\Delta n_t = 0$ at the beginning of the applied pulse V_p. In phase II the bias is returned from V_l to V_b and the depletion width also returns from W_l to W_b. In the region between W_l and W_b, electrons are emitted with an emission rate e_n by the deep levels to the conduction band and then swept out of the depletion region. In this phase Δn_t changes with time t as

$$\Delta n_t(t) = N_t(1 - e^{-C_n \tau})e^{-e_n t}, \tag{8}$$

where we have assumed $t = 0$ to be the moment the filling pulse is turned off. Substituting Eq. (8) into Eq. (5), the corresponding change ΔC in the junction capacitance C is

$$\frac{\Delta C}{C} = -\frac{N_t(1 - e^{-C_n \tau})e^{-e_n t}}{2N_d}. \tag{9}$$

Using Eq. (9) we can determine the capture rate C_n by measuring ΔC as a function of the filling pulse width τ. Similarly, the thermal emission rate e_n can be deduced from the time dependence of $\Delta C/C$. The thermal emission and capture rates are further related to each other by the principle of detailed balance [10, 15],

$$e_n = N_c C_n e^{-(E_c - E_t)/kT}, \tag{10}$$

where E_c and N_c are, respectively, the conduction band edge energy and effective density of states, and k is the Boltzmann constant. The energy $E_c - E_t$ is the thermal ionization energy of the deep level. In the literature it is a common practice to define the probability of capture in terms of the capture cross-section σ_n. It is related to the capture rate C_n by

$$C_n = n_0 \langle v \rangle \sigma_n, \tag{11}$$

5.2 STUDY OF DX CENTERS USING CAPACITANCE TECHNIQUES   **465**

where $\langle v \rangle$ and n_0 are, respectively, the thermal velocity and concentration of the free carriers. The capture cross-section is usually assumed to be thermally activated and has the temperature dependence

$$\sigma_n(T) = \sigma_n(\infty)e^{-E_B/kT},\tag{12}$$

where E_B is defined as the capture barrier height. Substituting Eqs. (12) and (11) into Eq. (10), we obtain

$$e_n(T) = n_o \langle v \rangle \sigma_n(\infty) N_c e^{-(E_c-E_t+E_B)/kT}.\tag{13}$$

Equations (12) and (13) are commonly used in the literature for determining the capture and emission barrier heights

2. CAPACITANCE TRANSIENT WHEN SCANNING TEMPERATURE — DEEP-LEVEL TRANSIENT SPECTROSCOPY

According to Eq. (13), one can in principle determine the emission barrier height by measuring the temperature dependence of the emission rate. The latter can be obtained by measuring the capacitance transient after a filling pulse as a function of time while keeping the temperature T constant. An alternative to scanning the time at fixed temperature is to scan the temperature while keeping a "time window" constant [12]. In this approach, a filling pulse is applied first and then the capacitance is measured only at two preset times t_1 and t_2, which are said to define a "time window." The corresponding capacitances are denoted by C_1 and C_2, respectively. Their difference $(C_1 - C_2)$ is then measured as a function of temperature. The idea is that $(C_1 - C_2)$ is nonzero only when there is significant change in the deep occupation during the time window. If t_1 and t_2 are both $\ll 1/e_n(T)$, then both C_1 and C_2 are unchanged so their difference $(C_1 - C_2) \sim 0$. On the other hand, if t_1 and t_2 are both $\gg 1/e_n(T)$, then C_1 and C_2 are both almost equal to the equilibrium value and so again $(C_1 - C_2) \sim 0$. Thus, a plot of $(C_1 - C_2)$ versus temperature will exhibit a peak when $1/e_n(T)$ is approximately equal to the time window $(t_2 - t_1)$. Such a $(C_1 - C_2)$ versus temperature curve is known as a *DLTS spectrum* [12]. The temperature T_m where the DLTS curve shows a maximum allows the emission rate at T_m to be determined:

$$e_n(T_m) \approx 1/(t_2 - t_1).\tag{14}$$

The DLTS technique takes advantage of the fact that $e_n(T)$ depends on T exponentially so it takes a relatively small change in T to vary $e_n(T)$ by many orders of magnitude. As may be expected, a slow scanning of temperature is desirable in obtaining DLTS spectra. Using different time windows and measuring the corresponding values of T_m, one can obtain the temperature dependence of e_n and hence the *DLTS activation energy* E_{DLTS}, defined as

$$E_{DLTS} = [E_c - E_t] + E_B. \tag{15}$$

From Eq. (13) it is clear that we can also interpret the energy E_{DLTS} as an emission barrier height.

3.　Photocapacitance Transient Measurements

From the preceding discussions we see that the emission rate of deep levels can be made negligibly small by lowering the sample temperature so that $kT \ll E_{DLTS}$. Under this condition it may become possible to photo-ionize the deep level. To carry out such photoionization measurement one fills the deep levels by setting the bias voltage $V_b = 0$ at room temperature so that $n_t = N_t$. The sample is then cooled to low temperature in the dark. A reverse bias voltage V_b is then applied to the junction. As the thermal emission rate e_n is extremely small and negligible at low temperature, a metastable state is created. If the sample is now illuminated with radiation with sufficient energy to optically excite carriers out of the deep level, the photoinduced change of the junction capacitance is a measure of the photoionization of the deep levels. Let $h\nu$ be the incident photon energy and $\Phi(h\nu)$ be its flux density. σ_n^o and σ_p^o are, respectively, the electron and hole photoionization cross-sections for the deep centers. Then dn_t/dt is given by [16]

$$dn_t/dt = \Phi(h\nu)[\sigma_p^o(h\nu)(N_t - n_t) - \sigma_n^o(h\nu)n_t]. \tag{16}$$

Let $t = 0$ be the point when the light is turned on. Since most of the deep centers are occupied before the light is turned on, $n_t(0) = N_t$ and (16) can be approximated at $t \geq 0$ by

$$dn_t/dt \approx -\Phi(h\nu)\sigma_n^o(h\nu)N_t. \tag{17}$$

5.2 STUDY OF DX CENTERS USING CAPACITANCE TECHNIQUES **467**

Substituting Eq. (17) into Eq. (6), we obtain

$$dC/dt \approx (CN_t/2N_d)\Phi(h\nu)\sigma_n^0(h\nu). \tag{18}$$

Thus, by measuring the initial rate of change $dC/dt(t \sim 0)$ and the incident photon flux density, one can measure the electron photoionization cross-section $\sigma_n^0(h\nu)$.

The capacitance transient technique is suitable for measuring samples inside the DAC. First, the DAC allows electric and optical perturbations to be applied simultaneously to the sample. Second, the advantage of such transient experiments is that a small change in the junction capacitance can be more precisely determined on top of a large constant background stray capacitance associated with a sample inside the DAC. In DLTS measurements, the sample temperature is monitored by a calibrated Si diode thermometer attached to the diamond anvil nearest to the sample. To minimize the temperature difference between the sample and the diode sensor, the cell temperature is changed very slowly. To achieve a short equilibration time between the sample and thermometer, the thermal mass of the cell should be minimal.

IV. Experimental Studies of DX Centers

1. INTRODUCTION

The deep trap known as the DX center was discovered in 1979 by Lang and co-workers [17] in n-type $Al_xGa_{1-x}As$ with $x > 0.22$. During the past two decades the properties of the DX center have been studied extensively [18–21]. Lang et $al.$ named this defect the DX center because they thought that it involved a complex consisting of a donor atom D and an unknown constituent X. Since this center was first observed in alloys of GaAlAs only, it was believed that X is an intrinsic defect found in alloys. Some key characteristics of the DX center which distinguish it from other deep centers are (1) its optical ionization energy E_{op} (~ 1 eV) is an order of magnitude larger than its thermal ionization energy ($E_c - E_t$); and (2) its capture barrier height is on the order of 0.2 eV, and it has a very small capture cross-section for electrons at 77 K, resulting in a very long lifetime for free electrons. This give rise to the phenomenon known as persistent photoconductivity (PPC). The electron capture cross-section of a number of deep levels in GaAs and GaP has been studied extensively by Henry and Lang [22], as summarized in Fig. 2. These data can be fitted by the multiphonon emission theory (MPT) [22, 23] with a capture barrier height E_B as defined in Eq. (12) when kT is larger than the phonon energy. On the other hand, there is a lower limit of 10^{-21} cm^2 for the capture cross-section at low temperatures due to optical capture. However, the capture cross-section of the DX center can be several orders of magnitude lower than 10^{-21} cm^2 as shown in Fig. 2. This implies that optical capture into the DX center is almost prohibited. Lang explained these properties of the DX centers by the existence of a large lattice relaxation in these centers [17, 18]. Since Lang *et al.*'s original paper, great progress has been made toward developing a microscopic model of the DX center. In this review, we mainly focus on some transient capacitance experiments performed inside the diamond anvil high pressure cells which have helped to elucidate the nature of the DX center.

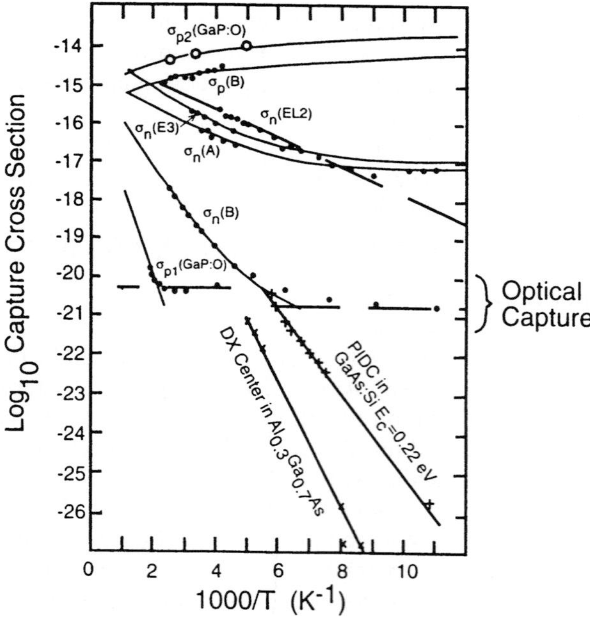

FIG. 2. The capture cross-section of PIDC (+) in GaAs:Si under 2.9 GPa [27] compared to that of the DX center (×) in Al$_{0.3}$Ga$_{0.7}$As:Si [35]. Notice how both of them behave differently from the capture behaviors of most other deep centers in GaAs and GaP. Reproduced from Ref. 22.

5.2 STUDY OF DX CENTERS USING CAPACITANCE TECHNIQUES   **469**

2. ESTABLISHMENT OF THE DX CENTER AS DUE TO SUBSTITUTIONAL DONORS

Lifshitz *et al.* [24] were the first group to notice the existence of a correlation between the effect of pressure and of Al alloying on the band structure properties of AlGaAs. They noticed that 0.1 GPa of pressure has approximately the same effect on the conduction band of GaAs as increasing the Al concentration by 1%. Since Lang *et al.* [17] have found that the DX center appeared in $Al_xGa_{1-x}As$ only when $x > 0.22$, it was natural to ask whether the DX center will appear in GaAs under a pressure of more than 2 GPa, assuming that the properties of the DX center is determined entirely by the conduction band structure. In 1985, Mizuta *et al.* [25] applied pressure to n-type GaAs doped with Si using a Bridgman anvil device and discovered that at pressures exceeding 2.4 GPa, a peak appeared in the DLTS spectrum (as shown in Fig. 3), with a DLTS activation energy of 0.31–0.33 eV. Similar results were obtained for n-type GaAs doped with Sn. Mizuta *et al.* identified this DLTS peak with the DX center peak found by Lang *et al.* in $Al_xGa_{1-x}As$ for $x > 0.2$. In a subsequent paper, Tachikawa *et al.* found evidence of PPC in GaAs:Si under pressure at 77 K [26] using an LED loaded into the high-pressure cell as the light source. Their result is shown in the inset of Fig. 3. If this pressure-induced deep center (PIDC) were indeed identical to the DX center, then the pressure experiments would have invalidated Lang's proposal that the DX center in AlGaAs involves an unknown constituent X introduced by alloying. Instead, one has to conclude that the DX center is the result of a shallow-to-deep transformation of substitutional Si donors in GaAs *induced by changes in the conduction-band structure only.* Such changes can be produced either by alloying or by pressure.

A crucial test of whether the PIDC in GaAs:Si and the DX center in AlGaAs are identical is the determination of the photoionization energy E_{op} and the thermal ionization energy $E_c - E_t$ of the PIDC and comparing them with those of the DX centers. Figure 4 shows the capture behavior of the PIDC in GaAs under 2.5–2.9 GPa of pressure. By analyzing the experimental result using Eqs. (10)–(13), the PIDC was found to obey the MPT with a capture barrier height $E_B = 0.22$ eV. Furthermore, the pressure coefficient of the capture barrier height (dE_B/dP) was determined to be -21 meV/GPa. This value of E_B is comparable to the value $E_B = 0.33 \pm 0.05$ eV for Si-doped AlGaAs measured by Lang [18]. Similarly, the values of $E_{DLTS} = 0.30$ eV and the pressure coefficient $dE_{DLTS}/dP = -13$ meV/GPa were obtained by DLTS measurement inside the DAC. These values are in good agreement with the results of Mizuta *et al.* The thermal ionization energy $E_c - E_t$ obtained from these energies using Eq. (15) was found

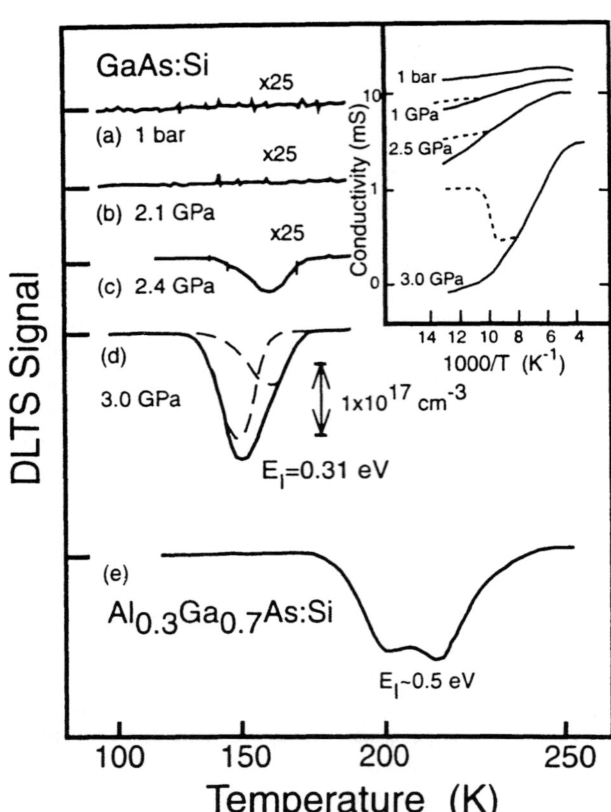

FIG. 3.   (a)–(d) The DLTS spectra of GaAs : Si measured at different
pressures. The rate window used is 66 sec^{-1} with t_2/t_1 = 2. Reproduced from [25]. (e) The corresponding spectrum for the DX center in
Al$_{0.3}$Ga$_{0.7}$As : Si. The inset shows persistent photoconductivity effect of
GaAs : Si under different pressures. Reproduced from [26]. The solid
curves were measured in the dark while the broken curves were measured after light illumination.

to be 0.08 eV, which is comparable with the value of 0.10 ± 0.05 eV
measured by Lang for Si-doped AlGaAs [18].

Photoionization experiments are difficult to perform with the Bridgman
anvil device used by Mizuta *et al.* [25] since these anvils are made from
sintered diamond and are opaque. To overcome this difficulty, we have
instead used a DAC to perform photocapacitance transient measurements
on this PIDC in Si-doped GaAs [27, 28]. The transparent diamond anvils

5.2   STUDY OF DX CENTERS USING CAPACITANCE TECHNIQUES   **471**

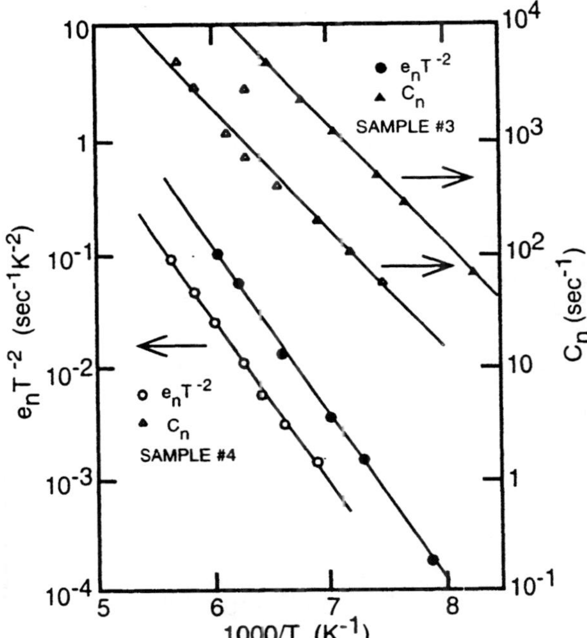

FIG. 4. Plots of the electron emission and capture rates versus 1/temperature for PIDC in two samples of GaAs under 2.9 GPa of pressure. Reproduced from [27].

allowed us to measure the photoionization spectra of the PIDC by applying Eq. (18). The results are shown in Fig. 5. By fitting our experimental data with the theory of Lang, Logan, and Jaros [17], we determined the photoionization threshold energy E_{op} of the PIDC in Si-doped GaAs to be 1.4 eV. This value is larger than its thermal ionization energy 0.08 eV by an order of magnitude. As seen from Fig. 5, there is also good agreement with the DX center's photoionization spectra measured by Legros et al. [29] in Si-doped AlGaAs. These results, together with those of Mizuta et al. [25] in GaAs:Si, thus show conclusively that the PIDC found in GaAs has all the important attributes of the DX centers in AlGaAs. To our knowledge this is the first time pressure has played such a crucial role in revealing the nature of a deep center in semiconductors.

It is interesting to note that on close examination there are actually quantitative differences between the DX centers in GaAs under pressure and those found in AlGaAs at ambient pressure [30]. The origin of this

472   Ming-Fu Li and Peter Y. Yu

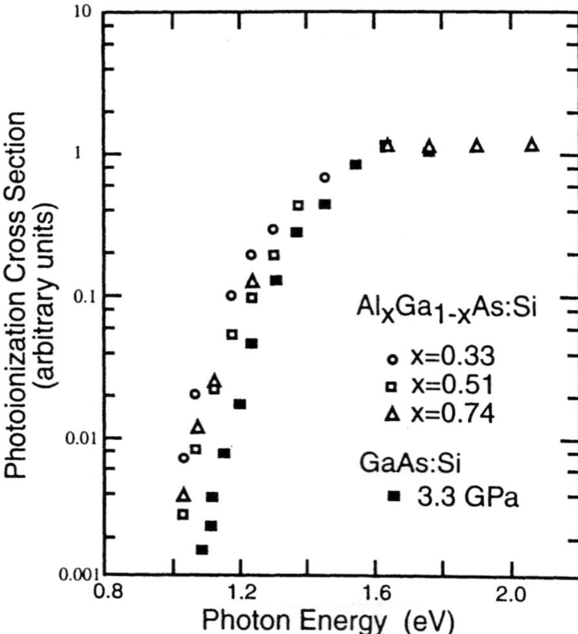

Fig. 5.   Photoionization spectrum of the PIDC in GaAs:Si under
a pressure of 3.3 GPa [28] at 77 K, compared with that of the DX
center in Al$_x$Ga$_{1-x}$As:Si at 84 K for three different values of x [29].

difference has been explained by Mooney *et al.* [31] to be due to the
different local environments of the DX center in the alloy. Although a Si
DX center has only one possible local environment, the analogous center
in AlGaAs can have one, two, or three Al atoms as its neighbors. The
convincing experiments in support of this explanation again involve the
application of pressure. The experiments were performed by Calleja *et al.*
[32] and by Baba *et al.* [33]. The results of Calleja *et al.* [32] are shown in
Fig. 6. They measured the DLTS spectra of Al$_x$Ga$_{1-x}$As samples with x =
0, 0.04, 0.08, respectively, under pressure. In Fig. 6, the lower temperature
DLTS peak with an activation energy of 0.34 eV is attributed to a DX
center without Al atoms as nearest neighbors. This assignment is consistent
with high-pressure data on GaAs:Si [27, 34]. The higher DLTS peak with
activation energy 0.44 eV corresponds to a DX center with Al atoms as
nearest neighbors and is also consistent with previous results in AlGaAs:Si
[35]. The 0.1-eV difference in DLTS activation energies between DX cen-

5.2 Study of DX Centers Using Capacitance Techniques **473**

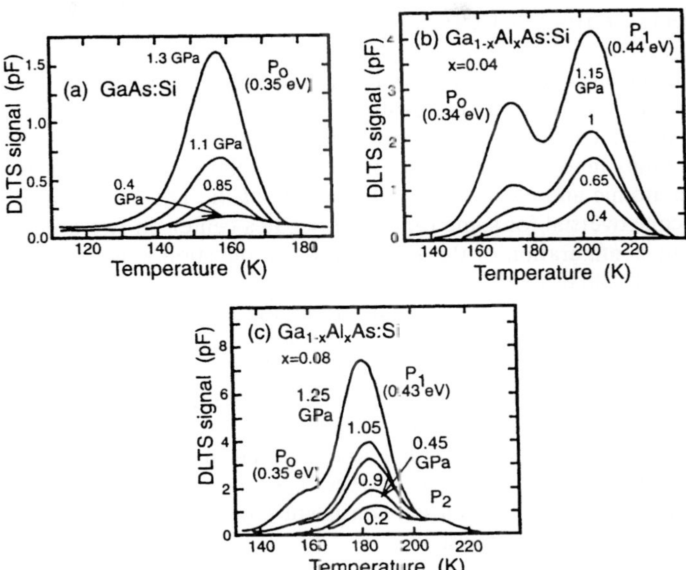

FIG. 6. DLTS spectra of DX centers under various applied pressures: (a) GaAs;Si, (b)Al$_x$Ga$_{1-x}$As ($x = 0.04$), and (c) Al$_x$Ga$_{1-x}$As ($x = 0.08$). Reproduced from [32].

ters with and without Al atoms as nearest neighbors has been confirmed by self-consistent theoretical calculations [36].

In addition to DX centers produced by group IV donors, such as Si, the pressure dependence of DX centers formed by group VI donors, such as Te, has also been studied [37]. The experiments were performed on Al$_x$Ga$_{1-x}$As epilayers, with $x = 0.15, 0.25$, and 0.35 and doped with 5×10^{16} cm^{-3} of Te. The DLTS peak emerges, respectively, at 1.6 GPa, 0.7 GPa, and 1 bar for the $x = 0.15, 0.25$, and 0.35 samples. These experiments showed that Liftshitz *et al.*'s result that 0.1 GPa of pressure has approximately the same effect on the conduction band of GaAs as increasing the Al concentration by 1% is correct for predicting the energy of the DX level relative to the conduction band edge. Together with the results in GaAs:Si they show quite convincingly that at ambient pressure the DX level associated with donor atoms in GaAs is actually a *resonance state* above the conduction band. As a result of the change in the conduction-band structure caused either by alloying or by pressure, the DX center emerges from the conduction band into the energy gap and becomes the stable ground state of the donor.

474   Ming-fu Li and Peter Y. Yu

3. Models of the DX Center

Since it became clear that the DX center is a simple substitutional donor
in GaAs which exhibits a shallow-to-deep transformation as a result of
changes in the conduction-band structure induced by either pressure or
alloying, many models have been proposed to explain its properties. However, the atomic and electronic configurations of the DX center have remained controversial for some time. The debates center on two areas. On
the atomic configuration of the DX center, the question is whether there
is large [17, 38, 39] or small lattice relaxation [40–43]. As far as the electronic
configuration of the DX center is concerned, the issue is whether the DX
center has a negative on-site Coulomb interaction U (abbreviated as $-U$)
or a positive U between the two electrons localized on the same impurity.
If the former case is correct, then the ground state of the DX center contains
two electrons [38, 43], whereas in the latter case it will contain only one
electron [41, 42]. Now it is generally accepted that the model proposed by
Chadi and Chang in 1988 [38] is correct. The important features of this
model, based on their supercell self-consistent pseudopotential calculation,
can be summarized as follows:

1. The DX center is a $-U$ center resulting from the reaction

$$2d^0 \rightarrow d^+ + DX^-, \tag{19}$$

   where d^0 and d^+ represent fourfold-coordinated substitutional donors
   in the neutral and ionized state, respectively. DX^- is a negative charged
   donor that has captured two electrons.
2. The DX^- defect formation involves a large bond-rupturing displacement of the host lattice atoms. For donors on cation sites, such as
   Si_{Ga}, the donor atom is displaced as depicted in Figs. 7a and b. In the
   case of donors located on anion sites, such as S_{As}, one of its nearest-
   neighbor Ga (or Al) atoms along a bond axes is displaced as shown
   in Figs. 7c and d. In other words, the local symmetry of a donor is
   charge dependent. When the donor electron occupancy is 0 or 1,
   corresponding to the positively charged d^+ or neutral charge d^0 states,
   the donor atom symmetry is T_d and there is no lattice relaxation.
   When the donor electron occupancy is 2, corresponding to a negatively
   charged DX^- state, the defect symmetry is reduced to C_{3v} as a result
   of bond-breaking relaxation.

One prediction of the $-U$ model is that the DX center should produce
no electron paramagnetic resonance (EPR) signal. The reason is because

5.2 STUDY OF DX CENTERS USING CAPACITANCE TECHNIQUES **475**

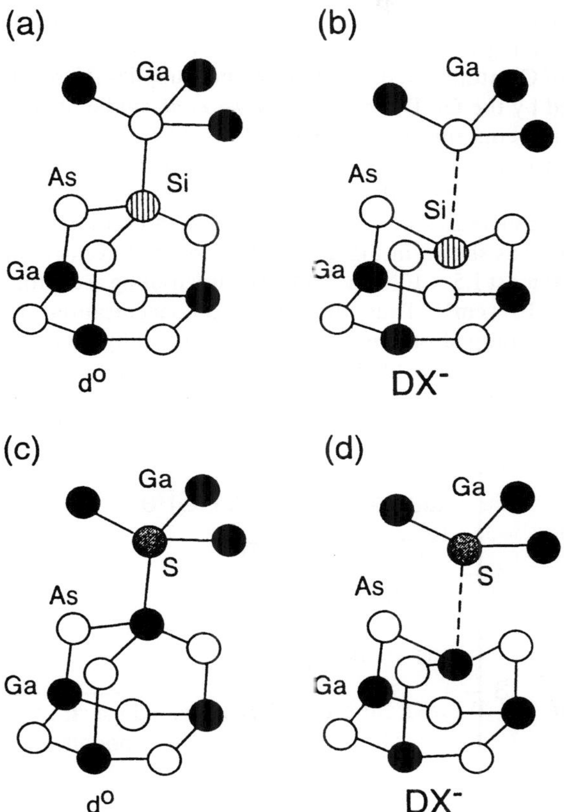

FIG. 7. Schematic diagrams of the normal substitutional d° (a and c) and the broken-bond DX⁻ configurations (b and d) of Si and S donors in GaAs. Reproduced from [38].

these two electrons should have opposite spin in order to satisfy the Pauli Exclusion Principle. Unfortunately, two experimental attempts to test the validity of the $-U$ model turned out to be contradictory [43, 44]. The strongest confirmation of the $-U$ property of the DX center comes from codoping experiments carried out by different groups using a variety of sample sources and measurement techniques.

Fujisawa *et al.* [45] performed the first successful codoping test of the $-U$ model by applying pressure to GaAs codoped with two donors: Ge and Si with different binding energies. At a pressure of 2.2 GPa, Ge in GaAs is converted into DX centers while Si remains as a shallow donor.

476   MING-FU LI AND PETER Y. YU

If the $-U$ model is correct, then the number of Ge DX centers can be varied by changing the concentration of shallow Si donors while keeping the number of Ge atoms fixed, since the former supplies the second electron to be trapped by the Ge DX state. Fujisawa *et al.* studied several samples in which the Ge concentration is fixed at 1×10^{17} cm^{-3} while the Si doping concentration is varied up to 2.6×10^{17} cm^{-3}. From the DLTS spectra, Fujisawa *et al.* found that the concentration of electrons trapped at the Ge DX centers increases with Si concentration and saturates at a value of 2.3×10^{17} cm^{-3} as shown in Fig. 8. The Ge donor concentration was estimated to be at most 1.5×10^{17} cm^{-3} with a compensating acceptor concentration of 0.5×10^{17} cm^{-3}. Thus, the saturated concentration of 2.3×10^{17} cm^{-3} electrons trapped on Ge cannot be explained by a Ge ground state with only one electron. Instead, one has to assume that each Ge atom can trap two electrons with a concentration of 1.15×10^{17} cm^{-3} Ge atoms.

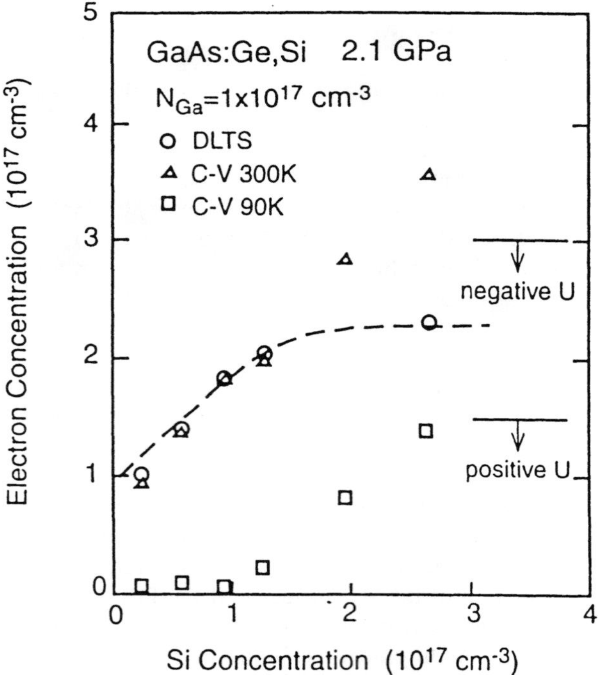

FIG. 8.   Plots of the electron concentration at Ge DX centers determined by DLTS (O) and of the free carrier concentration (△) measured by CV at two different temperatures in GaAs codoped with Ge and Si at 2.1 GPa as a function of Si donor concentration. Reproduced from [45].

5.2 STUDY OF DX CENTERS USING CAPACITANCE TECHNIQUES **477**

Unfortunately, Fujisawa *et al.*'s experiment is not unambiguous. First, if the true Ge concentration in their samples is 2.3×10^{17} cm^{-3} while the concentration of compensating acceptors is 1.3×10^{17} cm^{-3}, then their experimental result is consistent with a $+U$ center model. Second, the concentration of trapped electrons estimated by the DLTS method may not be reliable enough because of the high concentration of deep defects and the very large edge region effect [17]. Third, Fujisawa *et al.* tried to maintain the Ge doping concentration constant but, due to fluctuation in the growth conditions, the precision in controlling the doping concentration is low.

Baj *et al.* [46] avoided the difficulties in interpreting the codoping experiment of Fujisawa *et al.* by using a single GaAs sample codoped with Te and Ge instead. In addition to the large lattice relaxation DX levels, the Ge impurities in GaAs form also a small lattice relaxation A_1 level. These levels lie in the conduction band at ambient pressure but move into the gap at pressures exceeding 1.0 GPa. On the other hand, Te remains a shallow donor level in the gap at pressure less than 1.5 GPa. Thus, the idea behind the experiment of Baj *et al.* is to use pressure to convert Ge first into the positive U A_1 level impurities (labeled as the D^0 state in some literatures [47]) and then into the deep DX$^-$ center while shallow Te levels provide the electrons to be trapped on the DX levels. Furthermore, instead of measuring the concentration of electrons trapped at the DX centers by DLTS, the free carrier concentration is determined by the Hall effect, which is more precise than DLTS. A combination of control methods, such as irradiating the sample with light and changing its temperature, allow the number of electrons trapped on the DX centers to be varied via PPC. Figure 9 shows the Hall carrier concentration after light illumination measured by Baj *et al.* as a function of pressure at 77 and 100 K, respectively. Both curves show a step at pressure between 0.5 GPa to 1.0 GPa. The step in the 77 K curve is smaller and has a magnitude 1×10^{17} cm^{-3}. This step is explained by the trapping of electrons from the conduction band into the shallower A_1 level of the Ge centers. Thus, the concentration of Ge impurity is determined accurately to be 1×10^{17} cm^{-3} since each A_1 state captures only one electron. The deeper DX level associated with the Ge impurities does not capture electrons at 77 K because of its large capture barrier height. However, at 100 K the capture rate of the Ge DX center becomes much faster, so if the DX center is a $-U$ center, one expects to see a bigger drop in the carrier concentration due to trapping into the DX state. Indeed, Baj *et al.* found that the step in the 100 K curve in Fig. 9 is 2×10^{17} cm^{-3}, or exactly twice the concentration of the Ge impurity. This experiment unambiguously demonstrates that each DX level of the Ge impurity in GaAs captures two electrons. The beauty of this experiment is that the

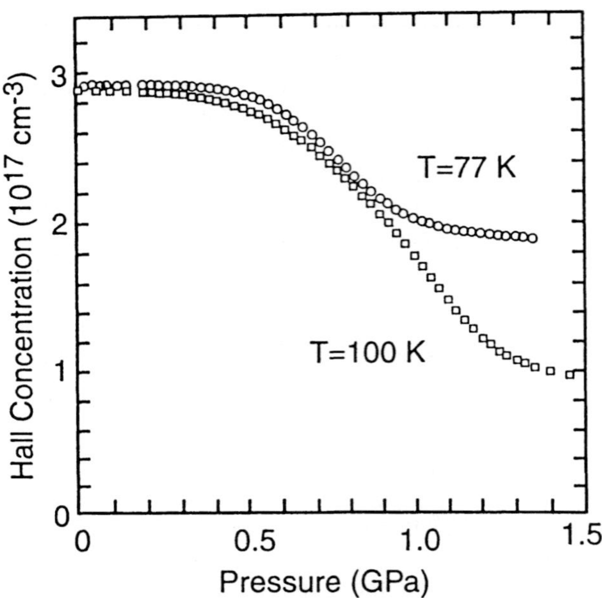

FIG. 9.   Plots of the Hall carrier concentration in GaAs codoped with Ge and Te measured as a function of hydrostatic pressure at $T = 77$ and 100 K. Reproduced from [46].

result is independent of any compensating acceptors. Also, no prior information on doping concentrations is needed. The only requirement is that the concentration of Te shallow donor be higher than that of the Ge donors so that Te can provide enough electrons to fill the Ge DX states.

Willke *et al.* [48] also reported a codoping experiment using $Al_xGa_{1-x}As$ samples which contain Si and Sn instead of Ge and Te. Under hydrostatic pressure, both Si and Sn DX levels move into the energy gap. The idea behind their experiment is to use light to selectively photoionize the two deep levels, since the photoionization thresholds of Sn and Si in AlGaAs are different, 0.8 and 1.1 eV, respectively. Figure 10 shows the phototransient carrier concentration (n) of their sample at 4.2 K as measured by Hall and Shubnikov–de Haas effects. In Fig. 10a the sample is first illuminated with 1 eV radiation at $t = 0$ sec. Then the illumination is switched to 1.4-eV radiation starting at $t = 7000$ sec. The transient step at $t = 0$ sec is accounted for by photoionization of the Sn centers and suggests a Sn concentration of 5.6×10^{17} cm^{-3}. The transient step at $t = 7000$ sec is due to Si and indicates that its concentration is 1.21×10^{18} cm^{-3}. The sample is then subjected to a thermal cycling in the dark to 70 K in order to allow

5.2   STUDY OF DX CENTERS USING CAPACITANCE TECHNIQUES   **479**

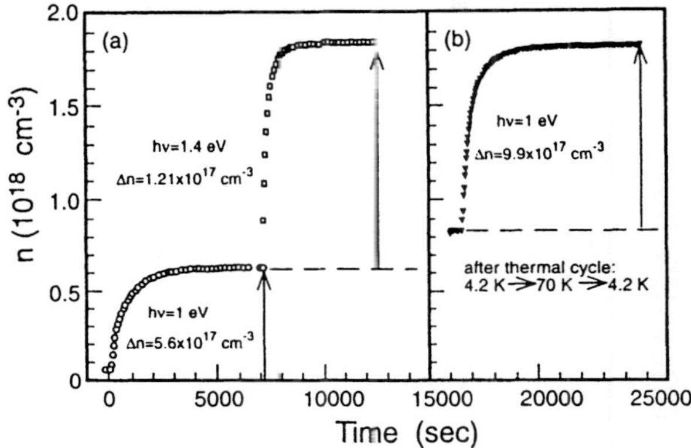

FIG. 10.   Phototransients in the carrier concentration (n) measurement at 4.2 K in $Al_xGa_{1-x}As$ sample codoped with Sn and Si. (a) The sample was first slowly cooled down in the dark and then illuminated with 1 eV radiation at $t = 0$ sec followed by irradiation with 1.4-eV light at $t = 7000$ sec. (b) The sample was thermally cycled in the dark to 70 K and then illuminated with 1-eV radiation at 4.2 K. Reproduced from [48].

carriers to be recaptured into the shallower Sn DX centers only. The idea is that PPC of the deeper Si DX center now provides the free carriers to be trapped on the Sn centers in case the latter has a negative U. When the measurement is repeated with 1-eV radiation (result shown in Fig. 10b, the step in the phototransient is now found to be larger, 9.9×10^{17} cm^{-3}. This is almost double the Sn concentration of 5.6×10^{17} cm^{-3} as predicted by the $-U$ model for the DX center. Thus, the evidence again supports that the Sn DX centers have $-U$.

Finally, we briefly describe another codoping experiment based on similar ideas but not using GaAs. In this experiment [49], $GaAs_{0.6}P_{0.4}$ samples with a uniformly doped background of Te with concentration N_{Te} are used. A Gaussian distribution of S with concentration $N_s(x)$, where x is the depth, is introduced by ion implantation. It is well known that Te is a shallow donor while S forms a DX center ground state in GaAsp [49, 50]. The carrier concentration (n) is measured by the CV method. When the sample is illuminated at 77 K by light, the carrier concentration n_{op} is given by $N_{Te} - N_A + N_s$ (shown in Fig. 11), where N_A is the compensating acceptor concentration. If n is measured in the dark instead, the resultant carrier concentration n_{dark} will depend on whether the DX centers trap one or two electrons. If the DX centers trap two electrons, then

$$n_{dark} = N_{Te} - N_A - N_s \tag{20}$$

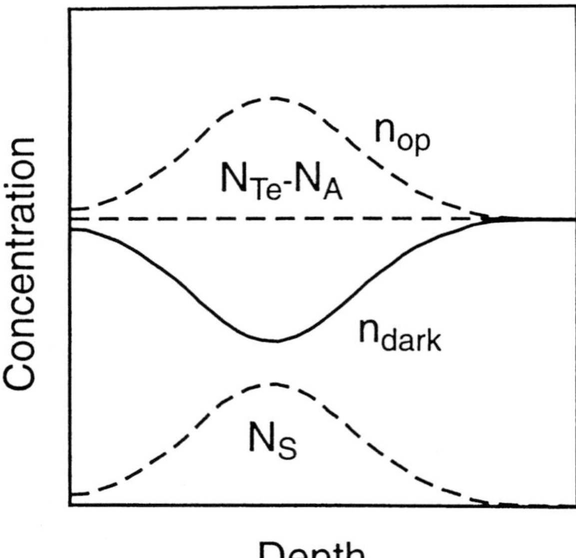

FIG. 11.   Schematic diagram of donor distribution as a function of depth in a GaAs$_{0.6}$P$_{0.4}$ sample codoped with Te and S. N_s represents the implanted S ion spatial profile. $N_{Te} - N_A$ is the free carrier profile in the sample before implantation of S. n_{dark} is the predicted free-carrier profile if the S DX centers capture two electrons. n_{op} is the predicted profile after all the S DX centers have been emptied of electrons via PPC. Reproduced from [49].

will have the shape shown in Fig. 11. This curve should be a mirror image of the curve n_{op} with respect to the horizontal line $N_{Te} - N_A$. On the other hand, if the DX centers trap only one electron, then $n_{dark} = N_{Te} - N_A$ will be a horizonal line as in the unimplanted sample. Other defects produced by ion implantation can complicate this scheme. Fortunately, these defects have no PPC effect and can be distinguished from the DX center signal. Figure 12 shows the carrier spatial profiles measured by the CV technique. In case of the dark profile, n_{dark}, the curve has been corrected for the effects due to ion-implantation-induced defects. The fact that the profiles n_{op} and n_{dark} are roughly mirror images of each other with respect to the unimplanted curve clearly shows that the sulfur DX center traps two electrons. Finally, by taking into consideration the fact that the DLTS spectra in GaAsP exhibit only one peak [51], one can conclude that the two-electron DX state of S in GaAsP is indeed the ground state, and therefore the S DX center in GaAsP is a $-U$ system. If this were not the case, then

5.2   Study of DX Centers Using Capacitance Techniques   **481**

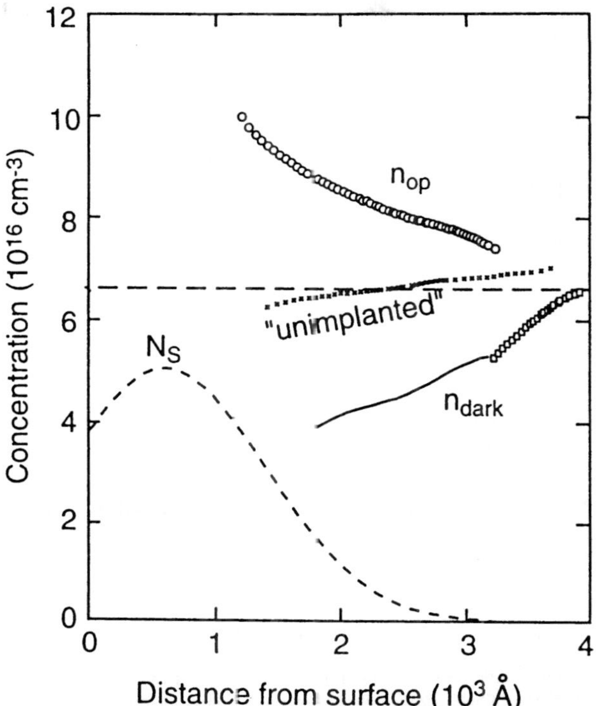

FIG. 12.   The spatial profiles of free carrier concentration in the GaAs$_{0.4}$P$_{0.6}$ sample codoped with Te and S measured at 77 K using CV techniques. The broken curve N_s is the calculated S profile based on the ion implantation parameters. The curve labeled "unimplanted" represents the free carrier profile due to the shallow Te donors before the sample was implanted with S. The curve labeled n_{dark} is the free carrier profile measured with the sample in the dark. This curve consists of two parts. The points are the measured data, while the solid curve is obtained after correction for the effects of the ion-implantation-induced defects. The curve labeled n_{op} is the spatial profile obtained under light illumination. Reproduced from [49].

the DLTS specta would show a second peak at a lower temperature, in disagreement with experiment [48, 50].

V.   Concluding Remarks

In this chapter we have described how to perform capacitance measurements on samples subjected to high pressure inside the DAC. As an illustra-

482   MING-FU LI AND PETER Y. YU

tion of the usefulness of this technique, we have discussed its application to determine the nature of the DX centers in GaAs and related alloys. Because of space limitations, it is impossible to summarize all the contributions made by high-pressure techniques to our understanding of DX centers. For example, we have omitted mentioning many important high-pressure optical experiments performed on DX centers inside the DAC [52]. We should also point out that the DX center is not the only deep center that capacitance experiments inside the DAC have made significant contributions to understanding [53].

Acknowledgments

The part of this work performed in Singapore was supported by the Singapore NSTB RIC-University Research Funding Project 681305. The work at Berkeley was supported by the Director, Office of Energy Research, Office of Basic Energy Sciences, Materials Sciences Division, of the U.S. Department of Energy under Contract No. DE-AC03-76SF00098.

REFERENCES

1. W. Paul, *J. Appl. Phys.* **32**, 2082 (1961).
2. P. Y. Yu and B. Welber, *Solid State Commun.* **25**, 209 (1978).
3. See, for example, review article by A. Jayaraman in *Rev. Modern Phys.* **57**, 1013 (1986).
4. H. K. Mao and P. M. Bell, *Rev. Sci. Instrum.* **52**, 615 (1981).
5. R. L. Reichlin, *Rev. Sci. Instrum.* **54**, 1674 (1983).
6. D. Erskine, P. Y. Yu, and G. Martinez. *Rev. Sci. Instrum.* **58**, 406 (1987).
7. D. Patel, T. C. Crumbaker, J. R. Sites, and I. L. Spain, *Rev. Sci. Instrum.* **57**, 2795 (1986).
8. D. Patel and I. L. Spain, *Rev. Sci. Instrum.* **58**, 1317 (1987).
9. H. Huiberts, Ph.D. Thesis, Vrije Universiteit, Amsterdam (1997).
10. C. T. Sah, L. Forbes, L. L. Rosier and A. F. Tasch, *Solid State Electron.* **13**, 759 (1970).
11. C. T. Sah, *Solid State Electron.* **19**, 975 (1975).
12. D. V. Lang, *J. Appl. Phys.* **45**, 3023 (1974).
13. G. L. Miller, D. V. Lang, and L. C. Kimerling, *Ann. Rev. Mater. Sci.* **7**, 377 (1977).
14. H. G. Grimmeis, *Ann. Rev. Mater. Sci.* **7**, 343 (1977).
15. M. F. Li, *Modern Semiconductor Quantum Physics*, World Scientific, Singapore, 1994, Section 360.
16. A. Chantre, G. Vincent, and D. Bois, *Phys. Rev. B* **23**, 5335 (1981).
17. D. V. Lang, R. A. Logan, and M. Jaros, *Phys. Rev. B.* **19**, 1015 (1979).
18. D. V. Lang, in *Deep Centers in Semiconductors*, ed. by S. T. Pantelides. Gordon and Breach, New York, 1985, p. 489.
19. P. M. Mooney, *J. Appl. Phys.* **67**, R1 (1990).
20. J. C. Bourgoin (ed.), *Physics of the DX Centers in AlGaAs Alloys.* Sci. Tech. Publications, Vaduz, 1990.

5.2 Study of DX Centers Using Capacitance Techniques **483**

21. K. J. Malloy and K. Khachaturyan, in *Imperfections in III/V Materials,* Vol. 38, *Semiconductors and Semimetals.* Academic Press, Boston, 1993, p. 235.
22. C. H. Henry and D. V. Lang, *Phys. Rev. B* **15,** 989 (1977).
23. K. Huang and R. Rhys, *Proc. Roy. Soc.* **204,** 406 (1950).
24. N. Lifshitz, A. Jayaraman, and R. A. Logan, *Phys. Rev. B* **21,** 670 (1980).
25. M. Mizuta, M. Tachikawa, H. Kukimoto, and S. Minomura, *Jpn. J. Appl. Phys.* **24,** L143 (1985).
26. M. Tachikawa, T. Fujisawa, H. Kukimoto, A Shibata, G. Oomi, and S. Minomura, *Jpn. J. Appl. Phys.* **24,** L893 (1985).
27. M. F. Li, P. Y. Yu, E. R. Weber, and W. L. Hansen, *Appl. Phys. Lett.* **51,** 349 (1987).
28. M. F. Li, P. Y. Yu, R. Weber, and W. L. Hansen, *Phys. Rev. B* **35,** 7505 (1987).
29. R. Legros, P. M. Mooney, and S. L. Wright, *Phys. Rev. B* **35,** 7505 (1987).
30. M. F. Li, P. Y. Yu, W. Shan, W. L. Hansen, and E. R. Weber, *Proc. 19th Int. Conf. Physics of Semiconductors,* ed. by W. Zawadzki. Institute of Physics, Polish Academy of Sciences, 1988, p. 1051.
31. P. M. Mooney, T. N. Theis, and S. L. Wright, *Appl. Phys. Lett.* **53,** 2546 (1988).
32. E. Calleja, F. Garcia, A. Gomez, E. Munoz, P. M. Mooney, T. N. Morgan, and S. L. Wright, *Appl. Phys. Lett.* **56,** 934 (1990).
33. T. Baba, M. Mizuta, T. Fujisawa, J. Yoshino, and H. Kukimoto, *Jpn. J. Appl. Phys.* **28,** L891 (1989).
34. T. N. Theis, P. M. Mooney, and S. L. Wright, *Phys. Rev. Lett.* **60,** 3619 (1988).
35. B. L. Zhou, K. Ploog, E. Gmelin, X. O. Zheng, and M. Schultz, *Appl. Phys. A* **28,** 223 (1982).
36. Z. B. Zhang, *Phys. Rev. B* **44,** 3417 (1991).
37. W. Shan, P. Y. Yu, M. F. Li, W. L. Hansen, and E. Bauser, *Phys. Rev. B* **40,** 7831 (1989).
38. D. J. Chadi and K. J. Chang, *Phys. Rev. Lett.* **61,** 873 (1988); *Phys. Rev. B* **39,** 10366 (1989).
39. K. A. Khachaturyan, E. R. Weber, and M. Kaminska, in *Defects in Semiconductors, 15.* Trans. Tech., Switzerland, 1989, p. 1067.
40. J. C. M. Henning and J. P. M. Ansems, *Semicond. Sci. Tech.* **2,** 1 (1987).
41. H. P. Hjarmarson and T. J. Drumond, *Appl. Phys. Lett.* **48,** 656 (1986).
42. J. C. Bourgoin, *Solid State Phenomena* **10,** 253 (1989).
43. E. Yamaguchi, K. Sheraishi, and T. Ohno, *J Phys. Soc. Japan* **60,** 3093 (1990).
44. K. A. Khachaturyan, D. D. Awschalom, J. R. Rosen, and E. R. Weber, *Phys. Rev. Lett.* **63,** 1311 (1989).
45. T. Fujisawa, J. Yoshino, and H. Kukimoto, *Jpn. J. Appl. Phys.* **29,** L388 (1990); and in *Proc. 20th Int. Conf. Phys. Semiconductors,* ed. by E. M. Anastassakis and J. D. Joannopoulos. World Scientific, Singapore, 1990, Vol. 1, p. 509.
46. M. Baj, L. H. Dmowski, and T. Stupinski, *Phys. Rev. Lett.* **71,** 3529 (1993); M. Baj and L. H. Dmowski, *J. Phys. Chem. Solids* **56,** 589 (1995).
47. For a discussion on the theory of the D° state see, for example, J. Dabrowski and M. Scheffler, in *Defects in Semiconductors 16, Materials Science Forum* **83–87** (Trans Tech Publ., Switzerland, 1992), p. 735. For discussions on experimental results see, for example, T. Suski in *Defects in Semiconductors 17, Materials Science Forum* **143–147** (Trans Tech Publ., Switzerland, 1994), p. 975.
48. U. Willke, M. L. Fille, D. K. Mauce, J. C. Portal, and P. Gibart, in *Proc. 22nd Int. Conf. Phys. Semiconductors,* ed. by D. J. Lockwood. World Scientific, Singapore, 1994, p. 2295.
49. M. F. Li, Y. Y. Luo, P. Y. Yu, E. R. Weber H. Fujioka, A. Y. Du, S. J. Chua and Y. T. Lim, *Phys. Rev. B* **50,** 7996 (1994); and in *Proc. 22nd Int. Conf. Phys. Semiconductors,* ed. by D. J. Lockwood. World Scientific, Singapore, 1994, p. 2303.
50. M. G. Craford, G. E. Stillman, N. Holonyak, Jr., and J. A. Rossi, *J. Electronic Materials* **20,** 3 (1991); *Phys. Rev.* **168,** 867 (1968).

484   MING-FU LI AND PETER Y. YU

51. R. A. Craven and D. Finn, *J. Appl. Phys.* **50,** 6334 (1979).
52. See, for example, J. A. Wolk, W. Walukiewicz, M. L. Thewalt, and E. E. Haller, *Phys. Rev. Lett.* **68,** 3619 (1992) for a description of the infrared vibrational study and J. Zeman, M. Zigone, and G. Martinez, *Phys. Rev. B* **51,** 1755 (1995) for a review of the Raman investigation of the DX center, both under pressure inside the DAC.
53. See, for example, N. M. Johnson, W. Shan, and P. Y. Yu, *Phys. Rev. B* **39,** 3431 (1989) for a study of the pressure dependence of the P_b center at the $\langle 111 \rangle$ Si–SiO$_2$ interface.

Chapter 2

Semiconductor Band Structures

857

Spin-Orbit Interaction Effects in Zincblende Semiconductors: <u>Ab</u> <u>Initio</u> Pseudopotential Calculations

Ming-Fu Li*, Michael P. Surh[†], and Steven G. Louie[†]

*University of Science and Technology, *Academia Sinica*
Beijing, People's Republic of China

[†] Department of Physics, University of California, and
Materials and Chemical Sciences Division, Lawrence Berkeley Laboratory,
Berkeley, CA 94720, U.S.A.

Ab initio band structure calculations have been performed for the spin-orbit interaction effects at the top of the valence bands for GaAs and InSb. Relativistic, norm-conserving pseudopotentials are used with no correction made for the gaps from the local density approximation. The spin-orbit splitting at Γ and linear terms in the $\vec{k}$ dependence of the splitting are found to be in excellent agreement with existing experiments and previous theoretical results. The effective mass and the cubic splitting terms are also examined.

We report results from an *ab initio* calculation of the effect of spin-orbit interaction at the top of the valence bands for GaAs and InSb. The method employed is an extension of the analysis of Hybertsen and Louie [1] to systems without inversion symmetry. In this approach, the relativistic pseudopotentials [2,3] are first used to obtain a scalar relativistic band structure in a planewave local density functional (LDA) calculation, yielding both the unsplit eigenvalues and wavefunctions. The spin-orbit interaction is then added as a perturbation on the self-consistent scalar bands, and the fully relativistic eigenvalues and wavefunctions are obtained at each $\vec{k}$ by diagonalizing the Hamiltonian matrix in the basis of the scalar relativistic wavefunctions. In a planewave basis the perturbation Hamiltonian is given by:

$$H^{so}_{s,\vec{K};s',\vec{K}'} = -i <s|\,\vec{S}\,|s'> \frac{\vec{K}\times\vec{K}'}{KK'} \sum_{n=1,2} e^{i(\vec{K}-\vec{K}')\cdot\vec{\tau}_n}$$

$$\times \{12\pi V^{so;n}_{l=1}(\vec{K},\vec{K}') + 60\pi \frac{\vec{K}\cdot\vec{K}'}{KK'} V^{so;n}_{l=2}(\vec{K},\vec{K}')\} \qquad (1)$$

The notation is equivalent to that in Ref. 1, except that here the basis $\vec{\tau}_n$ yields complex structure factors.

The scalar bands are calculated with a planewave basis extending up to a kinetic energy of 16 Ry for GaAs and 14 Ry for InSb. The calculated LDA direct gaps are 0.485 eV for GaAs and -0.201 eV for InSb without spin-orbit splitting. The spin-orbit perturbation is included for the first 8 scalar relativistic bands (4 valence and 4 conduction). Essentially the same results are obtained near Γ if only the top 3 valence and lowest conduction band are included. The $E(\vec{k})$ are then fitted in the vicinity of Γ for the valence states Γ_7 and Γ_8 to Mth order polynomials in each of the high symmetry directions Δ, Λ, Σ (M is typically 20). The results are summarized in Fig. 1 and Tables 1 and 2 and discussed below:

1) The spin-orbit splitting between the states Γ_7 and Γ_8 are in excellent agreement with experimental results for GaAs and good agreement with experimental InSb results.

2) The linear spin-orbit splittings which arise from the lack of inversion symmetry are precisely consistent with the group theoretical relations in Ref. 4 which have the ratios for different $\vec{k}$ directions (Fig. 1):

$$\frac{C^1_{\Sigma a}}{\sqrt{1 + \frac{\sqrt{3}}{2}}} = \frac{C^1_{\Sigma b}}{\sqrt{1 - \frac{\sqrt{3}}{2}}} = \frac{C^1_{\Lambda a}}{\sqrt{2}} = C^1_{\Delta ab} = C^1 \tag{2}$$

$$C^1_{\Lambda bc} = 0 \; ; \quad -C^1_{\Delta cd} = C^1_{\Delta ab} \; ; \quad -C^1_{\Sigma d} = C^1_{\Sigma a} \; ; \quad -C^1_{\Sigma c} = C^1_{\Sigma b} \tag{3}$$

where C^1 is the linear splitting coefficient in the [100] direction [4,5]. The only available experimental data for InSb is in very good agreement with our calculation. In addition, our results for GaAs are in close agreement with the LMTO calculation of Cardona $et.$ $al.$ [6]. This agreement comes despite the small calculated LDA direct gaps. These results indicate that the valence band linear terms are mainly due to the interaction among the Γ_{15} valence bands.

3) The quadratic coefficients and cubic splittings for the Γ_8 bands are listed in Table 2. The quadratic coefficients are strongly anisotropic due, in part, to a mixing of the light and heavy holes in the [100] and [110] directions. The cubic coefficients differ from those of Ref. 6 by two or three orders of magnitude, since the region examined here is close to Γ, so that the light and heavy holes are nearly degenerate and intermix. This is in contrast to Ref. 6, where the analysis was performed for a larger region of $\vec{k}$, so the holes are split by their different effective

859

masses and do not mix. Additionally, the LDA gaps have not been empirically corrected to the experimental values here.

4) The split-off hole band is very isotropic and parabolic with quadratic coefficients of -37 eVÅ^2 for GaAs and -91 eVÅ^2 for InSb.

We wish to acknowledge valuable discussions with P. Y. Yu, M. Cardona, and M. S. Hybertsen. This work was supported by NSF Grant No. DMR8319024 and by the Director, Office of Energy Research, Office of Basic Energy Sciences, Materials Science Division of the U.S Department of Energy under Contract No. DE-AC03-76SF00098. CRAY computer time was provided by the Office of Energy Research of the Department of Energy.

[1] M.S. Hybertsen and S.G. Louie, Phys. Rev. B34, 2920 (1986).
[2] L. Kleinman, Phys. Rev. B21, 2630 (1980).
[3] G.B. Bachelet and M. Schluter, Phys. Rev. B25, 2103 (1982).
[4] G. Dresselhaus, Phys. Rev. 100, 580 (1955).
[5] E.O. Kane, in Semiconductors and Semimetals, ed. R.K. Willardson and A.C. Beer (Academic, 1966) Vol. 1, p.75.
[6] M. Cardona, N.E. Christensen, and G. Fasol, Phys. Rev. Lett. 56, 2831 (1986); M. Cardona, N.E. Christensen, and G. Fasol, to be published.

Table 1. Comparison of present calculation (pseud.) of spin-orbit splittings and linear terms to experiment and LMTO calculations of Ref. 6. Units are eV and meVÅ for the splitting and linear terms, respectively.

	GaAs			InSb		
	Pseud.	LMTO	Exp.	Pseud.	LMTO	Exp.
s-o splitting	0.345	0.351	0.341[a] 0.350[b]	0.758	0.799	0.81 (1.5K)[c] 0.813 (4K)[d]
linear term	-3.42	-3.4,-3.6	——	-9.85	-9.2	|9.3|[e]

a)   D.E. Aspnes, A.A. Studna, Phys. Rev. B7, 4605 (1973).
b)   T. Nishino, M. Okuyama, Y. Hamakawa, J. Phys. Chem. Solids 30, 2671 (1969).
c)   C.R. Pidgeon, S.H. Grove, J. Feinleib, Sol. St. Comm. 5, 677 (1969).
d)   R.L. Aggarwal, in Semiconductors and Semimetals, Vol. 9, eds. R.K. Willardson and A.C. Beer, (Academic, 1972) p·151.
e)   Landoldt-Börnstein Tables, Vol. 17a, eds. O. Madelburg, M. Schultz, and H. Weiss (Springer, Berlin, 1982).

860

Table 2. Quadratic (C^2) and cubic (C^3) coefficients for the Γ_8 bands. The units are eVÅ^2 and eVÅ^3 for the C^2 and C^3 coefficients, respectively.

	GaAs	InSb
$C^2_{\Delta ab} = C^2_{\Delta cd}$	-6.9×10^1	2.2×10^1
$C^2_{\Lambda a} = C^2_{\Lambda d}$	-4.8	-5.6
$C^2_{\Lambda bc}$	-1.33×10^2	5.5×10^1
$C^2_{\Sigma a} = C^2_{\Sigma d}$	-1.32×10^1	-1.6
$C^2_{\Sigma b} = C^2_{\Sigma c}$	-1.24×10^2	5.1×10^1
$C^3_{\Delta ab} = C^3_{\Delta cd}$	-4.7×10^5	-8.0×10^4
$C^3_{\Lambda a} = C^3_{\Lambda d}$	absolute value $< 10^{-1}$	-1.5×10^3
$C^3_{\Lambda bc}$	0.00	0.00
$C^3_{\Sigma a} = C^3_{\Sigma d}$	-2.4×10^5	-3.6×10^4
$C^3_{\Sigma b} = C^3_{\Sigma c}$	2.6×10^5	3.7×10^4

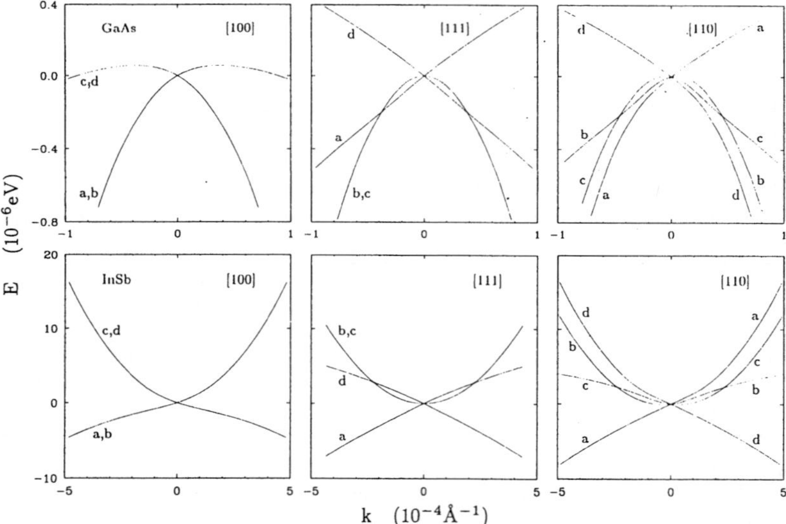

Figure 1. Band structures at the top of the Γ_8 valence bands for GaAs and InSb in the [100], [111], and [110] symmetry directions.

PHYSICAL REVIEW B   VOLUME 41, NUMBER 12   15 APRIL 1990-II

Deformation potentials at the top of valence bands in semiconductors: *Ab initio* pseudopotential calculations

Zong-Quan Gu
Institute of Semiconductors, Academia Sinica, P.O. Box 912, Beijing 100 083, China

Ming-Fu L
*Graduate School, University of Science and Technology of China, Academia Sinica, P.O. Box 3908, Beijing, China;
Institute of Semiconductors, Academia Sinica, P.O. Box 912, Beijing 100 083, China;
Department of Physics, Fudan University, Shanghai 200 433, China*

Jian-Qing Wang
Graduate School, University of Science and Technology of China, Academia Sinica, P.O. Box 3908, Beijing, China

Bing-Sing Wang
Institute of Semiconductors, Academia Sinica, P.O. Box 912, Beijing 100 083, China
(Received 25 September 1989)

Ab initio pseudopotential calculations of optical and acoustic deformation potentials for nine typical semiconductors, Si, AlSb, GaAs, GaP, GaSb, InP, InAs, InSb, and AlAs, are carried out systematically. The optical-deformation-potential values are, overall, larger than previous self-consistent calculations by the linear combination of muffin-tin orbitals (LMTO) method and are closer to experiments. The acoustic-deformation-potential values have the same sign as LMTO calculations. They are found to be sensitive to potential perturbed by strain, and therefore the previous, simpler empirical-pseudopotential calculations based on rigid-pseudopotential form factors generate incorrect sign and magnitude for acoustic deformation potentials. The internal-strain parameters for these semiconductors are found to have the same order of magnitude and same chemical trends as LMTO calculations, with, however, different values.

I. INTRODUCTION

The deformation potential is an important electron-lattice interaction parameter related to many physical phenomena in semiconductors, including electrical conductivity,[1] Raman[2] and Brillouin scattering,[3] phonon renormalization due to free carriers,[4] splittings and shifts of excitonic energies and band edges under stress,[5] temperature dependence of band edges,[6] stress dependence of effective mass,[7] etc. Recent theoretical calculations of deformation potentials in semiconductors have been carried out along two different lines: (1) Empirical calculations, of either the linear combination of atomic orbitals (LCAO) method and/or the empirical-pseudopotential method (EPM). (2) Self-consistent first-principles calculations, i.e., the relativistic linear combination of muffin-tin orbitals (LMTO) method and the pseudopotential method.

The LCAO method[8,9] is in principle less reliable than other sophisticated methods and the results are strongly dependent on the basis atomic orbitals selected. The EPM calculation[8,9] suffers from the rigid-ionic-pseudopotential approximation, since the redistribution of valence-band electrons under strain has significant influence to the values of deformation potentials.[0] Recently, the self-consistent LMTO method was used to calculate the optical (ODP) and acoustic (ADP) deformation potentials systematically for a variety of semiconductors.[11–13] The earlier ODP values obtained by LMTO calculations on the basis of spherically symmetric potentials are, in general, considerably smaller than those determined experimentally.[11,12] Results of Brey, Christensen, and Cardona[13] have shown that the strain-induced changes in nonspherical parts of the crystal potentials are significant. When the nonspherical potential correction is made, ODP's thus obtained increase. However, the values are still smaller than the experimental data systematically.

On the other hand, the deduction of deformation potentials from experiments are quite involved. It is difficult to distinguish strictly between "experimental" and "theoretical" values, since the experimental results are deduced by adjusting values of deformation potentials to fit the experimental data using the theoretical formula under certain approximation and with other parameters involved (see references cited in Refs. 8 and 13). In other words, the experimental values are also in less reliability and accuracy. From the current status of divergency of data extracted from different experimental and theoretical methods, the most sophisticated *ab initio* pseudopotential method seems desirable to be used in the calculation of deformation potentials. Nielsen and Martin[14] have made *ab initio* pseudopotential calculations on stress properties of Si, Ge, and GaAs. However, no systematic data of deformation potentials except Si were reported.

In this paper, we provide careful and systematic *ab initio* pseudopotential calculations of ODP's and ADP's for nine typical semiconductor crystals. The host band structures with and without spin-orbit-splitting effects are both calculated and the results are compared. Numerical convergency and linearity are carefully tested to increase the reliability of the results. The remainder of the paper is organized as follows. In Sec. II we outline the definitions of ODP's and ADP's as they have appeared in this paper, since the different definitions in the literature often result in confusion. In Sec. III, we describe the computational details of *ab initio* pseudopotential calculations. Finally, the results are presented and discussed in Sec. IV.

II. ODP AND ADP AT THE TOP OF THE VALENCE BANDS

For illustration purposes, we first consider the host Hamiltonian of zinc-blende-structure semiconductors without the spin-orbit-splitting term. At the top of the valence bands at the Γ point are threefold-degenerate states of Γ_{15} representation of T_d symmetry. Following Kane, the strain Hamiltonian coupled to Γ_{15} electronic states can be written as[15]

$$\hat{H}^{\text{str}} = D_5(\varepsilon_{xy}\tilde{\gamma}_{5xy} + \varepsilon_{xz}\tilde{\gamma}_{5xz} + \varepsilon_{yz}\tilde{\gamma}_{5yz}) . \tag{1}$$

For simplicity, we have dropped the terms in $\hat{H}^{\text{str}}$ uncoupled to Γ_{15} electronic states. ε_{ij} is the strain tensor, $\tilde{\gamma}$ is Clebsch-Gordan Matrix

$$\tilde{\gamma}_{5xy} = \sqrt{\tfrac{3}{2}}\begin{bmatrix} 0 & 1 & 0 \\ 1 & 0 & 0 \\ 0 & 0 & 0 \end{bmatrix}, \quad \tilde{\gamma}_{5xz} = \sqrt{\tfrac{3}{2}}\begin{bmatrix} 0 & 0 & 1 \\ 0 & 0 & 0 \\ 1 & 0 & 0 \end{bmatrix},$$

$$\tilde{\gamma}_{5yz} = \sqrt{\tfrac{3}{2}}\begin{bmatrix} 0 & 0 & 0 \\ 0 & 0 & 1 \\ 0 & 1 & 0 \end{bmatrix},$$

and D_5 is the deformation-potential constant defined by Kane. For Γ_{15} acoustic phonon, the strain tensor corresponds to a pure shear and is represented by

$$\tilde{\varepsilon} = \varepsilon\begin{bmatrix} 0 & 1 & 1 \\ 1 & 0 & 1 \\ 1 & 1 & 0 \end{bmatrix} . \tag{2}$$

Substituting Eq. (2) into Eq. (1) the strain Hamiltonian matrix is written as

$$|\langle\psi_i^{\Gamma_{15}}|H^{\text{str}}|\psi_j^{\Gamma_{15}}\rangle| = D_5\varepsilon\sqrt{\tfrac{3}{2}}\begin{bmatrix} 0 & 1 & 1 \\ 1 & 0 & 1 \\ 1 & 1 & 0 \end{bmatrix} . \tag{3}$$

When diagonalizing the matrix of Eq. (3), we obtain three eigenvalues:

$$\Delta E_1^\varepsilon = 2\sqrt{\tfrac{3}{2}}D_5\varepsilon , \tag{4}$$

$$\Delta E_2^\varepsilon = \Delta E_3^\varepsilon = -\tfrac{1}{2}\Delta E_1^\varepsilon . \tag{5}$$

In this paper, we use notations similar to Refs. 9 and 13.

Introducing the ADP constant

$$d'(\Gamma_{15}) = \frac{D_5}{\sqrt{2}} , \tag{6}$$

Eq. (4) can be rewritten as

$$\Delta E_1^\varepsilon = 2\sqrt{3}\varepsilon d'(\Gamma_{15}) . \tag{7}$$

Equations (5) and (7) show that strain of Γ_{15} acoustic phonon gives rise to the energy splitting of degenerated Γ_{15} electronic states with singlet component shifting upwards and doublet components shifting downwards if both ε and d' are positive.

Similarly, the interaction between long-wavelength optical phonons and Γ_{15} electron states can be described by[10]

$$\hat{H}^{\text{ODP}} = \frac{D_5^0}{a}(u_x^{\text{rel}}\tilde{\gamma}_{5yz} + u_y^{\text{rel}}\tilde{\gamma}_{5xz} + u_z^{\text{rel}}\tilde{\gamma}_{5xy}) , \tag{8}$$

where a is the lattice constant of cubic crystal. $\mathbf{u}_{\text{rel}} = (u_x^{\text{rel}}, u_y^{\text{rel}}, u_z^{\text{rel}})$ is the relative optical displacement vector between two sublattices A and B. For zinc-blende crystals, we define $\mathbf{u}^{\text{rel}} = \delta_0/\sqrt{3}(1,1,1)$. Comparing Eq. (8) with Eqs. (1) and (2), $\hat{H}^{\text{ODP}}$ is equivalent to $\hat{H}^{\text{str}}$ with $D_5\varepsilon$ replaced by $D_5^0\delta_0 a/\sqrt{3}$. Similarly introducing ODP's constant

$$d_0(\Gamma_{15}) = \sqrt{2}D_5^0 , \tag{9}$$

the optical displacement δ_0 induces energy splitting for degenerate Γ_{15} electronic states, as in the following:

$$\Delta E_1^0 = \frac{\delta_0}{a}d_0(\Gamma_{15}) , \tag{10}$$

$$\Delta E_2^0 = \Delta E_3^0 = -\tfrac{1}{2}\Delta E_1^0 . \tag{11}$$

In fact, the acoustic deformation is further complicated by internal strain.[16,17] The stress tensor $\tilde{\varepsilon}$ corresponds to a macroscopic or external strain which only describes the relative displacement of atoms belonging to one sublattice. Accompany the external strain there exists a microscopic or internal strain corresponding the relative displacement between different sublattices. For Γ_{15} acoustic phonons of zinc-blend semiconductors, the internal strain accompanied to the external strain $\tilde{\varepsilon}$ of Eq. (2) is an optical displacement $\mathbf{u}^{\text{rel}}$ between two sublattices A and B. Both external and internal strains transform the crystal symmetry from T_d to C_{3v} and elongate the [111] bond length by $\delta_\varepsilon = (a_0\varepsilon\sqrt{3}/2)$ and $\delta_0 = -\zeta\delta_\varepsilon$, respectively. Here ζ is the internal-strain parameter defined by Kleinman.[17] The total enlongation of the [111] bond is the sum of external and internal contributions, $\delta = \delta_\varepsilon + \delta_0 = (1-\zeta)\delta\varepsilon$. Total energy shift of singlet component of Γ_{15} electronic state is accordingly, $\Delta E_1 = \Delta E_1^\varepsilon + \Delta E_1^0$, which can also be written as

$$\Delta E_1 = 2\sqrt{3}\varepsilon d(\Gamma_{15}) , \tag{12}$$

$$d = d' - \frac{\zeta}{4}d_0 . \tag{13}$$

When the spin-orbit interaction is included, the Γ_{15}

electronic states at the top of valence bands split into a fourfold degenerate Γ_8 band and a twofold Γ_7 band. The deformation potentials $d'(\Gamma_8)$, $d_0(\Gamma_8)$ and $d(\Gamma_8)$ are similarly defined by Eqs. (7), (10), and (13) except for the notation Γ_{15}, which is replaced by Γ_8 and the energy shift ΔE_1 of the singlet component of Γ_{15} is replaced by the energy splitting ΔE of Γ_8:

$$\Delta E^\varepsilon = 2\sqrt{3}\,\varepsilon d'(\Gamma_8) \, , \tag{7'}$$

$$\Delta E^0 = \frac{\delta_0}{a} d_0(\Gamma_8) \, . \tag{10'}$$

Further, the above analysis can be extended to diamond structure semiconductors if we replace the notation Γ_{15} by Γ'_{25} and Γ_8 by Γ_8^+, respectively. Equations (7), (7'), (10), (10'), and (13) are the key relations upon which this work is based to determine the signs and magnitudes of deformation potentials. For comparison purposes, $d_0(\Gamma_8)$, $d'(\Gamma_8)$, and $d(\Gamma_8)$ defined in this paper are the same as d_0, d', and d defined in Ref. 13. $d_0(\Gamma_{15})$, $d'(\Gamma_{15})$ defined in this paper are the same as d_0, d' defined in Ref. 9. Note that $d_0(\Gamma_{15})$ given here differ by a factor of $\frac{2}{3}$ from $d_0''^v$ in Ref. 10.

III. AB INITIO PSEUDOPOTENTIAL CALCULATIONS AND SPIN-ORBIT SPLITTING

We have calculated the ODP's and ADP's of Γ electronic states at the top of the valence bands for semiconductor crystals Si, AlSb, GaAs, GaP, GaSb, InAs, InP, InSb, and AlAs. *Ab initio* pseudopotentials are generated by the method of Hamann, Schlüter, and Chiang[18] with the modification introduced by Kleinman[19] for relativistic correction. The Ceperly-Alder method[20] and Perdew-Zunger parametrization[21] are used for correlation energy calculation. The spin-orbit splitting at the top of valence bands is calculated by the method developed by Hybertsen and Louie for diamond structure[22] and extended by Li, Surh, and Louie for zinc-blende-structure semiconductors.[23]

The scalar relativistic bands are obtained by a plane-wave expansion in the calculation by the Hohenberg-Kohn-Sham local-density approximation (LDA). The plane-wave basis extends up to a kinetic energy of 16 Ry. The spin-orbit interaction is added as a perturbation on the self-consistent scalar bands, and the fully relativistic eigenvalues and wave functions are obtained at $\mathbf{k}=0$ by diagonalizing the matrix Hamiltonian in the scalar relativistic basis. The perturbation Hamiltonian in a plane-wave representation is expressed by

$$\langle \mathbf{k},s|H^{\text{s.o.}}|\mathbf{k}',s'\rangle = -i\langle s|\hat{S}|s'\rangle \frac{\mathbf{K}\times\mathbf{K}'}{KK'} \sum_n S_n(\mathbf{K}-\mathbf{K}')\left[12\pi V_{l=1}^{\text{s.o.}}{}^n(\mathbf{K},\mathbf{K}') + 60\pi \frac{\mathbf{K}\cdot\mathbf{K}'}{KK'} V_{l=2}^{\text{s.o.}}{}^n(\mathbf{K},\mathbf{K}')\right] \, . \tag{14}$$

Here $\mathbf{K},\mathbf{K}'$ and s,s' represent the plane waves and spinors, with $V_l^{\text{s.o.}}$ being the nonlocal perturbation in a plane-wave basis, and S_n the complex structure factor for each atomic species n $(=A,B)$ in a cell. The exact form of $V_l^{\text{s.o.}}{}^n(\mathbf{K},\mathbf{K}')$ is

$$V_l^{\text{s.o.}}(\mathbf{K},\mathbf{K}') = \frac{1}{\Omega_c} \int_0^{+\infty} dr\, r^2 j_l(Kr)\hat{V}_l^{\text{s.o.}}(r) j_l(K'r) \, , \tag{14'}$$

where $\hat{V}_l^{\text{s.o.}}$ is the spin-orbit part of the ionic pseudopotential, j_l is the spherical Bessel functions, and Ω_c is unit-cell volume. The perturbation is applied to a 15×16 matrix including four valence bands and four lowest conduction bands. The calculated values of spin-orbit splittings $\Delta_{\text{s.o.}} = E_{\Gamma_8} - E_{\Gamma_7}$ of the valence and lowest conduction bands for different semiconductors are listed in Table I. Experimental results for spin-orbit splittings are also listed in Table I for comparison. The calculated spin-orbit splittings $\Delta_{\text{s.o.}}$ for valence bands are in excellent agreement with experiments.

The deformation potentials were calculated for both cases of with and without spin-orbit interactions. Careful numerical convergence tests in computations were carried out for improved reliability of the data. In general, ADP's values are 1 order of magnitude smaller than ODP's and the following tests are more important for ADP's.

(1) Self-consistent convergence test of crystal potentials: In self-consistent calculations the convergent parameter δE sets the criterion for the difference dV between input and output crystal potential energies. The iteration procedure stops when dV is less than the small value of δE. For usual band-structure calculations $\delta E \sim 10^{-4}$ Ry is small enough to obtain accurate band energies, and 3–4 iterations are required to achieve consistency. This criterion for δE is also good enough for ODP calculations. However, $\delta E \sim 10^{-5}$ or even 10^{-6} Ry is necessary with 6–8 iteration steps needed to obtain reliable ADP values. Here an example is given for the importance of δE values selected in ADP calculations for the case of Si. When $\delta E = 1\times 10^{-4}$ Ry is chosen, there is an apparent nonlinearity in the deformation range of $\varepsilon a_0 = 0 - 2\times 10^{-3}$ a.u., with linear terms of $d'(\Gamma'_{25}) = -2.26$ eV. However, when $\delta E \leq 1\times 10^{-5}$ Ry are selected, very good linearity is achieved in the same range of εa_0 with quite different linear terms of $d'(\Gamma'_{25}) = -1.65$ eV (-1.66 eV for $\delta E = 1\times 10^{-6}$ Ry). This reveals an interesting fact that although ADP's are 1 order of magnitude smaller than ODP's, it does not mean that ADP's have a weaker dependence on the potential perturbation induced by acoustic strain. Actually, the ADP values are sensitive to the potential perturbation and a more accurate description of crystal potential is needed in order to obtain reliable results.

(2) Selection of deformation range: For ODP's calculation, excellent linearity were found for the deformation range of $(\delta/\sqrt{3}a_0) = (0-2)\times 10^{-4}$. The nonlinear terms

8336   GU, LI, WANG, AND WANG   41

TABLE I. Spin-orbit splittings $\Delta_{s.o.}$ (eV). The experimental data are taken from Ref. 24; an asterisk represents no experimental data available. The previous theoretical calculations reported $\Delta_{s.o.}(\Gamma_{15c})$ to be 0.03–0.04 eV. See Ref. 26.

Compound	$\Gamma_{8v}-\Gamma_{7v}$ ($\Gamma_{8v}^{+}-\Gamma_{7v}^{+}$ for Si) Theory (this work)	Experiment	$\Gamma_{8c}-\Gamma_{7c}$ Theory (this work)	Experiment
Si	0.05	0.0441 (1.8 K)[a] 0.045 (10 K)[b] 0.0435 (50 K)[c] 0.044 (200 K)[d]	0.03	*
AlSb	0.68	0.75 (86 K)[e] 0.645 (80 K)[f]	0.06	0.3[g]
GaAs	0.35	0.33 (80 K)[h] 0.341 (4.2 K)[b] 0.350 (2.5–300)[b] 0.340 (300 K)[b]	0.19	0.171 (4.2 K)[b]
GaP	0.09	0.08 (2–300 K)[i] 0.082 (1.6–25 K)[j] 0.08 (80 K)[b]	0.17	
GaSb	0.67	0.752 (4.2 K)[k] 0.82 (4.2 K)[l] 0.756 (10 K)[b] 0.749 (30 K)[m]	0.22	0.34 (5 K)[a] 0.213 (10 K)[b] 0.29 (300 K)[b]
InP	0.11	0.108 (4.2 K)[n] 0.108 (5 K)[a] 0.11 (300 K)[b]	0.45	0.07 (300 K)[b]
InAs	0.37	0.38 (1.5 K)[o] 0.371 (295 K)[b] 0.41 (300 K)	0.47	
InSb	0.76	0.81 (1.5 K)[o] 0.803 (4 K)[p] 0.9 (77 K)[q]	0.43	0.39 (5 K)[a] 0.4 (77 K)[g] 0.37 (80 K)[r] 0.33 (300 K)[b]
AlAs	0.30	0.30 (300 K)[b]	0.035	0.2 (300 K)[b]

[a]Wavelength modulated absorption.
[b]Electroreflectance.
[c]Electroabsorption.
[d]Optical transmission.
[e]Intervalence band absorption.
[f]Piezotransmission.
[g]Thermoreflectance.
[h]Subvalence-band transition absorption.
[i]Energy derivative reflectance.
[j]Transmission.
[k]Auger recombination.
[l]Angle-resolved photoelectron spectroscopy.
[m]Stress modulated magnetoreflectance.
[n]Photovoltaic effect.
[o]Magneto-electro reflectance.
[p]Magnetoabsorption.
[q]Cyclotron resonance.
[r]Differential magnetoreflectance.

are negligibly small and accurate converged results can be extracted. For the ADP calculation, the situation is serious since ADP values are an order of magnitude smaller. The range of acoustic deformation is decided by balancing between the random numerical fluctuation and nonlinearity. Incorrect choices for acoustic deformation range and potential convergence parameter δE might give rise to false values or even opposite signs for d' due to numerical error. Fortunately, when $\delta E = 1 \times 10^{-6}$ Ry is selected, ADP values can be extracted from 3 to 4 data points within the acoustic deformation range of $\varepsilon a_0 = 0 - 2 \times 10^{-3}$ a.u. with numerical error estimated to be less than 5% for Si, AlSb, GaAs, GaP, GaSb, InAs, and InSb, and 10% for InP.

(3) Convergence test for energy cuts E_{cut}: We have calculated ODP's and ADP's for several different E_{cut} up to 18 Ry. The error due to finite energy cut of 16 Ry is estimated to be less than 5%.

(4) Convergence test for number of special k points[25] used in Brilloin-zone integration: Under optical or acoustic deformation, point-group symmetry of the crystals is reduced from T_d to C_{3v} for zinc-blende structure and from O_h to D_{3d} for diamond-type crystals. When we calculated band energies for $\delta = 0$ (or $\varepsilon = 0$) in C_{3v} or D_{3d} point-group symmetry, we observed a finite energy offset as compared to the calculations with T_d or O_h symmetry. As many as 60 special k points in an irreducible wedge of the Brilloin zone are needed for the offset to vanish, and convergent ODP's and ADP's result. However, when the energy offset is subtracted off for the curve fitting, 10 special k points are sufficient to obtain convergent results for ODP's and ADP's within the accuracy of 2%.

IV. ADP AND ODP RESULTS AND DISCUSSION

In Tables II and III, we list the results of our *ab initio* pseudopotential calculations of ODP's and ADP's for Si, AlSb, GaAs, GaP, GaSb, InAs, InP, InSb and AlAs. The internal-strain parameter ζ defined by

$$\zeta = 4[d'(\Gamma_8) - d(\Gamma_8)]/d_0(\Gamma_8) \tag{13'}$$

is also calculated by using values of $d'(\Gamma_8)$ and $d_0(\Gamma_8)$ obtained in this work and the experimental values of $d(\Gamma_8)$ listed in Table I of Ref. 13. The data of previous self-consistent LMTO calculations and the experimental values of $d_0(\Gamma_8)$, $d(\Gamma_8)$, and ζ collected in Ref. 13 are also listed for comparison. We summarize our results in the following respects.

(1) The independent calculations of deformation potentials for two cases of both with and without spin-orbit interaction are generally consistent. The only exception is ODP of GaSb. The reason of this deviation for GaSb is presently not clear.

(2) ODP's obtained in our *ab initio* pseudopotential calculation are in overall greater than those calculated by self-consistent LMTO, and are closer to and yet still smaller than the experimental data. However, there are two exceptions: d_0 of Si in our calculation is larger than the experimental determined result. The other case is the $d_0(\Gamma_8)$ of GaSb which in our calculation is abnormally smaller than the LMTO result and is far less than the experimental value. Note, however, that $d_0(\Gamma_{15})$ of GaSb in our calculation is close to experiment.

(3) From Table III, the data of ADP's from different calculations are much more scattered than ODP's. In comparing our *ab initio* pseudopotential calculations with empirical pseudopotential calculations by Blacha *et al.* which we list in Table III, we notice that even the signs of d' of the two calculations are very different. We ascribe the difference to the effect of valence electron redistribution induced by acoustic deformation. As indicated in Sec. III, ADP's are very sensitive to the potential perturbation caused by acoustic strain, and more accurate self-consistent crystal potentials are needed than in case of the usual self-consistent band-structure calculations. The rigid-ionic-pseudopotential model in empirical-pseudopotential calculations completely neglect the valence electron redistribution effect around atomic sites,

TABLE II. Optical-deformation potentials (eV).

Compound		Si	AlSb	GaAs	GaP	GaSb	InAs	InP	InSb	AlAs
$d_0(\Gamma_{15})$ theory (this work)		53.3	31.7	38.0[a]	36.2[a]	39.5	27.1	26.2	29.4	31.6
$d_0(\Gamma_8)$	Theory (this work)	46.6	32.3	37.1[a]	33.1[a]	19.9	25.3	22.6	28.6	31.2
	Theory[b] (LMTO calc.)	27.1	21.3	25.0	24.3	23.4	20.8	20.1	19.7	22.0
	Expt.[c]	40		48[d]	44	32	42	35	39	
		27[e,f]	37[f]	41[f]	47[f]					

[a]The values reported in this work are slightly different from those reported in Ref. 10. This work ensures better convergence and reliability.
[b]Reference 13.
[c]All other experimental data are taken from Ref. 24.
[d]Reference 27.
[e]Reference 28.
[f]Reference 29.

TABLE III. Acoustic deformation potentials (eV) and internal-strain parameters.

Compound		Si	AlSb	GaAs	GaP	GaSb	InAs	InP	InSb	AlAs
$d'(\Gamma_{15})$	this work	-1.11	-1.75	-1.39	-1.51	-1.08	-1.11	-1.29	-1.38	-2.08
	EPM calc.[a]	-2.3	0.8	2.7	-1.1	0.1	1.2	0.6	0.5	
$d'(\Gamma_8)$	This work	-1.15	-1.84	-1.43	-1.46	-0.97	-1.20	-1.26	-1.53	-2.21
	LMTO calc.[b]	-2.28	-2.25	-0.99	-0.82	-1.17	-0.57	-0.65	-0.55	-2.86
$d(\Gamma_8)$expt.[c]		-5.3	-4.3	-4.5	-4.5	-4.6	-3.6	-5.0	-5.0	
ζ	this work	0.38	0.31	0.33	0.37	0.73 (0.37)[d]	0.38	0.66	0.49	
	LMTO calc.[b]	0.45	0.38	0.56	0.61[e]	0.59[e]	0.58	0.87	0.90	
	N.M. calc.[f]	0.53		0.48						
	expt.	0.73[g],0.54[h]		0.76[i]						

[a]Reference 9.
[b]Reference 13.
[c]Reference 24.
[d]Derived from $d_0(\Gamma_{15})$.
[e]ζ in Table II of Ref. 13 corrected by d_0, d' and d reported there.
[f]Reference 14.
[g]Reference 30.
[h]Reference 31.
[i]Reference 32.

and is obviously too crude for accurately determining ADP's, and in some cases even generate the wrong sign. The self-consistent LMTO method after nonspherical potential correction, however, gives the same sign for all d' as our *ab initio* pseudopotential calculations. Our calculations show that all the d' for the nine semiconductors studied are negative as listed in Table III. The magnitudes, however, between two calculations are quite different. A possible cause of this difference is due to the different treatments of potential perturbations under stress. Since the value of d' is sensitive to the potential perturbation in detail, and the nonspherical corrections are of a large fraction of the zeroth-order values as indicated in Ref. 13, a first-order perturbation treatment in Ref. 13 might not be enough to obtain accurate results for d'.

The internal-strain parameters ζ calculated by Eq. (13') are quite different by using the same experimental values of $d(\Gamma_8)$ listed in Table III but different $d_0(\Gamma_8)$ and $d'(\Gamma_8)$ obtained by our calculations and LMTO calculations, respectively, as shown in Table III. They do not agree with experimental results very well. However, our calculated values are of the same order of magnitude as LMTO results. Because of the possible underestimated $d_0(\Gamma_8)$ for GaSb our calculated ζ for this material is abnormally large. The value in parenthesis in the table is derived from the calculated $d_0(\Gamma_{15})$ and the result is much more reasonable. It is very interesting to observe that our calculations have roughly produced the same chemical trends as LMTO calculations for ζ of zinc-blende semiconductors studied. Finally, from the recent investigation of Zhu, Fahy, and Louie,[33] it was shown that the pressure dependence of the LDA calculation presented in this work is expected to yield reliable deformation potential not only for valence bands but also for conduction bands.

In conclusion, *ab initio* pseudopotential calculations of deformation potentials for nine typical semiconductor materials of Si, AlSb, GaAs, GaP, GaSb, InAs, InP, InSb, and AlAs are carried out. After careful convergence tests for various important convergency parameters, reliable results have been obtained. The optical-deformation-potential values d_0 thus obtained are generally larger than the previous LMTO calculations and are in better agreement with experiments. The calculated values of acoustic deformation potential d' have negative signs for all eight semiconductors, in agreement with LMTO calculations but with differing magnitudes. We found that values of d' are very sensitive to the potential perturbation induced by strain. Our *ab initio* pseudopotential calculations with enhanced potential self-consistency criterion stands on a more rigorous theoretical base than previous calculations and is, therefore, valuable to assess the existing and forthcoming experimental and theoretical data.

ACKNOWLEDGMENTS

We would like to acknowledge Professor Steven G. Louie for providing the *ab initio* pseudopotential calculation programs and parameters of atomic pseudopotentials used in this work. M. F. Li has profited from many valuable discussions with Michael P. Surh. The authors are grateful to Professor Kun Huang for his encouragement in this work. We also acknowledge Dr. M. Cardona and Dr. N. E. Christensen for a comment on the definition of $d_0(\Gamma_{15})$ in Eq. (10). This work is supported by the National Science Foundation of China. Digital Equipment Corporation VAX computer times are provided by Peking Astronomy Observatory and by the Institute of Semiconductors, Academia Sinica.

[1] J. D. Wiley, in *Semiconductors and Semimetals*, edited by R. K. Willardson and A. C. Beer (Academic, New York, 1975), Vol. 10, p. 91.

[2] M. Cardona, in *Light Scattering in Solids II*, edited by M. Cardona and G. Güntherodt (Springer-Verlag, Heidelberg, 1982).

[3] C. Hamaguchi, Phys. Rev. B 11, 3876 (1975).

[4] L. Pintschovius, J. A. Verges, and M. Cardona, Phys. Rev. B 26, 5658 (1982).

[5] M. Chandrasekhar and F. H. Pollak, Phys. Rev. B 15, 2127 (1977).

[6] P. B. Allen and M. Cardona, Phys. Rev. B 27, 4760 (1983).

[7] D. C. Aspnes and M. Cardona, Phys. Rev. B 17, 726 (1978); 17, 741 (1978).

[8] W. Pötz and P. Vogl, Phys. Rev. B 24, 2025 (1981).

[9] A. Blacha, H. Presting, and M. Cardona, Phys. Status Solidi B 126, 11 (1984).

[10] B. S. Wang, Z. Q. Gu, J. Q. Wang, and M. F. Li, Phys. Rev. B 39, 12 789 (1989).

[11] N. E. Christensen, Phys. Rev. B 30, 5753 (1984).

[12] N. E. Christensen, Solid State Commun. 50, 177 (1984).

[13] L. Brey, N. E. Christensen, and M. Cardona, Phys. Rev. B 36, 2638 (1987).

[14] O. H. Nielsen and R. M. Martin, Phys. Rev. B 32, 3792 (1985).

[15] E. O. Kane, Phys. Rev. 178, 1368 (1969).

[16] M. Born and K. Huang, *Dynamical Theory of Crystal Lattices* (Clarendon, Oxford, 1985).

[17] L. K. Kleinman, Phys. Rev. 128, 2614 (1962).

[18] D. R. Hamann, M. Schlüter, and C. Chiang, Phys. Rev. Lett. 43, 1494 (1979).

[19] L. K. Kleinman, Phys. Rev. B 21, 2630 (1982); G. B. Bachelet

[20] and M. Shlüter, *ibid.* 25, 2103 (1982).

[20] D. M. Ceperly and B. J. Alder, Phys. Rev. Lett. 45, 566 (1980).

[21] J. P. Perdew and A. Zunger, Phys. Rev. B 23, 5075 (1981).

[22] M. S. Hybertsen and S. G. Louie, Phys. Rev. B 34, 2920 (1986).

[23] M. F. Li, M. P. Surh, and S. G. Louie, in *Proceedings of the 19th International Conference on the Physics of Semiconductors*, edited by W. Zawadzki (Institute of Physics, Polish Academy of Sciences, Warsaw, 1988), p. 857.

[24] *Landolt-Börnstein, Numerical Data and Functional Relationships in Science and Technology* (Springer-Verlag, Berlin, 1982), Vol. 17a.

[25] D. J. Chadi and M. L. Cohen, Phys. Rev. B 8, 5747 (1973).

[26] G. G. Wepfer, T. C. Collins, and R. N. Euwema, Phys. Rev. B 4, 1296 (1971).

[27] M. H. G. Grimsditch, D. Olego, and M. Cardona, Phys. Rev. B 20, 1758 (1979).

[28] J. Jacoboni, G. Gagliani, L. Reggiani, and V. Turci, Solid State Electron. 21, 315 (1978).

[29] P. Lawaetz, Ph.D thesis, The Technical University of Denmark, 1978.

[30] H. d'Amour, W. Denner, H. Schulz, and M. Cardona, J. Appl. Crystallogr. 15, 148 (1982).

[31] C. S. G. Cousins, L. Gerward, J. Staun Olsen, B. Selsmark, and B. J. Sheldon, J. Phys. C 20, 29 (1987).

[32] C. N. Koumelis, G. E. Zardas, C. A. Loudos, and D. K. Leventuri, Acta Crystallogr. Sec. A 32, 84 (1975).

[33] X. Zhu, S. Fahy, and S. G. Louie, Phys. Rev. B 39, 7840 (1989).

PHYSICAL REVIEW B   VOLUME 43, NUMBER 17   15 JUNE 1991-I

Shear-deformation-potential constant of the conduction-band minima of Si: Experimental determination by the deep-level capacitance transient method

Ming-Fu Li

Graduate School, University of Science and Technology of China, Beijing China;
Institute of Semiconductors, Academia Sinica, Beijing, China;
and Fudan University, Shanghai, China

Xue-Shu Zhao and Zong-Quan Gu

Institute of Semiconductors, Academia Sinica, P.O. Box 912, Beijing, China

Jian-Xin Chen

Polytechnic University of Beijing, Beijing, China

Yan-Jin Li and Jian-Qing Wang

Graduate School, University of Science and Technology of China, Beijing, China
(Received 15 October 1990)

The shear-deformation-potential constant Ξ_u of the conduction-band minima of Si has been measured by a method which we called deep-level capacitance transient under uniaxial stress. The uniaxial-stress (F) dependence of the electron emission rate e_n from deep levels to the split conduction-band minima of Si has been analyzed. Theoretical curves are in good agreement with experimental data for the S^0 and S^+ deep levels in Si. The values of Ξ_u obtained by the method are 11.1 ± 0.3 eV at 148.9 K and 11.3 ± 0.3 eV at 223.6 K. The analysis and the Ξ_u values obtained are also valuable for symmetry determination of deep electron traps in Si.

I. INTRODUCTION

The shear-deformation-potential constant Ξ_u of the conduction band was introduced as early as in 1958 by Herring and Vogt,[1] and it has been determined experimentally by a variety of different techniques[2–14] because of its importance in semiconductor physics. Unfortunately, the values of Ξ_u determined by different techniques are quite different, ranging from 7 to 11 eV. On the other hand, most of these methods are quite involved or indirect with possible uncertainties as will be discussed in Sec. V.

The purpose of this work is to provide a method to determine the parameter Ξ_u. A short report[15] has been presented and we will describe it in detail in this paper. The method is based on the constant-temperature deep-level capacitance transient technique.[16] A deep level is used only as a reference level in the gap and the value of the deformation potential constant Ξ_u is obtained straightforwardly from the change of emission rate e_n of the electrons from the deep traps to the conduction-band minima due to applied uniaxial stress. Since the method is simple and direct with no ambiguous parameters, the results should be reliable.

The Ξ_u value obtained by this method is particularly valuable for uniaxial-stress deep-level transient measurements for symmetry determination of deep centers in semiconductors, where the splitting of the minima of the conduction bands plays a main obstacle for precise determination of deep-level shift and splitting under uniaxial

stress.[17]

We will divide the paper as follows: In Sec. II, the principle of the deep-level capacitance transient method for determining Ξ_u will be discussed. In Sec. III, experimental details and measurements will be outlined. In Sec. IV, experimental results will be given. In Sec. V we will compare our method to various existing methods. In Sec. VI we will present some concluding remarks.

II. PRINCIPLE OF THE METHOD— RECOMBINATION KINETICS OF DEEP LEVELS UNDER UNIAXIAL STRESS

In this section we discuss the principle of deep-level transient measurements as a means to determine Ξ_u. We first generalize the Shockley-Read-Hall recombination kinetic theory of deep traps[18] to the case when the minima of the conduction bands $E_c^{(i)}$ ($i=1-6$ for Si) are lifted from degeneracy under uniaxial stress. This problem was first discussed by Yao *et al.*[19] Here we follow the approach by Li,[17] which is most appropriate for our purpose.

The original argument of Li includes the general case where the conduction-band minima E_c and deep level E_T are both degenerate. In this work, a deep level E_T is introduced by sulfur impurities. Ludwig[20] has identified the symmetry of the deep-level wave function of an isolated S^+ center to be A_1, by electron spin resonance techniques. Therefore, E_T is nondegenerate and the analysis can be simplified.

The rate of emission per unit volume of the electrons from the deep levels E_T to the jth conduction-band minimum is $e_n^{(j)} n_T$. Here $e_n^{(j)}$ is the emission rate and n_T is the electron concentration of E_T levels. The rate of capture of the electrons from the jth conduction-band minimum to the deep levels E_T is $C_n^{(j)} n^{(j)} P_T$. Here $C_n^{(j)}$ is the capture rate in $cm^3 s^{-1}$, $n^{(j)}$ is the electron concentration in the jth conduction-band minimum, $P_T = (N_{TT} - n_T)$ is the hole concentration of E_T levels, and N_{TT} is the S impurity concentration. At thermal equilibrium, we have

$$\sum_{j=1}^{l} C_n^{(j)} n^{(j)} P_T = \sum_{j=1}^{l} e_n^{(j)} n_T . \tag{1}$$

We introduce

$$\sum_{j=1}^{l} e_n^{(j)} = e_n , \tag{2a}$$

$$\overline{C_n} = \sum_{j=1}^{l} C_n^{(j)} n^{(j)} \Big/ \sum_{j=1}^{l} n^{(j)} , \tag{2b}$$

and

$$E_c^{(j)} = \overline{E_c} + \delta E_c^{(j)} , \tag{3a}$$

$$\overline{E_c} = \sum_{j=1}^{l} E_c^{(j)} / l , \tag{3b}$$

$$g_c^{-1} = \sum_{j=1}^{l} e^{-\delta E_e^{(j)}/k_B} . \tag{3c}$$

Here $l = 6$ for the case of Si. On the other hand, we have

$$n_T = N_{TT} f_T ,$$

$$f_T = \frac{1}{1 + e^{(E_T - E_F)/k_B}} , \tag{4}$$

$$n^{(j)} = N_c e^{-(E_c^{(j)} - E_F)/k_B} , \tag{5}$$

and

$$N_c = \frac{2(2\pi m_c^* k_B)^{3/2}}{h^3} . \tag{6}$$

Here E_F is the Fermi energy and m_c^* is the effective mass at conduction-band minima. From previous piezoresistance experiments, the stress dependence of the effective mass of the conduction band for Si is very weak.[21] From the data of Ref. 21, the estimated change of $(1/m_c^*)(dm_c^*/dF)$ is less than 10^{-3}/kbar. Therefore, it is justified to neglect the stress dependence of m_c^* in our experiment.

Substituting Eqs. (2)–(6) into Eq. (1) we obtain

$$e_n = g_c^{-1} \overline{C_n} N_c e^{-(\overline{E_c} - E_T)/k_B} . \tag{7}$$

The rate equation of n_T is modified to

$$\frac{dn_T}{dt} = -(e_n + \overline{C_n} n) n_T + \overline{C_n} n N_{TT} , \tag{8}$$

where $n = \sum_{j=1}^{l} n^{(j)}$ is the carrier concentration.

Equation (8) is exactly the same rate equation for the case without stress.[16] Therefore, e_n and $\overline{C_n}$ can be measured by well-known constant-temperature capacitance transient experiments with emission-time constant e_n^{-1} (Ref. 16) and capture-time constant $(e_n + \overline{C_n} n)^{-1} \approx (\overline{C_n} n)^{-1}$.[22]

Following Herring's deformation-potential analysis,[1] the shift in energy of the jth minimum for Si is given by

$$\Delta E_c^{(j)} = \sum_{\alpha,\beta} (\Xi_d \delta_{\alpha\beta} + \Xi_u K_\alpha^{(j)} K_\beta^{(j)}) u_{\alpha\beta} , \tag{9}$$

where $K_\alpha^{(j)}$ and $K_\beta^{(j)}$ are components of a unit vector pointing from the center of the Brillouin zone towards the position in k space of the jth conduction-band minimum. The subindex α or β designates a component along one of the cubic axes of the crystal, and $u_{\alpha\beta}$ are components of the strain tensor. The symbols Ξ_d and Ξ_u are the volume and shear deformation potential constants, respectively.

$\Delta E_c^{(j)}$ in Eq. (9) can be divided into two parts:

$$\Delta E_c^{(j)} = \Delta \overline{E_c} + \delta E_c^{(j)} , \tag{10}$$

where $\overline{E_c}$ and $\delta E_c^{(j)}$ are defined by Eqs. (3b) and (3a). By Eqs. (9) and (10) the shift of the average energy $\overline{E_c}$ is given by

$$\Delta \overline{E_c} = (\Xi_d + \tfrac{1}{3}\Xi_u) \mathrm{Tr}\, \vec{u} . \tag{11}$$

The shift of the jth minimum with respect to $\overline{E_c}$ is given by

$$\delta E_c^{(j)} = \Xi_u \left[\left(\sum_{\alpha,\beta} K_\alpha^{(j)} K_\beta^{(j)} u_{\alpha\beta} \right) - \tfrac{1}{3} \mathrm{Tr}\, \vec{u} \right] . \tag{12}$$

The strain components $u_{\alpha\beta}$ are related to stress components $\sigma_{\alpha\beta}$ by the elastic constants $S_{\alpha\beta}$ for cubic crystals as follows:[23]

$$\begin{bmatrix} u_{xx} \\ u_{yy} \\ u_{zz} \\ u_{yz} \\ u_{zx} \\ u_{xy} \end{bmatrix} = \begin{bmatrix} S_{11} & S_{12} & S_{12} & 0 & 0 & 0 \\ S_{12} & S_{11} & S_{12} & 0 & 0 & 0 \\ S_{12} & S_{12} & S_{11} & 0 & 0 & 0 \\ 0 & 0 & 0 & \tfrac{1}{2}S_{44} & 0 & 0 \\ 0 & 0 & 0 & 0 & \tfrac{1}{2}S_{44} & 0 \\ 0 & 0 & 0 & 0 & 0 & \tfrac{1}{2}S_{44} \end{bmatrix} \begin{bmatrix} \sigma_{xx} \\ \sigma_{yy} \\ \sigma_{zz} \\ \sigma_{yz} \\ \sigma_{zx} \\ \sigma_{xy} \end{bmatrix} . \tag{13}$$

For uniaxial compression along the [100] direction, the stress tensor $\sigma_{\alpha\beta}$ is expressed by

$$(\sigma_{\alpha\beta}) = -F \begin{bmatrix} 1 & 0 & 0 \\ 0 & 0 & 0 \\ 0 & 0 & 0 \end{bmatrix} , \tag{14}$$

where F is the compression force per unit cross-sectional area. Combining Eqs. (12)–(14), we obtain

$$\delta E_c^{(j)} = \begin{cases} -\tfrac{2}{3} F \Xi_u (S_{11} - S_{12}) , & j = 1, 2 \\ \tfrac{1}{3} F \Xi_u (S_{11} - S_{12}) , & j = 3, 4, 5, 6 , \end{cases} \tag{15}$$

where $j=1,2$ correspond to two minima with $\mathbf{K}^{(j)}$ parallel to the stress direction while the rest correspond to the other four perpendicular directions.

From Eqs. (3c), (7), and (15), we have

$$\ln e_n = \ln g_c^{-1} + \ln \overline{C_n} + \ln N_c - \frac{1}{k_B T}(\overline{E_c} - E_T) \ . \tag{16}$$

Defining

$$x = \tfrac{1}{3} F \Xi_u (S_{11} - S_{12})/k_B T \ , \tag{17}$$

we get

$$g_c^{-1} = 4e^{-x} + 2e^{2x} \ . \tag{18}$$

We notice that $3x$ is just the splitting of the conduction-band minima in unit of $k_B T$. In Eq. (16) the capture rate $\overline{C_n}$ of the electrons is independent of the stress within experimental uncertainty for the case of the S deep level, as will be discussed in Sec. IV.

From these considerations the shift of $\ln e_n$ under stress is obtained by

$$\Delta(\ln e_n) = \Delta(\ln g_c^{-1}) - \frac{1}{k_B T}\Delta(\overline{E_c} - E_T) \ . \tag{16'}$$

For the low-stress limit $x \ll 1$, Eq. (16) reduces to

$$\Delta e_n \doteq \frac{\partial \ln e_n}{\partial F}\bigg|_{F \to 0} F + \frac{\Xi_u^2 (S_{11} - S_{12})^2}{3(k_B T)^2} F^2 \ , \tag{19}$$

with[17]

$$\frac{\partial \ln e_n}{\partial F}\bigg|_{F \to 0} = \frac{1}{k_B T}\frac{\partial(E_T - \overline{E_c})}{\partial F} \ . \tag{19'}$$

The physical meaning of Eq. (19') is apparent. When the splitting between conduction-band minima is small in comparison to $k_B T$, the Boltzmann factor e^x in Eq. (4) is linearly dependent on x. Therefore, the deep level sees the conduction-band minima in average distance. The variation of $\ln e_n$ under stress is mainly determined by the variation of $\overline{E_c} - E_T$; the splitting of the conduction band can be treated as a small perturbation proportional to F^2.

In large-stress limit, i.e., $3x \gg 1$, Eq. (16') reduces to

$$\frac{\partial \ln e_n}{\partial F}\bigg|_{F \to \infty} = -\frac{1}{k_B T}\left[\frac{\partial(E_c^{(1)} - E_T)}{\partial F}\right] \ . \tag{20}$$

In other words, when the splitting of the conduction-band minima is large in comparison to $k_B T$, the deep level only sees the lower conduction-band minima $E_c^{(1)}$ and $E_c^{(2)}$, since almost no electrons occupy the higher minima. From Eqs. (15), (19'), and (20) one obtains

$$\Xi_u = \frac{3k_B T}{2(S_{11} - S_{12})}\left[\frac{\partial \ln e_n}{\partial F}\bigg|_{F \to \infty} - \frac{\partial \ln e_n}{\partial F}\bigg|_{F=0}\right] \ . \tag{21}$$

The first term in the large parentheses $F_1 = (\partial/\partial F)\ln e_n|_{F \to \infty}$ can be determined from experimental data of $\ln e_n$ versus F by taking the high limit. The second term $F_2 = (\partial/\partial F)\ln e_n|_{F=0}$ is obtained by the following iteration procedure: As F_1 is an order of magnitude larger than F_2, Eqs. (19) and (21) are a pair of weakly coupled equations with two parameters F_2 and Ξ_u. For the first iteration, we neglect F_2 in Eq. (21) to obtain the value for Ξ_u. Using the Ξ_u thus obtained and Eq. (19) to fit the experimental curve of $\ln e_n$ versus F, the value of F_2 is obtained. For the second iteration, we put the value of F_2 from the first iteration into Eq. (21), and a new Ξ_u is obtained. The process continues until converged F_2 and Ξ_u are achieved.

III. EXPERIMENTAL DETAILS AND MEASUREMENTS

A. Sample preparation

$N_d = 3 \times 10^{15}$ cm^{-3} phosphorous-doped Si single crystals prepared by the Czochralski method were x-ray-oriented in the [100] direction within 0.1° and cut in samples with 1×0.7-mm^2 cross section and 7 mm in length. P^+-n junctions with $\Phi = 0.6$ mm located at the middle of the 7×1-mm^2 side surface were made by boron diffusion. The samples were checked by deep-level transient spectroscopy (DLTS) measurements with no detectable deep impurities. The samples were subsequently placed in a quartz ampoule with 2 mg of 99.999% sulfur and evacuated to 1×10^{-5} Torr and sealed. The diffusion temperature was 1180 °C for 2 h. Typical DLTS spectra with S^0 and S^+ peaks were obtained for the samples with capacitance transient amplitude rate $\Delta C/C \sim 0.01$ at 10-V bias. By $\delta C/C \sim \frac{1}{2}(N_{TT}/N_d)$, the S concentration was estimated to be 6×10^{13} cm^{-3}.

B. Stress apparatus

Figure 1 shows the schematic diagram of the stress apparatus. It consists of a sample holder (A), a semi-ball-shaped top block (B) with the surface adjustable to fit the top surface of the sample (C), and a bottom rod (D). The bottom rod is constrained to move along the stress direction and is lifted by means of a lever (E) through the rod (F) and frame (G). Weights are hung on the lever arm. The lever is mounted on a platform (H) which also supports the sample holder (A) by a pipe (P). The force applied to the sample can be precisely calibrated by this stress apparatus. The cross-sectional area of the sample was measured with a depth gauge; this was checked by a scaled optical microscope. The accuracy of the measurement of the cross-sectional area was estimated to be about 1%.

C. Measurement

The emission rate e_n was determined by measuring the capacitance transient signal at constant temperature. The measurements were carried out by a computer-controlled data-acquisition system, as shown in Fig. 2. The junction capacitance was measured by a Boonton 72 B capacitance meter at 1-MHz frequency. For every cycle of capacitance transient, 15 capacitance data were read sequentially by a HP3456A digital voltmeter and then were fed into the computer. The temperature stabil-

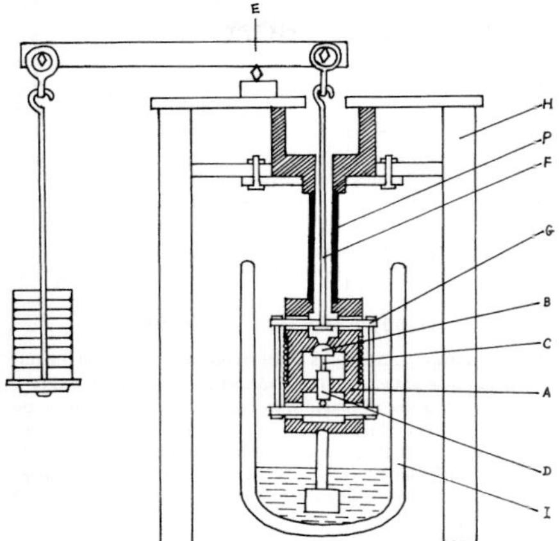

FIG. 1. Uniaxial stress apparatus: (A) Sample holder; (B) semi-ball-shaped top block; (C) sample; (D) bottom rod; (E) lever; (F) stainless steel rod; (G) frame; (H) support platform; (I) liquid-nitrogen Dewar; (P) pipe.

ity is extremely important, since the emission rate ϵ_n is very sensitive to the temperature. According to Eq. (7),

$$\frac{de_n}{e_n} \doteq \frac{\overline{E_c} - E_T}{k_B T}\frac{dT}{T} , \tag{22}$$

if we neglect the weak temperature dependences of g_c^{-1}, $\overline{C_n}$, and N_c. In our system, temperature fluctuation was less than ± 0.03 K. For S^+ and S^0 levels, as listed in Table I, the corresponding relative changes de_n/e_n were ~ 3.7 and 4.7×10^{-3}, respectively. $2-5 \times 10^2$ repeated measurements were taken and averaged to further reduce the fluctuations of e_n due to electrical and temperature fluctuations.

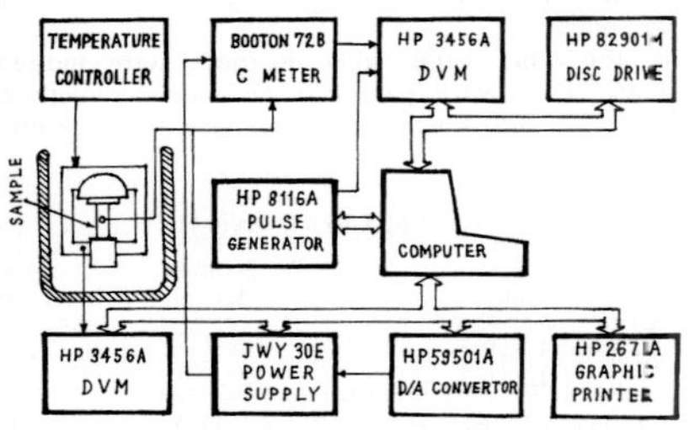

FIG. 2. Data-acquisition system for constant capacitance transient measurement under stress.

IV. EXPERIMENTAL RESULTS

DLTS spectra for typical S^+ and S^0 levels in Si were obtained.[24] From an Arrhenius plot, we obtain $\overline{E_c} - E_T = 0.53$ eV for the S^+ and 0.30 eV for the S^0 levels, which are consistent with previous measurements.[16,24] For constant-temperature capacitance transient measurements, a transient signal with a single exponential time constant e_n^{-1} was recorded within the whole stress range of our measurement. The stress dependence of $\ln e_n$ is shown in Fig. 3 for the S^+ level and Fig. 4 for the S^0 level at 223.6 and 148.9 K, respectively. Figure 5 is a typical measurement that describes the relation between the initial capacitance transient amplitude $\Delta C(\tau)$ and the stress F for the S^0 level. The capture rate $\overline{C_n}$ can be evaluated from variation of $\Delta C(\tau)$ with τ. Here τ is the filling pulse length.[22] From Fig. 5, $\overline{C_n}$ is confirmed to be almost stress independent within the experimental accuracy. The previous hydrostatic experiments[25,26] have reported similar results that the electron capture rates of deep levels including S^0 and S^+ in Si are insensitive to hydrostatic pressure.

In practice, the curve-fitting procedure of Figs. 3 and 4 was as follows: Because the highest stress we could apply to the sample before sample breaking was not large enough to satisfy the condition $3x \gg 1$ of Sec. II, we were not able to obtain the precise value of $(\partial/\partial F)\ln e_n|_{F \to \infty}$ in Eq (20) from the experiment. An alternative procedure was used. We used Eqs. (16)–(18) and Ξ_u as an adjustable parameter to fit the curve. As indicated in Figs. 3

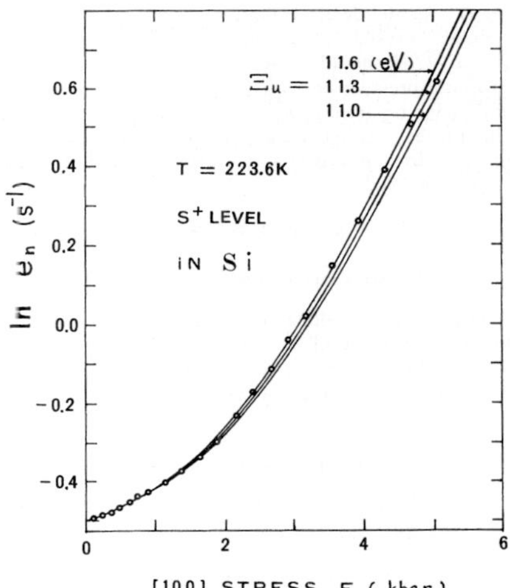

FIG. 3. Variation of emission rate of electrons $\overline{C_n}$ from S^+ levels to the conduction band of Si under [100] stress at 223.9 K.

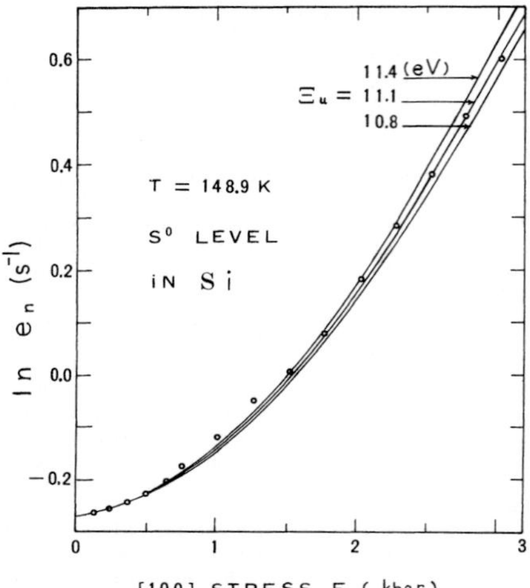

FIG. 4. Variation of emission rate of electrons $\overline{C_n}$ from S^0 levels to the conduction band of Si under [100] stres at 148.9 K.

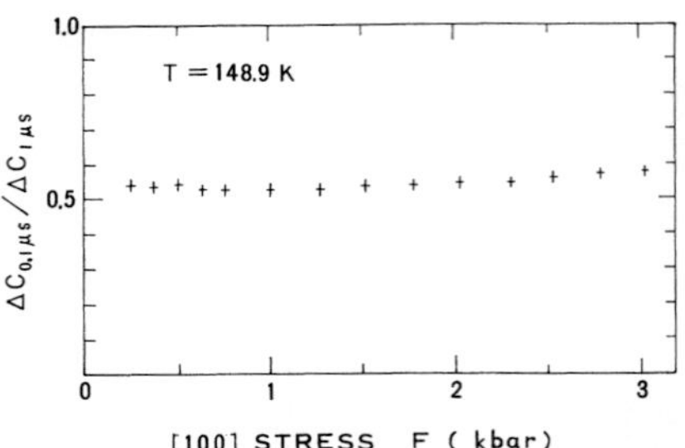

FIG. 5. $\Delta C_{0.1\,\mu s}/\Delta C_{1\,\mu s}$ vs [100] stress plot. $\Delta C_{1\,\mu s}$ is the initial capacitance transient amplitude with filling pulse length 1 μs. 1 μs is an order of magnitude larger than the capture-time constant τ_c. Therefore $\Delta C_{1\,\mu s}$ is saturated with this pulse length. $\Delta C_{0.1\,\mu s}$ is the initial capacitance transient amplitude with filling pulse length 0.1 μs. 0.1 μs is the same order of magnitude as τ_c. Any change of τ_c gives rise to change of $\Delta C_{0.1\,\mu s}/\Delta C_{1\,\mu s}$.

and 4, the curve fitting is sensitive to the value of Ξ_u.

The values of Ξ_u and $d(\overline{E_c}-E_T)/dF$ obtained by the iteration procedure are listed in Table I. By our previous analysis[26] for the defect Hamiltonian with T_d symmetry in cubic crystal, the uniaxial stress derivative $d(\overline{E_c}-E_T)/dF$ should be isotropic and equal to $\frac{1}{3}$ of the hydrostatic pressure derivative $d(\overline{E_c}-E_T)/dP$. Jantsch et al.[25] have reported $d(\overline{E_c}-E_T)/dP$ of S^0 and S^+ levels in Si to be -1.7 ± 0.1 and -2.05 ± 0.1 meV/kbar, respectively. In comparison with our results in Table I, the 1:3 rule is satisfied.

V. DISCUSSION

A. Assessment of existing experiments

A large variety of different methods has been used to determine the value of Ξ_u. Table II is a list of these methods. The values for Ξ_u determined by different methods lie in the range of 7–11 eV. Many of these methods were quite involved and less reliable. For instance, the method of number 12 in Table II measured the linewidth of cyclotron resonance to determine the carrier scattering relaxation time τ, and then used the theoretical formula of Herring and Vogt (HV) to obtain Ξ_u from τ. The equations of HV are rather complicated. The method numbered 14 was based on the effect of carrier concentration on elastic constants. Three equations were used to fit three parameters Ξ_u, Fermi energy E_F, and carrier concentration n. However, the values of E_F and n fitted from the experimental curves were not consistent with each other when Fermi statistics was used for calculating n. The piezoresistance method of number 13 was based on a hypothesis which has never been confirmed by experiments; there scattering mechanisms and mobilities were taken to be independent of stress. The piezo-optic method of number 2 suffers from uncertainties in determination of the carrier density by Hall-effect measurements due to lack of an exact knowledge of μ_H/μ. However, values of Ξ_u determined by the piezo-optic method are very close to those obtained in the present work.

On the other hand, some methods were indirect methods. The electric paramagnetic resonance methods numbered 3 and 4 were indirect, since the measured

TABLE I. Shear-deformation-potential constant Ξ_u and [100] stress derivative of S deep levels in Si.

Deep level	$\overline{E_c}-E_T$ (eV)	$d(\overline{E_c}-E_T)/dF$ (meV/kbar)	$\Xi_u{}^a$ (eV)	Temperature (K)
S^0	0.30	−0.57	11.1±0.3	148.9
S^+	0.53	−0.68	11.3±0.3	223.6

^{a}The values of $(C_{11}-C_{12})=(S_{11}-S_{12})^{-1}=1.024\times10^{12}$ and 1.018×10^{12} dyn/cm^2 at 148.9 and 223.6 K are used, respectively, to deduce Ξ_u, as illustrated in the text. The values of C_{11} and C_{12} are taken from Ref. 27.

TABLE II. Comparison of Ξ_u's for Si determined by various methods.

Method	Ξ_u (eV)	Ref. no.
1. Deep-level transient method	11.1 ± 0.3 (149 K) 11.3 ± 0.3 (224 K)	this work
2. Piezo-optic effect	11.3 ± 0.3 (300 K)	2
3. Electron paramagnetic resonance	11 ± 1 (1.25 K)	3
4. Same as no. 3	11.4 ± 1.1 (1.3–4 K)	4
5. Piezospectroscopy of indirect exciton absorption	8.6 ± 0.2 (80 K) 9.2 ± 0.3 (295 K)	5
6. Same as no. 5	8.3 ± 0.4 (77 K)	6
7. Piezospectroscopy of indirect exciton spectrum	8.6 ± 0.4 (77 K)	7
8. Same as no. 7	8.1 ± 0.3	8
9. Piezospectroscopy of shallow-donor excited states (Sb,P,As,Mg)	8.77 ± 0.07 (10 K)	13
10. Same as no. 9; (S)	7.9 ± 0.2 (low K)	10
11. Same as no. 9; (P)	7.9 ± 0.2 (10 K)	11
12. Linewidth of cyclotron resonance	8.5 ± 0.1 (2.5–5 K)	12
13. Piezoresistance	8.3 ± 0.3 (300 K)	13
14. Effect of carrier concentration on elastic constants	8.6 ± 0.4 (298 K)	14

quantity is Ξ_u/E_{12}, where E_{12} is the splitting between the singlet and doublet of donor levels in Si. The piezospectroscopy methods of numbers 5–11 were also indirect methods, which relied upon the validity of effective-mass theory. Methods 5–8 measured the splitting of the indirect exciton lines under stress. This included not only the splittings due to conduction-band minima, but also possible splittings of binding energies of excitons attached to different valleys, and splittings of energies of phonons involved in indirect exciton absorption. Methods numbered 9–11 measured the splittings of the excited donor states under stress. This implicitly assumed that the ionization energies of excited donor states to the conduction-band minima were stress independent or had the same stress dependence.

Since the values of Ξ_u determined by different methods are quite diverse and most of these methods have different kinds of uncertainties and ambiguities, independent techniques based on different principles are still desirable and significant in resolving the discrepancies that still exist in the measurements of deformation potentials in semiconductors. The technique of the present work is an electrical measurement in nature and incorporates simple theory and equations. The method measures the splitting of conduction-band minima directly. Thus it should prove valuable for clarifying the controversy of the different values of Ξ_u obtained by different authors.

B. Possible temperature dependence of Ξ_u

It is indicated in Table I that the shear deformation potential of Si conduction-band minima Ξ_u is temperature dependent. In the linear temperature approximation,

$d\Xi_u/dT \sim +3$ meV/K. The interesting fact to note is that from data of piezospectroscopy of Balslev,[5] a temperature coefficient of $d\Xi_u/dT \sim +3$ meV/K is also obtained. As has been pointed out by Brooks,[28] the energy term $E_c^{(j)}$ of Eq. (4) is actually the free energy, which is temperature dependent. From this context, Ξ_u, which measures the splitting of $E_c^{(j)}$ between different j, is reasonably temperature dependent. Furthermore, Van Vechten[29] has verified that the energy difference between two electronic states measured by no-phonon optical measurements should be equal to the free-energy difference between the same two states measured by thermal experiments.[24] Therefore, the temperature dependence of measured Ξ_u by methods 9–11 of Table II should be similar to the present work.

The arguments can also be used for methods 5–8 if the temperature dependences of the phonons participating in the absorption of indirect excitons attached to different minima are the same. Although the data of the present work and those given in Ref. 5 both indicate a temperature dependence of $\sim +3$ meV/K for Ξ_u, we feel that the result is not fully developed, and further experiments with higher accuracy and more temperature points are needed to confirm it.

VI. CONCLUDING DISCUSSIONS

We have developed a method—deep-level capacitance transient under uniaxial stress—to measure the shear deformation potential constant Ξ_u of the conduction band of Si. The technique is advantageous in its simplicity of the basic principle and formulation with parameters that can be measured precisely. The central theme of this method is the direct measurement of the splitting of the

14 046 LI, ZHAO, GU, CHEN, LI, AND WANG 43

conduction-band minima, which yields the values of $\Xi_u = 11.1 \pm 0.3$ eV at 148.9 K and 11.3 ± 0.3 eV at 223.6 K. The result is particularly valuable for symmetry determination of deep electron traps in Si by stress experiments.

ACKNOWLEDGMENTS

The authors wish to thank Professor Kun Huang for several stimulating discussions. This work was supported by the National Science Foundation of China.

[1] C. Herring and E. Vogt, Phys. Rev. **101**, 944 (1958).

[2] K. J. Schmidt-Tiedemann, *Proceedings of the International Conference on the Physics of Semiconductors, Exeter 1962* (Institute of Physics and the Physical Society, Adlard & Ltd., Dorking, England, 1962), p. 191.

[3] D. K. Wilson and G. Feher, Phys. Rev. **124**, 1068 (1961).

[4] G. D. Watkins and F. S. Ham, Phys. Rev. B **1**, 4071 (1970).

[5] I. Balslev, Phys. Rev. **143**, 636 (1966).

[6] L. D. Laude, F. H. Pollak, and M. Cardona, Phys. Rev. **133**, 2623 (1971).

[7] I. P. Akimchenko and V. A. Vdovenkov, Fiz. Tverd. Tela (Leningrad) **11**, 658 (1968) [Sov. Phys.—Solid State **11**, 528 (1969)].

[8] J. C. Merle, M. Capizzi, P. Fiorini, and A. Frova, Phys. Rev. B **17**, 4821 (1978).

[9] V. J. Tekippe, H. R. Chandrasekhar, P. Fisher, and A. K. Ramdas, Phys. Rev. B **6**, 2348 (1972).

[10] W. E. Krag, W. H. Kleiner, H. J. Zeiger, and S. Fischler, J. Phys. Soc. Jpn. Suppl. **21**, 230 (1966).

[11] W. E. Krag, W. H. Kleiner, and H. J. Zeiger, in *Proceedings of the 10th International Conference on the Physics of Semiconductors, Cambridge, Mass., 1970*, edited by S. P. Keller, J. C. Hensel, and F. Stern (U.S. Atomic Energy Commission, Division of Technical Information, Washington, D.C., 1970), p. 271.

[12] R. Ito, H. Kawamura, and M. Fukai, Phys. Lett. **13**, 26 (1964).

[13] J. E. Aubrey, W. Gubler, T. Henningson, and S. H. Koenig, Phys. Rev. **130**, 1667 (1963).

[14] J. J. Hall, Phys. Rev. **161**, 756 (1967).

[15] M. F. Li, J. X. Chen, X. S. Zhao, and Y. J. Li, in *Defects in Semiconductors*, Vols. 10–12 of *Materials Science Forum*, edited by H. J. vonBardeleben (Trans Tech , Switzerland, 1986), p. 469.

[16] C. T. Sah, L. Forbes, L. L. Rosier, and A. F. Tasch, Solid State Electron. **13**, 759 (1970); L. L. Rosier and C. T. Sah, *ibid.* **14**, 41 (1971).

[17] M. F. Li, Acta Phys. Sin. **34**, 1549 (1985).

[18] W. Schockley and W. T. Read, Phys. Rev. **87**, 835 (1952); R. N. Hall, *ibid.* **83**, 228 (1951); C. T. Sah, Proc. IEEE **55**, 654 (1967).

[19] X. C. Yao, G. G. Qin, S. R. Zeng, and M. H. Yuan, Acta Phys. Sin. **33**, 377 (1984).

[20] G. W. Ludwig, Phys. Rev. **137**, A1520 (1965).

[21] C. C. Koroluke, Sov. J. Semicond. Phys. Tech. **15**, 784 (1981).

[22] A. Zylbersztejn, Appl. Phys. Lett. **33**, 200 (1978); G. L. Miller, D. V. Lang, and L. C. Kimerling, Ann. Rev. Mater. Sci. **7**, 377 (1977).

[23] J. F. Nye, *Physical Properties of Crystals* (Oxford University, Oxford, London, 1964).

[24] H. G. Grimmeiss, E. Janzen, and B. Skarstam, J. Appl. Phys. **51**, 4212 (1980).

[25] W. Jantsch, K. Wunstel, O. Kumagai, and P. Vogl, Phys. Rev. B **25**, 5515 (1982); Physica B+C **117&118B**, 188 (1983).

[26] M. F. Li, J. X. Chen, Y. S. Yao, and G. Bai, J. Appl. Phys. **58**, 2589 (1985).

[27] H. J. McSkimin, J. Appl. Phys. **24**, 988 (1953).

[28] H. Brooks, in *Advances in Electronics and Electron Physics*, edited by L. Marton (Academic, New York, 1955), Vol. 7, p. 118.

[29] J. A. Van Vechten, in *Handbook on Semiconductors*, edited by S. P. Keller (North-Holland, Oxford, 1980), Vol. 3, p. 3.

PHYSICAL REVIEW B   VOLUME 44, NUMBER 16   15 OCTOBER 1991-II

First-principles calculations for quasiparticle energies of GaP and GaAs

Jian-Qing Wang* and Zong-Quan Gu
Department of Physics, Institute of Semiconductors, P.O. Box 912, Beijing, China

Ming-Fu Li
Department of Physics, Graduate School, University of Science and Technology of China, P.O. Box 3908, Beijing, China
(Received 18 January 1991)

We have applied the Green-function method in the GW approximation to calculate quasiparticle energies for the semiconductors GaP and GaAs. Good agreement between the calculated excitation energies and the experimental results was achieved. We obtained calculated direct band gaps of GaP and GaAs of 2.93 and 1.42 eV, respectively, in comparison with the experimental values of 2.90 and 1.52 eV, respectively. An *ab initio* pseudopotential method has been used to generate basis wave functions and charge densities for calculating the dielectric matrix elements and self-energies. To evaluate the dynamical effects of the screened interaction, the generalized-plasma-pole model has been utilized to extend the dielectric matrix elements from static results to finite frequencies. We present the calculated quasiparticle energies at various high-symmetry points of the Brillouin zone and compare them with the experimental results and other calculations.

I. INTRODUCTION

Although the theoretical framework for studying quasiparticle excitations in solids by the Green-function approach was presented decades ago,[1,2] accurate first-principles calculations for the quasiparticle energies in semiconductors and insulators have only become possible recently[3–5] due to the development of computer capacities and computational techniques. The traditional density-functional theory (DFT),[6,7] when applied to these many-electron systems in conjunction with the local-density approximation (LDA), has produced excellent results for the ground-state properties (e.g., bulk structural and dynamical properties[8–12] and problems of surfaces[13,14] and defects[15]). But it is well known that when DFT band structures of these systems are calculated, the band gaps generated and the excitation energies are largely underestimated.[16–18] A fundamental difficulty is the fact that the Kohn-Sham one-electron equation, though a powerful tool in calculations for ground-state properties, does not correspond to real situations where there is an elementary excitation in the system. Sham and Schlüter[19] and Perdew and Levy,[20] in their analyses of energy-band gaps, have demonstrated that for a system with empty conduction bands separated from filled valence bands by an energy gap, the true band gap of single-particle excitations deviates from the Kohn-Sham gap by a large amount. The difference can be of the same order of magnitude as the DFT band gap itself.[21] A later study showed[4,22] that the LDA is a very good approximation to the DFT exchange-correlation functional in semiconductors, contributing 80%. Thus most of the band-gap difference of the experimental and DFT's should be attributed to the incapabilities of Kohn-Sham equations to account for elementary excitations in many-electron systems.

Quasiparticle description, on the other hand, provides more complete characterizations for elementary excitations in many-electron systems. One technique often resorted to is the Green-function method, which is a powerful method for solving quasiparticle problems.[1] The spectral function defined by the single-particle Green function contains complete information on single-particle excitations in a system. The Green-function characterizations of quasiparticle states for metals are relatively simple since the dielectric functions of these systems can be diagonalized and free-electron-gas results[23] can be used.

Quantitative calculations on quasiparticle energies for semiconductors and insulators are, however, more difficult due to the complexity of the dielectric functions in these materials. Off-diagonal elements of the dielectric function cannot be ignored.[24] Hybertsen and Louie have done a first-principles pseudopotential calculation of elementary excitation energies for C, Si, Ge, and LiCl.[3] In their calculations, vertex corrections to polarizability were neglected. Good quantitative results were obtained in this lowest-order approximation compared with experimental values. In particular, the band gaps and excitation energies for these materials are within 5% of the experimental values of the corresponding excitations, representing a great improvement over the traditional DFT approach. A recent study has shown that the vertex corrections to the GW approximation in Si are indeed negligible.[25]

In this work we have carried out first-principles pseudopotential calculations for quasiparticle energies of the semiconductors GaP and GaAs. The calculated results are in good agreement with experimental values. In these calculations the dynamical behavior of the dielectric function in the screened interaction is approximated by the generalized-plasma-pole (GPP) model, in which the imaginary part has a δ-function frequency dependence.[3,5] The dominant features of the true dielectric function are depicted by this model where strong absorption prevails and the model is simple to apply.

We organize the paper as follows. The general framework with which we have calculated the quasiparticle energies and the approximations involved is outlined in Sec. II. In Sec. III, we describe the technical details of the first-principles calculations of self-energies. We present, in Sec. IV, the calculated quasiparticle energies at various high-symmetry points and at the conduction-band minimum of GaP in the Brillouin zone (BZ). We compare our calculated results with experiment and find good agreement.

II. GREEN-FUNCTION METHOD AND THE GPP MODEL

The elementary excitations in many-electron systems are readily expressed by a Green function. The spectral representation of the Green function reveals complicated structure, with peaks in its energy dependence corresponding to quasiparticle states. The energies with which the peaks are located are generally complex with the real parts corresponding to the quasiparticle energies and the imaginary parts to the lifetimes of these excitation states.

The concept of quasiparticle self-energy Σ is introduced by the one-electron Green function in order to specify the interactions between the quasiparticle excitations and their environments. The Green function satisfies the equation

$$[\omega - h(\mathbf{x}) - V(\mathbf{x})]G(\mathbf{x}, \mathbf{x}'; \omega)$$
$$- \int \Sigma(\mathbf{x}, \mathbf{x}''; \omega)G(\mathbf{x}'', \mathbf{x}'; \omega)d\mathbf{x}'' = \delta(\mathbf{x} - \mathbf{x}') , \quad (1)$$

where $h(\mathbf{x})$ is the one-electron Hamiltonian operator and $V(\mathbf{x})$ is the one-particle potential term. Quasiparticle states correspond to solutions of the homogeneous equations

$$[E_{n\mathbf{k}} - h(\mathbf{x}) - V(\mathbf{x})]\Phi_{n\mathbf{k}}(\mathbf{x})$$
$$- \int \Sigma(\mathbf{x}, \mathbf{x}'; E_{n\mathbf{k}})\Phi_{n\mathbf{k}}(\mathbf{x}')d\mathbf{x}' = 0 , \quad (2)$$

with $E_{n\mathbf{k}}$ the quasiparticle energy of state $|n\mathbf{k}\rangle$ and $\Phi_{n\mathbf{k}}(\mathbf{x})$ the corresponding quasiparticle wave function. Unlike the *real* potential operator in a quantum-mechanical system, the self-energy operator $\Sigma(\mathbf{x}, \mathbf{x}'; E)$ is nonlocal, energy dependent, and non-Hermitian. Thus the eigenenergies of quasiparticle states can be complex.

The set of coupled equations[1–3] for self-energy Σ, the Green function G, and the dynamical screened interaction

$$W(\mathbf{x}t, \mathbf{x}'t') = \int v_c(\mathbf{x}t, \mathbf{x}''t'')\epsilon^{-1}(\mathbf{x}'', t'', \mathbf{x}'t')d(\mathbf{x}''t'')$$

allows the self-energy to be expressed formally as a series expansion in W instead of the bare Coulomb potential v_c, where ϵ^{-1} is the dielectric function of the system. This has the advantage of avoiding slow converging series due to the relatively large contribution from v_c. In the lowest-order approximation the vertex corrections are neglected, thus leading to the simple GW approximation

for self-energy:

$$\Sigma(\mathbf{r}, \mathbf{r}'; E) = i \int \frac{dE'}{2\pi} e^{-i\delta E'} G(\mathbf{r}, \mathbf{r}'; E - E')W(\mathbf{r}, \mathbf{r}'; E') ,$$
$$(3)$$

where $\delta = 0^+$.

In evaluating W in our calculations, the dynamical behavior of the dielectric function $\epsilon_{\mathbf{GG}'}^{-1}(\mathbf{q}, \omega)$ in reciprocal space and energy representation was approximated by the GPP model. In this model the frequency dependence of ϵ^{-1} is depicted by that of a single plasmon with an effective frequency $\widetilde{\omega}_{\mathbf{GG}'}(\mathbf{q})$ and an amplitude $A_{\mathbf{GG}'} = -(\pi/2)\Omega_{\mathbf{GG}'}^2(\mathbf{q})/\widetilde{\omega}_{\mathbf{GG}'}(\mathbf{q})$, where $\Omega_{\mathbf{GG}'}^2(\mathbf{q})$ is the effective bare plasma frequency of the plasmon.[3,5] The general properties of the many-electron system set up certain constraints to the parameters in the model. As has been discussed in Refs. 3 and 5, the GPP model has two constraining criteria that are of universal validity: (1) the dielectric function approaches its static limit at $\omega = 0$; (2) the generalized f sum rule relates the first frequency moment of the time Fourier transformed imaginary (second) part $\epsilon_{2,\mathbf{GG}'}^{-1}(\mathbf{q}, \omega)$ of the dielectric function to the crystal charge density $\rho(\mathbf{G})$ in reciprocal space.[3]

For systems without a center of inversion symmetry as in GaP and GaAs, the situation is more complicated than the systems with a center of inversion symmetry. We have used the relation for the parameters as in Ref. 5,

$$\widetilde{\omega}_{\mathbf{GG}'}(\mathbf{q}) = \left[\frac{\lambda_{\mathbf{GG}'}(\mathbf{q})}{\cos(\phi_{\mathbf{GG}'}(\mathbf{q}))} \right]^{1/2} , \quad (4a)$$

$$\lambda_{\mathbf{GG}'}(\mathbf{q})e^{i\phi_{\mathbf{GG}'}(\mathbf{q})} = \frac{\Omega_{\mathbf{GG}'}^2(\mathbf{q})}{\delta_{\mathbf{GG}'} - \epsilon_{1,\mathbf{GG}'}^{-1}(\mathbf{q}, \omega=0)} , \quad (4b)$$

where $\epsilon_{1,\mathbf{GG}'}^{-1}(\mathbf{q}, \omega=0)$ is the time Fourier transformed real (first) part of the dielectric function in the static limit $\omega = 0$. All the quantities on the right-hand side of Eq. (4b) could be evaluated by the first-principles pseudopotential calculations, thus the model becomes truly adjustable parameter free.

III. *Ab Initio* QUASIPARTICLE ENERGY CALCULATIONS

We used the *ab initio* pseudopotential method to perform first-principles calculations of the inversed dielectric-function matrix $\epsilon_{\mathbf{GG}'}^{-1}(\mathbf{q})$ and self-energy Σ. The dielectric function ϵ^{-1} was evaluated in the random-phase approximation (RPA) with the irreducible polarizability $P_{\mathbf{GG}'}^0(\mathbf{q})$ represented in the standard Alder-Wiser expression.[26] The pseudopotentials were generated by the method of Hamann, Schlüter, and Chiang.[27,28] The scalar relativistic bands were obtained by a plane-wave expansion in the local-density approximation. The plane-wave basis extended up to a kinetic energy of 16 Ry.

In order to achieve convergence in matrix inversion for ϵ^{-1} and in calculating the quasiparticle self-energies, large matrix sizes are needed for $\epsilon_{\mathbf{GG}'}^{-1}(\mathbf{q})$. The matrix size we chose was 137×137. The large G elements of the plane-wave matrix require the inclusion of high conduction bands. Thus $n_c = 200$ bands were included in calculating $P^0_{GG'}(\mathbf{q})$. Convergence tests indicated an average convergency of within 1%. Ten special Chadi $\mathbf{k}$ points[29] were used in the full symmetry irreducible Brillouin zone (IBZ) for $\mathbf{k}$-point integration.[24]

The GW quasiparticle self-energy Eq. (3) is separated into two parts $\Sigma = \Sigma_{SE} + \Sigma_{CH}$. The first part Σ_{SE} is the screened-exchange (SE) self-energy, which originates from poles of the Green function G. When the time-delayed effect becomes zero, i.e., when $\omega = 0$ in ϵ^{-1}, Σ_{SE} becomes the approximation of Hedin.[1] The poles of the screened Coulomb interaction generate the Coulomb-hole (CH) self-energy Σ_{CH}. Expanding the screened interaction in plane waves and using the GPP model to approximate the dielectric function, one obtains the matrix elements for Σ_{SE} and Σ_{CH} taken between eigenvectors $|n\mathbf{k}\rangle, |n'\mathbf{k}\rangle$. Considering the rotation properties for the dielectric function $\epsilon^{-1}_{GG'}(\mathbf{q})$, the effective bare plasma frequency $\Omega^2_{GG'}(\mathbf{q})$, and the effective frequency $\tilde{\omega}_{GG'}(\mathbf{q})$, we have the following expressions:

$$\langle n\mathbf{k}|\Sigma_{SE}(\mathbf{r},\mathbf{r}';E)|n'\mathbf{k}\rangle = -\frac{1}{\Omega}\sum_{n_1 GG'}^{occ}\sum_{\mathbf{q}\in IBZ_k^{(+)}}\sum_R [f_{k-q,q}^{n_1 n *}(\mathbf{G})f_{k-q,q}^{n_1 n'}(\mathbf{G}')\epsilon^{-1}_{GG'}(\mathbf{q},E-\varepsilon_{n_1,\mathbf{k}-\mathbf{q}})$$

$$+ f_{kq}^{nn_1}(\mathbf{G})f_{kq}^{n'n_1 *}(\mathbf{G}')\epsilon^{-1}_{GG'}(\mathbf{q},E-\varepsilon_{n_1,\mathbf{k}+\mathbf{q}})]v_c(\mathbf{q}+\mathbf{G}') \quad , \tag{5a}$$

$$\langle n\mathbf{k}|\Sigma_{CH}(\mathbf{r},\mathbf{r}';E)|n'\mathbf{k}\rangle = \frac{1}{\Omega}\sum_{n_1 GG'}\sum_{\mathbf{q}\in IBZ_k^{(+)}}\sum_R \left[f_{k-q,q}^{n_1 n *}(\mathbf{G})f_{k-q,q}^{n_1 n'}(\mathbf{G}')\frac{1}{[E-\varepsilon_{n_1,\mathbf{k}-\mathbf{q}}-\tilde{\omega}_{GG'}(\mathbf{q})]} \right.$$

$$\left. + f_{kq}^{nn_1}(\mathbf{G})f_{kq}^{n'n_1 *}(\mathbf{G})\frac{1}{[E-\varepsilon_{n_1,\mathbf{k}+\mathbf{q}}-\tilde{\omega}_{GG'}(\mathbf{q})]} \right]\frac{\Omega^2_{GG'}(\mathbf{q})}{2\tilde{\omega}_{GG'}(\mathbf{q})}v_c(\mathbf{q}+\mathbf{G}') \quad . \tag{5b}$$

In the above expressions, the summation of $\mathbf{q}$'s is over the little $\mathbf{k}$ group irreducible Brillouin zone $IBZ_k^{(+)}$ for the state vector $\mathbf{k}$. The plus sign on $IBZ_k^{(+)}$ indicates that it is the IBZ when time-reversal symmetry is explicitly included, and the subscript k denotes that the IBZ is generated by rotations in the little k group, i.e., rotations that satisfy

$$R\mathbf{k}+\mathbf{G}_R = \mathbf{k} \quad , \tag{5c}$$

$\mathbf{G}_R$ being a reciprocal translation vector. The rotations R in the last summation are operations that bring irreducible $+\mathbf{q}$'s to their corresponding stars (of $\mathbf{k}$ vector). The function $f_{k,q}^{n_1 n} = \langle n_1\mathbf{k}|e^{-i(\mathbf{q}+\mathbf{G})\cdot\mathbf{r}}|n\mathbf{k}+\mathbf{q}\rangle$ and ε_{nk} is the eigenenergy of state $|n\mathbf{k}\rangle$. The energy-band summation in Σ_{SE} only includes occupied states, whereas the sum in Σ_{CH} is over all the states.

The $\mathbf{q}=0$ term in Eq. (5) deserves special attention because of the presence of the Coulomb potential $v_c(\mathbf{q}+\mathbf{G}')$, which produces a $1/q^2$ singularity for wing elements $W_{G0}(\mathbf{q}\to 0)$ and head element $W_{00}(\mathbf{q}\to 0)$ for the screened interaction W. Furthermore, the plane-wave matrix $f_{kq}^{n_1 n}(\mathbf{G}=0)|_{q\to 0}$, although being normal when $\mathbf{G}\neq 0$, also presents a $\mathbf{q}\cdot\mathbf{A}$ type of singularity as $\mathbf{q}\to 0$ unless $n_1 = n$, where $\mathbf{A}$ is a vector dependent on eigenvectors $|n\mathbf{k}\rangle$ and $|n_1\mathbf{k}\rangle$. $f_{kq}^{nn}(\mathbf{G}=0)_{q\to 0}$ is, on the other hand, equal to unity. Singularities of the type $\mathbf{q}\cdot\mathbf{A}$ are also produced[30] by heads and wings of $\epsilon^{-1}_{GG'}(\mathbf{q}\to 0)$ and $\Omega^2_{GG'}(\mathbf{q}\to 0)$. The contributions of $\mathbf{q}=0$ point to quasiparticle self-energies mainly coming from the head term $(\mathbf{G}=\mathbf{G}'=0)$ and the $n_1 = n$ band. The other terms contribute only on the order of $O(q_{sz}^3)$, which can be neglected, where $q_{sz} = (6\pi^2/N\Omega_c)^{1/3}$ is the radius of the small sphere occupied by a $\mathbf{q}$ point, $\Omega = N\Omega_c$ being the total volume of the crystal. The final expressions for the $\mathbf{q}=0$ contribution to the screened exchange and the Coulomb hole self-energies are

$$\langle n\mathbf{k}|\Sigma_{SE}|n\mathbf{k}\rangle = -\Theta(\mu-E)$$

$$\times\Omega\frac{2}{\pi}e^2\left[1+\frac{\omega_p^2}{(E-\varepsilon_{nk})-\tilde{\omega}_{00}^2}\right]q_{sz} \quad , \tag{6a}$$

$$\langle n\mathbf{k}|\Sigma_{CH}|n\mathbf{k}\rangle = \Omega\frac{1}{\pi}\frac{e^2\omega_p^2}{\tilde{\omega}_{00}(E-\varepsilon_{nk}-\tilde{\omega}_{00})}q_{sz} \quad . \tag{6b}$$

In the above expressions, $\Theta(E)$ is the unit-step function, μ is the Fermi energy, $\tilde{\omega}_{00}^2 = \omega_p^2/[1-\epsilon^{-1}_{00}(\mathbf{q}\to 0)]$ with $\epsilon^{-1}_{00}(\mathbf{q}\to 0) = 1/\epsilon_M$ being the inverse of the macroscopic dielectric constant ϵ_M of the system, and ω_p is the free-electron plasma frequency. We have calculated for GaP and GaAs the values of ϵ_M in a previous work,[31] and obtained the values 10.71 and 12.55 in comparison with the experimental results of 10.86 and 12.4, respectively.

The quasiparticle self-energy $\langle n\mathbf{k}|\Sigma(\mathbf{r},\mathbf{r}';E)|n\mathbf{k}\rangle$ is energy dependent. It must be evaluated at the quasiparticle energy E_{nk} for Eq. (2) to be self-contained. From the fact[3] of near-perfect overlap of $\Phi_{nk}(\mathbf{r})$ with the LDA one-particle wave function $\phi_{nk}(\mathbf{r})$, one arrives at

$$E_{nk} = \varepsilon_{nk} - \langle n\mathbf{k}|V_{xc}^{LDA}|n\mathbf{k}\rangle + \langle n\mathbf{k}|\Sigma(E_{nk})|n\mathbf{k}\rangle \quad , \tag{7}$$

where V_{xc}^{LDA} is the LDA exchange-correlation potential. We started to evaluate the self-energy in Eq. (7) at the LDA eigenenergies ε_{nk} and expanded it about these LDA values to the first order in $O(E-\varepsilon)$. The energy derivative $d\Sigma/dE$ was evaluated by a finite energy difference of the order of 1.0 eV. Varying the size of dE in this energy range did not change the calculated values of $d\Sigma/dE$, indicating a very good linearity of Σ versus E. The calculated E_{nk}'s from the first iteration were used as the starting energies of the second iteration. But in our calculations the converged quasiparticle energies were already achieved in the first iteration. One hardly needed to go to the second step.

IV. RESULTS AND DISCUSSIONS

The results of our *ab initio* pseudopotential calculations of the quasiparticle energies at the high-symmetry points Γ, X, L and the lowest conduction-band minima for GaP and GaAs are listed in Tables I and II (column II), respectively. The excitation energies are referenced with respect to the tops of valence bands at Γ. In column III of these tables we list the LDA energies of the corresponding states. In column IV the results of the calculations from other works are presented. In column V we list the results derived from experimental measurements for comparison. We observe that the direct band gaps for these materials from our calculations agree within a few percent with those of the experiments. And the calculated quasiparticle energies are also in good agreement with the experiments. In the calculations, the Ceperly-Alder form for the correlation potential[32] had been used. The

energy cutoff for crystal charge density was 64 Ry for both materials. Convergent results were obtained with 100 energy bands in the sum for Σ_{CH}.

In evaluating self-energy Eq. (5), the integration of $\mathbf{q}$ is carried over IBZ_k, which is in general larger than (or equal to) and includes the full symmetry IBZ. In the case of the $4\times4\times4$ uniformly divided BZ (including $\mathbf{q}=0$), the full symmetry IBZ consists of 8 $\mathbf{q}$ points. For state $|n\mathbf{k}\rangle$ with $\mathbf{k}$ in the lower-symmetry point in the BZ, one needs to use a larger irreducible $\mathbf{q}$-point set. Either the X or L point, e.g., has a 13-$\mathbf{q}$-point set in the respective IBZ_k, although the two sets are not identical to each other. The dielectric matrix elements $\epsilon_{\mathbf{GG'}}^{-1}(\mathbf{q},\omega)$ for these lower-symmetry $\mathbf{k}$ points are only evaluated at the seven full symmetry irreducible $\mathbf{q}$ points (excluding $\mathbf{q}=0$). In this case the dielectric matrices at other $\mathbf{q}$ points in the IBZ_k can be obtained from those at the IBZ $\mathbf{q}$'s by symmetry properties.[24] In doing so, however, some matrix elements are lost in the $\mathbf{G}$ summations in Eq. (5) by rotations $R\mathbf{G}-\mathbf{G}_R$ because these components are being operated out of the finite matrix size. The reciprocal translation vectors $\mathbf{G}$ and $\mathbf{G'}$ in the first summation run over the whole reciprocal space. But in our calculations, $\mathbf{G}$'s are limited to a certain subshell of relatively large radius. A $\mathbf{G}$-vector subshell is defined as those $\mathbf{G}$'s with the same magnitude. The $\mathbf{G}$'s in the summations are limited to 137. When $\mathbf{k}\neq0$, the reciprocal translation vector $\mathbf{G}_R$ in Eq. (5c) can be nonzero for some rotations. Thus when mapping $\mathbf{q}$ points in $IBZ_k^{(+)}$ by rotation R into the full Brillouin zone within the first BZ, some of the additional reciprocal vectors $R\mathbf{G}-\mathbf{G}_R$ for $\mathbf{G}$'s near but under the limiting subshell can become beyond the limiting sub-

TABLE I. Quasiparticle energies for GaP (eV).

Excitation	This work	LDA	Others[a]	Expt.[b]
Γ_{15v}	0.0	0.0	0.0	0.0
Γ_{1v}	-12.637	-12.430	-12.99	$-12.3, -13.2$
Γ_{1c}	2.933	1.804	2.88	2.895
Γ_{15c}	4.499	3.901	5.24	4.6, 4.87
X_{1v}	-9.683	-9.573	-9.46	-9.6
X_{3v}	-7.451	-6.748	-7.07	$-6.8, -6.9$
X_{5v}	-3.076	-2.671	-2.73	$-3.0, -2.7$
X_{1c}	1.825	1.553	2.16	2.354
X_{3c}	2.157	1.789	2.71	
L_{1v}	-10.809	-10.473	-10.60	$-10.8, -10.6$
L_{2v}	-7.364	-6.863	-6.84	$-6.8, -6.9$
L_{3v}	-1.287	-1.108	-1.10	$-0.9, -1.2$
L_{1c}	2.331	1.179	2.79	2.637
L_{3v}-L_{1c}	3.618	2.287		3.91
X_{3c}-X_{1c}	0.332	0.236		0.355[c]
Δ_{min}	1.811	1.518		2.350
CB	0.014	0.035		0.0035

[a]Reference 33.
[b]Reference 34.
[c]Reference 35.

<u>44</u>   FIRST-PRINCIPLES CALCULATIONS FOR QUASIPARTICLE . . .   8711

TABLE II. Quasiparticle energies for GaAs (eV).

Excitation	This work	LDA	Others[a,b]	Expt.[c]
Γ_{15v}	0.0	0.0	0.0	0.0
Γ_{1v}	-13.06	-12.70		-13.1
Γ_{1c}	1.42	0.46	1.58, 1.29	1.522
Γ_{15c}	4.55	3.73		4.716
X_{1v}	-10.33	-10.35		-10.75
X_{3v}	-7.12	-6.80		-6.70
X_{5v}	-2.82	-2.65	$-2.64, -2.79$	-2.80
X_{1c}	1.98	1.34	2.19, 2.05	2.08
X_{3c}	2.26	1.57	2.41	2.58
X_{5c}	11.06			
L_{1v}	-11.29	-11.07		-11.24
L_{1v}	-6.87	-6.63		-6.70
L_{3v}	-1.21	-1.12	$-1.11, -1.19$	-1.30
L_{1c}	1.72	0.94	1.93, 1.69	1.85
L_{3c}	5.40	4.61		

[a]Reference 4.
[b]Reference 5.
[c]Reference 34.

shell. These **G** vectors should have been excluded in the calculations. Of course, some other **G**'s beyond the limiting subshell can be translated to the inside, hence they should have been contributing to the summations of Eqs. (5a) and (5b). This induces some errors in the calculated self-energies on the order of 0.1 eV. Nevertheless, utilizing this symmetry property has great advantages since calculating dielectric matrices for different **q**'s consists of the dominating computation efforts.

For GaP in Table I, as we compare the quasiparticle calculation results with those of the LDA calculations in reference to the experimental results, we observe an overall improvement for the calculated elementary excitation energies. In particular, the calculated energy gaps in direct transitions at these high-symmetry points are in very good agreement with the experimental values, within a few percent, whereas the LDA calcalutions produce results that deviate from the experimental values by large fractions. For example, for the minimum direct energy gap at Γ our calculated value 2.93 eV agrees excellently with the measured value 2.90 eV. The calculated values for the L_{3v}-L_{1c} transition, 3.62 eV, and for the X_{3c}-X_{1c} splitting, 0.332 eV, are also in good agreement with the measured values of 3.91 and 0.355 eV, respectively.

The LDA calculations have produced for the height of the barrier at X_{1c} with respect to the Δ_1 conduction-band minimum (CB) a value 35 meV that is much too large compared with the experimental value of 3.5 meV. The value of our quasiparticle calculations, 14 meV, has improved the theoretical value, in closer agreement with the experiments.

We also observe that our quasiparticle results for the excitation energies at X and L are in better agreement with the experiments than the LDA results. For example, the value of $E(L_{1c})$ is increased to 2.33 eV from the LDA value 1.18 eV, in good agreement with the experimental value 2.64 eV. But there exist certain discrepancies. The indirect energy gap at the conduction-band minimum from our calculations differs from the experimental value by 25%, deviating by 0.54 eV. Deviations on this order can also be seen on several other states. These discrepancies are partially attributed to the computation errors introduced by indirectly obtaining ϵ^{-1} for the **q** points outside the full symmetry IBZ by symmetry transformation in the **q**-point integration in Eq. (5). The calculated results in column IV of Table I are from empirical-pseudopotential-method calculations in Ref. 35.

The results of the calculated quasiparticle energies for GaAs are listed in Table II. We also list the calculation results from Refs. 4 and 5. For GaAs, the results are very good. The calculated quasiparticle energies are in very good agreement with the experimental values within a few percent. The calculated direct fundamental energy gap is 1.42 eV, which agrees well with the measured value of 1.52 eV. Our calculated results also agree with other calculations.[4,5]

In conclusion, we have carried out first-principles pseudopotential calculations of quasiparticle energies for the semiconductors GaP and GaAs. Very good agreement between our calculated excitation energies and direct band gaps with experiment is found. The fundamental direct energy gaps for GaP and GaAs from our work are 2.93 and 1.42 eV, respectively, in comparison with the experimental values of 2.90 and 1.52 eV, respectively.

ACKNOWLEDGMENTS

The authors would like to thank Professor Steven G. Louie for generously providing the *ab initio* pseudopotential band-structure calculation program. One of us

8712   JIAN-QING WANG, ZONG-QUAN GU, AND MING-FU LI   <u>44</u>

(J.-Q.W.) would like to acknowledge Dr. S.-B. Zhang for sending us detailed calculation results for GaAs and their paper *prior* to publication. We are grateful to Professor K. Huang for his interest and encouragement on this problem. Digital Equipment Corporation Vax computer times were provided by the Peking Astronomy Observatory and the Institute of Semiconductors, Academia Sinica, respectively.

*Present address: Group 412, Institute of Physics, Academia Sinica, P.O. Box 603, Beijing 100 080, China.

[1] L. Hedin, Phys. Rev. **139**, A796 (1965).

[2] L. Hedin and S. Lundqvist, in *Solid State Physics,* edited by H. Ehrenreich, F. Seitz, and D. Turnbull (Academic, New York, 1969), Vol. 23, p. 1.

[3] M. S. Hybertsen and S. G. Louie, Phys. Rev B **34**, 5390 (1986).

[4] R. W. Godby, M. Schlüter, and L. J. Sham, Phys. Rev. Lett. **56**, 2415 (1986); Phys. Rev. B **36**, 6497 (1987); **37**, 10 159 (1988).

[5] S. B. Zhang, D. Tománek, M. L. Cohen, S. G. Louie, and M. S. Hybertsen, Phys. Rev. B **40**, 3162 (1989).

[6] P. Hohenberg and W. Kohn, Phys. Rev. **136**, B864 (1964).

[7] W. Kohn and L. J. Sham, Phys. Rev. **140**, A1133 (1965).

[8] M. T. Yin and M. L. Cohen, Phys. Rev. B **26**, 5668 (1982).

[9] K. Kunc and R. M. Martin, Phys. Rev. B **24**, 2311 (1981).

[10] M. Schlüter and L. J. Sham, Phys. Today **35**(2), 30 (1982).

[11] S. Froyen and M. L. Cohen, Phys. Rev. B **28**, 3258 (1983).

[12] C. O. Rodriguez, R. A. Casali, E. L. Peltzer, O. M. Cappannini, and M. Methfessel, Phys. Rev. B **40**, 3975 (1989).

[13] *Many Body Phenomena at Surfaces,* edited by D. C. Langreth and H. Suhl (Academic, New York, 1984).

[14] G.-X. Qian, R. M. Martin, and D. J. Chadi, Phys. Rev. Lett **60**, 1962 (1988).

[15] K. J. Chang and D. J. Chadi, Phys. Rev. Lett. **60**, 1422 (1988); D. J. Chadi and K. J. Chang, *ibid.* **60**, 2187 (1988).

[16] D. R. Hamann, Phys. Rev. Lett. **42**, 662 (1979).

[17] *Theory of the Inhomogeneous Electron Gas,* edited by S. Lundqvist and N. H. March (Plenum, New York, 1983).

[18] G. B. Bachelet and N.E. Christensen, Phys. Rev. B **31**, 879 (1985).

[19] L. J. Sham and M. Schlüter, Phys. Rev. Lett. **51**, 1888 (1983); Phys. Rev. B **32**, 3883 (1985).

[20] P. Perdew and M. Levy, Phys. Rev. Lett. **51**, 1884 (1985).

[21] M. Lannoo, M. Schlüter, and L. J. Sham, Phys. Rev. B **32**, 3890 (1985).

[22] F. Gygi and A. Baldereschi, Phys. Rev. Lett. **62**, 2160 (1989).

[23] J. K. Lindhard, Dan. Vidensk. Selsk. Mat.–Fys. Medd. **8**, 28 (1954).

[24] M. S. Hybertsen and S. G. Louie, Phys. Rev. B **35**, 5585 (1987).

[25] R. Daling and N. van Haeringen, Phys. Rev. B **40**, 11 659 (1989).

[26] S. L. Alder, Phys. Rev. **126**, 413 (1962); N. Wiser, *ibid.* **129**, 62 (1963).

[27] D. R. Hamann, M. Schlüter, and C. Chiang, Phys. Rev. Lett. **43**, 1494 (1979).

[28] L. Kleinman, Phys. Rev. B **21**, 2630 (1980); G. B. Bachelet and M. Schlüter, *ibid.* **25**, 2103 (1982).

[29] D. J. Chadi and M. L. Cohen, Phys. Rev. B **8**, 5747 (1973).

[30] R. M. Pick, M. H. Cohen, and R. M. Martin, Phys. Rev. B **1**, 910 (1970).

[31] J.-Q. Wang, Z.-Q. Gu, and M.-F. Li, Chin. Phys. Lett. **8**, 21 (1991).

[32] D. M. Ceperley and B. I. Alder, Phys. Rev. Lett. **45**, 566 (1980); J. P. Perdew and A. Zunger, Phys. Rev. B **23**, 5048 (1982).

[33] J. R. Chelikowsky and M. L. Cohen, Phys. Rev. B **14**, 556 (1976).

[34] *Landolt-Börnstein, Numerical Data and Functional Relationships in Science and Technology, Vol. 17a/22a,* edited by O. Madelung (Springer-Verlag, Berlin, 1982/1987).

[35] A. Onton, Phys. Rev. B **4**, 4449 (1971).

Reprinted with permission from W.J. Fan, M.F. Li, T.C. Chong, J.B. Xia, J. Appl. Phys.,
Vol.79, pp.188–194 (1996). Copyright 1996, American Institute of Physics.

Electronic properties of zinc-blende GaN, AlN, and their alloys Ga$_{1-x}$Al$_x$N

W. J. Fan, M. F. Li, and T. C. Chong
*Center for Optoelectronics, Department of Electrical Engineering, National University of Singapore,
Singapore 0511*

J. B. Xia
*National Laboratory for Superlattices and Microstructures, Institute of Semiconductors, Academia Sinica,
Beijing 100083, People's Republic of China*

(Received 12 May 1995; accepted for publication 21 September 1995)

The electronic properties of wide-energy gap zinc-blende structure GaN, AlN, and their alloys
Ga$_{1-x}$Al$_x$N are investigated using the empirical pseudopotential method. Electron and hole effective
mass parameters, hydrostatic and shear deformation potential constants of the valence band at Γ and
those of the conduction band at Γ and X are obtained for GaN and AlN, respectively. The energies
of Γ, X, L conduction valleys of Ga$_{1-x}$Al$_x$N alloy versus Al fraction x are also calculated. The
information will be useful for the design of lattice mismatched heterostructure optoelectronic
devices based on these materials in the blue light range application. © *1996 American Institute of
Physics.* [S0021-8979(96)05701-0]

I. INTRODUCTION

The wide-energy gap III–V nitride semiconductors GaN
and AlN have received considerable attention in both
experiment[1-8] and theory[9-13] for their device applications in
the blue and ultraviolet wavelengths. A number of reviews
on GaN and AlN were given by Strite and Morkoc,[14] Davis
et al.,[15] and Pankove.[16] The vast majority of research on
III–V nitrides has been focused on the wurtzite crystal phase.
The reason is that most of III–V nitrides have been grown on
sapphire substrates which generally transfer their hexagonal
symmetry to the nitride film. Nevertheless, interest in zinc-
blende nitrides has been growing recently. The zinc-blende
GaN has a higher saturated electron drift velocity[14] and a
somewhat lower energy band than wurtzite GaN. Mizuta
et al.[17] first reported bulk zinc-blende GaN grown on (001)
GaAs. There have been several recent studies of the zinc-
blende GaN.[6,7,18,19] AlN in zinc-blende structure has an indirect gap at X point with 5.11 eV,[20] while zinc-blende GaN
with direct gap of 3.5 eV has been reported by Bloom.[9]

To provide a basis for understanding future wide-energy
gap device concepts and applications based on zinc-blende
III–V nitride semiconductors, particularly Ga$_{1-x}$Al$_x$N/GaN
lattice mismatched heterostructure devices, we have computed the electronic band structure of zinc-blende
Ga$_{1-x}$Al$_x$N alloys, and the deformation potential constants
for zinc-blende GaN and AlN using the empirical pseudopotential method (EPM).[22] In comparison, the self-consistent
pseudopotential method in the local-density approximation
(LDA) usually underestimates the energy band.[21] The quasiparticle method[21] is reliable but it is very time consuming for
computation. The empirical pseudopotential method is
simple and expected to give quick and reasonably reliable
results. However, in the literature, there is a lack of zinc-
blende GaN and AlN experimental data regarding the band
structures and it is hard to extract empirical form factor parameters for the EPM calculation. Recently, Rubio *et al.*[12]
calculated the band structures of zinc-blende GaN and AlN
using the quasiparticle method. In our work, we adjust the
EPM parameters to fit the band energies of AlN and GaN

obtained by the quasiparticle calculation. We then use these
EPM parameters and the method in Ref. 23 to calculate the
deformation potential constants of zinc-blende GaN and
AlN, and use virtual crystal approximation (VCA) to calculate the zinc-blende Ga$_{1-x}$Al$_x$N alloys band structure.

II. BAND STRUCTURES OF ZINC-BLENDE GaN AND AlN

The EPM method in Ref. 22 is used in the calculation of
the zinc-blende GaN and AlN band structures. The experimental measured energy gap of zinc-blende GaN E_g^Γ is 3.2
eV.[5,6] Rubio *et al.*[12] reported that the energy gaps of GaN E_g^Γ
and AlN E_g^X are 3.1 and 4.9 eV, respectively, using the quasiparticle method. We adjust the symmetric and antisymmetric pseudopotential form factors to fit the band energies of
AlN and GaN obtained by Rubio *et al.*'s quasiparticle calculation. The lattice constants a of GaN and AlN used are 4.50
and 4.35 Å, respectively, the same values used by Rubio
et al. in their quasiparticle calculation. The final adjusted
symmetric and antisymmetric form factors of GaN and AlN
are given in Table I. The energies calculated using the EPM
method for zinc-blende GaN and AlN are listed in Table II
for the high-symmetry points Γ, X, and L in the Brillouin
zone. All energies are with reference to the top of the valence
band Γ_8^v. The band structures of GaN and AlN are shown in
Figs. 1 and 2, respectively. The results show that GaN is a
direct-gap semiconductor with the minimum of conduction
band at Γ point, and AlN has an indirect gap with the minimum of conduction band at X point. The calculated energy

TABLE I. Symmetric and antisymmetric form factors in Ry of zinc-blende
GaN and AlN.

$\|G\|^2$ $(2\pi/a)^2$	3	4	8	11	12
V_{GaN}^S	−0.300		0.060	0.070	
V_{GaN}^A	0.280	0.200		0.040	0.020
V_{AlN}^S	−0.300		0.080	0.110	
V_{AlN}^A	0.280	0.330		0.015	0.013

TABLE II. Zinc-blende GaN and AlN energies in eV at high symmetry points. All values refer to the top of the valence band.

			GaN		
Γ_7^c	10.300	X_7^c	6.805	$L_{4,5}^c$	9.916
Γ_6^c	3.383	X_6^c	4.571	L_6^c	5.636
Γ_8^v	0.000	X_7^v	−2.693	$L_{4,5}^v$	−0.931
Γ_8^v	0.000	X_6^v	−2.699	L_6^v	−0.938
Γ_7^v	−0.011	X_6^v	−6.149	L_6^v	−6.743
			AlN		
Γ_7^c	13.406	X_7^c	10.661	$L_{4,5}^c$	12.014
Γ_6^c	5.936	X_6^c	5.102	L_6^c	9.423
Γ_8^v	0.000	X_7^v	−2.337	$L_{4,5}^v$	−0.728
Γ_8^v	0.000	X_6^v	−2.343	L_6^v	−0.735
Γ_7^v	−0.011	X_6^v	−5.262	L_6^v	−6.179

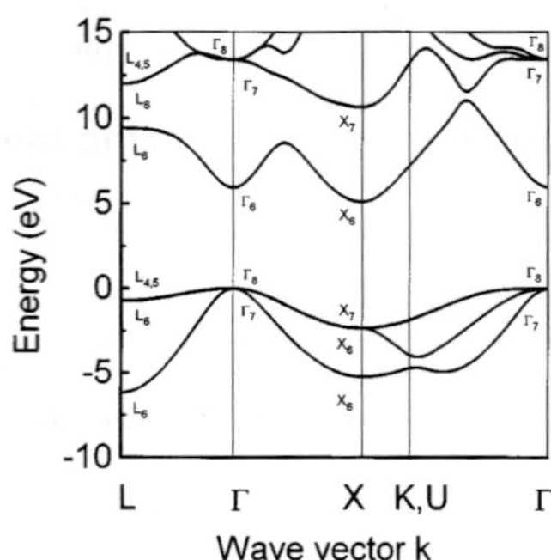

FIG. 2. The band structure of zinc-blende AlN.

gaps of GaN E_g^Γ and AlN E_g^X are 3.38 and 5.10 eV, respectively, which are in good agreement (within 8% discrepancy) with the experimental results[6] and quasiparticle results, as listed in Table III.

The spin-orbit splitting is included in our band structure calculations. The method was explained in Refs. 24 and 25. There is no existing experimental measurement nor theoretical calculation for spin-orbit splitting energies of zinc-blende GaN and AlN. Only the spin-orbit splitting energy at the top of the valence band Δ_{so} of wurtzite GaN is available and reported to be 11 meV.[26] However, it is noted that the values of Δ_{so} of wurtzite and zinc-blende crystals are very closed to each other for many other semiconductors in Ref. 26. Therefore, we take the Δ_{so} value of zinc-blende GaN to be 11 meV. On the other hand, in zinc-blende crystals the spin-orbit splitting at Γ is determined primarily by the anion atomic spin-orbit splitting in the valence band and by the cation atomic spin-orbit splitting in the conduction band.[27] Therefore, it is reasonable to assume that Δ_{so} of GaN and AlN are about the same. The corresponding spin-orbit parameters α, μ defined in Ref. 25 and spin-orbit splitting energies Δ_{so} are listed in Table IV.

We calculate the bottom of the conduction band and the heavy hole, light hole, and spin-orbit splitting bands at 11 k-points with $\mathbf{k}$ from $-0.005(2\pi/a)$ to $0.005(2\pi/a)$. The corresponding effective masses can be fitted using the formula $E = \hbar^2 k^2/2m^*$. Heavy hole, light hole, and spin-orbit splitting effective masses m_h^*, m_l^*, m_{so}^* along [100], [111], and [110] directions and electron effective masses at Γ point $m_e^*(\Gamma)$, longitudinal and transverse electron effective masses at X point $m_l^*(X)$ and $m_t^*(X)$ for GaN and AlN are shown in Table V. Using

$$m_h^{[100]} = m_0/(\gamma_1 - 2\gamma_2), \quad m_l^{[100]} = m_0/(\gamma_1 + 2\gamma_2),$$
$$m_h^{[111]} = m_0/(\gamma_1 - 2\gamma_3), \quad m_l^{[111]} = m_0/(\gamma_1 + 2\gamma_3), \quad (1)$$

we obtain the Luttinger parameter $\gamma_1 = 3.07$, $\gamma_2 = 0.86$, and $\gamma_3 = 1.26$ for GaN; and $\gamma_1 = 1.92$, $\gamma_2 = 0.47$, and $\gamma_3 = 0.75$ for AlN. For a reference, the effective masses of wurtzite GaN and AlN reported in the literature are also listed in Table V for comparison. On the other hand, the electron effective masses at Γ and X points for the zinc-blende GaN and AlN can be compared with the existing experimental and theoretical data with good agreement. For instance, $m_e^*(\Gamma)$ for GaN is $0.13m_0$ in our calculation, $0.15m_0$ obtained by electron-spin-resonance experiment,[38] and $0.21m_0$ by first-principle pseudopotential calculation.[39] The longitudinal and transverse electron effective masses at X point $m_l^*(X)$, and $m_t^*(X)$ for AlN are $0.53m_0$ and $0.31m_0$, respectively, in our calculation, and are $0.51m_0$ and $0.31m_0$, respectively, obtained by the first-principle calculation.[39]

TABLE III. The energy bands in eV of zinc-blende GaN and AlN.

	GaN			AlN	
E_g^Γ	3.38	3.1[a]	3.2[b]	5.94	6.0[a]
E_g^X	4.57	4.7[a]		5.10	4.9[a]
E_g^L	5.64	6.2[a]		9.42	9.3[a]

[a]Quasiparticle results from Ref. 12.
[b]Experimental results from Ref. 6.

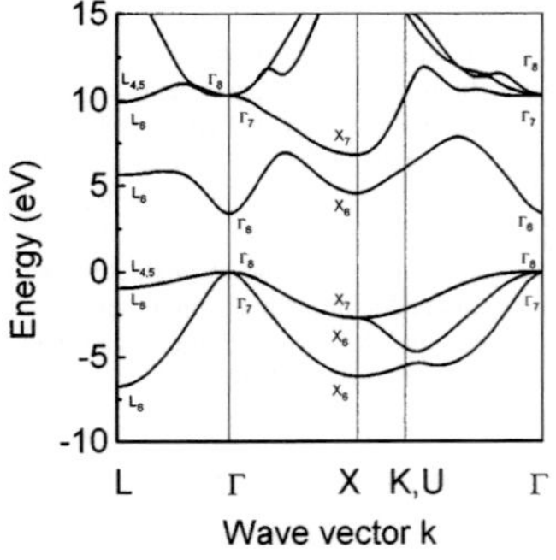

FIG. 1. The band structure of zinc-blende GaN.

TABLE IV. Spin-orbit parameters α, μ defined in Ref. 25 and calculated spin-orbit splitting energies at the top of the valence band Δ_{so} for zinc-blende GaN and AlN.

Compound	α	μ	Δ_{so}(meV)
GaN	0.410	0.083	11
AlN	1.170	0.030	11

III. DEFORMATION POTENTIAL CONSTANTS OF ZINC-BLENDE GaN AND AlN

A. Pseudopotential calculation of the band structure of a deformed crystal

Zinc-blende AlN and GaN have different lattice constants of 4.35 and 4.50 Å, respectively. When they form heterostructure, there is a large strain. The change of energy-band structure due to the strain is thus important and cannot be neglected.

We use the method similar to Ref. 23 to calculate the band structure of a deformed crystal. The first step is to find the analytical formula for the atomic pseudopotential form factor

$$V_i(\mathbf{G}) = \frac{1}{\Omega} \int V_i(\mathbf{r}) e^{-i\mathbf{G} \cdot \mathbf{r}} \, d\mathbf{r}. \tag{2}$$

The subscript $i = 1,2$ denotes the cation or anion of the zinc-blende lattice. Ω is the unit-cell volume and $\mathbf{G}$ is the reciprocal lattice vector. In our calculation, a modified empty-core model[24] is used for the atomic pseudopotential based on a local approximation for both the cation and anion atoms

$$V_i(r) = \begin{cases} -A_i & r < R_i \\ -Z_i r^{-1} e^{-\alpha_i r} & r > R_i \end{cases}, \tag{3}$$

where A_i is a constant representing a finite well depth and R_i is approximately the radius of the physical atomic core. The

additional screening factor $e^{-\alpha_i r}$ serves to obtain a better curve fitting for our form factors. The Fourier transform of Eq. (3) is

$$V_i(\mathbf{G}) = -\frac{8\pi}{\Omega} \left[\frac{Z_i e^{-\alpha_i R_i}}{G^2 + \alpha_i^2} \left(\cos GR_i + \frac{\alpha_i}{G} \sin GR_i \right) + \frac{A_i}{G^3} (\sin GR_i - GR_i \cos GR_i) \right]. \tag{4}$$

Symmetric and antisymmetric pseudopotential form factor can be written as[22]

$$V^S(\mathbf{G}) = \tfrac{1}{2}[V_1(\mathbf{G}) + V_2(\mathbf{G})], \tag{5}$$

$$V^A(\mathbf{G}) = \tfrac{1}{2}[V_1(\mathbf{G}) - V_2(\mathbf{G})]. \tag{6}$$

Substituting Eq. (4) into Eq. (5) and Eq. (6), the $\mathbf{G}$ dependence of $V^S(\mathbf{G})$ and $V^A(\mathbf{G})$ can be obtained. The parameters A_i, Z_i, R_i, and α_i are then adjusted to fit the discrete form factors listed in Table I using Eq. (4). The results obtained for GaN and AlN are shown in Figs. 3, 4 and Table VI. When the crystal lattice is deformed by a strain tensor $\overset{\leftrightarrow}{\mathbf{e}}$, the reciprocal lattice vector changes from $\mathbf{G}$ to $\mathbf{G}_\epsilon$ while the unit cell volume changes from Ω to $\Omega_\epsilon = \Omega + \Delta\Omega$. In the rigid-ion approximation, the atomic pseudopotential $V_i(\mathbf{r})$ in the real space remains unchanged. The change of the symmetric or antisymmetric pseudopotential form factor of the deformed crystal is only due to the change in the magnitude of $|\mathbf{G}_\epsilon|$ and the change in volume of the unit cell, therefore we have

$$V_\epsilon^{S,A}(\mathbf{G}_\epsilon) = \frac{\Omega}{\Omega_\epsilon} V^{S,A}(|\mathbf{G}_\epsilon|), \tag{7}$$

TABLE V. Calculated light-hole, heavy-hole and spin-orbit splitting effective masses along [100], [111], and [110] directions and electron masses at Γ point $m_e^*(\Gamma)$, longitudinal and transverse electron effective masses at X point $m_l^*(X)$ and $m_t^*(X)$ of zinc-blende GaN and AlN. Some existing results of the Wurtzite structure are also listed for comparison (in the unit of free electron mass m_0).

Valence band	m_h^*	m_l^*	m_{so}^*	m^* for wurtzite structure
GaN [100]	0.74	0.21	0.33	$(\Gamma \to K)1.58^a$, $(\Gamma \to A)2.03^a$
				$(\Gamma \to M)1.93^a$, 0.4^c, 0.8^c, 1.0^c
[111]	1.82	0.18	0.33	
[110]	1.51	0.19	0.33	
AlN [100]	1.02	0.35	0.51	$(\Gamma \to K)3.40^a$, $(\Gamma \to A)0.30^a$
				$(\Gamma \to M)3.52^a$
[111]	2.85	0.30	0.51	
[110]	2.16	0.31	0.51	
Conduction band	$m_e^*(\Gamma)$	$m_l^*(X)$	$m_t^*(X)$	m_e^* for wurtzite structure
GaN	$0.13(0.15^d, 0.21^e)$	0.58	0.30	$(\Gamma \to K)0.36^a$, $(\Gamma \to A)0.27^a$
				$(\Gamma \to M)0.33^a$, 0.20^b, 0.1^c, 0.2^c, 0.28^c
AlN	0.21	$0.53(0.51^e)$	$0.31(0.31^e)$	$(\Gamma \to K)0.42^a$, $(\Gamma \to A)0.33^a$
				$(\Gamma \to M)0.40^a$

[a]Reference 31.
[b]Reference 14.
[c]Reference 26.
[d]Reference 38.
[e]Reference 39.

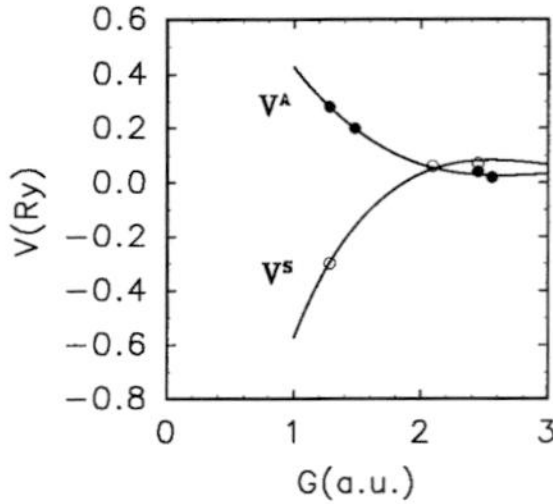

FIG. 3. The form factors of zinc-blende GaN vs reciprocal vector G.

TABLE VI. Fitted parameters of the pseudopotentials. The subscript 1(2) denotes the parameter associated with the atomic potential of the cation (anion) atom. All the quantities are in atomic units.

Parameters	GaN	AlN
A_1	−0.201	7.303
A_2	12.277	9.894
R_1	0.864	1.617
R_2	2.588	2.929
Z_1	3.508	14.210
Z_2	7.126	4.738
α_1	−0.571	−0.131
α_2	−0.575	−0.675

where $V_\epsilon^S(\mathbf{G}_\epsilon)$ is the symmetric or antisymmetric pseudopotential form factor of the deformed crystal and $V^{S,A}(\mathbf{G})$ is obtained by Eqs. (5) and (6). The band structure of deformed crystal can be calculated using pseudopotential given in Eq. (7) and the deformed lattice. The range of ϵ is from 0 to 1.2×10^{-3} in our calculation.

B. Deformation potential constants

Since for a heterostructure with growth direction along [001] direction, the lattice mismatch takes place in the (001) plane. The strain is biaxial, which may be decomposed into hydrostatic and pure-shear strains. The hydrostatic strain is given by the following strain tensor:

$$\overset{\leftrightarrow}{\mathbf{e}}_h = \epsilon \begin{pmatrix} 1 & 0 & 0 \\ 0 & 1 & 0 \\ 0 & 0 & 1 \end{pmatrix} \tag{8}$$

and a Γ_{12} component of the pure-shear strain is given by

$$\overset{\leftrightarrow}{\mathbf{e}}_s = \epsilon \begin{pmatrix} -1 & 0 & 0 \\ 0 & -1 & 0 \\ 0 & 0 & 2 \end{pmatrix}. \tag{9}$$

Under the shear strain $\overset{\leftrightarrow}{\mathbf{e}}_s$ of Eq. (9), the internal strain parameter ξ defined by Kleinman[28] equals zero. This corre-

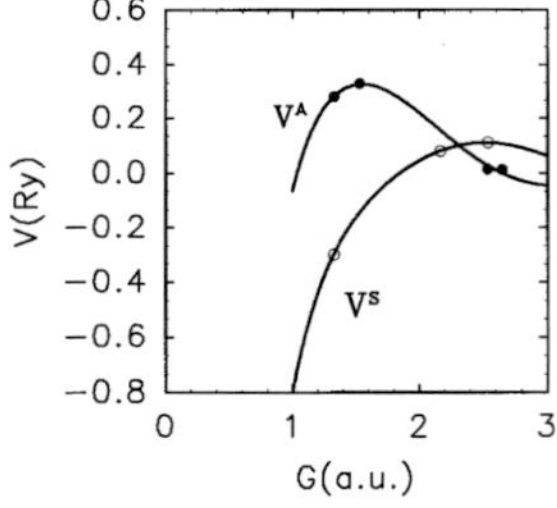

FIG. 4. The form factors of zinc-blende AlN vs reciprocal vector G.

sponds to intercellular distances transforming the macroscopic strain tensor $\overset{\leftrightarrow}{\mathbf{e}}$ in our empirical pseudopotential calculation.

1. Top of the valence band at k=0

We follow the notation and discussion in Ref. 29. The total Hamiltonian for a p-like multiplet can be written as

$$H = H_{so} + H_{strain}, \tag{10}$$

where H_{so} is the spin-orbit Hamiltonian, and

$$H_{strain} = -a(\epsilon_{xx} + \epsilon_{yy} + \epsilon_{zz}) - 3b[(L_x^2 - L^2/3)\epsilon_{xx} + c.p.]$$
$$- \sqrt{3}d[(L_x L_y + L_y L_x)\epsilon_{xy} + c.p.], \tag{11}$$

where L is the angular momentum operator, and $c.p.$ denotes cyclic permutation with respect to the indices x, y, z. We neglect the strain-dependent spin-orbit Hamiltonian. For the strain of $\overset{\leftrightarrow}{\mathbf{e}}_h + \overset{\leftrightarrow}{\mathbf{e}}_s$, taking the valence band wave functions in the $|J, M_j\rangle$ representation, the Hamiltonian matrix of Eq. (10) becomes

$$\begin{array}{ccc} |3/2,3/2\rangle & |3/2,1/2\rangle & |1/2,1/2\rangle \end{array}$$
$$\begin{vmatrix} -\delta E_h - \delta E_s & 0 & 0 \\ 0 & -\delta E_h + \delta E_s & \sqrt{2}\,\delta E_s \\ 0 & \sqrt{2}\,\delta E_s & -\delta E_h - \Delta_{so} \end{vmatrix}, \tag{12}$$

where Δ_{so} is the spin-orbit splitting energy, and

$$\delta E_h = 3a\epsilon \tag{13}$$

$$\delta E_s = 3b\epsilon. \tag{14}$$

The resulting three eigenvalues of Eq. (12) are

$$E_{v1} = -\delta E_h - \delta E_s \tag{15a}$$

$$E_{v2} = -\delta E_h - (\Delta_{so} - \delta E_s)/2 + [\Delta_{so}^2 + 2\delta E_s \Delta_{so} + 9\delta E_s^2]^{1/2} \tag{15b}$$

$$E_{v3} = -\delta E_h - (\Delta_{so} - \delta E_s)/2 - [\Delta_{so}^2 + 2\delta E_s \Delta_{so} + 9\delta E_s^2]^{1/2}. \tag{15c}$$

Next we put $\delta E_h = 0$, and only consider the shear strain. For the case of $|\delta E_s| \ll \Delta_{so}$, neglecting $(\delta E_s/\Delta_{so})^2$ and higher order terms, Eq. (15) reduces to

	GaN			AlN	
	Our results	Experiment	Calculation	Our results	Calculation
Valence band					
a	13.6			1.9	
b	-1.9		-1.6^{a}	-2.2	
Conduction band					
$a(\Gamma_6^c)$	-21.3			-11.7	
$\Xi_d(X_6^c)$	-9.2			-7.7	
$\Xi_u(X_6^c)$	7.1		6.9^{a}	6.6	
Inter band					
$\Gamma^v\to\Gamma^c$, $a(\Gamma_6^c)+a$	-7.7	-8.8^{b}	$-4.5^{\mathrm{c}},-7.3^{\mathrm{d}}$	-9.8	-9.0^{d}
$\Gamma^v\to X^c$, $\Xi_d(X_6^c)+a$	4.4		-0.05^{e}	-5.8	-0.4^{e}

[a] Using $d_3=-2.8$, $d_1^3=5.6$ in Ref. 34, which are related to our parameters by $d_3=\sqrt{3}b$, $d_1^3=\sqrt{\tfrac{2}{3}}\Xi_u$.

[b] Using $dE/dP=4.0$ meV kbar^{-1} in Ref. 32, $B_s=199$ GPa in Ref. 36 and Murnaghan equation at low pressure $p=-3B_0\epsilon$.

[c] Using $dE/dP=1.46$ meV kbar^{-1}, $B_0=3.09$ Mbar in Ref 35, and Murnaghan equation at low pressure $p=-3B_0\epsilon$.

[d] $\alpha=-22$ eV for GaN and -27 eV for AlN in Ref. 33, α is related to our parameter by $\alpha=3[a(\Gamma_6^c)+a]$.

[e] $\alpha=-0.14$ eV for GaN and -1.1 eV for AlN in Ref. 33, α is related to our parameter by $\alpha=3[\Xi_d(X_6^c)+a]$.

$$E_{\mathrm{hh}}=-\delta E_s, \qquad (16a)$$

$$E_{\mathrm{lh}}=+\delta E_s, \qquad (16b)$$

$$E_{\mathrm{so}}=-\Delta_{\mathrm{so}}. \qquad (16c)$$

When δE_s is a positive (negative) value, the heavy-hole energy E_{hh} is lower (higher) then the light-hole energy E_{lh}. We use this method to determine the sign of b in Eq. (14). We calculate the heavy-hole and light-hole energies under the shear strains $\epsilon=0$, 1×10^{-4}, 4×10^{-4}, 8×10^{-4}, 1.2×10^{-3} and then fit them with Eqs. (15a) and (15b) to obtain b. The values of a and b for GaN and AlN are listed in Table VII.

2. Conduction band at Γ and X

The energy shift under strain for the Γ_6^c conduction band is characterized by the deformation potential constant $c(\Gamma_6^c)$

$$\Delta E(\Gamma_6^c)=a(\Gamma_6^c)(\epsilon_{xx}+\epsilon_{yy}+\epsilon_{zz}). \qquad (17)$$

Therefore, the interband hydrostatic pressure deformation potential at Γ is $a(\Gamma_6^c)+a$. From Table VII, the values of $a(\Gamma_6^c)+a$ for GaN and AlN are -7.7 and -9.8 eV, respectively.

For 6 conduction valleys at X, the energy shift under strain can be expressed by[30]

$$\Delta E^{(i)}=\Xi_d(\epsilon_{xx}+\epsilon_{yy}+\epsilon_{zz})+\Xi_u\sum_{\mu\nu}a_\mu^{(i)}a_\nu^{(i)}\epsilon_{\mu\nu}, \qquad (18)$$

where $\mathbf{a}^{(i)}=\{a_x^{(i)},a_y^{(i)},a_z^{(i)}\}$ is the unit vector from Γ to the ith X point in the Brillouin zone, Ξ_d is the hydrostatic deformation potential constant, and Ξ_u is the shear deformation potential constant at X valleys. For strain $\vec{\mathbf{e}}_h+\vec{\mathbf{e}}_s$, Eq. (18) can be rewritten as

$$\Delta E^{(i)}=\begin{cases} 3\epsilon_1\Xi_d-\epsilon_2\Xi_u & i=1,2,3,4 \\ 3\epsilon_1\Xi_d+2\epsilon_2\Xi_u & i=5,6 \end{cases}, \qquad (19)$$

where $i=5$, 6 valleys located at $\mathbf{k}=2\pi/a(0,0,\pm1)$, while $i=1$, 2, 3, 4 valleys located at $\mathbf{k}=2\pi/a(\pm1,0,0)$ and

$\mathbf{k}=2\pi/a(0,\pm1,0)$. The calculated values of Ξ_d, Ξ_u, and $a(\Gamma_6^c)$ for GaN and AlN are listed in Table VII. Our results in Table VII are in over all agreement with previous experimental or theoretical data available in the literature, except the $\Gamma^v\to X^c$ deformation potential $\Xi_d(X_6^c)+a$. Our results are 4.4 eV for GaN and -5.8 eV for AlN, compared with -0.05 eV for GaN and -0.4 eV for AlN in Ref. 33. This discrepancy should be further checked by experiments.

C Band structures of Ga$_{1-x}$Al$_x$N alloys

The method described in Ref. 24 is used to calculate the band structures of Ga$_{1-x}$Al$_x$N alloys. With virtual crystal approximation, the symmetric and antisymmetric form factors for Ga$_{1-x}$Al$_x$N alloys can be expressed as

$$V^S=[(1-x)\Omega_{\mathrm{GaN}}V_{\mathrm{GaN}}^S+x\Omega_{\mathrm{AlN}}V_{\mathrm{AlN}}^S]/\Omega$$

$$V^A=[(1-x)\Omega_{\mathrm{GaN}}V_{\mathrm{GaN}}^A+x\Omega_{\mathrm{AlN}}V_{\mathrm{AlN}}^A]/\Omega, \qquad (20)$$

where, Ω, Ω_{GaN}, and Ω_{AlN} are the volumes of the primitive cells of Ga$_{1-x}$Al$_x$N, GaN, and AlN, respectively. The lattice constant of alloy Ga$_{1-x}$Al$_x$N is given by

$$a=(1-x)a_{\mathrm{GaN}}+xa_{\mathrm{AlN}}. \qquad (21)$$

Using this lattice constant and the corresponding reciprocal lattice vectors and the pseudopotential form factors $V^S(\mathbf{G})$ and $V^A(\mathbf{G})$ of Eq. (20), we can calculate the band structures of Ga$_{1-x}$Al$_x$N alloys.

The curves for conduction Γ, X, L valleys of Ga$_{1-x}$Al$_x$N alloys vs Al mole fraction x are given in Fig. 5, the reference energy level being the top of the valence band. These curves can be fitted by the following polynomials:

$$E_g^\Gamma=3.38+2.50x+0.05x^2, \qquad (22)$$

$$E_g^X=4.57-0.08x+0.61x^2, \qquad (23)$$

$$E_g^L=5.64+2.99x+0.80x^2. \qquad (24)$$

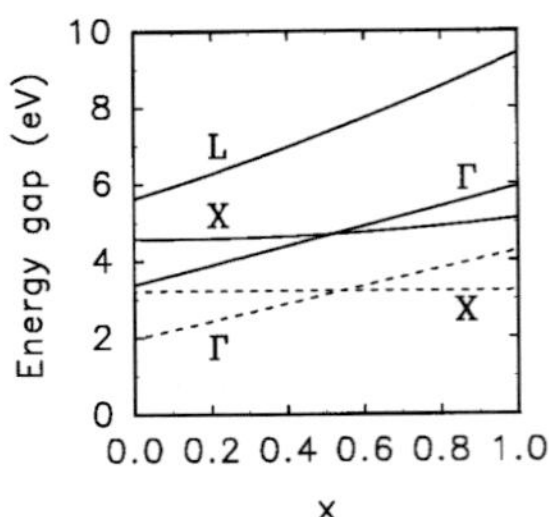

FIG. 5. The energy gaps of zinc-blende $Ga_{1-x}Al_xN$ alloys vs Al mole fraction x. The solid lines are our calculation results. The dashed lines are the calculation results of Albanesi et al. for the ideal zinc-blende structure using self-consistent LMTO method in LDA approximation, which underestimates the energy gap.

The unit of the energy is eV. The band gap bowing factors are $b_\Gamma = 0.05$ eV and $b_X = 0.61$ eV in Eqs. (22) and (23), respectively. It is well known that the bowing factors obtained by virtual crystal approximation may deviate from experiments.[37] Recent investigation of semiconductor alloys[37] shows that the bowing factor deviation may be due to (1) partial ordering of alloy structures and (2) bond-length relaxation effect in the alloy crystal structures. Since, in our empirical calculation, the existing experimental data are not available to take these effects into consideration, our bowing factors are only for reference. Albanesi et al. have calculated the band gap variation with Al fraction for the ideal structure and for the distorted structure including bond-length relaxation by self-consistent linear muffin-tin orbital (LMTO) method in LDA. The self-consistent LDA calculation suffers from underestimating the energy gap.[21] However, the tendency of the variation of energy gap versus Al fraction x is still valuable for reference. In Fig. 5, Albanesi et al.'s ideal structure results are also plotted for comparison. Both our results and Albanesi et al.'s ideal structure results show small bowing factors. The bowing factors obtained by Albanesi et al. when including bond length relaxation are $b_\Gamma = -0.40$ eV and $b_X = -0.92$ eV. In a very recent first principle calculation of Wright and Nelson,[40] they obtained the bowing factors for zinc-blende $Ga_{1-x}Al_xN$ to be $b_\Gamma = 0.53$ eV and $b_X = -0.10$ eV. There are a lack of experimental bowing factor values for zinc-blende $Ga_{1-x}Al_xN$. The experimental value for wurtzite $Ga_{1-x}Al_xN$ is $b_\Gamma = 0.98$ eV.[14] The Γ to X energy gap transition occurs at $x = 0.52$ in our calculation and $x = 0.57$ for the ideal structure in Albanesi et al.'s calculation.

IV. CONCLUSION

Electronic band structures of zinc-blende GaN, AlN, and their alloys $Ga_{1-x}Al_xN$ are calculated using the empirical pseudopotential method. It is shown that GaN is a direct-gap semiconductor, with energy gap 3.38 eV. AlN has an indirect gap of 5.10 eV with the conduction band minimum at X point. Electron, hole, and spin-orbit splitting effective masses are calculated. Hydrostatic and shear deformation potential constants of valence bands at Γ and those of conduction bands at Γ and X are calculated for GaN and AlN, respectively. The energy levels for $Ga_{1-x}Al_xN$ alloys are calculated with virtual crystal approximation. A direct to indirect energy-gap transition is predicted at $x = 0.52$.

ACKNOWLEDGMENTS

W. J. Fan would like to acknowledge K. L. Teo for discussion on the calculation of the symmetry and antisymmetry form factors. This work was supported by National University of Singapore research Grant No. RP920621.

[1] W. C. Johnson, J. B. Parsons, and M. C. Crew, J. Phys. Chem. 36, 2561 (1932).

[2] E. Tiede, M. Thimann, and K. Sensse, Chem. Berichte 61, 1568 (1928).

[3] H. P. Maruska and J. J. Tietjen, Appl. Phys. Lett. 15, 327 (1969).

[4] Proceedings of the 7th trieste ICTP-IUAP Semiconductor Symposium, edited by C. G. van de Walle [Physica B 185, R9 (1993)].

[5] T. Lei, T. D. Moustakas, R. J. Graham, Y. He, and S. J. Berkowitz, J. Appl. Phys. 71, 4933 (1992).

[6] T. Lei, M. Fanciulli, R. J. Molnar, T. D. Moustakas, R. J. Graham, and J. Scanlon, Appl. Phys. Lett. 95, 944 (1991).

[7] M. J. Paisley, Z. Sitar, J. B. Posthill, and R. F. Davis, J. Vac. Sci. Technol. A 7, 701 (1989).

[8] Z. Sitar, M. J. Paisley, J. Rnan, J. W. Choyke, and R. F. Davis, J. Mater. Sci. Lett. 11, 261 (1992).

[9] S. Bloom, G. Harbeke, E. Meier, and I. B. Ortenburger, Phys. Status Solidi B 66, 161 (1974).

[10] P. E. Van Camp, V. E. Van Doren, and J. T. Deverse, Phys. Rev. B 44, 9056 (1991).

[11] A. Munoz and K. Kunc, Phys. Rev. B 44, 10372 (1991).

[12] A. Rubio, J. L. Corkill, M. L. Cohen, E. L. Shirley, and S. G. Louie, Phys. Rev. B 48, 11810 (1993).

[13] A. F. Wright and J. S. Nelson, Phys. Rev. B 50, 2159 (1994).

[14] S. Strite and H. Morkoc, J. Vac. Sci. Technol. B 10, 1237 (1192).

[15] R. F. Davis et al., Mater. Sci. Eng. B 1, 77 (1988).

[16] J. I. Pankove, Mater. Res. Soc. Symp. Proc. 97, 409 (1987); 162, 515 (1990).

[17] M. Mizuta, S. Fujieda, Y. Matsumoto, and T. Kawamara, Jpn. J. Appl. Phys. 25, L945 (1986).

[18] Z. Sitar, M. J. Paisley, B. Yan, and R. F. Davis, Mater. Res. Soc. Symp. Proc. 162, 537 (1990).

[19] T. P. Humphreys, C. A. Sukow, R. J. Nemanich, J. B. Posthill, R. A. Radder, S. V. Hattangady, and R. J. Markunas, Mater. Res. Soc. Symp. Proc. 162, 53 (1990).

[20] W. R. L. Lambrecht and B. Segall, Phys. Rev. B 43, 7070 (1991).

[21] M. S. Hybertsen and S. G. Louie, Phys. Rev. Lett. 55, 1418 (1985); Phys. Rev. B 32, 7005 (1985); B 34, 5390 (1986).

[22] M. L. Cohen and T. K. Bergstresser, Phys. Rev. 141, 789 (1966).

[23] Ming-Fu Li, Zong-Quan Gu, and Jian-Qing Wang, Phys. Rev. B 42, 5714 (1990).

[24] K. L. Teo, Y. P. Feng, M. F. Li, T. C. Chong, and J. B. Xia, Semicond. Sci. Technol. 9, 349 (1994).

[25] J. R. Chelikowsky and M. L. Cohen, Phys. Rev. B 14, 556 (1976).

[26] Landolt-Bornstein, Numerical Data and Functional Relationships in Science and Technology, edited by Harbeke, Madelung, and Rossler (Springer, Berlin, 1982), Vol. 17a.

[27] G. G. Wepfer, T. C. Collins, and R. N. Euwema, Phys. Rev. B 4, 1296 (1971).

[28] I. Goroff and L. Kleinman, Phys. Rev. 132, 1080 (1963).

[29] F. H. Pollak, Surf. Sci. 37, 863 (1973).

[30] C. Herring and E. Vogt, Phys. Rev. 101, 944 (1956).

[31] Y. N. Xu and W. Y. Ching, Phys. Rev. B 48, 4335 (1993).

[32] S. J. Hwang, W. Shan, R. J. Hauenstein, and J. J. Song, Appl. Phys. Lett. 64, 2928 (1994).

[33] N. E. Christensen and I. Gorczyca, Phys. Rev. B 50, 4397 (1994).

[34] K. Kim, W. R. L. Lambrecht, and B. Segall, Phys. Rev. B **50**, 1502 (1994).

[35] Lu Wenchang, Zhang Kaiming, and Xie Xide, J. Phys.: Condens. Matter **5**, 875 (1993).

[36] E. A. Albanesi, W. R. L. Lambrecht, and B. Segall, Phys. Rev. B **48**, 17841 (1993).

[37] A. Zunger and S. Mahajan, in *Handbook on Semiconductors*, edited by S. Mahajan (Elsevier, New York, 1994), Vol. 3, p. 1399.

[38] M. Fanciulli, T. Lei, and T. D. Moustakas, Phys. Rev. B **48**, 15144 (1993).

[39] E. Miwa and A. Fukumoto, Phys. Rev. B **48**, 7897 (1993).

[40] A. F. Wright and J. S. Nelson, Appl. Phys. Lett. **66**, 3051 (1995).

Valence hole subbands and optical gain spectra of GaN/Ga$_{1-x}$Al$_x$N strained quantum wells

W. J. Fan, M. F. Li,[a)] and T. C. Chong
Center for Optoelectronics, Department of Electrical Engineering, National University of Singapore, Singapore 119260

J. B. Xia
National Laboratory for Superlattices and Microstructures, Institute of Semiconductors, Academia Sinica, Beijing 100083, People's Republic of China

(Received 4 March 1996; accepted for publication 30 May 1996)

The valence hole subbands, TE and TM mode optical gains, transparency carrier density, and radiative current density of the zinc-blende GaN/Ga$_{0.85}$Al$_{0.15}$N strained quantum well (100 Å well width) have been investigated using a 6×6 Hamiltonian model including the heavy hole, light hole, and spin-orbit split-off bands. At the $k=0$ point, it is found that the light hole strongly couples with the spin-orbit split-off hole, resulting in the so+lh hybrid states. The heavy hole does not couple with the light hole and the spin-orbit split-off hole. Optical transitions between the valence subbands and the conduction subbands obey the $\Delta n=0$ selection rule. At the $k \neq 0$ points, there is strong band mixing among the heavy hole, light hole, and spin-orbit split-off hole. The optical transitions do not obey the $\Delta n=0$ selection rule. The compressive strain in the GaN well region increases the energy separation between the **so1**+lh1 energy level and the hh1 energy level. Consequently, the compressive strain enhances the TE mode optical gain, and strongly depresses the TM mode optical gain. Even when the carrier density is as large as 10^{19} cm^{-3}, there is no positive TM mode optical gain. The TE mode optical gain spectrum has a peak at around 3.26 eV. The transparency carrier density is 6.5×10^{18} cm^{-3}, which is larger than that of GaAs quantum well. The compressive strain overall reduces the transparency carrier density. The J_{rad} is 0.53 kA/cm^2 for the zero optical gain. The results obtained in this work will be useful in designing quantum well GaN laser diodes and detectors. © *1996 American Institute of Physics.* [S0021-8979(96)08417-4]

I. INTRODUCTION

The wide energy gap III-V nitride semiconductors GaN, AlN, and their quantum well structures have received considerable attention for their device applications in the blue and ultraviolet wavelengths.[1-18] Recently, the successful fabrication of the blue light III-V nitride semiconductor laser was first demonstrated by Nakamura.[1] The vast majority of research on III-V nitrides has been focused on the wurtzite crystal phase. The reason is that most of III-V nitrides have been grown on sapphire substrates which generally transfer their hexagonal symmetry to the nitride film. Nevertheless, interest in zinc-blende nitrides has been growing recently.[2-6] The zinc-blende GaN has a higher saturated electron drift velocity and a somewhat lower energy gap than wurtzite GaN.[7]

In this work, the effective mass approximation is used to calculate valence hole subbands and the TE and TM mode optical gains of the zinc-blende GaN/Ga$_{0.85}$Al$_{0.15}$N strained quantum well. In bulk zinc-blende GaN and AlN, the spin-orbit split-off at the top of the valence band is around 17 meV.[19] Therefore, the 4×4 valence band effective mass Hamiltonian of Broido and Sham[20] is not suitable for the calculation and the spin-orbit split-off band must also be taken into consideration. People *et al.*[21] developed a 6×6 valence-band Hamiltonian in the calculation of strained bulk semiconductor layers. Meney *et al.*[22] compared the valence

band structure calculated by 4×4, 6×6, and 8×8 models, respectively. The influence from the conduction band to the valence band can be neglected because III-V nitride semiconductors have larger energy gaps. We modify the 6×6 Hamiltonian of People *et al.* to investigate the valence hole subband structures, optical gain, and radiative current density of the zinc-blende GaN/Ga$_{0.85}$Al$_{0.15}$N strained quantum well. In Sec. II, the 6×6 Hamiltonian is used to calculate the hole subband energies, wavefunctions, and density of states. The hole subbands are found to be far from parabolic and very anisotropic. Therefore, the axial approximation and the simple parabolic band approximation are not suitable for this case. We use numerical calculation methods and the Cray J916 supercomputer with very long computation time to obtain reliable and accurate results, rather than using any analytical approximation. In Sec. III, the method and the numerical calculation results of optical transition matrix elements between the conduction subbands and the valence subbands are described. In Sec. IV, the numerical calculation method of carrier density is given. In Sec. V, the TE and TM mode optical gains are calculated respectively. Discussion of the results and comparison with other work are given.

II. CALCULATION METHOD AND VALENCE HOLE SUBBANDS

The 6×6 Hamiltonian in Ref. 21 is modified to investigate the valence hole subband structures of zinc-blende

---

[a)]Electronic mail: elelimf@nus.sg

GaN/Ga$_{0.85}$Al$_{0.15}$N strained multiquantum wells (MQW) grown on (001)-oriented substrates. Following Luttinger,[23] we shall deal with the negative of valence electron Hamiltonian in our analysis. The Hamiltonian of the MQW is therefore given by

$$
\begin{array}{cccccc}
j=1, & 2, & 3, & 4, & 5, & 6, \\
(3/2,3/2), & (3/2,1/2), & (3/2,-1/2), & (3/2,-3/2), & (1/2,1/2), & (1/2,-1/2),
\end{array}
$$

$$
H_{SL}^{v}=
\begin{bmatrix}
H+V(z) & \alpha & \beta & 0 & i\alpha/\sqrt{2} & -i\sqrt{2}\beta \\
\alpha^{*} & L+V(z) & 0 & \beta & -i[D/\sqrt{2}-\sqrt{2}E] & i\sqrt{3/2}\alpha \\
\beta^{*} & 0 & L+V(z) & -\alpha & -i\sqrt{3/2}\alpha^{*} & -i[D/\sqrt{2}-\sqrt{2}E] \\
0 & \beta^{*} & -\alpha^{*} & H+V(z) & -i\sqrt{2}\beta^{*} & -i\alpha^{*}/\sqrt{2} \\
-i\alpha^{*}/\sqrt{2} & i[D/\sqrt{2}-\sqrt{2}E] & i\sqrt{3/2}\alpha & i\sqrt{2}\beta & S+V(z) & 0 \\
i\sqrt{2}\beta^{*} & -i\sqrt{3/2}\alpha^{*} & i[D/\sqrt{2}-\sqrt{2}E] & i\alpha/\sqrt{2} & 0 & S+V(z)
\end{bmatrix},
\qquad (1)
$$

where

$$
H=\frac{\hbar^{2}}{2m_{0}}[(k_{x}^{2}+k_{y}^{2})(\gamma_{1}+\gamma_{2})+k_{z}^{2}(\gamma_{1}-2\gamma_{2})]-E(z).
$$

$$
L=\frac{\hbar^{2}}{2m_{0}}[(k_{x}^{2}+k_{y}^{2})(\gamma_{1}-\gamma_{2})+k_{z}^{2}(\gamma_{1}+2\gamma_{2})]+E(z),
$$

$$
\alpha=\frac{\hbar^{2}}{2m_{0}}2\sqrt{3}[k_{z}(ik_{y}-k_{x})\gamma_{3}],
$$

$$
\beta=\frac{\hbar^{2}}{2m_{0}}\sqrt{3}[2ik_{x}k_{y}\gamma_{3}-(k_{x}^{2}k_{y}^{2})\gamma_{2}],
$$

$$
D=\frac{\hbar^{2}}{2m_{0}}[2(k_{x}^{2}+k_{y}^{2})\gamma_{2}-4k_{z}^{2}\gamma_{2}], \qquad (2)
$$

$$
S=\frac{\hbar^{2}}{2m_{0}}[(k_{x}^{2}+k_{y}^{2}+k_{z}^{2})\gamma_{1}]+\Delta_{0},
$$

$$
E(z)=
\begin{cases}
-\frac{2}{3}D_{u}(1+2c_{12}/c_{11})e_{xx}, & in\ well, \\
0, & in\ barrier,
\end{cases}
$$

$$
V(z)=
\begin{cases}
V_{0}, & in\ barrier, \\
0, & in\ well.
\end{cases}
$$

$V(z)$ is the periodic potential of MQW, $\frac{2}{3}D_{u}$ is the valence-band uniaxial deformation potential, Δ_{0} is the spin-orbit split-off energy, m_{0} is the free-electron mass, and γ_{1}, γ_{2}, γ_{3} are the Luttinger parameters. We assume that the MQW is grown on a thick Ga$_{0.85}$Al$_{0.15}$N buffer layer, so that a compressive strain exits only in the GaN well region. The in-plane strain $e_{xx}=(a-a_{0})/a_{0}$, where a is the unstrained lattice constant of Ga$_{0.85}$Al$_{0.15}$N and a_{0} the unstrained lattice constant of GaN. c_{11} and c_{12} are elastic stiffness constants.

The six dimensional hole envelope wave function for the MQW can be expanded as

$$
\phi_{n_{v},k}=\{\phi_{n_{v},k}^{j}\}, \quad (j=1,2,...6.), \qquad (3)
$$

where

$$
\phi_{n_{v},k}^{j}=\exp[i(k_{x}x+k_{y}y)]\sum_{m}a_{n_{v},k,m}^{j}\frac{1}{\sqrt{L}}\exp\left[i\left(k_{z}\right.\right.
$$

$$
\left.\left.+m\frac{2\pi}{L}\right)z\right], \qquad (4)
$$

and $L=l+d$ is the period of the MQW, where l and d are the widths of the wells and the barriers, respectively.

In order to distinguish the heavy hole (hh), light hole (lh), and spin-orbit band (so) components in the MQW wave function, we introduce the following probability functions:

$$
P^{hh}(n_{v},k)=\sum_{j=1,4}\sum_{m}a_{n_{v},k,m}^{j*}a_{n_{v},k,m}^{j}
$$

$$
P^{lh}(n_{v},k)=\sum_{j=2,3}\sum_{m}a_{n_{v},k,m}^{j*}a_{n_{v},k,m}^{j} \qquad (5)
$$

$$
P^{so}(n_{v},k)=\sum_{j=5,6}\sum_{m}a_{n_{v},k,m}^{j*}a_{n_{v},k,m}^{j}.
$$

From $P^{hh}(n_{v},k),P^{lh}(n_{v},k),P^{so}(n_{v},k)$, we can estimate the respective components of the heavy hole, light hole and spin-orbit split-off states in the MQW state $\phi_{n_{v},k}$. Note that the following sum rule holds,

$$
\sum_{i}P_{(n_{v},k)}^{i}=1, \quad i=\mathrm{hh,lh,so.} \qquad (6)
$$

We use Eqs. (1) and (2) to calculate the valence band structures of GaN/Ga$_{0.85}$Al$_{0.15}$N strained quantum well. The lattice constants of GaN and AlN are 4.50 and 4.35 Å, respectively.[4] The lattice constant of Ga$_{0.85}$Al$_{0.15}$N is 4.4775 Å obtained by linear interpolation. The corresponding compressive strain is 0.5% in the well region. The well width is 100 Å, the barrier width is 200 Å. We use the spin-orbit split-off Δ_{0} value of 17 meV[19] for the well and barrier. The experimental measured energy gap at the Γ point of zinc-bende GaN is 3.2 eV.[2] We take the calculated energy gap at the Γ point of zinc-blende AlN to be 5.94 eV,[24,25] and the band offset $\Delta E_{c}/\Delta E_{v}=70/30$,[26] so the $\Delta E_{c}=288$ meV and $\Delta E_{v}=123$ meV for GaN/Ga$_{0.85}$Al$_{0.15}$N quantum well using linear interpolation. Other parameters used in this work are listed in Table I. The conduction subbands of the quantum

TABLE I. Parameters used in the calculation.

Luttinger parameters γ_1, γ_2, γ_3 (Ref. 24)	3.07, 0.80, 1.26
Elastic stiffness c_{11}, c_{12} (Gpa) (Ref. 8)	296, 154
Deformation potential D_d^{cv}, $\frac{2}{3}D_u^v$ (eV) (Ref. 24)	−7.7, 1.6
Electron effective mass m_e for GaN (m_0) (Ref. 27)	0.15
Refractive index n (Ref. 9)	2.2
Intraband relaxation time τ (ps) (Ref. 28)	0.1
Lattice constants for GaN, AlN(Å) (Ref. 4)	4.50, 4.35
Spin-orbit split-off Δ_0 (meV) (Ref. 19)	17
Energy gaps E_g (Γ) for GaN (Ref. 2), AlN (eV) (Ref. 24)	3.2, 5.94
Band offset $\Delta E_c / \Delta E_v$ (Ref. 26)	70/30

well are calculated by the simple parabolic band approximation using electron effective mass in Table I.

Figure 1 shows the in-plane dispersion curves of the valence hole subbands of the GaN/Ga$_{0.85}$Al$_{0.15}$N compressive strained quantum well calculated by the 6×6 model. Figure 2 shows the corresponding wave functions of the energy levels at $k=0$ in Fig. 1. Figure 3 shows the variation of hh, lh, and so components in each subband state versus the in-plane wave vector. From Fig. 3, we can see that the hh does not couple with lh and so at $k=0$. Therefore, the states of $n=1$, 3, 5 at $k=0$ can be identified as hh1, hh2, and hh3 states, respectively. On the other hand, the lh state strongly couples with so state even at $k=0$.[29] According to Figs. 2 and 3, the states of $n=2$, 4, 6 at $k=0$ can be identified as **so1**+lh1, **so2**+lh2, and **so3**+lh3 states, respectively. At the $k\neq0$ points, there is band mixing among the heavy hole, light hole, and spin-orbit split-off states due to the contribution of the nondiagonal parts in Eq. (1). Figure 4 shows the in-plane dispersion curves of the valence hole subbands of GaN/Ga$_{0.85}$Al$_{0.15}$N quantum well where strain effect is neglected [$D_u=0$ in Eq. (2)]. Comparing Fig. 1 with Fig. 4, the hybrid state **lh1**+so1 (with **lh1** dominant) changes into the **so1**+lh1 state (with **so1** dominant) when the strain effect is taken into consideration.

We have compared that in the case of 200 Å barrier width, there is almost no interaction between the neighboring well and no energy dispersion along the z direction. Therefore, the result obtained can also be used for the case of single quantum well structure.

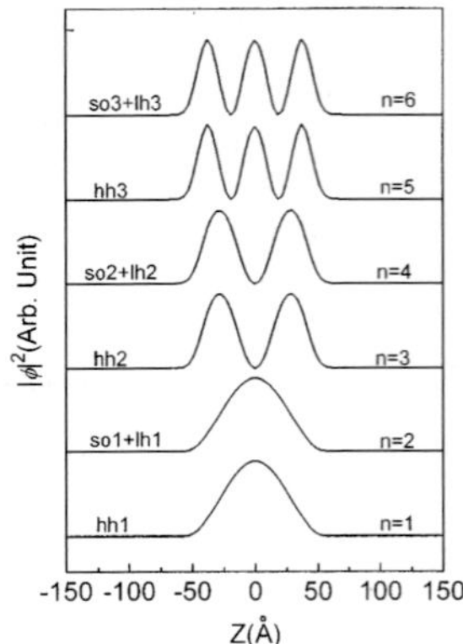

FIG. 2. The wave functions of the hole subband energy levels at $k=0$ of GaN/Ga$_{0.85}$Al$_{0.15}$N strained quantum well.

III. OPTICAL TRANSITION MATRIX ELEMENTS

The optical transition matrix elements for transitions between the valence hole subbands and the conduction electron subbands are given by:[30,31]

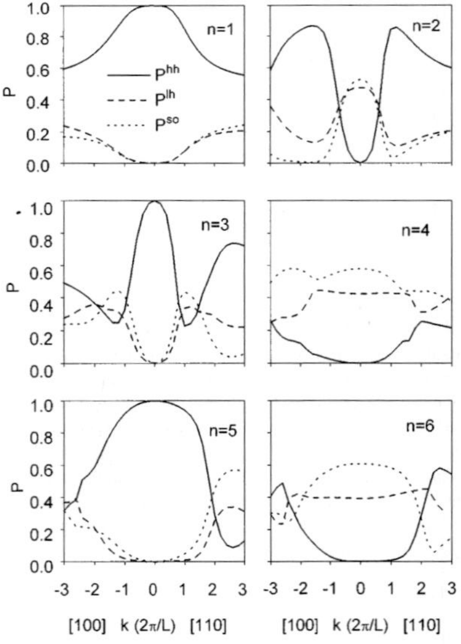

FIG. 3. The variation of the hh, lh, and so components in each state of GaN/Ga$_{0.85}$Al$_{0.15}$N strained quantum well vs the in-plane wave vector.

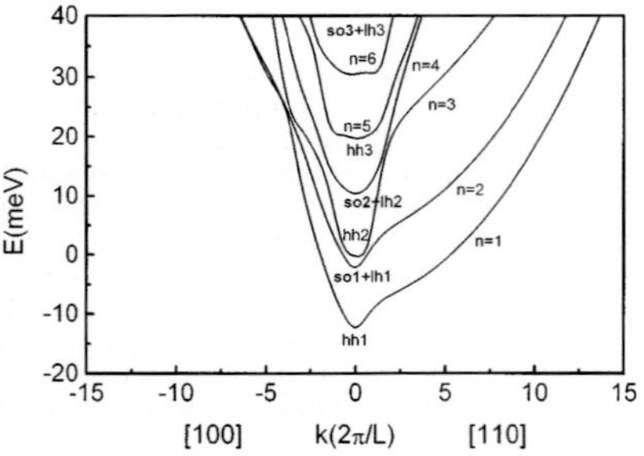

FIG. 1. The in-plane dispersion curves of the valence hole subbands of GaN/Ga$_{0.85}$Al$_{0.15}$N compressively strained quantum well.

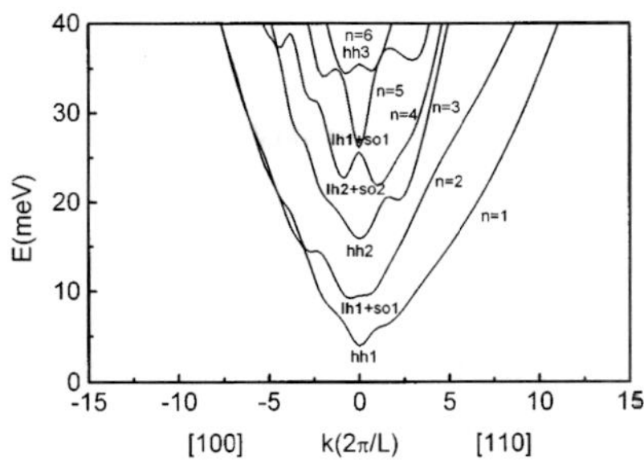

FIG. 4. The in-plane dispersion curves of the valence hole subbands of GaN/Ga$_{0.85}$Al$_{0.15}$N quantum well without strain.

$$M_i^{n_v n_c} = \langle \Psi_{n_v,k} | \hat{p}_i | \Phi_{n_c,k} \rangle, \quad i = x, y, z, \tag{7}$$

where, $\hat{p}_i$ is the momentum operator, $\Phi_{n_c,k}$ and $\Psi_{n_v,k}$ are the actual electron and hole wave functions, respectively. The spin-orbit-coupling wave functions of the valence band at the Γ point are[21]

$$\left| \frac{3}{2}, \frac{3}{2} \right\rangle = \frac{1}{\sqrt{2}} |(X+iY)\uparrow\rangle,$$

$$\left| \frac{3}{2}, \frac{1}{2} \right\rangle = \frac{i}{\sqrt{6}} |[(X+iY)\downarrow - 2Z\uparrow]\rangle,$$

$$\left| \frac{3}{2}, -\frac{1}{2} \right\rangle = \frac{1}{\sqrt{6}} |[(X-iY)\uparrow + 2Z\downarrow]\rangle, \tag{8}$$

$$\left| \frac{3}{2}, -\frac{3}{2} \right\rangle = \frac{i}{\sqrt{2}} |(X-iY)\downarrow\rangle,$$

$$\left| \frac{1}{2}, \frac{1}{2} \right\rangle = \frac{1}{\sqrt{3}} |[(X+iY)\downarrow + Z\uparrow]\rangle,$$

$$\left| \frac{1}{2}, -\frac{1}{2} \right\rangle = \frac{i}{\sqrt{3}} |[-(X-iY)\uparrow + Z\downarrow]\rangle,$$

where, $|X\rangle$, $|Y\rangle$, $|Z\rangle$, and $|s\rangle$ are the orbital wave functions of the top of the valence band and the bottom of the conduction band, respectively. $\uparrow$ and $\downarrow$ denote spin-up or spin-down components. Therefore, the actual hole wave function $\Psi_{n_v,k}$ is the product of the hole envelope wave function in Eq. (3) and the spin-orbit-coupling wave functions of the top of the valence band in Eq. (8). The actual wave function of the conduction electron subands $\Phi_{n_c,k}$ can be expanded as:

$$\Phi_{n_c,k} = \exp[i(k_x x + k_y y)] \sum_m d_{n_c,k,m} \frac{1}{\sqrt{L}}$$

$$\times \exp\left[i\left(k_z + m \frac{2\pi}{L}\right) z \right] |s\rangle. \tag{9}$$

Substituting Eqs. (3), (8), (9), into (7), we have

$$M_x^{n_v n_c} = \langle \Psi_{n_v,k} | \hat{p}_x | \Phi_{n_c,k} \uparrow \rangle + \langle \Psi_{n_v,k} | \hat{p}_x | \Phi_{n_c,k} \downarrow \rangle$$

$$= p_0 \sum_m \left(\frac{1}{\sqrt{2}} a_{n_v,k,m}^1 + \frac{i}{\sqrt{6}} a_{n_v,k,m}^2 \right.$$

$$+ \frac{1}{\sqrt{6}} a_{n_v,k,m}^3 + \frac{i}{\sqrt{2}} a_{n_v,k,m}^4$$

$$\left. + \frac{1}{\sqrt{3}} a_{n_v,k,m}^5 - \frac{i}{\sqrt{3}} a_{n_v,k,m}^6 \right) d_{n_c,k,m}^*, \tag{10}$$

$$M_y^{n_v n_c} = \langle \Psi_{n_v,k} | \hat{p}_y | \Phi_{n_c,k} \uparrow \rangle + \langle \Psi_{n_v,k} | \hat{p}_y | \Phi_{n_c,k} \downarrow \rangle$$

$$= p_0 \sum_m \left(\frac{i}{\sqrt{2}} a_{n_v,k,m}^1 - \frac{1}{\sqrt{6}} a_{n_v,k,m}^2 \right.$$

$$- \frac{i}{\sqrt{6}} a_{n_v,k,m}^3 + \frac{1}{\sqrt{2}} a_{n_v,k,m}^4$$

$$\left. + \frac{i}{\sqrt{3}} a_{n_v,k,m}^5 - \frac{1}{\sqrt{3}} a_{n_v,k,m}^6 \right) d_{n_c,k,m}^*, \tag{11}$$

$$M_z^{n_v n_c} = \langle \Psi_{n_v,k} | \hat{p}_z | \Phi_{n_c,k} \uparrow \rangle + \langle \Psi_{n_v,k} | \hat{p}_z | \Phi_{n_c,k} \downarrow \rangle$$

$$= p_0 \sum_m \left(-\frac{2i}{\sqrt{6}} a_{n_v,k,m}^2 + \frac{2}{\sqrt{6}} a_{n_v,k,m}^3 \right.$$

$$\left. + \frac{1}{\sqrt{3}} a_{n_v,k,m}^5 + \frac{i}{\sqrt{3}} a_{n_v,k,m}^6 \right) d_{n_c,k,m}^*, \tag{12}$$

where, $p_0 = \langle s | p_x | X \rangle = \langle s | p_y | Y \rangle = \langle s | p_z | Z \rangle$. p_0 is given by[32,33]

$$|p_0|^2 = \left(\frac{m_0}{m_e} - 1 \right) \frac{(E_g + \Delta_0)}{2(E_g + \frac{2}{3}\Delta_0)} m_0 E_g. \tag{13}$$

E_g is the unstrained energy gap of GaN. m_e is the electron effective mass of GaN.

For a quantum well laser with well layer growth direction along z, M_z corresponds to the TM mode optical transition and M_x (or M_y) corresponds to the TE mode optical transition.[34] Figures 5 and 6 show the squared optical transition matrix elements for TE and TM modes transitions from the four valence subband energy levels ($n=1$–4) to the first conduction subband energy level as a function of k along the [100] direction for the GaN/Ga$_{0.85}$Al$_{0.15}$N compressively strained quantum well. These results are obtained by numerical calculation of Eqs. (10)–(12). As indicated in Figs. 5 and 6 at the $k=0$ point, the optical transitions from the valence subband states to the conduction subband states obey the selection rule $\Delta n = 0$. The squared optical transition matrix elements for the TE mode are the sum of the major contribution from the electron to heavy hole transition and the minor contribution from the electron to hybrid state of light hole and spin-orbit split-off hole transition (Fig. 5). For the TM mode, the contribution comes from the electron to hybrid state of light hole and spin-orbit split-off hole transition only (Fig. 6). This is because M_x in Eq. (10) includes heavy

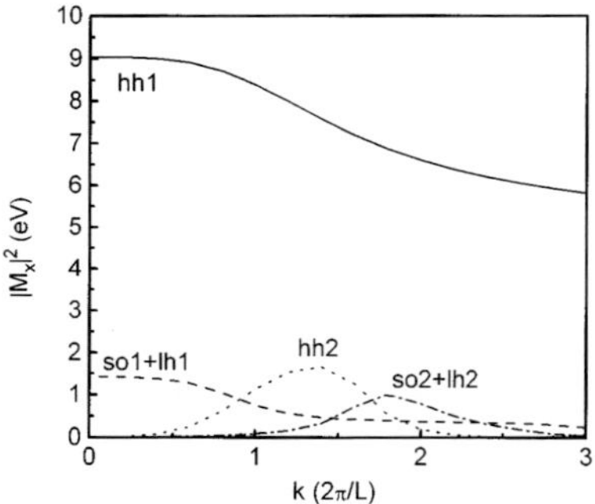

FIG. 5. The squared optical transition matrix elements for TE mode from the first conduction subband to the valence subbands as a function of k along the [100] direction for the GaN/Ga$_{0.85}$Al$_{0.15}$N compressively strained quantum well.

hole, light hole, and spin-orbit split-off hole wave functions. However, M_z in Eq. (12) does not include heavy hole wavefunction. At the $k \neq 0$ points, the $\Delta n = 0$ selection rule does not hold. There is band mixing between heavy hole, light hole, and spin-orbit split-off hole. Therefore, the squared optical transition matrix elements are the sum of the contribution from both electron to heavy hole and electron to hybrid state of light hole and spin-orbit split-off states.

IV. CARRIER DENSITY

The carrier density in a given band can be found for a given quasi-Fermi level by integrating the density of states

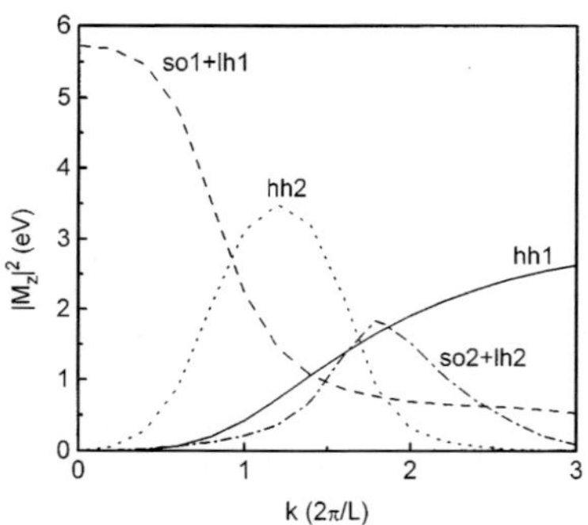

FIG. 6. The squared optical transition matrix elements for TM mode from the first conduction subband to the valence subbands as a function of k along the [100] direction for the GaN/Ga$_{0.85}$Al$_{0.15}$N compressively strained quantum well.

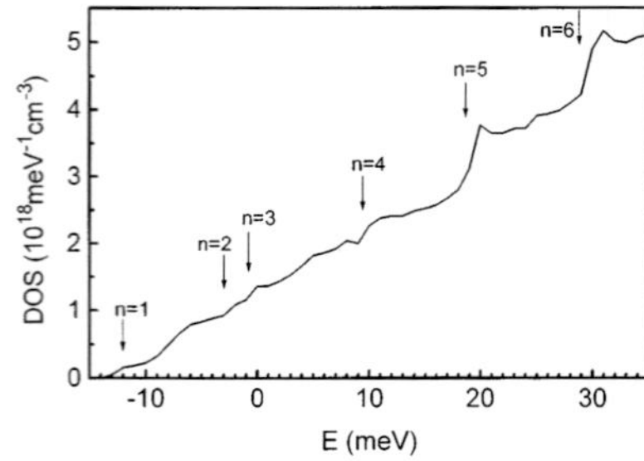

FIG. 7. The density of state for the valence hole subbands of the GaN/Ga$_{0.85}$Al$_{0.15}$N compressively strained quantum well.

multiplied by the occupation probability over the entire band. For the parabolic subband in the conduction band, the electron concentration can be written as[32]

$$N = \frac{m_e k_B T}{\pi \hbar^2 l} \sum_{n_c} \ln\{1 + \exp[-(E_{en_c} - E_{fc})/k_B T]\}, \quad (14)$$

where k_B is the Boltzmann's constant, T is the temperature, m_e is the electron effective mass, E_{fc} is the electron quasi-Fermi level. The sum is over all quantized subbands within the conduction band of the quantum well, and the E_{en_c} are the quantized energy levels. For the valence band, the subband structure is far from parabolic, as indicated in Fig. 1. Equation (14) is no longer valid. Thus, for this case, it is more appropriate to find the carrier density by numerically integrating over k space. We have

$$P = \sum_{n_v} \int \frac{1}{4\pi^2 l} f_v[E_{hn_v}(k_x,k_y)]dk_x dk_y, \quad (15)$$

where E_{hn_v} is the hole energy (not the electron energy) in the valence subband. The Fermi–Dirac distributions for electrons in the conduction bands f_c and for holes (not for electron) in the valence subbands f_v are defined as

$$f_c = \frac{1}{1 + \exp[(E_{en_c} - E_{fc})/k_B T]}, \quad (16)$$

$$f_v = \frac{1}{1 + \exp[(E_{hn_v} - E_{fv})/k_B T]}, \quad (17)$$

where E_{fv} are the hole quasi-Fermi level in the valence band.

Figure 7 is the density of states (DOS) of the valence subbands for a quantum well obtained by numerical calculation. In Fig. 7, the density of states of the valence subbands is quite different from a step function predicted by simple parabolic subbands in a quantum well. This result is expected since the valence subbands are far from parabolic case and are highly anisotropy, as shown in Fig. 1. To check the reliability of our numerical calculation program, we have used our program to calculate the GaAs/AlGaAs 50 Å quantum well density of states, and compared with the result of Szmulowicz et al.[35] The density of state curves obtained by

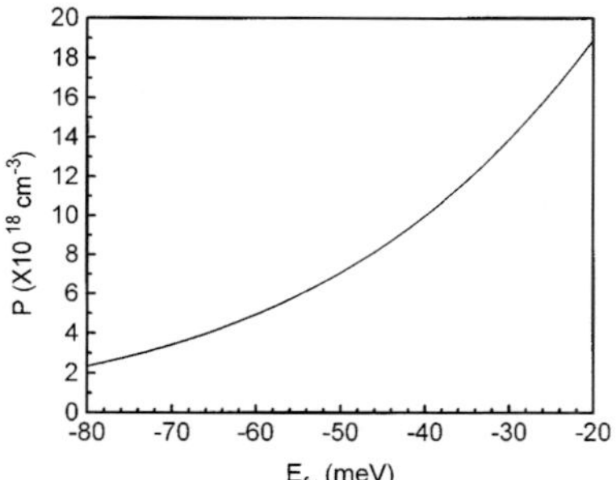

FIG. 8. The calculated hole concentration as a function of Fermi energy level for the GaN/Ga$_{0.85}$Al$_{0.15}$N compressively strained quantum well at $T=300$ K.

us and obtained by Szmulowicz *et al.* are almost identical. Figure 8 shows the calculated hole concentration as a function of Fermi energy level E_{fv} for the GaN/Ga$_{0.85}$Al$_{0.15}$N compressively strained quantum well at $T=300$ K.

V. OPTICAL GAIN AND RADIATIVE CURRENT DENSITY

According to the formula given in Refs. 32 and 36 and the definition of f_c, f_v in Eqs. (16) and (17), the optical gain can be written as

$$g(E)=\frac{\pi e^2 \hbar}{m_0^2 \varepsilon_0 n c E}\sum_{n_c,n_v}\int\int\frac{|M^{cv}|^2}{4\pi^2 l}[f_c+f_v-1]$$
$$\times\frac{1}{\pi}\frac{\hbar/\tau}{(E_{eh}-E)^2+(\hbar/\tau)^2}dk_x dk_y,\tag{18}$$

where E is the photon energy, ϵ_0 is the free-space dielectric constant, n is a refractive index, c is the light velocity, τ is the intraband relaxation time. The transition energy E_{eh} is given by

$$E_{eh}=E_{en_c}+E_{hn_v}+E_g^s,\tag{19}$$

where

$$E_g^s=E_g+2D_d^{cv}(1-c_{12}/c_{11})e_{xx},\tag{20}$$

D_d^{cv} is the interband hydrostatic deformation potential. $g(E)$ in Eq. (18) is also a function of carrier density. Using a specific electron density N, and charge neutrality condition $N=P$, from $N(E_{fc})$ and $P(E_{fv})$ curves obtained in Sec. IV, we obtain the values of D_{fc} and E_{fv}. Substituting Eqs. (16) and (17) into Eq. (18) and do the numerical calculation of Eq. (18), we obtain the value of $g(E)$.

The spontaneous emission rate can be given by[37]

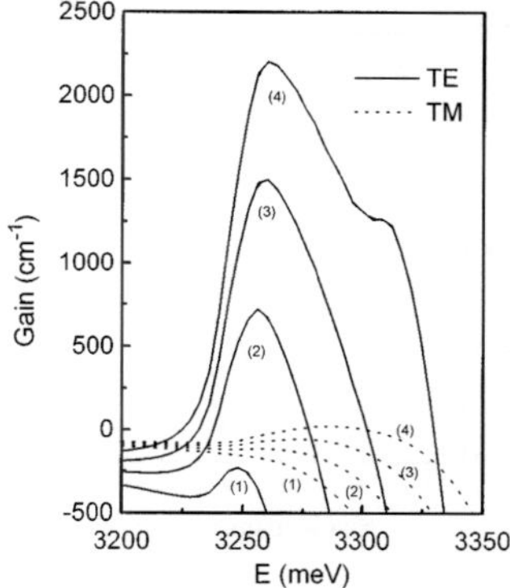

FIG. 9. Optical gain spectra at $T=300$ K for the GaN/Ga$_{0.85}$Al$_{0.15}$N compressively strained quantum well at carrier densities, (1) $N=6\times10^{18}$ cm^{-3}, (2) 8×10^{18} cm^{-3}, (3) 10×10^{18} cm^{-3}, (4) 12×10^{18} cm^{-3}.

$$R_{sp}(E)=\frac{ne^2 E}{\pi m_0^2 \varepsilon_0 \hbar^2 c^3}\sum_{n_c,n_v}\int\int\frac{|M^{cv}|^2}{4\pi^2 l}f_c f_v\frac{1}{\pi}$$
$$\times\frac{\hbar/\tau}{(E_{eh}-E)^2+(\hbar/\tau)^2}dk_x dk_y.\tag{21}$$

The radiative current density J_{rad} can be calculated from the spontaneous emission spectrum using[32]

$$J_{rad}=el\int R_{sp}(E)dE.\tag{22}$$

A. Polarization dependence of laser light output

Figures 9 and 10 show the optical gain spectra at $T=300$ K for GaN/Ga$_{0.85}$Al$_{0.15}$N quantum well with strain and without strain, respectively. Comparing Fig. 9 with Fig. 10, we

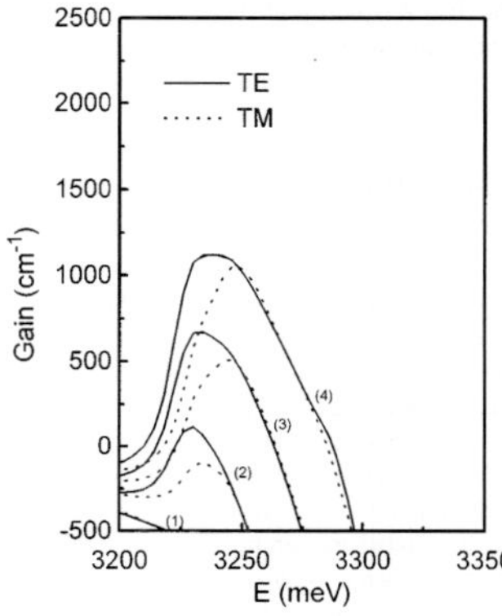

FIG. 10. Optical gain spectra at $T=300$ K for the GaN/Ga$_{0.85}$Al$_{0.15}$N quantum well without strain at carrier densities, (1) $N=6\times10^{18}$ cm^{-3}, (2) 8×10^{18} cm^{-3}, (3) 10×10^{18} cm^{-3}, (4) 12×10^{18} cm^{-3}.

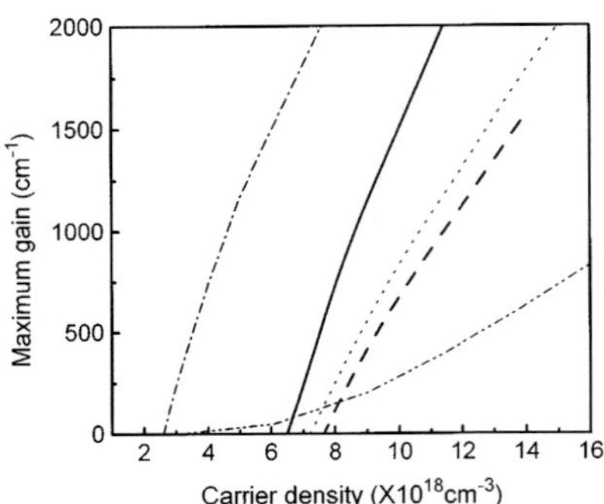

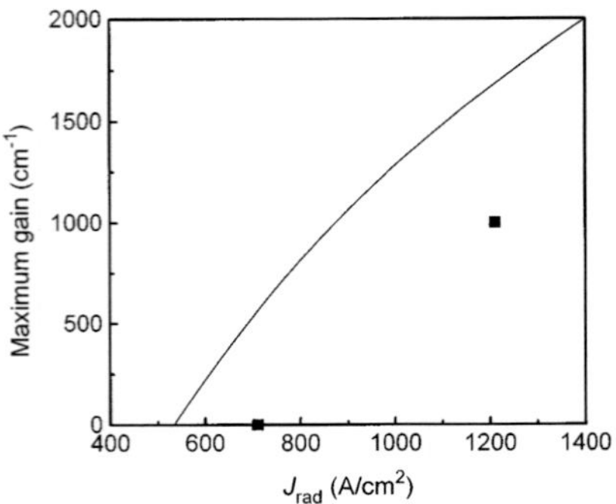

FIG. 11. Maximum optical gain of TE mode as a function of carrier density for the GaN/Ga$_{0.85}$Al$_{0.15}$N compressively strained quantum well at $T=300$ K (solid line) and for the quantum well without strain (dashed line). We also show for comparison the maximum gain for an unstrained 100 Å GaN quantum well between Al$_{0.26}$Ga$_{0.74}$N barriers (dot line) from Ref. 38, for a 100 Å GaAs/Al$_{0.4}$Ga$_{0.6}$As quantum well (dot-dashed line) from Ref. 38 and for bulk GaN material (dot-dot-dashed line) from Ref. 39.

FIG. 12. Maximum optical gain of TE mode as a function of radiative current density for the GaN/Ga$_{0.85}$Al$_{0.15}$N quantum well at $T=300$ K.

find that the compressive strain in the GaN well region strongly depresses the TM mode optical gain and enhances the TE mode optical gain. This can be understood as follows: comparing Fig. 1 with Fig. 4 shows that under compressive strain the energy separation between the **so1**+lh1 energy level and hh1 energy level increases. Therefore, the ratio of the number of **so1**+lh1 holes to the number of hh1 holes decreases. Consequently, the TM mode optical gain is depressed, because the TM mode optical gain is mainly contributed by the optical transitions between the **so1**+lh1 holes and the conduction subbands, as indicated in Fig. 6. Figure 9 shows that there is no positive TM optical mode gain even when the carrier density is as large as 10^{19} cm^{-3}. Consequently, the 100 Å well width GaN/Ga$_{0.85}$Al$_{0.15}$N compressively strained quantum well laser diode is expected to only have a TE mode laser light output, if the injection carrier is not far beyond the above carrier density.

B. Maximum optical gain as a function of carrier density and J_{rad}

Figure 11 shows the maximum optical gain of TE mode as a function of carrier density for the GaN/Ga$_{0.85}$Al$_{0.15}$N quantum well laser diode. Figure 12 shows the maximum optical gain of TE mode as a function of J_{rad} for the same laser diode. Meney et al. have also briefly reported in a letter[38] the optical gain calculation of 100 Å well width GaN quantum well with Ga$_{0.74}$Al$_{0.26}$N barrier, using an 8×8 valence Hamiltonian and some assumed effective mass parameters. Their calculated maximum TE optical gain as a function of carrier density is also indicated in Fig. 11. The consistency between our work and Meney et al.'s work is reasonably good if we note that Meney et al. used a slightly different barrier material and they have not considered the strain effect in their curve. For comparison, we also show in Fig. 11 the maximum gain for a 100 Å GaAs/Al$_{0.4}$Ga$_{0.6}$As quantum well.[38] As explained by Meney et al., due to the larger effective mass of electron and hole in GaN, the transparency carrier density of GaN is larger than that of GaAs. In Fig. 12, two squares are also obtained from Meney et al.'s calculation for comparison. Fang et al. have calculated the optical gain of bulk GaN material.[39] Their calculation used simplified two-band model, neglecting the complication of the valence band structure and only heavy hole band is taken into account. Therefore, their result is qualitatively significant but quantitatively unreliable. In Fig. 11, Fang et al.'s result is also plotted for comparison.

The following points are noted in our work: (1) Owing to the lack of experimental data for GaN, we use the intraband relaxation time $\tau=0.1$ ps in Eq. (18), as usually utilized in GaAs materials.[28] However, due to the large structural defect density which may introduce a large number of scattering centers, the intraband relaxation time is expected to be reduced and this in turn reduces the quantum efficiency. (2) The large structural defect density also introduces a high density of deep trap levels in the energy gap. Many of these deep traps are nonradiative deep defects, which cause additional leakage current due to the nonradiative recombination between the electrons and holes. This has not been considered in our letter. (3) Chow et al. indicated that the excitonic effect due to the Coulomb interaction between electrons and holes contributes significantly to the magnitude of the optical gain.[40] We have not considered the excitonic effect in our calculation. However, their calculation is only for the bulk materials, and for simple parabolic heavy hole valence band. A more elaborate theoretical method which takes into consideration the coupling between heavy hole, light hole, and spin-orbit split-off hole in the quantum well, the strain effect and Coulomb interaction effect between the electrons and holes will be necessary to obtain more reliable quantitative results for the optical gain of an ideal GaN quantum well laser. On the other hand, improving growth techniques to reduce the defect density is essential for fabricating high quality GaN laser diodes.

VI. CONCLUSION

The valence hole subbands, TE and TM mode optical gains, transparency carrier density and radiative current density of the zinc-blende $GaN/Ga_{0.85}Al_{0.15}N$ strained quantum well (100 Å well width) have been investigated using a 6 $\times$6 Hamiltonian model including the heavy hole, light hole, and spin-orbit split-off bands. At the $k=0$ point, it is found that the light hole strongly couples with the spin-orbit split-off hole, resulting in the so+lh hybrid states. The heavy hole does not couple with the light hole and the spin-orbit split-off hole. The optical transitions between the valence subbands and the conduction subbands obey the $\Delta n=0$ selection rule. At the $k \neq 0$ points, there are strong band mixing among the heavy hole, light hole, and spin-orbit split-off hole. The optical transitions do not obey the $\Delta n=0$ selection rule. The compressive strain in the GaN well region increases the energy separation between the so1+lh1 energy level and the hh1 energy level. Consequently, the compressive strain enhances the TE mode optical gain, and strongly depresses the TM mode optical gain. There is no positive TM mode optical gain even when the carrier density is as large as 10^{19} cm^{-3}. The TE mode optical gain spectrum has a peak at around 3.26 eV. The transparency carrier density is 6.5×10^{18} cm^{-3}, which is larger than that of GaAs quantum well. The compressive strain overall reduces the transparency carrier density. The J_{rad} is 0.53 kA/cm^2 for the zero optical gain. The results obtained in this work will be useful in designing quantum well GaN laser diodes and detectors.

ACKNOWLEDGMENTS

We would like to acknowledge the Computer Centre, National University of Singapore, and National Supercomputing Research Center, National University of Singapore for the use of supercomputers. This work was supported by National University of Singapore Research Grant No. RP 920621 and Singapore National Science and Technology Board RIC-University Research Project 681305.

[1] S. Nakamura, M. Senoh, S. Nagahma, N. Iwasa, T. Yamada, T. Matsushita, H. Kiyoku, and Y. Sugimoto, Jpn. J. Appl. Phys. **35**, 174 (1996).

[2] T. Lei, T. D. Moustakas, R. J. Graham, Y. He, and S. J. Berkowitz, J. Appl. Phys. **71**, 4933 (1992).

[3] S. J. Hwang, W. Shan, R. J. Hauenstein, and J. J. Song, Appl. Phys. Lett. **64**, 2928 (1994).

[4] A. Rubio, J. L. Corkill, M. L. Cohen, E. L. Shirley, and S. G. Louie, Phys. Rev. B **48**, 11810 (1993).

[5] E. A. Albanesi, W. R. L. Lambrecht, and B. Segall, Phys. Rev. B **48**, 17841 (1993).

[6] A. F. Wright and J. S. Nelson, Phys. Rev. B **50**, 2159 (1994).

[7] S. Strite and H. Morkoç, J. Vac. Sci. Technol. B **10**, 1237 (1992).

[8] K. Kim, W. R. L. Lambrecht, and B. Segall, Phys. Rev. B **50**, 1502 (1994).

[9] N. E. Christensen and I. Gorczyca, Phys. Rev. B **50**, 4397 (1994).

[10] *Proceedings of the 7th Trieste ICTP-IUAP Semiconductor Symposium*, edited by C. G. van de Walle [Physica B **185**, R9 (1993)].

[11] T. Lei, M. Fanciulli, R. J. Molnar, T. D. Moustakas, R. J. Graham, and J. Scanlon, Appl. Phys. Lett. **95**, 944 (1991).

[12] Z. Sitar, M. J. Paisley, J. Rnan, J. W. Choyke, and R. F. Davis, J. Mater. Sci. Lett. **11**, 261 (1992).

[13] P. E. Van Camp, V. E. Van Doren, and J. T. Deverse, Phys. Rev. B **44**, 9056 (1991).

[14] A. Munoz and K. Kunc, Phys. Rev. B **44**, 10372 (1991).

[15] M. A. Khan, R. A. Skogman, and J. M. Van Hove, Appl. Phys. Lett. **56**, 1257 (1990).

[16] S.-H. Ke, M.-C. Huang, and R.-Z. Wang, Solid State Commun. **89**, 105 (1994).

[17] S. Krishnankutty, R. M. Kolbas, M. A. Khan, J. N. Kuznia, J. M. Van Hove, and D. T. Olson, J. Electron. Mater. **21**, 437 (1992), *ibid.* **21**, 609 (1992).

[18] H. Crille and F. Bechstedt, Superlattices Microstruct. **16**, 29 (1994).

[19] G. Ramirez-Flores, H. Navarro-Contreras, A. Lastras-Martinez, R. C. Powell, and J. E. Greene, Phys. Rev. B **50**, 8433 (1994).

[20] D. A. Broido and L. J. Sham, Phys. Rev. B **31**, 883 (1995); *ibid.* **34**, 3917 (1985).

[21] R. People and S. K. Sputz, Phys. Rev. B **41**, 8431 (1990).

[22] A. T. Meney, B. Gonul, and E. P. O'Reilly, Phys. Rev. B **50**, 10893 (1994).

[23] J. M. Luttinger, Phys. Rev. **102**, 1030 (1956).

[24] W. J. Fan, M. F. Li, T. C. Chong, and J. B. Xia, J. Appl. Phys. **79**, 188 (1996).

[25] W. J. Fan, M. F. Li, T. C. Chong, and J. B. Xia, Solid State Commun. **97**, 381 (1996).

[26] E. A. Albanesi, W. R. L. Lambrecht, and B. Segall, J. Vac. Sci. Technol. B **12**, 2470 (1994).

[27] M. Fanciulli, T. Lei, and T. D. Moustakas, Phys. Rev. B **48**, 15144 (1993).

[28] M. Yamada and Y. Suematsu, J. Appl. Phys. **52**, 2653 (1981).

[29] W. J. Fan, M. F. Li, T. C. Chong, and J. B. Xia, Superlattices Microstruct. (to be published).

[30] J. N. Schulmann and Y. C. Chang, Phys. Rev. B **24**, 4445 (1981).

[31] J. B. Xia, Phys. Rev. B **38**, 8358 (1988).

[32] S. W. Corzine, R.-H. Yan, and L. A. Coldren, in *Quantum Well Lasers*, edited by P. S. Zory, Jr. (Academic, New York, 1993), p. 49.

[33] E. O. Kane, J. Phys. Chem. Solids **1**, 249 (1957).

[34] T. C. Chong and C. G. Fonstad, IEEE J. Quantum Electron. **25**, 171 (1989).

[35] F. Szmulowicz and G. J. Brown, Phys. Rev. B **51**, 13203 (1995).

[36] M. Asada and Y. Suematsu, IEEE J. Quantum Electron. **21**, 434 (1985).

[37] G. Lascher and F. Stern, Phys. Rev. **133**, A553 (1964).

[38] A. T. Meney and E. P. O'Reilly, Appl. Phys. Lett. **67**, 3013 (1995).

[39] W. Fang and S. L. Chuang, Appl. Phys. Lett. **67**, 751 (1995).

[40] W. W. Chow, A. Knorr, and S. W. Koch, Appl. Phys. Lett. **67**, 754 (1995).

Reprinted with permission from Y.C. Yeo, T.C. Chong, M.F. Li, J. Appl. Phys., Vol.83, pp.1429–1436 (1998). Copyright 1998, American Institute of Physics.

Electronic band structures and effective-mass parameters of wurtzite GaN and InN

Y. C. Yeo, T. C. Chong, and M. F. Li
Centre for Optoelectronics, Department of Electrical Engineering, National University of Singapore, 10 Kent Ridge Crescent, S119260, Singapore

(Received 16 July 1997; accepted for publication 22 October 1997)

The electronic band structures of wurtzite GaN and InN are calculated by the empirical pseudopotential method (EPM) with the form factors adjusted to reproduce band features which agree with recent experimental data and accurate first-principles calculations. The electron and hole effective masses at the Γ point are obtained using a parabolic line fit. Further, using the effective-mass Hamiltonian and the cubic approximation for wurtzite semiconductors, band edge dispersion at the Γ point obtained using the k.p method is fitted to that calculated using the EPM by adjusting the effective-mass parameters. Thus, we derived important band structure parameters such as the Luttinger-like parameters for GaN and InN which will be useful for material design in wide-gap nitride-based semiconductor lasers employing InGaN. The results also showed that the cubic approximation is fairly successful in the analysis of valence band structures for wurtzite nitrides. © *1998 American Institute of Physics.* [S0021-8979(98)05503-0]

I. INTRODUCTION

The group-III wurtzite nitride semiconductors, AlN, GaN, InN, and their alloys, have direct energy band gaps corresponding to a wide spectrum of wavelengths from the red to the ultraviolet (UV). Optoelectronic devices active in this entire range can potentially be fabricated using these materials. In particular, the blue and UV regions bear great technological importance. Blue GaN-based lasers and light-emitting diodes (LEDs) are highly demanded as short-wavelength laser sources in high-density optical data storage systems and as a source of the third primary color in display applications. Over last few years, intensive research activities have led to the commercial production of high-brightness blue/green LEDs,[1] and the demonstration of room-temperature (RT) blue/violet laser emission in the InGaN/GaN/AlGaN-based heterostructures under pulsed currents[2–4] and continuous-wave (CW) operation.[5,6] Recently, RT CW operation of the InGaN multiple quantum well (MQW) LEDs achieved a lifetime of 35 h.[6] This is expected to improve further through the growth of better quality crystals and the reduction of operating current densities.

Fundamental studies in an attempt to attain superior laser characteristics such as lower threshold current density and higher differential quantum efficiency rely on the knowledge of electronic band structures. Therefore, theoretical calculation of the band structure of the group-III nitrides has been a subject of intensive study. For GaN, first-principles band structure calculations were performed by many investigators.[7–19] In Ref. 7, the quasiparticle correction using the Green's function with screened Coulomb interaction (GW) approach was employed to correct the band gap underestimation caused by the local-density approximation (LDA). In Ref. 8, the overestimation of the band gap due to the use of a minimal basis set offsets the LDA error to a certain extent. Reference 9 used the addition of external potentials sharply peaked at the atomic sites to correct for the

small band gap. All other self-consistent calculations also give reliable valence band structures but the band gaps are significantly underestimated. For InN, self-consistent calculations were also performed.[16–21] Empirical and semiempirical calculations done more than a decade ago for both GaN[22–24] and InN[24–26] obtained accurate band gaps but did not harness the advantage of the empirical adjustment due to a lack of experimental data and the availability of reliable valence band features from first-principles calculations.

The purpose of this article is to clarify the electronic properties of wurtzite GaN and InN using the empirical pseudopotential method (EPM) and derive important band structure parameters describing the heavy-hole (HH), the light-hole (LH), and the crystal-field split-hole (CH) bands in the vicinity of the Brillouin zone (BZ) center. In the work of Suzuki *et al.*,[13] the effective-mass or Luttinger-like parameters were extracted for GaN and AlN by fitting band edge dispersion curves with those calculated from a self-consistent full-potential linearized-augmented plane-wave (FP-LAPW) method within the LDA. To date, no such parameters are available for InN. Material parameters for InN will provide useful reference in the context of material design for the InGaN MQW laser diodes. In this work, we apply the EPM[27–30] and adjust the form factors of GaN and InN to reproduce band features which agree with recent experimental data and the valence bands of accurate first-principles calculations in literature. The effective masses for electrons and holes at the Γ point are obtained by a parabolic line fit. In addition, using the effective-mass Hamiltonian and cubic approximation for wurtzite semiconductors,[31,32] we derive the effective-mass parameters, $A_i s$, for GaN and InN by fitting the band edge dispersion obtained by the k.p method to that calculated by the EPM. These parameters will be useful for the investigation of InGaN quantum well (QW) structures. The organization of this article is as follows. The calculation of electronic band structures using the EPM is described in Sec. II. In Sec. III, we show the derivation of the effective-mass parameters. Section IV summarizes the results.

II. ELECTRONIC BAND STRUCTURE

A. The empirical pseudopotential method

The use of EPM[27–30] on wurtzite GaN and InN will be described. We improve on the results of EPM calculations more than ten years ago[22–25] by tuning the pseudopotential form factors to obtain valence band structures which agree with self-consistent computations and experimental data. In addition, like all other calculations based on the EPM,[22–25] we accurately reproduce the fundamental band gap to give reliable conduction band and estimates of energy gaps at other symmetry points. The matrix element of the crystal pseudopotential V^{PS} in the spin-free pseudopotential Hamiltonian appears as[29]

$$\langle PW,\mathbf{G}_i|\hat{V}^{PS}|PW,\mathbf{G}_j\rangle = V^S(\mathbf{G}_i-\mathbf{G}_j)\cdot S^S(\mathbf{G}_i-\mathbf{G}_j)$$
$$+ iV^A(\mathbf{G}_i-\mathbf{G}_j)\cdot S^A(\mathbf{G}_i-\mathbf{G}_j), \tag{1}$$

where $|PW,\mathbf{G}\rangle$ is a plane-wave state indexed by the reciprocal lattice vector $\mathbf{G}$, $V^S(\mathbf{G}_i-\mathbf{G}_j)$ and $V^A(\mathbf{G}_i-\mathbf{G}_j)$ are the symmetric and asymmetric form factors, respectively, and $S^S(\mathbf{G}_i-\mathbf{G}_j)$ and $S^A(\mathbf{G}_i-\mathbf{G}_j)$ are, respectively, the symmetric and asymmetric structure factors specific to the wurtzite crystal.[30] The basis set of plane waves was chosen with the criterion that $|\mathbf{G}+\mathbf{k}|$ lie within a sphere defined by a kinetic energy E_1 equal to 13.5 Ry. Second-order contribution from all vectors $\mathbf{G}$, such that $E_1<|\mathbf{G}+\mathbf{k}|<E_2$, where $E_2=20$ Ry, are added to the matrix according to a perturbation method due to Löwdin.[33] Convergence tests have been performed by increasing E_1 and E_2 with no noticeable change in the eigenenergies. The lattice parameters used are $a=3.189$ Å, $c=5.185$ Å for GaN, and $a=3.54$ Å, $c=5.70$ Å for InN.[34] The internal structural parameter u describing the relative displacement of the Group III and Group V sublattices in the c direction follows from the results of Wei and Zunger[16] using the FP-LAPW method within the LDA. The values of u for GaN and InN are, respectively, 0.3768 and 0.3790 which are in excellent agreement with 0.3768 and 0.3792 by Majewski *et al.*,[17] and with 0.3770 and 0.3784 by Wright and Nelson,[35] both calculated using self-consistent pseudopotentials based on the LDA. Spin-orbit interaction is considered with the addition of the following spin-orbit matrix element:[36–38]

$$\langle PW,\mathbf{K}_i,s|\hat{H}_{so}|PW,\mathbf{K}_j,s'\rangle$$
$$= (\mathbf{K}_i\times\mathbf{K}_j)\cdot\sigma_{s,s'}[-i\lambda^S\cdot S^S(\mathbf{G}_i-\mathbf{G}_j)$$
$$+\lambda^A\cdot S^A(\mathbf{G}_i-\mathbf{G}_j)], \tag{2}$$

where $\mathbf{K}_i=\mathbf{G}_i+\mathbf{k}$, $\mathbf{K}_j=\mathbf{G}_j+\mathbf{k}$, and $\sigma_{s,s'}=\pm1$ is the usual Pauli spin index denoting either spin up or down. The symmetric and asymmetric contributions to the spin-orbit Hamiltonian, λ^S and λ^A, respectively, are related to the cationic and anionic spin-orbit contributions, λ_c and λ_a, respectively:

$$\lambda^S=(\lambda_c+\lambda_a)/2, \quad \lambda^A=(\lambda_c-\lambda_a)/2, \tag{3}$$

TABLE I. Wurtzite reciprocal lattice vectors, structure factors, and adjusted form factors (in Rydbergs) for GaN and InN.

$|G|^2$ (in units of $2\pi/a$)	$|S^S(G)|$	$|S^A(G)|$	GaN $V^S(G)$	GaN $V^A(G)$	InN $V^S(G)$	InN $V^A(G)$
0	1	0	0	0	0	0
3/8	0	0	0	0	0	0
1 1/3	0.50	0	−0.348	0	−0.324	0
1 1/2	0.71	0.71	−0.371	0.276	−0.255	0.209
1 17/24	0.33	0.80	−0.280	0.225	−0.217	0.174
2 5/6	0.35	0.35	−0.053	0.110	−0.040	0.119
3 3/8	0	0	0	0	0	0
4	1	0	0.059	0	0.044	0
4 3/8	0	0	0	0	0	0
4 17/24	0.80	0.33	0.083	0.051	0.056	0.032
5 1/3	0.50	0	0.056	0	0.045	0
5 1/2	0.71	0.71	0.066	0.036	0.045	0.029
5 17/24	0.33	0.80	0.067	0.033	0.036	0.0205
6	0	1	0	0.0065	0	0.0125
6 5/6	0.35	0.35	0.021	0.005	−0.006	0.016
7 1/3	0	0.50	0	0	0	−0.060

$$\lambda_c=\mu B_{nl}^c(K_i)B_{nl}^c(K_j), \quad \lambda_a=\alpha\mu B_{nl}^a(K_i)B_{nl}^a(K_j). \tag{4}$$

In Eq. (4), α is the ratio of spin splitting of the free anion atom to that of the free cation atom, μ is an adjustable spin-orbit parameter, and $B_{nl}(K)$ is an orthogonalization integral given by

$$B_{nl}(K)=\beta\int_0^\infty j_{nl}(Kr)R_{nl}(r)r^2dr, \tag{5}$$

where $j_{nl}(Kr)$ is the spherical Bessel's function of the l th angular momentum, $R_{nl}(r)$ is the radial part of the core wavefunction which can be found from tabulated Hartree–Fock–Slater orbitals,[39] and β is a normalization constant as defined in Ref. 40. The parameter μ can be varied to give the correct spin splitting of the valence band at the Γ point.

B. Results of EPM calculations

Initial calculations were performed using the spin-free pseudopotential Hamiltonian by adjusting the pseudopotential form factors to obtain desirable features at high symmetry points. These form factors (Table I) are then used to

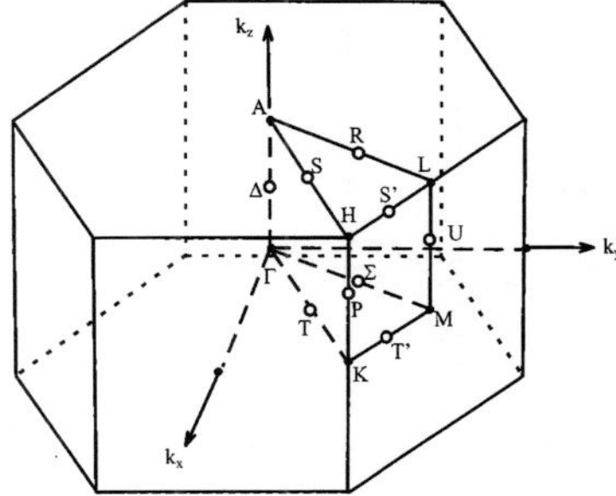

FIG. 1. The first Brillouin zone for the wurtzite-type crystals.

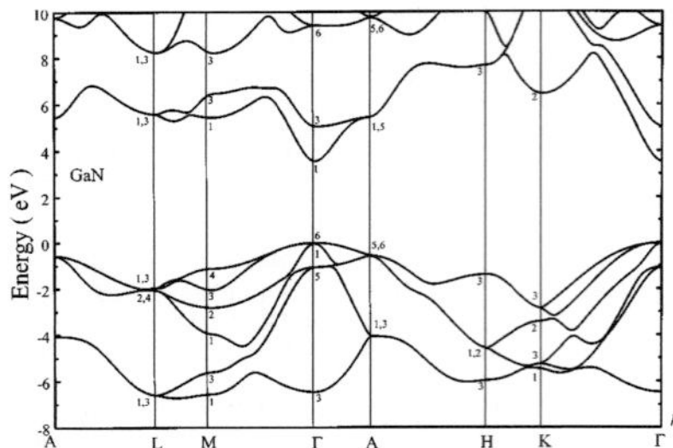

FIG. 2. The electronic band structure of wurtzite GaN along the high symmetry lines of the first Brillouin zone. Spin-orbit interaction is neglected.

determine the electronic energy bands on a fine grid of points along the high symmetry lines of the BZ depicted in Fig. 1. In Figs. 2 and 3, we plot the calculated electronic band structures for GaN and InN, respectively, without the inclusion of the spin-orbit interaction.

The direct band gap of GaN is adjusted to 3.50 eV, in agreement with experimental data[22,41,42] and elaborate self-consistent calculations with corrections[7–9] (Table II). The width of the top valence band (arising from N $2p$ orbitals) is calculated to be 6.8 eV, near the experimental value of 7.0 eV as determined by Hunt et $al.$[43] The crystal-field splitting Δ_{cr} of 21 meV is consistent with 22 meV as determined by Dingle et $al.$[44] and the interpretation[45] of recent experimental data.[46] Fine details of the band structure in the vicinity of the Γ point valence band maximum (VBM) are investigated with the inclusion of the spin-orbit interaction. In Fig. 4(a), we show the energy dispersion for GaN at the Γ point with and without the spin-orbit interaction. The spin-obit interaction and the noncubic crystal-field split the VBM of wurtzite semiconductors into three states, Γ_9^{v}, $\Gamma_7^{v(1)}$, and $\Gamma_7^{v(2)}$, which would otherwise be a doubly degenerate Γ_6^{v} state and a singly degenerate Γ_1^{v} state in the absence of spin-orbit interaction. The energies of the top three states, $E(\Gamma_9^{v})$, $E(\Gamma_7^{v(1)})$, and $E(\Gamma_7^{v(2)})$, are related to the spin-orbit splitting Δ_{so} and crystal-field splitting Δ_{cr} by[31]

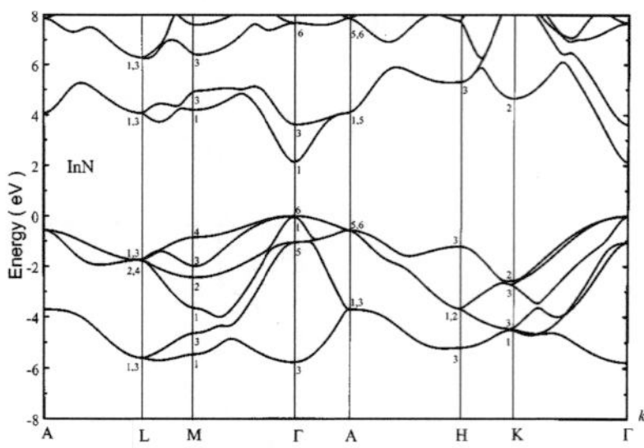

FIG. 3. The electronic band structure of wurtzite InN along the high symmetry lines of the first Brillouin zone. Spin-orbit interaction is neglected.

$$E(\Gamma_9^{v}) = 0,$$

$$E(\Gamma_7^{v(1)}) = -\frac{1}{2}\,(\Delta_{\mathrm{so}}+\Delta_{\mathrm{cr}})$$

$$+\frac{1}{2}\,\sqrt{(\Delta_{\mathrm{so}}+\Delta_{\mathrm{cr}})^2 - \frac{8}{3}\,\Delta_{\mathrm{so}}\Delta_{\mathrm{cr}}},$$

$$E(\Gamma_7^{v(1)}) = -\frac{1}{2}\,(\Delta_{\mathrm{so}}+\Delta_{\mathrm{cr}})$$

$$-\frac{1}{2}\,\sqrt{(\Delta_{\mathrm{so}}+\Delta_{\mathrm{cr}})^2 - \frac{8}{3}\,\Delta_{\mathrm{so}}\Delta_{\mathrm{cr}}}, \tag{6}$$

where reference is taken with respect to the VBM. The parameter μ is adjusted to give a Δ_{so} of 11 meV in agreement with Ref. 44. This corresponds to $E(\Gamma_9^{v})-E(\Gamma_7^{v(1)})$ = 5.84 meV (6 meV[42,44,47,48]), and $E(\Gamma_9^{v})-E(\Gamma_7^{v(2)})$ = 26.1 meV (22 meV,[47] 28 meV,[42] 18 meV,[44] 24 meV[48]) according to Eq. (6) and is in good agreement with the experimental data given in parentheses. In Table II, we also show critical point transition energies in comparison with those of other calculations and experimental data. An indisputable assignment of experimental features to specific electronic transition is difficult since the BZ of the wurtzite lattice involves more high symmetry points than in the zinc-blende lattice. In the spectroscopic ellipsometry studies of Logothetidis et $al.$,[49] three peaks at 7.0, 7.9, and 9.0 eV in the dielectric-function spectrum of wurtzite GaN may be associated, respectively, with primary transitions at the L and M points, with the $H_3^{v}-H_3^{c}$ and $M_2^{v}-M_1^{c}$ excitations, and with the onset of transitions at the K point.[49] This deduction is based on the converging results of Huang and Ching,[8] Gorczyca and Christensen,[9] and Xu and Ching.[18] Our results compare reasonably well with the assigned electronic transitions and with these calculations. The final part of Table II illustrates the agreement between eigenenergies in this work and those of other calculations.

For InN, the energy band gap is adjusted to 2.04 eV which agrees with the experimental results of Refs. 50–52 (Table III). The valence band dispersion resembles those of self-consistent calculations in Refs. 18–21. Only the eigenenergies of Ref. 19 are available for comparison in Table III. Our valence band width of 5.8 eV compares commendably with 5.73 eV calculated by the linear muffin-tin orbital (LMTO)-LDA method,[19] 5.98 eV by the orthogonalized linear combination of atomic orbitals (OLCAO)-LDA method,[18] and 6 eV by a semiempirical tight-binding (TB) method.[53] The pseudofunction method obtained a valence band width of 6.8 eV,[21] and a TB method gave an underestimated value of 3.5 eV.[26] Earlier EPM calculations[25] overestimated the valence band width to be 9.6 eV and this is expected to lead to lighter-than-expected hole effective masses. For Δ_{cr}, we obtained a value of 17 meV which is smaller than 21 meV for GaN. This is qualitatively similar to the results of Majewski et $al.$[17] which reported 35.3 meV and 27.8 meV for GaN and InN, respectively, and to the results of Wei and Zunger[16] which reported 42 and 41 meV for GaN and InN, respectively. The detailed energy dispersions at the Γ point are shown in Fig. 4(b) for InN with and without the

TABLE II. Comparison of eigenenergies of GaN with experimental and theoretical results.

	Experiment/eV	Other calculations/eV	This work/eV
Γ point features			
$\Gamma_6^v - \Gamma_1^c$ (E_g)	3.44,[a] 3.50,[b] 3.60[c]	2.3(3.5),[d] 3.50,[e] 2.59(3.44),[f] 2.89,[g] 2.71,[h] 3.0,[i] 2.76[j]	3.50
$\Gamma_3^v - \Gamma_6^v$ (top valence band width)	7.0[k]	7.4(8.0),[d] 6.78,[e] 7.18,[j] 7.21[l]	6.8
$\Gamma_1^v - \Gamma_6^v$ (Δ_{cr})	0.022[m]	0.072,[l] 0.042,[n] 0.0353[o]	0.021
$\Gamma_6^v - \Gamma_3^c$		4.6(5.9),[d] 4.86,[e] 4.76[j]	4.97
Critical-point transition energies			
$L_{2,4}^v - L_{1,3}^c$		6.4(8.2),[d] 7.29,[c] 7.05[h]	7.57
$M_4^v - M_1^c$	7.0[p]	6.1(7.6),[d] 7.05,[e] 6.7[h]	6.61
$M_4^v - M_3^c$		6.7(8.5),[d] 7.3,[e] 6.9(7.4),[f] 7.25[h]	7.69
$H_3^v - H_3^c$		8.1(9.9),[d] 8.4,[e] 8.25(8.9)[f]	9.0
$M_2^v - M_1^c$	7.9[p]	7.9(9.6),[d] 8.5,[e] 7.85(8.7)[f]	8.26
$K_3^v - K_2^c$		8.1(10.1),[d] 9.45,[e] 8.41(9.0)[f]	9.43
$K_2^v - K_2^c$	9.0[p]	7.9(9.8),[d] 9.05,[e] 9.82[q]	10.1
Eigenenergies w.r.t. Γ_{6v}			
K_1^v		$-5.6(-6.1)$,[d] -5.37,[j] -5.47[q]	-5.43
K_3^v		$-5.5(-6.1)$,[d] -5.29,[j] -5.24[q]	-5.25
K_2^v		$-3.0(-3.2)$,[d] -2.80,[j] -2.85[q]	-3.44
K_3^v		$-3.2(-3.5)$,[d] -3.02,[j] -2.96[q]	-2.77
K_2^c		4.9(6.6),[d] 4.93,[j] 5.74(6.97),[q] 6.35[r]	6.70
M_1^v		$-6.8(-7.4)$,[d] -6.56,[j] -6.49[q]	-6.52
M_3^v		$-5.6(-6.1)$,[d] -5.35,[j] -5.32[q]	-5.59
M_1^v		$-4.4(-4.9)$,[d] -4.25,[j] -4.18[q]	-3.88
M_2^v		$-2.8(-3.2)$,[d] -2.57,[j] -2.75[q]	-2.79
M_3^v		$-2.4(-2.6)$,[d] -2.29,[j] -2.14[q]	-2.01
M_4^v		$-1.0(-1.1)$,[d] -0.88,[j] -0.93[q]	-1.12
M_1^c		5.1(6.5),[d] 5.02,[j] 5.22(6.45),[q] 5.94[r]	5.49
$A_{1,3}^v$		$-4.1(-4.6)$,[d] -4.06,[j] -3.86[e]	-3.97
$A_{5,6}^v$		$-0.5(-0.6)$,[d] -0.49,[j] -0.52[e]	-0.55
$A_{1,3}^c$		4.6(6.1),[d] 5.00,[j] 5.58,[e] 5.31[r]	5.31
H_3^v		$-6.4(-7.1)$,[d] -6.22,[j] -5.98[e]	-5.93
H1,2v		$-4.6(-4.9)$,[d] -4.35,[j] -3.85[e]	-4.54
H_3^v		$-1.5(-1.6)$,[d] -1.35,[j] -1.36[e]	-1.40
H_3^c		6.6(8.3),[d] 6.62,[j] 7.47[r]	7.89
$L_{1,3}^v$		$-7.0(-7.6)$,[d] -6.75,[j] -6.34[e]	-6.57
$L_{2,4}^v$		$-2.0(-2.2)$,[d] -1.84,[j] -1.76[e]	-2.06
$L_{1,3}^v$		$-2.0(-2.2)$,[d] -1.81,[j] -1.71[e]	-1.95
$L_{1,3}^c$		4.4(6.0),[d] 4.54,[j] 5.53,[e] 5.49[r]	5.66

[a]Room temperature measurement, Perlin et al., Ref. 41.
[b]Photoluminescence excitation spectra, Monemar, Ref. 42.
[c]Reflectivity, Bloom et al., Ref. 22.
[d]Ab initio pseudopotential-LDA, GW corrected values given in parentheses, Rubio et al., Ref. 7.
[e]Minimal basis semi-ab initio OLCAO, Huang and Ching, Ref. 8.
[f]LMTO-LDA, with corrected value in parentheses, Gorczyca and Christensen, Ref. 9.
[g]Pseudopotential-LDA, Miwa and Fukumoto, Ref. 10.
[h]LCAO-LDA, Xu and Ching, Ref. 18.
[i]Norm-conserving pseudopotential, Min, Chan and Ho, Ref. 11.
[j]Norm-conserving pseudopotential, Palummo et al., Ref. 14.
[k]Synchrotron radiation photoemission spectroscopy, Hunt et al., Ref. 43.
[l]Full-potential LAPW-LDA, Suzuki, Uenoyama and Yanase, Ref. 13.
[m]Dingle et al., Ref. 44.
[n]LAPW-LDA, Wei and Zunger, Ref. 16.
[o]Total-energy pseudopotential-LDA, Majewski, Städele, and Vogl, Ref. 17.
[p]Ellipsometry, Logothetidis et al., Ref. 49.
[q]LMTO-LDA, with corrected value in parentheses, Lambrecht and Segall, Ref. 12.
[r]LMTO-LDA with adjusted gap, Christensen and Gorczyza, Ref. 19.

spin-orbit interaction. The Δ_{so} is adjusted to 3 meV to match the result of Wei and Zunger.[16] The smaller Δ_{so} of InN in comparison with that of GaN deviates from the usual trend in which Δ_{so} increases with the atomic number of the cation, i.e., $\Delta_{so}(InX) > \Delta_{so}(GaX) > \Delta_{so}(AlX)$ for $X = $ P, As, and Sb where the increase of the predominant p bonding is reflected as the cation becomes heavier. In the nitrides, the N $2p$ orbital is so deep in energy that it effects a substantial hy-

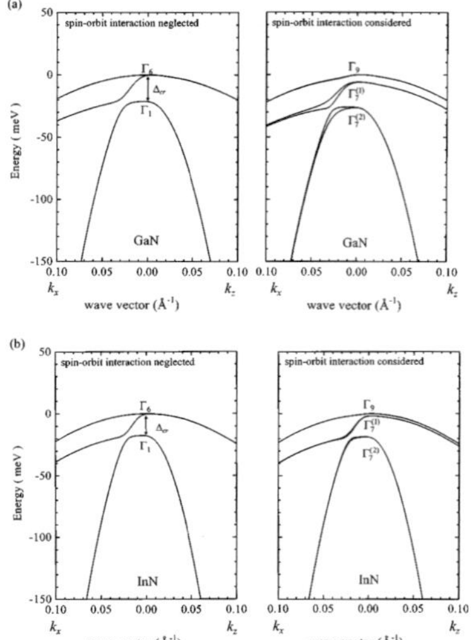

FIG. 4. The Γ point energy band dispersion with and without the spin-orbit interaction for (a) GaN and (b) InN.

TABLE III. Comparison of eigenenergies of InN with experimental and theoretical results.

	Experiment/eV	Other calculations/eV	This work/eV
Γ point features			
$\Gamma_6^v - \Gamma_1^c$ (E_g)	1.89,[a] 2.05,[b] 2.11[c]	2.05,[d] 1.02,[e] 2,[f] 2.04[g]	2.04
$\Gamma_3^v - \Gamma_6^v$ (valence band width)		5.73,[d] 5.98,[e] 9.6,[f] 6.8,[h] 3.5,[i] 6[j]	5.77
$\Gamma_1^v - \Gamma_6^v$ (Δ_{cr})		0.041,[k] 0.0278[l]	0.017
$\Gamma_5^v - \Gamma_6^v$		0.80[d]	1.05
Critical-point transition energies			
$\Gamma_5^v - \Gamma_3^c$	4.7[f]	4[f]	4.65
$\Gamma_5^v - \Gamma_6^c$	8.9[f]	8.4[f]	8.74
$M_2^v - M_1^c$	4.95[f]	5.1[f]	6.65
$M_4^v - M_1^c$			5.05
$M_4^v - M_3^c$			5.80
$U_4^v - U_3^c$		5.9[f]	5.3
$H_3^v - H_3^c$	5.4[f]	5.4[f]	6.51
$K_3^v - K_2^c$	7.2[f]	7.5[f]	7.38
$K_2^v - K_2^c$		8.1[f]	7.20
$L_{2,4}^v - L_{1,3}^c$			5.83
Eigenenergies w.r.t. Γ_{6v}			
K_1^v		-4.45[d]	-4.50
K_3^v		-4.42[d]	-4.48
K_3^v		-2.35[d]	-2.75
K_2^v		-2.18[d]	-2.57
K_2^c		$5.41(7.16)$[d]	4.64
M_1^v		-5.38[d]	-5.46
M_3^v		-4.44[d]	-4.64
M_1^v		-3.39[d]	-3.67
M_2^v		-2.31[d]	-2.45
M_3^v		-1.66[d]	-2.01
M_4^v		-0.76[d]	-0.85
M_1^c		$4.18(5.93)$[d]	4.20
$A_{1,3}^v$			-3.71
$A_{5,6}^v$			-0.57
$A_{1,3}^c$			4.06
H_3^v			-5.21
$H_{1,2}^v$			-3.70
H_3^v			-1.22
H_3^c			5.29
$L_{1,3}^v$			-6.57
$L_{2,4}^v$			-2.06
$L_{1,3}^v$			-1.95
$L_{1,3}^c$			4.06

[a]Room temperature optical absorption, Tansley and Foley, Ref. 50.
[b]Absorption edge at 300 K, Tyagai *et al.*, Ref. 51.
[c]Absorption edge at 78 K, Osamura *et al.*, Ref. 52.
[d]LMTO-LDA, band gap corrected values given in parentheses, Lambrecht and Segall, Ref. 20.
[e]LCAO-LDA, Xu and Ching, Ref. 18.
[f]EPM calculation and experimental reflectivity, Foley and Tansley, Ref. 25.
[g]Model pseudopotential, Grinyaev *et al.*, Ref. 24.
[h]Pseudofunction-LDA, Tsai *et al.*, Ref. 21.
[i]Empirical tight-binding method, Jenkins, Hong and Dow, Ref. 26.
[j]Semiempirical tight-binding method, Yang, Nakajima, and Sakai, Ref. 53.
[k]LAPW-LDA, Wei and Zunger, Ref. 16.
[l]Total-energy pseudopotential-LDA, Majewski, Städele, and Vogl, Ref. 17.

bridization with the cation d orbitals leading to a significant mixing of d character at the VBM.[16] The d hybridization has a negative contribution to Δ_{so} and is more effective in the heavier In than in the lighter Ga. This causes the reversal of the trend in Δ_{so} of nitrides in contrast with the phosphides, arsenides, and antimonides as shown in the LDA-based self-consistent calculations of Refs. 16 and 17. While this trend is noteworthy, we must be aware that the LDA also overestimates the d-p hybridization. This leads to an underestimation of Δ_{so} in the nitrides which is more significant in InN where the hybridization is more pronounced. Therefore, we believe the actual Δ_{so} of InN should be larger than 3 meV, but smaller than 11 meV (that of GaN). At present, experimental data on Δ_{so} of InN is not available for verification. Next, we present the critical-point transition energies of InN. The experimental reflectivity and EPM results of Foley and Tansley[25] are used for comparison. The experimental peak at 4.7 eV was assigned to the $\Gamma_5^v - \Gamma_3^c$ transition and this agrees with our value of 4.65 eV. Experimental peak at 4.95 eV could be due to the $M_4^v - M_1^c$ transition. We relate features at 5.4 eV to transitions involving $U_4^v - U_3^c$, $M_4^v - M_3^c$, and $L_{2,4}^v - L_{1,3}^c$. For higher energy transitions, our calculations agree with those of Ref. 26 in which the 7.2 eV peak is associated with transitions at K, and that the 8.9 eV peak is probably due to the $\Gamma_5^v - \Gamma_6^c$ transition. Further experimental work is needed to achieve an unambiguous assignment of energies to critical-point transitions.

III. DERIVATION OF EFFECTIVE MASS PARAMETERS

A. Effective-mass approximation

The use of the effective-mass Hamiltonian and the cubic approximation for wurtzite semiconductors will be described here. We adopt the effective-mass Hamiltonian derived by

Chuang and Chang[12] using the k.p method. The full Hamiltonian for an unstrained bulk wurtzite semiconductor is described by[32]

$$
H=\begin{bmatrix}
F & -K^* & -H^* & 0 & 0 & 0 \\
-K & G & H & 0 & 0 & \Delta \\
-H & -H^* & \lambda & 0 & \Delta & 0 \\
0 & 0 & 0 & F & -K & H \\
0 & 0 & \Delta & -K^* & G & -H^* \\
0 & \Delta & 0 & H^* & -H & \lambda
\end{bmatrix}, \quad (7)
$$

where

$$
F=\Delta_1+\Delta_2+\lambda+\theta, \quad G=\Delta_1-\Delta_2+\lambda+\theta,
$$

$$
\Delta=\sqrt{2}\Delta_3,
$$

$$
\lambda=\frac{\hbar^2}{2m_0}[A_1 k_z^2+A_2(k_x^2+k_y^2)], \quad (8)
$$

$$
\theta=\frac{\hbar^2}{2m_0}[A_3 k_z^2+A_4(k_x^2+k_y^2)],
$$

$$
K=\frac{\hbar^2}{2m_0}A_5(k_x+ik_y)^2, \quad H=\frac{\hbar^2}{2m_0}A_6(k_x+ik_y)k_z.
$$

In Eq. (8), Δ_1 is the crystal-field split energy, Δ_2 and Δ_3 account for the spin-orbit interaction, k_i is the wave vector, and $A_i s$ are the effective-mass parameters. For $\Delta_1>\Delta_2>0$, as in the case of GaN and InN, the three bands from top to bottom are labeled as HH, LH, and CH, respectively. At the Γ point, these correspond to the Γ_9^v, $\Gamma_7^{v(1)}$, and $\Gamma_7^{v(2)}$ states, respectively. Under the cubic approximation[13,31,32] which exploits the similarity between the c (0001) axis in the wurtzite crystal and the (111) direction in the cubic crystal, we have

$$
A_1-A_2=-A_3=2A_4, \quad A_1+4A_5=\sqrt{2}A_6, \quad \Delta_2=\Delta_3. \quad (9)
$$

According to Eq. (9), we have three independent $A_i s$ and two independent $\Delta_i s$. Thus, five independent parameters have to be determined to reproduce the EPM calculated band structure at the VBM. In addition, we note the relationship between $A_i s$ and the hole masses in the k_z direction ($m^{\parallel}$) and in the $k_x k_y$ plane ($m^{\perp}$). The superscripts $\parallel$ and $\perp$ specifically denote the **k**-directional dependence parallel and perpendicular to the k_z direction (c axis), respectively. In the k_z direction ($k_x=k_y=0$),

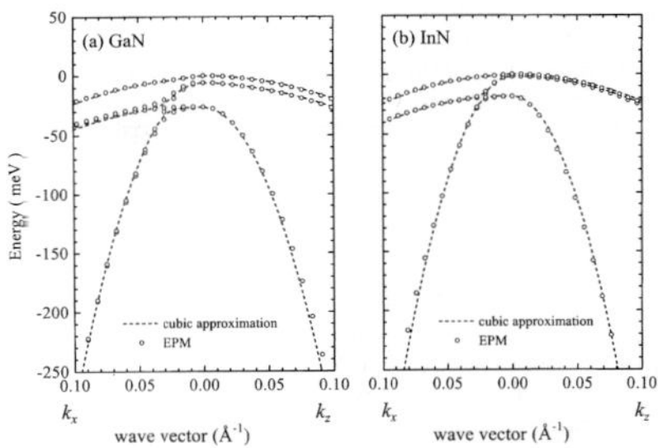

FIG. 5. The fitted band structures for (a) GaN and (b) InN at the Γ point using the effective-mass theory within the cubic approximation (dashed lines) with the spin-orbit interaction taken into account. The results of EPM calculation are plotted with open circles.

$$
m_{HH}^{\parallel}/m_0=-(A_1+A_3)^{-1},
$$

$$
m_{LH}^{\parallel}/m_0=-(A_1+A_3)^{-1}, \quad (10)
$$

$$
m_{CH}^{\parallel}/m_0=-A_1^{-1},
$$

and in the $k_x k_y$ plane ($k_z=0$),

$$
m_{HH}^{\perp}/m_0=-(A_2+A_4-A_5)^{-1},
$$

$$
m_{LH}^{\perp}/m_0=-(A_2+A_4+A_5)^{-1}, \quad (11)
$$

$$
m_{CH}^{\perp}/m_0=-A_2^{-1}.
$$

The energy dispersion $E(\mathbf{k})$ in the vicinity of the Γ point can be obtained by diagonalizing

$$
\det|H(\mathbf{k})-E(\mathbf{k})\mathbf{I}|=0, \quad (12)
$$

where $\mathbf{I}$ is a 6×6 unit matrix.

B. Results of numerical fitting

A parabolic line fit to the conduction band dispersion was used to determine the effective masses of the electrons. The lowermost conduction band shows slight anisotropy for **k** directions. For GaN, $m_c^{\parallel}=0.19\,m_0$ in the k_z direction, and $m_c^{\perp}=0.17\,m_0$ in the $k_x k_y$ plane. For InN, $m_c^{\parallel}=0.11\,m_0$ in the k_z direction, and $m_c^{\perp}=0.10\,m_0$ in the $k_x k_y$ plane.

For the valence bands, we derive the effective-mass parameters using two approaches. In the first approach, a parabolic line fit to each of the HH, LH, and CH bands is used to obtain their effective-masses in the k_z and in-plane ($k_x k_y$) directions at the Γ point. Using these effective masses, the parameters $A_i s$ for $i=1-5$ are then obtained from Eqs. (10) and (11). In the second approach, the Hamiltonian given in Eq. (7) within the cubic approximation is used in the k.p method to obtain a band dispersion which fit the EPM results in a three-dimensional (3D) mesh of **k** points near the Γ point. The $A_i s$ are treated as adjustable parameters and the values of $\Delta_i s$ are fixed according to $\Delta_1=\Delta_{cr}$ and $\Delta_2=\Delta_3=\Delta_{so}/3$. In the fitting process, we adopted a least-square fit by minimizing the total squared error between the fitted eigenenergies of Eq. (12) and the results of EPM calculation in which the total error is summed over the mesh-sampled **k**

TABLE IV. Valence band effective-mass parameters of GaN and InN obtained using a simple parabolic line fit and using a three-dimensional fit.

	GaN effective-mass parameters		InN effective-mass parameters	
	Line fit	3D fit	Line fit	3D fit
A_1	−7.14	−7.24	−9.62	−9.28
A_2	−0.57	−0.51	−0.72	−0.60
A_3	6.57	6.73	8.97	8.68
A_4	−3.30	−3.36	−4.22	−4.34
A_5	−3.28	−3.35	−4.35	−4.32
A_6	—	−4.72	—	−6.08

TABLE V. The hole effective masses for the HH, LH, and CH bands of GaN and InN obtained using a simple parabolic line fit and using a 3D fitting procedure. The superscripts $\parallel$ and $\perp$ denote the **k**-directional dependence in the k_z and the in-plane $(k_x - k_y)$ directions, respectively.

	GaN effective masses (m_0)		InN effective masses (m_0)	
	Line fit	3D fit	Line fit	3D fit
$m_{HH}^{\parallel}$	1.76	1.96	1.56	1.67
$m_{LH}^{\parallel}$	1.76	1.96	1.56	1.67
$m_{CH}^{\parallel}$	0.14	0.14	0.10	0.10
$m_{HH}^{\perp}$	1.69	1.87	1.68	1.61
$m_{LH}^{\perp}$	0.14	0.14	0.11	0.11
$m_{CH}^{\perp}$	1.76	1.96	1.39	1.67

points. This fitting scheme is termed the 3D fitting and is similar to the procedure done by Suzuki *et al.* in Ref. 13. In Fig. 5, we plot the fitted band structures for GaN and InN using dashed lines against the corresponding EPM results which are shown using open circles. The 3D-fitted dispersion reproduces the EPM results at the Γ point very well. In Table IV, the effective-mass parameters used in the 3D fitting and those derived using a parabolic line fit are summarized. For both GaN and InN, corresponding parameters obtained using the line fit and the 3D fit agree with each other. In addition, parameters of InN are generally larger than those of GaN due to the lighter hole masses of InN.

Next, we compare the hole effective masses obtained using the two approaches. The extraction of effective masses using the parabolic line fit was previously described. For the 3D fit, we can obtain the hole effective masses from the adjusted $A_i s$ by a substitution of the parameters into Eqs. (10) and (11). Table V summarizes the hole effective masses for the HH, LH, and CH bands of GaN and InN obtained using a simple parabolic line fit and using a 3D fit. From the results of Tables IV and V, and Fig. 5, it is evident that the 3D fitting of the effective-mass parameters employing the cubic approximation is rather successful in reproducing accurate valence band dispersions in the vicinity of the Γ point.

IV. CONCLUSION

In conclusion, we have presented a detailed study of the electronic band structures of wurtzite GaN and InN. The valence band structures from the EPM calculation are attested by a comparison with experimental data and first-principles calculations. We showed critical-point transition energies for GaN and InN which could be useful for future experimental studies of these materials. The effective-mass parameters are derived using two approaches: first, using a parabolic line fit; and second, using a 3D fitting of the band dispersion obtained by the k.p method to that of the EPM results. Both approaches yield effective-mass parameters that are in reasonable agreement with each other. We also obtained the effective masses of electrons and holes at the Γ point in the k_z and in the in-plane $(k_x k_y$ plane) directions. The effective-mass Hamiltonian together with the cubic approximation is shown to reproduce accurate band dispersions in the vicinity

of the Γ point. We also extracted valence band parameters for GaN and InN which could provide useful reference for the study of InGaN MQW structures.

ACKNOWLEDGMENTS

We appreciate fruitful discussions with Dr. S.-H. Wei of the National Renewable Energy Laboratory, Colorado, U.S., and Dr. J. A. Majewski of the Walter Schottky Institut, Technische Universität München, Germany, on the spin-orbit splitting in the nitrides. Gratitude is expressed to Dr. M. Suzuki and Dr. S. Kamiyama of the Central Research Laboratories and the Semiconductor Research Center, respectively, Matsushita Electric Industrial Co. Ltd., Japan, for their helpful information. We thank the support of the Singapore National Science and Technology Board RIC-university research grant for project 681305, and the computing facilities from the NUS Computer Center.

[1] S. Nakamura, M. Senoh, N. Iwasa, S. Nagahama, T. Yamada, and T. Mukai, Jpn. J. Appl. Phys., Part 2 **34**, L1332 (1995).

[2] I. Akasaki, H. Amano, S. Sota, H. Sakai, T. Tanaka, and M. Kaike, Jpn. J. Appl. Phys., Part 2 **34**, L1517 (1995).

[3] S. Nakamura, M. Senoh, S. Nagahama, N. Iwasa, T. Yamada, T. Matsushita, H. Kiyoku, and Y. Sugimoto, Jpn. J. Appl. Phys., Part 2 **35**, L74 (1996).

[4] S. Nakamura, M. Senoh, S. Nagahama, N. Iwasa, T. Yamada, T. Matsushita, H. Kiyoku, and Y. Sugimoto, Appl. Phys. Lett. **68**, 3269 (1996).

[5] S. Nakamura, M. Senoh, S. Nagahama, N. Iwasa, T. Yamada, T. Matsushita, Y. Sugimoto, and H. Kiyoku, Appl. Phys. Lett. **69**, 3034 (1996).

[6] S. Nakamura, Materials Research Society Internet J. of Nitride Semicond. Research **2**, 5 (1997).

[7] A. Rubio, J. L. Corkill, M. L. Cohen, E. L. Shirley, and S. G. Louie, Phys. Rev. B **48**, 11 810 (1993).

[8] M. Z. Huang and W. Y. Ching, J. Phys. Chem. Solids **46**, 977 (1985).

[9] I. Gorczya and N. E. Christensen, Solid State Commun. **80**, 335 (1991).

[10] K. Miwa and A. Fukumoto, Phys. Rev. B **48**, 7897 (1993).

[11] B. J. Min, C. T. Chan, and K. M. Ho, Phys. Rev. B **45**, 1159 (1992).

[12] W. R. L. Lambrecht and B. Segall, *Properties of Group III Nitrides*, edited by J. E. Edgar (INSPEC, IEE, London, 1994), p. 141.

[13] M. Suzuki, T. Uenoyama, and A. Yanase, Phys. Rev. B **52**, 8132 (1995); M. Suzuki and T. Uenoyama, Jpn. J. Appl. Phys., Part 1 **34**, 3442 (1995).

[14] M. Palummo, C. M. Bertoni, L. Reining, and F. Finocchi, Physica B **185**, 404 (1993).

[15] G. D. Chen, M. Smith, J. Y. Lin, H. X. Jiang, S. H. Wei, M. A. Khan, and C. J. Sun, Appl. Phys. Lett. **68**, 2784 (1996).

[16] S. H. Wei and A. Zunger, Appl. Phys. Lett. **69**, 2719 (1996); and private communication.

[17] J. A. Majewski, M. Städele, and P. Vogl, Materials Research Society Internet J. Nitride Semicond. Research **1**, 30 (1996).

[18] Y. N. Xu and W. Y. Ching, Phys. Rev. B **48**, 4335 (1993).

[19] N. E. Christensen and I. Gorczyca, Phys. Rev. B **50**, 4397 (1994).

[20] W. R. L. Lambrecht and B. Segall, *Properties of Group III Nitrides*, edited by J. E. Edgar (INSPEC, IEE, London, 1994), p. 151.

[21] M. H. Tsai, D. W. Jenkins, and J. D. Dow, Phys. Rev. B **38**, 1541 (1988).

[22] S. Bloom, G. Harbeke, E. Meier, and I. B. Ortenburger, Phys. Status Solidi B **66**, 161 (1974).

[23] S. Bloom, J. Phys. Chem. Solids **32**, 2027 (1971).

[24] S. N. Grinyaev, V. Ya Malakhov, and V. A. Chaldyshev, Sov. Phys. J. **29**, 311 (1986).

[25] C. P. Foley and T. L. Tansley, Phys. Rev. B **33**, 1430 (1986).

[26] D. W. Jenkins, J.-D. Hong, and J. D. Dow, Superlattices Microstruct. **3**, 365 (1987).

[27] J. C. Phillips and L. Kleinman, Phys. Rev. **116**, 287 (1959).

[28] M. L. Cohen and T. K. Bergstresser, Phys. Rev. **141**, 789 (1966).

[29] M. L. Cohen and H. Heine, *Solid State Physics: Advances in Research and Applications*, Vol. 24, edited by H. Ehrenreich, F. Seitz, and D. Turnbull (Academic, New York & London, 1970).

[30] M. L. Cohen and J. R. Chelikowsky, *Electronic Structure and Optical Properties of Semiconductors* (Springer, New York, 1989).

[31] G. L. Bir and G. E. Pikus, *Symmetry and Strain-Induced Effects in Semiconductor* (Wiley, New York, 1972).

[32] S. L. Chuang and C. S. Chang, Phys. Rev. B **54**, 2491 (1996).

[33] P. Löwdin, J. Chem. Phys. **19**, 1396 (1951).

[34] R. F. Davis, Proc. IEEE **79**, 702 (1991).

[35] A. F. Wright and J. S. Nelson, Phys. Rev. B **50**, 2159 (1994); Phys. Rev. B **51**, 7866 (1995).

[36] G. Weisz, Phys. Rev. **149**, 504 (1966).

[37] S. Bloom and T. K. Bergstresser, Solid State Commun. **6**, 465 (1970).

[38] J. R. Chelikowsky and M. L. Cohen, Phys. Rev. B **14**, 556 (1976).

[39] F. Herman and S. Skillman, *Atomic Structure Calculations* (Prentice Hall, Englewood Cliffs, NJ, 1966).

[40] J. P. Walter, M. L. Cohen, Y. Petroff, and M. Balkanski, Phys. Rev. B **1**, 2661 (1970).

[41] P. Perlin, I. Gorczyca, S. Porowski, T. Suski, N. E. Christensen, and A. Polian, Jpn. J. Appl. Phys., Part 1 **32**, 334 (1993).

[42] B. Monemar, Phys. Rev. B **10**, 676 (1974).

[43] R. W. Hunt, L. Vanzetti, T. Castro, K. M. Chen, L. Sorba, P. I. Cohen, W. Gladfelter, J. M. Van Hove, J. N. Kuznia, M. A. Khan, and A. Franciosi, Physica B **185**, 415 (1993).

[44] R. Dingle, D. D. Sell, S. E. Stokowski, and M. Ilegems, Phys. Rev. B **4**, 1211 (1971).

[45] S. L. Chuang and C. S. Chang, Phys. Rev. B **54**, 2491 (1996).

[46] B. Gil, O. Briot, and R. L. Aulombard, Phys. Rev. B **52**, R17 028 (1995).

[47] D. Volm, K. Oettinger, T. Streibl, D. Kovalev, M. Ben-Chorin, J. Diener, B. K. Meyer, J. Majewski, L. Eckey, A. Hoffman, H. Amano, I. Akasaki, K. Hiramatsu, and T. Detchprohm, Phys. Rev. B **53**, 16 543 (1996).

[48] K. Pakula, A. Wysmolek, K. P. Korona, J. M. Baranowski, R. Stepniewski, I. Grzegory, M. Bockowski, J. Jun, S. Krukowski, M. Wroblewski, and S. Porowski, Solid State Commun. **97**, 919 (1996).

[49] S. Logothetidis, J. Petalas, M. Cardona, and T. D. Moustakas, Phys. Rev. B **50**, 18 017 (1994).

[50] T. L. Tansley and C. P. Foley, J. Appl. Phys. **59**, 3241 (1986).

[51] V. A. Tyagai, A. M. Evstigneev, A. N. Krasiko, A. F. Adreeva, and V. Ya Malakhov, Sov. Phys. Semicond. **11**, 1257 (1977).

[52] K. Osamura, S. Naka, and Y. Murakami, J. Appl. Phys. **46**, 3432 (1975).

[53] T. Yang, S. Nakajima, and S. Sakai, Jpn. J. Appl. Phys., Part 1 **34**, 5912 (1995).

JOURNAL OF APPLIED PHYSICS   VOLUME 84, NUMBER 4   15 AUGUST 1998

Analysis of optical gain and threshold current density of wurtzite InGaN/GaN/AlGaN quantum well lasers

Y. C. Yeo, T. C. Chong,[a] M. F. Li, and W. J. Fan
Department of Electrical Engineering, National University of Singapore, 10 Kent Ridge Crescent, S119260, Singapore

(Received 20 January 1998; accepted for publication 8 May 1998)

The valence subband structures, density-of-states, and optical gain of (0001) wurtzite $In_xGa_{1-x}N$/GaN quantum wells (QWs) are studied using a numerical approach. We used the effective-mass parameters of GaN and InN derived using the empirical pseudopotential method. By varying the well width and mole fraction of In in the well material, the effects of quantum confinement and compressive strain are examined. A narrower well width and a higher In mole fraction in the well lead to transverse electric enhancement and transverse magnetic suppression of the optical gain. From the relationship between the optical gain and the radiative current density, we obtain the transparent current density for a single QW to be 200 A/cm^2. The InGaN/GaN/AlGaN separate confinement heterostructure multiple QW (MQW) laser structure is then analyzed. It is shown that a suitable combination of well width and number of QWs should be selected in optimizing the threshold current density in such MQW lasers. © *1998 American Institute of Physics.* [S0021-8979(98)00416-2]

I. INTRODUCTION

The interest in wurtzite (WZ) GaN-based semiconductors for the fabrication of blue light-emitting diodes (LEDs) and lasers stems from their prospective applications in full color displays and high-density data storage systems. Over the last few years, intensive research has led to the demonstration of room-temperature (RT) blue/violet laser emission in the InGaN/GaN/AlGaN-based heterostructures under pulsed currents[1,2] and continuous-wave operation.[3] Recently, InGaN multiple QW (MQW) structure laser diodes with high power and long lifetime were reported.[4] The threshold current density was 4.2 kA/cm^2,[4] a significant reduction from 8.8 kA/cm^2 reported earlier.[5] From a device standpoint, it is important to optimize the InGaN/GaN/AlGaN laser structure to achieve even lower threshold current density and higher differential quantum efficiency. Apart from improving the crystalline quality of the material, a study of the optical gain with varying strain and quantum confinement in the QW, and with varying device parameters in a MQW structure, is necessary. For a single QW (SQW), the material composition in the well (in the InGaN/GaN SQW) or barrier (in the GaN/AlGaN SQW) can be adjusted to control the amount of lattice mismatch and the barrier height. The barrier height and well width of the QW determine the magnitude of quantum confinement, and directly modify the subband structures and the optical gain properties. For a MQW structure laser diode, parameters such as the number of QWs used, the thicknesses of the barrier and cladding layers, and the material used for these layers, can be varied to give an optimal structure with low threshold current density. On the experimental side, Ref. 6 reported the optimization of the well structure of InGaN MQW laser diodes using the results of the well number de-

pendence of the optical pumping threshold power for stimulated emission. The threshold optical power for a device having three InGaN QWs was 33 kW/cm^2, the lowest reported to date.[6] RT pulsed operation was achieved for a laser diode having five periods of $In_{0.14}Ga_{0.86}N$ (2 nm)/$In_{0.05}Ga_{0.95}N$ (4 nm) QWs. The threshold current density was 9.5 kA/cm^2.[6] On the theoretical side, the most widely studied structure is the GaN/AlGaN SQW as only the effective-mass parameters of GaN and AlN are available.[7,8] The biaxial strain effect, the effect of a varying well width and Al mole fraction in the barrier, and the many-body Coulomb effects have been subjects of intensive study for the WZ (0001) GaN/AlGaN SQW.[9–13] For MQW structures, the dependence of the threshold current density on the well number has not been well investigated. This relationship for MQW structure laser diodes containing InGaN would be useful because most devices[1–6] employed InGaN as the active layers. Recently, we derived the effective-mass parameters for GaN and InN using the empirical pseudopotential method.[14] Such material parameters for InN were previously unknown. Thus, material design in wide-gap nitride-based semiconductor lasers containing InGaN can be examined.

In this article, we investigate the optical gain and the threshold performance of the InGaN/GaN SQW and the InGaN/GaN/AlGaN separate confinement heterostructure (SCH) MQW structure. The effect of varying the In mole fraction in the InGaN QW is taken into account. For the SQW, we also analyze the valence subband structures and the density-of-states where the effects of quantum confinement and compressive strain are studied by varying the well width, L_w, and the mole fraction of In, x, in the well material ($In_xGa_{1-x}N$). For the InGaN/GaN/AlGaN SCH MQW, we vary the device structure parameters and show that a suitable combination of L_w and number of QWs, n_w, should be selected to obtain low threshold currents. The organization of

---

[a]Electronic mail: electc@leonis.nus.edu.sg

0021-8979/98/84(4)/1813/7/$15.00   1813   © 1998 American Institute of Physics

1814   J. Appl. Phys., Vol. 84, No. 4, 15 August 1998   Yeo *et al.*

this article is as follows. The calculation of valence subband structures is shown in Sec. II A. In Sec. II B, we show the calculation of the optical gain spectra based on a numerical integration over a large $k_x - k_y$ space without the use of analytical approximations. The results are documented and discussed in Sec. III. Section IV concludes the findings of this work. Our results could be useful in the design of MQW lasers based on the WZ nitride-based semiconductors.

II. THEORY

A. The valence subband structures

The 6×6 effective-mass Hamiltonian, H, for (0001) WZ crystals is given by[15]

$$H=\begin{bmatrix} F & -K^* & -H^* & 0 & 0 & 0 \\ -K & G & H & 0 & 0 & \Delta \\ -H & -H^* & \lambda & 0 & \Delta & 0 \\ 0 & 0 & 0 & F & -K & H \\ 0 & 0 & \Delta & -K^* & G & -H^* \\ 0 & \Delta & 0 & H^* & -H & \lambda \end{bmatrix} \begin{array}{l} |u_1\rangle=-|(X-iY)\uparrow\rangle/\sqrt{2} \\ |u_2\rangle=|(X-iY)\uparrow\rangle/\sqrt{2} \\ |u_3\rangle=|Z\uparrow\rangle \\ \\ |u_4\rangle=|(X-iY)\downarrow\rangle/\sqrt{2} \\ |u_5\rangle=-|(X-iY)\downarrow\rangle/\sqrt{2} \\ |u_6\rangle=|Z\downarrow\rangle \end{array} , \qquad (1)$$

where

$$F=\Delta_1+\Delta_2+\lambda+\theta, \quad G=\Delta_1-\Delta_2+\lambda+\theta, \quad \Delta=\sqrt{2}\Delta_3,$$

$$\lambda=\frac{\hbar^2}{2m_0}[A_1 k_z^2+A_2(k_x^2+k_y^2)]+D_1\epsilon_{zz}+D_2(\epsilon_{xx}+\epsilon_{yy}),$$

$$\theta=\frac{\hbar^2}{2m_0}[A_3 k_z^2+A_4(k_x^2+k_y^2)]+D_3\epsilon_{zz}+D_4(\epsilon_{xx}+\epsilon_{yy}), \qquad (2)$$

$$K=\frac{\hbar^2}{2m_0}A_5(k_x+ik_y)^2+D_5\epsilon_+, \quad H=\frac{\hbar^2}{2m_0}A_6(k_x+ik_y)k_z+D_5\epsilon_{z+}.$$

In Eqs. (1)–(2), Δ_1 is the crystal-field split energy, Δ_2 and Δ_3 account for the spin-orbit interaction, k_i is the wave vector, ϵ_{ij} is an element of the strain tensor, $\epsilon_+=\epsilon_{xx}+2i\epsilon_{xy}-\epsilon_{yy}$, $\epsilon_{z+}=\epsilon_{zx}+i\epsilon_{yz}$, A_i's are the effective-mass parameters, and D_i's are the deformation potentials. For $\Delta_1>\Delta_2>0$, the three bands from top to bottom are labeled as HH, LH, and CH, respectively[15] in the bulk crystal. The basis functions in (1) representing the HH, LH, and CH bands at the Γ point are $(|u_1\rangle,|u_4\rangle)$, $(|u_2\rangle,|u_5\rangle)$, and $(|u_3\rangle,|u_6\rangle)$, respectively. The u_i's are composed using $|X\rangle$, $|Y\rangle$, and $|Z\rangle$ which are the p_x, p_y, and p_z wave functions with their dipoles along the $[10\bar{1}0]$, $[11\bar{2}0]$, and $[0001]$ directions respectively. For the biaxial-strained (0001) InGaN/GaN SQW, the strain tensor in the well region contains

$$\epsilon_{xx}=\epsilon_{yy}=\frac{a_0-a}{a}, \quad \epsilon_{zz}=-\frac{2C_{13}}{C_{33}}\epsilon_{xx},$$

$$\epsilon_{xy}=\epsilon_{yz}=\epsilon_{zx}=0, \qquad (3)$$

where a_0 and a are the lattice constants of the GaN barrier and the InGaN well layers, respectively, and C_{13} and C_{33} are the stiffness constants of the InGaN well layer. The material parameters for GaN and InN are shown in Table I. The cubic approximation[7,12,22] where

$$A_1-A_2=-A_3=2A_4, \quad A_1+4A_5=\sqrt{2}A_6, \quad \Delta_2=\Delta_3, \qquad (4)$$

$$D_1-D_2=-D_3=2D_4, \quad D_1+4D_5=\sqrt{2}D_6,$$

has been used. The valence subband structures of a MQW are evaluated by diagonalizing

$$\sum_{j=1}^{6}[H_{ij}+\delta_{ij}E_0^v(z)]\phi_m^{(j)}(z,\mathbf{k})=E_m^v(\mathbf{k})\phi_m^{(i)}(z,\mathbf{k}),$$

$$i=1,2,..6, \qquad (5)$$

where m indexes the valence subbands, and $E_0^v(z)$ is the periodic MQW profile of the unstrained valence band energy which varies in the [0001] direction. Strain-induced band-edge shifts are accounted for by the nondiagonal elements of H_{ij}. The six-dimensional envelope function, $\phi_m^{(j)}(z,\mathbf{k})$, in (5) is described by

$$\phi_m^{(j)}(z,\mathbf{k})=e^{i(k_x x+k_y y)}\sum_p a_{m,p,k_x,k_y}^{(j)}\frac{1}{\sqrt{L}}e^{i(k_z+p\cdot2\pi/L)z},$$

$$j=1,2,...,6, \qquad L=L_w+L_b, \qquad (6)$$

where L_w is the well width, L_b is the barrier width, and p is an integer running through the plane waves that compose the z-dependent envelope function. $a_{m,p,k_x,k_y}^{(j)}$ is the coefficient of each plane-wave. When the barrier width is large enough, the QWs are not coupled together and the energy dispersion,

J. Appl. Phys., Vol. 84, No. 4, 15 August 1998

TABLE I. Material parameters for GaN and InN.

Parameters	GaN	InN
Lattice constants[a–c] (Å)		
a	3.189	3.54
c	5.185	5.70
Energy parameters[a–e]		
E_g(eV) at 300 K	3.50	2.04
$\Delta_1(=\Delta_{cr})$ (meV)	21[d]	17[e]
Δ_{so}(meV)	11[d]	3[e]
$\Delta_2=\Delta_3=\Delta_{so}/3$ (meV)	3.67	1
Conduction-band effective-masses[f]		
m_e^z/m_0	0.19	0.11
m_e^t/m_0	0.17	0.10
Valence-band effective-mass parameters[f]		
A_1	−7.24	−9.28
A_2	−0.51	−0.60
A_3	6.73	8.68
A_4	−3.36	−4.34
A_5	−3.35	−4.32
A_6	−4.72	−6.08
Deformation potentials[g] (eV)		
a_c	−4.08	
D_1	0.7	
D_2	2.1	
D_3	1.4	
D_4	−0.7	
Elastic stiffness constants (10^{11} dyn/cm^2)[c,h]		
C_{13}	15.8	12.4
C_{33}	26.7	18.2

[a]See Ref. 16.   [e]See Ref. 20.
[b]See Ref. 17.   [f]See Ref. 14.
[c]See Ref. 18.   [g]See Ref. 15.
[d]See Ref. 19.   [h]See Ref. 21.

$E_m^v(\mathbf{k})$, in the k_z direction is negligible. This gives the energy subband structure for a single QW, $E_m^v(k_x,k_y)$. The wave function for the mth valence subband is described by

$$\Psi_{m,k_x,k_y}^v(z)=\sum_j \phi_m^{(j)}(z,k_x,k_y)|u_j\rangle. \tag{7}$$

B. The optical gain spectrum

The optical gain, $g(\hbar\omega)$, is evaluated from[12]

$$g(\hbar\omega)=g_{sp}^e(\hbar\omega)\left[1-\exp\left(\frac{\hbar\omega-(F_c-F_v)}{k_BT}\right)\right], \tag{8}$$

$$g_{sp}^e(\hbar\omega)=\frac{q^2\pi}{n_rc\epsilon_0m_0^2\omega L_w}\sum_{n,m}\int\int|\hat{e}.M_{nm}(k_x,k_y)|^2$$

$$\times\frac{1}{4\pi^2}\cdot\frac{f_n^c(k_x,k_y)[1-f_m^v(k_x,k_y)](\hbar\gamma/\pi)}{[E_{nm}^{cv}(k_x,k_y)-\hbar\omega]^2+(\hbar\gamma)^2}$$

$$\times dk_x\, dk_y. \tag{9}$$

In (8), ω is the photon angular frequency, F_c and F_v are the quasi-Fermi levels for the electrons and holes, respectively, and k_B is Boltzmann's constant. In (9), q is the electronic charge, m_0 is the free electron rest mass, c is the free space velocity of light, ϵ_0 is the free space permittivity, n_r is the refractive index, $f_n^c(k_x,k_y)$ and $f_m^v(k_x,k_y)$ are the Fermi-Dirac distributions for electrons in the conduction and valence subbands respectively, and $\hat{e}$ is the polarization vector

of the optical electric field. The intraband relaxation time, γ^{-1}, is assumed to be 0.1 ps. Summation over electron spins is implicit in (9). $E_{nm}^{cv}(k_x,k_y)$ denotes the interband energy between the mth valence subband, $E_m^v(k_x,k_y)$, and the nth conduction subband, $E_n^c(k_x,k_y)$. The conduction subband structures are solved from

$$[H^c+E_0^c(z)]\varphi_n(z,\mathbf{k})=E_n^c(\mathbf{k})\varphi_n(z,\mathbf{k}), \tag{10}$$

$$H^c=\frac{\hbar^2}{2}\left(\frac{k_x^2+k_y^2}{m_e^t}+\frac{k_z^2}{m_e^z}\right)+P^c(z), \tag{11}$$

$$\varphi_n(z,\mathbf{k})=e^{i(k_xx+k_yy)}\sum_p g_{n,p,k_x,k_y}\frac{1}{\sqrt{L}}\,e^{i(k_z+p\cdot2\pi/L)z}. \tag{12}$$

In (10)–(12), $E_0^c(z)$ is the MQW profile of the unstrained conduction band energy, m_e^t and m_e^z are the electron effective masses perpendicular and parallel to the growth direction respectively, and $P^c(z)$ accounts for the hydrostatic energy shift in the conduction band which is equal to $a_c(\epsilon_{xx}+\epsilon_{yy}+\epsilon_{zz})$ in the well and zero in the barrier region. a_c is the conduction band deformation potential. The z-dependent envelope function, $\varphi_n(z,\mathbf{k})$, of the conduction subband state in (12) is constituted by plane waves, each weighed by the coefficient g_{n,p,k_xk_y}.[23] The conduction subband wave function, $\Psi_{n,k_x,k_y}^c(z)$, is given by

$$\Psi_{m,k_x,k_y}^c(z)=\sum_{\eta=\uparrow,\downarrow}\varphi_n(z,k_x,k_y)|S,\eta\rangle, \tag{13}$$

where η is electron spin. Thus, using (7) and (13), we can evaluate the momentum matrix element, $M_{nm}(k_x,k_y)=\langle\Psi_{m,k_x,k_y}^v|\hat{p}|\Psi_{n,k_x,k_y}^c\rangle$, for transitions between $\Psi_{n,k_x,k_y}^c(z)$ and $\Psi_{m,k_x,k_y}^v(z)$ where $\hat{p}$ is the momentum operator. The k-selection rule is observed. The band edge momentum matrix elements $\langle S|\hat{p}_x|X\rangle$, $\langle S|\hat{p}_y|Y\rangle$, and $\langle S|\hat{p}_z|Z\rangle$, given by[12,15]

$$|\langle S|\hat{p}_x|X\rangle|^2$$
$$=|\langle S|\hat{p}_y|Y\rangle|^2$$
$$=\frac{m_0}{2}\left(\frac{m_0}{m_e^r}-1\right)$$
$$\times\frac{E_g[(E_g+\Delta_1+\Delta_2)(E_g+2\Delta_2)-2\Delta_3^2]}{(E_g+\Delta_1+\Delta_2)(E_g+\Delta_2)-\Delta_3^2}, \tag{14}$$

$$|\langle S|\hat{p}_z|Z\rangle|^2=\frac{m_0}{2}\left(\frac{m_0}{m_e^z}-1\right)$$
$$\times\frac{(E_g+\Delta_1+\Delta_2)(E_g+2\Delta_2)-2\Delta_3^2}{(E_g+2\Delta_2)},$$

have been used.

1816   J. Appl. Phys., Vol. 84, No. 4, 15 August 1998   Yeo *et al.*

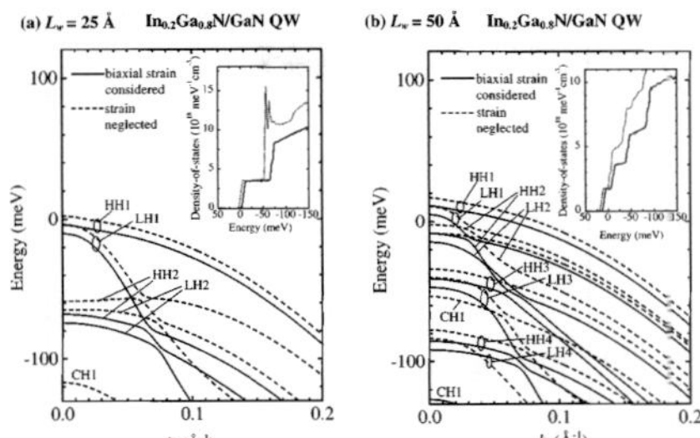

FIG. 1. The in-plane dispersion of valence subbands of a $In_{0.2}Ga_{0.8}N$/GaN SQW with (solid curve) and without (dashed curve) consideration of the biaxial compressive strain for two different well widths (a) 25 Å and (b) 50 Å. The corresponding densities-of-states are plotted in the inset.

FIG. 2. The in-plane dispersion of valence subbands of a $In_xGa_{1-x}N$/GaN SQW for $x=0.1$ (dashed) and 0.2 (solid) two different well widths (a) 25 Å and (b) 50 Å. Biaxial compressive strain is considered. The corresponding densities-of-states are plotted in the inset.

III. RESULTS AND DISCUSSION

A. Valence subband structures and optical gain of $In_xGa_{1-x}N$/GaN SQWs

The in-plane valence band dispersions of the $In_xGa_{1-x}N$/GaN SQWs are calculated from (1) and (5) using the parameters of GaN and InN shown in Table I. The effective mass parameters are from Ref. 14. Alloy properties of $In_xGa_{1-x}N$ are obtained by a linear interpolation except for the gap energy which is taken from the optical measurement of Ref. 24. We have performed alloy band structure calculations using the empirical pseudopotential method for $In_xGa_{1-x}N$ where $x=0.1$ and 0.2, and obtained close agreement with the results of Ref. 24. For the InN–GaN interface, the band-gap difference, ΔE_g, divides according to $\Delta E_v : \Delta E_c = 70:30$.[25] The deformation potentials for GaN are from Ref. 15 which are obtained from a fit to the experimental data of Ref. 26. Deformation potentials for InGaN are approximated to be equal to those of GaN. In Fig. 1, the valence subband structures of the $In_{0.2}Ga_{0.8}N$/GaN SQW are shown for (a) $L_w=25$ Å and (b) $L_w=50$ Å with (solid) and without (dashed) strain accounted for. The energy dispersions are plotted against the in-plane wave vector, k_t, since they are independent of the azimuthal angle in the k_x-k_y plane. The labeling of the subbands follows from the predominant composition of the wave function at the Γ point in terms of the HH, LH, and CH bases. The inset shows the corresponding densities-of-states. The densities-of-states are obtained by integrating the states of the valence structures over a large **k** space. A fine mesh size of 0.001 25 $Å^{-1}$ is used for the computation and convergence tests have been performed by varying the mesh size and limits of integration with negligible change in the results. The effect of the biaxial strain is seen by comparing the dashed and solid curves. Inclusion of the biaxial strain does not change the subband structures remarkably and the valence band-edge density-of-states remains relatively unaltered. This is because the C_{6v} symmetry of the WZ crystal is not reduced by the biaxial strain. The energies of $|X\rangle$ and $|Y\rangle$ constituting the HHi and LHi subbands at the Γ point are not differentiated and these subbands (with the same i) remain closely spaced. Only the

CHi subbands (which are constituted by $|Z\rangle$ at the Γ point) are distinguished[27] and they move towards lower electron energy when the biaxial compressive strain is considered [see Fig. 1(b)]. The effect of increasing quantum confinement with narrower well width can be observed by comparing Figs. 1(a) with 1(b). We see that a larger quantum size effect in a narrower well does not effectively separate HHi and LHi. This is because the quantum size effect is isotropic in the in-plane directions and does not effectively separate the energies of LHi and HHi. In Fig. 2, we plot the valence subband dispersion for two In mole fractions [$x=0.1$ (dashed) and 0.2 (solid)] for (a) $L_w=25$ Å and (b) $L_w=50$ Å. An increase in x introduces a larger strain and a larger QW potential, both of which generally push the valence subbands downward. Features of the density-of-states shift deeper into the valence band with an increase in x. In general, it is observed that the decrease of L_w, increase of x, or inclusion of larger biaxial compressive strain would bring the quasi-Fermi level towards lower electron energy at each carrier concentration. This enhances the contribution of the topmost subband to the optical gain. It is noted that the HH and LH components of each subband gives rise to momentum matrix elements associated with the transverse electric (TE) polarization, and the CH component contributes to those for the transverse magnetic (TM) polarization.

The peak TE and TM gains for the $In_xGa_{1-x}N$/GaN SQW are plotted in Fig. 3 to illustrate the effects of a reduction in well width [$L_w=25$ Å (open symbols) and 50 Å (solid symbols)] and a change in x [$x=0.1$ (triangle) and 0.2 (circle)]. TE and TM optical gains are plotted using solid and dashed lines, respectively. Figure 3 reveals that the thinner well has a higher transparent carrier density but a higher differential gain above transparency. For the $In_{0.1}Ga_{0.9}N$/GaN SQW with $L_w=50$ Å, the $C2$ subband becomes appreciably populated at about $n=3.5\times10^{19}$ cm^{-3}. From there, the secondary TE peak gain in the gain spectrum becomes the dominant peak and contributes to an increase in the differential gain. For the $In_{0.2}Ga_{0.8}N$/GaN SQW with the same well width ($L_w=50$ Å), the secondary TE peak becomes dominant at a higher carrier density of $n=4$

J. Appl. Phys., Vol. 84, No. 4, 15 August 1998

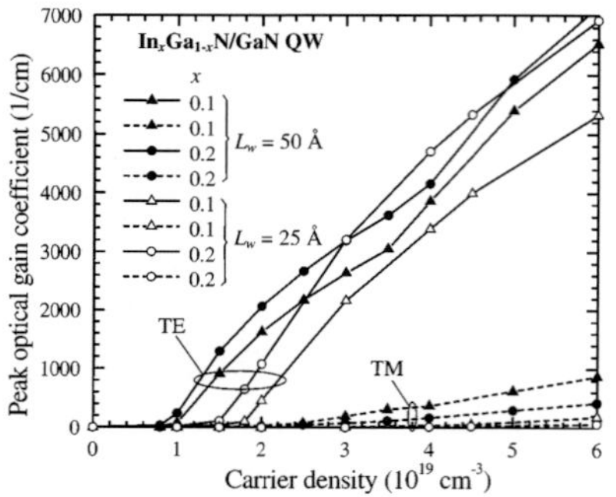

FIG. 3. Peak TE and TM optical gain coefficient as a function of carrier density for a compressively strained $In_xGa_{1-x}N/GaN$ QW for $x=0.1$ (triangle) and 0.2 (circle). Solid symbols are used for $L_w=50$ Å and open symbols are for $L_w=25$ Å.

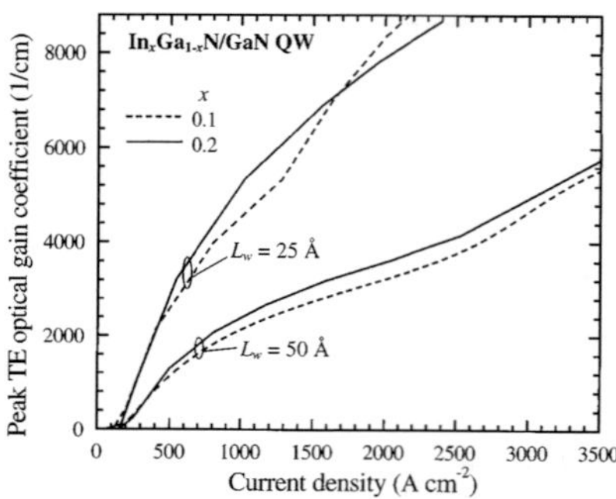

FIG. 5. Peak TE optical gain coefficient as a function of the radiative current density for a compressively strained $In_xGa_{1-x}N/GaN$ QW for $x=0.1$ (dashed) and 0.2 (solid) for well widths of $L_w=50$ and 25 Å.

$\times 10^{19}$ cm^{-3} as the separation between the $C1$ and $C2$ subbands is slightly larger. For $L_w=25$ Å, the primary TE peak remains dominant for all carrier densities plotted in Fig. 3. With the increase of x from 0.1 to 0.2, the transparent carrier density is reduced from 1.0×10^{19} to 0.9×10^{19} cm^{-3} for $L_w=50$ Å and from 1.8×10^{19} to 1.5×10^{19} cm^{-3} for $L_w=25$ Å.

The effect of the biaxial strain on the peak optical gain is illustrated next. In Fig. 4, the peak optical gain is plotted against the injected carrier density for the 50 Å $In_xGa_{1-x}N/GaN$ SQW with strain accounted (solid curves) and neglected (dashed curves). The effect of the biaxial strain on the peak optical gain is regarded as the difference between the corresponding curves with and without strain considered. It is noted that the magnitude of the strain is higher or more compressive in a well with a higher In content. In general, a higher In mole fraction in the well, or a

higher compressive strain would lead to TE enhancement and TM suppression of the optical gain. In Fig. 5, the peak TE gain is plotted against the radiative current density for a compressively strained $In_xGa_{1-x}N/GaN$ QW for $x=0.1$ (dashed) and 0.2 (solid) for $L_w=25$ and 50 Å. The transparent current density is about 200 A/cm^2.

B. The InGaN/GaN/AlGaN SCH MQW

In the following analysis, the threshold performance of a SCH MQW laser diode (Fig. 6) is investigated. The optical guiding layers sandwiching the MQW are 0.1 μm thick GaN layers, and the cladding layers are taken to be $Al_{0.1}Ga_{0.9}N$. The GaN barrier layers dividing the $In_{0.2}Ga_{0.8}N$ QWs are each 70 Å thick. Two well widths of 25 and 50 Å are considered. The optical confinement factor, Γ, is calculated from the electric-field profile of the TE_0 mode solved from Maxwell's equations. In solving for the optical field in the MQW structure of Fig. 6, an approximation is adopted where the refractive index of the $In_{0.2}Ga_{0.8}N$ active layers is taken to be the same as that of GaN. It is assumed that the thin $In_{0.2}Ga_{0.8}N$ active layers do not significantly change the optical guiding properties of the GaN layers. This simplifies the calculation since the MQW structure can be analyzed as a

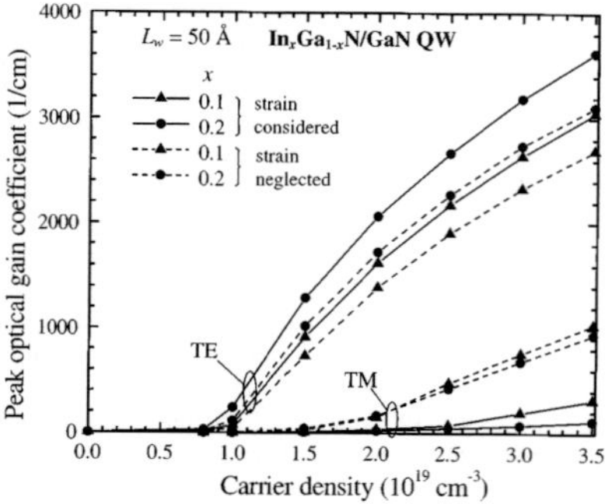

FIG. 4. Peak TE and TM optical gain as a function of the injected carrier density for a $L_w=50$Å compressively strained $In_xGa_{1-x}N/GaN$ QW for $x=0.1$ (triangle) and 0.2 (circle) with (solid curve) and without (dashed curve) strain taken into consideration.

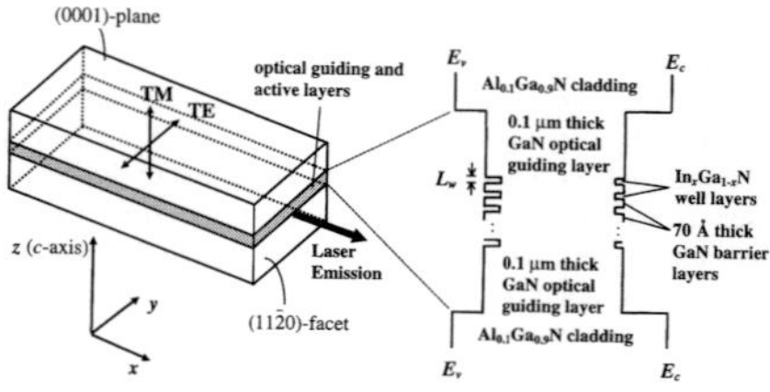

FIG. 6. Schematic of the InGaN/GaN/AlGaN separate-confinement heterostructure (SCH) multiple quantum well (MQW). The potential profile of the MQW is shown in detail.

1818   J. Appl. Phys., Vol. 84, No. 4, 15 August 1998   Yeo *et al.*

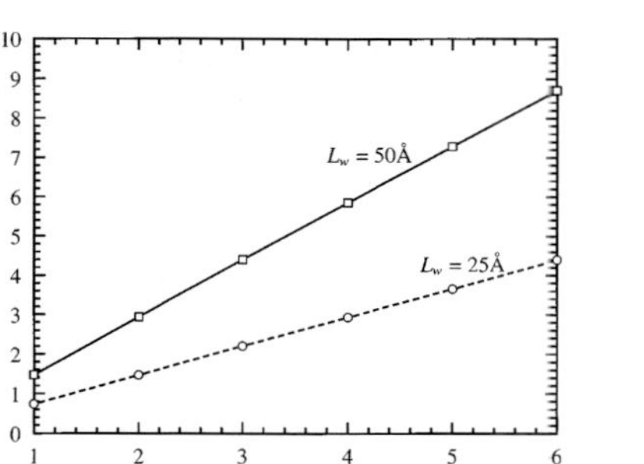

FIG. 7. Optical confinement factor, Γ, as a function of the number of quantum wells for a SCH-MQW laser structure consisting of $In_{0.2}Ga_{0.8}N$ well layers ($L_w = 50$ or 25 Å), 70 Å thick barrier layers, and 0.1 μm thick GaN optical guiding layers sandwiched by the $Al_{0.1}Ga_{0.9}N$ cladding layers.

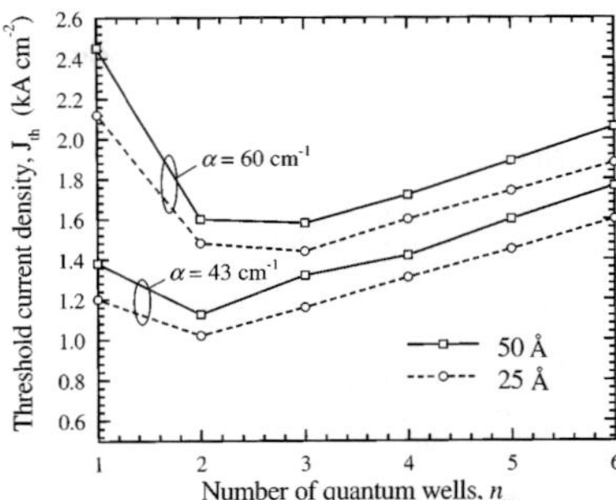

FIG. 9. Threshold current density as a function of the number of wells for a SCH-MQW laser consisting of $In_{0.2}Ga_{0.8}N$ well layers ($L_w = 50$ or 25 Å), 70 Å thick barrier layers, and 0.1 μm thick GaN optical guiding layers sandwiched by $Al_{0.1}Ga_{0.9}N$ cladding layers.

three-layer slab waveguide, where GaN is the guiding layer and AlGaN is the cladding. The optical confinement factor is then computed using

$$\Gamma = \frac{\frac{1}{2}\int_{inside}Re(\mathbf{E}\times\mathbf{H}^*)\cdot\hat{x}dz}{\frac{1}{2}\int_{total}Re(\mathbf{E}\times\mathbf{H}^*)\cdot\hat{x}dz}, \tag{15}$$

in which $\int_{inside}Re(\mathbf{E}\times\mathbf{H}^*)\cdot\hat{x}dz$ is evaluated over regions where the active layers exist. The calculated optical confinement factors of the SCH-MQW lasers are plotted in Fig. 7 as a function of the number of wells. In Figs. 8(a) and 8(b), the modal gain, Γ_g, is plotted against the current density for $L_w = 25$ and 50 Å, respectively. Homogeneous injection of carriers in the various wells is assumed. In Fig. 9, the threshold current density is plotted as a function of the number of wells, n_w, in the MQW for a given absorption loss. Considering only the absorption loss, α_i, of 43 cm^{-1},[5] the lowest threshold current density of 1.02 kA/cm^2 is obtained using $L_w = 25$ Å and $n_w = 2$. For a total loss, α, of 60 cm^{-1} (of which 43 cm^{-1} is the absorption loss and 17 cm^{-1} is the

mirror loss, α_m) the lowest threshold current density of 1.24 kA/cm^2 is obtained using $L_w = 25$ Å and $n_w = 3$. A mirror loss of 17 cm^{-1} would result from a 0.040 cm long laser cavity with 50% reflectivity at both mirror facets. Figure 9 also shows that the MQW with a narrower well width has lower threshold current densities for all n_w from 1 to 6 and that the optimal number of QWs for better threshold performance will be increased with higher losses. For α beyond 80 cm^{-1}, the optimal number of quantum wells should be greater than 3.

For $\alpha = 90$ cm^{-1}, the optimal threshold current density is obtained to be 2.1 kA/cm^2 using $n_w = 4$ and $L_w = 25$ Å. Nakamura *et al.*[5] reported a threshold current density of 8.8 kA/cm^2 for a MQW with $L_w = 35$ Å and $n_w = 4$. The laser structure in Ref. 5 used a $In_{0.15}Ga_{0.85}N/In_{0.05}Ga_{0.95}N$ MQW, and $Al_{0.08}Ga_{0.92}N$ cladding layers. The mirror loss was 46 cm^{-1},[5] giving $\alpha = 89$ cm^{-1}. Our predicted threshold current density of 2.1 kA/cm^2 is smaller than that reported in Ref. 5 considering the difference in L_w and the ideal crystalline quality assumed. Defect states in the QW of the real device would lead to nonradiative recombination and is expected to increase the threshold current density. A more recent report by Nakamura *et al.*[4] using a similar structure obtained a threshold current density of 4.2 kA/cm^2 which gave closer agreement with our result, probably due to better crystalline quality. It should also be noted that many-body effects, inhomogeneous broadening due to spatial variations in QW thickness or composition, as well as the leakage current due to device structure were not taken into account in this work. For a more accurate modeling of GaN-based QW lasers, such effects should be considered.[28]

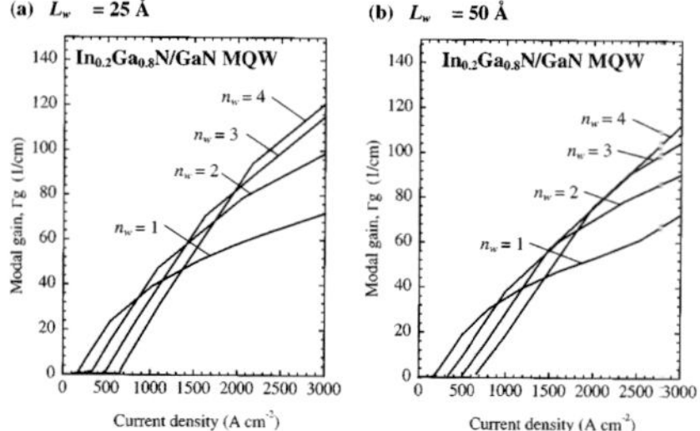

FIG. 8. Modal gain, Γ_g as a function of the current density for the SCH-MQW laser structure consisting of $In_{0.2}Ga_{0.8}N$ well layers of widths (a) 50 Å and (b) 25 Å, 70 Å thick barrier layers, and 0.1 μm thick GaN optical guiding layers sandwiched by the $Al_{0.1}Ga_{0.9}N$ cladding layers. n_w takes values from 1 to 4.

IV CONCLUSION

We have conducted a study on the electronic and optical properties of the InGaN/GaN SQW and the InGaN/GaN/AlGaN SCH MQW. For the InGaN/GaN SQW, a thinner well width offers higher TE gain. The threshold current density for a InGaN/GaN/AlGaN SCH-MQW was also ana-

J. Appl. Phys., Vol. 84, No. 4, 15 August 1998   Yeo *et al.*   1819

lyzed. A suitable optimal number of quantum wells, depending on the absorption loss, should be selected in the design of the device structure to reduce the threshold current density.

ACKNOWLEDGMENTS

Gratitude is expressed to Dr. M. Suzuki and Dr. S. Kamiyama of the Central Research Laboratories and the Semiconductor Research Center respectively, Matsushita Electric Industrial Co. Ltd., Japan, for their helpful information. We are thankful for the support of the Singapore National Science and Technology Board RIC-University Research Grant Project No. 681305, and the computing facilities from the NUS Computer Centre.

[1] I. Akasaki, H. Amano, S. Sota, H. Sakai, T. Tanaka, and M. Kaike, Jpn. J. Appl. Phys., Part 2 **34**, L1517 (1995).

[2] S. Nakamura, M. Senoh, S. Nagahama, N. Iwasa, T. Yamada, T. Matsushita, H. Kiyoku, and Y. Sugimoto, Jpn. J. Appl. Phys., Part 2 **35**, L74 (1996).

[3] S. Nakamura, M. Senoh, S. Nagahama, N. Iwasa, T. Yamada, T. Matsushita, Y. Sugimoto, and H. Kiyoku, Appl. Phys. Lett. **69**, 3034 (1996).

[4] S. Nakamura, M. Senoh, S. Nagahama, N. Iwasa, T. Yamada, T. Matsushita, Y. Sugimoto, and H. Kiyoku, Jpn. J. Appl. Phys., Part 2 **36**, L1059 (1997).

[5] S. Nakamura, MRS Internet J. Nitride Semicond. Res. **2**, article 5 (1997).

[6] H. Kawai, F. Nakamura, T. Kobayashi, K. Funato, and M. Ikeda, presented at the Late News Session, Symposium D, MRS Fall 1997 Meeting Boston, MA, Dec 1–5, 1997.

[7] M. Suzuki, T. Uenoyama, and A. Yanase, Phys. Rev. B **52**, 8132 (1995).

[8] K. Kim, W. R. L. Lambrecdt, B. Segall, and M. van Schilfgaarde, Phys. Rev. B **56**, 7363 (1997).

[9] A. T. Meney and E. P. O'Reilly, Appl. Phys. Lett. **67**, 3013 (1995).

[10] M. Suzuki and T. Uenoyama, Jpn. J. Appl. Phys. **35**, 1420 (1996).

[11] Yu. M. Sirenko, J.-B. Jeon, K. W. Kim, M. A. Littlejohn, and M. A. Stroscio, Appl. Phys. Lett. **69**, 2504 (1996).

[12] S. L. Chuang, IEEE J. Quantum Electron. **32**, 1791 (1996).

[13] W. W. Chow, A. F. Wright, and J. S. Nelson, Appl. Phys. Lett. **68**, 296 (1996).

[14] Y. C. Yeo, T. C. Chong, and M. F. Li, J. Appl. Phys. **83**, 1429 (1998).

[15] S. L. Chuang and C. S. Chang, Phys. Rev. B **54**, 2491 (1996).

[16] R. F. Davis, Proc. IEEE **79**, 702 (1991).

[17] S. Strite and H. Morkoc, J. Vac. Sci. Technol. B **10**, 1237 (1992).

[18] *Properties of Group III Nitrides* edited by J. H. Edgar (INSPEC, IEE, London, 1994).

[19] R. Dingle and M. Ilegems, Solid State Commun. **9**, 175 (1971); R. Dingle, D. D. Sell, S. E. Stokowski, and M. Ilegems, Phys. Rev. B **4**, 1211 (1971).

[20] S. H. Wei and A. Zunger, Appl. Phys. Lett. **69**, 2719 (1996), and private communication.

[21] V. A. Savastenko and A. U. Sheleg, Phys. Status Solidi A **48**, K135 (1978).

[22] G. L. Bir and G. E. Pikus, *Symmetry and Strain-Induced Effects in Semiconductor* (Wiley, New York, 1974.)

[23] Y. C. Yeo, T. C. Chong, M. F. Li, and W. J. Fan, IEEE J. Quantum Electron. **34**, 526 (1998).

[24] K. Osamura, S. Naka, and Y. Murakami, J. Appl. Phys. **46**, 3432 (1975).

[25] G. Martin, A. Botchkarev, A. Rockett, and H. Mockoc, Appl. Phys. Lett. **68**, 2541 (1996).

[26] B. Gil, O. Briot, and R. L. Aulombard, Phys. Rev. B **52**, R17028 (1995).

[27] K. Domen, K. Horino, A. Kuramata, and T. Tanahashi, IEEE J. Sel. Top. Quantum Electron. **3**, 450 (1997).

[28] W. W. Chow, A. F. Wright, A. Girndt, F. Jahnke, and S. W. Koch, Appl. Phys. Lett. **71**, 2608 (1997).

Chapter 3

Analog Integrated Circuit Design

LINEARITY IMPROVEMENT OF CMOS TRANSCONDUCTORS FOR LOW SUPPLY APPLICATIONS

M. F. Li, X. Chen and Y. C. Lim

Indexing terms: Circuit theory and design, Active filters, Operational transconductance amplifiers

A new CMOS transconductor circuit is proposed. It incorporates a current bootstrapping loop to improve the linearity without using a large source follower and high supply voltage.

The design of CMOS transconductor circuits has attracted much attention in recent years because of their suitability for the implementation of continuous-time active filters. Several methods have been developed for improving the linearity and the input signal dynamic range of CMOS transconductors [1–3]. Almost all of these methods are based on the MOS transistor behaviour using the following approximation:

$$I_D = k(V_{GS} - V_T)^2 \qquad (1)$$

where I_D is the drain current, V_T is the threshold voltage, $k = k_0 W/L$ is the transconductance parameter, W and L are, respectively, the width and length of the channel, and $k_0 = \mu C_{ox}/2$, where μ is the effective surface carrier mobility and C_{ox} is the gate oxide capacitance per unit area. Nedungadi *et al.* [1] have suggested using a crosscoupled quad cell (CCQC) as depicted in Fig. 1 to linearise the voltage–current transfer function. Two matched input MOS transistors M1 and M2

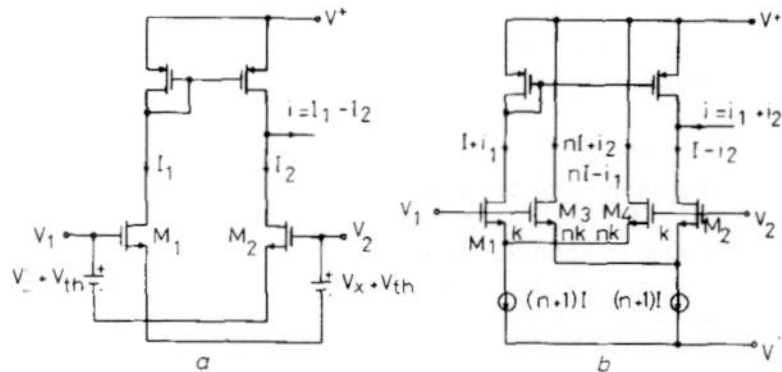

Fig. 1 *CCQC linear transconductor* [1]

a Schematic diagram
b Circuit diagram

are coupled by two voltage sources of equal value $V_T + V_x$ as shown in Fig. 1*a*. Representing $V_1 - V_2$ by v and using the relationship of eqn. 1 gives

$$I_1 = k(V_x + v)^2 \qquad\qquad I_2 = k(V_x - v)^2 \quad (2)$$

The output current i is given by

$$i = I_1 - I_2 = 4kV_x v \qquad (3)$$

The linear input signal range is bounded by

$$|v| < V_x \qquad (4)$$

The linear output current range is bounded by

$$|i| < 4kV_x^2 \qquad (5)$$

When $v = 0$, the quiescent current I_0 for each transistor is equal to kV_x^2. The circuit of Fig. 1*a* is implemented using the CCQC [1] shown in Fig. 1*b*. Let V_{GS3} and V_{GS4} be the gate-to-source voltages of transistors M3 and M4, respectively, of Fig. 1*b*. Each of the source followers (M3 or M4) provides a voltage shift of

$$V_{GS3} \simeq V_{GS4} \simeq V_T + (I_0/k)^{1/2} \qquad (6)$$

Comparing with Fig. 1a, the equivalent V_x in Fig. 1b is

$$V_x \simeq V_{xo} = (I_o/k)^{1/2} \tag{7}$$

The circuit in Fig. 1 has the advantage of simplicity and wide differential and common mode input ranges. However, within the linear range, the drain current of M1 (or M2) may vary from zero to $4kV_x^2$. To keep V_{GS3} and V_{GS4} constant, M3 and M4 must have large quiescent currents nI_o, where n may be as large as 20. This results in high power consumption. M3 and M4 also occupy a very large chip area. A valuable improved version of this circuit has been suggested by Seevinck *et al.* [2]. They replaced the single nMOS transistors M3 and M4 in Fig. 1b by CMOS pairs. Although this is an effective method and no large DC quiescent current is needed, the reduction in dynamic range due to the V_{GS} of a CMOS pair is twice that due to an nMOS transistor. This renders the transconductor unsuitable for use in low power situations. This problem can be serious because low voltage supply is preferred to avoid hot electron problems in MOS transistors [4].

We propose a new method to improve the linearity of the transconductor in Fig. 1 with small quiescent current and without sacrificing dynamic range. Thus it is suitable for low supply applications. Fig. 2 shows the improved transconductor circuit. M1, M2 are input transistors. The role of

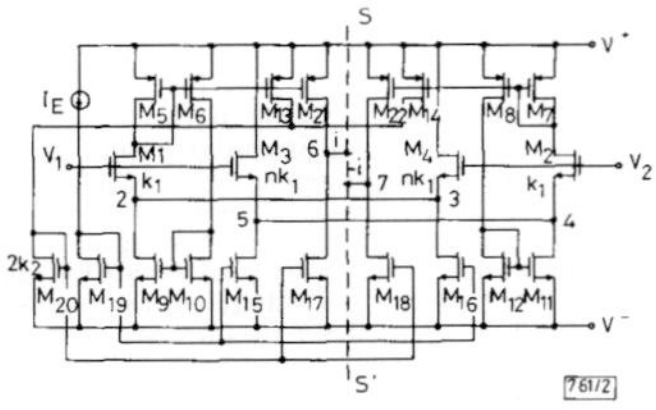

Fig. 2 *Improved CCQC linear transconductor incorporating current bootstrapping loops (M6, M9, M10, and M8, M11, M12)*

M3 and M4 is the same as that in Fig. 1b. Two current bootstrapping loops consisting of M1, M5, M6, M9, M10 and M2, M7, M8, M11, M12, respectively, are added. Neglecting the channel modulation effect, current $I_9(I_{11})$ of M9 (M11) is equal to current $I_1(I_2)$ of M1 (M2); there is no current through node 2(4) and node 3(5). The effective load impedance seen by the voltage follower M4 (M3) is thus infinite. $I_4(I_3)$ and hence $V_{GS4}(V_{GS3})$ becomes invariant with respect to changes in $I_1(I_2)$. The values of k for some of the transistors are indicated in Fig. 2. The quiescent currents of all the transistors are controlled by the bias current I_E. Note that $I_{17} = I_{18} = (I_1 + I_2)/2 = k(V_x^2 + v^2)$. The output current at node 6 is $I_1 - I_{17} = 2kV_x v$ and that at node 7 is $-2kV_x v$. It should be mentioned that the circuit of Fig. 2 is symmetrical about the line ss'. As a result, the common mode rejection is high. The output voltage dynamic range is large and is equal to the supply voltage minus $2(V_{GS13} - V_t)$.

If the channel modulation (λ) effect and other higher order effects are neglected, the circuit of Fig. 2 implements a perfectly linear transconductor for any value of n. When the channel modulation effect is considered, eqn. 1 should be replaced by

$$I_D = k(V_{GS} - V_T)^2(1 + \lambda V_{DS}) \tag{8}$$

where λ is a constant representing the channel modulation effect. We have simulated our circuit by taking the first order channel modulation effect into consideration. Our findings may be summarised as follows:

(*a*) The channel modulation effect of most transistors modifies the transconductance value slightly.

(*b*) The transconductance value is slightly dependent on common mode input and output voltages.

(*c*) The most serious consequence of the channel modulation effect is caused by imperfection of the current bootstrapping loop. This introduces a small third order term into the transconductance as a result of slight changes in $V_{GS3}(V_{GS4})$ due to a small current flowing between node 2(4) and node 3(5). This third order term can be reduced by increasing n.

The SPICE simulation result is shown in Fig. 3. Significant improvement in linearity is predicted when a bootstrapping loop is used.

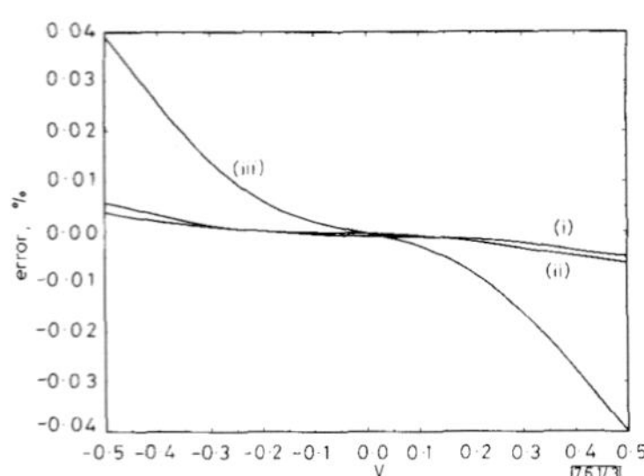

Fig. 3 *SPICE simulation results of circuit shown in Fig. 2*

Deviation from linearity for:
(i) $\lambda = 0.004$ and $n = 2$
(ii) $\lambda = 0.01$ and $n = 2$
(iii) $\lambda = 0.01$ and $n = 20$
Bootstrapping loops are removed for case (iii)

Acknowledgments: This work is supported by the National University of Singapore Research Grant RP 3920621.

© IEE 1993   *19th April 1993*

M. F. Li, X. Chen and Y. C. Lim (*Department of Electrical Engineering, National University of Singaore, Singapore 0511, Singapore*)

References

1 NEDUNGADI, A., and VISWANATHAN, T. R.: 'Design of linear CMOS transconductance elements', *IEEE Trans.*, 1984, **CAS-31**, pp. 891–894
2 SEEVINCK, E., and WASSENAAR, R. F.: 'A versatile CMOS linear transconductor/square-law function circuit', *IEEE J. Solid-State Circuits*, 1987, **SC-22**, pp. 366–377
3 KARDONTCHIK, J. E.: 'Introduction to the design of transconductor capacitor filters' (Kluwer Academic Publishers, Dordrecht, 1992), Chap. 9
4 HU, C.: 'VLSI electronics microstructure science', *in*: 'Advanced MOS device physics. Vol. 18' (Academic Press, San Diego, 1989)

LOW-VOLTAGE LINEAR OTA WITH RAIL-TO-RAIL DIFFERENTIAL MODE INPUT SIGNAL CAPABILITY

*X.W.Zhang, M F.Li and Uday Dasgupta**

Department of Electrical Engineering
National University of Singapore
Singapore 119260
* Tritech Microeletronics Ltd, Singapore 768442
email: elelimf@nus.edu.sg

Abstract

A low-voltage linear operational transconductance
amplifier (OTA) with rail-to-rail differential mode input
signal capability is designed and fabricated by $0.8\mu m$
CMOS techonlogy. The circuit is capable of operating at
supply voltage larger than two times the MOS transistor
threshold voltage V_T. When the supply voltage is $=1V$,
the total harmonic distortion (THD) is less than 1.8% for
2V peak-to-peak (V_{pp}) input.

1. Introduction

It is a trend to use low-voltage supplies in CMOS integrated circuits because of the reliability issue of small
size MOSFET transistors[1], and the increasing use of
low-weight long life battery-operated portable electronic
systems. However, reduction of power supply voltage
reduces the signal dynamic range and signal to noise ratio. In order to solve such problems, many recent designs
used rail-to-rail common mode input architecture in low-
voltage operational amplifiers[2, 3]. In rail-to-rail operational amplifier circuit, the common mode input voltage
extends to rail-to-rail, however, the differential input signal is very small.

When used in active filter circuit, the OTA has more
stringent input requirement which requires as large as
possible linear differential input range. We present in
this paper a linear OTA circuit operating at supply voltage larger than $2V_T$ with rail-to-rail differential mode input signal capability.

2. Principle of Operation

The new designed OTA architecture is shown in Fig. 1.
It consists of two complementary building blocks OTAN
and OTAP connected in parallel. When $V_{in} > 0$ only
OTAN operates, and when $V_{in} < 0$, only OTAP operates. The OTAN circuit is shown in Fig. 2.

The left part of the vertical dashed line is the input
stage of OTAN. It is similar to a circuit structure pro-

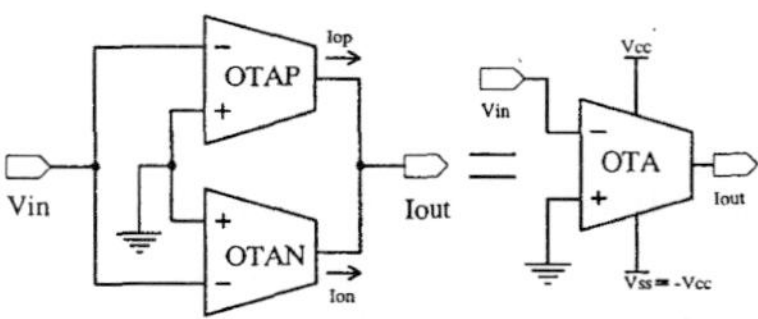

Fig. 1. Configuration of OTA (an OTAN and an OTAP in
parallel connection).

posed in[4]. The input transistors N1 and N2 are conducting elements operating in the non-saturation region.
The amplifiers AN1, AN2 and transistors N3, N4 constitute two feed back loops. From the virtual ground
principle, the drain voltages of N1 and N2 are fixed at
V_{DN}[4, 5, 6]. The current mirror consisting of P5 and P6
works as the load of the input stage and the differential-
to-single-end converter. The drain current of N1 and N2
operating in the non-saturation region can be expressed
as[7]:

$$I_D = K_N \cdot (V_{GS} - V_{TN}) \cdot V_{DSN} - (1/2) \cdot K_N \cdot V_{DSN}^2$$

Where I_D is the drain current, V_{GS} is the gate to
source voltage. V_{DSN} is the drain to source voltage of
N1 and N2($V_{DSN} = V_{DN} - V_{SS}$), V_{TN} is the threshold
voltage of the n-MOS transistor, and

$$K_N = (\mu_n \cdot C_{ox}) \cdot (W/L)_N$$

Here μ_n is the electron mobility, and C_{ox} is the capac-
itance(per unit area) of the gate oxide. W and L are the
channel width and channel length of the MOS transistors
N1 and N2. The non-saturation condition of the n-MOS
transistor is :

$$V_{DSN} < (V_{GS} - V_{TN}) \tag{1}$$

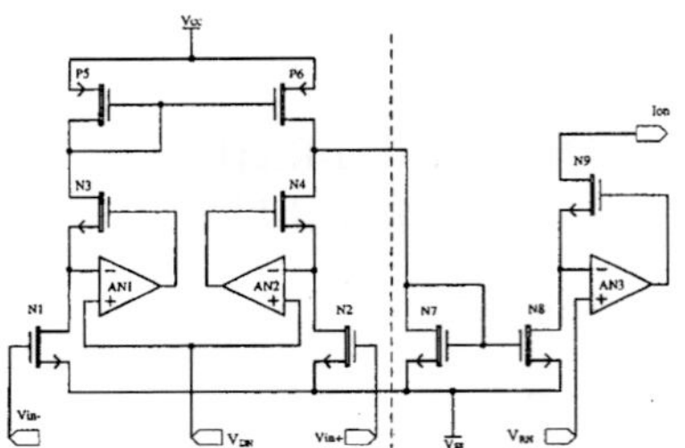

Fig. 2. The circuit diagram of OTAN.

Assume two transistors in each pair of N1 and N2, N3 and N4, P5 and P6 perfectly match each other. Then the current difference of the two input transistors N1 and N2 is:

$$I_{DN1} - I_{DN2} = K_N \cdot V_{DSN} \cdot (V_{GN1} - V_{GN2}) \quad (2)$$

Where V_{GN1} and V_{GN2} are the gate voltages of N1 and N2, respectively. It can be seen from Eq. (2) that I_{DN1}-I_{DN2} is linear with respect to $(V_{GN1}$-$V_{GN2})$ provided that V_{DSN} is a constant. From Eq. (1), it is obvious that the following relations must hold to ensure the validity of Eq. (2) :

$$\begin{aligned} V_{GN1} &> V_{TN} + V_{DN} \\ V_{GN2} &> V_{TN} + V_{DN} \end{aligned} \quad (3)$$

In Fig. 2, the right part of the vertical dashed line is the output stage of OTAN. The current mirror N7-N8 has two functions. First, owing to the current mirror, OTAN output can only sink current when $V_{GN1} > V_{GN2}$. When $V_{GN1} \leq V_{GN2}$, the output current is zero. Second, the mirror may have a current multiplication factor M_N which is the ratio of the channel width of N8 to N7 for the same channel length. A large M_N value improves the efficiency of the power consumption of the circuit. N9 and AN3 constitute an improved low-voltage regulated cascode circuit[8, 9], stacked on N8 to increase the output impedance of OTAN. AN1-AN3 have the same circuit as shown in Fig. 3.

Although in Fig. 2 the circuit structure of N8, N9 and AN3 is similar to that of N1, N3 and AN1 in the input stage, it should be noted that N8 must operate in the saturation region. The control voltage V_{RN} of AN3 should satisfy:

$$V_{RN} \geq \sqrt{\frac{2 \cdot L \cdot I_{DN8MAX}}{K_N \cdot W}} + V_{SS} \quad (4)$$

Here I_{DN8MAX} is the maximum output current of N8. L and W are the channel length and width of N8 respectively.

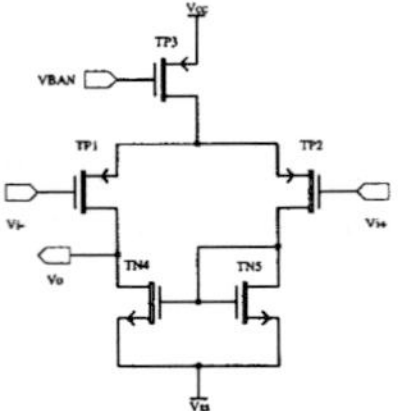

Fig. 3. The circuit diagram of amplifier AN1-AN3.

From Eq. (2), taking M_N into consideration, the output current I_{on} of OTAN can thus be expressed as:

$$I_{on} = \begin{cases} (M_N \cdot K_N \cdot V_{DSN}) \cdot (V_{GN2} - V_{GN1}), \\ \qquad\qquad \text{if } V_{GN1} \geq V_{GN2} \\ 0, \qquad\qquad \text{if } V_{GN1} \leq V_{GN2} \end{cases} \quad (5)$$

Owing to the bounds imposed by Eq. (3), OTAN does not show rail-to-rail input performance. In order to extend the input range to rail-to-rail, a complementary OTAP circuit as shown in Fig. 4 is used in conjunction with OTAN.

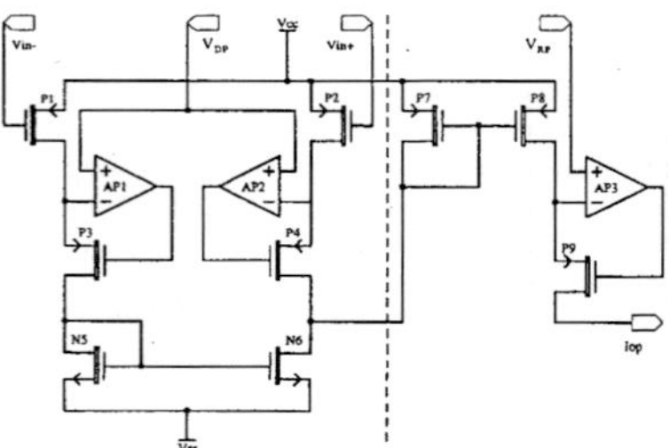

Fig. 4. The circuit diagram of OTAP.

OTAP is derived from OTAN by replacing all n-type transistors by p-type transistors and vice versa, and swapping the polarities of the supply rails. The voltage and current quantities of OTAP satisfy the following equations:

$$I_{op} = \begin{cases} (M_P \cdot K_P \cdot V_{DSP}) \cdot (V_{GP2} - V_{GP1}), \\ \qquad\qquad \text{if } V_{GP1} \leq V_{GP2} \\ 0, \qquad\qquad \text{if } V_{GP1} \geq V_{GP2} \end{cases} \quad (6)$$

$$\begin{aligned} V_{GP1} &< V_{DP} - |V_{TP}| \\ V_{GP2} &< V_{DP} - |V_{TP}| \\ K_P &= (\mu_P \cdot C_{ox}) \cdot (W/L)_P \end{aligned} \quad (7)$$

Where V_{GP1} and V_{GP2} are the gate voltages of the two input p-MOS transistors P1 and P2, respectively. V_{TP} is the threshold voltage of p-MOS transistors. μ_p is the hole mobility. V_{DSP} is the source to drain voltage of input transistors P1 and P2($V_{DSP} = V_{CC} - V_{DP}$). $(W/L)_P$ is the channel width over length ratio of the input transistors P1 and P2. M_P is the multiplication factor of current mirror(P7,P8) as M_N defined in OTAN. OTAP can only source current when $V_{GP1} < V_{GP2}$.

When OTAN and OTAP are connected in parallel, as shown in Fig. 1, according to Eq. (5) and Eq. (6), the output total current I_{out} of the OTA is expressed by:

$$I_{out} = \begin{cases} I_{on} = -(M_N \cdot K_N \cdot V_{DSN}) \cdot V_{in}, \\ \qquad \text{when } V_{in} \geq 0 \\ I_{op} = -(M_P \cdot K_P \cdot V_{DSP}) \cdot V_{in}, \\ \qquad \text{when } V_{in} \leq 0 \end{cases} \quad (8)$$

The bounds imposed by Eqs. (3) (7) can now be replaced by a unified bound as:

$$\begin{aligned} V_{CC} - V_{SS} &= 2V_{CC} \\ &> V_{TN} + |V_{TP}| + V_{DSN} + V_{DSP} \end{aligned}$$
$$(9)$$

In order to ensure linearity over the entire range of V_{in}, the transconductances of OTAN and OTAP must be equal. Denote this transconductance by GM. Hence,

$$GM = M_N \cdot K_N \cdot V_{DSN} = M_P \cdot K_P \cdot V_{DSP} \quad (10)$$

Taking note of Eq. (10), Eq. (8) can now be written as:

$$I_{out} = -GM \cdot V_{in} \quad (11)$$

for the entire rail-to-rail input voltage range.

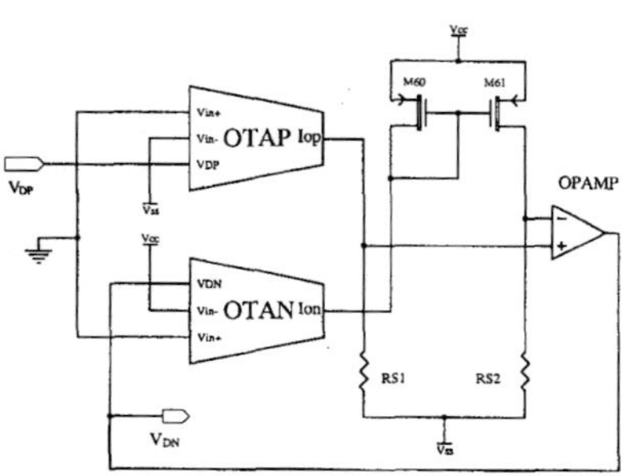

Fig. 5. The circuit diagram of master-bias.

The biasing voltages V_{DN} and V_{DP} cannot be selected independently if Eq. (10) is to be satisfied. Fig. 5 shows a circuit which will produce an output voltage V_{DN} satisfying Eq. (10) when V_{DP} is given. In Fig. 5, $I_{op} = M_P \cdot K_P \cdot V_{DSP} \cdot |V_{SS}|$ and the voltage across the resistor RS1 is $V_P = RS1 \cdot M_P \cdot K_P \cdot V_{DSP} \cdot |V_{SS}|$. The voltage across the resistor RS2 is $V_N = RS2 \cdot M_N \cdot K_N \cdot V_{DSN} \cdot V_{CC}$. For RS1=RS2 and $|V_{SS}| = V_{CC}$, we have $M_P \cdot K_P \cdot V_{DSP} = M_N \cdot K_N \cdot V_{DSN}$ satisfying Eq. (10). Since in an analog or mixed signal VLSI system, many identical OTAs are needed, the circuit in Fig. 5 can now be used as a master-bias V_{DN} generator to control all OTAs.

Fig. 6. The microscopic photograph of an OTA cell including master-bias circuit on the chip.

3. Experimental Result

The OTA has been simulated by HSPICE, using Chartered Semiconductor Manufacturing Ltd(CSM) 0.8μm BSIM28 model, and fabricated by the 0.8μm standard CMOS technology. Fig. 6 shows the microscopic photograph of an OTA cell including the master-bias circuit. The layout size of an OTA cell is 850μm × 108μm. The transconductance(GM) can be adjusted by changing the control voltage V_{DP}. Fig. 7 gives the experimental dc transfer curve measured by HP4156 semiconductor parameter analyser. It can be seen that the linear input range is from rail to rail.

The ac frequency response of the OTA is given in Fig. 8. The -3db bandwidth is 450kHz which is fit to the filter design with cut off frequency of 3.5KHz in an audio system.

The THD is measured by HP3589 spectrum analyser. The spectrum of the OTA output for a 2V_{pp} sinusoidal input signal at 1kHz is shown in Fig. 9. The Total Harmonic Distortion is less than 1.8%.

4. Conclusion

The design of a rail-to-rail low voltage single ended input OTA is demonstrated. It can be extended to differential input straightforwardly. The transconductance with respect to the input voltage is constant from rail-to-rail and the circuit is capable of operating at low-voltage

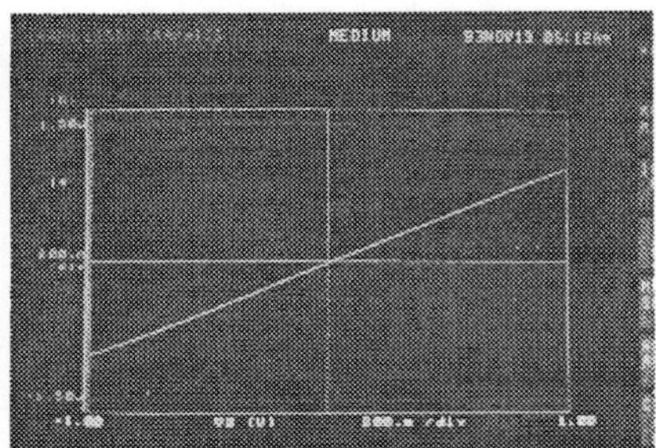

Fig. 7. The dc transfer curve of the OTA measured by
HP4156 semiconductor parameter analyser
($V_{CC} = -V_{SS} = 1V$). It corresponds to transconductance
GM=$1\mu A/V$. GM can be changed in a wide range by
changing the control voltage V_{DP}, or by changing the ratio of
W/L of the input transistors or M_N and M_P in Eq. 8.

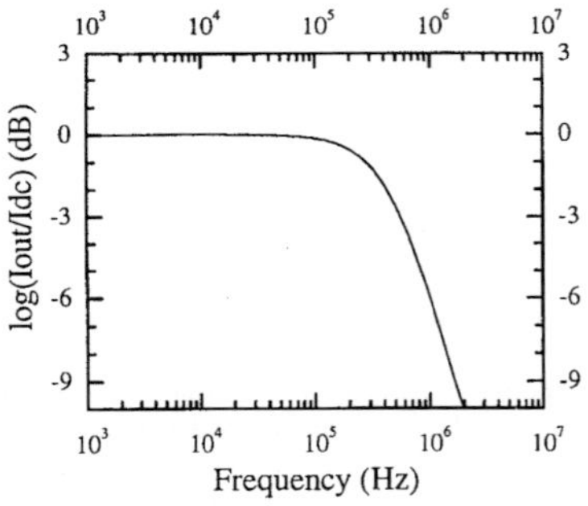

Fig. 8. The measured ac frequency response of the OTA.

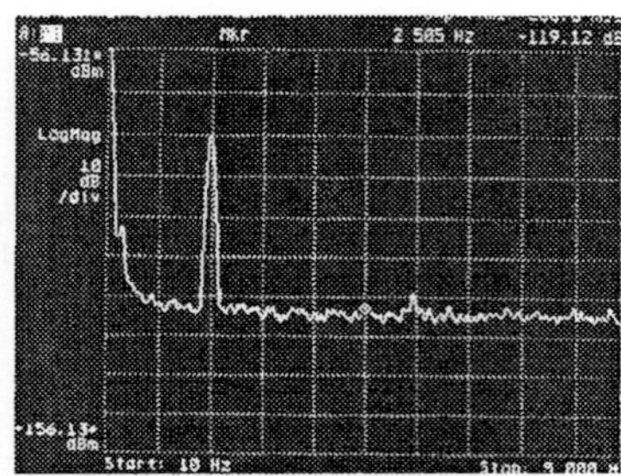

Fig. 9. The spectrum of the OTA output for a $2V_{pp}$ sinusoidal
input at 1KHz.($V_{CC} = -V_{SS} = 1V$)

provided that the supply voltage $2V_{CC}$ is larger than two
times the threshold voltage of the MOS transistors. The
THD is less than 1.8% for rail-to-rail(V_{pp}) input. This
kind of OTA will be a very useful building block in low-
voltage active filter design.

Acknowledgement We would like to thank Dr. Lian Yong
for assistance in this work. This work is supported by the
Singapore National Science and Technology Board Research
Grant NSTB /17/2/3.

5. References

[1] E. Takeda, C. Y. Yang and A Miura-Hamada, *Hot-Carrier Effects in MOS Devices*, Academic Press, San Diego, 1995.

[2] R. Hongervorst, R. J. Wiegerink P. A. J. Jong, J. Fongderie, R. F. Wassenaar and J. H. Huijising, "CMOS low-voltage operational amplifiers with constant-gm rail-to-rail input stage," *Analog Integrated Circuits and Signal Processing* , Vol. 5, pp. 135-146, 1994.

[3] S. Sakurai and M. Ismail, *Low-Voltage CMOS Operational Amplifiers*, Kluwer Academic, Boston, 1995.

[4] Veikko R. Sarri, "CMOS Transconductance Circuit with Triode Mode Input," United States Patent number 4,656,436, Apr. 7,1987.

[5] J. Pennock, P. Frith and R. G. Barker, "CMOS Triode Transconductor Continuous Time Filters," *Proc. IEEE Custom Integrated Circuits Conference*, pp. 231-234, 1996.

[6] A. L. Coban and P. E. Allen, "Low-voltage CMOS transconductance cell based on parallel operation of triode and saturation transconductors," *Electronic Letters*, Vol. 30, pp. 1124, 1994.

[7] P. E. Gray and R. G. Meyer, *Analog Integrated Circuits*, Wiley & Sons, New York, 1993.

[8] E. Sackinger and W. Guggenbuhl, "A High-Swing, High-Impedance MOS Cascode Circuit," *IEEE J. Solid Stage Circuits*, Vol. 25, pp. 289-297, 1990.

[9] K. Bult and G. J. G. M. Geelen, "A Fast-Settling CMOS Op Amp for SC Circuits with 90-dB DC Gain," *IEEE J. Solid State Circuits*, Vol. 25, pp. 1379-

Reprinted paper with permission from Y.J. Ha, M.F. Li and A.Q. Liu, Analog Integrated Circuits and Signal Processing, Vol.27, pp.7–17 (2001). Copyright © 2001, Springer Netherlands.

Analog Integrated Circuits and Signal Processing, 27, 7–17, 2001
© 2001 Kluwer Academic Publishers. Manufactured in The Netherlands.

A New CMOS Buffer Amplifier Design Used in Low Voltage MEMS Interface Circuits

YAJUN HA,[1,*] M. F. LI[1] AND AI QUN LIU[2]

[1]*Department of Electrical Engineering, National University of Singapore, Singapore 119260*
[2]*Institute of Materials Research and Engineering, 3 Research Link, Singapore 117602*
E-mail: yjha@imec.be; elelimf@nus.edu.sg

Received December 8, 1999; Accepted July 5, 2000

Abstract. To achieve low voltage high driving capability with quiescent current control, a class-AB CMOS buffer amplifier using improved quasi-complementary output stage and error amplifiers with adaptive loads is developed. Improved quasi-complementary output stage enables it more suitable for low voltage applications, while adaptive load in error amplifier is used to increase the driving capability and reduce the sensitivity of the quiescent current to fabrication process variation. The circuit has been fabricated in 0.8 μm CMOS process. With 300 Ω load in a ± 1.5 V supply, its output swing is 2.42 V. The mean value of quiescent current for eight samples is 204 μA, with the worst deviation of 17%.

Key Words: CMOS buffer amplifier, MEMS, low voltage

Introduction

Buffer amplifier is an important building block in MEMS interface circuit design. It couples the processed signals from previous on-chip stages with high output impedance to off-chip loads as shown in Fig. 1. The buffer amplifier comprises two stages in our design, namely input and output stages. The conceptual structure of a buffer is shown in Fig. 2, where output stage is further divided into two parts, the driver stage, and the output transistors MO_1 and MO_2. The input stage of the buffer amplifier is realized by an operational transconductance amplifier (OTA). The output buffers able to drive low impedance off-chip loads to voltage near the power supply while achieving good linearity, low quiescent power dissipation and controlled bias have been extensively investigated in recent years. Buffers with class-AB configuration employing quasi-complementary output stage (as shown in Fig. 3(a)) [1,2] is well-suited for meeting these demands, and has been widely used. However, two of its inherent drawbacks make it less attractive when driving a heavy load with low voltage supply.

First, the open loop gain of the error amplifiers A and B in Fig. 3(a) cannot be too high, generally not more

than 10 [2,3]. Otherwise, the quiescent current of the output buffer may has a large variation from the design target value due to random input offset voltage of the error amplifier. This low open loop gain of error amplifier limits the driving capability of the buffer amplifier.

Second, almost all the previous error amplifiers are designed to work with power supply voltage of 5 V or higher [2–7]. Under low power supply of 3 V or 1.5 V, when the common mode voltage V_o changes to one rail of the power supply, one of the error amplifiers may not work properly. Consequently, one of two output transistors will lose control of its gate voltage.

A new buffer amplifier (See Fig. 3(b)) is proposed in this work. To solve the first problem, the error amplifiers A and B use an adaptive load scheme introduced in [8] with further improvement. The adaptive load in error amplifier is used to obtain an adaptive gain. In the quiescent state, the open loop gains of the error amplifiers A and B are very low to achieve stable quiescent current. While in the driving state, the open loop gains of the error amplifiers will be changed to very high so as to get an improved driving capability. To solve the second problem, the proposed new quasi-complementary output stage uses two differential-output error amplifiers (as shown in Fig. 3b). Amplifier A uses p-MOS input transistor pair while amplifier B uses n-MOS input transistor pair. When V_o in Fig. 3(b) changes to

*Yajun Ha is now a Ph.D. candidate in IMEC, Belgium.

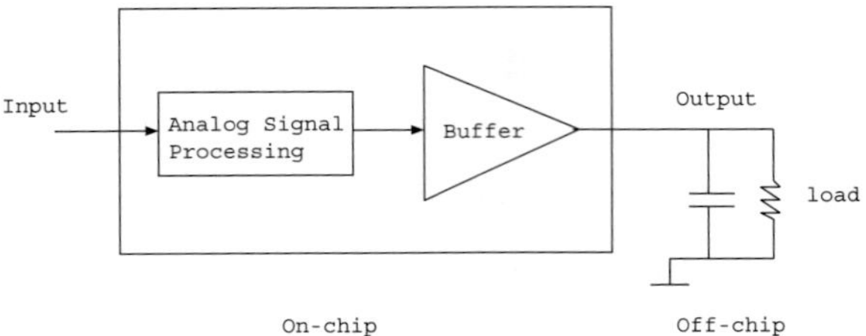

Fig. 1. Buffer amplifier in analog signal processing system.

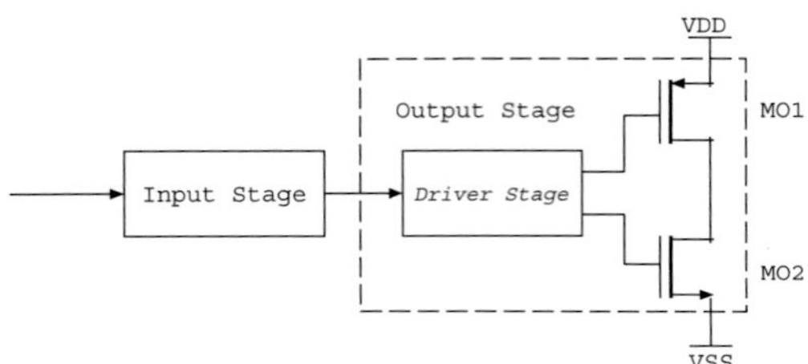

Fig. 2. Structure of buffer amplifier.

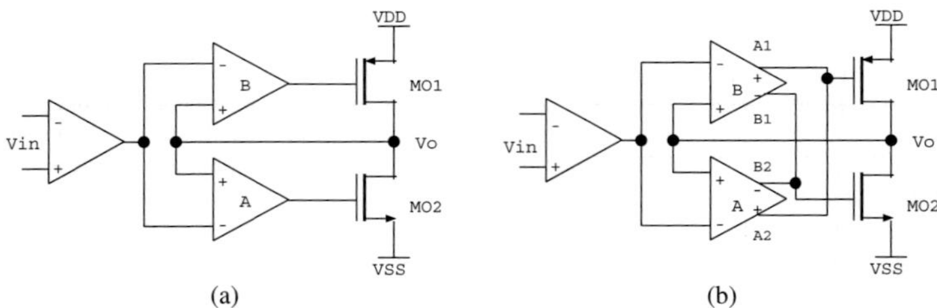

Fig. 3. Configuration of buffer with quasi-complementary output stage (a) traditional version, (b) improved version.

one rail of the power supply, one error amplifier will lose control while the other error amplifier will work properly to control the gates of two output transistors.

The proposed buffer amplifier has been successfully realized and used in the interface circuit for an optical MEMS accelerometer [9].

In the next section, detailed circuit description of the buffer amplifier is given. Simulation and chip measurement results are summarized and discussed in Section 3.

Circuit Description

A. *Adaptive Load*

Adaptive load was first introduced in [8]. Adaptive load is a resistive load which can be changed, under certain conditions, from one value of resistance to another. In this work, we use an improved adaptive load in the design of error amplifier in Fig. 3(b).

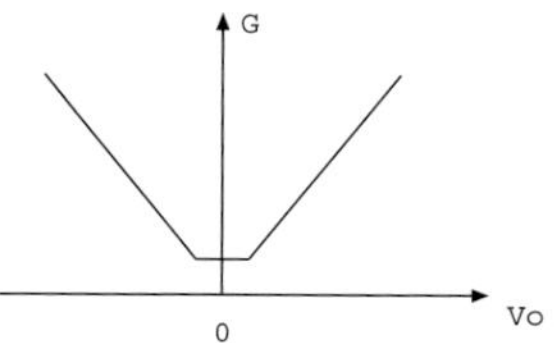

Fig. 4. Expected characteristics of gains of error amplifiers *A* and *B*.

As discussed in Section 1, in the quasi complementary output stage, the gains (G) of the error amplifiers *A* and *B* influence two sides of circuit performance. On the one hand, gains of error amplifiers decide the driving capability of the output stage. The higher the gain of error amplifier, the smaller the output resistance of output node, and larger driving capability at the output node. On the other hand, G influence the fluctuation of quiescent current of the buffer (the higher the gain of error amplifier, the easier the quiescent current deviates from its designed target value, due to random offset of error amplifiers).

To obtain the high driving capability and steady quiescent current at the same time, the gains G of error amplifiers *A* and *B* should behave as shown in Fig. 4. G is low in the region near the quiescent operating point (output voltage $V_o = 0$), and becomes high when V_o is far from quiescent point.

Fortunately, characteristics of G as shown in Fig. 4 can be implemented using adaptive load concept, in Fig. 5, by triggering low conductance loads (current source I_1 and I_2) or high conductance loads (diode-connected transistors MA_{12} and MA_{20}). In the region near the quiescent operation point, the adaptive switches MA_{13} and MA_{21} turn on, high conductance loads (MA_{12} and MA_{20}) are used to achieve low G. In the region far away from quiescent point, MA_{13} and MA_{21} turn off, low conductance loads (I_1 and I_2) are used to achieve a high G.

The implementation of such adaptive switches and thus adaptive loads used in error amplifier will be described in the following subsection.

B. Error Amplifier with Adaptive Load

In Fig. 5, MA_{12}/MA_{13} and MA_{21}/MA_{20} are used as adaptive load. The primary feature of the new adaptive load is to use variable bias for the two switches MA_{13}

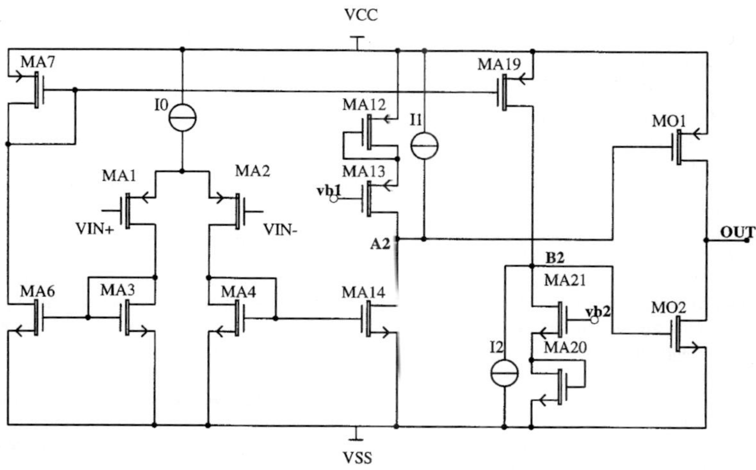

Fig. 5. Simplified schematic of error amplifier *A* in the improved quasi-complementary output stage as shown in Fig. 3(b).

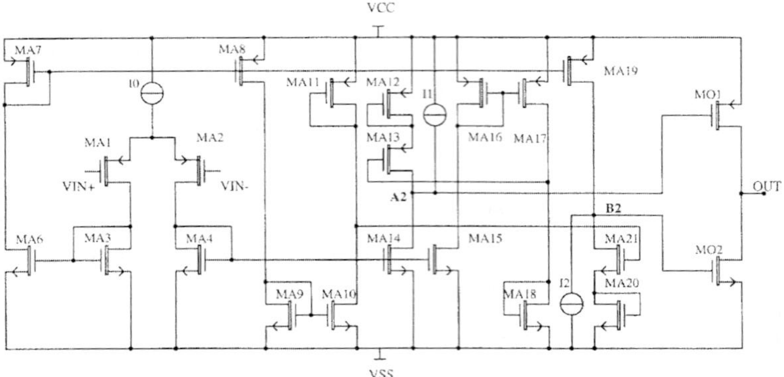

Fig. 6. Complete schematic of error amplifier *A* in improved quasi-complementary output stage.

and MA_{21} to avoid the troublesome determination of α in [8]. The variable biases V_{b1} and V_{b2} enable switches MA_{13} and MA_{21} turn on in quiescent state, while turn off in driving mode.

Fig. 6 gives the implementation circuit of the V_{b1} and V_{b2}. The work principle is as follows. Under quiescent conditions, the conductances at nodes A_2 (B_2) is very large, because switch MA_{13} (MA_{21}) is on, and MA_{12} (MA_{20}) is connected as a diode. Therefore, the gain of the error amplifier is low and sensitivity at nodes A_2 and B_2 is reduced, thus a stable quiescent current is obtained.

In driving mode, when $(V_{in+} - V_{in-})$ increases, node A_2 and drain voltage of MA_{15} are pulled down, large currents go through MA_{13}, MA_{12}, MA_{16}, MA_{17} and MA_{18}. Before MA_{13} turns off, source-to-gate voltage of MA_{12} becomes larger with the increase of its drain current. Correspondingly, the source voltage of MA_{13} decreases. On the other hand, gate voltage of MA_{13} will increase, because of the increasing gate-to-source voltage of MA_{18} created by the increasing currents of MA_{16} and MA_{17}. The different changing direction for the gate voltage and source voltage of MA_{13} eventually turns off MA_{13} and causes the overall resistance of the adaptive load to increase, therefore, leads to an improved driving capability. Similar analysis can be done in driving mode when $(V_{in+} - V_{in-})$ decreases.

Amplifier *B* in Fig. 3(b) is derived from amplifier *A* by replacing all n-MOS transistors by p-MOS transistors and vice versa, and swapping the polarities of the supply rails.

C. *Amplifier Using the Proposed Output Stage*

To test the performance and functionality of the proposed output buffer stage, it was incorporated into a two-stage opamp. Complete circuit schematic of the proposed buffer amplifier is given in Fig. 7. The designed opamp follows the structure of Fig. 3(b). It consists of a constant transconductance rail-to-rail input stage [10] (M_1–M_{40}), and the proposed class AB output stage. Class AB output stage is comprised by error amplifiers *A* (MA_1–MA_{21}) and *B* (MB_1–MB_{14}), as well as two output transistors MO_1 and MO_2.

Simulation and Chip Measurement Results

Fig. 8 and Fig. 9 show the output transistor drain currents of MO_1 and MO_2 versus output voltage V_o when A_2 and B_1 in Fig. 3(b) are disconnected or connected from the gates of output transistors. In Fig. 8, output transistors drain currents are measured when node A_2 and B_1 in Fig. 3(b) are disconnected. When $V_o > 0.4$ V, amplifier *A* in Fig. 3(b) does not work properly. The drain current of MO_2 loses proper control and becomes very large. When $V_o < -0.75$ V, amplifier *B* does not work properly. The drain current of MO_1 loses proper control and becomes very large. Such facts are reflected in Fig. 8. In Fig. 9, drain currents of the output transistor are measured when nodes A_2 and B_1 are connected as in Fig. 3(b). The drain current of MO_1 or MO_2 are

A New CMOS Buffer Amplifier Design 11

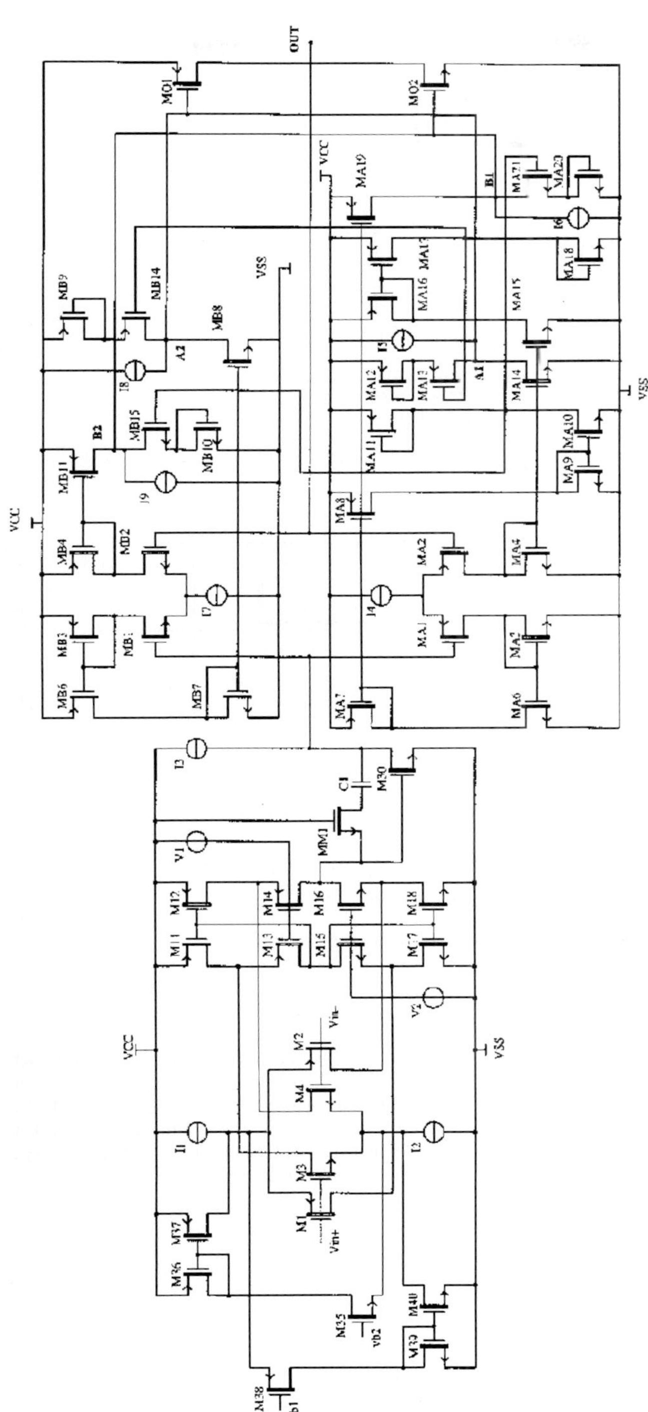

Fig. 7. Complete circuit schematic for the proposed buffer amplifier.

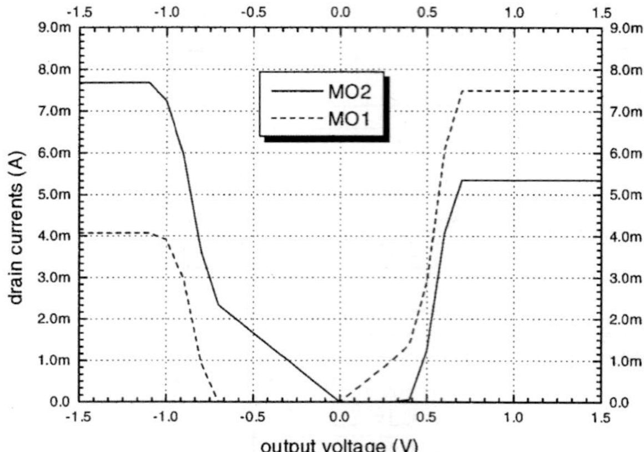

Fig. 8. Simulation of drain current of output transistors when nodes A_2 and B_1 in Fig. 3(b) are disconnected ($V_{DD} = 1.5$ V).

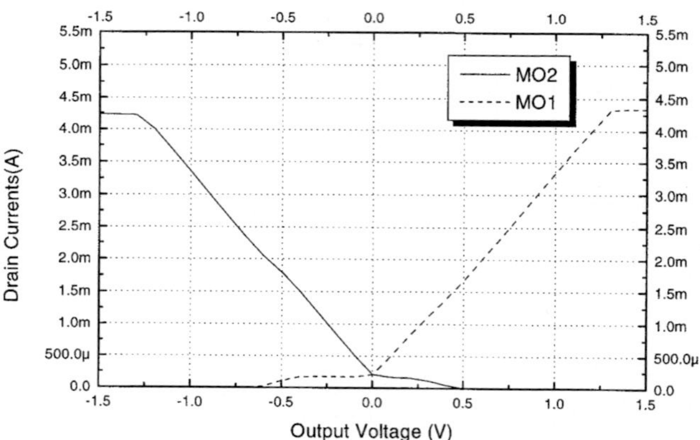

Fig. 9. Simulation of drain currents of output transistors in Fig. 3 (b) ($V_{DD} = 1.5$ V).

controlled properly in the whole output range, as expected in our design.

Fig. 10 shows the die photo of two fabricated buffer amplifiers. Fig. 11 shows the measured DC transfer characteristics when the buffer is connected in an unity-gain configuration under different loads. With 300 Ω load in a ± 1.5 V supply, the output swing of the buffer amplifier is $+1.16$ V to -1.26 V. The mean value of quiescent current for eight samples is 204 μA, with the worst deviation of 17%. The amplifier has a DC gain of 74 dB, and unity gain-bandwidth of 1 MHz (as shown in Fig. 13) with phase margin of 62°.

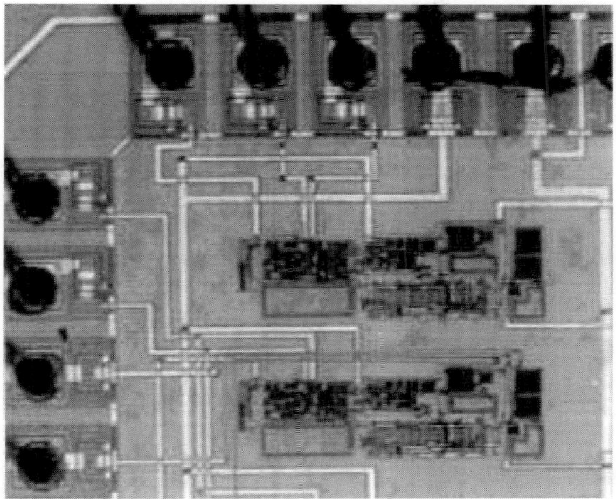

Fig. 10. Die photo of two buffer amplifiers.

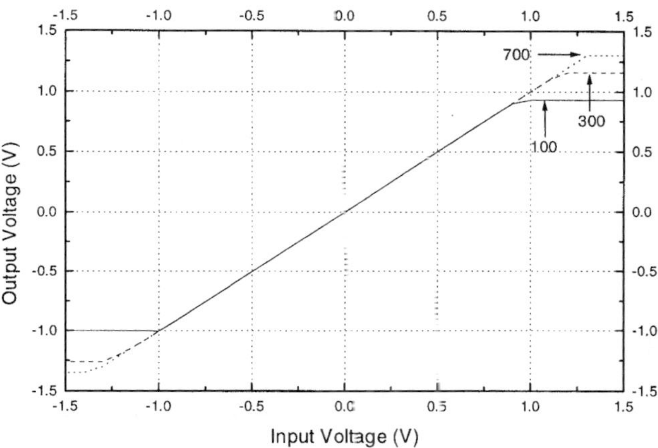

Fig. 11. Measured DC transfer function for the buffer amplifier connected as an unity–gain follower with 100 Ω, 300 Ω, and 700 Ω loads respectively.

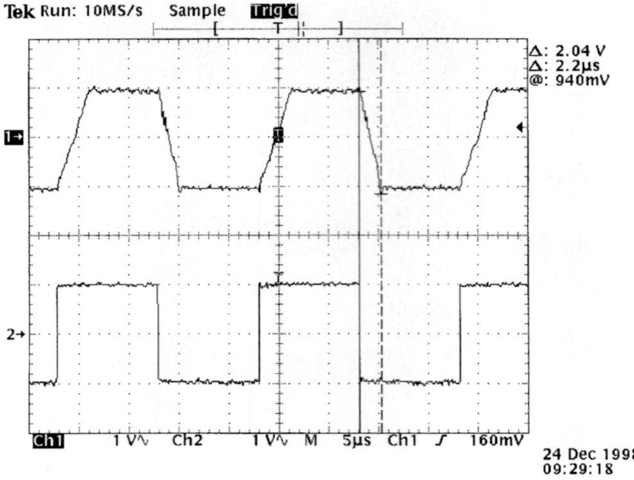

Fig. 12. Measured large signal response of proposed buffer amplifier with a $2V_{pp}/100$ kHz step. The buffer amplifier is connected as an unity-gain follower. (Lower curve—input signal, upper curve—output signal).

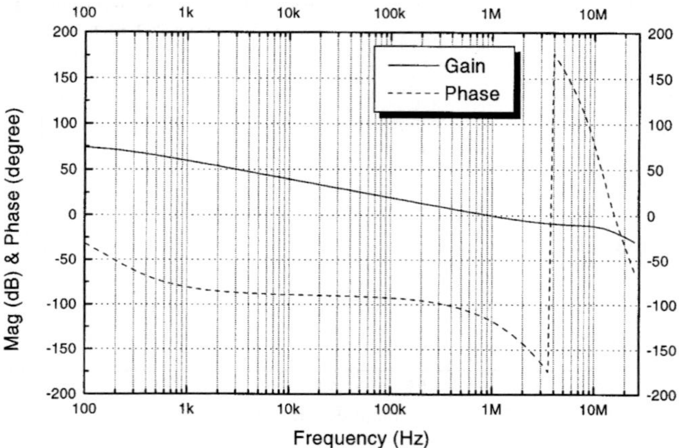

Fig. 13. AC response of the buffer amplifier.

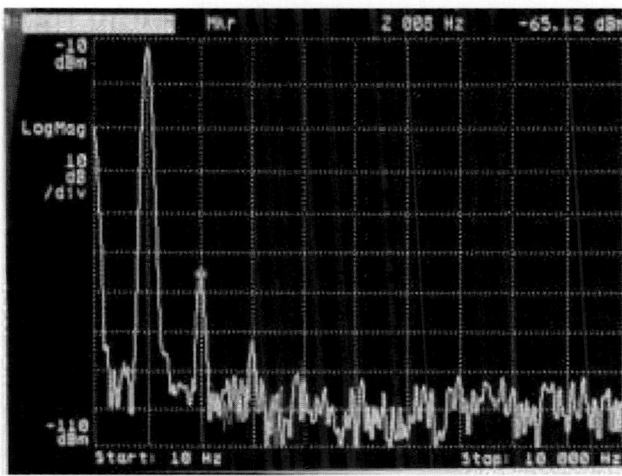

Fig. 14. Magnitude spectrum of the proposed buffer amplifier when connected in an unity-gain follower with 1 kHz sine wave input.

Large signal transient response of the proposed buffer amplifier with a $2V_{pp}/50$ kHz step input is shown in Fig. 12 while the measured magnitude spectrum of the proposed buffer amplifier when connected in an unity-gain follower with 1 kHz sine wave input is plotted in Fig. 14. Performance of the buffer amplifier is summarized in Table 1.

Table 1. Performance summary of the proposed buffer amplifier ($V_{supply} = \pm1.5$ V, $R_L = 300$ Ω, $C_L = 33$ pf)

Parameters		Measured Results
A_{vol}		74 dB
F_u		1 MHz
Phase Margin		62°
PSRR +	(DC)	87 dB
	(1 kHz)	80 dB
PSRR −	(DC)	92 dB
	(1 kHz)	81 dB
THD @ 1 kHz, $2V_{pp}$, 300 Ω load		54 dB
Slew Rate		positive 0.6 V/μs, negative 0.9 V/μs
V_{offset}		2.52 mV
Quiescent Power Dissipation		1.43 mW
Output Swing		2.42 V
Input Noise Density @ 1 kHz		95 nV/$\sqrt{\text{Hz}}$

Conclusion

In this paper, a low-voltage high driving capability CMOS buffer with quiescent current control is developed. By applying adaptive loads and improved error amplifiers, the buffer amplifier achieves both excellent load driving capability and stable quiescent current. Improved quasi-complementary output stage enables the proposed buffer to work at power supply below ±1 V.

Acknowledgment

The authors would like to thank Dr. Lian Yong and Ms. Zhang Xiwen for assistance in this work. This work is supported by the Singapore National Science and Technology Board Research Grant NSTB/17/2/3. Yajun Ha would like to thank IMEC, Belgium for kindly supporting him to present partial results of this work in ICECS'99, Cyprus.

References

1. Gray, P. R. and Meyer, R. G., *Analysis and Design of Analog Integrated Circuits,* John Wiley & Sons, Inc, 1993.

2. Ahuja, B. K., Gray, P. R., Baxter, W. M. and Uehara, G. T., "A programmable CMOS dual channel interface processor for telecommunications applications." *IEEE J. Solid-State Circuits* 19, pp. 892–899, December 1984.

3. Kih, J., Chang, B., Jeong, D. K. and Kim, W., "Class-AB large-swing CMOS buffer amplifier with controlled bias current." *IEEE J. Solid-State Circuits* 28, pp. 1350–1353, December 1993.

4. Brehmer, K. E. and Weiser, J. B., "Large swing CMOS power amplifier." *IEEE J. Solid-State Circuits* 29, pp. 624–629, December 1983.

5. Fisher, J. A., "A high-performance CMOS power amplifier." *IEEE J. Solid-State Circuits* 20, pp. 1200–1205, December 1985.

6. Nagaraj, K., "Large-swing CMOS buffer amplifier." *IEEE J. Solid-State Circuits* 24, pp. 181–183, February 1989.

7. Mistlberger, F. and Koch, R., "Class-AB high-swing CMOS power amplifier." *IEEE J. Solid-State Circuits* 27, pp. 1089–1092, July 1992.

8. You, F., Embabi, S. H. K. and Sinencio, E. S., "Low-voltage class AB buffers with quiescent current control." *IEEE J. Solid-State Circuits* 33, pp. 915–919, June 1998.

9. Chollet, F., Tang, X. S., Liu, A. Q., Ha, Y. and Li, M. F., "Micromachined shutter and low-voltage electronics for optical displacement/acceleration sensing," in *Proceedings of the 10th Int. Conf. on Solid-State Sensors and Actuators,* 1999.

10. Hogervorst, R. and Huijsing, J. H., *Design of Low-Voltage, Low-Power Operational Amplifier Cells,* Kluwer Academic Publishers, 1996.

11. Ha, Y., Li, M. F. and Liu, A. Q., "Low voltage high driving capability CMOS buffer used in MEMS interface circuits," in *Proceedings of the 6th Int. Conf. on Electronics, Circuits and Systems,* 1999.

Yajun Ha received the Bachelor of Engineering degree from the Department of Information and Electronics at the Zhejiang University, China, in 1996. From 1996 to 1997, he was a research engineer with the Shanghai Aerospace Bureau. In 1999, he received the Master of Engineering degree in electrical engineering from the National University of Singapore. He is currently working toward the Ph.D. degree at the Katholieke University of Leuven (K. U. Leuven), Belgium, and supported by the Inter-university Microelectronics Center (IMEC), Belgium. His research interests include analog VLSI circuit design, reconfigurable systems, VLSI systems design, and design automation.

Associate Professor **A. Q. Liu** received his Ph.D. in Applied Mechanics from National University of Singapore (NUS) in 1994. His M.S. degree was in Applied Physics, and B. Eng. Degree was in Mechanical Engineering from Xi'an Jiaotong University. He started to explore MEMS technology in 1995 when he had worked in the DSO National Laboratory. In 1997, he joined Institute of Materials Research & Engineering (IMRE), National University of Singapore, as a senior research fellow, to establish and drive the MEMS program, and build up MEMS core technology. Currently, he is an associate professor of Division of Microelectronics, School of Electrical & Electronic Engineering, Nanyang Technological University (NTU). His research interest is optical and RF MEMS technology in infocomm applications. He has implemented MEMS technology in a number of devices related to positive optical network (PON) systems, such as OXCs and add/drop multiplexers. Integration fabrication process, RF devices and electronic interface circuitry are also his major contribution areas.

Li Ming-Fu graduated from the Department of Physics, Fudan University, Shanghai, China in 1960. After graduation he joined the Department of Applied Physics, University of Science and Technology of China (USTC) as a teaching assistant and lecturer. In 1978, he joined the graduate school faculty, Chinese Academy of Sciences, Beijing, first as an associate professor and, in 1986, a professor. He has also served as adjunct professor at the Institute of Semiconductors, Chinese Academy of Sciences, Fudan University, and USTC, Hefei.

He was a visiting scholar at Case Western Reserve University, Cleveland, OH, in 1979, University of Illinois at Urbana-Champain from 1979 to 1981, and was a visiting scientist at University of California at

Berkeley and Lawrence Berkeley National Laboratories from 1986 to 1987, 1990 to 1991, and 1993. He joined the Department of Electrical Engineering, National University of Singapore, as an associate professor in 1991, and a professor in 1996. His current research interests are in the areas of reliability physics in deep sub-micron CMOS devices, analog IC design, and wide energy gap group III nitride. He has published over 140 research papers and two books, including *Modern Semiconductor Quantum Physics* (World Scientific, 1994). He has served on several international program committees and advisory committees in international semiconductor conferences in China, Japan, Canada, Germany, and Singapore.

IEEE TRANSACTIONS ON CIRCUITS AND SYSTEMS PART I: FUNDEMENTAL THEORY AND APPLICATIONS, VOL. 47, NO. 1, JANUARY 2000   1

A Low-Voltage CMOS OTA with Rail-to-Rail Differential Input Range

M. F. Li, Uday Dasgupta, X. W. Zhang, and Yong Ching Lim, *Senior Member, IEEE*

Abstract—This paper describes a novel circuit design technique for a low-voltage CMOS operational transconductance amplifier (OTA) where the output current versus the input voltage relationship is linear for differential input voltage range extended from rail to rail. The input transconducting CMOS transistors operate in the nonsaturation region. A pair of complementary n-type input OTAN and p-type OTAP circuits in conjunction with n-MOS and p-MOS output current mirrors are connected in parallel to implement the rail-to-rail voltage input and push–pull current output. The transconductances g_{mn} and g_{mp} of the OTAN and the OTAP are tuned by biasing voltages $VDSN$ and $VDSP$, respectively. A masterbias generator (with $VDSP$ as input) is used to generate $VDSN$ as output so as to satisfy $g_{mn} = g_{mp}$. The circuit is capable of operating at supply voltage larger than two times the MOS transistor threshold voltage V_T. When the supply voltage is 2 V and the MOS transistor threshold voltage is 0.7 V, using SPICE level 3-2 μm CMOS technology device parameters, the simulation result of the output current deviation from perfect linearity is less than 0.3% for rail-to-rail differential input voltage range.

I. Introduction

T HE TREND in the use of low-voltage power supplies for CMOS integrated circuits is set by the reliability issue of small size MOSFET transistors[1] and the increasing use of low-weight long life battery-operated portable electronic systems. However, reduction of power supply voltage reduces the signal dynamic range and signal to noise ratio. Therefore, recent designs attempt to maximize the dynamic range by incorporating rail-to-rail input/output capability. There have been several reports on the design of rail-to-rail low-voltage operational amplifiers [2], [3] but none on the OTA's of rail-to-rail linear differential input specification.

In rail-to-rail differential input operational amplifier circuits, the common mode input voltage extends to rail-to-rail, however, the differential input voltage is very small. Common mode rail-to-rail capability is achieved usually by using an n-MOS differential pair with transconductance g_{mn} and a p-MOS differential pair with transconductance g_{mp} in parallel [2], [3]. The total transconductance of the input stage is

$$g_{mT} = g_{mN} + g_{mP}.$$

In the standard CMOS technology, only enhancement-mode MOS transistors are used. When the common mode input voltage is near the negative power supply V_{SS}, only the p-MOS

pair operates. When the common mode input voltage is near the positive power supply V_{DD}, only the n-MOS pair operates. An auxiliary circuit is used to adjust the biasing currents of the n-MOS pair and p-MOS pair such that the value of g_{mT} remains constant in the whole common mode input range [3]. When used in active filter applications, the OTA with rail-to-rail input has more stringent input requirements than an operational amplifier because the differential mode input is also desired to extend to rail-to-rail. In this paper, in order to simplify discussion, we define $V_{SS} = -V_{DD}$. There are four possible combinations for the two input voltages V_{i+} and V_{i-} of a differential input OTA: the signs of V_{i+} and V_{i-} may be $+.+$, $+.-$, $-.+$, $-.-$, respectively. The OTA with rail-to-rail input capability should guarantee a constant g_m for all possible V_{i+} and V_{i-} input combinations. Obviously, the circuit architecture in a rail- to-rail operational amplifier cannot be used in the OTA with rail-to-rail input capability because the parameters g_{mn} and g_{mp} cannot be adjusted to satisfy four different cases. Therefore, new circuit architecture should be considered.

II. Single Ended Input Rail-to-Rail OTA Architecture

We first consider the single ended input rail-to-rail OTA circuit. It is a very useful building block in many active filter designs. The sign of input V_i may be $+$ or $-$ and the circuit architecture includes two adjustable parameters ($VDSN$ and $VDSP$, as defined in (4) and (9P)). The architecture of an OTA is shown in Fig. 1. The OTA consists of two complementary building blocks, the OTAN and OTAP, connected in parallel. When the sign of V_i is $+$, only the OTAN operates and when the sign of V_i is $-$, only the OTAP operates. The input stage of the OTAN consists of $N1$–$N4$, $P5$, $P6$, AN$_1$, and AN$_2$, as shown in Fig. 2. It is similar to a circuit structure proposed in [4]. The input n-MOS transistors $N1$ and $N2$ operate in the nonsaturation region [4]–[7] with fixed drain-to-source voltage V_{DS}. They are the transconducting elements. The I–V equation of an n-MOS transistor in the nonsaturation region is expressed by [8]

$$I_D = K_n.(V_{GS} - V_{TN}).V_{DS} - (1/2).K_n.V_{DS}^2 \quad (1)$$

where

I_D drain current;
V_{GS} gate to source voltage;
V_{DS} drain to source voltage;
V_{TN} threshold voltage of the n-MOS transistor.

$$K_n = (\mu_n \varepsilon / t_{ox})(W/L), \quad (2N)$$

Manuscript received May 5, 1997; revised April 28, 1999. This work was supported inpart by the Singapore National Science and Technology Board under Grant NSTB/17/2/3. This paper was recommended by Associate Editor J. E. Franca.
The authors are with the Department of Electrical Engineering, National University of Singapore, Singapore, 119260.
Publisher Item Identifier S 1057-7122(00)00719-4.

2   IEEE TRANSACTIONS ON CIRCUITS AND SYSTEMS PART I: FUNDEMENTAL THEORY AND APPLICATIONS, VOL. 47, NO. 1, JANUARY 2000

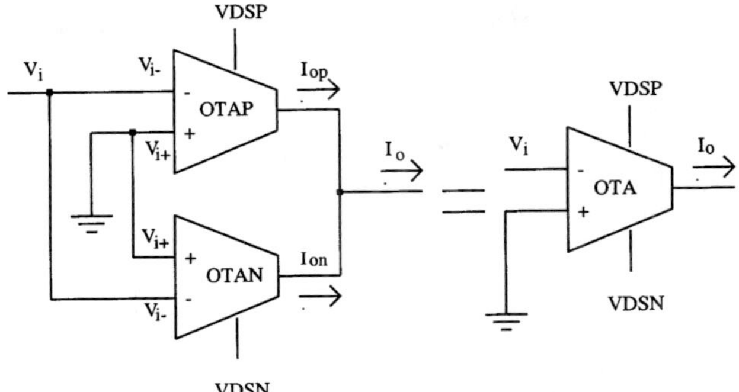

Fig. 1. An OTA consists of the OTAN and OTAP in parallel connection. The circuit diagrams of the OTAN and OTAP are shown in Fig. 2 and Fig. 5, respectively. $VDSN$ is used to control the transconductance g_{mn} of the OTAN. $VDSP$ is used to control the transconductance g_{mp} of the OTAP.

where:

μ_n	electron mobility;
ε and t_{ox}	permittivity and thickness of the gate oxide, respectively;
W and L	channel width and channel length of the MOS transistor, respectively.

The nonsaturation condition of the n-MOS transistor is

$$V_{DS} < (V_{GS} - V_{TN}). \tag{3}$$

In Fig. 2, N_1 and N_2 are input transistors. Two amplifiers AN_1 and AN_2 and two transistors N_3 and N_4 constitute two negative feedback loops. If the gains of the amplifiers AN_1 and AN_2 are large enough, from the virtual short principle, we have

$$V_{DSN1} = V_{DSN2} = VDSN. \tag{4}$$

Using (1), the difference between the drain currents of N_1 and N_2 is

$$I_{DN1} - I_{DN2} = (K_n.VDSN).(V_{GN1} - V_{GN2}) \tag{5}$$

where V_{GN1} and V_{GN2} are the gate voltages of input transistors N_1 and N_2, respectively. It can be seen from (5) that $I_{DN1} - I_{DN2}$ is linear with respect to $(V_{GN1} - V_{GN2})$ when $VDSN$ is a constant. From (3) it can be easily shown that the following relations must hold to ensure the validity of (5):

$$V_{GN1} > V_{TN} + VDSN + V_{SS} \tag{6N}$$

$$V_{GN2} > V_{TN} + VDSN + V_{SS}. \tag{7N}$$

The current mirror $N_7 - N_8$ in Fig. 2 has two functions. First, owing to the current mirror, the OTAN output can only sink current when $V_{GN1} > V_{GN2}$. When V_{GN1} is less than or equal V_{GN2}, the output current is zero. This is a desired characteristic of the OTAN, as will be explained later. Second, the mirror may have a current multiplication factor M_N where M_N is the ratio of the channel widths of N_7 to N_8 for the same channel length. A large M_N value improves the power efficiency of the circuit and the output voltage range. N_9 and AN_3 constitute an improved low-voltage regulated cascode circuit [9], [10] stacked on N_8 to increase the output impedance of the OTAN. AN_1–AN_3 have the same circuit as shown in Fig. 3. Although the circuit structure of (N_8, N_9, AN_3) is similar to the input circuit structure of (N_1, N_3, AN), N_8 operates in the saturation region to maintain the high output impedance of the OTAN. The control voltage V_{RN} of AN_3 should satisfy

$$V_{RN} \geq \sqrt{\frac{2LI_{DN_8\mathrm{MAX}}}{K_N W}}. \tag{8}$$

Here $I_{DN_8\mathrm{MAX}}$ is the maximum output current of N_8 and L and W are the channel length and channel width of N_8, respectively. V_{RN} should be as small as possible in order to increase the output voltage swing.

From (5), and taking M_N into consideration, the output current I_{ON} of the OTAN can thus be expressed as

$$\begin{aligned}
I_{ON} &= M_N.(I_{DN2} - I_{DN1}) \\
&= (M_N.Kn.VDSN).(V_{GN2} - V_{GN1}), \\
&\quad \text{when } V_{GN1} > V_{GN2}, \\
&= 0, \\
&\quad \text{when } V_{GN1} < V_{GN2}.
\end{aligned} \tag{9N}$$

Owing to the bounds imposed by (6N) and (7N), the OTAN does not show rail-to-rail input performance. In order to extend the input range to rail to rail, an OTAP circuit, as shown in Fig. 4, is used. The OTAP is the complementary equivalent circuit for the

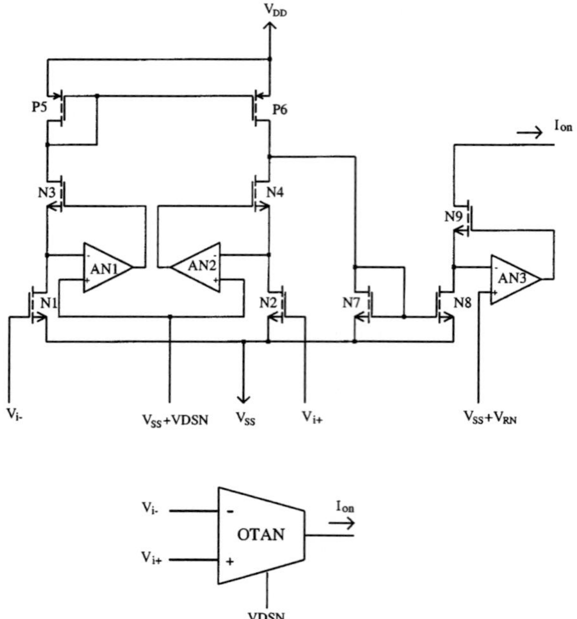

Fig. 2.   The circuit diagram of the OTAN and its symbol.

whereOTAN. In Fig. 4, AP_1 and AP_2 are the complementary amplifier circuit for AN_1 and AN_2 in Fig. 3. The voltage and current quantities of the OTAP satisfy the following equations:

$$I_{OP} = (M_P.K_P.VDSP)(V_{GP2} - V_{GP1}),$$
$$\text{when } V_{GP1} < V_{GP2},$$
$$= 0, \qquad\qquad (9P)$$
$$\text{when } V_{GP1} > V_{GP2}$$

$$V_{GP1} < V_{DD} - |V_{TP}| - VDSP \qquad (6P)$$

$$V_{GP2} < V_{DD} - |V_{TP}| - VDSP \qquad (7P)$$

$$K_P = (\mu_P \varepsilon / t_{ox})(W/L) \qquad (2P)$$

V_{GP1} and V_{GP2} gate voltages of two input p-MOS transistors;

V_{TP} threshold voltage of p-MOS transistors;

μ_p hole mobility;

$VDSP$ source to drain voltage of input transistors P_1 and P_2 in the OTAP.

The OTAP can only source current when $V_{GP1} < V_{GP2}$. When the OTAN and OTAP are connected in parallel, as shown in Fig. 1, V_{GN1} in the OTAN and V_{GP1} in the OTAP are connected together to the input voltage V_i of the OTA. The output current of the OTA is equal to the sum of the currents from the OTAN and OTAP. When $V_i > 0$, the OTAN output sinks current at the same time the OTAP output current is zero. When $V_i < 0$, the output current is sourced by the OTAP and OTAN output current is zero. In other words, the output is a push-pull circuit. According to (9N) and (9P), the output current I_O of the OTA is expressed by

$$I_O = I_{ON} = -(M_N.K_n.VDSN).V_i,$$
$$\text{when } V_i > \text{ or } = 0,$$
$$= I_{OP} = -(M_P.K_P.VDSP).V_i,$$
$$\text{when } V_i < \text{ or } = 0. \qquad (10)$$

The bounds imposed by (6N), (6P), (7N), (7P), can now be replaced by a unified bound of power supply voltage

$$V_{DD} - V_{SS} = 2V_{DD} > V_{TN} + |V_{TP}| + VDSN + VDSP. \qquad (11)$$

4   IEEE TRANSACTIONS ON CIRCUITS AND SYSTEMS PART I: FUNDEMENTAL THEORY AND APPLICATIONS, VOL. 47, NO. 1, JANUARY 2000

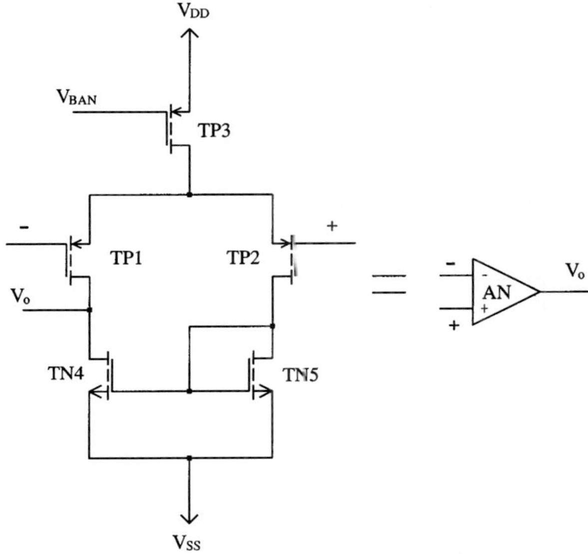

Fig. 3.   The circuit diagram of the amplifier AN_1–A_4 and its symbol.

It can be shown by considering Figs. 2–4 that the magnitudes of $VDSN$ and $VDSP$ may be very small if the aspect ratios (W/L) of TN_5, TN_4 in Fig. 3 and the complementary transistors in AP in Fig. 4 are very large. The values of $VDSN$ and $VDSP$ are determined by the transconductance values depicted in (10).

It can be seen from (10) that in order to ensure linearity over the entire range of V_i, the transconductances of the OTAN and OTAP must be equal. Denote this transconductance by GM. Hence

$$M_N > K_n.VDSN = M_P.K_P.VDSP = GM. \qquad (12)$$

Taking note of (12), (10) can now be rewritten as

$$I_O = -GM.V_i \qquad (13)$$

for the entire rail-to-rail input voltage range.

The biasing voltages $VDSN$ and $VDSP$ cannot be selected independently if (12) is to be satisfied. Fig. 5 shows a circuit which will produce an output voltage $VDSN$ satisfying (12), given $VDSP$. In Fig. 5 $I_{OP} = M_P.K_P.VDSP.V_{DD}$ and the voltage across the resistor R_{S1} is $V_P = R_{S1}.M_P.K_P.VDSP.V_{DD}$. The voltage across the resistor R_{S2} is $V_N = R_{S2}.M_N.K_N.VDSN.V_{DD}$. For $R_{S1} = R_{S2}$ we have $M_P.K_P.VDSP = M_N.K_N.VDSN$, satisfying (12).

Since in an analog or mixed signal VLSI system, many identical OTA's are needed, the circuit in Fig. 7 can now be used

as a masterbias $VDSN$ generator which can be shared among numerous OTA's within a VLSI system. $VDSP$ adjusts the transconductance value of all the OTA's simultaneously. It can be used for the tuning of the time constant in filter applications.

III Differential Input Rail-to-Rail OTA Architecture

The OTA described so far is single-ended input. Nevertheless, it can be extended to a differential input operational transconductance amplifier DOTA, as shown in Fig. 6. The DOTA consists of two complementary building blocks, the DOTAN and DOTAP, connected in parallel. Fig. 7 shows the circuit of the DOTAN. N_1–N_9, P_5, P_6 and AN_1–AN_3 constitute an OTAN circuit with V_{GN2} connected to zero. By (9N) we have

$$\begin{aligned} I_{DN9} &= (M_N.K_N.VDSN).V_{GN1}, & \text{when } V_{GN1} > 0 \\ &= 0, & \text{when } V_{GN1} < 0. \end{aligned} \qquad (14)$$

N_{10}–N_{13}, P_{14}–P_{18}, and AN_4, AP' constitute an OTAN' circuit, which is similar to an OTAN except that: 1) the inverting input V_{GN10} is connected to ground while the noninverting input V_{GN11} is connected to the V_{i+} input terminal of the DOTAN. 2) The drain to source voltages V_{DSN10} and V_{DSN2} are stabilized by the same amplifier AN_2 feedback loop. 3) The output current mirror is replaced by p-MOS transistors P_{16}, P_{17} with the same current multiplication factor M_N and

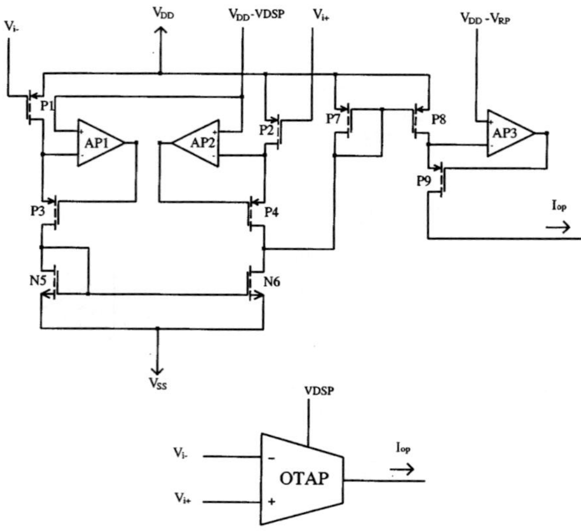

Fig. 4.   The circuit diagram of the OTAP and its symbol.

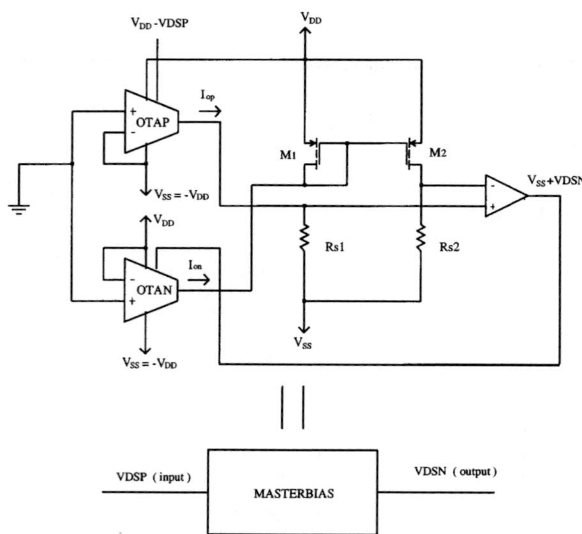

Fig. 5.   Masterbias circuit. The input voltage is $VDSP$. The output voltage is $VDSN$.

6   IEEE TRANSACTIONS ON CIRCUITS AND SYSTEMS PART I: FUNDAMENTAL THEORY AND APPLICATIONS, VOL. 47, NO. 1, JANUARY 2000

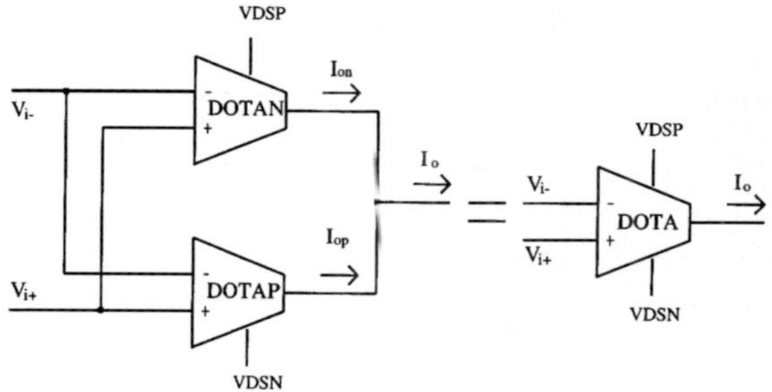

Fig. 6.   Differential input operational transconductance amplifier DOTA.

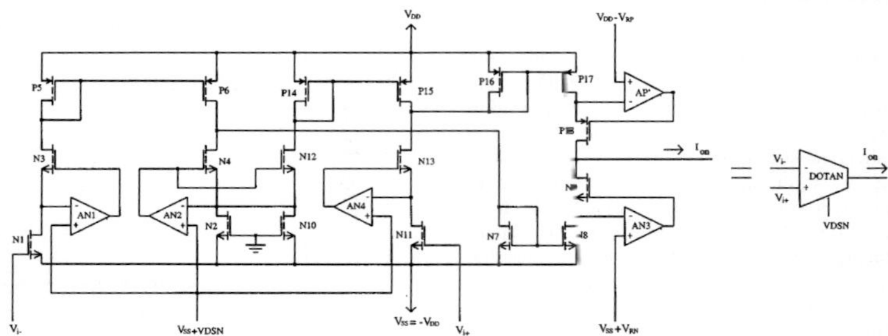

Fig. 7.   The circuit diagram of the DOTAN and its symbol.

related regulated cascode circuit P_{18} and AP'. The output current of (14) should be replaced by

$$I_{DP18} = (M_N.K_N.V_{DSN}).V_{GN11}, \quad \text{when } V_{GN11} > 0$$
$$= 0, \qquad\qquad\qquad\quad \text{when } V_{GN11} < 0. \tag{15}$$

The DOTAP in Fig. 6 is the complementary equivalent circuit for the DOTAN. For similar reasons as those of the DOTAN, the output current of the DOTAP can be written as

$$I_{DP9} = -(M_P.K_P.VDSP).V_{GP1}, \quad \text{when } V_{GP1} < 0$$
$$= 0, \qquad\qquad\qquad\qquad \text{when } V_{GP1} > 0 \tag{16}$$

$$I_{DN18} = -(M_P.K_P.VDSP).V_{GP11}, \quad \text{when } V_{GP11} < 0$$
$$= 0, \qquad\qquad\qquad\qquad\quad \text{when } V_{GP11} > 0. \tag{17}$$

Comparing Figs. 7, 8, and (12) and 14–(17), the output current of the DOTA is

$$I_O = I_{DP18} - I_{DN9} + I_{DP9} - I_{DN18} = GM.(V_{i+} - V_{i-}) \tag{18}$$

for all possible rail-to-rail input V_{i+} and V_{i-}. Furthermore, we have

$$V_{i+} = V_{GP11} = V_{GN11} \tag{19}$$

$$V_{i-} = V_{GN1} = V_{GP1}. \tag{20}$$

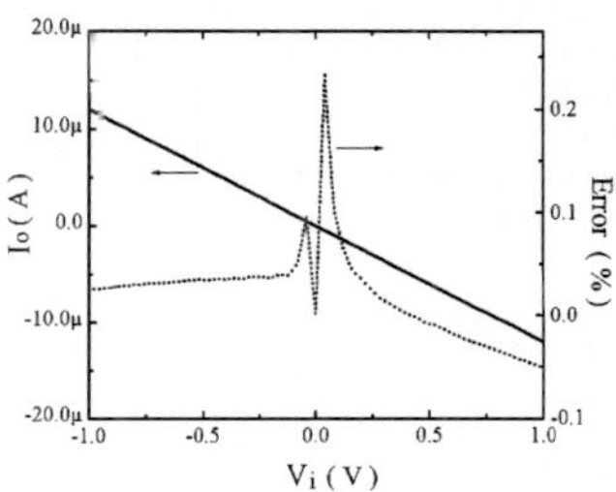

Fig. 8.   Simulation result of the OTA in Fig. 1. DC transfer curve of the output current I_O and linearity error defined by (21) ($V_{DD} = -V_{SS} = 1$ V).

The biasing circuit of the DOTA's in an analog or mixed signal VLSI system is identical to that of the OTA's, as shown in Fig. 5. Only the OTAN and OTAP in Fig. 5 are replaced by the DOTAN and DOTAP respectively.

IV. SIMULATION RESULTS AND DISCUSSION

The OTA and DOTA circuit have been simulated by HSPICE, using SPICE level-three 2-µCMOS technology device parameters. The supply voltage $2V_{DD}$ is 2 V. The MOS transistor

TABLE I
LIST OF TRANSISTOR SIZES IN THE OTAN
AND OTAP CIRCUITS

CELL	MOSFET NO.	W (um)	L (um)
OTAN	N1, N2	15	5
	N3, N4	300	2
	P5, P6	60	3
	N7	62	2
	N8	adjustable	2
	N9	160	2
OTAP	P1, P2	18	5
	P3, P4	700	2
	N5, N6	40	3
	P7	75	2
	P8	adjustable	2
	P9	360	2
AN	TP1, TP2	100	3
	TP3	30	3
	TN4, TN5	100	3

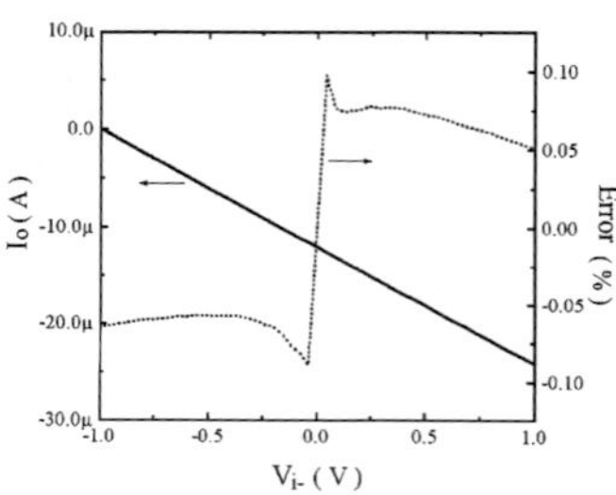

Fig. 9.   Simulation results of the DOTA in Fig. 6, with V_{i+} connected to V_{SS} while V_{i-} is varied from rail-to-rail,—dc transfer curve of the output current I_O and linearity error defined by (21) ($V_{DD} = -V_{SS} = 1$ V).

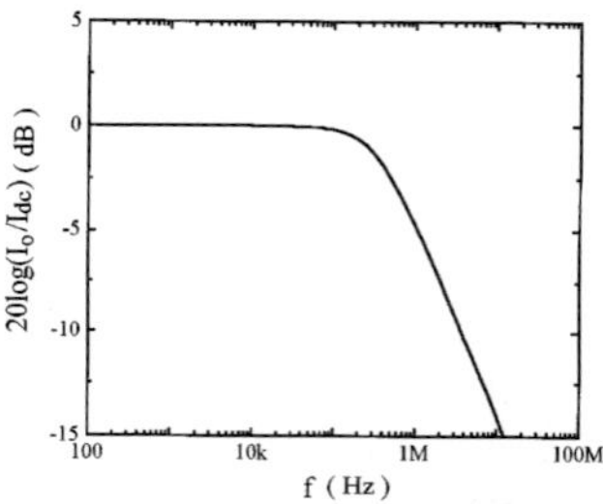

Fig. 10.   Simulation result of the frequency response of the DOTA.

threshold voltage $V_{TN} = -V_{TP} = 0.7$ V (V_{TN} can be different from $-V_{TP}$). V_{RN} in Figs. 2 and 7 and V_{RP} in Figs. 4 and 7 are 0.05 V. In obtaining a high linearity I–V transfer function, the high voltage gain of the amplifiers AN's and AP's (around 10^2 in our design) and large aspect ratios W/L (around 10^2 in our design) of N_3, N_4 in Fig. 2 and P_3, P_4 in Fig. 4 are desired to obtain very low load impedance observed at the drains of the input transistors. The transistor sizes of the OTA circuit are listed in Table I. The channel lengths of the input transistors N_1, N_2 in Fig. 2 and P_1, P_2 in Fig. 4 are 5 µm. Smaller channel lengths of the input transistors are not recommended, due to the short channel effect (which will degrade the OTA linearity), mismatch effect, and noise performance degradation. The layout of the OTA circuit occupies 0.10 mm^2. The quiescent power dissipation of one OTA is 0.21 mW. This is comparable to a recently reported fifth-order GM-C filter design using seven OTA's [11] with the total power dissipation of 2.5 mW (supply voltage 3 V) and 1.6-mm^2 cell size.

The simulation dc transfer curve $I_O = I_O(V_\text{in})$ and the linearity error defined as

$$\varepsilon = \frac{I_O(V_\text{in}) - I_O(0) - GM(0)V_\text{in}}{GM(0)V_\text{in}} 100\% \qquad (21)$$

are shown in Figs. 8 and 9. Fig. 9 is the most stringent differential input range test when the V_{i+} (or V_{i-}) of the DOTA connected to V_{DD} and V_{i-} (or V_{i-}) varies from V_{DD} to V_{SS}. The error is seen to be less than 0.3% in the rail-to-rail differential input range. This error simulation is based on the perfect match of transistor pairs in the OTA and DOTA circuits. Mismatch between components may cause larger distortion of linearity, particularly at the transition point of the OTAN and OTAP. The common mode rejection ratio (CMRR) and the power supply rejection ratio (PSRR) will be poorer than long tail differential pairs, due to mismatch of components. Quantitative estimation of the effect of mismatch depends on the fabrication technology and will be reported else where.

The simulation result of the frequency response of the DOTA is shown in Fig. 10. A –3-db cutoff frequency of over 0.55 MHz is obtained. For large signal distortion simulation, the total harmonic distortion (THD) is less than 0.3% for 2-V peak-to-peak differential input at 1 KHz.

V. CONCLUSION

The design of a rail-to-rail low voltage single ended input OTA and a rail-to-rail low voltage differential input DOTA has been demonstrated for the first time. The transconductance characteristic with respect to input voltage is linear from rail to rail and the circuit is capable of operating at low voltage provided that the supply voltage $2V_{DD}$ is larger than twice the threshold voltage of the MOS transistors. Simulation results using 2-µm technology SPICE level-three parameters show that the OTA or DOTA peak deviation from perfect linearity is less than 0.3% for rail-to-rail differential input voltage with $2V_{DD} = 2$ V.

REFERENCES

[1] T. H. Ning, P. W. Cook, R. H. Dennard, C. M. Osburn, S. E. Schuster, and H. N. Yu, "1 µm MOSFET VLSI technology, Part IV—Hot-electron design constrains," *IEEE Trans. Electron. Devices*, vol. 26, pp. 346–353, 1979.

[2] R. Hongervorst, R. J. Wiegerink, P. A. J. Jong, J. Fonderie, R. F. Wassenaar, and J. H. Huijsing, "CMOS low-voltage operational amplifiers with constant-gm rail-to-rail input stage," *Analog Integrated Circuits and Signal Processing*, vol. 5, pp. 135–146, 1994.

[3] S. Sakurai and M. Ismail, *Low-Voltage CMOS Operational Amplifiers*. Boston, MA: Kluwer Academic, 1995.

IEEE TRANSACTIONS ON CIRCUITS AND SYSTEMS PART I: FUNDAMENTAL THEORY AND APPLICATIONS, VOL. 47, NO. 1, JANUARY 2000

[4] A. Wyzynski, "Low-voltage CMOS and BiCMOS triode transconductors and integrators with gain-enhanced linearity and output impedance," *Electron. Lett.*, vol. 30, pp. 211–213, 1994.

[5] "U. S. Patent," 4 656 436, Apr. 7, 1987.

[6] J. Pennock, P. Frith, and R. G. Barker, "CMOS triode transconductor continuous time filters," in *Proc. IEEE Custom Integrated Circuits Conf.*, 1996, pp. 231–234.

[7] A. L. Coban and P. E. Allen, "Low-voltage CMOS transconductance cell based on parallel operation of triode and saturation transconductors," *Electron. Lett.*, vol. 30, p. 1124, 1994.

[8] P. R. Gray and R. G. Meyer, *Analog Integrated Circuits.* New York, NY: Wiley, 1993.

[9] E. Sackinger and W. Guggenbuhl, "A high-swing, high-impedance MOS cascode circuit," *IEEE J. Solid-State Circuits*, vol. 25, pp. 289–297, 1990.

[10] K. Bult and G. J. G. M. Geelen, "A fast-settling CMOS op amp for SC circuits with 90-dB DC gain," *IEEE J. Solid-State Circuits*, vol. 25, pp. 1379–1384, 1990.

[11] C. C. Hung, K. A. I. Halonen, M. Ismail, V. Porra, and A. Hyogo, "A low-voltage, low-power CMOS fifth-order elliptic GM-C filter for baseband mobile, wireless communication," *IEEE Trans. Circuits Syst. Video Technol.*, vol. 7, pp. 584–592, 1997.

M. F. Li graduated from the Department of Physics, Fudan University, Shanghai, China, in 1960.

After graduation, he joined the Department of Applied Physics, University of Science and Technology of China (USTC) as a Teaching Assistant and Lecturer. In 1978, he joined the Graduate School Faculty, Chinese Academy of Sciences, Beijing, China, first as an Associate Professor and, in 1986, a Professor. He has also served as Adjunct Professor at the Institute of Semiconductors, Chinese Academy of Sciences, Fudan University, and USTC, Hefei, China. He was a Visiting Scientist at Case Western Reserve University, Cleveland, OH, in 1979, the University of Illinois, Urbana-Champaign from 1979 to 1981, the University of California, Berkeley and Lawrence Berkeley National Laboratories from 1986 to 1987, 1990 to 1991, and 1993. He joined the Department of Electrical Engineering, National University of Singapore, as an Associate Professor in 1991 and a Professor in 1996. His current research interests are in the areas of reliability physics in deep submicron CMOS devices, analog IC design, and wide energy gap group III nitrides. He has published over 140 research papers and two books, including *Modern Semiconductor Quantum Physics* (World Scientific, 1994). He has served on several international program committees and advisory committees in semiconductor conferences in China, Japan, Canada, Germany, and Singapore.

Uday Dasgupta received the B.Tech. degree in electronics and electrical communication engineering from Indian Institute of Technology, Kharagpur, India, in 1976 and the M.Eng. degree from Department of Electrical Engineering, National University of Singapore in May 1999.

From 1977 to 1983 he worked as a Research and Development Engineer for Sonodyne Electronics, Calcutta, India, designing board-level high-fidelity stereo systems and heavy-duty regulated power supplies using discrete semiconductor devices. From 1983 to 1992 he was with Semiconductor Complex, Punjab, India as a CMOS analog IC design engineer working on telecom integrated circuits. From 1992 to 1995 he was with S.G.S. Thomson, Singapore, using BiCMOS process. From 1995 to 1999 he was with TriTech Microelectronics where he worked on low noise and low distortion CMOS audio IC products. Presently he is with the Institute of Microelectronics, Singapore as a Member of the Technical Staff and is working on the design of high-frequency CMOS analog IC's.

X. W. Zhang was born in Zhe Jiang, China, in 1970. She received the B.Eng. degree from the Institute of Microelectronics, Tsinghua University, China, in 1993.

From 1993 to 1996 she was a Circuit Designer in Tailai Electronic Pte., Ltd., Guang Zhou, China. She is now an analog IC designer and also a Professional Officer with the National University of Singapore. Her research interests are low-voltage low-power analog IC design.

Yong Ching Lim (S'80–M'80–SM'92) received the A.C.G.I. and the B.Sc., D.I.C. and Ph.D. degrees in electrical engineering from Imperial College, University of London, U.K., in 1977 and 1980, respectively.

From 1980 to 1982, he was a National Research Council Research Associate in the Naval Postgraduate School, Monterey, CA. Since 1982, he has been with the Department of Electrical Engineering, National University of Singapore. His research interests include VLSI circuits and systems design and digital signal processing.

Dr. Lim was selected to receive the 1996 IEEE Circuits and Systems Society's Guillemin–Cauer Award and the 1990 IREE (Australia) Norman Hayes Award. He served as an Associate Editor for the IEEE TRANSACTIONS ON CIRCUITS AND SYSTEMS from 1991 to 1993 and has been an Associate Editor for IEEE TRANSACTIONS ON CIRCUITS, SYSTEMS, AND SIGNAL PROCESSING since 1993. He served in the DSP Technical Committee of the Circuits and Systems Society as Secretary from 1996 to 1998 and as Chairman from 1998 to 2000. He is a member of Eta Kappa Nu.

A 1.2 V Rail-to-Rail Differential Mode Input Linear CMOS Transconductor

Aimin Xu and M.F.Li

Signal Processing and VLSI Design Lab, Department of Electrical and Computer Engineering
National University of Singapore, 10 Kent Ridge Crescent, Singapore 119260
Email : elelimf@nus.edu.sg

Abstract

Rail-to-rail differential mode input capability in low voltage CMOS transconductor design is implemented by a pull-down and a pull-up follower in an input buffer. By generating three intermediate voltages in the buffer stage , three voltages are converted to three currents and subsequently summed using MOSFETSs operating in the triode region . The nonlinear current components can be cancelled completely. The lowest supply voltage V_{dd} is constrained by $2V_{th}$ of the MOS transistors . A single 1.2V operational transconductor amplifier (OTA) was designed using 0.35μm CMOS technology and the simulated $I\text{-}V$ linearity error is less than 1% within the rail-to-rail differential input range. The achieved THD is less than 0.6% for a 1 KHz, 1.2 V_{pp} input signal.

1. Introduction

In low voltage IC design, it has become very important to maximize the input/output dynamic range in order to increase the signal to noise ratio [1]. In low voltage operational amplifier design, the common mode input voltage is large but differential mode voltage is small. A different requirement exists for OTA used in g_m-C active filters where the differential mode input range is of primary concern. [2-5]. In this paper, we present an method and circuit implementation using an input buffer to generate three intermediate voltages. The voltages are subsequently converted to three currents and summed up using MOSFETs operating in the triode region. The method allows rail-to-rail differential mode input in the low voltage OTA design with excellent V-I linearity conversion. The proposed OTA circuit is simulated using AMS CMOS 0.35μm process.

2. Method and circuit implementation

Fig.1 shows the conventional input topology of an OTA with the input n-MOSFET working in the triode region [2-5]., A gain-boosting amplifier feedback is then used to improve the linearity . As shown in Fig.1(a), the

amplifier feed back loop forces V_{ds} of M1 to be at a constant voltage and is approximately equal to the control voltage V_{bias}. The current of the OTA in Fig.1(b) can be written as [3]:

$$I = I(v_{in}) = \begin{cases} G_m(v_{in} - V_t - 0.5V_{bias}), & v_{in} \geq V_t + V_{bias} \\ I_{NL}(v_{in}), & v_{in} < V_t + V_{bias} \end{cases} \quad (1)$$

$$G_m = \mu C_{ox}(W/L)V_{bias} \quad (1a)$$

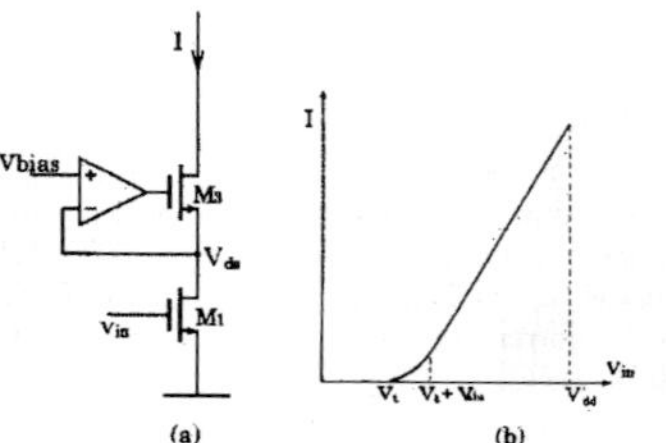

Figure 1. A gain-boosting cascode transconductor block. The input transistor operates at the triode region. The output current is linear versus input voltage when v_{in} is larger than V_t+V_{ds} . (a) The circuit diagram . (b) I versus v_{in} curve expressed by equation (1) .

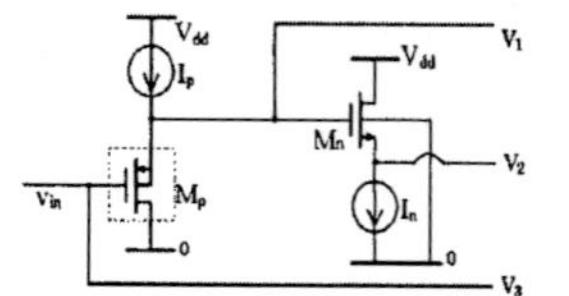

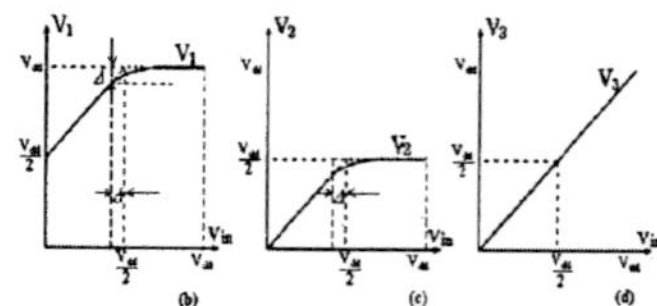

Fig.2 (a) An input buffer generates three intermediate voltages V_1, V_2 and V_3. V_1, V_2 and V_3 are functions of v_{in} as illustrated by (b)-(d) .

where $I_{NL}(v_{in})$ is a nonlinear function since the n-MOSFET operates in saturation and subthreshold regions when $v_{in} < V_t + V_{bias}$.

In our design, an input buffer as shown in Fig.2 is used to generate three buffer output voltages as a function of v_{in}. The pull-up follower M_p in Fig.2(a) provides a voltage shift V_p of v_{in} in the lower half of the input range as shown in Fig.2(b) . A constraint is set by adjusting the current source I_p and L/W of M_p to satisfy:

$$V_p = (V_{dd} / 2) > V_t + V_{bias} + \Delta \qquad (2)$$

Δ is a small voltage drop which will be defined in (5). Next, we have:

$$V_1(v_{in}) = V^{ideal} (v_{in}) + \delta (v_{in}) \qquad (3)$$

$$V^{ideal}(v_{in}) = \begin{cases} v_{in} + (0.5V_{dd}) & v_{in} \le 0.5V_{dd} \\ V_{dd} & 0.5V_{dd} < v_{in} \le V_{dd} \end{cases} \qquad (3a)$$

$\delta(v_{in})$ is a small voltage correction when the current source I_p is implemented by a p-MOSFET with source connected to V_{dd}. $\delta(v_{in})$ is not zero only when V_1 is very close to V_{dd} so that the p-MOSFET no longer operates in the saturation region. In other words,

$$\delta (v_{in}) = 0, \quad \text{when } v_{in} < (0.5V_{dd} - \Delta) \qquad (4)$$

Δ can be estimated by I_p and the size of the p-MOSFET which implementing the current source I_p. The pull-down follower M_n in Fig.2(a) provides a DC shift of $-V_n$ on V_1 . Again, the current source I_n and L/W of M_n is adjusted to satisfy:

$$V_n = V_p = 0.5 \, V_{dd} \qquad (5)$$

we have:

$$V_2(v_{in}) = V_1(v_{in}) - 0.5V_{dd} \qquad (6)$$

Finally, we introduce V_3 which is v_{in} itself:

$$V_3(v_{in}) = v_{in} \qquad (7)$$

In the second step, we convert the three output voltages in Fig.1(a) into three currents which are subsequently summed up. (for simplicity, $I_i(v_{in})$ is equivalent to $I(V_i(v_{in}))$, i = 1,2,3):

$$I_{total}(v_{in}) = I_1(v_{in}) - I_2(v_{in}) + I_3(v_{in}) \qquad (8)$$

Combining equations (1) - (7) , we can compute $I_i (v_{in})$, i = 1, 2, 3 based on two v_{in} ranges:

$$0 < v_{in} < (0.5V_{dd} - \Delta) \qquad (9a)$$

$$(0.5V_{dd} - \Delta) \le v_{in} < V_{dd} \qquad (9b)$$

We find that $I_{total}(v_{in})$ satisfies the same linear equation in the rail-to-rail range $0 \le v_{in} \le V_{dd}$:

$$I_{total}(v_{in}) = G_m [v_{in} + V_{dd}/2 - V_t - 0.5V_{bias}], \qquad (10)$$

This result is illustrated in Fig.2(b)(c)(d). In the input range of (9a), V_1 is a linear function of v_{in} and the V-I conversion as defined by equation (1) is also linear. Therefore I_1 is a linear function of v_{in} . On the other hand, $V_2 = V_3$ is in the nonlinear region of V-I conversion. However the nonlinear components of I_2 and I_3 cancel out each other completely. In the input range of (9b), the V-I conversion of V_1, V_2 and V_3 are all linear according to equations (1) and (2). Both V_1 and V_2 are nonlinear with v_{in} however $V_1 - V_2$ is a constant . Therefore the nonlinear components of I_1 and I_2 cancel out each other completely and therefore $I_1 - I_2$ is a constant equivalent to $G_m V_{dd}/2$. Fig.3 shows the simulated results of I_1, I_2, I_3 and $I_1 - I_2 + I_3$.

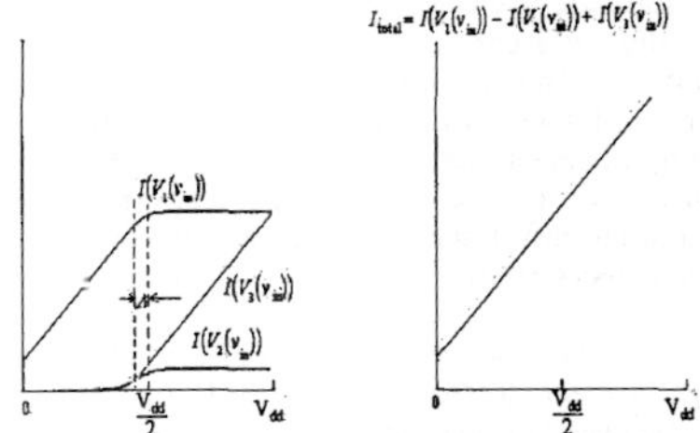

Fig.3 Simulation results of I_1, I_2, I_3 and $I_1 - I_2 + I_3$.

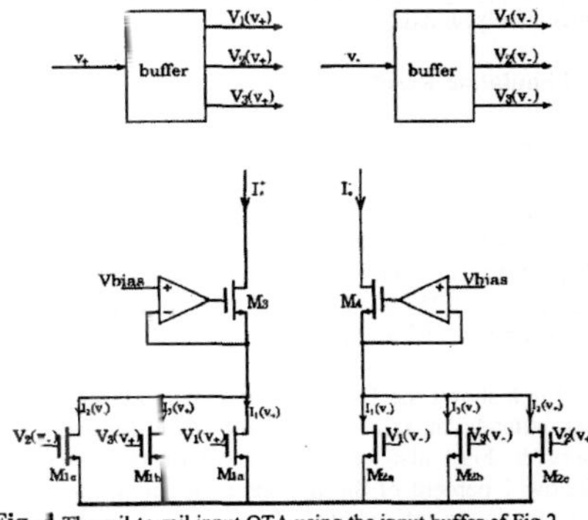

Fig. 4 The rail-to-rail input OTA using the input buffer of Fig.2

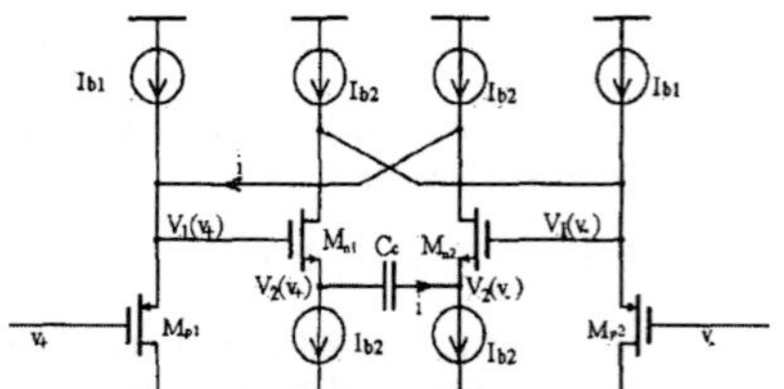

Fig.5 The improved buffer circuit . The parasitic capacitance compensation technique is used .

Fig.4 is the circuit implementation of the proposed linear OTA. From Fig.4 and (8) and (10), we have:

$$I_{out} = I_o^+ - I_o^- =$$
$$(I_1(v_+) + I_2(v_-) + I_3(v_+)) - I_1(v_-) + I_2(v_+) + I_3(v_-))$$
$$= I_{total}(v_+) - I_{total}(v_-) = G_m(v_+ - v_-) \qquad (11)$$

M_n substrate is connected to ground and therefore V_n is actually not a constant due to the body effect of the n-MOSFET. This will introduce some linearity error in (6) but it will not significantly affect the final linearity of the OTA because in the $V_{in} < V_{dd}/2$ region when this body effect should be considered, the transistor current is closed to zero. Therefore the current linearity error is a second order effect.

Fig.5 is an improved version of the buffer circuit in Fig.2 realizing the functions $V_1(v_+)$, $V_2(v_+)$, $V_1(v_-)$, $V_2(v_-)$. A compensation capacitor C_c is used to cancel a pole introduced by the follower in the buffer, so as to improve the buffer frequency response. Detailed analysis will be published elsewhere .

3. Simulation Results

A complete OTA (Fig. 9) based on 0.35-μm n-well CMOS process was designed and simulated using Cadence spectreS. A wide-swing current mirror [3] is used to generate I_{out} from I_o^+ and I_o^-. The lowest supply voltage for this circuit is determined by (2). For $V_t \approx 0.5$ V, a $V_{dd} = 1.2V$ is used and the simulated power dissipation was 193.4 uW.

Fig.6 shows the DC I-V curve when the differential input extends to rail-to-rail. Very good linearity was observed. Fig.6 also shows the nonlinearity error which is below 1 percent of the full-scale output current . The error function is defined by:

$$error = \frac{I_{out} - G_m(0)(v_+ - v_-)}{G_m(0)V_{dd}} \qquad (12)$$

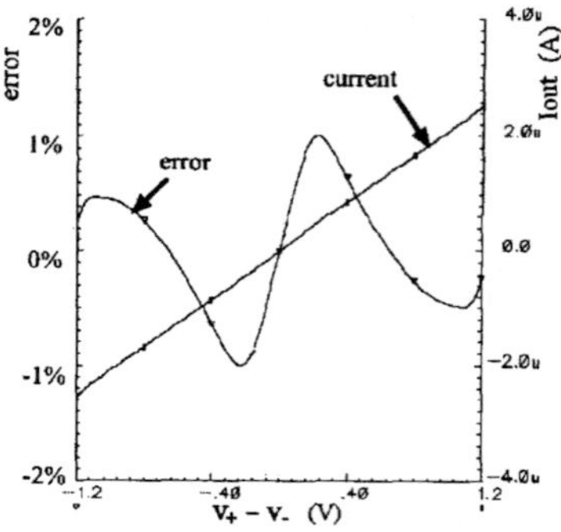

Fig.6 Simulation result of the differential input voltage versus output current of the OTA , and the linearity error defined by equation (12) . V_{dd}=1.2 V .

The THD was simulated using Hspice and the result is shown in Fig.7. From the figure, the THD is below 0.6% for the rail-to-rail input voltage at 1 KHz signal frequency .

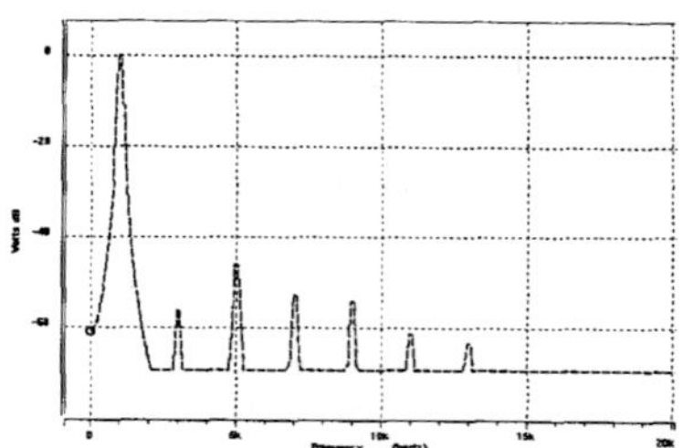

Fig.7 The FFT of the OTA output when the input is a sine wave with 1KHz frequency and 1.2 V$_{p-p}$. The harmonic distortion is below 40dB, and the total harmonic distortion (THD) is below 0.6%.

Fig. 8 shows the frequency response of an integrator composed of the OTA and a 10pF capacitor load. Table 1 shows the specification of the OTA .

Supply voltage V_{dd} (V)	1.2
Linear input voltage range	Rail-to-rail
Linearity error (Rail-to-Rail)	< 1%
THD (input V_{p-p}= 1.2V,1KHz)	0.6%
Power consummation (W)	194.6u
-3dB Bandwidth	4M
CMOS Technology	0.35µ

Table 1. Specification of the OTA.

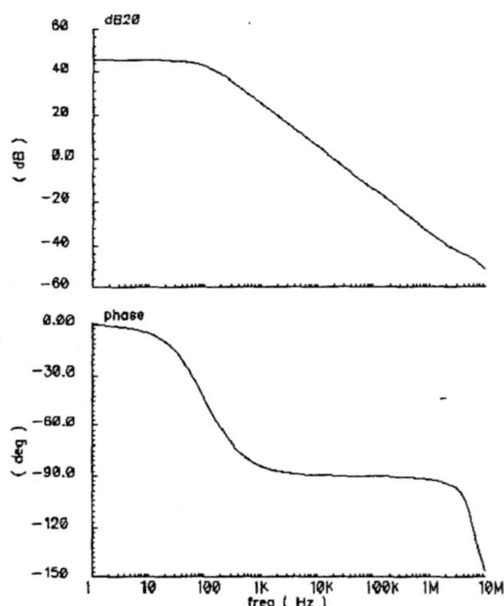

Fig. 8 Frequency response of an integrator composed of the OTA and a 10p capacitor.

4.Conclusion

A novel input structure with the capability of rail-to-rail differential mode input range in a low voltage linear CMOS OTA design is demonstrated. The input structure consists of a pull-up and a pull-down follower to produce three intermediate input voltages and subsequently converted to three currents and summed up. Using this method, the nonlinear current components are cancelled and good output current linearity is achieved. An OTA with this input structure was designed using 0.35µm CMOS technology. Using a low supply voltage of 1.2 V, the linearity of the OTA is about 1% over the rail-to-rail input range and the THD is below 0.6% for a 1.2Vp-p input at 1KHz.

Acknowledgement

This work was supported by the National University of Singapore Academic Research Fund R263-000-160-112 .

References

[1] S.Sakurai and M. Ismail , *Low-Voltage CMOS Operational Amplifiers.* Boston, MA : Kluwer Academic, 1995 .
[2] M.F.Li Uday Dasgupta, X.W.Zhang and Y.C.Lim, " A Low-Voltage CMOS OTA with Rail-to-Rail Differential Input Range " , IEEE Trans C & S Part 1 , vol. 47, no.1, pp.1-8 (2000)
[3] D.A. Johns and Ken Martin. Analog Integrated Circuit Design, 1997.
[4] A.L.Coban and P.E.Allen, " Low-voltage CMOS transconductance cell based on parallel operation of triode and saturation transconductors ", Electronic Lett., vol.30, pp.1124-1125, 1994.
[5] P. Likittamapong, A. Worapishet and C. Toumazou "Linear CMOS Triode Transconductor for low-voltage applications". *IEEE Electronics Letters*, vol. 34, no. 12, pp. 1124--1125, Jun. 1998.

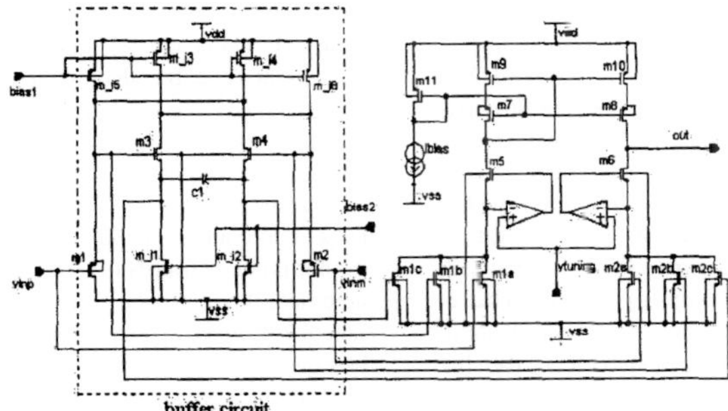

Fig.9 The complete circuit diagram of the OTA

Reprinted paper with permission from Luo Zhenying, M.F. Li, Yong Lian and S.C. Rustagi, Analog Integrated Circuits and Signal Processing, Vol.37, pp.233342 (2003). Copyright © 2003, Springer Netherlands.

Analog Integrated Circuits and Signal Processing, 37, 233–242, 2003
© 2003 Kluwer Academic Publishers. Manufactured in The Netherlands.

A New Low Voltage CMOS Transconductor for VHF Filtering Applications

LUO ZHENYING,[1] M.F. LI,[1*] YONG LIAN[1] AND S.C. RUSTAGI[2]

[1] *Signal Processing and VLSI Design Lab, Department of Electrical and Computer Engineering, National University of Singapore, Singapore 119260*
[2] *Institute of Microelectronics, Singapore 117685*
E-mail: elelimf@nus.edu.sg

Received February 14, 2002; Revised October 25, 2002; Accepted December 2, 2002

Abstract.   In this paper, a new differential input CMOS transconductor circuit for VHF filtering application is introduced. The new circuit has a very high frequency bandwidth, large linear differential mode input range and good common mode signal rejection capability. Using 0.35 μm CMOS technology with 3 V power supply, the transconductor has a ±0.9 V linear differential input range with a −54 dB total harmonic distortion (THD) and more than 1 GHz − 3 dB bandwidth. The large signal DC analysis and small signal ac analysis derived by compact equations are in line with SpectreS simulation. A 3rd order elliptic low pass g_m-C filter with a cutoff frequency of 150 MHz is demonstrated as an application of the new transconductor.

Key Words: low voltage, CMOS, transconductor, VHF, filter

1. Introduction

CMOS transconductor is a useful building block for the design of Analog and mixed signal integrated circuit systems, particularly for the design of continuous-time g_m-C filters. Over the pass few years, a few CMOS transconductor designs have been reported for high frequency continuous time signal processing applications [1–3]. In this paper, a new structure to realize the low voltage CMOS VHF transconductor is proposed. The 0.35 μm CMOS BSIM3v3 model is used in SpectreS simulation. In Section 2, using a power supply voltage of 3 V, the DC analysis shows that the linear V-I conversion of the transconductor can be achieved with a high common mode rejection and a large linear differential mode input voltage range of ±0.9 V. In Section 3, the small signal frequency analysis is derived. The result shows that a high frequency bandwidth of more than 1 GHz is achieved, with good agreement with the SpectreS simulation. An auxiliary circuit controlling the output DC voltage level is introduced in Section 4. Finally, the SpectreS simulation results of the transconductor and a 3rd order elliptic low pass g_m-C filter is presented in Sections 5 and 6.

*To whom correspondence should be addressed.

2. DC Analysis of the Transconductor

The proposed transconductor circuit is shown in Fig. 1. The idea is to create a circuit structure with minimum number of internal nodes so that the circuit structure is suitable for high frequency operation. In addition, the circuit should have a high common mode input rejection. The circuit structure in Fig. 1 is reflection symmetric about the SS′ line. When the differential mode input $V_{id} = 0$ with only the common mode input V_{cm} is applied, the input does not change the circuit symmetry. If all current mirrors are ideal with unity current reflection, it is clear from Fig. 1 that the output current $I_{out+} = I_{out-} = 0$. The circuit inherently has a good common mode rejection. Actually, checking the input at transistors M_2 and M_3, when V_{cm} is increased, the increased current through M_2 compensates the decreased current through M_3 and therefore their current summation, I_1 changes little. However, if the differential mode input V_{id} is increased, both currents through M_2 and M_3 increase and therefore their sum I_1 changes significantly. On the other hand, the differential mode input V_{id} destroys the symmetry of the circuit about the SS′ line and leads to the current sum I_2 also changes significantly in the opposite sign of I_1. Therefore $I_{out+} = -I_{out-}$, and the output current $I_{out} = I_{out+} - I_{out-}$ is increased.

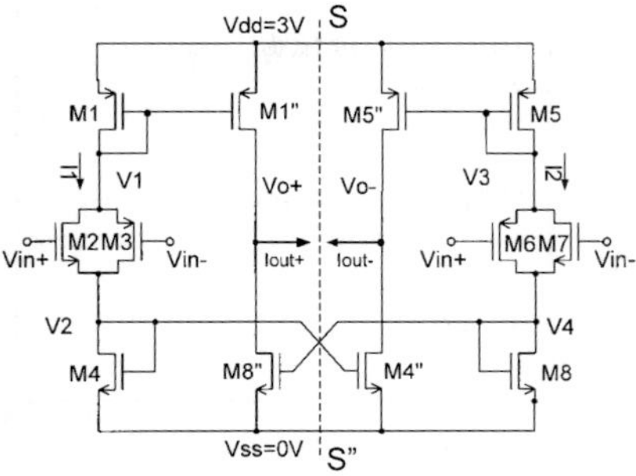

Fig. 1.   The proposed transconductor circuit.

Detailed analysis shows that I_{out} changes almost linearly with V_{id} with a transconductance coefficient almost independent of V_{cm} within a certain range. This is analyzed as follows where the long channel CMOS device I-V equations for the saturation mode operation are used [4] as a first approximation:

For *n*MOS transistors:

$$I_{\text{ds}} = K_n(V_{\text{gs}} - V_{\text{tn}})^2 \tag{1}$$

$$K_n = \frac{\mu_n C_{\text{ox}}}{2}\left(\frac{W}{L}\right)_n \tag{1A}$$

and for *p*MOS transistors:

$$I_{\text{sd}} = K_p(V_{\text{sg}} - V_{\text{tp}})^2 \tag{2}$$

$$K_p = \frac{\mu_p C_{\text{ox}}}{2}\left(\frac{W}{L}\right)_p \tag{2A}$$

where V_{tn} and V_{tp} are the *absolute value* of the *n*MOS and *p*MOS transistor threshold voltages respectively. Adjusting the W/L ratio of the *n*MOS and *p*MOS transistors to fit the following relationship:

$$K_n = K_p = \frac{\mu_{n,p} C_{\text{ox}}}{2}\left(\frac{W}{L}\right)_{n,p} = K \tag{3}$$

or $W_p/W_n = \mu_n/\mu_p \tag{3A}$

Re-writing (1) and (2) using normalized drain current:

$$I_i = (I_{\text{sd}}/K)_i \tag{4}$$

For drain current of M_1, we have:

$$I_1 = (V_{\text{dd}} - V_1 - V_{\text{tp}})^2 \tag{5}$$

for the sum of the drain currents of M_2 and M_3, we have:

$$I_1 = \left(V_{\text{cm}} + \frac{V_{\text{id}}}{2} - V_2 - V_{\text{tn}}\right)^2$$
$$+ \left(V_1 - V_{\text{cm}} + \frac{V_{\text{id}}}{2} - V_{\text{tp}}\right)^2 \tag{6}$$

for the drain current of M_4, we have:

$$I_1 = (V_2 - V_{\text{ss}} - V_{\text{tn}})^2 \tag{7}$$

Similarly for M_5, M_6, M_7 and M_8, we have:

$$I_2 = (V_{\text{dd}} - V_3 - V_{\text{tp}})^2 \tag{8}$$

$$I_2 = \left(V_{\text{cm}} - \frac{V_{\text{id}}}{2} - V_4 - V_{\text{tn}}\right)^2$$
$$+ \left(V_3 - V_{\text{cm}} - \frac{V_{\text{id}}}{2} - V_{\text{tp}}\right)^2 \tag{9}$$

$$I_2 = (V_4 - V_{\text{ss}} - V_{\text{tn}})^2 \tag{10}$$

From (5) to (10), we obtain the following result:

$$I_{1,2} = \left(-2V_{\text{tp}} - 2V_{\text{tn}} - V_{\text{ss}} + V_{\text{dd}} \pm V_{\text{id}} - \frac{\sqrt{2 \cdot (2V_{\text{cm}} \pm V_{\text{id}} - 4V_{\text{tn}} - 2V_{\text{ss}})(-2V_{\text{cm}} \pm V_{\text{id}} - 4V_{\text{tp}} + 2V_{\text{dd}})}}{2}\right)^2 \tag{11}$$

Each current mirror in Fig. 1 has a pair of identical transistors. We can easily obtain:

$$I_{\text{out}+} = -I_{\text{out}-} = I_1 - I_2 \tag{12}$$

$$\text{and}\quad I_{\text{out}} = I_{\text{out}+} - I_{\text{out}-} = 2(I_1 - I_2) \tag{13}$$

giving 0.35 μm CMOS technology typical values to V_{tn} and V_{tp} and substituting $V_{\text{dd}} = 3$ V, $V_{\text{ss}} = 0$ V into (11)

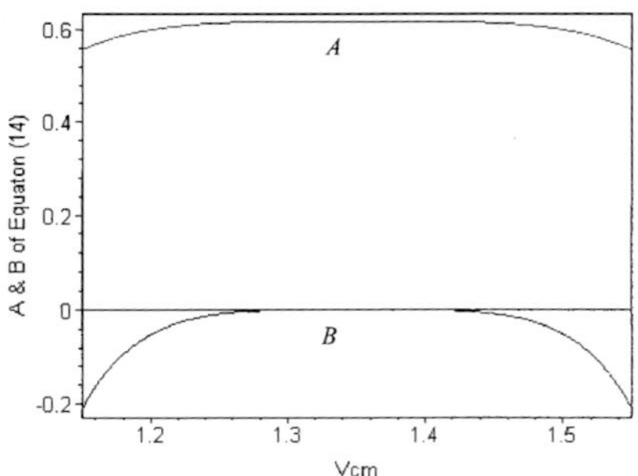

Fig. 2.   *A* in (14) is almost constant versus V_{cm} for V_{cm} from 1.2 V to 1.5 V. *B* in (14) is much smaller than *A* (less than 0.1) in this V_{cm} range. $V_{cm\text{-}ground} = (1.2 + 1.5)/2 = 1.35$ V is designated as "*common mode ground voltage*".

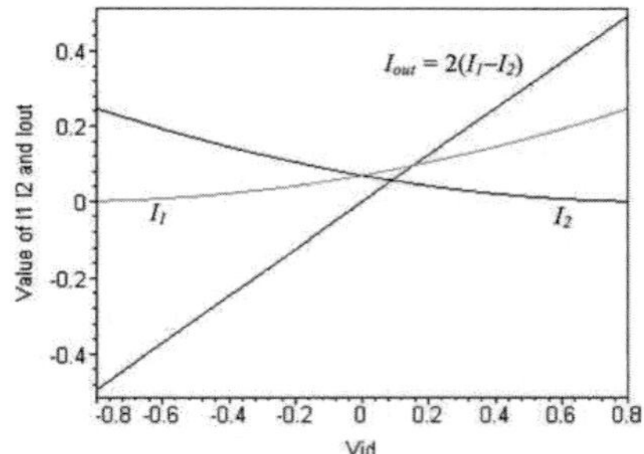

Fig. 3. I_1, I_2 and $I_{out} = 2(I_1 - I_2)$ versus V_{id} ($V_{cm} = V_{cm\text{-}ground}$).

and (12), we obtain the following Taylor expansion of I_{out}:

$$I_{out} = A \cdot V_{id} + B \cdot V_{id}^3 + O\left(V_{id}^5\right) \qquad (14)$$

Both *A* and *B* in (14) are functions of V_{cm} as plotted in Fig. 2. The analytical expressions of *A* and *B* in (14) can be found in the Appendix. As indicated in Fig. 2, the transconductance value *A* is almost a constant within the V_{cm} range:

$$1.2 \text{ V} < V_{cm} < 1.5 \text{ V} \qquad (15)$$

and *B* is very close to 0 in this range. From (15), we designate $(1.2 + 1.5)/2 = 1.35$ V as the "*common mode ground voltage*" $V_{cm\text{-}ground}$. In the system design, a common mode feedback control is used to force the output common mode voltage approaching 1.35 V.

In the above analysis, all MOS transistors in Fig. 1 operate in the saturation region and strong inversion. The following conditions must be satisfied by the input MOS transistors M_2, M_3 and M_6, M_7:

for M_2:

$$(V_{ss} + V_{tn}) + V_{tn} \leq V_{in+} \leq (V_{dd} - V_{tp}) + V_{tn}$$

$$\therefore 0.94 \text{ V} \leq V_{in+} \leq 2.85 \text{ V} \qquad (16a)$$

for M_3:

$$(V_{ss} + V_{tn}) - V_{tp} \leq V_{in-} \leq (V_{dd} - V_{tp}) - V_{tp}$$

$$\therefore -0.15 \text{ V} \leq V_{in-} \leq 1.77 \text{ V} \qquad (16b)$$

similarly, for M_6 and M_7:

$$-0.15 \text{ V} \leq V_{in+} \leq 1.77 \text{ V}, 0.94 \text{ V} \leq V_{in-} \leq 2.85 \text{ V} \qquad (16c)$$

combining (16a)–(16c), we obtain:

$$0.94 \text{ V} \leq V_{in\pm} \leq 1.77 \text{ V} \qquad (17)$$

(17) is another constraint condition for the input signal. Since $V_{cm\text{-}ground} = 1.35$ V is almost at the middle of the range defined in (17), when the common mode voltage is at $V_{cm\text{-}ground}$, the differential mode input will have a maximum AC input range. Figure 3 is the plot for (11)–(13) which shows an almost linear output current I_{out} versus the input differential voltage V_{id} while the input common mode voltage is kept on $V_{cm\text{-}ground}$.

Although the above analysis based on (1) (2) neglected the following effects: the finite output impedance [4], body effect of input nMOSs [4] and short channel effects [5], the overall specification is predicted fairly well compared with more accurate circuit simulators such as SpectreS and using the 0.35 μm BSIM3v3 model.

3.   Small Signal AC Analysis of the Transconductor

In the AC analysis of the transconductor circuit, the following approximations are used:

1. The small signal equivalent circuits as shown in Fig. 4 are used for all MOS transistors.
2. Using the same scaling factor to characterize the parasitic capacitances of nMOS and pMOS transistors. Or $C_i = \alpha_i W \cdot W$ is the channel width (while

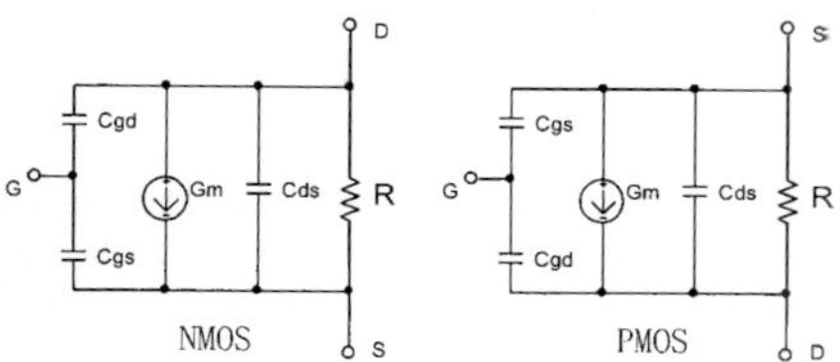

Fig. 4. *n*MOS and *p*MOS transistors small signal equivalent circuits.

the channel lengths of all transistors are the same). The index *i* specifies C_{gs}, or C_{gd} or C_{ds}. Therefore, according to (3A), C_{gs} (or C_{gd}, C_{ds}) of *p*MOS transistor is $\mu_n/\mu_p \approx 3$ times as that of *n*MOS transistor.

3. According to (1) and (2), the transconductance of the transistor is $G_m = \sqrt{2K \cdot W \cdot I_{ds}/L}$. When the common mode input voltage is at $V_{\text{cm-ground}}$, the currents through M_2 and M_3 (M_6 and M_7) are nearly equal and are half of the current through M_1 (or M_4), so the G_m of M_2 and M_3 (M_6 and M_7) is $1/\sqrt{2}$ times the G_m of M_1 and M_4 (M_5 and M_8).

4. For *n*MOS transistors and *p*MOS transistors, the output resistance is $R = V_E L/I_{\text{DSsat}}$ [4] and it roughly neglects the difference of Early voltage per unit-channel length V_E between the *n*MOS and *p*MOS transistors.

5. The output voltage is clamped to a constant voltage level when simulating the V-I response. In

other word, it is grounded during the small signal analysis. Otherwise the output node will introduce more poles or zeros depending on the load condition and cause the mathematical analysis to be too complex.

Under these approximations, the small signal equivalent circuit of the g_m-Cell is shown in Fig. 5.

Using *Kirchoff's Current Law* (KCL):

At node V_1:

$$-V_1 \cdot s(C_{ds} + 2C_{gs}) - \frac{V_1}{R} - V_1 G_m - 3V_1 s C_{gd}$$
$$+ 3(V_{\text{in\_n}} - V_1)s C_{gs} - (V_1 - V_{\text{in\_n}})\frac{G_m}{\sqrt{2}} - \frac{(V_1 - V_2)}{R}$$
$$- (V_1 - V_2)s \cdot 4C_{ds} - (V_{\text{in\_p}} - V_2)\frac{G_m}{\sqrt{2}}$$
$$- (V_1 - V_{\text{in\_p}})s C_{gd} = 0 \qquad (18)$$

At node V_2:

$$(V_{\text{in\_p}} - V_2)s C_{gs} + (V_{\text{in\_p}} - V_2)\frac{G_m}{\sqrt{2}} + (V_1 - V_2)s$$
$$\cdot 4C_{ds} + \frac{V_1 - V_2}{R} + (V_1 - V_{\text{in\_n}})\frac{G_m}{\sqrt{2}}$$
$$+ (V_{\text{n\_n}} - V_2)s \cdot 3C_{gd} - V_2 s C_{gd}$$
$$- V_2 s(C_{ds} + 2C_{gs}) - \frac{V_2}{R} - V_2 G_m = 0 \qquad (19)$$

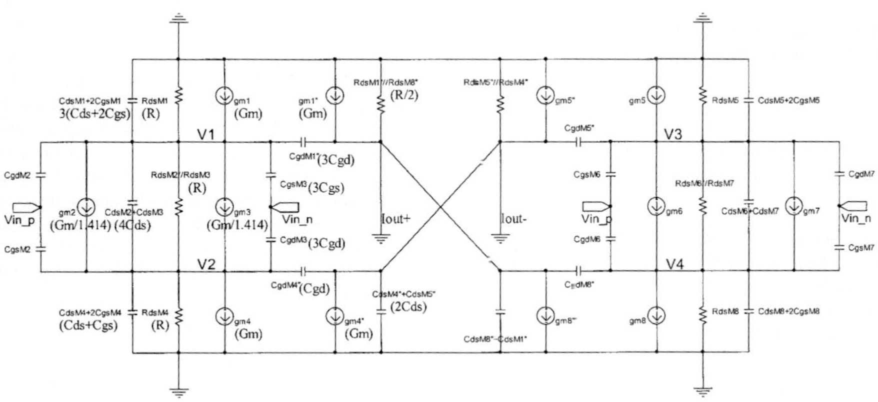

Fig. 5.   Small signal equivalent circuit of the proposed transconductor cell.

substituting $V_{\text{in}\_p} = -V_{\text{in}\_n} = V_{\text{id}}$ into (18) and (19) and solving these two equations, we obtain:

$$V_1 = V_{\text{id}}\frac{a_{21}s^2 + a_{11}s + a_{01}}{b_{21}s^2 + b_{11}s + b_{01}},$$

$$V_2 = V_{\text{id}}\frac{a_{22}s^2 + a_{12}s + a_{02}}{b_{22}s^2 + b_{12}s + b_{02}}, \quad (19a)$$

the expression for parameters a_{ij} and b_{ij} can be found in the Appendix. Since the g_m-Cell structure is reflection symmetric about line SS′, therefore:

$$\therefore V_4 = V_2|_{V_{\text{id}}\Rightarrow -V_{\text{id}}} = -V_{\text{id}}\frac{a_{22}s^2 + a_{12}s + a_{02}}{b_{22}s^2 + b_{12}s + b_{02}} \quad (19b)$$

the output current:

$$I_{\text{out}} = I_{\text{out}+} - I_{\text{out}-} = 2I_{\text{out}+} = 2\cdot(V_1 s C_{\text{gd}} - V_1 G_m + V_4 s C_{\text{gd}} - V_4 G_m)$$

$$I_{\text{out}} = K\frac{(s + z_1)(s + z_2)(s + z_3)}{(s + p_1)(s + p_2)} \cdot V_{\text{id}} \quad (20)$$

substituting the typical values of the following parameters into (20):

$$C_{\text{gd}} = 2 \times 10^{-15}\ F, C_{\text{gs}} = 11 \times 10^{-15}\ F,$$

$$C_{\text{ds}} = 18 \times 10^{-18}\ F, G_m = 400 \times 10^{-6}\ A/V, \quad (21)$$

$$R = 180 \times 10^3\ \Omega$$

after some approximation and simplification, we obtain the expression of the two poles as:

$$p_{1,2} \approx$$

$$-\frac{\left(10C_{\text{gs}} + 6.8C_{\text{gd}} \pm \sqrt{40C_{\text{gs}}^2 + 24C_{\text{gd}}C_{\text{gs}} + 8C_{\text{gd}}^2}\right) \cdot G_m}{(4C_{\text{gd}} + 9C_{\text{gs}})(4C_{\text{gd}} + 3C_{\text{gs}})}$$

Since the three zeros have very complicated expressions, which will not be shown here. After calculation, we obtain:

$$K \approx -4.0 \times 10^{-15}$$

$$p_1 \approx -0.49 \times 10^{10}, \quad p_2 \approx -1.8 \times 10^{10}$$

$$z_1 \approx -0.52 \times 10^{10}, \quad z_2 \approx -4.5 \times 10^{10},$$

$$z_3 \approx 9.1 \times 10^{10}$$

and their relationship:

$$|p_1| \cong |z_1| < |p_2| < |z_2| < |z_3|$$

here pole p_1 and zero z_1 is very closed together and can roughly be cancelled each other. Substituting the typical parameter values above, we obtain the numerical expression of I_{out}:

$$I_{\text{out}} = -4.0 \times 10^{-15}$$

$$\times \frac{(s + 4.5 \times 10^{10})(s - 9.1 \times 10^{10})}{s + 1.8 \times 10^{10}} \cdot V_{\text{id}} \quad (22)$$

The Bode plot of transfer function in (22) is shown in Fig. 6. It shows a large −3 dB bandwidth of 2.9 GHz ($1.8 \times 10^{10}/2\pi \approx 2.9$ GHz). It is in good agreement with the SpectreS simulation result in Fig. 9.

4.   Output Common Mode DC Level Stability

The output common voltage in Fig. 1 may not be at the desired level $V_{\text{cm-ground}}$ and is sensitive to process variations. Therefore, an auxiliary circuit is used to control the output common mode dc level as shown in the right half circuit of Fig. 7.

The circuit consists of N_1–N_4, N_1'', N_8'' is a copy of half of the tranconductor circuit M_1–M_4, M_1'', M_8''. N_5–N_9 is an auxiliary differential amplifier with the input of N_8 connected to the desired common mode voltage $V_{\text{cm-ground}}$ and the input of N_7 connected to the output V_{sample} (the drain of N_1'' and N_8''). N_{10} is parallel to N_8'' and is controlled by the output V_0 of the auxiliary amplifier which creates a negative feedback ensuring V_{sample} equals to $V_{\text{cm-ground}}$. M_{11}, M_{12} are the replica of N_{10} and are parallel to M_8'' and M_4'' respectively. This ensures that the output V_{o+} and V_{o-} equal to $V_{\text{cm-ground}}$ while the input of the transconductor is also set to $V_{\text{cm-ground}}$. One of the merits of this auxiliary circuit is that it does not introduce any additional internal node into the signal path, and thus will not affect the frequency response of the transconductor. On the other hand, this output dc voltage control scheme is not sensitive to the device parameter variation as has been verified by the SpectreS simulation.

5.   SpectreS Simulation Results

The following are the simulation results using SpectreS BSIM3v3 model with the device

238   *Zhenying et al.*

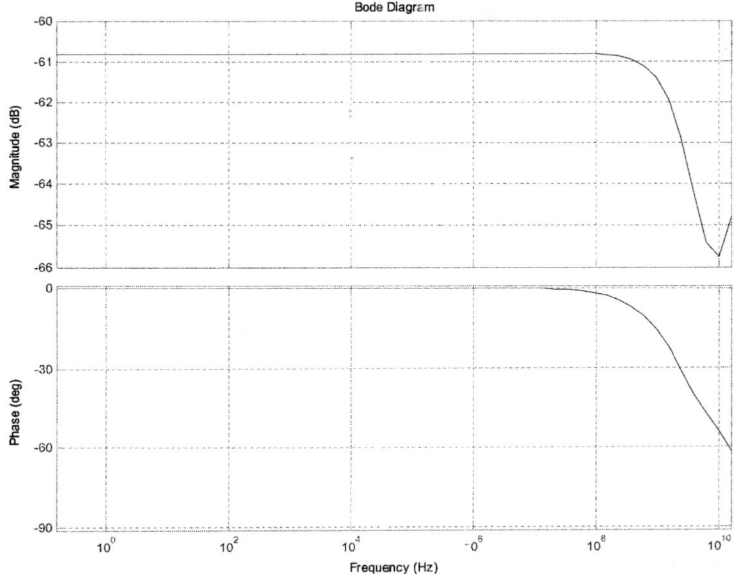

Fig. 6.   Bode plot of I_{out} versus frequency using (22). It exhibits only one pole and two zeros in the whole frequency range.

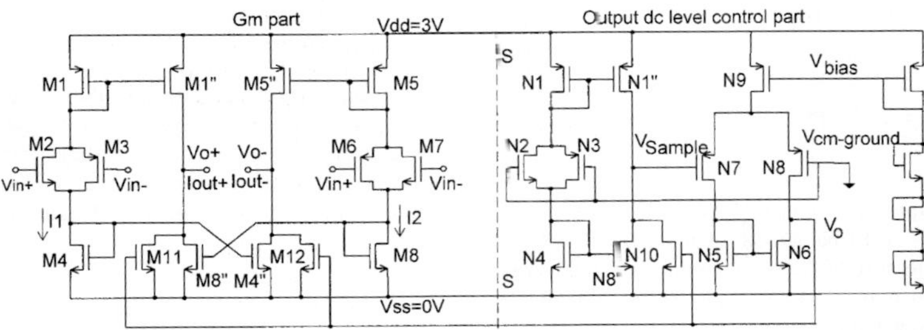

Fig. 7.   A complete schematic of the proposed transconductor. W/L (M_1, M_1'', M_5, M_5'', M_3, M_6, N_1, N_1'', N_3) = 34.7 μm/0.3 μm; W/L (M_2, M_4, M_7, M_8, M_{11}, M_{12}, M_4'', M_8'', N_2, N_4, N_{10}, N_8'') = 10 μm/0.3 μm.

parameters using 0.35 μm CMOS technology. The extracted device parameters are around the same as in (21).

Figure 8 shows the simulation of I_{out} versus V_{id}. The linear V-I conversion characteristic highlights the validity of the theoretical analysis. The transconductance can be tuned by means of the power supply voltage V_{dd}. Though it's not easy for implementation, this tuning method is applied by some designs [1, 3].

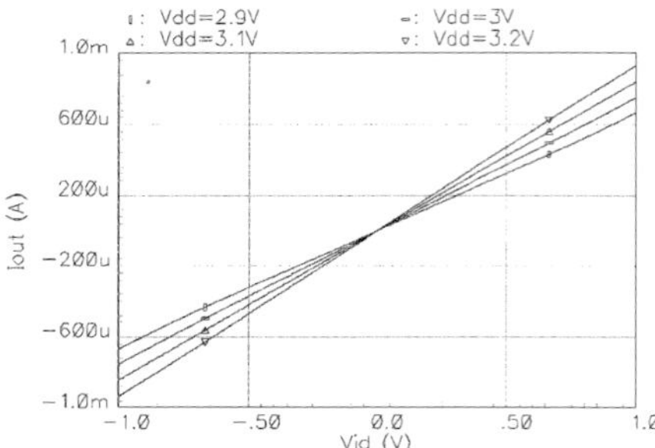

Fig. 8. SpectreS simulation of I_{out}, versus V_{id} of the transconductor. The g_m can be tuned by changing the power supply.

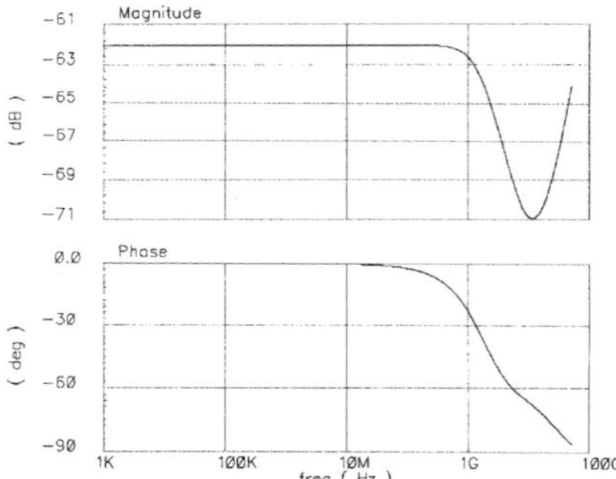

Fig. 9. Frequency response of the g_m-Cell.

The frequency response of the g_m-Cell is shown in Fig. 9. A -3 dB bandwidth of more than 1 GHz is obtained because of the simplicity of the circuit structure, and it is in good agreement with the analytical result obtained in Fig. 6.

Most of the previous analyses are based on the premise that the nMOS and pMOS are matched by (3). Since the ratio k_n/k_p of the transconductance parameters for nMOS(k_n) and pMOS (k_p) can vary within a range larger than 10% [7], an inspection of the performance of the proposed gm-Cell due to nMOS and pMOS mismatch is given. In Fig. 10, the channel width of pMOSs (Wp) in the gm-Cell changes from 30 to 40 μm, which represents the variation of parameter

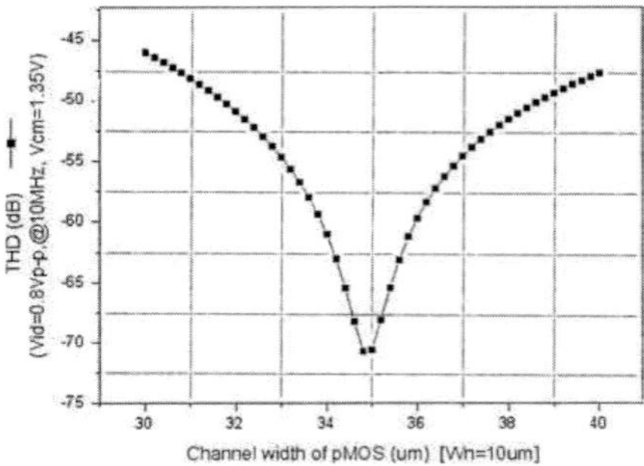

Fig. 10. Change of THD of the transconductor circuit, when channel width of pMOSs (Wp) in the g_m-Cell is changing while the channel width of nMOS (W_n) is a constant of 10 μm, which represents the mismatch of parameters during process.

Table 1. Specification of the transconductor.

Supply voltage V_{dd} and V_{ss}	3 V and 0 V
Linear input voltage range	-0.9 V $< V_{id} < 0.9$ V
THD($V_{id} = 0.8\,V_{p\text{-}p}$, @10 MHz, $V_{cm} = 1.35$ V)	-54 dB
-3 dB Bandwidth	>1 GHz
CMOS technology	$0.35\ \mu$m
Power consumption	<0.8 mW

values during process. If the pMOS is designed with a 34.7 μm channel width, the THD of the gm-Cell will be at its best value—less than -70 dB (0.032%). If a tolerance of ±10% is introduced (20% variation, Wp varies from 31.3 to 38.1 μm), Fig. 10 indicates that even

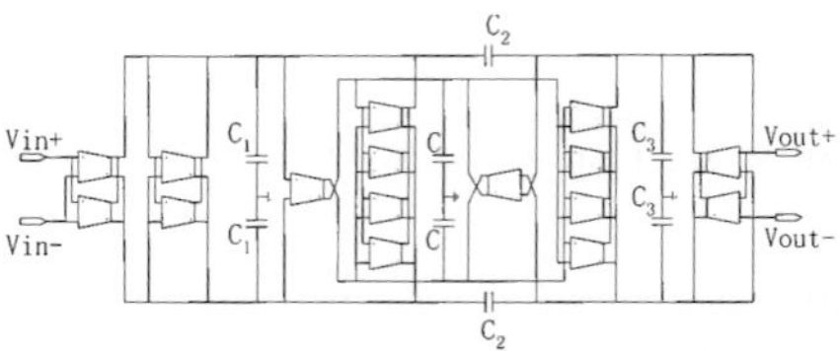

Fig. 11. 3rd order elliptic low pass filter using the proposed transconductor. $g_m = 750\ \mu$A/V, $C_1 = C_3 = 6.56$ pF, $C_2 = 400$ fF, $C = 1.38$ pF.

240   *Zhenying et al.*

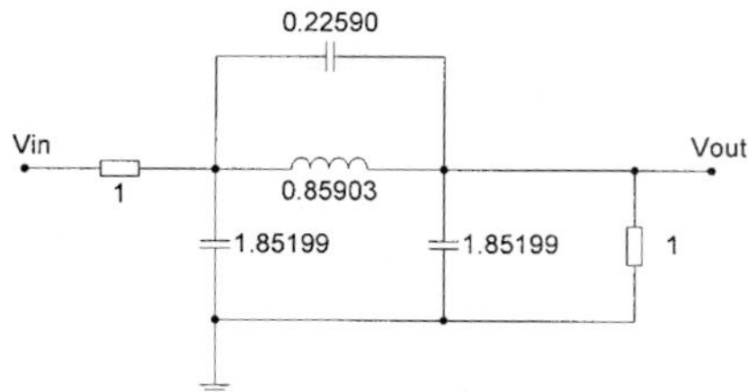

Fig. 12.   3rd order elliptic low-pass LC ladder filter.

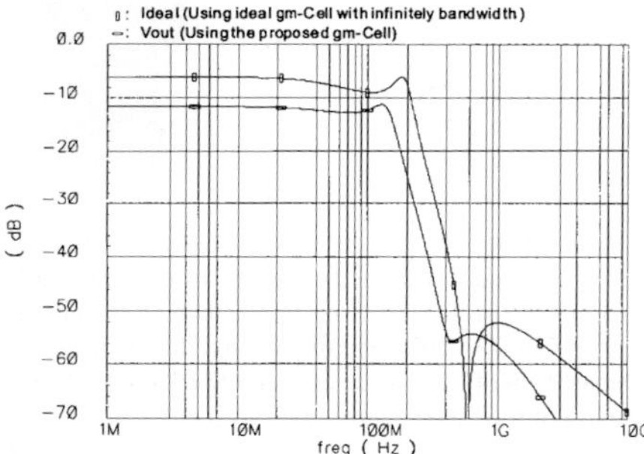

Fig. 13. Simulation result of the filter. A cutoff frequency of 150 MHz is obtained.

$$I_{1,2} = \left(-2V_{tp} - 2V_{tn} - V_{ss} + V_{dd} \pm V_{id} - \frac{\sqrt{2 \cdot (2V_{cm} \pm V_{id} - 4V_{tn} - 2V_{ss})(-2V_{cm} \pm V_{id} - 4V_{tp} + 2V_{dd})}}{2}\right)^2$$

$$I_{out} = 2(I_1 - I_2) = A \cdot V_{id} + B \cdot V_{id}^3 + O(V_{id}^5)$$

in this worse case, the THD can be achieved less than -48 dB (0.4%). Normally, if the variation range is narrow to $\pm 5\%$, the THD will be less than -54 dB.

The achieved specification of the transconductor is listed in Table 1.

6. g_m-C Filter Application

Figure 11 is an active implementation of 3rd order elliptic low-pass filter using the proposed transconductor. This filter is derived from a passive ladder filter since ladder filters have good sensitivity and dynamic range properties. The normalized passive prototype filter [6] is given in Fig. 12. Using the gyrator approach, both resistors and inductors are replaced by the proposed transconductor element. The simulation result together with the theoretical frequency response of this kind of filter is shown in Fig. 13. A cutoff frequency of 150 MHz is obtained using the proposed g_m-Cell.

7. Conclusion

A new high frequency low voltage transconductor circuit which is suitable for VHF g_m-C filter application is proposed. The transconductor inherently has a good common mode rejection ability and very high cutoff frequency. Using 0.35 μm CMOS technology with 3 V power supply, the transconductor has a ± 0.9 V linear differential input range with a -54 dB total harmonic distortion (THD) and greater than 1GHz bandwidth. The transconductor used in a 3rd order elliptic low-pass g_m-C filter with a cutoff frequency of 150 MHz is also demonstrated.

Appendix

A. Calculation of the Coefficients A and B of I_{out} in (14)

From (11) and (12), we have:

substituting the following values to the parameters:

$$V_{tn} = 0.466 \text{ V}, \quad V_{tp} = 0.617 \text{ V},$$
$$V_{dd} = 3 \text{ V}, \quad V_{ss} = 0 \text{ V}.$$

and let:

$$t = -(10V_{cm} - 9)(5V_{cm} - 9) \quad \text{and} \quad s = V_{cm} - 1.35.$$

We can derive the following results:

$$A \approx 0.2 \frac{(9\sqrt{t} - 4t)(2\sqrt{t} - 9)}{t}$$

$$B \approx -24960 \frac{s^4}{t^{5/2}}$$

B. The Detailed Expression of a_{ij} and b_{ij} in (19a) and (19b)

From the typical parameter value shown in (21), we make the following two approximations:

1. $G_m R = 400 \times 10^{-6} \times 180 \times 10^3 = 72 \gg 1$;
2. $C_{ds} \ll C_{gs}$ and C_{gd}, so that C_{ds} is neglected all the time during the approximation while comparing with C_{gs} or C_{gd}.

For V_1:

$$V_1 = V_{id} \frac{a_{21}s^2 + a_{11}s + a_{01}}{b_{21}s^2 + b_{11}s + b_{01}}$$

$$a_{21} \approx \left(-4C_{gd}^2 + 9C_{gs}C_{gd} + 9C_{gs}^2\right)$$

$$a_{11} \approx (6.1C_{gd} + 8.7C_{gs})G_m$$

$$a_{01} \approx 1.4G_m^2$$

$$b_{21} \approx \left(16C_{gd}^2 + 48C_{gs}C_{gd} + 27C_{gs}^2\right)$$

$$b_{11} \approx (13.7C_{gd} + 20.5C_{gs})G_m$$

$$b_{01} \approx 2.4G_m^2$$

For V_4:

$$V_4 = V_2|_{V_{id} \Rightarrow -V_{id}} = -V_{id} \frac{a_{22}s^2 + a_{12}s + a_{02}}{b_{22}s^2 + b_{12}s + b_{02}}$$

$$a_{22} \approx \left(-12C_{gd}^2 - 23C_{gs}C_{gd} + 9C_{gs}^2\right)$$

$$a_{12} \approx (1.2C_{gd} + 12.3C_{gs})G_m$$

$$a_{02} \approx 1.4G_m^2$$

$$b_{22} \approx \left(16C_{gd}^2 + 48C_{gs}C_{gd} + 27C_{gs}^2\right)$$

$$b_{12} \approx (13.7C_{gd} + 20.5C_{gs})G_m$$

$$b_{02} \approx 2.4G_m^2$$

References

1. B. Nauta, "A CMOS transconductance-C filter technique for very high frequencies." *IEEE Journal of Solid-State Circuits*, vol. 27, no. 2, pp. 142–153, 1992.
2. A. Assi, M. Sawan, and R. Raut, "A new CMOS tunable transconductor dedicated to VHF continuous-time filters." *VLSI, 1997. Proceedings Seventh Great Lakes Symposium*, 1997, pp. 143–148.
3. S. Szczepanski, J. Jakusz, and R. Schaumann, "A linear fully balanced CMOS OTA for VHF filtering applications." *IEEE Transactions on Circuits and System*, vol. 44, no. 3, pp. 174–187, 1997.
4. K.R. Laker and W.M.C. Sansen, *Design of Analog Integrated Circuits and Systems*. McGraw-Hill: New York, 1994.
5. P.K. Ko, "Approaches to scaling," in *VLSI Electronics Microstructure Science*, vol. 18, *Advanced MOS Device Physics*, N.G. Einspruch and G.Sh. Gildenblat, (Eds.). Academic Press: San Diego, 1989, p. 1.
6. J.E. Kardontchik *Introduction to the Design of Transconductor-Capacitor Filters*. Kluwer Academic publishers.
7. S. Koziel and S. Szczepanski, "Design of highly linear tunable CMOS OTA for continuous-time filteres." *IEEE Trans. Circuits and Systems-II*, vol. 49, no. 2, pp. 110–122, 2002.

Luo Zhenying was born in Guangdong, China in 1978. He received the B.Sc. degree from the Department of Physics, University of Science and Technology of China in 2001 and the M.Eng. degree from the Department of Electrical Engineering of National University of Singapore, Singapore, in 2003.

His research interests are low-voltage low-power analog IC design and RF IC design.

Ming-Fu Li graduated from the Department of Physics, Fudan University, Shanghai, in 1960.

After graduation, he joined the University of Science and Technology of China (USTC) as a Teaching assistant and then lecturer. In 1978, he joined the Graduate School faculty, Chinese Academy of Sciences, Beijing, and became a professor in 1986. He has also served as Adjunct Professor at the Institute of Semiconductors, Chinese Academy of Sciences, Fudan University, and USTC, Hefei.

He was a visiting scholar at Case Western Reserve University, Cleveland, OH in 1979, and at the University of Illinois at Urbana-Champain from 1979 to

81, and was a visiting scientist at the University of California at Berkeley and Lawrence Berkeley National Laboratories from 1986 to 1987, 1990 to 1991, respectively. He joined the Department of Electrical Engineering, National University of Singapore in 1991, and became a Professor in 1996. His current research interests are in the areas of CMOS device technology, reliability, quantum modeling, and Analog IC design. He has published over 200 research papers and two books, including Modern Semiconductor Quantum Physics (World Scientific, 1994). He has served on several international program committees and advisory committees in international semiconductor conferences in China, Japan, Canada, Germany and Singapore.

Yong Lian received the B.Sc degree from the School of Management of Shanghai Jiao Tong University, China, in 1984 and the Ph.D degree from the Department of Electrical Engineering of National University of Singapore, Singapore, in 1994. He was with the Institute of Microcomputer Research of Shanghai Jiao Tong University, Brighten Information Technology Ltd, SyQuest Technology International, and Xyplex Inc. from 1984 to 1996. He joined the National

University of Singapore in 1996 where he is currently an Associate Professor at the Department of Electrical and Computer Engineering. His research interests include digital filter design, VLSI implementation of signal processing systems, and RF IC design.

He received the 1996 IEEE Guillemin-Cauer Award for the best paper published in the IEEE Transactions on Circuits and Systems. He currently serves as an Associate Editor for the IEEE Transactions on Circuits and Systems Part II and has been an Associate Editor for Circuits, Systems and Signal Processing since 2000.

Subhash C. Rustagi received his M.Sc. and Ph.D. degrees in Physics from Kurukshetra University, India in the years 1975 and 1981 respectively. His Ph.D. work was on the analysis of p-n junction and high frequency silicon bipolar transistor structures. Subhash is presently the Member of Technical Staff, at Institute of Microelectronics, Singapore. His research interests are in the area of design, modeling and characterization of active and passive devices for RF integrated circuits, analog and RF circuit design.

Subhash is a Senior Member of IEEE.

Chapter 4

CMOS Transistors (I) (Reliability)

IEEE ELECTRON DEVICE LETTERS, VOL. 18, NO. 12, DECEMBER 1997

Investigation of Interface Traps in LDD pMOST's by the DCIV Method

B. B. Jie, M. F. Li, C. L. Lou, W. K. Chim, D. S. H. Chan, and K. F. Lo

Abstract—Interface traps in submicron buried-channel LDD pMOST's, generated under different stress conditions, are investigated by the direct-current current–voltage (DCIV) technique. Two peaks C and D in the DCIV spectrum are found corresponding to interface traps generated in the channel region and in the LDD region respectively. The new DCIV results clarify certain issues of the underlying mechanisms involved on hot-carrier degradation in LDD pMOST's. Under channel hot-carrier stress conditions, the hot electron injection and electron trapping in the oxide occurs for all stressing gate voltage. However, the electron injection induced interface trap spatial location changes from the LDD region to the channel region when the stressing gate voltage changes from low to high.

I. INTRODUCTION

HOT-CARRIER degradation is a bottleneck of the reliability of submicron metal-oxide-semiconductor field effect transistors (MOST's) [1]–[5]. In this letter, we briefly report new results of interface traps and oxide charge generation by hot-carrier injection in LDD pMOST's using the direct-current current-voltage (DCIV) method recently developed by Neugroschel and Sah and co-workers [6]. Our clear-cut results show that DCIV method can really have an impact to hot-carrier degradation studies and clarify certain issues of the underlying mechanisms involved in hot-carrier degradation [2]–[5].

II. MEASUREMENT TECHNIQUE

The principle of the DCIV method was described in [6]. In our experimental measurement, buried-channel LDD pMOST's in n wells with 0.6 μm channel length, 40 μm channel width, and gate oxide thickness of 125 Å were used. The base (bulk) current I_b of the p/n/p-BJT in Fig. 1 is used to measure the recombination current via the interface traps distributed over the channel region (region C; interface traps C, base current component I_{bC}), or interface traps distributed in the LDD drain region or drain-to-bulk pn junction depletion region (region D; interface traps D, base current component I_{bD} [7]). Either I_{bC} or I_{bD} is a function of the gate voltage V_{gb}, showing a sharp peak at a certain V_{gb}. It is easy to derive from Shockley–Read–Hall theory [9] that the peak V_{gb}

Manuscript received March 31, 1997; revised July 2, 1997. This work was supported by the Singapore National Science and Technology Board Research Grant NSTB/17/2/3 and RIC-University Research Project 681 305.

B. B. Jie, M. F. Li, W. K. Chim, and D. S. H. Chan are with the Center for Integrated Circuit Failure Analysis and Reliability, Faculty of Engineering, National University of Singapore, Singapore 119260.

C. L. Lou and K. F. Lo are with the Chartered Semiconductor Manufacturing Pte. Ltd., Singapore 738406.

Publisher Item Identifier S 0741-3106(97)08900-3.

value corresponds to the case whereby the surface potential at the trap site results in the intrinsic Fermi energy E_i at the surface coinciding with the average of the quasi-Fermi energy of electrons and holes $(E_{FE} + E_{FH})/2$. The DCIV method has the advantage that the $I_b - V_{gb}$ spectrum (denoted as the DCIV spectrum) can distinguish the interface traps located at different sites by the different V_{gb} peaks. It is commonly recognized that the interface traps above E_i are acceptor traps and those below E_i are donor traps [10]. Therefore, the total net charge of the interface traps C or D is nearly zero at the peak V_{gb}. The circuit connected as in the inset of Fig. 1 is defined as the normal mode. We also define a reverse mode by interchanging the source and drain terminals in Fig. 1. The I_{bC} peak amplitude should be almost identical for the normal mode and reverse mode measurements. On the other hand, under the reverse mode measurement, if the V_{gb} at the I_{bD} peak does not induce a conducting p-channel between the source and the drain, then the floating drain voltage $V_{db} < V_{sb}$ (which is confirmed by measurement). Correspondingly, the interface traps D at the drain side induce a much smaller recombination current I_{bD} under the reverse mode measurement than IbD under the normal mode measurement.

III. EXPERIMENTAL RESULTS AND DISCUSSION

The DCIV spectra under normal and reverse mode measurements after different stress conditions are illustrated in Figs. 1–3.

Case (1), substrate-hot-carrier (SHC) injection condition. (Fig. 1)

The holes injected by the collector junction are accelerated vertically to the interface, and surmount the Si/SiO$_2$ barrier in the channel region and be injected into the gate oxide. Interface traps C are generated homogeneously in the channel region. As expected, the amplitude of the DCIV spectrum peak C for the normal mode measurement in Fig. 1(a) are almost identical to that for the reverse mode measurement in Fig. 1(b). Peak C shifts left to more negative V_{gb} when increasing the stress time. This left shift is due to positive charge of the trapped holes in the gate oxide above the channel region, since at the peak position, the net charge of the C interface traps is nearly zero as explained in Section II. Consequently, the absolute value of the pMOST threshold voltage increases, which is also confirmed by the measurement.

Case (2), channel hot-carrier injection at large $|V_{ds}|$ and small $|V_{gs}|$ (Fig. 2).

The holes are accelerated in the channel and attain high kinetic energies close to the drain. Electrons are generated

584   IEEE ELECTRON DEVICE LETTERS, VOL. 18, NO. 12, DECEMBER 1997

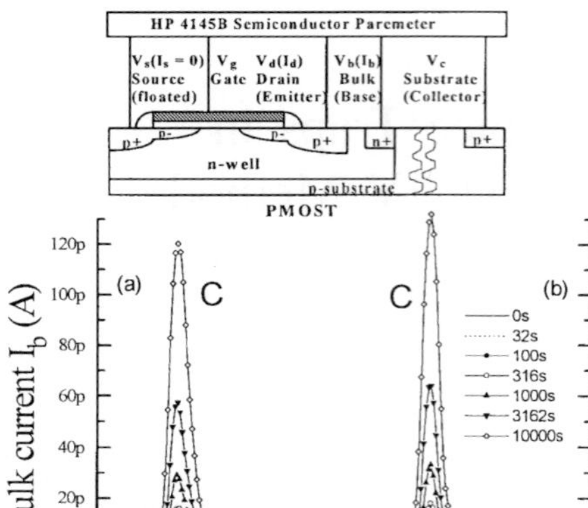

Fig. 1. The normal (a) or reverse (b) mode DCIV spectra after substrate-hot-carrier injection stress ($V_{sb} = V_{db} = -4.0$ V, reverse biased source and drain junctions, $V_{cb} = 0.4$ V, forward biased bulk/substrate junction, $V_{gb} = -13$ V). The inset shows the cross-sectional schematic of the LDD pMOST and the DCIV measurement set-up. A vertical parasitic p/n/p-BJT is used with the condition: $V_{eb} = +0.3$ V (forward biased emitter-base), $V_{bc} = 0.0$ V, $I_s = 0.0$ A (source open).

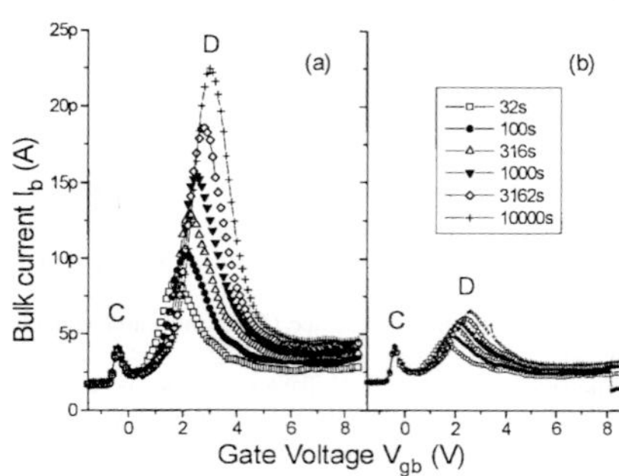

Fig. 2. The normal (a) and reverse (b) mode DCIV spectra after channel hot-carrier injection stress ($V_{sb} = 0.0$ V, $V_{db} = -9.5$ V, $V_{gb} = -1.0$ V, $V_{cb} = -7.0$ V reverse biased bulk/substrate junction).

by hot hole impact ionization. The electrons in region D are injected into the oxide and generate interface traps D. The amplitude of Peak D of the D traps' current in the DCIV spectrum under the reverse mode measurement [Fig. 2(b)] is greatly reduced compared with the normal mode measurement [Fig. 2(a)]. This is expected because the V_{gb} bias at the peak D corresponds to an accumulation region in the channel and no p-channel between the source and the drain. Furthermore,

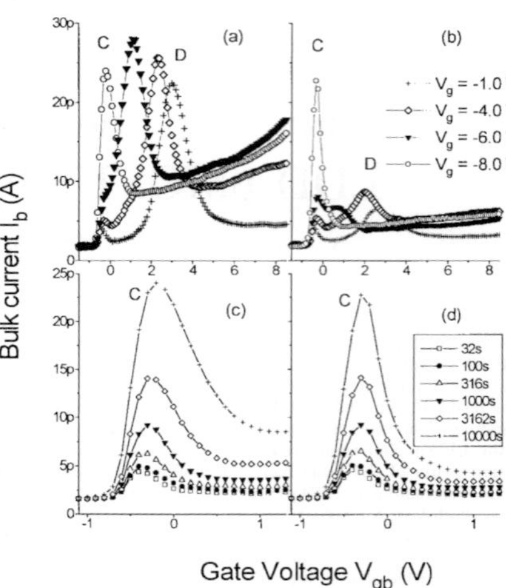

Fig. 3. The normal (a), (c) and reverse (b), (d) modes DCIV spectra after channel hot-carrier injection stress with the same stress conditions of Fig. 2, however changing V_{gb} from low value to high value. For (a) and (b), $V_{gb} = -1.0$ to -8.0 V, respectively, after stress time = 10 000 s; for (c) and (d), $V_{gb} = -8.0$ V after different stress times.

peak D shifts right to more positive V_{gb} with increasing stress time. Following the similar argument as in case (1), the right shift is due to negative charges of the trapped electrons in the gate oxide. Consequently, the absolute value of the pMOST threshold voltage decreases, which is also confirmed by the measurement.

Case (3), channel hot-carrier (CHC) injection at different V_{gs} (Fig. 3).

Case (3), same as stress condition (2); however $|V_{gs}|$ gradually increases from the case (2) value to $|V_{ds}|$. As indicated in Fig. 3(a), when V_{gs} gradually increases from -1.0 to -8.0 V, the peak D gradually shifts from the peak V_{gb} value in Fig. 2 to the peak V_{gb} value in Fig. 1. This indicates that the interface states generated by hot electron injection change their spatial location from the D region to the C region. We rule out the possibility that peak D shifts from D to C is due to positive charge trapping in the oxide, since peak D still shifts to C even if the stress time reduces to 32 s and the injected charge in the oxide can be neglected. This is further confirmed by observing the peak amplitude under the reverse mode measurement as indicated in Fig. 3(b). In the reverse mode measurement, the amplitudes of the D peaks are greatly reduced compared with the normal mode peaks, at small $|V_{gs}|$ stress condition. However at the stress condition of $V_{gs} = -8.0$ V, the reverse mode peak amplitude suddenly increases, indicating again that the interface traps change their space to the C region. The C traps in Fig. 3 are also due to the electron injection. At least a part of the electrons are trapped in the oxide, causing the DCIV peak V_{gb} value to shift to right side when the stress time is increased, as indicated in Fig. 3(c). This is different from case (1) in Fig. 1.

In conclusion, we found two peaks, C and D, in the DCIV spectrum of submicron buried-channel LDD pMOST's. Peak C corresponds to channel interface traps while peak D corresponds to interface traps located in the LDD region or the drain-bulk pn junction depletion region. Under CHC stress condition, the hot electron injection and electron trapping in the oxide occur for all stressing gate voltage. However, the generated interface traps change their spatial location from the D region to the C region when the stressing gate voltage changes from low to high.

ACKNOWLEDGMENT

M. F. Li would like to thank C. T. Sah for informing him of this new DCIV method.

REFERENCES

[1] T. H. Ning, P. W. Cook, R. H. Dennard, C. M. Osburn, S. E. Schuster, and H. N. Yu, "1 μm MOST VLSI technology—Part IV: Hot-electron design constraints," *IEEE Trans. Electron Devices*, vol. ED-26, pp. 346–353, 1979.

[2] P. Heremans, R. Bellens, G. Groeseneken, A. V. Schwerin, W. Weber, B. Brox, and H. E. Maes, "The mechanisms of hot-carrier degradation," in *Hot-Carrier Design Considerations for MOS Devices and Circuits*, C. T. Wang, Ed. New York: Van Nostrand Reinhold, 1992, pp. 1–119.

[3] C. Fu, "Hot-carrier effects," in *VLSI Electronics Microstructure Science, Vol. 18, Advanced MOS Device Physics*, N. G. Einspruch, G. Sh. Gildenblat, Eds. San Diego, CA: Academic, 1989, pp. 119–160.

[4] R. Woltjer, A. Hamada, and E. Takeda, "Time dependence of p-MOSFET hot-carrier degradation measured and interpreted consistently over en orders of magnitude," *IEEE Trans. Electron Devices*, vol. 40, pp. 392–400, Feb. 1993.

[5] F. Matsuoka, H. Iwai, H. Hayashida, K. Hama, Y. Toyoshima, and K. Maeguchi, "Analysis of hot-carrier induced degradation mode on pMOSFET's," *IEEE Trans. Electron Devices*, vol. 37, pp. 1487–1495, 1990.

[6] A. Neugroschel, C.T. Sah, K. M. Han, M. S. Carroll, T. Nishida, J. T. Kavalieros, and Y. Lu, "Direct-current measurements of oxide and traps on oxidized silicon," *IEEE Trans. Electron Devices*, vol. 42, pp. 1657–1662, 1995.

[7] Different peaks corresponding to the different recombination current components in the channel region and the pn junction transition region was first discussed by C. T. Sah, "A new semiconductor tetrode—The surface-potential controlled transistor," *Proc. IRE*, vol. 49, pp. 1523–1639, 1961.

[8] W. Shockley and W. T. Read, "Statistics of recombination of holes and electrons," *Phys. Rev.*, vol. 87, pp. 835–842, 1952.

[9] R. N. Hall, "Germanium rectifier characteristics," *Phys. Rev.*, vol. 86, p. 228, 1951.

[10] G. A. Scoggan and T. P. Ma, "Effects of electron-beam radiation on MOS structures as influenced by the silicon dopant," *J. Appl. Phys.*, vol. 48, pp. 294–300, 1977.

238   IEEE ELECTRON DEVICE LETTERS, VOL. 20, NO. 5, MAY 1999

Nondestructive DCIV Method to Evaluate Plasma Charging Damage in Ultrathin Gate Oxides

Hao Guan, Yaohui Zhang, B. B. Jie, Y. D. He, Ming-Fu Li, *Member, IEEE,* Zhong Dong, Joseph Xie, J. L. F. Wang, Andrew C. Yen, George T. T. Sheng, and Weidan Li

Abstract— Understanding and minimizing plasma charging damage to ultrathin gate oxides become a growing concern during the fabrication of deep submicron MOS devices. Reliable detecting techniques are essential to understand its impact on device reliability. As the gate oxide thickness of MOST's rapidly scales down, the conventional nondestructive methods such as capacitor C–V and threshold voltage and subthreshold swing of MOST's are no longer effective for evaluating this damage in gate oxide. In this paper, the new developed direct-current current–voltage (DCIV) technique is reported as an effective monitor for plasma charging damage in ultrathin oxide. The DCIV measurements for p-MOST's with both 50- and 37-Å gate oxides clearly show the plasma charging damage region on the wafers and are consistent with the results of charge-to-breakdown measurements. In comparing with charge-to-breakdown measurement and other conventional methods, the DCIV technique has the advantages of nondestructiveness, high sensitivity and rapid evaluation.

I. Introduction

PLASMA charging damage to gate oxide is a growing concern during the fabrication of MOS device [1]–[5]. Conventionally, capacitor C–V and the shift of MOST device parameters, such as threshold voltage V_{th}, subthreshold swing S_S, and transconductance g_m were suggested as a monitor for the plasma charging damage [6]. However, as gate oxide is scaled below 40 Å, these conventional methods have been demonstrated to be insensitive and therefore not suitable for monitoring the plasma charging damage [6], [7]. Consequently, some destructive methods, such as the charge-to-breakdown measurements, were proposed to evaluate plasma damage in the ultrathin oxides [7]. In this paper, we demonstrate that the newly developed direct-current current–voltage (DCIV) technique [8] is an effective monitor for plasma charging damage in ultrathin oxides with the advantages of nondestructiveness, high sensitivity and fast evaluation.

Manuscript received October 7, 1998; revised January 5, 1999. This work was supported by the Singapore National Science and Technology Board Research Grant NSTB/17/2/3 and RIC-University Research Project 681305.

H. Guan, B. B. Jie, Y. D. He, and M. F. Li are with the Department of Electrical Engineering, National University of Singapore, Singapore 119260 (e-mail: elelimf@leonis.nus.edu.sg).

Y. Zhang, J. Xie, J. L. F. Wang, A. C. Yen, and G. T. T. Sheng are with the Department of Deep Sub-Micron Integrated Circuits, Institute of Microelectronics, Singapore Science Park II, Singapore 117684.

Z. Dong is with Chartered Semiconductor Manufacturing Ltd., Singapore 738406.

W. Li is with LSI Logic Co., Santa Clara, CA 95054 USA.

Publisher Item Identifier S 0741-3106(99)04418-3.

II. Experiments

The 50- and 37-Å gate oxide thickness MOST's were fabricated on 6- and 8-in p-type (100) silicon wafers respectively. A 50-Å gate oxide in p-MOST's was grown after LOCOS isolation, followed by n^+ poly-Si deposition, and 37 Å gate oxide in p-MOST's was grown after shallow trench isolation, followed by p^+ poly-Si deposition. (It is not accurate comparison of 50- and 37-Å gate oxide samples using different isolation technology). Oxide thickness was determined by Fowler–Nordheim I–V characteristic [9], by optical method and further verified by TEM. P-MOST's with metal antenna structures attached to the gate were used to monitor plasma charging damage. Metal etch was performed in a commercial reactor using BCl_3/Cl_2 chemistry by optimal recipe. After definition of metal patterns, the photoresist is stripped off in an asher. Finally, wafers were annealed in forming gas at 400 °C for 30 min. The schematic diagram of our DCIV measurement was shown in Fig. 1(a) which has been described in [10]. The gate voltage sweeping speed in DCIV measurement is 0.15 V/s. Plasma charging damages were carefully examined by DCIV measurements in 50-Å gate oxides by using W/L = 20 μm/0.4 μm p-MOST's with metal antenna connected to the poly-gate, and 37 Å gate oxides by using W/L = 50 μm/0.24 μm p-MOST's with metal antenna connected to poly-gate. The same p-MOST's with no antenna located in the same cells were used as references. All electrical measurements were made by a HP4156 semiconductor parameter analyzer.

III. Results and Discussion

Fig. 1(b) shows the typical DCIV spectra of two p-MOST's after uniform and constant gate current stresses to simulate the plasma charging. The stress current used in our experiment is 12 mA/cm^2. The DCIV current peak C at peak gate voltage V_{pc} (near 0 V) is due to electron-hole recombination via the Si/SiO$_2$ interface traps in the MOST channel region [8], [10] and the DCIV peak amplitude is proportional to the interface trap density. The DCIV current far apart from the peak voltage V_{pc} is due to damage in other region [10] and can be disregarded in this work. Fig. 1(b) shows effectiveness of the DCIV technique to measure the interface traps in MOST's with ultrathin gate oxides under constant gate current stress. Since plasma charging is a kind of constant gate current stress [11]–[14], we should expect the same DCIV technique can be used for evaluating plasma charging damage.

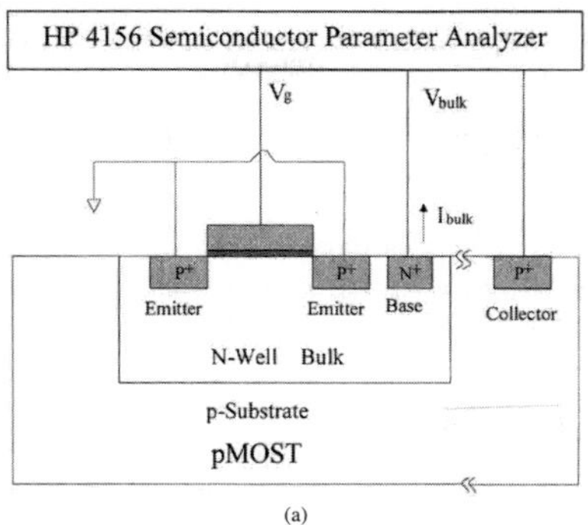

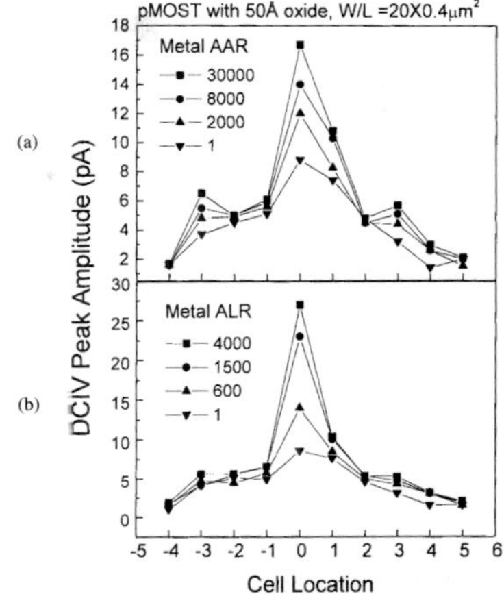

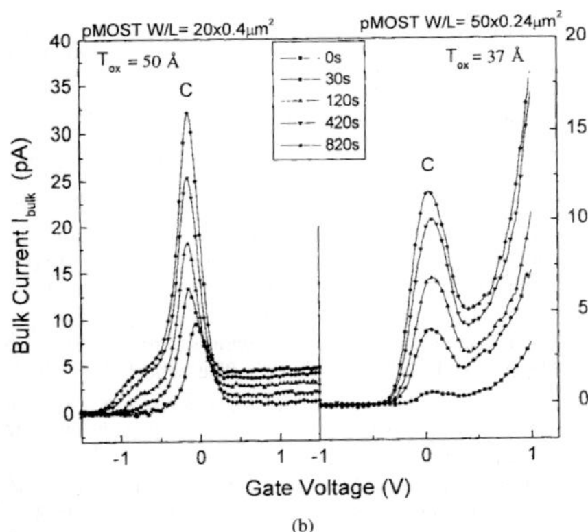

Fig. 1. (a) Cross-sectional schematic of the DCIV measurement set-up. A vertical parasitic p/n/p-BJT is used with the condition: $V_{eb} = +0.3$ V (forward biased) and $V_{bc} = 0$ V, (b) DCIV spectra after a constant-current gate injection (gate current density $= -12$ mA/cm^2, s, d, bulk, substrate all connected to ground while $V_g < 0$) in a 50-Å gate oxide, W/L $= 20$ μm/0.4 μm p-MOST and a 37 Å gate oxide, W/L $= 50$ μm/0.24 μm p-MOST. The DCIV peak corresponds to the recombination current via the interface traps located at the channel region.

Fig. 2. DCIV peak amplitude of 50-Å gate oxide, W/L $= 20$ μm/0.4 μm p-MOST's with various metal AAR [Fig. 2(a)] or ALR [Fig. 2(b)] ratio as functions of cell position.

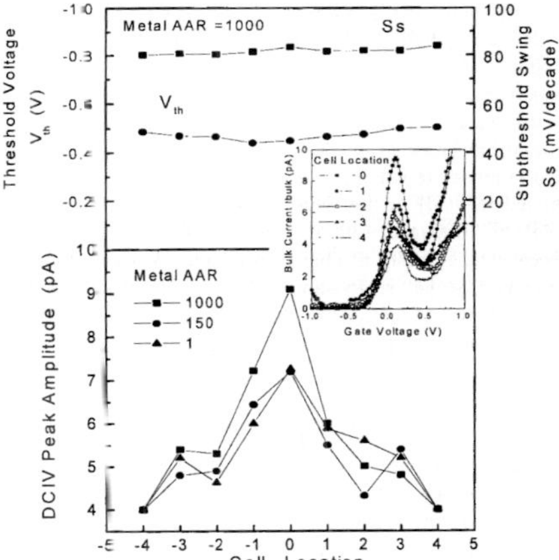

Fig. 3. (a) Threshold voltage V_{th} or subthreshold swing S_S as a function of cell position. (b) The peak amplitudes of DCIV spectra as functions of cell position. Both (a) and (b) are measured for the same 37-Å gate oxide, W/L $= 50$ μm/0.24 μm p-MOST's with various metal AAR. DCIV spectra are shown in the insect.

Fig. 2(a) and (b) shows the DCIV peak amplitudes of 50-Å gate oxide, using W/L $= 20$ μm/0.4 μm p-MOST's with various metal antenna area ratio (AAR) or antenna length ratio (ALR), as a function of cell location. The measurements were performed on ten cells along a central line across the wafer. It is seen that the DCIV peak amplitudes for the devices located at the wafer center are larger than those for the devices located at the edge of the wafer. Moreover, the DCIV peak amplitude increases with increasing AAR and ALR in the wafer center. MOST's with larger antenna ratio or located at the wafer center have higher rate of interface state generation during plasma processing, indicating more serious plasma charging damage occurred.

In our 37-Å oxide p-MOST's with AAR $= 1000$, the device parameters V_{th} and S_S change very slightly across the wafer, as illustrated in Fig. 3(a). This result is consistent with previous report [6], [7] and can not be used as a reliable guide to predict the plasma charging damage [7]. In order to identify

IEEE ELECTRON DEVICE LETTERS, VOL. 20, NO. 5, MAY 1999

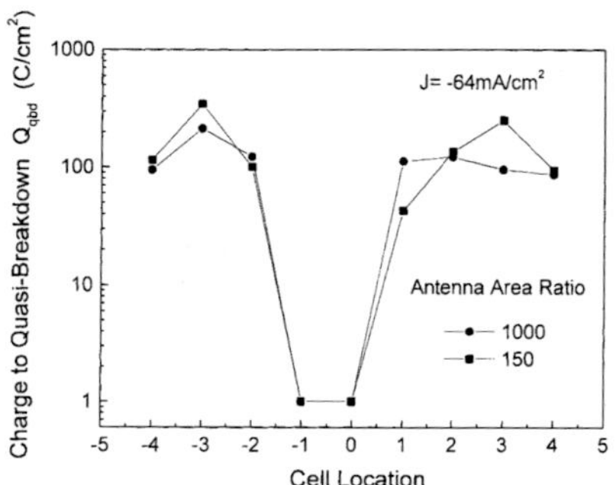

Fig. 4. Charge-to-quasibreakdown Q_{qbd} as a function of cell location, measured for the same 37-Å gate oxide devices as in Fig. 3.

the real plasma charging damage in this ultrathin gate oxide, we performed the DCIV measurements for the same devices on nine cells along a central line across the wafer. As illustrated in Fig. 3(b), the DCIV peak amplitudes indicate large difference at different cell location as well as different AAR. This difference demonstrates that real damages have occurred in the ultrathin oxide at the wafer center during ashing treatment. It is unusual that plasma charging damage could be induced in devices with AAR = 1, as shown in Fig. 3(b). There are probably other damage events (e.g., radiation) which could be responsible for the oxide degradation. The plasma damages in the wafer center were further identified by the charge-to-quasibreakdown (or soft breakdown) measurements [15], [16] (a moderate voltage drop under constant current stress as defined in [15]) in the same 37-Å gate oxide devices with AAR of 1000 and 150, respectively. Fig. 4 illustrates the experimental results of charge-to-quasibreakdown (Q_{qbd}) as a function of the cell location. It is clear that the Q_{qbd} at the wafer center is much lower than the wafer edge, indicating more serious plasma charging damages in the wafer center than in wafer edge. This is consistent with DCIV results.

IV. CONCLUSION

DCIV technique is a powerful method in monitoring the plasma charging damage in ultrathin gate oxide. It is demonstrated by using the DCIV technique to p-MOST's with gate oxide thicknesses 50- and 37-Å and with different metal antenna structures. From the DCIV measurements, we found that if the cell is located at the wafer center or the AAR is large, the plasma charging damages are serious for both

50- and 37-Å thick oxides. This conclusion is also verified by the charge-to-quasibreakdown measurement. This DCIV technique shows advantages such as nondestructiveness, high sensitivity, and rapid evaluation. It can be an reliable method for evaluating plasma charging damage in deep sub-micron device fabrication.

ACKNOWLEDGMENT

The authors would like to thank Prof. C. T. Sah for informing us of this new DCIV method.

REFERENCES

[1] J. P. McVittie, "Process charging in ULSI: Mechanisms, Impact and Solutions," in *IEDM Tech. Dig.*, 1997 (invited paper).
[2] F. Shone, K. Wu, J. Shaw, E. Hokelet, S. Mittal, and A. Haranahalli, "Gate oxide charging and elimination for metal antenna capacitor and transistor in VLSI CMOS double-layer metal technology," *VLSI Technol.*, 1989, p. 73.
[3] H. Shin, C. C. King, T. Horiuchi, and C. Hu, "Thin oxide charging current during plasma etching of aluminum," *IEEE Electron Device Lett.*, vol. 12, p. 404, Aug. 1991.
[4] A. Joshi, L. Chung, B. W. Min, and D. L. Kwong, "Gate oxide thickness dependence of RIE-induced damage on N-channel MOSFET reliability," in *Int. Reliab. Phys. Symp.*, 1996, p. 300.
[5] K. P. Cheung and C. P. Chang, "Plasma-charging damage: A physics model," *J. Appl. Phys.*, vol. 75, p. 4415, 1994.
[6] D. Park and C. Hu, "Plasma charging damage on ultrathin gate oxides," *IEEE Electron Device Lett.*, vol. 19, p. 1, Jan. 1998.
[7] H. C. Lin, C. C. Chen, C. H. Chien, S. K. Hsein, M. F. Wang, T. S. Chao, T. Y. Huang, and C. Y. Chang, "Evaluation of plasma charging in ultrathin gate oxides," *IEEE Electron Device Lett.*, vol. 19, p. 68, Mar. 1998.
[8] A. Neugroschel, C. T. Sah, K. M. Han, M. S. Carroll, T. Nishida, J. T. Kavalieros, and Y. Lu, "Direct-current measurements of oxide and traps on oxidized silicon," *IEEE Trans. Electron Devices*, vol. 42, p. 1657–1662, 1995.
[9] A. Gupta, P. Fang, M. Song, M. R. Lin, D. Wollesen, K. Chen, and C. Hu, "Accurate determination of ultrathin gate oxide thickness and effective polysilicon doping of CMOS devices," *IEEE Trans. Electron Device Lett.*, vol. 18, pp. 580–582, 1997.
[10] B. B. Jie, M. F. Li, C. L. Lou, W. K. Chim, D. S. H. Chan, and K. F. Lo, "Investigation of interface traps in LDD pMOST's by the DCIV method," *IEEE Electron Device Lett.*, vol. 18, p. 583, Dec. 1997.
[11] W. Lukaszek, "Understanding and controlling wafer charging damage," *Solid State Technol.*, June 1998, p. 101.
[12] S. Fang, S. Murakawa, and J. P. McVittie, "Modeling of oxide breakdown from gate charging during resist ashing," *IEEE Trans. Electron Devices*, vol. 41, p. 1848, Oct. 1994.
[13] S. Ma, J. P. McVittie, and K. C. Saraswat, "Prediction of plasma charging induced gate oxide damage by plasma charging probe," *IEEE Electron Device Lett.*, vol. 18, p. 468, Oct. 1997.
[14] C. H. Chien, C. Y. Chang, H. C. Lin, T. F. Chang, S. G. Chiou, L. P. Chen, and T. Y. Huang, "Resist-related damage on ultrathin gate oxide during ashing," *IEEE Electron Device Lett.*, vol. 18, p. 33, 1997.
[15] S. H. Lee, B. J. Cho, J. C. Kim, and S. H. Choi, " Quasibreakdown of ultrathin gate oxide under high field stress," in *IEDM Tech. Dig.*, 1994, pp. 605–608.
[16] B. E. Weir, P. J. Silverman, D. Monroe, K. S. Kirsch, M. A. Alam, G. B. Alers, T. W. Sorsch, G. L. Timp, F. Baumann, C. T. Liu, Y. Ma, and D. Hwang, "Ultrathin gate dielectrics: They break down, but do they fail?," in *IEDM Tech. Dig.*, 1997, pp. 73–76.

586   IEEE ELECTRON DEVICE LETTERS, VOL. 20, NO. 11, NOVEMBER 1999

Role of Hole Fluence in Gate Oxide Breakdown

M. F. Li, Y. D. He, S. G. Ma, B.-J. Cho, K. F. Lo, and M. Z. Xu

Abstract—A simple model which links the primary hole and Fowler–Nordheim (FN) electron injections to oxide breakdown is established and the calculation based on this model is in good agreement with our experiments. When the sum of the active trap density D^{pri} due to primary hole injection and the active trap density D^n due to FN electron injection reaches a critical value D_{cri}, the oxide breaks down. The hole is two orders of magnitude more effective than FN electron in causing breakdown. These new findings are imperative in predicting oxide reliability and device lifetime.

I. Introduction

GATE OXIDE breakdown is one of the most significant bottle-necks in scaling metal-oxide-semiconductor field-effect transistors (MOST's). Early work on breakdown was attributed to electron injection and electron trapping [1], while subsequent works [2], [3] concluded that breakdown was due to Fowler–Nordheim (FN) electron-induced hole trapping. In this letter, we call the latter "secondary hole"[1]. Although this secondary hole trapping model has been widely accepted, there are disagreements [4]–[7]. Recently, Kamakura *et al.* [7] investigated oxide breakdown using substrate hole (we name it "primary hole") injection, however, they overestimated the primary hole fluence, as will be explained in the following section. The objective of this work is to establish a simple quantitative model to link the effect of electron and hole injections in gate oxide breakdown. This new finding lays the groundwork for predicting the oxide reliability and device lifetime.

II. Measurements

Using a well-known structure, shown in Fig. 1 [6]–[9], we have carefully designed a testing scheme which is able to control and measure the FN tunneling electron fluence Q_n, and the primary hole injection fluence Q_p^{pri}, separately. An

Manuscript received December 3, 1998; revised June 25 1999. This work was supported by the Singapore National Science and Technology Board Research Grant NSTB/17/2/3 and RIC-University Research Project 681305.

M. F. Li, Y. D. He, S. G. Ma, and B.-J. Cho are with the Center for Integrated Circuit Failure Analysis and Reliability, Department of Electrical Engineering, National University of Singapore, Singapore 119260.

K. F. Lo is with Chartered Semiconductors Manufacturing Pte. Ltd, Singapore 738406.

M. Z. Xu is with the Institute of Microelectronics, Peking University, Beijing 100871, China.

Publisher Item Identifier S 0741-3106(99)08995-8.

[1] The FN electron induced hole current in MOST was first observed in 1980 as illustrated in Weinberg and Fischetti, *J. Appl. Phys.*, vol. 57, p. 443, 1985; Sah *et al.* has given a comprehensive analysis of the physical original of this FN electron induced hole current in C.-T. Sah, *Fundamentals of Solid-State Electronics, Study Guide*, Appendix B, World Scientific, 1993, and Y. Lu and C. T. Sah, *J. Appl. Phys.*, vol. 76, p. 4724, 1994; *Phys. Rev.*, vol. 52, p. 5657, 1995, and they used the terminology "secondary hole."

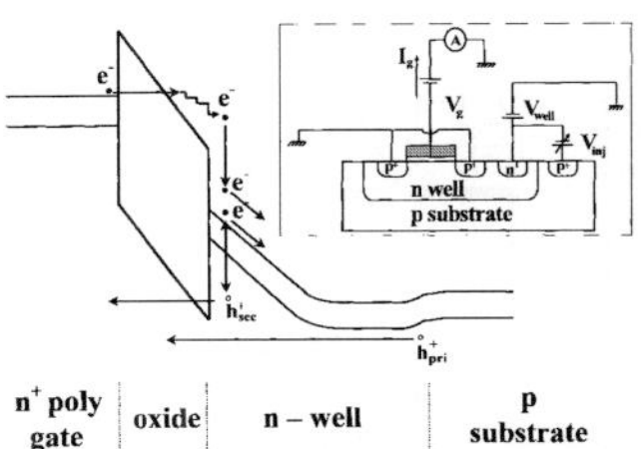

Fig. 1. Energy band diagram of pMOST during FN electron and primary hole injections. The inset shows pMOST test structure. V_{inj} switches on and off periodically with an adjusted duty cycle.

n⁺ poly-gate pMOST in n-well (gate area: 1×50 $\mu\mathrm{m}^2$, gate oxide thickness: 6 nm) is used with a switching forward bias V_{inj} applying to the substrate-well pn junction. As shown in Fig. 2, when $V_{\mathrm{inj}} = 0$ (switch-off), the gate current is mainly the FN tunneling electron current J_n. When $V_{\mathrm{inj}} > 0$ (switch-on), the gate current is $J_n + \Delta J$, where $\Delta J = J_p^{\mathrm{pri}} + \Delta J_n$. Here, J_p^{pri} is the primary hole current injected from the forward-biased p-substrate to the n-well junction [8]. ΔJ_n is the increase of FN electron current when the switch turns on[2]. Since there are pre-existed high-density fast hole traps (probably oxygen vacancies) in the thermal oxide with a high trapping and detrapping (more accurately, electron-"trapped hole" recombination) rates [10], [5], when the switch turns on, hole injection will rapidly cause hole trapping in the oxide, and thereby change the oxide field and cause the increase in the FN electron current ΔJ_n. When the switch turns off, the fast hole traps detrapped rapidly and ΔJ_n disappears. During stress, V_{inj} switches on and off periodically with a duty cycle $t_{\mathrm{on}}/[t_{\mathrm{on}} + t_{\mathrm{off}}]$, as shown in Figs. 1 and 2; the primary hole current J_p^{pri} $(=\lambda \Delta J)$ and the increase of FN electron current ΔJ_n $(=(1-\lambda)\Delta J)$ can be distinguished by a transient measurement, as illustrated in the inset of Fig. 2, which depicts a falling edge in Fig. 2 at a higher time resolution. Here, λ (<1) is a fraction number. We chose falling edge rather than rising edge because the hole trap detrapping rate is slower than trapping rate, so a clearer distinction can be made. At $t'' = 0$ in the inset of Fig. 2, V_{inj} changes from "on" to "off" state. Gate current suddenly drops at $t'' = 0$, which corresponds to the disappearance of J_p^{pri} component, which is then followed by a gradual decrease of current within a few

[2] The authors are indebted to the anonymous reviewer who referred to [5], [10] and pointed out the ΔJ_n current component and suggested the transient current experiment as, indicated in Fig. 3. In [6] and [7], the authors have not taken ΔJ_n $(=(1-\lambda)\Delta J)$ component into consideration; therefore they overestimated J_p^{pri} term.

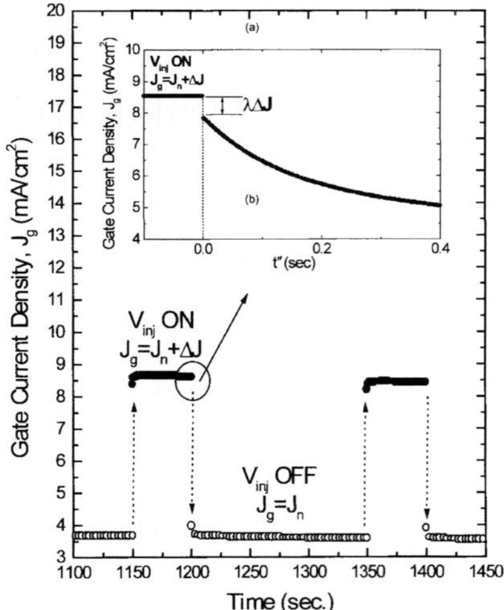

Fig. 2. Typical FN electron and primary hole injection current versus time diagram, with V_{inj} switches on ($\bullet$) ($V_{\text{inj}} = 2$ V) and switches off ($\circ$) ($V_{\text{inj}} = 0$ V) periodically with a fixed duty cycle $t_{\text{on}}/[t_{\text{on}} + t_{\text{off}}]$. Here, $V_g = -9.2$ V, $V_{\text{well}} = 7.5$ V. Only two cycles are shown in the figure. When the switch turns on, the current increases. ΔJ consists of two components, as explained in the text. The inset shows one of the falling edges of injection current versus time diagram at higher time resolution. At $t'' = 0$, V_{inj} changes from on to off in less than 1 μs. A sudden drop of current $\lambda\Delta J$ corresponds to the disappearance of the J^{pri} component, then is followed by a gradual decrease of FN current due to detrapping of fast hole traps in the oxide.

seconds, which corresponds to the decrease of the FN electron current due to the detrapping of fast hole traps in the oxide. From the inset of Fig. 2, we can estimate $\lambda = 0.13 \pm 0.02$ when $V_g = -9.2$ V. From our measurements, it was found that λ was roughly a constant during constant voltage stress until breakdown. However, λ is V_g-dependent. From Fig. 2, we can calculate the electron fluence Q_n and the primary hole fluence Q_p^{pri} using the following equations:

$$Q_n(t) = \int_0^t \{J_n(t') + (1 - \lambda)\Delta J(t')\}\, dt' \tag{1}$$

$$Q_p^{\text{pri}}(t) = \lambda \int_0^t \Delta J(t')\, dt'. \tag{2}$$

Notice that $\Delta J(t') = 0$ when the switch is off.

III. Results and Discussion

Fig. 3 shows Q_n and Q_p^{pri} at breakdown point when the primary holes are injected by $V_{\text{inj}} = 2$ V with a duty cycle $t_{\text{on}}/[t_{\text{on}} + t_{\text{off}}] \approx 1$. When V_g is in the range of -6 V to -9.5 V, Q_n changes more than one order of magnitude while Q_p^{pri} remains in the range of 0.5 ± 0.2 C/cm^2, which is nearly a constant. It indicates that under this range of stress conditions, breakdown is mainly controlled by the hole fluence.

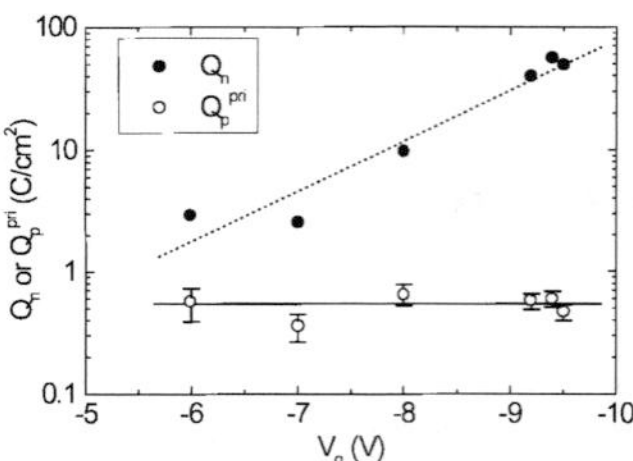

Fig. 3. $Q_p^{\text{pri}}(\circ)$ and $Q_n(\bullet)$ at breakdown for different gate voltage V_g. $V_{\text{well}} = 7.5$ V, $V_{\text{inj}} = 2$ V when switch-on, with duty cycle $t_{\text{on}}/[t_{\text{on}} + t_{\text{off}}] \approx 1$.

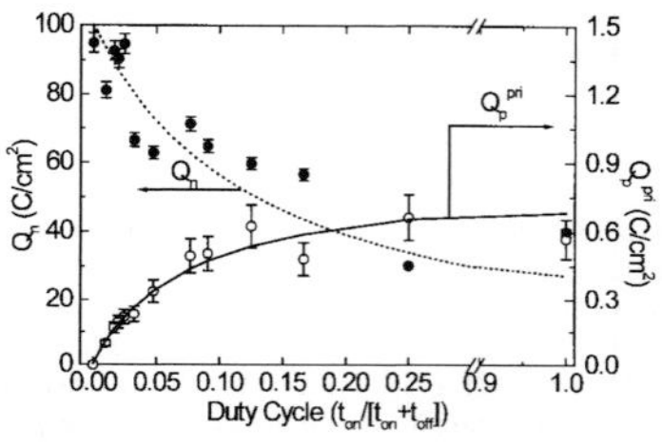

(a)

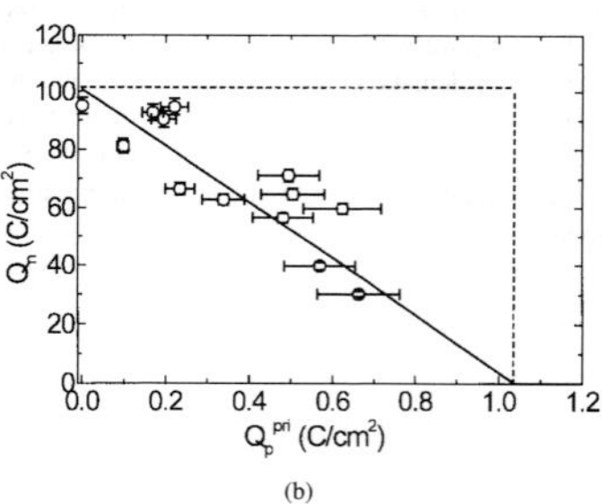

(b)

Fig. 4 (a) Breakdown Q_p^{pri} and Q_n as a function of V_{inj} switch-on duty cycle. ($V_g = -9.2$ V, $V_{\text{well}} = 7.5$ V, $V_{\text{inj}} = 2$ V). (b) Same data of (a) are re-plotted to show the relationship between Q_p^{pri} and Q_n. The solid line is a least square fit of the data which supports (3) (Case I, from the text). The dashed line represents Case II.

Fig. 4(a) shows Q_n and Q_p^{pri} at breakdown as a function of V_{inj} switch-on duty cycle. The results can be quantitatively explained by the following simple model. It has been widely accepted that a critical density of oxide traps activated[3] during stress is required to trigger oxide breakdown [11], [12]. We define a trap density D^{pri} ($=\eta^{\text{pri}}Q_p^{\text{pri}}$) activated by primary hole fluence and a trap density D^n ($=\eta^n Q_n$) activated by FN

---

[3] The authors are indebted to Prof. C. T. Sah, who pointed out to us that the kinetic energy of holes in the oxide may not be high enough to generate new oxide traps. They therefore use "activate" to replace "generate" traps.

588   IEEE ELECTRON DEVICE LETTERS, VOL. 20, NO. 11, NOVEMBER 1999

electron fluence, where η is the probability of trap activation by each hole (or electron) with charge q (or $-q$) per unit length [13]. A question is naturally raised whether trap D^{pri} and D^n belong to the same type of trap (Case I) or belong to two different types of trap (Case II). If Case I, oxide breakdown will occur when $D^{\text{pri}} + D^n$ reaches a critical value D_{cri}. Then, at breakdown point

$$Q_p^{\text{pri}} + (\eta^n/\eta^{\text{pri}}) \times Q_n = D_{\text{cri}}/\eta^{\text{pri}} = Q_{p\,\text{cri}}^{\text{pri}}. \qquad (3)$$

$Q_{p\,\text{cri}}^{\text{pri}}$ is the Q_p^{pri} value at breakdown when Q_n is zero. Fig. 4(b) uses the same data of Fig. 4(a), but re-plots Q_n as a function of Q_p^{pri}. The result shows a linear correlation which clearly supports (3). The solid line in Fig. 4(b) is the least square fit of the data. From this curve, we found $(\eta^{\text{pri}}/\eta^n) = Q_{n\,\text{cri}}/Q_{p\,\text{cri}}^{\text{pri}} = 100/1.03 = 98$, where $Q_{n\,\text{cri}} = 100$ C/cm^2 is the Q_n at breakdown point when Q_p^{pri} is zero. In other words, the primary hole is two orders of magnitude more effective than FN electron in activating the trap D. Finally, the dashed line in Fig. 4(b) corresponds to Case II that D^n and D^{pri} belong to two different types of trap and breakdown is a race process between D^n trap and D^{pri} trap activations. When either D^n or D^{pri} reaches the critical value $D_{\text{cri}}^{\text{pri}} = \eta^{\text{pri}} Q_{p\,\text{cri}}^{\text{pri}}$, or $D_{\text{cri}}^n = \eta^n Q_{n\,\text{cri}}$, oxide breakdown will occur. The experimental data obviously do not support this case.

In summary, our experiments strongly support that oxide breakdown is stimulated by a combined effect when the sum of the active trap density D^{pri} activated by primary holes and the activated trap density D^n activated by FN electrons reaches a critical value D_{cri}. Primary hole is two orders of magnitude more effective than FN electron in activating traps which cause breakdown.

ACKNOWLEDGMENT

The authors thank Profs. C. T. Sah and C. Yang for their invaluable discussions. They also thank the anonymous reviewer who pointed out to us the possible ΔJ_n contribution and suggested the transient current experiment at higher time resolution.

REFERENCES

[1] E. Harari, "Dielectric breakdown in electrically stressed thin films of thermal SiO$_2$," *J. Appl. Phys.*, vol. 49, pp. 2478–2489, 1978.

[2] I. C. Chen, S. E. Holland, and C. Hu, "Electrical breakdown in thin gate and tunneling oxides," *IEEE Trans. Electron Devices*, vol. ED-32, pp. 413–422, 1985.

[3] K. F. Schuegraf and C. Hu, "Hole injection SiO$_2$ breakdown model for very low voltage lifetime extrapolation," *IEEE Trans. Electron Devices*, vol. 41, pp. 761–767, 1994.

[4] Z. A. Weinberg and T. N. Nguyen, "The relation between positive charge and breakdown in metal-oxide-silicon structures," *J. Appl. Phys.*, vol. 61, pp. 1947–1956, 1987.

[5] D. Arnold, E. Cartier, and D. J. DiMaria, "Theory of high-field electron transport and impact ionization in silicon dioxide," *Phys. Rev. B*, vol. 49, pp. 10278–10297, 1994.

[6] H. Satake, S. Takagi, and A. Toriumi, "Evidence of electron-hole cooperation in SiO$_2$ dielectric breakdown," in *Proc. IRPS*, 1997, pp. 156–163.

[7] Y. Kamakura, H. Utsunomiya, T. Tomita, K. Umeda, and K. Taniguchi, "Investigations of hot-carrier-induced breakdown of thin oxides," in *IEDM Tech. Dig.*, 1997, pp. 81–84.

[8] C. T. Sah and T. Nishida, "Mechanisms of electronic trapping in SiO$_2$ on Si," in *21st Int. Conf. Phys. Semiconductors*, 1992, vol. 1, pp. 28–40.

[9] T. H. Ning and H. N. Yu, "Optical induced injection of hot electrons into SiO$_2$," *J. Appl. Phys.*, vol. 45, pp. 5373–5378, 1974.

[10] D. J. DiMaria, E. Cartier, and D. A. Buchanan, "Anode hole injection and trapping in silicon dioxide," *J. Appl. Phys.*, vol. 80, pp. 304–317, 1996.

[11] D. J. Dumin, J. R. Madux, R. S. Scott, and R. Subramonioum, "A model relating wearout to breakdown in thin oxides," *IEEE Trans. Electron Devices*, vol. 41, pp. 1570–1580, 1994.

[12] P. P. Apte and K. C. Saraswat, "Correlation of trap generation to charge-to-breakdown: A physical-damage model of dielectric breakdown," *IEEE Trans. Electron Devices*, vol. 41, p. 1595, 1994.

[13] Y. Nissan-Cohen, J. Shappir, and D. Frohman-Bentchkowsky, "Trap generation and occupation dynamics in SiO$_2$ under charge injection stress," *J. Appl. Phys.*, vol. 60, pp. 2024–2035, 1986.

Jpn. J. Appl. Phys. Vol. 38 (1999) pp. 4696–4698
Part 1, No. 8, August 1999
©1999 Publication Board, Japanese Journal of Applied Physics

Correlation between Charge Pumping Method and Direct-Current Current Voltage Method in p-Type Metal-Oxide-Semiconductor Field-Effect Transistors

Bin-Bin JIE, Kok-Hooi NG, Ming-Fu LI and Keng-Foo LO[1]

Centre for Integrated Circuit Failure Analysis and Reliability, Faculty of Engineering, National University of Singapore, Singapore 119260
[1]*Chartered Semiconductor Manufacturing, 60 Woodlands Industrial Park D, Street 2, Singapore 738406*

(Received March 11, 1999; accepted for publication May 13, 1999)

In this study, the correlation between the charge pumping (CP) method and the direct-current current voltage (DCIV) method in p-type metal-oxide-semiconductor field-effect transistors (pMOSFETs) is investigated. We found that the two techniques probe essentially the same interface traps in the channel (C) region as well as in the drain (D) region. A good linear relationship is observed between the DCIV signal and the CP signal. One obvious advantage of the DCIV method is that it can clearly and directly distinguish the interface traps in the C region and in the D region, respectively. This is possible in the CP method only after complicated manipulation of raw experimental data. The shift of V_{cp} shows the net effect of compensating oxide charge and interface trapped charge, while the shift of $V_{gb,max}$ mainly shows the effect of the oxide charge.

KEYWORDS: charge pumping, DCIV, interface trap, oxide charge, pMOSFET

1. Introduction

The charge pumping (CP) method[1-3] is a very popular dynamic method to characterize interface traps and oxide charges in metal-oxide-semiconductor field-effect transistors (MOSFETs). Recently, a novel static method—direct-current current voltage (DCIV) method[4,5]—has been developed to measure the interface traps and oxide charges in MOSFETs. In this work, we investigate whether the dynamic CP method and the static DCIV method measure the same interface traps and oxide charges. The similarity and difference in the results obtained from the two methods on the same stressed devices are reported and analyzed.

2. Experimental Details

The first test device structure (device #1) is a $0.6 \times 40\,\mu\mathrm{m}^2$ buried-channel lightly doped drain (LDD) pMOSFET with 12.5 nm gate oxide thickness. The device was stressed under the drain avalanche hot-carrier stress (DAHC) with a gate voltage of -8.0 V and a drain voltage of -2.0 V. The second test device structure (device #2) is a $1.0 \times 20\,\mu\mathrm{m}^2$ surface-channel pMOSFET with 4.0 nm gate oxide thickness. The device was stressed under the uniform Fowler-Nordheim tunneling (FN) stress with a gate voltage of -6.5 V, while the source, drain and well were connected to the ground.

For DCIV measurement, we used the normal mode test configuration reported in ref. 6 with p-drain to n-well junction forward biased with bias voltage $V_{eb} = 0.4$ V for device #1 and 0.3 V for device #2, and p-source to n-well junction floated. This is equivalent to the top-E configuration DCIV measurement proposed in ref. 4. For CP measurement, we used the popular version of the CP method—variable base-level CP measurement.[1-3] A 100 kHz frequency square pulse train was applied to the gate. The pulse train has a constant pulse amplitude (6 V for device #1 and 3 V for device #2, peak-to-peak) and its base level V_{base} is swept from an accumulation level to an inversion level.

3. Results and Discussion

Figure 1 shows the DCIV spectra (a) and the CP curves (b, c) of device #1 under DAHC stresses.

For the fresh device, there is a sharp DCIV peak C at $V_{gb} \approx -0.1$ V (V_{gb} is the gate to n-well bias voltage) as shown in Fig. 1(a), corresponding to the interface traps in the channel region (C region).[6] As demonstrated in ref. 5, process-residue interface traps in the channel region generally exist and can be tested in fresh MOSFETs. The CP curve for C region interface traps in the same fresh device is a bell-shaped curve covering a wide range of V_{base}, from -7 V to $+0.5$ V, as shown in Fig. 1(b).

Under DAHC stress, interface traps were generated in the LDD drain region or the drain-to-bulk pn junction depletion region (D region). Figure 1(a) shows two clearly distinguished DCIV peaks C and D corresponding to the interface traps located in the C region and the D region, respectively.[6] The D peak height, I_{bmax}^{D}, increases monotonically with the stress time, indicating that DAHC stress generates interface traps in the D region. However, the C peak height, I_{bmax}^{C}, reduces to a lower level after DAHC stress and then becomes saturated, as shown in Fig. 1(a). This reduction of channel interface traps in pMOSFETs during DAHC is probably caused by hydrogenation of the residual interfacial bonds by hydrogen.[7] On the other hand, the measured CP curve after DAHC stress in Fig. 1(b) is the sum of the C component that originates from the interface traps in the C region and the D component from those in the D region.[2] The negative fixed charges are generated in the D region after DAHC stress,[8] and thus the left transition edge of the CP current at around $V_{base} \approx -7$ V to -5 V is mainly the C component of the CP current, as indicated in Fig. 22 of ref. 3. We assume that the C component of the CP curve keeps the same shape although the magnitude is scaled down after DAHC stress. [This assumption is supported by the experimental results shown in Fig. 3(b).] Consequently, we obtained the C component of the CP current after DAHC stress using the scaling factor of $I_{cp,\,after\,stress}/I_{cp,\,fresh}(V_{base} = -6.2\,\mathrm{V})$, as indicated in Fig. 1(b). $I_{cp,\,after\,stress}/I_{cp,\,fresh}(V_{base} = -6.2\,\mathrm{V})$ means the ratio of the total CP current after DAHC stress to that of the fresh device when $V_{base} = -6.2$ V. The D component of CP current can then be acquired by subtracting the C component CP current from the measured CP current. It is clearly shown in Fig. 1(b) that the two components of CP current after DAHC stress overlap in a large V_{base} region. Figure 1(c) shows the D component of CP current after different DAHC stressing times.

It is therefore obvious that the DCIV method can clearly

Jpn. J. Appl. Phys. Vol. 38 (1999) Pt. 1, No. 8   B.-B. Jie *et al.*   4697

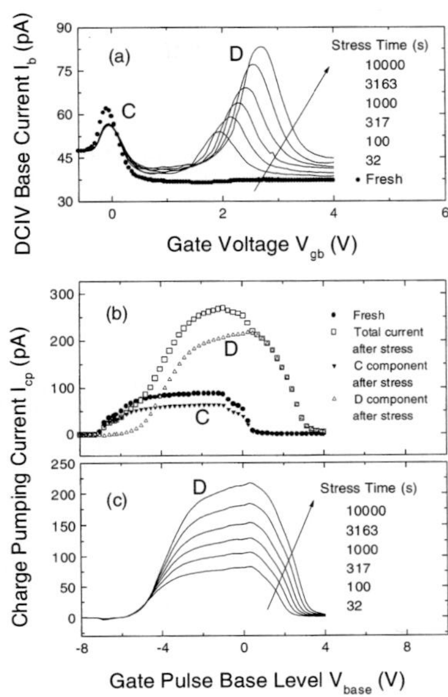

Fig. 1. CP current of a buried-channel pMOSFET and DCIV base current of the same pMOSFET after drain avalanche hot-carrier stress with a stressing gate voltage of -2 V, and a stressing drain voltage of -8 V. In (a), the DCIV base current I_b is plotted against the gate voltage V_{gb} with the stress time as a parameter. In (b), the total CP current after stress is decomposed into the C component and the D component. In (c), the D component of CP current is plotted against the gate pulse base level V_{base} with the stress time as a parameter.

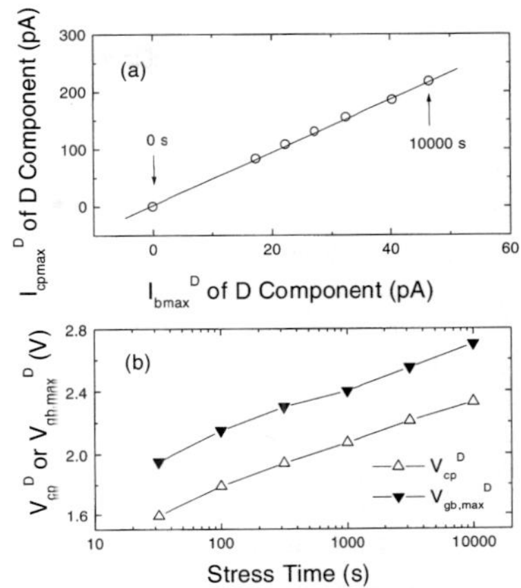

Fig. 2. Correlation between D component of CP current and D component of DCIV base current of the pMOSFET in Fig. 1. (a) The maximum DCIV current, I^D_{bmax}, of the D component against the corresponding maximum CP current, I^D_{cpmax}, of the D component. (b) V^D_{cp} of the D component of CP current and $V^D_{gb,max}$ of the D component of DCIV current against the stress time.

and directly distinguish the interface traps in the C region and the D region. This is possible in the CP method only after complicated manipulation of the raw experimental data.

Figure 2(a) plots the DCIV peak current amplitude I^D_{bmax} of the D peak in the DCIV spectra in Fig. 1(a) against the maximum CP current I^D_{cpmax} of the D component of CP curves in Fig. 1(c). A good linear relationship between I^D_{bmax} and I^D_{cpmax} of the D component is exhibited after DAHC stress. As the density of the stress-induced interface traps is directly proportional to both the CP current I^D_{cpmax} and the DCIV current I^D_{bmax}, the results indicate that both the DCIV and CP methods measure the same stress induced interface traps in the D region.

The shift of the edges of the CP curves implies the presence of charges in the gate oxide or at the interface. The threshold voltage V_{cp} of the CP technique was obtained from the base level of the falling edge in the CP curves at 50% of the maximum CP current.[3] Similarly, the shift of the peak in the DCIV spectrum indicates the presence of stress-induced oxide and interface trap charges. The peak gate voltage $V_{gb,max}$ was obtained from the gate voltage of the DCIV spectrum at the maximum point of the peak. In Fig. 2(b), we plot both V^D_{cp}

and $V^D_{gb,max}$ of the D component signals against the DAHC stress time in Fig. 1. The V^D_{cp} curve is almost parallel to the $V^D_{gb,max}$ curve.

Figure 3 shows the DCIV spectra (a) and the CP curves (b) of device #2 under uniform Fowler-Nordheim (FN) stress. The interface traps were generated mainly in the C region after uniform FN stress.[6] Figure 3(a) shows the sharp C peak of the DCIV spectra while Fig. 3(b) shows the measured CP curves corresponding to the C component of CP curve. Both the shapes of the CP curves and DCIV spectra remain almost unchanged after the stress. Figure 4(a) plots the increment of maximum CP current ΔI^C_{cpmax} against the increment of maximum DCIV current ΔI^C_{bmax} after different stress times (0–10000 s) in Fig. 3. The good linear relationship again demonstrates that the CP and DCIV methods probe essentially the same interface traps in the C region after uniform FN stress. Figure 4(b) plots the shift of V^C_{cp} and $V^C_{gb,max}$ (compared with the fresh device), ΔV^C_{cp} and $\Delta V^C_{gb,max}$, against the FN stress time. ΔV^C_{cp} and $\Delta V^C_{gb,max}$ show the same trend, while $\Delta V^C_{gb,max}$ is about two times larger than ΔV^C_{cp} at 10000 s. The zero value of the first three points of $\Delta V_{gb,max}$ does not indicate no shift of $V_{gb,max}$. In our DCIV measurement, the sweep step of V_{gb} was set at the value of 0.03 V, and thus any $V_{gb,max}$ shift less than 0.03 V cannot be detected accurately.

It is interesting to note that the shifts of both V^D_{cp} and $V^D_{gb,max}$ of the D component interface trap signals are the same after DAHC stress, as shown in Fig. 2(b). The C component interface trap signals, $\Delta V^C_{gb,max}$, however, are about twofold of ΔV^C_{cp} after FN stress, as shown in Fig. 4(b). This is expected due to the following consideration: interface trapped charge has an influence on the edges of the CP curves.[3] The shift

4698   Jpn. J. Appl. Phys. Vol. 38 (1999) Pt. 1, No. 8

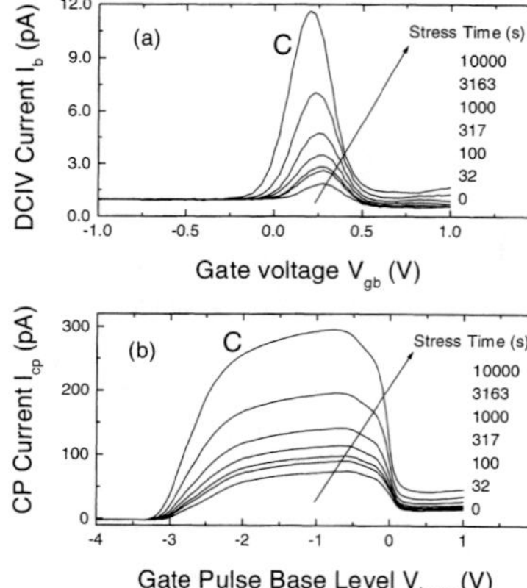

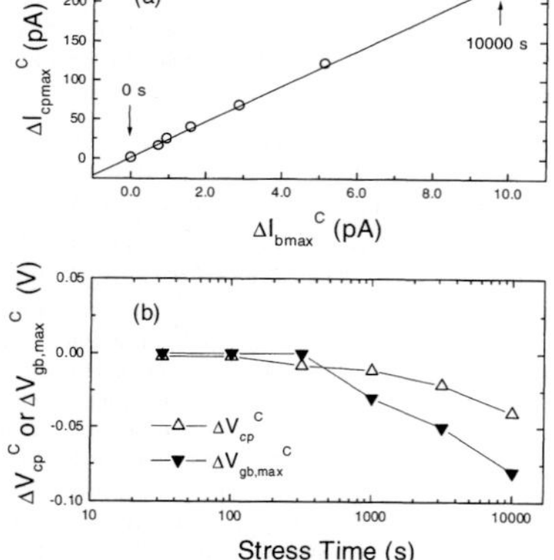

Fig. 3. CP current and DCIV base current of the same surface-channel pMOSFET after Fowler-Nordheim tunneling stress with a stressing gate voltage of -6.5 V. (a) DCIV base current against the gate voltage with the stress time as a parameter. (b) CP current against the gate pulse base level with the stress time as a parameter.

Fig. 4. Correlation between CP current and DCIV base current of the pMOSFET in Fig. 3. (a) The increment of the maximum CP current, $\Delta I^C_{\text{cpmax}}$, against the increment of the corresponding maximum DCIV current, ΔI^C_{bmax}. (b) $\Delta V^C_{\text{cp}} = V^C_{\text{cp}} - V^C_{\text{cp,fresh}}$ of CP current, and $\Delta V^C_{\text{gb,max}} = V^C_{\text{gb,max}} - V^C_{\text{gb,max,fresh}}$ of DCIV current against the stress time.

of V_{cp} is caused by both oxide charge and interface trapped charge. However, the shift of $V_{\text{gb,max}}$ is mainly due to the effect of the oxide charge, since the total net charge of the interface traps is nearly zero at the peak $V_{\text{gb,max}}$.[6] In the case of DAHC stress applied to a pMOSFET, negative oxide charge generated near the drain side is the dominant degradation mechanism compared to the interface trap generation.[8] Thus, the effect of interface trapped charge on the shift of V_{cp} can be neglected. As a result, $V_{\text{gb,max}}$ and V_{cp} have almost the same shift. However, in the case of uniform FN stress, interface trap generation is comparable to oxide trapped charge. Thus, the interface trapped charge compensates part of the oxide charge in the shift of V_{cp}, while the shift of $V_{\text{gb,max}}$ mainly measures the oxide charge.

4. Conclusion

We have compared the DCIV method with the variable base-level CP technique and found that: (1) the two techniques probe essentially the same interface traps in the C region as well as in the D region. There is a good linear relationship between the DCIV signal and CP signal; (2) the DCIV method has the advantage of simplicity in analysis of results. It can clearly and directly distinguish the interface traps in the C region and the D region. The variable base-level CP method can achieve this only after complicated manipulation of raw experimental data; (3) the shift of V_{cp} shows the net effect of compensating oxide charge and interface trapped charge, while the shift of $V_{\text{gb,max}}$ shows mainly the effect of the oxide charge.

Acknowledgments

This work was supported by the Singapore NSTB Research Grant NSTB/17/2/3 and RIC-university research fund project 681305.

1) G. Groeseneken, H. E. Maes, N. Beltran and R. F. De Keersmaecker: IEEE Trans. Electron Devices **31** (1984) 42.
2) P. Heremans, R. Bellens, G. Groeseneken, A. V. Schwerin, W. Weber, B. Brox and H. E. Maes: *Hot Carrier Design Considerations for MOS Devices and Circuits,* ed. C. T. Wang (Van Nostrand Reinhold, New York, 1992) Chap. 1, p. 1.
3) P. Heremans, J. Witters, G. Groeseneken and H. E. Maes: IEEE Trans. Electron Devices **36** (1989) 1318.
4) A. Neugroschel, C. T. Sah, K. M. Han, M. S. Carroll, T. Nishida, J. T. Kavalieros and Y. Lu: IEEE Trans. Electron Devices **42** (1995) 1657.
5) J. Cai and C. T. Sah: IEEE Electron Device Lett. **20** (1999) 60.
6) B. B. Jie, M. F. Li, C. L. Lou, W. K. Chim, D. S. H. Chan and K. F. Lo: IEEE Electron Device Lett. **18** (1997) 583.
7) K. M. Han and C. T. Sah: IEEE Trans. Electron Devices **45** (1998) 1380.
8) R. Woltjer, A. Hamada and E. Takeda: IEEE Trans. Electron Devices **40** (1993) 392.

1608   IEEE TRANSACTIONS ON ELECTRON DEVICES, VOL. 47, NO. 8, AUGUST 2000

A Thorough Study of Quasi-Breakdown Phenomenon of Thin Gate Oxide in Dual-Gate CMOSFET's

Hao Guan, *Student Member, IEEE*, Ming-Fu Li, *Senior Member, IEEE*, Yandong He, Byung Jin Cho, *Member, IEEE*, and Zhong Dong, *Member, IEEE*

Abstract—The conduction mechanism of quasibreakdown (QB) mode for thin gate oxide has been studied in dual-gate CMOSFET with a 3.7-nm thick gate oxide. Systematic carrier separation experiments were conducted to investigate the evolutions of gate, source/drain, and substrate currents before and after gate oxide quasibreakdown (QB). Our experimental results clearly show that QB is due to the formation of a local physically-damaged-region (LPDR) at Si/SiO$_2$ interface [1]. At this region, the effective oxide thickness is reduced to the direct tunneling (DT) regime. The observed high gate leakage current is due to DT electron or hole currents [14] through the region where the LPDR is generated. Twelve V_g, I_{sub}, $I_{s/d}$ versus time curves and forty eight I–V curves of carrier separation measurements have been demonstrated. All curves can be explained in a unified way by the LPDR QB model and the proper interpretation of the carrier separation measurements. Particularly, under substrate injection stress condition, there is several orders of magnitude increase of $I_{sub}(I_{s/d})$ at the onset point of QB for n(p) - MOSFET, which mainly corresponds to valence electrons DT from the substrate to the gate. Consequently, cold holes are left in the substrate and measured as substrate current. These cold holes have no contribution to the oxide breakdown and thus the lifetime of oxide after QB is very long. Under gate injection stress condition, there is sudden drop and even change of sign of $I_{sub}(I_{s/d})$ at the onset point of QB for n(p)-MOSFET, which corresponds to the disappearance of impact ionization and the appearance of hole DT current from the substrate to the gate.

Index Terms—Direct tunneling, MOS devices, oxide breakdown.

I. INTRODUCTION

THE INTEGRITY of thin gate oxide is one of the most crucial reliability issues for ULSI (ultralarge scale integrated) circuits. When gate oxide is thinner than 5 nm, a new anomalous degradation mode has been reported over the past few years, referred to as quasibreakdown (QB) [1], or B-mode SILC [2], [4], or soft-breakdown [3] indistinctly. Typical features of quasibreakdown are high gate leakage current at low oxide field, the large gate signal fluctuations [1]–[9], and several orders of magnitude increase of substrate current in n-MOSFET after QB [1], [7]. Several models have been proposed to explain the QB conduction mechanism, such as physical damage model near to the anode interface [1], the multiple trap assisted tunneling model [3], the variable range hopping conduction model [4], and so on. Although there is consensus that QB is a localized phenomenon in a very small area [1]–[9], however, there is no generally accepted model of the QB conduction mechanism, giving an overall explanation of major observations under the QB. We consider the QB mechanism from the following two important facts: 1) Why QB is thickness-dependent and is observed only in the oxide thinner than 5 nm, and 2) Why there is a correlated sudden increase of substrate current at the onset of QB as has been observed in nMOSFET's [1], [7]. In this paper, a systematic and thorough investigation has been made, using constant current stress and carrier separation measurements [10]–[12], [14] to both n- and p-MOSFET's. All experimental results convincingly support that QB is due to the formation of a local physically-damaged-region (LPDR) at Si/SiO$_2$ interface, as first proposed by Lee and Cho [1]. At this region, the effective oxide thickness is reduced to the direct tunneling (DT) regime. The observed high gate leakage current and correlated high substrate current at low field are due to DT electron or hole currents at the LPDR. Using this model, we can interpret all of our experimental curves in a unified way and also clarify some muddles which have never been understood in the previous works.

II. EXPERIMENTAL AND NOTATIONS

The MOSFET's used in our experiments are dual gate CMOS devices (p$^+$ polysilicon gate for p-MOSFET's and n$^+$ polysilicon gate for n-MOSFET's) fabricated by 0.18 μm design rule technology with a 3.7 nm gate oxide and channel width and length $W/L = 50$ μm/0.5 μm. Constant current stresses for both polarities, i.e., gate and substrate side injections, were applied to n- and p-MOSFET's when source/drain and substrate connected to the ground. Static current–voltage (I–V) characteristics were monitored using an HP4156A semiconductor parameter analyzer. The carrier separation experiments were conducted to measure the gate current I_g, the sum of the source and drain currents $I_{s/d}$, and the substrate current I_{sub} separately before and after QB, applying both polarity of V_g. All currents flowing into the device are taken as positive and $I_g + I_{s/d} + I_{sub} = 0$ if there is no other leakage channel. There are totally 48 I–V curves (p or n MOSFET's, fresh or after QB, gate injection stress or substrate injection stress, positive or negative V_g). In this paper, I_1 denotes valence hole tunneling current through the oxide barrier; I_2 denotes valence electron tunneling current through the oxide barrier; I_3 denotes conduction electron tunneling current through the oxide barrier; I_4 denotes current generated by electron-hole impact ionization [13]. γ is the impact

Manuscript received October 4, 1999; revised March 6, 2000. This work was supported by the Singapore NSTB Research Grant NSTB/17/2/3 and NUS University Research Project RP3982754. The review of this paper was arranged by Editor C. Y. Yang.

H. Guan, M.-F. Li, Y. He, and B. J. Cho are with the Department of Electrical Engineering, National University of Singapore, Singapore 119260, Republic of Singapore (e-mail: elelimf@leonis.nus.edu.sg).

Z. Dong is with Chartered Semiconductor Manufacturing Ltd., Singapore 738406, Republic of Singapore.

Publisher Item Identifier S 0018-9383(00)06038-X.

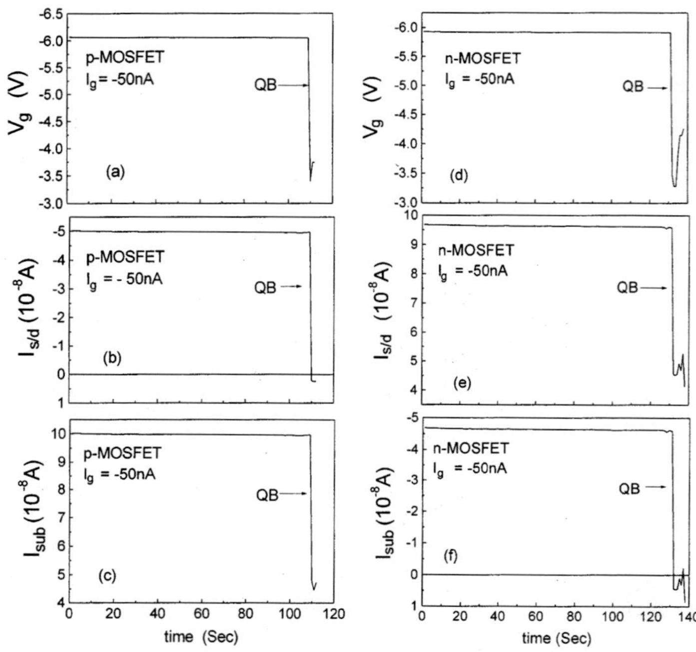

Fig. 1. Evolutions of V_g, $I_{d/s}$, and I_{sub} subjected to negative constant current stress (gate injection) $I_g = -50$ nA (-200 mA/cm^2); (a)–(c) p-MOSFET, and (d)–(f) n-MOSFET. (W/L : $50\,\mu$m/0.5 μm, oxide thickness: 3.7 nm).

ionization quantum yield; η is the probability of the impact-ionized hot hole crossing over the oxide barrier and reaching the opposite side.

III. RESULTS AND DISCUSSIONS

A. Constant Current Stress

Twelve curves in Figs. 1 and 2 illustrate the evolutions of gate voltage V_g, source/drain current $I_{s/d}$ and substrate current I_{sub} of both p- and n-MOSFET's during constant current stress. Fig. 1 is the case of gate side injection with a stress current of -50 nA, (-200 mA/cm^2) and Fig. 2 is the case of substrate side injection with a stress current of $+100$ nA ($+400$ mA/cm^2). The current stresses were removed right after the oxide QB and followed by carrier separation measurements to measure I_g, $I_{s/d}$, and I_{sub} for both polarity of V_g. The carrier separation $I–V$ curves for the fresh devices are plotted together for comparison.

B. Proper Interpretation of Carrier Separation Measurements

The carrier separation measurements conducted in this work are more comprehensive than those conducted in [10]–[12], [14]

and the following two points should be clarified. 1) When the device operates in the inversion mode, [applying large enough positive (negative) V_g to the n-MOSFET (p-MOSFET)] , the device acts as an effective MOS transistor and $I_{s/d}$ measures the electron (hole) current and I_{sub} measures the hole (electron) current, as explained in [10]–[12], [14]. When the device operates in the accumulation mode [applying negative (positive) V_g to the n(p)-MOSFET), or noninversion mode (V_g is lower than the threshold voltage of the MOSFET)], we can consider the device acts as an quasibipolar npn (pnp) transistor with a nearly unity minority carrier base transport factor [15]. The source and drain act as the collector and $I_{s/d}$ measures the minority current in the base, i.e., electron (hole) current. The substrate acts as the base and I_{sub} measures the majority current in the base, i.e., hole (electron) current. In other words, in both polarity of V_g, in the carrier separation measurements in an n(p)-MOSFET, $I_{s/d}$ measures the electron (hole) current and I_{sub} measures the hole (electron) current. 2) The above mentioned arguments are based on absence of electron-hole recombination. However, when the oxide is in the QB state, there is very high density of interface traps [16]. These interface traps are effective recombination centers when $|V_g|$ is very low [17], [18] (lower than about 1 V).

IEEE TRANSACTIONS ON ELECTRON DEVICES, VOL. 47, NO. 8, AUGUST 2000

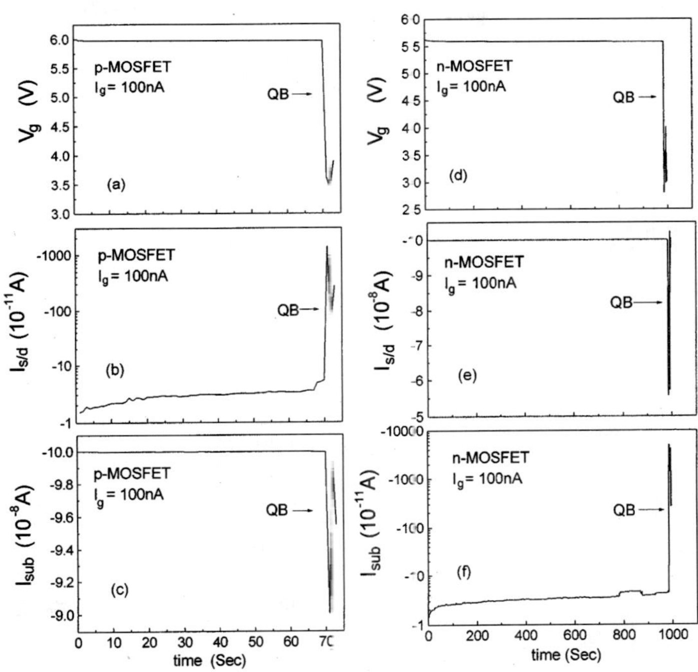

Fig. 2. Same as Fig. 1, but subjected to positive constant current stress (substrate injection) $I_g = +100$ nA ($+400$ mA/cm^2).

Consequently, the minority carriers have extremely short diffusion length L along the interface direction. If the current path is longer than L, the minority carrier current will transfer to majority carrier current before reaching the electrode. More quantitative analysis will be reported elsewhere.

C. Analysis for Inversion Mode Carrier Separation Measurements

1) p-MOSFET's (Inversion Mode): Fig. 3 shows I_g, I_{sub} and $I_{s/d}$ versus V_g, in fresh p-MOSFET and after QB generated by a) negative (gate injection) and b) positive (substrate injection) constant current stresses. The corresponding band diagrams of the devices are illustrated in Fig. 4.

Fresh Device: The conduction mechanisms of three current components of fresh device in Fig. 3 are indicated in Fig. 4(b). Fig. 1(b) and (c) shows that the magnitude of $I_{s/d}$ before QB is nearly identical to the gate current [$I_{s/d} \approx I_g = -50$ nA, $I_{sub} = -(I_g + I_{s/d}) = +100nA$]. It indicates that the electron currents I_3 and I_2, when tunneled to the substrate, produce electron-pairs by impact ionization with a quantum yield $\gamma \approx 1$

[13], as shown in Fig. 4(b). The impact ionized holes flow out through the source/drain.

Device After QB: Fig. 1(b) shows that before and after QB, $I_{s/d}$ is changed from -50 nA to $+2$ nA. In Fig. 3, comparing the $I–V$ curves after QB for the cases of gate injection [Fig. 3(a)] and substrate injection [Fig. 3(b)], the $I–V$ curves after QB for both gate and substrate injections look similar, indicating carrier transport process in the oxide after QB is independent on stress polarity. We therefore only focus our discussion on one of them. Three curves after QB in Fig. 3(b) show surprising analogy with three curves of fresh p$^+$ polysilicon gate p-MOSFET with 2.5 nm gate oxide which is in the direct tunneling (DT) regime, reported by Shi *et al.* [14] (see [14, Fig. 7]). Only voltage and current scaling are slightly different, which can be ascribed to the different oxide thickness and area. (For a fixed V_g, the DT current is determined by oxide thickness and area of the LPDR region, while the $I_{s/d}/I_{sub}$ ratio is determined by the oxide thickness). This analogy gives a very strong support of the QB model proposed in [1] that after QB, a LPDR region at the Si/SiO$_2$ interface reduces the effective SiO$_2$ thickness at that local region and gives rise to DT current. Using this model, we can explain

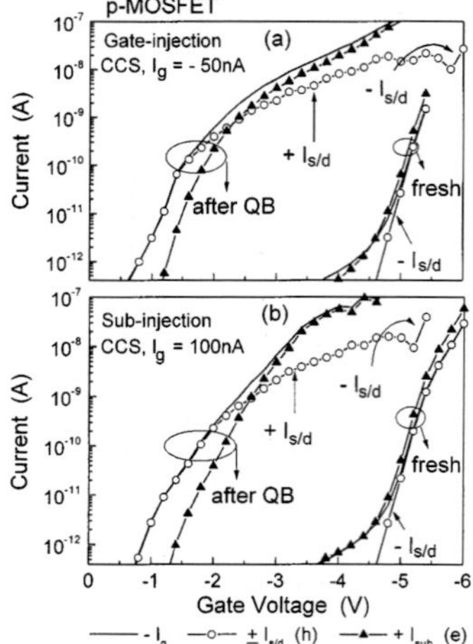

Fig. 3. Carrier separation *I–V* curves for p-MOSFET's (W/L : 50 μm/0.5 μm, oxide thickness : 3.7 nm) measured in inversion mode: (a) fresh and QB after gate injection stress; (b) fresh and QB after substrate injection stress. [The data points in Fig. 3(a) for the fresh device with $|V_g| > 5.5$ V have been deleted.].

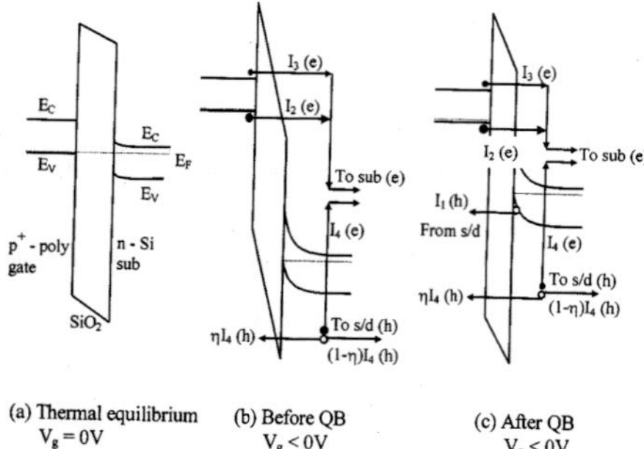

Fig. 4. Schematic band diagram of p-MOSFET in inversion mode measurement. (c) illustrates the band diagram at LPDR region after QB. The effective oxide thickness at LPDR is reduced, the FN current components I_2, I_3 in (b) change into DT current components. After QB, another DT current component I_1 in (c) should also be taken into account.

all of our 48 *I–V* curves and 8 *I–t* curves in this paper in a unified way.

In Fig. 3(b), in very low voltage regime ($|V_g| < 1V$), $I_{s/d}$ is positive and $I_{s/d} \approx -I_g$, implying that holes DT from the LPDR to the gate (I_1 in Fig. 4) is the dominating process. In this very low voltage regime, there is no measurable I_{sub}, which means that there is almost no electron tunneling from the gate conduction band (I_3) or valence band (I_2) to the substrate. This

can be explained by Fig. 4(a). When the V_g is close to zero, there are no enough conduction electrons in the p$^+$ gate to tunnel through the barrier (I_3). For the valence electrons in the gate (I_2), due to the band alignment as shown in Fig. 4(a), the elastic tunneling through the barrier to the substrate is forbidden at very low V_g. In the low voltage regime $1V < |V_g| < 2.5V$, as shown in Fig. 3(b), I_{sub} becomes measurable and increases when $|V_g|$ increases. This I_{sub} corresponds to DT of valence electrons (I_2), since when applying negative V_g, the elastic electron DT from the gate valence band to the substrate conduction band becomes allowed, as shown in Fig. 4(c). The argument that I_{sub} is mainly contributed from valence band electron DT in low voltage regime has been further analyzed in Shi *et al.* work in 2.5-nm fresh oxide [14]. When $|V_g| > 2.5$ V, I_{sub} starts to supersede $I_{s/d}$ and becomes the dominant component of I_g. I_{sub} consists of electrons tunneling from the gate ($I_3 + I_2$), and impact ionization induced electron current (I_4). Further, as shown in Fig. 3(b), when $|V_g|$ increases from 2 to 5 V, the increase of $I_{s/d}$ tends to slow down and even turns to decrease, with a downward U-turn of $I_{s/d}$ curve, and finally changes the sign from positive to negative. These findings are completely consistent with DT of electrons and holes in fresh p-MOSFET in [14]. In this region, when $|V_g|$ increases, the number and the quantum yield γ of injected electrons increase [13], therefore more electron-hole pairs [I_4 in Fig. 4(c)] are generated in the substrate region. When the major portion of generated holes flow out to the source/drain [$[1 - \eta]I_4$ in Fig. 4(c)], the corresponding source/drain current becomes negative, as illustrated in Fig. 4(c). This negative source/drain current will compensate the positive source/drain current I_1, causing a downward U-turn of $I_{s/d}$ curve and even changing the sign of $I_{s/d}$ from positive to negative, when $[1-\eta]I_4$ over-compensates the hole tunneling current I_1.

2) n-MOSFET's (Inversion Mode): Fig. 5(a) shows I_g, I_{sub} and $I_{s/d}$ versus positive V_g, in fresh n-MOSFET and after QB generated by negative (gate injection) constant current stress. Similar curves were obtained by applying positive (substrate injection) constant current stress as well. The corresponding band diagrams of the devices are illustrated in Fig. 5(b)–(d).

Fresh Device: The conduction mechanisms of three current components of fresh device in Fig. 5(a) are indicated in Fig. 5(c). Fig. 2(e) shows that the magnitude of $I_{s/d}$ before QB is nearly identical to the gate current however with opposite polarity ($I_{s/d} \approx -I_g = -100$ nA), which can be explained by Fig. 5(c) that electrons flow in from the source/drain (I_3) and tunnel through the barrier to the gate. The impact ionized electron current I_4 has almost no contribution to the gate current since the hole current $[1 - \eta]I_4 \approx I_4$ compensates the electron current I_4 in the gate current as shown in Fig. 5(c). A very small portion of holes ηI_4 tunnels through the oxide barrier to the substrate, giving rise to a very small negative substrate current I_{sub} as shown in Fig. 5(a) and Fig. 2(f). From the data in Fig. 5(a), the value of η can be estimated as $\eta \approx I_{sub}/I_{s/d} \approx 3 \times 10^{-11}A/10^{-7}A = 3 \times 10^{-5}$, (at $V_g = 5.6V$), if $\gamma \approx 1$ is assumed.

Device After QB: Fig. 2(d)–(f) shows that V_g suddenly drops from 5.6 V to 3.5 V, $|I_{s/d}|$ suddenly reduces, and $|I_{sub}|$ correspondingly increases three orders of magnitude, from 10^{-11} to

1612   IEEE TRANSACTIONS ON ELECTRON DEVICES, VOL. 47, NO. 8, AUGUST 2000

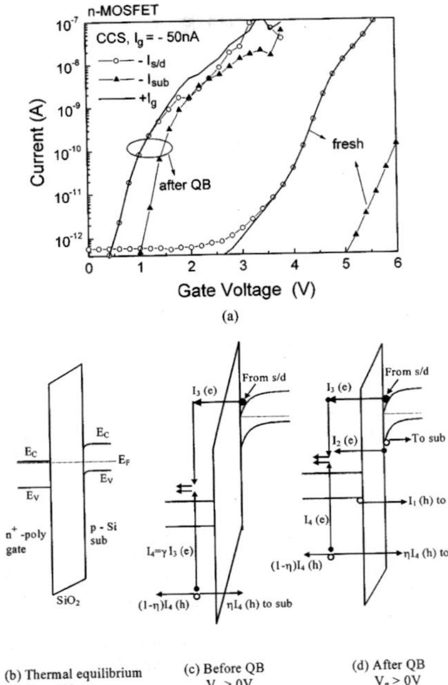

(b) Thermal equilibrium   (c) Before QB   (d) After QB
    V_g = 0V   V_g > 0V   V_g > 0V

Fig. 5.   (a) Carrier separation I–V curves for n-MOSFET's (W/L : 50 μm/0.5 μm, oxide thickness: 3.7 nm) measured in inversion mode: fresh and QB after gate injection stress. (similar curves were obtained after substrate injection stress). (b)–(d) Schematic band diagram of n-MOSFET in inversion mode measurement. Fig. 5(d) illustrates the band diagram at LPDR region after QB. The effective oxide thickness at LPDR is reduced, the FN current components I_3 in (c) changes into DT current component. After QB, DT current components I_1 and I_2 in (d) should also be taken into account.

10^{-8} A at the onset of QB. This has been reported but not clearly understood in [1], [7]. In Fig. 5(a), when $V_g < 1V$, only electrons supplied from the source and drain, DT through the oxide barrier (I_3) to the gate side. This corresponds to the negative $I_{s/d}$ current measured at V_g lower than 1 V. The valence electrons can not tunnel to the gate side (I_2) at very low V_g. Since at very low V_g, due to the band alignment as shown in Fig. 5(b), the valence electron elastic DT from the substrate to the gate is forbidden. The hole current I_1 by DT from the gate to the substrate is also negligible since the hole concentration in the n$^+$ gate is extremely low. When V_g is increased, I_{sub} is increased and can be measured, as shown in Fig. 5(a) when V_g is higher than 1 V. This is mainly due to the valence electrons DT from the substrate to the gate (I_2), as shown in Fig. 5(d). When valence electrons tunnel through the oxide, "cold" holes are left in the substrate and move out, giving rise to the measurable negative substrate current I_{sub}. We thus give a natural explanation of sudden increase of $|I_{sub}|$ after QB in Fig. 2(f). We have also considered

two possible hole current components ηI_4 and I_1 as shown in Fig. 5(d). ηI_4 is hot hole component originated from the impact ionization $I_4 = \gamma I_3$. Similar to the case of p-MOSFET, the impact ionization quantum yield γ is very small at $V_g \approx 3.5$ V after QB and therefore ηI_4 hot hole component is expected to be very small. I_1 is due to DT of holes from gate valence band to the substrate. I_1 is also very small and can be neglected because hole concentration in the n$^+$ gate is very low and the oxide barrier is high. We thus argue that the sudden increase of I_{sub} after QB is mainly due to cold hole current component I_2, not hot hole current. Hot hole injection is known to be very effective to reduce the oxide lifetime [19]. However, after QB, the major hole current component is cold hole. We believe this is the reason why after QB, the oxide usually can endure for a very long time under stress.

D. Analysis for Accumulation Mode Carrier Separation Measurements

1) p-MOSFET's (Accumulation Mode): Fig. 6(a) shows I_g, I_{sub} and $I_{s/d}$ versus positive V_g, in fresh p-MOSFET and after QB generated by negative (gate injection) constant current stress. Similar curves were obtained by applying positive constant current stress as well. The corresponding band diagrams of the devices are illustrated in Fig. 6(b)–(d).

Fresh Device: The conduction mechanisms of three current components of fresh device are indicated in Fig. 6(c) when V_g is high. The band diagram in Fig. 6(c) is quite similar to the case of inversion mode n-MOSFET shown in Fig. 5(c). However, in Fig. 5(c), $I_{s/d}$ measures the electron current I_3 and the I_{sub} measures the hole current ηI_4, while in Fig. 6(c), I_{sub} measures the electron current I_3 and $I_{s/d}$ measures the hole current ηI_4. We can therefore expect that when positive V_g is applied to both n-MOSFET and p-MOSFET fresh devices, three I–V curves for p-MOSFET (in accumulation mode) can be roughly derived from three I–V curves for n-MOSFET (in inversion mode) with swapping $I_{s/d}$ and I_{sub}. This is confirmed by the experimental results of the I–V curves for fresh devices in Fig. 5(a) and Fig. 6(a) and Fig. 2(b), (c), (e), and (f) before QB.

Device After QB: The interpretation of the I–V curves after QB in Fig. 6(a) is more complicated in the low V_g regime. Three I–V curves in Fig. 6(a) should not be derived from three I–V curves for n-MOSFET in Fig. 5(a) with swapping $I_{s/d}$ and I_{sub}, due to following two facts: first, when V_g is close to zero, the band diagram of p-MOSFET in Fig. 6(b) is very different from that of n-MOSFET in Fig. 5(b). Second, there is a very high density of Si/SiO$_2$ interface traps when the oxide is in QB state [16]. Correspondingly, the quasibipolar pnp transistor has a nearly zero minority carrier base transport factor. In this case, both electron and hole currents are measured as I_{sub}. Taking these two facts into account, let us turn to interpret three I–V curves after QB in Fig. 6(a). Comparing three curves after QB in Fig. 6(a) and the band diagrams in Fig. 6(b) and (d), there are three current components: the substrate conduction electrons DT to the gate (I_3); the gate valence holes DT to the substrate (I_1); and the substrate valence electrons DT to the gate (I_2), leaving "cold" holes behind. In the low V_g regime, the hole cur-

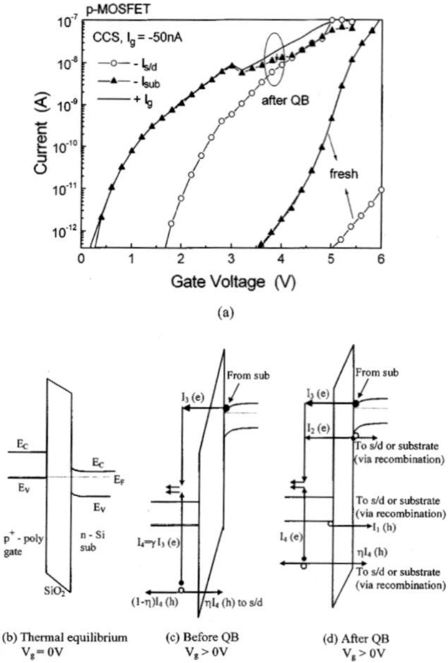

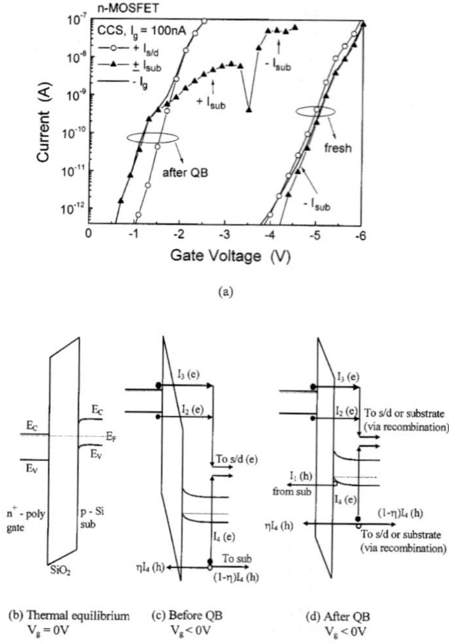

Fig. 6. (a) Carrier separation I–V curves for p-MOSFET's (W/L : 50 μm/0.5 μm, oxide thickness: 3.7 nm) measured in accumulation mode: fresh and QB after gate injection stress. (similar curves were obtained after substrate injection stress). (b)–(d) Schematic band diagram of p-MOSFET in accumulation mode measurement.(d) illustrates the band diagram at LPDR region after QB. The effective oxide thickness at LPDR is reduced, the FN current components I_3 in (c) changes into DT current component. After QB, DT current components I_1 and I_2 in (d) should also be taken into account.

Fig. 7. (a) Carrier separation I–V curves for n-MOSFET's (W/L : 50 μm/0.5 μm, oxide thickness: 3.7 nm) measured in accumulation mode: fresh and QB after substrate injection stress. (similar curves were obtained after gate injection stress). (b)–(d) Schematic band diagram of n-MOSFET in accumulation mode measurement. (d) illustrates the band diagram at LPDR region after QB. The effective oxide thickness at LPDR is reduced, the FN current components I_2 and I_3 in (c) change into DT current components. After QB, DT current components I_1 in (d) should also be taken into account.

rent components transfer to electron current by recombination and are measured as substrate current. In other words, I_{sub} measures both hole current and electron current and $I_{s/d} \approx 0$ when V_g is low. When V_g is increased, the interface state recombination rate decreases and the minority carrier base transport factor increases toward one. Consequently, the I_{sub} measures the electron current and $I_{s/d}$ measures the hole current. When V_g is large, the band diagram in Fig. 6(d) is similar to the band diagram in Fig. 5(d). Therefore we expect that three I–V curves in Fig. 6(a) can be derived from three I–V curves in Fig. 5(a) with swapping $I_{s/d}$ and I_{sub}, and two I–t curves in Fig. 2(b) and (c) can be derived from those in Fig. 2(e) and (f) with swapping $I_{s/d}$ and I_{sub}. These have been confirmed as seen in the figures. The sudden changes of I_{sub} and $I_{s/d}$ in I–t curves for p-MOSFET at the onset point of QB have never been reported before.

2) n-MOSFET's (Accumulation Mode): Fig. 7(a) shows I_g, I_{sub} and $I_{s/d}$ versus negative V_g, in fresh n-MOSFET and after QB generated by the positive (substrate injection) constant current stress. Similar curves were obtained by applying negative

constant current stress as well. The corresponding band diagrams of the devices are illustrated in Fig. 7(b)–(d).

Fresh Device: The conduction mechanisms of three current components of fresh device are indicated in Fig. 7(c). The band diagram in Fig. 7(c) is quite similar to the case of Fig. 4(b). However, in Fig. 4(b) I_{sub} measures the electron currents (I_3, I_2 and I_4) and $I_{s/d}$ measures the hole current ($[1 - \eta]I_4$), while in Fig. 7(c) $I_{s/d}$ measures the electron current (I_3, I_2 and I_4) and I_{sub} measures the hole current ($[1 - \eta]I_4$). We can therefore expect, when negative V_g is applied to both n-MOSFET and p-MOSFET fresh devices, three I–V curves for n-MOSFET can be roughly derived from three I–V curves for p-MOSFET with swapping $I_{s/d}$ and I_{sub}. This is confirmed by the experimental results of I–V curves for the fresh devices in Fig. 3 and Fig. 7(a), and Fig. 1(b), (c), (e), and (f) before QB.

Device After QB: From the above argument, when V_g is high, the band diagram in Fig. 7(b)–(d) is similar to that in

1614   IEEE TRANSACTIONS ON ELECTRON DEVICES, VOL. 47, NO. 8, AUGUST 2000

Fig. 4. We then expect to derive three I–V curves in Fig. 7(a) from those in Fig. 3, and derive two I–t curves in Fig. 1(e) and (f) from those in Fig. 1(b) and (c) with swapping $I_{s/d}$ and I_{sub}. This is confirmed in the experiments, including the change of sign from $+$ to $-$ for I_{sub} in Fig. 7(a) and Fig. 1(f) as for $I_{s/d}$ in Fig. 3 and Fig. 1(b). This is a very strong support of the whole argument in this paper. When V_g is low, the band diagram in Fig. 7(b) is similar to that in Fig. 6(b), if swapping the substrate and the gate and swapping the V_g polarity. On the other hand, the quasibipolar npn transistor in Fig. 7 has a nearly zero minority carrier base transport factor, since after QB, the SiO$_2$/Si interface has a very high density of interface traps which act as effective recombination centers when V_g is low. Although the gate electrons tunnel through the oxide barrier (I_3) is a major current component, however it is measured as substrate current when V_g is low. Therefore in this case I_{sub} measures both electron current and hole current. We expect that at low V_g regime, only I_{sub} can be measured with a shape similar to the case of p-MOSFET accumulation mode. This was confirmed in the experiments in Figs. 6(a) and 7(a).

E. Random Telegraph Switching Noise (RTSN)

Based on the LPDR-DT model, it is natural to establish a possible explanation of random telegraph switching noise (RTSN) after oxide QB [6]. In the LPDR region, the damaged structure has two or more metastable states corresponding to different effective oxide thickness, as shown in Fig. 8(a). The thermal transition between two (or multi) metastable states leads to RTSN noise fluctuation between two (or multi) levels. This RTSN model predicts a correlation between $I_{s/d}$ RTSN and I_{sub} RTSN This has been confirmed by experiment as shown in Fig. 8(b).

IV. SUMMARY AND CONCLUSIONS

Systematic carrier separation experiments were conducted to investigate the evolutions of gate, source/drain, and substrate currents before and after gate oxide quasibreakdown (QB) in dual-gate CMOSFET's.

Proper Interpretation of Carrier Separation Measurements: When the device operates in the inversion mode, the n(p)-MOSFET acts as an effective MOSFET. When the device operates in the accumulation mode, or noninversion mode, the device acts as a quasibipolar npn (pnp) transistor with a nearly unity minority carrier base transport factor. In both polarity of V_g, in the carrier separation measurements in an n(p)-MOSFET, $I_{s/d}$ measures the electron (hole) current and I_{sub} measures the hole (electron) current. However, when the oxide is in the QB state, there is very high density of interface traps [16]. These interface traps are effective recombination centers when $|V_g|$ is very low [17], [18]. Consequently, the minority carrier diffusion length L along the interface direction is extremely short. If the minority current path is longer than L, the minority current will transfer to the majority current before reaching the electrode.

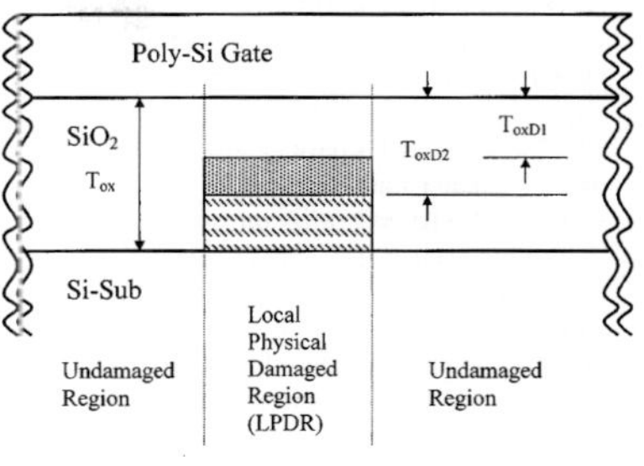

(a)

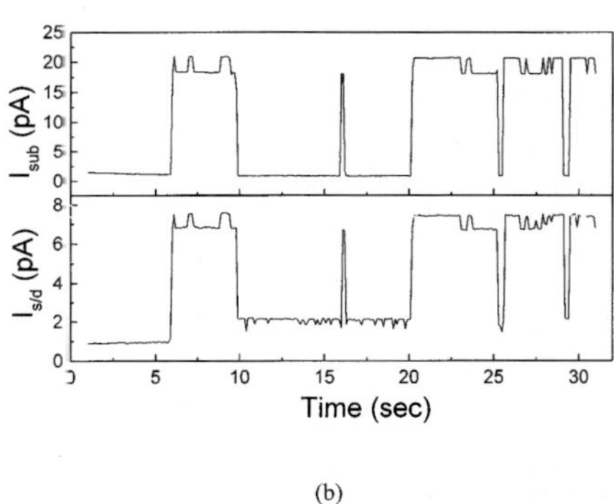

(b)

Fig. 8. (a) Schematic drawing for a LPDR region at the Si/SiO$_2$ interface with two metastable states, corresponding to two different effective oxide thickness T_{OXD1} and T_{OXD2} in a MOSFET with original gate oxide thickness T_{OX}. (b) RTSN signals after QB observed in a p-MOSFET, stressed under constant current with a density of -40 mA/cm^2. The RTSN signal was monitored at a constant gate voltage of -4 V.

QB Mechanism:

1) QB is due to the formation of a local physically-damaged-region (LPDR) at Si/SiO$_2$ interface. At this region, the effective oxide thickness is reduced to the direct tunneling regime [1]. The observed high gate leakage current is due to direct tunneling electron or hole currents at the LPDR. Twelve V_g, I_{sub}, $I_{s/d}$ versus time curves (Figs. 1 and 2) and forty eight I–V curves (Figs. 3, 5, 6, 7 and similar curves after the opposite polarity stress.) of carrier separation measurements (p- or n-MOSFET's, fresh or after QB, gate injection stress or substrate injection stress, positive or negative V_g) have been demonstrated. All curves can be explained in a unified way with the LPDR QB model.

2) Under **substrate injection stress condition**, there is several orders of magnitude increase of $I_{sub}(I_{s/d})$ at the

GUAN *et al.*: QUASI-BREAKDOWN PHENOMENON OF THIN GATE OXIDE 1615

onset point of QB for n(p) -MOSFET, [Fig. 2(b) and (f)] which mainly corresponds to valence electrons DT from the substrate to the gate [I_2 in Figs. 5(d) and 6(d)]. Consequently, cold holes are left in the substrate and move out. These cold holes have almost no contribution to the oxide breakdown and the lifetime of oxide after QB is very long. Under *gate injection stress condition*, there is sudden drop and even change of sign of $I_{sub}(I_{s/d})$ at the onset point of QB for n(p)-MOSFET, [Fig. 1(b) and (f)] which corresponds to the disappearance (or reduction) of impact ionization induced hole current [$(1 - \eta)I_4$ in Figs. 4(c) and 7(d)] and the appearance of hole DT current from the substrate to the gate [I_1 in Figs. 4(c) and 7(d)].

3) All *I–V* curves after QB generated by gate injection resemble to the corresponding *I–V* curves after QB generated by substrate injection . The conduction mechanism after QB is independent on stress polarity.

4) When V_g is high, all *I–V* curves of p and n-MOSFET's (fresh or after QB) [Figs. 3, 5(a), and 7(a)] can be derived from each other with swapping $I_{s/d}$ and I_{sub}. Only after QB and when V_g is low, two facts should be considered: the band alignment in low V_g regime and the recombination via interface traps.

REFERENCES

[1] S. H. Lee, B. J. Cho, J. C. Kim, and S. H. Choi, "Quasibreakdown of ultrathin gate oxide under high field stress," in *IEDM Tech. Dig.*, 1994, pp. 605–608.

[2] K. Okada, S. Kawasaki, and Y. Hirofuji, "New experimental findings on stress induced leakage current of ultra thin silicon dioxides," in *Ext. Abst. of 1994 Int. Conf. Solid State Devices and Materials*, 1994, pp. 565–567.

[3] M. Depas, T. Nigam, and M. M. Heyns, "Soft breakdown of ultra-thin gate oxide layers," *IEEE Trans. Electron Devices*, vol. 43, pp. 1499–1504, 1996.

[4] K. Okada and K. Taniguchi, "Electrical stress-induced variable range hopping conduction in ultra-thin silicon dioxides," *Appl. Phys. Lett.*, vol. 70, pp. 351–353, 1997.

[5] T. Yoshida, S. Miyazaki, and M. Hirose, "Analytical modeling of quasibreakdown of ultra-thin gate oxides under constant current stress," in *Ext. Abst. 1996 Int. Conf. Solid State Devices and Materials*, 1996, pp. 539–541.

[6] T. Tomita *et al.*, "A new soft breakdown model for thin thermal SiO$_2$ films under constant current stress," *IEEE Trans. Electron Devices*, vol. 46, pp. 159–164, 1999.

[7] F. Crupi *et al.*, "On the properties of the gate and substrate current after soft breakdown in ultrathin oxide layers," *IEEE Trans. Electron Devices*, vol. 45, pp. 2329–2334, 1998.

[8] M. Houssa, T. Nigam, P. W. Mertens, and M. M. Heyns, "Soft breakdown in ultrathin gate oxide: Correlation with the percolation theory or nonlinear conductors," *Appl. Phys. Lett.*, vol. 73, pp. 514–516, 1998.

[9] E. Miranda *et al.*, "Soft Breakdown Fluctuation Events in Ultrathin SiO$_2$ Laysers," *Appl. Phys. Lett.*, vol. 73, pp. 490–492, 1998.

[10] A. S. Ginovker, V. A. Gristsenko, and S. P. Sinitsa, "Two-band conduction of amorphous silicon nitride," *Phys. Stat. Sol. A*, vol. 26, pp. 489–496, 1974.

[11] B. Eitan and K. Kolodny, "Two components of tunneling current in metal-oxide-semiconductor structure," *Appl. Phys. Lett.*, vol. 43, pp. 106–108, 1983.

[12] L. D. Yao, F. T. Liou, and S. Chen, "Hole current in dual dielectric under positive gate voltage," *IEEE Electron Device Lett.*, vol. EDL-4, p. 261, 1983.

[13] C. Chang, C. Hu, and R. W. Brodersen, "Quantum yield of electron impact ionization in silicon," *J. Appl. Phys.*, vol. 57, pp. 302–309, 1985.

[14] Y. Shi, T. P. Ma, S. Prasad, and S. Dhanda, "Polarity dependent gate tunneling currents in dual-gate CMOSFET's," *IEEE Trans. Electron Devices*, vol. 45, pp. 2355–2360, 1998.

[15] S. M. Sze, *Physics of Semiconductor Devices*. New York: Wiley, 1981, ch. 3.

[16] H. Guan, B. J. Cho, M. F. Li, Y. D. He, and Z. Dong, "A study of quasibreakdown mechanism in ultra-thin gate oxide by using DCIV technique," in *Proc. IEEE 7th Int. Symp. on Physical and Failure Analysis of Integrated Circuits*, 1999, pp. 81–84.

[17] A. Neugroschel *et al.*, "Direct-current measurements of oxide and traps on oxidized silicon," *IEEE Trans. Electron Devices*, vol. 42, pp. 1657–1662, 1995.

[18] B. B. Jie *et al.*, "Investigatioon of interface traps in LDD pMOST's by the DCIV method," *IEEE Electron Device Lett.*, vol. 18, pp. 583–585, 1997.

[19] K. F. Schuegraf and C. Hu, "Hole injection SiO$_2$ breakdown model for very low voltage lifetime extrapolation," *IEEE Trans. Electron Devices*, vol. 41, pp. 761–767, 1994.

Hao Guan (S'99) was born in Nanjing, China, in 1969. He received the B.S. and M.S. degrees from the Department of Physics, Nanjing University, in 1991 and 1994, respectively. Since 1997, he has been pursuing the Ph.D. degree in the Department of Electrical Engineering, National University of Singapore.

In 1994, he joined Shenzhen State LCD Displaying Engineering Company, China, as an Engineer. Currently, he is with the QRA Department, TECH Semiconductor (Singapore) Pte, Ltd., as a Senior Engineer, working on device characterization and reliability. His general research interests include thin oxide reliability, conduction mechanism of thin oxide, device characterization, and modeling.

Ming-Fu Li (M'91–SM'99) graduated from the Department of Physics, Fudan University, Shanghai, China, in 1960.

After graduation, he joined the Department of Applied Physics, University of Science and Technology of China (USTC), as a Teaching Assistant and Lecturer. In 1978, he joined the Graduate School Faculty, Chinese Academy of Sciences, Beijing, first as an Associate Professor and in 1986, as Professor. He has also served as Adjunct Professor at the Institute of Semiconductors, Chinese Academy of Sciences, Fudan University, and USTC, Hefei, China . He was a Visiting Scholar at Case Western Reserve University, Cleveland, OH, in 1979, University of Illinois at Urbana-Champain from 1979 to 1981, and was a visiting scientist at University of California, Berkeley and Lawrence Berkeley National Laboratories from 1986 to 1987, 1990 to 1991, and 1993. He joined the Department of Electrical Engineering, National University of Singapore, as an Associate Professor in 1991, and became a Professor in 1996. His current research interests are in the areas of reliability physics in deep submicron CMOS devices, analog IC design, and wide energy gap group III nitride. He has published over 140 research papers and two books., including *Modern Semiconductor Quantum Physics* (Singapore: World Scientific, 1994).

Mr. Li has served on several international program committees and advisory committees in international semiconductor conferences in China, Japan, Canada, Germany, and Singapore.

Yandong He received the B.S. degree in radio electronics and the M.S. degree in microelectronics from Peking University, China, in 1990 and 1993, respectively.

From 1993 to 1997, she was with the Institute of Microelectronics (IME), Peking University. She joined the Department of Electrical Engineering, National University of Singapore, in 1998 as a Post-Master Fellow, where she has done research in deep submicron CMOS devices and integrated circuit technology. Her research interests include process-related problems in submicron CMOS devices, ultrathin gate oxide degradation and quasi/breakdown mechanism, hot-carrier degradation effects and electrical characterization techniques, circuit design, simulation and verification.

1616   IEEE TRANSACTIONS ON ELECTRON DEVICES, VOL. 47, NO. 8, AUGUST 2000

Byung Jin Cho (M'97), received the B. S. degree in electrical engineering from Korea University, Seoul, Korea, in 1985, and the M. S. and Ph.D. degrees in electrical engineering from Korea Advanced Institute of Science and Technology, Taejon, in 1987 and 1991, respectively.

From 1991 to 1993, he was with the Inter-University Microelectronic Center), Leuven, Belgium, as a Research Fellow, where he worked on advanced silicon processing. In 1993, he joined the Memory R&D Division, Hyundai Electronics Ind. Co., Korea, as a Section Manager, where he led a research team for the process development for 256M and 1G DRAM and flash EEPROM. Since 1997, he has been with the Department of Electrical Engineering, National University of Singapore, where he is currently an Associate Professor. His research interests include CMOS process integration, ultrathin gate oxide, device isolation technology, junction formation and thermal processing, and MOS device physics. He was the first to use the termin "quasi-breakdown" of thin oxide in his paper at IEDM'94. He has published over 50 technical papers and holds over 30 patents.

Dr. Cho is an active member of the Electrochemical Society and is listed in *Who's Who In The World*.

Zhong Dong (M'99) was born in Sichuan, China, in 1963. He received the B.S degree in physics from Sichuan University, Chengdu, 1983, and the M.S. degree in Microelectronics from the University of Electronic Science and Technology of China in 1986.

He has worked in the diffusion area for 13 years. Since 1995, he has been with Chartered Semiconductor Manufacturing. As a Principal engineer in TD, his focus is on ultrathin gate dielectric and the gate module processes for 0.18μm and beyond, specializing in the nitrided gate oxide formed by rapid thermal oxidation and nitridation.

IEEE ELECTRON DEVICE LETTERS, VOL. 22, NO. 5, MAY 2001   233

Interface Traps at High Doping Drain Extension Region in Sub-0.25-μm MOSTs

Gang Chen, *Student Member, IEEE*, M. F. Li, *Senior Member, IEEE*, and Xing Yu

Abstract—A huge bulk (or drain) current I_b (or I_d) peak versus gate voltage was observed for the 0.25-μm or sub-0.25-μm metal-oxide-semiconductor field effect transistors (MOSTs) with high doping concentration source/drain extension, when the drain-bulk pn junction is forward biased. This current is increased under Fowler–Nordheim (FN) or channel hot carrier (CHC) stress and is identified as thermal-trap-tunneling electron current at the drain extension-gate overlap region. It is extremely sensitive that one interface trap will induce 0.1 pA current increment of peak I_b (or I_d).

Index Terms—CMOSFETs, hot carriers, reliability, semiconductor–insulator interfaces, tunneling.

I. INTRODUCTION

HOT-CARRIER effect has been an interesting subject for the metal-oxide-semiconductor field effect transistors (MOSTs) [1]–[4]. Many researchers have studied the stress-induced interface traps in the basewell-channel (BC) region [3]–[11], the drain-base junction space-charge (JSC) region [8], and the lightly-doped drain-extension region (DE) [8]. However, the frequently used charge-pumping [5] and capacitance–voltage (CV) [11] methods cannot detect and distinguish the interface traps in the highly doped DE region [12] in 0.25-μm or sub-0.25-μm MOS devices. Actually interface traps in DE region may have serious impact on MOS device degradation such as increases of drain series resistance, turn-off leakage current etc. Therefore the detection and estimation of interface traps in the DE region are important. Using the newly developed direct-durrent–current–voltage ($DCIV$) technique, which measures two current peaks corresponding to the interface traps in the BC and JSC regions respectively [7], [8], a third huge peak is detected [13] and is tentatively identified as thermal-trap-tunneling current via the interface traps located at the gate-DE overlap region.

II. ANALYSIS

The p-MOSTs are used for demonstration. In 0.25-μm or sub-0.25-μm MOS technology, the doping concentration of the

Manuscript received December 19, 2000; revised February 13, 2001. This work was supported by the Singapore National Science and Technology Board Research Grant NSTB/17/2/3 and National University of Singapore Research Grant RP3982754. The review of this paper was arranged by Editor K. De Meyer.

G. Chen and M. F. Li are with SNDL, Department of Electrical and Computer Engineering, National University of Singapore, Singapore 119 260 (e-mail: elelimf@nus.edu.sg).

X. Yu is with Chartered Semiconductor Manufacturing, Ltd., Singapore 738 406.

Publisher Item Identifier S 0741-3106(01)03706-5.

P$^+$ Drain Extension

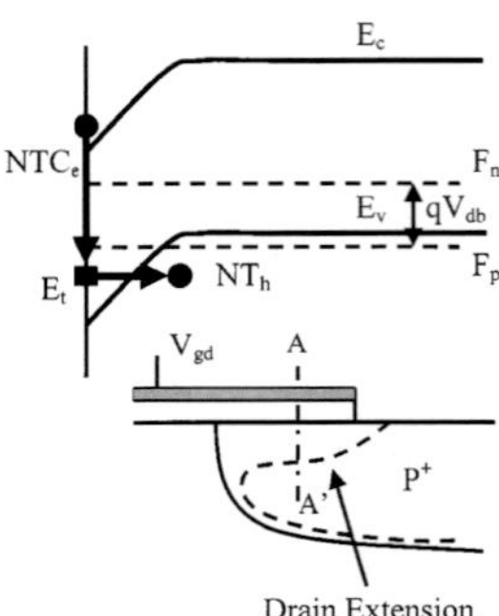

Fig. 1. Band bending in the gate-to-drain overlap region of DE of p-MOSFET under positive V_{gd} and forward biased drain-to-substrate pn junction V_{db}. When the DE doping concentration is high, t_{ox} is thin, and V_{gd} is large, the vertical electric field (along A–A′ profile) is strong enough to induce the thermal-trap-tunneling process of electron transition from the conduction band to the valence band. The electron quasi-Fermi energy F_n is above the hole quasi-Fermi energy F_p by qV_{db} approximately. The arrow NTC_e indicates the net process of electrons thermally captured by the interface traps from the conduction band, and the arrow NT_h indicates the net process of electron tunneling from traps to the valence band (thermal-trap-tunneling).

source/drain extension (SDE) is very high and the gate oxide thickness is very thin. When applying high gate-drain voltage V_{gd}, the surface electric field under the gate-drain overlap region becomes very strong and electron tunneling may occur. If the drain-bulk (well) p–n junction is reverse biased, it causes gate induced drain leakage (GIDL), which is due to band-to-band tunneling [14] or band-trap-band tunneling [15]. In the $DCIV$ drain-emitter measurement mode [8], [10], where the drain-bulk p–n junction is forward biased, the surface electric field parallel to the interface is very small, while that vertical to the interface is very strong. In this case, band-to-band tunneling is forbidden and only two-step process [15], [16] may occur. Namely, electrons from the Si conduction band are thermally captured by the interface traps and subsequently tunneling from the traps to the valence band (TTT), as indicated in Fig. 1. The net thermal capture rate NTC_e is given by [17]

$$NTC_e = C_n[n_1 f_t - n_s(1 - f_t)] \tag{1}$$

where
C_n electron capture rate;

234   IEEE ELECTRON DEVICE LETTERS, VOL. 22, NO. 5, MAY 2001

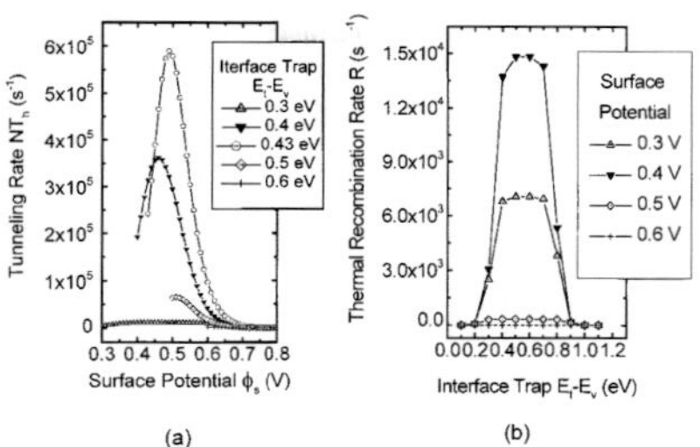

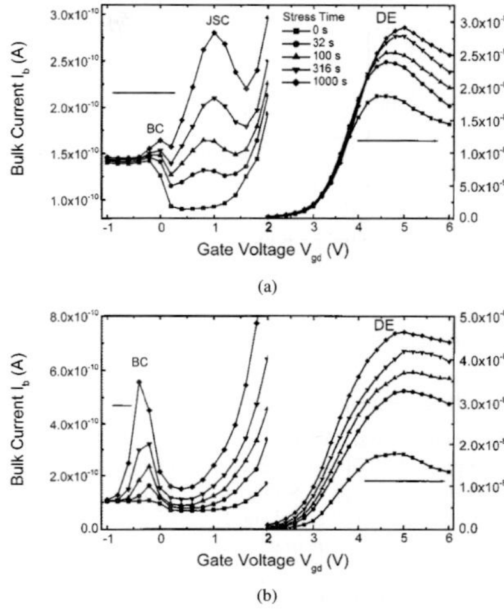

Fig. 2. (a) Simulation results of NT_h versus surface potential Φ_s by (4) for 0.25 μm p-MOST. DE doping concentration is 2×10^{19} cm^{-3}. $V_{db} = 0.4$ V. In (4), $C_n = 2.3 \times 10^8$ s/cm^3 [6]. $m_p = 0.2m_o$ where m_o is the free electron mass and $\tau_{ov} = 0.1$ ps [15]. Tunneling via interface trap with $(E_t - E_v) = 0.43$ eV is most effective and actually dominates the tunneling current with a peak at $\Phi_s \approx 0.48$ V. (b) Simulation of thermal recombination rates R versus interface state position E_t by (5) with the same parameters as in (a).

$n_1 \qquad n_i e^{(E_t - E_i)/kT}$;

$n_i \qquad$ Si intrinsic carrier concentration;

$E_i \qquad$ intrinsic energy level;

$f_t \qquad$ electron occupation factor at interface trap energy E_t;

$n_s \qquad$ conduction electron concentration at Si surface.

The net tunneling rate NT_h indicated in Fig. 1 can be expressed as [18]

$$NT_h = \frac{f_t - f_v}{\tau_h} \qquad (2)$$

where f_v is the electron occupation factor of the valence band state with energy coincided with the trap energy E_t. τ_h is the tunneling time constant, which can be expressed as [15], [19]

$$\tau_h = \tau_{0v} \exp\left[\frac{4(2m_p)^{1/2}(E_t - E_v)^{3/2}}{3q\hbar F}\right] \qquad (3)$$

where

$\tau_{0v} \qquad$ effective transit time in valence band;

$\hbar \qquad$ Plank's constant;

$m_p \qquad$ effective mass of a hole;

$E_v \qquad$ top of the valence band of Si at the Si–SiO$_2$ interface;

$q \qquad$ charge of hole;

$F \qquad$ electric field.

In the steady state, $NTC_e = NT_h$. Combining (1)–(3) and neglecting n_1, which is much smaller than n_s, we can derive

$$NT_h(E_t) = \frac{[1 - f_v(E_t)]n_s}{C_n^{-1} + n_s \tau_h(E_t)}. \qquad (4)$$

Fig. 2(a) is the simulation results of NT_h (E_t) against the surface potential Φ_s based on (4). The average electric field is roughly estimated using Φ_s/d where d is the surface depletion layer width. Fig. 2(a) shows that interface traps with energy level $E_{teff} \approx E_v + 0.43$ eV have a peak tunneling rate when $\Phi_s \approx 0.48$ V.

As a comparison, the Shockley–Read–Hall (SRH) thermal recombination rate [8] through the interface traps is also simulated for the same device and same electron thermal capture rate C_n.

Fig. 3. (a) I_b versus V_{gd} before and after channel hot carrier (CHC) stresses. Three peaks (BC, JSC, and DE) are observed, corresponding to the interface traps at BC, JSC, and DE regions respectively. Peaks BC and JSC are shown clearly only when the current scale is two orders of magnitude magnified. Measurements were implemented at $V_{db} = 0.4$ V, $V_{sub} = V_{bulk}$, while source was floated. CHC stresses were implemented by grounding source and bulk, while $V_{gs} = -2$ V, and $V_{ds} = -7$ V. (b) I_b versus V_{gd} before and after FN stresses. Two peaks (BC and DE) are observed, corresponding to the interface traps at BC and DE regions respectively. Peaks BC is shown clearly only when the current scale is two orders of magnitude magnified. Measurements as in (a). FN stresses were implemented by grounding source, drain, and bulk, while $I_g = -3$ A.

assuming $C_n = C_p$, where C_p is the hole capture rate. The steady state thermal recombination rate R is given by [17]

$$R(E_t) = C_n \frac{p_s n_s - n_i^2}{n_s + p_s + 2n_i ch\left(\dfrac{E_t - E_i}{kT}\right)} \qquad (5)$$

where p_s is the hole concentration at the surface. Fig. 2(b) shows R versus E_t. The peak R is about two orders of magnitude smaller than the peak NT_h, therefore can be neglected.

The measured current I_T is the sum of all TTT currents in space and trap energy E_t. Since the interface traps at E_{teff} play a dominant role in tunneling, we can roughly estimate I_T by

$$I_T = N_{it}[qNT_h(E_{teff})]W \qquad (6)$$

where W is the channel width, N_{it} is the equivalent effective number of interface traps per unit channel width in the DE region. From Fig. 2(a), $NT_h(E_{teff}) \approx 6 \times 10^5$/s, or one interface trap will induce 0.1 pA of current. This is extremely sensitive. I_T can be measured either as bulk current I_b or drain current I_d. The advantage of I_b measurement is that it can separate from pn junction Shockley diffusion current in an n-well as explained in the DCIV method [6]. However, when the gate oxide is very thin and V_{gd} is high, I_b also measures the gate current due to FN tunneling, while I_d does not.

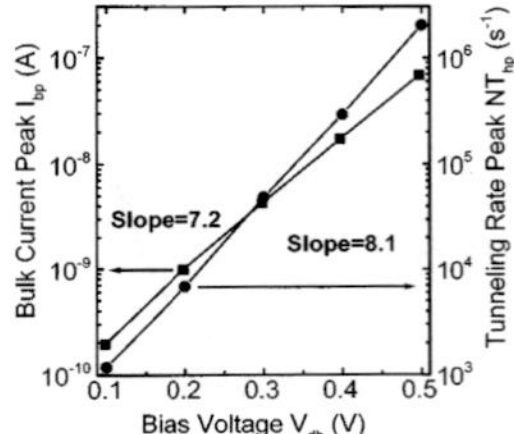

Fig. 4.   Simulated peak NT_h value NT_{hp} by (4) versus different forward bias V_{db} is compared with the experimental peak I_b value I_{bp} versus V_{db}.

III. EXPERIMENTAL RESULTS AND DISCUSSION

The devices used in this work were 0.25 (or 0.18) μm technology p-MOSTs with channel width/length of 50/0.5 μm, oxide thickness $t_{ox} \approx 5.8$ nm (or 3.7 nm), and DE doping concentration 2×10^{19} cm^{-3} (or 4×10^{19} cm^{-3}). The DCIV test configuration is illustrated in [10]. Fig. 3(a) and (b) show the measured results of 0.25 μm technology p-MOS transistors. For 0.18 μm technology devices, the results are similar. There are three I_b current peaks corresponding to interface traps in three different regions. Two small peaks BC and JSC at $V_{gd} \approx -0.1$ V and $+1$ V are due to SRH thermal recombination via interface traps located at BC region and JSC region, respectively [7]–[10]. The BC peak current increases only when applying FN stress when $V_{ds} = 0$ [9], [10]. The JSC peak current increases only when applying channel hot carrier (CHC) stress when V_{ds} is high [9], [10]. The third peak current (denoted by DE) at $V_{gd} \approx 4.5 \sim 5$ V, two orders of magnitude higher than the peak currents of BC and JSC, increases when applying either FN or CHC stress as indicated in Fig. 3(a) and (b). The DE peak is identified as due to TTT process in the DE-gate overlap region as illustrated in Fig. 1 [20]. For DE doping concentration of 2×10^{19} cm^{-3} and $t_{ox} = 5.8$ nm, using depletion layer approximation, the simulated peak NT_h at $\Phi_s \approx 0.48$ V in Fig. 2(a) corresponds to a $V_{gd} \approx 4$ V, which is comparable to the experimental DE peak at $V_{gd} \approx 4.5 \sim 5$ V in Fig. 3. The gate current I_g in this V_{gd} range is lower than 0.1 pA which can be neglected.

In Fig. 4, the simulated peak NT_h value by using (4) versus different forward bias V_{db} is compared with the experimental peak I_b value versus V_{db}. There is a good agreement between the simulations and experimental results.

From Fig. 4, when the stress time is increased, the magnitude of the DE peak also increases but is now shifted to a higher V_{gd} value. This implies that a negative charge is trapped in the oxide at the top of the gate-DE overlap region during stress. When detrapping oxide charge by applying negative V_{gd}, DE peak shifts back to lower V_{gd}.

In conclusion, interface traps at high doping DE region of 0.25-μm or sub-0.25-μm MOS transistors can be measured and characterized by forward biased bulk (or drain) current peak at high gate voltage. The current is identified as TTT current at the DE-gate overlap region [20]. The method is extremely sensitive as one interface trap is able to induce about 0.1 pA increment of peak bulk (or drain) current.

ACKNOWLEDGMENT

The authors would like to thank Dr. C. T. Sah for his invaluable comments and Dr. Samudra and his students for providing device simulation tools.

REFERENCES

[1]   T. H. Ning, P. W. Cook, R. H. Dennard, C. M. Osburn, S. E. Schuster, and H. N. Yu, "1 μm MOST VLSI technology—Part IV: Hot-electron design constraints," *IEEE Trans. Electron Devices*, vol. ED-26, pp. 346–353, Apr. 1979.

[2]   C. Hu, "Hot-carrier effects," in *VLSI Electronics Microstructure Science*, N. G. Einspruch and G. Sh. Gildenblat, Eds.   San Diego, CA: Academic, 1989, vol. 18, Advanced MOS Device Physics, pp. 119–160.

[3]   P. Heremans, R. Bellens, G. Groeseneken, A. V. Schwerin, W. Weber, B. Brox, and H. E. Maes, "The mechanisms of hot-carrier degradation," in *Hot-Carrier Design Considerations for MOS Devices and Circuits*, C. T. Wang, Ed.   New York: Van Nostrand Reinhold, 1992, pp. 1–119.

[4]   E. Takeda, C. Y. Tang, and A. Miura-Hamada, *Hot-Carrier Effects in MOS Devices*.   San Diego, CA: Academic, 1995.

[5]   G. Groeseneken, H. E. Maes, N. Beltran, and R. F. Keersmaecker, "A reliable approach to charge-pumping measurements in MOS transistors," *IEEE Trans. Electron Devices*, vol. ED-31, pp. 42–53, Jan. 1984.

[6]   A. Neugroschel, C. T. Sah, K. M. Han, M. S. Carroll, T. Nishida, J. T. Kavalieros, and Y. Liu, "Direct-current measurement of oxide and interface traps on oxidized silicon," *IEEE Trans. Electron Devices*, vol. 42, pp. 1657–1662, Sept. 1995.

[7]   K. M. Han and C. T. Sah, "Positive oxide charge from hot hole injection during channel-hot-electron stress," *IEEE Trans. Electron Devices*, vol. 45, pp. 1624–1627, July 1998.

[8]   J. Cai and C. T. Sah, "Interface electronic traps in surface controlled transistors," *IEEE Trans. Electron Devices*, vol. 47, pp. 576–583, Mar. 2000.

[9]   B. B. Jie, K. H. Ng, M. F. Li, and K. F. Lo, "Correlation between charge pumping method and direct-current current voltage method in p-type metal-oxide-semiconductor field-effect transistors," *Jpn. J. Appl. Phys.*, vol. 38, pp. 4696–4698, 1999.

[10]   B. B. Jie, M. F. Li, C. L. Lou, W. K. Chim, D. S. H. Chan, and K. F. Lo, "Investigation of interface traps in LDD pMOSTs by DCIV methods," *Electron Device Lett.*, vol. 18, pp. 583–585, 1997.

[11]   S. M. Sze, *Semiconductor Device Physics*.   New York: Wiley, 1981.

[12]   S. Thompson, P. Packan, and M. Bohr, "MOS scaling: Transistor challenges for the 21st century," *Intel Technol. J.*, pp. 1–19, 1998.

[13]   Similar experimental results and theoretical explanations using the TTT model were reported in the Ph.D thesis of Dr. Jin Cai at the University of Florida in May 2000. Private communication from Professor Chih-Tang Sah.

[14]   J. Chen, T. Y. Chan, I. C. Chen, P. K. Ko, and C. Hu, "Subbreakdown drain leakage current in MOSFETs," *IEEE Electron Device Lett.*, vol. EDL-8, pp. 515–517, Nov. 1987.

[15]   T. Tse-En, C. Huang, and T. Wang, "Mechanisms of interface trap-induced drain leakage current in off-state n-MOSFETs," *IEEE Trans. Electron Devices*, vol. 42, pp. 738–743, Apr. 1995.

[16]   P. Speckbacher, J. Berger, A. Asenov, F. Koch, and W. Weber, "The gated diode configuration in MOSFETs, a sensitive tool for characterizing hot-carrier degradation," *IEEE Trans. Electron Devices*, vol. 42, pp. 738–743, July 1995.

[17]   M. F. Li, *Modern Semiconductor Quantum Physics*.   Singapore: World Scientific, 1994, sec. 351.

[18]   ——, *Modern Semiconductor Quantum Physics*.   Singapore: World Scientific, 1994, sec. 570.

[19]   I. Lundstrom and C. Svensson, "Tunneling to traps in insulators," *J. Appl. Phys.*, vol. 43, pp. 5045–5047, 1972.

[20]   Numerical simulation by Dr. Cai in his thesis [13], using the analytical theory [8], with TTT transition included as described by (5), could not account for the very broad experimental I_b-V_{gb} lineshape. Thus, further study is needed.

734   IEEE ELECTRON DEVICE LETTERS, VOL. 23, NO. 12, DECEMBER 2002

Dynamic NBTI of p-MOS Transistors and Its Impact on MOSFET Scaling

G. Chen, M. F. Li, C. H. Ang, J. Z. Zheng, and D. L. Kwong

Abstract—For the first time, a dynamic negative bias temperature instability (DNBTI) effect in p-MOSFETs with ultrathin gate oxide (1.3 nm) has been studied. The interface traps generated under NBTI stressing corresponding to p-MOSFET operating condition of the "high" output state in a CMOS inverter, are subsequently passivated when the gate to drain voltage switches to positive corresponding to the p-MOSFET operating condition of the "low" output state in the CMOS inverter. Consequently, this DNBTI effect significantly prolongs the lifetime of p-MOSFETs operating in a digital circuit, and the conventional static NBTI (SNBTI) measurement underestimates the p-MOSFET lifetime. A physical model is presented to explain the DNBTI. This finding has significant impact on future scaling of CMOS devices.

Index Terms—Annealing, CMOSFETs, negative bias temperature instability (NBTI), semiconductor–insulator interfaces, ultrathin gate oxide.

I. INTRODUCTION

WITH the continuous shrinking of the transistor dimensions, generation of interface traps during negative bias temperature instability (NBTI) stress in p-MOS transistors has become one of the most critical reliability issues that ultimately determine the lifetime of CMOS devices [1]–[3], [11]. In conventional NBTI study, a constant negative bias is applied to the gate electrode of a p-MOS transistor at high temperatures with S/D grounded [1]–[3], [11]. However, during the operation of a p-MOSFET in a CMOS inverter, the applied gate bias is switching between "high" and "low" voltages, while the drain bias is alternating between "low" and "high" voltages, correspondingly. Therefore, it is critically important to investigate NBTI under such dynamic stress conditions. As we will show in this letter, the conventional static NBTI measurement has neglected the passivation effects of the interface traps during the operation of p-MOSFETs in digital CMOS circuits, and therefore overestimates the degradations of p-MOS devices. It is found for the first time that a large portion of the interface traps generated under the NBTI stressing, corresponding to p-MOSFET operating condition of the "high" output state in a CMOS inverter, are passivated electrically when the gate to drain voltage switches to positive corresponding to the

Manuscript received August 15, 2002; revised September 27, 2002. This work was supported by the Singapore A*STAR Research Grant EMT/TP/00/001.2 and the National University of Singapore Research Grant R263-000-077-112. The review of this letter was recommended by Editor E. Sangiorgi.

G. Chen and M. F. Li are with the Silicon Nano Device Lab and CICFAR, Department of Electrical and Computer Engineering, National University of Singapore, Singapore 119260 (e-mail: elelimf@nus.edu.sg).

C. H. Ang and J. Z. Zheng are with Chartered Semiconductor Manufacturing Ltd., Singapore 738406.

D. L. Kwong is with the Department of Electrical and Computer Engineering, University of Texas at Austin, Austin, TX 78741 USA.

Digital Object Identifier 10.1109/LED.2002.805750

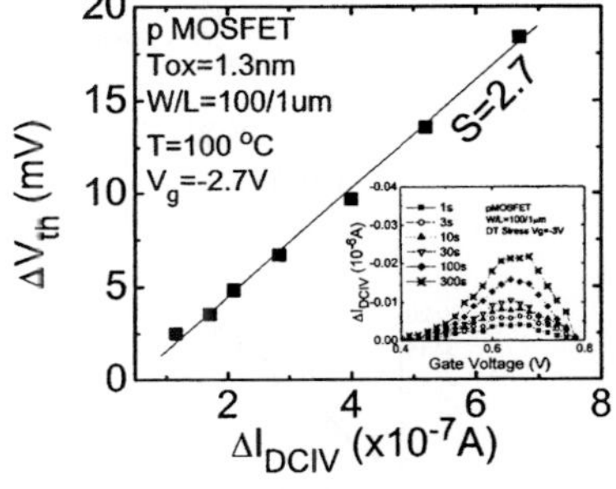

Fig. 1. Correlation between DCIV current and threshold voltage changes. Interface trap generation plays a major role in NBTI. The inset shows the typical DCIV spectra of the same p-MOSFET.

p-MOSFET operating condition of the "low" output state in a CMOS inverter. As a result, this dynamic NBTI (DNBTI) effect greatly prolongs the lifetime of p-MOSFETs operating in a digital circuit, while the conventional static NBTI (SNBTI) measurement underestimates the p-MOSFET lifetime.

Although electric passivation (EP) of interface traps has been reported before in MOSFETs during hot-carrier stress [4]–[6], its effects on NBTI and device lifetime have not been investigated. In this work, careful EP experiments and its frequency dependence have been conducted. We show that due to this EP effect, the lifetime of p-MOSFETs under DNBTI stress corresponding to a realistic operation condition in a digital circuit is approximately one order of magnitude longer than that under conventional SNBTI stress.

II. DEVICE FABRICATION AND INTERFACE TRAPS MEASUREMENT

CMOS devices were fabricated using standard dual-gate CMOS technology. Gate oxide of 1.3 nm thickness was grown by rapid thermal oxidation followed by an exposure to high-density nitrogen plasma. TEM was used to measure the gate oxide thickness. The NBTI temperature was set at 100 °C. Direct current current–voltage (DCIV) technique, which detects interface traps by measuring the bulk current due to Shockley–Read–Hall recombination through interface traps [7], [8], was used to accurately monitor the interface trap density in the p-MOSFETs during NBTI stressing. Typical DCIV spectra as a function of NBTI stress time are shown in the inset of Fig. 1.

0741-3106/02$17.00 © 2002 IEEE

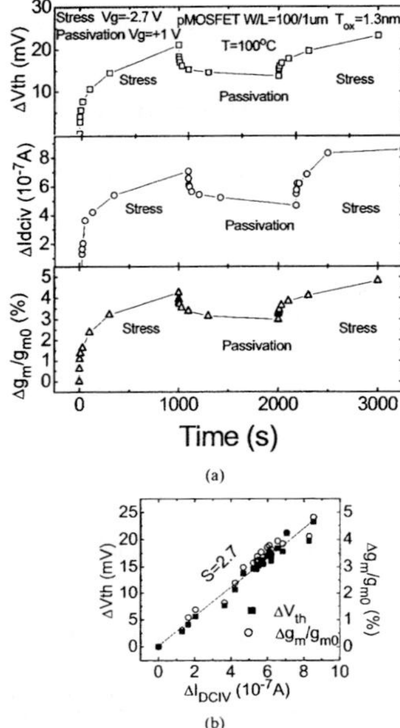

Fig. 2.　(a) Variations of $V_{\rm th}$, DCIV current, and g_m under uniform dynamic NBTI stressing–passivation–stressing process. The gate voltage is switched to negative and positive, while the S/D is connected to ground. (b) Correlation between $V_{\rm th}$ shift, DCIV, and g_m variations in (a). The $\Delta V_{\rm th}$–$\Delta I_{\rm DCIV}$ slope is exactly the same as in Fig. 1.

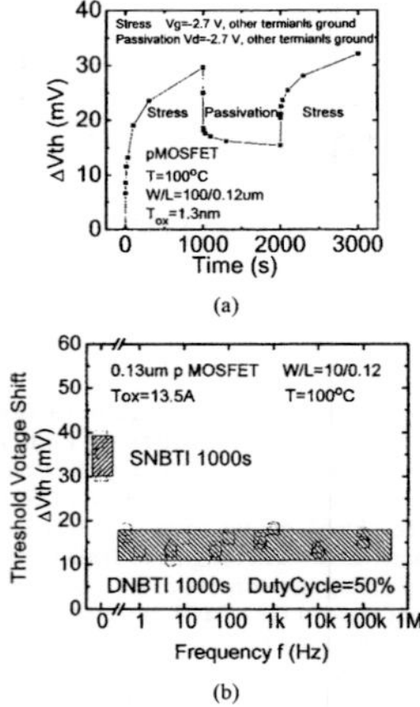

Fig. 3.　(a) Illustration of nonuniform dynamic NBTI, which simulates high-low operation of a p-MOSFET in a digital inverter. Due to the electric-passivation effect, the conventional (static) NBTI overestimates the p-MOSFET degradation. (b) The frequency dependency of DNBTI. A 50% duty cycle square wave of 0 to -2.5 V was applied to the gate. A same square wave with 180° phase shift was applied to the drain. Other terminals were grounded.

III. STATIC AND DYNAMIC NBTI EFFECTS

Under conventional SNBTI stress with negative gate-to-S/D bias on a p-MOSFET at high temperature, the major degradation is the shift of threshold voltage $\Delta V_{\rm th}$ [1]. Fig. 1 shows the excellent correlation between $\Delta V_{\rm th}$ and $\Delta N_{\rm it}(\Delta I_{\rm DCIV})$ under SNBTI stress. The DNBTI results are shown in Figs. 2 and 3(a). As can be seen clearly, when reversing the gate-to-S/D electric field to the opposite polarity (positive) during DNBTI stressing, a reduction (passivation) of $\Delta N_{\rm it}$ and thus $\Delta V_{\rm th}$ is observed. Fig. 2(a) shows the DNBTI effect under uniform stress condition where S/D are grounded. $\Delta N_{\rm it}$, $\Delta V_{\rm th}$, and $-\Delta g_m$ increase during the stress period of a negative gate voltage, as expected for SNBTI. However, they all decrease over the "passivation" period when the gate voltage is switched to positive. The relationship among $\Delta V_{\rm th}$ versus $\Delta N_{\rm it}$ and $\Delta g_m\%$ is plotted in Fig. 2(b) which shows a perfect linearity, with $\Delta V_{\rm th}$ versus $\Delta N_{\rm it}$ slope identical to that of the SNBTI results shown in Fig. 1. Fig. 3(a) shows the DNBTI effect under nonuniform stress condition. During the stress period, a uniform SNBTI stress is applied, i.e., a negative gate voltage is applied with S/D grounded. This stress condition corresponds to the p-MOSFET operation condition in a CMOS inverter when the output is "high." During the passivation period, a negative drain voltage is applied while the source and the gate are grounded, resulting a positive gate-to-drain voltage. This nonuniform stressing simulates the p-MOSFET operation condition in a CMOS inverter when the output is "low." As shown in Fig. 3(a), a passivation effect, i.e., $\Delta V_{\rm th}$ reduction, is seen when the p-MOS is operated with the output "low." The passivation effect is found frequency independent, as shown in Fig. 3(b).

Based on the results shown in Figs. 1–3, we conclude that due to the DNBTI effect, the generated interface traps and the accompanying degradations of $V_{\rm th}$ and g_m during the typical SNBTI stress period are partially "annealed" during the subsequent passivation stress period. As a result, the device lifetime under DNBTI stress becomes much longer than that projected under the conventional SNBTI stress. The lifetime measurements were made on p-MOSFETs with gate oxide thickness of 1.3 nm and channel length of 0.12 μm under both DNBTI and SNBTI stressing. As shown in Fig. 4, for device operating at 100 °C and the lifetime defined by the time when $\Delta V_{\rm th}$ reaches 30 mV, the lifetime for DNBTI is one order of magnitude longer than that for SNBTI. The supply voltage for 10-year (10-y) projected lifetime V_{10Y} is 0.9 V under SNBTI while it is 1.2 V under DNBTI stressing.

IV. DISCUSSION

The passivation effect or dynamic NBIT can be explained by extending the previous Hydrogen diffusion–reaction model [2], [9]–[11]. The interface trap generation is ascribed to hydrogen release from a hydrogen terminated silicon-dangling bond (Si ≡ Si–H), first proposed by Balk in 1965 [9]. Under

IEEE ELECTRON DEVICE LETTERS, VOL. 23, NO. 12, DECEMBER 2002

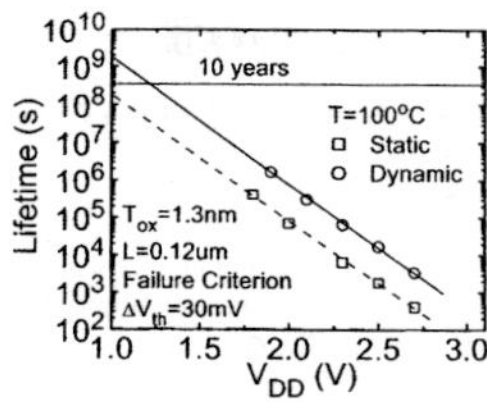

Fig. 4. Lifetime projection for p-MOSFETs with gate oxide thickness 1.3 nm, under static and dynamic NBTI stress. The lifetime is defined by the time when $\Delta V_{th} = 30$ mV. The projected 10-y lifetime operating voltage V_{10y} is 1.2 V for dynamic NBTI stress, and is 0.9 V for static NBTI stress which underestimates the lifetime in real digital operation.

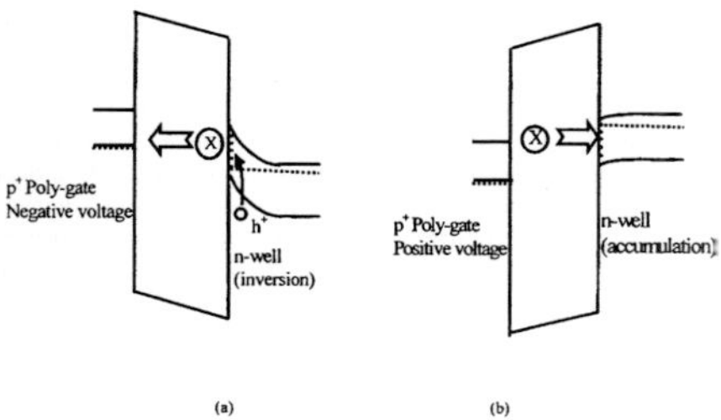

Fig. 5. (a) N_{it} generation: Si–H bond breaking at the interface and hydrogen diffusion toward the gate electrode during negative gate bias stressing. (b) N_{it} passivation: hydrogen moves back to the interface and passivates the Si dangling bonds during positive gate bias. X stands for hydrogen species.

high-temperature and negative gate bias stress conditions, the holes from the induced inversion layer react with the interface trap precursors (Si $\equiv$ Si–H), breaking the H–Si bond and resulting in interface traps N_{it} (Si dangling bonds). The produced hydrogen species, denoted as X in Fig. 5, diffuse/drift to the gate electrode. During this stress period, the interface acts as a hydrogen source. The EP effect can be readily interpreted by the reverse reaction between N_{it} and X species, as shown in Fig. 5(b). The interface trap passivation by hydrogen in SiO_2 was first explained in a comprehensive study by Sah et al. in 1984 [10]. When the bias polarity is reversed (positive or zero gate bias), the channel inversion layer disappears. The breaking of Si–H bond is interrupted due to a lack of holes, and at the same time, X moves back to the SiO_2/Si interface under the influence of positive gate voltage and passivates the Si dangling bond, resulting in N_{it} reduction. In this period, the interface acts as a hydrogen sink. Further investigation of this interface trap generation/passivation mechanism in DNBTI is in progress and will be reported elsewhere.

V. CONCLUSION

NBTI of a p-MOSFET with 1.3-nm thick gate oxide operating in a CMOS inverter in a digital circuit was investigated. A dynamic NBTI effect is observed, in which the generated interface traps and the accompanying degradations of V_{th} and g_m during the typical steady NBTI stress period (gate to drain voltage is negative, when the inverter output is "high") are partially "repaired" during the subsequent passivation period (gate to drain voltage is positive, when the inverter output is "low"). As compared to traditional static NBTI, the dynamic NBTI suppresses high-temperature degradation to a large extent and significantly prolongs the p-MOSFET lifetime or increase the 10-y operation voltage. The p-MOSFET operating in a CMOS inverter actually sustains a dynamic NBTI stress rather than static NBTI stress, which has been identified as a scaling-limiting factor. Therefore, this finding has significant impact for future device scaling. The dynamic NBTI effect can be interpreted by extending the model of static NBTI, based on the interaction between hydrogen species and silicon dangling bonds.

REFERENCES

[1] N. Kimizuka, T. Yamamoto, T. Mogami, K. Yamaguchi, K. Imai, and T. Horiuchi, "Impact of bias temperature instability for direct-tunneling ultrathin gate oxide on MOSFET scaling," in *Dig. Tech. Papers—Symp. VLSI Technology*, 1999, pp. 73–74.

[2] S. Ogawa, M. Shimaya, and N. Shiono, "Interface-trap generation at ultrathin SiO_2–Si interfaces during negative-bias temperature aging," *J. Appl. Phys.*, vol. 77, pp. 1137–1148, 1995.

[3] T. Yamamoto, K. Uwasawa, and T. Mogami, "Bias temperature instability in scaled p⁺ polysilicon gate p-MOSFETs," *IEEE Trans. Electron Devices*, vol. 46, pp. 921–926, 1999.

[4] N. C. Das and V. Nathan, "Hot-carrier-induced interface trap annealing in silicon field effect transistors," *J. Appl. Phys.*, vol. 74, no. 12, pp. 7596–7599, 1993.

[5] K. M. Han and C. T. Sah, "Reduction of interface traps in p-channel MOS transistors during channel-hot-hole stress," *IEEE Trans. Electron Devices*, vol. 45, pp. 1380–1382, 1998.

[6] G. Chen, M. F. Li, and Y. Jin, "Electric passivation of interface traps at drain junction space charge region in p-MOS transistors," *Microelectron. Reliab.*, vol. 41, pp. 1427–1431, 2001.

[7] A. Neugroschel, C. T. Sah, K. M. Han, M. S. Carroll, T. Nishida, J. T. Kavalieros, and Y. Lu, "Direct-current measurement of oxide and interface traps on oxidized silicon," *IEEE Trans. Electron Devices*, vol. 42, pp. 1657–1662, 1995.

[8] G. Chen, M. F. Li, W. Y. Loh, B. J. Cho, D. S. H. Chan, C. H. Ang, J. Z. Zheng, and D. L. Kwong, "Direct monitoring of interface trap generation in MOSFETs with tunneling (1.3 nm) gate Ooxide using DCIV method", to be published.

[9] P. Balk, "Effects of hydrogen annealing on silicon surfaces," in *Proc. Electro Chemical Society Meeting*, vol. 14, 1965, p. 237.

[10] C.-T. Sah, J. Y. C. Sun, and J. J. T. Tzou, "Study of the atomic models of three donor-like traps on oxidized silicon with aluminum gate from their processing dependences," *J. Appl. Phys.*, vol. 55, pp. 1525–1545, 1984.

[11] S. Ogawa and N. Shiono, "Generalized diffusion-reaction model for the low–field charge-buildup instability at the Si–SiO_2 interface," *Phys. Rev. B. Condens. Matter*, vol. 51, pp. 4218–4230, 1995.

DYNAMIC NBTI OF PMOS TRANSISTORS AND ITS IMPACT ON DEVICE LIFETIME

G. Chen, K. Y. Chuah, M. F. Li, Daniel SH Chan, C. H. Ang[1], J. Z. Zheng[1], Y. Jin[1] and D. L. Kwong[2]

SNDL and CICFAR, Dept. of Electrical & Computer Engineering, National University of Singapore, Singapore 119260

Tel: 65 6874 2559, Fax: 65 6779 1103, E-mail: **elelimf@nus.edu.sg**

[1] Chartered Semiconductor Manufacturing Ltd, Singapore 738406

[2] Dept. of Electrical & Computer Engineering, The University of Texas at Austin, TX 78752, USA

ABSTRACT

We report a new NBTI phenomenon for the first time for the p-MOSFETs with ultra thin gate oxides. We demonstrate that in a CMOS inverter circuit, the interface traps generated under the NBTI stressing in a p-MOSFET (corresponding to the "high" output state of the inverter) are subsequently passivated when the gate to drain voltage switches to positive (corresponding to the "low" output state of the inverter). As a result, it was found that this "Dynamic" NBTI (DNBTI) operating in a CMOS inverter circuit prolongs significantly the device lifetime while the conventional "static" NBTI (SNBTI) underestimates the device lifetime. Furthermore, the DNBTI effect is dependent of temperature and gate oxide thickness , however independent of operation frequency. A physical model is proposed for DNBTI that involves the interaction between hydrogen and silicon dangling bonds. This finding has significant impact on the determination of maximum operation voltage as well as lifetime projection for future scaling of CMOS devices.

INTRODUCTION

With the continuous shrinking of the transistor dimensions, new reliability issues emerge, among which negative bias temperature instability (NBTI) of p-MOSFET has been identified as a critical limiting factor that ultimately determines the lifetime of the devices [1-3]. However, as will be shown, the conventional NBTI based on static experimental data, overestimate the degradation of the p-MOSFET in digital CMOS circuits by overlooking the electric passivation (EP) effect during normal operations of digital circuits. During the operation of a p-MOSFET in a CMOS inverter, the applied gate bias (input signal) is switching between "high" and "low" voltages, while the drain bias (output signal) is alternating between "low" and "high" voltages, correspondingly. Therefore, it is critically important to investigate NBTI under such dynamic stress conditions. During "low" (low output) phase of an inverter, the EP effect effectively reduces the interface traps generated during "high" (high output) phase, and consequently, recovers the degradations of device parameters to a certain degree.

Although the electric passivation or annealing of interface traps N_i has been studied for hot carrier degradations [4-6], there has been no study on its effects on p-MOS NBTI under realistic circuit operation. In this paper, we will show that the DNBTI effect greatly prolongs the lifetime of p-MOSFETs operating in a digital circuit, while the conventional SNBTI measurement underestimates the p-MOSFET lifetime.

DEVICE FABRICATION AND DEGRADATION CHARACTERIZATION

P-MOSFETs were fabricated using standard dual-gate CMOS technology. Gate oxides with thicknesses of 1.3nm, 1.4nm, 1.7nm and 2.1nm were grown by Rapid Thermal Oxidation followed by an exposure to high-density nitrogen plasma. TEM and CV measurement and simulation [7] confirmed the thickness of these ultrathin gate oxides. Carrier separation measurement fits well to tunneling simulation [8], indicating that the leakage is mainly caused by direct tunneling. DCIV technique [9,10] was improved in this work (Appendix) to accurately monitor interface trap formation during both stressing and passivation phases. By proper treatment of the measured bulk current in a gate controlled bipolar-junction-transistor configuration, the electron-hole-recombination current I_{DCIV} that is proportional to the interface trap concentration N_{it}, is successfully obtained for devices with gate oxide as thin as 1.3nm .

EXPERIMENTAL RESULTS: THE DNBTI PHENOMENON

The SNBTI stress was first applied to p-MOSFETs at a temperature of 100 °C. The threshold voltage shift ΔV_{th} and DCIV current I_{DCIV} were monitored and the relationship between them was investigated. Fig.1 shows an excellent correlation between ΔV_{th} and ΔN_{it} (ΔI_{DCIV}) under SNBTI stress.

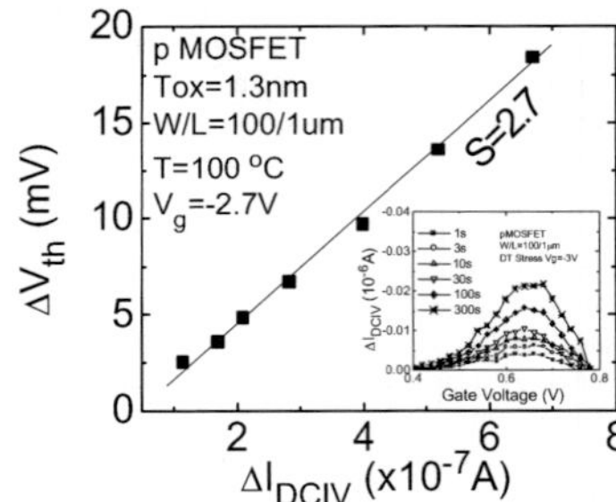

FIGURE 1. CORRELATION BETWEEN DCIV CURRENT AND THRESHOLD VOLTAGE CHANGES. INTERFACE TRAP GENERATION PLAYS A MAJOR ROLE IN NBTI. THE INSET SHOWS THE TYPICAL DCIV SPECTRA FOR 1.3 NM GATE OXIDE P-MOSFET.

Interestingly, when reversing the electric field to the opposite polarity (the passivation mode) during SNBTI stressing, a reduction of ΔN_{it} and thus ΔV_{th} and $-\Delta g_m\%$ was observed (Fig.2). It is shown in Fig.2 that the ΔN_{it}, ΔV_{th} and $-\Delta g_m\%$ are increasing and decreasing simultaneously during the stress-passivation-stress sequence. Therefore, the degradations of N_{it}, V_{th} and g_m are probably due to the same origin of interface trap generation. The relations between ΔV_{th} vs. ΔN_{it} and $\Delta g_m\%$ are plotted in Fig.2 (b). It is shown that the linearity among these three parameters is excellent and the slope of ΔV_{th} vs. ΔN_{it} plot is exactly the same as that in Fig.1.

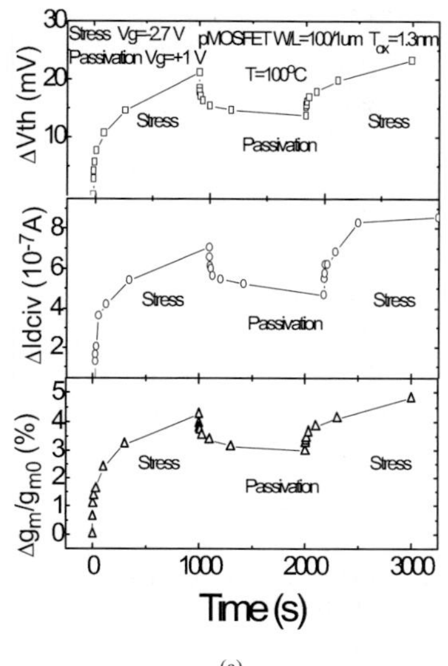

(a)

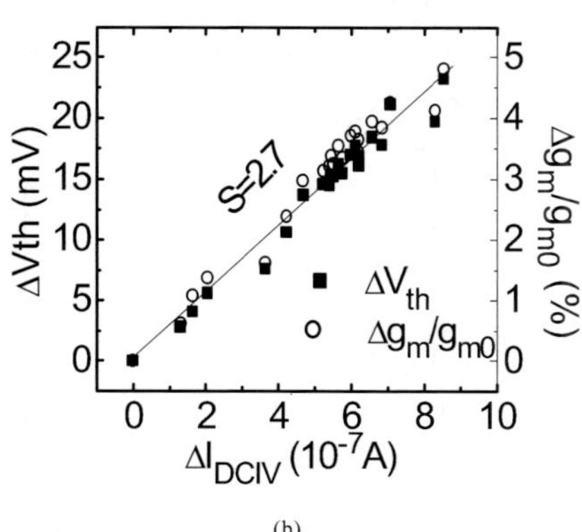

(b)

FIGURE 2. (a) VARIATIONS OF V_{TH}, DCIV CURRENT, AND g_m UNDER STRESS-PASSIVATION-STRESS PROCESS. (b) CORRELATION BETWEEN V_{TH} SHIFT, DCIV CURRENT AND g_m VARIATIONS IN (a). THE ? V_{TH}-? I_{DCIV} SLOPE IS EXACTLY THE SAME AS IN FIG.1.

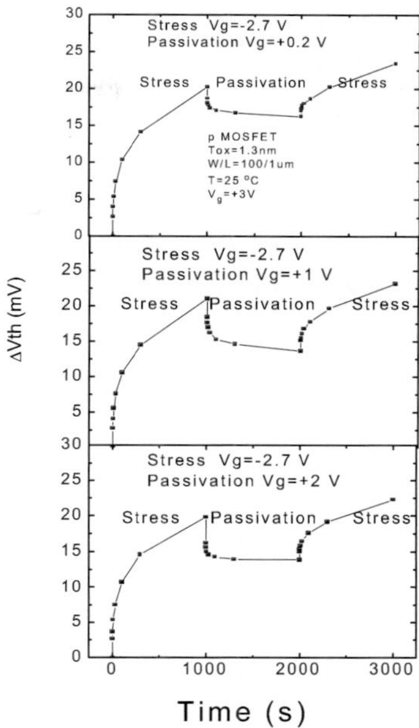

FIGURE 3. VARIATION OF V_{TH} UNDER STRESS-PASSIVATION-STRESS. THE STRESS CONDITIONS ARE SIMILAR AS IN FIG.2 . PASSIVATION RATE INCREASES WITH INCREASING OF PSSIVATION GATE VOLTAGE (AND THE ELECTRIC FIELD).

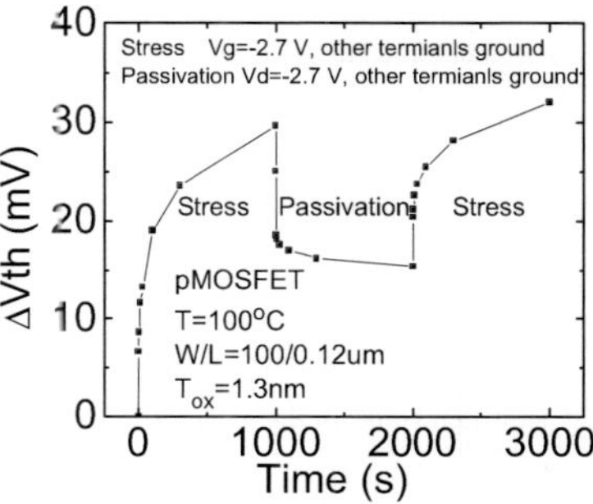

FIGURE 4. ILLUSTRATION OF DYNAMIC NBTI, DURING HIGH-LOW OPERATION OF A P-MOSFET IN A DIGITAL CMOS INVERTER. DUE TO THE ELECTRIC-PASSIVATION EFFECT, THE CONVENTIONAL SNBTI OVERESTIMATES THE DEGRADATION.

We also found that the passivation rate is electric field dependent, i.e., the larger the field applied, the more rapid the passivation. Fig.3 shows ΔV_{th} under the stress-passivation-stress sequence where the stress conditions remain the same except changing the passivation gate voltage. The data indicates that a higher electric field favors the passivation process.

To simulate a p-MOSFET during the "low" phase in a CMOS inverter, a non-uniform electric field is applied during the "passivation" period, i.e., a negative drain bias is applied while the other terminals are kept grounded. As shown in Fig.4, the EP effect is similar to that seen under uniform electric field passivation.

In digital circuits, the devices are operated under dynamic conditions. For a p-MOSFET in an inverter (Fig.5), the input signal V_{in} and the output signal V_{out} are opposite in terms of phase. To simulate the stressing condition, we deliberately apply a train of square wave to the gate and an opposite phase signal to the drain, as shown in Fig.6.

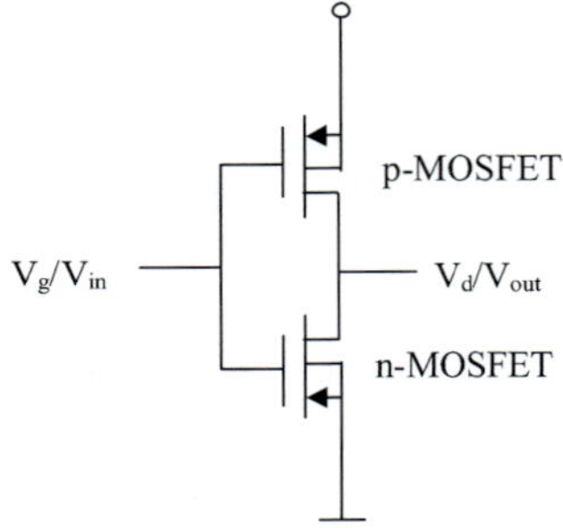

FIGURE 5. A P-MOSFET IN A CMOS INVERTER ACTUALLY UNDERGOES DYNAMIC STRESSING.

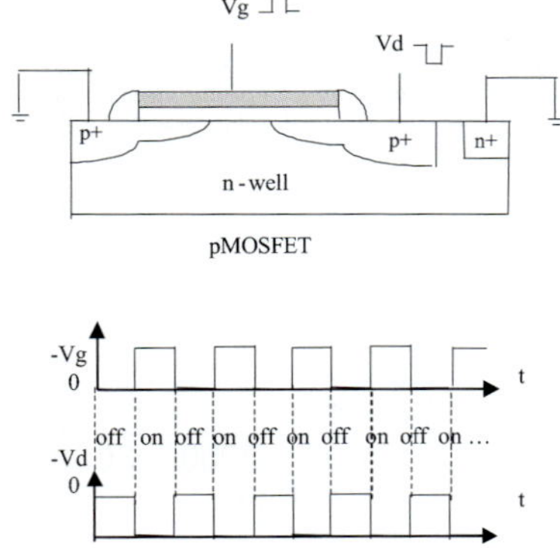

FIGURE 6. TO SIMULATE THE P-MOSFET STRESSING CONDITION, TWO OPPOSITE PHASE SQUARE WAVE TRAINS ARE APPLIED TO ITS GATE AND DRAIN ELECTRODES, RESPECTIVELY.

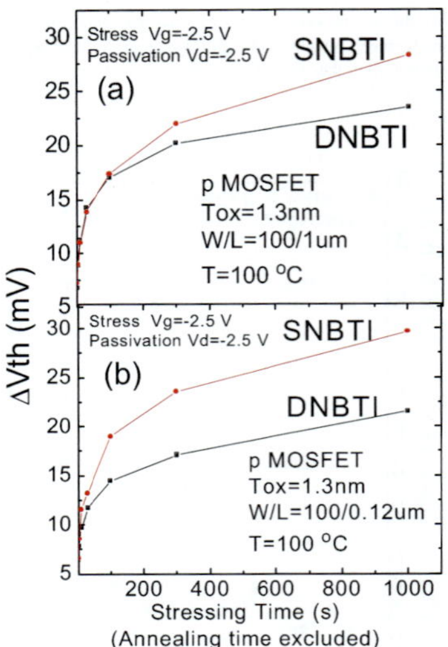

FIGURE 7. COMPARISON BETWEEN SNBTI AND DNBTI FOR (a) LON CHANNEL AND (b) SHORT CHANNEL P-MOSFETS. FOR DNBTI, TH STRESSING TIME IS HALF THE NOMINAL STRESS TIME (OPERATIC TIME), GIVEN THE 50% DUTY CYCLE. THE DNBTI STRESS FREQUENC ~ 0.5 Hz.

Under such dynamic stress conditions, it is found that th interface trap generation and passivation take place mainly near th drain-channel corner, and therefore devices with shorter chann benefit more from the EP effect than longer channel ones, confirmed by Figs.7 (a) and (b).

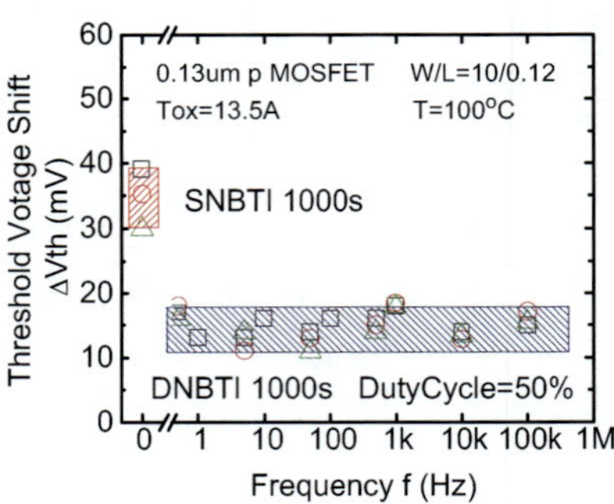

FIGURE 8. THE FREQUENCY DEPENDENCY OF DNBTI, WITH THE SAME STRESS AND PASSIVATION VOLTAGES AS IN FIG.7. DIFFERENT SYMPOLS FOR DIFFERENT SAMPLES.

The frequency dependency of DNBTI was studied in the range of 0.5 to 100k Hz, as shown in Fig. 8. As can be seen, the DNBTI is frequency independent in this frequency range with significantly lower degradation than SNBTI.

The V_{th} degradations under the same stress condition but different temperatures were plotted in Fig.9. It is shown that although both SNBTI and DNBTI degradations are smaller at lower temperatures, the EP effect is obvious even under room temperature, indicating that the passivation is not due to thermal activation effect.

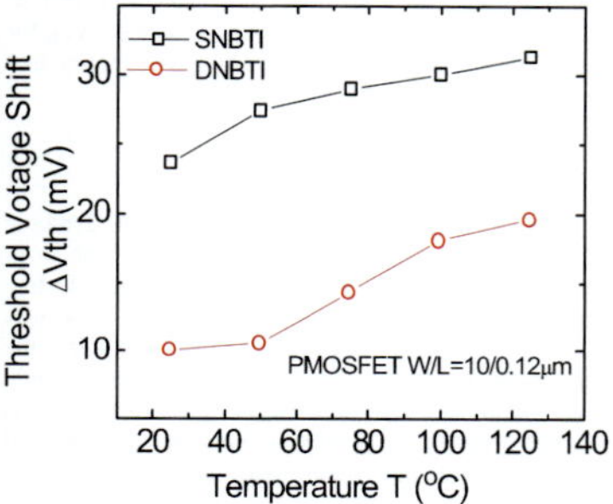

FIGURE 9. THE TEMPERATURE DEPENDENCY OF SNBTI AND DNBTI, WITH THE STRESS VOLTAGE SAME AS IN FIG.7.

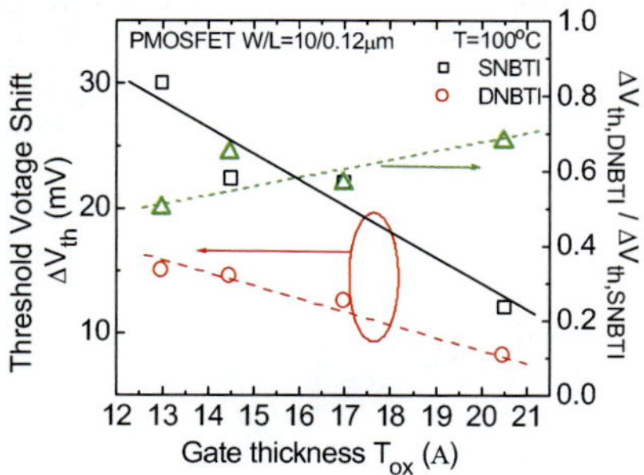

FIGURE 10. THE OXIDE-THICKNESS DEPENDENCY OF SNBTI AND DNBTI UNDER THE SAME STRESS VOLTAGE AS IN FIG.7.

The device degradations for different oxide thicknesses were also compared. Fig. 10 shows that for thinner gate oxide, the passivation effect in DNBTI is larger.

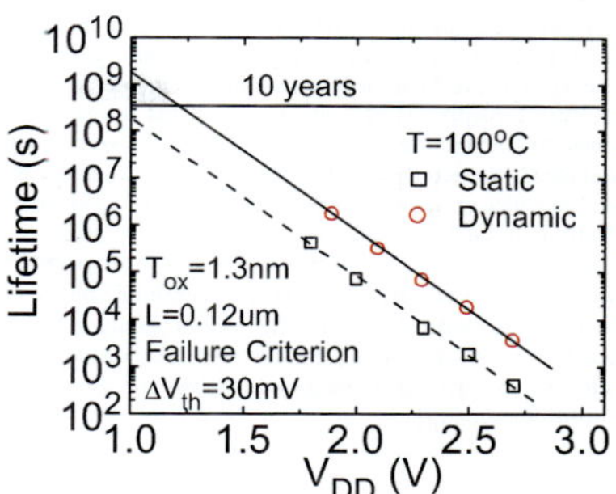

FIGURE 11. SNBTI AND DNBTI LIFETIME (30 MV V_{TH} SHIFT) PROJECTIONS FOR P-MOSFETS. THE PROJECTED 10-YEAR LIFETIME OPERATING VOLTAGE V_{10Y} IS 1.2V FOR DNBTI STRESS, AND IS 0.9V FOR SNBTI STRESS WHICH OVERESTIMATES THE DEGRADATION IN REAL DIGITAL OPERATION.

Because of the significant EP effect of interface traps on p-MOSFET operating in a CMOS inverter, the device lifetime under DNBTI stress can be much longer than that projected under the conventional SNBTI stress. The lifetime projection for both DNBTI and SNBTI was compared and the 10-year lifetime operation voltages were extracted in Fig.11. It is shown that V_{10Y} is 0.9 V for SNBTI degradation while it is 1.2V for DNBTI degradation. Under the same stress voltage, the lifetime predicted by DNBTI is almost one order of magnitude longer than predicted by SNBTI.

DISCUSSION

As first proposed in 1965 by P. Balk at IBM [11], and now widely accepted [12-14], interface traps are related to Si dangling bonds when hydrogen is released from a Si-H bond. Sah et al have proposed in 1983 the reverse process: passivation of interface trap by the absorbtion of hydrogen at the dangling bond site [15]. The excellent correlation among ΔV_{th}, Δg_m and ΔN_{it} (ΔI_{DCIV}) under NBTI stress, as shown in Figs.1-3, indicates that for ultrathin gate oxide devices, the device parameter degradation is mainly caused by the interface trap generation. Based on the interface trap generation and passivation mechanisms proposed in [11] and [15], and the recent diffusion-reaction model proposed for the electrochemical reaction of interface trap generation during NBTI degradation [16], we propose the following reactions for interface trap generation and passivation in DNBTI (eqs (1) and (2), and Fig.12):

$$(Si_3 \equiv Si\text{-}H) + h^+ \leftrightarrow Si_3 \equiv Si^* + X_{interface} \qquad (1)$$

and

$$X_{interface} \xleftrightarrow{\quad diffusion \quad} X_{bulk} \qquad (2)$$

Here $Si_3 \equiv Si\text{-}H$ is the precursor for the Si-H bond. When it interacts with a hole h^+ in the inversion layer or the source/drain extension region under the NBTI stress, the hole breaks the Si-H bond and creates an interface trap by releasing hydrogen species

$X_{interface}$ at the Si/SiO$_2$ interface. One thing remaining uncertain is how the holes accelerated by operating voltage as low as 2.7 V in this work can break the Si-H bond and generate the interface traps. This is a big unsolved issue and probably more complicated processes such as the Auger mechanism may be involved [17, 18]. We therefore consider (1) as a more generalized reaction in that some detailed reaction mechanism has not been shown explicitly . The produced hydrogen species denoted as X in eq (1), either in the form of molecules or neutral atoms or ions, will diffuse/drift to the gate electrode through the bulk gate oxide, as expressed by eq (2) (Fig.12a). In this process, the interface acts as a hydrogen source. The major symptom of NBTI is the shift of threshold voltage ΔV_{th}. For ultrathin gate oxide MOSFETs, the ΔV_{th} is mainly induced by interface traps building up along the silicon-gate oxide interface, as oxide charges are easily detrapped by tunneling and thus makes a smaller contribution [19].

The EP effect is interpreted by the reverse interaction between N_{it} and hydrogen species as shown in Fig. 12b. When the gate bias polarity is reversed from the negative to positive, the channel inversion layer disappears and depletion layers are formed at source/drain. The breaking of Si-H bond stops due to lack of holes. On the other hand, the reverse reaction occurs when the hydrogen moves back to the SiO$_2$/Si interface and passivates the Si dangling bonds, resulting in ΔN_{it} reduction. In this period, the interface acts as a hydrogen sink.

Finally, as proposed in [16], the NBTI degradation is diffusion-controlled rather than reaction-controlled. As can be seen in Fig. 13, a slope of 0.25 is observed in the time dependence of ΔV_{th} for both stressing and passivation modes, supporting the theory that both processes under DNBTI are diffusion-controlled.

CONCLUSION

Negartive bias temperature instability (NBTI) under dynamic operation that simulates a practical stress condition for a p-MOSFET in a CMOS inverter was investigated and the electric passivation effect of interface traps during positive bias was demonstrated for the first time. A physical model involving the interactions between hydrogen species and silicon dangling bonds is proposed to explain this Dynamic NBTI phenomenon. It is shown that DNBTI suppresses p-MOS device degradation and significantly prolongs the device lifetime and increase the 10-year operation voltage. This finding has a significant impact on future CMOS device scaling projections.

Appendix

INTERFACE TRAPS MEASUREMENT BY IMPROVED DCIV METHOD FOR TUNNELLING OXIDE DEVICES

By proper biasing and signal processing, the DCIV method [9,10] is able to effectively monitor the interface traps and oxide charges in MOSFETs with gate oxide thicknesses down to 1.3 nm . For such tunneling gate oxide in DCIV measurement, the measured bulk current I_b consists of the following components (Fig.A1) :

$$I_b (V_e, V_g, N_{it}) = I_{BG} (V_e, V_g) + I_{DCIV} (V_e, V_g, N_{it}) + I_{SUB}(V_g) + I_{GIT}(V_g, N_{it}) + I_{TTT}(V_e, V_g, N_{it}) \tag{A1}$$

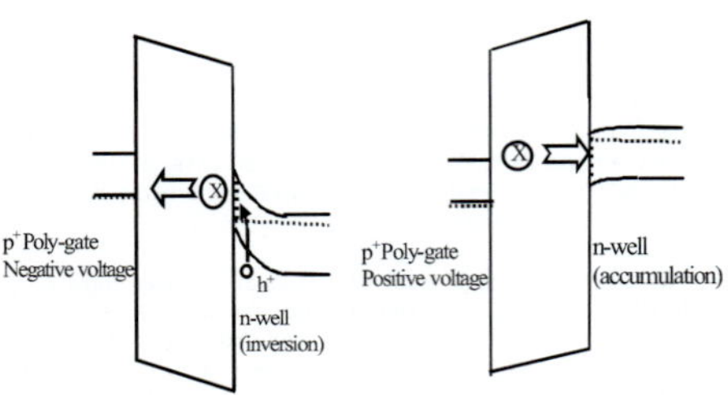

FIGURE 12. (a) N_{IT} FORMATION AND HYDROGEN DIFFUSION TOWARDS THE GATE ELECTRODE DURING NEGATIVE GATE BIAS STRESSING. (b) HYDROGEN RETURNING TO THE INTERFACE AND N_{IT} PASSIVATION DURING POSITIVE GATE BIAS . X STANDS FOR HYDROGEN SPECIES.

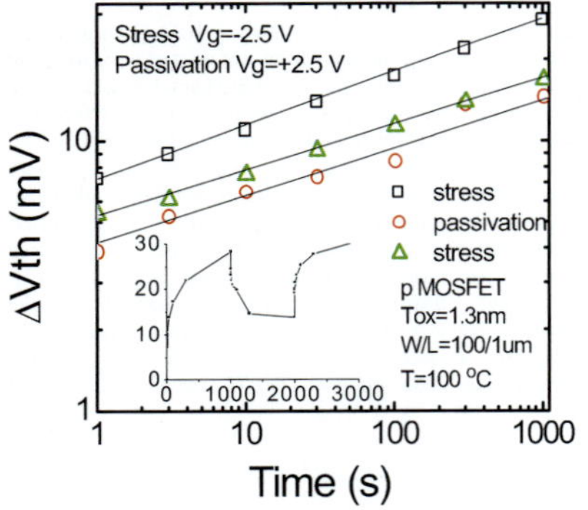

FIGURE 13. THE LOG-LOG VTH SHIFT VERSUS TIME CURVES FOR THE STRESS -PASSIVATION-STRESS PROCEDURE. THE~ 0.25 SLOPES INDICATE THAT THE PROCESSES ARE DIFFUSION -CONTROLLED.

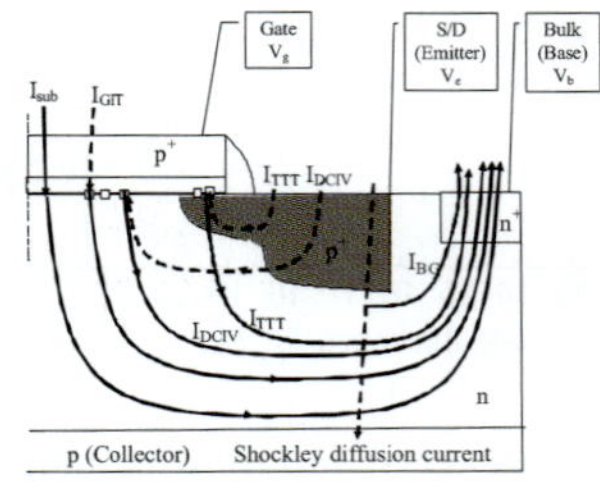

FIGURE A1 . DIFFERENT CURRENT COMPONENTS IN EQ(A1) . ELECTRON CURENTS MARKED BY SOLID LINES, HOLE CURRENTS MARKED BY DASHED LINES .

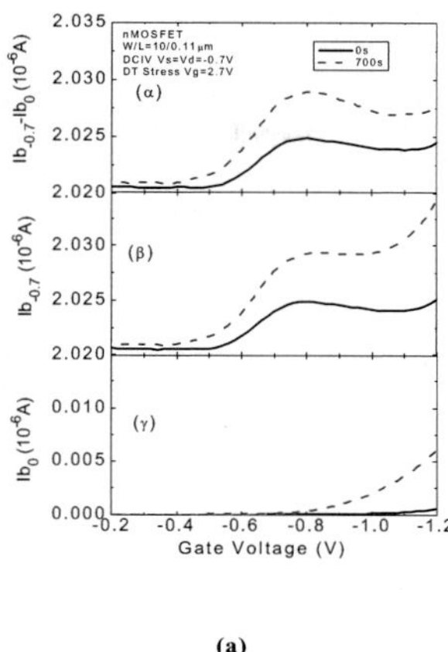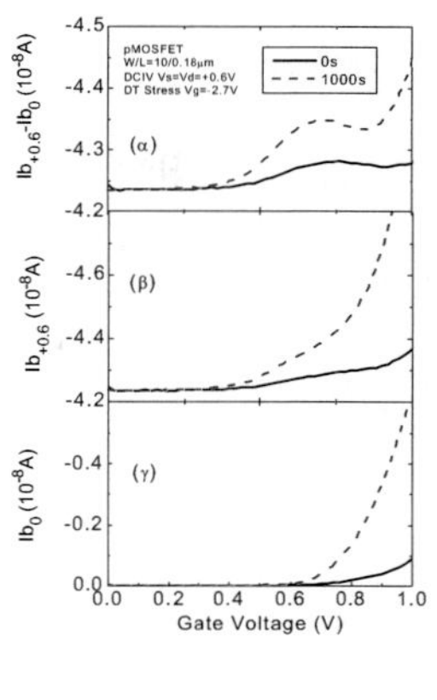

(a)   (b)

FIGURE A2. *Illustration of extracting recombination current I_{DCIV} from measured bulk current I_b. (a) nMOSFET, and (b) pMOSFET. (a) Extracted I_{DCIV}, (β) measured bulk current with positive emitter bias condition (0.7 V for nMOSFET and 0.6 V for pMOSFET) , and (?) measured bulk current with zero biased emitter.*

In eq.(A1) ,the Shockley minority carrier diffusion current from the forward biased emitter (biasing voltage V_e) partially converts to majority carrier bulk current I_{BG} due to electron-hole recombination in the bulk. I_{DCIV} is caused by electron-hole recombination via the interface traps [9,10], and can be modulated by V_g to reach a peak value I_{PDCIV}, when V_g adjusting the interface trap energy in the middle of the Si energy gap to be coincident with the Fermi energy ,with the peak amplitude proportional to the effective number of interface traps N_{it} . For n (p)-MOSFET with ultrathin gate oxide, V_g of I_{PDCIV} are found to be in the range of 0 to -1(+1) V, as shown in Fig.A2. The rest of terms in eq.(A1) may seriously interfere with I_{DCIV} when the oxide is ultrathin. I_{SUB} (V_g) is the intrinsic gate to bulk direct tunneling majority carrier current which is much smaller than the total gate leakage current $I_g = (I_{SUB}+I_{S/D})$ [21,22] in the DCIV peak V_g region since the holes (electrons) face 1.1 eV Si energy gap in the $n^+(p^-)$ poly-Si gate in n(p)-MOSFET [20]. This is the major reason why the DCIV method is applicable to ultrathin tunnel gate oxide device. I_{GIT} is the gate to bulk tunneling currents via interface traps [20] and I_{TTT} is the Thermal-Trap-Tunneling current at s/d extension [23]. We have developed a simple approach to remove the interfering current components in equation (1) in order to obtain clear DCIV spectra by optimizing the bias condition and proper processing of the measured signal Since I_{SUB} and I_{GIT} are tunneling currents through the channel, they are nearly independent of emitter forward bias voltage, V_e. Therefore, when $V_e=0V$, I_{BG}, I_{DCIV}, and I_{TTT} can be ignored. Equation (A1) becomes:

$$I_b (0, V_g, N_{it}) = I_{SUB} (V_g) + I_{GIT} (V_g, N_{it}) \qquad (A2)$$

By measuring I_b under bias $V_e > 0$ and $V_e = 0$, respectively, one finds the difference in I_b:

$$I_b (V_e, V_g, N_{it}) - I_b(0, V_g, N_{it}) = I_{BG} (V_e, V_g) + I_{DCIV} (V_e, V_g, N_{it}) + I_{TTT}(V_e, V_g, N_{it}) \qquad (A3)$$

The background recombination current I_{BG} is independent of N_{it} and remains the same (no peak) before and after stress, so it can be easily removed by comparing I_b before and after stress. Also, I_{TTT} increases with V_e at a slower rate than that of I_{DCIV} .Therefore, by increasing V_e, the impact of I_{TTT} on I_{DCIV} is suppressed significantly. Since the shape of I_{TTT} is known in [22], it changes monotonically in the range of V_g where I_{DCIV} has a peak. Therefore I_{TTT} can be easily de-convoluted out of I_{DCIV}. When increasing V_e, the ratio I_{TTT}/I_{DCIV} and therefore the error from de-convolution will be suppressed. The improved DCIV measurement scheme is illustrated in Fig.A2. Curve γ shows I_b $(0, V_g, N_{it})$. Curve β shows $I_b(V_e, V_g, N_{it})$ when $V_e =0.6 \sim 0.7$ V. Curve α is the difference between β and γ, representing $I_{DCIV} + I_{TTT}$ (plus a background I_{BG}). The peak current at around $|V_g| \sim 0.7$-0.8 V represents I_{PDCIV} superposed by the current component I_{TTT} monotonically increasing with $|V_g|$ in the range of I_{PDCIV} [23]. The oxide charge can be detected by the DCIV peak shift [24].

As reported in [20], an SILC gate current of a capacitor (or bulk current for a transistor) at voltage near flat-band (LV-SLIC) was interpreted as tunneling currents via interface traps. Fig.A3 shows the excellent correlation between DCIV measurements and LV-SILC measurements on our devices under uniform direct tunneling (DT) stressing. The good linearity is a strong support to both analysis of [20] and this work, showing evidence that both measurements give effective detection of interface traps. However, DCIV peak mainly detects the interface traps in the mid gap of Si while the relative increment of LV-SILC peaks mainly detects the interface traps nearby the conduction/valence band edge for n/p MOS.

Acknowledgement

This work was supported by Singapore A*STAR EMT/TP/00/001,2 research grant and the National University of Singapore R263-000-077-112 research grant .

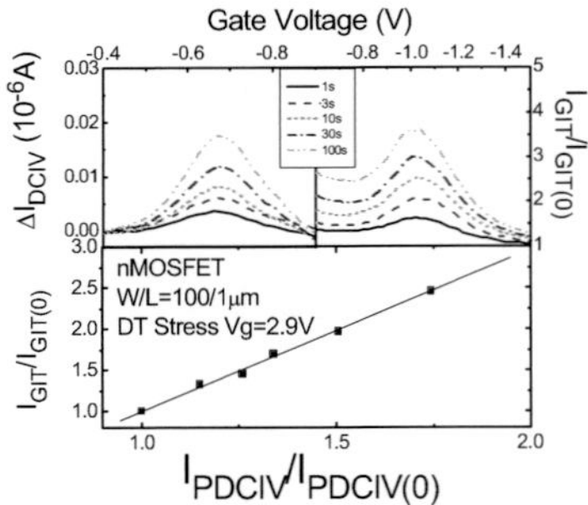

FIGURE A3. CORRELATION OF N_{IT} INCREAMENTS MEASURED BY DCIV AND LV-SILC, BOTH EXTRACED AT PEAK VALUES. THE UPPER TWO CURVES SHOW THE DCIV CURRENT INCREAMENTS AND LV-SILC (I_{GIT}) RELATIVE INCREAMENTS, RESPECTIVELY.

REFERENCES

[1] N. Kimizuka, T. Yamamoto, T. Mogami, K. Yamaguchi, K. Imai, T. Horiuchi, "Impact of bias temperature instability for direct-tunneling ultra-thin gate oxide on MOSFET scaling", *Digest of Technical Papers, Symposium on VLSI Technology1999*, p. 73.

[2] C. H. Liu, M. T. Lee, C. Y. Lin, J. Chen, K. Schruefer, J. Brighten, N. Rovedo, T. B. Hook, M.V. Khare, S. F. Huang, C. Wann, T. C. Chen, T. H. Ning, "Mechanism and process dependence of negative bias temperature instability (NBTI) for pMOSFETs with ultrathin gate dielectrics", *IEDM Tech Digest 2001*, p.861.

[3] T. Yamamoto, K. Uwasawa, T. Mogami, "Bias temperature instability in scaled p^+ polysilicon gate p-MOSFET's", *IEEE Trans. ED* Vol. 46, p.921, 1999.

[4] N.C. Das and V. Nathan, "Hot carrier induced interface trap annealing in silicon field effect transistors", *J. Appl. Phys.* Vol.74 ,p.7596, 1993.

[5] KM Han, CT Sah, "Reduction of interface traps in p-channel MOS transistors during channel-hot-hole stress", *IEEE Trans ED* Vol.45,p.1380, 1998.

[6] G. Chen, M. F. Li and Y. Jin, "Electric passivation of interface traps at drain junction space charge region in p-MOS transistors",*Microelectronics Reliability*, Vol. 41, p. 1427,2001.

[7] W.K. Henson, K.Z. Ahmed, E.M. Vogel, J.R. Hauser, J.J. Wortman, R.D. Venables, M. Xu, and D. Venables, "Estimating oxide thickness of tunnel oxides down to 1.4 nm using conventional capacitance-voltage measurements on MOS capacitors", *IEEE Electron Device Letters*, Vol.20, p.179,1999.

[8] Y.T. Hou, M. F. Li, Y. Jin, and W.H. Lai, "Direct tunnelling hole current through ultrathin gate oxides in metal-oxide-semiconductor devices", *Journal of Applied Physics*, Vol.91, p.258,2002.

[9] A Neugroschel, CT Sah, KM Han, M S Carroll, T Nishida, JT Kavalieros, Y Lu, "Direct-current measurement of oxide and interface traps on oxidized silicon", *IEEE Trans ED* Vol.42, p.1657,1995.

[10] Chih-Tang Sah , "DCIV diagnosis for submicron MOS transistor : Design, Process, Reliability and Manufacturing", 2001 6th International Conference on Solid-State and Integrated Circuit Technology , Proceedings, Vol.1, p.1 , Shanghai .

[11] P. Balk, "Effects of hydrogen annealing on Silicon surfaces", *Extended Abstracts of Electronics Division, Electrochemical Society Spring Meeting*, 1965, vol. 14, p.237.

[12] C. T. Sah, Fundamentals of Sold state Electronics, Solution Manual, World Scientific, 1996, p.101.

[13] C. Hu, S.C. Tam, F. C. Hsu, P. K. Ko, T. Y. Chan, and K.W. Terril, "Hot-electron-induced MOSFET degradation—model, monitor, and improvement." *IEEE Trans. ED*, Vol. 32, p. 375, 1985.

[14] E.H. Poindexter, "Chemical reactions of hydrogenous species in the Si/SiO_2 system", *J.Non-Crystalline Solids*, vol. 187, p.257,1995.

[15] Chih-Tang Sah, J.Y.C. Sun and J.J.T. Tzou, J. Appl. Phys., "Study of the atomic models of three donor-like traps on oxidized silicon with aluminum gate from their processing dependences", Vol. 54, p.5864,1983.

[16] S. Ogawa, M.Shimaya, N. Shiono, "Interface –trap generation at ultrathin SiO_2 –Si interfaces during negative-bias temperature aging", J. Appl. Phys., Vol.77, p.1137,1995; S. Ogawa, N. Shiono, "Generalized diffusion-reaction model for the low-field charge-buildup instability at the Si-SiO_2 interface", Phys. Rev. B, Vol. 51, p.4218,1995.

[17] Yi Lu and Chih-Tang Sah, "Energy and momentum conservation during energetic-carrier generation and recombination in silicon", Phys. Rev. B, vol. 52, p.5657. 1995.

[18] C.W.Tsai et al, "Valence-Band Tunneling Enhanced Hot Carrier Degradation in Ultra-Thin Oxide nMOSFETs", 2000 IEDM Tech.Digest, p.139.

[19] R. Thewes, R. Brederlow, C. Schlünder, P. Wieczorek, B. Ankele, A. Hesener, J. Holz, S. Kessel and W. Weber "MOS Transistor Reliability under Analog operation", *Microelectronics Reliability*, Vol. 40, p.1545,2000.

[20] A. Ghetti, E. Sangiorgi, J. Bude, T.W. Sorsch, and G. Weber, "Tunneling into interface states as reliability monitor for ultrathin oxides", *IEEE Trans. ED*, v.47, p. 2358, 2000.

[21] Yin Shi, T.P.Ma et al,"Polarity dependent gate tunneling currents in dual-gate CMOSFET's", *IEEE Trans. ED* ,Vol.45,p.2355,1998.

[22] Y.T. Hou, M.F.Li et al, *IEDM 2002*, " Quantum tunneling and scalability of HfO_2 and HfAlO gate stacks" , p.731.

[23] G. Chen, M.F. Li, and X.Yu, "Interface traps at high doping drain extension region in sub-0.25- μm MOSTs", *IEEE Electron Device Letters*, vol. 22, p. 233, 2001.

[24] B.B.Jie, W.K.Chim, M.F.Li, K.F.Lo, "Investigation of interface traps located at different regions in p-MOS transistors using DCIV technique", *IEEE Trans.ED*, Vol. 48, p.913, 2001.

NEW RELIABILITY ISSUES OF CMOS TRANSISTORS
WITH 1.3 nm THICK GATE OXIDE

M. F. Li, B. J. Cho, G. Chen, and W.Y. Loh

Silicon Nano Device Laboratory, Department of Electrical and Computer Engineering,
National University of Singapore, 4 Engineering Drive 3, Singapore 117576

D. L. Kwong

Microelectronics Research Center, Department of Electrical and Computer Engineering
The University of Texas at Austin, Austin, TX 78712

Abstract: Several new reliability issues facing CMOS transistors with $t_{ox}=1.3$ nm thick gate oxide and their impacts on projection of operation voltage V_{10Y} for 10-year lifetime are discussed.
1). Oxide lifetime is defined by an event taking place much earlier than oxide breakdown: a strongly transistor-size dependent increment in gate leakage current. Contrary to T_{BD} [1], the newly defined oxide lifetime is shorter when the transistor size is smaller.
2). Contrary to previous reports on thicker gate oxide [2], $V_g=V_d$ is the worst-case hot carrier degradation condition for both n-MOS and p-MOS, with p-MOS showing smaller V_{10Y} than n-MOS.
3). A "dynamic" NBTI of p MOS devices during AC stressing is reported. It is clearly demonstrated that the conventional (static) NBTI [3] underestimates p-MOS device lifetime. The DNBTI determines the overall CMOS device lifetime and will have significant impact on projection of maximum operating voltage in practical operation of digital circuits.

INTRODUCTION

Over the past 30 years, MOSFETs (metal-oxide-semiconductor field-effect transistor) dimensions have shrunk from gate length of 5 micron in the early 1970s to 0.1 micron today, and are expected to reach 0.065 micron within next 2 years [4]. Currently, the scaling of the device geometry in CMOS technology requires the fabrication of gate oxide with thickness t_{ox} below 2 nm, and gate length below ~100 nm for high density and high-speed circuits. On the other hand, while the device sizes are shrinking, the supply voltage is also getting lower. For instance, according to the ITRS2001 road map [4], for the printed gate length 65 nm devices, the equivalent gate oxide thickness is around 1.1-1.6 nm and the nominal power supply voltage V_{dd} is around 1.0 V. The reliability issues for such devices should be re-examined and may be very different from those with thicker gate oxides and longer channel lengths. This is because, first, in such devices, gate leakage current becomes extremely high due to direct tunneling of electrons and holes through the gate oxide. Second, the accumulation increment of electron or hole kinetic energy obtained from field acceleration in the device should never exceeds 1 2V, which is likely not enough for breaking the atomic bonds to create interface traps or oxide traps [5]. A systematic investigation of reliability issues for CMOS devices with oxide thickness less than 1.5

nm is very necessary, however not much research work has been done in this area. This paper summarizes our recent investigation of device degradations and new reliability issues for advanced CMOS devices with t_{ox}=1.3 nm under stress conditions that are relevant to device operation. The emphasis of the work is focused on new experimental findings rather than physical explanations of the degradation mechanism.

CMOS TRANSISTORS

CMOS devices were fabricated using standard dual-gate CMOS technology. Gate oxide of 1.3 nm thickness was grown by Rapid Thermal Oxidation (RTO) followed by an exposure to high-density nitrogen plasma. Fig. 1 is the TEM section of 1.3 nm gate oxide. Fig. 2 is the C-V measurements and simulation results fitting to 1.3 nm oxide thickness. The C-V simulation method for high tunneling leakage capacitors is according to [6]. Fig. 3 shows carrier separation measurements which are in good agreement with our tunneling simulation [7], indicating that the leakage is mainly caused by the intrinsic direct tunneling mechanism.

Fig. 1 *TEM cross section of 1.3 nm gate oxide.*

Fig. 2 *C-V measurements and simulation results (solid lines) fitting to 1.3nm oxide thickness by the method of [6]*

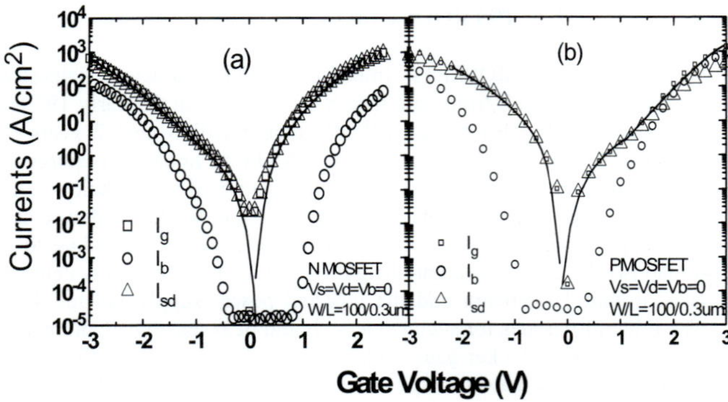

Fig.3 *Carrier separation measurements and simulation results of gate leakages (solid lines) for (a) an n-MOSFET and (b) a p-MOSFET. $V_b=V_s=V_d=0$. Gate oxide thickness t_{ox} =1.3 nm.*

V_{10Y} PROJECTIONS WITH DIFFERENT STRESS CONDITIONS

The device degradations and the projections of operation voltage V_{10Y} for 10-year lifetime under different stress conditions have been systematically investigated for CMOS devices with 1.3nm thick gate oxide.

(i) Oxide degradation:

For oxide thicker than 3 nm, dielectric breakdown can be clearly differentiated into quasi-breakdown (QB) [8-10] or conventional breakdown depending on the severity of the degradation [1,11,12]. However, as oxide thickness shrinks, gate leakage current increases significantly especially in the direct tunneling regime. Wu *et al.* [13] has proposed a new failure criterion, using a dual voltage time-dependent dielectric wearout (TDDW) to characterize and monitor device failure for 1.8-2.7 nm oxide. For ultrathin oxides, Monsieur *et al* [14] have further observed that current increase in 1.7-2.4 nm oxides is progressive and shows characteristics dissimilar to QB. In their detailed study, it has been observed that different device areas have almost identical wear-out current thus leading them to conclude that progressive breakdown (PBD) dynamics is independent on device area.

In this study, we report that the progressive wear-out behavior and its area dependence in 1.3 nm oxides are quite different from previous reports for thicker oxides. In particular, it is observed that the progressive wear-out and gate leakage current is highly localized and area dependent. Using this localized gate leakage current as oxide failure criterion, it is observed that smaller devices actually have shorter lifetime, which is opposite to the trend observed in conventional area dependence of time-dependent dielectric breakdown (T_{DB}) [1,11].

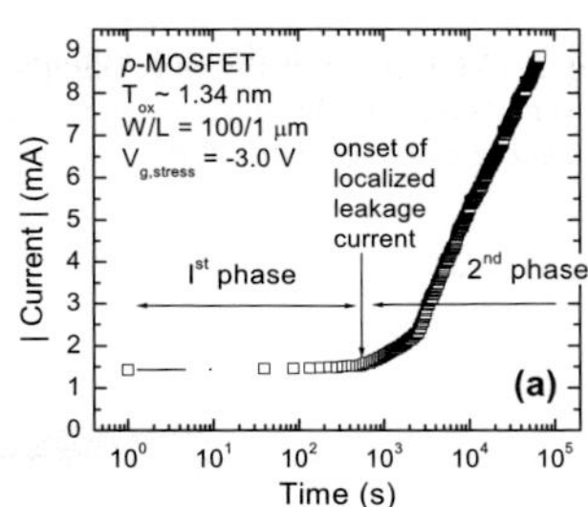

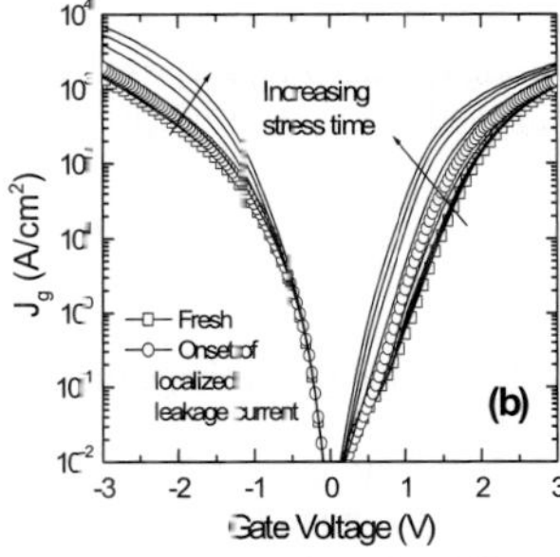

Fig.4 **(a)** *Evolution of gate leakage current of 1.3 nm gate oxide on p-MOSFET during the stress with constant gate voltage of -3 V.* **(b)** *Associated I-V characteristics of the oxide at fresh and post-stressed conditions.*

Figure 4(a) shows the evolution of gate leakage current under constant gate voltage stressing. Gate leakage current remains relatively constant in the initial phase of voltage stressing, and increases significantly only after about 600s of stressing. In the second phase, gate leakage current increases almost linearly with a logarithmic time scale. The trend is similar to the wear-out current and PBD reported in other papers [13,14]. Fig. 4(b) shows the associated current-voltage (*I-V*) characteristics of the p-MOSFETs during the stress. Under the negative gate bias of V_g < -1 V, gate leakage current increases significantly after the time corresponding to the onset of the second phase in Fig.4 (a). Under the accumulation condition (positive gate bias), the gate leakage current increase is observed even at low gate bias. This localized gate leakage is critical to device performance due to its impact on low voltage standby leakage current.

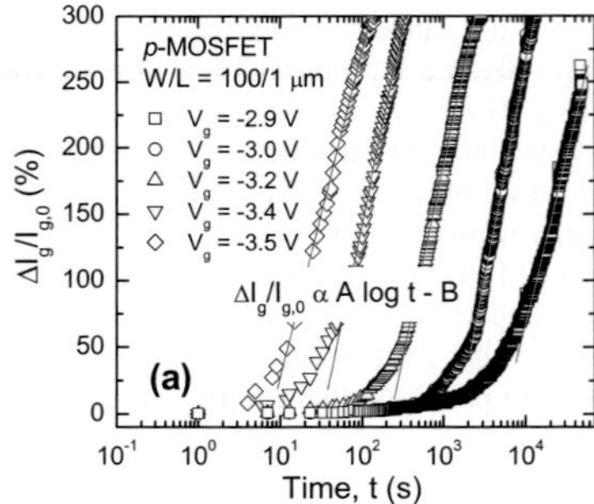

Fig.5 *Percentage change in gate leakage current when stressed under different gate voltages. The stressing and measuring voltage is the same.*

Figure 5 shows the evolution of normalized gate leakage current under electrical stressing with different constant gate voltages. Normalized gate leakage current is defined as $\Delta I_g/I_{g,0}$ where $\Delta I_g = I_g - I_{g,0}$ and $I_{g,0}$ and I_g are the gate current before and after stress respectively. A steady increase in the normalized gate leakage current is observed after the onset of the localized gate leakage. An almost parallel shift is observed in normalized gate leakage currents for different gate stress voltages. Form Fig. 5, the normalized gate leakage current can be described by the following equations where $A = 2.7$ and $B = C$ -12.85 $|V_g|$:

$$\frac{\Delta I_g}{I_{g,0}} \approx 2.7\log t + 12.85\left|V_g\right| - C \tag{1}$$

where V_g is the applied gate voltage (in volt) and C is a constant.

Figure 6 shows the current density J_g for different sample areas when a constant voltage of –3 V is applied to the gate. (Negative sign for J_g indicates that electron is flowing into the gate electrode). It can be observed that the density of gate current increases much faster in smaller area samples although the actual gate leakage current is lower. The differing slope or rate of increase for the gate current density of different gate areas shows that the degradation is not uniformly distributed. Unlike the experimental data obtained by Monsieur *et al.* for thicker oxides, Fig. 6 shows that the leakage current density in ultra-thin oxide of 1.3 nm is not the same for different device areas. This can be reconciled if the degradation is highly localized. This is similar to the gate current and current density observed after the occurrence of QB or breakdown, which is well known as a local phenomenon [8].

Figure 7 shows the rate of gate current density increase with respect to injected electron fluency for different gate areas. The effective localized spot generation probability, which is directly proportional to the rate of increase in the normalized leakage current density, shows that different gate areas have different degradation rates. This is obtained from the slope of graphs in Fig. 6 and shows that the localized spot generation rate is area-dependent and is not uniformly distributed like stress-induced leakage current (SILC)[15]. From Fig. 7, an almost constant degradation rate is observed at both very small and very large gate areas. The saturation in degradation rate values for very large sample area may explain the differing observations by other researchers. For the intermediate gate areas of 1 μm^2 to 100 μm^2, the rate of gate current density increase is slowly decreasing with sample area. The implication of the result is that for ultra-thin gate oxides, smaller gate area sample will have faster rate of current density increase and this will lead to a shorter lifetime, contrary to that suggested by conventional breakdown statistics.

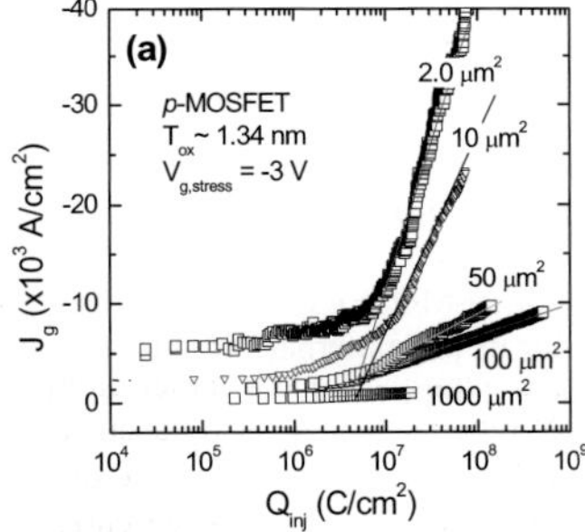

Fig.6 Evolution of *gate leakage current density stressed under constant voltage stress of -3V for different channel areas.*

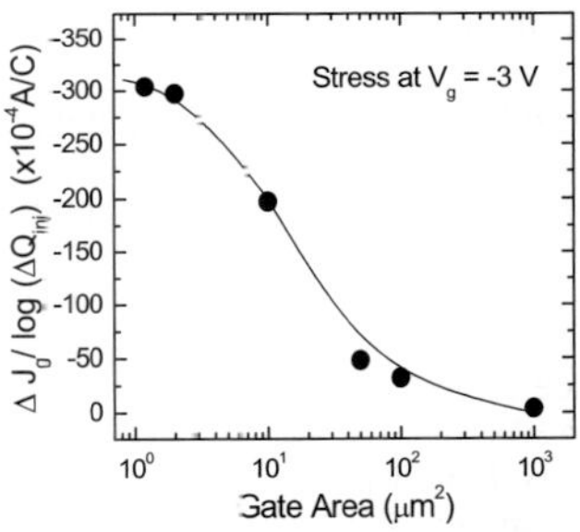

Fig.7 *Effective localized spot generation rates as a function of gate area .*

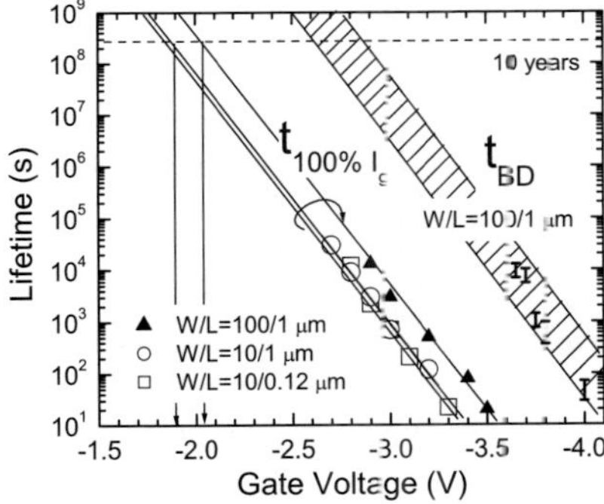

Fig.8 *Maximum allowable operating voltage for 10 year operating lifetime $T_{100\% \, Ig}$ using 100% increase in gate leakage current as a criterion. For comparison, time-to-breakdown T_{BD}, is also evaluated for 1.3 nm gate oxides.*

Figure 8 shows the maximum allowable operating voltage for 10 years lifetime projected using gate current density as a criterion for device failure. In this case, we have adopted a 100% increase in gate leakage current as the criterion. Previous breakdown studies in thin oxides have shown that oxide degradation is a gate-voltage-driven phenomenon [16]. By assuming the same degradation dynamics for localized gate leakage current, the projected lifetimes of 1.3 nm gate oxide with different gate areas are obtained as shown in Fig. 8. For comparison, the projected lifetime based on time-to-breakdown using higher voltage stressing on the same type of devices is also evaluated. Due to the long stressing time required for breakdown or quasi-breakdown to occur at lower stress voltage range, only a limited range of gate voltage can be performed. Nevertheless, it can be observed that projected lifetime based on time-to-breakdown T_{BD} is much larger than that based on the localized gate current increase. This shows that T_{BD} is not a significant problem in ultra-thin oxide. Instead, time-to-100% increase in localized gate current $T_{100\% \ Ig}$ becomes the limiting factor, which is dependent on gate area. Larger area sample shows longer $T_{100\% \ Ig}$. From Fig. 8, the operating voltage with projected 10-year lifetime V_{10Y} is about 2.03 V for 100 μm^2 gate area and 1.9 V for 10 μm^2 gate area.

(ii) Hot-carrier (HC) degradations:

Worst stressing conditions Estimated by High Voltage Stressing

Traditionally, worst HC stress condition for n-MOSFET is under the bias condition where I_b peaks, and when I_g maximizes for its p counterpart. In ultra thin gate oxide devices, however, the conditions become more complicated. As shown in Fig. 9, I_b peaks are still very clear for both n- and p-MOSFETs, which is opposite to the observation of [17] for thicker oxide where the authors claimed that the I_b peaks had become very flat and it was hard to locate the peak position. Due to the large direct tunneling gate leakage currents, no I_g peak was found under any of the stressing conditions.

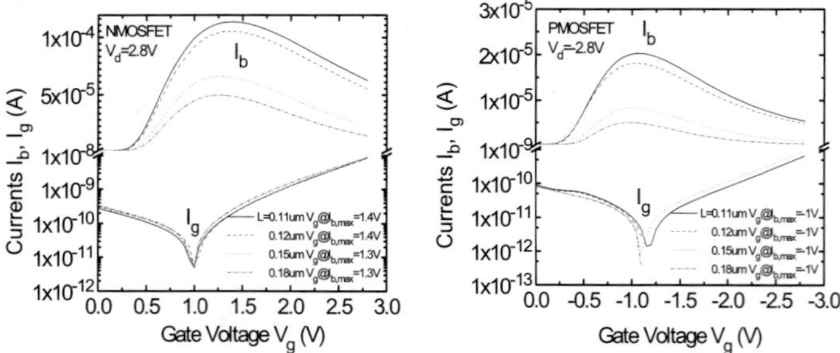

Fig.9 *Gate and bulk current during HC stressing for (a) n and (b) p MOSFETs. $T_{ox}=2.0$ nm. For $T_{ox}=1.3nm$, the result is similar.*

Two typical stressing conditions, i.e., $V_g=V_d$ and $V_g@I_{b,peak}$ were applied to n-MOSFETs of different gate lengths (Figs.10 (a) and (b)). It was found that for shorter gate n-MOSFETs, the worst case is $V_g=V_d$, while for longer gate devices, the worst case is when I_b peaks. For p-MOSFETs, $V_g=V_d$ is the common worst stressing case for both long and short channel devices (Figs. 10(c) and (d)). DCIV measurement [18,19] and transistor parameters degradation reveal the same information. The findings are in agreement with those in [17, 20]. In [17], the reason for the worst case for shorter gate n-MOSFETs from $V_g@I_{b,peak}$ to $V_g=V_d$ is attributed to the difference in the ionization energy and the energy to create interface traps. Under the condition of $V_g@I_{b,peak}$, although the ionization reaches its maximum, yet the rate of interface trap generation has not maximized. Moreover, the scattering between the carriers and the interface is more severe when $V_g=V_d$. In our experiment, we really detected more interface trap generated under the condition of $V_g=V_d$, which also support above opinion. However, there is also another very important fact : the series resistance under the spacer region, which is a critical factor in deciding g_m , and also related with other parameters such as $I_{d,linear}$, V_{th}, degrade much more under the $V_g=V_d$ stress, for the carrier trajectory in this condition is closer to drain extension and thus the carriers are more readily to induce damages and increase series resistance.

Worst Stressing Conditions Estimated by Lifetime Prediction

Failure criterion is chosen as the 10% degradation of g_m. The lifetime projections for n- and p-MOSFETs with gate oxide thickness of 2 and 1.3 nm were conducted at room temperature. First reported in [2], CMOS devices of 2 nm gate oxide suffer the most degradation at different V_d-V_g combinations under different voltage ranges. In our experiment, the devices with 2 nm oxide show the same tendency as reported in [2], i.e., at high voltage area, the worst stressing condition is $V_g=V_d$, while it transits to $I_{b,max}$ as the operation voltage is lowered. The worst-case transition voltage is 2.86 V for n-MOSFET and 1.5 V for p-MOSFET (Fig. 11 (a)). n-MOSFET degrades faster than p-MOSFET. The CMOSFET overall lifetime is limited by n-MOSFET under $I_{b,max}$ working condition. For the devices of 1.3 nm oxide, the behavior is totally different. As can be found in Fig.11 (b), for n-MOSFET, the worst stressing condition is $V_d=V_g$ if operation voltage is below 1.41 V and changed to $I_{b,max}$ when operation voltage exceeds 1.41V. For p-MOSFET devices, $V_g=V_d$ remains the worst stress condition for the whole voltage range less than 3 V. Moreover, the eventual lifetime is determined by p-MOSFET transistor, other than n-MOSFET. Therefore, CMOSFET HC worst stressing conditions are not only working voltage and device type dependent, but also gate oxide thickness dependent.

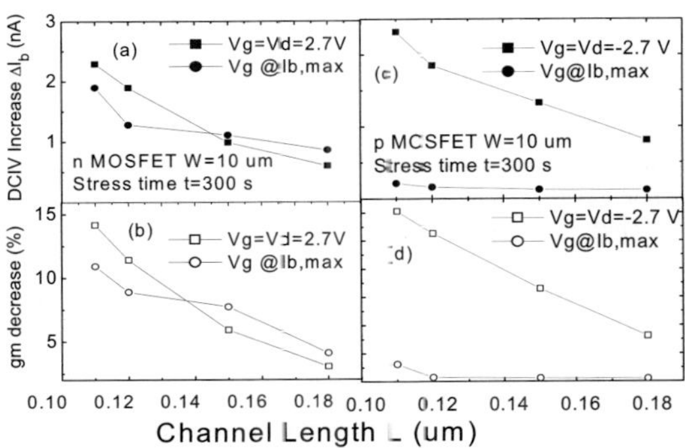

Fig.10 *DCIV and transistor parameters changes under different types of stresses for n (a), (b) and p MOSFETs (c) (d). $t_{ox}=1.3nm.$.*

(iii) Dynamic Negative Bias Temperature Instability (DNBTI) for p-MOS:

With the continuous shrinking of the transistor dimensions, generation of interface traps during NBTI stress in p-MOSFET has become one of the most critical reliability issues that ultimately determine the lifetime of CMOS devices [3,21,22]. In conventional NBTI study, a constant negative bias is applied to the gate electrode of a p-MOSFET at high temperatures with S/D grounded [3,22]. However, during the operation of a p-MOSFET in a CMOS inverter, the applied gate bias is switching between "high" and "low" voltages, while the drain bias is alternating between "low" and "high" voltages, correspondingly. Therefore, it is critical to investigate NBTI under such dynamic stress conditions. We found that a large portion of the interface traps generated under the NBTI stressing, corresponding to p-MOSFET operating condition of the "high" output state in a CMOS inverter, are passivated electrically when the gate to drain voltage switches to positive, corresponding to the p-MOSFET operating condition of the "low" output state in a CMOS inverter. As a result, this dynamic NBTI (DNBTI) effect greatly prolongs the lifetime of p-MOSFETs operating in a digital circuit, while the conventional static NBTI (SNBTI) measurement underestimates the p-MOSFET lifetime .

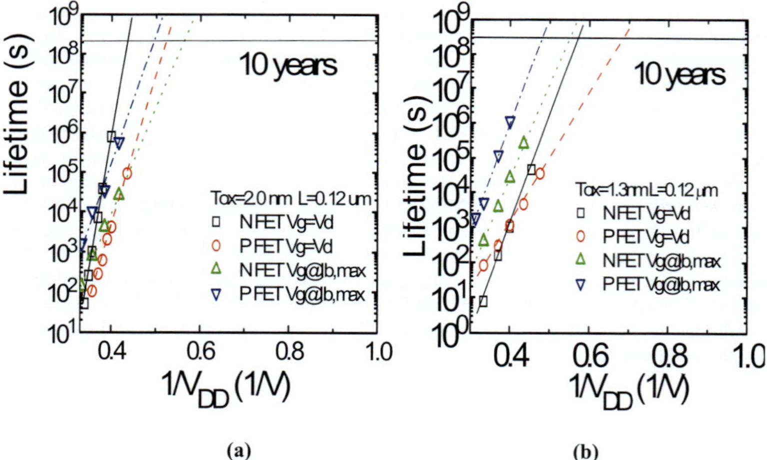

Fig.11 *Lifetime projection of MOSFETs under two typical HC stressing conditions. Worst stress conditions for different gate oxide, supply voltage and device type show different V_g and V_d combination. (a) T_{ox}=2.0nm, n MOS under $I_{b,max}$ limits V_{DD} scaling. (b) T_{ox}=1.3nm, p MOS under $V_g=V_d$ limits V_{DD} scaling.*

Under conventional SNBTI stress with negative gate-to-s/d bias on a p-MOSFET at high temperature, the major degradation is the shift of threshold voltage ΔV_{th} [3]. Fig.12 shows the excellent correlation between ΔV_{th} and ΔN_{it} (ΔI_{DCIV}) under SNBTI stress. The DNBTI results are shown in Figs.13-16 . As can be seen clearly, when reversing the gate-to-s/d electric field to the opposite polarity (positive) during DNBTI stressing, a reduction (passivation) of ΔN_{it} and thus ΔV_{th} is observed. Fig.13 shows the DNBTI effect under uniform stress condition where S/D are grounded. ΔN_{it}, ΔV_{th} and -

Δg_m increase during the stress period of a negative gate voltage, as expected for SNBTI. However, they all decrease over the "passivation" period when the gate voltage is switched to positive. The relationship among ΔV_{th} vs. ΔN_{it} and $\Delta g_m\%$ is plotted in Fig. 13(b) which shows a perfect linearity, with ΔV_{th} vs. ΔN_{it} slope identical to that of the SNBTI results shown in Fig. 12. Figure 14 shows the DNBTI effect under non-uniform stress condition. During the stress period, a uniform SNBTI stress is applied, i.e., a negative gate voltage is applied with S/D grounded. This stress condition corresponds to the p-MOSFET operation condition in a CMOS inverter when the output is "High" (Fig. 15). During the passivation period, a negative drain voltage is applied while the source and the gate are grounded, resulting a positive gate-to-drain voltage. This non-uniform stressing simulates the p-MOSFET operation condition in a CMOS inverter when the output is "Low". As shown in Fig. 14, a passivation effect, i.e., ΔV_{th} reduction, is seen when the p-MOSFET is operated with the output "Low".

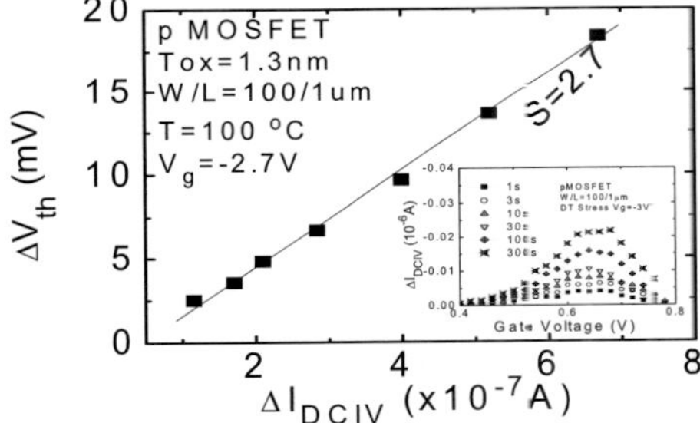

Fig. 12 *Correlation between DCIV current and threshold voltage changes. Interface trap generation plays a major role in NBTI. The inset shows the typical DCIV spectra for 1.3 nm gate oxide p-MOSFET.*

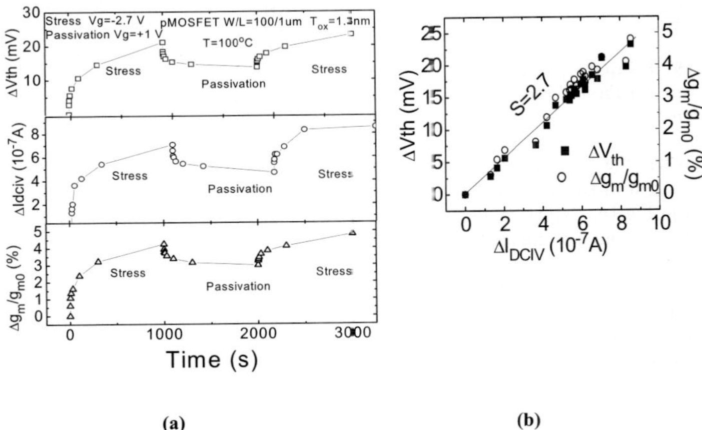

(a)   **(b)**

Fig.13 *(a) VARIATIONS OF V_{TH}, DCIV CURRENT, AND g_m UNDER STRESS-PASSIVATION-STRESS PROCESS. (B)CORRELATION BETWEEN V_{TH} SHIFT, DCIV CURRENT AND g_m VARIATIONS IN (a).*

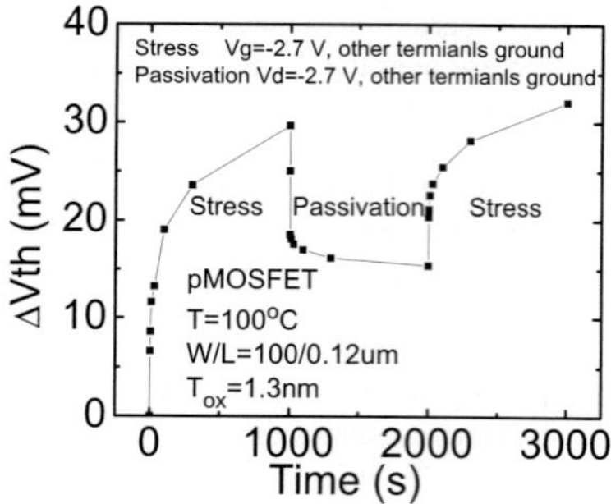

Fig.14. *Illustration of Dynamic NBTI, during high-low operation of a p-MOSFET in a digital CMOS inverter. Due to the electric-passivation effect, the conventional SNBTI overestimates the degradation.*

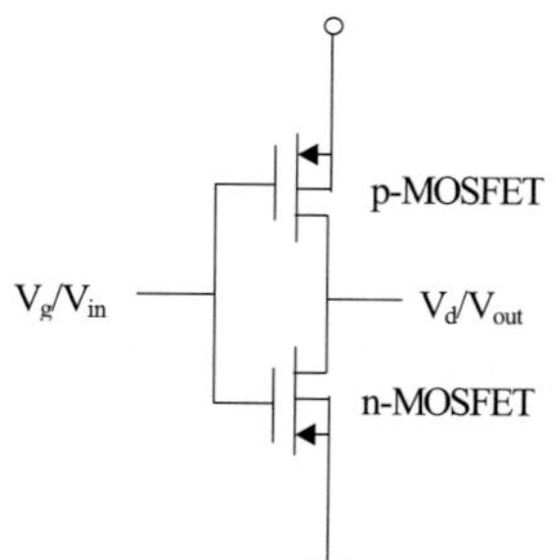

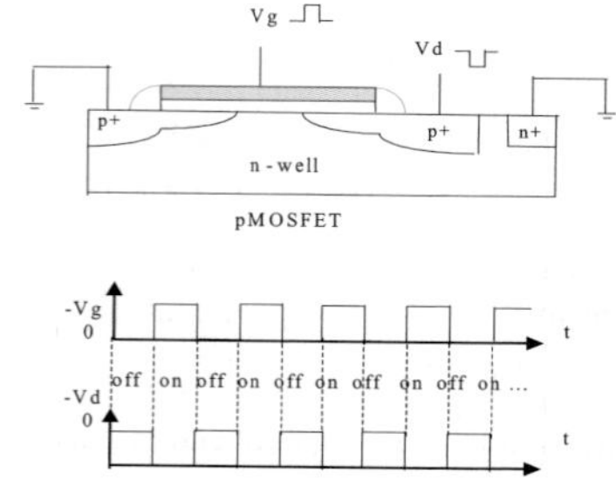

Fig.15 *A p-MOSFET in a CMOS inverter actually undergoes dynamic stressing.*

Fig.16 *To simulate the p-MOSFET stressing condition, two opposite phase square wave trains are applied to its gate and drain , respectively.*

To test the frequency dependence of this dynamic NBTI effect, we deliberately apply a train of square wave to the gate and an opposite phase signal to the drain, as shown in Fig.16. NBTI degradations under such square wave stressing with different frequencies were monitored. The results are shown in Fig. 17.

Based on the results shown in Figs. 13-17, we conclude that due to the DNBTI effect, the generated interface traps and the accompanying degradations of V_{th} and g_m during the typical SNBTI stress period are partially "annealed" during the subsequent passivation stress period. As a result, the device lifetime under DNBTI stress becomes much longer than that projected under the conventional SNBTI stress. The lifetime estimations were made on p-MOSFETs with gate oxide thickness of 1.3 nm and channel length of 0.12 μm under both DNBTI and SNBTI stressing. As shown in Fig.18, for device operating at 100°C and the lifetime defined by the time when ΔV_{th} reaches 30 mV, the lifetime for DNBTI is one order of magnitude longer than that for

10

SNBTI. The supply voltage for 10-year projected lifetime V_{10Y} is 0.9 V under SNBTI while it is 1.2V under DNBTI stressing.

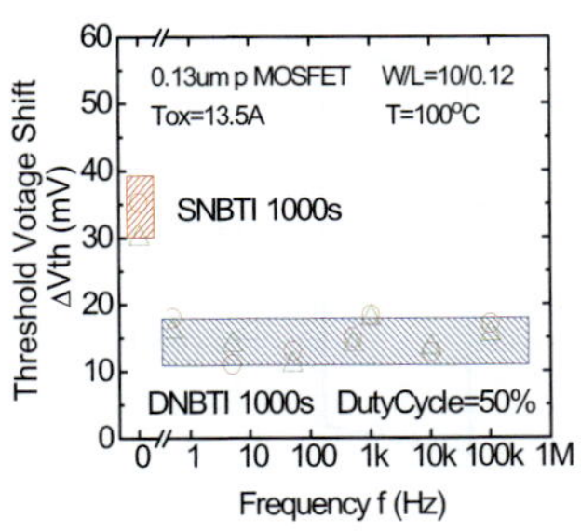

FIG.17 *THE FREQUENCY DEPENDENCY OF DNBTI, WITH THE STRESSING VOLTAGE 2.5V IN FIG.16.*

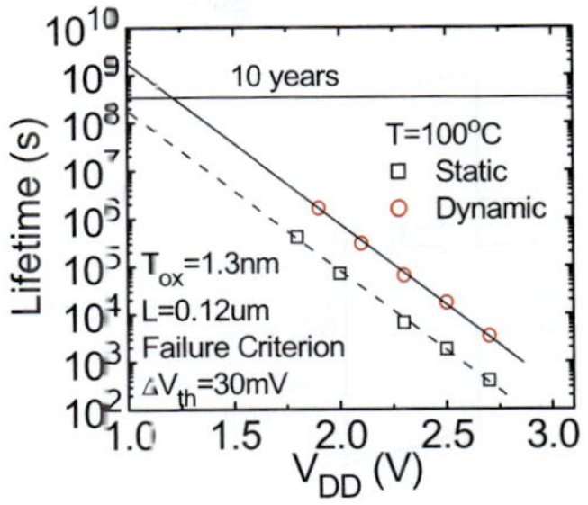

FIG.18 *SNBTI AND DNBTI LIFETIME (30 MV V_{TH} SHIFT) PROJECTIONS FOR P-MOSFETs. V_{10Y} IS 1.2V FOR DNBTI STRESS, AND IS 0.9V FOR SNBTI STRESS.*

As first proposed in 1965 by P. Balk at IBM [23], and now widely accepted [5,24,25], interface trap is related to Si dangling bond when hydrogen is released from a Si-H bond. Sah et al have proposed in 1983 the reverse process: passivation of interface trap by absorbing hydrogen at the dangling bond site [26]. The excellent correlation among ΔV_{th}, Δg_m and ΔN_{it} (ΔI_{DCIV}) under NBTI stress, as shown in Figs.12-13, indicates that for ultrathin gate oxide devices, the device parameter degradation is mainly caused by the interface trap generation. Based on the interface trap generation and passivation mechanisms proposed in [23] and [26], and the recent diffusion-reaction model proposed for the electrochemical reaction of interface trap generation during NBTI degradation [27], we propose the following reactions for interface trap generation and passivation in DNBTI (Eqs. (2) and (3), and Fig. 19):

$$(Si_3 \equiv Si\text{-}H) + h^+ \leftrightarrow Si_3 \equiv Si^* + X_{interface} \tag{2}$$

and

$$X_{interface} \xleftrightarrow{diffusion} X_{bulk} \tag{3}$$

Here $Si_3 \equiv Si\text{-}H$ is the precursor for the Si-H bond. When it interacts with a hole h^+ in the inversion layer or the source/drain extension region under the NBTI stress, the hole breaks the Si-H bond and creates an interface trap by releasing hydrogen species $X_{interface}$ at the Si/SiO$_2$ interface. One thing remaining uncertain is how the holes accelerated by operating voltage as low as 2.5-2.7 V in this work can break the Si-H bond and generate the interface traps. This is a big unsolved issue and probably more complicated processes such as Auger mechanism may be involved [28,29]. We therefore consider Eq. (2) as a more generalized reaction that some detailed reaction mechanism has not been shown explicitly in Eq. (2). The produced hydrogen species denoted as X in Eq. (2), either in the form of molecules or neutral atoms or ions, will diffuse/drift to the gate electrode through the bulk gate oxide, as expressed by Eq. (3) (Fig. 19a). In this process, the interface acts as a hydrogen source. The major symptom of NBTI is the shift of threshold voltage ΔV_{th}. For ultrathin gate oxide MOSFETs, the

ΔV_{th} is mainly induced by interface traps build up along the silicon-gate oxide interface, as oxide charges are easily detrapped by tunneling and thus have less contribution.

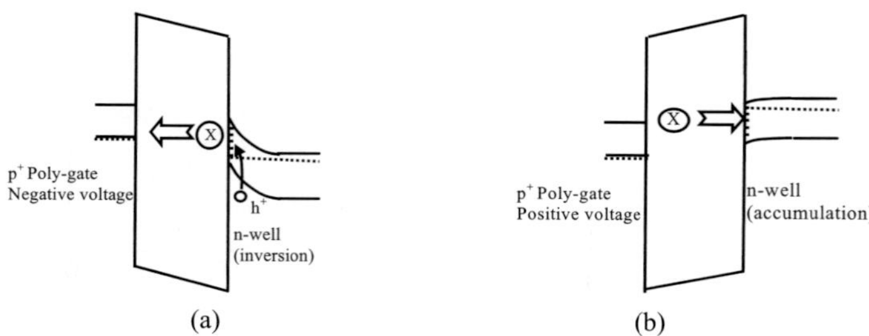

FIG.19. *(a) N_{IT} FORMATION AND HYDROGEN DIFFUSION TOWARDS THE GATE ELECTRODE DURING NEGATIVE GATE BIAS STRESSING. (b) HYDROGEN RETURNING TO THE INTERFACE AND N_{IT} PASSIVATION DURING POSITIVE GATE BIAS. X STANDS FOR HYDROGEN SPECIES.*

The electrical passivation (EP) effect is interpreted by the reverse interaction between N_{it} and hydrogen species as shown in Fig.19b. When the gate bias polarity is reversed from negative to positive, the channel inversion layer disappears and depletion layers are formed at source/drain. The breaking of Si-H bond stops due to lack of holes. On the other hand, the reverse reaction occurs when the hydrogen moves back to the SiO_2/Si interface and passivates the Si dangling bonds, resulting in ΔN_{it} reduction. In this period, the interface acts as a hydrogen sink.

CONCLUSION

Some new reliability issues of CMOS transistors with $t_{ox}=1.3\ nm$ thick gate oxide and their impacts on projection of operation voltage V_{10Y} for 10-year lifetime are discussed. Oxide breakdown is no longer a critical issue of oxide reliability because t_{BD} under operating voltage is extremely long. Oxide lifetime is defined by an event taking place much earlier than oxide breakdown: a strongly transistor-size dependent time-to-100% increment in gate leakage current $T_{100\%\ Ig}$. Contrary to t_{DB}, the newly defined oxide lifetime is shorter when the transistor size is smaller. For life time of $T_{100\%\ Ig}$, V_{10Y} is about *2.03 V* for *100 μm^2* gate area and *1.9 V* for *10 μm^2* gate area .

Contrary to previous reports on thicker gate oxide, $V_g=V_d$ is the worst-case hot carrier degradation condition for both n-MOSFET and p-MOSFET. When failure criterion is chosen as the 10% degradation of g_m, for $L=0.12\ \mu m$ devices, V_{10Y} is *1.5V* for p-MOSFET and is *1.8 V* for n- MOSFET.

A "dynamic" NBTI during AC stressing is reported. It is clearly demonstrated that the conventional (static) NBTI underestimates p-MOSFET lifetime because it has overlooked the electric passivation effect when p-MOSFET is operated in an inverter in the digital circuit . The DNBTI determines the overall CMOS device lifetime and will have significant impact on projection of maximum operating voltage in practical dynamic operation of digital circuits. When *30 mV V_{th}* shift is taken as failure criterion , V_{10Y} is only *0.9V* for SNBTI stress , and *1.2 V* for DNBTI. SNBTI actually underestimate the device lifetime and DNBTI appropriately determines the overall CMOS device lifetime under digital operation.

ACKNOWLEDGEMENT

We thank Drs. Ang Chew Hoe and Zheng Jia Zhen with Chartered Semiconductor Manufacturing Ltd, Singapore for providing devices used in this work. This work was supported by Singapore A*STAR EMT/TP/00/001.2 Research Grant.

REFERENCES

[1]J.H.Stathis, *IRPS* 2001, p.132.

[2] E.Li et al, *IEEE TED*, **48**, p.671 (2001).

[3] N.Kimizuka et al, *VLSI Tech.* 1999, p.73.

[4] ITRS 2001, http://public.itrs.net/Files/2001ITRS/

[5] Chih-Tang Sah, Fundamentals *of Solid-State Electronics –Solution Manual* , Apendix- Transistor Reliability , p.101 ,World Scientific, Singapore (1996).

[6] W.K.Henson , K.Z.Ahmed, E.M.Vogel, J.R.Hauser , J.J. Wortman, R.D. Venables , M.Xu and D. Venables, *IEEE EDL*, **20**, p.179 (1999).

[7] Y.T.Hu, M.F.Li, Y. Jin and W.H.Lai, *J.Appl. Phys.* **91**, p.258 (2002).

[8] S. H . Lee, B. J. Cho, J.C. Kim, S.H. Choi, *IEDM Tech. Dig.*1994, p. 605.

[9] M. Depas, T. Nigam, M. M. Heyns, *IEEE TED*, **43**, p.1499 (1996).

[10] K. Okada, K. Taniguchi, *Appl. Phys.Lett.,***70**, p.351 (1997).

[11] J. Sune, E. Y. Wu, D. Jimenez, R. P. Vollertsen and E. Miranda, *IEDM Tech. Dig. 2001*, p.117.

[12] M. A. Alam, B. E. Weir, and P. J. Silverman, *IEEE TED*, **49**, 239 (2002)

[13] Y.Wu et al, *IEEE EDL* , **20**, 262 (1999)

[14] F. Monsieur, E. Vincent, D. Roy, S. Bruyere, J. C. Vildeuil, G. Pananakakis, G. Ghibodo , *IRPS* 2002, p. 45.

[15] D.J. Dumin et al, *IEEE ICMTS* 1997, p. 61.

[16] P. E. Nicollian, W. R. Hunter and C. Hu, *IRPS* 2000, p.7.

[17] E. Li, E Rosenbaum, J Tao, G C-F Yeap, M.R Lin, P Fang, *IRPS 1999*, p.253.

[18] A.Neugroschel, C.T.Sah, K.M.Han, M.S.Carrol, T. Nishida, J.T.Kavalieros, and Y.Lu, *IEEE TED*, **42**, 1657 (1995).

[19] Chih-Tang Sah, *6th Int. Conf. Solid-State and Integrated Circuit Technology* , *Proceedings,* 2001, vol.1, p.1.

[20] C. Lin, S Biesemans, L.K Han, K Houlihan, T Schiml, K Schruefer, C Wann, J Chen, R Mahnkopf, *IEDM Tech Digest 2000*, p.135.

[21] C. H. Liu, M. T. Lee, C. Y. Lin, J. Chen, K. Schruefer, J. Brighten, N. Rovedo, T. B. Hook, M.V. Khare, S. F. Huang, C. Wann, T. C. Chen, T. H. Ning, *IEDM Tech Digest 2001*, p.861.

[22] T. Yamamoto, K. Uwasawa, T. Mogami, *IEEE TED* **46**, 921 (1999).

[23] P. Balk, *Extended Abstracts of Electronics Division, Electrochemical Society Spring Meeting*, vol. 14, p.237 (1965).

[24] C. Hu, S.C. Tam, F. C. Hsu, P. K. Kc, T. Y. Chan. and K.W. Terril, *IEEE TED*, **32**, 375 (1985).

[25] E.H. Poindexter, *J.Non-Crystalline Solids*, **187**, 257 (1995).

[26] Chih-Tang Sah, J.Y.C. Sun and J.J.T. Tzou, *J. Appl. Phys.,* **54**, 5864 (1983).

[27] S. Ogawa, M.Shimaya, N. Shiono, J. Appl. Phys., **77**,1137 (1995); S. Ogawa, N. Shiono, *Phys. Rev. B*, **51**, 4218 (1995).

[28] Yi Lu and Chih-Tang Sah, *Phys. Rev. B*, **52**, 5657 (1995).

[29] C.W.Tsai et al, "Valence-Band Tunneling Enhanced Hot Carrier Degradation in Ultra-Thin Oxide nMOSFETs", *IEDM Tech.Digest 2000*, p.139.

Japanese Journal of Applied Physics
Vol. 43, No. 11B, 2004, pp. 7807–7814
©2004 The Japan Society of Applied Physics

Review Paper

Dynamic Bias-Temperature Instability in Ultrathin SiO$_2$ and HfO$_2$ Metal-Oxide-Semiconductor Field Effect Transistors and Its Impact on Device Lifetime

Ming Fu L$_I$[1,2,*], Gang C$_{HEN}$[1], Chen S$_{HEN}$[1], Xin Peng W$_{ANG}$[1], Hong Yu Y$_U$[1], Yee-Chia Y$_{EO}$[1] and Dim Lee K$_{WONG}$[3]

[1]*Silicon Nano Device Lab (SNDL), ECE Department, National University of Singapore, 119260 Singapore*
[2]*Institute of Microelectronics, 117685 Singapore*
[3]*Microelectronics Research Center, Department of ECE, University of Texas, Austin, TX 78712, USA*

(Received June 7, 2004; accepted July 6, 2004; published November 15, 2004)

In this paper, we review our recent work on the dynamic bias-temperature instability (BTI) in metal-oxide-semiconductor field effect transistors (MOSFETs) with ultrathin SiO$_2$ and high-K gate dielectrics, operating in a digital inverter circuit. Key findings are: (1) For p-MOSFETs with ultrathin SiO$_2$ gate dielectrics, negative BTI (NBTI) is mainly due to the generation of interface traps. Under dynamic NBTI stress, the interface traps generated in the stressing phase are subsequently passivated in the passivation phase with a zero gate bias. As a result, p-MOSFET lifetime is significantly enhanced and the enhanced lifetime is frequency-independent up to 100 kHz. (2) For n- and p-MOSFETs with ultrathin HfO$_2$ gate dielectrics, BTI is mainly caused by charge trapping in HfO$_2$. Similar lifetime enhancements in both n- and p-MOSFETs are observed under dynamic BTI stress. However, in contrast to SiO$_2$ devices, dynamic BTI in HfO$_2$ is strongly frequency-dependent, i.e., a high frequency results in a slight device degradation and hence a long lifetime. A physical model that accounts for two-step trapping and detrapping in HfO$_2$ is proposed to explain frequency-dependent BTI in HfO$_2$ gate dielectrics. Simulation results based on the new model shows excellent agreement with all experimental data.   [DOI: 10.1143/JJAP.43.7807]

KEYWORDS: CMOS, FET, reliability

1. Introduction

With the continuous decrease in transistor dimensions, new reliability issues emerge. For metal-oxide-semiconductor field effect transistors (MOSFETs) with ultrathin SiO$_2$ gate dielectrics, negative bias-temperature instability (NBTI) has been identified as a critical limiting factor that ultimately determines their lifetimes.[1–3] For MOSFETs with ultrathin high-κ HfO$_2$ gate dielectrics, it has been reported that the devices suffer from serious charge trapping problems, resulting in positive bias threshold voltage instability (PBTI) in n-MOSFETs.[4–7] However, in a practical complementary metal-oxide-semiconductor (CMOS) inverter in a digital circuit, both n- and p-MOSFETs operate under dynamic stress conditions. The input signal V_{in} and output signal V_{out} are opposite in phase, switching alternatively between high and low voltages as indicated in Fig. 1. In this paper, we will review some of our recent investigations into dynamic BTI degradation.

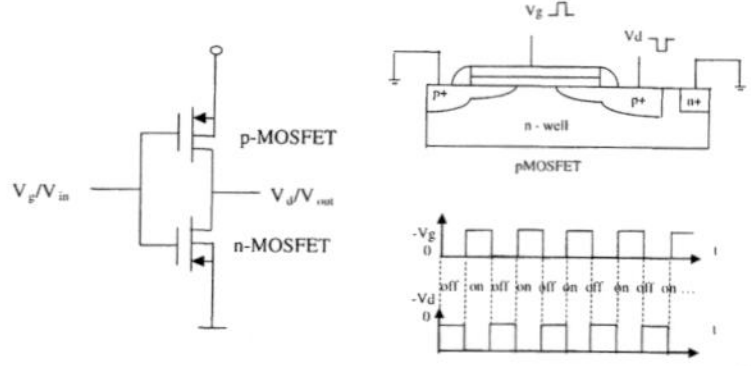

Fig. 1. To simulate the MOSFET stressing condition in an inverter, two opposite-phase square-wave trains are applied to its gate and drain electrodes, respectively. The figure shows p-MOSFET stress voltage only. The n-MOSFET operates at a complementary stress voltage.

*E-mail address: elelimf@nus.edu.sg

2. Dynamic NBTI Degradation in p-MOSFETs with SiO$_2$ Gate Dielectrics

2.1 Device fabrication

p-MOSFETs were fabricated using standard dual-gate CMOS technology. Gate oxides with thicknesses of 1.3 nm, 1.4 nm, 1.7 nm and 2.1 nm were grown by rapid thermal oxidation followed by exposure to high-density nitrogen plasma. A DCIV technique[8–11] was used to monitor interface trap generation and passivation. Oxide thicknesses were determined by C–V measurement and XTEM.

2.2 Interface trap generation and passivation

Static NBTI stress was first applied to p-MOSFETs at a temperature of 100°C. The threshold voltage shift ΔV_{th} and DCIV current I_{DCIV} were monitored and their relationship was investigated. Figure 2 shows an excellent correlation between ΔV_{th} and ΔN_{it} (ΔI_{DCIV}) under static NBTI stress. When the gate voltage was reversed of its polarity or was decreased to zero (the passivation phase), a decrease in ΔN_{it} and thus in ΔV_{th} and $-\Delta g_m\%$ was observed (Fig. 3). It is shown in Fig. 3 that ΔN_{it}, ΔV_{th}, and $-\Delta g_m\%$ have the same trend of increase and decrease during the stress-passivation-stress sequence. Therefore, the degradations of N_{it}, V_{th} and g_m are probably due to interface trap generation. The relationships of ΔV_{th} with ΔN_{it} and $\Delta g_m\%$ are plotted in Fig. 3(b). A similar observation of a decrease in ΔV_{th} at the passivation phase has been recently reported.[12–16]

2.3 Dynamic NBTI in p-MOSFET in inverter circuit[10,11]

To simulate p-MOSFET during the low (passivation) phase in a CMOS inverter (Fig. 1), a nonuniform electric field is applied, i.e., a negative bias is applied to the drain while keeping the other terminals grounded. As shown in Fig. 4, the electric field passivation (EP) effect is similar to that observed in uniform electric field passivation (Fig. 3).

7808   Jpn. J. Appl. Phys., Vol. 43, No. 11B (2004)

M. F. Li *et al.*

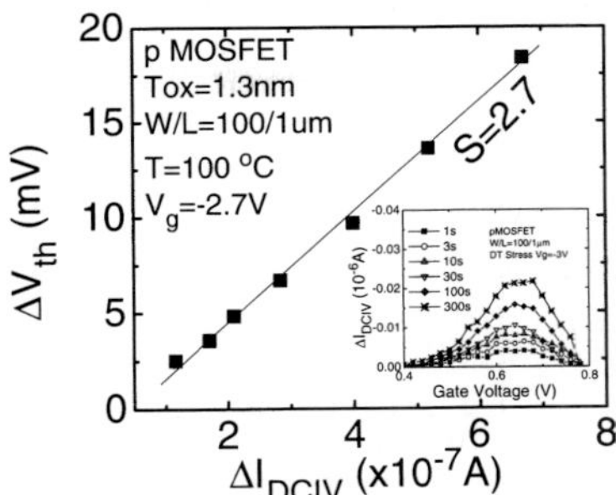

Fig. 2.   Correlation between DCIV current and threshold voltage changes. Interface trap generation plays a major role in NBTI in SiO_2/MOSFET. The inset shows typical DCIV spectra for 1.3 nm gate oxide p-MOSFET.

Under such dynamic stress conditions, it is found that interface trap generation and passivation occur mainly near the drain-channel corner therefore, devices with short channels benefit more from the EP effect than those with long channels, as shown in Fig. 5.

The frequency dependence of dynamic NBTI was studied using Fig. 6 in the range of 0.5 to 100 kHz. As can be seen in the figure, the dynamic NBTI (DNBTI) is frequency-independent in this frequency range with a significantly lower degree of degradation than static NBTI (SNBTI). V_{th} degradations under the same stress condition but different temperatures were plotted in Fig. 7. It is shown that although both SNBTI and DNBTI degradations are small at low temperatures, the EP effect is evident even at room temperature, indicating that the EP is not due to the thermal activation effect. p- and n-MOSFET degradations for different oxide thicknesses were also compared. Figure 8 shows that for thinner gate oxides, the passivation effect in DNBTI is larger.

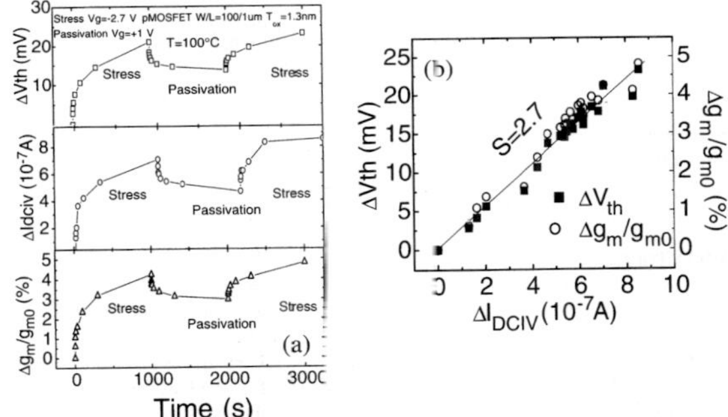

Fig. 3.   (a) Variations of V_{th}, DCIV current and g_m under stress-passivation-stress process. (b) Correlation between V_{th} shift, DCIV current and g_m variations in (a). The ΔV_{th}–ΔI_{DCIV} slope is exactly the same as in Fig. 2.

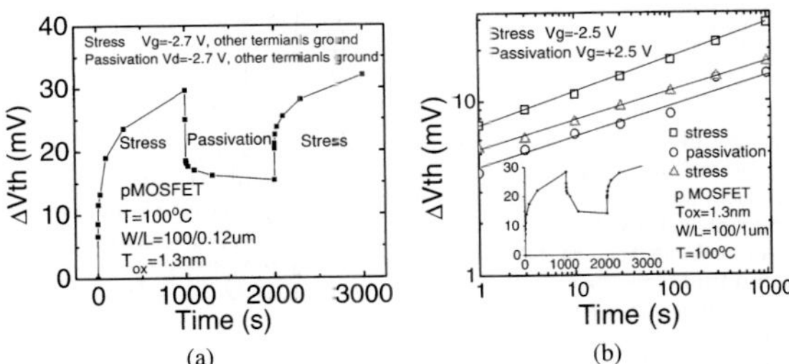

Fig. 4.   Illustration of dynamic NBTI, during alternate stress and passivation operation of a p-SiO_2/MOSFET in a digital CMOS inverter. Due to the EP effect, the conventional static NBTI overestimates the degradation. (a) Linear plot. (b) log-log plot indicates power law time evolution with a 0.25 slope for both stress and passivation phases.

Jpn. J. Appl. Phys., Vol. 43, No. 11B (2004)

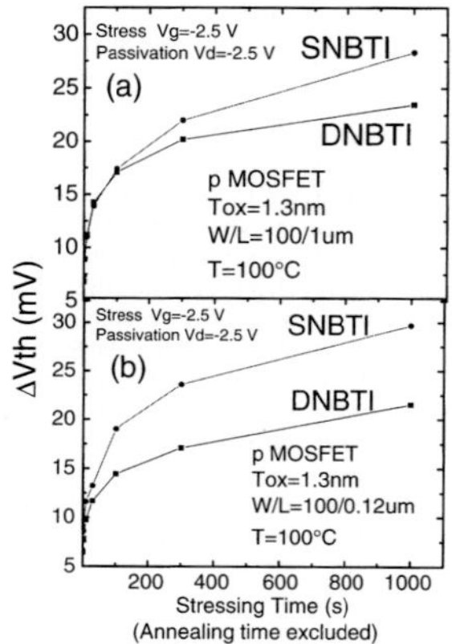

Fig. 5. Comparison between SNBTI and DNBTI for (a) long-channel and (b) short-channel p-MOSFETs. For DNBTI, the stressing time is half the nominal stress time (operation time), given a 50% duty cycle. The DNBTI stress frequency ∼0.5 Hz.

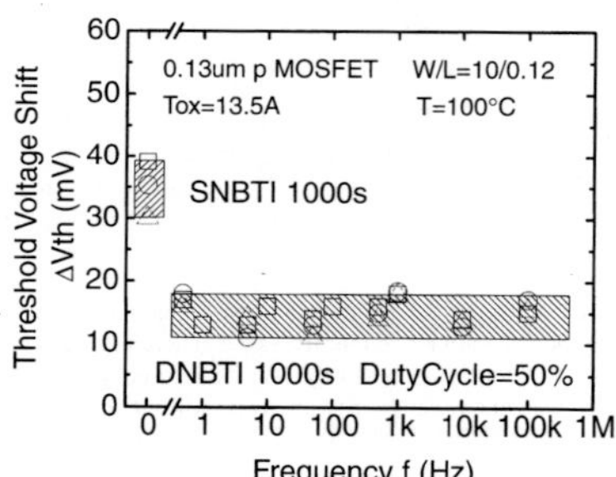

Fig. 6. The frequency dependency of dynamic NBTI for p-SiO$_2$/MOSFETs, with the same stress and passivation voltages as those in Fig. 5. Different symbols are used for denoting different samples.

2.4 Device lifetime

Because of the significant EP effect of interface traps on p-MOSFET operating in a CMOS inverter, p-MOSFET lifetime under DNBTI stress can be much longer than that projected based on conventional SNBTI stress data. The lifetime projections for both SNBTI and DNBTI were compared and 10-year lifetime operation voltages were extracted and shown in Fig. 9. It is shown that the operation voltages for a 10 year lifetime V_{10Y} are 0.9 V for SNBTI degradation and 1.2 V for DNBTI degradation when a 30 mV V_{th} shift is used as a failure criterion. Under the same stress voltage, the lifetime predicted by DNBTI is almost one order of magnitude longer than that predicted by SNBTI.

M. F. Li *et al.*   7809

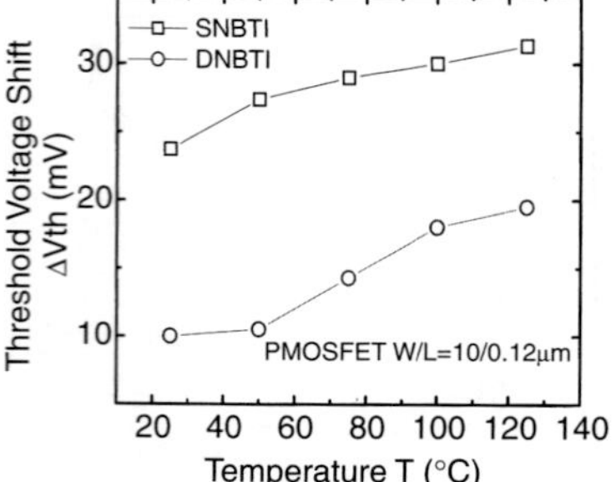

Fig. 7. Temperature dependency for SNBTI and DNBTI for SiO$_2$/MOSFETs after 1000 s stress, at the same stress voltage as that in Fig. 5.

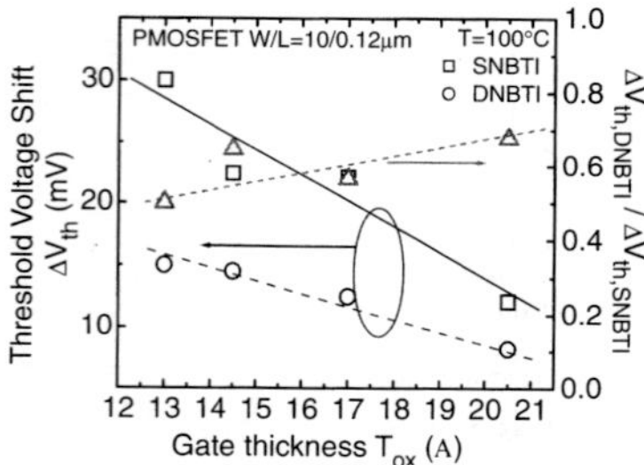

Fig. 8. Oxide-thickness dependency for SNBTI and DNBTI after 1000 s stress, at the same stress voltage as that in Fig. 5 for p-SiO$_2$/MOSFETs.

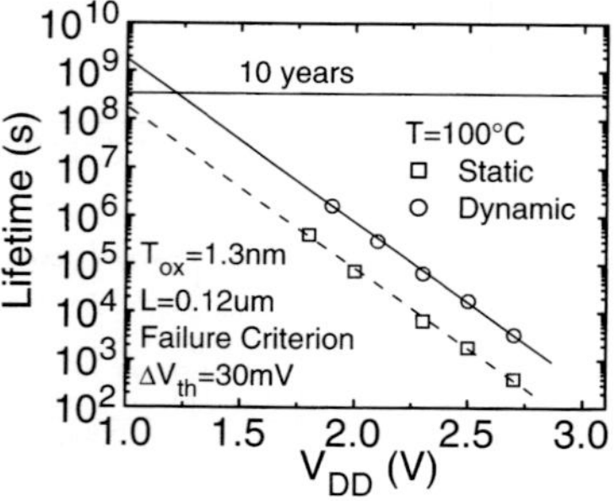

Fig. 9. SNBTI and DNBTI lifetime (30 mV V_{th} shift) projections for p-SiO$_2$/MOSFETs. The projected 10 year lifetime operation voltages V_{10Y} are 1.2 V for DNBTI stress and 0.9 V for SNBTI stress which overestimates degradation in real digital operation.

7810   Jpn. J. Appl. Phys., Vol. 43, No. 11B (2004)

M. F. Li *et al.*

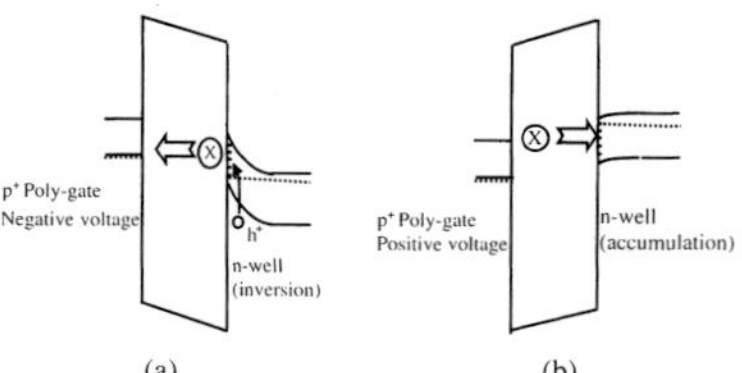

Fig. 10. (a) N_{it} formation and hydrogen diffusion towards the gate electrode during negative gate bias stressing. (b) Hydrogen species returning to the interface and N_{it} passivation during positive gate bias. X stands for hydrogen species.

2.5 Modeling of dynamic NBTI

As first proposed in 1965 by Balk at IBM,[17] and now widely accepted,[18–20] interface traps are related to Si dangling bonds when hydrogen is released from a Si–H bond. Sah *et al.* proposed in 1983 the reverse process, that is, the passivation of interface traps by the absorption of hydrogen at a dangling bond site.[21] The excellent correlations among ΔV_{th}, Δg_m and ΔN_{it} (ΔI_{DCIV}) under NBTI stress, as shown in Figs. 2 and 3, indicate that for ultrathin-gate-oxide devices, device degradation is mainly caused by interface trap generation, ΔN_{it}. On the basis of interface trap generation and passivation mechanisms proposed in refs. 17 and 21 and the recent reaction-diffusion (R-D) model of the electrochemical reaction of interface trap generation during NBTI degradation,[22,23] we propose the following reactions for interface trap generation and passivation in dynamic NBTI (eqs. (1) and (2), and Fig. 10)

$$(Si_3 \equiv Si\text{-}H) + h^+ \leftrightarrow Si_3 \equiv Si^* + X_{interface} \qquad (1)$$

$$X_{interface} \overset{diffusion}{\longleftrightarrow} X_{bulk} \qquad (2)$$

Here, $Si_3 \equiv Si\text{-}H$ is the precursor of the Si-H bond. When it interacts with a hole h^+ in the inversion layer or the source/drain extension region under the NBTI stress, the hole breaks the Si-H bond and generates an interface trap by releasing hydrogen species $X_{interface}$ at the Si/SiO$_2$ interface. The produced hydrogen species denoted as X in eq. (1), in the form of either molecules or neutral atoms or ions, will diffuse/drift to the gate electrode through the bulk gate oxide, as expressed by eq. (2) (Fig. 10(a)). In this process, the interface trap acts as a hydrogen source.

The EP effect is interpreted by the reverse interaction between N_{it} and hydrogen species, as shown in Fig. 10(b). When the gate bias polarity is reversed, i.e., from negative to positive, the channel inversion layer disappears and depletion layers are formed at the source/drain. The breaking of Si-H bonds stops due to lack of holes. On the other hand, the reverse reaction occurs when hydrogen species move back to the SiO$_2$/Si interface and passivate the Si dangling bonds, resulting in ΔN_{it} decrease. In this period, the interface trap acts as a hydrogen sink.

2.6 Recent development

Alam[24] has recently examined dynamic BTI problems using numerical and analytical solutions of hydrogen R-D

equations. His simulation results are in very good agreement with our experimental data shown in Figs. 3 and 5. His simulation also supports our result shown in Fig. 6, indicating that dynamic BTI degradation is frequency-independent or at best weakly frequency-dependent. Chakravarthi *et al.* have performed a similar simulation and they proposed that NBTI is dominated by the diffusion of neutral atomic and molecular-hydrogen-related defects.[25] However, different mechanisms of DNBTI degradations have been proposed by other groups. Huard *et al.* argued that DNBTI degradation in SiO$_2$ MOSFET is mainly caused by hole trapping/detrapping on pre-existing traps, and not by interface trap passivation by hydrogen atoms diffusing back to the interface.[26] Moreover, Abadeer and Ellis[12] and Chakravarthi[25] reported frequency-dependent dynamic BTI degradation. However their data show only a weak frequency-dependence. Other recent studies on dynamic NBTI in SiO$_2$ MOSFETs can be found in IRPS 2004.[27,28]

3. Dynamic BTI Degradation in n- and p-MOSFETs with HfO$_2$ Gate Dielectrics

3.1 Device fabrication

Both n- and p-MOSFETs with HfO$_2$ gate dielectrics and HfN metal gate electrodes were fabricated. HfO$_2$ gate dielectrics were grown in a MOCVD chamber. The details of the fabrication can be found in ref. 29. An equivalent oxide thickness (EOT) of $\sim$1.3 nm was obtained for both n- and p-MOSFETs using C–V curves with quantum mechanical correction.

3.2 Charge trapping and detrapping in dynamic BTI

V_{th} instability was first investigated under static stress for both n- and p-channel MOSFETs at room temperature (Fig. 11). Subthreshold swing (ss) was monitored during the stress to evaluate the correlation between V_{th} shift and interface state density. For both n- and p-transistors, we observed that ss does not show any observable change with stress time, indicating that interface state generation plays no significant role in V_{th} instability. This is also confirmed by the DCIV method which showed no significant increase in interface trap density under stress. These suggest that the BTI issue is primarily due to charge trapping in the bulk dielectric.[30–32] This observation is radically different from the case of SiO$_2$ dielectrics.

In the case of dynamic BTI, although HfO$_2$ dielectrics exhibit severe threshold voltage shifts after stress, a large portion of the V_{th} shift can be recovered during the electrical passivation phase due to charge detrapping, as shown in Fig. 12. Figure 12 also shows that for both n- and p-MOSFETs, there may be a fast component in trapping and detrapping, appearing as a sharp rising or dropping edge of the ΔV_{th} evolution in the initial stage of either the stress or passivation phase, respectively. The sharp rising or dropping edge is followed by a slow component in trapping or detrapping. Also, V_{th} shift in n-MOSFET is much greater than that in p-MOSFET.

As for the frequency dependence of BTI degradation, Fig. 13 shows that V_{th} shift is strongly dependent on the frequency of dynamic stress. For both n- and p-MOSFETs, BTI degradation under dynamic stress is improved as compared with that under static stress. This improvement

Jpn. J. Appl. Phys., Vol. 43, No. 11B (2004)   M. F. Li *et al.*   7811

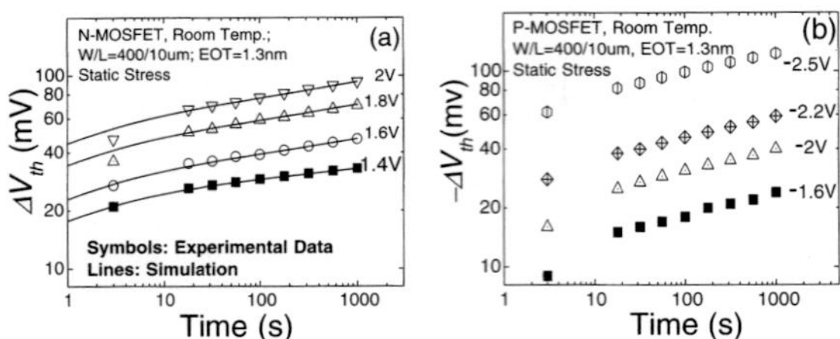

Fig. 11. Time dependence of V_{th} shift for (a) n-HfO$_2$/MOSFETs, and (b) p-HfO$_2$/MOSFETs, under various static inversion biases. There are two different slopes before and after 10 s stress, indicating two different types of traps with high and low trapping rates.

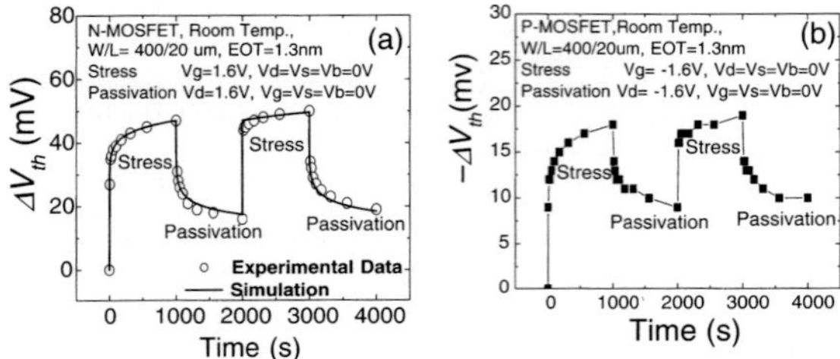

Fig. 12. Time evolution of V_{th} shift for (a) n-, and (b) p-HfO$_2$/MOSFETs, under dynamic stress with a duty cycle of 0.5. The V_{th} degradation/recovery is due to charge trapping and detrapping of two different types of traps in the HfO$_2$ gate dielectric, that is, fast and slow traps, with different capture and emission rates of carriers. Fast traps appear as a sharp rising or dropping edge of ΔV_{th} evolution in the initial stage of each phase. In Fig. 12(a), the simulation results for n-MOSFETs V_{th} degradation/recovery based on the model proposed in this work are shown.

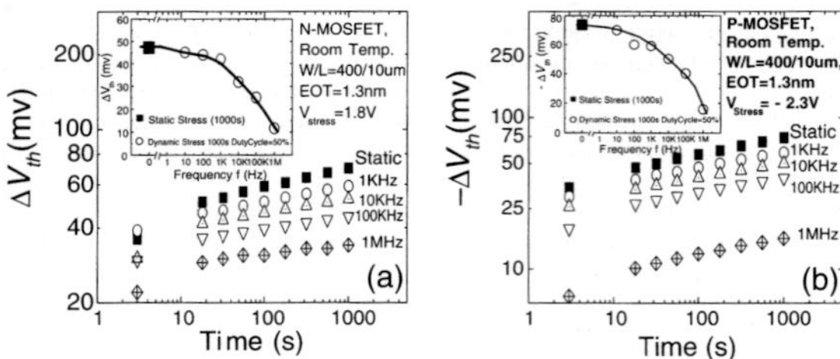

Fig. 13. V_{th} shift time evolution for (a) n-HfO$_2$/MOSFETs and (b) p-HfO$_2$/MOSFETs, under static and dynamic stresses of different frequencies. The insets of Figs. 13(a) and 13(b) show ΔV_{th} at the 1000th second stress time under both static and dynamic stresses of different frequencies for n- and p-MOSFETs, respectively.

is enhanced as the stress frequency is increased (up to 1 MHz as demonstrated in this work).[33] Rhee *et al.*[34] have reported similar frequency-dependent dynamic BTI phenomena for n-MOSFETs with PVD-grown HfO$_2$ gate dielectrics. In contrast, dynamic NBTI is almost frequency-independent in p-MOSFETs with SiO$_2$ gate dielectrics, as indicated in Fig. 6.

3.3 Device lifetime

The 10-year-lifetime projection based on ΔV_{th} is shown in Figs. 14 and 15 for various stress frequencies. As can be seen, device lifetime significantly increases with stress frequency at the same stress voltage. When $\Delta V_{th} = 30$ mV is used as the failure criterion, the operation voltages for a 10 year lifetime V_{10Y} are 0.8 V (and -0.9 V) for SNBTI

7812   Jpn. J. Appl. Phys., Vol. 43, No. 11B (2004)   M. F. Li *et al.*

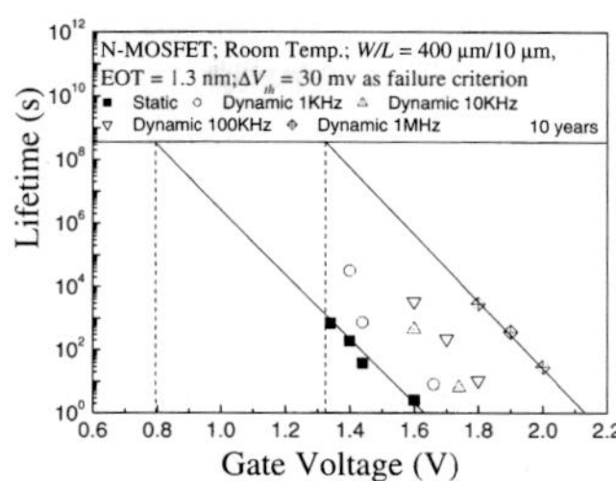

Fig. 14. 10 year lifetime projection for n-HfO$_2$/MOSFETs based on V_{th} shift. $\Delta V_{th} = 30$ mV is set as the device failure criterion. $V_{10year} = 0.8$ V for static BTI and 1.3 V for dynamic BTI at 1 MHz.

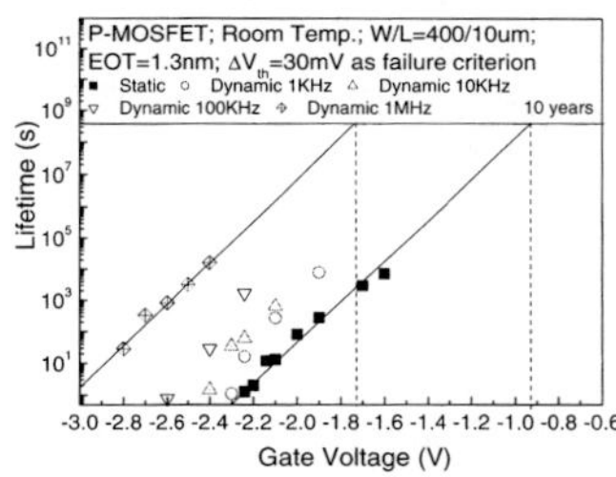

Fig. 15. 10 year lifetime projection for p-HfO$_2$/MOSFETs based on V_{th} shift. $\Delta V_{th} = 30$ mV is set as the device failure criterion. $V_{10year} = -0.9$ V for static BTI and -1.7 V for dynamic BTI at 1 MHz.

degradation for n- (and p-) MOSFETs and 1.3 V (and -1.7 V) for DNBTI degradation at 1 MHz. n-MOSFET has a shorter lifetime than p-MOSFET when the stress condition is the same.

3.4 Modeling frequency dependency of dynamic BTI

Here we will only consider n-MOSFETs. The approach can be easily extended to p-MOSFETs. To explain the frequency-dependence of V_{th} shift in dynamic BTI, we consider the number of trapped electrons Δn in the time interval Δt during one stress cycle in the dynamic BTI experiment, as shown in Fig. 16. When the stress cycle increases from T (as given $T = 1/f$, where f is the stress frequency) to 2T, the number of trapped electrons increases from Δn_T to Δn_{2T}. If the relationship between Δn and Δt is linear (curve L), then $\Delta n_{2T} = 2\Delta n_T$. Therefore, the number of trapped electrons during the same stress time would be the same for two frequencies and dynamic BTI would be frequency-independent. Because of the frequency-dependence observed in dynamic BTI, Δn_{2T} must be larger than $2\Delta n_T$, which means that the Δn-Δt relationship should be described by a concave curve (CC), as shown in Fig. 16. Based on the above-mentioned analysis, we propose a two-step procedure. Here, we assume that electron traps in the dielectric must be activated in the stress phase before it can

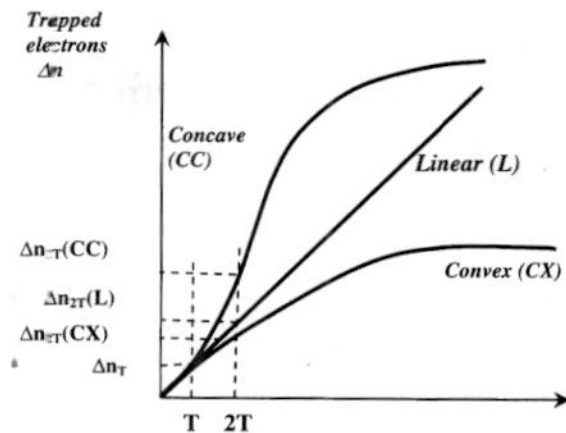

Fig. 16. Three possible cases of trapped electrons Δn vs stress time Δt in one cycle of stress phase. Only curve CC can explain the observed frequency dependence of dynamic BTI in HfO$_2$/MOSFETs.

capture electrons, and when the electric field is removed in the passivation phase, an empty trap can be deactivated. Thus we can describe the dynamics of trapping by eqs. (3) and (4), and detrapping by eqs. (5) and (6).

$$\frac{d\rho}{dt} = \frac{1}{\tau_A}(N - \rho);$$

trap activation during stress phase)   (3)

$$\frac{dn}{dt} = \frac{1}{\tau_C}(\rho - n);$$

(trapping procedure during stress phase)   (4)

$$\frac{d\rho}{dt} = \frac{-1}{\tau_D}(\rho - n);$$

(trap deactivation during passivation phase)   (5)

$$\frac{dn}{dt} = \frac{1}{\tau_E}n.$$

(detrapping procedure during passivation phase)   (6)

Where, τ_A, τ_C, τ_D and τ_E are the time constants for trap activation, capture (trapping), deactivation and emission (detrapping), respectively, n is the trapped-electron density, ρ is the active trap density, and N is the total (active and inactive) trap density. The deactivation time constant τ_D is so small that the empty traps are deactivated almost completely at each passivation phase and therefore the emptied activation trap density (ρ-n) is almost zero at the start of each stress phase and increases with time. This is the key finding in our study. The time evolution of V_{th} can be calculated for both trapping and detrapping by solving these equations. With the same trap distribution $N(\tau_C)$ used in the static stress simulation and adding the time constant of τ_E, τ_A and τ_D (Table I), eqs. (3)–(6) can simulate all static and dynamic time evolutions under different frequencies and different stress voltages. The simulation is in complete agreement with all experiment data, as shown in Figs. 12(a) and 17.

4. Conclusions

In this paper, we review our recent works on the dynamic bias temperature instability (BTI) in MOSFETs with ultra-thin SiO$_2$ and high-κ gate dielectrics, operating in a digital inverter circuit. For SiO$_2$ MOSFETs, the electric passivation effect of interface traps during a positive or zero bias was

Jpn. J. Appl. Phys., Vol. 43, No. 11B (2004)   M. F. Li *et al.*   7813

Table I. HfO$_2$ trapping, detrapping, activation and deactivation time constants used in calculations using eqs. (3)–(6).

Symbol	Description	Value		
$N(\tau_C)$	Distribution function of capture time constant τ_C	Sum of two log-normal distributions.		
		peak	40 µs	0.1 s
		width σ	3.1	3.5
τ_E	Emission time constant	$100 \cdot \tau_C$		
τ_A	Activation time constant	80 µs		
τ_D	Deactivation time constant	5 µs		

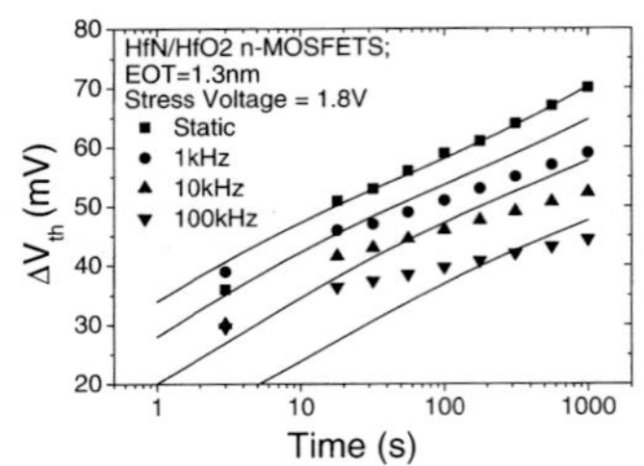

Fig. 17. Calculated time evolutions of V_{th} in n-HfO$_2$/MOSFET under static and dynamic stresses at different frequencies using eqs. (3)–(6) are plotted using lines, showing good agreement with experimental data (symbols).

demonstrated in p-MOSFET NBTI. In addition, V_{th} evolution satisfies the power law dependence on time with a power index of 0.25 at both stress and passivation phases. A physical model involving the interactions between hydrogen species and silicon dangling bonds is proposed to explain dynamic NBTI. It is shown that DNBTI suppresses p-MOSFET degradation and significantly increases MOSFET lifetime and 10 year operation voltage. This effect is oxide-thickness and channel-length dependent. The difference in suppressing degradation of devices between DNBTI and SNBTI will be significant only for ultrathin gate oxides. For 1.3 nm SiO$_2$ gate oxide p-MOSFETs with a channel length of 0.12 µm, using $\Delta V_{th} = 30$ mV as a failure criterion, the 10 year operation voltages V_{10year} are -0.9 V for SNBTI operation and -1.2 V for DNBTI operation.

For HfO$_2$ MOSFETs, we have shown that V_{th} evolution satisfies the power law dependence on time with an initial fast component followed by a slow component. The frequency dependence of dynamic BTI is observed in the experiment. For a dynamic stress of a given gate voltage amplitude, a decrease in the extent of BTI degradation for both n- and p-MOSFETs is observed with an increase of stress frequency. Accordingly, device lifetime significantly increases with an increase in stress frequency. For EOT $= 1.3$ nm HfO$_2$ gate dielectric and long channel (10 µm) devices, using $\Delta V_{th} = 30$ mV as a failure criterion, V_{10year} are 0.8 V and 1.3 V for static and 1 MHz dynamic BTI operations of an n-MOSFET, and -0.9 V and -1.7 V for static and 1 MHz dynamic BTI operations for that of a

p-MOSFET, respectively. A two-step model that accounts for carrier capture/emission and trap activation/deactivation in HfO$_2$ dielectrics under stress is proposed to explain frequency-dependent BTI. The calculation results show excellent agreement with all experimental data.

Acknowledgements

This work was supported by the Singapore A-STAR research grant R263-000-267-305. The authors wish to thank Dr. Ang Chew Hoe of Chartered Semiconductor for providing the CMOS devices with 1.3 nm SiO$_2$ gate oxide. M. F. Li would like to thank Dr. J. H. Stathis for helpful discussion on device lifetime projection.

1) N. Kimizuka, T. Yamamoto, T. Mogami, K. Yamaguchi, K. Imai and T. Horiuchi: *Dig. Tech. Pap. Symp. VLSI Technology 1999*, p. 73.
2) C. H. Liu, M. T. Lee, C. Y. Lin, J. Chen, K. Schruefer, J. Brighten, N. Rovedo, T. B. Hook, M. V. Khare, S. F. Huang, C. Wann, T. C. Chen and T. H. Ning: *Int. Electron Devices Meet. Tech. Dig. 2001*, p. 861.
3) T. Yamamoto, K. Uwasawa and T. Mogami: IEEE Trans. ED **46** (1999) 921.
4) A. Kerber, E. Cartier, L. Pantisano, M. Rosmeulen, R. Degraeve, T. Kauerauf, G. Groeseneken, H. E. Maes and U. Schwalke: *Int. Reliability Physics Symp. Proc. 2003*, p. 41.
5) S. Zafar, A. Callegari, E. Gusev and M. V. Fishetti: *Int. Electron Devices Meet. Tech. Dig. 2002*, p. 517.
6) K. Onishi, R. Choi, C. S. Kang, H. J. Cho, Y. H. Kim, R. E. Nieh, J. Han, S. A. Krishnan, M. S. Akbar and J. C. Lee: IEEE Trans. ED **50** (2003) 1517.
7) A. Shanware, M. R. Visokay, J. J. Chambers, A. L. P. Rotondaro, H. Bu, M. J. Bevan, R. Khamankar, S. Aur, P. E. Nicollian, J. McPherson and L. Colombo: *Int. Reliability Physics Symp. Proc. 2004*, p. 208.
8) A. Neugroschel, C. T. Sah, K. M. Han, M. S. Carroll, T. Nishida, J. T. Kavalieros and Y. Lu: IEEE Trans. ED **42** (1995) 1657.
9) C. T. Sah: *Proc. 2001 6th Int. Conf. Solid-State and Integrated Circuit Technology, Shanghai*, Vol. 1, p. 1.
10) G. Chen, M. F. Li, D. S. H. Chan, C. H. Ang, J. Z. Zheng and D. L. Kwong: IEEE Electron Device Lett. **23** (2002) 734.
11) G. Chen, K. Y. Chuah, M. F. Li, D. S. H. Chan, C. H. Ang, J. Z. Zheng, Y. Jin and D. L. Kwong: *Int. Reliability Physics Symp. Proc. 2003*, p. 196.
12) W. Abadeer and W. Ellis: *Int. Reliability Physics Symp. Proc. 2003*, p. 17.
13) S. Rangan, N. Mielke and E. C. C. Yeh: *Int. Electron Devices Meet. Tech. Dig. 2003*, p. 341.
14) S. Tsujikawa, T. Mine, K. Watanabe, Y. Shimamoto, R. Tsuchiya, K. Ohnishi, T. Onai, J. Yugami and S. Kimura: *Int. Reliability Physics Symp. Proc. 2003*, p. 183.
15) M. Ershov, R. Lindley, S. Saxena, A. Shibkov, S. Minehane, J. Babcock, S. Winters, H. Karbasi, T. Yamashita, P. Clifton and M. Redford: *Int. Reliability Physics Symp. Proc. 2003*, p. 606.
16) H. Usui, M. Kanno and T. Morikawa: *Int. Reliability Physics Symp. Proc. 2003*, p. 610.
17) P. Balk: Ext. Abstr. Electronics Division, Electrochemical Society Spring Meet., 1965, Vol. 14, p. 237.
18) C. T. Sah: *Fundamentals of Solid State Electronics, Solution Manual* (World Scientific, 1996) p. 101.
19) C. Hu, S. C. Tam, F. C. Hsu, P. K. Ko, T. Y. Chan and K. W. Terril: IEEE Trans. ED **32** (1985) 375.
20) E. H. Poindexter: J. Non-Cryst. Solids **187** (1995) 257.
21) C. T. Sah, J. Y. C. Sun and J. J. T. Tzou: J. Appl. Phys. **54** (1983) 5864.
22) S. Ogawa, M. Shimaya and N. Shiono: J. Appl. Phys. **77** (1995) 1137.
23) S. Ogawa and N. Shiono: Phys. Rev. B **51** (1995) 4218.
24) M. A. Alam: *Int. Electron Devices Meet. Tech. Dig. 2003*, p. 345.
25) S. Chakravarthi, A. T. Krishnan, V. Reddy, C. F. Machala and S. Krishnan: *Int. Reliability Physics Symp. Proc. 2004*, p. 273.

7814   Jpn. J. Appl. Phys., Vol. 43, No. 11B (2004)   M. F. Li *et al.*

26) V. Huard *et al.*: *Int. Reliability Physics Symp. Proc. 2004*, p. 40.
27) S. S. Tan, T. P. Chen, C. H. Ang and L. Chan: *Int. Reliability Physics Symp. Proc. 2004*, p. 35.
28) B. Zhu, J. S. Suehle and J. B. Bernstein: *Int. Reliability Physics Symp. Proc. 2004*, p. 689.
29) H. Y. Yu *et al.*: *Int. Electron Devices Meet. Tech. Dig. 2003*, p. 99.
30) A. Kerber, E. Cartier, L. Pantisano, M. Rosmeulen, R. Degraeve, T. Kauerauf, G. Groeseneken, H. E. Maes and U. Schwalke: *Int. Reliability Physics Symp. Proc. 2003*, p. 41.
31) S. Zafar, A. Callegari, E. Gusev and M. V. Fischetti: *Int. Electron Devices Meet. Tech. Dig. 2002*, p. 517.
32) A. Shanware, M. R. Visokay, J. J. Chambers, A. L. P Rotondaro, H. Bu, M. J. Bevan, R. Khamankar, S. Aur, P. E. Nicollian, J. McPherson and L. Colombo: *Int. Reliability Physics Symp. Proc. 2004*, p. 208.
33) C. Shen H. Y. Yu, X. P. Wang, M. F. Li, Y. C. Yeo, D. S. H. Chan, K. L. Bera and D. L. Kwong: *Int. Reliability Physics Symp. Proc. 2004*, p. 401.
34) S. J. Rhee, Y. H. Kim, C. Y. Kang, H. J. Cho, R. Cho, C. H. Cho, M. S. Akbar and J. C. Lee: *Int. Reliability Physics Symp. Proc. 2004*, p. 269.

Negative U Traps in HfO$_2$ Gate Dielectrics and Frequency Dependence of Dynamic BTI in MOSFETs

C.Shen[1,2], M.F.Li[1,2], X.P.Wang[1], H.Y.Yu[1], Y. P. Feng[3], A. T.-L. Lim[3], Y.C.Yeo[1], D.S.H.Chan[1], D.L.Kwong[4]

[1] Silicon Nano Device Lab (SNDL) , ECE Dept, National University of Singapore, Singapore 119260 Email: elelimf@nus.edu.sg
[2] Institute of Microelectronics , Singapore 117685 , [3] Physics Dept, National University of Singapore, Singapore 119260
[4] ECE Dept, University of Texas, Austin, TX 78712, USA

Abstract

We report for the first time the following new findings in charge trapping in HfO$_2$ gate dielectrics: 1) Using an ultra-fast electronic method to measure the MOSFET V_{th}, two different traps (fast and slow) are identified, each leading to charge trapping characteristics with different dependence on the frequency of dynamic stress. 2) First-principle calculation of oxygen vacancy related traps in HfO$_2$ reveals the negative U (–U) property of traps in the strongly ionized HfO$_2$ dielectric. Each trap can trap two electrons and lower the trap energy due to a large lattice relaxation. 3) A physic-based model for the frequency dependence of dynamic BTI is established. The BTI frequency dependence of the slow traps is explained by the –U property of these traps. Excellent agreement between the simulation and experimental data was achieved.

Introduction

HfO$_2$ has been demonstrated as one of the most promising high-K gate dielectrics (1). Recently, it was observed that charge trapping occurs in HfO$_2$ gate dielectric, leading to severe BTI (2-6). IMEC group recently developed a pulsed I_d-V_g method to measure V_{th} and extract charge trapping characteristics, demonstrating that the conventional DC measurement method underestimates the charge trapping effect as a result of fast de-trapping occurring during V_{th} measurement (2). We have improved the method as shown in Fig.1 to reduce the V_{th} measurement time t_M to $1\mu s$. Further reducing t_M to $0.1\mu s$ is possible if the transmission line effect is considered.

Experimental results

Both n- and p-MOSFETs with MOCVD HfO$_2$ gate dielectric (EOT=1.3 nm) and HfN/TaN metal gate stack were fabricated, with the process flow detailed in (7). V_{th} is measured using the falling or rising edge of a pulse as illustrated in Fig.2. Fig.3 shows the measured V_{th} shift (ΔV_{th}) after a fixed 1sec DC stress but with different measurement time t_m at falling edge of the stress voltage. The smaller ΔV_{th} obtained using a longer t_m is due to additional de-trapping during a longer t_m. Under static BTI stress, Fig.4 highlights the significantly larger V_{th} degradation when V_{th} is monitored by the fast measurement method, as compared to DC (3-6) measurement method.

For dynamic BTI stress, a rectangular pulse with 50% duty cycle and a fall and rise time of t_m is applied to the gate of the MOSFET. Fig.5 shows the transient response for very low frequency dynamic BTI stress. A large portion of the ΔV_{th} can be recovered during the electrical passivation phase as the result of charge de-trapping. The DC measurement (open symbols in Fig.5) allows all charges in the fast traps to de-trap during measurement and the measured charges are contributed only from traps that de-trap slowly (slow traps). Figs.6 &7 show the frequency dependence of dynamic BTI degradation results using the DC method, i.e. contributed by the slow traps. Increasing the stress frequency reduces the degradation.

On the other hand, in Fig.8, fast measurements are used to measure the total of fast and slow traps. No clear frequency dependence of the accumulation charge can be observed within the measurement error. We interpret this result as follows. The fast trap charge trapping is increased when the frequency is increased, and the frequency dependence effects of fast and slow traps compensates to each other when measured by the fast measurement. The results indicate that the slow and fast traps belong to two kinds of traps with different frequency dependency.

Negative U properties of traps in HfO$_2$

The idea of –U trap was proposed by Anderson (8). For a –U trap, capturing the first electron with trap energy E(-/0) leads to a meta-stable state. Capturing the second electron with trap energy E(--/-) leads to a stable state with a negative value of U=E(--/-)- E(-/0) , due to the lattice relaxation (9,10). –U traps are mostly found in strong ionic semiconductors (11) due to large lattice-electron interaction and recently also found in SiO$_2$ (12, 13). Since HfO$_2$ is a more ionic material than SiO$_2$, it is therefore reasonable to expect that certain traps in HfO$_2$ are –U traps. Very recently, Kang et al (14) showed that H in monoclinic HfO$_2$ is a –U trap, capturing two electrons with U=-1.6 eV. It is known that HfO$_2$ gate dielectric has a monoclinic or cubic structure by TEM analysis (15). In this work, we investigate the oxygen vacancy in cubic HfO$_2$ by first-principles calculation in a 23-atom super-cell of cubic HfO$_2$, as illustrated in Fig.9. The calculation is based on the density functional theory and generalized gradient approximation using the Perdew-Burke-Ernzerhof exchange-correlation potential (16). One oxygen atom is removed in the center of the tetrahedron to create the oxygen vacancy. Our calculation gave the required accuracy and well reproduced the bulk properties of HfO$_2$ reported elsewhere (17, 18).

30.6.1

Five different charge states for oxygen vacancy in cubic HfO_2, namely V^{++} ← (capture the second hole) V^+ ←(capture the first hole) V^O (capture the first electron) → V^- (capture the second electron) →V^{--} were investigated. The electron-lattice interaction will cause different lattice relaxations for different oxygen vacancy states which are illustrated in Fig.10. When the oxygen vacancy is in neutral V^O state, all 4 Hf atoms relax inward by 0.029 Å, as shown in Fig. 10a. Removing one or two electrons from V^O results in the formation of V^+ or V^{++} states. The 4 Hf atoms relax outward by 0.036 Å for V^+ or 0.094 Å for V^{++} state, as shown in Fig. 10b. In all the V^O, V^+ and V^{2+} charged states, the LR does not change the T_d^2 symmetry. Finally, adding one or two electrons to V^O results in formation of V^- or V^{--}. The symmetry changes from T_d^2 to C_{2c}^2, as shown in Fig. 10c. Two of the Hf atoms relax by more than 0.2 Å for V^- or 0.3 Å for V^{2-}, moving toward each other. Another pair of Hf atoms relax relatively less by 0.10 Å for V^- or 0.11 Å for V^{--}, moving apart from each other. The lattice relaxation reduces the trap energy to a minimum.

The energies of the five charge states under lattice relaxation are calculated and illustrated in Fig. 11. Due to the −U property of the trap, it is energetically more favorable to have two electrons (holes) trapped at this defect. When the electrons are injected into HfO_2 of the n-MOSFET (Fig.12), V^O will first capture one electron, which is meta-stable. Therefore it subsequently captures the second electron. Conversely, when the holes are injected into the HfO_2 in the p-MOSFET (Fig.12), V^O will first capture one hole (release electron from trap state E(+/0), and subsequently captures the second hole (release electron from trap state E(++/+)) . The capturing of two holes by −U traps of oxygen vacancy in SiO2 was demonstrated by Chadi (13). The calculation also showed that the total energy of two neutral oxygen vacancy traps $2V^O$ is lower than the total energy of V^{++} + V^{--} . It indicates that if no electron or hole injection, all oxygen vacancy traps keep neutral.

Modeling of frequency dependent BTI degradation

Specifically, we analyze the case of the n-MOSFET. The power-law dependence in $\Delta V_{th} \sim t$ can be obtained by assuming that the electron traps have a distribution $N(\tau_C)$ in the trapping time constant domain (3). In Fig.13, if the number of electrons being trapped during time Δt in one cycle (stress and passivation) in the dynamic stressing experiment is Δn, then the frequency dependence of BTI degradation can be explained by the $\Delta n \sim \Delta t$ curve in three different cases. **For the slow traps**, the curve must have a concave-up shape as explained in Fig.13. This can be interpreted using the −U property of traps. The rate equations describing the trapping and de-trapping behavior are shown in tables I and II. Solving equations (1)-(4) in table I yields the static and dynamic time evolutions under different frequencies, with the time constants listed in Table II. Since the de-trapping of the first electron is fast, almost all traps with only one electron

emit that electron when the stress is removed during the passivation phase. Only those traps with two captured electrons retain the electrons. Therefore, only n_2 is responsible for the cumulative V_{th} shift in measurement. On the other hand, at each initial time of the stress phase [$\Delta t = 0$ in Fig.13] n_1 is almost zero and then increases versus time. Using this initial condition, the solution of n_2 from eq. (2) gives a concave-up curve, and therefore, the cumulative increase of n_2 is reduced when the frequency increases. Results of calculations using eq. (1)-(4) are in excellent agreement with all DC experimental data under static and dynamic stress, as shown in Figs. 14 &15. For p-MOSFET, we can use similar but complementary description to that for n-MOSFET, using the -U trap property of capturing two holes as illustrated in Fig.10. **For the fast traps**, the curve must have a concave-down shape as explained in Fig.13. This can be obtained by normal trap first order trapping (3, 19) and de-trapping equations.

Conclusions

An ultra-fast method is developed to measure the MOSFET V_{th}. Fast and slow traps with different frequency dependence of charge trapping were found in HfO_2. When the dynamic stress frequency is increased, charge trapping by slow traps is reduced while that by fast traps is increased. The frequency dependency of slow-trap-charge is interpreted by a set of two-step equations, based on the −U property of slow traps, with excellent agreement with all experimental results. The −U property of oxygen vacancy in cubic HfO_2 is confirmed by first-principle calculation. The fast traps are conventional traps and the frequency dependency of charge trapping can be explained by conventional first order trapping de-trapping equations.

Acknowledgement: This work was supported by Singapore A*STAR R263-000-267-305 grant. MF Li gratefully acknowledge invaluable discussion with Prof. P.Y.Yu of UC Berkeley on −U traps, and the HfO_2 TEM structural information provided by Chih-Hang Tung of IME, Singapore.

References

(1) S.Datta et al, IEDM2003, p.653. (2) A. Kerber et al, IRPS 2003, p.41 (3) S. Zafar et al, IEDM 2002, p.517. (4) K. Onishi *et al*, TED 50, p.1517 (2003). (5) C. Shen et al, IRPS2004, p.601. (6) S. J. Rhee et al, IRPS 2004, p.269. (7) H.Y.Yu et al, IEDM 2003, p.99. (8) P.W.Anderson, PRL 34, p.953 (1975). (9) Baraff et al, PR. B21, p.5662 (1980). (10) M.F.Li, Modern Semiconductor Quantum Physics, World Scientific (1994) , p.316. (11) D. J. Chadi, 22nd Int. Conf. on the Physics of Semiconductors, p.2331. (12) P.E.Blochl et al, PRL 83, p.372 (1999). (13) D.J.Chadi, APL 83, p.437 (2003). (14) J. Kang et al, APL 84, p.3894 (2004). (15) J.Aarik et al, Apply Surface Sci. 173, p.15 (2001). (16) J.P. Perdew et al, PRL 77, p.3865 (1996). (17) A. S. Foster et al, PR. B 65, 174117 (2002). (18) Ximyuan Zhao et al, PR. B 65, 233106 (2002). (19) A.Shanware et al, IRPS 2003, p.208.

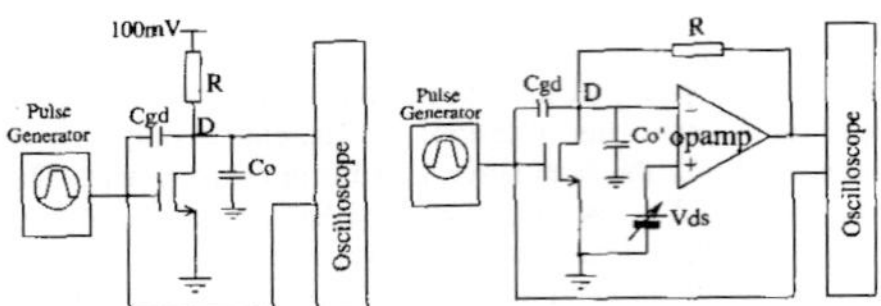

Fig.1 (a) The pulsed I_d-V_g measurement developed by IMEC (2). The measurement speed is limited by the charging current of C_{gd} and C_O, where C_O includes C_{ds} of the MOSFET, the co-axial cable capacitance (~100pF), and the input capacitance of the oscilloscope (10-20 pF), (b) The improved method used in this work . Voltage at D is fixed to V_{ds} due to the virtue ground principle of the Op.Amp , and there is no charging current through C_O'. The charging current through C_{gd} can be deducted when C_{gd}-V_g is known.

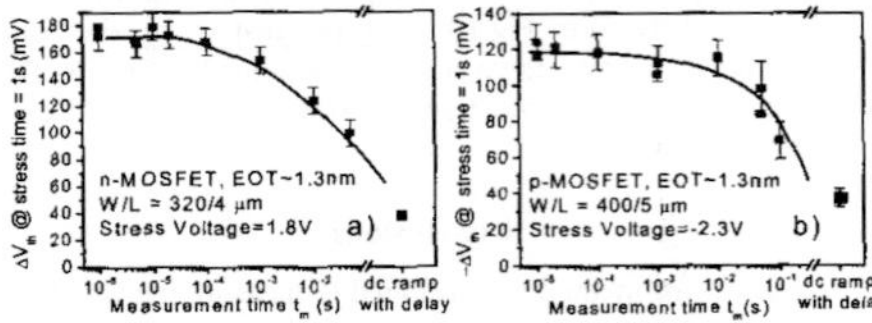

Fig. 3 Measured ΔV_{th} after 1 sec stress, using different pulse fall time t_m as illustrated in Fig.2. Reduction of ΔV_{th} with increasing t_m is due to additional de-trapping during a longer t_m. ΔV_{th} measured by the conventional DC measurement [3-6] only detects the charge trapped by slow traps (slow de-trapping) as described in this work.

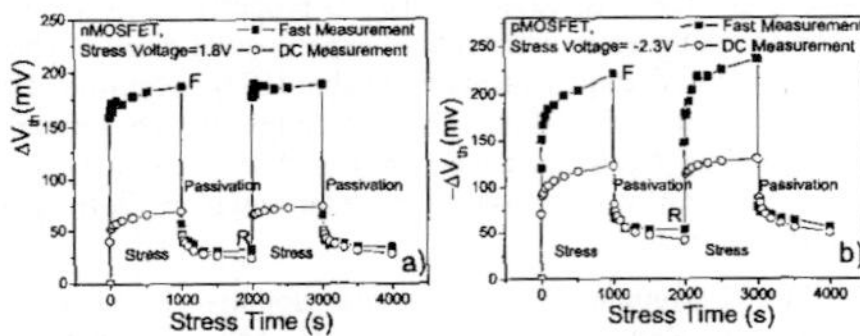

Fig. 5 ΔV_{th} transient variation under dynamic stress with f=0.0005 Hz, measured by DC (open circles) and fast measurement (t_m=5µs) (solid squares). DC measurement only measures the slow-trap-charge.

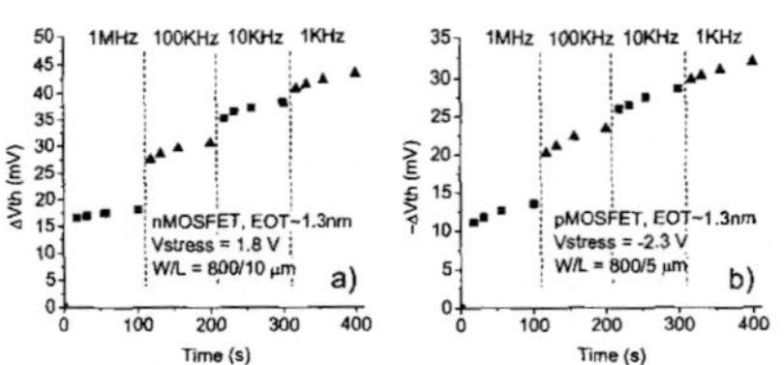

Fig.7 One sample is stressed by different frequencies in sequence, the stress time at each frequency is 100s, ΔV_{th} is measured by the DC method for (a) n-MOSFET, and (b) p-MOSFET. The frequency dependency of BTI degradation of slow-trap-charge is clearly shown, and is in consistent with the results in Fig.6

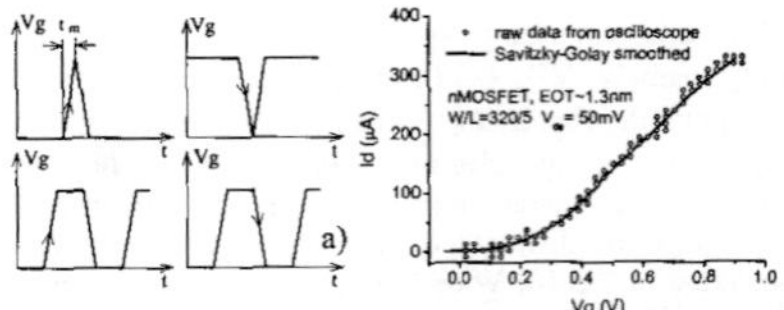

Fig. 2 (a) Various V_g waveforms used for sensing V_{th}, during static and dynamic stress. V_{th} can be measured at falling (or rising) edge of the pulse. (b) The measured Id-Vg curve using falling edge t_m =1 µs. Shorter t_m induces larger noise and error.

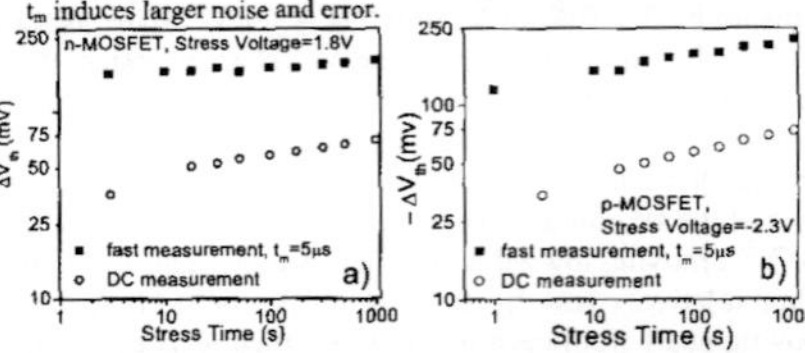

Fig. 4 Time evolution of ΔV_{th} under static stress. DC measurement measured the slow-trap-charge only and the fast measurement (fall edge t_m=5µs) measured the total of fast and slow trap-charge. Both curves follow the power law.

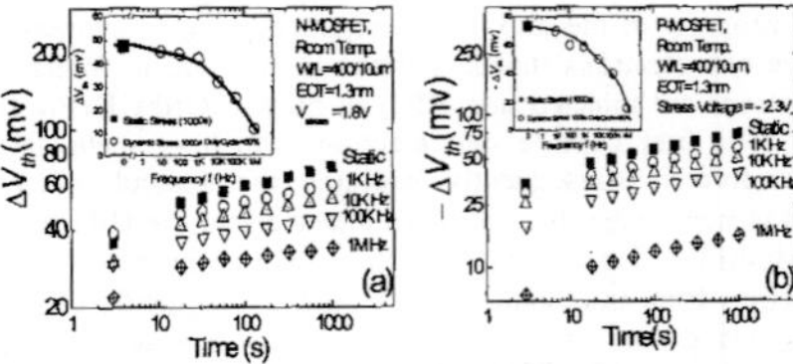

Fig. 6 ΔV_{th} time evolution for (a) n-MOSFETs, and (b) p-MOSFETs, under static and dynamic stresses of different frequencies, measured by the DC method, the measured ΔV_t is due to slow traps. The insets show the ΔV_{th} at the 1000th second, stressed under both static and dynamic stress of different frequencies. With increasing stress frequency, the BTI degradation is reduced.

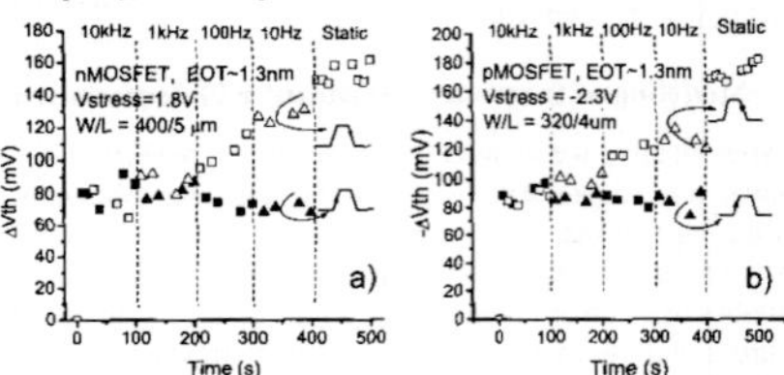

Fig. 8 Same as Fig.7, but measured by the fast measurement method (t_m=5µs). Solid symbols are obtained from measurement at rising edge of the pulse to measure the accumulative charge of the total of fast and slow traps under dynamic stress. Open symbols are obtained from measurement at the falling edge of the pulse. The difference between solid and open symbols reflects the amplitude of transient response in Fig.5 (R and F in Fig.5). It is the V_{th} shift at R point that reflects accumulative charge (solid symbols) under dynamic stress, which shows no clear frequency dependence.

30.6.3

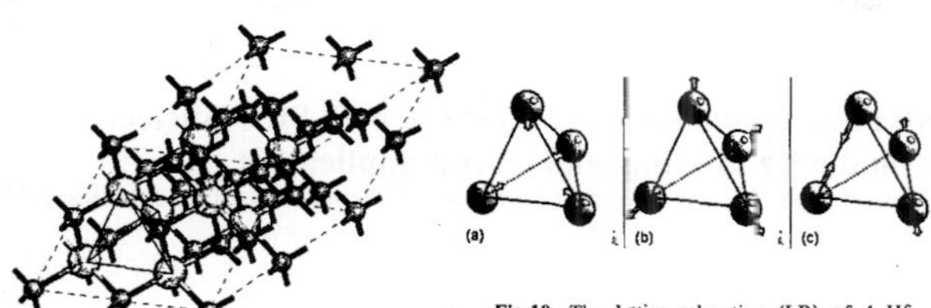

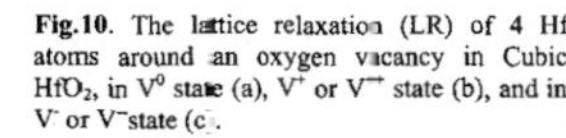

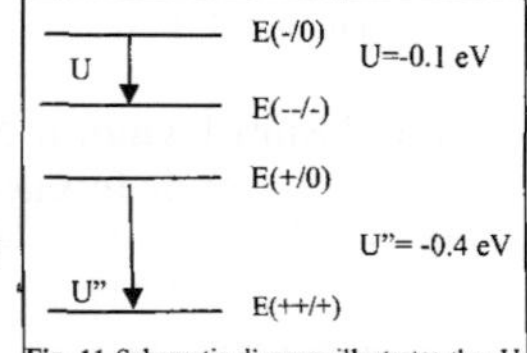

Fig .9 The 23-atom super-cell of the cubic HfO$_2$ used in the first-principles calculation.

Fig.10. The lattice relaxation (LR) of 4 Hf atoms around an oxygen vacancy in Cubic HfO$_2$, in V^0 state (a), V$^+$ or V^{++} state (b), and in V$^-$ or V^{--}state (c).

Fig. 11 Schematic diagram illustrates the -U property obtained from the first principles calculation of oxygen vacancy in cubic HfO$_2$. A neutral vacancy captures one electron at energy level E(-/0) and captures the second electron at E(--/-) with U=E(--/-) -E(-/0)= -0.1 eV. An V^{++} capture one electron at E(+/++) and capture the second electron at E(0/+) with U"=E(++/+)-E(+/0)=-0.4 eV.

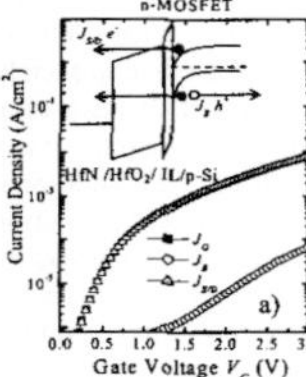

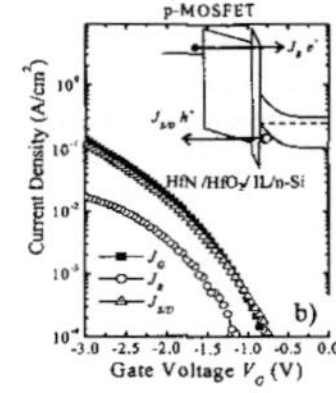

Fig. 12 (Left) Carrier separation results under gate bias stressing: **(a)** n-MOSFETs. $J_{S/D}$ measured the conduction electron tunneling and J_B measured the valence electron tunneling (see inset) , **(b)** p-MOSFETs, J_B measured the electron tunneling from the gate, $J_{S/D}$ measured hole tunneling from Si substrate (see inset). In n (p) MOSFET, electron (hole) current is the predominant injection current under stress.

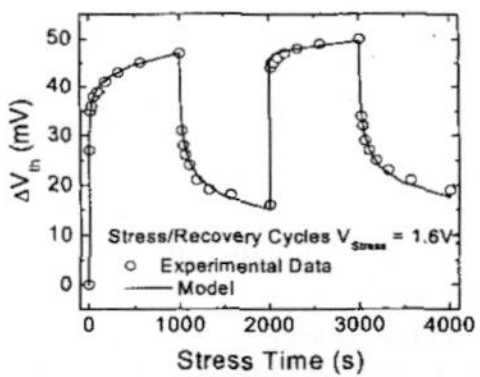

Fig.14 Comparison between the experimental data of transient ΔV_{th} measured by the DC method and the simulation results by eqs (1-4) in tables I and II (for n-MOSFET). DC measurement time of 0.5 sec is considered in the simulation.

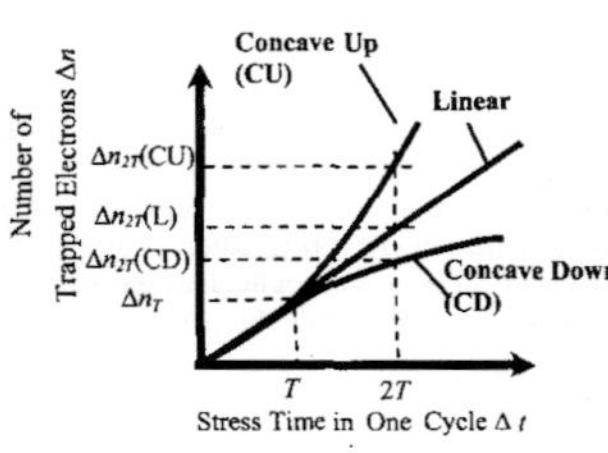

Fig.13 (Left) Three possible cases for the relationship between the number of trapped electrons Δn versus stress time Δt in one cycle of the dynamic stress. When Δt increases from T (where $T = 1/f$, and f is the stress frequency) to $2T$, the number of trapped electrons increases from Δn_T to Δn_{2T}. If $\Delta n \sim \Delta t$ is linear, then $\Delta n_{2T} = 2\Delta n_T$. Therefore, the number of trapped electrons during the same stressing time would be the same for two frequencies and the accumulative V_{th} shift would be frequency independent. When Δn_{2T} is larger (concave up) or smaller (concave down) than $2\Delta n_T$, the dynamic degradation is reduced or increased when the stress frequency is increased.

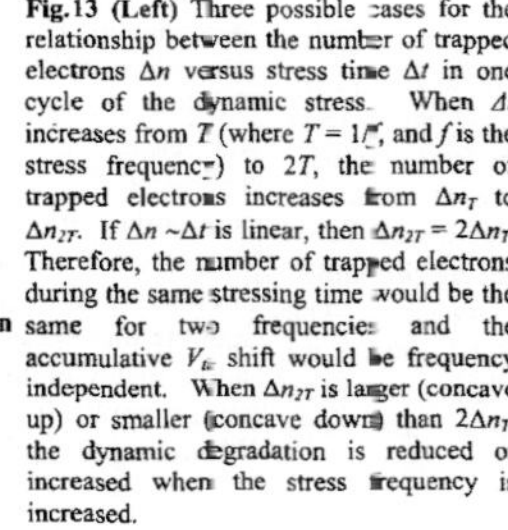

Table I. Equations (1-4) illustrate the trapping and de-trapping processes of -U traps (slow traps) in an n-MOSFET

$$\frac{dn_1}{dt}=\frac{1}{\tau_{C1}}(N-n_1-n_2)-\frac{dn_2}{dt} \quad \text{(1)} \quad \text{Capturing the first electron during stress}$$

$$\frac{dn_2}{dt}=\frac{1}{\tau_{C2}}n_1 \quad \text{(2)} \quad \text{Capturing the second electron during stress}$$

$$\frac{dn_1}{dt}=-\frac{1}{\tau_{E1}}n_1-\frac{dn_2}{dt} \quad \text{(3)} \quad \text{Emitting the first electron during passivation}$$

$$\frac{dn_2}{dt}=-\frac{1}{\tau_{E2}}n_2 \quad \text{(4)} \quad \text{Emitting the second electron during passivation}$$

N: total trap concentration, n_1: concentration of traps occupied by one electron, n_2: concentration of traps occupied by two electrons. Other parameters are defined in Table II. The de-trapping term in (3) does not appear in eq. (1). This is interpreted as follows. In a band diagram where the traps are mainly located at the high-k dielectric near to the SiO$_2$ interfacial layer, de-trapping is due to electron tunneling from the trap to the Si substrate through the SiO$_2$ interfacial layer. When a positive gate voltage is applied in the stress phase, the one-electron trap energy in the high-k dielectric moves into energy gap region of the Si substrate, and tunneling becomes forbidden.

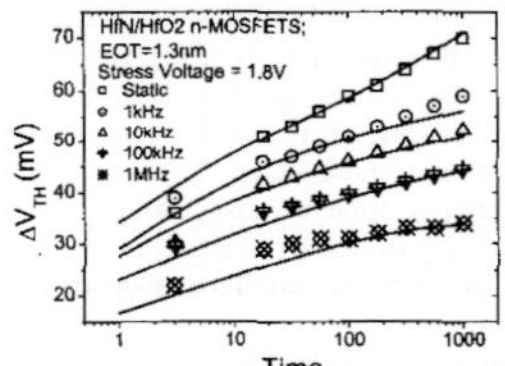

Fig.15 Same as Fig.14, but for time evolution of ΔV_{th} under static and dynamic stress with different frequency.

Table II. Definitions and values of parameters used in Table I

Symbol	Description	Value
$N(\tau_C)$	Distribution function of capture time constant τ_{C2} of the second electron. It is given by a sum of two log-normal distributions.	Two log-normal pdf: peak 40 µs 0.1 s width σ 3.1 3.5
τ_{E2}	Emission time constant of the second electron	$100\cdot\tau_{C2}$
τ_{C1}	Capture time constant of the first electron	$10^{-5}\cdot\tau_{C2}$
τ_{E1}	Emission time constant of the first electron	0.1 µs

30.6.4

Fast and Slow Dynamic NBTI components in p-MOSFET with SiON dielectric and their impact on device life-time and circuit application

T. Yang[1,2], M.F.Li[1,2], C.Shen[1], C.H.Ang[3], Chunxiang Zhu[1], Y.C.Yeo[1], G.Samudra[1], Subhash C. Rustagi[2], M.B.Yu[2], and D.L.Kwong[2,4]

[1]SNDL and CICFAR, ECE Dept, National University of Singapore, Singapore 119260,
Tel:65-68742559,Fax:65-67791103,Email:elelimf@nus.edu.sg
[2]Institute of Microelectronics, Singapore, 117685, [3]Chartered Semiconductor Manufacturing, Singapore, 738406
[4] Microelectronics Research Center, Dept. of ECE, The University of Texas, Austin, TX 78712,USA

Abstract: For the first time, we perform a systematic investigation of the fast and slow components of dynamic NBTI (DNBTI)in p-MOSFET with SiON gate dielectric. The new findings are: (1)The recent debate in the slow DNBTI component measured by conventional DC method [1-5] is clarified. We show evidence that the slow DNBTI is due to interface traps N_{it} generation and passivation. The conventional methods used over the past years seriously underestimate N_{it} due to passivation of N_{it} during measurement. (2) The fast DNBTI component measured by the fast method [7] is due to trapping and de-trapping of hole traps N_{ot} in SiON. The accumulative degradation increases with increasing stress frequency. Model simulations are in excellent agreement with all experiments. (3)We re-evaluate the impact of DNBTI on device lifetime and circuit applications in the light of this new finding.

Introduction: DNBTI is a critical reliability issue for CMOS transistors with silicon oxynitride SiON gate dielectric [1-6]. Using the fast measurement method developed in [7], the DNBTI degradation in ultra-thin SiON gate dielectrics is re-investigated.

Device fabrication and measurement: Devices were fabricated using 0.11 μm CMOS technology. The gate oxides with EOT of 1.3nm were grown by thermal oxidation followed by decoupled plasma nitridation and post-deposition thermal annealing. The fast V_{th} measurement method [7] is illustrated in Fig.1. Pulses are inserted in gate stress and I_d-V_g is measured at falling or rising edge of the pulse (Fig. 1 inset). Fig.2 shows that 50us measurement time is fast enough. Interface trap density is measured by an improved DCIV method[1].

What DC and fast methods measure in DNBTI: Fig.3 shows the DNBTI [1,7] response of a p-FET, measured by conventional DC and fast methods respectively. **DC method** :Figs.4-6 clearly demonstrate that the slow components in the DNBTI measured by the DC method is mainly due to generation and passivation of the interface traps N_{it} [1-4], and not due to trapping and de-trapping in SiON hole traps N_{ot}as proposed in [5]. Figs.4,5 show that the transient result depends on the order of measurements of N_{it} and V_{th} , indicating both measured N_{it} and ΔV_{th} are underestimated due to passivation of N_{it} during measurement. Fig.6 demonstrates the misleading interpretation by [5] in the passivation phase due to the wrong curve alignment. **Fast method**: In Fig.8, using the pulse falling and rising edges to measure points S$_f$ and P in Fig.3, the transient amplitude can be measured. In Fig.9, rising edge measures the accumulative degradation under DNBTI. Figs.10-12 show the static and dynamic NBTI measured by fast and DC methods respectively, showing different time, V_g and temperature dependences. The slope of time dependence of the fast method result is much smaller than the value of 0.25 anticipated in interface trap diffusion-reaction model [2]. Actually the fast NBTI component, defined as the difference between ΔV_{th} measured by fast and slow method, can be interpreted by trapping and de-trapping of pre-existing SiON hole traps N_{ot} to be explained as follows.

Does oxide traps affect the device lifetime and circuit applications: Fig.8 shows that the transient amplitude of ΔV_{th} under dynamic stress is reduced and approaches zero when the frequency is increased. Therefore no transient effect is expected in the fast digital circuit applications. As for the device lifetime, if the charge accumulated in N_{ot} in the stress phase is completely de-trapped in the passivation phase, there is no net accumulation and no effect on the DNBTI device life time. Otherwise there is a net charge accumulation and N_{ot} will contribute to additional DNBTI degradation. Fig.9 clearly shows that the later case is true. Fig.8 shows that fast DNBTI accumulated degradation (measured at P) is increased when increasing the frequency. Notice, however, that the additional degradation due to N_{ot} has a slower time evolution than the slow NBTI component (Fig.10). The longer stress time, the smaller effect of N_{ot} on the total NBTI degradation. Therefore N_{ot} only seriously affects the high voltage operation. When the device operates at low voltage with 10-year lifetime, the effect of N_{ot} degradation can be almost neglected. In this case, the operation voltage with a 10-year lifetime V_{10y}, estimated by the conventional DC measurement [1], is still appropriate (Fig.13). Fig.14 shows the ΔV_{th} evolution under a sine wave V_g stress in an analog circuit, measured by the fast method. Due to the exponential ΔV_{th}~V_g relation (Fig.11), ΔV_{th} and therefore $(V_g$-$V_{th})$ is no longer a sine wave. This introduces an additional nonlinear distortion in ultra-low frequency large signal analog applications.

Modeling for SiON hole traps: The fast NBTI component can be simulated by the following equations of trapping and de-trapping of the pre-existing hole traps N_{ot} [8]:

$$\frac{dp}{dt} = \frac{1}{\tau_C}(N_{ot} - p) - \frac{1}{\tau_{E1}}p \tag{1}$$

$$\frac{dp}{dt} = -\frac{1}{\tau_{E2}}p \tag{2}$$

where p is the hole concentration in N_{ot}. N_{ot} has wide spectrum of different trapping and de-trapping time constants τ_C and τ_E (Fig.8(b)). Using (1) & (2), the simulated time evolutions of fast NBTI degradations are in excellent agreement with all static and dynamic stress condition as shown in Figs 7,8,10.

Conclusions: We clarify that the debate in the slow dynamic NBTI component in p-MOSFETs with SiON gate dielectrics, measured by DC method, mainly due to generation and passivation of interface traps N_{it}. We report and model a new fast DNBTI component due to trapping and de-trapping of hole traps in SiON. The fast DNBTI component affects the device lifetime at high voltage and introduces non-linear distortion in large signal ultra-low frequency analog applications.

This work was supported by Singapore A*STAR grants R263-000-267-305 and R-398-000-019-305.

References [1]G.Chen et al, IRPS,p.196 (2003), [2]M.Alam, IEDM, p. 345 (2003), [3] S.Mahapatra et al, IEDM,p.105 (2004), [4] S.Charkravarthi et al., IRPS,p.273 (2004), [5]V.Huard et al, IRPS,p.40 (2004), [6]W.Abadeer et al, IRPS,p.17 (2003), [7] C.Shen et al, IEDM,p.733 (2004), [8]Y.Nissan-Cohen et al, JAP,v.58, p.2252 (1985), [9]H.Aono et al, IRPS, p.23 (2004)

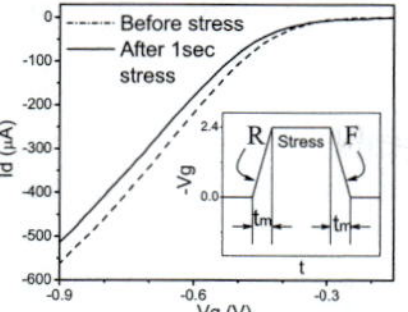

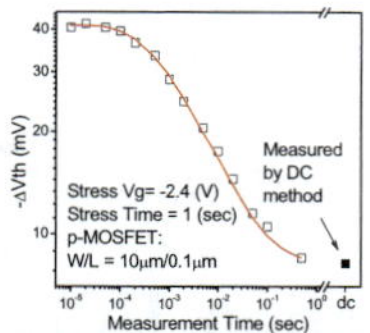

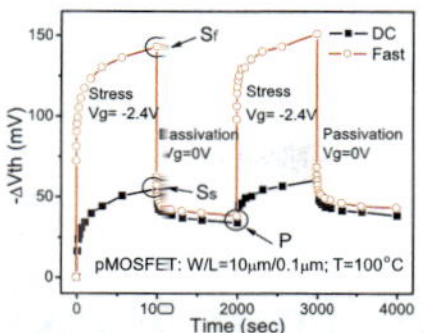

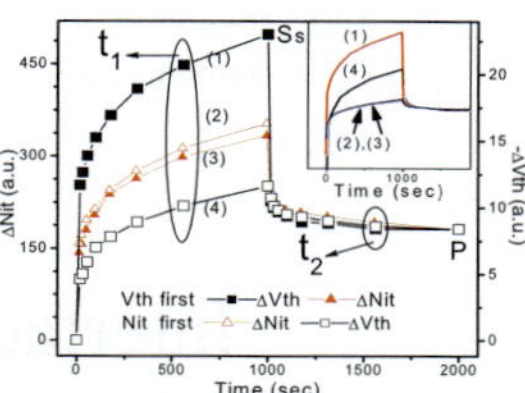

Fig.1 Fast measurement of I_d-V_g using the rising (R) or falling (F) edge of a V_g pulse [7]. Waveform with a negative bias stress (V_g= -2.4V) was used. Measurement time t_m may be varied.

Fig.2 ΔV_{th} as measured using different measurement time t_m (Fig.1 inset), after 1sec stress. ΔV_{th} measured using a conventional DC method is also shown .

Fig.3 ΔV_{th} under dynamic stress as measured by fast and DC methods. Sf and Ss are the end of stress phase measured by fast or DC method respectively; P is the end of the passivation phase.

Fig.4 The DNBTI measured by DC method can be explained mainly by generation and passivation of interface traps N_{it}. (Figs.4 & 5) Two characterization approaches: In the first, V_{th} was measured (curve 1) followed by N_{it} measurement (curve 3). In the second, N_{it} was measured (curve 2) followed by V_{th} measurement (curve 4). All curves are normalized at the P point, because no passivation of N_{it} during measurement at P so the true N_{it} and ΔV_{th} are measured (Fig.5 right). During stress phase, the measured ΔN_{it} and ΔV_{th} depends on the measurement sequence because N_{it} passivation during measurement (Fig.5 left). If there is no N_{it} passivation , only oxide charge de-trapping in the passivation phase as proposed in [5],curves 1-4 should be ordered as shown in the right upper inset.

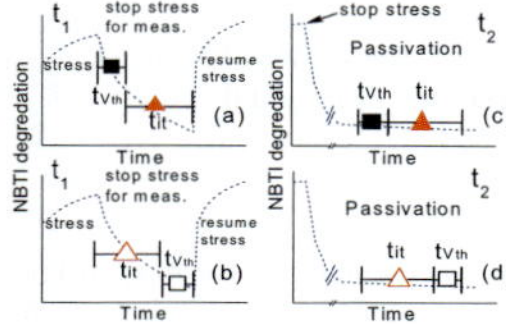

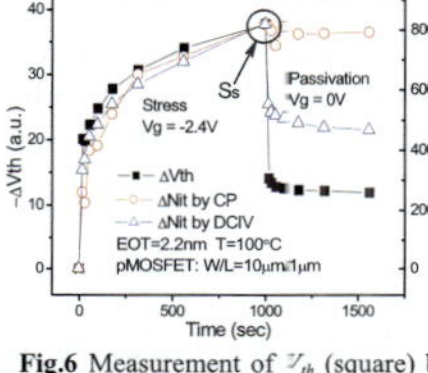

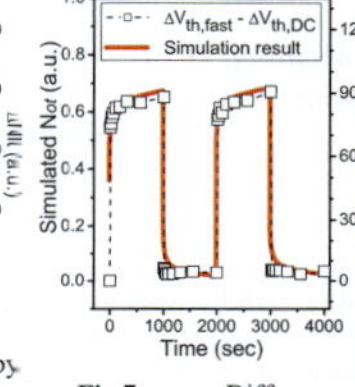

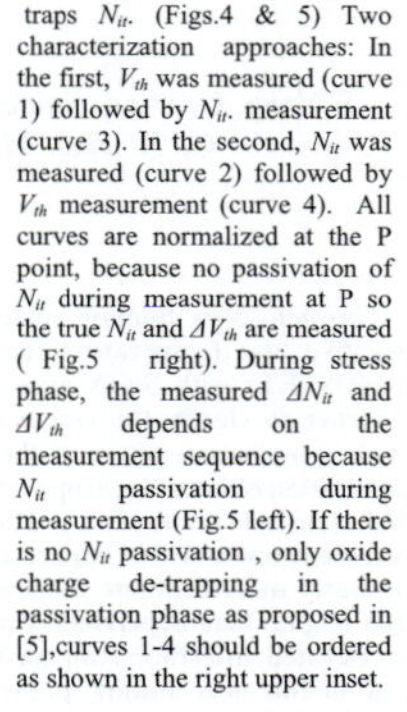

Fig.5 Zoom-in times t_1 and t_2 in Fig.4. Once stopping stress for measurements, NBTI degradation recovers. (a) ΔV_{th} first measured during t_{vth}, followed by ΔN_{it} measured during t_{Vth}~$t_{Vth}+t_{it}$. (b), the reverse sequence is used. Putting (a), (b) together explains the four data points at time t_1 in Fig.4. Note that N_{it} measurement is slower than V_{th} measurement. c) and d), explains data t_2 in Fig.4.

Fig.6 Measurement of V_{th} (square) by DC method followed by N_{it} using the CP and DCIV methods. Curves are aligned at point S_S [5]. Due to this wrong alignment, N_{it} is lifted up at the passivation phase. N_{it} is lifted more using CP measurement, consistent with Fig.1 of [5], due to the longer measurement time t_{it} compared to DCIV measurement.

Fig.7 Difference between ΔV_{th}'s from fast and DC methods reflect the ΔV_{th} component due to fast traps (square). The ΔV_{th} trend agrees well with simulated N_{ot} (solid line).

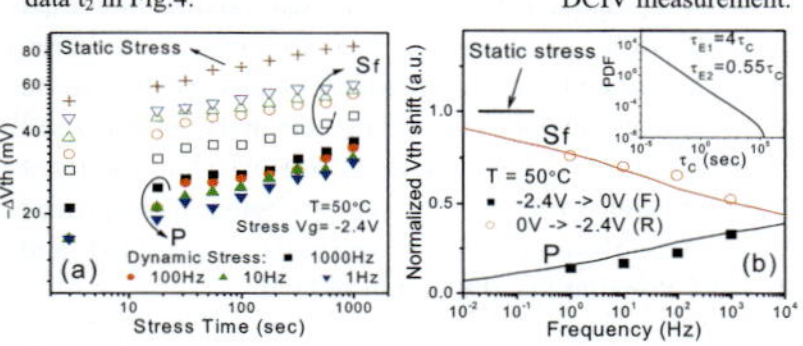

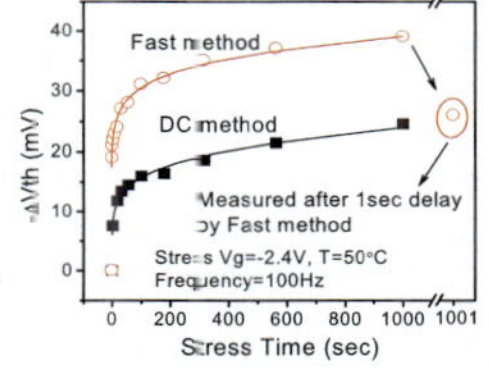

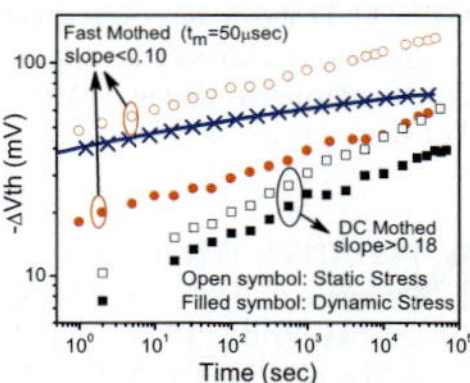

Fig.8 a) Frequency f dependence of ΔV_{th} due to DNBTI as measured by fast method. Falling (or Rising) edge measure S_f (or P) in Fig.3. When increasing f, the accumulated degradation P increases while the transient amplitude (S_f – P) decreases and approaches zero. b) Frequency dependence of the fast DNBTI component ($\Delta V_{th,fast}$ – $\Delta V_{th,slow}$ @1000sec). Simulated results are plotted using solid lines. The inset shows the spectrum of τ_C and τ_E, employed in eqs(1)(2).

Fig.9 Measured ΔV_{th} due to DNBTI using DC method and fast method (using R edge to measure the accumulation degradation). The point in the circle is the result of fast measurement stop at 1000 s and measured after 1 s delay. De-trapping of accumulated charge in the fast traps under DNBTI is observed when the stress is interrupted.

Fig.10 Measured ΔV_{th} under dynamic and static stresses by fast and DC methods. After 5×10^5 s stress with ΔV_{th} =130 (mV), no degradation enhancement [9] or saturation [10] was observed. The experiment fit data of $\Delta V_{th,fast}$ – $\Delta V_{th,slow}$ ("X" symbols) matches well with N_{ot} simulation result (line).

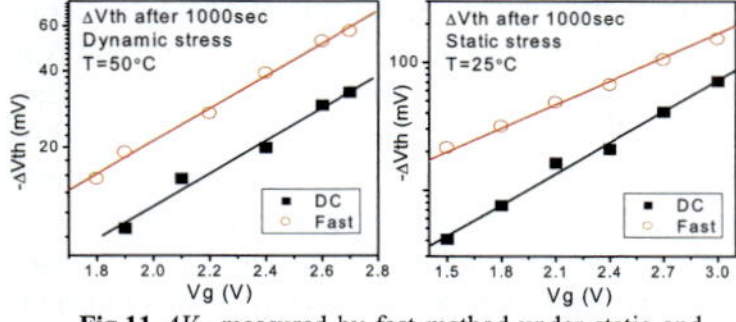

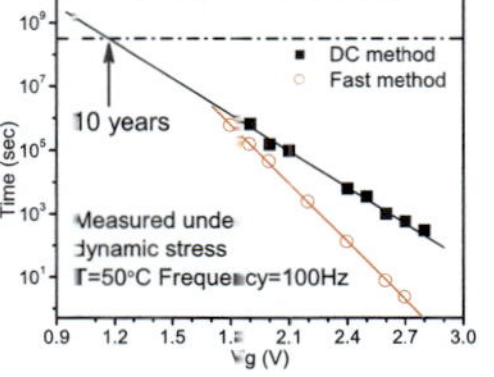

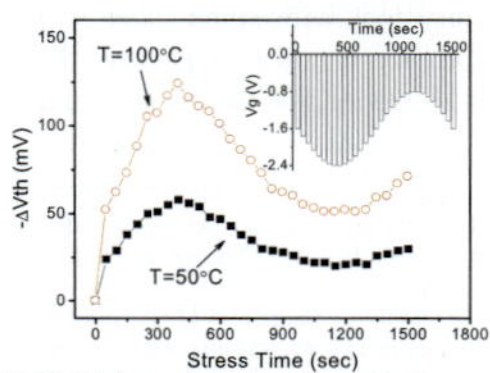

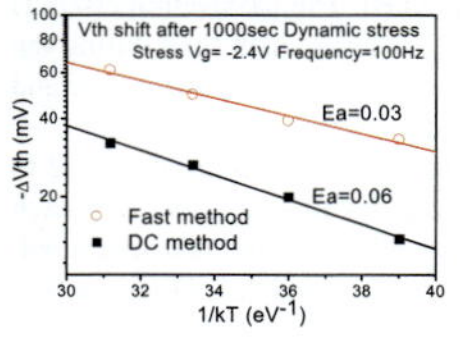

Fig.11 ΔV_{th} measured by fast method under static and dynamic stress shows exponential dependence on V_g.

Fig.12 Temperature dependence of DNBTI measured by fast and DC methods.

Fig.13 Device lifetime (ΔV_{th}=30mV as failure criterion) evaluated under dynamic stress. The DC method over-estimates the device lifetime at high stress voltage V_g.

Fig.14 Under a sine wave V_g stress input, the measured ΔV_{th} is not a sine function of time, due to exponential ΔV_{th}-V_g relationship (Fig.10). This would induce a non-linear signal distortion in ultra-low frequency large signal analog applications.

758   IEEE ELECTRON DEVICE LETTERS, VOL. 26, NO. 10, OCTOBER 2005

Interface Trap Passivation Effect in NBTI Measurement for p-MOSFET With SiON Gate Dielectric

T. Yang, C. Shen, *Student Member, IEEE*, M. F. Li, *Senior Member, IEEE*, C. H. Ang, *Senior Member, IEEE*, C. X. Zhu, *Member, IEEE*, Y.-C. Yeo, *Member, IEEE*, G. Samudra, *Member, IEEE*, and D.-L. Kwong, *Senior Member, IEEE*

Abstract—New findings of interface trap passivation effect in negative bias temperature instability (NBTI) measurement for p-MOSFETs with SiON gate dielectric are reported. We show evidence to clarify the recent debate: the recovery of V_{th} shift in the passivation phase of the dynamic NBTI is mainly due to passivation of interface traps (N_{it}), not due to hole de-trapping in dielectric hole traps (N_{ot}). The conventional interface trap measurement methods, dc capacitance–voltage and charge pumping, seriously underestimate the trap density N_{it}. This underestimation is gate bias dependent during measurement, because of the accelerated interface trap passivation under positive gate bias. Due to this new finding, many of previous reliability studies of p-MOSFETs should be re-investigated.

Index Terms—MOSFETs, negative bias temperature instability (NBTI), Silicon oxynitride (SiON).

I. Introduction

NEGATIVE bias temperature instability (NBTI) has become a critical issue for the p-MOSFETs with ultrathin Silicon oxynitride (SiON) gate dielectric. Under static NBTI stress, the p-MOSFET shows a significant negative threshold voltage shift. By adding a passivation phase (applying a positive or zero gate bias) after the NBTI stress, recovery of NBTI degradation has been observed. Recently, several studies on the origin of this NBTI recovery effect have been reported [1], [14], [2]–[7]. G. Chen *et al.* [1], [14], by using dc current–voltage (DCIV) method [8], found that the V_{th} recovery in the passivation phase is due to the passivation of SiO_2/Si interface traps, consistent with some other works [2]–[4], [9], [10]. On the contrary, Huard *et al.* [5] used the charge pumping (CP) method [11], argued that the interface trap density remains as a constant

Manuscript received April 27, 2005; revised June 30, 2005. This work was supported by Singapore A*STAR under Grant R-263-000-267-305 and Grant R-398-000-019-305. The review of this letter was arranged by Editor M. Ostling.

T. Yang, C. Shen, and M. F. Li are with the Institute of Microelectronics, Singapore 117685, and also with the SNDL and CICFAR, Department of Electrical and Computer Engineering, National University of Singapore, Singapore 119260 (e-mail: elelimf@nus.edu.sg).

C. H. Ang is with Chartered Semiconductor Manufacturing, Singapore 738406.

C. X. Zhu, Y.-C.Yeo, and G. Samudra are with the SNDL and CICFAR, Department of Electrical and Computer Engineering, National University of Singapore, Singapore 119260.

D.-L. Kwong is with the Institute of Microelectronics, Singapore 117685 and also with the Microelectronics Research Center, Department of Electrical and Computer Engineering, University of Texas, Austin, TX 78712 USA.

Digital Object Identifier 10.1109/LED.2005.855419

during the passivation phase, and the transient recovery of V_{th} in the passivation phase is due to de-trapping of hole traps in the dielectric. In this letter, we clarify the contradicting views on the V_{th} recovery mechanism in the passivation phase and then discuss the effect of interface trap passivation during measurement.

II. Experiment

Transistors were fabricated using 0.11-μm CMOS technology. The gate dielectrics with equivalent oxide thickness (EOT) = 1.3 and 4.5 nm) were grown by thermal oxidation followed by decoupled plasma nitridation and post-deposition thermal annealing. A HP4155C parameter analyzer was used to measure the device characteristics. For N_{it} measurement of p-MOSFET with EOT = 1.3 nm, we used an improved DCIV method as illustrated in the Appendix of [1], [14], and an improved CP method as illustrated in [12].

III. Results and Discussion

For the NBTI characterization, devices with EOT = 1.3 nm were stressed under a constant negative gate voltage followed by a passivation phase ($V_g = 0$ V), while the source, drain, and bulk were grounded. Stress in the stress phase was periodically interrupted for V_{th} and N_{it} measurement. During each interruption, both V_{th} (extracted by I_d-V_g measurement) and N_{it} (extracted by DCIV measurement) were measured (Fig. 1). The results depend on the order of V_{th} and N_{it} measurements. This can be interpreted as follows. During each interruption of stress for measurement, the negative gate voltage is reduced (for I_d-V_g measurement), or even turned positive (for DCIV or CP measurement). The interface trap density reduces at this moment due to passivation of interface traps. The measured values of V_{th} and N_{it} are therefore underestimated. The passivation effect is more pronounced during the N_{it} measurement than during the V_{th} measurement, because of the longer measurement time and the more positive bias employed.

The following experiments in Figs. 2 and 3 are designed to further illustrate the interface trap passivation effect. For each fresh device in Fig. 2, the V_{th} and N_{it} were first measured, giving the initial threshold voltage $V_{th,0}$ and the initial interface trap density $N_{it,0}$. A stress ($V_g = -2.4$ V for EOT = 1.3 nm devices, -4.5 V for EOT = 4.5 nm devices) was then applied for 500 s. Threshold voltage shift $\Delta V_{th1,S}$ was measured at the end

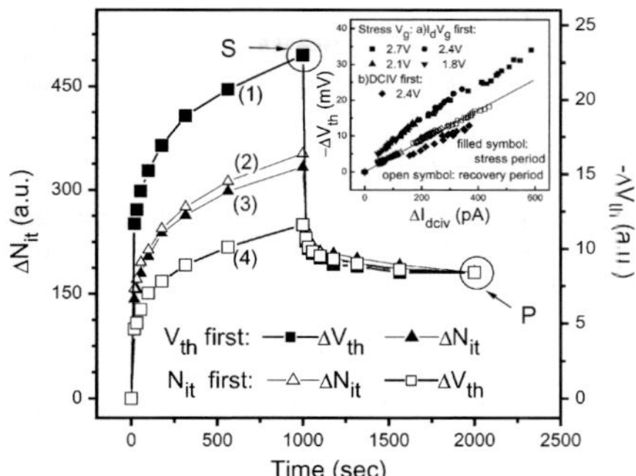

Fig. 1.   NBTI degradation of pMOSFET (SiON EOT = 1.3 nm) characterized by threshold voltage shift (ΔV_{th}) and change of interface trap density (ΔN_{it}) under a stress phase (t = 0 to 1000s) and a passivation phase (t = 1000 to 2000 s). The stress phase is intermittently interrupted for I_d-V_g measurement (for V_{th} extraction) and DCIV measurement (for N_{it} extraction). Two characterization approaches: In the first, V_{th} was measured (solid square symbols on curve 1), followed by N_{it} measurement (solid triangle symbols on curve 3) during each interruption of stress. In the second, N_{it} was measured (open triangle symbols on curve 2), followed by V_{th} measurement (open square symbols on curve 4). The inset shows the correlation of ΔV_{th} and ΔN_{it} for NBTI stress under various stress voltage, in the stress and passivation phases, using both measurement approaches. In the stress phase, the two measurement approaches (I_d-V_g first or DCIV first) yield different correlation (different slopes in ΔV_{th}-ΔN_{it} plot). However, in the passivation phase, two approaches yield the same correlation. The real correlation between ΔV_{th} and ΔN_{it} is shown in the passivation phase only because no interface trap passivation during measurement. (see also Figs. 2 and 3). Therefore the normalization of four curves should be made at the P point at the end of passivation phase.

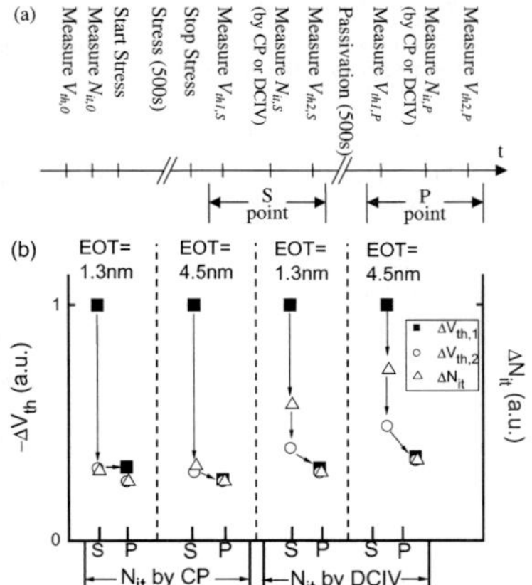

Fig. 2.   (a) Measurement sequences at both S and P points (indicated in Fig. 1) (b) Measured ΔV_{th1}, ΔN_{it}, and ΔV_{th2} data at points S and P. ΔN_{it} and ΔV_{th2} data are normalized at P because no further passivation of N_{it} occurs at P.

of the stress (point S as indicated in Fig. 1), followed by a measurement of the change in interface trap density $\Delta N_{it,S}$ using

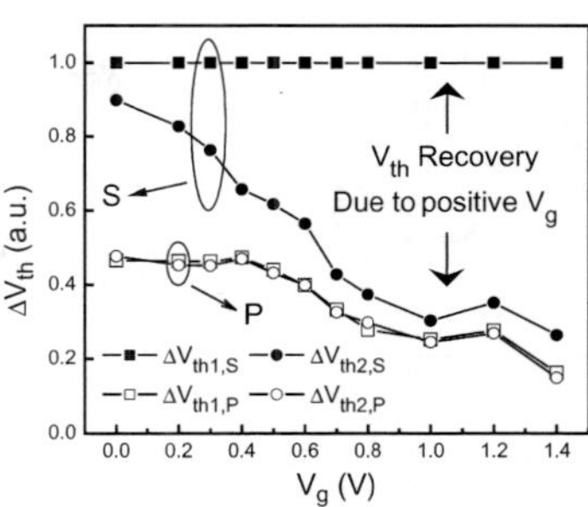

Fig. 3.   In the measurement sequence of Fig. 2(a), the measurement of N_{it} is replaced by a 0.5-s positive gate bias stress V_g (for both points S and P). Plot $\Delta V_{th,2}$ as a function of different V_g, normalized by $\Delta V_{th,1}$ at point S. The results show clearly that the positive gate bias accelerates ΔV_{th} passivation.

CP or DCIV method, and another measurement of the threshold voltage shift $\Delta V_{th2,S}$. After a 500 s lapse in the passivation phase ($V_g = 0$), $\Delta V_{th1,P}$, $\Delta N_{it,P}$, and $\Delta V_{th2,P}$ were measured sequentially again at the end of passivation phase (point P as indicated in Fig. 1). In Fig. 2(b), ΔN_{it} and ΔV_{th} data are normalized at P because no further passivation of N_{it} occurs at P.

All data from devices with thin (EOT = 1.3 nm) and thick (EOT = 4.5 nm) gate dielectrics show the same trend: at point S, ΔV_{th2} after the N_{it} measurement is much smaller than ΔV_{th1} before the N_{it} measurement. For the thick dielectric with EOT = 4.5 nm, the measured FN gate tunneling current rises significantly when $|V_g|$ is above 5 V. Therefore at a stress voltage of -4.5 V, the gate tunneling current is negligible and no charge trapping/detrapping in the dielectric. All variation of V_{th} is due to variation of N_{it} (generation and passivation). The reduction of ΔV_{th} is due to the passivation of interface traps. The N_{it} measurement accelerates the reduction of ΔV_{th}, due to the positive gate bias applied to the device during N_{it} measurement, as further illustrated by Fig. 3. Fig. 2(b) also clearly shows that no obvious recovery of ΔN_{it} can be observed in the passivation phase using CP measurement [5], because most of the interface traps have already been passivated during the CP measurement. Since the thin dielectric device has the same trend as the thick dielectric device, we believe that the V_{th} recovery is dominated by N_{it} passivation [1], [14] rather than N_{ot} detrapping [5].

Fig. 3 illustrates the acceleration effect of interface trap passivation under a positive gate bias. The same measurement sequence as that in Fig. 2 was used, however the measurement of N_{it} was replaced by a 0.5 s positive gate bias stress V_g. Fig. 3 plots $\Delta V_{th,2}$ as a function of different positive V_g, normalized by $\Delta V_{th,1}$ at point S. The results show clearly that the positive bias accelerates ΔV_{th} passivation. This explains the different results obtained by DCIV and CP measurements in Fig. 2(b). In DCIV measurement, a recombination current I_{DCIV} through the interface traps shows a peak when the Fermi level coincides with the Si mid gap at surface [13]. In our measurement, the maximum V_g applied to the device to show the peak is around $+0.5 \sim 0.6$ V. In the stress phase measurement, the interface

IEEE ELECTRON DEVICE LETTERS, VOL. 26, NO. 10, OCTOBER 2005

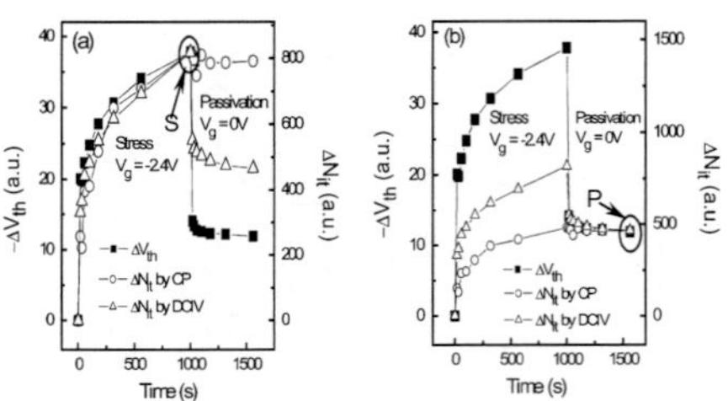

Fig. 4. The data of ΔV_{th} and ΔN_{it} measured by both CP and DCIV, plotted versus stress and passivation time: (a) all data aligned at point S [5]; (b) all data aligned at P point (this work).

traps have already passivated to some extend before reaching the I_{DCIV} peak gate voltage (Fig. 3 data measured at point S), so the result is always underestimated. However after a time period of passivation in the passivation phase at point P, the passivation rate is almost zero (Fig. 3 data measured at point P) and therefore DCIV method measures the real interface trap density. The underestimation is more serious when using CP technique. In CP measurement, the device changes from inversion to accumulation to pump the electrons between conduction/valence bands through the interface states [11]. Therefore a more positive V_g than that in DCIV measurement should be applied. In our CP measurement, the maximum positive V_g applied to the device is +1.2 V. On the other hand, CP measurement uses longer time than DCIV measurement. This explains why $\Delta V_{\mathrm{th},2,S}$ and ΔN_{it} measured by CP is smaller than that measured by DCIV as indicated in Fig. 2(b), because more interface traps are passivated during the CP measurement.

In Fig. 4 (a), all curves are aligned at point S, according to the discussion in [5]. However, this alignment is wrong because the measured N_{it} is underestimated during the stress phase. Due to this incorrect alignment, N_{it} is lifted up at the passivation phase. N_{it} is lifted more by using CP measurement, consistent with Fig. 1 of [5]. Thus, our work clarifies the contradiction between two different interpretations of the V_{th} recovery phenomena in the passivation phase.

IV. Conclusion

From our experiments, we showed evidence that the recovery of V_{th} shift in the passivation phase after NBTI stress of the p-MOSFETs with SiON gate dielectric is mainly due to passivation of interface traps (N_{it}), not due to hole de-trapping in dielectric hole traps. The passivation effect is accelerated by an applied positive gate bias. The widely used CP and DCIV methods underestimate the value of N_{it} significantly because of this passivation effect.

Acknowledgment

One of the authors, T. Yang, would like to thank IME/NUS JML for supporting her scholarship for her M.Eng. study at NUS.

References

[1] G. Chen, M. F. Li, C. H. Ang, J. Z. Zheng, and D. L. Kwong, "Dynamic NBTI of pMOS transistors and its impact on MOSFET scaling," *IEEE Electron Device Lett.*, vol. 23, no. 7, pp. 734–736, Jul. 2002.

[2] M. A. Alam, "A critical examination of the mechanics of dynamic NBTI for PMOSFETs," *IEDM Tech. Dig.*, pp. 345–348, 2003.

[3] S. Mahapatra, M. A. Alam, P. B. Kumar, T. R. Dalei, and D. Saha, "Mechanism of negative bias temperature instability in CMOS devices: Degradation, decovery, and impact of nitrogen," in *IEDM Tech. Dig.*, 2004, pp. 105–108.

[4] S. Chakravarthi, A. T. Krishnan, V. Reddy, C. F. Machala, and S. Krishnan, "A comprehensive framework for predictive modeling of negative bias temperature instability," in *Proc. IEEE Reliab. Phys. Symp.*, 2004, pp. 273–282.

[5] V. Huard and M. Denais, "Hole trapping effect on methodology for DC and AC negative bias temperature instability measurements in PMOS," in *Proc. IEEE Reliab. Phys. Symp.*, 2004, pp. 40–45.

[6] W. Abadeer and W. Ellis, "Behavior of NBTI under AC dynamic circuit conditions," in *Proc. IEEE Reliab. Phys. Symp.*, 2003, pp. 17–22.

[7] S. Tsujikawa, T. Mone, K. Watanabe, Y. Shinamoto, R. Tsuchiya, K. Ohnishi, T. Onai, J. Yugami, and S. Kimura, "Negative bias temperature instability of p-MOSFETs with ultrathin SiON gate dielectrics," in *Proc. IEEE Reliab. Phys. Symp.*, 2003, pp. 183–187.

[8] A. Neugroschel, C. T. Sah, K. M. Han, M. S. Caroll, T. Nishida, J. T. Kavalieros, and Y. Lu, "Direct-current measurement of oxide and interface traps on oxidized silicon," *IEEE Trans Electron Devices*, vol. 47, no. 11, pp. 1657–1662, Nov. 1995.

[9] S. Ogawa, M. Shimiya, and N. Shionon, "Interface trap generation at ulthathin SiO_2-Si interfaces during negative-bias temperature aging," *J. Appl. Phys.*, vol. 77, pp. 1137–1148, 1995.

[10] N. Kimizuka, K. Yamaguchi, K. Imai, T. Lizuka, C. T. Liu, R. C. Keller, and T. Horiuchi, "NBTI enhancement by nitrogen incorporation into ultrathin gate oxide for 0.10 μm gate CMOS generaion," in *Symp. VLSI Tech. Dig.*, 2000, pp. 91–92.

[11] G. Groeseneken, H. E. Maes, N. Beltran, and R. F. De Keersmaecker, "A reliable approach to charge-pumping measurements in MOS transistors," *IEEE Trans. Electron Devices*, vol. ED-31, no. 1, pp. 42–53, Jan. 1984.

[12] S. S. Chung, S.-J. Chen, C.-K. Yang, S.-M. Cheng, S.-H. Lin, Y.-C. Sheng, H.-S. Lin, K.-T. Hung, D.-Y. Wu, T.-R. Yew, S.-C. Chien, F.-T. Liou, and F. Wen, "A novel and direct determination of the interface traps in sub-100 nm CMOS devices with direct tunneling regime (12–16 A) gate oxide," in *Symp. VLSI Tech. Dig.*, 2002, pp. 74–75.

[13] J. Cai and C. T. Sah, "Interfacial electronic traps in surface controlled transistors," *IEEE Trans. Electron Devices*, vol. 47, no. 5, pp. 576–583, May 2000.

[14] G. Chen et al., "Dynamic NBTI of PMOS transistors," in *Proc. IEEE Reliab. Phys. Symp.*, 2003, pp. 196–202.

IEEE ELECTRON DEVICE LETTERS, VOL. 27, NO. 1, JANUARY 2006   55

A Fast Measurement Technique of MOSFET I_d–V_g Characteristics

C. Shen, *Student Member, IEEE*, M.-F. Li, *Senior Member, IEEE*, X. P. Wang, Yee-Chia Yeo, *Member, IEEE*, and D.-L. Kwong, *Senior Member, IEEE*

Abstract—In this letter, we developed an improved ultrafast measurement method for threshold voltage V_{th} measurement of MOSFETs. We demonstrate I_d–V_g curve measurement within 1 μs to extract the threshold voltage of MOSFET. Errors arising from MOSFET parasitics and measurement setup are analyzed quantatatively. The ultrafast V_{th} measurement is highly needed in the investigation of gate dielectric charge trapping effect when traps with short detrapping time constants are present. Application in charge trapping measurement on HfO$_2$ gate dielectric is demonstrated.

Index Terms—CMOSFETs, reliability, semiconductor device measurements, trapping.

I. INTRODUCTION

STUDIES on the stress induced threshold voltage (V_{th}) shift of MOSFET transistors, notably positive/negative bias temperature instability conventionally evaluate the V_{th} degradation in measure-stress-measure cycles with dc parametric measurement tools such as the HP4156 semiconductor parameter analyzer. The time delay between the end of stress and the I_d–V_g measurement for V_{th} extraction is typically in the order of 0.1–1 s, and the degraded V_{th} will possibly recover during this delay. The recovery during the short delay has long been thought negligible until recently. Studies on charge trapping in high-κ and SiON dielectrics in recent years showed that the recovery in V_{th} is significant even within an 1 ms delay [1]–[4]. Therefore, a fast and accurate V_{th} measurement technique is required to capture all the fast transient trapping/detrapping phenomenon. Kerber *et al.* developed a pulsed I_d–V_g method to measure hysteresis in I_d–V_g within 10–100 μs, which has been applied to charge trapping studies in high-κ gate dielectrics [1]. In this letter, we present a systematic discussion on an improved fast I_d–V_g measurement technique [2], which can achieve 1 μs measurement time without using expensive RF measurement setup.

Manuscript received July 27, 2005; revised October 10, 2005. This work was supported by the Singapore A*STAR under Research Grants R-263-000-267-305 and R-398-000-019-305. The review of this letter was arranged by Editor M. Ostling.

C. Shen, X. P. Wang and Y.-C. Yeo are with the Silicon Nano Device Laboratory and Center for Integrated Circuit Failure Analysis and Reliability, Department of Electrical and Computer Engineering, National University of Singapore, Singapore 119260.

M.-F. Li is with the Silicon Nano Device Laboratory and Center for Integrated Circuit Failure Analysis and Reliability, Department of Electrical and Computer Engineering, National University of Singapore, Singapore 119260, and also with the Institute of Microelectronics, Singapore 117685 (e-mail: elelimf@nus.edu.sg).

D.-L. Kwong is with the Institute of Microelectronics, Singapore 117685.

Digital Object Identifier 10.1109/LED.2005.861025

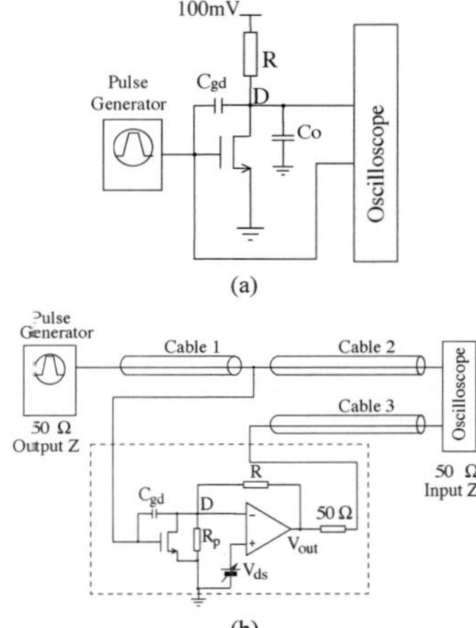

Fig. 1. Schematics illustrating the fast I_d–V_g measurement setup developed (a) by Kerber [1] and (b) in this letter. Resistor R and R_p are both 1 kΩ. Coaxial cables 2 and 3 have the same length, so the signal propagation delays are the same. The V_{gs} and V_{out} signals reach the oscilloscope input at the same time.

II. MEASUREMENT METHOD

The measurement method developed by Kerber *et al.* has been widely used to evaluate the fast charging trapping in high-κ dielectrics [1]. The method is shown in Fig. 1(a). A short trapezoidal pulse is applied to the gate of MOSFET. The oscilloscope measures the voltage drop across the sense resistor R at the drain, and hence the drain current. However, this method suffers from the following limitation for ultrafast measurement. In this method, the drain voltage of the MOSFET under test is not a constant, but changes with changing drain current. The parasitic capacitor C_0 and C_{gd} must be charged or discharged as the drain voltage and gate voltage change, and the charging current distorts the measured drain current. The parasitic capacitor C_0 consists of the C_{db} of the MOSFET, the input capacitance of the oscilloscope and the cable capacitance, and ranges from 20 to over 100 pF depending on the length of the cable. The distortion is more serious when the voltage pulses rise or fall time becomes shorter.

To circumvent the above mentioned problems, we developed an improved pulsed I_d–V_g measurement technique, as shown

IEEE ELECTRON DEVICE LETTERS, VOL. 27, NO. 1, JANUARY 2006

in Fig. 1(b). By virtual short circuit property of op-amp (the voltage of two input terminals forced to be equal), the drain voltage of the MOSFET is fixed at V_{ds} supplied by the voltage source, so no charging and discharging current flows through C_0. A high-speed op-amp (OPA655) with 400-MHz unity gain bandwidth is used to achieve fast measurement. As samples are measured in probe station environment, the op-amp circuit (enclosed by the dashed line in figure) is mounted immediately above the probe holder. Probe holders are modified and wire connection to the transistor source and drain terminal is made less than 10 cm to minimize parasitics. The drain current is measured by the sense resistor R. Resistors ranged from 1–10 $k\Omega$ are used in this letter for different transimpedance gain. Resistor $R_p = R$ is used to ensure circuit stability when the MOSFET is off. The output voltage measured at the oscilloscope is related to the MOSFET drain current by

$$V_{\mathrm{out}} = (I_D - I_{\mathrm{gd}}) \cdot R + V_{\mathrm{ds}} \tag{1}$$

where R is the sense resistance, V_{ds} is the drain voltage, and I_{gd} is the current from gate to drain through the parasitic capacitor C_{gd}. The current I_{gd} is caused by the fast transient at the gate and is given by

$$I_{\mathrm{gd}} = C_{\mathrm{gd}} \cdot \frac{dV_{\mathrm{gd}}}{dt} = C_{\mathrm{gd}} \cdot \frac{dV_{\mathrm{gs}}}{dt} \tag{2}$$

In the measurement, the MOSFET is biased in linear region in I_d–V_g measurements, and C_{gd} is given by

$$C_{\mathrm{gd}} = C_{\mathrm{overlap},d} + \frac{1}{2}C_{\mathrm{inv}} \tag{3}$$

where $C_{overlap,d}$ and C_{inv} are the capacitance of the drain overlap region and the inversion capacitance, respectively.

For short-channel devices, C_{gd} is small, and the corresponding I_{gd} is much smaller than the drain current. Therefore, the charging current through C_{gd} can be ignored. When a symmetric triangular pulse is applied at the gate as shown in Fig. 2 inset, I_d–V_g curve can be measured at both the up-trace and down-trace of the pulse. In the two cases, dV_g/dt are of the same magnitude but of opposite sign. For nMOSFET with short channel length $L = 0.1~\mu$m, the I_d–V_g curves measured in the up-trace and down-trace of V_{gs} both coincide with that from dc measurement, as shown in Fig. 2, which indicates negligible effect of charging current through C_{gd}. V_{out} waveform must be synchronous to V_{gs} waveform in order to get the correct I_d–V_g curve. A delay difference δt between V_{gs} and V_{out} waveforms will generate approximately a horizontal shift of I_d–V_g curve, and is given by

$$\delta V_{\mathrm{gs}} = \frac{dV_{\mathrm{gs}}}{dt} \cdot \delta t. \tag{4}$$

When measurement time (rise/fall time) is reduced, both δt and dV_{gs}/dt increases, and the distortion of I_d–V_g curve worsens quickly. In order to reduce the delay difference δt, 1) length of coaxial cables 2 and 3 in Fig. 1(b) should be approximately same to minimize the delay difference between V_{gs} and V_{out} signal due to signal propagation in the cables. One meter of cable length difference causes 5 ns delay time difference. If in the fast measurement V_{gs} ramps from 0 to 1 V in 1 μs, 5 ns

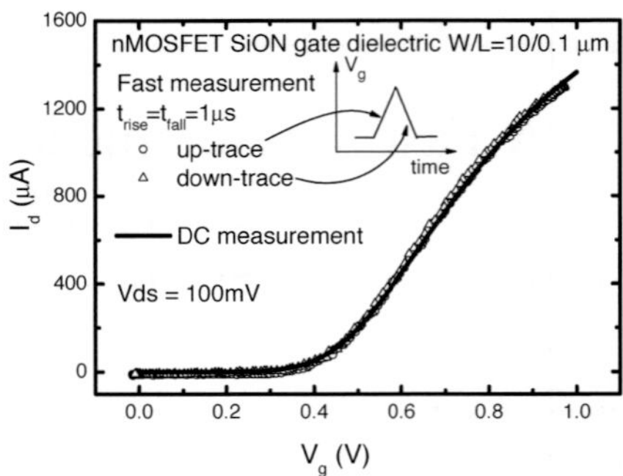

Fig. 2. I_d–V_g characteristics measured from a short-channel nMOSFET with SiON gate dielectric, with the V_g waveform shown in the inset. Fast I_d–V_g measurement (1 μs measurement time) result is identical to the conventional dc-ramp measurement result in both up trace and down trace.

introduces 5 mV shift in I_d–V_g curve. 2) very high speed op-amp must be used to reduce the delay caused by the op-amp circuit. The measurement speed is limited to 1 μs in this letter due to the frequency response of the op-amp.

For long channel devices, the charging current through the large C_{gd} can be a significant portion of the drain current, and need to be corrected. Fig. 3(a) shows that for nMOSFET with $L_g = 10~\mu$m, the I_d–V_g curve measured in the up-trace of V_{gs} does not coincide with that from the down-trace. The difference between the two I_d–V_g curves is inversely proportional the rise time t_{rise} of V_{gs}. This indicates a capacitive branch and is attributed to C_{gd}. This gate-to-drain capacitance can be measured with a slightly modified split-CV measurement. The source and body of the MOSFET are grounded, and capacitance between gate and drain is recorded, as shown in the inset of Fig. 3(b). This measured C–V curve provides good approximation to C_{gd} when the MOSFET is operating in linear region when V_{ds} is small. The effect of charging and discharging of C_{gd} to the drain current measurement can be corrected using (1) and (2). Fig. 3(b) shows the raw and corrected I_d–V_g curve obtained from a MOSFET with channel length $L = 10~\mu$m. After correction, the up-trace and down-trace I_d–V_g curves coincide, verifying the above analysis. A first order analysis suggests that the error in drain current due to C_{gd} scales with the channel length as

$$\frac{\delta I_d}{I_d} \propto L^2 \cdot \frac{dV_{\mathrm{gs}}}{dt}. \tag{5}$$

Therefore, the effect of C_{gd} quickly vanishes for gate length less than 1 to 2 μm for 1 μs measurement time.

III. APPLICATIONS TO CHARGE TRAPPING IN HIGH-κ GATE DIELECTRICS

We applied this improved fast measurement technique to study the charge trapping in high-κ dielectrics. The MOSFET transistors used in this experiment have MOCVD deposited HfO_2 gate dielectric with $EOT \approx 1.3$ nm [5]. After an initial

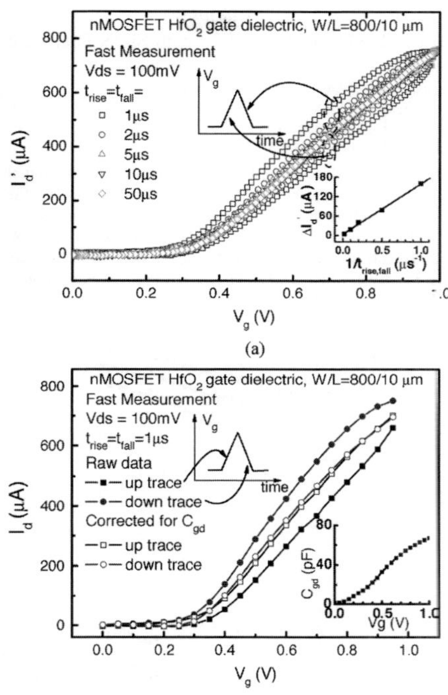

(a)

(b)

Fig. 3. (a) I_d–V_g characteristics measured from a long-channel nMOSFET with HfO₂ gate dielectric, with the V_g waveform as in Fig. 2. The effect of charging current through C_{gd} is not deducted, so the obtained current is labeled I'_d. As the rise time and fall time of the waveform decreases, the I_d–V_g curves from up-trace and down-trace move apart from each other, due to the charging and discharging of C_{gd} capacitance. The difference between the two curves is inversely proportional to the rise time and fall time of the V_g waveform as shown in the inset. (b) The effect of C_{gd} is corrected using (1) and (2), for the $t_{\mathrm{rise}} = 1\ \mu$s case. The gate to drain capacitance is measured using modified split-CV method, and is shown in the inset.

I_d–V_g measurement, MOSFETs are stressed for 1 s, and I_d–V_g is measured again at the falling edge of the stress voltage. Threshold voltage shift is extracted from the horizontal shift of the I_d–V_g curve before and after the stress. In Fig. 4, the V_{th} shift of an nMOSFET is plotted with varying falling edge time, i.e., I_d–V_g measurement time, at the end of the stress. It is observed that as measurement time is increased above 100 μs, the measured V_{th} drops due to the significant detrapping during the measurement. The V_{th} shift measured by conventional dc measurement is only 20% of the true value, and gives a

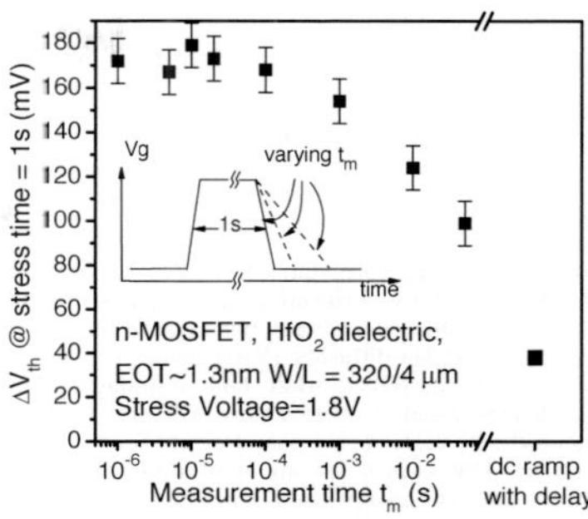

Fig. 4. NMOSFET transistors with HfO₂ gate dielectrics are stressed for 1 s with $V_g = 1.8$ V, after which I_d–V_g is measured with different measurement time t_m at the falling edge of the stress voltage. The threshold voltage shift before and after stress (ΔV_{th}) is plotted against measurement time, and the threshold voltage shift obtained with dc measurement is shown for comparison. Slow measurements leads to serious underestimation of the threshold voltage shift caused by charge trapping.

very poor under-estimation of the trapped charge in high-κ dielectrics. On the other hand, from this figure, measurement time of 5 μs to 10 μs is fast enough to evaluate the charge trapping in this particular device adequately.

IV. Conclusion

We have developed a fast measurement technique to obtain MOSFET I_d–V_g characteristics in 1 μs without using expensive RF measurement setup. This method is used to evaluate the charge trapping in HfO₂ dielectric layer, which distinguished the fast and slow traps in HfO₂. The 1-μs measurement time is shown to be enough for this particular application.

References

[1] A. Kerber, E. Cartier, L. A. Ragnarsson, M. Rosmeulen, L. Pantisano, R. Degraeve, T. Kauerauf, G. Groeseneken, H. E. Maes, and U. Schwalke, "Characterization of the VT-instability in SiO₂/ HfO₂ gate dielectrics," in *Proc. IEEE IRPS*, 2003, pp. 41–45.

[2] C. Shen, M. F. Li, X. P. Wang, H. Y. Yu, Y. P. Feng, A. T.-L. Lim, Y.-C. Yeo, D. S. H. Chan, and D. L. Kwong, "Negative-U traps in HfO₂ gate dielectrics and frequency dependence of dynamic BTI in MOSFETs," in *IEDM Tech. Dig.*, 2004, pp. 733–736.

[3] M. Denais, A. Bravaix, V. Huard, C. Parthasarathy, G. Ribes, F. Perrier, Y. Rey-Tauriac, and N. Revil, "On-the-fly characterization of NBTI in ultrathin gate oxide PMOSFETs," in *IEDM Tech. Dig.*, 2004, pp. 109–112.

[4] T. Yang, M. F. Li, C. Shen, C. H. Ang, C. Zhu, Y.-C. Yeo, G. Samudra, S. C. Rustagi, M. B. Yu, and D. L. Kwong, "Fast and slow dynamic NBTI components in p-MOSFET with sion dielectric and their impact on device life-time and circuit application," in *VLSI Symp. Tech. Dig.*, 2005, pp. 92–93.

[5] H. Y. Yu, J. F. Kang, J. D. Chen, C. Ren, Y. T. Hou, S. J. Whang, M. F. Li, D. S. H. Chan, K. L. Bera, C. H. Tung, A. Du, and D.-L. Kwong, "Thermally robust high quality HfN/HfO₂ gate stack for advanced CMOS devices," in *IEDM Tech. Dig.*, 2003, pp. 99–102.

Characterization and Physical Origin of Fast V_{th} Transient in NBTI of pMOSFETs with SiON Dielectric

C. Shen[*†], M.-F. Li[*†‡], C. E. Foo[*], T. Yang[*], D. M. Huang[‡], A. Yap[§], G. S. Samudra[*], Y.-C. Yeo[*]
* SNDL and CICFAR, ECE Dept., National University of Singapore, Singapore 119260. Email: elelimf@nus.edu.sg
† Institue of Microelectronics, Singapore 117685.
‡ Microelectronics Dept., Fudan University, Shanghai 201203, China.
§Chartered Semiconductor Manufacturing, Singapore, 738406

Abstract—**Highly reliable characterization of fast transient in NBTI is achieved by performing initial and stressed $I-V$ measurements in ultra-short time (100 ns). We further provide evidences that reaction-diffusion (R-D) model can not explain the fast transient in NBTI, while hole trapping (HT) model explains all experimental observations. We also establish that previous on-the-fly methods are sound except for the slow initial measurement. This caused the apparent disagreements among results from different groups using on-the-fly methods, which is resolved in this work by the fast on-the-fly technique.**

I. INTRODUCTION

In recent years, the effect of fast recovery of NBTI characterization in SiON gate dielectric has received increasing attention, and various groups proposed new measurement schemes to measure the true V_{th} shift free from the recovery effect [1]–[5]. However, controversy arises as different groups, using their respective measurement setup, reported different NBTI characteristics. Varghese *et al.* proposed a universal power-law exponent of around 0.16 [5], while Denais *et al.* reported logarithmic dependence [1] based on a slightly different measurement procedure. However, both results are again inconsistent with the alternative fast $I_d - V_g$ measurement, which yields a much smaller power-law slope (often < 0.1) [3], [4]. In this work, we provide a critical comparison and resolve the inconsistency between different measurement techniques. Measurement results shown in this work are from p-MOSFET samples with DPN SiON of 1.3nm EOT, while samples with thicker RTN SiON (1.8-2.1nm EOT) showed consistent results. These results are then examined against the predictions from both the reaction-diffusion (R-D) model and the hole-trapping (H-T) model.

II. REVIEW ON MEASURMENT TECHNIQUES

Conventional (slow) on-the-fly characterization has been claimed to be free from the fast recovery of V_{th} [1], [2], [5]. It soon became popular as it can be performed on common semiconductor parameter analyzers such as HP4156. However, the initial pre-stress measurement of I_{d0} takes 10-100 ms with a typical setup on HP4156, during which some V_{th} degradation occur under the high gate voltage ($= V_{g,stress}$). The measured I_{d0} is lower than the actual value, thus the V_{th} shift after the subsequent stress is underestimated. At short stress time, the underestimation is more serious, which results in a kink before 10 sec as observed in the log-log plot as shown in Fig. 1(a). We observed this kink as well as the power-law slope of 0.156 as reported in Ref. [5] if the initial measurement is slow. However, when the initial measurement time is reduced from 100 ms to 1 ms, the kink becomes less obvious, and the slope decreases. Furthermore, assuming $\Delta V_{th} = 40\,\mathrm{mV}$ in the 1 ms initial measurement at $V_g = -2.4\,\mathrm{V}$, one can compensate this initial degradation by adding $40\,\mathrm{mV}$ to the subsequent ΔV_{th} values. As a result, the kink is further weakened and the slope

reduced. The reduction of V_{th} degradation during the initial pre-stress measurement is therefore the key to improve the on-the-fly technique. Slow measurement underestimates V_{th} degradation and yields an errorneous slope in the log-log plot. Alternatively, the same data can be plotted on a semi-log scale as in Fig. 1(b), where a surprising logarithmic relatioin is seen, as Ref. [1] reported.

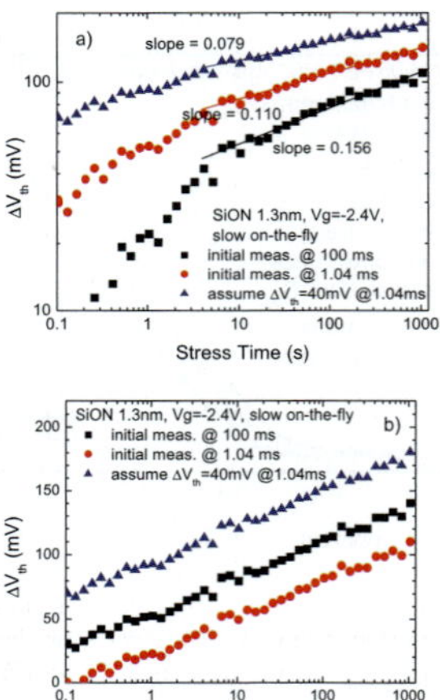

Fig. 1. (a) Conventional (slow) on-the-fly suffers from degradation during the measurement of the inital I_{d0}, and estimates ΔV_{th}. When plotted in log-log scale, a characteristic kink is observed in short time. The slower the initial (pre-stress) measurement is, the stronger the kink is. If one compensates for the degradation during the initial (pre-stress) measurement by adding an assumed amount to the value of ΔV_{th}, the kink weakens (triangle symbols). (b) When the same data is plotted in semi-log scale, straight lines are observed showing logarithmic relationship.

To circumvent this problem with slow initial measurement, the pre-stress I_{d0} and V_{th0} is measured by the fast $I_d - V_g$ technique [3], [4], [6] within $1\mu s$ to minimize I_d degradation. The drain current I_d is then monitored on-the-fly during stress. The measured ΔV_{th} by this *fast on-the-fly* technique is plotted in Fig. 2, where results from a few other techniques described in Table I are also plotted for comparison. Note that the fast measurement setup used here is an improved version based on ref. [6], and is able to measure the $I_d - V_g$ characteristics within 100 ns. As shown, the fast on-the-fly measurement as in this work, and the fast $I_d - V_g$ measurement as in ref. [6] yield comparable results, while all other techniques show underestimation of varying amount.

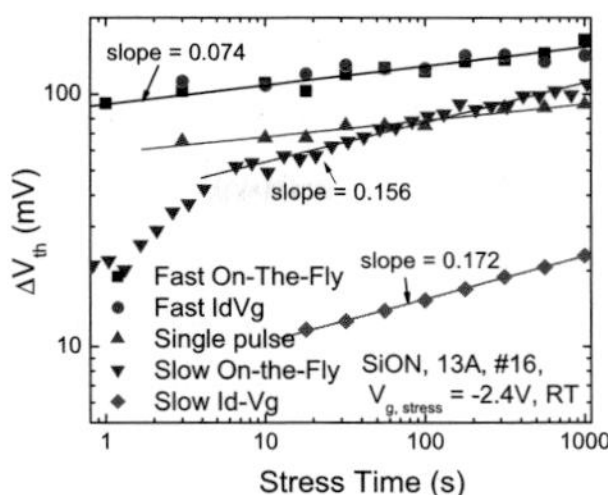

Fig. 2. ΔV_{th} for identical pMOSFETs is measured with five different measurement techniques, which yield very different NBTI results. The fast $I_d - V_g$ (100 ns measurement time) and the fast on-the-flay measurements produce the same and reliable results. All other methods show underestimated ΔV_{th}. The single pulse measurement [7] used in this work requires a 300 ns pulse due to long settling time, which can be improved with proper (but costly) RF measurement setup.

TABLE I

SUMMARY OF MEASUREMENT TECHNIQUES

Method	Source of error	Effect on measured ΔV_{th}
Slow on-the-fly	Initial pre-stress I_d measurement ($V_g = V_{stress}$) is too slow, during which I_d degrades.	large under-estimation
Fast on-the-fly	Initial pre-stress I_d measurement (100-200 ns) induces small stress and thus small I_d degradation.	small under-estimation
Fast $I_d - V_g$	Small recovery after stress is removed, during the V_g sweep measurement (100-200 ns).	small under-estimation
Single pulse ΔI_d	Long settling time is required for good accuracy (< 1%), during which recovery occurs.	small-medium under-estimation

In short, the apparent inconsistency in the NBTI literature is a result of imperfect measurement setups. The proposed power-law exponent of 0.16 is specific to a particular measurement setup instead of universal. The fast on-the-fly method and the fast $I_d - V_g$ method are thus far least affected by the V_{th} transient artifacts, and are the better choices for NBTI characterization.

III. PHYSICAL ORIGIN OF THE FAST NBTI COMPONENT

Using experimental data from the reliable characterization techniques, the physical origin of the fast NBTI component is discussed next. We have shown that the power-law exponent of 0.16 is not reliable, and the data from the more careful measurements show a much less exponent ($\sim$0.07-0.10) if plotted in log-log scale. The R-D theory, which predicts an exponent of 0.25, or 0.16 when an additional reaction step is introduced ($2H^0 \leftrightarrow H_2$) [5], [8], [9], therefore faces serious challenges. In principle, one could introduce even higher order reactions to further reduce the exponent to fit to the experimental power-law slope. We attempt this problem beyond the phenomenal power-law slope, but from a few fundamental characteristics of a reaction-diffusion process. We present the following arguments to demonstrate the incompatibility of the R-D model with the fast V_{th} transient observed experimentally.

This incompatibility arises from the basis of the R-D model, which assumes that the process is diffusion-limited.

A. Fast Recovery and Dependence on Stress Time

We first noticed that, even after long time stress (1000s), the V_{th} shift still shows dramatically fast recovery (more than 60% within the first second after the stress is removed, as shown in Fig. 3. This fast recovery is however not expected in the reaction-diffusion model. As illustrated in Fig. 4, an approximate triangular hydrogen concentration profile is gradually set-up during the 1000 sec stress, and the total amount of the released hydrogen species (area under the curve) must equal to the total amount of interface traps, which is in turn proportional to the V_{th} shift. After the stress is removed, if the ΔV_{th} were to recover by 60% within the first 1 sec, 60% of the hydrogen species must have diffused back to the interface in this short time to react with the interface traps. In other words, one must assume a much faster backward diffusion than the forward diffusion, which is against one's understanding about a diffusion process.

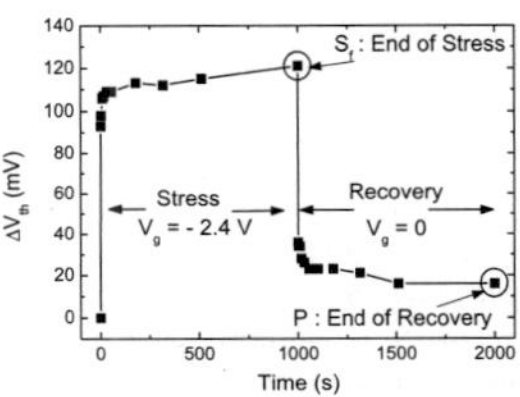

Fig. 3. Using fast $I_d - V_g$ technique, the dynamics of ΔV_{th} under stress/recovery cycles is studied ($f = 1/2000$ Hz is shown). S_f denotes the ΔV_{th} at the end of the stress half-cycle, and P denotes that at the end of recovery half-cycle.

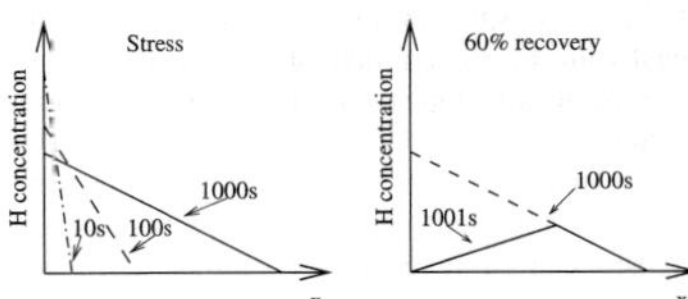

Fig. 4. Hypothetical hydrogen profile necessary to explain the fast V_{th} recovering after long time (1000 s) stress within the reaction-diffusion framework. For 60% of the ΔV_{th} to recover within 1 sec, 60% of the hydrogen species would have to diffuse back to the interface in this short time, traveling a distance that took $\sim$1000 sec in out-diffusion.

In fact, in the reaction-diffusion model, as hydrogen species diffuse out to a greater distance from the interface during a longer stress time, it accordingly takes longer time to diffuse back after stress is removed. Therefore, the recovery after a 1000 sec stress is expected to be appreciably slower than the recovery after a 1 sec stress. However, the observed fast recovery does not show dependence on the stress time, as shown in Fig. 5, for the case of both V_{th} recovery and $I_{d,lin}$ recovery.

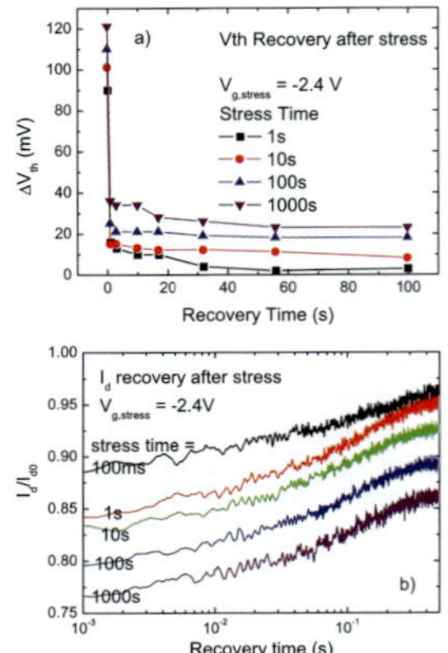

Fig. 5.   a) After stress is removed, the majority of ΔV_{th} recovers in a very short time. Even after 1000 sec stress, >60% of the ΔV_{th} recovers within 1 sec after the stress is removed. b) Similar fast recovery in drain current is seen after stress is removed, despite of the long stress time.

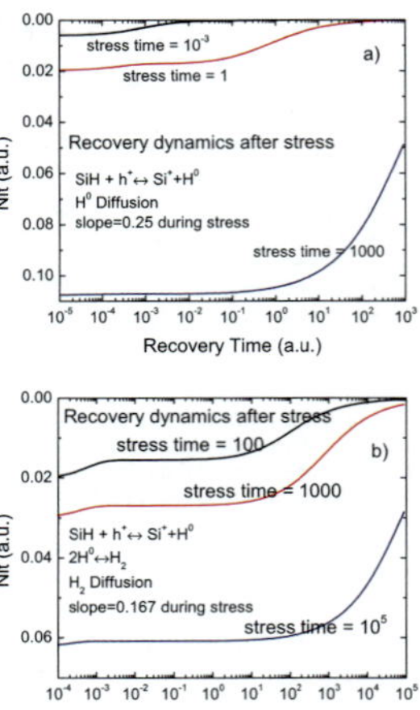

Fig. 6.   According to the reaction-diffusion model ((a)H^0 diffusion and (b) H_2 diffusion), after stressed for a time t long enough so that diffusion dominates, recovery occurs mainly between $0.1t$ and $10t$ after stress is removed.

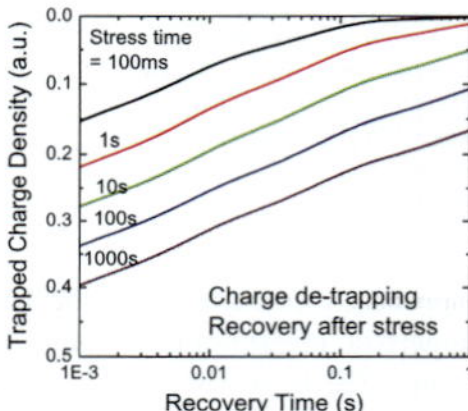

Fig. 7.   Simulated recovery process according to the hole trapping/de-trapping model. Fast V_{th} recovery is expected even for long time stress, which qualitatively agrees with Fig.5.

To verify the above arguments, the reaction-diffusion equations are solved numerically for both cases of H^0 and H_2 diffusion [8], [10]. It is found that after being stressed for long enough time, where hydrogen diffusion is the limiting process, the time for recovery is closely related to the stress time t_{stress}. As shown in Fig. 6, recovery mainly occurs between $0.1 \times t_{stress}$ to $10 \times t_{stress}$, and it takes a time approximately equal to t_{stress} for 50% recovery. This observation on the simulatioin results confirms the above arguments, but obviously contradicts with the fast V_{th} recovery observed experimentally.

On the other hand, simulation with the hole-trapping model equations [3], [4] is qualitatively consistent with the observed fast recovery for all stress times, as shown in Fig. 7. When holes get trapped in a certain kind of trap states during stress, their de-trapping time-constant during recovery is determined by the trap states, but does not depend on how long holes have been trapped in the hole traps. This marks a fundamental difference between the dynamics of trapping/de-trapping process and the reaction-diffusion process, which suggests that the experimental data favors the charge trapping model on a fundamental instead of coincidental basis.

B. Effect of Measurement Delay

The R-D model simulation predicts that the relative error caused by measurement delay diminishes after long time stress, as shown in Fig. 8. When the stress time exceeds 10 times the measurement delay, the relative error due to delay becomes negligible. This is in contrast with the experimental data shown in Fig. 9, where a short delay causes large error even after long-time stress. The simulation with the hole trapping model, on the other hand, agrees with the experimentas qualitatively as shown in Fig. 10.

C. Frequency Dependence Under Dynamic Stress

The reaction-diffusion equations are solved repeatedly to simulate the V_{th} degradation under dynamic stress. The amplitude of V_{th} transient between the end of stress half-cycle (S_f point in Fig.3) and the end of recovery half-cycle (P point) would diminish when dynamic stress time exceeds $100\times$ of the period of one stress/recovery cycle (Fig. 11). After long time stress, V_{th} shift under dynamic stress shows very little oscillation, and converges to a trend-line that is 1) parallel to that under static stress, and 2) frequency independent. However, as shown in Fig.12, we experimentally observe significant transient amplitude after a 1000 sec, 100 kHz dynamic stress, which is drastically different from the predictions from the R-D model. On the other hand, the hole-trapping model shows agreement with frequency-dependence experiments as detailed in the previous work [3], [4], and shown in Fig. 13.

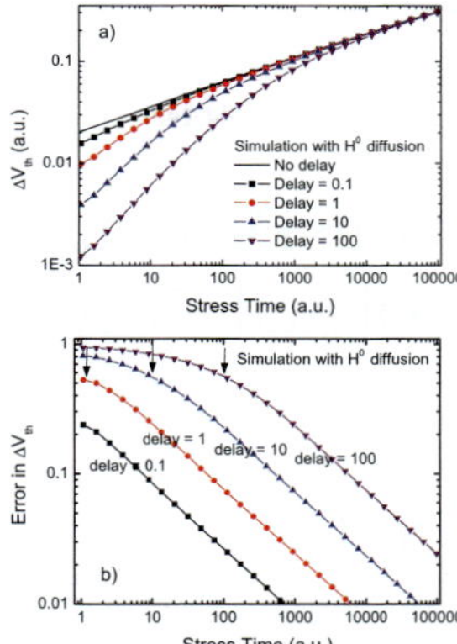

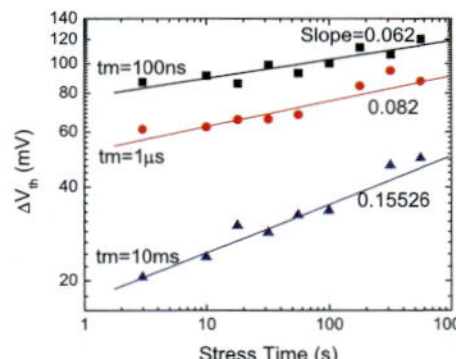

Fig. 8. (a) From the R-D model, ΔV_{th} with measurement delay is simulated. (b) The error due to delay diminishes when stress time is much greater than the delay time.

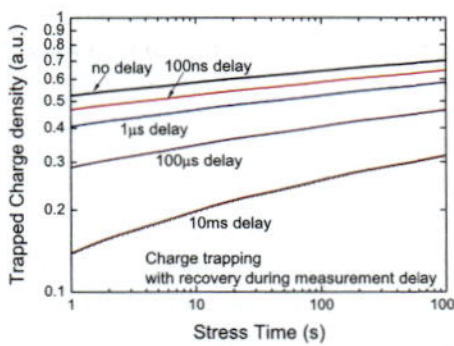

Fig. 9. As measurement time t_m increases, V_{th} shift is underestimated, and the power-law exponent increased. A short delay leads to large underestimation of V_{th} even after 1000 sec of stress.

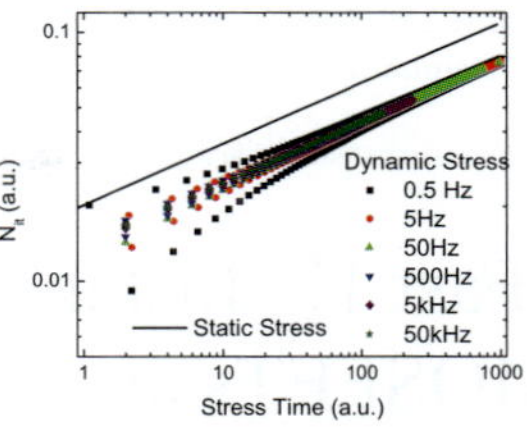

Fig. 10. From the trapping/de-trapping model, the trapped charge (and thus ΔV_{th}) is simulated with delay. A short delay causes large error, which persists even after long-time stress.

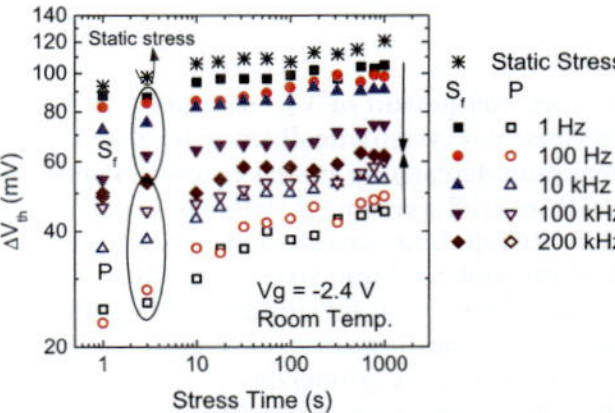

Fig. 11. From R-D model, ΔV_{th} under dynamic stress is simulated. The S_f point and P points (see Fig. 3) should merge to a frequency-independent trend-line after long time stress.

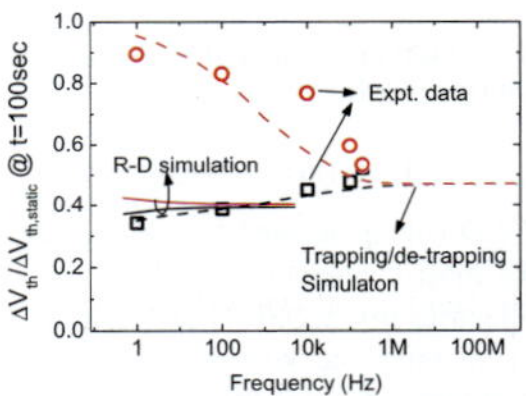

Fig. 12. Experimental ΔV_{th} data under dynamic stress. The difference between S_f and P points is large and not closing after long time stress.

Fig. 13. The trapping/de-trapping model predicts that, under dynamic stress, ΔV_{th} at both S_f and P points are frequency dependent, and the difference between the two is large.

reasonable explanation to all experimental observations.

V. ACNOWLEDGEMENT

This work is supported by A*STAR Grant R398000019305, and by Applied Materials Inc. through research grant R263C00385720. C.S. would like to thank H. Reisinger of Infineon and Y.G. Zhao of Keithley for inspiring discussions on fast measurement techniques.

IV. CONCLUSIONS

The inconsistency in the existing reports on the fast components of NBTI is resolved with a review on the major techniques. The conventional on-the-fly measurement is shown to be unreliable unless the initial pre-stress measurement is done using fast method to minimize initial stress. R-D model is shown to be fundamentally incompatible with the fast NBTI data, while charge trapping/de-trapping model provides

REFERENCES

[1] M. Denais, *et al.*, IEDM Tech. Digest 2004, pp. 109-112.
[2] S. Rangan *et al.*, IEDM Tech. Digest 2003, pp. 341-344.
[3] T. Yang *et al.*, VLSI Symp. 2005, pp. 92-93.
[4] T. Yang *et al.*, EDL 26, pp. 826-828 (2005).
[5] D. Varghese *et al.*, IEDM Tech. Digest 2005, pp. 701-704.
[6] C. Shen *et al.*, EDL 27, pp. 55-57 (2006).
[7] C. D. Young *et al.*, SSDM 2004, pp. 216-217.
[8] S. Chakravarthi *et al.*, IRPS Proceedings 2004, pp. 273-282.
[9] A. T. Krishnan *et al.*, IEDM Tech. Digest 2005, pp. 705-708.
[10] M. Alam *et al.*, IEDM Tech. Digest 2003, p.345.

IEEE TRANSACTIONS ON ELECTRON DEVICES, VOL. 53, NO. 12, DECEMBER 2006   3001

Fast V_{th} Instability in HfO$_2$ Gate Dielectric MOSFETs and Its Impact on Digital Circuits

Chen Shen, *Student Member, IEEE*, Tian Yang, Ming-Fu Li, *Senior Member, IEEE*, Xinpeng Wang, C. E. Foo, Ganesh S. Samudra, Yee-Chia Yeo, *Member, IEEE*, and Dim-Lee Kwong, *Senior Member, IEEE*

Abstract—**Fast component of V_{th} instability in MOSFET with HfO$_2$ gate dielectric is systematically measured and characterized. A charge-trapping/detrapping model is used to simulate the V_{th} instability with overall agreement with the experiments. Experimental and modeling data provide and predict the fast V_{th} shift under both static and dynamic stress conditions. These data are incorporated into HSpice circuit simulation to evaluate the impact of V_{th} shift on the performance of digital circuit in realistic situations. Considering the properties of the fast V_{th} instability, circuit performance can be optimized by circuit design in addition to process improvements. This should be included to the guideline of process development and circuit design for future CMOSFET digital systems.**

Index Terms—**CMOSFETs, digital circuits, reliability, static random access memory (SRAM), trapping.**

I. INTRODUCTION

THRESHOLD voltage instability due to charge trapping is one of the most challenging problems for incorporating high-κ gate dielectrics in CMOSFET technology [1]–[8]. Early studies [1]–[3] on threshold voltage instability in HfO$_2$ and other high-κ gate dielectric used the traditional measure-stress-measure scheme to characterize the evolution of threshold voltage shift, with dc parametric analyzers such as HP4156. These studies suggested that the V_{th} shift is due to electron and hole trapping in high-κ film and is moderate, although much larger than that in SiO$_2$. However, Kerber *et al.* revealed that there is significant V_{th} shift within the time frame of tens of

microseconds [4], using a fast I_d–V_g measurement technique. When the stress is removed, the V_{th} degradation quickly recovers. This fast degradation/recovery in threshold voltage shift has a magnitude of a few hundred millivolts, which is much larger than previously observed with slow measurement.

Most authors agree that in HfO$_2$, electron/hole trapping and detrapping in preexisting traps are responsible for the threshold voltage shift observed in both fast and slow measurements [1]–[8]. Shen *et al.* suggested that distinctive fast and slow charge traps exist in HfO$_2$ gate dielectric, based on the observation that charge-trapping measurement using fast and slow methods yields opposite dependence on the frequency of stress voltage signal [7]. One may infer that the fast and slow charge-trapping components may be associated with different defects or trapping mechanisms, and can be studied separately. The fast charge trapping causes threshold voltage to shift over 100 mV, and leads unacceptably large drive current degradation, which renders the fast charge trapping the main show-stopper for HfO$_2$ as gate dielectric material. The slow charge-trapping component is a reliability problem, while the fast charge trapping is both a time-zero and a reliability problem.

Vast effort have been dedicated to reducing the charge trapping in HfO$_2$, and several approaches have been demonstrated to reduce the slow charge-trapping component [2], [9]–[15], or to reduce the fast charge trapping [5], [8]. However, as significant fast hole trapping was also recently discovered in the SiON dielectric [16]–[18], we may not be able to totally eliminate the fast charge trapping in high-κ film. We likely have to accept a certain amount of fast V_{th} instability in MOSFETs with high-κ dielectric. As a result, accurate modeling of the fast charge-trapping component is mandatory to assess its impact on circuit performance, and to determine the maximum allowable fast charge trapping as the target of process improvement effort. In this paper, we systematically characterized the fast charge-trapping component in HfO$_2$ gate dielectric, propose a model and assess its impact on digital circuits.

Manuscript received April 7, 2006; revised July 11, 2006. This work was supported by the Singapore A*STAR Nanoelectronics Research Program under Grant R398-000-019-305. The review of this paper was arranged by Editor G. Groeseneken.

C. Shen, X. Wang, C. E. Foo, and G. S. Samudra are with the Silicon Nano-Device Laboratory and the Center for Integrated Circuit Failure Analysis and Reliability, Department of Electrical and Computer Engineering, National University of Singapore, Kent Ridge 119260, Singapore.

T. Yang was with Silicon Nano-Device Laboratory, National University of Singapore, Kent Ridge 119260, Singapore and was with Agere Systems, Singapore 118256.

M.-F. Li is with the Silicon Nano-Device Laboratory and the Center for Integrated Circuit Failure Analysis and Reliability, Department of Electrical and Computer Engineering, National University of Singapore, Kent Ridge 119260 Singapore and also with the Institute of Microelectronics, Singapore 117685.

Y.-C. Yeo is with the Silicon Nano-Device Laboratory and Center for Integrated Circuit Failure Analysis and Reliability, Department of Electrical and Computer Engineering, National University of Singapore, Kent Ridge 119260, Singapore and also with the Agency for Science, Technology, and Research (A*STAR), Singapore 138668.

D.-L. Kwong is with the Microelectronics Research Center, Department of Electrical and Computer Engineering, The University of Texas at Austin, Austin, TX 78758 USA.

Digital Object Identifier 10.1109/TED.2006.885680

II. SAMPLE PREPARATION AND MEASUREMENT TECHNIQUE

Both n- and p-MOSFETs with 4.0-nm metal–organic chemical vapor deposition (MOCVD) HfO$_2$ gate dielectric and HfN/TaN metal gate stack were fabricated, with the process flow detailed in [19]. The equivalent electrical thickness of the HfO$_2$ gate dielectric is 1.3 nm after source/drain (S/D) anneal at 950 °C.

As mentioned earlier, a fast characterization technique is required to study the fast charge trapping in HfO$_2$. The

3002   IEEE TRANSACTIONS ON ELECTRON DEVICES, VOL. 53, NO. 12, DECEMBER 2006

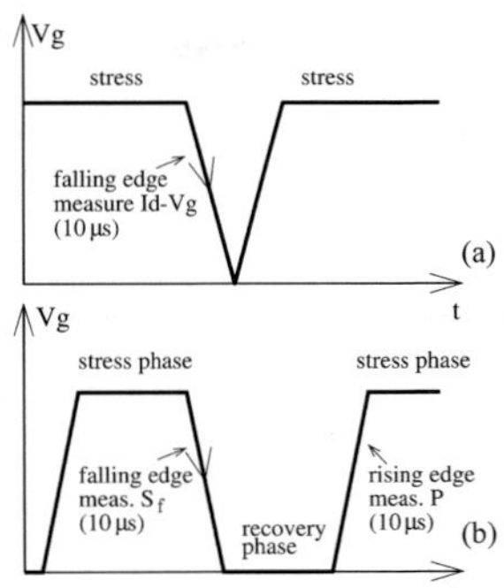

Fig. 1. Waveform of stress voltage used in (a) static stress and (b) dynamic stress.

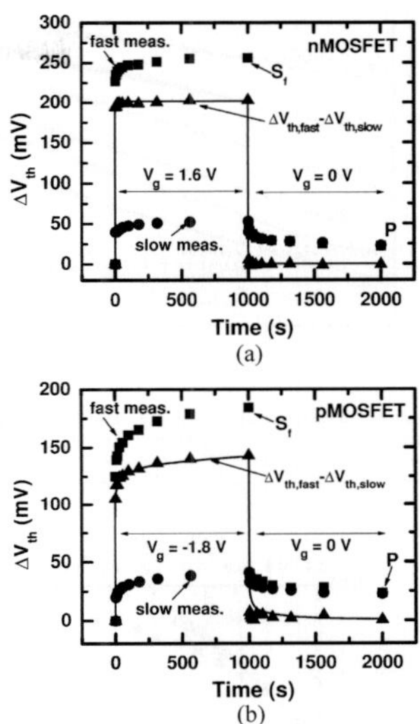

Fig. 2. Threshold voltage shift under stress/recovery cycles with frequency = 1/2000 Hz. Results of fast ($\Delta V_{\mathrm{th,fast}}$, squares) and slow ($\Delta V_{\mathrm{th,slow}}$, dots) measurements are compared. The difference between fast and slow measurement is plotted with triangles, representing the fast trap contribution to the total V_{th} degradation. Solid line shows simulation data as described in Section III-F. (a) nMOSFET. (b) pMOSFET.

single-pulse measurement originally proposed by Kerber *et al.* [4] was not sufficiently quantitative, and has difficulty in short measurement time down to 1 μs. Young *et al.* used RF measurement technique and a multipulse approach [20] to measure the intrinsic properties of high-κ MOSFET with ultrashort pulses (35 ns each). However, the multipulse scheme needs a series of pulses to obtain the I_d–V_g characteristics, which takes long time and is not suitable in studying the charge-trapping characteristics. In this paper, we use an improved single-pulse scheme with the best measurement speed reduced to 1 μs, noise suppressed, and sources of error analyzed [21]. Measurement time in this paper is 10 μs, which has been shown to be fast enough for evaluating the fast detrapping states in these samples [7]. Both static and dynamic stress are possible with this setup. In the case of static stress, dc stress voltage is applied to the gate of the MOSFET, with pulses down to 0 V intermittently inserted, as shown in Fig. 1(a). I_d–V_g curve is measured as the gate voltage is dropping from stress voltage to zero, and the threshold voltage is extracted. In the case of dynamic stress, square-wave stress voltage is applied on the gate of the MOSFET, and I_d–V_g characteristics can be measured at both the rising-edge or the falling-edge of the waveform. In the case of nMOSFET, the threshold voltage extracted from the I_d–V_g curve obtained at the falling-edge represents the V_{th} degradation right after the stress phase (S_f point), while the V_{th} measured at the rising-edge represents the partially recovered V_{th} after the recovery phase (P point).

Fig. 2 shows time evolution of V_{th} degradation in n- and p-MOSFETs under 1000 s of static stress followed by 1000 s of recovery, using the V_g waveform shown in Fig. 1. In a way, this can be viewed as the first stress/recovery period of a dynamic stress with frequency = 1/2000 Hz. As a comparison, slow measurement with HP4156A parametric analyzer is also used in this paper to study the slow component of charge trapping in HfO$_2$. A large difference in ΔV_{th} measured by fast technique ($\Delta V_{\mathrm{th,fast}}$) and slow technique ($\Delta V_{\mathrm{th,slow}}$) is observed. This difference ($\Delta V_{\mathrm{th,fast} } - \Delta V_{\mathrm{th,slow}}$) is plotted with triangles in the figure, and reflects the contribution of fast charge trap com-

ponent to the total V_{th} degradation. However, the time constants of detrapping for fast and slow traps do not have a clear-cut boundary. Therefore, the value of $\Delta V_{\mathrm{th,slow}}$ from slow measurement on one hand is the slow charge-trapping component with some amount of recovery (detrapping), and on the other hand may contain some fast charge-trapping component. As the fast component is observed to be much larger than the slow component, we decide to use the threshold voltage shift measured by fast measurement ($\Delta V_{\mathrm{th,fast}}$) to estimate the fast component in this paper, unless otherwise stated. Comments on the effect of subtracting the contribution of the slow component is provided when it is necessary. The fast charge-trapping component is huge after only 1 s of stress, but the growth of ΔV_{th} after that is slow. When stress voltage is removed, the fast component of threshold voltage shift quickly recovers to almost zero.

Alternatively, Chan *et al.* proposed to monitor the recovery of the linear-region drain–current after stress voltage is removed [22], [23]. After high voltage stress is removed, a voltage slightly higher than the threshold voltage is applied on the gate, and the drain–current is plotted against time. The recovery characteristics of an nMOSFET is plotted in Fig. 3. The linear-region drain–current is proportional to the gate overdrive ($V_g - V_{\mathrm{th}}$), and the threshold voltage shift is approximated from the relation $\Delta I_d = g_m \cdot \Delta V_{\mathrm{th}}$. In the initial period of the recovery, the slope the recovery characteristics are similar. As the recovery progresses, the slope reduces and recovery saturates with a turning point that depends on the time of stress.

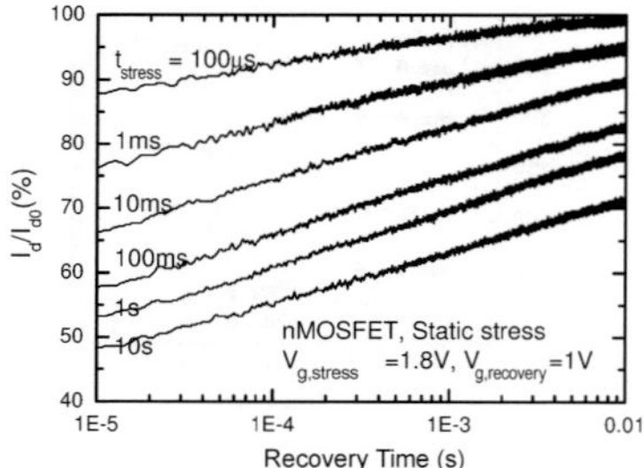

Fig. 3. Recovery of linear region drain–current I_d with respect to the prestress I_{d0}, after stress voltage is removed from the gate of the nMOSFET. Recovery after different stress time is shown. The drain–current during recovery is measured by the fast measurement, at $V_g = 1.0$ V, and $V_d = 0.1$ V.

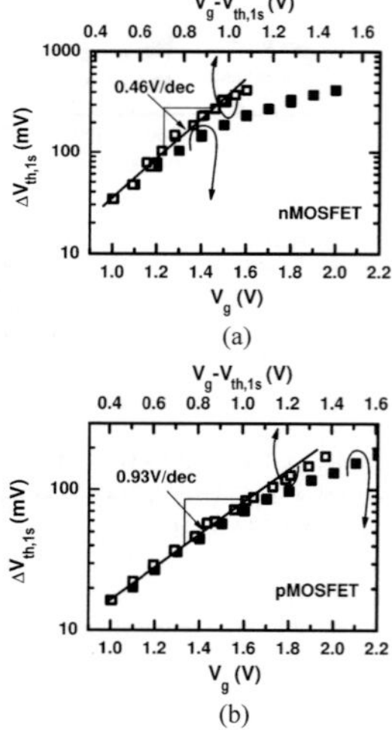

Fig. 4. V_{th} shift dependence on static stress voltage and gate overdrive at the end of a 1-s stress $(V_g$-$V_{\text{th,1s}})$, measured by fast technique. Exponential dependence of $\Delta V_{\text{th,1s}}$ on the gate voltage overdrive V_g-$V_{\text{th,1s}}$ is observed.

III. CHARACTERIZATION OF FAST CHARGE TRAPS

A. Voltage Dependence

To investigate the voltage dependence of the fast charge-trapping component, transistors are stressed at different voltage levels for 1 s (static stress), and the change in V_{th} compared to the fresh device is plotted against the stress voltage, or alternatively against the gate overdrive voltage at the end of the stress, as shown in Fig. 4. As the fast charge-trapping component builds up in very short time, the ΔV_{th} after 1 s provides a good estimate on the steady-state ΔV_{th}.

It is proposed that the observed exponential dependence of $\Delta V_{\text{th,1s}}$ on the gate overdrive is characteristic of the fast charge-trapping component. In the case of nMOSFET, assume that the fast electron traps in HfO$_2$ are shallow with respect to the conduction band of HfO$_2$, and are distributed in space scale [24] or in energy scale. These preexisting trap sites can be occupied by electrons only if the trap level is below or near the energy of electron injection. Qualitatively, under larger positive gate stress voltage, trap levels in HfO$_2$ are moved downwards with respect to the substrate Fermi level, therefore more traps become available for charge trapping. However, as the V_{th} increases under stress while V_g is kept constant, the electric field across dielectric decreases, and fewer fast traps are available. Therefore, the amount of fast charge trapping is dependent on the gate overdrive (or electric field) at the end of the stress $(V_g$-$V_{\text{th,1s}})$, as shown in Fig. 4. The voltage dependence of $\Delta V_{\text{th,1s}}$ in nMOSFET has a larger slope (0.46 V/dec) than that of pMOSFET (0.93 V/dec), showing a stronger voltage dependence. The voltage dependence obtained from slow measurement is weaker (close to 1.2 V/dec) for both n- and p-MOSFET. As a result, if the slow component were to be subtracted off from the voltage dependence shown in Fig. 4, the actual slope would be slightly steeper.

It should be noted that, although the amount of trapped charge (or ΔV_{th}) shows exponential dependence on electric field, the density of available trap sites does not necessarily have the same field dependence. As Nissan-Cohen *et al.* pointed out [25], the density of trapped charge in steady state is determined by the balance between charge trapping and detrapping processes, the ratio of trapped charge density to trap density depends on the relative strength of trapping versus detrapping, which in turn may be field dependent. This gives rise to a new dimension of complications, and a quantitative physical model of the exponential voltage dependence could not be reached in this paper. Nevertheless, the empirical relationship obtained above is used in later sections.

B. Frequency Dependence

Figs. 5 and 6 show the frequency dependence of V_{th} shift under dynamic stress. As an example with extremely low frequency, in Fig. 2, the ΔV_{th} after stress phase is labeled S_f point, and the ΔV_{th} after recovery phase is labeled P point. As shown in Figs. 5 and 6, when the stress frequency is increased, the maximum dynamic V_{th} degradation (S_f point) decreases, while the cumulative dynamic V_{th} degradation (P point) measured by the fast technique increases.

The frequency dependence of the slow component, which was discussed in [6], [7], is opposite to the frequency dependence of the P point described above. Therefore, even after subtracting the slow charge-trapping component from the total V_{th} shift measured here, the V_{th} degradation at P point still increases with frequency. The opposite frequency dependence of the fast and slow components in charge trapping is one important clue suggesting that the two components are distinctive and may have different physical origins. In previous work, we have shown that a two-step trapping/detrapping model is required to explain the peculiar frequency dependence of

3004

IEEE TRANSACTIONS ON ELECTRON DEVICES, VOL. 53, NO. 12, DECEMBER 2006

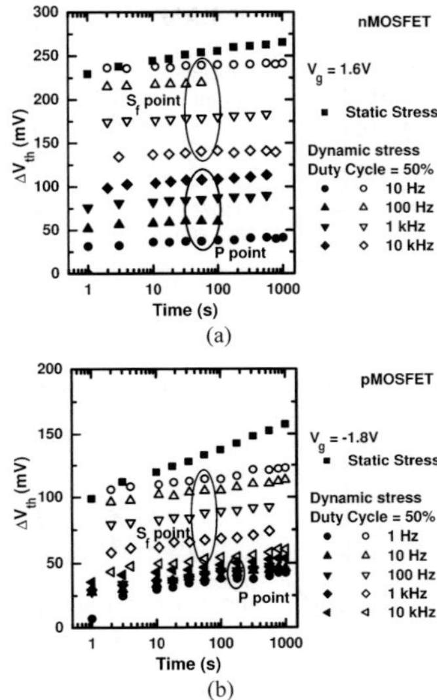

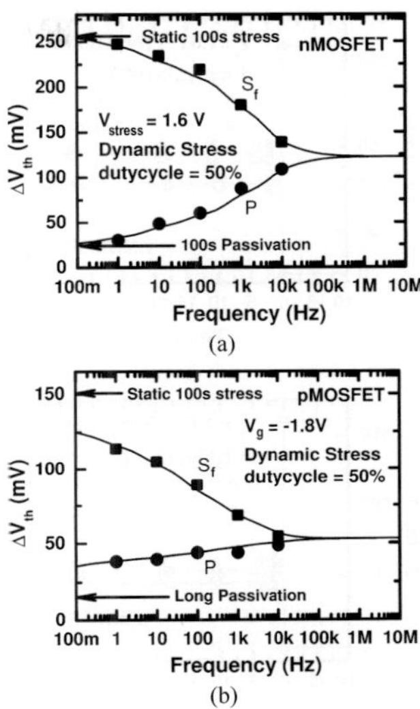

Fig. 5. Threshold voltage shift under static stress and dynamic stress of different frequencies, measured by fast technique. For dynamic stress, both the ΔV_{th} at the end of stress phase (S_f) point and at the end of recovery phase (P) point are plotted. (a) nMOSFET. (b) pMOSFET.

Fig. 6. Frequency dependence of ΔV_{th} after 100 s of dynamic stress, using fast measurement. Squares represents the ΔV_{th} at the end of the stress phase in a stress/recovery cycle, while dots represents the ΔV_{th} at the end of the recovery phase. As frequency increases, the amplitude of ΔV_{th} in a stress/recovery cycle (difference between S_f point and P point) reduces, and the accumulated ΔV_{th} (P point) increases. Solid lines shows simulation result. (a) nMOSFET. (b) pMOSFET.

the slow component [6], [26]. The fast component, on the other hand, can be explained with conventional charge-trapping dynamics, as will be discussed in Section III-F.

The difference between the ΔV_{th} of S_f point and P point reflects the transient amplitude as illustrated in Fig. 2. In the high-frequency limit, S_f point and P point converges, which is most evident in Fig. 6. As frequency increases, in one period of dynamic stress, the degradation in the stress phase and the recovery in recovery phase both decreases, so the threshold voltage transient amplitude reduces, and approaches zero at high frequency. As a result, V_{th} of the MOSFET with HfO_2 dielectric, although exhibiting large instability, can be predicted when it is switching at high-frequency, if the operation voltage and duty cycle is known. This property may have important implication to digital circuits, as we attempt to explore in Section IV-B.

C. Stress History

Fig. 7 shows that evolution of ΔV_{th} under dynamic stress with two different prior conditions: (1) 100-s static stress and (2) zero stress applied, on two identical transistors. It is obvious that prior condition or stress history does not affect the steady-state V_{th} shift of dynamic stress. This independence on prior stress history agrees with the assumption that all the traps in the dielectric are preexisting and are not generated during the stress. The steady-state ΔV_{th} is determined by stress condition

(voltage, duty cycle, etc.), but not by stress history. The transition from previous steady-state ΔV_{th} to a new equilibrium takes less than 1 s as observed in Fig. 7.

D. Duty-Cycle Dependence

Figs. 8 and 9 show the dependence of ΔV_{th} on the duty cycle of the dynamic stress. In Fig. 8, the duty cycle dependence is rather weak in the range of 20%–80%. Due the limitation of the measurement scheme, it is difficult to investigate the situation for duty cycle < 20% or > 80%. We are not yet able to understand the weak duty cycle dependence in the middle range and the steep dependence at the low and high ends. We suspect that the contribution from the slow charge component may have distorted the duty cycle dependence. From Fig. 9, the voltage dependence of the dynamic stress of different duty cycle is identical (same slope) to that observed in static stress.

E. Temperature Dependence

Fig. 10 shows the temperature dependence of $\Delta V_{th,1s}$. Negative temperature dependence is observed in the V_{th} shift of nMOSFETs, while pMOSFETs show almost no temperature dependence. Therefore, discussion based on room temperature situation represents the worst case.

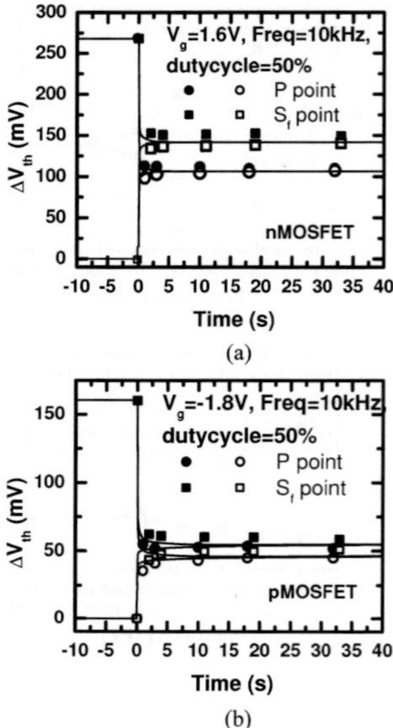

Fig. 7. Evolution of ΔV_{th} during transition from static stress to dynamic stress (filled symbols), and ΔV_{th} of fresh device under dynamic stress (open symbols). The steady-state V_{th} shift of dynamic stress does not depend on the prior stress history, which is expected if all the traps are preexisting. Transition time from static to dynamic stress is in the 100-ms time scale. Solid lines show simulation result. (a) nMOSFET. (b) pMOSFET.

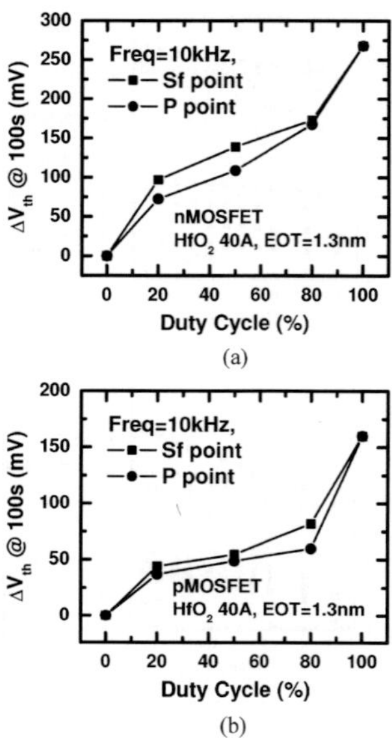

Fig. 8. ΔV_{th} after 100-s dynamic stress of different duty cycle, but same stress voltage, frequency, and rise/fall time. (a) nMOSFET. (b) pMOSFET.

F. Modeling of the Fast V_{th} Instability

Empirical models of the voltage and temperature dependence of charge-trapping/detrapping process could be directly obtained from the characterization described above. However, more physical insights are required to model the dynamic behavior of the fast charge trapping, specifically its dependence on frequency. We follow the dynamic model proposed by Nissan-Cohen [25], and write the equations for electron trapping, electron detrapping, hole trapping, and hole detrapping, respectively

$$\frac{dn}{dt} = \frac{1}{\tau_{C,n}}(N_{\text{ot},n} - n) - \frac{1}{\tau_{E1,n}}n \tag{1}$$

$$\frac{dn}{dt} = -\frac{1}{\tau_{E2,n}}n \tag{2}$$

$$\frac{dp}{dt} = \frac{1}{\tau_{C,p}}(N_{\text{ot},p} - p) - \frac{1}{\tau_{E1,p}}p \tag{3}$$

$$\frac{dp}{dt} = \frac{1}{\tau_{E2,p}}p \tag{4}$$

where $n(p)$ is the trapped electron (hole) concentration in oxide traps, which has a total density $N_{\text{ot},n}(N_{\text{ot},p})$, respectively. A series of time constants τ are used to characterize the rate of individual processes, with subscript "C" denoting time constant for trapping process, subscript "E" for detrapping, "n" for electrons, and "p" for holes, respectively. For n-MOSFET, only electron trapping/detrapping are considered, while for p-MOSFET, only holes are considered.

In principle, with given initial condition, successively solving (1) and (2) for each stress/recovery cycle, and repeating the procedure for m cycles yields the trapping/detrapping behavior in n-MOSFET. However, as millions of such cycles are involved in this paper, numerical error can accumulate to an unacceptable level with iterative calculation. Instead the close-form expression for the density of trapped electrons after m cycles is derived (see Appendix A), assuming that the period is T and the duty cycle is γ

$$n = \left[\frac{\lambda(1-A)AB}{1-AB}[1-(AB)^m] + \lambda(1-A)\right]N + A(AB)^m n_0 \tag{5}$$

with

$$\lambda = \frac{\tau_{E1,n}}{\tau_{C,n} + \tau_{E1,n}} \tag{6}$$

$$A = \exp\left(-\gamma T\left(\frac{1}{\tau_{C,n}} + \frac{1}{\tau_{E1,n}}\right)\right) \tag{7}$$

$$B = \exp(-\gamma T/\tau_{E2,n}). \tag{8}$$

3006   IEEE TRANSACTIONS ON ELECTRON DEVICES, VOL. 53, NO. 12, DECEMBER 2006

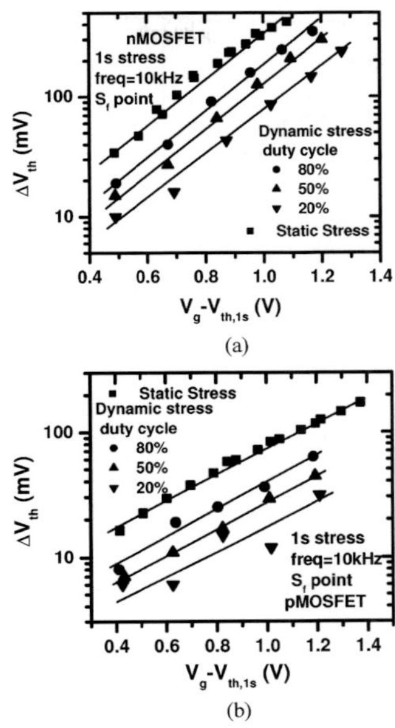

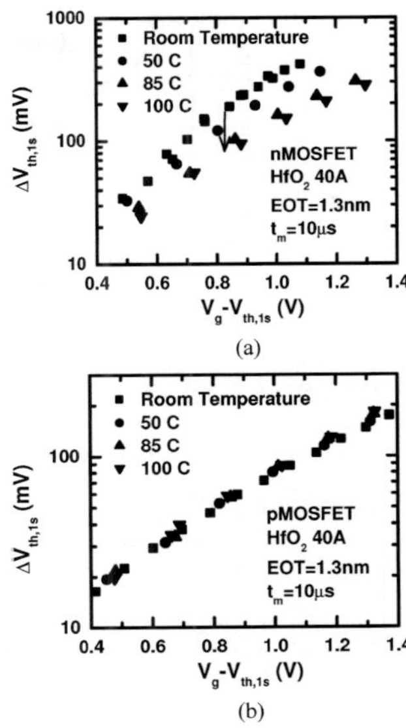

Fig. 9.   Stress voltage dependence of ΔV_{th} under static and dynamic stress of different duty cycle. The slope in voltage dependence is the same for static stress and dynamic stress of different duty cycles. (a) nMOSFET. (b) pMOSFET.

Fig. 10.   Stress voltage dependence of ΔV_{th}, stressed under different temperatures. nMOSFET shows negative temperature dependence, while pMOSFET shows zero temperature dependence. (a) nMOSFET. (b) pMOSFET.

The equations for hole trapping and detrapping in p-MOSFET can be derived in a complementary procedure.

We assumed that N_{ot} has a widespreading spectrum of trapping and detrapping constants τ_C and τ_E [1], as shown in Fig. 11. This distribution is obtained by fitting of simulation result to the experimental results. Such broad distribution on time constants must be used to explain the gradual frequency dependence observed in Fig. 6. Note that the correlation between τ_{E1}, τ_{E2}, and τ_C can be quite complicated in reality. To reduce the number of fitting parameters, we chose the simplest linear correlation. It is assumed that for a trap with trapping time constant τ_C, the detrapping time constants during stress and recovery are $\tau_{E1} = a \cdot \tau_C$ and $\tau_{E2} = b \cdot \tau_C$, respectively, where a and b are constants. This simple relationship has enable us to reproduce most of the characteristics observed in experiments. The correct frequency dependence is obtained from solving (1)–(4) and is in good agreement with experiments, as shown in Figs. 2, 6, and 7 as solid lines.

Under very high frequency, fast charge trapping is predicted to be frequency independent, as shown in Fig. 6. This is because in the high-frequency limit, $T \to 0$ and the solution (5)–(8) approaches constant. This prediction enables us to use the experimental data obtained at low frequency to assess the effect of charge trapping in real digital circuit, which operates at much higher frequency.

The recovery in drain–current, after removal of stress, is also simulated with the detrapping time constant distribution obtained above, and is shown in Fig. 12, which also agrees with the experimental observations qualitatively (Fig. 3). Note that in Fig. 3, the gate voltage during recovery is 1 V, while the time constants used here is obtained from the case of 0-V recovery voltage. This difference may explain the apparent faster recovery in simulation than in experiment.

IV. Impacts on Digital Circuits

After careful characterization and modeling of the fast traps, we are ready to look at its impact on digital circuits. Our discussion is based on both the experimentally measured ΔV_{th} and the high-frequency dynamic stress data projected by the charge-trapping model. We used HSpice and a 65-nm predictive technology model (vth0 = 0.22 V) [27] for circuit simulation. In the simulation, the parameter vth0 is changed to reflect the V_{th} shift due to trapping [28]. However, the effect of mobility degradation is ignored.

A. Static Random Access Memory (SRAM)

In the widely used dynamic voltage scaled system, V_{DD} can be regulated to lower values for low power operation [29]. Static noise margin (SNM) of SRAM can limit the minimum allowed operation voltage. In a 6T SRAM cell shown in Fig. 13, the worst case for SNM occurs when the storage node x

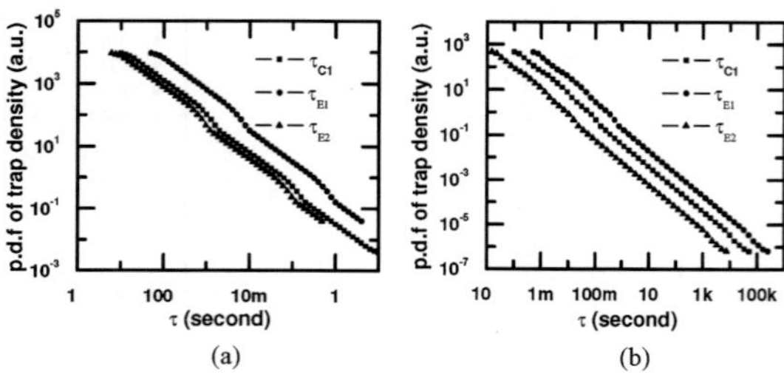

Fig. 11.   Spectrum of trapping (τ_C) and detrapping time constants (τ_{E1}, τ_{E2}) used in calculation from eqs. (1) and (4), for (a) electrons and (b) holes. Broad distribution on time constants must be used to explain the gradual frequency dependence observed in Fig. 6.

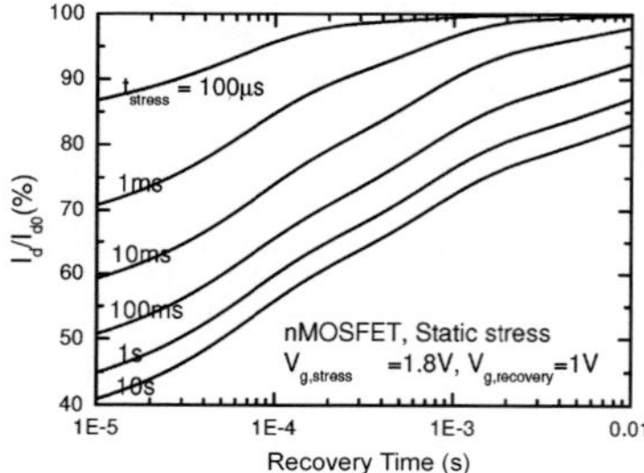

Fig. 12.   Simulation of recovery of linear region drain–current I_d after stress voltage is removed from the gate of the nMOSFET, as shown in Fig. 3.

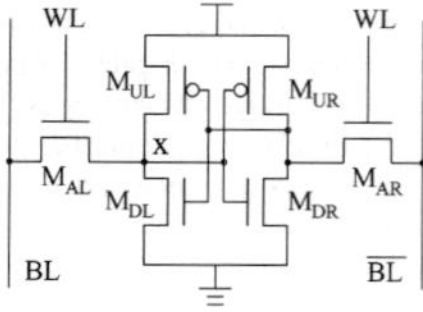

Fig. 13.   Schematics of 6T SRAM cell.

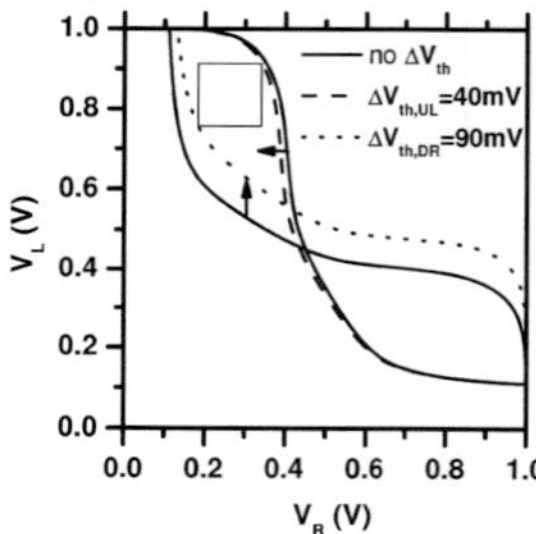

Fig. 14.   Butterfly plot of the SRAM cell showing the transfer characteristics between voltage at the left storage node (V_L) and that at the right storage node (V_R). V_{th} increase of M_{UL} causes the V_L-V_R curve to shift to the left, and V_{th} increase of M_{DR} causes the V_R-V_L curve to shift up. The SNM measured by the maximum square enclosed in the butterfly plot, is therefore reduced.

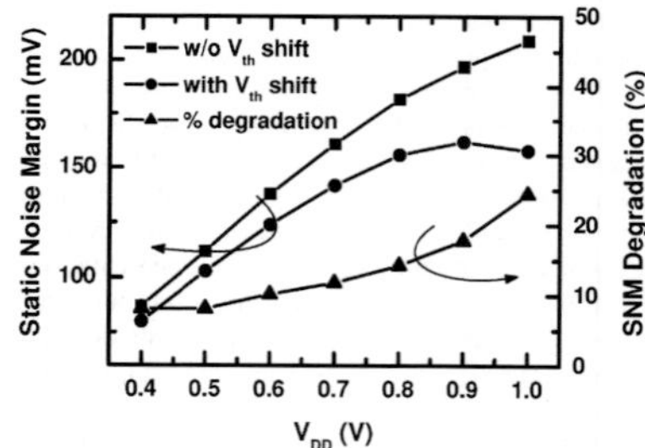

Fig. 15.   SNM of SRAM cell with cell ratio $\beta = 2$. Square symbols show the SNM with no V_{th} degradation, dots show the case with worst case V_{th} degradation, triangles show the percentage loss in SNM due to worst case V_{th} degradation. Under low V_{DD}, the SNM degradation due to fast V_{th} instability is much reduced.

previously stored "1" for long time, and is reading the stored "1." Therefore, the gate voltage for transistors M_{DR} and M_{UL} have been at V_{DD} for long time (statically stressed), and have high V_{th}. The other four transistors in the cell have been in recovery for long time, and thus have normal V_{th}.

Using the V_{th} shift value extracted from Fig. 4, SNM can be obtained from circuit simulation. In the butterfly plot shown in Fig. 14, the increased V_{th} of M_{DR} causes one curve to shift up, and the increased V_{th} of M_{UL} causes one curve to shift to the left. The SNM, measured by the maximum square enclosed by the butterfly plot, decreases when the worst case threshold voltage shift is considered.

In Fig. 15, SNM calculated with and without considering V_{th} degradation is compared, for an SRAM cell with cell ratio $\beta = 2$. The dimensions (W/L) for the cell nMOSFET is 140/70 nm, pMOSFET is 90/70 nm, and access transistor is 90/90 nm. As V_{DD} increases, the percentage loss in SNM increases, due to the exponential voltage increase of V_{th} shift. Under reduced V_{DD}, the much reduced fast V_{th} instability

3008   IEEE TRANSACTIONS ON ELECTRON DEVICES, VOL. 53, NO. 12, DECEMBER 2006

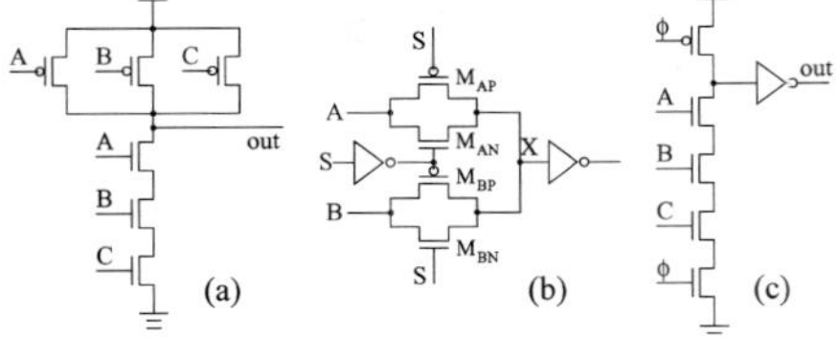

Fig. 16.   Schematics of (a) NAND3 gate implemented with static logic, (b) two-input multiplexer implemented with CMOS transmission gate, and (c) NAND3 gate implemented with dynamic logic.

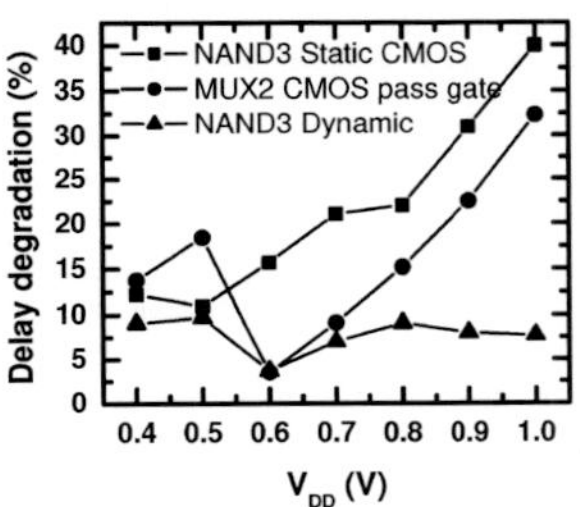

Fig. 17.   Percentage increase in gate propagation delay due to V_{th} degradation as function of supply voltage. Logic gates implemented with transmission gate and dynamic logic shows much reduced degradation in delay.

causes little SNM degradation. In this low V_{DD} situation, other factors, such as random dopant fluctuation and process variation, will dominate the SNM degradation. Therefore, the fast V_{th} instability may not be the limiting factor to the minimum operating voltage of SRAM cell.

B. Logic Circuits

Considering the impact of fast V_{th} shift on ring oscillators, the circuit enjoys the benefit of being dynamically stressed (less ΔV_{th}), and shows improved performance at high frequency [30]. However, in the most general case in logic circuits, ΔV_{th} under static stress should be used to determine the worst case propagation delay. This is because 1) certain circuit input may be static and 2) When one switches from static stress to dynamic stress, it takes relatively long time for the V_{th} of the MOSFET to transit from the steady-state value under static stress to that under dynamic stress. This transition time is estimated to be about 100 ms from the simulation performed in Fig. 7), which is many orders of magnitude slower compared with the switching frequency in circuits.

For the three-input NAND gate (NAND3) implemented with static CMOS logic shown in Fig. 16(a), worst case delay degradation occurs when all three inputs were previously high for long time, all switched to low briefly, and are switching to high at the same time. The V_{th} recovery during the brief low inputs period is negligible. Therefore, V_{th} shift under static stress must be used to determine the propagation delay degradation. Delay

degradation is plotted as a function of the supply voltage in Fig. 17. Delay degradation can be well approximated by the degradation in the $I_{d,\text{sat}}$ (due to ΔV_{th}) of the nMOSFETs [28]. Therefore, other logic gates and circuits implemented with the static logic style would have similar amount of degradation as long as ΔV_{th} of static stress determines the performance.

For CMOS transmission gate, at most one transistor in the pair is under stress at any time, while both transistors conduct when switched on [Fig. 16(b)]. For example, M_{AP} is under static stress when $S = 0$, $A = X = 1$, but in this situation $\bar{S} = A = X = 1$, so M_{AN} is not under stress. When the transmission gate turns on, both M_{AN} and M_{AP} contribute to the current conduction. If M_{AN} and M_{AP} have equal contribution to the total conductance, the percentage delay degradation is about half of the drive current degradation of M_{AP}. However, the static CMOS inverter for complementary control signal $\bar{S}$ generation makes the delay degradation closer to the static logic case. Overall, CMOS transmission gate logic (MUX2) shows less delay degradation than NAND3 (Fig. 17), although ΔV_{th} under static stress is considered.

For dynamic logic [Fig. 16(c)], transistors are all dynamically stressed at the clock frequency with a predefined duty cycle. In precharge phase, clock ϕ is low, all the four nMOSFETs are cutoff (detrapping), regardless of the values at the three inputs. Therefore, the nMOSFETs in the pull-down network are stressed dynamically with duty cycle not greater than the clock duty cycle. In the buffer inverter that feeds to the next stage, only low-to-high transition need to be optimized, so the pull-up pMOSFET is more important. The pMOSFET is also stressed dynamically, and the stress duty cycle is no worse than the clock duty cycle. Therefore, V_{th} shift under dynamic stress represents the worst case for dynamic logic circuits in terms of propagation delay. With the much lower V_{th} shift under high-frequency dynamic stress (Fig. 6), the delay degradation of the dynamic NAND3 gate is much lower than the static logic. As the most delay sensitive parts of the circuits are often implemented with dynamic logic, the degraded delay of static CMOS logic in noncritical path might not be the limiting factor of overall circuit speed.

It is not possible to exhaustively examine all building blocks in logic circuits in this paper. However, it is demonstrated that the use of certain circuit designs can take advantage of the dynamic nature of charge-trapping effect in a deterministic manner. This potentially allow the performance of logic circuit to be less affected by fast V_{th} instability, and thus impose less stringent target to the optimization of high-κ dielectrics.

V. SUMMARY

The fast component of V_{th} in MOSFET with HfO$_2$ gate dielectric is systematically characterized and modeled. ΔV_{th} due to fast charge trapping is shown to be predictable in both static and dynamic stress situations. HSpice circuit simulation is performed based on experimental ΔV_{th} to evaluate the impact of V_{th} shift on the performance of digital circuits in more realistic operating conditions. Circuit performance should be optimized with both process improvement and circuit design techniques.

APPENDIX A
DERIVATION OF THE DYNAMIC TRAPPING EQUATIONS

The electron trapping (1) can be solved analytically, and the solution can be written in matrix form

$$\begin{pmatrix} n \\ N \end{pmatrix} = \begin{pmatrix} A & \lambda(1-A) \\ 0 & 1 \end{pmatrix} \begin{pmatrix} n_0 \\ N \end{pmatrix} = T \begin{pmatrix} n_0 \\ N \end{pmatrix} \quad (\text{A.1})$$

with

$$\lambda = \frac{\tau_{E1,n}}{\tau_{C,n} + \tau_{E1,n}} \quad (\text{A.2})$$

$$A = \exp\left(-t_s\left(\frac{1}{\tau_{C,n}} + \frac{1}{\tau_{E1,n}}\right)\right) \quad (\text{A.3})$$

t_s the stress time, and n_0 the initial density of trapped electrons before stress.

Similarly the solution to the detrapping, (2) is written as

$$\begin{pmatrix} n \\ N \end{pmatrix} = \begin{pmatrix} B & 0 \\ 0 & 1 \end{pmatrix} \begin{pmatrix} n_0 \\ N \end{pmatrix} = D \begin{pmatrix} n_0 \\ N \end{pmatrix} \quad (\text{A.4})$$

with

$$B = \exp(-t_r/\tau_{E2,n}). \quad (\text{A.5})$$

t_r the recovery time, and n_0 the density of trapped electron before recovery.

With the trapping and detrapping matrices T and D, the trapped charge after m stress/recovery cycle can be obtained from successively applying T and D to the initial condition

$$\begin{pmatrix} n \\ N \end{pmatrix} = (D \cdot T)^m \begin{pmatrix} n_0 \\ N \end{pmatrix}. \quad (\text{A.6})$$

The matrix $(D \cdot T)$ can be diagonalized to

$$(D \cdot T) = P \cdot G \cdot P^{-1}. \quad (\text{A.7})$$

with

$$P = \begin{pmatrix} \frac{\lambda(1-A)B}{1-AB} & 1 \\ 1 & 0 \end{pmatrix} \quad (\text{A.8})$$

$$P^{-1} = \begin{pmatrix} 0 & 1 \\ 1 & -\frac{\lambda(1-A)B}{1-AB} \end{pmatrix} \quad (\text{A.9})$$

$$G = \begin{pmatrix} 1 & 0 \\ 0 & AB \end{pmatrix}. \quad (\text{A.10})$$

With the diagonalization, (A.6) can be evaluated from

$$\begin{pmatrix} n \\ N \end{pmatrix} = PG^m P^{-1} \quad (\text{A.11})$$

and (5) is arrived.

ACKNOWLEDGMENT

The authors would like to thank P. Yan for valuable discussions.

REFERENCES

[1] S. Zafar, A. Callegari, E. Gusev, and V. Fischetti, "Charge trapping in high k gate dielectric stacks," in *IEDM Tech. Dig.*, 2002, pp. 517–520.

[2] A. Shanware, M. Visokay, J. Chambers, A. Rotondaro, H. Bu, M. Bevan, R. Khamankar, S. Aur, P. Nicollian, J. McPherson, and L. Colombo, "Evaluation of the positive biased temperature stress stability in HfSiON gate dielectrics," in *Proc. IRPS*, Dallas, TX, 2003, pp. 208–213.

[3] K. Onishi, R. Choi, C. S. Kang, H.-J. Cho, Y. H. Kim, R. Nieh, J. Han, S. Krishnan, M. Akbar, and J. Lee, "Bias-temperature instabilities of polysilicon gate HfO$_2$ MOSFETs," *IEEE Trans. Electron Devices*, vol. 50, no. 6, pp. 1517–1524, Jun. 2003.

[4] A. Kerber, E. Cartier, L. A. Ragnarsson, M. Rosmeulen, L. Pantisano, R. Degraeve, T. Kauerauf, G. Groeseneken, H. E. Maes, and U. Schwalke, "Characterization of the VT-instability in SiO$_2$/HfO$_2$ gate dielectrics," in *Proc. IEEE IRPS*, 2003, pp. 41–45.

[5] A. Shanware, M. Visokay, J. Chambers, A. Rotondaro, J. McPherson, L. Colombo, G. Brown, C. Lee, Y. Kim, M. Gardner, and R. Murto, "Characterization and comparison of the charge trapping in HfSiON and HfO$_2$ gate dielectrics," in *IEDM Tech. Dig.*, Washington, DC, 2003, pp. 939–942.

[6] C. Shen, H. Y. Yu, X. P. Wang, M.-F. Li, Y.-C. Yeo, D. S. H. Chan, K. L. Bera, and D. L. Kwong, "Frequency dependent dynamic charge trapping in HfO$_2$ and threshold voltage instability in MOSFETs," in *Proc. IRPS*, 2004, pp. 601–602.

[7] C. Shen, M. F. Li, X. P. Wang, H. Y. Yu, Y. P. Feng, A. T.-L. Lim, Y.-C. Yeo, H. Chan, and D. L. Kwong, "Negative-U traps in HfO$_2$ gate dielectrics and frequency dependence of dynamic BTI in MOSFETs," in *IEDM Tech. Dig.*, 2004, pp. 733–736.

[8] B. Lee, C. Young, R. Choi, J. Sim, G. Bersuker, C. Kang, R. Harris, G. Brown, K. Matthews, S. Song, N. Moumen, J. Barnett, P. Lysaght, K. Choi, H. Wen, C. Huffman, H. Alshareef, P. Majhi, S. Gopalan, J. Peterson, P. Kirsh, H.-J. Li, J. Gutt, M. Gardner, H. Huff, P. Zeitzoff, R. Murto, L. Larson, and C. Ramiller, "Intrinsic characteristics of high-κ devices and implications of fast transient charging effects (FTCE)," in *IEDM Tech. Dig.*, 2004, pp. 859–862.

[9] A. Morioka, H. Watanabe, M. Miyamura, T. Tatsumi, M. Saitoh, T. Ogura, T. Iwamoto, T. Ikarashi, Y. Saito, Y. Okada, H. Watanabe, Y. Mochiduki, and T. Mogami, "High mobility MISFET with low trapped charge in HfSiO films," in *VLSI Symp. Tech. Dig.*, 2003, pp. 165–166.

[10] X. Yu, C. Zhu, X. Wang, M. Li, A. Chin, A. Du, W. Wang, and D.-L. Kwong, "High mobility and excellent electrical stability of MOSFETs using a novel HfTaO gate dielectric," in *VLSI Symp. Tech. Dig.*, Honolulu, HI, 2004, pp. 110–111.

[11] X. P. Wang, M.-F. Li, C. Ren, X. F. Yu, C. Shen, H. H. Ma, A. Chin, C. X. Zhu, J. Ning, M. B. Yu, and D.-L. Kwong, "Tuning effective metal gate work function by a novel gate dielectric HfLaO for nMOSFETs," *IEEE Electron Device Lett.*, vol. 27, no. 1, pp. 31–33, Jan. 2006.

[12] E. Gusev, V. Narayanan, S. Zafar, C. Cabral, Jr., E. Cartier, N. Bojarczuk, A. Callegari, R. Carruthers, M. Chudzik, C. D'Emic, E. Duch, P. Jamison, P. Kozlowski, D. LaTulipe, K. Maitra, F. McFeely, J. Newbury, V. Paruchuri, and M. Steen, "Charge trapping in aggressively scaled metal gate/high-high-κ stacks," in *IEDM Tech. Dig.*, 2004, pp. 729–732.

[13] H.-H. Tseng, M. Ramon, L. Hebert, P. Tobin, D. Triyoso, J. Grant, Z. Jiang, D. Roan, S. Samavedam, D. Gilmer, S. Kalpat, C. Hobbs, W. Taylor, O. Adetutu, and B. White, "ALD HfO$_2$ using heavy water (D$_2$O) for improved MOSFET stability," in *IEDM Tech. Dig.*, 2003, pp. 4.1.1–4.1.4.

[14] H. H. Tseng, P. J. Tobin, E. A. Hebert, S. Kalpat, M. E. Ramon, L. Fonseca, Z. X. Jiang, J. K. Schaeffer, R. I. Hegde, D. H. Triyoso, D. C. Gilmer, W. J. Taylor, C. C. Capasso, O. Adetutu, D. Sing, J. Conner, E. Luckowski, B. W. Chan, A. Haggag, S. Backer, R. Noble, M. Jahanbani, Y. H. Chiu, and B. E. White, "Defect passivation with fluorine in a Ta$_x$C$_y$/high-κ gate stack for enhanced device threshold voltage stability and performance," in *IEDM Tech. Dig.*, 2005, pp. 696–699. no. 29.4.

[15] M. Inoue, S. Tsujikawa, M. Mizutani, K. Nomura, T. Hayashi, K. Shiga, J. Yugami, J. Tsuchimoto, Y. Ohno, and M. Yoneda, "Fluorine incorporation into HfSiON dielectric for Vth control and its impact on reliability for poly-Si gate pFET," in *IEDM Tech. Dig.*, 2005, p. 17.1.

[16] T. Yang, M. F. Li, C. Shen, C. H. Ang, C. Zhu, Y.-C. Yeo, G. Samudra, S. C. Rustagi, M. B. Yu, and D. L. Kwong, "Fast and slow dynamic NBTI components in p-MOSFET with SiON dielectric and their impact on device life-time and circuit application," in *VLSI Symp. Tech. Dig.*, 2005, pp. 92–93.

[17] T. Yang, C. Shen, M.-F. Li, C. H. Ang, C. X. Zhu, Y.-C. Yeo, G. Samudra, S. C. Rustagi, M. B. Yu, and D.-L. Kwong, "Fast DNBTI component in

3010   IEEE TRANSACTIONS ON ELECTRON DEVICES, VOL. 53, NO. 12, DECEMBER 2006

p-MOSFET with SiON dielectric," *IEEE Electron. Device Lett.*, vol. 26, no. 11, pp. 826–828, Nov. 2005.

[18] M. Denais, A. Bravaix, V. Huard, C. Parthasarathy, G. Ribes, F. Perrier, Y. Rey-Tauriac, and N. Revil, "On-the-fly characterization of NBTI in ultra-thin gate oxide PMOSFET's," in *IEDM Tech. Dig.*, 2004, pp. 109–112.

[19] H. Y. Yu, J. F. Kang, J. D. Chen, C. Ren, Y. T. Hou, S. J. Whang, M. F. Li, D. S. H. Chan, K. L. Bera, C. H. Tung, A. Du, and D.-L. Kwong, "Thermally robust high quality HfN/HfO$_2$ gate stack for advanced CMOS devices," in *IEDM Tech. Dig.*, 2003, pp. 99–102.

[20] C. D. Young, Y. G. Zhao, M. Pendley, B. H. Lee, K. Matthews, J. H. Sim, R. Choi, G. Bersuker, and G. A. Brown, "Ultra-short pulse I–V characterization of the intrinsic behavior of high-κ devices," in *Proc. SSDM*, 2004, pp. 216–217.

[21] C. Shen, M.-F. Li, X. P. Wang, Y.-C. Yeo, and D.-L. Kwong, "A fast measurement technique for MOSFET I_d–V_g characteristics," *IEEE Electron Device Lett.*, vol. 27, no. 1, pp. 55–57, Jan. 2006.

[22] C. T. Chan, C. J. Tang, C. H. Kuo, H. C. Ma, C. W. Tsai, H. C.-H. Wang, M. H. Chi, and T. Wang, "Single-electron emission of traps in HfSiON as high-κ gate dielectric for MOSFETs," in *Proc. IRPS*, 2005, pp. 41–44.

[23] C. T. Chan, C. J. Tang, T. Wang, H. C.-H. Wang, and D. D. Tang, "Positive bias and temperature stress induced two-stage drain–current degradation in HfSiON nMOSFETs," in *IEDM Tech. Dig.*, 2005, p. 22.7.

[24] A. Kerber, E. Cartier, L. Pantisano, R. Degraeve, T. Kauerauf, Y. Kim, A. Hou, G. Groeseneken, H. Maes, and U. Schwalke, "Origin of the threshold voltage instability in SiO$_2$/HfO$_2$ dual layer gate dielectrics," *IEEE Electron Device Lett.*, vol. 24, no. 2, pp. 87–89, Feb. 2003.

[25] Y. Nissan-Cohen, J. Shappir, and D. Frohman-Bentchkowsky, "Dynamic model of trapping-detrapping in SiO$_2$," *J. Appl. Phys. (USA)*, vol. 58, no. 6, pp. 2252–2261, Sep. 15, 1985.

[26] C. Shen, M. F. Li, H. Y. Yu, X. P. Wang, Y.-C. Yeo, D. S. H. Chan, and D.-L. Kwong, "Physical model for frequency-dependent dynamic charge trapping in metal-oxide-semiconductor field effect transistors with HfO$_2$ gate dielectric," *Appl. Phys. Lett.*, vol. 86, no. 9, p. 093510, 2005.

[27] Y. Cao, T. Sato, D. Sylvester, M. Orshansky, and C. Hu, "New paradigm of predictive MOSFET and interconnect modeling for early circuit design," in *Proc. CICC*, pp. 201–204. [Online]. Available: http://www.eas.asu.edu/~ptm/

[28] A. Krishnan, V. Reddy, S. Chakravarthi, J. Rodriguez, S. John, and S. Krishnan, "NBTI impact on transistor and circuit: Models, mechanisms and scaling effects," in *IEDM Tech. Dig.*, Washington, DC, 2003, pp. 14–15.

[29] T. Burd, T. Pering, A. Stratakos, and R. Brodersen, "A dynamic voltage scaled microprocessor system," *IEEE J. Solid-State Circuits*, vol. 35, no. 11, pp. 1571–1580, Nov. 2000.

[30] C. Kang, J. Lee, R. Choi, J. Sim, C. Young, B. Lee, and G. Bersuker, "Charge trapping effects in HfSiON dielectrics on the ring oscillator circuit and the single stage inverter operation," in *IEDM Tech. Dig.*, San Francisco, CA, 2004, pp. 485–488.

Tian Yang received the B.S. degree from Fudan University, Shanghai, China, and the M.Eng. degree from National University of Singapore (NUS), Kent Ridge, Singapore, in 2003 and 2006, respectively, both in electrical engineering. She is currently working toward the Ph.D. degree in electrical engineering at Purdue University, West Lafayette, IN.

From 2003 to 2005, she was with the Silicon Nano-Device Laboratory (SNDL) and the Center for Integrated Circuit Failure Analysis and Reliability, Department of Electrical and Computer Engineering, NUS where she was involved in the research of reliability of MOS transistors. From 2005 to 2006, she was with Agere System.

Ming-Fu Li (M'91–SM'99) received a degree from the Department of Physics, Fudan University, Shanghai, China, in 1960.

After graduation, he joined the University of Science and Technology of China (USTC), Hefei, China, as a Teaching Assistant and then as a Lecturer. In 1978, he joined the Graduate School Faculty, Chinese Academy of Sciences, Beijing, China, and was promoted to Professor in 1986. He was also an Adjunct Professor with the Institute of Semiconductors, Chinese Academy of Sciences, Fudan University, and USTC. He was a Visiting Scholar with Case Western Reserve University, Cleveland, OH, in 1979, and with the University of Illinois, Urbana-Champaign, from 1979 to 1981, and was a Visiting Scientist with the University of California, Berkeley, and with the Lawrence Berkeley National Laboratories from 1986 to 1987 and 1990 to 1991, respectively. He then joined the Department of Electrical Engineering, National University of Singapore, in 1991, and was promoted to Professor in 1996. He is the author of over 200 research papers and two books, including *Modern Semiconductor Quantum Physics* (Singapore: World Scientific, 1994). His current research interests include CMOS device technology, reliability, quantum modeling, and analog IC design.

Dr. Li has served on several international program committees and advisory committees in international semiconductor conferences in China, Japan, Canada, Germany, and Singapore.

Xinpeng Wang received the B.Eng. and M.Eng. degrees from Tsinghua University, Beijing, China, in 1999 and 2002, respectively. He is currently working toward the Ph.D. degree at Silicon Nano-Device Laboratory (SNDL), Department of Electrical and Computer Engineering, National University of Singapore (NUS), Kent Ridge, Singapore.

His research interests include processing, characterization, and reliability of advanced gate dielectrics and metal gates for future generation of devices.

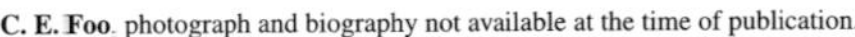

C. E. Foo. photograph and biography not available at the time of publication.

Chen Shen (S'00) received the B.Eng. degree (first class honors) in electrical engineering from the National University of Singapore (NUS), Kent Ridge, Singapore, in 2003, where he is currently working toward the Ph.D. degree at Silicon Nano-Device Laboratory (SNDL).

He received the NUS President Graduate Fellowship in 2006. His research interests include reliability physics of SiO$_2$ and high-κ gate dielectrics in CMOS device, electrical characterization techniques, and nonvolatile memory devices.

Ganesh S. Samudra received the Ph.D. degree from Purdue University, West Lafayette, IN, in 1985.

He is currently an Associate Professor with National University of Singapore (NUS), Kent Ridge, Singapore, Department of Electrical and Computer Engineering, and has been teaching for about fifteen years. He was a Visiting Professor with the Massachusetts Institute of Technology in 2001, and worked for three years at Texas Instruments (TI) before joining NUS. At TI, he worked on the development of technology computer-aided design tools linking device and circuit simulator and defining three-dimensional structures for simulation. At NUS, he is mostly involved in research projects in the area of simulation and novel devices. He has published about 100 technical papers in journals and conferences.

Yee-Chia Yeo (S'97–M'02) received the B.Eng. (with first class honors) and M.Eng. degrees in electrical engineering from the National University of Singapore (NUS), Kent Ridge, Singapore, and the M.S. and Ph.D. degrees in electrical engineering and computer sciences from the University of California, Berkeley.

He had worked on optoelectronic devices with the British Telecommunications Laboratories, U.K., and on CMOS technology at the University of California, Berkeley. During 2001-2003, he worked on exploratory transistor technologies at Taiwan Semiconductor Manufacturing Company. He is currently with the Silicon Nano-Device Laboratory (SNDL) and the Center for Integrated Circuit Failure Analysis and Reliability, Department of Electrical and Computer Engineering, NUS, where he is an Assistant Professor of electrical and computer engineering and leads research efforts in strained transistor technology and sub-30-nm device work. He is also with the Agency for Science, Technology, and Research (A*STAR), Singapore, where he is a Research Program Manager, managing a nanoelectronics research program. He has authored or coauthored more than 120 journal and conference papers, and a book chapter. He has 41 U.S. patents, and more than 50 U.S. patents pending.

Prof. Yeo served on the International Electron Devices Meeting, Subcommittee on CMOS Devices during 2005-2006. He received the 1995 Institution of Electrical Engineers (IEE) Prize, the 1996 Lee Kuan Yew Gold Medal, the 1996 Institution of Engineers Singapore Gold Medal, 1997–2001 NUS Overseas Graduate Scholarship Award, the 2001 IEEE Electron Devices Society Graduate Student Fellowship Award, the 2002 IEEE Paul Rappaport Award, and 2003 TSMC Invention Awards. In 2006, he received the Singapore Young Scientist Award and the Singapore Youth Award in Science and Technology, the highest national accolade conferred on youth in Singapore.

Dim-Lee Kwong (A'84–SM'90) received the B.S. degree in physics and the M.S. degree in nuclear engineering, both from National Tsing Hua University, Taiwan, R.O.C., in 1977 and 1979, respectively, and the Ph.D. degree in electrical engineering from Rice University, Houston, TX, in 1982.

He was an Assistant Professor of Electrical Engineering Department at the University of Notre Dame during the years 1982–1985. He was a Visiting Scientist at the IBM General Technology Division, Essex Junction, Vermont during the summer of 1985 working on 4-MB DRAM technology. He joined The University of Texas at Austin, Microelectronics Research Center and Department of Electrical and Computer Engineering in 1985 as an Assistant Professor. He was promoted to Associate Professor in 1986 and to Full Professor in 1990. He is the author of more than 360 referred archival journal and 320 referred archival conference proceedings publications, has presented more than 50 invited talks at international conferences, and has been awarded with more than 25 U.S. patents. His current areas of research interests include high-κ gate dielectrics and dual metal gate electrodes for both Si- and Ge-channel MOSFETs, nonvolatile memory devices based on Si/Ge/metal nanocrystal, high-κ, and polymer, Si photonics, Si-based bioelectronics, and Si nanowire devices. Fifty students received their Ph.D. degrees under his supervision.

Dr. Kwong received numerous awards including Best Dissertation Award in 1982, the IBM Faculty Award in 1984, the Engineering Foundation Teaching Award from The University of Texas at Austin in 1994, and holds the Earl N. and Margaret Brasfield Endowed Fellowship.

On-The-Fly Interface Trap Measurement and Its Impact on the Understanding of NBTI Mechanism for p-MOSFETs with SiON Gate Dielectric

W.J.Liu[1], Z.Y.Liu[1], Daming Huang[1], C.C.Liao[2], L.F.Zhang[2], Z.H.Gan[2], Waisum Wong[2], C.Shen[3], and Ming-Fu Li[1,3]

[1]State Key Lab ASIC & Syst., Dept. Microelectronics, Fudan University, Shanghai 201203, China,
Email: mfli@fudan.edu.cn; dmhuang@fudan.edu.cn
[2]Semiconductor Manufacturing International Corp., Shanghai 201203 China,
[3]SNDL, ECE Dept., National University of Singapore, Singapore 117576

Abstract

For the first time, we developed an on-the-fly method OFIT to measure the interface trap density N_{IT} without recovery during measurement. The OFIT produces the most reliable experimental data of the interface trap generation dynamics under stress and therefore provides a solid ground to check various modeling work. The slope n of t^n time evolution of ΔN_{IT} under stress is temperature dependent, supporting dispersive Hydrogen transport in the oxide. Comparing OFIT data with the data measured by ultra-fast pulsed V_{th} measurement, we successfully decompose the NBTI ΔV_{TH} into interface trap component $\Delta V_{TH}{}^{IT}$ and oxide charge component $\Delta V_{TH}{}^{OX}$ quantitatively for the p-MOSFETs with SiON gate dielectric.

Introduction

Recent investigations (1-7) showed that recovery during measurement delay plays a key role in introducing NBTI measurement error. Since the existing interface traps measurement methods, charge pumping (CP) (8) and DCIV (9), all interrupt the stress during measurement, until now there are no reliable data of ΔN_{IT} time evolution under stress, and its field and temperature dependence, due to recovery during measurement (1,3). On the other hand, the physical origin of NBTI V_{TH} shift, due to interface traps $q\Delta N_{IT}$ (10,12,13), or oxide charge $q\Delta N_{OX}$ (11), or both (1,14,15), is under debate. Therefore, on-the-fly ΔV_{TH} measurement (2,5-7) is doubtful for inspecting the N_{IT} R-D model due to possible N_{OX} signal interference. In this work, we have developed a novel on-the-fly N_{IT} (OFIT) measurement method with no recovery during measurement. The experiments provide the most reliable data to inspect the N_{IT} R-D model directly. Combining with the ΔV_{TH} data measured by the fast pulsed method (16), the contributions of NBTI ΔV_{TH} by ΔN_{IT} and ΔN_{OX} can also be decomposed unambiguously.

Device Fabrication

Poly-Si/SiON gate stack p-MOSFETs were fabricated. SiON was grown by thermal oxidation followed by plasma nitridation and post-deposition thermal annealing. Thicker EOT of *2.8 nm* and *3.5 nm* are used to reduce the gate tunneling current and oxide/poly gate interface effects, and therefore increase the measurement accuracy. The main results are thickness independent for the *2.8nm* and *3.5nm* samples.

On-The-Fly interface Trap measurement

A. Re-examination of the conventional CP (CCP) method

In the conventional CP (CCP) measurement (8), the gate stress V_{NBTI} is interrupted during CP measurement and a series of pulses with V_{HIGH} =+1.1V (accumulation) and V_{LOW}= -0.9V (inversion) are applied for the p-MOSFET measurements. In the early CP theory (8, eq.19), when keeping the same voltage rising (falling) rate (not rising (falling) time) dV/dt, I_{CP} should be independent of the duty cycle of the pulse and the V_{LOW} magnitude, when V_{LOW} is low enough causing channel inversion (Fig.1).

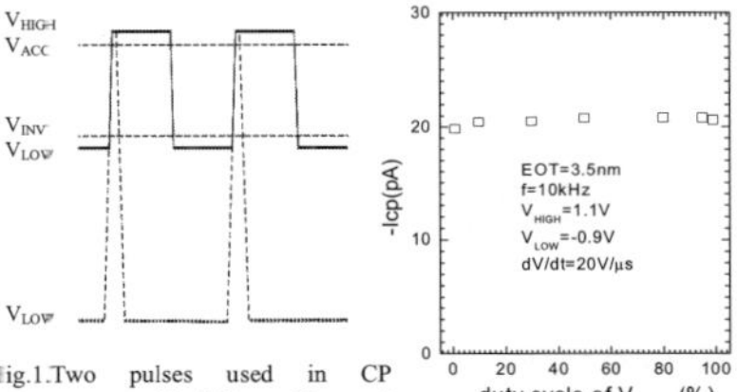

Fig.1.Two pulses used in CP measurement with different V_{LOW} and duty cycle (solid and dashed lines), while keeping V_{HIGH}, rising and falling voltage rates dV/dt, and frequency the same. According to the early CP theory (8), two pulses should measure the same CP current, if the interface trap number is fixed.

Fig.2. I_{CP} is independent of pulse duty cycle when $|V_{LOW}|$ is small (0.9V) and therefore there is no interface trap generation during measurement.

In our work, Figs. 2-5 give the more accurate experimental results of this aspect. Fig.2 shows that when $|V_{LOW}|$ is small enough (*0.9V*) that no generation of interface traps during measurement, the measured I_{CP} keeps the same when changing the duty cycle from 1% to 99%. However, in Fig.3, when $|V_{LOW}|$ is large (*4V*), changing V_{HIGH} duty cycle from 99% to 96%, I_{CP} increases accordingly. This is due to generation of the interface traps under single I_{CP} measurement when V_G goes to V_{LOW}, even the duty cycle of V_{LOW} is only 1% to 4%. Using least square fit of these data, one can obtain the I_{CP} value at 100% duty cycle of V_{HIGH}. It is the I_{CP} value without new generation of interface traps during measurement. Fig.4 shows that when raising the $|V_{LOW}|$ value, and changing the duty cycle while keeping the same voltage rising (falling) rate, I_{CP} increases slightly by increasing

$|V_{LOW}|$ or reducing V_{HIGH} duty cycle. When the same least square fit procedure in Fig.3 is applied for data in Fig.4 at $|V_{LOW}| = 3V$ and 4V respectively, we obtain the I_{CP} versus $|V_{LOW}|$ curve as shown in Fig.5, demonstrate the I_{CP} weak dependence of $|V_{LOW}|$ for a fixed number of interface traps, and same voltage rising (falling) rate of the pulse under CP measurements.

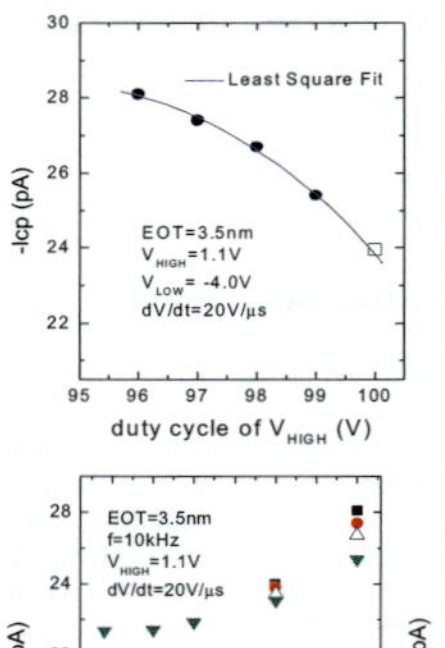

Fig.3. When $|V_{LOW}|$ is large (4V), I_{CP} increases with decreasing duty cycle of V_{HIGH}, due to generation of interface traps during single measurement when V_G going to $|V_{LOW}|$, even V_{LOW} duty cycle is very low (1% to 4%). Least square fitting the data (solid line) and extending to 100% V_{HIGH} duty cycle gives the I_{CP} value (open square symbol) with no interface trap generation during measurement.

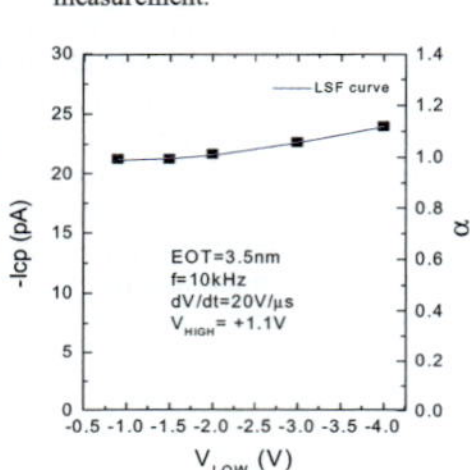

Fig.4. The change of I_{CP} current when changing V_{LOW} and the duty cycle.

Fig.5. The I_{CP} versus V_{LOW} curve, while fix the rising and falling dV/dt rate, the frequency, and the number of interface traps. This curve is used for calibration in our OFIT method. The calibration factor α is normalized (=1) at V_{LOW}=-0.9V.

B. The OFIT method

In the new OFIT measurement, the change of V_g is shown in Fig.6(b). *At the initial stage before stress and in the recovery phase, V_{NBTI} is off and the measurement pulses are the same as in the CCP. In the stress phase, V_{NBTI} is always on. During I_{CP} measurement, the CP pulses are applied.

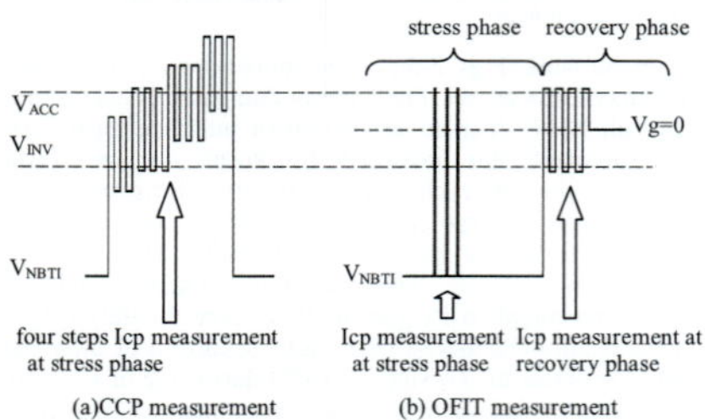

Fig.6. The gate voltage pulse under stress phase and recovery phase for (a) conventional CP (CCP) method, (b) On-the-fly (OFIT) method.

When pulse is *"on"*, $V_{HIGH} = +1.1V$ with as small as possible V_{HIGH} duty cycle (1%) and when pulse is *"off"*, $V_{LOW}=V_{NBTI}$. The pulse rising and falling rates are the same for stress phase and recovery phase measurements. Comparing to the CCP method, the OFIT reduces the V_{NBTI} interruption time by a factor of 10^{-2} to 10^{-3}. The ΔN_{IT} recovery during measurement is therefore negligible as will be verified later. In the OFIT measurement, since V_{NBTI} is always on at the stress phase, there is a constant substrate current due to gate-substrate tunneling. It should be deducted from the measured I_{SUB}. For the samples with EOT=2.8nm, I_{CP} was measured by $I_{S/D}$ rather than I_{SUB} to reduce the constant tunneling background from the gate. Finally, the measured I_{CP} in the stress phase should be divided by a calibrating factor α shown in Fig.5, to compare with the I_{CP} in the recovery phase and in the initial stage.

Results and Discussions

A. R-D model

Figs.7,8 show the ΔN_{IT} time evolution under stress and recovery phase, measured by CCP and OFIT respectively.

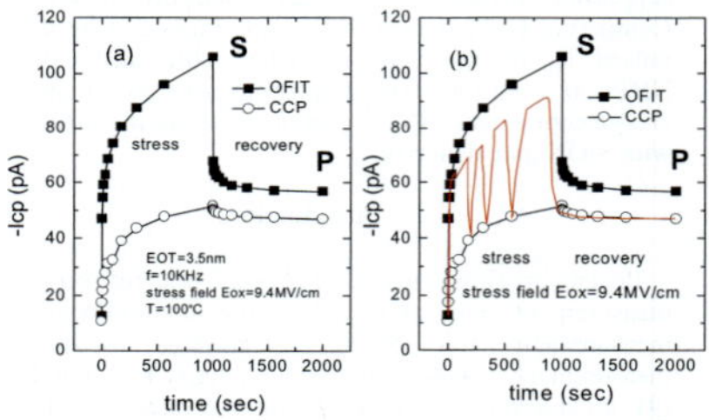

Fig.7. (a) The time evolution of ΔN_{it} under stress phase and recovery phase, measured by CCP and OFIT respectively. (b) Illustration that smaller I_{CP} measured by CCP is due to recovery during measurement in the stress phase. Note that in the recovery phase, the I_{CP} measured by CCP is also lower than that measured by OFIT. It is because that the number of cumulative interface traps generated in the stress phase (at S point) is smaller in the CCP measurement than in the OFIT measurement, due to the recovery in the CCP measurement at the stress phase.

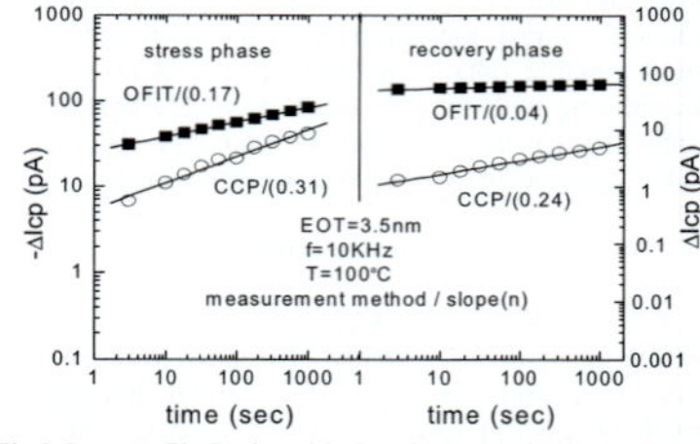

Fig.8 Same as Fig.7, plotted in Log-Log scale in the stress phase (a) and recovery phase (b) respectively. The slope n=0.17≈1/6 by OFIT measurement supports the H_2 diffusion in the R-D model [12]. CCP measurement overestimates the n value.

The CCP method seriously underestimates the ΔN_{IT} values at the stress phase due to the recovery during measurements. Fig.8 shows the $N_{IT}=A\ t^n$ time evolution under stress with slope $n=0.17 \cong 1/6$ at $T=100^{\circ}C$ for OFIT data.

In Fig.9, when varying the V_{HIGH} duty cycle of $10^4\ Hz$ OFIT pulses from 1% to 50%, the results are the same, indicating that ΔN_{it} recovery has a time constant longer than $50\ \mu s$ and our OFIT measurements are free from recovery and are reliable. In the contrary, using CCP measurement, the obtained slope is different for different measurement delay.

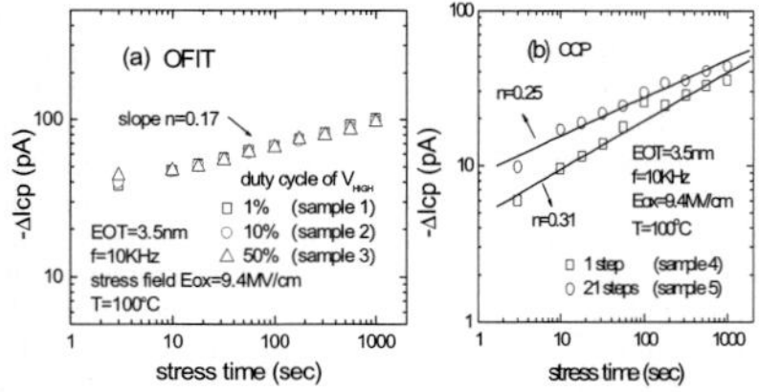

Fig.9 (a) OFIT measurements with different V_{HIGH} duty cycle, varying from 1% to 50%, corresponding to the V_{NBTI} interruption time duration from 1µs to 50µs when $f=10KHz$. The results (n=0.17 at 100°C) are the same, indicating recovery free during OFIT measurements. (b) For CCP measurements with different measurement delay, the results are different. Longer measurement delay, larger overestimated n value, and then the results are not reliable.

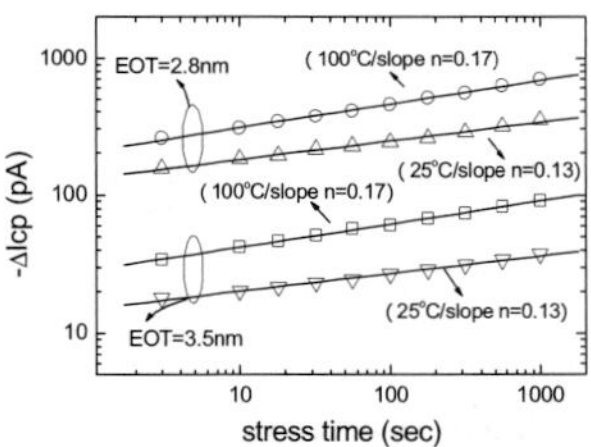

Fig.10 n in OFIT measurement is independent of oxide thickness within the measurement error. For $EOT=2.8nm$ samples, pulse frequency = 500KHz, and $I_{S/D}$ rather than I_{SUB} was measured for I_{CP}, to reduce the gate tunneling current background.

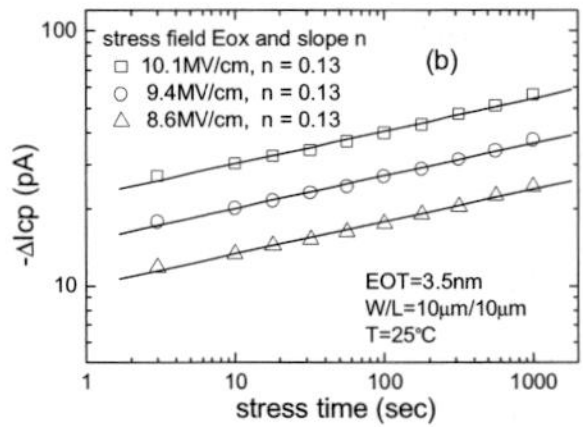

Fig.11 OFIT measurement under different oxide field stress E_{OX}. The slope n is independent of E_{OX} within the measurement error.

Figs.10,11,12 show that the slope n of the ΔN_{IT} degradation is independent of the oxide EOT and field E_{OX} within the measurement error, and linearly temperature dependent when $n<1/6$. These results support the R-D model of H_2 diffusion (12) with dispersive (13), not Arrhenius transport. Using n(T) function in the dispersive transport theory, $n=kT/6E_O$ with the characteristic DOS width of localized hydrogen states in the oxide (13) $E_O\cong32\ meV$, the experimental temperature dependence of n (Fig.12b) can be explained.

B. Decomposition of ΔV_{TH} by Interface traps and oxide charge components in the stress phase

In Fig.13, we combine the ΔN_{IT} curve using the OFIT measurement and the ΔV_{TH} curve using the fast pulsed I-V measurement (16) by aligning two curves at the end of the recovery phase P point. At this point, the ΔV_{TH} is caused by both $q\Delta N_{IT}$ charge and $q\Delta N_{OX}$ charge with $\eta_P = [\Delta N_{IT}/(\Delta N_{IT}+\Delta N_{OX})]_{at\ P} \leq 1$. The ΔV_{TH} component contributed by $q\Delta N_{IT}$, denoted by ΔV_{TH}^{IT}, is thus represented by the OFIT curve if $\eta_P =1$, or below the OFIT curve if $\eta_P<1$. The ΔV_{TH} component contributed by $q\Delta N_{OX}$, denoted by ΔV_{TH}^{OX}, is the difference between two curves if $\eta_P =1$ or more if $\eta_P <1$, as shown in Fig. 14. Plotting ΔV_{TH}^{OX} in Log-Log scale in the stress phase results a slope of $n=0.05$ at $100^{\circ}C$, much smaller than the slope $n=0.17$ of ΔV_{TH}^{IT} at the same temperature.

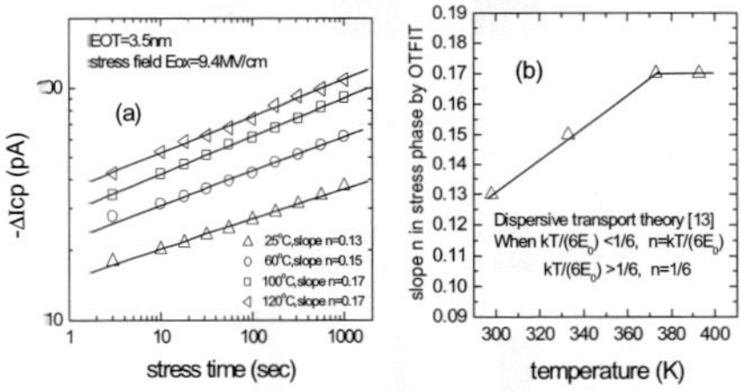

Fig.12. The linear temperature dependence of slope n when n<1/6, supports the H_2 dispersive transport (13) with hydrogen trap energy depth of 32 meV.

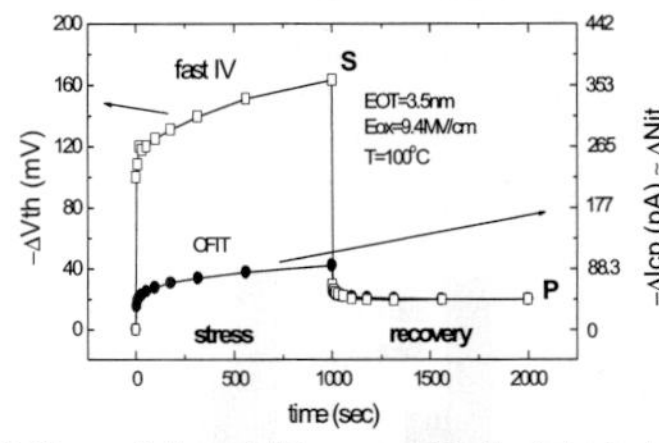

Fig.13 Time evolutions of ΔV_{TH} measured by the fast-pulsed method (14), and ΔN_{IT} measured by OFIT. ΔN_{IT} curve is scaled to align the ΔV_{TH} curve at the end of recovery phase P point. The ΔV_{TH} contribution by $q\Delta N_{IT}$ (ΔV_{TH}^{IT}) is represented by the OFIT curve if $\eta_P =1$, or below the OFIT curve if $\eta_P <1$. The ΔV_{TH} contribution by $q\Delta_{OX}$ (ΔV_{TH}^{OX}) is the difference between two curves if $\eta_P =1$ or more if $\eta_P <1$. Here $\eta_P=[\Delta N_{IT}/(\Delta N_{IT}+\Delta N_{OX})]_{at\ P}$.

Figs.15,16,17 show the oxide EOT, oxide field E_{OX}, and temperature dependences of $q\Delta N_{IT}$ and $q\Delta N_{OX}$ contributions at the end of the stress phase S point, assuming $\eta_P \cong 1$. In all cases, ΔN_{OX} component is a major component and ΔN_{IT} component is a minor component in the stress phase. It

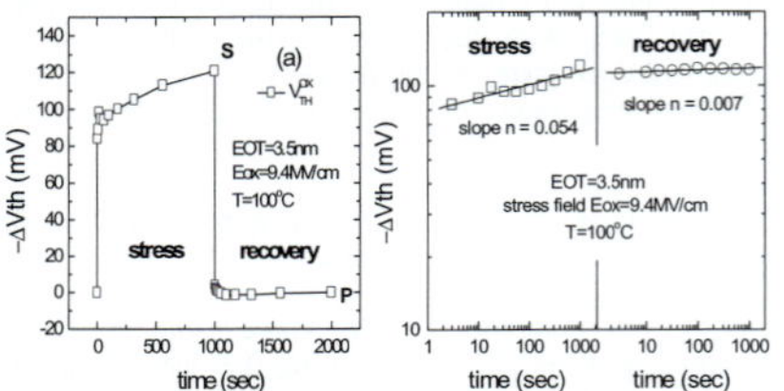

Fig.14 The time evolution of $\Delta V_{TH}{}^{OX}$ extracting from the difference of two curves in Fig.13, and assuming $\eta_P \cong 1$. $\Delta V_{TH}{}^{OX} = B\ t^n$ with $n=0.054$ at stress phase at $T = 100^{\circ}C$.

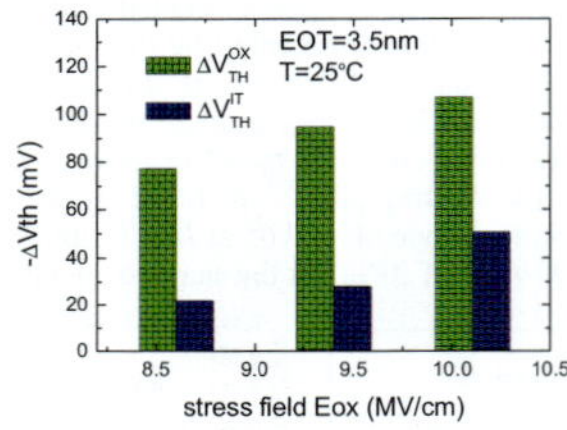

Fig.15 Oxide field E_{OX} dependence of $\Delta V_{th}{}^{IT}$ and $\Delta V_{th}{}^{OX}$ at S point of 1000 sec stress, assuming $\eta_P \cong 1$.

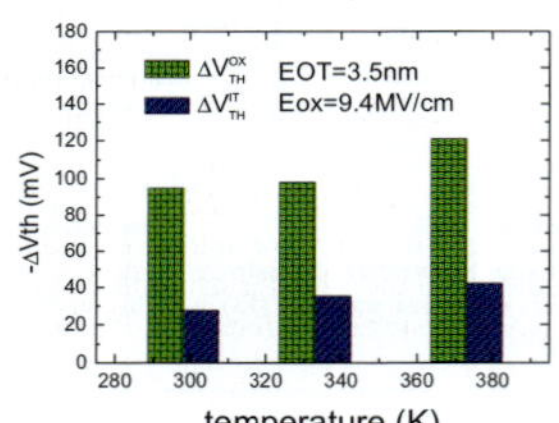

Fig.16 Temperature dependence of $\Delta V_{th}{}^{IT}$ and $\Delta V_{th}{}^{OX}$ at S point of 1000 sec stress, assuming $\eta_P \cong 1$.

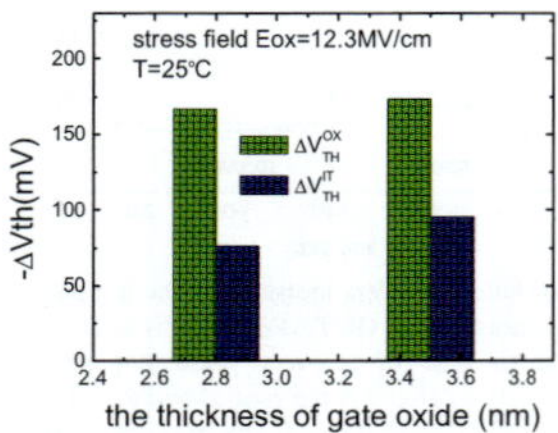

Fig.17 Oxide thickness dependence of $\Delta V_{th}{}^{IT}$ and $\Delta V_{th}{}^{OX}$ at S point of 1000 sec stress, assuming $\eta_P \cong 1$.

explains why previous works using ΔV_{TH} measurement (I-V or on-the-fly) to measure the slope n have different values and temperature dependence comparing with our results in this work. ΔV_{TH} is not suitable to check the N_{IT} R-D model due to the N_{OX} signal interference.

Conclusions

A novel on-the-fly Interface trap N_{IT} (OFIT) measurement method is developed with no recovery during measurement. Based on the OFIT results, all major issues on the NBTI mechanism can be solved: **(1)** At $T=100^{\circ}C$, the t^n time evolution of ΔN_{IT} under stress with $n=0.17\cong 1/6$ is obtained, and is independent of the oxide EOT and field E_{OX}, supporting the H_2 diffusion R-D model (12). The slope n is linearly temperature dependent, when $n<1/6$, supporting dispersive (13), not Arrhenius transport, with the characteristic DOS width of localized hydrogen states in the oxide $E_O \cong 32\ meV$. **(2)** Combining OFIT measurements with the fast pulsed I-V method for ΔV_{TH} measurements, decomposition of ΔV_{TH} into ΔN_{IT} component ($\Delta V_{TH}{}^{IT}$) and oxide charge ΔN_{OX} component ($\Delta V_{TH}{}^{OX}$) can be obtained quantitatively. In the stress phase, the ΔN_{IT} component is a minor component with slow recovery time $\geq 50\ \mu s$. The ΔN_{OX} component is a major component with much faster recovery time (2). $\Delta V_{TH}{}^{OX}$ has t^n time evolution with much smaller slope $n=0.05$. Thus, previous works using ΔV_{TH} (I-V or on-the-fly) measurement to inspect the N_{IT} R-D model is not reliable. It underestimates the slope n, due to the interference of the ΔN_{OX} component.

Acknowledgement

This work is supported by the Micro/Nano-electronics Science and Technology Innovation Platform of Fudan University.

References

(1) T. Yang et al, Sym VLSI, p. 92, 2005.

(2) C. Shen et al, IEDM, p. 333, 2006.

(3) T. Yang et al, EDL 26, p. 826, 2005.

(4) M. Ershov et al, APL v. 83, p.1647, 2003.

(5) M. Denais et al, IEDM, p.109, 2004.

(6) S. Rangan, N. Mielke, E. C. C. Yehl, IEDM, p. 341, 2003.

(7) D. Varghese et al, IEDM, p. 701, 2005.

(8) G. Groeseneken, H. E. Maes, N. Beltrán, R. F. De Keersmaecker, TED, v. 31, p. 42, 1984.

(9) A. Neugroschel et al, TED, v. 42, p. 1657, 1995.

(10) A. Alam, IEDM, p. 346, 2003.

(11) V. Huard, M. Denais, IRPS, p. 40, 2004.

(12) S. Chacravarthi, A. T. Krishnan, V. Reddy, C. F. Machala and S. Krishnanl, IRPS, p. 273, 2004.

(13) B. Kazer et al, IRPS, p. 381, 2005.

(14) H. Reisinger et al, IRPS, p. 448, 2006.

(15) T. Grasser et al, IRPS, p. 265, 2007.

(16) C. Shen, M.-F. Li, X. P. Wang, Yee-Chia Yeo, D.-L.Kwong, EDL, v.27, p. 55, 2006.

62 IEEE TRANSACTIONS ON DEVICE AND MATERIALS RELIABILITY, VOL. 8, NO. 1, MARCH 2008

Understand NBTI Mechanism by Developing Novel Measurement Techniques

Ming-Fu Li, Daming Huang, Chen Shen, *Student Member, IEEE*, T. Yang, W. J. Liu, and Zhiying Liu

(Invited Paper)

Abstract—Our recent investigations and understanding of the negative bias temperature instability (NBTI) degradation in p-MOSFETs with ultrathin SiON gate dielectric are reviewed. The progressive understanding of NBTI mechanism is mainly related to the novel measurement techniques we developed. We show in this paper the following: 1) For the conventional charge pumping and direct–current current–voltage interface trap measurement, the interface trap density N_{it} is underestimated due to the recovery during measurement delay. The existing N_{it} data should be reexamined; 2) an ultrafast pulsed $I-V$ method [fast pulsed measurement (FPM)] is developed to measure ΔV_{th} with measurement time $t_M = 100$ ns. It can be considered as free from recovery during measurement; 3) due to the degradation during the initial threshold voltage measurement, the existing slow on-the-fly (OTF) ΔV_{th} measurement distorts (overestimates) the slope and induces a kink at early stress time in the Log-Log curve of the time evolution of NBTI degradation. A fast OTF ΔV_{th} measurement method is developed to overcome this problem; 4) a novel OTF interface trap (OFIT) measurement method is developed which is free from interface trap recovery during measurement. The OFIT measurement provides the most reliable data to inspect the interface trap R-D model; 5) combining the OFIT and FPM measurements, we decompose the NBTI ΔV_{th} into two components: A slow ΔV_{th}^{it} component contributed by ΔN_{it} with a slow recovery time longer than 50 μs and a fast ΔV_{th}^{ox} component contributed by ΔN_{ox} with a broad spectrum of recovery time, including a component with very fast recovery time (100 ns); and 6) the dynamic degradation by ΔV_{th}^{it} component is frequency-independent and can be measured by a dc method, whereas the dynamic degradation by ΔV_{th}^{ox} component measured by FPM is increased by increasing frequency. The ten-year lifetime of the p-MOSFETs is mainly determined by the degradation of the ΔV_{th}^{it} component.

Index Terms—CMOS, negative bias temperature instability (NBTI), reliability.

I. Introduction

THE NEGATIVE bias temperature instability (NBTI) of p-MOSFET is a crucial reliability problem which determines the lifetime of modern CMOS transistors with SiON gate dielectric [1]–[6]. In the past several years, there were difficulties and confusions to explore the NBTI mechanism due to the unsuitable measurement techniques which misled the understanding of the NBTI mechanism. In 2002, we first reported the recovery effect of NBTI degradation in p-MOSFETs with an ultrathin SiON gate oxide when the negative gate bias is stopped, and the device lifetime under dynamic NBTI (DNBTI) degradation should be longer than that under static NBTI degradation [7]. The NBTI recovery effect is now widely reported and investigated [8]–[12]. In 2003, Ershov *et al.* [12] first reported that due to the recovery in the measurement delay, the slope n and the magnitude a in the at^n time evolution of the NBTI degradation are distorted. In literature, the main issues of NBTI degradation can be ascribed to the following: 1) There are diverse experimental data on the slope n of the NBTI degradation reported by different labs and interpreted by different NBTI mechanisms; 2) the physical origin of NBTI degradation and recovery in p-MOSFETs with SiON gate dielectrics has different interpretations. Whether the NBTI degradation and recovery are due to reaction–diffusion of hydrogen species interacted with Si dangling-bond-induced interface traps at the $Si-SiO_2$ interface (R-D model) or due to trapping/detrapping of oxide traps is under debate [5], [13]–[17]. These two issues are related to each other and are all related to some unsuitable measurement techniques, which may mislead the understanding of the NBTI mechanism. We have proposed that there are two NBTI components: a slow component which is contributed by the interface trap degradation and a fast component which is contributed by trapping/detrapping of oxide charge [17]. In this paper, we will summarize some measurement techniques developed in our group, and our understanding of the NBTI degradation and recovery mechanism, and clarify the issues and misinterpretations in the previous works. We shall also verify our two-component NBTI proposal by rigorous experiments.

The devices used in this paper are poly-Si/SiON gate stack p-MOSFETs. The gate oxynitride SiON with EOTs of 1.3, 2.8,

Manuscript received July 4, 2007; revised August 30, 2007. This work was supported in part by the Micro/Nanoelectronics Science and Technology Innovation Platform, Fudan University, and in part by Singapore A*STAR Research Grant at Silicon Nano Device Laboratory, Department of Electrical and Computer Engineering, National University of Singapore.

M.-F. Li is with the State Key Lab of ASIC & System, Microelectronics Department, Fudan University, Shanghai 201203, China and also with the Silicon Nano Device Laboratory, Department of Electrical and Computer Engineering, National University of Singapore, Singapore 117576 (e-mail: mfli@fudan.edu.cn).

D. Huang, W. J. Liu, and Z. Y. Liu are with the State Key Lab of ASIC & System, Microelectronics Department, Fudan University, Shanghai 201203, China.

C. Shen and T. Yang are with the Silicon Nano Device Laboratory, Department of Electrical and Computer Engineering, National University of Singapore, Singapore 117576.

Color versions of one or more of the figures in this paper are available online at http://ieeexplore.ieee.org.

Digital Object Identifier 10.1109/TDMR.2007.912273

1530-4383/$25.00 © 2008 IEEE

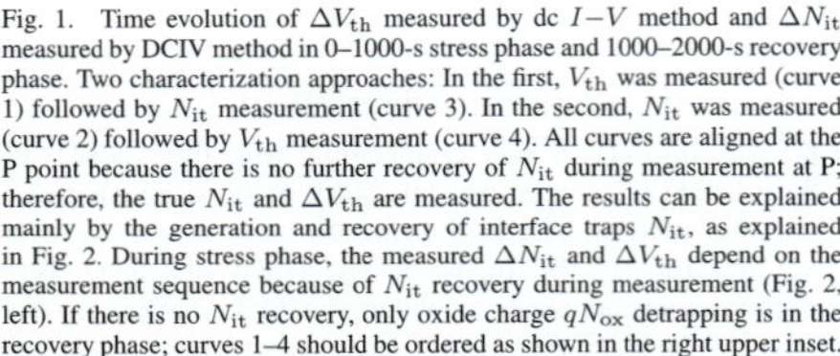

Fig. 1.   Time evolution of ΔV_{th} measured by dc $I{-}V$ method and ΔN_{it} measured by DCIV method in 0–1000-s stress phase and 1000–2000-s recovery phase. Two characterization approaches: In the first, V_{th} was measured (curve 1) followed by N_{it} measurement (curve 3). In the second, N_{it} was measured (curve 2) followed by V_{th} measurement (curve 4). All curves are aligned at the P point because there is no further recovery of N_{it} during measurement at P; therefore, the true N_{it} and ΔV_{th} are measured. The results can be explained mainly by the generation and recovery of interface traps N_{it}, as explained in Fig. 2. During stress phase, the measured ΔN_{it} and ΔV_{th} depend on the measurement sequence because of N_{it} recovery during measurement (Fig. 2, left). If there is no N_{it} recovery, only oxide charge qN_{ox} detrapping is in the recovery phase; curves 1–4 should be ordered as shown in the right upper inset.

3.5, and 4.5 nm were grown by thermal oxidation followed by plasma nitridation and postdeposition thermal annealing.

II. RECOVERY OF INTERFACE TRAPS DURING DC MEASUREMENT DELAY

In this section, we first illustrate the problems of NBTI characterization by conventional dc V_{th} measurement and interface trap density N_{it} measurement by the conventional direct–current current–voltage (DCIV) [18], [19] or charge pumping (CP) [20] method. For the NBTI characterization shown in Fig. 1, p-MOSFETs were stressed 1000 s under a constant negative gate voltage, followed by a recovery phase ($V_g = 0$ V) of another 1000 s, while the source, drain, and bulk were grounded. Stress in the stress phase was periodically interrupted for V_{th} and N_{it} measurements. During each interruption, both V_{th} (extracted by dc $I_d{-}V_g$ measurement) and N_{it} (extracted by DCIV measurement) were measured. The results depend on the order of V_{th} and N_{it} measurements [17], [21]. This is because that during each interruption of stress for measurement, the negative gate voltage is reduced (for dc $I_d{-}V_g$ measurement) or even turned to positive (for DCIV or CP measurement). The interface trap density is reduced at this moment due to the recovery of interface traps, as shown in Fig. 2. The measured values of ΔV_{th} and ΔN_{it} are therefore underestimated. The recovery is more serious during the ΔN_{it} measurement than during the ΔV_{th} measurement because of the longer measurement time and the more positive bias employed.

There are some common misinterpretations of these curves: If curve 4 in Fig. 1 is only observed, or CP or DCIV curve in Fig. 3(a), one may misleadingly conclude that the recovery of the interface traps is pretty small because the difference of

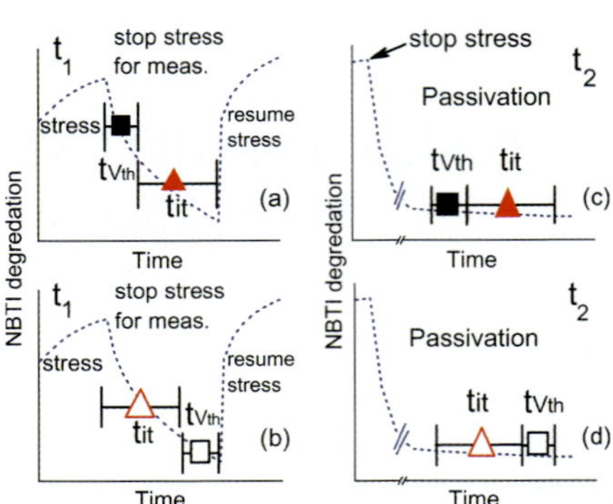

Fig. 2.   Zoom-in times t_1 and t_2 in Fig. 1. Once interrupting the stress for measurements, NBTI degradation recovers. (a) ΔV_{th} first measured during $t_{V\mathrm{th}}$, followed by ΔN_{it} measured during $t_{V\mathrm{th}} \sim t_{V\mathrm{th}} + t_{\mathrm{it}}$. (b) The reverse sequence is used. Putting (a) and (b) together explains the four data points at time t_1 in Fig. 1. Note that N_{it} measurement is slower than V_{th} measurement. (c) and (d) explain data t_2 in Fig. 1.

ΔV_{th} at S and P points is small (Fig. 3(a), curves aligned at S point). However, the real situation is that the recovery is very large at stress phase; therefore, the measured ΔV_{th} at stress phase is much lower than the real ΔV_{th} due to the recovery in measurement delay, as shown in Fig. 2 (left). This is clear if we align the curves at P point rather than at S point [Fig. 3(b)]. Only the alignment at P point is reasonable because there is no further recovery during measurement at P point. This will be more clearly illustrated in Fig. 10.

From our experiments, we show evidence that the widely used CP and DCIV methods underestimate the value of N_{it} significantly because of this recovery effect. All previously reported data of interface traps based on CP and DCIV measurements should be reassessed. Fig. 3(b) shows that CP has more serious recovery than DCIV during measurement in the stress phase. In DCIV measurement, a recombination current I_{DCIV} through the interface traps shows a peak when the Fermi level coincides with the Si midgap at surface [19]. In our measurement, in order to show a clear peak, the maximum V_g applied to the device is around $+0.5{-}0.6$ V. In the stress phase measurement, the interface traps already recovered to some extent before reaching the I_{DCIV} peak gate voltage; therefore, the measured ΔN_{it} value is underestimated. The underestimation is more serious when using the CP technique. In CP measurement, the device changes from inversion to accumulation to pump the electrons between conduction/valence bands through the interface traps [20]. Therefore, a more positive V_g than that used in DCIV measurement should be applied. In our CP measurement, the maximum positive V_g applied to the device is $+1.2$ V. On the other hand, the CP measurement uses longer time than the DCIV measurement. This explains why ΔN_{it} measured by CP is smaller than that measured by DCIV in the stress phase, as shown in Fig. 3(b), because more interface traps are recovered during the CP measurement.

64   IEEE TRANSACTIONS ON DEVICE AND MATERIALS RELIABILITY, VOL. 8, NO. 1, MARCH 2008

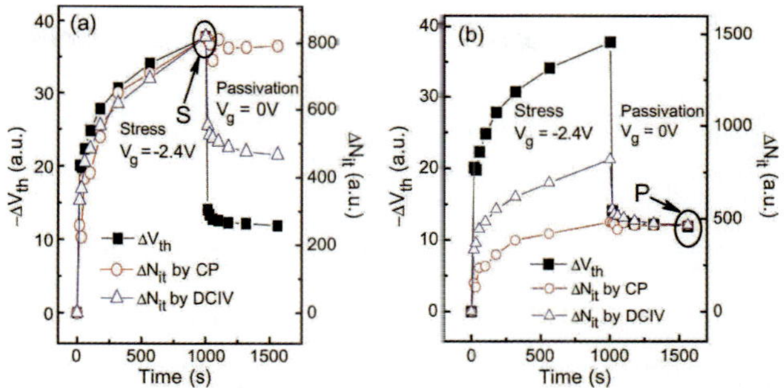

Fig. 3. Data of ΔV_{th} and ΔN_{it} measured by both CP and DCIV are plotted versus stress and recovery time. (a) All data are aligned at point S [15]. Since, in the stress phase, the recovery in ΔN_{it} measurement is more serious than in ΔV_{th} measurement, as shown in Fig. 2, all data aligned at S point will overestimate the ΔN_{it} value, leading to misinterpretation of the NBTI recovery mechanism. This will be more clearly shown in Fig. 10. (b) All data are aligned at P point (our work). At the P point, there is no recovery in ΔN_{it} measurement or ΔV_{th} measurement. The measured data reflect the real ΔN_{it} and ΔV_{th} values at this point.

However, our experiments cannot completely exclude the recovery of ΔV_{th} partially due to hole detrapping in dielectric hole traps. This will further be discussed in Section VI.

III. FPM METHOD

Since there is recovery effect during the dc measurement delay which distorts the NBTI degradation signal, it is natural to ask if we can design a fast measurement method to reduce the measurement delay and reduce the recovery effect. How fast should the measurement take to be completely free from recovery during measurement? Our investigation showed that there is a fast NBTI component with very fast recovery time. Only a fast measurement method with measurement time $t_M = 100$ ns to measure the ΔV_{th} can be considered as free from recovery. It will be shown later.

IMEC group has developed a pulsed I_d-V_g method to measure the threshold voltage V_{th} and extract the charge trapping characteristics in HfO$_2$ high-κ gate dielectric [22]. It is our first thought as to why not use this pulsed I_d-V_g method to measure the NBTI degradation in p-MOSFETs with SiON gate dielectric since all previous experiments were only reported by dc measurement. On the other hand, we thought that the pulsed I_d-V_g measurement can further be improved by inserting an amplifier in the measurement loop. We have improved the fast measurement method with the V_{th} measurement time t_M as short as 1 μs in 2004 [23], [24] and further reduced to 100 ns in 2006 [25]. The method is shown in Fig. 4. For the circuit in Fig. 4(a) used by the IMEC group, the measurement speed is limited by the charging current of capacitors C_{gd} and C_O, when voltage at D is varied. C_O includes C_{ds} of the MOSFET, the coaxial cable capacitance, and the input capacitance of the oscilloscope. When the measurement is faster, the capacitance charging current is larger, which finally distorts the real I_d current with an unreliable V_{th} measurement. The improved method [denoted by the fast pulsed measurement method (FPM)] used in our measurement employs a circuit shown in Fig. 4(b). Voltage at D is fixed to V_{ds} due to the virtual ground principle of the Op.

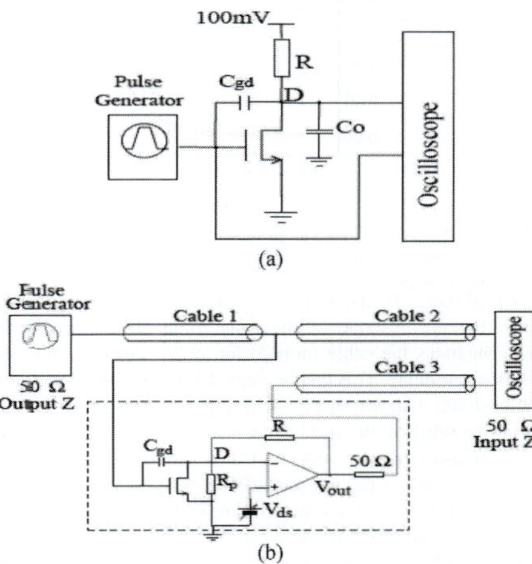

Fig. 4. Circuit diagrams of the fast I_d-V_g measurement. (a) Developed by IMEC [22]. (b) Developed in this paper. The length of the coaxiable cables 2 and 3 should be adjusted; therefore, the propagation delays of V_g and I_g should be the same. With improved probe setup, the I_d-V_g and, therefore, V_{th} can be measured by $t_m = 100$ ns.

Amp, and there is no charging current through C_O'. The charging current through C_{gd} can be deducted when $C_{gd}-V_g$ is measured. When the propagation delay in the transmission cables is carefully shortened and the reflection at the terminal is reduced, the measurement time t_M can be reduced to 100 ns [25].

Fig. 5 shows the I_d-V_g curve measured by the FPM using voltage scan of the rising (up) or falling (down) edge of the stress pulse applied to the gate. The V_{th} can be determined by the I_d-V_g curve.

When the measurement time t_M is in the range of 1 μs or shorter, it is important that the voltage V_g signal and the current

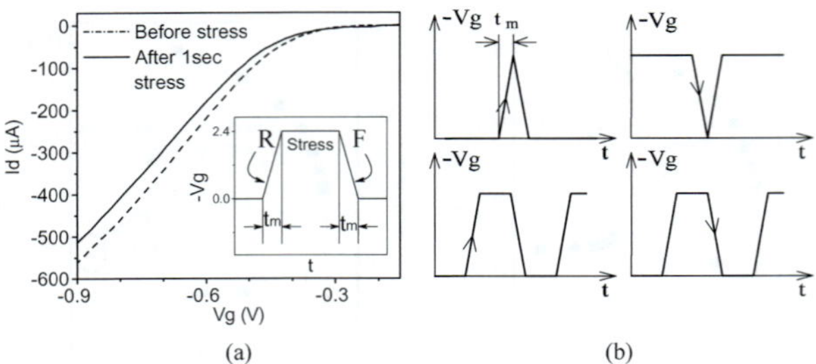

Fig. 5. (a) I_d–V_g curve measured by FPM using the rising (up) or falling (down) edge of a V_g pulse as indicated in the inset (note that the vertical scale $-V_g$ rather than $+V_g$ is employed). Measurement time t_m can be varied. (b) Various V_g waveforms are used for V_{th} measurement during static and dynamic stresses.

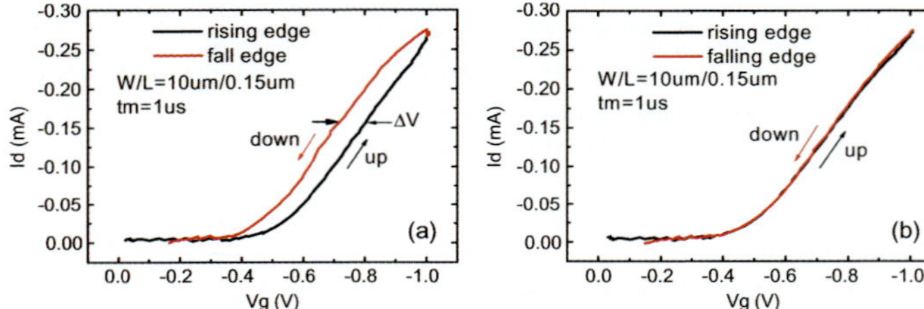

Fig. 6. I_d–V_g curves recorded by the oscilloscope using FPM with $t_M = 1$ μs. (a) $\Delta t_{\text{delay}} \cong 36$ ns corresponding to the propagation delay of the cable and amplifier with an equivalent cable length of 5.4 M. (b) Adjusting to $\Delta t_{\text{delay}} \cong 0$ ns.

I_g signal (actually, the voltage output of the amplifier) should reach the oscilloscope at the same time. The propagation delay includes the cable propagation delay and the amplifier propagation delay. This can be adjusted by the cable length of cables 2 and 3 shown in Fig. 4(b). Fig. 6(a) shows the I_d–V_g curve recorded by the oscilloscope for an FPM measurement using a triangular pulse with $t_M = 1$ μs and peak height $= 1$ V, as shown in the upper left part of Fig. 5(b). When measuring I–V by up (rising) and down (falling) edges of the triangular pulse, a loop appears, as shown in Fig. 6(a), when there is a difference of propagation delay Δt_{delay} between V_g signal and I_g signal. The loop width ΔV can be estimated by

$$\Delta V = 2\Delta t_{\text{delay}} \left| \frac{dV}{dt} \right|_{\text{Pulse}} . \tag{1}$$

When $\Delta t_{\text{delay}} = 0$ by adjusting the cable lengths of cables 2 and 3, the loop curve is replaced by a coincide I_d–V_g curve, as shown in Fig. 6(b).

Fig. 7 shows the measured ΔV_{th} after a 1-s stress, which is measured by the FPM method with different measurement time t_M as defined in the inset of Fig. 7. The results show that there is a very fast recovery component that is as fast as 100-ns (or shorter) recovery time. However, in the next section, we shall show that $t_M = 100$ ns is fast enough to be considered as recovery free measurement.

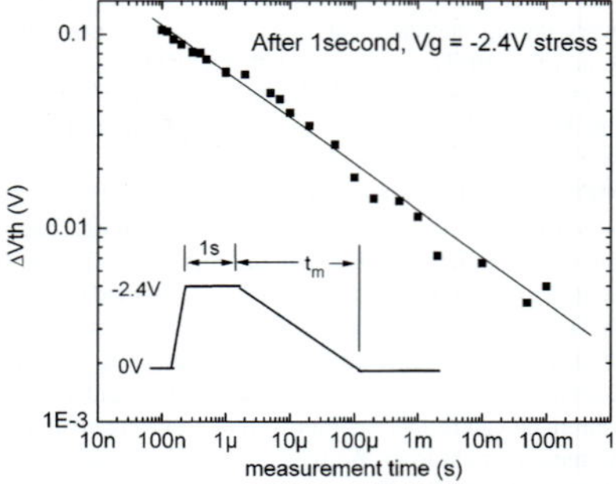

Fig. 7. Measured ΔV_{th} after 1-s stress, which is measured by the FPM method with different measurement time t_M as defined in the inset of Fig. 7. Recovery starts as short as 100 ns after stress is removed, indicating that there is a very fast component with recovery time of 100 ns or shorter.

IV. DISTORTION INDUCED BY THE SOTF METHOD AND THE NEW FOTF METHOD

In the existing literature, the effect of recovery in the NBTI characterization has been investigated from another point of view. Ershov *et al.* [12] have shown the distortion of the time evolution curve of ΔV_{th} under a static NBTI stress due to the time delay under V_{th} measurement. By inserting longer

66

IEEE TRANSACTIONS ON DEVICE AND MATERIALS RELIABILITY, VOL. 8, NO. 1, MARCH 2008

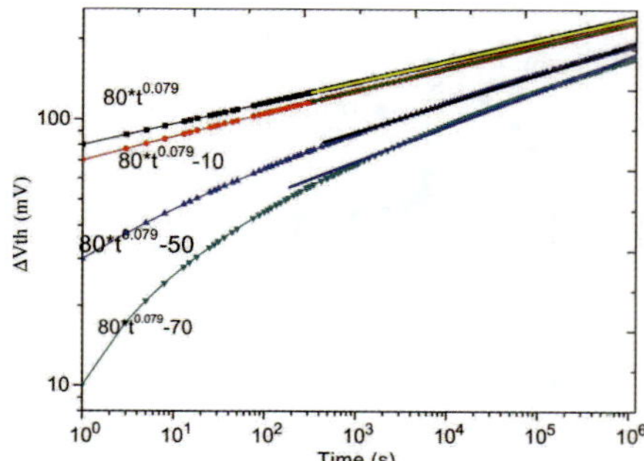

Fig. 8. Simulation curves of $\Delta V_{\mathrm{th}}^{M} = at^{n} - \Delta V$(mV), with $a = 80$, $n = 0.079$, and $\Delta V = 0$, 10, 50, and 70, respectively.

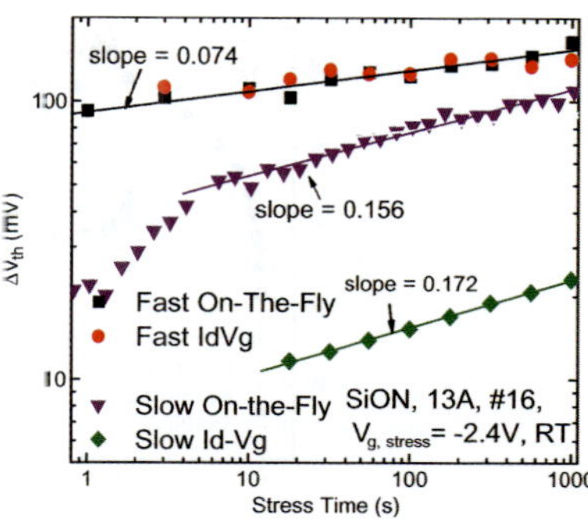

Fig. 9. ΔV_{th} for identical p-MOSFETs is measured with different measurement methods, which yield very different NBTI results. The FPM (fast $I_d - V_g$) method and the FOTF method with $t_M = 100$ ns measure the same reliable results. All other methods show underestimated ΔV_{th}. The curve measured by SOFT method has an artificial kink at 6 s due to the degradation of initial $V_{\mathrm{th}0}$ measurement. It also underestimates the ΔV_{th} magnitude and overestimates the slope n.

time delay, the measured Log ΔV_{th} versus $\log t$ curve not only shows a smaller ΔV_{th} amplitude but also a larger slope n. This is correctly interpreted as due to the recovery of NBTI degradation under measurement delay [12]. To avoid unintentional recovery during measurement, an on-the-fly (OTF) method of ΔV_{th} measurement has been developed recently [8], [26]. The change of V_{th} was calculated by measuring the change of I_d of MOSFET at the linear range of a small V_d bias, without interruption of the NBTI stress. Varghese et al. [27] have taken comprehensive measurements based on the OTF ΔV_{th} measurement to discuss the origin of NBTI degradation.

Conventional slow OTF (SOTF) measurement has been claimed to be free from the fast recovery of V_{th} [8], [26]. However, the initial prestress measurement of I_{do} takes 10–100 ms with a typical setup on HP4156 parameter analyzer. During the initial prestress measurement, some V_{th} degradation occurs under the stress gate voltage. Due to this degradation, the measured initial threshold voltage $V_{\mathrm{th}0}^{M}$ by SOTF measurement is actually distorted from the real prestress threshold voltage $V_{\mathrm{th}0}$ by

$$V_{\mathrm{th}0}^{M} = V_{\mathrm{th}0} + \Delta V \qquad (2)$$

and therefore, all the measured threshold voltage shift by SOTF measurement $\Delta V_{\mathrm{th}}^{M}$ is distorted from the real threshold voltage shift ΔV_{th} by

$$\Delta V_{\mathrm{th}}^{M} = \Delta V_{\mathrm{th}} - \Delta V. \qquad (3)$$

Fig. 8 shows the simulation results of $\log \Delta V_{\mathrm{th}}$ versus $\log t$ plots for the real ΔV_{th} with a power law index n: $\Delta V_{\mathrm{th}} = ct^{n}$, and the measured $\Delta V_{\mathrm{th}}^{M}$.

Fig. 8 clearly shows the following: 1) The SOTF method shifts down the real curve (underestimate ΔV_{th}) and distorts (overestimate) the slope (the power law index n); 2) at short stress time when $\Delta V_{\mathrm{th}}^{M} < \Delta V$, the slope of $\Delta V_{\mathrm{th}}^{M}$ is seriously distorted (overestimated), whereas at middle stress time when $\Delta V_{\mathrm{th}}^{M} > \Delta V$, the slope of $\Delta V_{\mathrm{th}}^{M}$ is lightly overestimated, which makes an artificial kink around the stress time of $\Delta V_{\mathrm{th}}^{M} = \Delta V$; and 3) at very long time when $\Delta V_{\mathrm{th}}^{M} \gg \Delta V$, the

effect of ΔV can be neglected, and the $\Delta V_{\mathrm{th}}^{M}$ curve measured by SOFT approaches to the real slope and real amplitude of ΔV_{th}. To improve this, we developed a fast OTF (FOTF) ΔV_{th} measurement method, replacing the HP4156 measurement in the SOFT by our fast FPM measurement to measure the initial $V_{\mathrm{th}0}^{M}$. Thus, the initial degradation ΔV in (2) reduces to zero.

Fig. 9 shows our measurement results by the SOTF and FOTF, respectively. For the curve measured by SOFT, we observed a kink at around 5 s and the power law slope $n = 0.156$, which is consistent with the data reported in [27] and is consistent with the simulation curve in Fig. 8 with nonzero ΔV. For the curve measured by FOFT and by FPM with $t_M = 100$ ns, excellent agreement between the two methods is reached, demonstrating that both methods give the reliable results of real ΔV_{th} degradation. In both curves, the kink at around 5 s disappears, with the slope $n = 0.074$. It can be well explained by Fig. 8 and by (2) and (3) with zero ΔV. It also indicates that FPM measurement with $t_M = 100$ ns can be considered as free from recovery effect.

We therefore conclude that all previous publications based on the SOTF method should be reexamined. The power law index has been overestimated. The threshold voltage amplitude has been underestimated. On the other hand, whether OTF ΔV_{th} can be used for inspecting the interface trap R-D model is doubtful because we do not know whether the measured ΔV_{th} is completely due to the contribution of the interface traps ΔN_{it} or due to the contribution of oxide trap charge ΔN_{ox} with different time evolution. This will be clarified in Section VI.

V. OFIT MEASUREMENT

To solve all the aforementioned problems, we have developed a novel OTF interface trap (OFIT) measurement method with no recovery during measurement. The experiments provide the most reliable data to inspect the N_{it} R-D model

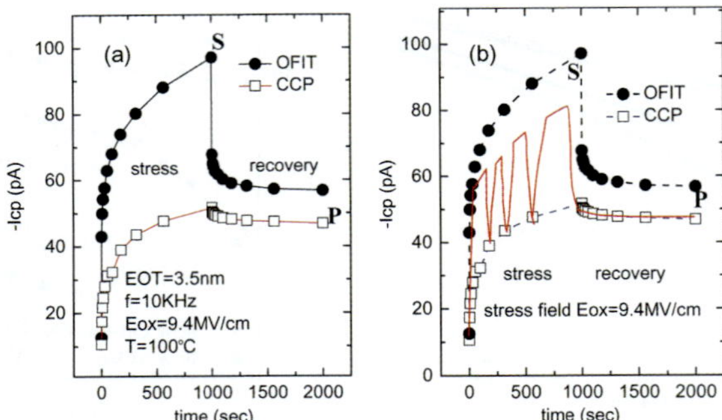

Fig. 10. (a) The time evolution of ΔN_{it} under stress phase and recovery phase measured by CCP and OFIT, respectively. (b) Illustration that smaller ΔN_{it} measured by CCP is due to the recovery during measurement in the stress phase. Note that in the recovery phase (1000–2000 s), the I_{CP} measured by CCP is also lower than that measured by OFIT. It is because that the number of cumulative interface traps generated in the stress phase (at S point) is smaller in the CCP measurement than in the OFIT measurement, which is due to the recovery in the CCP measurement, as shown in (b).

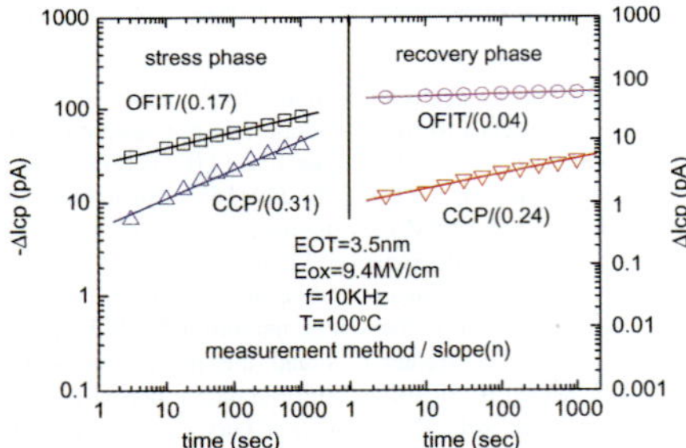

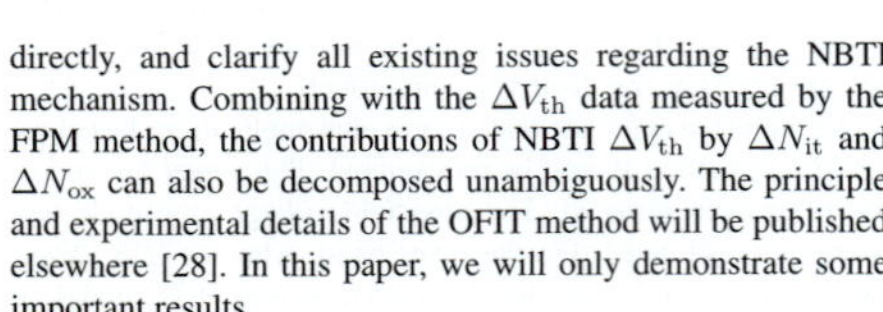

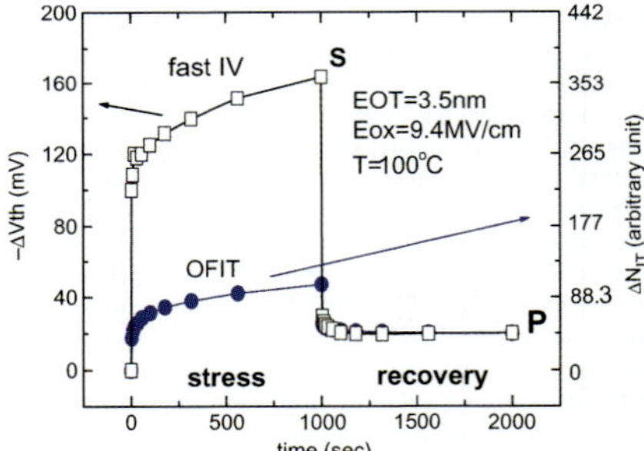

Fig. 11. Same as in Fig. 10, plotted in Log-Log scale in the (a) stress phase and (b) recovery phase, respectively. The slope $n = 0.17 \approx 1/6$ by OFIT measurement supports the H_2 diffusion in the R-D model [6]. CCP measurement overestimates the n value due to the recovery in the measurement delay.

Fig. 12. Time evolutions of ΔV_{th} measured by FPM and of ΔN_{it} measured by OFIT. ΔN_{it} curve is scaled to align the ΔV_{th} curve at P point at the end of recovery phase. The ΔV_{th} contributed by $q\Delta N_{it}(\Delta V_{th}^{it})$ is represented by the OFIT curve if $\eta_P = 1$, or below the OFIT curve if $\eta_P < 1$. The ΔV_{th} contributed by $q\Delta_{ox}(\Delta V_{th}^{ox})$ is the difference between two curves if $\eta_P = 1$ or more if $\eta_P < 1$. Here, $\eta_P = [\Delta N_{it}/(\Delta N_{it} + \Delta N_{ox})]_{at\ P}$.

directly, and clarify all existing issues regarding the NBTI mechanism. Combining with the ΔV_{th} data measured by the FPM method, the contributions of NBTI ΔV_{th} by ΔN_{it} and ΔN_{ox} can also be decomposed unambiguously. The principle and experimental details of the OFIT method will be published elsewhere [28]. In this paper, we will only demonstrate some important results.

Fig. 10 shows the time evolution of ΔN_{it} under 1000-s stress phase and 1000-s recovery phase measured by the OFIT method and the conventional CP (CCP) method, respectively. It is observed that the CCP method underestimates the ΔN_{it} value by around 50% due to the recovery during measurement delay, as shown in Fig. 10(b). The CCP method also distorts and much reduces the difference of ΔI_{CP} between S and P points, as also shown in the CP curve in Fig. 3.

Fig. 11 shows the $\Delta N_{it} = A\, t^n$ time evolution under stress with slope $n = 0.17 \cong 1/6$ at $T = 100\ ^\circ$C for OFIT data. These results support the R-D model of H_2 diffusion [6]. Further

discussion will be published elsewhere. On the other hand, the CCP method overestimates an n value $(n = 0.31)$ due to the recovery in the measurement delay. The n value reported by different labs is quite different, obviously due to the different measurement delay time used for ΔN_{it} measurement.

We have made convergent test of our OFIT measurement. We obtained that the recovery time of interface traps ΔN_{it} is longer than 50 μs, as compared to the very fast recovery time (100 ns) component of ΔV_{th} demonstrated in Sections III and IV. The details will be reported elsewhere.

VI. DECOMPOSITION OF ΔV_{th} CONTRIBUTED BY THE INTERFACE TRAPS AND OXIDE CHARGE COMPONENTS IN THE STRESS PHASE

In Fig. 12, we combine the ΔN_{it} curve by the OFIT measurement and the ΔV_{th} curve by the FPM method by aligning

68   IEEE TRANSACTIONS ON DEVICE AND MATERIALS RELIABILITY, VOL. 8, NO. 1, MARCH 2008

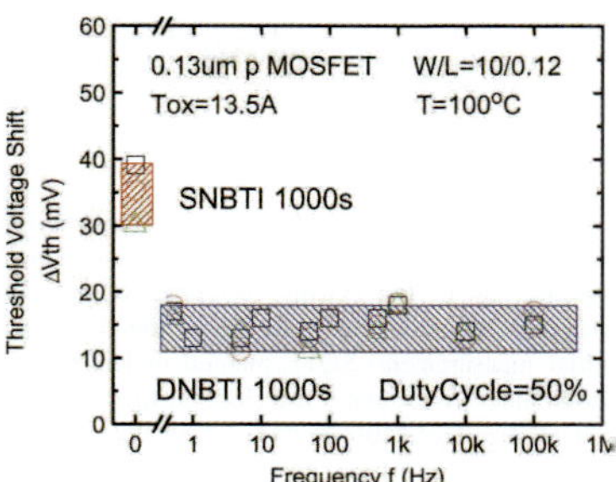

Fig. 13. Frequency dependence of slow component DNBTI, by dc measurement, with stress voltage $V_g = 2.5$ V. Different symbols for different samples. DNBTI degradation is frequency-independent and is less than the slow component static NBTI degradation, which is also measured by dc method [7].

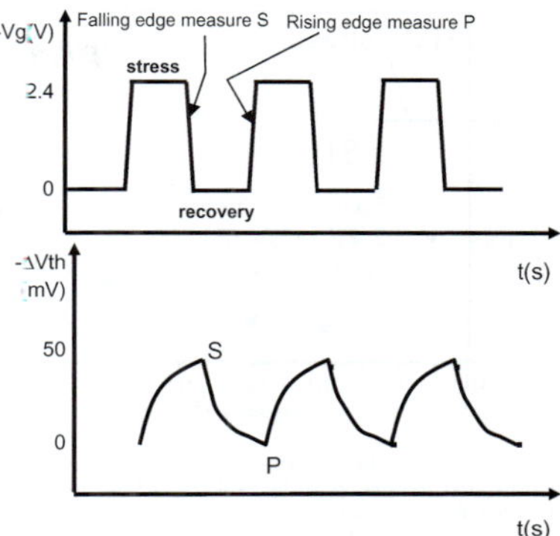

Fig. 14. FPM measures the ΔV_{th} at S and P points without interruption of the stress in the DNBTI stress condition.

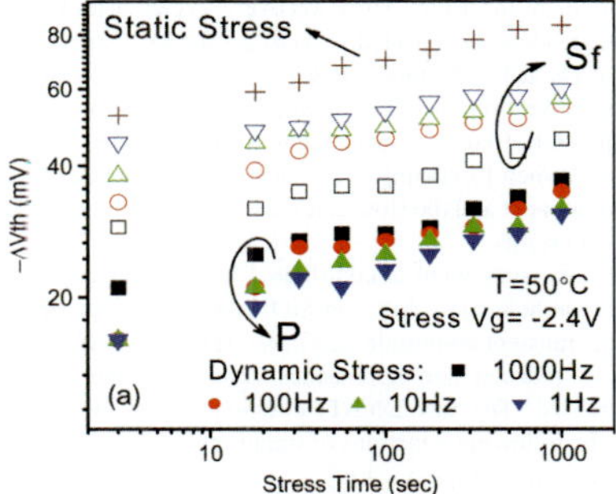

Fig. 15. Experimental ΔV_{th} data under DNBTI measured by FPM. The difference between S and P points is large and not closing after long stress time.

two curves at the P point at the end of the recovery phase. At this point, in the general case, ΔV_{th} may be caused by both $q\Delta N_{\text{it}}$ charge and the oxide trap charge $q\Delta N_{\text{ox}}$ with $\eta_P = [\Delta N_{\text{it}}/(\Delta N_{\text{it}} + \Delta N_{\text{ox}})]_{\text{at P}} \leq 1$. The ΔV_{th} component contributed by $q\Delta N_{\text{it}}$, which is denoted by $\Delta V_{\text{th}}^{\text{it}}$, is thus represented by the OFIT curve if $\eta_P = 1$ or below the OFIT curve if $\eta_P < 1$. The ΔV_{th} component contributed by $q\Delta N_{\text{ox}}$, which is denoted by $\Delta V_{\text{th}}^{\text{ox}}$, is the difference between two curves if $\eta_P = 1$ or more if $\eta_P < 1$. Plotting $\Delta V_{\text{th}}^{\text{ox}}$ in Log-Log scale in the stress phase results a slope of $n = 0.05$, which is much smaller than the slope $n = 0.17$ for $\Delta V_{\text{th}}^{\text{it}}$. We thus conclude that there are two components contributing to ΔV_{th} in NBTI degradation: a ΔN_{ox} component and a ΔN_{it} component in the stress phase. It explains why previous works that used ΔV_{th} measurement ($I-V$ or OTF) to measure the slope n have different values compared with our results in this paper. ΔV_{th} is not suitable to check the N_{it} R-D model due to the N_{ox} signal interference.

VII. Fast and Slow NBTI Components

In Section VI, we demonstrate that there are two components that contribute to ΔV_{th} in NBTI degradation. In Section V, we mentioned that the ΔN_{it} component has a recovery time longer than 50 μs. In Sections III and IV we demonstrated that ΔV_{th} has a very fast recovery component that is as fast as 100 ns. We therefore further conclude that there are two components in NBTI ΔV_{th} degradation: a slow component mainly contributed by ΔN_{it}, with slow recovery time that is longer than 50 μs, and a fast component mainly contributed by ΔN_{ox} with fast recovery time that is as fast as 100 ns. Therefore, we can expect that some NBTI behaviors measured by the slow dc method mainly reflect the behaviors of the slow ΔN_{it} component. They should be different from those measured by the FPM method which mainly reflect the behaviors of the fast ΔN_{ox} component.

A. Frequency Dependence of DNBTI Degradation

Fig. 13 shows the frequency dependence of DNBTI degradation measured by the dc method after 1000-s dynamic stress of

different frequency [7]. The dc measurement actually measures the cumulative degradation of slow DNBTI component due to interface trap degradation. By this measurement, it clearly shows that the slow component of DNBTI degradation is frequency independent in the frequency range of 1 Hz–100 kHz. Very recently, the IMEC group has conducted a very careful experiment to measure the DNBTI by an on-chip circuit to avoid the parasitic effect in the conventional DNBTI measurement through the probe when the frequency is very high. Their results confirm that DNBTI degradation is frequency independent in the frequency range of 1 Hz–2 GHz [29]. The frequency independence of the slow component DNBTI was explained by Alam [14] by an interface trap R-D theory.

However, the frequency dependence of DNBTI measured by the FPM has very different result. FPM can measure ΔV_{th} at S and P points, respectively, and does not interrupt the stress during measurement as explained in Fig. 14. Figs. 15 and 16 explore the frequency dependence of DNBTI degradation

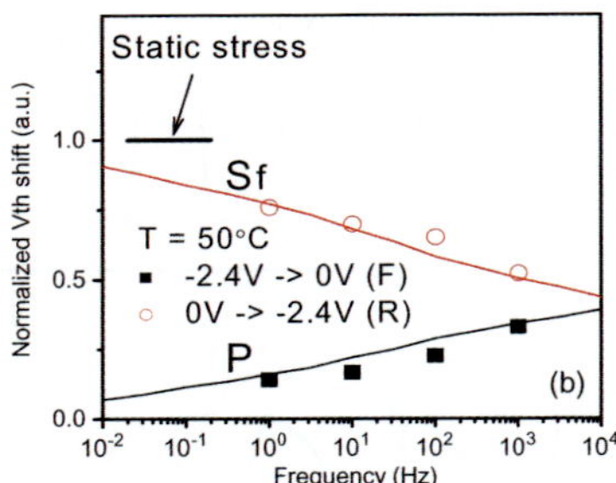

Fig. 16. Frequency dependence of DNBTI degradation measured at S (open circles) and P (solid squares) points by FPM after 1000-s stress time. The cumulative degradation measured at P is increased with increasing frequency. Solid lines are obtained by simulation (4) and (5). The simulation results predict that when the frequency is higher than 10^4 Hz, the difference between the P curve and S_f curve (the transient amplitude as shown in Fig. 14) approaches zero and the degradation becomes frequency-independent.

measured by the FPM. Stress-recovery cycles determined by the frequencies are repeated so as to get the total stress time of 1000 s. The ΔV_{th} degradation due to the fast component is also a cumulative process. The cumulation degradation measured at P point is frequency-dependent; increasing degradation is observed when increasing the frequency. This implies that the fast component and the slow component in NBTI have different physical origins.

The fast component is contributed by the trapping and de-trapping at hole traps ΔN_{ox} in SiON dielectric. Fig. 16 shows that the transient amplitude $(\Delta V_{\text{th,Sf}} - \Delta V_{\text{th,P}})$ under dynamic stress is reduced and approaches zero when the frequency is increased. The fast DNBTI component can be simulated using the following equations of trapping and detrapping of the preexisting hole traps N_{ot} [30]:

$$\frac{dp}{dt} = \frac{1}{\tau_C}(N_{\text{ox}} - p) - \frac{1}{\tau_{E1}}p \tag{4}$$

$$\frac{dp}{dt} = -\frac{1}{\tau_{E2}}p \tag{5}$$

where p is the trapped hole concentration; N_{ox} is the preex-isted trap concentration, which has a wide probability distrib-ution over trapping and detrapping time constants τ_C and τ_E [17], [31]. Fig. 16 shows the frequency dependence of the fast DNBTI component replotted on a normalized scale. Simulated results are plotted using solid lines, which are in good agree-ment with the experimental data.

B. Device Lifetime

Note that the degradation component $\Delta V_{\text{th}}^{\text{ox}}$ has a slower time evolution of $t^{0.05}$ than the degradation component $\Delta V_{\text{th}}^{\text{it}}$ with a fast time evolution of $t^{0.17}$; we can expect, for the longer stress time, the smaller effect of ΔN_{ox} on the total NBTI degra-dation. We have shown [17] that ΔN_{ox} only seriously affects the high-voltage operation. When the device operates at low voltage with a ten-year lifetime, N_{ox} degradation can almost be

neglected. The operation voltage with a ten-year lifetime V_{10y} should be mainly determined by ΔN_{it} degradation [17].

VIII. CONCLUSION

The conventional interface trap ΔN_{it} measurement methods CP and DCIV underestimate the ΔN_{it} value due to recovery during measurement. The existing SOTF ΔV_{th} measurement is actually not free from recovery due to the degradation at slow initial measurement. SOTF method underestimates the degradation and overestimates the slope in the Log-Log time evolution of the degradation curve and induces an artificial kink at the early stress time. We have developed some new measurement techniques: the ultrafast pulsed $I-V$ method (FPM), FOTF method, and OFIT measurement method to in-vestigate the NBTI mechanism. Both FOTF and FPM methods measure the NBTI ΔV_{th} degradation with excellent agreement, giving the reliable information. The OFIT measurement pro-vides the most reliable data to inspect the interface trap R-D model. Combining the OFIT and FPM methods, we demon-strate that the NBTI degradation in p-MOSFETs with SiON gate dielectric has two components with different physical origins. A slow $\Delta V_{\text{th}}^{\text{it}}$ component due to the contribution of interface trap ΔN_{it} has a recovery time longer than 50 μs. A fast $\Delta V_{\text{th}}^{\text{ox}}$ component due to the contribution of oxide trap charge $q\Delta N_{\text{ox}}$ has a broad spectrum of recovery time with very fast recovery time (100 ns) component, which can be well explained by oxide trapping/detrapping model. The cumulative degradation of $\Delta V_{\text{th}}^{\text{it}}$ component is frequency-independent in the DNBTI stress. The cumulative degradation of $\Delta V_{\text{th}}^{\text{ox}}$ com-ponent is increased with increasing frequency f, when $f < 10^4$ Hz, and approaches a constant when $f > 10^4$ Hz. The ten-year lifetime of the p-MOSFETs is mainly determined by the degradation of the $\Delta V_{\text{th}}^{\text{it}}$ component.

ACKNOWLEDGMENT

M.-F. Li would like to thank his former student Dr. G. Chen for his excellent work on NBTI degradation.

REFERENCES

[1] N. Kimizuka, T. Yamamoto, T. Mogami, K. Yamaguchi, K. Imai, and T. Horiuchi, "The impact of bias temperature instability for direct-tunneling ultra-thin gate oxide on MOSFET scaling," in *VLSI Symp. Tech. Dig.*, 1999, p. 73.
[2] G. La Rosa, F. Guarin, A. Acovic, J. Lukatis, and E. Crabbe, "NBTI-channel hot carrier effects in PMOSFETs in advanced CMOS technolo-gies," in *Proc. Int. Rel. Phys. Symp.*, 1997, p. 282.
[3] G. Chen, K. Y. Chuah, M. F. Li, C. H. Ang, J. Z. Zhen, and D. L. Kwong, "Dynamic NBTI of PMOS transistors and its impact on device lifetime," in *Proc. Int. Rel. Phys. Symp.*, 2003, p. 196.
[4] A. T. Krishnan, C. Chancellor, S. Chakravarthi, P. E. Nicollian, V. Reddy, A. Varghese, R. B. Khamankar, and S. Krishnam, "Material dependence of hydrogen diffusion: Implications for NBTI degradation," in *IEDM Tech. Dig.*, 2005, p. 705.
[5] M. A. Alam and S. Mahapatra, "A comprehensive model of PMOS NBTI degradation," *Microelectron. Reliab.*, vol. 45, no. 1, pp. 71–81, Jan. 2005.
[6] S. Chakravarthi, A. T. Krishnan, V. Reddy, C. F. Machala, and S. Krishnan, "A comprehensive framework for predictive modeling of negative bias temperature instability," in *Proc. Int. Rel. Phys. Symp.*, 2004, p. 273.

70   IEEE TRANSACTIONS ON DEVICE AND MATERIALS RELIABILITY, VOL. 8, NO. 1, MARCH 2008

[7] G. Chen, M. F. Li, C. H. Ang, J. Z. Zhen, and D. L. Kwong, "Dynamic NBTI of p-MOS transistors and its impact on MOSFET scaling," *IEEE Electron Device Lett.*, vol. 23, no. 12, pp. 734–736, Dec. 2002.

[8] S. Rangan, N. Mielke, and E. C. C. Yeh, "Universal recovery behavior of negative bias temperature instability," in *IEDM Tech. Dig.*, 2003, p. 341.

[9] S. Tsujikawa, T. Mine, K. Watanabe, Y. Shimamoto, R. Tsuchiya, K. Ohnishi, T. Onai, J. Yugami, and S. Kimura, "Negative bias temperature instability of pMOSFETs with ultra-thin SiON gate dielectrics," in *Proc. Int. Rel. Phys. Symp.*, 2003, p. 183.

[10] V. Huard, F. Monsieur, G. Ribes, and S. Bruyere, "Evidence for hydrogen-related defects during NBTI stress in p-MOSFETs," in *Proc. Int. Rel. Phys. Symp.*, 2003, p. 178.

[11] H. Usui, M. Kanno, and T. Morikawa, "Time and voltage dependence of degradation and recovery under pulsed negative bias temperature stress," in *Proc. Int. Rel. Phys. Symp.*, 2003, p. 610.

[12] M. Ershov, S. Saxena, H. Karbasi, S. Winters, S. Minehane, J. Babcock, R. Lindly, P. Clifton, M. Redford, and A. Shibkov, "Dynamic recovery of negative bias temperature instability in *p*-type metal–oxide–semiconductor field-effect transistors," *Appl. Phys. Lett.*, vol. 83, no. 8, pp. 1647–1649, Aug. 2003.

[13] K. O. Jepsson and C. M. Svenson, "Negative bias stress of MOS devices at high electric fields and degradation of MNOS devices," *J. Appl. Phys.*, vol. 48, no. 5, pp. 2004–2014, May 1977.

[14] M. A. Alam, "A critical examination of the mechanics of dynamic NBTI for PMOSFETs," in *IEDM Tech. Dig.*, 2003, p. 346.

[15] V. Huard and M. Denais, "Hole trapping effect on methodology for DC and AC negative bias temperature instability measurements in PMOS transistors," in *Proc. Int. Rel. Phys. Symp.*, 2004, p. 40.

[16] H. Reisinger, O. Blank, W. Heinrigs, A. Muhlhoff, W. Gustin, and C. Schlunder, "Analysis of NBTI deradation- and recovery-behavior based on ultra fast VT-measurements," in *Proc. Int. Rel. Phys. Symp.*, 2006, p. 448.

[17] T. Yang, M. F. Li, C. Shen, C. H. Ang, C. Zhu, Y.-C. Yeo, G. Samudra, S. C. Rustagi, M. B. Yu, and D. L. Kwong, "Fast and slow dynamic NBTI components in p-MOSFET with SiON dielectric and their impact on device life-time and circuit application," in *VLSI Symp. Tech. Dig.*, 2005, p. 92.

[18] A. Neugroschel, C. T. Sah, K. M. Han, M. S. Karoll, T. Nishida, J. T. Kavalieros, and Y. Lu, "Direct-current measurements of oxide and interface traps on oxidized silicon," *IEEE Trans. Electron Devices*, vol. 42, no. 9, pp. 1657–1662, Sep. 1995.

[19] J. Cai and C. T. Sah, "Interfacial electronic traps in surface controlled transistors," *IEEE Trans. Electron Devices*, vol. 47, no. 3, pp. 576–583, Mar. 2000.

[20] G. Groeseneken, H. E. Maes, N. Bertran, and R. F. Keersmaecker, "A reliable approach to charge-pumping measurements in MOS transistors," *IEEE Trans. Electron Devices*, vol. ED-31, no. 1, pp. 42–53, Jan. 1984.

[21] T. Yang, C. Shen, M. F. Li, C. H. Ang, C. X. Xue, Y. C. Yeo, G. Samudra, and D. L. Kwong, "Interface trap passivation effect in NBTI measurement for p-MOSFET with SiON gate dielectric," *IEEE Electron Device Lett.*, vol. 26, no. 10, pp. 758–760, Oct. 2005.

[22] A. Kerber, E. Cartier, L. A. Ragnarsson, M. Rosmeulen, L. Pantisano, R. Degraeve, T. Kauerauf, G. Groeseneken, H. E. Maes, and U. Schwalke, "Characterization of the V_T-instability in SiO_2/HfO_2 gate dielectrics," in *Proc. Int. Rel. Phys. Symp.*, 2003, p. 41.

[23] C. Shen, M. F. Li, X. P. Wang, H. Y. Yu, Y. P. Feng, A. T.-L. Lim, Y.-C. Yeo, D. S. H. Chan, and D. L. Kwong, "Negative U traps in HfO2 gate dielectrics and frequency dependence of dynamic BTI in MOSFETs," in *IEDM Tech. Dig.*, 2004, p. 733.

[24] C. Shen, M.-F. Li, X. P. Wang, Y.-C. Yeo, and D.-L. Kwong, "A fast measurement technique of MOSFET $I_d - V_g$ characteristics," *IEEE Electron Device Lett.*, vol. 27, no. 1, pp. 55–57, Jan. 2006.

[25] C. Shen, M. F. Li, C. E. Foo, T. Yang, D. M. Huang, G. S. Samudra, and Y. C. Yeo, "Characterization and physical origin of fast Vth transient in NBTI of pMOSFETs with SiON dielectric," in *IEDM Tech. Dig.*, 2006, p. 333.

[26] M. Denais, A. Bravaix, V. Huard, C. Parthasarathy, G. Ribes, F. Perrier, Y. R. Tauriac, and N. Revil, "On-the-fly characterization of NBTI in ultra-thin gate oxide PMOSFET's," in *IEDM Tech. Dig.*, 2004, p. 109.

[27] D. Varghese, D. Saha, S. Mahapatra, K. Ahmed, F. Nouri, and M. Alam, "Degradation and breakdown of 0.9 nm EOT SiO_2/ ALD HfO_2/metal gate stacks under positive constant voltage stress," in *IEDM Tech. Dig.*, 2005, p. 701.

[28] W. J. Wen *et al.*, "On-the-fly interface trap measurement and its impact on the understanding of NBTI mechanism for p-MOSFETs with SiON gate dielectric," in *IEDM Tech. Dig.*, 2007. to be published.

[29] R. Fernández, B. Kaczer, A. Nackaerts, S. Demuynck, R. Rodríguez, M. Nafría, and G. Groeseneken, "AC NBTI studied in the 1 Hz–2 GHz range on dedicated on-chip CMOS circuits," in *IEDM Tech. Dig.*, 2006, p. 237.

[30] Y. Nissan-Cohen, J. Shappir, and D. Frohman-Bentchkowsky, "Dynamic model of trapping-detrapping in SiO_2," *J. Appl. Phys.*, vol. 58, no. 6, pp. 2252–2261, Sep. 1985.

[31] S. Zafar, A. Callegari, E. Gusev, and M. V. Fischetti, "Charge trapping in high κ gate dielectric stacks," in *IEDM Tech. Dig.*, 2002, p. 517.

Ming-Fu Li received the degree from Fudan University, Shanghai, China, in 1960.

After graduation, he was with the University of Science and Technology of China (USTC), Hefei, China, first as a Teaching Assistant and then a Lecturer. In 1978, he was with the Graduate School, Chinese Academy of Sciences, Beijing, China, where he became a Professor in 1986. He was also an Adjunct Professor with the Institute of Semiconductors, Chinese Academy of Sciences, USTC, and Fudan University. He was a Visiting Scholar with the University of Illinois, Urbana, from 1979 to 1981, and a Visiting Scientist with the University of California, Berkeley, from 1986 to 1987 and 1990 to 1991. In 1991, he joined the National University of Singapore (NUS), Singapore, where he became a Professor with the Electrical and Computer Engineering Department. He was also an Adjunct Senior Member of Technical Staff with the Institute of Microelectronics, Singapore. He is currently a Professor with the Microelectronics Department, Fudan University, and an Adjunct Professorial Fellow with the NUS. He has published over 300 research papers and two books, including *Modern Semiconductor Quantum Physics* (World Scientific, 1994). His current research interests include CMOS device technology, and reliability and quantum modeling.

Mr. Li has served on numerous international program committees and advisory committees in semiconductor conferences in Canada, China, Germany, India, Japan, Singapore, Taiwan, China, and the USA.

Daming Huang received the B.S. degree in physics from Fudan University, Shanghai, China, in July 1982 and the Ph.D. degree in electrical engineering from the University of Illinois at Urbana-Champaign, in December 1989.

From December 1989 to May 1991, he was a Postdoctoral Research Associate with the University of South Florida, Tampa. Since June 1991, he has been a member of the faculty at Fudan University, where he became a Professor in 1995 in the Physics Department where he worked on the optical properties of semiconductor quantum wells, superlattices, and heterostructures. He was a Visiting Scientist with Virginia Commonwealth University, Richmond, from November 2000 to November 2002. In 2005, he joined the Microelectronics Department, Fudan University, as a Professor. His current research interest includes the reliability of the Si microelectronic devices.

Chen Shen (S'05) received the B.Eng. degree (first class honors) in electrical engineering from the National University of Singapore (NUS), Singapore, in 2003, where he is currently working toward the Ph.D. degree at the Silicon Nano Device Laboratory, Department of Electrical and Computer Engineering, NUS.

His research interests include reliability physics of SiO_2 and high-κ gate dielectrics in CMOS device, nonvolatile memory devices, and devices physics of novel I-MOS and T-FET transistors.

Mr. Shen received the NUS President Graduate Fellowship in 2006 and in 2007.

T. Yang, photograph and biography not available at the time of publication.

W. J. Liu received the B.Eng. degree in optoelectronics and the M.Eng. degree in optical engineering from Xi'an Technological University, Xi'an, China, in 2002 and 2005, respectively. He is currently working toward the Ph.D. degree in the State Key Lab of ASIC & System, Microelectronics Department, Fudan University, Shanghai, China.

He is focusing on semiconductor reliability issue—particularly negative bias temperature instability and the new measurement techniques for characterization of nano-CMOS devices.

Zhiying Liu is currently working toward the Ph.D. degree in microelectronics in the State Key Lab of ASIC & System, Microelectronics Department, Fudan University, Shanghai, China.

She joined the Laboratory of Reliability Research on Microelectronic Devices in 2005. Her current research interest includes the CMOS reliability issues.

IEEE TRANSACTIONS ON ELECTRON DEVICES, VOL. 56, NO. 2, FEBRUARY 2009   267

A Modified Charge-Pumping Method for the Characterization of Interface-Trap Generation in MOSFETs

Daming Huang, W. J. Liu, Zhiying Liu, C. C. Liao, Li-Fei Zhang,
Zhenghao Gan, Waisum Wong, and Ming-Fu Li

Abstract—A novel recovery-free interface-trap measurement method is presented in detail. This method is the modification of the conventional charge pumping (CP) by extending the pulse low level to the stress-bias and minimizing the pulse high-level duty cycle to suppress the recovery effect. The method is applied to study the negative-bias temperature instability in p-MOSFETs. As compared with the conventional CP, a much larger interface-trap generation under stress is observed by the new method. A power law time dependence ($\sim t^n$) of interface-trap generation is observed. The index n is less than that derived from conventional CP and increases with temperature, demonstrating a dispersive process involved in the trap generation dynamics.

Index Terms—Charge pumping (CP), interface traps, MOSFETs, negative-bias temperature instability (NBTI), reaction–diffusion model.

I. Introduction

NEGATIVE-BIAS temperature instability (NBTI) in p-MOSFETs has been investigated for four decades [1]. It has been one of the most critical reliability concerns for the present CMOS technologies when the device sizes shrink to nanometer scales [2]. The main effect of the NBTI is characterized by the threshold-voltage shift (ΔV_T) of a p-MOSFET under gate bias and temperature stress. It is widely believed that the ΔV_T is possibly induced by two microscopic components. One is the charge trapping in the gate oxide [3] and the other is the generation of the interface traps [4]. However, which component dominates the NBTI is still under debate

Manuscript received April 9, 2008; revised November 7, 2008. Current version published January 28, 2009. This work was supported by the Micro/Nanoelectronics Science and Technology Innovation Platform, Fudan University. The review of this paper was arranged by Editor C.-Y. Lu.

D. Huang and W. J. Liu are with the State Key Lab of ASIC and System, School of Microelectronics, Fudan University, Shanghai 201203, China (e-mail: dmhuang@fudan.edu.cn).

Z. Liu is with the State Key Lab of ASIC and System, School of Microelectronics, Fudan University, Shanghai 201203, China, and also with the School of Information and Communication Technology, Royal Institute of Technology, 100 44 Stockholm, Sweden.

C. C. Liao, L.-F. Zhang, Z. Gan, W. Wong are with the Semiconductor Manufacturing International Corporation, Shanghai 201203, China.

M.-F. Li is with the State Key Lab ASIC and System, School of Microelectronics, Fudan University, Shanghai 201203, China, and also with the Silicon Nano Device Laboratory, Department of Electrical and Computer Engineering, National University of Singapore, Singapore 117576, Singapore.

Color versions of one or more of the figures in this paper are available online at http://ieeexplore.ieee.org.

Digital Object Identifier 10.1109/TED.2008.2010585

[3]–[7]. The ΔV_T measured by the $I–V$ method cannot directly distinguish the different components, making the understanding of the NBTI difficult.

Another difficulty is the effect of recovery [3], [8]–[10] during the ΔV_T measurement when the stress is interrupted. This effect seriously distorts the information of the measured ΔV_T under stress [5], [10]. Similar to the static $I–V$ method for the ΔV_T measurement, the existing methods, which are the charge-pumping (CP) [11] and the direct-current $I–V$ (DCIV) [12] techniques for the interface trap (N_{IT}) measurement, both suffer from recovery effect [5]. Therefore, the experiment data thus obtained may not be reliable, and any quantitative comparison with the theoretical model may be irrelevant if the recovery effect is not properly considered.

To suppress the recovery effect, the static $I–V$ measurement for the ΔV_T has been improved in recent years. Pulsed $I–V$ techniques were developed to reduce the measurement delay to 1 μs or below [13], [14]. The new experimental results evidently show the delay dependence of the measured ΔV_T characteristics. Alternatively, an on-the-fly technique for the degradation in (g_m/I_D) and $I_{D,\mathrm{lin}}$ was proposed to obtain the ΔV_T [15], [16]. This method is recovery-free during the measurements immediately after the stress, although the initial value of I_D and g_m/I_D for the fresh device cannot be determined by this method correctly [14]. Despite these advantages, the quantity derived from both the pulsed $I–V$ and the on-the-fly techniques is the ΔV_T and provides no direct information on the mechanism of the NBTI, in case there are two components with different origins and different behaviors. To separate the two possible NBTI components of oxide charge and interface traps, it is desirable to have the techniques which can measure either of the components independently, as well as those that are recovery-free so that the available microscopic models can be tested quantitatively.

For this purpose, we have recently developed a modified CP (MCP) method for interface-trap measurement [17]. This method is based on the conventional CP but is recovery-free during the measurements. In this paper, we will present the MCP measurement method in detail and discuss the experimental results observed from p-MOSFETs under various biases and temperatures. After briefly describing the devices investigated in this paper in Section II, we will review the principle and the problems of the conventional CP associated with the NBTI characterization in Section III. In Section IV, we will present

0018-9383/$25.00 © 2009 IEEE

268

IEEE TRANSACTIONS ON ELECTRON DEVICES, VOL. 56, NO. 2, FEBRUARY 2009

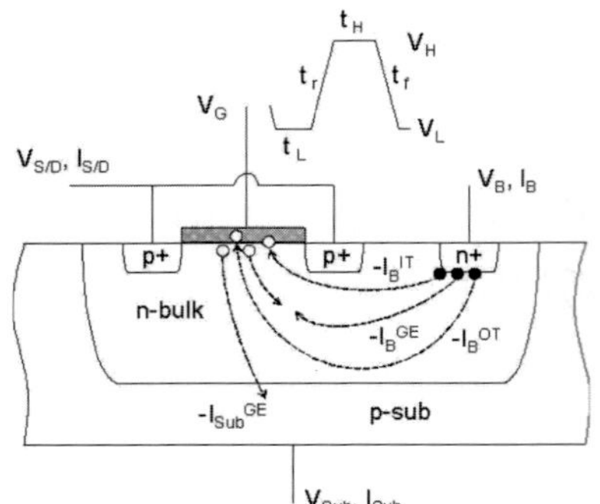

Fig. 1.   Schematic diagram of a p-MOSFET and a gate pulse used in the CP measurement, showing the different components of the CP current measured from the bulk and substrate. Solid circles represent electrons; open circles represent holes.

the MCP method and the experimental results in detail. We will conclude this paper in Section V.

II. DEVICES

The devices investigated in this paper are the p-MOSFETs fabricated with poly-Si/SiON gate stack. SiON (with 4×10^{15} cm^{-2} N) was grown by thermal oxidation followed by plasma nitridation and postdeposition thermal annealing. The device sizes (channel width/length) are 10 μm/10 μm, 4.0 μm/4.0 μm, and 10 μm/0.20 μm. A thick EOT of 3.5 nm is used to reduce the gate tunneling current and oxide/polygate interface effects, in order to increase the measurement accuracy.

III. REEXAMINE THE CONVENTIONAL CP METHOD

A. CP Current and Interface-Trap Density

The CP technique was first proposed by Brugler and Jespers [18]. It was extensively investigated by Groeseneken $et\ al.$ [11]. A schematic diagram of a p-MOSFET biased for the CP measurement is shown in Fig. 1, along with a gate pulse to show the notations. During the CP measurement, the source and drain are connected and grounded ($V_{S/D} = 0$). The substrate is also grounded ($V_{Sub} = 0$). A reverse bias V_R is applied to the bulk or well ($V_B = V_R$). When a series of trapezoidal pulses are applied to the gate with $V_H > V_{FB}$ (accumulation gate voltage) and $V_L < V_{INV}$ (inversion gate voltage), a CP current I_{CP} is measured from the bulk, as

$$I_{CP} \equiv I_B = I_B^{IT} + I_B^{GE} + I_B^{OT} \tag{1}$$

where I_B^{IT} is the contribution from interface traps, I_B^{GE} is from the geometric effect [11], [18], and I_B^{OT} is from the oxide traps [19].

The part of the interface traps detected by CP (N_{IT}^{CP}, in square centimeters) is proportional to the total trap density N_{IT}

$$N_{IT}^{CP}(t) = \alpha N_{IT}(t) \tag{2}$$

where $\alpha < 1$ depends on the device structure and the measurement conditions but is assumed to be independent of the stress time t. In this paper, the conventional CP is referred to as that no new interface traps are generated during the measurement. In this case, I_B^{IT} is measurement independently for a steady-state interface and simply related to N_{IT}^{CP} by [11]

$$I_B^{IT} = -qfA_G N_{IT}^{CP} \tag{3}$$

where q is the elementary charge, f is the pulse frequency, and A_G is the gate area. The condition for the conventional CP is satisfied when $V_H \sim 1.0$ V and $V_L \sim -1.0$ V, as typically used in the experiments.

From (1) and (3), we obtain

$$N_{IT}(t) = -\xi \left[I_B(t) - I_B^{GE}(t) - I_B^{OT}(t) \right] \tag{4}$$

where $\xi \equiv (\alpha q f A_G)^{-1}$. It shows that N_{IT} is proportional to I_B only when I_B^{GE} and I_B^{OT} are negligible ($\ll I_B$).

If the interface traps are uniformly distributed across the band gap E_g, then $\alpha = \Delta E / E_g$ and

$$\xi = \frac{1}{qfA_G} \frac{E_g}{\Delta E} \tag{5}$$

where ΔE is the bandwidth of the traps probed by CP [11]

$$\Delta E = -2kT \ln \left[\nu_{TH} n_i \sqrt{\sigma_n \cdot \sigma_p} \cdot \frac{|V_{FB} - V_{TH}|}{|\Delta V_G|} \cdot \sqrt{t_r \cdot t_f} \right] \tag{6}$$

where T is the temperature, k is the Boltzmann constant; ν_{TH} is the thermal velocity of conductive electrons; n_i is the intrinsic electron density; σ_n (σ_p) is the cross section of the electron (hole) traps; $\Delta V_G = V_H - V_L$ is the pulse amplitude; and t_r (t_f) is the pulse rising (falling) time.

In practice, ξ is not determined using (5) and (6) since the interface traps are not uniformly distributed across the band gap. Rather, the distribution has a U shape [20] and increases rapidly toward the conduction and valence edges.

B. Geometric Effect

The geometric component I_B^{GE} is the recombination current of the bulk electrons with the remaining channel holes during the rising phase of the gate pulses (see Fig. 1) [11]. This component can be eliminated, in principle, by reducing the gate length L_G or/and by increasing t_r [21]. However, a large L_G (i.e., a large gate area) and a small t_r (i.e., a high f) are often necessary to get enough high CP current to reduce the experimental error. Therefore, this component should be properly examined when the CP is used for the NBTI characterization.

We have applied the five-node CP method [21] to evaluate the I_B^{GE} for the devices used under different measurement conditions. In this method, a substrate current I_{Sub} is measured simultaneously with I_B. As shown in Fig. 1, I_{Sub} is fully due to the geometric effect I_{Sub}^{GE} and is related to I_B^{GE} by

$$I_{Sub} = I_{Sub}^{GE} = \beta I_B^{GE} \tag{7}$$

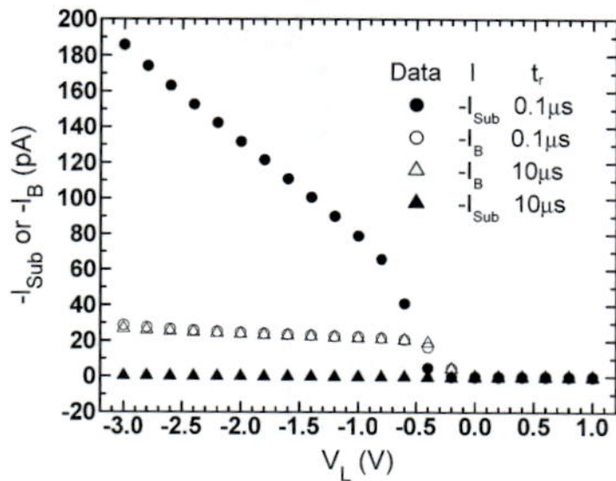

Fig. 2. CP current $-I_B$ and $-I_{\text{Sub}}$ as a function of V_L measured with t_r of 0.1 and 10 μs. The device is 10 μm × 10 μm. The other parameters are $V_H = 1.0$ V; $f = 10$ kHz; $t_f = t_r$; duty cycle is 50%; reverse bias $V_B = V_R = 0.3$ V; and $V_{S/D} = V_{\text{Sub}} = 0$ V.

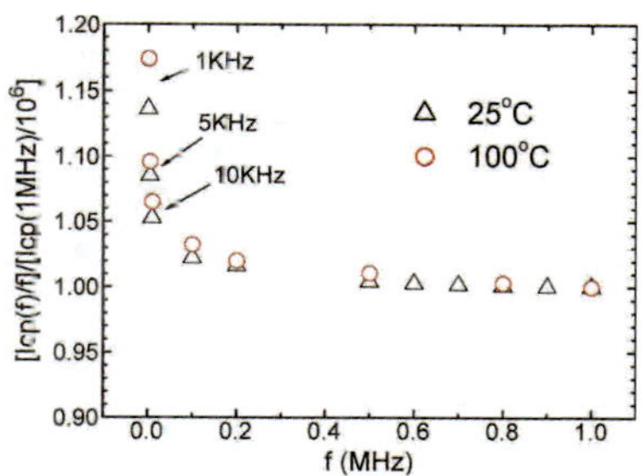

Fig. 3. CP current per cycle I_{CP}/f as a function of f, normalized to the data at 1 MHz ($I_{\text{CP}} = -I_B$), for 10 μm × 10 μm devices. When f decreases from 1 MHz to as low as 10 kHz, I_{CP}/f keeps a constant approximately with a variation less than 7% measured at both 100 °C and room temperature, indicating that, in this range, the oxide trap contribution to I_{CP} can be neglected.

where β can be estimated by measuring I_{Sub} and I_B twice with different t_r's, as shown in Fig. 2. Comparing two measurements, the change of I_{Sub} is purely due to the change of $I_{\text{Sub}}^{\text{GE}}$ denoted by $\Delta I_{\text{Sub}}^{\text{GE}}$; however, the change of I_B is the sum of the change of the CP current and the change of the geometric component. Therefore

$$\beta = \frac{\Delta I_{\text{Sub}}^{\text{GE}}}{\Delta I_B^{\text{GE}}} > \frac{\Delta I_{\text{Sub}}}{\Delta I_B}. \tag{8}$$

Fig. 2 shows the CP currents $-I_{\text{Sub}}$ and $-I_B$ as a function of V_L for a 10 μm × 10 μm device measured with $t_r = 0.1$ and 10 μs. Using (8) and the measurable ΔI_{Sub} and ΔI_B, β is estimated to be > 88 for $V_L = -3.0$ V. For the case of $t_r = 0.1$ μs, $-I_B^{\text{GE}} = -I_{\text{Sub}}/\beta < 2.1$ pA or $I_B^{\text{GE}}/I_B < 7.3\%$. Therefore, the geometric component of the CP current is negligible in the I_B measurement; however, it becomes very large and will seriously distort the CP current in the I_{SD} measurement. The geometric effect is less important for a lower $-V_L$, larger t_r, and for the devices with shorter L_G.

From the result shown in Fig. 2, the source/drain current $I_{\text{S/D}}$ may also be used as the probe for the CP current when $I_{\text{Sub}} \sim 0$ or $-I_B \sim I_{\text{S/D}}$ as in the case of the 10 μm × 10 μm device measured with $t_r = 10$ μs. Using $I_{\text{S/D}}$ instead of I_B to probe I_{CP} has the advantage of a more stable signal (less noisy) when the source and drain are better isolated to the substrate.

C. Oxide Trap Contribution

The oxide traps near the SiON–Si interface are detectable by CP via carrier tunneling at low gate-pulse frequencies [19]. The component I_B^{OT} can be extracted by plotting I_B/f as a function of f since I_B^{IT}/f and I_B^{GE}/f are both constants; however, I_B^{OT}/f increases when reducing f. Fig. 3 shows the measured $I_B(f)/f$ as a function of f for a 10 μm × 10 μm device. The data are normalized at 1 MHz. As shown, when f is reduced from 1 MHz, a slight increase of $I_B(f)/f$ is observed. The increment is less than 7% for f higher than 10 kHz. For f below 10 kHz, an accelerated increment is observed, and the contribution of I_B^{OT} to I_B may become significant.

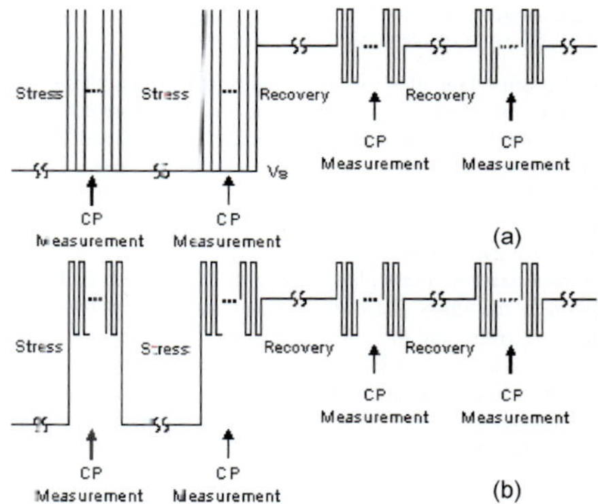

Fig. 4. (a) Schematic diagram showing the time evolution of the gate bias and pulses applied to p-MOSFETs for MCP measurements, including both stress and recovery phases. Each measurement in both stress and recovery phases takes 1.8 s. (b) Similar diagram for conventional CP measurements.

The results shown in Fig. 3 are also expected to be applicable to the shorter channel but otherwise identical devices. For shorter channel (small gate area) devices, a similar estimation of $I_{\text{CP}}^{\text{OT}}$ may be difficult due to the small I_{CP} measured at low f.

IV. MCP FOR INTERFACE-TRAP MEASUREMENT

A. Modification to Conventional CP

The MCP method reported here is the direct extension of the conventional CP measurement with the following modification to the gate pulses: 1) In the stress phase, V_L is set to the stress bias V_S (i.e., $V_L = V_S$) to make the measurement an almost on-the-fly one. The t_H (see Fig. 1) or the duty cycle for V_H is minimized to reduce the recovery if there is any during the measurement. Except if otherwise indicated, a duty cycle of 1%

270   IEEE TRANSACTIONS ON ELECTRON DEVICES, VOL. 56, NO. 2, FEBRUARY 2009

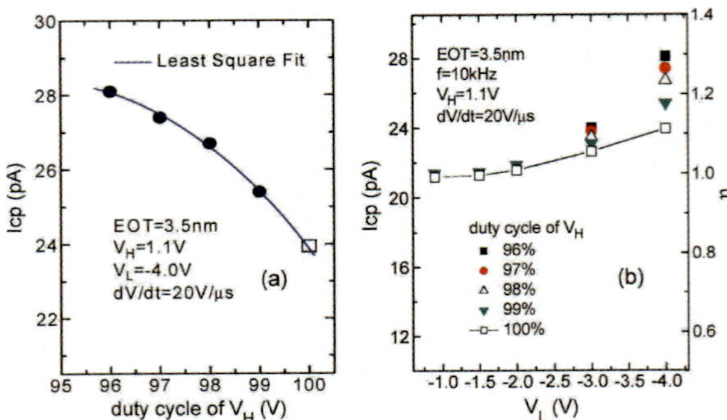

Fig. 5.   (a) Measured I_{CP} for four different duty cycles when $|V_L|$ is high (4 V). To suppress the new interface-trap generation during measurement, very high duty cycles (96%–99%) of V_H are used. The increase of I_{CP} with decreasing duty cycle (equivalent to longer stress time) is interpreted as new interface-trap generation during measurement. An extrapolation to near 100% duty cycle is made by second polynomial least square fitting the measured data, representing the measured I_{CP} without new interface-trap generation during measurement. (b) Measured I_{CP} as a function of V_L for four different V_H duty cycles. All measurements have the same pulse rising and falling rates dV/dt. The solid line connects the extrapolated data to near 100% duty cycle, producing the $\eta(V_L) = \xi(V_L)/\xi(V_L = -1 \text{ V})$ curve. SiON EOT = 3.5 nm.

for V_H is used for all of the MCP results in the stress phase, and 2) in the initial phase, the initial value $N_{IT}(0)$ before the stress phase and $N_{IT}(t)$ in the recovery phase are measured by the conventional CP with $V_L \cong -1.0$ V to avoid the stress effect. The pulse-voltage rising and falling rates dV/dt are adjusted to be the same as in the stress phase. Fig. 4(a) schematically shows the gate bias and pulses used for the stress/measurement characterization (stress phase) using the MCP method. In addition to the stress phase, the gate bias and the pulses used in the recovery and the initial phases are also presented. As a comparison, Fig. 4(b) shows the gate bias and pulses used for the NBTI characterization using conventional CP. In that case, a significant recovery may appear in the stress phase during the CP measurements when the stress is interrupted.

B. Rationale Behind the MCP Method

Some key points in the MCP method should be clarified.

1) In the MCP method, V_L is set to V_S in the stress phase and to -1.0 V in the recovery and initial phases. When the stress data are compared with the initial and recovery data, it is necessary to know the V_L dependence of the constant ξ in (4). In an early work, there was some discussion regarding the V_L dependence of I_B^{IT} [11]. The increase of I_B^{IT} was observed when increasing $|V_L|$. However, the early work did not distinguish whether the increase of I_B^{IT} is due to the decrease of ξ or the increase of N_{IT} during the measurement when $|V_L|$ is large. The accurate ξ value as a function of V_L can be obtained by the following experiment, as shown in Fig. 5. Fig. 5(a) shows the I_{CP} measured with the different duty cycles for V_H, from 96% to 99% with a large $|V_L| = 4.0$ V. Here, the duty cycle for V_H is defined as $t_H/(t_H + t_L)$, and the duty cycle for V_L is defined as $t_L/(t_H + t_L)$ assuming t_r and t_f are negligibly small. Increasing the duty cycle for V_H is equivalent to reducing the stress

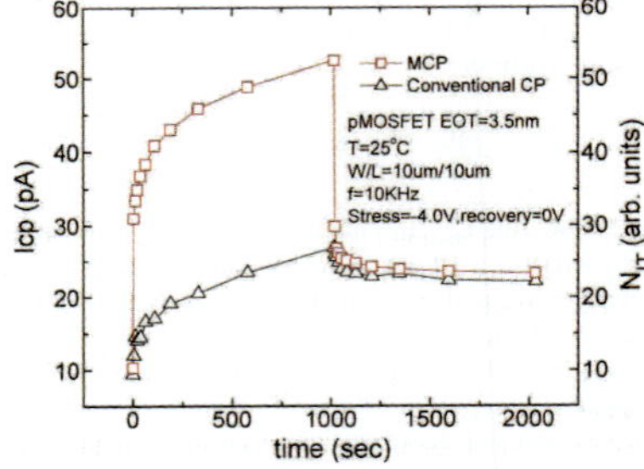

Fig. 6.   Time evolution of I_{CP} measured by (a) MCP and (b) conventional CP methods, including both stress and recovery phases plotted in linear scales.

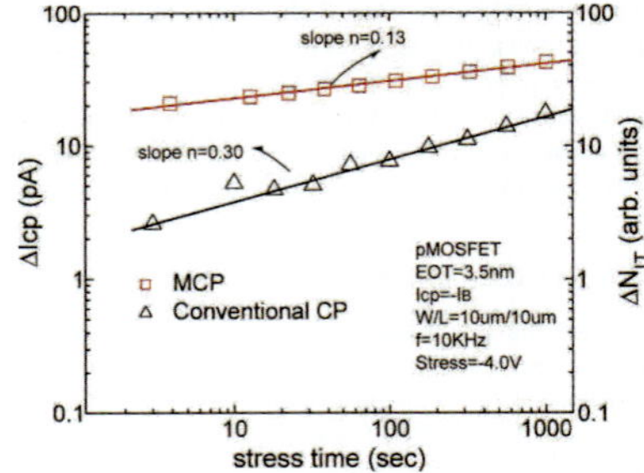

Fig. 7.   Log–Log plots of $\Delta I_{CP} \sim t$ characteristics measured using MCP and conventional CP for the stress phase.

time t_L during the measurement and, therefore, reducing the stress-induced new interface traps and the measured I_{CP}. By extrapolating the data to near 100% duty cycle

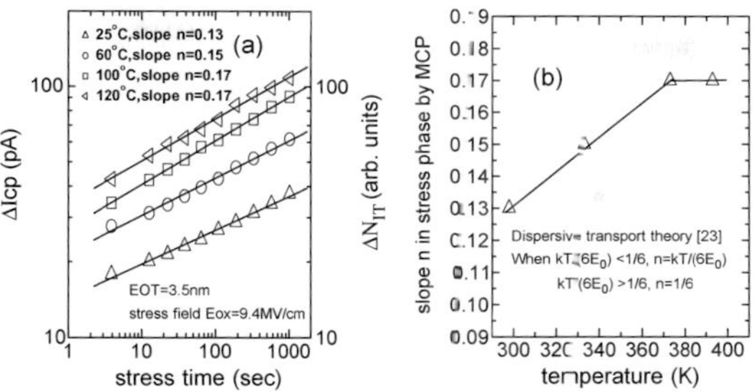

Fig. 8. (a) Time evolution of $\Delta I_{\rm CP}$ for the stress phase measured at different temperatures by the MCP method. (b) Linear temperature dependence of slope n when $n < 1/6$, supporting the H_2 dispersive transport [23] with a hydrogen trap energy depth of 32 meV.

by a least square fit, we obtain the $I_{\rm CP}$ with no stress effect, i.e., with no generation of new interface traps during the measurement. Fig. 5(b) shows the stress effect during the measurement and its correction for the MCP measurements with different V_L's. For $|V_L| \leq 2.0$ V, the $I_{\rm CP}$ is duty cycle independent up to 99%, and the stress effect is negligible. For $|V_L| \geq 3.0$ V, the $I_{\rm CP}$ reduces with the duty cycle, and the stress effect is not negligible. The data extrapolated to near 100% duty cycle for each V_L are also shown in Fig. 5(b) and connected by a solid curve. This line gives the weak V_L dependence of the $I_{\rm CP}$ for a fixed $N_{\rm IT}$ or the $[\xi(V_L)]^{-1}$ function which is needed in the MCP measurement to calibrate the measured $I_{\rm CP}$ in the stress phase. In the following, $I_{\rm CP}$ in the stress phase is the measured $I_{\rm CP}$ calibrated (divided) by a factor of $\eta = \xi(V_S)/\xi(V_L = -1$ V$)$ obtained from Fig. 5(b). The η value is ~ 1.1 for the devices used in the investigation when $V_S \sim -4.0$ V.

2) When V_L is extended to V_S, the gate tunneling current I_T is greatly enhanced since it is an exponential function of the gate voltage. In the recovery and initial phases, I_T is negligible when the oxide is thicker than 3 nm. In the stress phase, however, a tunneling contribution to the bulk current $I_{T,B} = I_T \times (1 - $ duty cycle of $V_H)$ may not be negligible and should be deducted from the measured CP current. In our experiments, $I_{T,B}$ was estimated by the carrier separation measurements. $I_{T,B}$ was measured both before and after the stress/recovery phases, and no stress-induced leakage current was observed.

C. Experimental Results

Fig. 6 shows the experimental results of the CP current ($I_{\rm CP} = -I_B$) as a function of time for both stress ($0 < t \leq 1000$ s) and recovery phases ($1000 < t \leq 2000$ s) measured from 10 μm $\times$ 10 μm devices at 25 °C. Before the stress is applied, the initial value $I_{\rm CP}(t = 0)$ is measured using the conventional CP. After that, a stress of -4.0V is applied on the gate. During the stress phase, the $I_{\rm CP}(0 < t \leq 1000$ s$)$ data are measured using the MCP method [Fig. 6(a)]. As a comparison, the data measured from a parallel device using the conventional CP are also shown [Fig. 6(b)]. Both sets of $I_{\rm CP}$ data in the

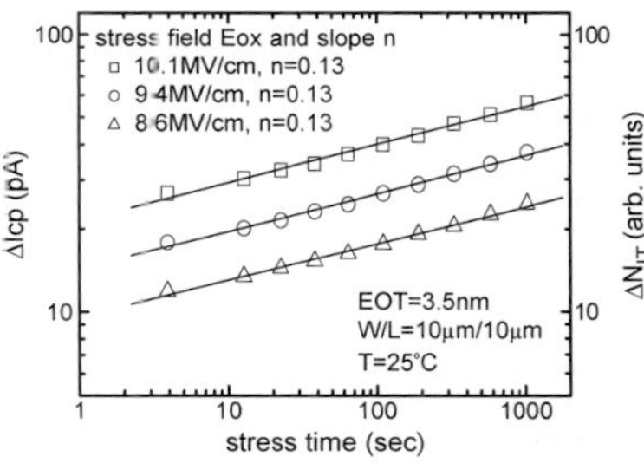

Fig. 9. Time evolution of $\Delta I_{\rm CP}$ for the stress phase measured at different oxide field stresses $E_{\rm OX}$ by the MCP method. The slope n is independent of $E_{\rm OX}$ within the measurement error.

recovery phases ($1000 < t \leq 2000$s) are measured with the conventional CP.

The most significant result shown in Fig. 6 is a much higher $I_{\rm CP}$ measured by the MCP method than that by the conventional CP in the stress phase. A much lower $I_{\rm CP}$ measured by the conventional CP is essentially due to the recovery effect, i.e., due to the recovery of the generated interface traps during the CP measurements.

D. Slopes in $\mathrm{Log}(\Delta N_{\rm it})-\mathrm{Log}(\Delta t)$ Plots

Using the method described earlier, $\Delta N_{\rm IT}$, as a function of Δt, is obtained. Fig. 7 shows the results plotted in Log–Log scales for the stress phase measured from a 10 μm $\times$ 10 μm device. The same $I_{\rm CP}(t)$ data plotted in linear scales have been shown in Fig. 6. The results demonstrate a power law $\Delta N_{\rm IT}(t) \sim t^n$ for the trap generation with stress time measured by both conventional CP and MCP methods. A slope n of 0.30 is derived from the conventional CP. However, it is much less (0.13) than that for the MCP measurement. The reduction in the slope demonstrates reduced recoveries during the $\Delta N_{\rm IT}$ measurements, similar to the cases of the pulsed I–V method for $\Delta V_{\rm TH}$ measurements with different measurement times [14].

272   IEEE TRANSACTIONS ON ELECTRON DEVICES, VOL. 56, NO. 2, FEBRUARY 2009

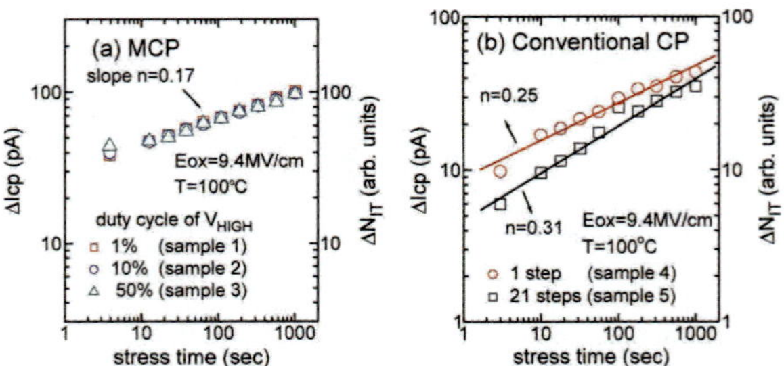

Fig. 10. Time evolution of ΔI_{CP} for the stress phase measured (a) by MCP method with different V_H duty cycles of 1%, 10%, and 50% and (b) by conventional CP with different measurement delays.

The MCP experiments are also performed on the 10 μm × 10 μm devices with different temperatures and different stress biases (field). Fig. 8 shows the results measured with a stress field of 9.4 MV/cm but at different temperatures. It shows that the slope n of the ΔN_{IT} degradation is linearly temperature dependent when $n < 1/6$. The results support the R-D model of H_2 diffusion [22] with dispersive [23] non-Arrhenius transport. The results shown in Fig. 3 also rule out the possibility that the change of n at different temperatures is due to the change of oxide trap contribution in the I_{CP} measurement. Using an $n(T)$ function in the dispersive transport theory, $n = kT/6E_0$ with the characteristic DOS width of localized hydrogen states in the oxide [23] $E_0 \cong 32$ meV, the experimental temperature dependence of n [Fig. 8(b)] can be explained.

Fig. 9 shows the MCP results measured at 25 °C but at different stress fields from 8.6 to 10.1 MV/cm. It is observed that the slope n does not change with the stress field.

E. Recovery Effect During Measurement

As compared with the conventional CP, a higher I_{CP} and a lower n are observed from the MCP measurements in the stress phase. The results demonstrate a lesser recovery suffered in the latter method. To test whether any recovery effect is still left to affect the result significantly in the MCP method, we show, in Fig. 10(a), the $\Delta I_{CP}(t)$ data measured in the stress phase with different duty cycles (1%, 10%, and 50%). For the pulse frequency used (10 kHz), the duty cycle for V_H from 1% to 50% corresponds to the interruption time t_H from 1 to 50 μs in each CP cycle. If there is any recovery for $t_H < 50$ μs, the measured $\Delta I_{CP}(t)$ data are expected to be different for the different duty cycles used. However, the observed results are the same, demonstrating a negligible recovery of the ΔN_{IT} during the MCP measurements. In other words, the recovery time for the interface-trap generation is > 50 μs.

Fig. 10(b) shows the similar results measured by the conventional CP method for the devices on the same wafer. A one-step measurement takes ~1 s to get each data point, while a 21-step measurement takes much longer (~20 s for each data point). As shown in Fig. 10(b), the longer measurement time leads to a lower ΔI_{CP} due to the recovery effect during the CP measurement. As a result, the slope n in the $\log \Delta I_{CP} - \log t$ plot is also increased from 0.25 to 0.31 when the measurement time is increased from ~1 to ~20 s.

V. CONCLUSION

We have presented, in detail, a novel MCP measurement method for the characterization of the interface-trap generation in p-MOSFETs under negative-bias stress. This method is the modification of the conventional CP by extending the pulse low level to stress bias and minimizing the pulse high-level duty cycle. The method has been shown to be recovery-free and has been applied to study the NBTI in p-MOSFETs. As compared with the conventional CP method, a much larger interface-trap generation under stress is observed by the MCP method. A power law time dependence ($\Delta N_{IT} \sim t^n$) is demonstrated, where the slope n increases with temperature. This result shows that a dispersive process must be involved in the trap generation process. The MCP method reported here can also be used for the other BTI configurations, such as the NBTI in n-MOSFETs and the PBTI in n- and p-MOSFETs.

REFERENCES

[1] D. K. Schroder and J. A. Babcock, "Negative bias temperature instability: Road to cross in deep submicron silicon semiconductor manufacturing," *J. Appl. Phys.*, vol. 94, no. 1, pp. 1–18, Jul. 2003.

[2] V. Reddy, A. T. Krishnan, A. Marshall, J. Rodriguez, S. Natarajan, T. Rost, and S. Krishnan, "Impact of negative bias temperature instability on digital circuit reliability," in *Proc. Int. Reliab. Phys. Symp.*, 2002, pp. 248–254.

[3] V. Huard and M. Denais, "Hole trapping effect on methodology for DC and AC negative bias temperature instability measurements in pMOS transistors," in *Proc. Int. Reliab. Phys. Symp.*, 2004, pp. 40–45.

[4] D. Varghese, D. Saha, S. Mahapatra, K. Ahmed, F. Nouri, and M. Alam, "On the dispersive versus Arrhenius temperature activation of NBTI time evolution in plasma nitrided gate oxides: Measurements, theory, and implications," in *IEDM Tech. Dig.*, 2005, pp. 684–687.

[5] T. Yang, M. F. Li, C. Shen, C. H. Ang, C. Zhu, Y. C. Yeo, G. Samudra, S. C. Rustagi, M. B. Yu, and D. L. Kwong, "Fast and slow dynamic NBTI components in p-MOSFET with SiON dielectric and their impact on device life-time and circuit application," in *VLSI Symp. Tech. Dig.*, 2005, pp. 92–93.

[6] H. Reisinger, O. Blank, W. Henrigs, A. Mühlhoff, W. Gustin, and C. Schlünder, "Analysis of NBTI degradation- and recovery-behavior based on ultra fast V_T measurements," in *Proc. Int. Reliab. Phys. Symp.*, 2006, pp. 448–453.

[7] T. Grasser, B. Kaczer, P. Hehenberger, W. Gös, R. O. Cobbor, H. Reisinger, W. Gustin, and C. Schlünder, "Simultaneous extraction of

recoverable and permanent components contributing to bias-temperature instability," in *IEDM Tech. Dig.*, 2007, pp. 801–804.

[8] G. Chen, M. F. Li, C. H. Ang, J. Z. Zheng, and D. L. Kwong, "Dynamic NBTI of p-MOS transistors and its impact on MOSFET scaling," *IEEE Electron Device Lett.*, vol. 23, no. 12, pp. 734–736, Dec. 2002.

[9] S. Tsujikawa, T. Mine, K. Watanabe, Y. Shimamoto, R. Tsuchiya, K. Ohnishi, T. Onai, J. Yugami, and S. Kimura, "Negative bias temperature instability of pMOSFETs with ultra-thin SiON gate dielectrics," in *Proc. Int. Reliab. Phys. Symp.*, 2003, pp. 183–188.

[10] M. Ershov, S. Saxena, H. Karbasi, S. Winters, S. Minehane, J. Babcock, R. Lindly, P. Clifton, M. Redford, and A. Shibkov, "Dynamic recovery of NBTI in p-type MOSFETs," *Appl. Phys. Lett.*, vol. 83, no. 8, pp. 1647–1649, Aug. 2003.

[11] G. Groeseneken, H. E. Maes, N. Beltran, and R. F. De Keersmaecker, "A reliable approach to charge-pumping measurements in MOS transistors," *IEEE Trans. Electron Devices*, vol. ED-31, no. 1, pp. 42–53, Jan. 1984.

[12] A. Neugroschel, C. T. Sah, K. M. Han, M. S. Carroll, T. Nishida, J. T. Kavalieros, and Y. Lu, "Direct-current measurements of oxide and interface traps on oxidized silicon," *IEEE Trans. Electron Devices*, vol. 42, no. 9, pp. 1657–1662, Sep. 1995.

[13] C. Shen, M. F. Li, X. P. Wang, Y. C. Yeo, and D. L. Kwong, "A fast measurement technique of MOSFET I_d–V_g characteristics," *IEEE Electron Device Lett.*, vol. 27, no. 1, pp. 55–57, Jan. 2006.

[14] C. Shen, M. F. Li, C. E. Foo, T. Yang, D. M. Huang, G. S. Samudra, and Y. C. Yeo, "Characterization and physical origin of fast V_{th} transient in NBTI of pMOSFETs with SiON dielectric," in *IEDM Tech. Dig.*, 2006, pp. 333–336.

[15] M. Denais, A. Bravaix, V. Huard, C. Parthasarathy, G. Ribes, F. Perrier, Y. R. Tauriac, and N. Revil, "On-the-fly characterization of NBTI in ultra-thin gate oxide PMOSFETs," in *IEDM Tech. Dig.*, 2004, pp. 109–112.

[16] S. Rangan, N. Mielke, and E. C. C. Yeh, "Universal recovery behavior of negative bias temperature instability," in *IEDM Tech. Dig.*, 2003, pp. 341–344.

[17] W. J. Liu, Z. Y. Liu, D. Huang, C. C. Liao, L. F. Zhang, Z. H. Gan, W. Wong, C. Shen, and M. F. Li, "On-the-fly interface trap measurement and its impact on the understanding of NBTI mechanism for p-MOSFETs with SiON gate dielectric," in *IEDM Tech. Dig.*, 2007, pp. 813–816.

[18] J. S. Brugler and P. G. A. Jespers, "Charge pumping in MOS devices," *IEEE Trans. Electron Devices*, vol. ED-16, no. 3, pp. 297–302, Mar. 1969.

[19] R. E. Paulsen and M. H. White, "Theory and application of charge pumping for the characterization of $Si-SiO_2$ interface and near-interface oxide-traps," *IEEE Trans. Electron Devices*, vol. 41, no. 7, pp. 1213–1216, Jul. 1994.

[20] M. H. White and J. R. Cricchi, "Characterization of thin-oxide MNOS memory transistors," *IEEE Trans. Electron Devices*, vol. ED-19, no. 12, pp. 1280–1288, Dec. 1972.

[21] G. Van den bosch, G. Groeseneken, and H. E. Maes, "On the geometric component of charge-pumping current in MOSFETs," *IEEE Electron Device Lett.*, vol. 14, no. 3, pp. 107–109, Mar. 1993.

[22] A. E. Islam, H. Kufluoglu, D. Varghese, and M. A. Alam, "Critical analysis of short-term negative bias temperature instability measurements: Explaining the effect of time-zero delay for on-the-fly measurements," *Appl. Phys. Lett.*, vol. 90, no. 8, p. 083 505, Feb. 2007.

[23] B. Kaczer, V. Arkhipov, R. Degraeve, N. Collaert, G. Groeseneken, and M. Goodwin, "Disorder-controlled-kinetics model for negative bias temperature instability and its experimental verification," in *Proc. Int. Reliab. Phys. Symp.*, 2005, pp. 381–387.

W. J. Liu is currently working toward the Ph.D. degree with the State Key Lab of ASIC and System, School of Microelectronics, Fudan University, Shanghai, China.

His interests include reliability physics of gate dielectrics in CMOS devices and characterization techniques and device physics of novel devices.

Zhiying Liu received the B.S. degree from Huazhong University of Science and Technology, Wuhan, China, in 2005. She is currently working toward the Ph.D. degree, as a joint Ph.D. student, in the State Key Lab of ASIC and System, School of Microelectronics, Fudan University, Shanghai, China, and in the School of Information and Communication Technology, Royal Institute of Technology, Stockholm, Sweden.

Her research interest includes nanodevices and their reliability issues.

C. C. Liao received the M. S. degree in electrical engineering from Chung Hua University, Hsinchu, Taiwan, in 1997 and the Ph.D. degree from National Chiao Tung University, Hsinchu, in 2005.

In 2001, he joined the Semiconductor Manufacturing International Corporation (SMIC), Shanghai, China. He is the author or coauthor of more than 40 technical papers on reliability, high-k, RF Si, ESD, and nano-CMOS. His current research interests include the development of high-k gate insulator processing for VLSI devices, the characterization of nano-MOS devices, and the physics of reliability in VLSI.

Li-Fei Zhang received the M.S. degree in physics from Fudan University, Shanghai, China, in 2003.

She then joined the Logic Technology Development Center, Semiconductor Manufacturing International Corporation (SMIC), Shanghai. She has more than five years of experience in process reliability testkey design, test methodology, and evaluation for mass production.

Daming Huang received the B.S. degree in Physics from Fudan University, Shanghai, China, in July 1982 and the Ph.D. degree in electrical engineering from the University of Illinois, Urbana, in December 1989.

From December 1989 to May 1991, he was a Postdoctoral Research Associate with the University of South Florida, Tampa. Since June 1991, he has been a member of the faculty at Fudan University, as a Professor, in 1995, with the Physics Department, where he worked on the optical properties of semiconductor quantum wells, superlattices, and heterostructures and where, since 2005, he has been with the State Key Lab of ASIC and System, School of Microelectronics as a Professor. He was a visiting scientist with Virginia Commonwealth University, Richmond, from November 2000 to November 2002. His current research interest includes the reliability of Si microelectronic devices.

Zhenghao Gan received the B.Eng. and M. Eng. degrees in materials science and engineering from Zhejiang University, Hangzhou, China, in 1995 and 1997, respectively, and the Ph.D. degree in mechanical engineering from Nanyang Technological University, Singapore, in 2002.

From 2001 to 2006, he was a Research Fellow with the School of Electrical and Electronic Engineering and School of Materials Science and Engineering, Nanyang Technological University, focusing on ULSI interconnect reliability research. In 2006, he joined the Semiconductor Manufacturing International Corporation, Shanghai, China, where he is currently a Technical Manager in LTD, heavily involved in reliability improvement and evaluation (both FEOL and BEOL) for the advanced technologies. He is the author of more than 30 technical papers on reliability in refereed journals. He is also the holder of several patents filed in China.

274   IEEE TRANSACTIONS ON ELECTRON DEVICES, VOL. 56, NO. 2, FEBRUARY 2009

Waisum Wong received B.S. and M.S. degrees in electrical engineering from Mississippi State University, Starkville, in 1986 and 1988, respectively, and the Ph.D. degree in electrical engineering from the University of Central Florida, Orlando, in 1992.

From 1992 to 1994, he was with Siliconix, where he worked on power device development and modeling. In 1994, he joined Exar to work on mixed-signal circuit design and MS SPICE modeling. From 1995 to 2005, he was with Intel, where he worked on mixed-signal/RF device modeling, analog circuit design, high-speed simulation support, and RF PDK development. In 2005, he joined the Semiconductor Manufacturing International Corporation, Shanghai, China, where he is currently in charge of the LTD/Device Group, which is responsible for device characterization, MS/RF SPICE modeling, TCAD, and reliability for advanced process development. He is the author or coauthor of more than 50 technical papers on spice modeling, reliability modeling, and circuit design. He is the holder of three U.S. and China patents, many of which are currently under review.

Ming-Fu Li received the degree from Fudan University, Shanghai, China, in 1960.

After graduation, he was with the University of Science and Technology of China (USTC), Hefei, China, first as a Teaching Assistant and then as a Lecturer. He was with the Graduate School, Chinese Academy of Sciences, Beijing, China, in 1978, where he was a Professor in 1986 and where he was also an Adjunct Professor with the Institute of Semiconductors, Chinese Academy of Sciences. He was also an Adjunct Professor with USTC and with Fudan University. He was a Visiting Scholar with the University of Illinois, Urbana, from 1979 to 1981, and a Visiting Scientist with the University of California, Berkeley from 1986 to 1987 and from 1990 to 1991. He was an Adjunct Senior Member of the Technical Staff with the Institute of Microelectronics, Singapore, Singapore. Being a Founding Member of the Silicon Nano Device Laboratory, he has been with the National University of Singapore (NUS) since 1991, first as a Professor with the Electrical and Computer Engineering Department and then, currently, as an Adjunct Professorial Fellow. He is also currently a Professor with the State Key Lab of ASIC and System, School of Microelectronics, Fudan University. He has published over 350 research papers and two books, including Modern Semiconductor Quantum Physics (Singapore; World Scientific, 1994). His current research interests include CMOS device technology and reliability and quantum modeling.

Prof. Li has served on numerous international program and advisory committees in semiconductor conferences in Canada, China, Germany, India, Japan, Singapore, Taiwan, China, and the USA.

Chapter 5

CMOS Transistors (II) (Technologies)

APPLIED PHYSICS LETTERS   VOLUME 81, NUMBER 19   4 NOVEMBER 2002

Thermal stability of $(HfO_2)_x(Al_2O_3)_{1-x}$ on Si

H. Y. Yu, N. Wu, M. F. Li,[a] Chunxiang Zhu, and B. J. Cho
Silicon Nano Device Lab, Department of Electrical and Computer Engineering, National University of Singapore, Singapore 119260

D.-L. Kwong
Department of Electrical and Computer Engineering, The University of Texas, Austin, Texas 78752

C. H. Tung
Institute of Microelectronics, Singapore 117685

J. S. Pan, J. W. Chai, W. D. Wang, and D. Z. Chi
Institute of Materials Research and Engineering, 3 Research Link Singapore 117602

C. H. Ang and J. Z. Zheng
Chartered Semiconductor Manufacturing Limited, Singapore 738406

S. Ramanathan
Genus Incorporated, Sunnyvale, California 94089

(Received 18 July 2002; accepted 13 September 2002)

The kinetics of the interfacial layer (IL) growth between Hf aluminates and the Si substrate during high-temperature rapid thermal annealing (RTA) in either N_2 ($\sim$10 Torr) or high vacuum ($\sim$2 $\times 10^{-5}$ Torr) is studied by high-resolution x-ray photoelectron spectroscopy and cross-sectional transmission electron microscopy. The significant difference of the IL growth observed between high vacuum and relatively oxygen-rich N_2 annealing (both at 1000 °C) is shown to be caused by the oxygen species from the annealing ambient. Our results also show that Hf aluminates exhibit much stronger resistance to oxygen diffusion than pure HfO_2 during RTA in N_2 ambient, and the resistance becomes stronger with more Al incorporated into HfO_2. This observation is explained by the combined effects of (i) smaller oxygen diffusion coefficient of Al_2O_3 than HfO_2, and (ii) higher crystallization temperature of the Hf aluminates. © 2002 American Institute of Physics. [DOI: 10.1063/1.1519733]

Recently, HfO_2-based high-k gate dielectrics including HfO_2,[1,2] Hf silicates,[3,4] and Hf aluminates,[5–9] as alternates to SiO_2 have been extensively studied for future generation of metal–oxide–semiconductor transistors to address the excessive high leakage current concern. Pure as-deposited amorphous HfO_2 crystallizes during postdeposition annealing (for example $\sim$400 °C),[6] which may induce grain-boundary leakage current and nonuniformity of the film thickness.[10] More importantly, HfO_2 is transparent to the oxygen diffusion.[3,11] Annealing in an oxygen-rich ambient will lead to fast diffusion of oxygen through the HfO_2, causing the growth of uncontrolled low-k interfacial layers (either SiO_x or SiO_x-containing layer).[10,12] The uncontrolled low-k layer poses a serious limitation to further scaling of the equivalent oxide thickness for HfO_2 gate dielectrics. Several research groups have demonstrated that alloying HfO_2 with Al results in the increase of crystallization temperature of HfO_2 films.[6–8] Furthermore, due to their reasonable k value and the band offset values to Si,[5,7,8] hafnium aluminates are thus regarded as a promising candidate for high-k gate dielectrics application. However, the impact of alloying Al into Hf on oxygen diffusivity through Hf aluminates has not been studied. In this letter, we demonstrate that the resistance to the oxygen diffusion in HfO_2 films can be greatly enhanced by the incorporation of Al.

[a]Electronic mail: elelimf@nus.edu.sg

$(HfO_2)_x(Al_2O_3)_{1-x}$ ($0 \leqslant x \leqslant 1$) films of four different Al_2O_3 compositions were deposited by atomic layer deposition at 300 °C onto 8 in. p-type (100) Si substrates. Trimethyl aluminum [TMA, $Al(CH)_3$], hafnium tetrachloride ($HfCl_4$), and water (H_2O) were used as precursors, and nitrogen was employed as carrier and purge gas. Hf–Al–O composite films were formed by switching between the metal precursors, TMA and $HfCl_4$, while an H_2O pulse follows every metal pulse.[8] Rapid thermal annealing (RTA) was conducted in either 10 Torr of N_2 or in a high vacuum ($\sim$2$\times 10^{-5}$ Torr) at several temperatures (800 °C–1000 °C) for 20 s. Thick films ($\sim$20 nm) were prepared for x-ray diffraction (XRD) and x-ray photoelectron spectroscopy (XPS) study. The XPS results show that the compositions of these samples are HfO_2, $(HfO_2)_{0.85}(Al_2O_3)_{0.15}$, $(HfO_2)_{0.67}(Al_2O_3)_{0.33}$, and Al_2O_3. Thin films were prepared using the same process recipe and characterized using XPS and cross-sectional transmission electron microscopy (XTEM). Thicknesses for the as-deposited high-k thin films determined by transmission electron microscopy are $\sim$5.5 nm (HfO_2), $\sim$4.4 nm $[(HfO_2)_{0.85}(Al_2O_3)_{0.15}]$, $\sim$4.2 nm $[(HfO_2)_{0.67}(Al_2O_3)_{0.33}]$, and $\sim$4.6 nm (Al_2O_3).

High-resolution XPS was used to quantitatively describe the growth of the interfacial layer (IL) between $(HfO_2)_x(Al_2O_3)_{1-x}$ films and Si substrate during RTA processing. All of the high-resolution scans were taken at a pho-

Appl. Phys. Lett., Vol. 81, No. 19, 4 November 2002   Yu *et al.*   3619

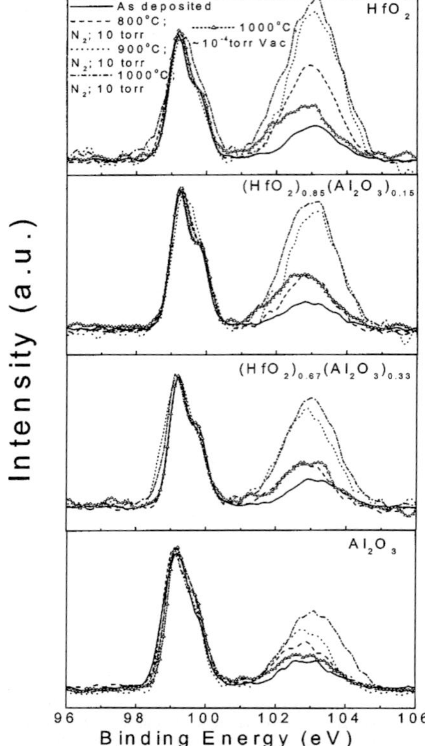

FIG. 1. XPS Si $2p$ core-level spectra recorded from various $(HfO_2)_x(Al_2O_3)_{1-x}$ samples (the thin ones) of as-deposited (solid lines), after 800 °C/N$_2$ annealing (dashed lines), 900 °C/N$_2$ annealing (dotted lines), 1000 °C/N$_2$ annealing (dashed dotted lines), and 1000 °C/high vacuum annealing (dashed lines with open triangles). The peak located at ~99.3 eV is assigned to Si—Si bonds from the substrates, and the one at ~103.0 eV to Si—O bonds from IL. The intensities for XPS peaks of Si—Si bonds have been normalized for comparison.

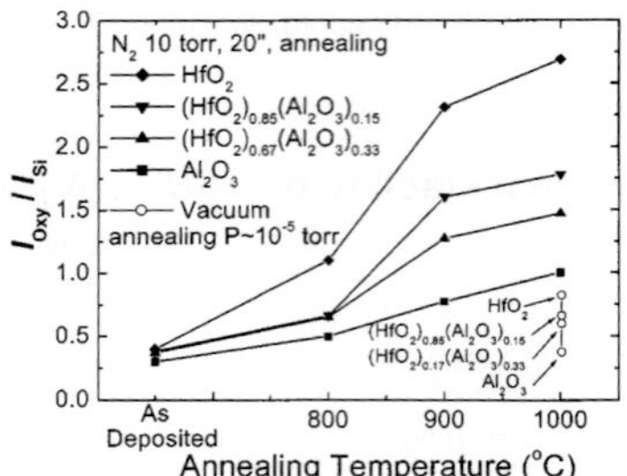

FIG. 2. The ratio of I_{Oxy}/I_{Si} for various $(HfO_2)_x(Al_2O_3)_{1-x}$ samples versus the different annealing conditions based on XPS spectra in Fig. 1. The change in this ratio directly correlates with the variation of the IL growth.

toelectron take-off angle of 90° and with a pass energy of 20 eV. The Si $2p$ core-level XPS spectra for the as-deposited and after various RTA annealed (800 °C/N$_2$, 900 °C/N$_2$, 1000 °C/N$_2$, and 1000 °C/high vacuum) samples are shown in Fig. 1. The peak located at ~99.3 eV is attributed to Si—Si bonds from Si substrates, and the one at ~103.0 eV to Si—O bonds from the interfacial layer.[13] For the sake of comparison, the intensities of XPS peaks of Si—Si bonds have been normalized. From Fig. 1, it is obvious that an IL exists for all of the as-deposited samples. For better understanding, the ratio of I_{Oxy}/I_{Si} for each sample with the different annealing conditions is plotted in Fig. 2. I_{Oxy} and I_{Si} are determined by integrating the Si—O and Si—Si peak area, respectively, after a Shirley background subtraction. The change in the ratio of I_{Oxy}/I_{Si} directly reflects the variation of the interfacial layer growth: The higher the ratio, the

thicker the IL. In the case of annealing in N$_2$, IL thickness increases with increasing temperature for all of the four samples. For a given annealing temperature, the extent of IL growth is determined versus Al%, with HfO$_2$ film (0% Al) showing the largest growth, and Al$_2$O$_3$ film the smallest. Doping of HfO$_2$ film with Al slows down the IL growth during annealing. Based on these XPS results, one can draw the conclusion that the ability to block oxygen diffusion through HfO$_2$ films is greatly enhanced by the incorporating of Al, and the ability becomes stronger when more Al is incorporated.

The ratio of I_{Oxy}/I_{Si} corresponding to a high vacuum ($\sim 2\times10^{-5}$ Torr) annealing at 1000 °C is also plotted in Fig. 2. Although the changes of I_{Oxy}/I_{Si} for these four samples show the same trend as that of annealing in N$_2$, these ratios are significantly smaller than those of annealing in N$_2$ at the same temperature. This result implies that the active source of oxygen in an N$_2$ ambient, not the oxygen species present in the high-k films themselves, is responsible for the IL growth during RTA.

It should also be noted that no Hf silicide (Hf—Si bonding) was detected from the Hf $4f$ core-level scans (data not shown) for all samples after annealing in a high vacuum at 1000 °C. This is because the pressure for annealing in our study is not low enough to trigger the formation of SiO species at the high-k/IL region, which is the key step for silicide formation.[14]

The high-resolution XTEM micrographs of two samples [HfO$_2$ and $(HfO_2)_{0.85}(Al_2O_3)_{0.15}$ films] before and after 900 °C annealing in N$_2$ are presented in Fig. 3. After annealing, it is observed that the growth of IL is greater for the HfO$_2$ sample compared to the Al-doped HfO$_2$, consistent with the XPS results shown in Fig. 1. The composition of the interfacial layer is likely to be Hf(Al) silicate,[12,15] which is supported by XPS measurements shown in Fig. 1. More evidence for silicate formation is the high-k film thickness decrease after annealing, as shown by XTEM images, which may be due to the consumption of high-k films through the reaction with IL[15] and/or film densification.[16] Further study of the IL composition by high spatial resolution electron energy loss spectroscopy is underway.

As reported previously,[6–8] HfO$_2$ film crystallization temperature is increased by alloying with Al. Our XRD study of

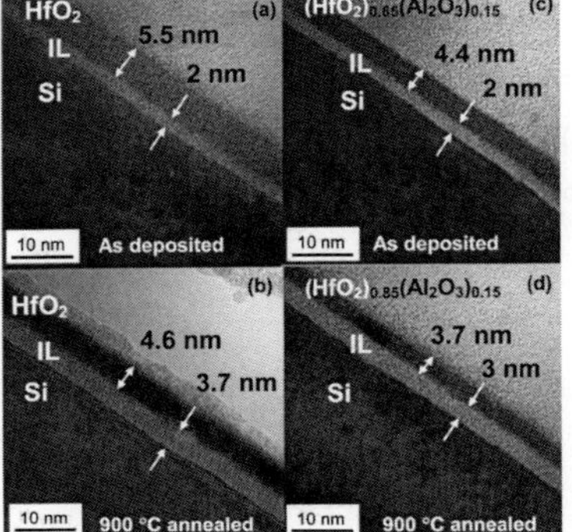
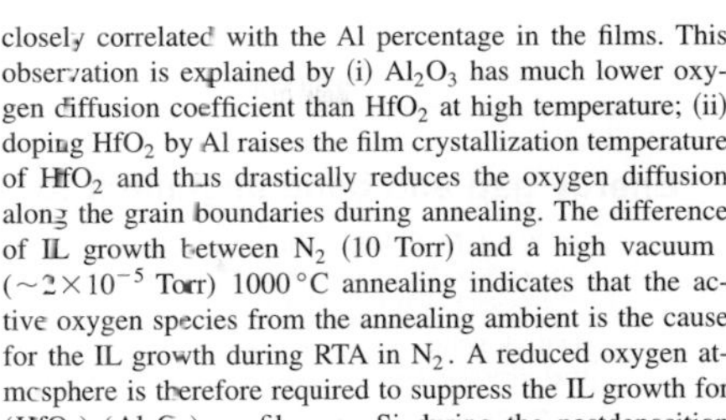

FIG. 3. HRTEM images for of (a) the as-is HfO$_2$ sample, (b) the 900 °C/N$_2$ annealed HfO$_2$ sample, (c) the as-is (HfO$_2$)$_{0.85}$(Al$_2$O$_3$)$_{0.15}$ sample, and (d) the 900 °C/N$_2$ annealed (HfO$_2$)$_{0.85}$(Al$_2$O$_3$)$_{0.15}$ sample.

20 nm thick (HfO$_2$)$_x$(Al$_2$O$_3$)$_{1-x}$ films after 900 °C annealing in N$_2$ also confirms this point, as depicted in Fig. 4. These spectra show that (HfO$_2$)$_{0.67}$(Al$_2$O$_3$)$_{0.33}$ remains amorphous after 900 °C annealing. In addition, the crystallization temperatures increases with more Al incorporated.[6–8] These results are well correlated with the experimental observation regarding oxygen diffusion through (HfO$_2$)$_x$(Al$_2$O$_3$)$_{1-x}$ films (Figs. 1 and 2): the higher Al concentration, the higher the crystallization temperature and, hence, the lower the oxygen diffusion along the grain boundaries of the high-k films which, in turn, reduces the IL growth. Another reason for the reduced rate of oxygen diffusion by the addition of Al$_2$O$_3$ is that the Al$_2$O$_3$ is known to have much lower oxygen diffusion coefficient compared to HfO$_2$ at high temperature.[17]

In conclusion, we report that both the thermal stability and the resistance to oxygen diffusion of HfO$_2$ are improved by adding Al to form Hf aluminates, and the improvement is closely correlated with the Al percentage in the films. This observation is explained by (i) Al$_2$O$_3$ has much lower oxygen diffusion coefficient than HfO$_2$ at high temperature; (ii) doping HfO$_2$ by Al raises the film crystallization temperature of HfO$_2$ and thus drastically reduces the oxygen diffusion along the grain boundaries during annealing. The difference of IL growth between N$_2$ (10 Torr) and a high vacuum ($\sim 2 \times 10^{-5}$ Torr) 1000 °C annealing indicates that the active oxygen species from the annealing ambient is the cause for the IL growth during RTA in N$_2$. A reduced oxygen atmosphere is therefore required to suppress the IL growth for (HfO$_2$)$_x$(Al$_2$O$_3$)$_{1-x}$ films on Si during the postdeposition process.

This work was supported by the Singapore grant (No. NSTB/EMT/TP/00/001.2) and the National University of Singapore grant (No. R263-000-221-112). The authors thank Professor J. Y. Lin of the Dept. of Physics at NUS for use of the XRD apparatus.

[1] Y. Harada, M. Niwa, S. J. Lee, and D. L. Kwong, Symp. VLSI Tech. Dig. 3-3-1, (2002).

[2] L. Kang, K. Onishi, Y. Jeon, B. H. Lee, C. Kang, W. J. Qi, R. Nieh, S. Gopalan, R. Choi, and J. C. Lee, Tech. Dig. - Int. Electron Devices Meet. **2000**, 35 (2000).

[3] G. D. Wilk, R. M. Wallace, and J. M. Anthony, J. Appl. Phys. **87**, 484 (2000).

[4] S. Gopalan, K. Onishi, R. Nieh, C. S. Kang, R. Choi, H. J. Cho, S. Krishna, and J. C. Lee, Appl. Phys. Lett. **80**, 4416 (2002).

[5] H. Y. Yu, M. F. Li, B. J. Cho, C. C. Yeo, M. S. Joo, D. L. Kwong, J. S. Pan, C. H. Ang, J. Z. Zheng, and S. Ramanathan, Appl. Phys. Lett. **81**, 376 (2002).

[6] W. Zhu, T. P. Ma, T. Tamagawa, Y. Di, J. Kim, R. Carruthers, M. Gibso, and T. Furukawa, Tech. Dig.-Int. Electron Devices Meet. **2001**, 20.4.1 (2001).

[7] G. D. Wilk, M. L. Green, M. Y. Ho, B. W. Busch, T. W. Scorsch, F. P. Klemers, B. Brijs, R. B. van Dover, A. Kornblit, T. Gustafsson, E. Garfunkel, S. Hillenius, D. Monroe, P. Kalavade, and J. M. Hergenrother, VLSI Tech. Dig. 9-4-1, (2002).

[8] A. R. Londergan, S. Ramanathan, K. Vu, S. Rassiga, R. Hiznay, J. Winkler, H. Velasco, L. Matthysse, T. E. Seidel, C. H. Ang, H. Y. Yu, and M. F. Li, (unpublished).

[9] Y. T. Hou, M. F. Li, H. Y. Yu, Y. Jin, and D. L. Kwong, Tech. Dig.–Int. Electron Meet. (to be published).

[10] G. D. Wilk, R. M. Wallace, and J. M. Anthony, J. Appl. Phys. **89**, 5243 (2001).

[11] A. Kumar, D. Rajdev, and D. L. Douglass, J. Am. Chem. Soc. **55**, 439 (1972).

[12] C. Hobbs, H. Tseng, K. Reid, B. Tayhor, L. Dip, L. Hebert, R. Garcia, R. Hedge, J. Grant, D. Gilmer, A. Frake, V. Dhandapani, M. Azrak, L. Prabhu, R. Rai, S. Bagchi, J. Conner, S. Backer, F. Dumbuya, B. Nguyen, and P. Tobin, Tech. Dig. - Int. Electron Devices Meet. **2001**, 30.1.1 (2001).

[13] J. F. Moulder, *Handbook of X-ray Photoelectron Spectroscopy*, 2nd ed. (Physical Electronics, Eden Prairie, MN, 1992).

[14] T. S. Jeon, J. M. White, and D. L. Kwong, Appl. Phys. Lett. **78**, 368 (2001).

[15] M. Cho, J. Park, H. B. Park, C. S. Hwang, J. Jeong, and K. S. Hyun, Appl. Phys. Lett. **81**, 334 (2002).

[16] Y. Kim, G. Gebara, M. Freiler, J. Barnett, D. Riley, J. Chen, K. Torres, J. Kim, B. Foran, F. Shaapur, A. Agarwal, P. Lysaght, G. Brown, C. Young, S. Borthakur, H. Li, B. Nguyen, P. Zeitzoff, G. Bersuker, D. Derro, R. Bergmann, R. Murto, A. Hou, H. Huss, E. Shero, C. Pomarede, M. Givens, M. Mazanec, and C. Werkhoven, Tech. Dig. - Int. Electron Devices Meet. **2001**, 20.2.1 (2001).

[17] M. L. Green, E. P. Gusev, R. Degraeve, and E. L. Garfunkel, J. Appl. Phys. **90**, 2057 (2001).

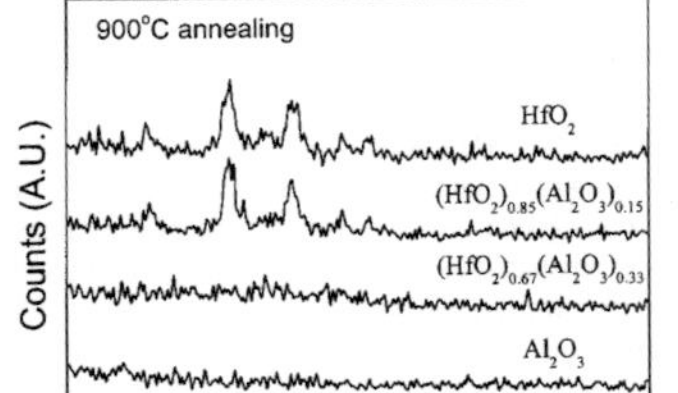

FIG. 4. XRD characteristics of various (HfO$_2$)$_x$(Al$_2$O$_3$)$_{1-x}$ samples (~ 20 nm) after 900 °C/N$_2$ annealing.

Reprinted with permission from H.Y. Yu, M.F. Li, B.J. Cho, C.C. Yeo, M. S. Joo, D.-L. Kwong, J.S.Pan, C.H. Ang, J.Z. Zheng and S. Ramanathan, Appl. Phys. Lett. Vol.81, pp.376–378, (2002). Copyright 2002, American Institute of Physics.

APPLIED PHYSICS LETTERS   VOLUME 81, NUMBER 2   8 JULY 2002

Energy gap and band alignment for $(HfO_2)_x(Al_2O_3)_{1-x}$ on (100) Si

H. Y. Yu, M. F. Li,[a] B. J. Cho, C. C. Yeo, and M. S. Joo
Silicon Nano Device Lab, Department of Electrical and Computer Engineering, National University of Singapore, Singapore 119260

D.-L. Kwong
Department of Electrical and Computer Engineering, The University of Texas, Austin, Texas 78752

J. S. Pan
Institute of Materials Research and Engineering, 3 Research Link, Singapore 117602

C. H. Ang and J. Z. Zheng
Chartered Semiconductor Manufacturing Ltd., Singapore 738406

S. Ramanathan
Genus Inc., Sunnyvale, California 94089

(Received 4 March 2002; accepted for publication 14 May 2002)

High-resolution x-ray photoelectron spectroscopy (XPS) was applied to characterize the electronic structures for a series of high-k materials $(HfO_2)_x(Al_2O_3)_{1-x}$ grown on (100) Si substrate with different HfO_2 mole fraction x. Al $2p$, Hf $4f$, O $1s$ core levels spectra, valence band spectra, and O $1s$ energy loss all show continuous changes with x in $(HfO_2)_x(Al_2O_3)_{1-x}$. These data are used to estimate the energy gap (E_g) for $(HfO_2)_x(Al_2O_3)_{1-x}$, the valence band offset (ΔE_v) and the conduction band offset (ΔE_c) between $(HfO_2)_x(Al_2O_3)_{1-x}$ and the (100) Si substrate. Our XPS results demonstrate that the values of E_g, ΔE_v, and ΔE_c for $(HfO_2)_x(Al_2O_3)_{1-x}$ change linearly with x. © 2002 American Institute of Physics. [DOI: 10.1063/1.1492024]

High-k gate dielectrics as alternates to SiO_2 have received tremendous attention due to the aggressive downscaling of complementary metal–oxide–semiconductor field effect transistor dimensions, which in turn results in increasing levels of tunneling current.[1] HfO_2 has emerged as one of the most promising high-k candidates due to its high dielectric constant, large energy gap, and compatibility with conventional complementary metal–oxide–semiconductor (CMOS) process.[2–4] However, it may suffer recrystallization at high temperature during postdeposition annealing, which in turn may induce higher leakage current and severe boron penetration issues. On the other hand, Al_2O_3 films grown directly on Si was reported to remain amorphous up to 1000 °C.[5] Recently Al has been proposed to alloy HfO_2 to raise the dielectric crystallization temperature, and encouraging results were demonstrated.[6] It was reported that when Al concentration is increased to 31.7%, the corresponding crystallization temperature increases to between 850 and 900 °C, which is about 400 °C higher than that for HfO_2. In this work, the energy gap (E_g) of $(HfO_2)_x(Al_2O_3)_{1-x}$, the valence band offset (ΔE_v) and the conduction band offset (ΔE_c) between $(HfO_2)_x(Al_2O_3)_{1-x}$ and the Si substrate as functions of x are obtained based on x-ray photoelectron spectroscopy (XPS) measurement. This information is of vital importance in assessing $(HfO_2)_x(Al_2O_3)_{1-x}$ as a most promising high-k gate dielectric in future CMOS device technology.

The principles of using high-resolution XPS to obtain both E_g and ΔE_v for the dielectrics including SiO_2 and various high-k materials can be found in Refs. 7–9. The conduction band offset (ΔE_c) can also be derived using the Si energy gap value of 1.12 eV.[10]

A total of five $(HfO_2)_x(Al_2O_3)_{1-x}$ samples of various x values prepared by atomic layer deposition (ALD), with low-doped p-type Si (100) wafers as substrates ($P \sim 10^{15}$ cm^{-3}) were studied in this work. A thin layer of oxide around 10 Å was thermally grown on each Si wafer after the pregate clean using HF last. The wafers were sent to GENUS for ALD and a pre-HF vapor clean to remove the oxide was conducted prior to the deposition of the dielectric films. The deposition temperature for the ALD process is 300 °C. The thicknesses of the $(HfO_2)_x(Al_2O_3)_{1-x}$ films are around 20 nm obtained by a spectroscopic ellipsometer (Woollam Model M-2000). The *ex situ* XPS measurements were carried out using a VG ESCALAB 220i-XL system,[11] equipped with a monochromatized Al $K\alpha$ source ($h\nu = 1486.6$ eV) for the excitation of photoelectrons. All of the high-resolution scans were taken at a photoelectron take-off angle of 90° and a pass energy of 20 eV. Under such configurations, the full width at half maximum of Si $2p_{5/2}$ core level recorded from H-terminated Si surface was measured as $\sim$0.45 eV, which gives an indication of the instrument energy resolution. Al $2p$, Hf $4f$, C $1s$, O $1s$, valence band maximum, and O $1s$ energy loss spectra were measured and analyzed. The intensities for all the XPS spectra reported here have been normalized for comparison and all of the spectra are calibrated against C $1s$ peak (285.0 eV) of adventitious carbon.

The chemical compositions of various $(HfO_2)_x(Al_2O_3)_{1-x}$ samples [change from HfO_2 ($x=1$) to Al_2O_3 ($x=0$)] can be determined by the intensities of the XPS lines. The five samples are denoted as HAO-1 to HAO-5, respectively, and their corresponding elemental

[a] Electronic mail: elelimf@nus.edu.sg

Appl. Phys. Lett., Vol. 81, No. 2, 8 July 2002

TABLE I. Elemental composition of various $(HfO_2)_x(Al_2O_3)_{1-x}$ samples (labeled as from HAO-1 to HAO-5) estimated by XPS. The HfO_2 mole fraction value x as in $(HfO_2)_x(Al_2O_3)_{1-x}$ are also given in the table. The Hf at. $\% = x/(5-2x)$ and the Al at. $\% = 2(1-x)/(5-2x)$ are determined by the intensities of XPS lines.

	HAO-1	HAO-2	HAO-3	HAO-4	HAO-5
Hafnium at. %	33.9%	25.8%	18.4%	9.6%	0
Aluminum at. %	0	9.2%	18.2%	27.7%	39.8%
Oxygen at. %	66.1%	65%	63.4%	62.7%	60.2%
HfO_2 mole fraction value x as in $(HfO_2)_x(Al_2O_3)_{1-x}$	1	~0.85	~0.67	~0.41	0

compositions as well as the value of x are given in Table I. All the samples show good stoichiometry and trace amounts of carbon are detected from all of the samples' surfaces.

XPS spectra for Hf $4f$, Al $2p$, and O $1s$ core levels are shown in Figs. 1(a), 1(b), and 1(c). It is observed that all the core level peak positions of Hf $4f$, Al $2p$, and O $1s$ experience a shift to higher binding energy with the increase of Al_2O_3 concentration in $(HfO_2)_x(Al_2O_3)_{1-x}$ system, and these changes are similar to the XPS chemical shifts in $ZrSiO_4$ vs SiO_2 and ZrO_2 as discussed in Ref. 12. The earlier shift is due to the fact that Hf is a more ionic cation than Al in $(HfO_2)_x(Al_2O_3)_{1-x}$,[13] and thus the charge transfer contribution changes with the increase of Al concentration.[12,13]

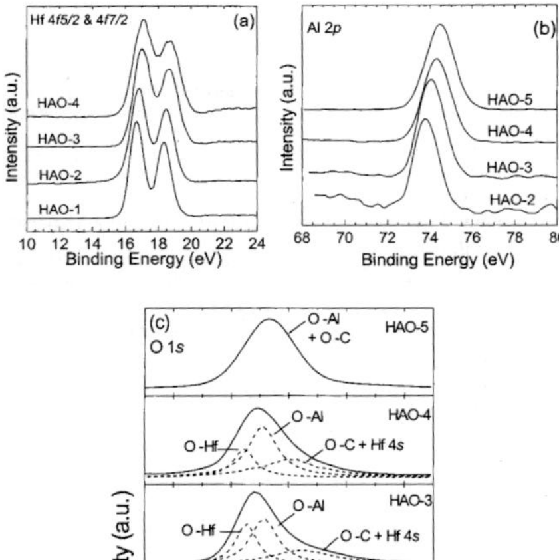

FIG. 1. XPS spectra for (a) Hf $4f$ core levels, (b) Al $2p$ core levels, and (c) O $1s$ core level taken from various $(HfO_2)_x(Al_2O_3)_{1-x}$ samples. The core level peak positions of Hf $4f$, Al $2p$, and O $1s$ shift continuously towards greater binding energy with increasing Al components. For O $1s$ spectra, solid lines are experimental data and dashed lines are the curve fitting results. From the curve fitting results, it is clearly shown that the Al–O bond (~531.2 eV) component increases with increasing Al composition.

For the O $1s$ core level spectra (the solid lines), a curve-fitting method (Gaussian fitting; the dashed lines) is applied to analyze the variation in O $1s$ spectra shape. For the samples HAO-2, HAO-3, and HAO-4, three peaks can be clearly resolved. The peak located at ~530.5 eV is attributed to Hf–O bonds, and another peak at ~531.2 eV to Al–O bonds. From the curve-fitting results as well as the O $1s$ spectra collected from HfO_2 (HAO-1) and from Al_2O_3 (HAO-5), it is obvious that Al–O components increase with increasing Al in $(HfO_2)_x(Al_2O_3)_{1-x}$. The shoulder at ~532.3 eV is generally interpreted as due to residual surface contaminants (i.e., C–O bonds)[14] and it is observed that this shoulder decreases with the decrease of Hf component in $(HfO_2)_x(Al_2O_3)_{1-x}$. However, Hf $4s$ photoelectron line is also located around this energy.[15] Therefore, it is suggested that both of the earlier-indicated sources contribute to the peak at ~532.3 eV.

Let us turn to focus on the major topic: energy band alignment for the $(HfO_2)_x(Al_2O_3)_{1-x}$. Figure 2(a) shows the O $1s$ energy-loss spectra, which are caused by the outgoing photoelectrons suffering inelastic losses to collective oscillations (plasmon) and single particle excitations (band to band transitions).[16] As is well known, the energy gap values for the dielectric materials can be determined by the onsets of energy loss from the energy-loss spectra.[8,16] By this mean, the energy gap value for HfO_2 (sample HAO-1) is measured as 5.25 ± 0.10 eV, and for Al_2O_3 (sample HAO-5) it is measured as 6.52 ± 0.10 eV. The energy gap value of Al_2O_3 is consistent with those reported by Itokawa *et al.*[8] (6.55 ± 0.05 eV) and Bender *et al.*[9] (6.7 ± 0.2 eV). From the results, a linear change of energy gap value with x in the $(HfO_2)_x(Al_2O_3)_{1-x}$ system is also observed.

The determination of valence band alignment of $(HfO_2)_x(Al_2O_3)_{1-x}$ on Si substrate was made by measuring the valence band maximum (VBMax)-difference between the $(HfO_2)_x(Al_2O_3)_{1-x}$ grown on p-Si(100) substrate samples and the H-terminated p-Si (100) substrate sample with the same substrate doping of $P \sim 10^{15}$ cm^{-3}, as demonstrated in Fig. 2(b).[17] The VBMax of each sample is determined by extrapolating the leading edge of valence band spectrum to the base line [the cross points in Fig. 2(b)] from its specific spectrum.[16] Thus, ΔE_v values of 3.03 ± 0.05 eV and 2.22 ± 0.05 eV are obtained for Al_2O_3 and HfO_2, respectively. The ΔE_v value of Al_2O_3 is consistent with the value 2.9 ± 0.2 eV reported by Bender *et al.*[9] A gradual change of the valence band density of states is also observed from sample HAO-1 to HAO-5, as indicated by the dashed arrow in Fig. 2(b).

With the knowledge of Si energy gap value of 1.12 eV,

378   Appl. Phys. Lett., Vol. 81, No. 2, 8 July 2002   Yu *et al.*

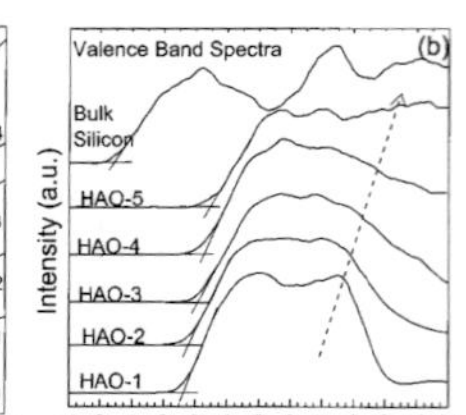

FIG. 2. (a) O $1s$ energy loss spectra for various $(HfO_2)_x(Al_2O_3)_{1-x}$ samples. The cross points (obtained by linearly extrapolating the segment of maximum negative slope to the base line) denote the energy gap E_g values. The dashed arrow shows the continuous change in the energy loss spectra contour from sample HAO-1 to HAO-5. (b) XPS valence band spectra taken from various $(HfO_2)_x(Al_2O_3)_{1-x}$ grown on (100) Si substrate samples and H-terminated (100) Si substrate sample. The cross point from each spectrum denotes the VBMax for that specific sample. The valence band alignment ΔE_v is obtained by the difference of VBMax between the $(HfO_2)_x(Al_2O_3)_{1-x}$ and the H-terminated Si. The dashed arrow indicates the gradual change in the valence band density of states from sample HAO-1 to HAO-5.

the ΔE_c values for $(HfO_2)_x(Al_2O_3)_{1-x}$ can be simply derived by the equation

$$\Delta E_c = E_g - \Delta E_v - 1.12 \ (eV). \tag{1}$$

Hence ΔE_c for HfO_2 is calculated as 1.91 ± 0.15 eV and for Al_2O_3, it is calculated as 2.37 ± 0.15 eV. Afanas'ev *et al.* reported 3.23 ± 0.08 eV for the (100) Si valence band to Al_2O_3 conduction band offset, measured by internal photoemission.[18] Using 1.12 eV energy gap for Si, the Si to Al_2O_3 conduction band offset ΔE_c is calculated to be 2.11 ± 0.08 eV, which in turn is in reasonable agreement with our XPS result.

The E_g, ΔE_v, and ΔE_c values obtained by XPS measurements and by Eq. (1) for samples HAO-1 to HAO-5 are plotted in Fig. 3. By linear least square fit, the following

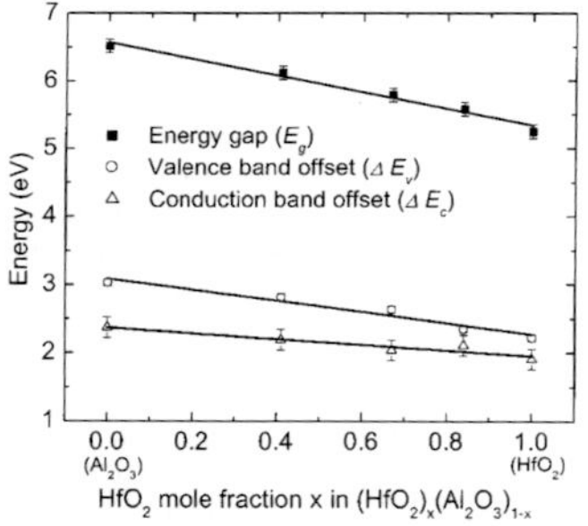

FIG. 3. Dependence of E_g, ΔE_v, and ΔE_c for $(HfO_2)_x(Al_2O_3)_{1-x}$ on HfO_2 mole fraction x. The E_g and ΔE_v data are obtained by XPS measurements. The ΔE_c data are calculated by Eq. (1). The solid lines are obtained by linear least square fits of the data points.

equations are obtained:

$$E_g = 6.52 - 1.27x \ (eV), \tag{2a}$$

$$\Delta E_v = 3.03 - 0.81x \ (eV), \tag{2b}$$

$$\Delta E_c = 2.37 - 0.46x \ (eV), \tag{2c}$$

where x stands for the mole fraction of HfO_2 in $(HfO_2)_x(Al_2O_3)_{1-x}$, as clearly demonstrated in Table I. Accordingly, the electrical properties of $(HfO_2)_x(Al_2O_3)_{1-x}$ gate dielectrics can be tuned by simply changing the HfO_2 mole fraction while keeping the stoichiometry of the materials.

In conclusion, high-resolution XPS measurements were performed to investigate E_g and ΔE_v of $(HfO_2)_x(Al_2O_3)_{1-x}$ high-k materials. Al $2p$, Hf $4f$, O $1s$ core levels spectra, valence band spectra, and O $1s$ energy loss spectra all show continuous changes with the variation of HfO_2 mole fraction x in $(HfO_2)_x(Al_2O_3)_{1-x}$. E_g, ΔE_v, and ΔE_c values for $(HfO_2)_x(Al_2O_3)_{1-x}$ on Si (100) are determined and can be expressed by $6.52 - 1.27x$ (eV), $3.03 - 0.81x$ (eV), and $2.37 - 0.46x$ (eV), respectively.

This work was supported by the Singapore NSTB/EMT/TP/00/001.2 Grant and the National University of Singapore R263-000-077-112 Grant.

[1] G. D. Wilk, R. M. Wallace, and J. M. Anthony, J. Appl. Phys. **89**, 5243 (2001).

[2] S. J. Lee, H. F. Luan, C. H. Lee, T. S. Jeon, W. P. Bai, Y. Senzaki, D. Roberts, and D. L. Kwong, 2001 Symp. VLSI Tech. Dig., 2001, p. 133.

[3] S. J. Lee, H. F. Luan, W. P. Bai, C. H. Lee, T. S. Jeon, Y. Senzaki, D. Roberts, D. L. Kwong, Tech. Dig. Int. Electron Devices Meet. **2000**, 31 (2000).

[4] L. Kang, K. Onishi, Y. Jeon, B. H. Lee, C. Kang, W. J. Qi, R. Nieh, S. Gopalan, R. Choi, and J. C. Lee, Tech. Dig. Int. Electron Devices Meet. **2000**, 35 (2000).

[5] A. Chin, C. C. Liao, C. H. Lu, W. J. Chen, and C. Tsai, Symp. VLSI Tech. Dig., 1999, p. 135.

[6] W. Zhu, T. P. Ma, T. Tamagawa, Y. Di, J. Kim, R. Carruthers, M. Gibso and T. Furukawa, Tech. Dig. Int. Electron Devices Meet. **2001**, 20.4.1 (2001).

[7] S. Miyazaki, H. Nishimura, M. Fukuda, L. Ley, and J. Ristein, Appl. Surf. Sci. **113/114**, 585 (1997).

[8] H. Itokawa, T. Maruyama, S. Miyazaki, and M. Hirose, Extended Abstracts of the 1999 Int. Conf. Solid State Devices and Materials, Tokyo, 1999, p. 158.

[9] H. Bender, T. Conard, H. Nohira, J. Petry, O. Richard, C. Zhao, B. Brijs, W. Besling, C. Detavernier, W. Vandervorst, M. Caymx, S. De Gendt, J. Chen, J. Kluth, W. Tsai, and J. W. Maes, International Workshop on Gate Isolation, Tokyo, 2001.

[10] S. M. Sze, *Physics of Semiconductor Devices* (Wiley, New York, 1981), p. 850.

[11] D. Mangelinck, J. Y. Dai, J. S. Pan, and S. K. Lahiri, Appl. Phys. Lett. **75**, 1736 (1999).

[12] M. J. Guittet, J. P. Crocombette, and M. Gautier-Soyer, Phys. Rev. B **63**, 125117 (2001).

[13] G. Lucovsky, International Workshop on Gate Isolator, Tokyo, 2001.

[14] H. Y. Yu, X. D. Feng, D. Grozea, Z. H. Lu, R. N. S. Sodhi, A.-M. Hor, and H. Aziz, Appl. Phys. Lett. **78**, 2595 (2001).

[15] J. F. Moulder *et al.*, *Handbook of X-ray Photoelectron Spectroscopy*, 2nd ed. (Physical Electronics, Eden Prairie, MN, 1992).

[16] F. G. Bell and L. Ley, Phys. Rev. B **37**, 8383 (1988).

[17] J. L. Alay and M. Hirose, J. Appl. Phys. **81**, 1606 (1997).

[18] V. V. Afanas'ev, M. Houssa, A. Stesmans, and M. M. Heyns, Appl. Phys. Lett. **78**, 3073 (2001).

Robust HfN Metal Gate Electrode for Advanced MOS Devices Application

H.Y. Yu, H.F. Lim, J.H. Chen, M.F. Li, C.X. Zhu, D.-L. Kwong#, C.H. Tung*, K.L. Bera*, and C.J. Leo*

Silicon Nano Device Lab, ECE Department, National University of Singapore, Singapore 119260 (elelimf@nus.edu.sg);
#Dept. of ECE, University of Texas at Austin, Austin, TX 78712, USA.; *Institute of Microelectronics, Singapore 117685

Abstract

A comprehensive study of HfN metal gate electrode for advanced MOS devices application is presented for the first time. It is found that HfN is an excellent barrier against oxygen diffusion, has a midgap work function (4.65 eV) on SiO_2, and exhibits superior thermal stability with underlying gate dielectric. Negligible degradation in EOT, work function, leakage current, and TDDB upon high-temperature treatments (up to 1000°C) has been observed in HfN gated MOS devices. These results suggest that HfN metal electrode is an ideal candidate for ultra thin body fully depleted SOI (FD-SOI) and symmetric double gate (SDG) MOS applications.

Introduction

Ultra thin body FD-SOI has received tremendous attention for future generation of low-power, high-speed MOS device applications. Metal gate with a midgap work function is highly desired for FD-SOI to make low channel doping viable, so as to eliminate channel mobility degradation and minimize the V_{th} variation [1]. While TaNx [2] and TiN [3] have been studied as midgap metal gates, these materials showed limited thermal stability [4] and thus are incompatible with conventional CMOS process. Due to the negatively larger heat of formation of HfN (also known as the most refractory metal nitride) compared to that of TiN and TaN (HfN:–88.2, TiN:–80.4, TaN:–60.3; kcal/mol) [5], it is expected that the thermal stability of HfN is superior to TiN and TaN. In this paper, the physical and electrical properties of PVD HfN metal gate electrode are reported for the first time. The low-resistivity, midgap work function HfN metal shows excellent resistance against high temperature treatments (up to 1000°C) in terms of EOT, work function, volume resistivity, and leakage current stability, making it an ideal gate electrode candidate for FD-SOI and SDG [6] MOS devices application.

Experimental

The MOS devices were fabricated using p-Si(100) substrates. For capacitors fabrication, after pre-gate cleaning, SiO_2 gate oxides with different thicknesses were thermally grown. HfN (~50nm) was then deposited by DC sputtering of Hf target in Ar+N_2 ambient. A TaN (~100nm) capping layer was sputtered on HfN to achieve a low gate sheet resistance (~10Ω/sq.). The TaN/HfN stack was RIE using Cl_2 based chemistry. The devices were then rapid thermal annealed (RTA) in N_2 (10torr) at 600°C, 800°C, and 1000°C for 20sec for thermal stability evaluation. All samples received back side Al metallization and forming gas sintering. NMOSFET with TaN/HfN gate were also fabricated, and process details could be found elsewhere [7].

Results and Discussion

As shown in Fig.1, the stoichiometric HfN (Hf:N~1:1)film exhibits the lowest volume resistivity compared to the N-rich HfN films, and thus is used in this work. Note that the HfN volume resistivity does not change under RTA treatments, as shown in Fig.2. Fig.3 and Fig.4 illustrate surface morphology of as-deposited HfN, and smooth surface is shown by AFM

image with RMS=1.481 nm. Fig.5 presents the AES depth profile of as-deposited and 1000°C RTA annealed HfN films. The HfN composition depth profile remains unchanged after 600°C RTA, and only the surface is oxidized into HfO_xN_y after 800°C and 1000°C RTA, suggesting HfN is a good oxygen diffusion barrier. This result correlates well with XTEM of the MCS capacitors (HfN/SiO_2/Si) with SiO_2~4.5nm before and after different RTA, as shown in Fig.6. No degradation of SiO_2 thickness and the HfN/SiO_2 interface were observed. The measured CV curves of all devices studied in this work match the simulation results, suggesting good interface quality. One such example is illustrated in Fig.7 for a 1000°C RTA device. Fig.8 plots V_{fb} versus Tox for TaN/HfN gated devices before and after RTA treatments to extract work function (Φ_M) of HfN. The superior electrical stability of the TaN/HfN gated devices under RTA treatments is demonstrated in Fig.9, where EOT and HfN Φ_M are plotted as a function of RTA temperature. Results show that EOT variations in all devices are negligible up to 1000°C RTA. Φ_M of as-deposited HfN is ~4.65eV. It remains constant after 600°C RTA, and slightly increases to ~4.7 eV after 1000°C RTA. The small amount of Φ_M increase (≤0.06eV) after 1000°C RTA is probably due to the change of HfN crystallization, as shown by XRD in Fig.10. Fig.11 demonstrates stability of gate leakage current of the TaN/HfN gated devices (SiO_2~3.12nm) after the various RTA treatments. Therefore, the mid-gap HfN metal electrode with superior thermal stability is highly suitable for the FD-SOI and/or SDG applications. Figs.12 and 13 show charge trapping and TDDB characteristics, respectively, of TaN/HfN gated devices under constant current stressing (CCS). As can be seen in Fig.13, TDDB is improved after RTA, and time-to-breakdown increases with increasing RTA temperature. This improvement is caused by the annealing of plasma damages generated during sputtering deposition of the metal electrodes. N-MOSFETs with TaN/HfN gate electrode are demonstrated with well-behaved I_{ds}-V_{ds} and I_{ds}-V_g and excellent subthreshold slope (68mV/dec) (Figs.14 & 15).

Conclusion

In summary, the material and electrical characteristics of PVD HfN as gate electrode material has been studied for the first time. It is found that HfN possesses a midgap work function (~4.65eV) with SiO_2 as gate dielectrics. MOS devices with HfN metal gate exhibit excellent thermal and electrical stability, with negligible degradation of EOT, work function, gate leakage, and TDDB upon RTA treatments up to 1000°C. These results suggest that HfN metal gate has great potential for FD-SOI and SDG MOS devices application.

Acknowledgements: This work is supported by Singapore A-STAR EMT/TP/00/001,2 grant.

References

[1] B. Cheng, et al., Int. SOI conf., p.91 (2001) [2] H. Shimada, et al. IEEE-TED, vol. 48, p.1619 (2001) [3] J. Chen, et al., Symp. VLSI Tech., p.25 (1999) [4] M. Wang, et al., Int. Symp. P2ID, p.36 (2001) [5] M. Wittmer, J. Vac. Sci. Techol. A, vol. 3, p.1797 (1985) [6] H.S.P. Wong, IBM J. Res. & Dev., vol. 46, p.133 (2002) [7] C.O. Chui, et al., Tech. Dig. IEDM, p.17.3.1 (2002)

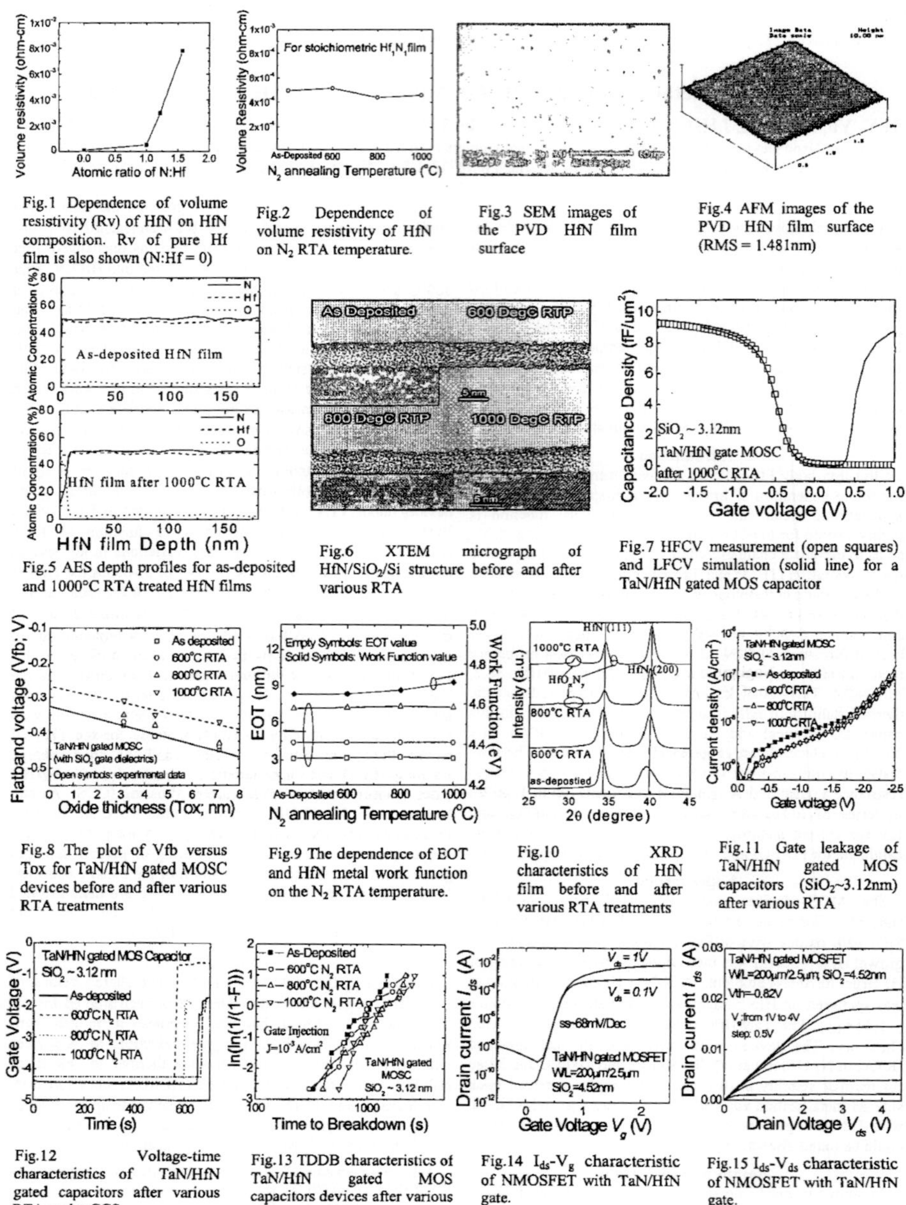

Fig.1 Dependence of volume resistivity (Rv) of HfN on HfN composition. Rv of pure Hf film is also shown (N:Hf = 0)

Fig.2 Dependence of volume resistivity of HfN on N_2 RTA temperature.

Fig.3 SEM images of the PVD HfN film surface

Fig.4 AFM images of the PVD HfN film surface (RMS = 1.481nm)

Fig.5 AES depth profiles for as-deposited and 1000°C RTA treated HfN films

Fig.6 XTEM micrograph of $HfN/SiO_2/Si$ structure before and after various RTA

Fig.7 HFCV measurement (open squares) and LFCV simulation (solid line) for a TaN/HfN gated MOS capacitor

Fig.8 The plot of Vfb versus Tox for TaN/HfN gated MOSC devices before and after various RTA treatments

Fig.9 The dependence of EOT and HfN metal work function on the N_2 RTA temperature.

Fig.10 XRD characteristics of HfN film before and after various RTA treatments

Fig.11 Gate leakage of TaN/HfN gated MOS capacitors ($SiO_2\sim3.12$nm) after various RTA

Fig.12 Voltage-time characteristics of TaN/HfN gated capacitors after various RTA under CCS

Fig.13 TDDB characteristics of TaN/HfN gated MOS capacitors devices after various RTA under CCS.

Fig.14 I_{ds}-V_g characteristic of NMOSFET with TaN/HfN gate.

Fig.15 I_{ds}-V_{ds} characteristic of NMOSFET with TaN/HfN gate.

Thermally Robust High Quality HfN/HfO$_2$ Gate Stack for Advanced CMOS Devices

H.Y. Yu[1], J.F. Kang[1,3], J.D. Chen[1], C. Ren[1], Y.T. Hou[1], S.J. Whang[1,4], M.-F. Li[1,2], D.S.H. Chan[1], K.L. Bera[2], C.H. Tung[2], A. Du[2], D.-L. Kwong[5]

[1]Silicon Nano Device Lab, Dept. of ECE, National University of Singapore, Singapore 119260
Tel:65-68742559, Fax:65-67791103, Email: elelimf@nus.edu.sg
[2]Institute of Microelectronics, Singapore 117685;
[3]Institute of Microelectronics, Peking University, Beijing 100871, P.R. China
[4]Jusung Engineering Co. Ltd, Korea 464-892
[5]Microelectronics Research Center, Dept. of ECE, The University of Texas, Austin, TX 78712, USA

ABSTRACT

We report for the first time a thermally stable and high quality HfN/HfO$_2$ gate stack for advanced CMOS applications. Due to the superior oxygen diffusion barrier of HfN as well as the thermal stability of HfN/HfO$_2$ interface, the EOT of HfN/HfO$_2$ gate stack has been successfully scaled down to less than 10Å with excellent leakage, boron penetration immunity, and long-term reliability even after 1000°C RTA treatment for 20sec., without using surface nitridation prior to HfO$_2$ deposition. The mobility is improved significantly for devices without surface nitridation. Negligible change in both EOT and the work function of HfN/HfO$_2$ gate stack are observed after 1000°C RTA.

INTRODUCTION

HfO$_2$ has emerged as one of the most promising high-K gate dielectrics (1,2). However, HfO$_2$ is poor barrier to oxygen diffusion, which caused the uncontrolled growth of low-K interfacial layer (IL) between HfO$_2$ and Si substrate during high-temperature post-processing (1). This imposes serious concern to EOT scalability. Although using surface nitridation or N-contained HfO$_2$ (i.e. HfOxNy) can minimize IL growth (3,4), they also cause severe carrier mobility degradation. Recently we have demonstrated that HfN gate electrode exhibits superior thermal stability with SiO$_2$ gate dielectrics and excellent barrier against oxygen diffusion (5). In this paper, we present for the first time the thermally robust high quality HfN/HfO$_2$ gate stack with excellent EOT scalability against high temperature treatments, without using surface nitridation, and investigate its performance and reliability for CMOS applications.

EXPERIMENTAL

For MOSCAPs fabrication, after the active area definition with 4000Å field oxide, and a standard DHF-last RCA pre-gate clean process, HfO$_2$ films were deposited at 400°C using a MOCVD cluster tool. Some samples received in-situ surface nitridation (SN) treatment in NH$_3$ at 700°C prior to CVD HfO$_2$ deposition. All samples were then subjected to post-deposition annealing (PDA) at 700°C in N$_2$ ambient, followed by PVD deposition of HfN(~50nm)/TaN(as capping layer on HfN; ~100nm) (5). Devices were then rapid thermal annealed (RTA) in N$_2$ at 900°C-1000°C for 20sec for thermal stability evaluation. For MOSFETs fabrication, S/D implantations of phosphorus for n-MOS and BF$_2$ for p-MOS with a dose of 5×10^{15} cm^{-2} were performed followed by RTA activation in N$_2$ at 950°C for 30s. EOT and flat band voltage (V_{fb}) were simulated by taking into account the quantum mechanical correction.

RESULTS AND DISCUSSION

MOSCAPs:

Fig.1 compares measured and simulated CV curves of HfN/HfO$_2$ devices without SN after various RTA anneal, showing good agreement. The EOT of HfN/HfO$_2$ is as low as 8.2Å after FGA, and slightly increases to 8.8Å/ 9.1Å after 900°C/ 1000°C post-metal annealing (PMA). Insignificant variation of the gate leakage current is observed in these devices after RTA (Fig.2). Fig. 3 compares XTEM of these HfN/HfO$_2$ gate stacks after various RTA. Small change in the physical thickness of both HfO$_2$ and IL is seen, consistent with results in Figs.1 and 2. From the XTEM and CV data, it appears that the IL is not the pure SiO$_2$, and it has a K value of 7~8. The superior thermal stability of HfN/HfO$_2$ gate stack is

4.5.1

further demonstrated in Fig.4, where several devices with different EOT are plotted as a function of RTA temperatures. Negligible EOT variations in all HfN/HfO$_2$ devices (without SN) are demonstrated up to 1000°C RTA. For comparison, the EOT of TaN/HfO$_2$ gate stack shows significant increases after 800°C, 900°C and 1000°C RTA, even with SN, as shown in Fig.5. The inferior thermal stability of PVD TaN metal gate compared to HfN metal gate might be attributed to its negatively smaller heat of formation (6) and its inferior ability to block oxygen diffusion. Fig.6 depicts gate leakage at V$_{fb}$-1V as a function of EOT for HfN/HfO$_2$ without SN. Compared to poly/SiO$_2$ benchmark with the same EOT, HfN/HfO$_2$ shows more than 10^5 reduction in gate leakage. In Fig.7, the work function (Φ_M) of HfN on HfO$_2$, extracted from plots of V$_{fb}$ versus EOT, is shown as a function of RTA temperature. Φ_M of HfN after FGA is ~4.75eV and slightly increases to ~4.8 eV after 1000°C RTA. The small Φ_M variation after 1000°C RTA is related to the HfN crystallization change (5). Compared to Φ_M of HfN on SiO$_2$ dielectrics (~4.71eV after 1000°C RTA) (5), it is noted little dependence of Φ_M of HfN on the underlying gate dielectrics (HfO$_2$ or SiO$_2$).

Stress-induced leakage current (SILC) characteristics of the 1000°C RTA device (EOT=9.1Å) are shown in Fig.8. Setting failure criterion as 50% increment of Jg$_0$, operating voltage for 10-year lifetime for the HfN/HfO$_2$ device is projected as 2.2V. We have also studied the boron penetration effect in p-MOS with HfN/HfO$_2$ gate stack using BF$_2$ implantation (20keV; 5x10^{15}cm^{-2}) and the results are shown in Figs.9&10. There is no difference in V$_{fb}$ between the un-implanted (control devices) and BF$_2$-implanted devices after RTA at 900°C, 950°C, and 1000°C for 20s in N$_2$, indicating excellent boron penetration immunity of HfN/HO$_2$ gate stack. The use of SN results in larger CV hysteresis, as shown in Fig.11, which is due to the higher trapped charge density induced by NH$_3$ annealing treatments. Note that the high temperature PMA process can effectively reduce the hysteresis caused by the SN treatments.

MOSFETs:

Fig.12 shows HFCV from n-MOSFET (without SN treatment) with HfN/HfO$_2$ stack (EOT=1.18nm from the CV). Poly-depletion is eliminated as expected. Figs.13&14 show the typical electrical characteristics (Id-Vd & Id-Vg) of this n-

MOSFET with subthreshold slope (ss) of ~78mV/dec. Fig.15 compares effective electron mobility (μ_e) between n-MOSFETs devices with and without SN. Mobility is degraded for the SN device despite its larger EOT, which is attributed to the larger interface trap density (Dit) due to nitrogen, as shown in Fig.16, where Dit is measured by the direct-current current-voltage (DCIV) technique (7). Using interface trap capture cross section of 4.4 Å^2 (8), Dit is ~3x10^{11}/cm^2 for fresh SN nMOSFET (EOT~2.04nm), and ~7x10^{10}/cm^2 for fresh device W/O SN with EOT~1.18nm. Figs.17&18 show well-behaved Id-Vd, and Id-Vg characteristics of p-MOSFETs with HfN/HfO$_2$ W/O SN (EOT=1.28nm) with effective hole mobility shown in Fig. 19. Negative bias temperature instability (NBTI) for the p-MOSFET was also studied (Fig.20) at 100 °C. Using 50 mV shift of Vth as the failure criterion (9), the devices can satisfy the 10-year lifetime at an operation voltage of ~1V.

CONCLUSION

Thermally robust high quality HfN/HfO$_2$ gate stack is demonstrated for advanced CMOS applications. EOT of HfN/HfO$_2$ gate stack has been successfully scaled down to less than 10Å with excellent leakage, boron penetration immunity, and long-term reliability even after 1000°C annealing, without using surface nitridation prior to HfO$_2$ deposition. The mobility is improved significantly without surface nitridation. Negligible change in both EOT and the work function of HfN/HfO$_2$ gate stack are observed after 1000°C RTA.

Acknowledgement

This work was supported by Singapore A-STAR research grant EMT/TP/00/001,2.

Reference

(1) G.D. Wilk et al, *JAP.* **89**, p.5243, 2001
(2) H.Y. Yu et.al *APL.* **81**, p.376, 2002
(3) C.H. Choi, et al. *Tech Dig. of IEDM* p.857, 2002
(4) C.S. Kang, et al. *Tech Dig. of IEDM*, p.865, 2002
(5) H.Y. Yu, et.al *VLSI Tech*, p.151, 2003
(6) M. Wittmer, *JVST-A.* **3**, p.1797, 1985
(7) A.Neugroschel et al *T-ED.* **42**, p.1657, 1995
(8) G.Groeseneken et al, *T-ED.* **31**, p.42, 1984
(9) K. Onishi et al *Tech Dig. IEDM* p.30.3.1 2001

4.5.2

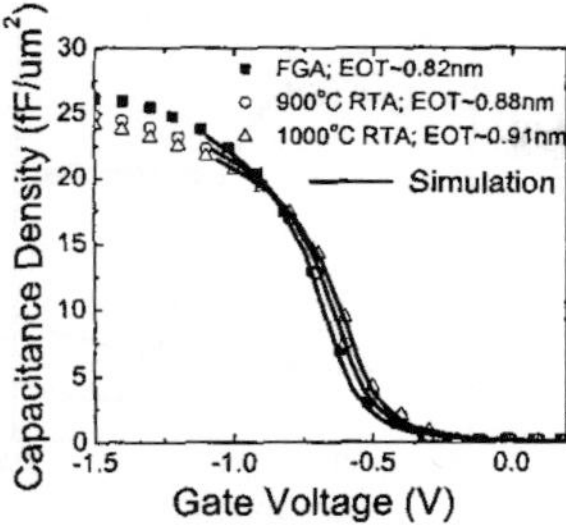

Fig.1. HFCV measurement (symbols) and LFCV simulation (solid lines) of n-MOSCAP (HfN gate on ultra-thin HfO_2) after different thermal treatment. No surface nitridation was performed before HfO_2 deposition.

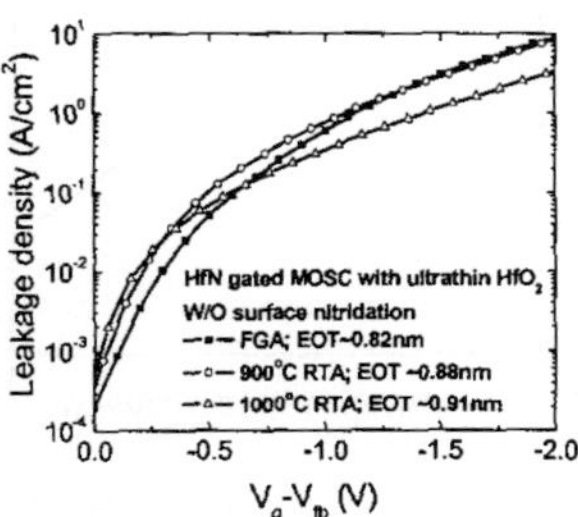

Fig.2. The leakage current measured from HfN/HfO_2 NMOSCAP after various thermal treatments (corresponding to Fig.1)

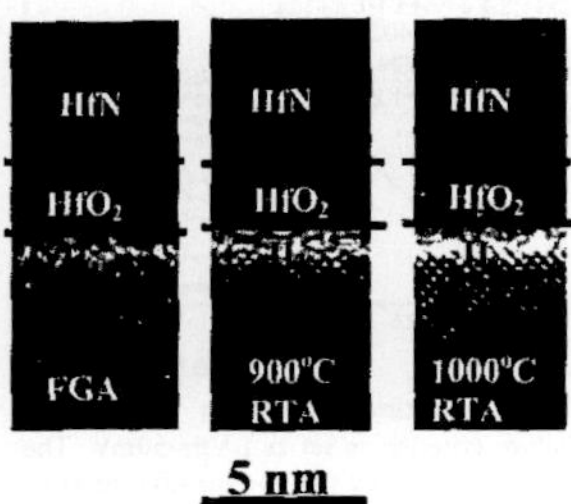

Fig.3. XTEM images of HfN/HfO_2 gate stack on (100)Si substrate after different thermal treatments (corresponding to Fig.1). For FGA sample, IL: 7~8Å & HfO_2: 22~23Å.

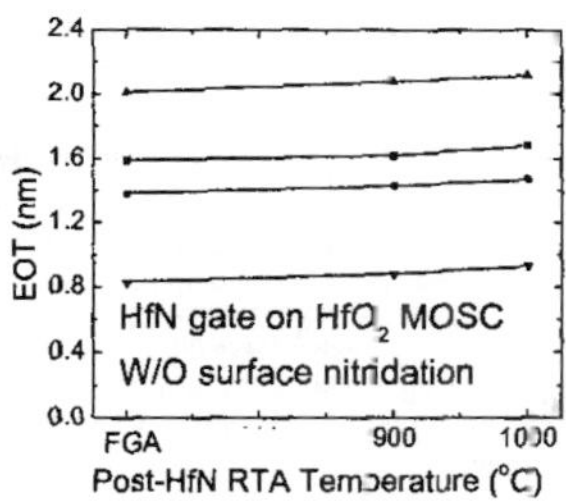

Fig.4. Dependence of EOT for HfN gated MOSCAP with HfO_2 gate dielectrics on the N_2 RTA temperature. No surface nitridation was done prior to HfO_2 deposition for all of the 4 samples.

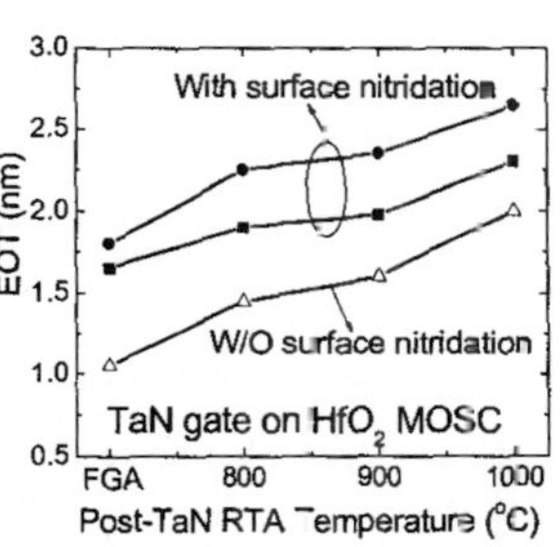

Fig.5. Dependence of EOT for TaN gated MOSCAP with HfO_2 gate dielectrics on the N_2 RTA temperature. Compared to HfN case, the EOT of TaN gated MOSCAP show significant increase after 800°C, 900°C & 1000°C RTA even with surface nitridation.

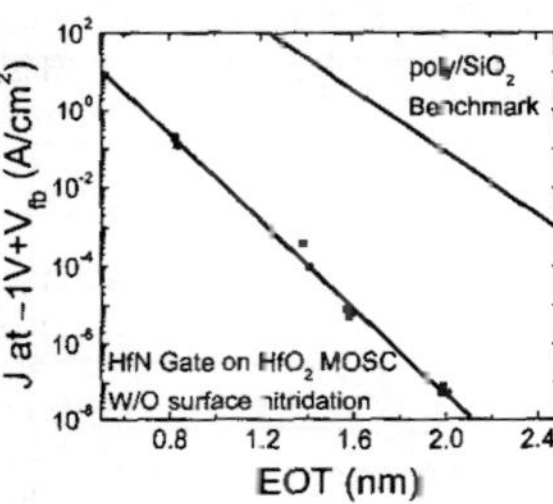

Fig.6. Leakage vs. EOT relationship for MOSCAP devices with HfN/HfO_2 gate stack. Compared to SiO_2 benchmark at the same EOT, HfO_2 provides more than 10^5 reduction in gate leakage current.

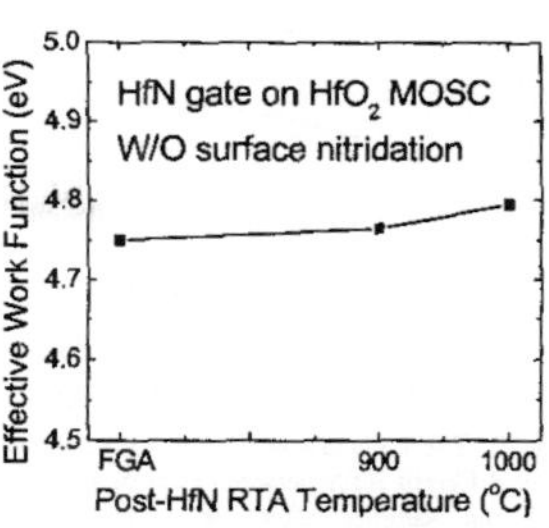

Fig.7. The dependence of effective work function of HfN metal gate on N_2 RTA temperature for HfN/HfO_2 MOSCAP.

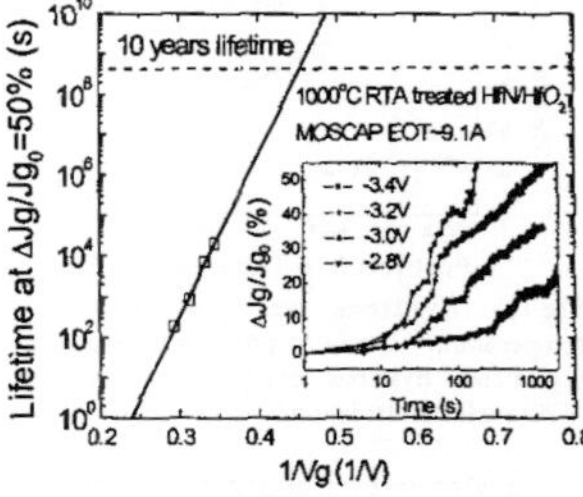

Fig.8. Lifetime projection based on SILC of HfN/HfO_2 MOSCAP after 1000°C RTA with EOT=9.1Å. Inset shows typical SILC time evolutions at four gate voltages. Failure criterion is set at 50% increment of Jg.

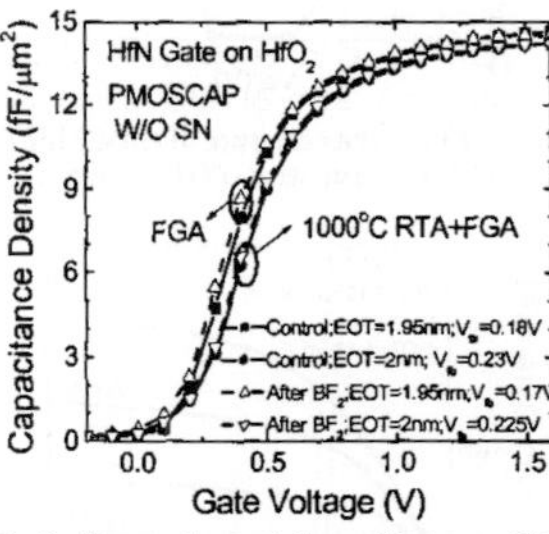

Fig.9. Boron implantation effect on CV characteristics of HfN/HfO_2 pMOSCAP. Solid symbols: control samples; empty symbols: samples after BF_2 implantation. Measured CV after FGA and 1000°C RTA dopant activation are shown for both kinds of samples.

4.5.3

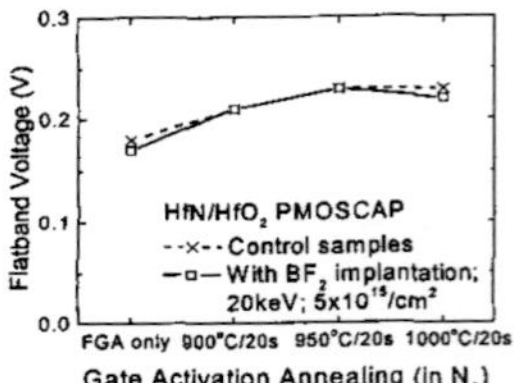

Fig.10. The comparison of the V_{fb} variation after different RTA between control samples and the samples after BF_2 implantation indicates HfN is an effective barrier to block boron penetration.

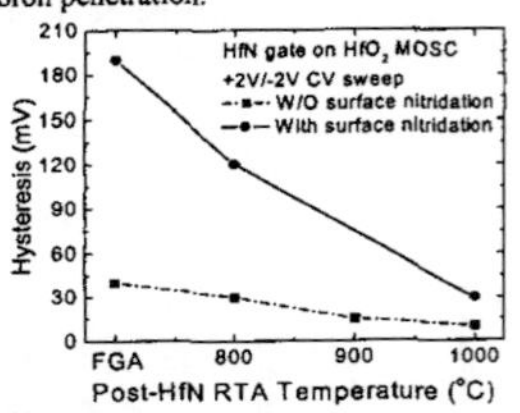

Fig.11. Hysteresis vs. post-HfN RTA temperature. Surface nitridation results in significant hystersis compared to samples W/O surface nitridation.

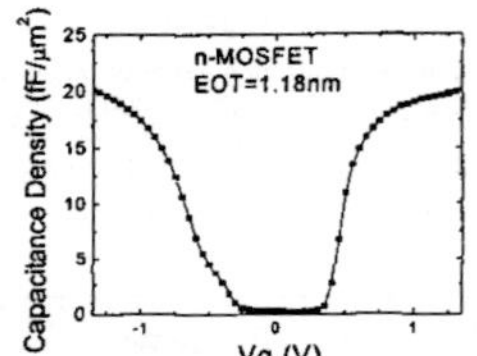

Fig.12. HFCV characteristics of a nMOSFET with HfN/HfO_2 gate stack (EOT=1.18nm)

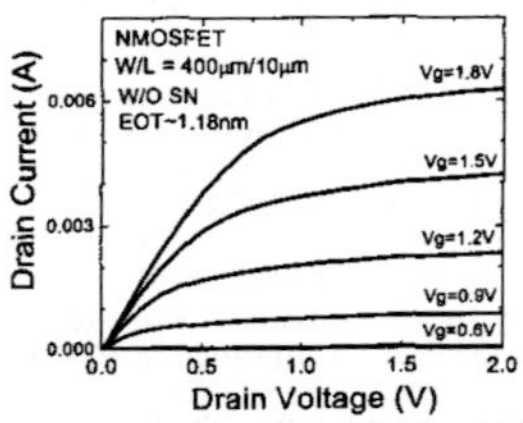

Fig.13. Id-Vd characteristics of a nMOSFET using HfN/HfO_2 gate stack (W/O SN) with EOT ~1.18nm.

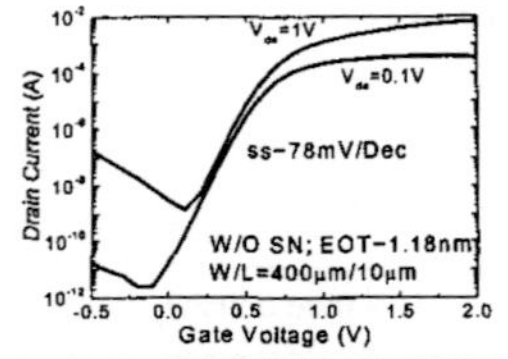

Fig.14. Id-Vg of nMOSFET with HfN/HfO_2 gate stack (no SN treatment) with EOT ~1.18nm

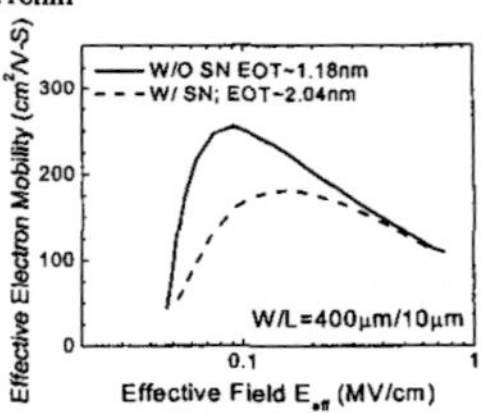

Fig.15. Effective electron mobility comparison between devices w/ and W/O SN treatment. Mobility is degraded for SN device despite the larger EOT.

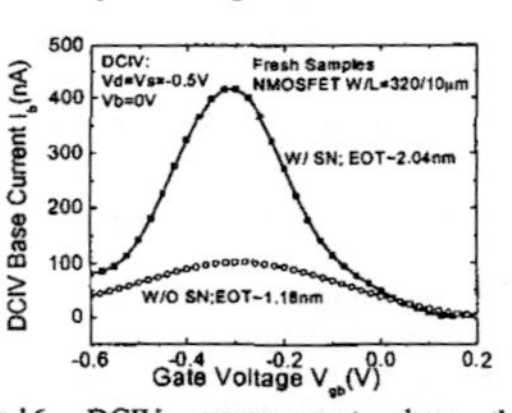

Fig.16. DCIV measurement show that interface trap density Dit is ~3×10^{11}/cm^2 for fresh SN nMOSFET (EOT~2.04nm), and Dit ~7×10^{10}/cm^2 for fresh device W/O SN with EOT~1.18nm. (using interface trap capture cross section of 4.4 Å^2 [8]).

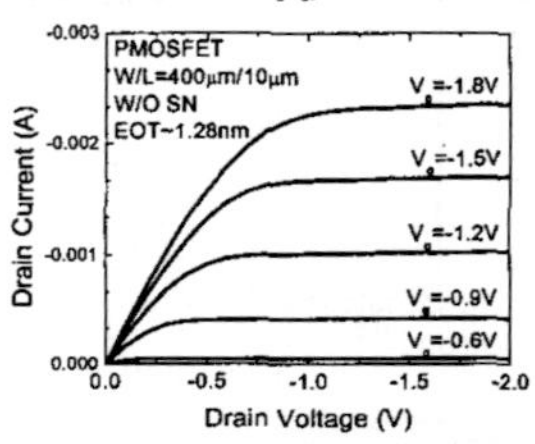

Fig.17. Id-Vd characteristics of pMOSFET with HfN/HfO_2 gate stack (W/O SN) with EOT~1.28nm

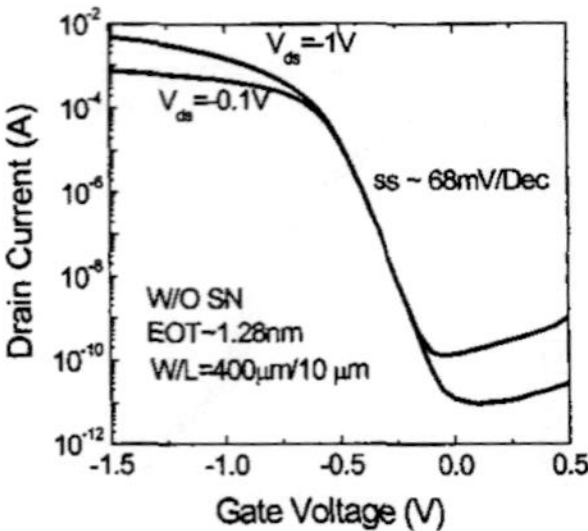

Fig.18. Id-Vg characteristics of pMOSFET with HfN/HfO_2 gate stack (W/O SN, with EOT=1.28nm)

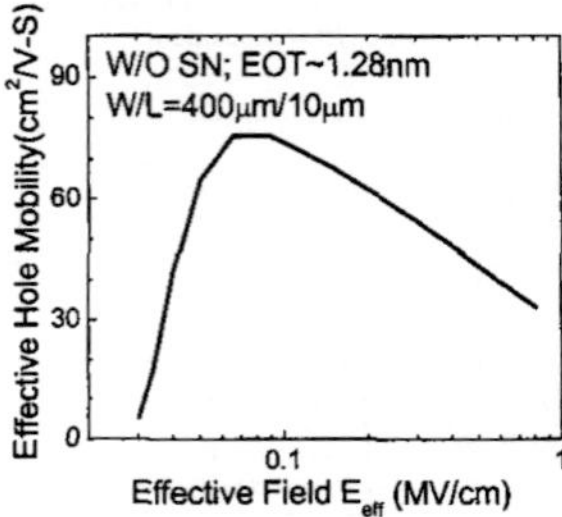

Fig.19. Effective hole mobility for the pMOSFET with HfN/HfO_2 stack with EOT=1.28nm (w/o SN)

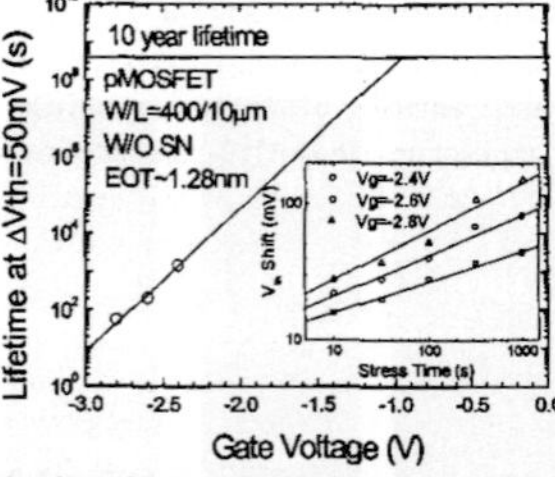

Fig.20. Lifetime projection for NBTI. Failure criterion is set as ΔVth=50mV. The devices can satisfy the 10-year lifetime at an operation voltage of ~1V. Inset shows Vth variation as a function of time during NBTI for three different negative gate biases at 100°C. (for pMOSFET W/O SN & with EOT=1.28nm)

4.5.4

High Performance ALD HfO$_2$-Al$_2$O$_3$ Laminate MIM Capacitors for RF and Mixed Signal IC Applications

Hang Hu[1], Shi-Jin Ding[1,2], HF Lim[1], Chunxiang Zhu[1], M.F. Li[1,2], S.J. Kim[1], XF Yu[1],
JH Chen[1], YF Yong[1], Byung Jin Cho[1], D.S.H. Chan[1], Subhash C Rustagi[2], MB Yu[2],
CH Tung[2], Anyan Du[2], Doan My[2], PD Foo[2], Albert Chin[3], Dim-Lee Kwong[4]

[1]SNDL, Dept. of ECE, National Univ. of Singapore, Singapore, 119260
Tel: 65-6874 8930, Fax: 65-6779 1103 Email: elezhucx@nus.edu.sg
[2]Institute of Microelectronics, Singapore, 117685, [3]Dept. of Electronics Eng., National Chiao Tung Univ., Taiwan
[4]Dept. of Electrical & Computer Eng., Univ. of Texas, Austin, TX 78712, USA

Abstract

In this paper, high performance ALD HfO$_2$-Al$_2$O$_3$ laminate metal-insulator-metal (MIM) capacitor is demonstrated for the first time with high capacitance density of 12.8 fF/μm^2 from 10 kHz to 20 GHz, low leakage current of 7.45×10^{-9} A/cm^2@2V, low VCC (voltage coefficients of capacitance), and excellent reliability. The superior electrical properties and reliability suggest that the ALD HfO$_2$-Al$_2$O$_3$ laminate is a very promising material for MIM capacitors for Si RF and mixed signal IC applications.

Introduction

Recently, MIM capacitors have attracted much attention because of its significant advantages for Si RF and mixed signal IC applications [1-3]. The challenge for MIM capacitor has been to achieve large capacitance density at RF frequencies with acceptable low leakage, RF performance, and reliability. Several high-κ materials have been employed to replace SiO$_2$ and Si$_3$N$_4$ [4-11]. In this work, we investigate the laminate of HfO$_2$ and Al$_2$O$_3$ for MIM capacitors application. HfO$_2$-Al$_2$O$_3$ laminate has been studied as DRAM dielectric [12]. In our work, we use HfO$_2$(5nm)/Al$_2$O$_3$(1nm) laminate with insertion of 1 nm Al$_2$O$_3$ layers as the starting and end layers. The use of 1 nm Al$_2$O$_3$ as the contacting layer to the top and bottom electrodes is to improve the meta/dielectric interface quality. As a result, we have successfully demonstrated high performance MIM capacitors fabricated by this ALD HfO$_2$-Al$_2$O$_3$ laminate layer structure. Our data suggest ALD HfO$_2$-Al$_2$O$_3$ laminate is a promising candidate for next generation MIM capacitor for RF and mixed signal IC applications.

Experiments

The MIM capacitors with TaN top and bottom electrodes were fabricated on top of 4 μm thick SiO$_2$ grown on Si substrate. High-κ dielectrics with alternatively deposited Al$_2$O$_3$ (1 nm) and HfO$_2$ (5 nm) laminate stacks are prepared by ALD at 320°C, with Al$_2$O$_3$ layers at beginning and end of ALD growth cycle. Three total thicknesses of 13, 31 and 43 nm were employed in the MIM capacitors. After top electrode

deposition, all samples were annealed in forming gas for 30 mins at 420°C. Al was deposited as the contact pads after TaN top electrode formation for a necessary good electric contact with the RF test probes.

Results and Discussion

A. RF Characteristics

Fig. 1 illustrates the fabricated 13 nm laminate structure and TEM photo together. Figs. 2 and 3 show the RF test structure and equivalent circuit for RF characterization. Dummy structures were also made for RF de-embedding purpose [13]. In Fig.4, S parameters were measured and shunt elements were de-embedded for three laminate MIM capacitors, using HP8510C Network Analyzer. The simulated two-ports S parameters were extracted by IC-CAP [14]. Using the equivalent circuit shown in Fig.3, the measured and simulated data over the entire frequency range from 50 MHz to 20 GHz are in very good agreement, which suggests the equivalent circuit model shown in Fig. 3 is suitable and reliable for the capacitor modeling and parameters extraction. Fig. 5 presents the single ended effective capacitance including the effect of parasitic inductances shown in Fig.3. These parasitic inductances are believed to be associated with the interconnect and are extrinsic to the MIM structure, and the useful operating frequency range for MIM capacitor could be further extended by reducing the parasitic from the interconnect. Fig.6 indicates the RF (50 MHz-20 GHz, extracted from S parameters) and low frequency ($\leq$1 MHz, measured by HP4284 LCR meter) capacitance density, showing a constant value from 10 kHz to 20 GHz. A capacitance density of 12.8 fF/μm^2 and a dielectric constant of ~19 are extracted for 13 nm laminate film, which is suitable for RF MIM capacitor application according to ITRS roadmap [15].

B. DC Properties

Fig. 7 presents the typical J-V characteristics of laminate MIM capacitors. Thin film of 13 nm exhibits an optimal leakage of 7.45×10^{-9} A/cm^2@2V, which is the lowest leakage

15.6.1

current ever reported at the similar capacitance density of MIM capacitors. The conduction mechanism of 13 nm film shows a transition from Schottky emission at low field to Pool-Frankel conduction at high field as illustrated in Fig.8. Fig.9 summarizes the VCCs of laminate MIM capacitors. The quadratic (α) and linear (β) VCC of 13 nm film are 1990 ppm/V^2 and 211 ppm/V, respectively. The small β is sufficient for RF application according to ITRS roadmap [15]. The quadratic (α) VCC is plotted as a function of high-κ thickness and measuring frequency, as shown in Figs.10 and 11, suggesting that α may have a logarithmic relationship with thickness [5, 6] and frequency [8], which may impose a limitation for certain precise MIM capacitors application where small α is needed and is useful for the design purpose. For the thickness dependence of α value, it is found to be intrinsic for many high-κ MIM capacitors [5, 6, 9]. Fig.12 shows the temperature dependence of leakage. The current decreases with measuring temperature till 100°C in 0~1.5V range, after that it increases with temperature at higher voltage (at 2V), which indicate the transitions of conduction mechanism move towards low voltage at high temperature. The TCC for laminate MIM capacitor is ~200 ppm/°C and is independent of high-κ thickness, as shown in Fig. 13.

C. Reliability and Lifetime

Fig. 14 shows the time evolution of VCCs under stress in which α value is decreased however β value is increased with stress time, the variation of α and β values corresponds to the "flatten-out" and the positive shift along the voltage axis of CV curves respectively. From Figs. 15 and 16, it is shown that both the leakage and VCCs will recover to pre-stress level after the stress is interrupted and held for sufficient time. The linear decrease of leakage with log stress time under constant voltage stress in Fig. 15 and the recovery in Fig.16 are probably due to traping and detraping in the dielectric. The correlation between the variations of leakage and VCC is shown in Fig.16. For the first time, it is shown that the origin of VCC may be related to traping and detraping in the dielectric during and after electrical stress. Fig. 17 shows hard breakdown characteristics of 13 nm laminate MIM capacitors under constant voltage stress. The cumulative TDDB for 13 nm laminate MIM capacitor is shown in Fig. 18. The extrapolated operating voltage for 10-year lifetime is 3.3 V, as illustrated in Fig. 19. The post-stress recovery phenomena may further prolong projected capacitor lifetime under AC condition. Table 1 compares our results on HfO$_2$-Al$_2$O$_3$ laminate MIM capacitors with other high capacitance density MIM capacitors reported recently [6-10], the overall superior electrical properties suggests high-κ laminate MIM capacitor is very useful for Si RF and mixed

signal IC applications.

Conclusions

In summary, high performance ALD HfO$_2$-Al$_2$O$_3$ laminate MIM capacitors have been demonstrated for the first time. Compared with various high-κ MIM capacitors reported so far, our data shows that the laminate MIM capacitor exhibits superior electrical performances such as high capacitance density up to 20 GHz, low leakage current and voltage coefficients, high breakdown field, as well as excellent device reliability. All these indicate that the HfO$_2$-Al$_2$O$_3$ laminate MIM capacitor is very promising for Si RF and mixed signal IC applications.

Acknowledgement

This work was supported by the Institute of Microelectronics Singapore and the National University of Singapore under Grant R-263-000-221-112 and Grant R-263-000-233-490.

Reference

[1]. M. Armacost et al., "A high reliability metal insulator metal capacitor for 0.18 μm copper technology", IEDM 2000, p.157.

[2]. C. H. Ng. et al., "Characterization and comparison of two metal-insulator-metal capacitor schemes in 0.13 μm copper dual damascene metallization process for mixed-mode and RF applications", IEDM 2002, p.241.

[3]. P.Zurcher et al., "Integration of thin film MIM capacitors and resistors into copper metallization based RF-CMOS and Bi-CMOS technologies", IEDM2000, p.153.

[4]. T. Yoshitomi et al., "High performance MIM capacitor for RF BiCMOS/CMOS LSIs", BCTM, 1999, p.133.

[5]. R. B. van Dover. et al., "Advanced dielectrics for gate oxide, DRAM and RF capacitors" IEDM 1998, p.823.

[6]. XF Yu et al., "A High density MIM capacitor (13 fF/μm^2) using ALD HfO$_2$ dielectrics", EDL. Vol. 24, 2003.

[7]. Tsuyoshi. I et al., "High-density Cu/Ta$_2$O$_5$/Cu MIM structure for SoC applications featuring a single-mask add-on process", IEDM 2002, p.940.

[8]. C. H. Huang, et al., "High density RF MIM capacitors using high-κ AlTaO$_x$ dielectrics", MTT-S. 2003. p.507.

[9]. Y. L. Tu. et al., "Characterization and comparison of high-k metal-insulator-metal (MiM) capacitors in 0.13 μm Cu BEOL for mixed-mode and RF applications", VLSI symp. 2003, p.79.

[10]. S.J. Kim et al., "HfO$_2$ and lanthanide-doped HfO$_2$ MIM capacitors for RF/Mixed IC applications", VLSI symp. 2003, p.77.

[11]. G.D.Wilk et al., "High-κ gate dielectrics: current status and materials properties considerations", JAP. 89, p.5243(2001).

[12]. J.H.Lee et al., "Mass production worthy HfO$_2$-Al$_2$O$_3$ Laminate capacitor technology using Hf liquid precursor for sub-100 nm DRAMs", IEDM, 2002, p.221

[13]. P.J.van Wijnen et al., IEEE BCTM 1987, p.70.

[14]. IC-CAP manual, Hewlett Packard, 1998.

[15]. ITRS roadmap 2002.

15.6.2

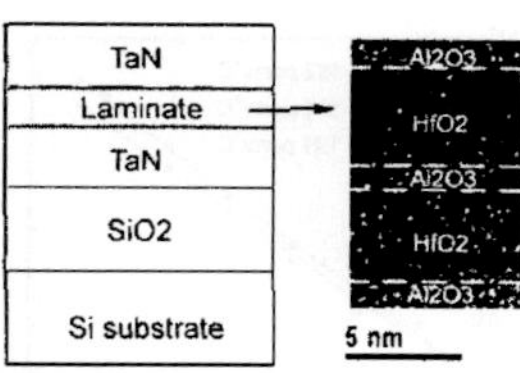

Fig. 1. TEM cross section of 13 nm laminate MIM capacitor.

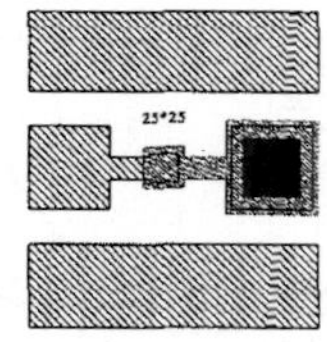

Fig. 2. The RF MIM capacitor layout.

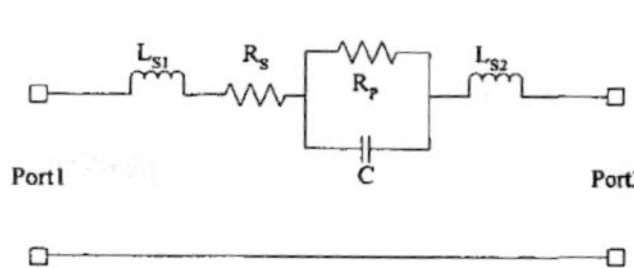

Fig. 3. The equivalent circuit model for capacitor simulation at RF regime.

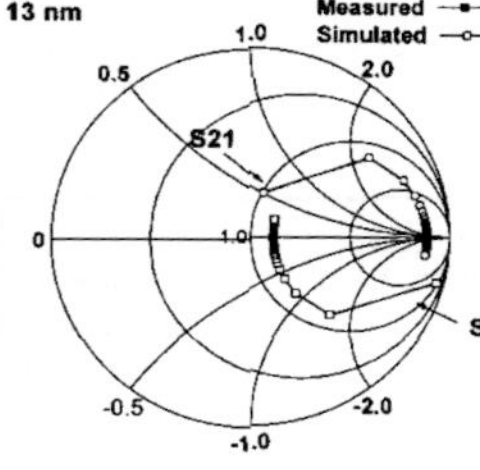

(a)

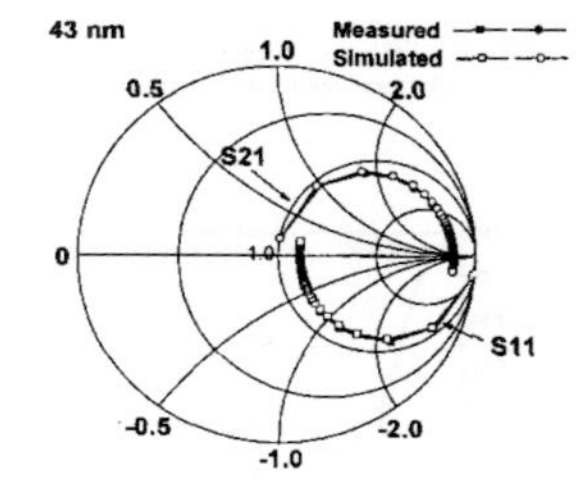

(b)

(c)

Fig. 4. Measured and simulated S-parameters for laminate MIM capacitors with three thicknesses. Simulations and parameter extractions were done by IC-CAP [14].

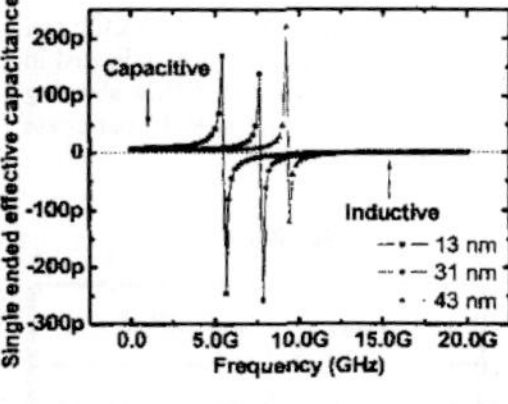

Fig. 5. High frequency response of laminate MIM capacitors from 50 MHz to 20 GHz.

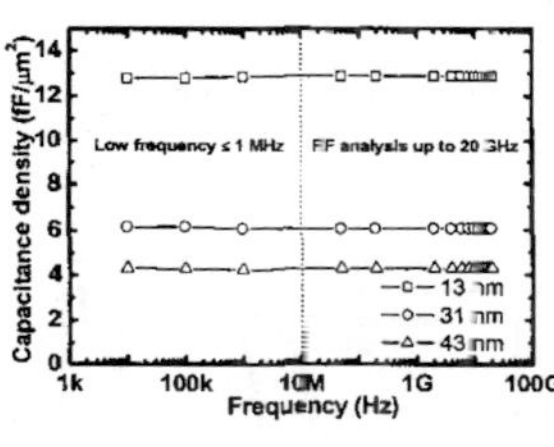

Fig. 6. The frequency dependence on capacitance density of laminate capacitors with three thicknesses.

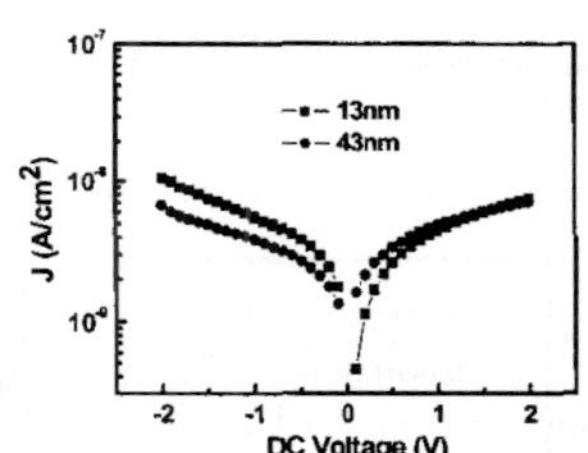

Fig.7. Typical J-V characteristics of laminate MIM capacitors.

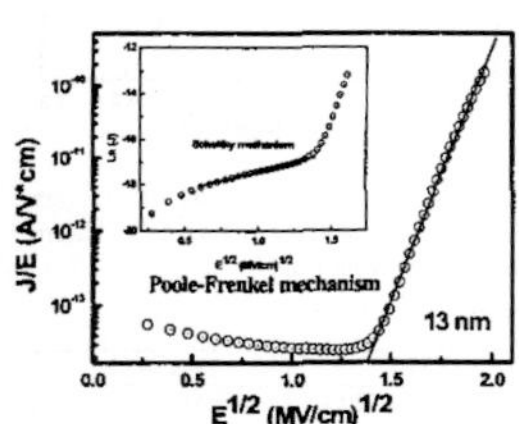

Fig. 8. Conduction mechanisms of 13 nm laminate MIM capacitor, showing Schottky emission at low field and Pool-Frenkel conduction at high field [15].

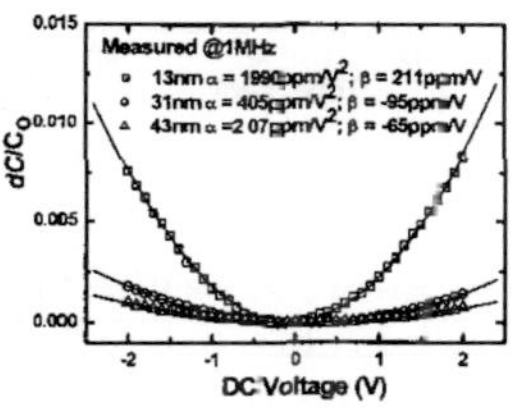

Fig. 9. Quadratic (α) and linear (β) VCCs of laminate MIM capacitors with thicknesses of 13, 31 and 43 nm.

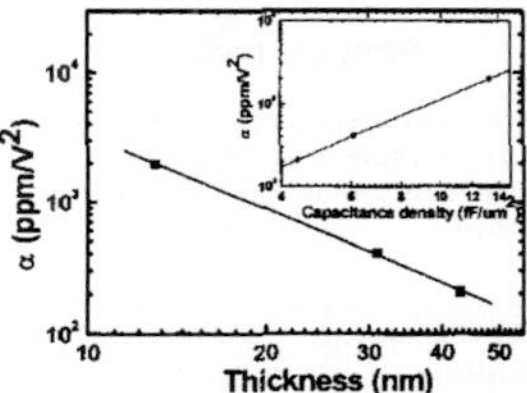

Fig. 10. Thickness dependence of quadratic VCC (α) for laminate MIM capacitors. The inset shows α dependence on capacitance density.

15.6.3

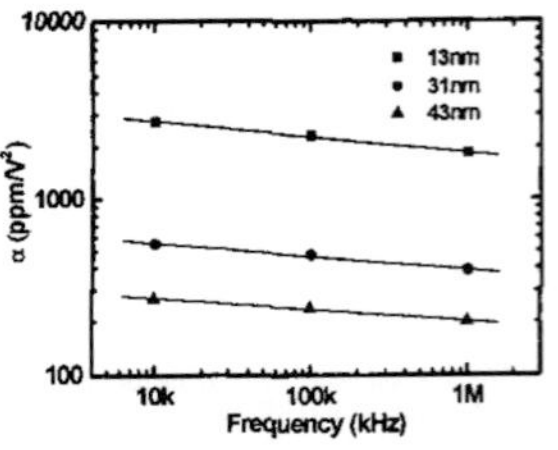

Fig. 11. Frequency dependence of quadratic VCC α for three laminate thicknesses.

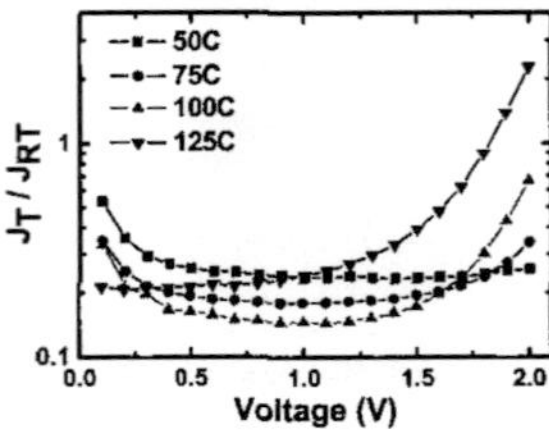

Fig. 12. Leakage obtained at different temperatures for 13 nm MIM capacitor (normalized to J_{RT}: leakage measured at room temperature).

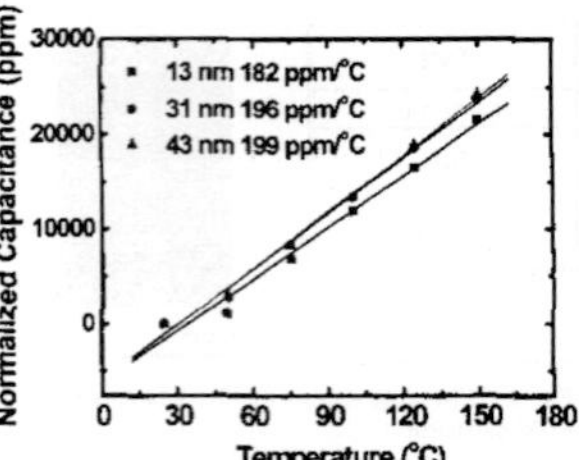

Fig. 13. TCC values for laminate MIM capacitors with three different thicknesses.

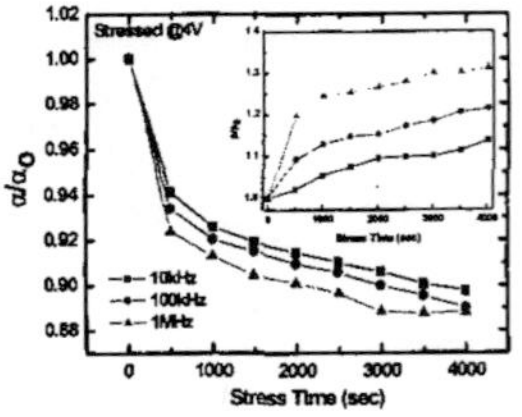

Fig. 14. Quadratic VCC α as a function of stress time. The inset shows time dependence of linear VCC β.

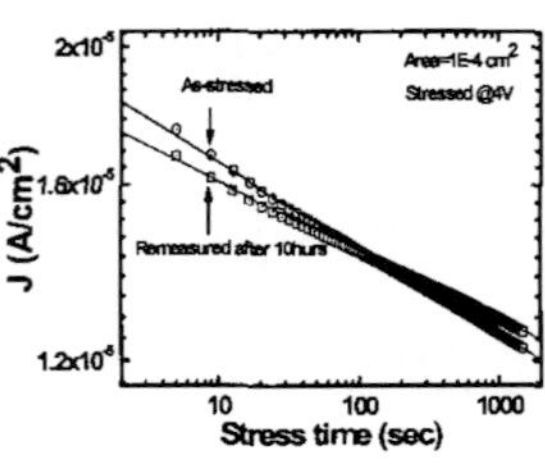

Fig. 15. Stress time dependence of leakage for a fresh device up to 2000s. The device was re-stressed and re-measured after interrupting stress for 10 hours.

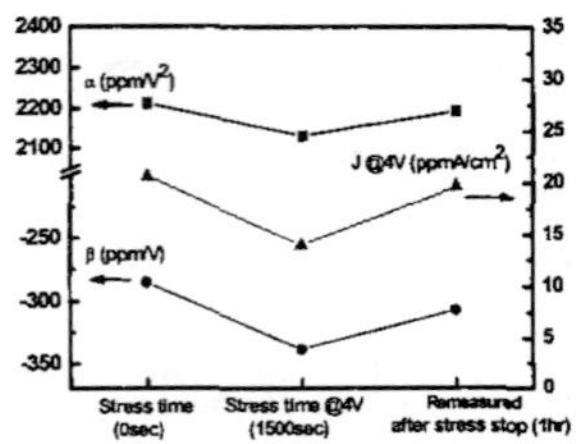

Fig. 16. Time dependence on VCCs and leakage, under stress condition indicated in Fig.15. Both leakage and VCCs showing recovery and correlation after 1 hour stress interruption.

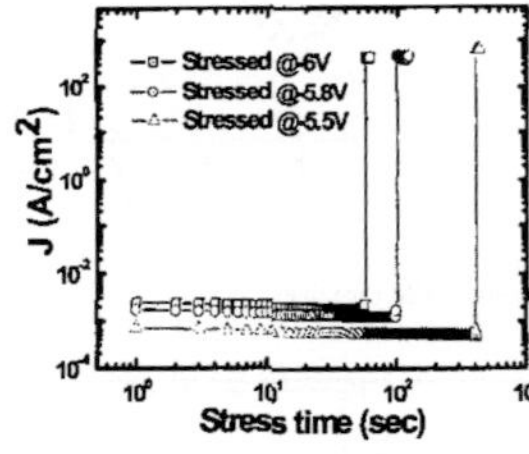

Fig. 17. Breakdown and leakage characteristics of 13 nm laminate MIM capacitors as a function of stress time under different constant voltage stress.

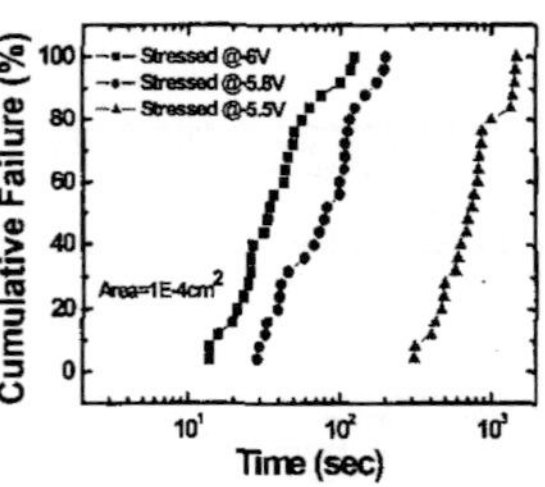

Fig. 18. Cumulative TDDB curves for 13 nm laminate MIM capacitor.

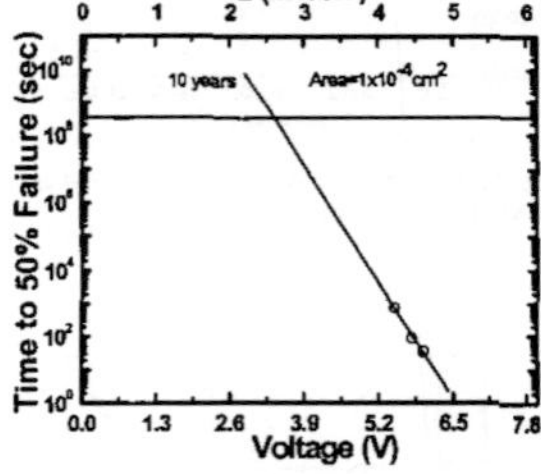

Fig. 19. Life time projection of 13 nm laminate MIM capacitor, using 50% failure time extracted from Fig.18.

Table 1: Comparison of various high-density high-κ MIM capacitors (2002-2003)

Reference	[6]	[7]	[8]	[9]	[10]	This work
Dielectric	HfO$_2$ (ALD)	Ta$_2$O$_5$	AlTaO$_x$ (PVD)	Ta$_2$O$_5$ (CVD)	Tb doped HfO$_2$ (PVD)	Hf/Al laminate (ALD)
Capacitance density (fF/μm^2)	13	9.2	10	9	13.3	12.8
Leakage (A/cm^2)	5.7×10^{-7}@2V	2×10^{-8}@1.5V	4.6×10^{-6}@1 V	—	1×10^{-7}@2V	7.45×10^{-9}@2V
VCC	607 ppm/V 853 ppm/V^2	2060 ppm/V 3580 ppm/V^2	2818 ppm/V^2	2050 ppm/V 475 ppm/V^2	332 ppm/V 2667 ppm/V^2	211 ppm/V 1990 ppm/V^2
TCC (ppm/°C)	—	~200	255	—	123	198

15.6.4

IEEE ELECTRON DEVICE LETTERS, VOL. 25, NO. 8, AUGUST 2004

565

N-Type Schottky Barrier Source/Drain MOSFET Using Ytterbium Silicide

Shiyang Zhu, *Member, IEEE*, Jingde Chen, M.-F. Li, *Senior Member, IEEE*, S. J. Lee, *Member, IEEE*, Jagar Singh, C. X. Zhu, *Member, IEEE*, Anyan Du, C. H. Tung, *Senior Member, IEEE*, Albert Chin, *Senior Member, IEEE*, and D. L. Kwong, *Senior Member, IEEE*

Abstract—Ytterbium silicide, for the first time, was used to form the Schottky barrier source/drain (S/D) of N-channel MOSFETs. The device fabrication was performed at low temperature, wich is highly preferred in the establishment of Schottky barrier S/D transistor (SSDT) technology, including the HfO_2 gate dielectric, and HaN/TaN metal gate. The $YbSi_{2-x}$ silicided N-SSDT has demonstrated a very promising characteristic with a recorded high I_{on}/I_{off} ratio of $\sim 10^7$ and a steep subthreshold slope of 75 mV/dec, which is attributed to the lower electron barrier height and better film morphology of the $YbSi_{2-x}/Si$ contact compared with other self-aligned rare earth metal-(Erbium, Terbium, Dysprosium) silicided Schottky junctions.

Index Terms—MOSFET, rare earth (RE) metal, Schottky, silicide.

I. INTRODUCTION

THE SCHOTTKY barrier source/drain transistor (SSDT) architecture [1] has been proposed to overcome the series resistance problem of ultrashallow S/D junction of sub50–nm MOSFETs [2]–[4], due to the abrupt silicide/Si interface and low resistance of silicide. The barrier height of the Schottky junction should be low enough to obtain a high-driving current [5] and to prevent two different slopes in the subthreshold region of the MOSFETs [4], [6]. P-channel SSDT (P-SSDT) with PtSi as Schottky S/D (hole barrier $\Phi_p = 0.24$–0.28 eV) has been fabricated with quite acceptable electrical performance with I_{on}/I_{off} ratio of $\sim 10^8$ [6], [7] and one subthreshold slope of 66 mV/dec [6]. However, the electrical performance of N-channel SSDT (N-SSDT) is still inferior mainly due to lack of suitable silicide material [4]–[6]. Erbium silicide has been widely adopted for N-SSDT, but its relatively high electron

barrier height and poor film morphology formed by solid-state reaction of deposited Er and substrate Si do not meet the device performance criteria. The I_{on}/I_{off} ratio of the recently reported N-SSDT with $ErSi_{2-x}$ is about 10^5 [8]. On the other hand, the log I_d versus V_g curve shows two slopes at the subthreshold region for long-channel devices [4], [6]. $DySi_{2-x}$ was reported recently [6]. However, it has the similar problems as that of Erbium silicide.

It is well known that the low work function metals usually have low Schottky electron barrier height [9]. Yb has lower photoelectric work function (2.59 eV) than Er (3.12 eV) and Dy (3.09 eV) [1], therefore, Yb silicide is expected to have a lower Schottky electron barrier height. This is the motivation for Yb silicide to be used for N-SSDT and eventually can lead to an excellent electrical device performance.

II. MOS DEVICE FABRICATION

A simplified low-temperature process, which has been described in our previous paper [6], [10] was used to fabricate N-SSDT with HfO_2 gate oxide and HfN/TaN metal gate. Starting substrates were p-type Si(100) wafers with resistivity of 4–8 $\Omega \cdot$cm. HfO_2 (4–6 nm) was deposited at 400 °C using $Hf[OC(CH_3)_3]_4$ and O_2 in a metal–organic CVD(MOCVD) system, followed by an *in situ* annealing in N_2 ambient at 700 °C. Then HfN (~ 50 nm) and TaN (~ 100 nm) were deposited sequentially in a sputtering system with a base pressure of $\sim 1.5 \times 10^{-7}$ torr. Wafers were patterned and subsequently etched using the standard photolithograph and dry etch processes. Immediately after the diluted hydrogen fluoride (DHF) solution dipping, the patterned wafer was loaded into the sputtering system again. Yb (or other RE metal: Er, Dy, Tb) (~ 100 nm) and HfN (~ 100 nm) were deposited in sequence. HfN was used as a capping layer to prevent RE metal oxidization during *ex situ* annealing. Silicidation was performed by rapid thermal anneal (RTA) at 600 °C for 1 min in N_2 ambient, followed by forming-gas anneal (FGA) at 420 °C for 1 h. The silicidation can be performed by only a FGA step, while the $YbSi_{2-x}$ film morphology is improved by the RTA step. The HfN capping layer and unreacted RE metal were selectively removed by wet etch in DHF (HF: $H_2O = 1 : 100$) and sulphuric-acid hydrogen peroxide mixed (SPM) solution ($H_2SO_4 : H_2O_2 = 3 : 1$ at 120 °C) sequentially. The square sheet resistance of $YbSi_{2-x}$ is ~ 3.8 Ω/sq. and the thickness is ~ 90 nm [measured by

Manuscript received April 14, 2004; revised May 11, 2004. The review of this letter was arranged by Editor C.-P. Chang.

S. Zhu is with the Silicon Nano Device Laboratory, Department of Electrical and Computer Engineering, National University of Singapore, Singapore 119260 (e-mail: elelimf@nus.edu.sg). He is also with the Department of Microelectronics, Fudan University, Shanghai 200433, China.

J. Chen, S. J. Lee, and C. X. Zhu are with the Silicon Nano Device Laboratory, Department of Electrical and Computer Engineering, National University of Singapore, Singapore 119260.

M.-F. Li is with the Silicon Nano Device Laboratory, Department of Electrical and Computer Engineering, National University of Singapore, Singapore 119260 and also with the Institute of Microelectronics, Singapore 117685.

J. Singh, A. Du, and C. H. Tung are with the Institute of Microelectronics, Singapore 117685.

A. Chin is with the Department of Electronics Engineering, National Chiao-Tung University, Hsinchu 300 Taiwan.

D. L. Kwong is with the Department of Electrical and Computer Engineering, The University of Texas, Austin, TX 78712 USA.

Digital Object Identifier 10.1109/LED.2004.831582

[1]The photoelectric work function data of various elements are from the software "ptable" (periodic table of the elements). E. L. Edgar, 1993.

566

IEEE ELECTRON DEVICE LETTERS, VOL. 25, NO. 8, AUGUST 2004

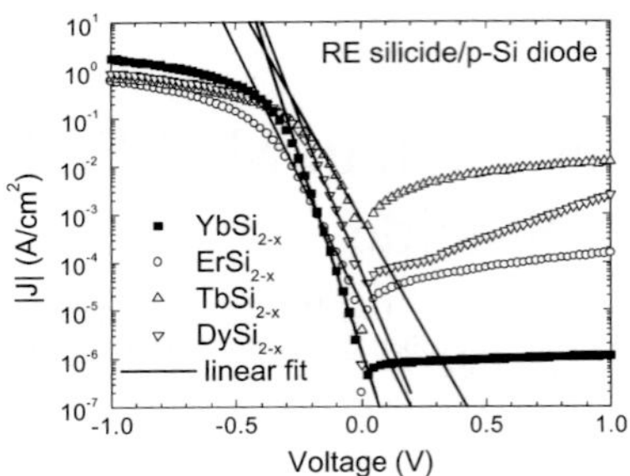

Fig. 1.   Room-temperature I–V curves of various RE silicide/p-Si(100) diodes and the linear fitting based on the thermal emission model. The deduced barrier heights and ideality factors are summarized in Table I.

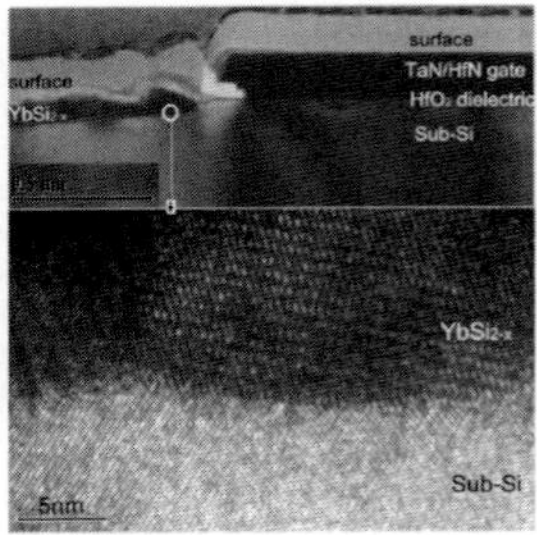

Fig. 2.   (Top) XTEM image of the final N-SSDT fabricated by the simplified one-mask process and a "hole" between the S/D and the gate acts as a sidewall spacer [10], and (bottom) a high resolution XTEM image of the polycrystalline YbSi$_{2-x}$/Si(100) contact. Even though the quite rough YbSi$_{2-x}$ surface, which is probably affected by the SPM solution during the selectively etching step, the polycrystalline YbSi$_{2-x}$/Si interface is quite smooth and flat.

cross-sectional transmission electron microscopy (XTEM)], the resistivity of YbSi$_{2-x}$ is calculated to be $\sim$34 $\mu\Omega\cdot$cm.

III. RESULTS AND DISCUSSION

Fig. 1 shows current–voltage (I–V) curves of various RE silicide/p-Si(100) Schottky diodes. The Schottky hole barrier height $\left(\Phi_P^{I-V}\right)$ and the ideality factor were deduced by linear fitting based on the thermal emission model [9]. The values are summarized in Table I. The hole barrier heights $\left(\Phi_P^{C-V}\right)$ deduced from capacitance–voltage (C–V) curves are also given. The YbSi$_{2-x}$/p-Si contact has the highest hole barrier height of 0.85 eV, lowest reverse bias leakage current, and the best rectifying property with near unity ideality factor. Other diodes have significantly higher leakage current at reverse bias, larger than unity ideality factors and larger difference between Φ_P^{I-V} and Φ_P^{C-V}, implying the unnegligible barrier height inhomogeneity [11]. Observing in the microscope, the surfaces of ErSi$_{2-x}$ and DySi$_{2-x}$ contain many square pits with micrometer size. While there are no such pits on the surface of YbSi$_{2-x}$ and TbSi$_{2-x}$ even though their surface roughness is still quite large

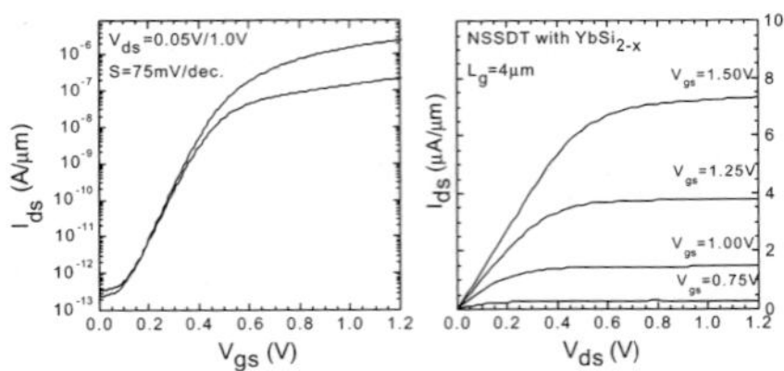

Fig. 3. I_d–V_d and I_d–V_g curves of N-SSDT with YbSi$_{2-x}$ S/D. The channel width and length are 400 and 4 μm respectively. Equivalent oxide thickness of HfO$_2$ is 2.5 nm as deduced from C–V and $V_{\text{th}} = 0.40$ V.

TABLE I

ELECTRICAL CHARACTERISTICS OF VARIOUS RE SILICIDE/p-Si(100) CONTACTS FORMED BY SOLID-STATE REACTION AND THE CORRESPONDING NSSDT PROPERTIES. BARRIER HEIGHTS DEDUCED FROM C–V HAVE RELATIVELY LARGE DEVIATION. DATA OF PtSi/n-Si AND P-SSDT [6] ARE ALSO INCLUDED FOR COMPARISON

Silicide/p-Si (100) diode	ErSi$_{2-x}$	TbSi$_{2-x}$	DySi$_{2-x}$	**YbSi$_{2-x}$**	*PtSi Silicide on n-Si(100)*
Hole barrier obtained by the I-V measurement $\Phi_p^{\text{I-V}}$ (eV)	0.71	0.60	0.67	**0.82**	*0.84 (electron barrier $\Phi_n^{I\text{-}V}$)*
Ideality factor in I-V	1.57	1.83	1.33	**1.04**	*1.02*
Hole Barrier obtained by C-V measurement $\Phi_p^{\text{C-V}}$ (eV)	~ 0.78	~ 0.87	~ 0.83	**~ 0.88**	*~0.86 (electron barrier $\Phi_n^{C\text{-}V}$)*
Averaged hole barrier $\Phi_p \equiv (\Phi_p^{\text{I-V}} + \Phi_p^{\text{I-V}})/2$ (eV)	0.75 [1]	0.74	0.75 [2]	**0.85**	*0.85 (electron barrier Φ_n)*
Electron Barrier $\Phi_n \cong 1.12\,\text{eV} - \Phi_p$ (eV) [3]	0.37	0.38	0.37	**0.27**	*0.27 (hole barrier Φ_p)*
I_{leakage} @ 1V (A/cm²)	1.5×10⁻⁴	1.2×10⁻²	2.3×10⁻³	**1.1×10⁻⁶**	*4.6×10⁷ @-1V*
SSDT properties: $I_{\text{on}}/I_{\text{off}}$ ratio	10³~10⁴	10³~10⁴	10⁴~10⁵	**~ 10⁷**	*~10⁸*
I_{ds} @ $|V_{ds}|$=V$_{gs}$-V$_{th}$ = 1V (μA/μm) (L$_g$ = 4 μm)	~ 1.4	~ 0.26	~ 2.5	**~ 3.4**	*~3.2*

[1] Compared to the reported value of *0.71 eV* for the ErSi$_{2-x}$ film from I-V [15] and *0.85 eV* from C-V [20]

[2] Compared to the reported value of *0.74 eV* for the DySi$_{2-x}$ film formed by UHV evaporation [18]

[3] Due to the difficulty to measure a low electron Schottky barrier height Φ_n directly, the high hole barrier height Φ_p is measured and the electron barrier height is calculated according to the approximation of $\Phi_n + \Phi_p \cong E_g$ (silicon bandgap).

as measured by atomic force microscopy. Fig. 2 (top) shows the cross-sectional XTEM image of the final N-SSDT fabricated by our simplified one-mask process.

Fig. 3 shows the I_d–V_d and I_d–V_g curves of N-SSDT with YbSi$_{2-x}$. The $I_{\text{on}}/I_{\text{off}}$ ratio reaches $\sim$10^7 with one subthreshold slope of $\sim$75 mV/dec, and its drivability is slightly larger than the corresponding P-SSDT with PtSi with the same device structure (Table I). To our knowledge, this is the best electrical performance for N-SSDT reported so far. For comparison, Fig. 4 shows the transfer characteristics of N-SSDT with the same device structure and technology, however using

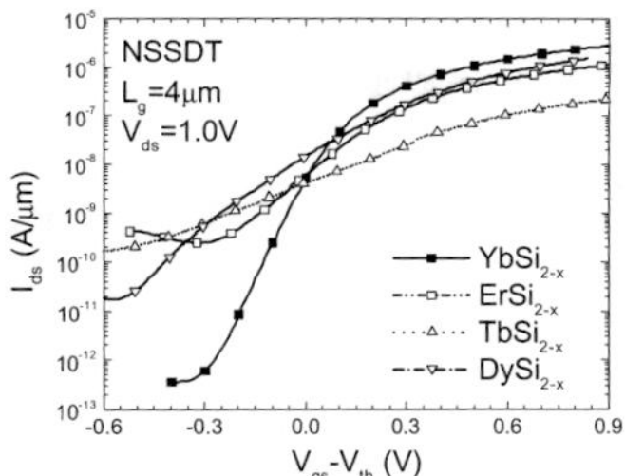

Fig. 4. Transfer characteristics of N-SSDTs with various RE silicides. All devices have the same size of $W/L = 400\ \mu$m$/4\ \mu$m, and were fabricated by the same process.

$ErSi_{2-x}$, $TbSi_{2-x}$ and $DySi_{2-x}$, respectively. The electrical results of these devices are summarized in Table I.

Besides the low electron barrier height, $YbSi_{2-x}$ has better film quality than other RE silicides. The growth of $ErSi_{2-x}$ or $DySi_{2-x}$ during solid-state reaction of deposited RE metal and substrate Si (100) is strongly nucleation preferred, resulting in a nonuniform, columnar growth of the layer with rough surface and interface [12], [13]. The formed silicide has been reported to be $ErSi_{1.7}$ or $DySi_{1.7}$ due to the Si vacancy in the silicide film [4]. In the case of $YbSi_{2-x}$, the formed silicide has been reported to be $YbSi_{1.8}$ [14]. Our x-ray diffraction and energy dispersive X-ray analysis (not shown here) also confirm that the formed film is $YbSi_{1.8}$. Less Si vacancy may cause the silicide more uniformly. From Fig. 2, the grain size of the polycrystalline $YbSi_{1.8}$ is about 5–10 nm and the grain growths approximately along Si[110] axis. Columnar growth, as in the cases of $ErSi_{2-x}$ and $DySi_{2-x}$, was not found.

The RE silicide property is sensitive to the oxygen contamination. For $ErSi_{2-x}$ silicide, $\Phi_N = 0.28$ eV when grown in ultrahigh-vacuum (UHV) condition [15], however Φ_N becomes higher when grown in normal vacuum level as reported in this work and other paper [16], [17]. Our result shows that $YbSi_{2-x}$ grown in normal vacuum condition has better rectifying characteristics than $ErSi_{2-x}$ grown in UHV condition. It implies that $YbSi_{2-x}$ is not so sensitive to oxygen as $ErSi_{2-x}$, or even better rectifying property of $YbSi_{2-x}$/Si contact may be obtained if it is grown in UHV condition. Very low barrier height (0.08 eV) of metal/n-Si contacts has been reported recently by surface passivation of a thin Se layer [18]. However, such method is infeasible for N-SSDT fabrication due to the requirement of self-aligned S/D formation.

IV. CONCLUSION

Several rare earth metals are investigated for silicide S/D. The $YbSi_{2-x}$ has been found to be a very promising candidate for N-SSDT as it provides a high drive current with a very low leakage current. It is probably due to the low electron barrier height of the $YbSi_{2-x}$/Si Schottky contact and smooth

$YbSi_{2-x}$/Si interface. It can be concluded that $YbSi_{2-x}$ is a much better silicide material than the usually used $ErSi_{2-x}$ for N-SSDT.

REFERENCES

[1] M. P. Lepselter and S. M. Sze, "SB-IGFET: An insulated-gate field-effect transistor using Schottky barrier contacts for source and drain," *Proc. IEEE*, vol. 56, pp. 1400–1401, 1968.

[2] M. Nishisaka, S. Matsumoto, and T. Asano, "Schottky source/drain SOI MOSFET with shallow doped-extension," in *Int. Conf. Solid-State Devices and Materials*, 2002, pp. 586–587.

[3] H.-C. Lin, M. F. Wang, F. J. Hou, H. N. Lin, C. Y. Lu, J. T. Liu, and T. Y. Huang "High-performance p-channel Schottky-barrier SOI FinFET featuring self-aligned PtSi source/drain and electrical junctions," *IEEE Electron Device Lett.*, vol. 24, pp. 102–104, Jan. 2003.

[4] J. Kedzierski, P. Xuan, E. H. Anderson, J. Bokor, T. J. King, and C. H. Hu, "Complementary silicide source/drain thin-body MOSFETs for the 20 nm gate length regime," in *IEDM Tech. Dig.*, 2000, pp. 57–60.

[5] W. Saitoh, A. Itoh, S. Yamagami, and M. Asada, "Analysis of short-channel Schottky source/drain metal–oxide–semiconductor field-effect transistor on silicon-on-insulator substrate and demonstration of sub50-nm n-type devices with metal gate," *Jpn. J. Appl. Phys.*, vol. 38, pp. 5226–6231, 1999.

[6] S. Y. Zhu, H. Y. Yu, S. J. Whang, J. H. Chen, C. Shen, C. Zhu, S. J. Lee, M. F. Li, D. S. H. Chan, W. J. Yoo, A. Du, C. H. Tung, J. Singh, A. Chin, and D. L. Kwong, "Schottky barrier source/drain MOSFETs with high-κ gate dielectrics and metal gate electrode," *IEEE Electron Device Lett.*, vol. 25, pp. 268–270, Mar. 2004.

[7] L. E. Calvet, H. Luebben, M. A. Reed, C. Wang, J. P. Snyder, and J. R. Tucker, "Suppression of leakage current in Schottky barrier metal–oxide–semiconductor field-effect transistors," *J. Appl. Phys.*, vol. 91, nc. 2, pp. 757–759, 2002.

[8] M. Jang, J. Oh, S. Maeng, W. Cho, S. Lee, K. Kang, and K. Park, "Characteristics of erbium-silicided n-type Schottky barrier tunnel transistors," *Appl. Phys. Lett.*, vol. 83, no. 13, pp. 2611–2613, 2003.

[9] R. T. Tung, "Recent advances in Schottky barrier concepts," *Mater. Sci. Eng.* vol. 35, pp. 1–138, 2001.

[10] S. Y. Zhu, H. Y. Hu, J. D. Chen, S. J. Whang, J. H. Chen, C. Shen, C. X. Zhu, S. J. Lee, M. F. Li, D. S. H. Chan, W. J. Yoo, A. Du, C. H. Tung, J. Singh, A. Chin, and D. L. Kwong, "Low temperature MOSFET technology with Schottky barrier source/drain, high-κ gate dielectric and metal gate electrode," *Solid State Electron.*, to be published.

[11] S. Y. Zhu, R. L. Van Meirhaeghe, C. Detavernier, F. Cardon, G. P. Ru, X. P. Qu, and B. Z. Li, "Barrier height inhomogeneities of epitaxial CoSi_2 Schottky contacts on n-Si(100) and (111)," *Solid State Electron.*, vol. 44, no. 4, pp. 663–671, 2000.

[12] C. H. Luo and L. J. Chen, "Growth kinetic of amorphous interlayers and formation of crystalline silicide phases in ultrahigh vacuum deposited polycrystalline Er and Tb thin films on (001)Si," *J. Appl. Phys.*, vol. 82, no. 8, pp. 3808–3814, 1997.

[13] A. Travlos, N. Salamouras, and N. Boukos, "Growth of rare earth silicides on silicon," *J. Phys. Chem. Sol.*, vol. 64, pp. 87–93, 2003.

[14] K. S. Chi and L. J. Chen, "Formation of ytterbium silicide on (111) and (001)Si by solid-state reactions," *Mater. Sci. Semicond. Processing*, vol. 4, pp. 269–272, 2001.

[15] F. Muret, T. A. N. Tan, N. Frangis, and J. van Landuyt, "Unpinning of the Fermi level at erbium silicide/silicon interface," *Phys. Rev. B, Condens. Matter*, vol. 56, no. 15, pp. 9286–9289, 1997.

[16] G. Kaltsas, A. Travlos, N. Salamouras, A. G. Nassiopoulos, P. Revva, and A. Traverse, "Erbium silicide films on (100) silicon, grown in high vacuum, fabrication and properties," *Thin Solid Films*, vol. 275, pp. 37–90, 1996.

[17] Z. Xu, *Properties of Metal Silicides*, K. Maex and M. Van Rossum, Eds. London, U.K.: Inspec, 1995, pp. 217–224.

[18] M. Tao, S. Agarwal, D. Udeshi, N. Basit, E. Maldonado, and W. P. Kirk, "Low Schottky barriers on n-type silicon (001)," *Appl. Phys. Lett.*, vol. 83, no. 13, pp. 2593–2595, 2003.

[19] M. Q. Huda and K. Sakamoto, "Use of ErSi_2 in source/drain contacts of ultrathin SOI MOSFETs," *Mater. Sci. Eng.*, vol. B89, pp. 378–381, 2002.

268   IEEE ELECTRON DEVICE LETTERS, VOL. 25, NO. 5, MAY 2004

Schottky-Barrier S/D MOSFETs With High-K Gate Dielectrics and Metal-Gate Electrode

Shiyang Zhu, *Member, IEEE*, H. Y. Yu, *Student Member, IEEE*, S. J. Whang, J. H. Chen, Chen Shen, Chunxiang Zhu, *Member, IEEE*, S. J. Lee, *Member, IEEE*, M. F. Li, *Senior Member, IEEE*, D. S. H. Chan, *Senior Member, IEEE*, W. J. Yoo, Anyan Du, C. H. Tung, *Senior Member, IEEE*, Jagar Singh, Albert Chin, *Senior Member, IEEE*, and D. L. Kwong, *Senior Member, IEEE*

Abstract—This letter presents a low-temperature process to fabricate Schottky-barrier silicide source/drain transistors (SSDTs) with high-κ gate dielectric and metal gate. For p-channel SSDTs (P-SSDT) using PtSi sourece/drain (S/D) , excellent electrical performance of $I_{on}/I_{off} \sim 10^7 - 10^8$ and subthreshold slope of 66 mV/dec have been achieved. For n-channel SSDTs (N-SSDTs) using DySi$_{2-x}$ S/D , I_{on}/I_{off} can reach $\sim 10^5$ at V_{ds} of 0.2 V with two subthreshold slopes of 80 and 340 mV/dec. The low-temperature process relaxes the thermal budget of high-κ dielectric and metal-gate materials to be used in the future generation CMOS technology.

Index Terms—High-κ, metal gate, MOSFET, Schottky.

I. INTRODUCTION

THE SERIES resistance of ultrashallow sourece/drain (S/D) junctions is a serious issue for future CMOS transistor scaling. Schottky-barrier silicide S/D (SSDT) structure has been suggested as a potential solution, due to the sharp silicide/silicon interface and low sheet resistance of silicide [1]–[5]. SSDT is particularly attractive when a metal-gate/high-κ gate stack is employed, as it avoids the use of a high-temperature annealing process required for implanted S/D junctions and poly gate, hence, eliminating the thermal stability issues associated with high-κ gate stack [6]. In this letter, we successfully demonstrate bulk SSDTs with HfO$_2$ high-κ dielectric, HfN metal gate, and PtSi (for pMOS) and DySi$_{2-x}$ (for nMOS) silicided S/D using a simplified low temperature process. The process can be easily extended to ultrathin body (UTB) silicon-on-insulator (SOI) structure to further improve the SSDT performance.

II. MOS DEVICE FABRICATION

Si (100) wafers of both n– and p-type with resistivity of 4–8 $\Omega \cdot$ cm were used as the starting substrate. First, $\sim$6-nm HfO$_2$ film was deposited on Si at 400 °C using Hf[OC(CH$_3$)$_3$]$_4$ and O$_2$ in a metal–organic chemical vapor deposition system, followed by an *in situ* annealing in N$_2$ ambient at 700 °C.Then HfN($\sim$50 nm) and TaN($\sim$100 nm) were deposited sequentially in a sputtering system with base pressure of 1.5×10^{-7} torr. TaN is used as a capping layer to reduce the gate sheet resistance ($\sim$ 10 Ω sq) [7]. The wafers were patterned using photolithography and reactive ion etching procedures. Immediately after dipping in a diluted hydrogen fluoride solution (DHF), the wafers were loaded into the sputtering system again for platinum ($\sim$100 nm) (for P-SSDT) or dysprosium ($\sim$100 nm)/HfN ($\sim$70 nm) stack (for N-SSDT) deposition. Since Dy is easily oxidized during *ex situ* anneal, a capping layer of thermally stable HfN [7] is used to prevent Dy oxidization. Silicidation of Pt or Dy was performed by forming gas anneal at 420 °C for 1 h. Then unreacted Pt was removed by hot diluted Aqua Regia etch. The HfN capping layer and unreacted Dy were etched by DHF and a diluted HNO$_3$ solution sequentially.

Although PtSi has flat surface, the surface of DySi$_{2-x}$ film is quite rough. This is because DySi$_{2-x}$, as well as other rare earth (RE) silicides, is not formed through layer-by-laye, but islanded-preferred during the solid-state reaction with substrate Si [8]. No improvement to the DySi$_{2-x}$ film morphology was found by using other capping layers such as Pt, Ti, Ta, Ru, and Al, etc. or by *in situ* vacuum anneal without a capping layer. HfN is found to be the most suitable capping layer in our study, due to its easy removal and no contamination in DySi$_{2-x}$.

III. DEVICE CHARACTERIZATION AND DISCUSSION

For the PtSi/n-Si Schottky contact, the electron barrier height (Φ_{bN}) was measured to be 0.85 and 0.86 eV from current–voltage (I–V) and capacitance–voltage (C–V), respectively. The corresponding hole barrier height ($\Phi_{bP} \cong E_g --\Phi_{bN}$) is about 0.26 eV, close to the reported value [9]. For the DySi$_{2-x}$/p–Si diode, Fig. 1 shows that Φ_{bP} are 0.66 eV (I–V) and 0.88 eV (C–V), compared to the reported value of $\sim$0.74 eV for the DySi$_{2-x}$ film formed by ultrahigh vacuum evaporation [9]. The large difference in Φ_{bP} from C–V and I–V, and the larger than

Manuscript received January 5, 2004. This work was supported by the Singapore A-STAR under Grant EMT/TP/00/001,2. The review of this letter was arranged by Editor A. Chatterjee.

S. Zhu is with the Department of Electrical and Computer Engineering, Silicon Nano Device Laboratory, National University of Singapore, 119260 Singapore, and also with the Department of Microelectronics, Fudan University, Shanghai, 200433 China.

H. Y. Yu, J. H. Chen, C. Shen, C. Zhu, S. J. Lee, D. S. H. Chan, and W. J. Yoo are with the Department of Electrical and Computer Engineering, Silicon Nano Device Laboratory, National University of Singapore, 119260 Singapore.

S. J. Whang is with the Jusung Engineering Co, Ltd., Gyeonggi 464-892 Korea.

M. F. Li is with the Department of Electrical and Computer Engineering, Silicon Nano Device Laboratory, National University of Singapore, Singapore 119260 and also with the Institute of Microelectronics, 117685 Singapore (e-mail: elelimf@nus.edu.sg).

A. Du, C. H. Tung, and J. Singh are with the Institute of Microelectronics, 117685 Singapore.

A. Chin is with the Department of Electronics Engineering, National Chiao Tung University, Hsinchu, 300 Taiwan, R.O.C.

D. L. Kwong is with the Department of Electrical and Computer Engineering, University of Texas, Austin, TX 78712 USA.

Digital Object Identifier 10.1109/LED.2004.826569

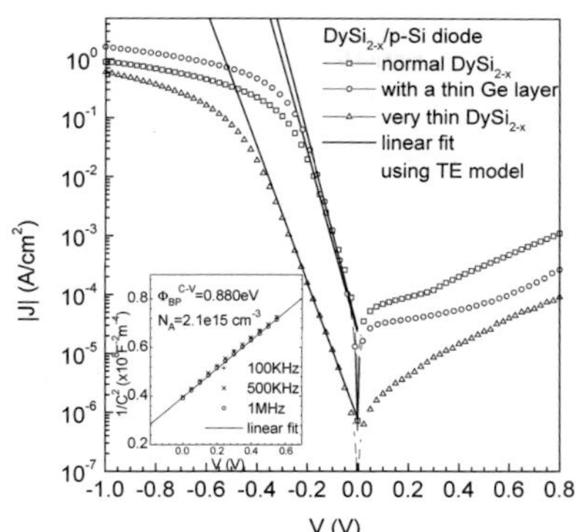

Fig. 1. *I–V* curves of three different $DySi_{2-x}$/p-Si diodes. (1) Normal $DySi_{2-x}$ formed by Dy/Si silicidation at 420 °C for 1 h. (2) Adding an ultrathin intermediate Ge layer (~1 nm). (3) A very thin $DySi_{2-x}$ with an Al capping layer. The barrier heights Φ_{bP} and ideality factors deduced from the thermal emission model are (0.66 eV, 1.10), (0.72 eV, 1.05), and (0.82 eV, 1.46), respectively. The inset is the *C–V* curves of diode 1. The deduced Φ_{bP} and N_A are 0.88 eV and 2.1×10^{15} cm^{-3}, respectively. Other two diodes have the similar *C–V* results within the experimental error.

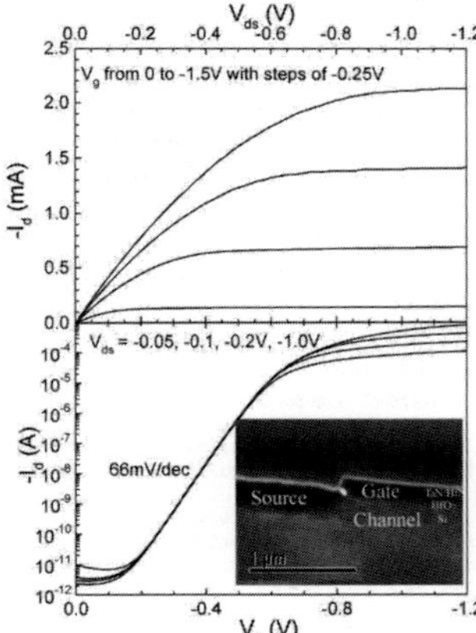

Fig. 2. I_d–V_{ds} and I_d–V_g curves of P-SSDT with PtSi S/D. The channel width and length are 400 and 4 μm. Equivalent oxide thickness (EOT) of HfO_2 gate dielectric is 2.0 nm, $V_t = -0.56$ V. The inset is the XTEM image of the device structure. A "hole" between S/D and channel acts as a sidewall spacer.

unity ideality factor (*n*) imply significant barrier height inhomogeneity of the $DySi_{2-x}$/Si interface [10], associated with the rough surface and interface of $DySi_{2-x}$/Si as observed by atomic force microscopes and cross-sectional transmission electron microscope (XTEM) measurements. Fig. 2 and the inset show the *I–V* characteristics of P-SSDT and the device structure. The drain current at small $V_g (I_{off})$ is between 1–10 pA

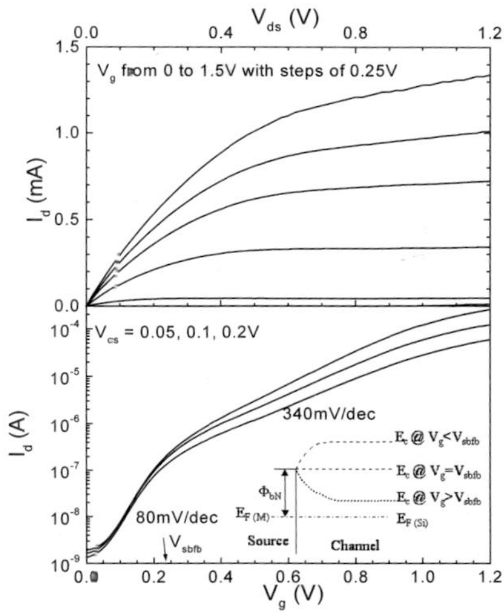

Fig. 3. I_d–V_{cs} and I_d–V_g curves of N-SSDT with $DySi_{2-x}$ S/D. The channel width and length are 400 and 4 μ m. EOT is 1.5 nm, $S = 80$ mV/dec at $V_g < 0.23$ V, and 340 mV/dec at $V_g > 0.23$ V, $V_t = 0.56$ V. The inset is the schematic band diagram of Schottky barrier at source/channel. If the device is already turned on before V_g reaches V_{sbfb}, it shows one subthreshold slope as conventional device. Otherwise, it shows two subthreshold slopes because the slope reduces dramatically when $V_g > V_{sbfb}$.

for devices with drain area of 1×10^{-4} cm^2, close to the theoretical value of the reverse current of the PtSi/n-Si diode with barrier height of 0.86 eV (~3.8 pA). I_{on} depends on process parameters more sensitively. In our simplified one-mask process, a "hole" between the S/D and gate is formed during DHF dipping because HfN can be etched by DHF while TaN cannot. The "hole" acts as a spacer to separate the S/D and the gate. The "hole" size can be controlled by DHF dipping time. For the $W/L = 200/2.5 - \mu$m devices, the I_{on} at $V_g = V_{ds} = -2.5$ V increases from 3.5 to 6.2 mA by reducing the DHF dipping time from 105 to 60 s. Our P-SSDTs have a I_{on}/I_{off} ratio ~ 10^8 at $V_g = V_{cs} = -1.0$ V, with the subthreshold slope of 66 mV/dec. Their performance is similar to or even slightly better than the best reported data for SiO_2/poly-Si gate P-SSDT [11]. N-SSDT has ~three orders of magnitude larger I_{off} than P-SSDT. Fig. 3 shows that the I_{on}/I_{off} ratio at a small V_{ds} (0.2 V) is about 10^5, close to the recently reported value of N-SSDT on SOI [12]. I_{off} increases significantly at large V_{ds} due to the quick increase of the $DySi_{2-x}$/p-Si diode leakage current with the increase of the reverse bias (see Fig. 1). The transfer I_d–V_g curves of N-SSDT show two different subthreshold slopes ~80 mV/dec at $V_g < 0.23$ V (V_{sbfb}: source-body flatband voltage) and ~340 mV/dec at $V_g > 0.23$ V. This behavior, as well as the large I_{off} observed in N-SSDT, can be explained by the relatively low hole barrier height Φ_{bP} of the $DySi_{2-x}$/p-Si contact as compared to the electron barrier height Φ_{bN} of the PtSi/n-Si contact in P-SSDT. As shown in the inset of Fig. 3, the tunneling

IEEE ELECTRON DEVICE LETTERS, VOL. 25, NO. 5, MAY 2004

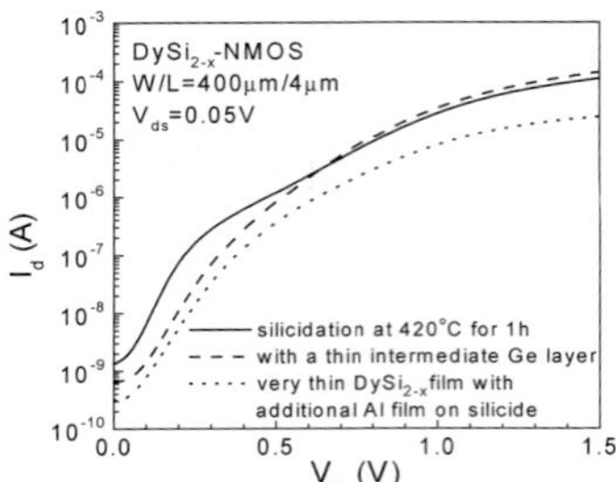

Fig. 4. Transfer curves of N-SSDTs with DySi$_{2-x}$ S/D formed at different conditions, (also see Fig. 1). The device with very thin DySi$_{2-x}$ has lower I_off and I_on. The $I_\mathrm{on}/I_\mathrm{off}$ ratio at $V_\mathrm{ds} = 50$ mV is improved about 4 to 5 times by adding a very thin ($\sim$1 nm) Ge intermediate layer.

current (J_TN) in N-SSDT through the electron Schottky barrier $\Phi_\mathrm{bN} \cong E_g - \Phi_\mathrm{bP} = 0.38$ eV between source and channel dominates in the $V_g > V_\mathrm{sbfb}$ region, resulting in two slopes below threshold [5], [13]. It exhibits only one subthreshold slope if V_sbfb is larger than the threshold voltage (V_th), which requires $\Phi_\mathrm{bN} < E_g/2 - (kT/q)\ln(N_A/n_i)$ ($\sim$0.26 eV in our case) and this condition is not satisfied for DySi$_{2-x}$/Si Schottky junction. On the other hand, PtSi–Si contact meets the requirement of $\Phi_\mathrm{bP} < E_g/2 - (kT/q)\ln(N_D/n_i)$ ($\sim$0.27 eV) for P-SSDT. Therefore, only one subthreshold slope of 66 mV/dec is observed in P-SSDT.

Low barrier height is a key issue for SSDT. RE silicides have the lowest electron barrier height among the known silicides [9]. However, their barrier height is not low enough to meet the N-SSDT requirement. Even worse, the large barrier height inhomogeneity of the RE silicide/Si contact causes I_off ($\sim$ 25 μA/cm^2) larger than the theoretical value of a uniform barrier height ($\sim$ 1.1 μA/cm^2 for DySi$_{2-x}$/p–Si diode with $\Phi_\mathrm{bP} = 0.74$ eV). Improving the RE silicide quality and reducing its barrier height are big challenges for N-SSDT. We found that adding a very thin ($\sim$1 nm) deposited amorphous Ge intermediate layer can improve the DySi$_{2-x}$ film morphology significantly, because the amorphous Ge may suppress the crystallization of DySi$_{2-x}$ during its growth. The better diode performance is shown in Fig. 1 with Φ_bP of 0.72 eV, n of 1.05, and about half order of magnitude lower reverse current than the normal ones. In Fig. 4, the transistor performance is also improved. The $I_\mathrm{on}/I_\mathrm{off}$ ratio at $V_\mathrm{ds} = 0.05$ V increases about 4 to 5 times. Very thin DySi$_{2-x}$ film has quite smooth surface, which can be formed by low temperature (300 °C, 1 h) or short time (400 °C, 1 min) anneal. In this case, an additional metal

(such as Al) film on the DySi$_{2-x}$ is necessary to reduce the series resistance. Fig. 1 shows that Φ_bP and n are 0.82 eVand 1.46, respectively. Fig. 4 shows that I_off at small V_g and V_ds is reduced by 8 to 9 times. The value of $\sim$ 2.0 μA/cm^2 is close to the theoretical prediction. However, I_on is also reduced due to the large S/D series resistance. Further improvement of the electrical performance of N-SSDT needs Φ_bN to be reduced to less than 0.2 eV with quite smooth interface. The use of UTB-SOI also helps due to the reduction of the Schottky contact area.

In conclusion, N- and P-SSDTs with high-κ dielectric and metal gate were demonstrated for the first time using a low temperature process. The transistors have comparable electrical performance to the best data of SSDT reported so far.

REFERENCES

[1] M. P. Lepselter and S. M. Sze, "SB-IGFET: An insulated-gate field-effect transistor using Schottky barrier contacts for source and drain," *Proc. IEEE*, vol. 56, pp. 1400–1401, 1968.

[2] M. Nishisaka, S. Matsumoto, and T. Asano, "Schottky source/drain SOI MOSFET with shallow doped-extension," in *Proc. Int. Conf. Solid State Devices and Materials*, 2002, pp. 586–587.

[3] J. P. Snyder, C. R. Helms, and Y. Nishi, "Experimental investigation of a PtSi source and drain field-emission transistor," *Appl. Phys. Lett.*, vol. 67, pp. 1420–1422, 1995.

[4] H.-C. Lin, M. F. Wang, F. J. Hou, H. N. Lin, C. Y. Lu, J. T. Liu, and T. Y. Huang, "High-performance P-channel Schottky-barrier SOI FinFET featuring self-aligned PtSi source/drain and electrical junctions," *IEEE Electron Device Lett.*, vol. 24, pp. 102–104, Feb. 2003.

[5] J. Kedzierski, P. Xuan, E. H. Anderson, J. Bokor, T. J. King, and C. H. Hu, "Complementary silicide source/drain thin-body MOSFETs for the 20-nm gate length regime," in *IEDM Tech. Dig.*, 2000, pp. 57–60.

[6] C. S. Kang, H. J. Cho, K. Onishi, R. Choi, R. Nieh, S. Goplan, S. Krishnan, and J. C. Lee, "Improved thermal stability and device performance of ultrathin (EOT< 10 Å) gate dielectric MOSFETs by using Hafnium Oxynitride (HfO$_x$N$_y$)," in *Symp. VLSI Tech. Dig.*, 2002, pp. 146–157.

[7] H. Y. Yu, J. F. Kang, J. D. Chen, C. Ren, Y. T. Hou, S. J. Whang, M. F. Li, D. S. H. Chan, K. L. Bera, C. H. Tung, A. Du, and D. L. Kwong, "Thermally robust high quality HfN/HfO$_2$ gate stack for advanced CMOS devices," *IEEE Electron Device Lett.*, vol. 25, pp. 70–72, Jan. 2004.

[8] B. Z. Liu and J. Nogami, "A scanning tunneling microscopy study of dysprosium silicide nanowire growth on Si(001)," *J. Appl. Phys.*, vol. 93, pp. 593–599, 2003.

[9] Z. Xu, *Properties of Metal Silicides*, K. Maex and M. Van Rossum, Eds. London, U.K.: IEEE-Inspec, 1995, p. 217.

[10] R. T. Tung, "Electron transport at metal-semiconductor interfaces: general theory," *Phys. Rev. B, Condens. Matter*, vol. 45, pp. 13 509–13 523, 1992.

[11] L. E. Calvet, H. Luebben, M. A. Reed, C. Wang, J. P. Snyder, and J. R. Tucker, "Suppression of leakage current in Schottky barrier metal-oxide-semiconductor field-effect transistors," *J. Appl. Phys.*, vol. 91, no. 2, pp. 757–759, 2002.

[12] M. Jang, J. Oh, S. Maeng, W. Cho, and S. Lee, "Characteristics of erbium-silicided n-type Schottky-barrier tunnel transistors," *Appl. Phys. Lett.*, vol. 83, no. 13, pp. 2611–2613, 2003.

[13] J. Knoch and J. Appenzeller, "Impact of the channel thickness on the performance of Schottky-barrier metal–oxide–semiconductor field-effect transistors," *Appl. Phys. Lett.*, vol. 81, no. 16, pp. 3082–3084, 2002.

IEEE ELECTRON DEVICE LETTERS, VOL. 25, NO. 5, MAY 2004

337

Fermi Pinning-Induced Thermal Instability of Metal-Gate Work Functions

H. Y. Yu, *Student Member, IEEE*, Chi Ren, Yee-Chia Yeo, *Member, IEEE*, J. F. Kang, X. P. Wang, H. H. H. Ma, Ming-Fu Li, *Senior Member, IEEE*, D. S. H. Chan, *Senior Member, IEEE*, and D.-L. Kwong, *Senior Member, IEEE*

Abstract—The dependence of the metal-gate work function on the annealing temperature is experimentally studied. We observe increased Fermi-level pinning of the metal-gate work function with increased annealing temperature. This effect is more significant for SiO_2 than for HfO_2 gate dielectric. A metal-dielectric interface model that takes the role of extrinsic states into account is proposed to explain the work function thermal instability. This letter provides new understanding on work function control for metal-gate transistors and on metal-dielectric interfaces.

Index Terms—Extrinsic states, Fermi pinning, metal gate, thermal stability, work function.

I. INTRODUCTION

METAL-GATE electrodes will be required for complementary metal–oxide–semiconductor (CMOS) transistors to eliminate the gate depletion and dopant penetration problems that are associated with the conventional polycrystalline silicon (poly-Si) gate electrode [1]. In selecting metal-gate materials for device integration, the metal work function (Φ_m) is an important consideration since it directly affects the threshold voltage and the performance of a transistor. The identification of metal-gate materials is a challenging task because metal-gate work functions are observed to be dependent on the underlying gate dielectric [2], [3] and on the fabrication process conditions [4]–[9]. The dependence of Φ_m on the gate dielectric material was explained by Yeo *et al.* [2], [3] to be due to dipole formation at the interface of the gate electrode and the gate dielectric. This model has been particularly successful for metal-dielectric interfaces where there is minimal interfacial reaction [2], [3], or where *intrinsic states* or *metal-induced gap states* (MIGS) dominate. On the other hand, the dependence of Φ_m on process conditions is not clearly understood or well explored, despite its importance.

Manuscript received February 23, 2004; revised March 15, 2004. This work was supported by the Singapore A-STAR under Grant EMT/TP/00/001,2. The review of this letter was arranged by Editor C.-P. Chang.

H. Y. Yu, C. Ren, Y.-C. Yeo, X. P. Wang, H. H. H. Ma, and D. S. H. Chan are with the Silicon Nano Device Laboratory, Department of Electrical and Computer Engineering, National University of Singapore, Singapore 119260.

J. F. Kang is with the Silicon Nano Device Laboratory, Department of Electrical and Computer Engineering, National University of Singapore, Singapore 119260 and with the Institute of Microelectronics, Peking University, Beijing 100871, China.

M.-F. Li is with the Silicon Nano Device Laboratory, Department of Electrical and Computer Engineering, National University of Singapore, Singapore 119260 and with the Institute of Microelectronics, Singapore 117685 (e-mail: elelimf@nus.edu.sg).

D.-L. Kwong is with the Department of Electrical and Computer Engineering, The University of Texas, Austin, TX 78712 USA.

Digital Object Identifier 10.1109/LED.2004.827643

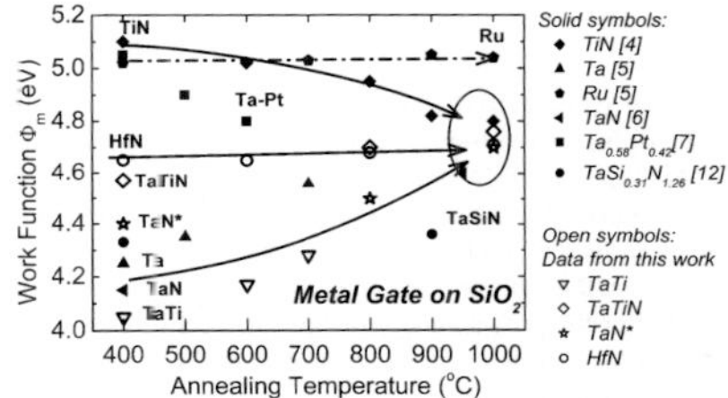

Fig. 1. Variation of metal-gate work function Φ_m with the annealing temperature. The gate dielectric is SiO_2. Intrinsic states at the interface of metals and SiO_2 do not play a very significant role in modifying the vacuum metal work function. Therefore, the change of Φ_m with increasing temperature is predominantly due to extrinsic states.

In this letter, we report experimental results on the dependence of the metal-gate work function on the process temperature, and present a model to explain the phenomenon of process-induced Φ_m thermal instability. We also show that high-temperature annealing could lead to the creation of *extrinsic states* at the metal-dielectric interface for particular combinations of metal-gate and gate-dielectric materials, and result in metal Fermi-level pinning.

II. FERMI-LEVEL PINNING INDUCED BY LOCALIZED EXTRINSIC STATES

Capacitors with HfN, TaN, TaTi, and TaTiN metal-gate electrodes were fabricated. The capacitor gate dielectric is silicon oxide (SiO_2) or hafnium oxide (HfO_2) with different thicknesses. Details of the fabrication were reported in [10]. The capacitors were annealed at different temperatures, and the work functions of the metal-gate electrodes were extracted from plots of the flatband voltage V_{fb} versus the equivalent SiO_2 thickness of the gate dielectric, with reference to the following:

$$\Phi_m = \Phi_{Si} + V_{fb} - \frac{Q_{ox}}{C_{ox}} \qquad (1)$$

where Q_{ox} is the equivalent oxide charge per unit area, C_{ox} is the oxide capacitance, and Φ_{Si} is the work function of the Si substrate [11].

Fig. 1 shows the dependence of Φ_m on the annealing temperature for various metal-gate materials formed on the SiO_2 gate dielectric. Experimental data from the literature [4]–[7], [12] are also included in this figure. The fabrication methods

338

IEEE ELECTRON DEVICE LETTERS, VOL. 25, NO. 5, MAY 2004

TABLE I
FABRICATION METHOD OF THE METAL-GATE ELECTRODES STUDIED IN FIG. 1

Metal	Brief description of the fabrication method	Reference
TiN	Atomic layer deposition at 350°C using TiCl$_4$/NH$_3$ precursors	[4]
Ta	Sputtering of Ta target	[5]
Ru	Sputtering of Ru target	[5]
TaN	Reactive sputtering of Ta target in Ar+N$_2$ mixed ambient	[6]
Ta$_{0.58}$Pt$_{0.42}$	Co-sputtering of Ta and Pt targets	[7]
TaSi$_{0.31}$N$_{1.26}$	Reactive co-sputtering of Ta and Si targets in Ar+N$_2$ mixed ambient	[12]
TaTi	Co-sputtering of Ta and Ti targets	This work
TaTiN	Reactive co-sputtering of Ta and Ti targets in Ar+N$_2$ mixed ambient	This work
TaN*	Reactive sputtering of Ta target in Ar+N$_2$ mixed ambient	This work
HfN	Reactive sputtering of Hf target in Ar+N$_2$ mixed ambient	This work

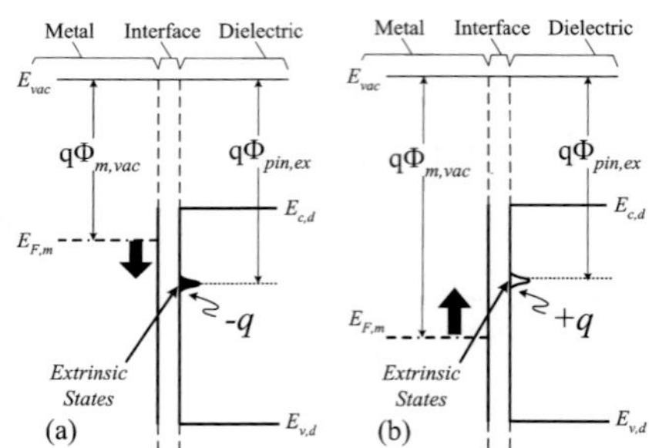

Fig. 2. Schematic energy band diagram for a metal gate on a dielectric, showing extrinsic states that pin the metal Fermi level. The energy level of the extrinsic states, i.e., pinning level, could be related to the interfacial bonding defects between the metal and the dielectric. (a) When the metal Fermi level $E_{f,m}$ is above the pinning level, i.e., $|q\Phi_m| < |q\Phi_{pin,ex}|$, a dipole is created that is charged negatively on the dielectric side. (b) When the metal Fermi level $E_{f,m}$ is below the extrinsic pinning level, i.e., $|q\Phi_m| > |q\Phi_{pin,ex}|$, a dipole is created that is charged positively on the dielectric side. For both cases, the interfacial dipole drives the metal Fermi level toward the pinning level. The conduction-band edge and the valence-band edge of the dielectric are denoted by $E_{c,d}$ and $E_{v,d}$, respectively.

of the metal-gate electrodes studied in Fig. 1 are summarized in Table I. A noteworthy trend is that the work function of metals like TiN, Ta, TaTi, TaN, TaPt, and TaTiN converge as the annealing temperature increases. This suggests metal-gate Fermi levels are pinned at about 4.7–4.8 eV below the vacuum level, assuming that the volume work function of the bulk metal is unchanged after the post-anneal. For SiO$_2$ dielectric, it is known that MIGS (or intrinsic states) at the metal–SiO$_2$ interface do not significantly modify the vacuum work function of the metal [3]. Therefore, the change in Φ_m with increasing temperature is likely to be due to the existence of a high density of localized extrinsic states. While the annealing time for these data may vary, it appears that the major factor determining the abovementioned Fermi pinning is the annealing temperature. Note that the charge neutrality level for SiO$_2$, which is used to explain the Fermi pinning induced by MIGS (or intrinsic states) [3], would not be expected to play an important role to determine the extrinsic states pinning level. Extrinsic states, usually associated with bonding defects, drive the Fermi pinning and the convergence of Φ_m. Given the chemical similarity of Ti, Hf, and Ta, it is plausible that extrinsic states with similar characteristics are formed between SiO$_2$ and these metals. It was recently reported that Fermi pinning occurs at the interface of poly-Si-HfO$_2$ (or Al$_2$O$_3$) and the interface of metal gate/metal oxide, resulting in high transistor threshold voltages [13], [14]. Our work suggests that Fermi pinning due to extrinsic states also occurs at the interface of metal gate and SiO$_2$. Additionally, we found that the extent of Fermi pinning increases with increasing annealing temperature. Elevation of the annealing temperature probably increases the density of extrinsic states and their effectiveness in pinning the Fermi level of the metal gate.

Fig. 2 shows our model for a metal-dielectric interface where extrinsic states dominate. Fig. 2(a) illustrates the case where the metal Fermi level $E_{f,m}$ is above the energy level of the extrinsic states, and, hence, the empty states at the pinning location are filled with electrons from the metal. This creates an interface dipole that is charged negatively on the dielectric side, driving $E_{f,m}$ toward the pinning position. Vice versa, for the case where the metal Fermi level $E_{f,m}$ is below the energy level of the extrinsic states, as shown in Fig. 2(b), the existing electrons at the pinning level tend to redistribute toward the metal side, resulting in an interface dipole that is charged positively at the dielectric side. The Fermi pinning effect will be less pronounced if $E_{f,m}$

is close to the pinning level of the extrinsic states. The extrinsic pinning level could be related to the interfacial bonding defects between the gate electrode and the gate dielectric, and is thus determined by both the gate electrode and the gate-dielectric materials. Pinning levels induced by Hf-Si and Al-O-Si bonds have been reported for poly-Si-HfO$_2$ and poly-Si—Al$_2$O$_3$ interfaces, respectively [13].

It should be noted that the creation of extrinsic states and the resulting Fermi pinning is not a universal phenomenon that occurs for all combinations of metal gates and gate dielectrics. Extrinsic states are absent at a defect-free interface where the metal work function is predominantly determined by intrinsic states [3]. In Fig. 1, the work function of HfN, Ru, and TaSi$_{0.31}$N$_{1.26}$ changes little with annealing temperature. This could be due to the absence of extrinsic states at the associated interfaces or, for the case of HfN, the close alignment between the $E_{f,m}$ and the Fermi pinning level.

Another way to show the impact of high-temperature process steps on the work function of metal gates is to plot metal work functions before and after high-temperature annealing on the horizontal and vertical axes, respectively. Fig. 3(a) shows the work function of metals on SiO$_2$ before and after annealing. When the work function of a metal gate does not change appreciably upon annealing at high temperature, it contributes a data point on the solid invariant line. Fig. 3(a) shows that most data points deviate from the invariant line, emphasizing the fact that the work function of most metal gates on SiO$_2$ change considerably upon annealing at high temperatures. In contrast, the work function of most metals on HfO$_2$ do not change significantly on annealing, as shown by the tight distribution of data points near the invariant line of Fig. 3(b). This may suggest that the generation of extrinsic states or creation of interfacial bonding defects upon annealing is less significant for metal gates on HfO$_2$ compared to metal gates on SiO$_2$. Considering the chemical sim-

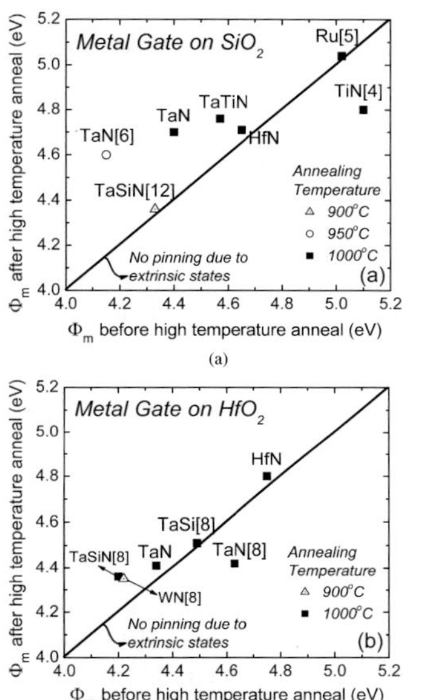

(a)

(b)

Fig. 3. Work function of metal gates on (a) SiO$_2$ and (b) HfO$_2$ before and after annealing at high temperatures. A 400 °C anneal was performed prior to the high-temperature anneal.

ilarity between the metal materials investigated in this letter, and the hafnium from the HfO$_2$ gate dielectric, little interfacial reaction is expected [8]. Therefore, the work function of the metal-gate materials on HfO$_2$ is determined by the *intrinsic states* or *MIGS* [2], [3], [14].

III. Conclusion

The dependence of the metal-gate work function on the annealing temperature was investigated. A metal-dielectric interface model that takes the role of extrinsic states into account was proposed to qualitatively explain the work function thermal instability. The creation of extrinsic states and the resulting Fermi-level pinning of the metal-gate work function is observed for certain combinations of metal-gate and gate-dielectric materials, particularly when the gate dielectric is SiO$_2$.

The effect appears to be thermodynamically driven, becoming more pronounced when the annealing temperature is higher. In general, the generation of extrinsic states upon annealing is less significant for metal gates on HfO$_2$ compared to metal gates on SiO$_2$.

References

[1] *International Technology Roadmap for Semiconductors*, (2002 Update), SIA, 2002.

[2] Y.-C. Yeo, P. Lanade, T.-J. King, and C. Hu, "Effects of high-κ gate-dielectric materials on metal and silicon gate workfunctions," *IEEE Electron Device Lett.*, vol. 23, pp. 342–344, June 2002.

[3] Y.-C. Yeo, T.-J. King, and C. Hu, "metal-dielectric band alignment and its implications for metal gate complementary metal–oxide-semiconductor technology," *J. Appl. Phys.*, vol. 92, pp. 7266–7271, 2002.

[4] J. Westlinder, T. Schram, L. Pantisano, E. Cartier, A. Kerber, G. S. Lujan, J. Olsson, and G. Groeseneken, "On the thermal stability of atomic layer deposited TiN as gate electrode in MOS devices," *IEEE Electron Device Lett.*, vol. 24, pp. 550–552, Sept. 2003.

[5] J. Lee, H. Zhong, Y. Suh, G. Heuss, J. Gurganus, B. Chen, and V. Misra, "Tunable work function dual metal gate technology for bulk and nonbulk CMOS," in *IEDM Tech. Dig.*, 2002, pp. 359–362.

[6] C. S. Kang, H. J. Cho, Y. H. Kim, R. Choi, K. Onishi, A. Shahriar, and J. C. Lee, "Characterization of resistivity and work function of sputtered-TaN film for gate electrode applications," *J. Vac. Sci. Technol. B, Microelectron. Process. Phenom.*, vol. 21, pp. 2026–2028, 2003.

[7] B. Y. Tsui and C. F. Huang, "Wide range work function modulation of binary alloys for MOSFET application," *IEEE Electron Device Lett.*, vol. 24, pp. 153–155, Mar. 2003.

[8] J. K. Schaeffer, S. B. Samavedam, D. C. Gilmer, V. Dhandapani, P. J. Tobin, J. Mogab, B. Y. Nguyen, B. E. White Jr, S. Dakshina-Murthy, R. S. Rai, Z. X. Jiang, R. Martin, M. V. Raymond, M. Zavala, L. B. La, J. A. Smith, R. Garcia, D. Roan, M. Koeeke, and R. B. Gregory, "Physical and electrical properties of metal gate electrodes on HfO$_2$ gate dielectrics," *J. Vac. Sci. Technol. B, Microelectron. Process. Phenom.*, vol. 21, pp. 11–17, 2002.

[9] C. Ren, H. Y. Yu, J. F. Kang, Y. T. Hou, D. S. H. Chan, M.-F. Li, W. D. Wang, and D.-L. Kwong, "Fermi-level pinning-induced effective work function thermal instability of TaN/SiO$_2$ gate stack," *IEEE Electron Device Lett.*, vol. 25, pp. 123–125, Mar. 2004.

[10] H. Y. Yu, J. F. Kang, C. Ren, J. D. Chen, Y. T. Hou, C. Shen, M. F. Li, D.S.H. Chan, K. L. Bera, A. C. H. Tung, and D. L. Kwong, "Robust high quality HfN–HfO$_2$ gate stack for advanced MOS device applications," *IEEE Electron Device Lett.*, vol. 25, pp. 70–72, Feb. 2004.

[11] Y. Taur and T. H. Ning, *Fundamentals of Modern VLSI Devices*. Cambridge, U.K.: Cambridge Univ. Press, 1998, p. 75.

[12] Y. S. Suh, G. P. Heuss, J. H. Lee, and V. Misra, "Effect of the composition on the electrical properties of TaSi$_x$N$_y$ metal gate electrodes," *IEEE Electron Device Lett.*, vol. 24, pp. 439–441, July 2003.

[13] C. Hobbs, L. Fonseca, V. Dhandapani, S. Samavedam, B. Taylor, J. Grant, L. Dip, D. Triyoso, R. Hegde, G. Gilmer, R. Garcia, D. Roan, L. Lovejoy, R. Rai, L. Hebert, H. Tseng, B. White, and P. Tobin, "Fermi-level pinning at the polySi/metal oxide interface," in *Symp. VLSI Tech. Dig.*, 2003, pp. 9–10.

[14] S. B. Samavedam, L. B. La, P. J. Tobin, B. White, C. Hobbs, L. Fonseaca, A. Demkov, J. Scheaffer, E. Luckowski, A. Martinez, M. Raymond, D. Triyoso, D. Roan, V. Dhandapani, R. Garcia, G. Anderson, K. Moore, H. Tseng, C. Capasso, O. Adetutu, D. Gilmer, W. Taylor, R. Hedge, and J. Grant, "Fermi-level pinning with sub-monolayer MeO$_x$ and metal gates," in *IEDM Tech. Dig.*, 2003, pp. 307–310.

New Developments in Schottky Source/Drain High-k/Metal Gate CMOS Transistors

Ming-Fu Li[1,2*] , Sungjoo Lee[1] , Shiyang Zhu[1,3] , Rui Li[1] , Jingde Chen[1] ,

Albert Chin[1] and D.L.Kwong[2,4]

[1] Silicon Nano Device Lab (SNDL), ECE Dept, National University of Singapore , Singapore 119260
[2] Institute of Microelectronics, Singapore 117685
[3] Department of Microelectronics, Fudan University, Shanghai, 200433, China
[4] ECE Dept , The University of Texas, Austin, TX 78712, USA

ABSTRACT

Recent developments in Schottky source/drain high-k/metal gate CMOS transistors (SSDT) will be presented. Bulk SSDTs with *1.5-2 nm* HfO_2 (or HfAlO) gate dielectric and HfN/TaN metal gate have been fabricated using a novel low temperature process. The Si N-SSDT using $YbSi_{2-x}$ silicide, due to the lower Schottky electron barrier of $YbSi_{2-x}$/Si, has demonstrated a record high I_{on}/I_{off} ratio of $\sim 10^7$ and a steep subthreshold slope of *75 mV/dec*. For P-SSDT, the Si SSDT using PtSi silicide S/D shows excellent I_{on}/I_{off} of $\sim 10^7 - 10^8$ and subthreshold slope of ~ 66 *mV/dec*, while the Ge SSDT using NiGe S/D shows $I_{on} \sim 5$ times larger than that of the Si counterpart with PtSi S/D, due to the lower hole Schottky barrier and the higher hole mobility of Ge channel. The implant-free low temperature process relaxes the thermal budget of high-k dielectric and metal gate Fermi pinning. More improved performances are expected by using ultra-thin-body (UTB) SOI or GOI structures, showing great potential of this low temperature process SSDTs for future sub-tenth micron CMOS technology.

I. Introduction

Future CMOS devices require metal-gate/high-K gate stacks, advanced source/drain engineering, and the use of UTB-SOI [1-3]. The series resistance of shallow source/drain junction is a serious issue for future scaling, and Schottky S/D transistor (SSDT) structure has been suggested as a potential solution [4-7]. However, the barrier height of the Schottky junction should be low enough to obtain high driving current [8,9] and to prevent two different slopes in the sub-threshold region of the MOSFETs[7,8]. This problem has not been well solved, particularly for N-SSDTs. SSDT is particularly attractive for metal gate/high-K gate stack as it avoids the use of high temperature

---

* Contact Author: email : elelimf@nus.edu.sg, Phone: 65-6874 2559

annealing process required for implanted S/D junctions, hence relaxes the thermal stability issues associated with high-K gate stack [10] and metal electrode Fermi pinning [11]. In this work, we successfully demonstrate bulk SSDTs with CVD HfO_2 (or HfAlO) high-K dielectric, PVD HfN/TaN metal gate and PtSi (for PMOS) and $YbSi_{2-x}$ (for NMOS) silicide S/D using a low temperature process. The highest temperature is *420°C* after high-K stack formation. The process can be easily extended to UTB-SOI structures.

II. Si N- and P- SSDTs , Devices Fabrication

A simplified low temperature process was used to fabricate SSDT with HfO_2 gate oxide and HfN/TaN metal gate as shown in Fig. 1.

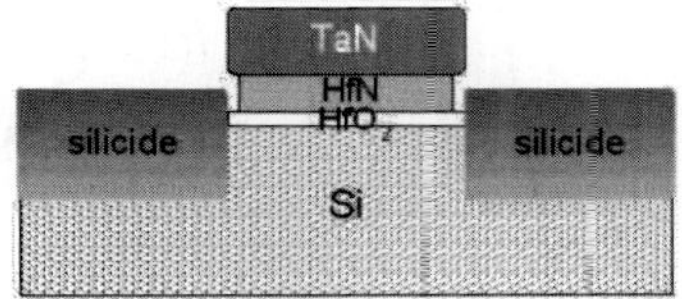

Figure 1.   Schematic of the SSDT structure

Starting substrates were n- or p- type Si(100) wafers with resistivity of *4 ~ 8 Ω·cm*. HfO_2 (*4 ~ 6 nm*) was deposited at 400°C using $Hf[OC(CH_3)_3]_4$ and O_2 in a MOCVD system, followed by an *in-situ* annealing in N_2 ambient at *700 °C*. Then HfN (*~50nm*) and TaN (*~100nm*) were deposited sequentially in a sputtering system with a base pressure of *~ 1.5×10^{-7} torr*. Wafers were patterned and subsequently etched using the standard photolithograph and dry etch processes. Immediately after the diluted HF solution (DHF) dipping, the patterned wafer was loaded into the sputtering system again. Pt (*~100nm*) for P-SSDT and Yb (or other rare earth (RE) metals: Er, Dy, Tb) (*~100nm*)/HfN (*~100nm*) stack (for N-SSDT) were deposited at room temperature. HfN was used as a capping layer to prevent RE metal oxidization during *ex-situ* annealing. Silicidation was performed by forming-gas anneal (FGA) at *420°C* for *1 hour*. For P-SSDT, un-reacted Pt is removed in hot diluted Aqua Regia solution. For N-SSDT, the HfN capping layer and un-reacted RE metal were selectively removed by wet etch in DHF (HF : H_2O = *1 : 100*) and SPM solution (H_2SO_4 : H_2O_2 = *3 : 1* at *120°C*) sequentially.

III. Si N- and P- SSDTs , Results and Discussion

Fig. 2 shows *I-V* and *C-V* curves of PtSi/n-Si Schottky diodes. Agreement of barrier height obtained from two measurement methods implies a good interface between Pt-silicide and Si.

Fig. 3 shows *I-V* characteristics of various RE silicide/p-Si(100) Schottky diodes. The Schottky hole barrier height (Φ_P^{I-V}) and the ideality factor were deduced by linear fitting of the forward *I-V* curves based on the thermal emission model [12]. Obtained values are summarized in Table 1. The hole barrier heights deduced from *C-V* curves (Φ_P^{C-V}) are also given. The YbSi$_{2-x}$/p-Si contact shows the highest hole barrier height of 0.85eV (and therefore lowest electron barrier), the lowest reverse bias leakage current, and the best rectifying property with near unity ideality factor. Other diodes have significantly higher leakage current at reverse bias, larger than unity ideality factors and larger difference between Φ_P^{I-V} and Φ_P^{C-V}, implying significant barrier height inhomogeneity [13].

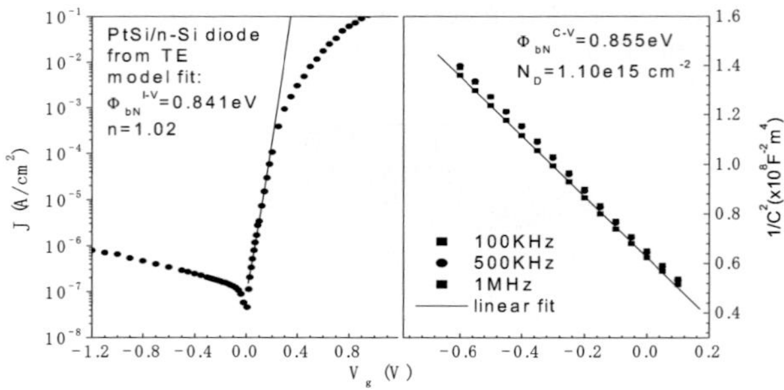

Figure 2   Room temperature *I-V* and *C-V* curves of the PtSi/n-Si(100) diodes. The deduced barrier heights are *0.84 eV (I-V)*, *0.86 eV (C-V)* for PtSi/n-Si

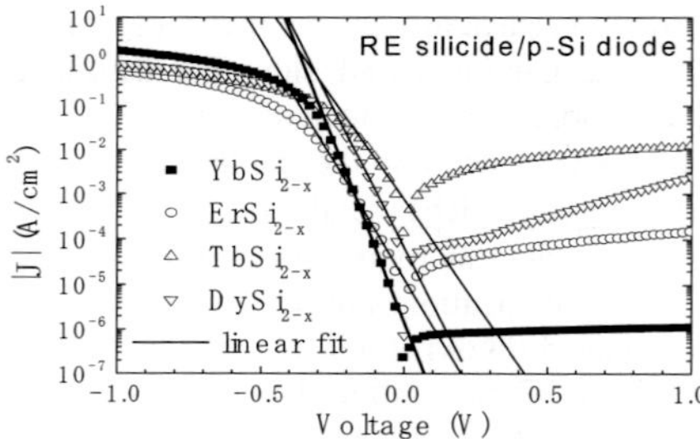

Figure 3.   Room temperature *I-V* curves of various RE silicide/p-Si(100) diodes and the linear fitting based on the thermal emission model. The deduced barrier heights and ideality factors are summarized in Table 1.

Figures 4 and 5 show the *I-V* characteristics of P- and N-SSDTs with W/L = *400 μm / 4 μm*. P-SSDT shows excellent electrical properties with I_{on}/I_{off} ~ *10⁷–10⁸*, and subthreshold slope of *66 mV/dec*. For N-SSDT, the I_{on}/I_{off} ratio reaches ~*10⁷* with one sub-threshold slope of ~ *75 mV/dec*, and its drivability is slightly larger than the corresponding P-SSDT with PtSi with the same device structure (Table 1). To our knowledge, this is the best electrical performance for N-SSDT reported so far. For comparison, Fig. 6 shows the transfer characteristics of N-SSDT with the same device structure and technology, using $ErSi_{2-x}$, $TbSi_{2-x}$ and $DySi_{2-x}$, respectively. The electrical results of these devices are summarized in Table 1.

Silicide/p-Si (100) diode	$ErSi_{2-x}$	$TbSi_{2-x}$	$DySi_{2-x}$	$YbSi_{2-x}$	*PtSi on n-Si(100)*
Hole barrier obtained by the I-V measurement Φ_p^{I-V} (eV)	0.71	0.60	0.67	**0.82**	*0.84 (electron barrier Φ_n^{I-V})*
Ideality factor in I-V	1.57	1.83	1.33	**1.04**	*1.02*
Hole Barrier obtained by C-V measurement Φ_p^{C-V} (eV)	~ 0.78	~ 0.87	~ 0.83	**~ 0.88**	*~ 0.86 (electron barrier Φ_n^{C-V})*
Averaged hole barrier $\Phi_p \cong (\Phi_p^{I-V} + \Phi_p^{I-V})/2$ (eV)	0.75	0.74	0.75	**0.85**	*0.85 (electron barrier Φ_n)*
Electron Barrier $\Phi_n \cong 1.12$ eV $- \Phi_p$ (eV)	0.37	0.38	0.37	**0.27**	*0.27 (hole barrier Φ_p)*
$I_{leakage}$ @ 1V (A/cm^2)	1.5×10^{-4}	1.2×10^{-2}	2.3×10^{-3}	**1.1×10^{-6}**	*4.6×10^{-7} @-1V*
SSDT properties: I_{on}/I_{off} ratio	$10^3 \sim 10^4$	$10^3 \sim 10^4$	$10^4 \sim 10^5$	**~ 10^7**	*~10^8*
I_{ds} @ $|V_{ds}|=V_{gs}-V_{th}= 1V$ ($\mu A/\mu m$) ($L_g = 4$ μm)	~ 1.4	~ 0.26	~ 2.5	**~ 3.4**	*~ 3.2*

Table 1. Electrical characteristics of various RE silicide/p-Si(100) contacts formed by solid-state reaction and the corresponding N-SSDT properties. The barrier heights deduced from *C-V* have relatively large deviation. The data of PtSi/n-Si and P-SSDT are also included for comparison.

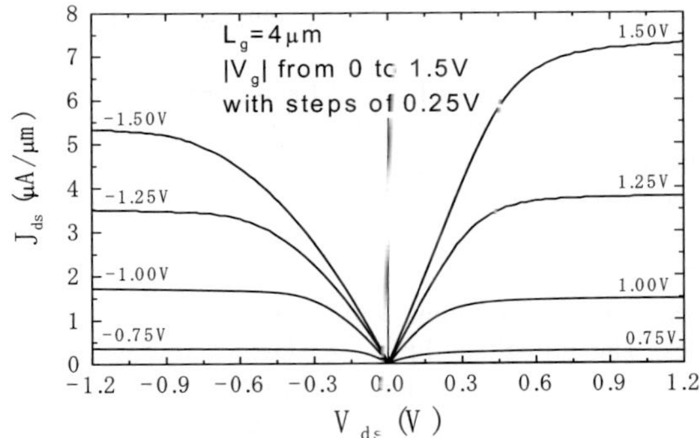

Figure 4. I_d-V_d of P-(left) and N-(right) SSDTs, (W/L) = *400/4 µm*. For P-SSDT, EOT = *2.0 nm*, V_t= *-0.56 V*; For N-SSDT, EOT = *2.5 nm*, V_t = *0.56 V*.

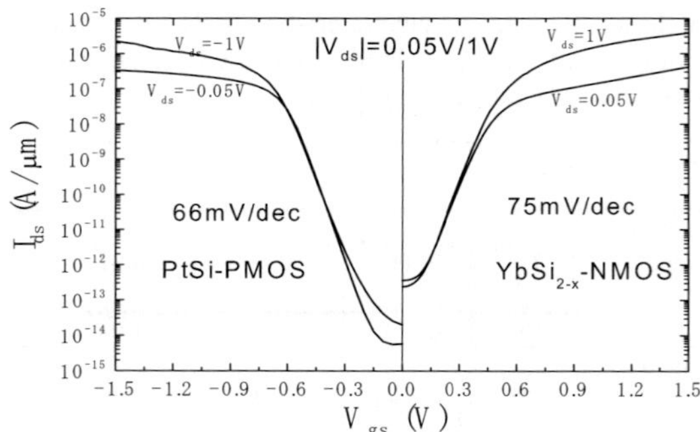

Figure 5.   Same as Figure 4, however I_d-V_g curves . For P-SSDT, $S = 66\ mV/dec$, I_{on}/I_{off} $\sim 10^7$- 10^8; For N-SSDT, $S = 75\ mV/dec$, I_{on}/I_{off} is around 10^7.

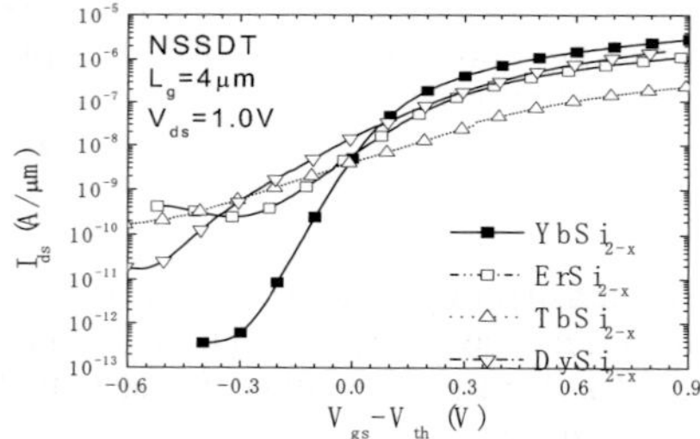

Figure 6.   The transfer characteristics of N-SSDTs with various RE silicides. All devices have the same size of $W/L=400\mu m/4\mu m$, and were fabricated by the same process.

Observing in the microscope, the surfaces of ErSi$_{2-x}$ and DySi$_{2-x}$ contain many square pits with micrometer size. While there are no such pits on the surface of YbSi$_{2-x}$ and TbSi$_{2-x}$ even though their surface roughness is still quite large as measured by AFM. Fig. 7 (top) shows the cross sectional XTEM image of the final N-SSDT fabricated by our simplified one-mask process.

Besides the low electron barrier height, YbSi$_{2-x}$ has better film quality than other rare earth metal silicides. The growth of ErSi$_{2-x}$ or DySi$_{2-x}$ during solid-state reaction of deposited rare earth metal and substrate Si (100) is strongly nucleation preferred, resulting in a non-uniform, columnar growth of the layer with rough surface and interface [14,15]. The formed silicide has been reported to be ErSi$_{1.7}$ or DySi$_{1.7}$ due to the Si vacancy in the silicide film [7]. In the case of YbSi$_{2-x}$, the formed silicide has been reported to be YbSi$_{1.8}$ [16]. Our XRD (x-ray diffraction) in Fig. 8 and EDX (energy dispersive X-ray) (not shown here) analysis also confirm that the formed film is $YbSi_{1.8}$. Less Si vacancy may cause more uniform silicide. From Fig. 7, the grain size of the polycrystalline $YbSi_{1.8}$ is about *5~10 nm* and the grain growths approximately along Si[110] axis. Columnar growth, as in the cases of $ErSi_{2-x}$ and $DySi_{2-x}$, was not found.

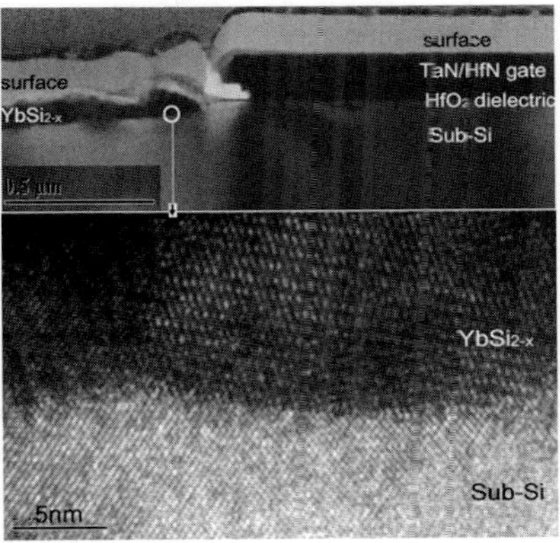

Figure 7. The cross sectional TEM image of the final N-SSDT fabricated by the simplified one-mask process (top), a "hole" between the S/D and the gate acts as a sidewall spacer [17], and the high resolution XTEM image of the polycrystalline $YbSi_{2-x}$ / Si(100) contact (bottom). Even though the quite rough $YbSi_{2-x}$ surface, which is probably affected by the SPM solution during the selective etching step, the polycrystalline $YbSi_{2-x}$ / Si interface is quite smooth and flat.

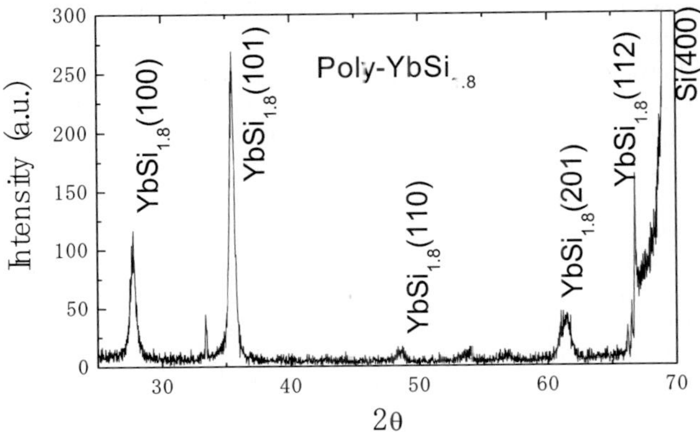

Figure 8. XRD analysis of Yb Silicide film

The RE silicide property is sensitive to the oxygen contamination. For $ErSi_{2-x}$ silicide, $\Phi_N = 0.28\ eV$ when grown in ultra-high-vacuum (UHV) condition [18], however Φ_N becomes higher when grown in normal vacuum level as reported in this work and other paper [19,20]. Our result shows that $YbSi_{2-x}$ grown in normal vacuum condition has better rectifying characteristics than $ErSi_{2-x}$ grown in UHV condition. It implies that $YbSi_{2-x}$ is not so sensitive to oxygen as $ErSi_{2-x}$, or even better rectifying property of $YbSi_{2-x}/Si$ contact may be obtained if it is grown in UHV condition.

IV. Ge P- SSDT

Using Ge substrate to replace Si substrate for P-SSDT has two-fold advantages of higher channel hole mobility and lower Schottky hole barrier of germanide/Ge contact due to the smaller ionization energy of Ge valence electrons compared with those of Si.

The starting substrates are n-type Ge (100) wafers with resistivity of $2 \sim 5\ \Omega\cdot cm$. After cleaning in a diluted HNO_3 solution and dipping in a diluted HF (DHF) solution, the wafers were loaded in a MOCVD system and annealed in pure NH_3 ambient at $600^{o}C$ for $30\ s$ for surface nitridation. Then $\sim 6\ nm$ HfAlO was deposited at $400^{o}C$, followed by an *in-situ* annealing in N_2 ambient at $600^{o}C$ for $1\ min$. The wafers were transferred into a sputtering system for HfN ($\sim 50\ nm$) and TaN ($\sim 100\ nm$) depositions with the same conditions as for metal gate of Si-SSDTs. The deposited wafers were patterned using conventional photolithography and reactive ion etching procedures. Immediately after dipping in the DHF solution to remove the remaining HfAlO film in the S/D region, the patterned wafers were loaded in the sputtering system again and a Ni film of $\sim 100\ nm$ was deposited. The Ni germanidation was performed by rapid thermal annealing (RTA) at $600^{o}C$ for $1\ min$. Then unreacted Ni was removed by wet etching in RCA1 (NH_4OH : H_2O_2 : $H_2O = 1 : 2 : 5$) solution. Because both NiGe and Ge substrate will be attacked by the RCA1 solution slowly, the selective etching time should be carefully optimized.

Fig. 9 shows the measured *I-V* curve of the Schottky diode with NiGe/n-Ge(100) contact. The traditional thermionic emission (TE) model [12] was used to fit the experimental forward current *I-V* data, from which apparent Schottky barrier height (Φ_n), ideality factor (n) and series resistance (R_s) were extracted to be $0.50\ eV$, 1.49 and $110\ \Omega$, respectively. The corresponding hole barrier height (Φ_p) can be calculated to be $\sim 0.16\ eV$, assuming that the sum of electron and hole barrier heights approximately equals to the Ge gap energy ($\sim 0.66\ eV$). Fig. 10 shows the I_d-V_d curves of Ge-PSSDT with channel width / length = $400\ \mu m$ / $8\ \mu m$. The threshold voltage is $\sim -0.41\ V$ from the linear fitting of the $I_{ds}^{1/2} \sim V_g$ curve at $V_{ds} = -0.1\ V$. The drain current of the device at $V_d = V_g$-$V_{th} = -1\ V$ is $\sim 6.5\ \mu A/\mu m$. For comparison, I_d-V_d curves of control silicon P-SSDTs with PtSi S/D and with the same device size are also shown in Fig. 10. For the control Si PSSDT, EOT = $2.0\ nm$, $V_{th} = -0.50\ V$ and the drain current at $V_d = V_g$-$V_{th} = -1\ V$ is $\sim 2.6\ \mu A/\mu m$. Therefore, the Ge device has $\sim 5\times$ larger drain drivability than the Si counterpart if they are scaled to the same EOT. Although the drive current improvement can be partly attributed to the fact that Ge has higher hole mobility than Si, the main reason is believed to be due to the smaller hole barrier between source and channel of Ge-PSSDT ($\sim 0.16\ eV$) than that of

Si-PSSDT (Φ_p = ~ *0.24 − 0.26 eV* for PtSi/Si [17,21]). Fig 10 (right) shows the I_d-V_g curves of Ge-PSSDT. It shows relatively large off-state current, I_{off}. The I_{on}/I_{off} ratio is *10^2 ~ 10^3*, about *5* orders of magnitude smaller than that of Si-PSSDT. The large I_{off} is mainly caused by the relatively low electron barrier height (Φ_n, ~ *0.5* eV in our experiment) at the drain/substrate contact that forms a reverse-biased NiGe/n-Ge diode. The reverse saturation current of the contact with an area of *1×10^{-4} cm²* is calculated to be ~ *4×10^{-6} A*, close to the value of I_{off} in Fig. 1C. Conventional Ge MOSFET with P-N junction S/D suffers from the same problem because the narrow gap energy of Ge also results in large PN junction leakage [22-25]. The large I_{cff} can be effectively reduced by using ultra-thin GOI (Germanium On Insulator) substrate because the contact area can be dramatically reduced.

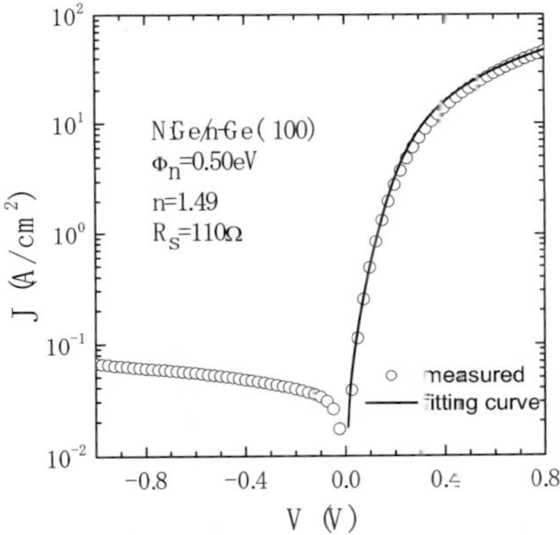

Figure 9.   I-V curve of the NiGe/n-Ge(100) diode, the solid line is a fitting curve based on the thermal emission (TE) model. Inset is the cross sectional TEM image of the NiGe/n-Ge(100) contact.

It has been pointed out that Schottky barrier heights of metals and germanides on n-Ge are pinned at between *0.54* and *0.61 eV* over a wide range of metal work function [26]. Erbium germanide (ErGe) was also used to fabricate Ge-PSSDTs in our experiment and show similar electrical characteristics as displayed in Fig.10. The quality of the NiGe film and the NiGe/Ge interface is sensitive to the annealing condition. Furnace annealing at *420°C* results in poorer rectifying property than that after RTA. However, very few data have been reported in the literature about the formation of germanide by solid-state reaction as well as the Schottky barrier properties of various metal/Ge or germanide/Ge contacts.   Systematic studies are still on-going in order to improve the quality of germanide/Ge contact by optimizing the fabrication parameters.

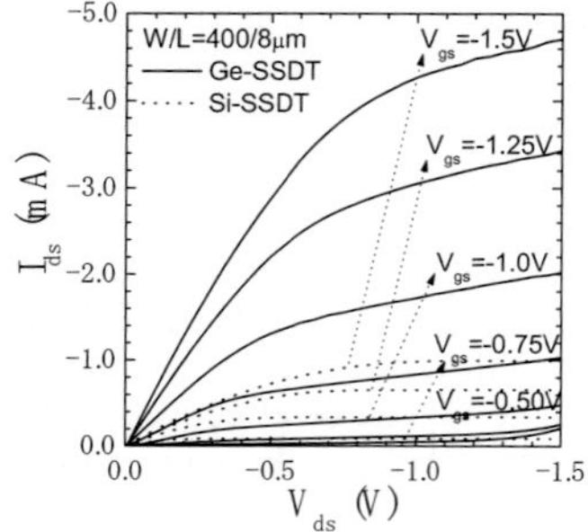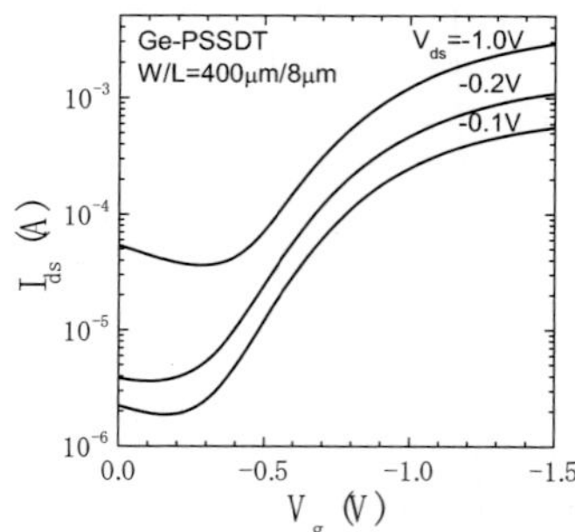

Figure 10. I_d-V_d (left) and I_d-V_g (right) curves of Ge-PSSDT with NiGe S/D. The channel width and length are *400 μm* and *8 μm*, EOT = ~ *3.8 nm*, V_{th} = ~ *-0.41 V*. For comparison, the I_d-V_d curves of corresponding Si-PSSDT with PtSi S/D are also shown (dotted lines), which has the same size and EOT = *2.0 nm*, V_{th} = ~ *-0.50 V*.

V. Conclusion

The YbSi$_{2-x}$ is a very promising candidate for Si N-SSDT as it provides a high drive current with a very low leakage current. It is probably due to the low electron barrier height of the YbSi$_{2-x}$/Si Schottky contact and smooth YbSi$_{2-x}$/Si interface. The NiGe was used to fabricate Ge P-SSDT with drive current 5 times larger than Si P-SSDT using Pt silicide . This is due to two-fold advantage of higher hole mobility and smaller valence electron ionization energy and therefore lower hole Schottky barrier in the Ge, comparing with Si. It can be expected that I_{off} can be reduced appreciably by using UTB-SOI and GOI due to the reduction of Schottky contact area. This low temperature process SSDTs show great potential in the future sub-tenth micron CMOS technology.

Acknowledgement : This project was supported by Singapore A-STAR R263-000-267-305 grant . We like to thank Chunxiang Zhu for useful discussion and Chih-Hang Tung and Anyan Du for TEM work.

References

1. H. Iwai, *IED Tech Dig.,* (2004)11.
2. C. Hu, *Symp. VLSI,* (2004) 4.
3. T. H. Ning, *Symp. VLSI,* (2003) 6.
4. M. P. Lepselter and S. M. Sze, *Proc. IEEE,* **56,** (1968) 1400.
5. M. Nishisaka, S. Matsumoto and T. Asano, *International Conference on Solid State Devices and Materials*, (2002) 586.
6. H-C. Lin, M. F. Wang, F. J. Hou, H. N. Lin, C. Y. Lu, J. T. Liu and T. Y. Huang, *IEEE EDL,* **24,** (2003) 102.
7. J. Kedzierski, P. Xuan, E. H. Anderson, J. Bokor, T. J. King, C. H. Hu, *IEDM Tech. Dig.,*(2000) 57.
8. W. Satoh, A. Itoh, S. Yamagami and M. Asada, Jpn. J. Appl. Phys. **38,** part 1, (1999) 6226.
9. S. Y. Zhu et al, *IEEE. EDL,* **25,**(2004) 268.
10. C. S. Kang, et al, *Symp. VLSI,* (2002) 146.
11. H.Y.Yu et al, *IEEE EDL,* **25,**(2004) 70.
12. E. H. Rhoderick, R. H. Williams, Metal-Semiconductor contacts, 2nd ed.,Clarendon Press, Oxford (1988)
13. S. Y. Zhu, R. L. Van Meirhaeghe, et al., Solid-State Electron **44,** (2000)663.
14. C. H. Luo, L. J. Chen, J. Appl. Phys., **82,** (1997)3803.
15. A. Travlos, N. Salamouras, N. Boukos, J. Physics and Chemistry of Solids, **64,** (2003) 87.
16. K. S. Chi, L. J. Chen, *Materials Science in Semiconductor Processing,* **4,**(2001) 269.
17. S. Y. Zhu, et al., *Solid-State Electronics,* **43,** (2004)1987.
18. P. Muret, T. A. N. Tan, N. Frangis, J. van Landuyt, *Phys. Review* B, **56,** (1997) 9286.
19. G. Kaltsas, A. Travlos, N. Salamouras, A. G. Nassiopoulos, P. Revva, A. Traverse, *Thin Solid Film,* **275,** (1996) 87.
20. Z. Xu, in *Properties of Metal Silicides,* edited K. Maex and M. Van Rossum, INSPEC, London, (1995) 217.
21. V. W. L. Chin, M. A. Green, J. W. V. Storey, *Solid-state Electronics,* **36,**(1993)1107.
22. S. C. Martin, L. M. Hitt, J. J. Rosenberg, *IEEE EDL,***10,** (1989) 325.
23. C. O. Chui, H. Kim, D. Chi, B. B. Triplett, P. C. McIntyre, K. C. Saraswat, *IEDM Tech. Dig.,* (2002) 437.
24. D. S. Yu, C. H. Huang, A. Chin, C. Zhu, M. F. Li, B. J. Cho, D. L. Kwong, *IEEE EDL,* **25,** (2004) 138.
25. A. Ritenour, S. Yu, M. L. Lee, N. Lu, W. Bai, A. Pitera, E. A. Fitzgerald, D. L. Kwong, D. A. Antoniadis, *IEDM Tech. Dig.,*(2003) 433.
26. C. C. Han, E. D. Marshall, F. Fang, L. C. Wang, S. S. Lau, D. Voreades, *J. Vac. Sci. Technol. B,* **6,** (1988)1662.

IEEE ELECTRON DEVICE LETTERS, VOL. 26, NO. 2, FEBRUARY 2005   81

Germanium pMOSFETs With Schottky-Barrier Germanide S/D, High-κ Gate Dielectric and Metal Gate

Shiyang Zhu, *Member, IEEE*, Rui Li, S. J. Lee, *Member, IEEE*, M. F. Li, *Senior Member, IEEE*, Anyan Du, Jagar Singh, Chunxiang Zhu, *Member, IEEE*, Albert Chin, *Senior Member, IEEE*, and D. L. Kwong, *Senior Member, IEEE*

Abstract—Schottky-barrier source/drain (S/D) germanium p-channel MOSFETs are demonstrated for the first time with HfAlO gate dielectric, HfN–TaN metal gate and self-aligned NiGe S/D. The drain drivability is improved over the silicon counterpart with PtSi S/D by as much as $\sim$5 times due to the lower hole Schottky barrier of the NiGe–Ge contact than that of PtSi–Si contact as well as the higher mobility of Ge channel than that of Si.

Index Terms—Germanium, high-κ, metal gate, MOSFET, Schottky.

I. Introduction

GERMANIUM is an attractive channel material due to its high low-field carrier mobility. Germanium p-channel MOSFETs with enhanced mobility have been demonstrated using germanium oxynitride, ZrO_2, Al_2O_3 and HfO_2 as the gate dielectric [1]–[4]. Another technology bottleneck for future scaling of MOSFET is the fabrication of ultra-shallow source/drain (S/D) with low series resistance [5]. A Schottky-barrier S/D transistor (SSDT) structure has been proposed to solve this problem [6], [7]. However, an SSDT is difficult to use to achieve high drive current due to the relatively high potential barrier (Schottky barrier) between the source and the channel [8]. This problem may be overcome, or at least alleviated, by using a Ge substrate because of the low Schottky-barrier height of germanide–Ge contact and the high carrier mobility of Ge. In this letter, p-channel SSDTs with HfAlO gate dielectrics, HfN–TaN metal gates, and NiGe S/Ds are demonstrated for the first time using a simplified low-temperature process. The highest temperature in the entire fabrication process was 600 °C.

II. MOS Device Fabrication

The starting substrates are N-type Ge (100) wafers with a resistivity of $2 \sim 5\ \Omega \cdot$cm. After cleaning in a diluted HNO_3 solution and dipping in a diluted HF (DHF) solution, the wafers were loaded in a metal–organic chemical vapor deposition system and annealed in pure NH_3 ambient at 600 °C for 30 s for surface nitridation. Then $\sim$6 nm HfAlO was deposited at 400 °C, followed by an *in situ* annealing in N_2 ambient at 600 °C for 1 min. The wafers were transferred into a sputtering system with a base pressure of $\sim 1.5 \times 10^{-7}$torr. HfN ($\sim$50 nm) and TaN ($\sim$100 nm) were deposited sequentially at room temperature as a metal gate electrode [9]. The deposited wafers were patterned using conventional photolithography and reactive ion etching procedures. Immediately after dipping in the DHF solution to remove the remaining HfAlO film in the S/D region, the patterned wafers were loaded in the sputtering system again and a Ni film of $\sim$100 nm was deposited. Because HfN can be etched by DHF, but TaN cannot, a "hole" between S/D and gate was formed due to the lateral etching of the HfN layer of the HfN–TaN gate stack during the DHF dipping, which acts as a spacer to separate the gate and S/D [10], [11]. Each transistor was surrounded by a guard ring, thus can be electrically separated from other devices, as shown in the inset of Fig. 4. The Ni germanidation was performed by rapid thermal annealing (RTA) at 600 °C for 1 min. Then, unreacted Ni was removed by wet etching in RCA1 ($NH_4OH : H_2O_2 : H_2O = 1 : 2 : 5$) solution. Because both the NiGe and Ge substrates will be attacked by the RCA1 solution slowly, the selective etching time should be carefully optimized.

III. Device Characterization and Discussion

Fig. 1 shows the gate capacitance–voltage (C–V) characteristics measured at 1 MHz as well as the cross-sectional transmission electron microscope (TEM) image of the TaN–HfN–HfAlO–n-Ge(100) gate stack of the final PSSDT. The TEM picture shows smooth interface between HfAlO and Ge substrate and the physical thickness of the amorphous HfAlO film is $\sim$5 nm. However, the equivalent SiO_2 thickness (EOT) extracted from the accumulation capacitances is $\sim$3.8 nm. The well-behaved C–V with small hysteresis of 18 mV was observed with no significant frequency dispersion, implying that HfAlO is a potential gate dielectric for Ge MOSFETs.

Manuscript received October 26, 2004; revised November 17, 2004. The review of this letter was arranged by Editor B. Yu.

S. Zhu is with the Silicon Nano Device Laboratory, Department of Electrical and Computer Engineering, National University of Singapore, Singapore 119260, and also with the Department of Microelectronics, Fudan University, Shanghai 200433, China.

R. Lui, S. J. Lee, and C. Zhu are with the Silicon Nano Device Laboratory, Department of Electrical and Computer Engineering, National University of Singapore, Singapore 119260.

M. F. Li is with the Silicon Nano Device Laboratory, Department of Electrical and Computer Engineering, National University of Singapore, Singapore 119260, and also with the Institute of Microelectronics, Singapore 117685 (e-mail: elelimf@nus.edu.sg).

A. Du and J. Singh are with the Institute of Microelectronics, Singapore 117685.

A. Chin is with the Department of Electronics Engineering, National Chiao-Tung University, Hsinchu 300, Taiwan, R.O.C.

D. L. Kwong is with the Department of Electrical and Computer Engineering, The University of Texas, Austin, TX 78712 USA.

Digital Object Identifier 10.1109/LED.2004.841462

0741-3106/$20.00 © 2005 IEEE

82   IEEE ELECTRON DEVICE LETTERS, VOL. 26, NO. 2, FEBRUARY 2005

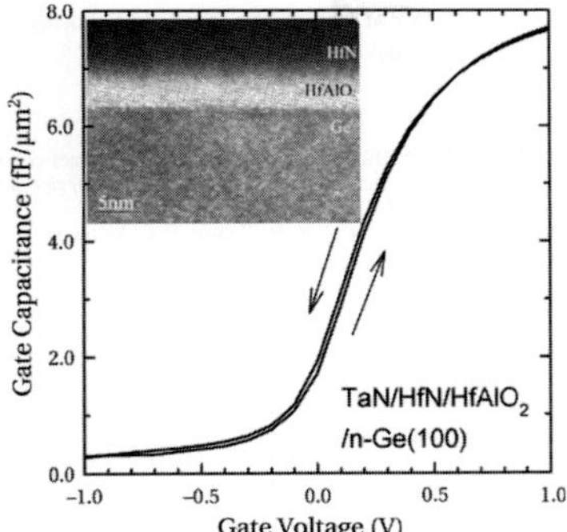

Fig. 1. Capacitance measured at 1 MHz on the fully processed Ge-PSSDT. Inset shows the cross-sectional TEM image of the TaN–HfN-HfAlO$_2$/n–Ge(100) stack.

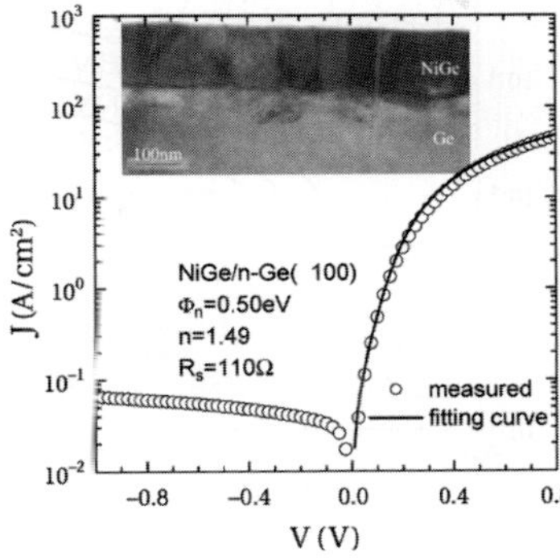

Fig. 2. *I–V* curves of the NiGe–n-Ge(100) diode, the solid line is a fitting curve based on the TE model. Inset is the cross-sectional TEM image of the NiGe–n-Ge(100) contact.

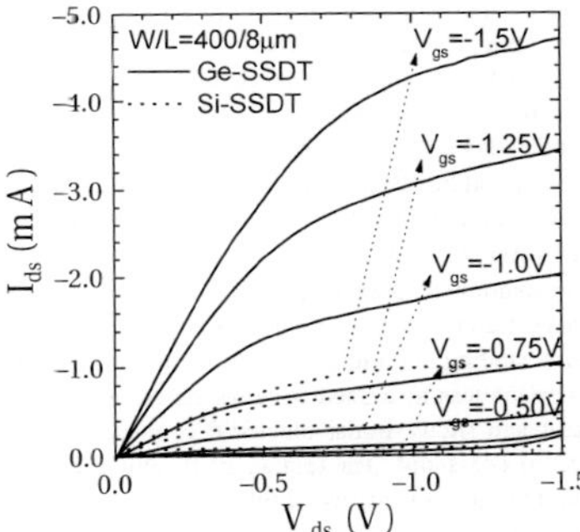

Fig. 3. I_d–V_d curves of the Ge-PSSDT with NiGe S/D. The channel width and length are 400 and 8 μm, EOT =∼ 3.8 nm, V_{th} =∼ −0.41 V. For comparison, the I_d–V_d curves of corresponding Si-PSSDT with PtSi S/D are also shown (dotted lines), which has the same size and EOT = 2.0 nm, V_{th} =∼ −0.50 V.

Fig. 2 shows the measured current–voltage (*I–V*) curve of the Schottky diode with NiGe–n-Ge(100) contact and its cross-sectional TEM image. The traditional thermionic emission (TE) model [12] was used to fit the experimental forward current *I–V* data, from which apparent Schottky-barrier height (Φ_n), ideality factor (n) and series resistance (R_s) were extracted to be 0.50 eV, 1.49 and 110 Ω, respectively. The corresponding hole barrier height (Φ_p) can be calculated to be ∼0.16 eV, assuming that the sum of electron and hole barrier heights approximately equals to the Ge gap energy (∼0.66 eV). The reverse leakage current of the Schottky diode at −1 V is ∼ 6×10^{-2} A/cm^2. The value is reasonable in view of germanide–Ge Schottky contact [13]. The NiGe layer has a thickness of ∼140 nm, thinner than the expected thickness of 250 nm if the deposited Ni (100 nm) is completely reacted with Ge to form NiGe due to the partially etching of NiGe by the RCA1 solution during the selective etching process. The TEM image shows that there is an intermediate layer between NiGe and Ge substrate with a quite rough interface between the intermediate layer and Ge substrate. Energy dispersive X-ray (EDX) analysis shows that this intermediate layer is a Ge-rich NiGe layer. This may be the main reason for the significantly large ideality factor, the relatively high reverse leakage current and the low apparent barrier height of the NiGe–n-Ge diode compared with the reported value [13]. It is expected that the rectifying property of the NiGe–Ge contact can be improved by reducing this intermediate layer.

Fig. 3 shows the I_d–V_d curves of Ge-PSSDT with channel width/length = $400/8$ μm. The threshold voltage is ∼ −0.41 V from the linear fitting of the $I_{ds}^{1/2}$ ∼ V_g curve at V_{ds} = −0.1 V. The drain current of the device at $V_d = V_g$–V_{th} = −1 V is ∼ 6.5 μA/μm. For comparison, control silicon PSSDTs with PtSi S/D were also fabricated using the similar process and same device size [10], [11], their I_d–V_d curves are also shown in Fig. 3. For the control Si PSSDT, EOT = 2.0 nm, V_{th} = −0.50 V and the drain current at $V_d = V_g$–V_{th} = −1 V is ∼ 2.6 μA/μm. Therefore, the Ge device has ∼ 4.8× larger drain drivability

than the Si counterpart if they are scaled to the same EOT. Although the drive current improvement can be partly attributed to the fact that Ge has higher hole mobility than Si, the main reason is believed due to the smaller hole barrier between source and channel of Ge-PSSDT (∼0.16 eV) than that of Si-PSSDT (Φ_p =∼ 0.24 − 0.26 eV for PtSi/Si [11], [14]). In the case of Si-PSSDT, it is probably difficult to reach the hole barrier height as low as that of Ge [15] because of the large hole band-offset (0.4 eV) or difference of valence electron ionization energy (0.2 eV) between Ge and Si [16]. This is one of the major motivations of Ge-PSSDT. Fig. 4 shows the I_d–V_g curves of Ge-PSSDT. It shows relatively large off-state current, I_{off}. The I_{on}/I_{off} ratio is $10^2 \sim 10^3$, about five orders of magnitude smaller than that of Si-PSSDT. The large I_{off} is mainly caused by the relatively low electron barrier height (Φ_n, ∼0.5 eV in our experiment) at the drain/substrate contact that forms a reverse-biased NiGe–n-Ge

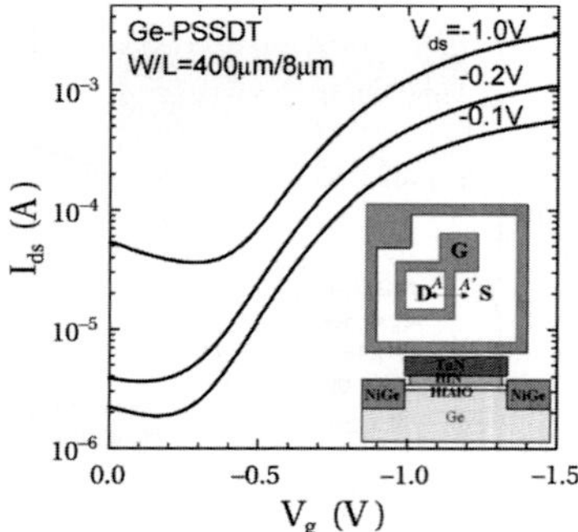

Fig. 4. I_d–V_g curves of the Ge-PSSDT with NiGe S/D, V_{ds} are -0.1, -0.2, and -1 V, respectively. The I_{on}/I_{off} ratio reaches $10^2 \sim 10^3$. The inset shows schematics of the device layer and cross-sectional structure.

diode. The reverse saturation current of the contact with an area of 1×10^{-4} cm^2 is calculated to be $\sim 4 \times 10^{-6}$ A, close to the value of I_{off} in Fig. 4. Conventional Ge MOSFET with p–n junction S/D suffers from the same problem because the narrow gap energy of Ge also results in large p–n junction leakage [1]–[4]. The large I_{off} can be effectively reduced by using ultrathin germanium on insulator substrate [3] because the contact area can be dramatically reduced.

It has been pointed out that Schottky-barrier heights of metals and germanides on n-Ge are pinned at between 0.54 and 0.61 eV over a wide range of metal work function [13]. Erbium germanide (ErGe) was also used to fabricate Ge-PSSDTs in our experiment and show similar electrical characteristics as displayed in Figs. 3 and 4. The quality of the NiGe film and the NiGe–Ge interface is sensitive to the annealing condition. Furnace annealing at 420 °C results in poorer rectifying property than that after RTA. However, very few data have been reported in the literature about the formation of germanide by solid-state reaction as well as the Schottky-barrier properties of various metal–Ge or germanide–Ge contacts. Systematically studies are still on-going in order to improve the quality of germanide–Ge contact by optimizing the fabrication parameters.

In conclusion, the first germanium PSSDT with HfAlO gate dielectric, HfN/TaN metal gate and NiGe S/D was fabricated using a simplified low temperature process. The drive current is about $4.8\times$ larger than that of the silicon counterpart due to the high hole mobility of Ge and the low hole Schottky-barrier height of the germanide–Ge contact.

REFERENCES

[1] S. C. Martin, L. M. Hitt, and J. J. Rosenberg, "P-channel germanium MOSFETs with high channel mobility," *IEEE Electron Device Lett.*, vol. 10, no. 7, pp. 325–326, Jul. 1989.

[2] C. O. Chui, H. Kim, D. Chi, B. B. Triplett, P. C. McIntyre, and K. C. Saraswat, "A sub-400 °C germanium MOSFET technology with high-κ dielectric and metal gate," in *IEDM Tech. Dig.*, 2002, pp. 437–440.

[3] D. S. Yu, C. H. Huang, A. Chin, C. Zhu, M. F. Li, B. J. Cho, and D. L. Kwong, "Al$_2$O$_3$–Ge-on-insulator n- and p-MOSFETs with filly NiSi and NiGe dual gates," *IEEE Electron Device Lett.*, vol. 25, no. 3, pp. 138–140, Mar. 2004.

[4] A. Ritenour, S. Yu, M. L. Lee, N. Lu, W. Bai, A. Pitera, E. A. Fitzgerald, D. L. Kwong, and D. A. Antoniadis, "Epitaxial strained germanium p-MOSFETs with HfO$_2$ gate dielectrics and TaN gate electrode," in *IEDM Tech. Dig.*, 2003, pp. 433–436.

[5] C. O. Chui, K. Gopalakrishnan, P. B. Griffin, J. D. Plummer, and K. C. Saraswat, "Activation and diffusion studies of ion-implanted p and n dopants in germanium," *Appl. Phys. Lett.*, vol. 83, pp. 3275–3277, 2003.

[6] M. P. Lepselter and S. M. Sze, "SB-IGFET: An insulated-gate field-effect transistor using Schottky-barrier contacts for source and drain," *Proc. IEEE*, vol. 56, pp. 1400–1401, 1968.

[7] J. Kedzierski, P. Xuan, E. H. Anderson, J. Bokor, T. J. King, and C. H. Hu, "Complementary silicide source/drain thin-body MOSFETs for the 20-nm gate length regime," in *IEDM Tech. Dig.*, 2000, pp. 57–60.

[8] W. Saitoh, A. Itoh, S. Yamagami, and M. Asada, "Analysis of short-channel Schottky source/drain metal-oxide-semiconductor field-effect transistor on silicon-on-insulator substrate and demonstration of sub-50-nm n-type devices with metal gate," *Jpn. J. Appl. Phys.*, vol. 38, pp. 6226–6231, 1999.

[9] H. Y. Yu, J. F. Kang, J. D. Chen, C. Ren, Y. T. Hou, S. J. Whang, M. F. Li, D. S. H. Chan, K. L. Bera, C. H. Tung, A. Du, and D. L. Kwong, "Robust high quality HfN/HfO$_2$ gate stack for advanced CMOS devices," *IEEE Electron Device Lett.*, vol. 25, no. 2, pp. 70–72, Feb. 2004.

[10] S. Y. Zhu, H. Y. Yu, S. J. Whang, J. H. Chen, C. Shen, C. Zhu, S. J. Lee, M. F. Li, D. S. H. Chan, W. J. Yoo, A. Du, C. H. Tung, J. Singh, A. Chin, and D. L. Kwong, "Schottky-barrier S/D MOSFETs with high-K gate dielectrics and metal-gate electrode," *IEEE Electron Device Lett.*, vol. 25, no. 5, pp. 268–270, May 2004.

[11] S. Y. Zhu, H. Y. Yu, J. D. Chen, S. J. Whang, J. H. Chen, C. Shen, C. Zhu, S. J. Lee, M. F. Li, D. S. H. Chan, W. J. Yoo, A. Du, C. H. Tung, J. Singh, A. Chin, and D. L. Kwong, "Low temperature MOSFET technology with Schottky-barrier source/drain, high-κ gate dielectric and metal gate electrode," *Solid State Electron.*, vol. 48, pp. 1987–1992, 2004.

[12] S. Y. Zhu, R. L. Van Meirhaeghe, G. P. Ru, and B. Z. Li, "Effects of the annealing temperature on Ni silicide–n-Si(100) Schottky contacts," *Solid State Electron.*, vol. 48, pp. 29–35, 2004.

[13] C. C. Han, E. D. Marshall, F. Fang, L. C. Wang, S. S. Lau, and D. Voreades, "Barrier height modification of metal/germanium Schottky diodes," *J. Vac. Sci. Technol. B, Microelectron. Process. Phenom.*, vol. 6, no. 6, pp. 1662–1666, 1988.

[14] V. W. L. Chin, M. A. Green, and J. W. V. Storey, "Current transport mechanisms studied by I–V–T and IR photoemission measurements on a p-doped PtSi Schottky diode," *Solid State Electron.*, vol. 36, no. 8, pp. 1107–1116, 1993.

[15] E. H. Rhoderick and R. H. Williams, *Metal-Semiconductor Contacts*, 2nd ed. Oxford, U.K.: Clarendon, 1988.

[16] W. A. Harrison, *Electronic Structure and the Properties of Solids*. San Francisco, CA: Freeman, 1980.

ECS Transactions, 1 (5) 717-730 (2006)
10.1149/1.2209318, copyright The Electrochemical Society

NEW INSIGHTS IN Hf BASED HIGH-k GATE DIELECTRICS IN MOSFETs

Ming-Fu Li[1,2*], Chunxiang Zhu[1], C.Shen[1], X.F.Yu[1], X.P.Wang[1], Y.P.Feng[3], A.Y.Du[2], Y.C.Yeo[1], G.Samudra[1], Albert Chin[1] and D.L.Kwong[1,2]

[1] Silicon Nano Device Lab (SNDL) , ECE Dept, National University of Singapore , Singapore 117576 elelimf@nus.edu.sg
[2] Institute of Microelectronics, Singapore 117685
[3] Department of Physics, National University of Singapore, 117542

ABSTARCT

I. Two different traps (fast and slow) in HfO_2 gate dielectric are identified. For the slow traps, the negative U (-U) property of trap is proposed and confirmed by the first principle calculation. Each trap can trap two electrons or two holes and lower the trap energy due to a large lattice relaxation. The observed experimental result of reduction in slow component of bias temperature instability (BTI) degradation with an increase in stress frequency can be simulated with excellent agreement, based on the concept of -U. A fast BTI component is also observed. The fast dynamic BTI degradation is increased with an increase in stress frequency. It is due to the existence of fast traps and can be simulated by the conventional trapping/de-trapping equations. The fast trap is a standard conventional trap.

II. Mixing Ta and La into HfO_2 to form HfTaO and HfLaO gate dielectric have been studied systematically. Comparing to the HfO_2 gate dielectric, the HfTaO and HfLaO have the advantages of much higher crystallization temperature, much lower charge trapping as well as the BTI degradation, and increased channel mobility. In addition, variation of La concentration in HfLaO/TaN or HfLaO/HfN gate stack can effectively tune the metal work function continuously from mid gap to 4eV. Possible physical explanation for these interesting properties are discussed.

1. INTRODUCTION

The serious problems in use of HfO_2 as a gate dielectric in future CMOS devices are mainly: (1) the trap charging effect and the corresponding BTI degradation in both n- and p- MOSFETs [1-6]; (2) the degradation of channel mobility, probably due to the remote Coulomb [7] and remote phonon[8] scattering from the dielectric; (3) Difficulty in tuning effective work function for HfO_2/metal gate stack, suitable for n- and p- MOSFETs, with high enough thermal budget [9-11]. In this paper, we present some new insights and property improvement of Hf based high-k gate dielectric to address these problems.

2. FAST AND SLOW TRAPS IN HfO_2 GATE DIELECTRICS

---

* Contact Author: email: elelimf@nus.edu.sg, phone: 65-687 42559

ECS Transactions, 1 (5) 717-730 (2006)

2.1 Fast V_{th} measurement

IMEC group has developed a pulsed I_d-V_g method to measure the threshold voltage V_{th} and extract charge trapping characteristics, demonstrating that the conventional DC measurement method underestimates the charge trapping effect as a result of fast de-trapping occurring during V_{th} measurement [1]. We have improved the method to further reduce the V_{th} measurement time t_M to 1μs or less. The method is illustrated in Fig.1. In the circuit Fig.1(a) used by IMEC group, the measurement speed is limited by the charging current of C_{gd} and C_O, when voltage at D is varied. C_O includes C_{ds} of the MOSFET, the co-axial cable capacitance, and the input capacitance of the oscilloscope. The improved method used in this work employs circuit shown in Fig.1(b). Voltage at D is fixed to V_{ds} due to the virtual ground principle of the Op. Amp , and there is no charging current through $C_O{}'$. The charging current through C_{gd} can be deducted when C_{gd}-V_g is known. Fig.2 shows the fast measurement I_d-V_g curve , using rising (R) or falling edge (F) of the stress pulse applied to the gate. The V_{th} can be determined by the I_d-V_g curve. Fig.3 shows the V_{th} shift under 1 sec stress, measured with different measurement time t_m illustrated in Fig.2. Fig.4 shows time evolutions of dynamic BTI degradation in n- and p-MOSFETs, measured by conventional DC measurement and the fast measurement respectively. In the fast measurement, V_{th} can be measured at the falling edge (S_f) or the rising edge (P) as indicated in Fig.2(b).

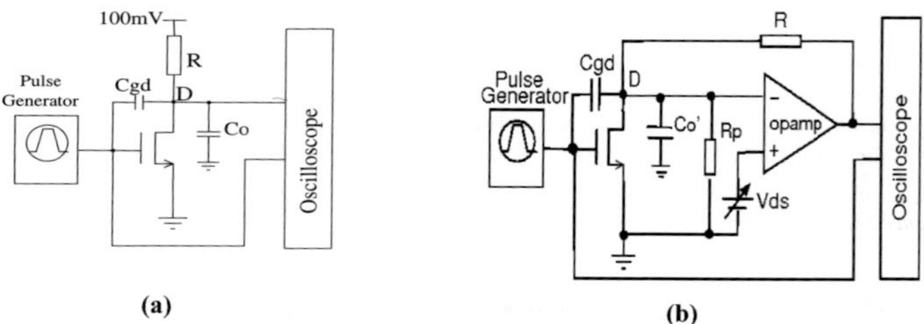

(a)   **(b)**

Figure 1. *(a) The pulsed I_d-V_g measurement developed by IMEC [1]. (b) The improved method used in this work . Voltage at D is fixed to V_{ds} due to the virtual ground principle of the Op.Amp, therefore no charging current from $C_0{}'$.The charge current from C_{gd} can be corrected. $R=R_P=1k\ \Omega$*

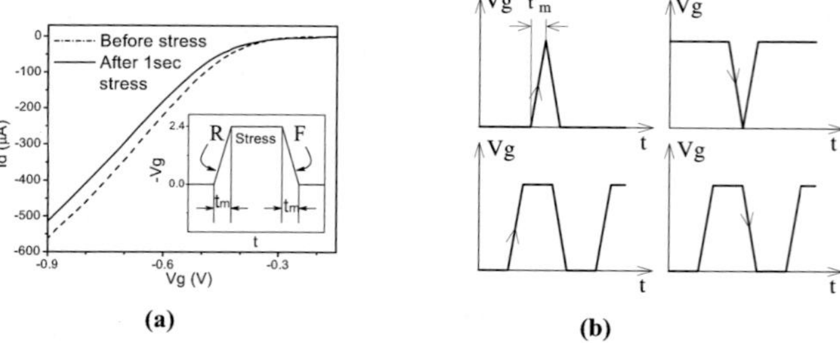

(a)   **(b)**

Figure 2 . *(a) Fast measurement of I_d-V_g using the rising (R) or falling (F) edge of a V_g pulse as indicated in the inset. Measurement time t_m can be varied. (b) Various V_g waveforms used for sensing V_{th}, during static and dynamic stress.*

ECS Transactions, 1 (5) 717-730 (2006)

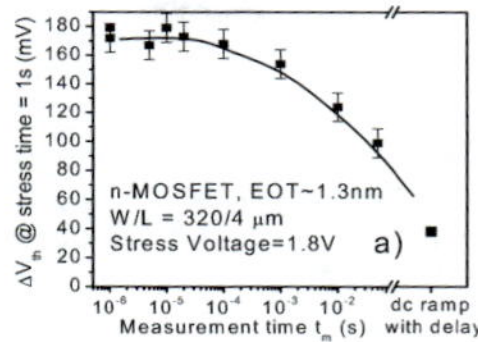
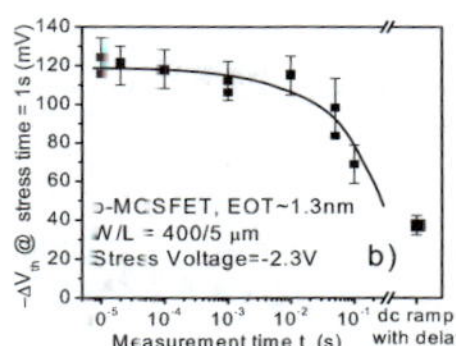

Figure 3. *Measured ΔV_{th} after 1 sec stress, using different pulse falling time t_m.. Reduction of ΔV_{th} with increasing t_m is due to additional de-trapping during a longer t_m. ΔV_{th} measured by the conventional DC measurement only detects the charge trapped by slow traps (slow de-trapping)[4] .*

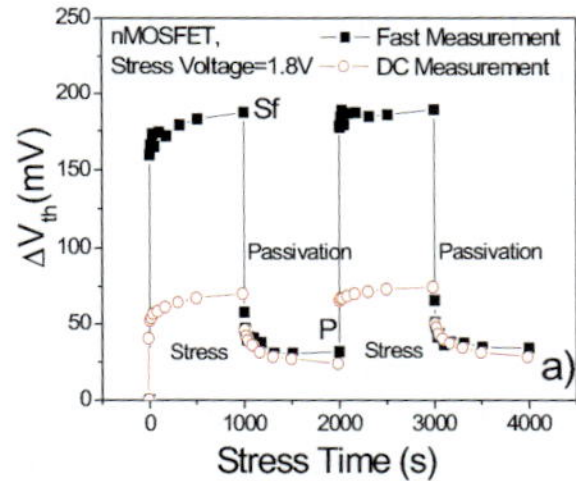
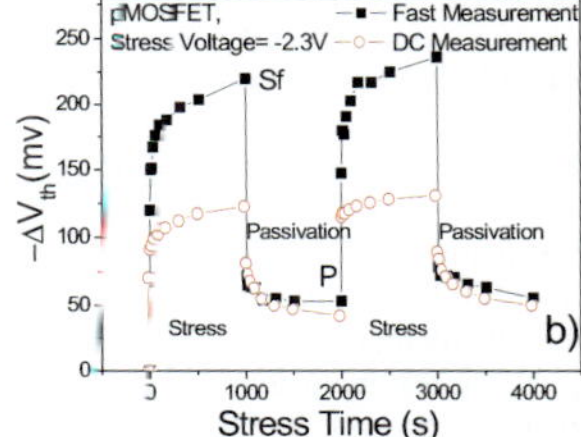

Figure 4. *ΔV_{th} shift under dynamic stress with frequency=(1/2000) Hz, measured by fast measurement (solid squares) and slow measurement (circles) respectively. The difference between fast and slow measurement reflects the fast trap contribution. n-MOSFET degradation is more severe, applying V_g=1.8 V after 1 sec gives rise ΔV_{th} of 150 mV.*

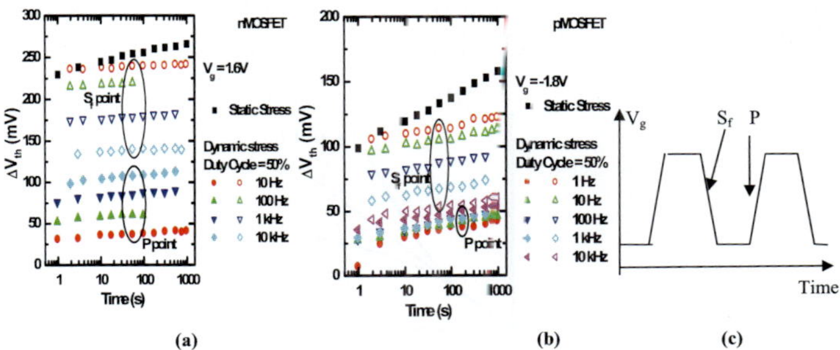

Figure 5. *Time evolution of ΔV_{th} degradation under static (solid squares) and dynamic BTI stresses with different frequencies measured by fast measurement. Under dynamic stress, both ΔV_{th} measured at the S_f point (open symbols) and the P point (solid symbols) are plotted. (a) for n-MOSFET, (b) for p-MOSFET. (c) S_f and P points measurement under a dynamic stress. S_f and P points are illustrated in Figure 4. With increasing stress frequency, the BTI degradation measured by fast measurement is increased.[4,13]*

ECS Transactions, 1 (5) 717-730 (2006)

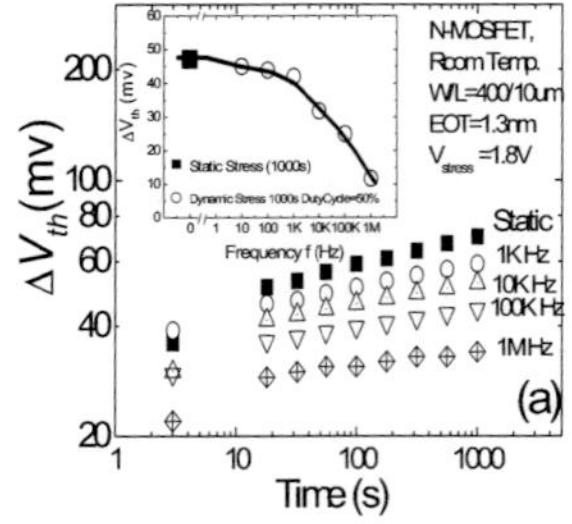
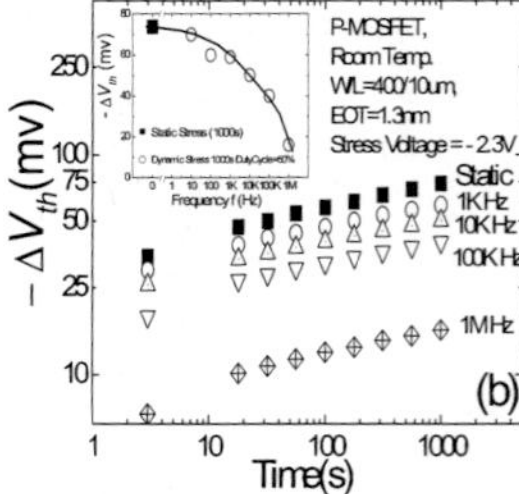

Figure 6. *Time evolution of ΔV_{th} degradation under static and dynamic BTI stresses with different frequencies measured by DC measurement.* **With increasing stress frequency, the BTI degradation measured by DC measurement is decreased.***[4]*

2.2 Frequency dependence of BTI - experimental

Figs.5 and 6 shows the frequency dependence of dynamic BTI degradation for both n- and p- MOSFETs, measured by the fast measurement (Fig.5) using the circuit in Fig.1b, and by DC method (Fig.6) using HP4156 parameter analyzer, respectively. The interesting thing is : by fast measurement, the BTI degradation is increased when increasing the stress frequency. However, by DC measurement, the BTI degradation is reduced when increasing the stress frequency [4,6]. We interpret this phenomena by the existence of two different kind of traps: fast traps and slow traps in HfO$_2$ dielectric. The fast measurement measures both fast traps and slow traps. $\Delta V_{th} = \Delta V_{th}^{F} + \Delta V_{th}^{S}$. The fast trap induced ΔV_{th}^{F} is much larger than the slow trap induced ΔV_{th}^{S} as indicated in Fig.4. Therefore the fast measurement in Fig.5 mainly measures the frequency dependence of the fast traps induced BTI degradation. The DC method measures the slow traps only because all fast traps de-trapped during DC measurement. Measurements in Fig.6 show the frequency dependence of the slow traps induced BTI degradation.

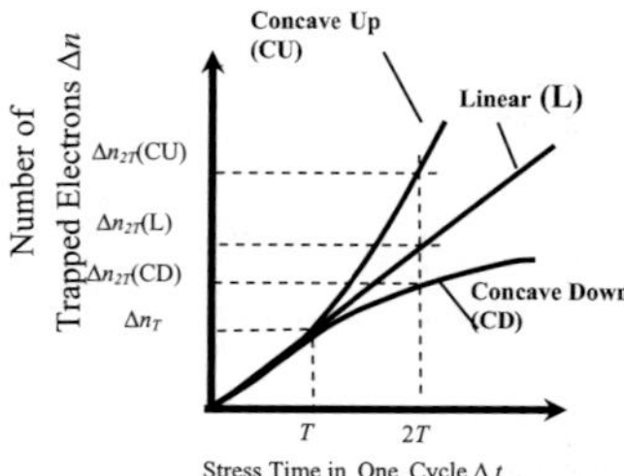

Figure 7. *Three possible cases for the relationship between the number of trapped electrons Δn versus stress time Δt in one cycle of the dynamic stress. When Δt increases from T (where T = 1/f, and f is the stress frequency) to 2T, the number of trapped electrons increases from Δn_T to Δn_{2T}. See explanation in the text.*

ECS Transactions, 1 (5) 717-730 (2006)

2.3 Frequency dependence of BTI –theoretical

Fig.7 is used to analyze three possible cases for the relationship between the number of trapped electrons Δn versus stress time Δt in one cycle of the dynamic stress. When Δt increases from T (where $T = 1/f$, and f is the stress frequency) to $2T$, the number of trapped electrons increases from Δn_T to Δn_{2T}. If $\Delta n \sim \Delta t$ relationship is linear, then $\Delta n_{2T} = 2\Delta n_T$. Therefore, the number of trapped electrons during the same stressing time would be the same for two frequencies and the accumulative V_{th} shift would be frequency independent. When Δn_{2T} is larger (concave up) or smaller (concave down) than $2\Delta n_T$, the dynamic degradation is reduced or increased when the stress frequency is increased.

Fast Traps : The V_{th} degradation due to fast traps leads to concave up curve in Fig.7, which can be modeled by the following conventional equations of one-hole (electron) trapping and de-trapping of the pre-existing traps N_{ot} [12]. For hole trapping in p-MOSFET, we have :

$$\frac{dp}{dt} = \frac{1}{\tau_C}(N_{ot} - p) - \frac{1}{\tau_{E1}}p \tag{1}$$

$$\frac{dp}{dt} = -\frac{1}{\tau_{E2}}p \tag{2}$$

where p is the hole (or electron) concentration in N_{ot} in p- (or n-) MOSFET. Solving (1) & (2) gives

$$\Delta n \propto 1 - \exp\left[-\Delta t(d/\tau_C + d/\tau_{E1} + (1-d)/\tau_{E2})\right], \tag{3}$$

where d is the duty cycle of the dynamic stress. This result clearly concaves down with Δt, corresponding to the frequency dependence observed for the fast traps: increased charge trapping accumulation under higher frequency. Using (1) & (2), with a wide spectrum of trapping and de-trapping time constants τ_C and τ_E [13] of N_{ot} , the simulated time evolutions of fast NBTI degradations are in good agreement with all static and dynamic stress condition shown in Figs 4,5,8.

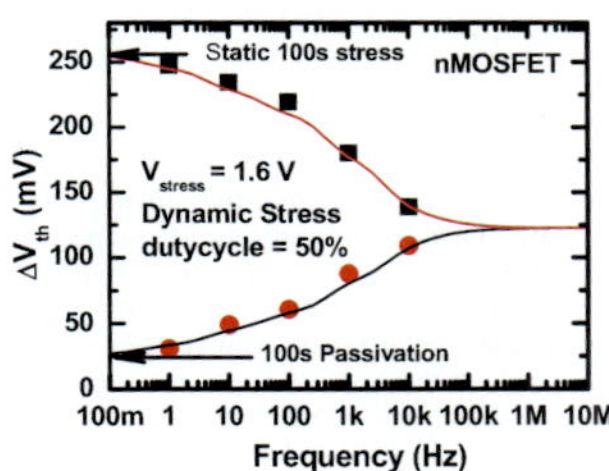

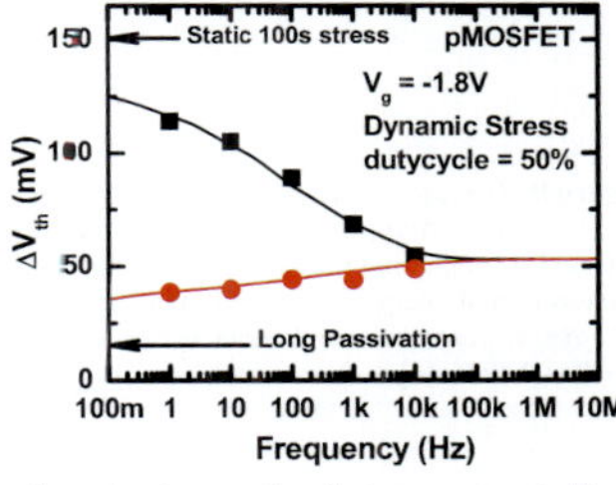

Figure 8. *Frequency dependence of ΔV_{th} after 100 sec dynamic stress, using fast measurement. The difference between S_f (square) and P (circle) reflects the transient amplitude indicated in Fig.4.. The solid lines show simulation data. P data reflect the cumulative degradation under dynamic stress. When the frequency is increased, the cumulative degradation is increased.*

ECS Transactions, 1 (5) 717-730 (2006)

Slow traps: The V_{th} degradation due to slow traps belong to concave up curve in Fig.7. We have carefully analyzed that the conventional one electron (or hole) trapping/de-trapping first order differential equation can never reach a concave up curve. It is well known that in the strongly ionized semiconductors, there are some traps with negative U (-U) property [14]. For a –U trap, capturing the first electron with trap energy E(-/0) leads to a meta-stable state. Capturing the second electron with trap energy E(--/-) leads to a stable state with a negative value of U=E(--/-)- E(-/0) , due to the large lattice relaxation (9,10)[15,16]. Since HfO$_2$ is a strongly ionic dielectric, it is therefore reasonable to expect that certain traps in HfO$_2$ are –U traps. It is known that HfO$_2$ gate dielectric has a monoclinic or cubic structure by TEM analysis [17]. In this work, we investigate the oxygen vacancy in cubic HfO$_2$ by first-principles calculation [4], as illustrated in Fig.9. Five different charge states for oxygen vacancy in cubic HfO$_2$, namely V^{++} ← (capture the second hole) V^+ ←(capture the first hole) V^O (capture the first electron) → V^- (capture the second electron) →V^{--} were investigated. The electron-lattice interaction will cause different lattice relaxations for different oxygen vacancy states which are illustrated in Fig.9a,b,c. The energies of the five charge states under lattice relaxation are calculated and illustrated in Fig.9d. Due to the –U property of the trap shown in Fig.9d, it is energetically more favorable to have two electrons (holes) trapped at this defect. When the electrons are injected into HfO$_2$ of the n-MOSFET, V^O will first capture one electron, which is meta-stable. Therefore it subsequently captures the second electron. Conversely, when the holes are injected into the HfO$_2$ in the p-MOSFET , V^O will first capture one hole (release electron from trap state E(+/0), and   subsequently captures the second hole (release electron from trap state E(++/+)) .

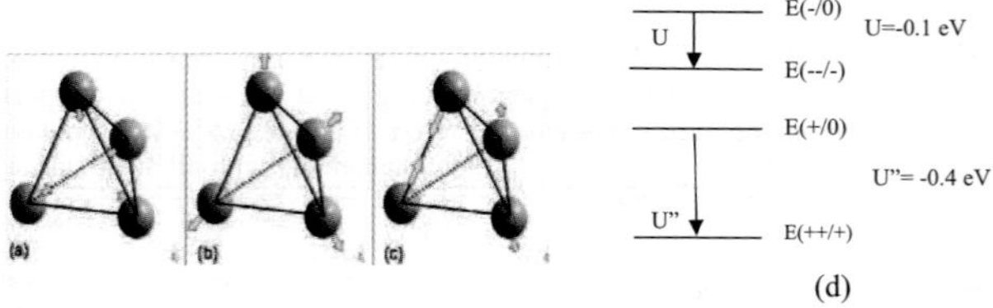

Figure 9. *The lattice relaxation (LR) of 4 Hf atoms around an oxygen vacancy in Cubic HfO$_2$, in V^0 state (a), V^+ or V^{++} state (b), and in V^- or V^-state (c). (d) is a* schematic diagram illustrates the -U property obtained from the first principles calculation of oxygen vacancy in cubic HfO$_2$. A neutral vacancy captures one electron at energy level E(-/0) and captures the second electron at E(--/-) with U=E(--/-) –E(-/0)= -0.1 eV. An V^{++} capture one electron at E(+/++) and capture the second electron at E(0/+) with U"=E(++/+)-E(+/0)=-0.4 eV. [4]

The concave-up curve in Fig.7 can be interpreted using the –U property of traps. Specifically, we analyze the case of the n-MOSFET .The rate equations describing the trapping and de-trapping behavior are :

ECS Transactions, 1 (5) 717-730 (2006)

$$\frac{dn_1}{dt} = \frac{1}{\tau_{C1}}\left(N - n_1 - n_2\right) - \frac{dn_2}{dt} \tag{4}$$

$$\frac{dn_2}{dt} = \frac{1}{\tau_{C2}} n_1 \tag{5}$$

$$\frac{dn_1}{dt} = -\frac{1}{\tau_{E1}} n_1 - \frac{dn_2}{dt} \tag{6}$$

$$\frac{dn_2}{dt} = -\frac{1}{\tau_{E2}} n_2 \tag{7}$$

where N is the total trap concentration, n_1 is the concentration of traps occupied by one electron, n_2 is the concentration of traps occupied by two electrons. Other parameters are defined in Table I.   The de-trapping term in (6) does not appear in eq. (4).   This is interpreted as follows. In a band diagram where the traps are mainly located at the high-k dielectric near to the SiO_2 interfacial layer, de-trapping is due to electron tunneling from the trap to the Si substrate through the SiO_2 interfacial layer. When a positive gate voltage is applied in the stress phase, the one-electron trap energy in the high-k dielectric moves into energy gap region of the Si substrate, and tunneling becomes forbidden. Solving equations (4)-(7) yields the static and dynamic time evolutions under different frequencies, with the time constants listed in Table I. Power-law dependence in $\Delta V_{th} \sim t$ can be obtained by assuming that the electron traps have a distribution $N(\tau_{C2})$ in the trapping time constant domain [3]. Results of calculations using eq. (4)-(7) are in excellent agreement with all DC experimental data under static and dynamic stress, as shown in Fig.10. For p-MOSFET, we can use similar but complementary description to that for n-MOSFET, using the $-U$ trap property of capturing two holes as illustrated in Fig.9.

Table I. Definitions and values of parameters used in eqs(4-7)

Symbol	Description	Value		
$N(\tau_{C2})$	Distribution function of capture time constant τ_{C2} of the second electron. It is given by a sum of two log-normal distributions.	Two log-normal pdf:		
			peak	40 µs	0.1 s
			width σ	3.1	3.5
τ_{E2}	Emission time constant of the second electron	$100 \cdot \tau_{C2}$		
τ_{C1}	Capture time constant of the first electron	$10^{-5} \cdot \tau_{C2}$		
τ_{E1}	Emission time constant of the first electron	0.1 µs		

ECS Transactions, 1 (5) 717-730 (2006)

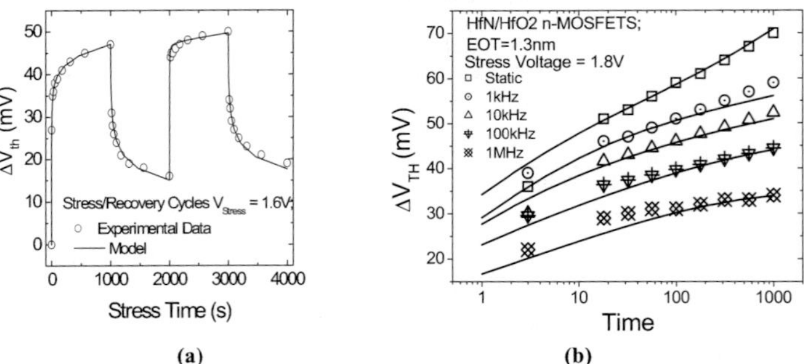

Figure 10.(a) *Comparison between the experimental data (symbols) of transient ΔV_{th} measured by the DC method and the simulation results by eqs (4-7) (solid lines) for n-MOSFET. DC measurement time of 0.5 sec is considered in the simulation.* *(b)* Same as (a), but for time evolution of ΔV_{th} under static and dynamic stress with different frequency.[4]

3. MIXING OF Ta AND La INTO HfO$_2$ GATE DIELECTRICS

Besides the charge trapping effect illustrated in session 2, HfO$_2$ as gate dielectric also suffers from low crystallization temperature [18], mobility degradation [7,8], and gate electrode Fermi pinning [9-11] problems. We have systematically investigated the mixing of Ta and La into HfO$_2$ to improve on the device performance degradation due to the abovementioned problems.

Both MOS-C and MOSFET devices were fabricated on (100) Si p-substrates. HfO$_2$, HfTaO [19] and HfLaO [20] were deposited using reactive sputtering techniques followed by post-deposit annealing in N$_2$ ambient at 700°C for 40 sec for HfTaO and 600°C for 30 s for HfLaO. For HfTaO, the composition of Ta was controlled by the ratio of applied power between Hf and Ta target. For HfLaO, HfLa target was used to avoid water absorption of La during exposure to air [21]. The composition of La was controlled by the power ratio between Hf and HfLa targets. TaN or HfN gate electrode was deposited using reactive sputtering. After gate patterning, phosphorus was implanted at *50 KeV* with a dose of *$5x10^{15}$ cm^{-2}* for HfTaO transistors and *100 KeV* with a dose of *$1x10^{15}$ cm^{-3}* for HfLaO transistors. Dopant activation annealing was done at various temperatures of *900~1000°C* in N$_2$ ambient for *30 sec*. After backside Al deposition, forming gas annealing was done at *420 °C* for *30 min*.

3.1 Raising crystallization temperature

Figures 11 and 12 illustrate that mixing Ta or La into HfO$_2$ can effectively raise the crystallization temperature from 400^0 C of pure HfO$_2$ [18]to more than 900^{0}C [19,20]. The phase transformation from the amorphous state to the crystal state depends on the free energies of both states, preferring the lowest free energy state. The energy part depends on the bond lengths, bond angles, and the bond-stretching and bond-bending force constant [22]. Incorporating Ta or La into HfO$_2$ will seriously change these

ECS Transactions, 1 (5) 717-730 (2006)

parameters because of the difference of Ta or La and Hf atomic sizes, the difference of co-ordination number of Ta_2O_5 or La_2O_3 and HfO_2 structures , and the difference of fractional ionic character of the bonds [23]. This is the possible physical reason for change of crystallization temperature. It will be an interesting topic for computer modeling using *ab initio* calculation, which we are conducting.

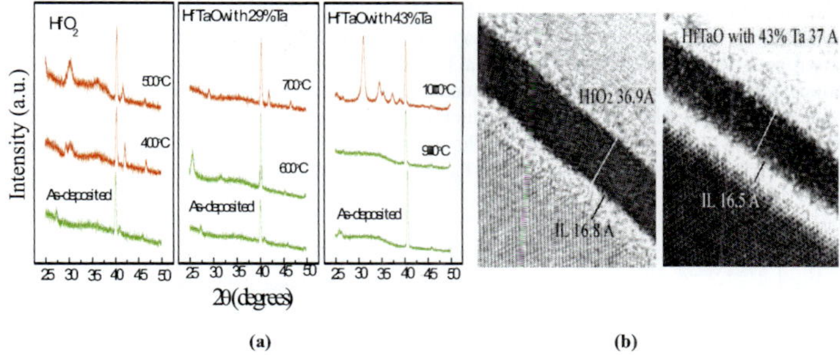

Figure 11. **(a)** *XRD spectra of HfO_2, HfTaO with 29% and 43% Ta after different temperature annealing for 30s in N_2. The Ta incorporated in HfO_2 films raises crystallization temperature significantly.* **(b)** *TEM image of HfO_2, HfTaO with 43% Ta after 950 °C annealing for 30 s. HfTaO remains amorphous structure. [19]*

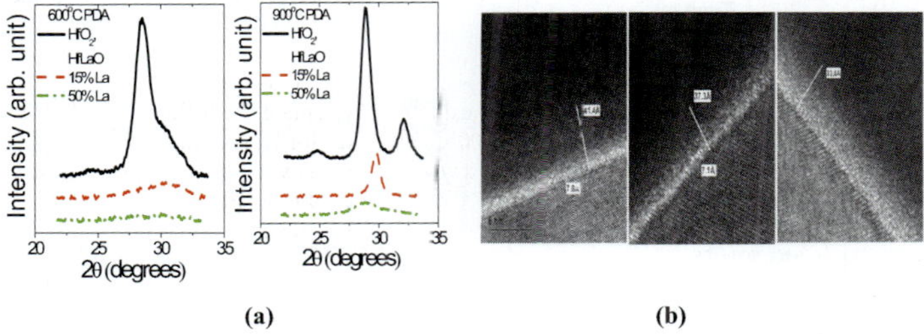

Figure 12. **(a)** *XRD spectra of HfO_2, HfLaO with 15% and 50% La after different temperature annealing for 30 sec in N_2. The La incorporated in HfO_2 films can raise the crystallization temperature.* **(b).** *TEM images of HfO_2 (left) and HfLaO with 15% (middle) and 50% (right) La after 900 °C annealing for 30 sec in N_2. The HfO_2 film is crystallized, however HfLaO films remain amorphous structures.[20]*

3.2 Reduction of charge trapping and improvement of channel mobility

Figures 13 and 14 illustrate the reduction of charge trapping and BTI degradation when mixing Ta or La into HfO_2 dielectric. For same device structure and BTI stress, there is 20

ECS Transactions, 1 (5) 717-730 (2006)

times reduction of V_{th} shift for devices with HfTaO gate dielectric and 10 times reduction for devices with HfLaO gate dielectric with 50% La , comparing with the devices with HfO$_2$ gate dielectric. Figures 15-18 show I_d –V_d curves and the electron mobility of n-MOSFETs with HfO$_2$, HfTaO and HfLaO gate dielectrics , respectively. Compared to HfO$_2$, HfTaO and HfLaO n-MOSFETs show higher drive current I_d scaling to the same device structure, and higher electron mobility. The peak electron mobility of n-MOSFETs with HfO$_2$, HfTaO with 43% Ta , and HfLaO with 50% La as gate dielectrics, are 140, 354, and 240 cm^2/V-s, respectively. The reduction of V_{th} shifty under BTI stress and improvement of channel mobility may be interpreted by the strengthen bonds in amorphous structured HfTaO and HfLaO than in crystallized HfO$_2$. The strengthened bonds raise the phonon vibration frequency and reduce phonon number and therefore reduce the remote phonon scattering, which is believed as a major scattering mechanism in MOSFETs with high-k dielectric[8]. The strengthened bonds also reduce the number of charged traps, and therefore reduce the V_{th} shift under BTI stress.

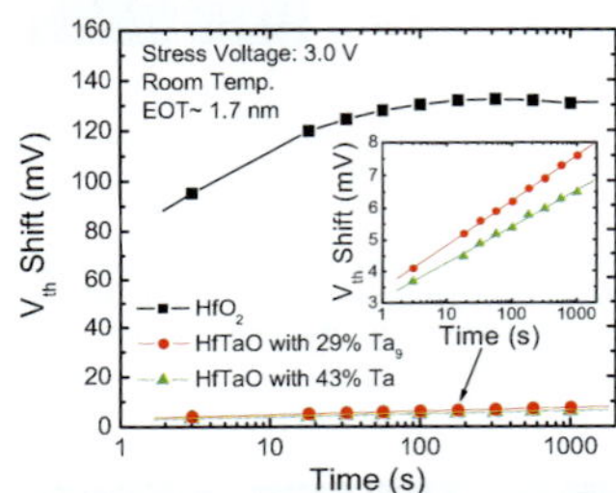

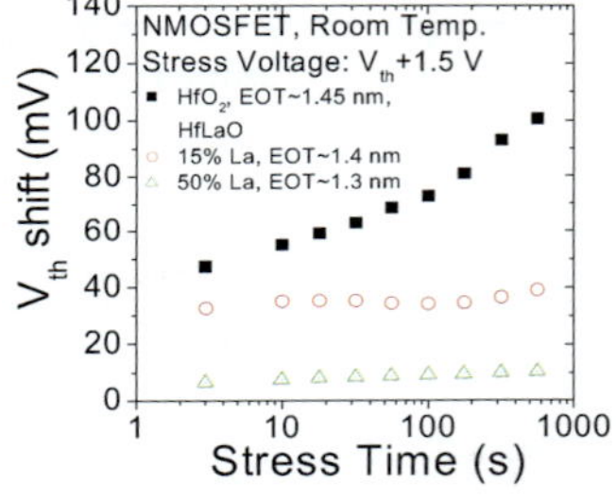

Figure 13. *BTI V_{th} shift under constant voltage stress in HfO$_2$, HfTaO with 29% and 43% Ta gate dielectric n-MOSFETs by DC measurement [19].*

Figure 14. *BTI V_{th} shift under constant voltage stress in HfO2, HfLaO with 15% and 50% La gate dielectric n-MOSFETs by DC measurement [20].*

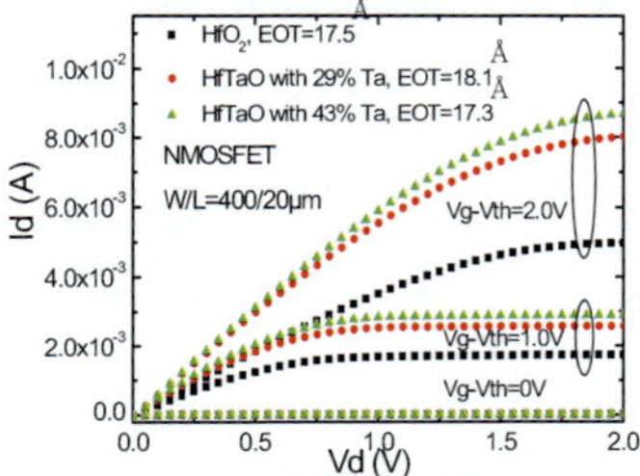

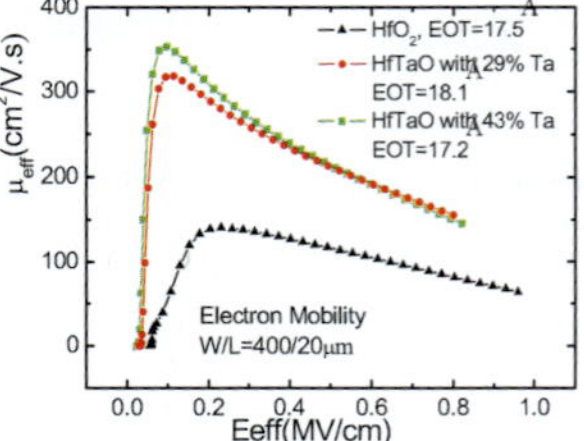

Figure 15 *Id-Vd curves for n-MOSFETs with HfO$_2$, HfTaO with 29% and 43% Ta gate dielectric [19]*

Figure 16.   *Effective electron mobility in n-MOSFETs with HfO$_2$, HfTaO with 29% and 43% Ta gate dielectric, extracted by split-CV method[19].*

ECS Transactions, 1 (5) 717-730 (2006)

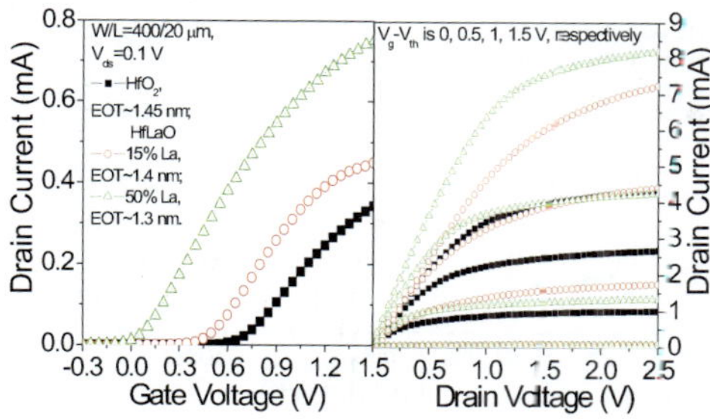

Figure 17. I_d-V_g (left) and I_d-V_d (right) curves of n-MOSFETs with HfO$_2$, and HfLaO with 15% and 50% La gate dielectrics. Scaling to the same EOT , I_d has 70% increment when using HfLaO (50% La) to replace HfO$_2$ as gate dielectric.[20].

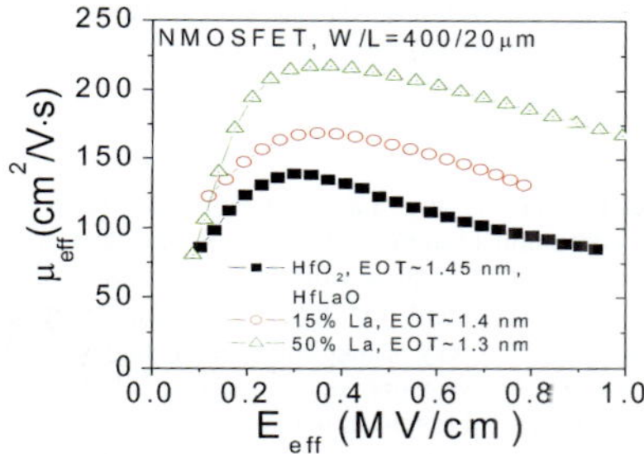

Figure 18. *Effective electron mobility in n-MOSFETs with HfO2, HfLaO with 15% and 50% La gate dielectric, extracted by split-CV method.[20]*

3.3 Metal gate work function tuning

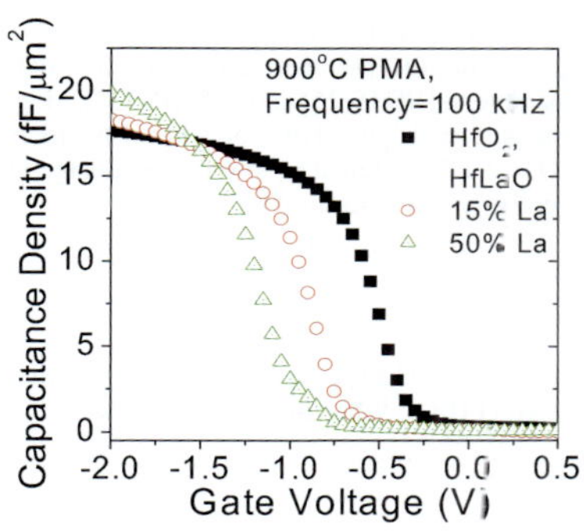

Figure 19. *Comparison of C-V curves for capacitors with HfO$_2$, HfLaO with 15% and 50% La gate dielectrics (with HfN gate).*

ECS Transactions, 1 (5) 717-730 (2006)

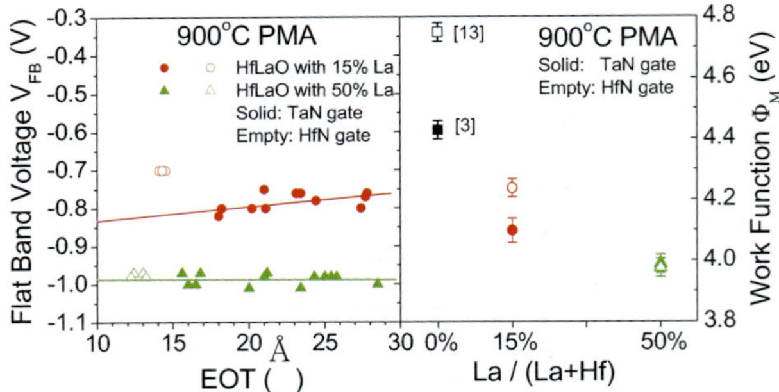

Figure 20. *Flat-band voltage V_{FB} (after 900 ^{0}C PMA) as a function of EOT (left), and the corresponding metal work function as a function of La composition in HfLaO films (right). For 50% La in HfLaO, both HfN and TaN have work functions of ~4.0 eV [20].*

Figs.19 and 20 show an interesting effect that varying La composition in HfLao can tune the effective work function of metal continuously in the HfLaO/metal gate stack from mid-gap (TaN : *4.44 eV* , HfN : *4.75 eV*) to *4 eV*, which will be very suitable for n-MOSFETs, in particular, it can meet the requirement for future ultra-thin body n-MOSFETs when the electron quantization effect is very strong and more stringent metal work function is required [24]. The reason of effective work function tuning by changing La composition in HfLaO is still not clear. One possible reason is that change of dielectric structure and atomic bonds in HfLaO may change HfLaO/metal interface states and Fermi pinning level, causing change of effective metal work function compared with HfO_2/metal.

3.4 Gate leakage current

Fig. 21 shows *Jg*-EOT characteristics of the HfO_2 and HfTaO gate dielectric MOS capacitors. It is noticed that the higher leakage current of HfTaO than HfO_2 is due to Ta_2O_5 lower electron barrier to Si, but it is still comparable to HfSiO [25] and HfSiON [27]. Fig. 22 shows that HfLaO gate leakage is comparable with pure HfO_2 and has 5-6 orders reduction compared with SiO_2 at the same EOT of 1.2-1.8 nm. The lower gate leakage of HfLaO than HfTaO is due to the higher electron barrier to Si for La_2O_3 .

4. CONCLUSION

The following new insights of Hf based high-k gate dielectrics are presented : **(I)** Two different traps, fast and slow traps were identified in HfO_2. When the dynamic stress

ECS Transactions, 1 (5) 717-730 (2006)

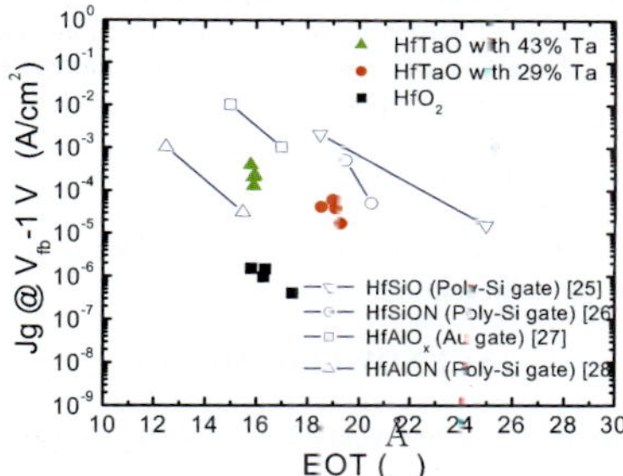

Fig.21 . *Comparison of gate current density Jg @V_fb-1V vs.EOT curves for HfO₂, HfTaO, and published data. [19,25-28]*

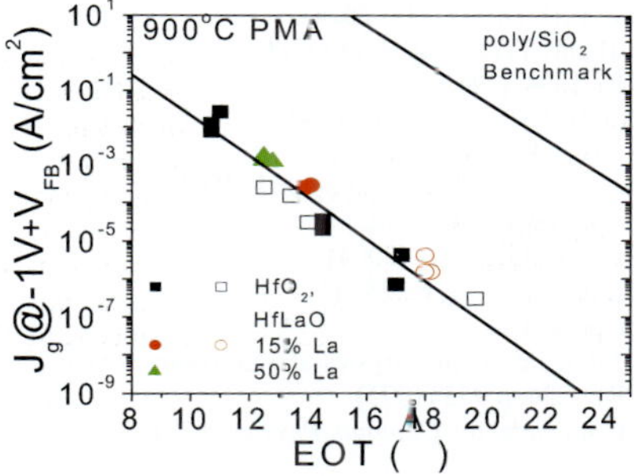

Figure 22. *Comparison of gate current density Jg@-1V + VFB for HfLaO with different EOT and La concentration. Solid symbols: HfN gate, empty symbols: TaN gate.*

frequency is increased, charge trapping by slow traps is reduced while that by fast traps is increased. The frequency dependency of slow-trap-charge is interpreted based on the –U property of slow traps, on excellent agreement with all the experimental results. The fast traps are conventional traps and the frequency dependency of charge trapping can be explained by the conventional first order trapping/de-trapping equations. **(II)** Mixing Ta and La into HfO₂ , the HfTaO and HfLaO as gate dielectrics have the following advantages comparing with the HfO₂ gate dielectric in following : (a) rise of the crystallization temperature from 400⁰C to higher than 900⁰C ,(b) Reduction of charge trapping and BTI

ECS Transactions, 1 (5) 717-730 (2006)

degradation by more than one order of magnitude, (c) Increase of the drive current I_{on} and the channel mobility by a factor of more than 2 for HfTaO and a factor of 1.6 for HfLaO transistors. The device performance improvements due to HfTaO and HfLaO gate dielectric are possibly because of hardening of bond structure , giving rise to (a) increment of phonon frequency and decrement of phonon number and therefore reduction of remote phonon scattering, (b) larger bond breaking energy and therefore higher crystallization temperature and reduction of traps creation by bond breaking. In addition, by varying the concentration of La in HfLaO, the effective work function of metal electrode in the HfLaO/TaN and HfLaO/HfN gate stacks can be tuned continuously from the mid gap to 4 eV, suitable for n-MOSFETs operation. This is possibly due to the change of extrinsic Fermi pinning state energy in the gap.

ACKNOWLEDGMENTS

This work was supported by Singapore A-STAR R263-000-267-305 and R398-000-019-305 grants.

REFERENCES

1. A. Kerber *et al*, IRPS Proceedings, p.41(2003).
2. K. Onishi *et al*, TED. v.50, p.1517 (2003),
3. S. Zafar *et al*, IEDM Tech. Dig., p.517 (2002).
4. C. Shen *et al*, IEDM Tech.Dig., p.733 (2004).
5. A. Shanware *et al*, IEDM Tech. Dig., p.939 (2003).
6. S. J. Rhee et al, IRPS 2004, p.269.
7. E.P.Gusev *et al.,* IEDM Tech. Dig., p.451(2001).
8. Z.Ren *et al.,* IEDM Tech.Dig., p.793 (2003).
9. S. B. Samavedam *et al.,* IEDM Tech. Dig., p.307 (2003).
10. Y. C. Yeo *et al.,* EDL. **23**, p. 342 (2002).
11. H. Y. Yu *et al.,* EDL, v. 25, p. 337 (2004).
12. Y.Nissan-Cohen *et al*, JAP,v.58, p.2252 (1985)
13. C.Shen *et al*, to be published.
14. D. J. Chadi, 24th Int. Conf. on the Physics of Semiconductors, p.2311(1995).
15. P.W.Anderson, PRL v.34, p.953 (1975).
16. M.F.Li, *Modern Semiconductor Quantum Physics*, p.316, World Scientific, Singapore (1994).
17. J.Aarik *et al,* Apply Surface Sci. **173**, 15 (2001).
18. W. Zhu *et al.* ,IEDM Tech.Dig., p. 463 (2001)
19. X.Yu *et al.,* Symp VLSI Tech.., p.110 (2004).
20. X.P.Wang *et al*, to be published.
21. D. S. Yu *et al.,* IEDM Tech Dig., p. 181 (2004).
22. N.F.Mott and E.A.Davis, *Electronic Processes in Non-Crystalline Materials*, Chapter 7, Clarendon Press, Oxford,(1979).
23. C.Kittel, *Introduction to Solid State Physics* , sixth edition, Chapter 3, John Wiley, New York (1991)
24. Tony Low et al, IEDM Tech. Dig., p.151 (2004).
25. A. Morioka et al., Symp VLSI Tech., p.165 (2003).
26. T. Nabatame et al., Symp VLSI Tech., p. 25 (2003).
27. M. Koyama et al., IEDM Tech. Dig., pp. 849-852 (2003).
28. H. Jung et al., IEDM Tech. Dig., p. 853 (2002).

160

IEEE ELECTRON DEVICE LETTERS, VOL. 27, NO. 3, MARCH 2006

Yb-Doped Ni FUSI for the n-MOSFETs Gate Electrode Application

J. D. Chen, H. Y. Yu, *Member, IEEE*, M. F. Li, *Senior Member, IEEE*, D.-L. Kwong, *Senior Member, IEEE*, M. J. H. van Dal, J. A. Kittl, A. Lauwers, P. Absil, M. Jurczak, and S. Biesemans

Abstract—In this letter, an n-type near-band edge fully silicided (FUSI) material–Yb-doped Ni FUSI is demonstrated for the first time. By doping Yb into Ni FUSI, it is shown that while maintaining the same equivalent oxide thickness and the similar device reliability, the work function of Ni FUSI (on SiON dielectrics) could be tuned from 4.72 to 4.22 eV. Yb-doped Ni FUSI is promising for the gate electrode application in n-MOSFETs.

Index Terms—Band edge workfunction, fully silicided (FUSI) , n-MOSFETs, Ni/Yb.

I. Introduction

METAL gate is expected, in the sub-45-nm CMOS technology nodes, to address the concerns associated with the poly-Si electrode such as poly-depletion [1]. Ni fully silicided FUSI technology attracted significant attention for this application due to its compatibility with conventional flows [2]–[7]. However, it is still a challenge for Ni FUSI to achieve the n-type bandedge work function (WF), which is required by bulk CMOS devices [8]. Addition of dopants such as As and Sb may lower the WF, but at the expense of introducing adhesion issues that impact manufacturability. Yb is known for a low WF (photoelectric WF $\sim$ 2.59 eV), and a low electron Schottky barrier height ($\sim$ 0.27 eV) has been reported for Yb silicide [9]. In this letter, by doping Yb into Ni FUSI, we show that, while maintaining the same equivalent oxide thickness (EOT) and the similar device reliability, WF of Ni FUSI (on SiON dielectrics) could be tuned from $\sim$ 4.72 eV to $\sim$ 4.22 eV. Yb-doped Ni FUSI is promising as a gate electrode for n-MOSFETs.

II. Experimental

The capacitors were fabricated using p-type Si substrate (with resistivity of 4–8 $\Omega \cdot$ cm). After active area definition, the gate stack of undoped poly-Si ($\sim$ 100 nm) and SiON dielectrics of different thickness was grown and patterned. Ni-Yb was then co-deposited in a plasma vapor deposition (PVD) tool. The ratio of Yb/Ni was controlled by adjusting the respective Yb and Ni

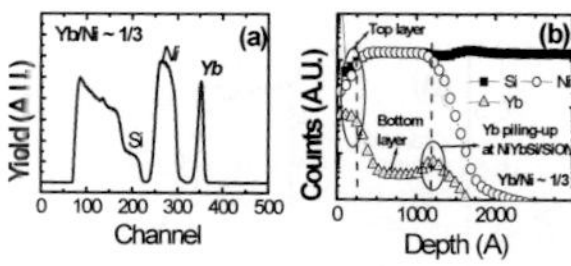

Fig. 1. (a) RBS spectrum and (b) SIMS spectrum of Yb-doped NiSi (Yb/Ni $\sim$ 1/3) show that Yb is mainly distributed at top layer of silicide. The piling-up of Yb at the NiYbSi/SiON interface is observed from SIMS data.

deposition rate and time. The total metal thickness was kept as $\sim$ 90 nm so that poly-Si could be completely silicided during FUSI process. The silicidation was done by one step annealing at 400 °C, 1 min. The selective etching was then carried out to remove the remaining unreacted Ni and Yb by dilute HNO_3. Ni FUSI control samples were made for comparison. Equivalent oxide thickness (EOT) and flat band voltage (V_{fb}) were simulated considering the quantum mechanical correction. Blank Yb-doped NiSi (or undoped NiSi) on SiON were prepared for the material characterization.

III. Results and Discussion

Both RBS and SIMS [Fig. 1(a) and (b)] analysis of the annealed samples imply that most of Yb in the Yb-doped NiSi film (with Yb/Ni$\sim$ 1/3, as defined by Yb/Ni deposition parameters) is distributed on the top layer after silicidation. Yb signal cannot be detected by RBS in the bottom layer as the Yb concentration is below the RBS detection limit ($<$ 1 at.%). The resulting Yb distribution is correlated with the fact that the dominant diffusion species during the respective Ni silicide and Yb silicide formation are Ni and Si. Ta was reported to have a similar distribution in NiTaSi during the silicide formation [10]. In addition, SIMS reveals that Yb is piling up at the NiYbSi/SiON interface after the silicidation process. The possible mechanism for Yb piling up is that Yb is not fully soluble inside the NiSi. Thus during the NiYb silicide formation, some Yb would be pushed to the interface between silicide/SiON. In Fig. 2, the XTEM micrograph shows that the bulk layer of Yb-doped NiSi (with Yb/Ni$\sim$ 1/3) is fully silicided (with a thickness of $\sim$ 120 nm), and it also confirms that there are two different layers in the NiYbSi as evidenced by RBS and SIMS data (Fig. 1). Moreover, a smooth NiYbSi/SiON interface is observed from the XTEM. Yb concentration in the NiYbSi top layer (Yb/ Ni}$\sim$ 1/3) is $\sim$ 12%, as determined by AES (data not shown). Note that the NiYbSi has a sheet resistance (Rs) of $\sim$ 2 $\Omega/\square$, which is comparable the Rs of NiSi with a similar thickness.

Manuscript received November 1, 2005; revised January 6, 2006. The review of this letter was arranged by Editor C.-P. Chang.

J. D. Chen and M. F. Li are with the Silicon Nano Device Laboratory, Department of Electrical and Electronic Engineering, National University of Singapore, Singapore 119260.

H. Y. Yu, A. Lauwers, P. Absil, M. Jurczak, and S. Biesemans are with the IMEC, Leuven B-3001, Belgium (e-mail: hongyu@imec.be).

D.-L. Kwong is with the Department of Electrical and Electronic Engineering, The University of Texas, Austin, TX 78712 USA.

M. J. H. van Dal is with Philips Research Leuven, Leuven B-3001, Belgium.

J. A. Kittl is an assignee to IMEC, Leuven B-3001, Belgium.

Digital Object Identifier 10.1109/LED.2006.870252

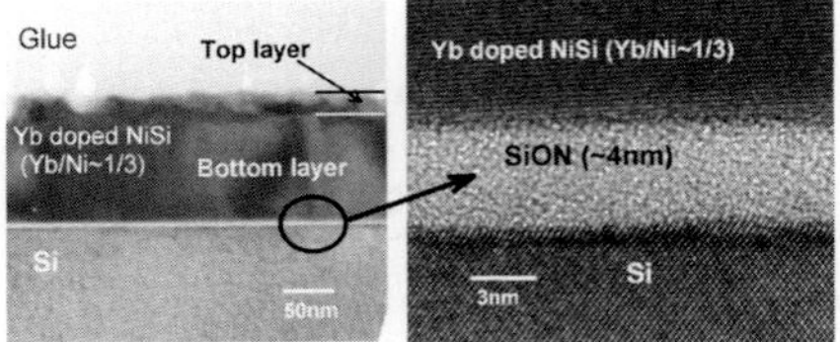

Fig. 2. XTEM shows that the bulk layer of Yb-doped NiSi (Yb/Ni ~ 1/3) is fully silicided, and the resulting silicide thickness is ~ 120 nm. Two different layers in the Yb-doped NiSi (corresponding to RBS and SIMS data in Fig. 1) are observed. The smooth NiYbSi/SiON interface is also revealed by XTEM.

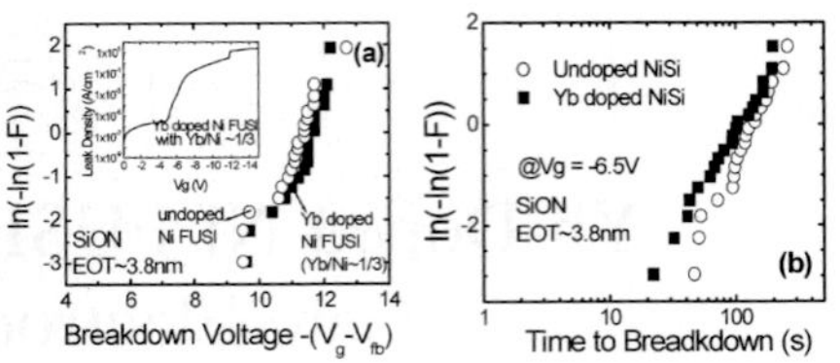

Fig. 4. (a) TZBD comparison between the devices with Yb-doped Ni FUSI and undoped Ni FUSI electrodes (on SiON dielectrics). Inset shows a typical J-V sweep for the device with Yb-doped Ni FUSI gate. (b) TDDB (under gate injection and constant voltage FN stress at -6.5 V) comparison between the devices with Yb-doped Ni FUSI and undoped Ni FUSI gate electrode (on SiON dielectrics).

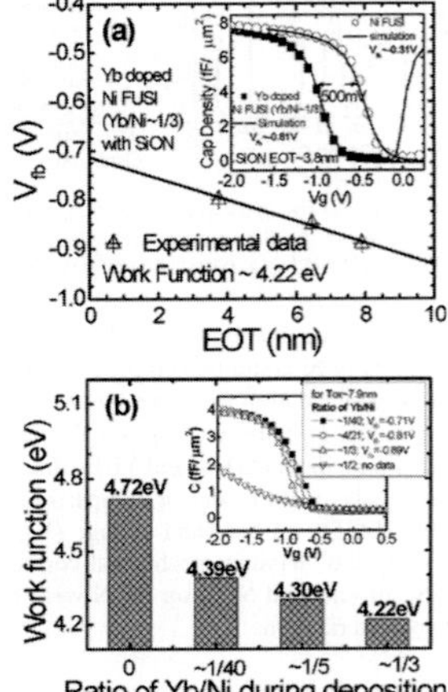

Fig. 3. (a) Plot of EOT versus V_{fb} for the devices with Yb-doped Ni FUSI (with Yb/Ni ~ 1/3) gate electrode. The inset compares the measured and simulated C–V data for capacitors with Yb-doped Ni FUSI (Yb/Ni ~ 1/3) and undoped Ni FUSI gate electrodes. No change in EOT is observed with addition of Yb. The extracted work function is 4.22 eV, with fixed charge of $Q_{ox}/q = 4.59 \times 10^{11}$ cm^{-2} (b) Work function of Yb-doped Ni FUSI is tunable by modifying Yb incorporation during deposition. It is noted that excessive Yb might degrade the device dielectric (e.g., Yb/Ni ~ 1/2) as shown in the inset. The simulated curves are generated by QMCV program, using EOT and V_{fb} as fitting parameters.

In the inset of Fig. 3(a), the simulated C–V data well match the measured one for the capacitors with both the Yb-doped NiSi (Yb/Ni ~ 1/3) and the undoped NiSi electrode, suggesting their negligible interface trap density D_{it}. Compared to the undoped Ni FUSI, while maintaining the same EOT, the Yb-doped Ni FUSI (Yb/Ni ~ 1/3) demonstrates a negative V_{fb} shift of ~ 500 mV. Considering that Yb itself has a low WF, and so does its silicide [9], the Yb piling up at the NiYbSi/SiON interface [Fig. 1(b)] is believed to be responsible for the V_{fb} shift. The WF of Yb-doped Ni FUSI (on SiON) is extracted based on the V_{fb} versus EOT dependence. In Fig. 3(a), for the case of

Yb/Ni ~ 1/3, the extracted WF is ~ 4.22 eV, which is desirable for n-MOSFETs application. Furthermore, it is found that WF of Yb-doped NiSi could be tuned by modifying the Yb incorporation during deposition [from 4.22 to 4.72 eV, as shown in Fig. 3(b)]. Note that excessive Yb might degrade the device dielectric integrity [e.g., Yb/Ni ~ 1/2 in this work as shown in the inset of Fig. 3(b)]. This might be due to that the excessive Yb reaching the interface consumes the dielectric [11]. Time zero breakdown (TZBD) and time dependant dielectrics breakdown (TDDB) characteristics (under gate injection, and constant voltage Fowler–Nordheim (FN) stress at -6.5 V) of the devices with Yb-doped Ni FUSI were studied and compared to those of the undoped Ni FUSI devices [Fig. 4(a) and (b), respectively]. Inset of Fig. 4(a) shows a typical J-V sweep for the device with Yb-doped Ni FUSI gate. It is seen that the Yb-doped Ni FUSI devices show comparable reliability characteristics to the undoped Ni FUSI. Finally, for the devices with Yb-doped Ni FUSI, we did not observe any interface adhesion issues found in other reports when WF is modulated by dopants such as As or Sb [12].

IV. Conclusion

We demonstrated for the first time the feasibility to modulate the work function of Ni FUSI gates from 4.72 to 4.22 eV by doping with Yb (an n-type bandedge FUSI). A systematic material study was conducted to characterize this novel FUSI material. The WF change is attributed to the piling up of Yb at the NiSi/SiON interface.

References

[1] *International Technology Roadmap for Semiconductor (ITRS)*, Semiconductor Industry Association, San Jose, CA, 2003.

[2] W. P. Maszara, Z. Krivokapic, P. King, J.-S. Goo, and M.-R Lin, "Transistors with dual work function metal gates by single full silicidation (FUSI) of polysilicon gates," in *IEDM Tech. Dig.*, 2002, pp. 367–370.

[3] J. Kedzierski, D. Boyd, P. Ronsheim, S. Zafar, J. Newbury, J. Ott, C. Cabral, M. Ieong, and W. Haensch, "Threshold voltage control in NiSi-gated MOSFETs through silicidation induced impurity segregation (SIIS)," in *IEDM Tech. Dig.*, 2003, pp. 315–318.

[4] K. G. Anil, A. Veloso, S. Kubicek, T. Schram, E. Augendre, J. F. de Marneffe, K. Devriendt, A. Lauwers, S. Brus, K. Henson, and S. Biesemans, "Demonstration of fully Ni-silicided metal gates on HfO₂ based high-k gate dielectrics as a candidate for low power applications," in *VLSI Symp. Tech. Dig.*, 2004, pp. 190–191.

[5] E. P. Gusev, C. Cabral, B. Linder, Y. H. Kim, K. Maitra, E. Cartier, H. Nayfeh, R. Amos, G. Biery, N. Bojarczuk, A. Callegari, R. Carruthers, S. A. Chohen, M. Copel, S. Fang, M. Frank, S. Guha, M. Gribelyuk, P. Jamison, R. Janny, M. Ieong, J. Kedzierski, P. Kozlowski, V. Ku, D. Lacey, D. Latulipe, V. Narayanan, H. Ng, P. Nguyen, J. Newbury, V. Paruchuir, R. Rengarajan, G. Shahidi, A. Steegen, M. Steen, S. Zafar, and Y. Zhang, "Advanded gated stacks with fully silicded (FUSI) gates and high-k dielectrics: Enahnced performance at reduced gate leakage," in *IEDM Tech. Dig.*, 2004, pp. 79–83.

[6] K. Hosaka, T. Kurahashi, K. Kawamura, T. Aoyama, Y. Mishima, K. Suzuki, and S. Sato, "A comprehensive study of fully-silicided gates to achieve wide-range work function differences (0.91 eV) for high-performance CMOS devices," in *Symp. VLSI Tech. Dig.*, pp. 66–67.

[7] J. A. Kittl, A. Veloso, A. Lauwers, K. G. Anil, C. Demeurisse, S. Kubicek, M. Niwa, M. J. H. van Dal, O. Richard, M. A. Pawlak, M. Jurczak, C. Vrancken, T. Chiarella, S. Brus, K. Maex, and S. Biesemans, "Scalability of Ni FUSI gate process: Phase and V_{th} control to 30 nm gate length," in *Symp. VLSI Tech. Dig.*, 2005, pp. 72–73.

[8] I. De, D. Hchri, A. Srivastava, and C. M. Osburn, "Impact of gate work function on device performance at the 50 nm technology node," *Solid State Electron.*, vol. 44, pp. 1077–1085, 2000.

[9] S. Y. Zhu, J D. Chen, M.-F Li, S. J. Lee, J. Singh, C. X. Zhu, A. Du, C. H. Tung, A. Chin, and D. L. Kwong, "N-type Schottky barrier source/drain MOSFET using ytterbium silicided," *IEEE Electron Device Lett.*, vol. 24, no. 5, pp. 565–567, May 2004.

[10] M. C. Sun, M. J. Kim, J. H. Ku, K. J. Roh, C. S. Kim, S. P. youn, S. W. Jung, S. Choi, N. I. Lee, H. K. Kang, and K. P. Suh, "Thermally robust Ta-doped Ni SALICIDE process promising for sub-50 nm CMOSFETs," in *VLSI Symp. Tech. Dig.*, 2003, pp. 81–82.

[11] O. M. Ndwandwe, Q. Y. Hlatshwayo, and R. Pretorius, "Thermodynamic stability of SiO2 in contact with thin metal films," *Mater. Chem. Phys.*, vol. 92, pp. 487–491, 2005.

[12] K. Sano, M. Hino, N. Ooishi, and K. Shibahara, "Workfunction tuning using various impurities for fully silicided NiSi gate," *Jpn J. Appl. Phys.*, vol. 44, no. 6A, pp. 3774–3777, 2005.

IEEE ELECTRON DEVICE LETTERS, VOL. 27, NO. 1, JANUARY 2006   31

Tuning Effective Metal Gate Work Function by a Novel Gate Dielectric HfLaO for nMOSFETs

X. P. Wang, Ming-Fu Li, *Senior Member, IEEE*, C. Ren, *Student Member, IEEE*, X. F. Yu, C. Shen, H. H. Ma, Albert Chin, *Senior Member, IEEE*, C. X. Zhu, *Member, IEEE*, Jiang Ning, M. B. Yu, and Dim-Lee Kwong, *Senior Member, IEEE*

Abstract—**Using a novel HfLaO gate dielectric for nMOSFETs with different La composition, we report for the first time that TaN (or HfN) effective metal gate work function can be tuned from Si mid-gap to the conduction band to fit the requirement of nMOS-FETs. This is explained by the change of interface states and Fermi pinning level by adding La into HfO_2. The superior performances of the nMOSFETs compared with those using pure HfO_2 gate dielectric are also reported, in terms of higher crystallization temperature and higher drive current I_d without sacrifice of very low gate leakage current, i.e. 5–6 orders reduction compared with SiO_2 at the same equivalent oxide thickness of $\sim$1.2–1.8 nm.**

Index Terms—**HfLaO, high-κ dielectric, MOSFET, metal gate.**

I. INTRODUCTION

THE HfO_2, as one of the most promising high-κ dielectrics, has been extensively investigated for replacing conventional SiO_2 and SiON gate dielectrics in MOSFETs [1]. We recently found that incorporating La into HfO_2 can greatly improve some serious problems of pure HfO_2 gate dielectric. The first problem is the low crystallization temperature (less than 500 °C) of HfO_2 [2]. The grain boundaries in the crystallized HfO_2 can serve as the fast paths for gate leakage current, or oxygen and some dopants atoms diffusion into the gate dielectric and even MOSFET channel region, which will induce nonuniform interfacial layer growths and κ values, threshold voltage (V_{th}) instability and defect generation [1], [3]. To increase the crystallization temperature, Si [4], Al [2], N [5]–[7] or Ta [8] have been incorporated into HfO_2 film to form HfO_2-based gate dielectrics. All of these materials exhibit high crystallization temperature and good thermal stability in contact with Si to withstand the conventional 900 °C–1000 °C activation annealing. However, their dielectric permittivities

[2], [4]–[7] or the electron barrier to Si [8] are reduced comparing with pure HfO_2, and therefore the gate leakage increases compared to HfO_2 with the same EOT. The second problem is the Fermi level pinning between the metal gate and the HfO_2 [9]. Due to Fermi level pinning, the metal/HfO_2 gate stack has difficulty to obtain effective work function of 4.1 eV, required by nMOSFETs after high temperature annealing for source/drain activation [10]. We have investigated the dielectric material HfLaO, which can improve both of the above-mentioned problems. The first problem will be discussed elsewhere in more detail [11] and this letter will mainly focus on the second problem. It is found that changing La composition in HfLaO films can effectively tune the work function of the metal gates continuously from Si midgap to around 4 eV, which can fit the requirement of nMOSFETs. Moreover, incorporation of HfO_2 with La also improves significantly the electrical performances of nMOSFETs in terms of drive current with no gate leakage degradation compared to pure HfO_2 gate dielectric.

II. EXPERIMENTAL

The MOS devices were fabricated using p-type (100) Si substrates with 6×10^{15} cm^{-3} boron doping. For capacitors fabrication, after field oxide growth ($\sim$400 nm) and area definition, DHF-last RCA pregate clean was performed. HfO_2 and HfLaO films with different La concentration and different thicknesses were then deposited using reactive sputtering with low oxygen concentration at room temperature, followed by an ex-situ post deposition anneal (PDA) in N_2 at 600°C for 30 s. Here, HfLa target (Hf:La= 1 : 1) was used to avoid water absorption of La during exposure to air [12]. The composition of La was controlled by the power ratio between Hf and HfLa targets, and detected by X-ray photoelectron spectroscopy (XPS). HfN ($\sim$50 nm) with a TaN capping layer ($\sim$100 nm) [13] or TaN gate electrodes ($\sim$150 nm) were then deposited using reactive sputtering, and then patterned using a Cl_2-based etchant. After that, the devices were rapid thermal annealed (RTA) in N_2 at 900°C for 30 s for thermal stability evaluation. For nMOSFETs fabrication, source/drain implantations of arsenic (100 keV, 1×10^{15} cm^{-2}) were performed, followed by RTA activation in N_2 at 900 °C for 30 s. In the end, all samples received back side Al metallization and forming gas sintering.

Electrical characteristics of the MOS devices were measured using HP4284A precision LCR meter and HP4156A parameter analyzer. The EOT and flatband voltage (V_{fb}) of the capacitors were simulated by taking into account quantum mechanical correction.

Manuscript received July 28, 2005; revised September 22, 2005. This work was supported by the Singapore A-STAR under Research Grant R263-000-267-305. The review of this letter was arranged by Editor C.-P. Chang.

X. P. Wang, C. Ren, X. F. Yu, C. Shen, H. H. Ma, A. Chin, and C. X. Zhu are with the Silicon Nano Device Laboratory, Department of Electrical and Computer Engineering, National University of Singapore, Singapore 119260.

M.-F. Li is with the Silicon Nano Device Laboratory, Department of Electrical and Computer Engineering, National University of Singapore, Singapore 119260, and also with the Institute of Microelectronics, Singapore 117685 (e-mail: elelimf@nus.edu.sg).

J. Ning and M. B. Yu are with the Institute of Microelectronics, Singapore 117685.

D.-L. Kwong is with the Institute of Microelectronics, Singapore 117685. and also with the Department of Electrical and Computer Engineering, University of Texas, Austin, TX 78712 USA.

Digital Object Identifier 10.1109/LED.2005.859950

0741-3106/$20.00 © 2005 IEEE

32   IEEE ELECTRON DEVICE LETTERS, VOL. 27, NO. 1, JANUARY 2006

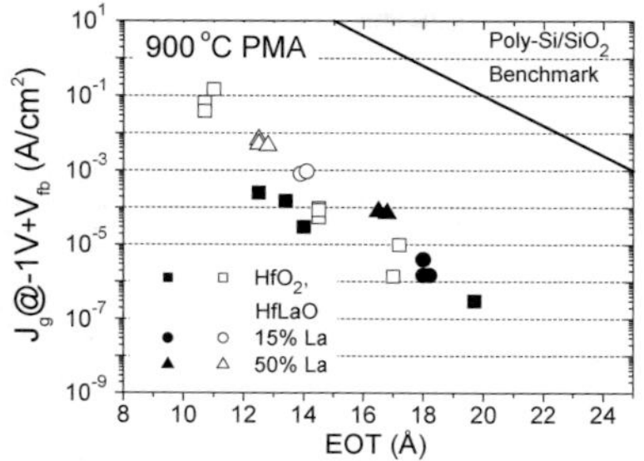

Fig. 1. Gate leakage current density vs. EOT relationship for MOS devices with HfO_2 and HfLaO (with 15% and 50% La) gate dielectrics and TaN (solid data point) or HfN (open data point) metal gate. Compared with poly/SiO_2 benchmark at the same EOT, both HfO_2 and HfLaO provide 5–6 orders reduction in gate leakage current.

III. RESULTS AND DISCUSSION

Fig. 1 shows that the gate leakage current densities of HfLaO films with 15% and 50% La are comparable with pure HfO_2 at the same EOT of $\sim$1.2–1.8 nm, and has around five orders reduction compared with poly-Si/SiO_2 benchmark.

It was reported that lanthanum oxide (La_2O_3) films can induce a large negative V_{fb} shift [14]. However, the origin of this shift is not understood. A possible reason for this phenomenon is due to the presence of high density positive fixed charge in the film, which is unacceptable for MOS devices application. In this work, V_{fb} was extracted by C–V curves and plotted as a function of EOT and La concentration in the HfLaO films as shown in Fig. 2. The results show that the amount of V_{fb} shift in HfLaO devices increases with increasing the concentration of La. However, for a fixed La concentration, the V_{fb} shift remains almost constant with the change of EOT, which indicates very low, if any, charge density in HfLaO films. Therefore the change of V_{fb} for HfLaO films is not due to oxide charges, but the change of effective metal gate work function. Recently, by FTIR measurement, we have reported that incorporated La atoms into HfO_2 can distort the monoclinic structure of HfO_2 due to the different bonding properties of La in terms of ionicity or coordination number [11]. This change of dielectric structure and atomic bonds may also change the energy of the interface states between HfLaO and the metal gate, causing the change of Fermi level pinning level, and the effective metal work function. Fig. 2(c) summarizes the V_{fb} as a function of La composition. From the figure, if neglecting the very weak dielectric charge effect, the work functions of both TaN and HfN can be tuned to around 4 eV in the case of 50% La, which matches the requirement of nMOSFETs. Fig. 3(a) and (b) shows the I_d–V_g and I_d–V_d characteristics of nMOSFETs with HfO_2 and HfLaO (with 15% and 50% La) gate dielectrics. As can be seen, there is a significant V_{th} shift between HfO_2 and HfLaO devices, corresponding to the V_{fb} data, and I_d increases with increasing concentration of La for the same device size and operation condition. The I_d improvement may be interpreted by the following: by adding La into HfO_2, the change of atomic

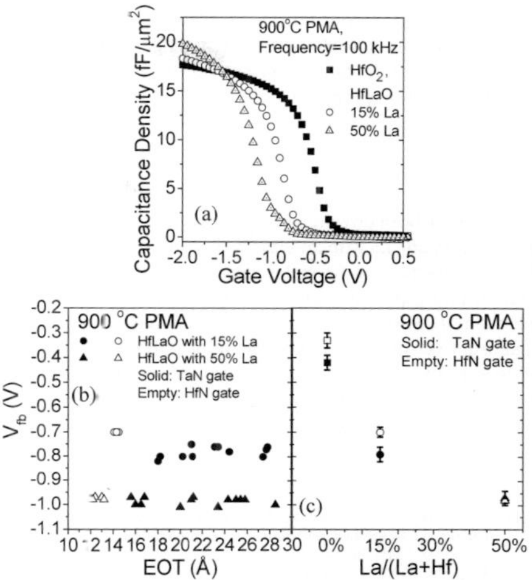

Fig. 2. (a) Typical C–V curves of capacitors with HfO_2 and HfLaO (with 15% and 50% La) gate dielectrics and HfN/TaN metal gate. (b) V_{fb} as a function of EOT, indicating no or very low, if any, oxide charge density in HfLaO films. (c) V_{fb} as a function of La composition in HfLaO films. For p-substrate of 6×10^{15} cm^{-3} doping, the metal work function = $(4.95 + qV_{fb})$ eV, neglecting the weak dielectric charge effect. For 50% La in HfLaO, both HfN and TaN have work functions of $\sim$4 eV.

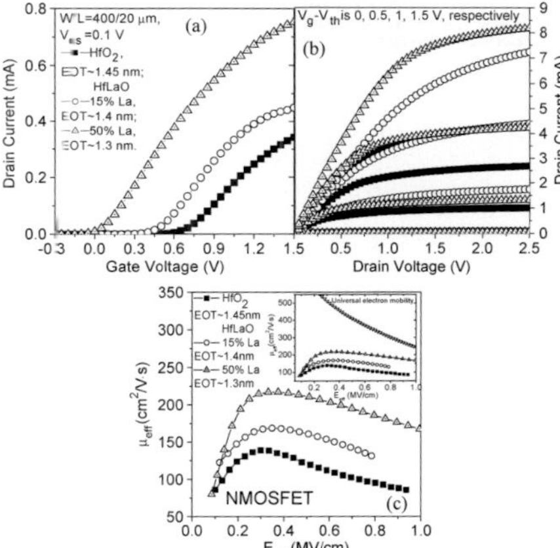

Fig. 3. (a) I_d–V_g and (b) I_d–V_d characteristics of nMOSFETs with HfO_2 and HfLaO (with 15% and 50% La) gate dielectrics. Scaling to the same EOT, I_d has improvement when using HfLaO to replace HfO_2 as gate dielectric. (c) Effective electron mobility extracted by split-C–V method and compared with universal electron mobility in the inset.

bonding and low energy phonon spectra observed in FTIR [11], may reduce the remote phonon scattering [15] and increase the electron channel mobility which is shown in Fig. 3(c). However, as compared with universal electron mobility as shown in the inset of Fig. 3(c), mobility improvement for these gate stacks requires further investigation.

IV. Conclusion

TaN (or HfN) metal-gate nMOSFETs with HfLaO gate dielectrics have been investigated systematically for the first time. By incorporating La with different composition into HfO_2 films, the work function of the metal gates can be effectively tuned from Si mid-gap to around 4 eV, which can meet the requirement of nMOSFETs. Incorporation of La into HfO_2 also shows significant performance improvement of nMOSFETs in terms of dielectric crystallization temperature and on drive current. At the same time, no gate leakage degradation was observed, compared with pure HfO_2 gate dielectric.

References

[1] G. D. Wilk, R. M. Wallace, and J. M. Anthony, "High-κ gate dielectrics: current status and materials properties considerations," *J. Appl. Phys.*, vol. 89, p. 5243, 2001.

[2] W. J. Zhu, T. Tamagawa, M. Gibson, T. Furukawa, and T. P. Ma, "Effect of Al inclusion in HfO_2 on the physical and electrical properties of the dielectrics," *IEEE Electron Device Lett.*, vol. 23, no. 11, p. 649, Nov. 2002.

[3] S. H. Bae, C. H. Lee, R. Clark, and D. L. Kwong, "MOS characteristics of ultrathin CVD HfAlO gate dielectrics," *IEEE Electron Devices Lett.*, vol. 24, no. 9, p. 556, Sep. 2003.

[4] G. D. Wilk, R. M. Wallace, and J. M. Anthony, "Hafnium and zirconium silicates for advanced gate dielectrics," *J. Appl. Phys.*, vol. 87, p. 484, 2000.

[5] C. H. Choi, S. J. Rhee, T. S. Jeon, N. Lu, J. H.Sim , R. Clark, M. Niwa, and D. L. Kwong, "Thermally stable CVD HfO_xN_y advanced gate dielectrics with poly-Si gate electrode," in *IEDM Tech. Dig.*, 2002, p. 857.

[6] A. L. P. Rotondaro, M. R. Visokay, J. J. Chambers, A. Shanware, R. Khamankar, H. Bu, R. T. Laaksonen, L. Tsung, M. Douglas, R. Kuan, M. J. Bevan, T. Grider, J. McPherson, and L. Colombo, "Advanced CMOS transistors with a novel HfSiON gate dielectric," in *Symp. VLSI Tech. Dig.*, 2002, p. 11.

[7] H. S. Jung, Y. S. Kim, J. P. Kim, J. H. Lee, J. H. Lee, N. I. Lee, H. K. Kang, K. P. Suh, H. J. Ryu, C. B. Oh, Y. W. Kim, K. H. Cho, H. S. Baik, Y. S. Chung, H. S. Chang, and D. W. Moon, "Improved current performance of CMOSFETs with nitrogen incorporated $HfO_2 - Al_2O_3$ laminate gate dielectric," in *IEDM. Tech. Dig.*, 2002, p. 853.

[8] X. F. Yu, C. X. Zhu, X. P. Wang, M. F. Li, A. Chin, A. Y. Du, W. D. Wang, and D. L. Kwong, "High mobility and excellent electrical stability of MOSFETs using a novel HfTaO gate dielectric," in *Symp. VLSI Tech. Dig.*, 2004, p. 110.

[9] S. B. Samavedam, L. B. La, P. J. Tobin, B. White, C. Hobbs, L. R. C. Fonseca, A. A. Demkov, J. Schaeffer, E. Luckowski, A. Martinez, M. Raymond, D. Triyoso, D. Roan, V. Dhandapani, R. Garcia, S. G. H. Anderson, K. Moore, H. H. Tseng, C. Capasso, O. Adetutu, D. C. Gilmer, W. J. Taylor, R. Hegde, and J. Grant, "Fermi level pinning with sub-monolayer MeO_x and metal gates," in *IEDM. Tech. Dig.*, 2003, p. 307.

[10] H. Y. Yu, C. Ren, Y.-C. Yeo, J. F. Kang, X. P. Wang, H. H. H. Ma, M.-F. Li, D. S. H. Chan, and D.-L. Kwong, "Fermi pinning induced thermal instability of metal gate work functions," *IEEE Electron Device Lett.*, vol. 25, no. 5, p. 337, May 2004.

[11] X. P. Wang, M. F. Li, A. Chin, C. X. Zhu, J. Shao, W. Lu, X. C. Shen, A. Y. Du, A. C. H. Huan, J. S. Pan, and D.-L. Kwong, "Investigation of material and electrical properties of MOS capacitors with HfLaO high-κ gate dielectric," J. Appl. Phys., to be published.

[12] H. Watanabe, N. Ikarashi, and F. Ito, "La-silicate gate dielectrics fabricated by solid phase reaction between La metal and SiO_2 underlayers," *Appl. Phys. Lett.*, vol. 83, p. 3546, 2003.

[13] H. Y. Yu, J. F. Kang, J. D. Chen, C. Ren, Y. T. Hou, S. J. Whang, M. F. Li, D. S. H. Chan, K. L. Bera, C. H. Tung, A. Y. Du, and D.-L. Kwong, "Robust high quality HfN/HfO_2 gate stack for advanced CMOS Devices," in *IEDM. Tech. Dig.*, 2003, p. 99.

[14] S. Guha, E. Cartier, M. A. Gribelyuk, N. A. Bojarczuk, and M. C. Copel, "Atomic beam deposition of lanthanum- and yttrium-based oxide thin films for gate dielectrics," *Appl. Phys. Lett.*, vol. 77, p. 2710, 2000.

[15] Z. Ren, M. V. Fischetti, E. P. Gusev, E. A. Cartier, and M. Chudzik, "Inversion channel mobility in high-κ high performance MOSFETs," in *IEDM. Tech. Dig.*, 2003, p. 793.

Dual Metal Gates with Band-Edge Work Functions on Novel HfLaO High-κ Gate Dielectric

X.P.Wang[1,2], C.Shen[1,2], Ming-Fu Li[1,2], H.Y.Yu[4], Yiyang Sun[2], Y.P.Feng[3], Andy Lim[1], Hwang Wan Sik[1], , Albert Chin[5],Y.C.Yeo[1], Patrick Lo[2] , D.L.Kwong[2]

[1] SNDL, ECE Dept, National University of Singapore, Singapore 117546, elelimf@nus.edu.sg
[2] Institute of Microelectronics, Singapore 117685, [3]Dept. Physics, National University of Singapore, Singapore 117540
[4] IMEC, Kapeldreef 75, B-3001 Leuven, Belgium , [5]Dept. of Electronics Eng., Nat'l Chiao-Tung Univ., Hsinchu, Taiwan ROC

Abstract

In this work, by using a novel HfLaO high-κ (HK) gate dielectric, we show for the first time that with a thermal budget of 1000 $^{\circ}C$, Fermi-Pinning in the HK-metal gate (MG) stack can be released. The effective metal work function (EWF) can be tuned by a wide range more than the requirement of bulk CMOSFETs, and also fits the future UTB-SOI CMOSFETs when Si body thickness is approaching 3 nm or less. As prototype examples, TaN gate with EWF ~3.9-4.4 eV and TaN/Pt gate with EWF ~5.5 eV are shown. In addition, by replacing HfO$_2$ with HfLaO, high κ value and low gate tunneling are maintained, BTI V_{th} instability is improved by one order. These new findings are correlated to the enhanced thermal stability and significantly reduced oxygen vacancy density in HfLaO compared to HfO$_2$ as estimated by the first-principles calculations.

Introduction

As outlined by ITRS, metal gate together with high-κ dielectrics would be required for sub-45 nm CMOS technology [ITRS, 1-6]. The required EWF of metal gate bulk CMOSFETs should be closed to those of highly doped poly-Si gate (~4.0 eV for n-FETs and ~5.2 eV for p-FETs). For ultra-thin body (UTB) SOI FETs, the EWF is even more stringent due to carrier quantization and surface roughness effects, as shown in Fig. 1 [7]. One of the key challenges for the integration of MG/HK (such as HfO$_2$ and HfSiO) into conventional CMOS technology is the high V_{th} of the FETs due to Fermi-Pinning effect between MG and HK [8,9]. Table I lists some latest technologies for metal EWF achievements. Till date, the metal EWF is not satisfied for p-FET with high thermal budget required in gate first CMOS technology, also not overall satisfied for UTB n-FET when Si body thickness is ≤3 nm with different surface orientation. Different mechanisms have been proposed to account for the high V_{th} observed in MG/HK FETs, such as Fermi Pinning [8,9], V_{fb} shift to midgap after annealing [10], reaction between metal and dielectric and oxygen vacancy at metal/dielectric interface [6,11].

In this work, we incorporate La to HfO$_2$ as a gate dielectric. As prototype examples, TaN (for n-FET) with EWF of ~3.9-4.4 eV and TaN/Pt (for p-FET) with EWF of ~5.5 eV are shown, with thermal budget up to 1000 $^{\circ}C$. Excellent dielectric properties for HfLaO include high crystallization temperature (up to 900 $^{\circ}C$), one order reduction of V_{th} shift under BTI stress and low leakage current comparable with HfO$_2$.

Experimental

(100) n & p doped (6x10^{15}cm^{-3}) Si substrates were used in MOS fabrication process: (1) DHF-last RCA pre-gate clear, (2) HfO$_2$ or HfLaO deposition using reactive sputtering with low oxygen concentration, followed by PDA in N$_2$ with a small amount of O$_2$ at 600 $^{\circ}C$ for 30 s. HfLa target was used to avoid water absorption of La during exposure to air [12]. The composition of La was controlled by the varying sputter power on Hf and HfLa targets, and was measured by XPS. (3) TaN or Pt (capped with TaN) gate electrode deposition using PVD, followed by gate patterning. (4) Devices were then annealed at different temperature up to 1000 $^{\circ}C$ with different ambient for thermal stability evaluation.

Results and Discussion

Experimental results: Figures 2 & 3 show the CV curves, the V_{fb} versus EOT plots and extracted EWF for TaN and TaN/Pt metal gates on HfLaO, all after RTA at 1000 $^{\circ}C$. Fig. 3 demonstrates that the dielectric charge effect is small and the shift of EWF is mainly due to change of EWF of the metal when adding La into HfO$_2$, This is probably due to release of Fermi Pinning at metal/dielectric interface. Fig. 4 shows the I_d –V_g characteristics of n- and p-FETs. The V_{th} in Fig. 4 are consistent with the V_{fb} obtained in Fig. 3. Incorporation of La in HfO$_2$ also improves the gate stack thermal stability with Pt electrode from less than 600 $^{\circ}C$ [6] (Fig. 2 only shows FGA result for HfO$_2$ dielectric) to 1000 $^{\circ}C$. This large Φ_M window (3.9-5.5 eV) between TaN and Pt electrodes shows the possibility of obtaining optimal metal gate work function for CMOSFETs on HfLaO dielectric. The XRD spectra (Fig. 5) show that HfLaO can raise the crystallization temperature to 900 $^{\circ}C$. Fig. 6 shows that HfLaO gate leakage is comparable with pure HfO$_2$ and ~5 orders reduction compared with SiO$_2$ at the same EOT. Fig. 7 shows one order reduction of BTI shift for HfLaO with 50% La, compared with HfO$_2$ gate dielectric.

***Ab initio* calculation of oxygen vacancy formation** :
Table II shows the first-principles calculation results (by VASP) of oxygen vacancy formation using monoclinic HfO$_2$ [13] and pyrochlore HfLaO [14] configurations shown in Fig. 8. By Table II, the oxygen vacancy density ratio between Hf$_2$La$_2$O$_7$ and HfO$_2$ with same volume was estimated to be 0.2 by

$$Ratio = \frac{D_{Td}\exp[-\dfrac{E_{Td}-E_{V4}}{k_B T}]+D_{C2v}\exp[-\dfrac{E_{c2v}-E_{v4}}{k_B T}]}{D_{v3}\exp[-\dfrac{E_{v3}-E_{v4}}{k_B T}]+D_{v4}}$$

Considering the non-fully thermal equilibrium under PMA annealing, the difference between the real amorphous phase and simulated crystal phase, the calculated ratio 0.2 gives a reasonable explanation of reduction of V_{th} shift under BTI stress for HfLaO device shown in Fig. 7. The reduction of oxygen vacancy density is probably also one of major facts to interpret the improved thermal stability and released Fermi pinning of the metal/HfLaO gate stack [6].

Conclusion

Using HfLaO as a high-κ gate dielectric, the Fermi-pinning between MG and HK can be released. TaN metal gate (for n-MOSFETs) with EWF of ~3.9-4.4 eV and TaN/Pt metal gate (for p-MOSFETs) with EWF of ~5.5 eV, with 1000 $^{\circ}C$ thermal budget are demonstrated. Comparing with HfO$_2$ gate dielectric, low gate current is maintained while BTI stress degradation is improved.

Acknowledgement: This work is supported by Singapore A-STAR research grant R263-000-267-305

Reference

[1] J.K.Schaeffer et al, *IEDM* 2004, p.287. [2] V.Narayanam et al, *Symp. VLSI Tech.* 2004, p.192. [3] S.Inumiya et al, *IEDM* 2005, p.27. [4] H.C.Wen et al *Symp.VLSI Tech.* 2005, p.46.[5] Z.B.Zhang et al, *Symp. VLSI Tech.* 2005, p.50.[6] E.Cartier et al, *Symp.VLSI Tech.* 2005, p.230.[7] Tony Low et al, *IEDM* 2004, p.151. [8] S.B. Samavedam *et al., IEDM*, p.307, 2003. [9] Y.C Yeo et al, *JAP* v.92,p.7266 (2002). [10] H.Y.Yu *et al., EDL*, v.25, p.337, (2004). [11] R.Jha et al, *IEDM* 2005, p.47.[12] D.S.Yu *et al., IEDM 2000* p.181. [13] J.Aarik et al, *Appl.Surface Sci.*, v.173,p.15 (2001).[14] A.W. Sleight, Inorganic Chemistry, v.7,p.1704(1968), Table IV.

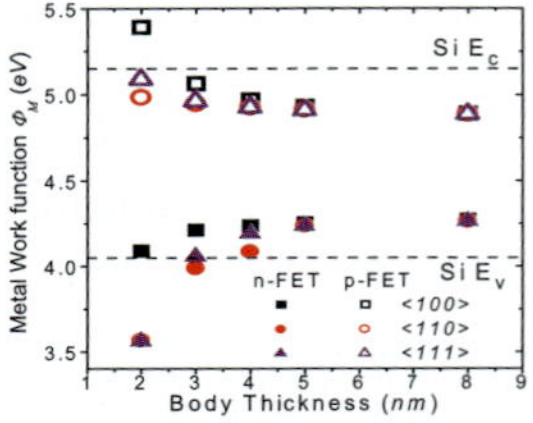

Fig. 1 The calculated optimal gate work function for Si UTB MOSFETs as a function of body thickness. Carrier quantization and surface roughness effect are considered [7].

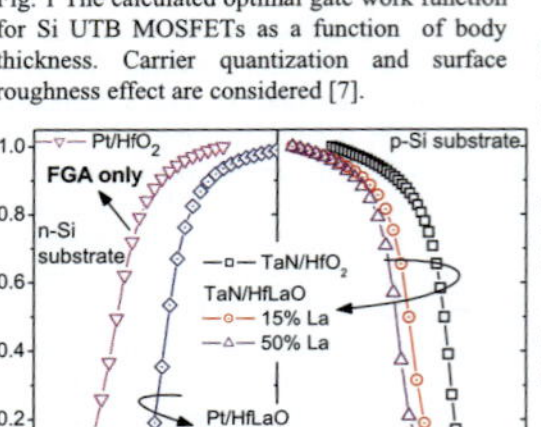

Fig. 2 Comparison of typical C-V curves for MOS HfLaO capacitors with different La concentration after 1000 °C PMA. Obvious V_{fb} shift can be seen after La incorporation.

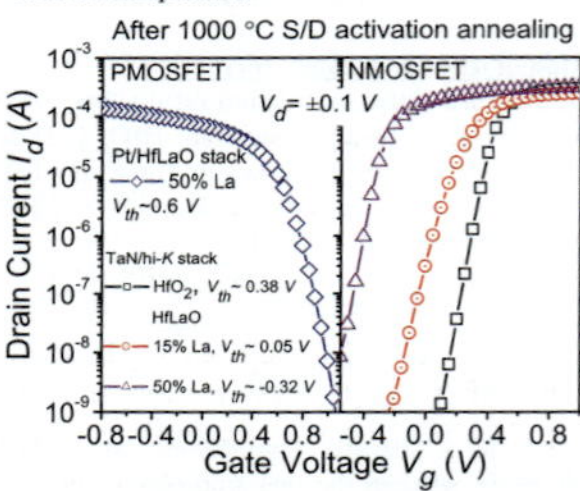

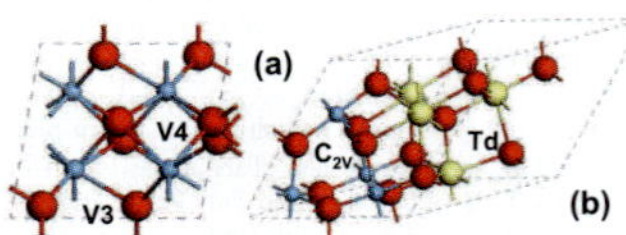

Fig. 4 Id-Vg curves of n- & p-MOSFETs using TaN for n-MOSFETs, TaN/Pt for p-MOSFETs, and HfLaO for gate dielectric. (p- & n-substrates doping concentration 6×10^{15} cm^{-3}).

Table I Comparison of some latest high-k/metal gate technologies in EWF

Gate Dielectric	Metal Gate	Effective WF (eV)	Bulk nFET	UTB nFET	Bulk pFET	UTB pFET	Thermal stability	Ref
HfO$_2$	Pt, Re				yes		500° C	[6]
HfO$_2$	Ru				high V$_{th}$			[5]
HfO$_2$	TaC	4.18	yes	no			1000° C	[1]
HfO$_2$	TaSiN W, Re		high V$_{th}$	no	high V$_{th}$	no	Gate last	[2]
HfSiO	Metal-Si-N	4.16-4.75	yes	no	high V$_{th}$	no	1000° C	[4]
HfSiON	TaSi$_x$		yes	no			1000° C	[3]

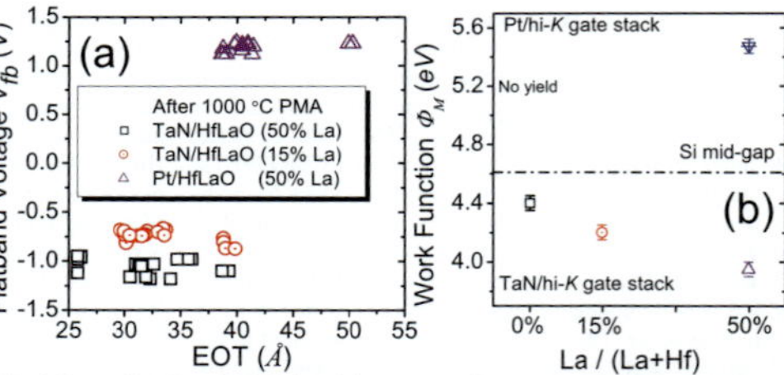

Fig. 3 V_{fb} as a function of EOT (a), and the corresponding metal work function as a function of La composition in HfLaO films (b). (p- & n-substrates doping concentration 6×10^{15} cm^{-3}).

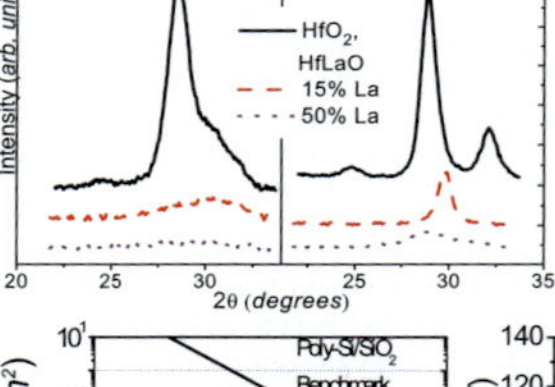

Fig. 5 XRD spectra of HfO$_2$, HfLaO films with 15% and 50% La after 600 °C and 900 °C annealing for 30 s in N$_2$. The La incorporated in HfO$_2$ film can increase the crystallization temperature up to 900 °C.

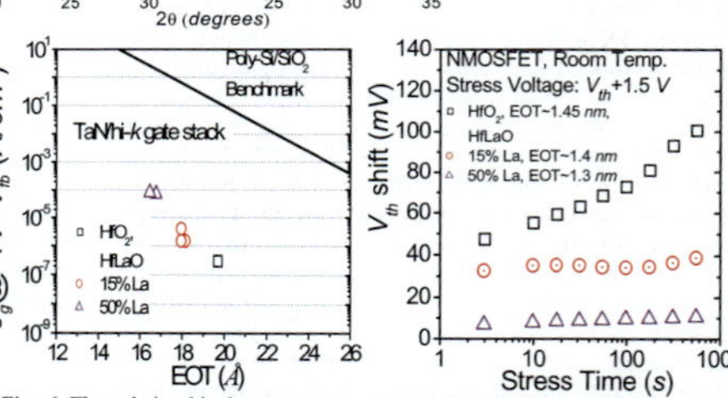

Fig. 6 The relationship between gate leakage current density and EOT for MOS devices with HfO$_2$ and HfLaO gate dielectrics and TaN metal gate.

Fig. 7 Charge trapping induced V_{th} shift under constant voltage stress for HfO$_2$, HfLaO with 15% and 50% La gate dielectric NMOSFETs.

Dark (red) large balls-oxygen atoms; Shallow (blue) small balls–Hf atoms; Shallow (yellow) large balls –La atoms.

Fig. 8 (a) Monoclinic HfO$_2$ primitive cell. Two possible Oxygen vacancy sites: V3 (3-fold coordinated) and V4 (4-fold coordinated); (b) Pyrochlore Hf$_2$La$_2$ O$_7$ primitive cell. Two possible Oxygen vacancy sites: Td (Td symmetry) and C$_{2v}$ (C$_{2v}$ symmetry).

Table II The formation energies of oxygen vacancies at varies sites in monoclinic HfO$_2$ and pyrochlore HfLaO (as shown in Fig. 8), calculated by first- principles calculations using VASP software.

Oxygen vacancy site	Formation energy E(eV)	Site density D(nm^{-3})
V3 site in HfO$_2$	6.51	28.6
V4 site in HfO$_2$	6.39	28.6
Td site in Hf$_2$La$_2$O$_7$	7.23	6.3
C$_{2v}$ site in Hf$_2$La$_2$O$_7$	6.51	38.0

862   IEEE ELECTRON DEVICE LETTERS, VOL. 28, NO. 10, OCTOBER 2007

NMOS Compatible Work Function of TaN Metal Gate With Erbium-Oxide-Doped Hafnium Oxide Gate Dielectric

Jingde Chen, *Student Member, IEEE*, X. P. Wang, Ming-Fu Li, *Senior Member, IEEE*, S. J. Lee, M. B. Yu, C. Shen, *Student Member, IEEE*, and Yee-Chia Yeo, *Member, IEEE*

Abstract—This letter demonstrates reduction in effective work function of tantalum-nitride (TaN) metal gate with erbium-oxide-doped hafnium oxide. We report that TaN effective metal-gate work function can be tuned from Si midgap to the conduction band to meet the work-function requirement of NMOSFETs by incorporating ErO in HfO_2 with an equivalent oxide thickness as low as 1.15 nm. Several other lanthanide-oxide doped hafnium oxides show similar characteristics.

Index Terms—Erbium, high-κ dielectric, lanthanide, metal gate, MOS capacitor, rare earth metal.

I. INTRODUCTION

AS ONE OF THE most promising high-κ dielectrics, HfO_2 has been extensively investigated to replace the conventional SiO_2 and SiON gate dielectrics in MOSFETs [1]. One major problem for HfO_2 is the Fermi-level pinning between the HfO_2 and a metal-gate material [2], [3]. Due to the Fermi-level pinning, difficulties are faced in obtaining an effective work function of 4.1 eV in the metal/HfO_2 gate stack as required by NMOSFETs [4]. It was previously reported that incorporating La into HfO_2 can tune the work function of tantalum-nitride (TaN) metal gate from midgap to Si conduction band edge [5]. In this letter, we found that several lanthanide metals have the same effect when incorporated into HfO_2, while erbium-oxide-doped hafnium oxide (HfErO) will be mainly discussed in this letter in detail. It is found that, when the atomic ratio of Hf : Er is 7 : 3 (30% Er) in the HfErO films, the effective work function (EWF) of the TaN metal gate can be tuned to around 4.10 eV, without degrading oxide integrity even after high temperature anneal.

Manuscript received June 8, 2007; revised July 5, 2007. This work was supported in part by the Applied Materials, Inc., Research Grant R263-000-385-720 and in part by the Singapore A-STAR Research Grant R263-000-267-305. The review of this letter was arranged by Editor A. Chatterjee.

J. Chen, X. P. Wang, S. J. Lee, C. Shen, and Y.-C. Yeo are with the Silicon Nano Device Laboratory, Department of Electrical and Computer Engineering, National University of Singapore, Singapore 119260.

M.-F. Li is with the Silicon Nano Device Laboratory, Department of Electrical and Computer Engineering, National University of Singapore, Singapore 119260, and also with the Department of Microelectronics, Fudan University, Shanghai 201203, China (e-mail: elelimf@nus.edu.sg; mfli@fudan.edu.cn).

M. B. Yu is with the Institute of Microelectronics, Singapore 117685.

Color versions of one or more of the figures in this letter are available online at http://ieeexplore.ieee.org.

Digital Object Identifier 10.1109/LED.2007.904210

II. EXPERIMENTAL SETUP

The MOS capacitors were fabricated using p-type (100) Si substrates with 6×10^{15} cm^{-3} boron doping. After field oxide growth ($\sim$400 nm) and active area definition, diluted hydrofluoric acid-last Radio Corporation of America pregate clean was performed. Control HfO_2 and also HfO_2 incorporating a lanthanide oxide were then deposited using reactive sputtering in N_2 ambient at room temperature. This was followed by an *ex situ* postdeposition anneal in N_2 at 600 °C for 30 s. The composition of lanthanide metal was controlled by the ratio of the power provided to the hafnium and lanthanide metal targets and determined using X-ray photoelectron spectroscopy. The samples are then immediately put into the sputter system (within 5 min) for gate-electrode deposition. The short time gap minimizes possible moisture absorption effect for the lanthanide oxide [6]. TaN gate electrodes ($\sim$150 nm) were deposited using reactive sputtering and, then, patterned using a Cl_2-based etchant. After that, the devices were rapid thermal annealed in N_2 at 1000 °C for 5 s for thermal stability evaluation. Finally, all samples received back side Al metallization and 420-°C forming gas sintering.

Electrical characteristics of the MOS capacitors were measured using HP4284A precision LCR meter and HP4156A parameter analyzer. The equivalent oxide thickness (EOT) and flatband voltage (V_{fb}) were obtained by fitting the high-frequency (100 KHz) capacitance–voltage (C–V) measurements with simulated C–V curves that account for quantum–mechanical effects. The simulator used is the QMCV program developed by UC Berkeley [7].

III. RESULTS AND DISCUSSION

Fig. 1 shows the normalized C–V curves for the TaN gated capacitors with HfO_2 and HfErO gate dielectrics with different Er concentrations after forming gas sintering at 420 °C [Fig. 1(a)] and after rapid thermal processing (RTP) at 1000 °C [Fig. 1(b)]. V_{fb} shift is observed for all HfErO samples at both temperatures compared to the control HfO_2 samples, while samples with 70% Er have more V_{fb} shift than those with 30% Er at both temperatures. A higher Er concentration leads to a larger V_{fb} shift.

Comparing Fig. 1(a) and (b), it is also observed that after 1000-°C anneal, the V_{fb} differential between HfErO and control

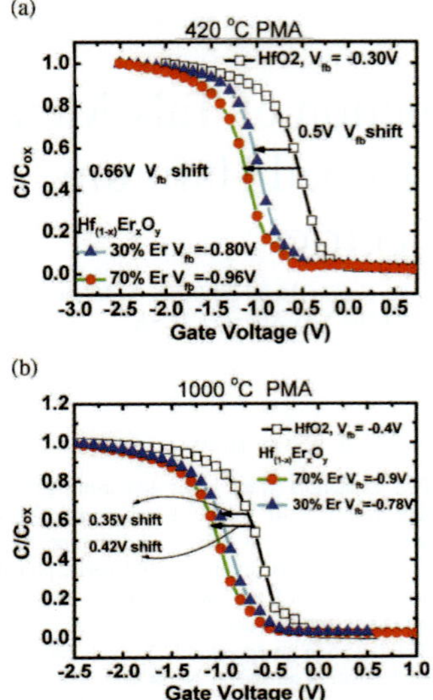

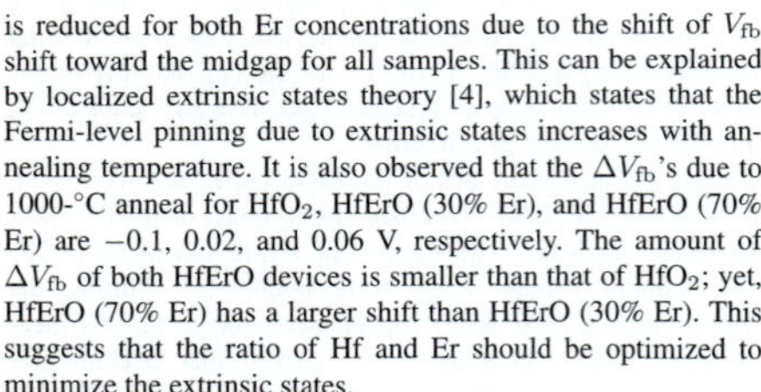

Fig. 1. (a) Typical C–V curves of capacitors with HfO$_2$ and HfErO (with 30% and 70% Er) gate dielectrics and TaN metal gate after 420-°C forming gas annealing. (b) Typical C–V curves of capacitors with HfO$_2$ and HfErO (with 30% and 70% Er) gate dielectrics and TaN metal gate after 1000-°C 5-s annealing.

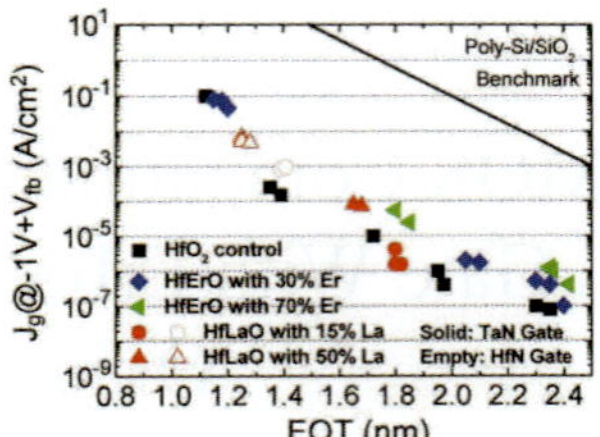

Fig. 2. Gate leakage current density versus EOT relationship for MOS devices with HfO$_2$ and HfLaO and HfErO gate dielectrics. The HfLaO data are from [5].

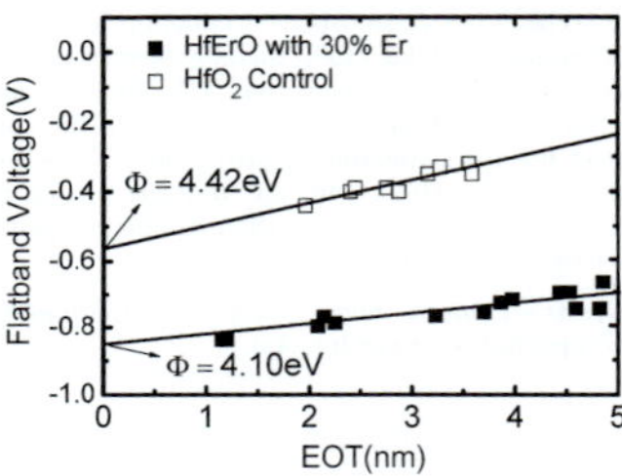

Fig. 3. V_{fb} versus EOT plots of capacitors with HfO$_2$ and HfErO (with 30% Er) gate dielectrics and TaN metal gate after 1000-°C 5-s annealing. The p-Si substrate doping is 6×10^{15} cm^{-3}. From these data, the EWF of 4.10 eV for TaN on HfErO is obtained.

is reduced for both Er concentrations due to the shift of V_{fb} shift toward the midgap for all samples. This can be explained by localized extrinsic states theory [4], which states that the Fermi-level pinning due to extrinsic states increases with annealing temperature. It is also observed that the ΔV_{fb}'s due to 1000-°C anneal for HfO$_2$, HfErO (30% Er), and HfErO (70% Er) are -0.1, 0.02, and 0.06 V, respectively. The amount of ΔV_{fb} of both HfErO devices is smaller than that of HfO$_2$; yet, HfErO (70% Er) has a larger shift than HfErO (30% Er). This suggests that the ratio of Hf and Er should be optimized to minimize the extrinsic states.

Fig. 2 shows the gate leakage current densities of HfO$_2$, HfLaO, and HfErO at different EOTs. The HfLaO data are from our previously published letter [5] but annealed at a lower temperature of 900 °C. The leakage current densities of HfErO locate on almost the same trend line as HfO$_2$ and HfLaO, and have around four orders of reduction compared with poly-Si/SiO$_2$ benchmark. The leakage is slightly higher than that of HfO$_2$. The leakage is also slightly higher for the HfErO with higher Er concentration. This can be partially explained by the lower permittivity of erbium oxide. The reported relative permittivity of Er$_2$O$_3$ is 14 [8], which is lower than that of HfO$_2$ (25) [9]. The incorporation of Er decreases the permittivity. To achieve the same EOT, the physical thickness of HfErO must be smaller than that of HfO$_2$, which leads to higher leakage

current, assuming the same band offset. It is reported that there is a substantial formation of silicate at the Er$_2$O$_3$/Si interface at high temperature (900 °C), and the silicate interfacial layer increases leakage current density [10]. This also explains the increase of leakage current density with an increase of erbium concentration. The other two factors for gate leakage are the oxide traps and band offset, which are yet to be studied.

Effective Φ_{ms} was extracted by the C–V curves and plotted as a function of EOT, as shown in Fig. 3. For 30% Er concentration HfErO after 1000-°C RTP, the extracted V_{fb} is 4.1 eV, which meets the NMOS requirement. The slope for HfErO is smaller than that for the HfO$_2$ control, which means that the fixed charge density in HfO$_2$ is reduced with the incorporation of erbium. This is similar to the previous published work on HfLaO [5]. As for the reason of the V_{fb} shift, the C–V–EOT plot rules out the V_{fb} shift due to the fixed charge in oxide. The Er is known to have an electronegativity of 1.24 on the Pauling scale, which is lower than those of Hf (1.3) and Ta (1.5) [11]. Due to the lower electronegativity of Er atoms, additional electron transfer from HfErO to TaN can be expected when Er atoms replace the Hf atoms. This effect compensates the electron transfer from the metal gate to the dielectric due to the Fermi-level pinning, giving rise to the partial release of Fermi-level pinning and reduction of $\Phi_{m,\mathrm{eff}}$ of TaN.

864   IEEE ELECTRON DEVICE LETTERS, VOL. 28, NO. 10, OCTOBER 2007

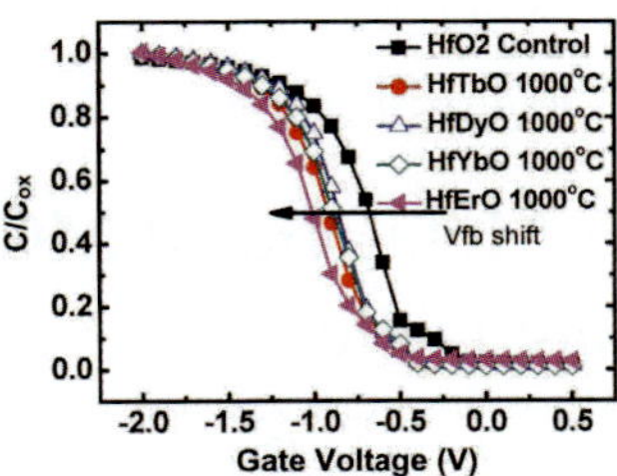

Fig. 4. $C–V$ curves of HfO$_2$ doped by ErO, TbO, YbO, and DyO after 1000-°C anneal. The ratios Hf : Er, Tb : Hf, Yb : Hf, and Dy : Hf are 3.0 : 7.0, 2.7 : 6.3, 1.7 : 8.3, and 2.1 : 7.9, respectively. All the curves show significant flatband voltage shift toward silicon conduction band.

Referring to a recent publication [12], it is also possible that the HfErO possesses lower oxygen-vacancy (VO) density than the HfO$_2$, which leads to the decrease of the amount of electron transfer induced by VO for the gate stacks with HfErO dielectric compared to those with HfO$_2$ and the partial release of Fermi-level pinning between the HfErO and the TaN.

The $C–V$ curves of MOS capacitors with different lanthanide-oxide doped HfO$_2$'s are shown in Fig. 4, namely, TbO, DyO, and YbO. All samples were fabricated with a 1000-°C 5-s RTP anneal and show significant $V_{\rm fb}$ shift compared to the control. The $V_{\rm fb}$ shift depends not only on the electronegativity of Er, Tb, Dy, and Yb but also on their atomic concentration in the dielectric. As their concentrations differ in Fig. 4, we do not correlate the effectiveness of work-function tuning with the electronegativity of each lanthanide metal. The leakage current data from thick dielectrics with EOT above 2.5 nm, e.g., from HfTbO, HfYbO, and HfDyO, are dominated by trap-assisted tunneling, not direct tunneling. Hence, they are not compared with the direct tunneling current data from the thin dielectrics shown in Fig. 2. The metals studied in this letter are all lanthanide (also rare earth metals) which exhibit similar physical and chemical properties. It was reported that the La-based oxide films can induce a large negative $V_{\rm fb}$ shift [13]. However, due to the moisture-absorption-induced oxide degradation [6], lanthanum and other lanthanide series metal oxides are not suitable to serve as gate oxide on their own. Recently, tuning the TaN work function by incorporating other lanthanide metals into HfO$_2$ by various methods has been reported. Incorporating LaO into HfO$_2$ is reported to tune the TaN work function to 4.0 eV [5]. There are other two recent works which reported that Gd$_2$O$_3$ and La$_2$O$_3$ interfacial layers on the Hf-based oxide are able to tune the HfO$_2$ work function to 4.2 eV [14], [15]. In fact, the interfacial layer eventually mixes up with the underlying Hf-based oxide due to the diffusion during thermal process [15], resulting in a similar structure to this letter. With the results in this experiment and in the literature mentioned, it is reasonable to claim that, probably, all lanthanide metals have the ability to tune the TaN metal-gate work function toward silicon band edge when incorporated into hafnium oxide, and the metal work function is tunable by adjusting the lanthanide concentration. This is an interesting and important result in high-κ/metal-gate technology development.

IV. CONCLUSION

The TaN metal-gate MOS capacitors with lanthanide-oxide doped gate dielectrics have been systematically investigated. By incorporating Er into HfO$_2$ films, the work function of the TaN metal gates can be effectively tuned from the Si midgap to 4.1 eV, near the silicon band edge, with an EOT as low as 1.15 nm. Similar results are obtained by using other lanthanide metals Tb, Dy, and Yb. Combined with our previous report of LaO doped HfO$_2$ [4], probably, all lanthanide metals share this property. This is important in the high-κ metal-gate technology development as well as the basic understanding of the Fermi-pinning release mechanism between the high-κ and the metal gate.

REFERENCES

[1] G. D. Wilk, R. M. Wallace, and J. M. Anthony, "High-κ gate dielectrics: Current status and materials properties considerations," *J. Appl. Phys.*, vol. 89, no. 10, pp. 5243–5275, May 2001.

[2] S. B. Samavedam, L. B. La, P. J. Tobin, B. White, C. Hobbs, L. R. C. Fonseca, A. A. Demkov, J. Schaeffer, E. Luckowski, A. Martinez, M. Raymond, D. Triyoso, D. Roan, V. Dhandapani, R. Garcia, S. G. H. Anderson, K. Moore, H. H. Tseng, C. Capasso, O. Adetutu, D. C. Gilmer, W. J. Taylor, R. Hegde, and J. Grant, "Fermi level pinning with sub-monolayer MeO$_x$ and metal gates," in *IEDM. Tech. Dig.*, 2003, pp. 307–310.

[3] Y.-C. Yeo, T.-J. King, and C. Hu, "Metal-dielectric band alignment and its implications for metal gate complementary metal–oxide–semiconductor technology," *J. Appl. Phys.*, vol. 92, no. 12, pp. 7266–7271, Dec. 2002.

[4] H. Y. Yu, C. Ren, Y.-C. Yeo, J. F. Kang, X. P. Wang, H. H. H. Ma, M.-F. Li, D. S. H. Chan, and D.-L. Kwong, "Fermi pinning induced thermal instability of metal gate work functions," *IEEE Electron Device Lett.*, vol. 25, no. 5, pp. 337–339, May 2004.

[5] X. P. Wang, M.-F. Li, C. Ren, X. F. Yu, C. Shen, H. H. Ma, A. Chin, C. X. Zhu, J. Ning, M. B. Yu, and D.-L. Kwong, "Tuning effective metal gate work function by a novel gate dielectric HfLaO for nMOSFETs," *IEEE Electron Device Lett.*, vol. 27, no. 1, pp. 31–33, Jan. 2006.

[6] Y. Zhao, M. Toyama, K. Kita, K. Kyuno, and A. Toriumi, "Moisture absorption-induced permittivity deterioration and surface roughness enhancement of lanthanum oxide films on silicon," in *Proc. SSDM Dig.*, 2005, pp. 546–547.

[7] C. A. Richter, A. R. Hefner, and E. M. Vogel, "A comparison of quantum–mechanical capacitance–voltage simulators," *IEEE Electron Device Lett.*, vol. 22, no. 1, pp. 35–37, Jan. 2001.

[8] H. R. Huff and D. C. Gilmer, *High Dielectric Constant Materials*. New York: Springer-Verlag, 2005, p. 246.

[9] J. Robertson, "Band offsets of wide-band-gap oxides and implications for future electronic devices," *J. Vac. Sci. Technol. B, Microelectron. Process. Phenom.*, vol. 18, no. 3, pp. 1785–1791, May 2000.

[10] T.-M. Pan, C.-L. Chen, W. W. Yeh, and S.-J. Hou, "Structural and electrical characteristics of thin oxide gate dielectrics," *Appl. Phys. Lett.*, vol. 89, no. 22, p. 222912, Nov. 2006.

[11] *CRC Handbook of Chemistry and Physics*, CRC Press, Boca Raton, FL, 2005, pp. 9–76.

[12] X. F. Wang, H. Y. Yu, M.-F. Li, C. X. Zhu, S. Biesemans, A. Chin, Y. Y. Sun, Y. P. Feng, A. Lim, Y.-C. Yeo, W. Y. Loh, Q. Q. Lo, and D.-L. Kwong, "Wide $V_{\rm fb}$ and $V_{\rm th}$ tunability for metal-gated MOS devices with HfLaO gate dielectrics," *IEEE Electron Device Lett.*, vol. 28, no. 4, pp. 258–260, Apr. 2007.

[13] S. Guha, E. Cartier, M. A. Gribelyuk, N. A. Bojarczuk, and M. C. Copel, "Atomic beam deposition of lanthanum- and yttrium-based oxide thin films for gate dielectrics," *Appl. Phys. Lett.*, vol. 77, no. 17, pp. 2710–2712, Oct. 2000.

[14] G. Thareja, H. C. Wen, R. Harris, P. Majhi, B. H. Lee, and J. C. Lee, "NMOS compatible work function of TaN metal gate with gadolinium oxide buffer layer on Hf-based dielectrics," *IEEE Electron Device Lett.*, vol. 27, no. 10, pp. 802–304, Oct. 2006.

[15] H. N. Alshareef, M. Quevedo-Lopez, H. C. Wen, R. Harris, P. Kirsch, P. Majhi, B. H. Lee, F. Jammy, D. J. Lichtenwalner, J. S. Jur, and A. I. Kingon, "Work function engineering using lanthanum oxide interfacial layers," *Appl. Phys. Lett.*, vol. 89, no. 23, p. 232103, Dec. 2006.

258   IEEE ELECTRON DEVICE LETTERS, VOL. 28, NO. 4, APRIL 2007

Wide V_{fb} and V_{th} Tunability for Metal-Gated MOS Devices With HfLaO Gate Dielectrics

X. P. Wang, *Student Member, IEEE*, H. Y. Yu, M.-F. Li, *Senior Member, IEEE*, C. X. Zhu, *Member, IEEE*, S. Biesemans, Albert Chin, *Senior Member, IEEE*, Y. Y. Sun, Y. P. Feng, Andy Lim, Yee-Chia Yeo, *Member, IEEE*, Wei Yip Loh, G. Q. Lo, and Dim-Lee Kwong, *Senior Member, IEEE*

Abstract—For the first time, we demonstrate experimentally that by using HfLaO high-κ gate dielectric, the flat-band voltage (V_{fb}) and the threshold voltage (V_{th}) of metal-electrode-gated MOS devices can be tuned effectively in a wide range (wider than that from the Si-conduction band edge to the Si-valence band edge) after a 1000-°C annealing required by a conventional CMOS source/drain activation process. As prototype examples shown in this letter, TaN gate with effective work function $\Phi_{m,\text{eff}} \sim 3.9$–4.2 eV and Pt gate with $\Phi_{m,\text{eff}} \sim 5.5$ eV are reported. A specific model based on the interfacial dipole between the metal gate and the HfLaO is proposed to interpret the results. This provides an additionally practical guideline for choosing the appropriate gate stacks and dielectric to meet the requirements of future CMOS devices.

Index Terms—CMOS, Fermi-level pinning, HfLaO, high-κ (HK) dielectric, interfacial dipole, metal gate (MG), work function.

I. Introduction

METAL gates (MGs) together with the high-κ (HK) dielectrics are considered as promising technology options for sub-45-nm CMOS technology [1]. The required effective work function of MGs for the bulk CMOS field-effect transistors (FETs) should be close to those of highly doped poly-Si gate (~ 4.1 eV for n-FETs and ~ 5.2 eV for p-FETs) [2]. One of the key challenges for the integration of MG/HK (such as HfO$_2$ and HfSiO) into the conventional CMOS technology is the high device V_{th} [3]–[12], which is believed to be due to either: 1) Fermi-level-pinning effect between the MG and the HK [9]–[11]; 2) metal/dielectric reaction [8], [12]; or 3) oxygen vacancy at metal/dielectric interface [8], [12].

Manuscript received November 16, 2006. This work was supported by the Singapore A-STAR Research Grant R263-000-267-305. The review of this letter was arranged by Editor A. Chatterjee.

X. P. Wang and M.-F. Li are with the Silicon Nano Device Laboratory, Department of Electrical and Computer Engineering, National University of Singapore, Singapore 119260, and also with the Institute of Microelectronics, Singapore 117685.

H. Y. Yu and S. Biesemans are with the Interuniversity Microelectronics Center, 3001 Leuven, Belgium.

C. X. Zhu, A. Lim, and Y.-C. Yeo are with the Silicon Nano Device Laboratory, Department of Electrical and Computer Engineering, National University of Singapore, Singapore 119260.

A. Chin is with the Department of Electronics Engineering, National Chiao-Tung University, Hsinchu, 300 Taiwan, R.O.C.

Y. Y. Sun and Y. P. Feng are with the Department of Physics, National University of Singapore, Singapore 117540.

W. Y. Loh, G. Q. Lo, and D.-L. Kwong are with the Institute of Microelectronics, Singapore 117685.

Color versions of one or more of the figures in this paper are available online at http://ieeexplore.ieee.org.

Digital Object Identifier 10.1109/LED.2007.891757

In our previous work, we have already demonstrated the superior performance of the nMOSFETs by using HfLaO as compared to those using the pure HfO$_2$ gate dielectric and reported that $\Phi_{m,\text{eff}}$ of TaN (or HfN) can be tuned from Si midgap to the conduction band by incorporating La into the HfO$_2$ gate dielectric [13]. The aforementioned results were obtained using a 900-°C annealing. In this letter, we further investigated the electrical properties of HfLaO after a 1000-°C annealing, which is required by a conventional CMOS source/drain activation process. We reported that HfLaO not only modulated $\Phi_{m,\text{eff}}$ toward the Si-conduction band edge, but also the Si-valence band edge, fulfilling the $\Phi_{m,\text{eff}}$ requirements for bulk CMOS devices. As prototype examples, TaN gate with $\Phi_{m,\text{eff}}$ of ~ 3.9–4.2 eV, Pt gate with $\Phi_{m,\text{eff}}$ of ~ 5.5 eV, and large $V_{\text{fb}}/V_{\text{th}}$ tunable range correspondingly are shown. It is believed that the interfacial dipole between the MG and the HfLaO plays a critical role for the data interpretation in this letter.

II. Experimental

(100) Si substrates with n-type or p-type doping concentration of 6×10^{15} cm^{-3} were used. The MOS-capacitor (MOSCAP) fabrication process comprises a 0.4-μm field oxide and active area definition, a DHF-last RCA pregate clean, a HfO$_2$ or HfLaO deposition with different physical thicknesses and different La concentrations using the reactive sputtering with low-oxygen concentration, followed by postdeposition anneal in N$_2$ with a small amount of O$_2$ at 600 °C for 30 s. Here, it should be noted that the HfLa target was used to reduce the moisture absorption of La during exposure to air [14]. We also tried to shorten the period between the MG deposition and the HfLaO deposition to avoid the possible moisture absorption. In addition, even though there was still some water absorbed during the device fabrication, it has been demonstrated that the absorbed water can be annealed out, even at relatively low temperatures [15]. The composition of La was controlled by the power ratio between the Hf and HfLa targets and was determined by X-ray photoelectron spectroscopy. TaN (~ 150 nm) or Pt (~ 50 nm) gate electrodes were formed by reactive sputtering and dc sputtering, respectively, followed by gate patterning. The devices then went through rapid thermal annealing (RTA) at 1000 °C for 5 s with different ambient for thermal-stability evaluation. For MOSFET fabrication, the source/drain implantations of arsenic (100 keV, 1×10^{15} cm^{-2})/BF$_2$(50 keV, 1×10^{15} cm^{-2}) were performed for nMOSFETs/pMOSFETs, respectively, followed

by the RTA activation at 1000 °C for 5 s. Finally, all samples received backside Al metallization and forming gas annealing at 420 °C for 30 min.

HP4284A precision LCR meter and HP4156A parameter analyzer were used to measure the electrical characteristics. The equivalent oxide thickness (EOT) and the $V_{\rm fb}$ of the capacitors were extracted using a simulator that accounts for the quantum–mechanical effects.

III. RESULTS AND DISCUSSION

Fig. 1(a) shows the typical capacitance–voltage ($C–V$) measurement (after normalization) of MOSCAPs with TaN and Pt MGs on either HfO_2 or HfLaO (with different La concentrations). Significant $V_{\rm fb}$ shift can be observed after the incorporation of La into the HfO_2. It should be mentioned that the incorporation of La into the HfO_2 improves the thermal stability of the gate stack with Pt electrode from less than 600 °C [8] to 1000 °C. The root reason for this improvement is still under investigation. To understand the significant $V_{\rm fb}$ shift, the $V_{\rm fb}$ values and the EOT measured from MOSCAPs with different dielectric thicknesses are plotted in Fig. 1(b). It turns out that the shift of $V_{\rm fb}$ for HfLaO dielectric is not caused by the fixed charges in the dielectric film but mainly by the change of $\Phi_{m,\rm eff}$ of TaN and Pt compared to the HfO_2 case. The corresponding $\Phi_{m,\rm eff}$s for TaN and Pt on HfLaO were obtained from Fig. 1(b) and summarized in Fig. 1(c). It is shown that dual MGs with band-edge $\Phi_{m,\rm eff}$ on HfLaO (with 50% La) after 1000 °C are achieved simultaneously (TaN gate with $\Phi_{m,\rm eff}$ of ~ 3.9 eV and Pt gate with $\Phi_{m,\rm eff}$ of ~ 5.5 eV). In Fig. 2, MOSFETs with the same gate stacks mentioned in Fig. 1 were also fabricated, and their transfer characteristics are shown. The $V_{\rm th}$ values for all devices are consistent with the $\Phi_{m,\rm eff}$ values obtained in Fig. 1(b), indicating that the wide $V_{\rm fb}$ and $V_{\rm th}$ tunability for these gate stacks stems from the modulation of the MG $\Phi_{m,\rm eff}$.

When metals and HK dielectrics are placed in contact, the MG/HK interface states would be created, and generally, the charge transfer could occur across the interface states at the dielectric side and the metal electrode side, leading to the formation of a dipole layer and Fermi-level pinning of $\Phi_{m,\rm eff}$ [10]. In this letter, two additional effects which influence the charge transfer, causing the release or partial release of Fermi-level pinning of $\Phi_{m,\rm eff}$, are considered. The first effect is charge transfer due to the electronegativity of the materials. La is known to have an electronegativity (the tendency of an atom in a molecule to attract electrons to it) of 1.1 on the Pauling scale, which is lower than those of Hf (1.3) and Ta (1.5) [16]. For the TaN gate, additional electron transfer from HK to MG would be expected when La atoms replace the Hf atoms due to the lower electronegativity of La atoms. This effect will compensate the electron transfer from MG to HK due to the Fermi-level pinning, giving rise to the partial release of Fermi-level pinning and reduction of $\Phi_{m,\rm eff}$ of TaN [Fig. 3(a)].

In the case of p-type MGs, the electron transfer from La to Pt due to the low electronegativity of La would not be feasible, because Pt, as a noble metal, generally shows intrinsically inert performances. However, a second effect should be considered. Oxygen-vacancy (V_O) creation in a dielectric is believed to in-

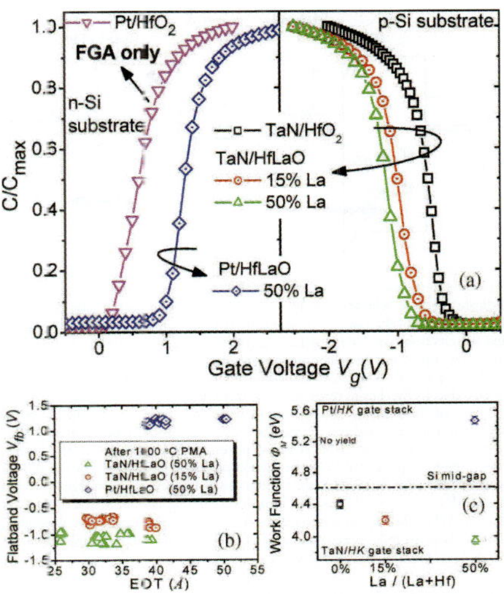

Fig. 1. (a) Comparison of typical $C–V$ curves for MOS HfO_2 and HfLaO capacitors with different La concentrations (La concentrations of 50%, 15%, and 0%, respectively) after a 1000-°C RTA. Obvious $V_{\rm fb}$ shift can be seen after La incorporation (b) $V_{\rm fb}$ as a function of EOT, indicating very low-oxide charge density in HfLaO films and (c) the corresponding $\Phi_{m,\rm eff}$ extracted as a function of La composition in HfLaO films. (Note: Doping concentration for both p- and n-substrates is 6×10^{15} cm^{-3}).

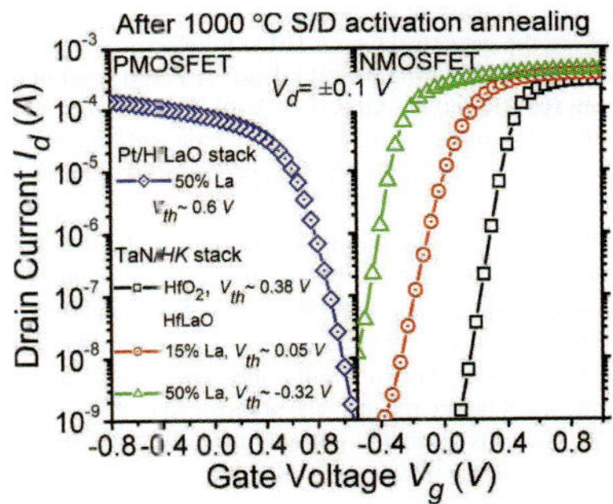

Fig. 2. Transfer characteristics ($I_d–V_g$) of MOSFETs with TaN (for nMOS) and Pt (for pMOS) on HfO_2 and HfLaO (with 15% and 50% La) gate dielectrics.

duce the electron transfer from V_O to the gate electrode, which would lead to a reduction of $\Phi_{m,\rm eff}$ of p-type MG such as Pt [8]. We have roughly estimated the V_O density ratio for pyrochlore $Hf_2La_2O_7$ to monoclinic HfO_2 by the Vienna *ab initio* simulation package to be 0.2. The details of this *ab initio* simulation will be published elsewhere. In addition, it is reported that the presence of V_O can serve as a trapping center [17], and the

260   IEEE ELECTRON DEVICE LETTERS, VOL. 28, NO. 4, APRIL 2007

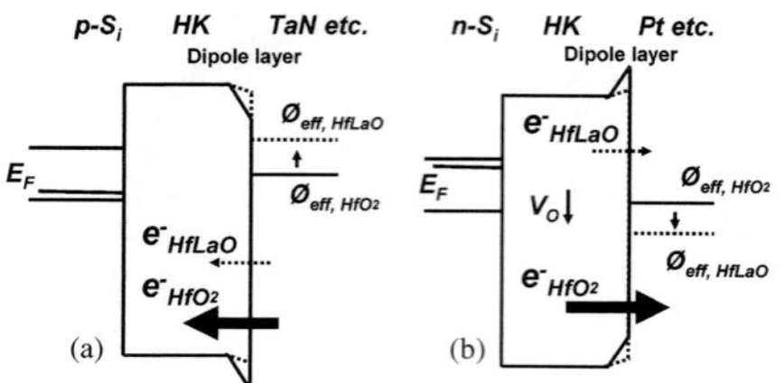

Fig. 3. Energy band diagram for MG/HK gate stacks. The dashed (solid) lines indicate the case whereby the HfLaO (HfO$_2$) is used as the gate dielectric. The subscripts of Φ_{eff} and e$^-$ (HfLaO and HfO$_2$) represent the $\Phi_{m,eff}$ of MGs and the electron transfer for HfLaO and HfO$_2$ case, respectively. The amount and direction of the net electron transfer are indicated by horizontal arrows. (a) For the nMOS with TaN/HfO$_2$ gate stack, the electron transfer from MG to HK due to the Fermi-level pinning shifts the $\Phi_{m,eff}$ to midgap [11]. In the case of TaN/HfLaO gate stack, additional electron transfer from HK to MG would be expected due to the lower electronegativity of La atoms. This effect compensates the electron transfer from MG to HK due to the Fermi-level pinning, giving rise to the reduction of TaN $\Phi_{m,eff}$. (b) For the pMOS with Pt gate, V_O in a dielectric is believed to induce electron transfer from V_O to MG, which would lead to Fermi-level pinning and reduction of the MG $\Phi_{m,eff}$. When the La atoms replace the Hf atoms, V_O concentration is reduced. This leads to the reduction of the amount of electron transfer and the release of Fermi-level pinning, increasing the $\Phi_{m,eff}$ of MG.

increase of V_O density will enhance the charge trapping effect, which will deteriorate the threshold instability and degrade the channel mobility accordingly. However, compared to the pure HfO$_2$ case, the devices show the reduced BTI degradation and the increased electron channel mobility with the incorporation of La into the HfO$_2$ [13], [18], which implies that the HfLaO possesses lower V_O density than the HfO$_2$. This result also leads to the decrease of the amount of electron transfer induced by V_O for the gate stacks with HfLaO dielectric compared to those with HfO$_2$ and the partial release of Fermi-level pinning between the HK and the MGs [Fig. 3(b)].

IV. CONCLUSION

In this letter, we demonstrate experimentally that by using the HfLaO HK gate dielectric, the TaN gate with $\Phi_{m,eff}$ $\sim$3.9–4.2 eV, the Pt gate with $\Phi_{m,eff}$ $\sim$ 5.5 eV, and the corresponding V_{fb} and V_{th} of MOS devices can be obtained after a 1000-°C annealing. By combining the experimental and first-principle calculation results, we propose that the change of the MG $\Phi_{m,eff}$ is attributed not only to the V_O density in HK layer but also to the difference of electronegativities among the materials involved in gate stacks.

REFERENCES

[1] International Technology Roadmap for Semiconductors (ITRS), San Jose, CA: Semiconductor Industry Assoc., 2003.
[2] I. De, D. Johri, A. Srivastava, and C. M. Osburn, "Impact of gate work-function on device performance at the 50 nm technology node," *Solid State Electron.*, vol. 44, no. 6, pp. 1077–1080, Jun. 2000.
[3] J. K. Schaeffer, C. Capasso, L. R. C. Fonseca, S. Samavedam, D. C. Gilmer, Y. Liang, S. Kalpat, B. Adetutu, H.-H. Tseng, Y. Shiho, A. Demkov, R. Hegde, W. J. Taylor, R. Gregory, J. Jiang, E. Luckowski, M. V. Raymond, K. Moore, D. Triyoso, D. Roan, B. E. White, Jr., and P. J. Tobin, "Challenges for the integration of metal gate electrodes," in *IEDM Tech. Dig.*, 2004, pp. 287–290.
[4] V. Narayanan, A. Callegari, F. R. McFeely, K. Nakamura, P. Jamison, S. Zafar, E. Cartier, A. Steegen, V. Ku, P. Nguyen, K. Milkove, C. Cabral, Jr., M. Gribelyuk, C. Wajda, Y. Kawano, D. Lacey, Y. Li, E. Sikorski, E. Duch, H. Ng, C. Wann, R. Jammy, M. Ieong, and G. Shahidi, "Dual work function metal gate CMOS using CVD metal electrodes," in *VLSI Symp. Tech. Dig.*, 2004, pp. 192–193.
[5] S. Inumiya, Y. Akasaka, T. Matsuki, F. Ootsuka, K. Torii, and Y. Nara, "A thermally-stable sub-0.9 nm EOT TaSix/HfSiON gate stack with high electron mobility, suitable for gate-first fabrication of hp45 LOP devices," in *IEDM Tech. Dig.*, 2005, pp. 27–30.
[6] H. C. Wen, H. N. Alshareef, H. Luan, K. Choi, P. Lysaght, H. R. Harris, C. Huffman, G. A. Brown, G. Bersuker, P. Zeitzoff, H. Huff, P. Majhi, and B. H. Lee, "Systematic investigation of amorphous transition-metal-silicon-nitride electrodes for metal gate CMOS applications," in *VLSI Symp. Tech. Dig.*, 2005, pp. 46–47.
[7] Z. B. Zhang, S. C. Song, C. Huffman, J. Barnett, N. Moumen, H. Alshareef, P. Majhi, M. Hussain, M. S. Akbar, J. H. Sim, S. H. Bae, B. Sassman, and B. H. Lee, "Integration of dual metal gate CMOS with TaSiN (NMOS) and Ru (PMOS) gate electrodes on HfO2 gate dielectric," in *VLSI Symp. Tech. Dig.*, 2005, pp. 50–51.
[8] E. Cartier, F. R. McFeely, V. Narayanan, P. Jamison, B. P. Linder, M. Copel, V. K. Paruchuri, V. S. Basker, R. Haight, D. Lim, R. Carruthers, T. Shaw, M. Steen, J. Sleight, J. Rubino, H. Deligianni, S. Guha, R. Jammy, and G. Shahidi, "Role of oxygen vacancies in V_{FB}/V_t stability of pFET metals on HfO2," in *VLSI Symp. Tech. Dig.*, 2005, pp. 230–231.
[9] S. B. Samavedam, L. B. La, P. J. Tobin, B. White, C. Hobbs, L. R. C. Fonseca, A. A. Demkov, J. Schaeffer, E. Luckowski, A. Martinez, M. Raymond, D. Triyoso, D. Roan, V. Dhandapani, R. Garcia, S. G. H. Anderson, K. Moore, H. H. Tseng, C. Capasso, O. Adetutu, D. C. Gilmer, W. J. Taylor, R. Hegde, and J. Grant, "Fermi level pinning with sub-monolayer MeOx and metal gates," in *IEDM Tech. Dig.*, 2003, pp. 307–310.
[10] Y.-C. Yeo, T.-J. King, and C. Hu, "Metal-dielectric band alignment and its implications for metal gate complementary metal-oxide-semiconductor technology," *J. Appl. Phys.*, vol. 92, no. 12, pp. 7266–7271, Dec. 2002.
[11] H. Y. Yu, C. Ren, Y.-C. Yeo, J. F. Kang, X. P. Wang, H. H. H. Ma, M.-F. Li, D. S. H. Chan, and D.-L. Kwong, "Fermi pinning induced thermal instability of metal gate work functions," *IEEE Electron Device Lett.*, vol. 25, no. 5, pp. 337–339, May 2004.
[12] R. Jha, B. Lee, B. Chen, S. Novak, P. Majhi, and V. Misra, "Dependence of PMOS metal work functions on surface conditions of high-K gate dielectrics," in *IEDM Tech. Dig.*, 2005, pp. 47–50.
[13] X. P. Wang, M.-F. Li, C. Ren, X. F. Yu, C. Shen, H. H. Ma, A. Chin, C. X. Zhu, J. Ning, M. B. Yu, and D.-L. Kwong, "Tuning effective metal gate work function by a novel gate dielectric HfLaO for NMOSFETs," *IEEE Electron Device Lett.*, vol. 27, no. 1, pp. 31–33, Jan. 2006.
[14] D. S. Yu, A. Chin, C. C. Laio, C. F. Lee, C. F. Cheng, W. J. Chen, C. X. Zhu, M.-F. Li, W. J. Yoo, S. P. McAlister, and D. L. Kwong, "3D GOI CMOSFETs with novel IrO2(Hf) Dual Gates and High-κ dielectric on 1P6M-0.18 μm-CMOS," in *IEDM Tech. Dig.*, 2004, pp. 181–184.
[15] H. Watanabe, N. Ikarashi, and F. Ito, "La-silicate gate dielectrics fabricated by solid phase reaction between La metal and SiO2 underlayers," *Appl. Phys. Lett.*, vol. 83, no. 17, pp. 3546–3548, Oct. 2003.
[16] [Online]. Available: http://en.wikipedia.org/wiki/Pauling_scale
[17] C. Shen, M. F. Li, X. P. Wang, H. Y. Yu, Y. P. Feng, A. T.-L. Lim, Y.-C. Yeo, D. S. H. Chan, and D.-L. Kwong, "Negative U traps in HfO2 gate dielectrics and frequency dependency of dynamic BTI in MOS-FETs," in *IEDM Tech. Dig.*, 2004, pp. 733–736.
[18] X. P. Wang, M. F. Li, A. Chin, C. X. Zhu, J. Shao, W. Lu, X. C. Shen, X. F. Yu, R. Chi, C. Shen, K. H. Huan, J. S. Pan, A. Y. Du, P. Lo, D. S. H. Chan, and D.-L. Kwong, "Physical and electrical characteristics of high-κ Gate Dielectric Hf(1-x)LaxOy," *Solid State Electron.*, vol. 50, no. 6, pp. 986–991, Jun. 2006.

Novel Hafnium-Based Compound Metal Oxide Gate Dielectrics for Advanced CMOS Technology

Ming-Fu Li[1,2], Chunxiang Zhu[1], Xin Peng Wang[1] and Xiongfei Yu[1]

[1]Silicon Nano-Device Lab (SNDL), ECE Dept, National University of Singapore, Singapore 119260

[2]Department of Microelectronics, Fudan University, Shanghai, China 201203

Abstract: *Improvement of Hf-based high-k gate dielectrics by incorporating Ta and La in HfO_2 are investigated systematically. The main issues of pure HfO_2 gate dielectric, including low crystallization temperature, channel mobility degradation, and bias temperature instability (BTI) degradation, can be effectively improved in HfTaO and HfLaO. Particularly, HfLaO with appropriate metal gate materials (TaN for N-FETs, Pt and Ru for P-FETs) can reduce the Fermi pinning between high-k and metal gate. The metal gate effective work function can be modulated by varying La percentage in HfLaO. Effective work function from 3.9 eV to 5.2 eV can be tuned continuously by adjusting the Ru thickness in a stacking multi-layer TaN/Ru metal gate, showing high potential for the application in next generation low threshold CMOS transistor technology.*

1. Introduction: HfO_2, as one of the most promising high-k gate dielectrics for the next generation CMOS devices [1,2], suffers from the following performance issues: *(I)*. Low crystallization temperature at around 400° C [3], *(II)*. Charge trapping and Bias temperature instability (BTI) [4,5], *(III)* Channel mobility degradation[6,7], *(IV)*. Fermi level pinning between the HfO_2 dielectric and gate electrode [8-10]. N- (P-) FET requires the gate electrode Fermi level to be aligned with the Si conduction (valence) band. However, due to Fermi pinning, the effective work function (EWF) of electrode (Poly-Si or metal) changes to midgap after high temperature annealing [10,11].

Incorporating Si or Al in HfO_2 to improve the performances *(I)*, *(II)* and *(III)* has been widely reported [12-16]. However this can not solve the Fermi pinning issue *(IV)*. On the other hand, since SiO_2 and Al_2O_3 has lower dielectric constant and higher leakage current, HfSiO or HfAlO is unavoidable to degrade the dielectric constant k and gate leakage, comparing to HfO_2. It is a natural thought to find a metal element with high k dielectric metal oxide, incorporate in HfO_2 to overcome this issue.

The first attempt in our team to improve the high k dielectric alone this line is to incorporate Ta [17,18] in HfO_2 films. Ta_2O_5 has k value of *26* [19]. We have successfully demonstrated that HfTaO can improve *(I)*, *(II)* and *(III)*. However, due to the low electron barrier of Ta_2O_5/Si (*1.5eV*) [19], HfTaO has higher gate leakage than HfO_2 with the same EOT. Our second attempt is to replace metal Ta by La [20-22]. La_2O_3 has high dielectric constant (*k~30*) and high electron barrier to Si (*2.3eV*) [19], therefore HfLaO is expected to have lower leakage than HfTaO. We incorporated La to HfO_2 and pleased to find that HfLaO not only improves *(I)*, *(II)* and *(III)*, but also solves the Fermi pinning issue *(IV)*.

In our work, both MOS-C and MOSFET devices were fabricated on (100) Si substrates with p or n doping concentration $6x10^{15}cm^{-3}$.

2. Incorporating Ta in HfO_2:

2.1 HfTaO gate dielectric with TaN metal gate [17]

Surface nitridation in NH_3 ambient was performed at 700°C for 10sec before high-*k* dielectric deposition. HfO_2 and HfTaO were deposited using sputtering techniques followed by post-deposition annealing in N_2 ambient at 700°C for 40sec. The composition of Ta was controlled by the power ratio between Hf and Ta target. The metal gate TaN (~160nm) was deposited using reactive sputtering.

Figs. 1 and 2 show XRD spectra and TEM picture of HfO_2 and HfTaO films. The crystallization temperature is increased from 400°C of HfO_2 to 1000°C of HfTaO with 43% Ta.

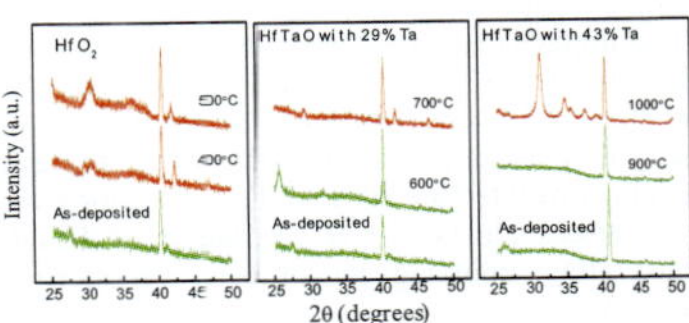

Fig. 1 XRD spectra of HfO_2, HfTaO films (40nm) with 29% and 43% Ta after different temperature annealing for 30s in N_2. The Ta incorporated in HfO_2 films raises the crystallization temperature significantly.

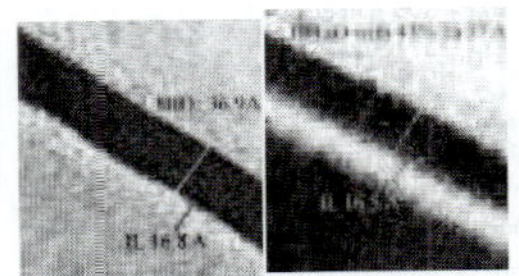

Fig. 2 TEM of HfO_2, HfTaO with 43%Ta after 950°C annealing for 30s. HfO_2 is crystallized however HfTaO remains amorphous structure.

Figure 3 shows Jg-EOT characteristics of the HfO_2 and HfTaO gate dielectric MOS capacitors. It is noticed that the higher leakage current of HfTaO than HfO_2 is due to lower band offset of Ta_2O_5, but it is still comparable to HfSiO [12] and HfSiON [13]. Figure 4(L) shows I_d-V_d of N-FETs with HfO_2 and HfTaO gate dielectric. Electron mobility of

N-FETs is compared in Fig. 4(R). It is observed that HfTaO can provide higher mobility than HfO_2. The peak mobility of HfO_2, HfTaO with 29% and 43% Ta are 140, 318 and 354 $cm^2/V·s$, respectively.

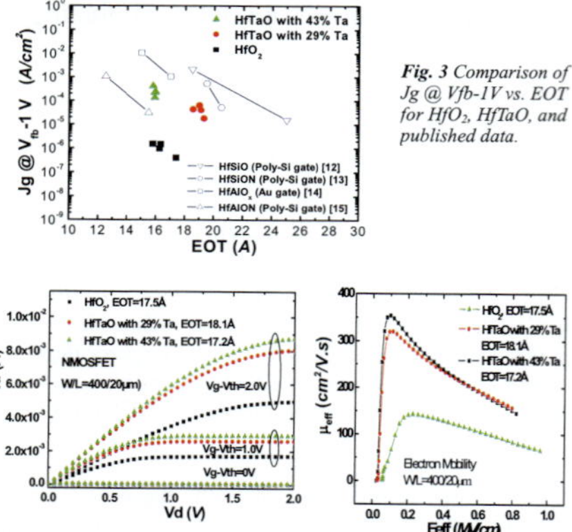

Fig. 3 Comparison of Jg @ Vfb-1V vs. EOT for HfO_2, HfTaO, and published data.

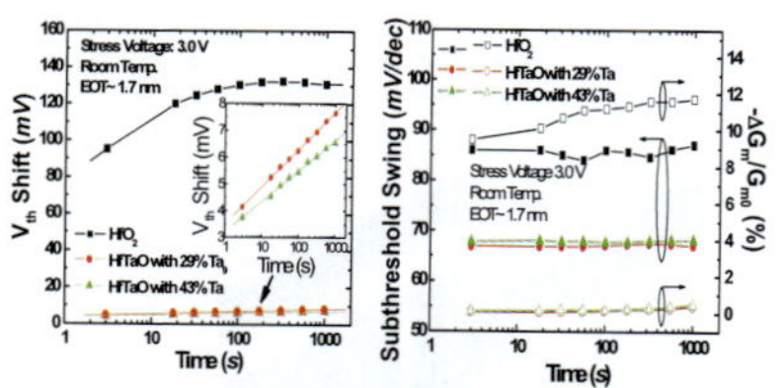

Fig. 4 (Left) Id-Vd curves for HfO2, HfTaO with 29% and 43% Ta gate dielectric N-FETs; *(Right)* Effective electron mobility of HfO_2, HfTaO with 29% and 43% Ta gate dielectric N-FETs extracted by split-CV method.

Fig. 5(L) shows comparison of the V_{th} shift under 3.0V constant voltage stress (CVS). The V_{th} shift in HfTaO (43% Ta) film is *6.2mV* after *1000sec* stress, which is 20 times lower than that of pure HfO_2 (*130mV*). Such a small V_{th} shift in HfTaO implies that Ta incorporated into HfO_2 films can significantly reduce charged traps in the bulk. No obvious sub-threshold swing change can be observed at the same voltage stress, as shown in Fig. 5(R), which indicates a negligible effect of interface state on V_{th} shift. Moreover, Fig. 5(R) illustrates almost no G_m variation of HfTaO compared to HfO_2, implying no obvious variation of mobility after CVS.

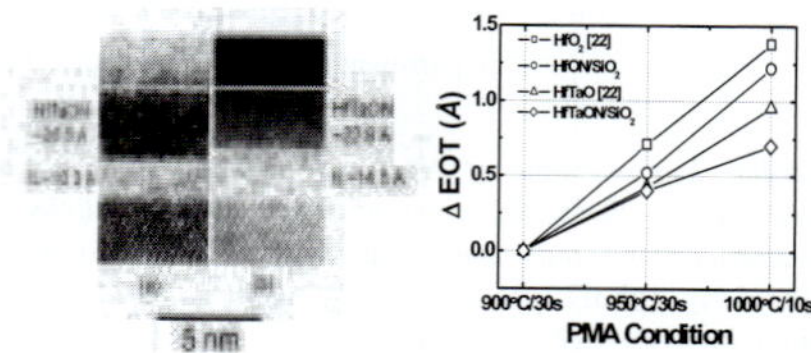

Fig. 5 (Left) Comparison of the V_{th} shift due to constant voltage stress in HfO_2, HfTaO with 29% and 43% Ta gate dielectric N-FETs; *(Right)* Sub-threshold swing and Gm variation as a function of constant voltage stress time.

2.2 HfTaON/SiO$_2$ with TaN metal gate [18]

To further improve thermal stability of HfTaO gate

dielectric, nitrogen was introduced. An ultra-thin layer of SiO_2 was also inserted between high-k and Si substrate for high carrier mobility. A 10Å SiO_2 was grown on the Si substrates as an interfacial layer (IL). HfON and HfTaON films with thickness of ~25Å were then deposited by reactive dc co-sputtering of Hf and Ta targets in $Ar/N_2/O_2$ ambient, followed by post-deposition annealing (PDA) in N_2 with 5% O_2 ambient at 700°C for 30sec.

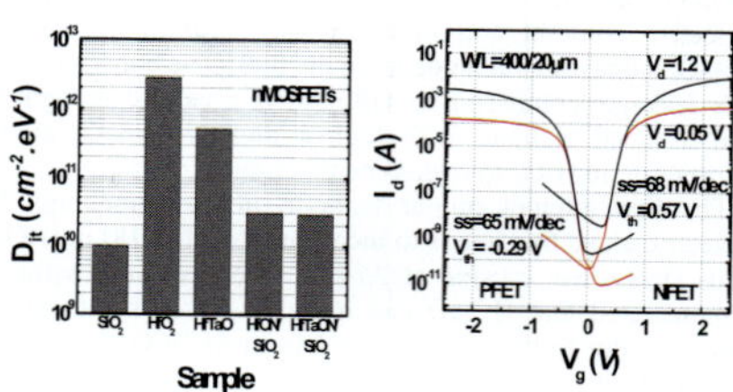

Fig. 6 (Left) TEM pictures of $HfTaON/SiO_2$ gate stacks without and with PMA at 1000°C for 10sec; *(Right)* The $HfTaON/SiO_2$ exhibits the lowest increase in EOT comparing to other gate stacks, which indicates that the $HfTaON/SiO_2$ shows the best thermal stability in those gate stacks.

The TEM pictures of $HfTaON/SiO_2$ gate stacks (a) without and (b) with the post-metal annealing (PMA) at 1000° C for 10sec are shown in Fig. 6(L). Fig. 6(R) exhibits that HfTaON provides the best thermal stability over other gate stacks.

The interface-state density (D_{it}) in N-FETs with different gate stacks is compared in Fig. 7(L). It was noted that the D_{it} is significantly reduced by inserting SiO_2 IL for both $HfON/SiO_2$ and $HfTaON/SiO_2$ films. Fig. 7(R) shows I_d-V_g characteristics of FETs with $HfTaON/SiO_2$ gate stacks after PMA at 1000°C for 10sec. As can be seen, both N- and P-FETs exhibited excellent sub-threshold swing, which are due to the good interface properties observed in $HfTaON/SiO_2$ gate stacks.

Fig. 7 (Left) Comparison of interface trap density D_{it} for n-FETs with different gate stacks; *(Right)* I_d-V_g characteristics for n- and p-FETs with $HfTaON/SiO_2$ gate stacks. The excellent ss (68 mV/dec for NFET, 65 mV/dec for PFET) can be achieved.

The I_d-V_d characteristics for the FETs with $HfTaON/SiO_2$ gate stack are shown in Fig. 8. Fig. 9 depicts electron and hole mobility, obtained by split C-V method, for FETs with

HfTaON/SiO$_2$ gate stack after PMA at 1000°C for 10sec. At the operation voltage of devices, whose effective field is about 0.8MV/cm, electron mobility of 100% and hole mobility of 96% of universal curves were obtained in the FETs with HfTaON/SiO$_2$ gate stacks.

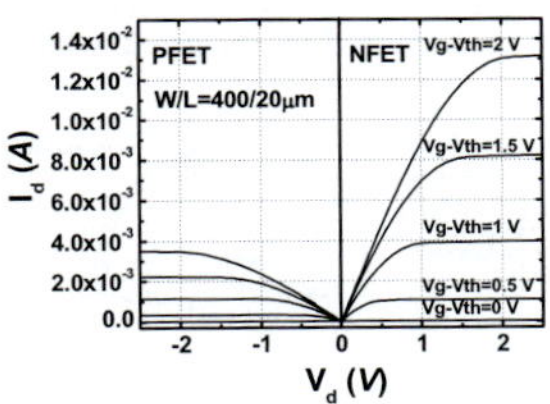

Fig. 8 I_d-V_d characteristics for p-FETs **(Left)** and n-FETs **(Right)** with HfTaON/SiO$_2$ gate stacks.

Fig. 10 shows comparison of the V_{th} instability in (a) N-FETs and (b) P-FETs with HfON/SiO$_2$ and HfTaON/SiO$_2$ gate stacks after PMA at 1000°C for 10sec. The constant voltage stresses of $V_{th} \pm 2V$ and $V_{th} \pm 2.5V$ were applied at the gate electrode and the conventional dc measurement was used to examine the V_{th} instability. As shown in Fig. 10, the V_{th} shifts in HfTaON/SiO$_2$ MOSFETs are much lower than those in HfON/SiO$_2$ under the same constant voltage stress. Fig. 11 shows the lifetime projection for V_{th} shifts in HfON/SiO$_2$ and HfTaON/SiO$_2$ gate stacks, with the failure criterion defined by $\Delta V_{th}=10mV$. The projected 10-year lifetime operating voltages for HfON/SiO$_2$ and HfTaON/SiO$_2$ were *1.1* and *2.1V* in N-FETs, and *-1.3* and *-1.9* V in P-FETs, respectively.

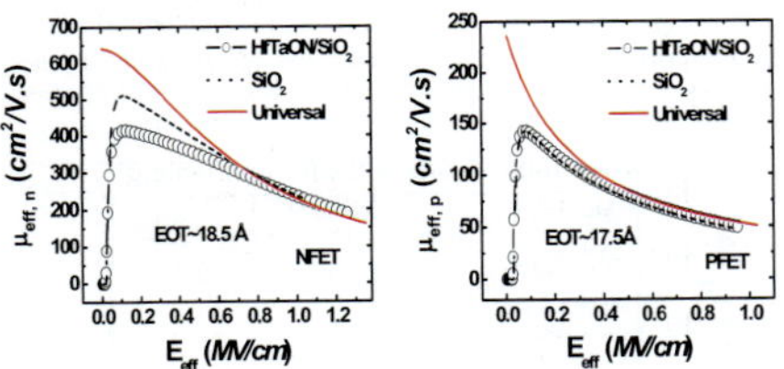

Fig. 9: Comparison of **(Left)** electron and **(Right)** hole mobility in SiO$_2$ and HfTaON/SiO$_2$ FETs.

Overall the HfTaON/SiO$_2$ gate stack provided much lower gate leakage current against SiO$_2$, good interface properties, excellent transistor characteristics, superior carrier mobility, excellent charge trapping induced V_{th} instability.

3. Incorporating La in HfO$_2$
3.1 HfLaO gate dielectric with TaN (or HfN) gate for NMOSFETs and with Pt gate for PMOSFETs [20-22]:
Reactive sputtering was used for HfLaO with low oxygen

concentration, followed by PDA in N$_2$ with a small amount of O$_2$ at 600°C for 30sec. HfLa target was used to avoid water absorption of La during exposure to air [23].

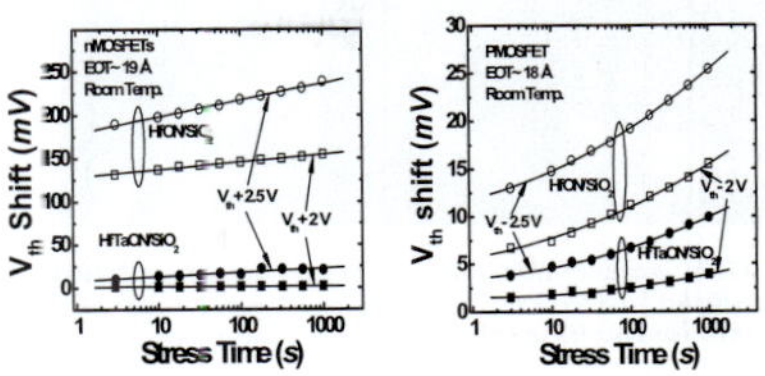

Fig. 10 Comparison of the V_{th} shift due to constant voltage stress in HfTaON/SiO$_2$ and HfON/SiO$_2$ NMOSFETs **(Left)** and PMOSFETs **(Right)** as a function of constant voltage stress time.

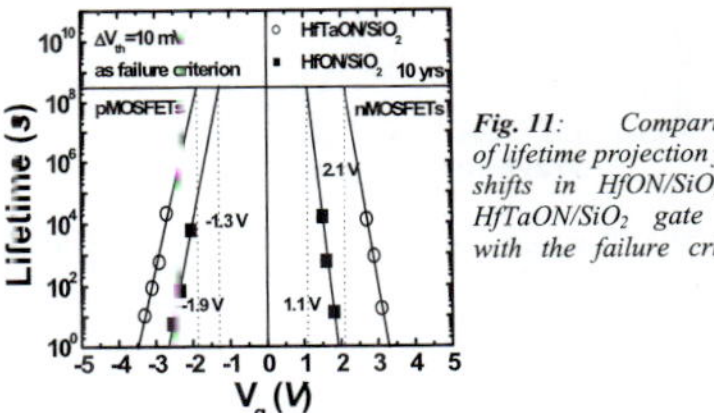

Fig. 11: Comparison of lifetime projection for V_{th} shifts in HfON/SiO$_2$ and HfTaON/SiO$_2$ gate stack with the failure criterion

The composition of La was controlled by the varying sputter power on Hf and HfLa targets, and was measured by XPS. TaN or Pt gate electrode was deposited using PVD, followed by gate patterning. Devices were then annealed at different temperature up to 1000°C with different ambient for thermal stability evaluation.

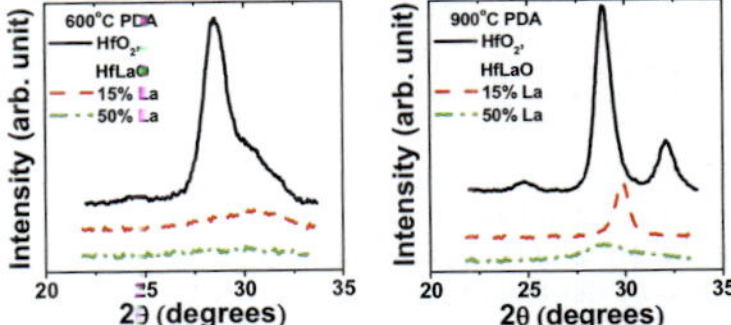

Fig.12 XRD spectra of HfO$_2$, HfLaO with 15% and 50% La after different temperature annealing in N$_2$. La incorporated in HfO$_2$ films raises the crystallization temperature.

In Fig. 12, the XRD spectra of HfO$_2$ and HfLaO films (~30nm) after different temperature annealing show clearly that incorporation of La in HfO$_2$ can raise the crystallization temperature. XTEM images of HfO$_2$ and HfLaO films (Fig. 13) confirm that HfO$_2$ is fully crystallized while HfLaO with 15% and 50% La can keep amorphous structure under 900°C annealing. In Fig. 14, the FTIR spectra (~200nm films on Si) shows that HfLaO keeps the amorphous structure after 900°C annealing and the phonon spectrum is changed from many narrow phonon peaks as shown in HfO_2 spectrum to a broader and weaker peak.

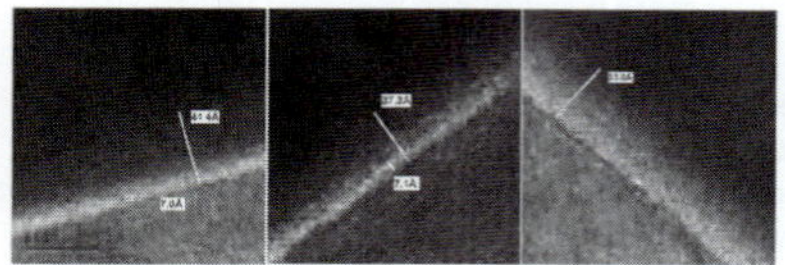

*Fig. 13 XTEM images of HfO₂ **(Left)** and HfLaO with 15% **(Middle)** and 50% **(Right)** La after 900°C annealing for 30sec in N₂. HfO₂ film is crystallized however HfLaO films remain amorphous structures.*

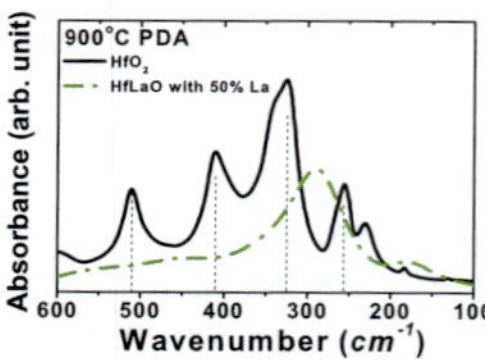

Fig. 14 The FTIR absorption spectra of HfO₂ and HfLaO with 50% La. Adding La in HfO₂ changes the spectra, indicating changes of atomic bonding and phonon energy.

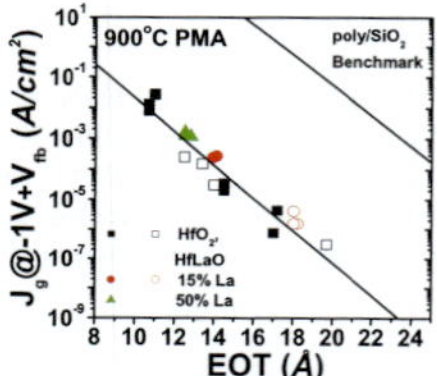

Fig. 15 Comparison of Jg@-1V +Vfb for HfLaO with different EOT and La concentration. Solid symbols: HfN gate; empty symbols: TaN gate.

Fig. 15 shows that HfLaO gate leakage is comparable with pure HfO_2 and has 5-6 orders reduction compared with SiO_2 at the same EOT of 1.2-1.8nm. Figs. 16 and 17 show the change of flat-band voltage V_{fb} versus La composition, all after 1000°C annealing. The EWF can be tuned continuously from the midgap to 3.9eV using TaN or HfN n-metal gate, and 5.5eV using Pt p-metal, as indicated in Fig. 17. **Figs. 16 and 17 clearly demonstrate that the shift of EWF is not due to charge effect in the dielectric, but due to change of EWF of the metal when adding La in HfO_2. Incorporating La in HfO_2 induces change of EWF to more n for n-metal and more p for p-metal.** Fig.18 shows the $I_d - V_g$ characteristics of N- and P-FETs. The V_{th} shifts in Fig. 18 are consistent with the V_{fb} shift obtained in Fig. 16. Incorporation of La in HfO_2 also improves the gate stack thermal stability with Pt electrode from less than 600^oC [24] (Fig. 16 only shows 420°C FGA result for Pt on

HfO_2 dielectric) to 1000^oC. This large EWF window (3.9-5.5eV) between TaN and Pt electrodes shows the possibility of obtaining optimal metal gate work function for CMOSFETs on HfLaO dielectric. Fig. 19 shows that using HfLaO with 50% La has 60-70% electron mobility improvement compared with using pure HfO_2 as gate dielectric. Fig. 20 shows one order reduction of BTI shift for HfLaO (with 50% La) compared with HfO_2 gate dielectric.

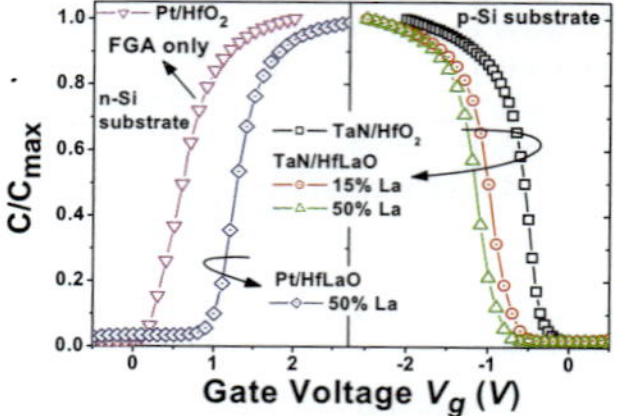

*Fig. 16 C-V curves for MOS HfLaO capacitors with different La concentration after 1000 °C PMA. **(Left)** p-metal Pt gate, n-Si substrate, **(Right)** n-metal TaN gate, p-Si substrate. By incorporating La in HfO₂, Vfb has positive (negative) shift, using p-metal gate Pt (n-metal gate TaN)*

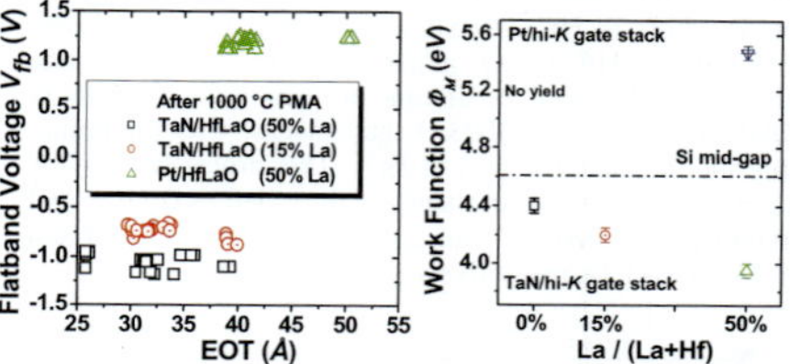

*Fig. 17 **(Left)** Vfb as a function of EOT (p- & n-substrates doping concentration 6x10¹⁵cm⁻³). **(Right)** the corresponding metal work function as a function of La composition in HfLaO films.*

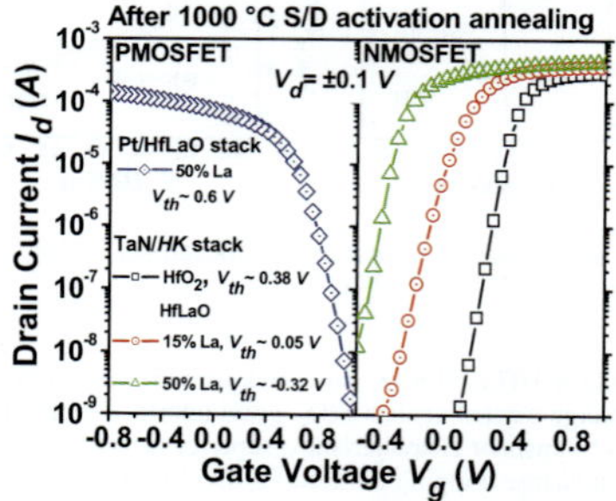

*Fig. 18 Id-Vg curves of N-**(Right)** & P-FETs **(Left)** using TaN for N-FETs, Pt for P-FETs, and HfLaO for gate dielectric. (p- & n-substrates doping concentration 6x10¹⁵cm⁻³).*

3.2. HfLaO gate dielectric with TaN/Ru stacking multi-layer metal gate for dual metal gate CMOSFETs.

Pt in Figs. 16-18 shows too large EWF which over-tune p-FET threshold voltage. On the other hand Pt is difficult for etching. Figure 21 shows the V_{fb} versus EOT of pure Ru metal on HfLaO (50% La atomic concentration) after

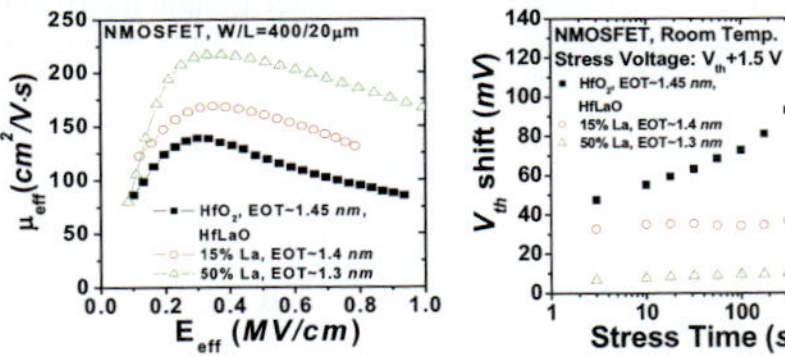

Fig. 19 Effective electron mobility of HfO_2, HfLaO with 15% and 50% La gate dielectric NMOSFETs extracted by split-CV method.

Fig. 20 BTI V_{th} shift under constant voltage stress in HfO_2, HfLaO with 15% and 50% La gate dielectric NMOSFETs.

1000°C annealing. This result shows small dielectric charge with $V_{fb}{\sim}0.91eV$ on $6{\times}10^{15}cm^{-3}$ n-doped Si. The corresponding EWF of Ru on HfLaO is $5.2eV$, indicating that Ru is a suitable metal on HfLaO for P-FET. Combining with the results of TaN on HfLaO as shown in Figs. 16 and 17, we show evidence that incorporating La to HfO_2 can release the Fermi pinning between metal and high-k dielectric, causing EWF shifts from midgap $4.64eV$ [25] to $5.2eV$ for p-metal Ru, and from $4.4eV$ to $3.9eV$ [20,21] for n-metal TaN respectively. Further, by controlling the thicknesses of stacking multi-layer of p-metal Ru and n-metal TaN, the metal gate EWF can be tuned continuously from $3.9eV$ to $5.2eV$ after 1000°C annealing. Ru and TaN stacks with different layer thicknesses were deposited by physical vapor deposition (PVD). After gate patterning by dry etch with the help of SiO_2 hard mask, BF_2 implantation and the post metal annealing (PMA) at 1000°C for 5sec were performed to activate s/d for P-FETs fabrication. Finally, sintering was done at 420°C in forming gas ambient for 30min after s/d and backside Al metallization. Fig. 22 shows the EDX results of TaN/Ru multi-layer as deposited and after 1000°C annealing. TaN was partially intermixed with Ru after 1000°C annealing. Fig. 23(L) depicts the CV curves with different Ru thicknesses for TaN/Ru stacks after 1000°C annealing. The trend of CV curves shift suggests that the EWF of gate electrode was increased by increasing the Ru thickness, a phenomena similar to those reported by [26] for stacking multi-layer on SiO_2. Fig. 23(R) shows a typical example for TaN/Ru gate stack with $100\mathring{A}$ Ru after different temperature annealing. The EWF decreased with the increase of annealing temperature. However the subsequent annealing at 1000°C does not affect the EWF of MGs, which indicates that the EWF changes after 1000°C annealing are stable and permanent. The I_d-V_g transfer curves of P-FETs with the TaN/Ru/HfLaO gate stack are shown in Fig. 24, demonstrating 1.3V V_{th} variation with the change of Ru thickness, consistent with their EWF data obtained from V_{fb}

measurement.

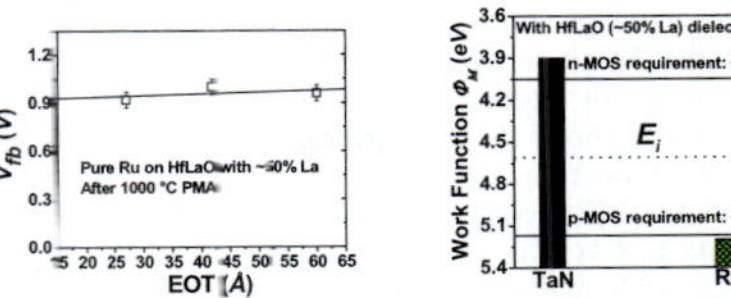

Fig. 21 (Left) V_{fb} as a function of EOT for pure Ru and HfLaO with ~50% La stacks after 1000°C annealing. (Right) EWFs of Ru and TaN on HfLaO (50% La) dielectric after 1000°C RTA.

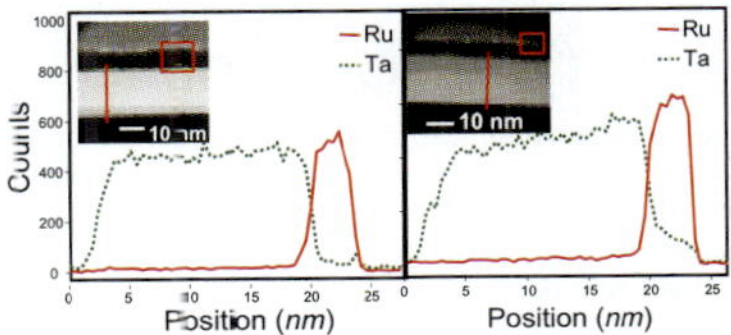

Fig. 22 EDX analysis for TaN and Ru stack, (Left) As deposited and (Right) After 1000°C 5sec RTA. Obvious Ta diffusion into Ru layer up to the interface between Ru and HfLaO was observed after 1000°C annealing.

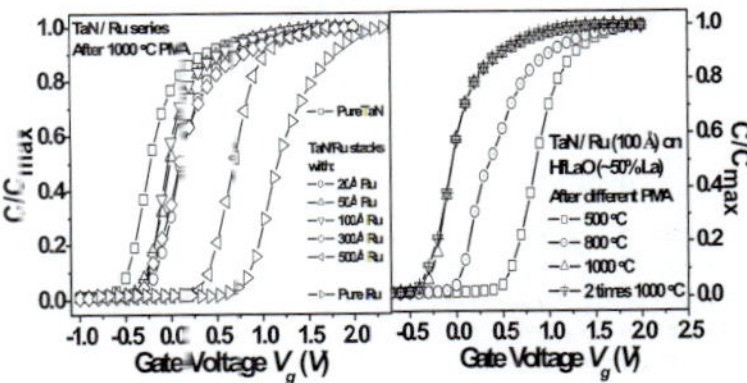

Fig. 23 (Left) CV curves for TaN/Ru series after high temperature annealing (1000°C). (Right) As the annealing temperature increases, there are more TaN diffusion into Ru up to the interface between MG and HK so as to decrease the EWF of metal gates.

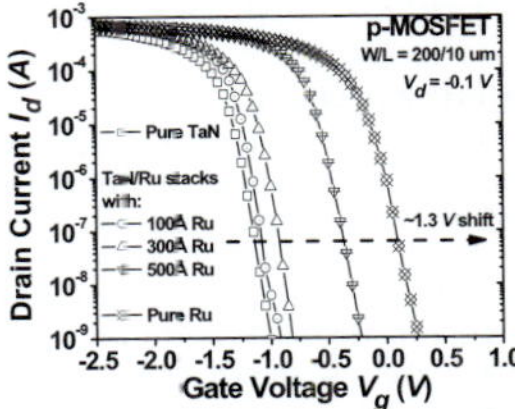

Fig. 24 Transfer characteristics of PMOSFETs with TaN/Ru gate stacks. Different Ru thicknesses introduce ~1.3V Vt shift.

3.3 Analysis and Discussion

Ab initio calculation of oxygen vacancy formation in HfLaO dielectric:

In order to explain the possible mechanism of lower BTI degradation and higher electron mobility in HfLaO than HfO$_2$, and the Fermi pinning release, we investigate the formation energy of oxygen vacancy V_o defect in HfLaO and HfO$_2$. Table I shows the *ab initio* calculation results (by VASP) of V_O formation using monoclinic HfO$_2$ [27] and pyrochlore HfLaO [28] configurations shown in Fig. 25. By the definition and results in Table I, the V_o density ratio between Hf$_2$La$_2$O$_7$ and HfO$_2$ with the same volume was estimated to be 0.2 by the following equation:

$$Ratio = \frac{D_{Td}\exp[-\dfrac{E_{Td}-E_{V4}}{k_BT}]+D_{C2v}\exp[-\dfrac{E_{c2v}-E_{v4}}{k_BT}]}{D_{v3}\exp[-\dfrac{E_{v3}-E_{v4}}{k_BT}]+D_{v4}}$$

It gives a reasonable explanation of reduction of BTI degradation (Fig. 20), probably also one of the facts to interpret the channel mobility and thermal stability improvement and released Fermi pinning of the metal/HfLaO gate stac]

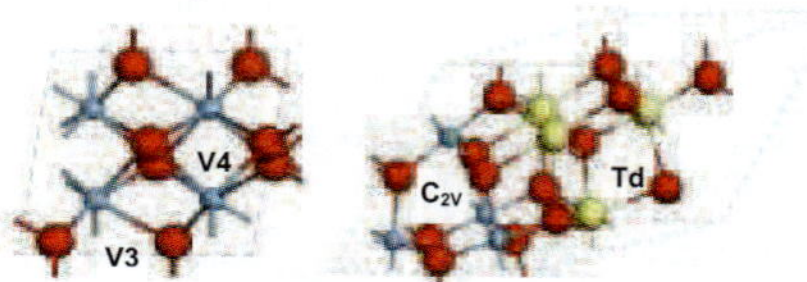

Dark (red) large balls-oxygen; atoms;

Shallow (blue) small balls–Hf atoms;

Shallow (yellow) large balls –La atoms.

Fig. 25 (Left) *Monoclinic HfO$_2$ primitive cell. Two possible Oxygen vacancy sites: V3 (3-fold coordinated) and V4 (4-fold coordinated); (Right) Pyrochlore Hf$_2$La$_2$O$_7$ primitive cell. Two possible V_o sites: Td (Td symmetry) and C$_{2v}$ (C$_{2v}$ symmetry).*

Table I The formation energies of V_o at varies sites in monoclinic HfO$_2$ and pyrochlore HfLaO, calculated by *ab* initio calculations.

Oxygen vacancy site	Formation energy E(eV)	Site density D(nm^{-3})
V3 site in HfO$_2$	6.51	28.6
V4 site in HfO$_2$	6.39	28.6
Td site in Hf$_2$La$_2$O$_7$	7.23	6.3
C$_{2v}$ site in Hf$_2$La$_2$O$_7$	6.51	38.0

Fermi pinning release mechanism

From our experimental data (not shown here), we believe that Fermi pinning release mechanism of HfLaO/metal gate is complicated and far from understanding now. Here we try to give a very preliminary explanation. Fermi pinning and EWF modulation is due to a dipole formation between the metal gate and high k dielectric (Fig. 26) [9]. For N-MOS with TaN/HfO$_2$ gate stack as shown in Fig. 26(Left), the dipole is formed by electron transfer from MG to HK. For TaN/HfLaO gate stack, additional electron transfer from HK to MG would be expected due to the lower electron negativity of La atom than Hf atom [29]. This effect compensates the electron transfer from MG to HK due to Fermi pinning, giving rise to the reduction of TaN EWF. For P-MOS with Ru gate as shown in Fig. 26(Right), electron transfer from La to Ru due to low electron negativity of La would not be feasible because Ru, as a noble metal, generally shows intrinsically inert performances. However, V_O in a dielectric are believed to induce electron transfer from V_O to MG, which would lead to Fermi pinning and reduction of the metal EWF [24,30]. Reduction of V_O concentration in HfLaO leads to the reduction of electron transfer and the release of Fermi pinning, increasing the EWF of MG.

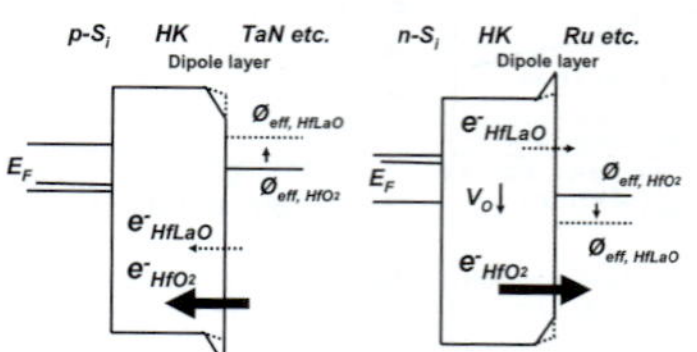

Fig. 26 *Energy band diagram for MG/HK gate stacks.*

4. Conclusion

Incorporating Ta and La in HfO$_2$ to improve the gate dielectric property is demonstrated in terms of crystallization temperature, BTI degradation, and channel mobility. Particularly, by increasing La composition in HfLaO, the effective work function EWF of n(p)-metal TaN and HfN(Pt and Ru) changes to more n(p) type from the midgap, after high temperature annealing (up to 1000°C). This is interpreted by Fermi pinning release between the high-k gate dielectric and metal gate. EWF from 3.9eV to 5.2eV can be tuned continuously by adjusting the Ru thickness in a stacking multi-layer TaN/Ru metal gate.

Acknowledgement: We thank IME Singapore for their strong technical support. Fruitful discussions with Professors Albert Chin , Dim-Lee Kwong and Dr H.Y.Yu are grateful. *Ab initio* calculation was conducted by Professor Feng Yuan Ping's group. This work was mainly supported by Singapore A*STAR project and Applied Materials Inc donated grant.

References

[1] S.Datta *et al.*, *IEDM*, p.653, 2003. [2]H.Y.Yu *et al.*, *IEDM*, p.99, 2003. [3] W. Zhu *et al.*, *IEDM*, p.463, 2001. [4] A.Kerber *et al.*, *IRPS*, p.41, 2003. [5]C.Shen et al, *IEDM*,p.733,2004,[6] E.P.Gusev *et al.*, *IEDM*, p.451, 2001. [7] Z.Ren *et al.*, *IEDM*, p.793, 2003. [8] S.B.Samavedam *et al.*, *IEDM*, p.307, 2003.[9]Y.C.Yeo *et al* JAP, v.92, p.7266,2002,[10] H.Y.Yu *et al.*, *EDL*, v.25, p.337, 2004,[11]C.Hobbs et al, *Symp VLSI Tech*, p.9 ,2003, [12] A. Morioka *et al.*, *Symp VLSI Tech.* p.165, 2003, [13] M. Koyama *et al.*, *IEDM*, p.849, 2002,. [14] T. Nabatame *et al.*, *SympVLSI Tech.* p. 25. 2003, [15] H. Jung *et al.*, *IEDM* p.853, 2002, [16] A.Shanware *et al.*, *IEDM*, p.939, 2003. [17] X. Yu *et al.*, *Symp VLSI Tech.* p.110,2004,[18] X. Yu *et al.*, *IEDM* p.31, 2005, [19] G.D.Wilk *et al.*, *JAP*, v.89, p.5243, 2001.[20] X.P.Wang *et al*, EDL, v.27, p.31,2006, [21] X.P.Wang SSE, to be published [22] X.P.Wang *et al*, *Symp VLSI Tech.* p.12, 2006, [23] D.S.Yu *et al.*, *IEDM* p.181, 2000. [24] E.Cartier *et al*, *Symp.VLSI Tech*, p.230,2005, [25] Z.B.Zhang *et al*, *Symp VLSI Tech.* p.50, 2005, [26] L.S.Jeon et al, *IEDM*, p.303, 2004, [27] J.Aarik *et al*, *Appl.Surface Sci.*, v.173,p.15 ,2001.[28] A.W. Sleight, Inorganic Chemistry, v.7,p.1704,1968, Table IV. [29] http://en.wikipedia.org/wiki/Pauling_scale,[30]R.Jia et al, *IEDM*,p.47,2005.

50   IEEE ELECTRON DEVICE LETTERS, VOL. 29, NO. 1, JANUARY 2008

Widely Tunable Work Function TaN/Ru Stacking Layer on HfLaO Gate Dielectric

X. P. Wang, *Student Member, IEEE*, M.-F. Li, *Senior Member IEEE*, H. Y. Yu, J. J. Yang, J. D. Chen,
C. X. Zhu, *Member, IEEE*, A. Y. Du, W. Y. Loh, S. Biesemans, Albert Chin, *Senior Member, IEEE*,
G. Q. Lo, and D.-L. Kwong, *Senior Member, IEEE*

Abstract—For the first time, we demonstrate experimentally that using HfLaO high-κ gate dielectric and vertical stacks of TaN/Ru metal layers, dual metal gates with continuously tunable work function over a very wide range from 3.9 to 5.2 eV, can be achieved after 1000 °C annealing required by a conventional CMOS source/drain activation process. The wide tunability of work function for this bilayer metal structure is attributed to metal interdiffusion during annealing and the release of Fermi level pinning between metal gates (Ru and TaN) and HfLaO. Moreover, this change is thermally stable and unaffected by a subsequent high temperature process.

Index Terms—CMOS, Fermi level pinning, HfLaO, high-κ dielectric, interdiffusion, metal gate, work function.

I. Introduction

THE GATE stack of metal gate/high-κ dielectric attracts immense interest for sub-45 nm CMOS technology [1]. Metal gates with work functions close to the conduction and valence band edges of Si are desired for the optimal design of bulk Si n- and p-MOSFETs, respectively [2]. However, it has been shown that these requirements are very difficult to reach due either to 1) the Fermi level pinning effect between metal gates and high-κ dielectrics [3], [4], 2) reaction between metal and dielectric, or 3) oxygen vacancy at the metal/dielectric interface [5], [6], especially for traditional gate first technology [7] which requires 1000 °C thermal annealing. Recently, MOS devices with lanthanum doped Hf-based oxide dielectric have shown

Manuscript received September 12, 2007; revised October 23, 2007. This work was supported by the Singapore A-STAR research Grant R263-000-267-305 and AMAT Grant R263-000-385-720. The review of this letter was arranged by Editor A. Chatterjee.

X. P. Wang is with the Silicon Nano Device Laboratory, Department of Electrical and Computer Engineering, National University of Singapore, Singapore 119260 and also with the Institute of Microelectronics, Singapore 117685.

M.-F. Li is with the Silicon Nano Device Laboratory, Department of Electrical and Computer Engineering, National University of Singapore, Singapore 119260. He is also with the Institute of Microelectronics, Singapore 117685 and also with the Department of Microelectronics, Fudan University, Shanghai 201203, China (e-mail: mfli@fudan.edu.cn).

H. Y. Yu and S. Biesemans are with the Interuniversity MicroElectronics Center, 3001 Leuven, Belgium.

J. J. Yang, J. D. Chen and C. X. Zhu are with the Silicon Nano Device Laboratory, Department of Electrical and Computer Engineering, National University of Singapore, Singapore 119260.

A. Y. Du, W. Y. Loh, G. Q. Lo and D.-L. Kwong are with the Institute of Microelectronics, Singapore 117685.

A. Chin is with the Department of Electronics Engineering, National Chiao-Tung University, Hsinchu 300, Taiwan, R.O.C.

Color versions of one or more of the figures in this letter are available online at http://ieeexplore.ieee.org.

Digital Object Identifier 10.1109/LED.2007.911608

superior performance compared to those using a pure HfO_2 gate dielectric. However, most of them are for n-MOSFETs [8]–[11]. Even though Pt has been demonstrated to show very high effective work function (EWF) on HfLaO for p-MOSFET [10], gate patterning for short-channel devices would be particularly challenging because Pt is very resistant to chemical or plasma etching.

In this letter, first we demonstrate experimentally that the EWF of Ru, an etching friendly p-metal, on HfLaO with $\sim$50% La is $\sim$5.2 eV. Moreover, by controlling thickness of the bottom metal layer Ru in TaN/Ru bilayer stack on HfLaO dielectric and intermixing TaN and Ru during annealing at 1000 °C, we show that the EWF of metal gates can be tuned continuously over a wide range, from n-type band edge (3.9 eV) to p-type band edge (5.2 eV).

II. Experimental

(100) Si substrates with n-type doping concentration of 5×10^{15} cm^{-3} were used. For MOS capacitor (MOSCAP), after a dilute hydrofluoric acid-last Radio Corporation of America pregate clean, HfLaO films ($\sim$50% La) with different physical thicknesses were deposited using reactive sputtering, followed by postdeposition annealing (PDA) in N_2 with a small amount of O_2 at 600 °C for 30 s. HfLa target (Hf : La = 1 : 1, atomic ratio) was used to reduce moisture absorption of La during exposure to air [12]. After PDA, the wafers were promptly transferred to a physical vapor deposition (PVD) tool to minimize exposure to moisture. The bottom Ru metal layer with different thicknesses (from 2 to 50 nm) was then deposited directly on top of the gate dielectric by dc sputtering. The top TaN metal layer (150 nm) was later deposited *in situ* by reactive sputtering. In addition, MOS devices with a single metal layer, pure Ru (50 or 100 nm) or TaN (150 nm), were fabricated for comparison. After gate patterning, all the devices went through rapid thermal annealing (RTA) at different temperatures up to 1000 °C for thermal stability evaluation. For MOSFET fabrication, source/drain implantations of BF_2 (50 keV, 1×10^{15} cm^{-2}) were performed, followed by RTA activation at 1000 °C for 5 s. Finally, all samples received backside Al metallization and forming gas annealing (FGA) at 420 °C for 30 min.

III. Results and Discussion

Fig. 1 shows the plot of flatband voltage (V_{fb}) versus equivalent oxide thickness (EOT) for pure Ru metal on HfLaO

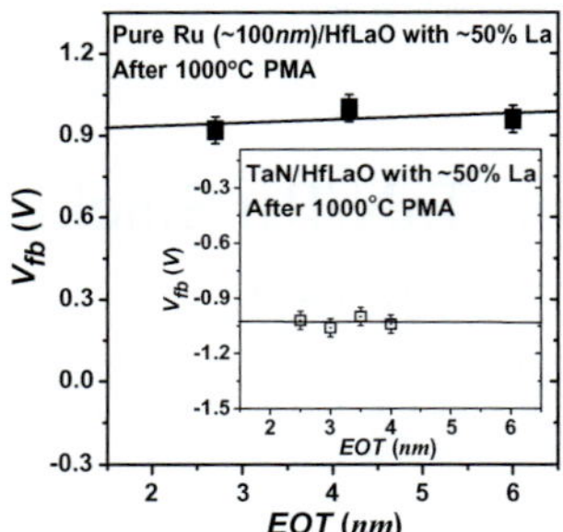

Fig. 1. Plot of V_{fb} versus EOT for devices with pure Ru metal on HfLaO dielectric on n-Si substrate (6×10^{15} cm^{-3} n-doped) after 1000 °C postmetal annealing (PMA). The case for devices with pure TaN gate on HfLaO dielectric on p-Si substrate (6×10^{15} cm^{-3} p-doped) is shown in the inset. Both V_{fb} and EOT were extracted from high-frequency C–V measurement. The corresponding EWF of Ru (TaN) on HfLaO is 5.2 (3.9) eV.

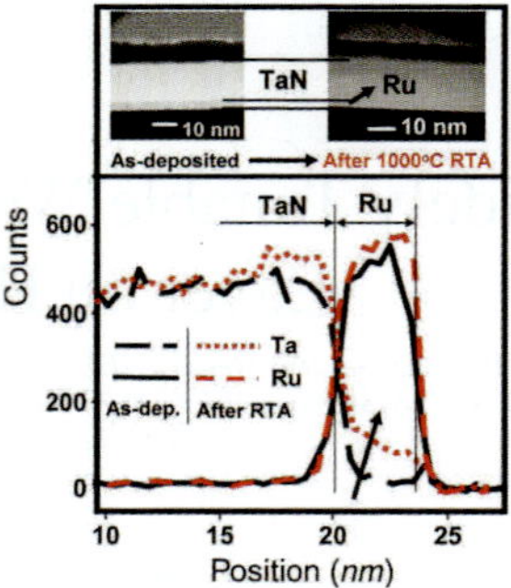

Fig. 2. EDX analysis for TaN/Ru/HfLaO stack as deposited and after 5 s RTA at 1000 °C. Ta diffusion into the Ru layer down to the interface between Ru and HfLaO after high-temperature annealing was detected. Moreover, the corresponding TEM pictures (insets) indicate good continuity and uniformity for different layers in the gate stack. Note that the top layers above the TaN layer were only used for the EDX analysis.

with ~50% La after 1000 °C annealing. Distribution of the data in Fig. 1 indicates low charge density near the HfLaO interface with the channel. The most likely explanation for the differences in V_{fb} is the EWF change of the gate electrode. However, a fixed charge in the HfLaO near the gate cannot be ruled out, although it is unlikely that the charge in HfLaO near the TaN gate would be different enough from that under the Ru gate to explain the observed difference in V_{fb} shifts. In addition, based on the V_{fb} values on 6×10^{15} cm^{-3} n-doped Si, a corresponding EWF of around 5.2 eV for Ru on HfLaO is extracted. Combined with the result of TaN on HfLaO as shown in the inset of Fig. 1, we show evidence that the incorporation of La can release Fermi level pinning between metal and HfO$_2$ dielectric, causing EWF shifts from midgap 4.64 [13] to 5.2 eV for p-metal Ru, and from 4.4 [8] to 3.9 eV for n-metal TaN, respectively. A specific model has been proposed to explain these phenomena [14], where the change of the metal EWF is attributed not only to the oxygen vacancy density in the high-κ layer, but also the difference in electronegativities of the materials involved in the gate stacks. Here, the EOT scale for the EWF extraction is from ~2.5 to ~6.0 nm, which does not show an obvious V_{fb} roll-off behavior [15], [16]. The suitability of these gate stacks for CMOS technologies that require EOT < 2.5 nm remains to be established.

Fig. 2 shows the energy dispersive X-ray spectroscopy (EDX) results of TaN and Ru (~5 nm) bilayer for as-deposited and annealed samples (1000 °C), respectively. It can be seen that the top metal layer TaN has little interdiffusion with the bottom metal layer Ru during the sputtering process, while more TaN has obviously diffused through Ru layer down to the Ru/HfLaO interface after high temperature annealing. Moreover, cross-sectional transmission electron microscopy (TEM) pictures in the insets indicate good continuity and uniformity of the different layers before and after high temperature annealing.

Fig. 3(a) shows the typical C–V curves of MOSCAPs with TaN, TaN/Ru bilayer structure (with different Ru thicknesses), and Ru metal gates on HfLaO after high temperature annealing

(1000 °C). The gradual V_{fb} shift from that of single TaN layer (~ -0.4 eV) to that of single Ru layer (~0.9 eV) can be clearly seen as the thickness of the bottom layer Ru increases from 0 to 50 nm. This results from an EWF difference for TaN and Ru, which can be explained by the observation from EDX shown above. Due to the diffusion of the top metal layer TaN through the bottom layer Ru to the Ru/HfLaO interface, the metal gate EWF can be modulated accordingly. The similar phenomena for multistacking-metal-layer on SiO$_2$ at low annealing temperature have been reported in the literature [17], [18]. Note that the capacitance variation between TaN and/or Ru gated devices is possibly due to different sputtering effects on the dielectrics during the respective TaN and Ru PVD process [19]. Fig. 3(b) shows a typical example for a TaN/Ru gate stack with 10 nm Ru after different temperature annealing. The V_{fb} decreased dramatically with higher annealing temperature due to the increase of TaN concentration at the interface between Ru and HfLaO. Moreover, the negligible V_{fb} and EOT variation after two subsequent annealings at 1000 °C suggests that the EWF changes for the gate stack after high temperature annealing are stable and permanent [20].

Fig. 4(a) shows the plots of the drain current (I_d) versus drain voltage (V_d) for long-channel p-MOSFETs with pure TaN, TaN/Ru stack with 10 nm Ru and pure Ru metal gates on an HfLaO dielectric. Compared with the case of pure TaN, the well-behaved and similar I_d–V_d characteristics for the cases of pure Ru and TaN/Ru stack indicate that the channel region of MOS devices has no obvious degradation due to the introduction of Ru and interdiffusion annealing for the TaN/Ru stack.

Fig. 4(b) shows plots of the drain current (I_d) versus gate voltage (V_g) for long-channel p-MOSFETs with pure TaN, TaN/Ru stack with different Ru thicknesses and pure Ru metal gates on an HfLaO dielectric. Very good transfer characteristics with a subthreshold slope around 75 mV/dec are demonstrated. In addition, considering the threshold voltage (V_{th}) values for

52   IEEE ELECTRON DEVICE LETTERS, VOL. 29, NO. 1, JANUARY 2008

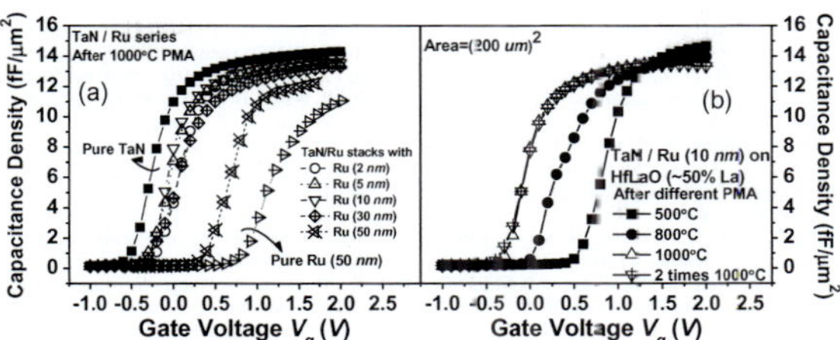

Fig. 3.   (a) Typical C–V curves for pure TaN, TaN/Ru stacks and pure Ru after 1000 °C PMA. The V_{fb} shifts toward the positive direction with the increase of the bottom Ru layer's thickness. (b) C–V curves for TaN/Ru stack with 10 nm Ru on HfLaO after different temperature PMA. The V_{fb} shifts toward the negative direction with the increase of annealing temperature up to 1000 °C.

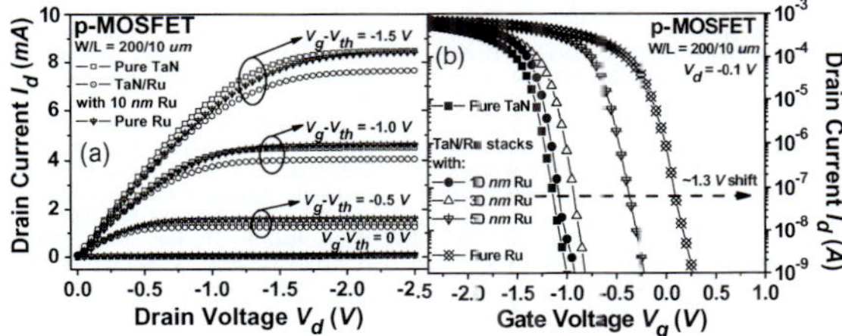

Fig. 4.   (a) I_d–V_d and (b) I_d–V_g characteristics of p-MOSFETs with pure TaN, TaN/Ru stack with 10 nm Ru and pure Ru (~50 nm) metal gate after 1000 °C annealing.

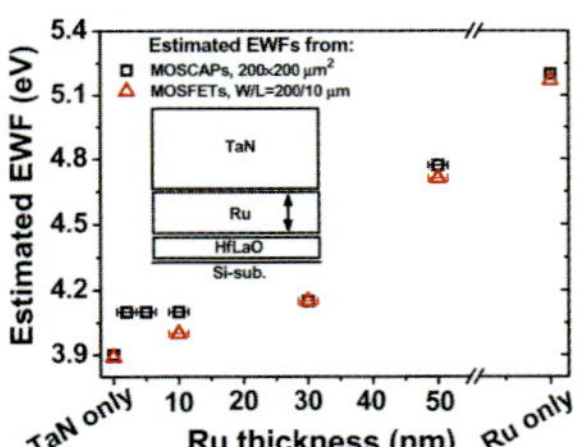

Fig. 5.   Summary of estimated EWFs from both MOSCAPs and MOSFETs for TaN/Ru stacks on HfLaO with different Ru thicknesses after 1000 °C PMA. For MOSCAPs, the EWF was obtained from the respective plot of V_{fb} versus EOT (Fig. 1), and V_{fb} shift for the devices with the different TaN/Ru electrodes [Fig. 3(a)]. While for MOSFETs, the EWF was estimated based on the device V_{th}, considering that the pure TaN gate EWF on HfLaO is ~3.9 eV.

the devices with single TaN and single Ru gate, EWF can be extracted as ~3.9 and ~5.2 eV, respectively, which are in good agreement with the results extracted from the MOSCAPs. In addition, the V_{th} shift for those devices with TaN/Ru stacks follows the same trend as the V_{fb} shift observed in MOSCAPs. The extracted EWFs from both MOSCAPs and MOSFETs are summarized in Fig. 5, which implies that a tuned EWF from n-type band edge to p-type band edge could be achieved by employing an HfLaO dielectric and TaN/Ru bilayer structure with different Ru thicknesses.

IV. CONCLUSION

A novel dual metal gate technology using a TaN/Ru bilayer metal structure on an HfLaO high-κ gate dielectric is demonstrated. Due to the interdiffusion between the top layer metal (TaN) and the bottom layer (Ru) during high temperature annealing (1000 °C) and the release of Fermi level pinning between metal gates (Ru and TaN) and HfLaO, we have demonstrated that in the EOT range of 2.5 nm and above, the EWF of these gate stacks has a widely tunable range from ~3.9 to ~5.2 eV.

REFERENCES

[1] *International Technology Roadmap for Semiconductors*. 2005 ed. [Online]. Available: http://www.itrs.net/
[2] I. De, D. Johri, A. Srivastava, and C. M. Osburn, "Impact of gate work-function on device performance at the 50 nm technology node," *Solid State Electron.*, vol. 44, no. 6, pp. 1077–1080, Jun. 2000.
[3] S. B. Samavedam, L. B. La, P. J. Tobin, B. White, C. Hobbs, L. R. C. Fonseca, A. A. Demkov, J. Schaeffer, E. Luckowski, A. Martinez, M. Raymond, D. Triyoso, D. Roan, V. Dhandapani, R. Garcia, S. G. H. Anderson, K. Moore, H. H. Tseng, C. Capasso, O. Adetutu, D. C. Gilmer, W. J. Taylor, R. Hegde, and J. Grant, "Fermi level pinning with sub-monolayer MeOx and metal gates," in *IEDM Tech. Dig.*, 2003, pp. 307–310.
[4] Y.-C. Yeo, T.-J. King, and C. Hu, "Metal-dielectric band alignment and its implications for metal gate complementary metal–oxide–semiconductor technology," *J. Appl. Phys.*, vol. 92, no. 12, pp. 7266–7271, Dec. 2002.
[5] E. Cartier, F. R. McFeely, V. Narayanan, P. Jamison, B. P. Linder, M. Copel, V. K. Paruchuri, V. S. Basker, R. Haight, D. Lim, R. Carruthers, T. Shaw, M. Steen, J. Sleight, J. Rubino, H. Deligianni, S. Guha, R. Jammy, and G. Shahidi, "Role of oxygen vacancies in V_{FB}/V_t

stability of pFET metals on HfO_2," in *VLSI Symp. Tech. Dig.*, 2005, pp. 230–231.

[6] R. Jha, B. Lee, B. Chen, S. Novak, P. Majhi, and V. Misra, "Dependence of PMOS metal work functions on surface conditions of high-k gate dielectrics," in *IEDM Tech. Dig.*, 2005, pp. 43–46.

[7] H. Y. Yu, C. Ren, Y.-C. Yeo, J. F. Kang, X. P. Wang, H. H. H. Ma, M.-F. Li, D. S. H. Chan, and D.-L. Kwong, "Fermi pinning-induced thermal instability of metal-gate work functions," *IEEE Electron Device Lett.*, vol. 25, no. 5, pp. 337–339, May 2004.

[8] X. P. Wang, M.-F. Li, C. Ren, X. F. Yu, C. Shen, H. H. Ma, A. Chin, C. X. Zhu, J. Ning, M. B. Yu, and D.-L. Kwong, "Tuning effective metal gate work function by a novel gate dielectric HfLaO for nMOSFETs," *IEEE Electron Device Lett.*, vol. 27, no. 1, pp. 31–33, Jan. 2006.

[9] H. N. Alshareef, H. R. Harris, H. C. Wen, C. S. Park, C. Huffman, K. Choi, H. F. Luan, P. Majhi, B. H. Lee, R. Jammy, D. J. Lichtenwalner, J. S. Jur, and A. I. Kingon, "Thermally stable N-metal gate MOSFETs using La-incorporated HfSiO dielectric," in *VLSI Symp. Tech. Dig.*, 2006, pp. 7–8.

[10] X. P. Wang, C. Shen, M. F. Li, H. Y. Yu, Y. Sun, Y. P. Feng, A. Lim, W. S. Hwang, A. Chin, Y.-C. Yeo, P. Lo, and D. L. Kwong, "Dual metal gates with band-edge work functions on novel HfLaO high-k gate dielectric," in *VLSI Symp. Tech. Dig.*, 2006, pp. 9–10.

[11] V. Narayanan, V. K. Paruchuri, N. A. Bojarczuk, B. P. Linder, B. Doris, Y. H. Kim, S. Zafar, J. Stathis, S. Brown, J. Arnold, M. Copel, M. Steen, E. Cartier, A. Callegari, P. Jamison, J.-P. Locquet, D. L. Lacey, Y. Wang, P. E. Batson, P. Ronsheim, R. Jammy, M. P. Chudzik, M. Ieong, S. Guha, G. Shahidi, and T. C. Chen, "Band-edge high-performance high-κ/metal gate n-MOSFETs using cap layers containing group IIA and IIIB elements with gate-first processing for 45 nm and beyond," in *VLSI Symp. Tech. Dig.*, 2006, pp. 178–179.

[12] H. Watanabe, N. Ikarashi, and F. Ito, "La-silicate gate dielectrics fabricated by solid phase reaction between La metal and SiO_2 underlayers," *Appl. Phys. Lett.*, vol. 83, no. 17, pp. 3546–3548, Oct. 2003.

[13] Z. B. Zhang, S. C. Song, C. Huffman, J. Barnett, N. Moumen, H. Alshareef, P. Majhi, M. Hussain, M. S. Akbar, J. H. Sim, S. H. Bae, B. Sassman, and B. H. Lee, "Integration of dual metal gate CMOS with TaSiN (NMOS) and Ru (PMOS) gate electrodes on HfO_2 gate dielectric," in *VLSI Symp. Tech. Dig.*, 2005, pp. 50–51.

[14] X. P. Wang, H. Y. Yu, M.-F. Li, C. X. Zhu, S. Biesemans, A. Chin, Y. Y. Sun, Y. P. Feng, A. Lim, Y.-C. Yeo, W. Y. Loh, P. Lo, and D.-L. Kwong, "Wide V_{fb} and V_{th} tunability for metal gated MOS devices with HfLaO gate dielectrics," *IEEE Electron Device Lett.*, vol. 28, no. 4, pp. 258–260, Apr. 2007.

[15] B. H. Lee, J. Oh, H. H. Tseng, R. Jammy, and H. Huff, "Gate stack technology for nanoscale devices," *Mater. Today*, vol. 9, no. 6, pp. 32–40, Jun. 2006.

[16] K. Akiyama, W. Wang, W. Mizubayashi, K. Salam, M. Ikeda, H. Ota, T. Nabatame, and A. Toriumi, "Vfb roll-off of metal gate/HfO_2/SiO_2/Si capacitors in thinner EOT regime," in *Proc. Extended Abstract IWDTF*, 2006, pp. 63–64.

[17] C. Lu, G. Wong, M. Deal, W. Tsai, P. Majhi, C. Chui, M. Visokay, J. Chambers, L. Colombo, B. Clemens, and Y. Nishi, "Characteristics and mechanism of tunable work function gate electrodes using a bilayer metal structure on SiO_2 and HfO_2," *IEEE Electron Device Lett.*, vol. 26, no. 7, pp. 445–447, Jul. 2005.

[18] A. E.-J. Lim, W. S. Hwang, X. P. Wang, D. M. Y. Lai, G. S. Samudra, D.-L. Kwong, and Y.-C. Yeo, "Metal-gate work function modulation using hafnium alloys obtained by the interdiffusion of thin metallic layers," *J. Electrochem. Soc.*, vol. 154, no. 4, pp. H309–H313, 2007.

[19] L.-Å. Ragnarsson, V. S. Chang, H. Y. Yu, H.-J. Cho, T. Conard, K. M. Yin, A. Delabie, J. Swerts, T. Schram, S. De Gendt, and S. Biesemans, "Achieving conduction band-edge effective work functions by La_2O_3 capping of hafnium silicates," *IEEE Electron Device Lett.*, vol. 28, no. 6, pp. 486–488, Jun. 2007.

[20] W. P. Bai, S. H. Bae, H. C. Wen, S. Mathew, L. K. Bera, N. Balasubramanian, N. Yamada, M. F. Li, and D.-L. Kwong, "Three-layer laminated metal gate electrodes with tunable work functions for CMOS applications," *IEEE Electron Device Lett.*, vol. 26, no. 4, pp. 231–233, Apr. 2005.

Chapter 6

CMOS Transistors (III)
(Quantum Simulations)

1188

IEEE TRANSACTIONS ON ELECTRON DEVICES, VOL. 48, NO. 6, JUNE 2001

Hole Quantization Effects and Threshold Voltage Shift in pMOSFET—Assessed by Improved One-Band Effective Mass Approximation

Y. T. Hou and Ming-Fu Li, *Senior Member, IEEE*

Abstract—Threshold voltage (V_T) shift due to quantum mechanical (QM) effects in pMOSFET is investigated based on a six-band effective mass approximation (EMA). Due to the valence band mixing, both subband energies and density of states (DOS) show remarkable difference from those derived from traditional one-band EMA using the bulk Si effective masses. In comparison with the experimental results, it is found that V_T shift in pMOSFET is significantly overestimated by the traditional one-band EMA, however it corresponds with our six-band EMA calculation. Based on the numerical results of our six-band EMA, new effective masses are determined empirically and their electric field dependence is also evaluated. Using these new effective masses instead of the bulk effective masses, one-band EMA still display effectiveness in describing hole quantization and V_T shift in an empirical manner. A set of constant energy quantizaiton/DOS effective masses ($0.29/1.14, 0.22/0.75, 0.24/0.66m_0$) for the first three subbands, neglecting their electric field dependence, is proposed for the modeling of QM effects in pMOSFET in this improved version of the one-band EMA formula. Computing time is minimized and results can be obtained with sufficient accuracy and correspond well with reported experimental data, thus the improved one-band EMA formula provide a firm ground in routine device simulation for deep submicron MOS devices.

Index Terms—CMOSFETs, inversion quantization, quantum mechanical effects, semiconductor device modeling, threshold voltage.

I. INTRODUCTION

THE two-dimensional (2-D) nature of carriers in MOSFETs has been known for a long time [1], [2], but it plays a more prominent role only recently when the scaling of MOSFET enters into deep submicron era. Due to the higher substrate doping concentration and thinner gate oxide in modern devices, the electric field near the Si/SiO$_2$ interface is strong enough even at the threshold region, so that the quantum mechanical (QM) effects becomes noticeable. In this situation, the classical three-dimensional (3-D) treatment becomes inadequate and carriers must be treated as quantum 2-D. The quantization of carriers will lead to a substantial change of the band structure and density of states (DOS) [1]–[3], and as a consequence, a

threshold voltage (V_T) shift in MOSFET. This phenomenon has been experimentally observed both in nMOSFETs and pMOSFETs [4]–[6]. Since V_T is a key parameter affecting the device performance, a physical insight into the QM effects and an accurate modeling of the V_T shift will be of great importance. For nMOSFET, the electron quantization as well as the V_T shift is well understood in one-band effective mass approximation (EMA) [4], [5], [7]–[9]. However, hole quantization in pMOSFET has not been studied in as much detail as electrons due to the complexity of the valence band structure of Si [10]–[15]. Previously, one-band EMA was also applied to hole quantization, using constant effective masses derived from bulk Si (brief as traditional one-band EMA) to calculate the hole subband energies and V_T shift in pMOSFET [3], [10], [13]. Such a procedure omits band mixing effect and the results obtained are not reliable as shown later. A few studies abandon the use of the EMA and used the pseudopotential full-band self-consistent formalism, accounting for the complicated valence band structure [11], [12]. Although these full-band self-consistent treatments may achieve the reliable results in principle, they are computational prohibitive for routine device simulations. We have developed a simple but reliable model based on six-band effective mass equation for hole quantization [16] Instead of the self-consistent method, the potential in the inversion layer is approximated by a periodic triangular well (zigzag) structure. An accurate modeling of 2-D hole DOS and the V_T shifts in pMOSFET by this model is reported. As a further simplification, a new set of empirical hole effective masses suitable for this hole quantization effect under electric field is also derived from the numerical results. With band-mixing effects included in these empirical hole effective masses, we have proven that V_T shift in pMOSFET can still be described properly under the framework of the one-band EMA (brief as improved one-band EMA). Compared to the traditional one-band EMA, the improved one-band EMA shows better agreement with the experiments.

II. THRESHOLD VOLTAGE SHIFT DUE TO QM EFFECTS— ONE BAND EMA

Although the actual potential profile within the inversion layer can be determined by solving self-consistently the coupled Schrödinger and Poisson equations [1], such a treatment is tedious and is not always necessary. It has been shown that, triangular well is a quite good approximation of the potential in

Manuscript received September 27, 2000; revised December 8, 2000. This work was supported by Singapore NSTB/EMT/TP/00/001,2 and the National University of Singapore RP3 982 754 Grants. The review of this paper was arranged by Editor C.-Y. Lu.

The authors are with SI Nano Device Laboratory, Department of Electrical and Computer Engineering, National University of Singapore, Singapore 119260 (e-mail: elelimf@nus.edu.sg).

Publisher Item Identifier S 0018-9383(01)04219-8.

weak inversion layer [3] and its validity in threshold region is also justified [9]. In triangular well approximation, the potential well is written as

$$\phi(z) = qF_s \cdot z \tag{1}$$

where F_s is the surface electric field in Si substrate and z is the axes perpendicular to the Si/SiO$_2$ interface. The devices are usually fabricated on (001) Si and thus we take $z = (001)$ here. Moreover, we take the top of the valence band at the substrate surface as the energy reference, and for convenience the hole energies are all expressed as positive values.

Under the depletion approximation, surface potential ϕ_s, and surface electric field F_s of the substrate with doping N_{sub} have the relation [17]

$$F_s = \sqrt{\frac{2qN_{\mathrm{sub}}}{\varepsilon_0\varepsilon_{\mathrm{Si}}}\phi_s}. \tag{2}$$

Here, we neglect the contribution from the inversion charge because it is much less than the depletion charge at threshold region (less than 1%). The 3-D inversion charge density at ϕ_s can be given by

$$\begin{aligned}
N_s^{\mathrm{class}} &= \int N_V e^{(E_f - \phi(z))/kT}\, dz \\
&= \frac{kT \cdot N_V}{F_s} e^{E_f/kT}(e^{\phi_s/kT} - 1)
\end{aligned} \tag{3}$$

where N_V is the effective DOS at the top of the valence band.

The classical definition of V_T is the gate voltage when surface potential $\phi_T^{\mathrm{class}} = 2\phi_B$, with $\phi_B = kT\ln(N_{\mathrm{sub}}/n_i)$ [17]. The 2-D V_T is determined by the gate voltage to populate the 2-D inversion layer to the same inversion charge density as the 3-D one, namely, $N_s^{\mathrm{class}}(\phi_T^{\mathrm{class}})$ [10]. In the 2-D QM description, if the subband dispersion E_n and DOS $D_n(E)$ are determined, the inversion charge density is

$$N_s^{\mathrm{QM}} = \sum_n \int f(E_n)D_n(E)\, dE_n. \tag{4}$$

At threshold region, $f(E)$ can be simplified to Maxwell-Boltzmann statistical function [9]. By equating $N_s^{\mathrm{QM}}(\phi_T^{\mathrm{QM}})$ in (4) to $N_s^{\mathrm{class}}(\phi_T^{\mathrm{class}})$ in (3), we obtain ϕ_T^{QM}, the surface potential at 2-D V_T. Finally, the V_T shift due to QM effect can be expressed as [10]

$$\begin{aligned}
\Delta V_T &= \left(\phi_T^{\mathrm{QM}} - \phi_T^{\mathrm{class}}\right) + T_{\mathrm{OX}} \\
&\cdot \frac{\varepsilon_{\mathrm{Si}}}{\varepsilon_{\mathrm{OX}}}\sqrt{\frac{2qN_{\mathrm{sub}}}{\varepsilon_{\mathrm{Si}}\varepsilon_0}}\left(\sqrt{\phi_T^{\mathrm{QM}}} - \sqrt{\phi_T^{\mathrm{class}}}\right).
\end{aligned} \tag{5}$$

Under traditional one-band EMA, as widely used for electrons, the derivation of V_T shift occurs in a simple and straightforward manner. In triangular well approximation, the energies of subband minima can be expressed analytically as [1], [18]

$$E_{ij} = \left(\frac{\hbar^2}{2m_{zi}}\right)^{1/3}\left[\frac{3}{2}\pi qF_s\left(j - \frac{1}{4}\right)\right]^{2/3} \tag{6}$$

where m_{zi} is the energy quantization effective mass of the bulk Si. The 2-D DOS $(g_i m_{di}/\pi\hbar^2)$ is independent of energy, where m_{di} and g_i ($g_i = 1$ for valence band of Si) are the DOS effective mass of the bulk Si and degenerate factor of the ith subband. Then, (4) has an analytical form as

$$N_s^{\mathrm{QM}} = \sum_{i,j}\frac{g_i m_{di}}{\pi\hbar^2}kT e^{(E_f - E_{ij})/kT}. \tag{7}$$

For complicated subband structure, (4) can only be evaluated numerically.

III. SUBBAND STRUCTURE AND DENSITY OF STATES BY SIX-BAND EMA

The above traditional one-band EMA has also been employed to calculate the hole subband energies as well as the V_T shift in pMOSFET [10], [13]. However, coupling of the valence bands casts doubts on the validity of this approximation [11], [12]. To account for band mixing effect, we have developed a model for hole quantization based on the multiband effective mass method. Six bulk bands are included in our model, they are heavy, light and spin-orbit split-off holes and their spin degenerate bands. The spin-orbit split-off energy for Si is only 44 meV [19] and thus the coupling of the split-off band to other bands is not negligible. From the $k \cdot p$ effective-mass theory, the Hamiltonian can be obtained by the unperturbed bulk band Hamiltonian adding a diagonal electric potential energy term $V(z)$ [19]. When a periodic triangle well potential (zigzag) with period L is used to replace the single triangle well as described in (1), the derivation of eigenenergies follows a similar treatment in III-V multiple quantum wells [20] with the eigenfunctions expanded by plane waves. L is selected to be sufficiently large so that the coupling between the neighboring wells is negligible and sufficiently small. The calculation of hole subband energies are efficient. Details of the model will be presented elsewhere [16]. The difference of the subband minimum energies at classical threshold region for varying substrate doping between traditional one-band EMA and our multiband model is given in Fig. 1(a) as a relative deviation in percentage. The electric fields versus N_D are determined from (2). In traditional one-band EMA, the energies are determined by (6). The energy quantization (m_z) and DOS (m_d) effective masses are deduced from bulk Si valence bands [3], [10], [13]. Their values are summarized in Table I. The energies for the first lowest subband $(n = 1)$ obtained from two methods are fully consistent. We have found that this subband is purely heavy hole at $k = 0$ [12], [16]. It is not surprising from the symmetric consideration of the Hamiltonian, where nondiagonal items for heavy hole all vanish at $k = 0$. For $n = 2$ and $n = 3$ subbands, band mixing occurs even at $k = 0$ [12], [16], which results in the inaccuracy of the traditional one-band EMA in the determination of subband energies, as shown in Fig. 1(a). With the increase of substrate doping N_D (surface electric field increases correspondingly), the relative deviation also increases. It is also found that the deviations of $n = 2$ and $n = 3$ subbands are in opposite signs. For $k \neq 0$, there exists band mixing for all subbands. Due to the strong band mixing effect, the dispersion of the hole subband is far from parabolic and also becomes dependent of electric filed.

1190   IEEE TRANSACTIONS ON ELECTRON DEVICES, VOL. 48, NO. 6, JUNE 2001

TABLE I
SUMMARY OF ENERGY QUANTIZATION AND
DOS EFFECTIVE MASSES IN UNIT OF FREE ELECTRON MASS m_0. BULK SI
VALUES ARE FROM [3]. THE FIELD DEPENDENT EMPIRICAL VALUES ARE
OBTAINED BY FITTING TO THE NUMERICAL DATA OF SIX-BAND EMA
CALCULATION AT DIFFERENT FIELD F_s (IN UNITS OF MV/cm)

		Bulk Si	Field dependent empirical	Constant empirical
Quantization mass m_z	n=1	0.29	0.2910	0.29
	n=2	0.20	$0.2038+0.0358Fs-0.0119Fs^2$	0.22
	n=3	0.29	$0.2502-0.0658Fs+0.0324Fs^2$	0.24
DOS mass m_d	n=1	0.65	$1.1825-0.1466Fs+0.0492Fs^2$	1.14
	n=2	0.25	$0.8031-0.1051Fs-0.0516Fs^2$	0.75
	n=3	0.29	$0.9105-0.8500Fs+0.4967Fs^2$	0.66

In Fig. 1(b), we plot the DOS at threshold voltage for a typical substrate doping density $N_D = 5 \times 10^{17}$ cm^{-3}. In traditional one-band EMA, the DOS profile is the well-known step-like function, shown as dashed line in Fig. 1(b). The difference is very clear on the DOS obtained by our six-band EMA. The DOS profile is deviated from the step-like function because the dispersion is no longer parabolic. Furthermore, the DOS from our model is overall much larger than that obtained from the traditional one-band EMA.

In order to obtain a greater insight of the DOS characteristics, we show in Fig. 2 the separate DOS profiles of the first three subbands at V_T for $N_D = 1 \times 10^{17}$ cm^{-3}, 5×10^{17} cm^{-3}, and 1×10^{18} cm^{-3}. The subband DOS profiles are all deviated from the step-like shape of 2-D carriers with parabolic dispersion. The first subband ($n = 1$) shows a much slower slope as the DOS rises from the subband minimum. At higher energies, the DOS approaches an almost constant value. Although DOS profiles of other two subbands remain a rapid increase near the subband minima, they show very different behavior at higher energies. The DOS profile of $n = 2$ subband forms two quasi-plateaus while two peaks appear near the subband minimum of the $n = 3$ subband. These peaks are caused by the camel back structures [21] near $k = 0$ on the subband dispersion curve [16]. From Fig. 2, the DOS profiles of hole subbands are generally irregular, especially near the subband minima. Since the extent of band mixing depends on the electric field, the DOS profile also becomes electric field dependent. We will not deal with other subbands of higher order because the first three subbands in total have accounted for almost all the inverted hole charges in our case. Even near threshold region, occupation of other higher subbands can be negligible especially at higher N_D, where QM effects are noticeable and is of interest in our discussion.

IV. THRESHOLD VOLTAGE SHIFT IN pMOSFET

Using the DOS profiles and subband energies obtained, the threshold voltage shift at various substrate doping can be determined numerically from (3)–(5), the results for 14-nm oxide

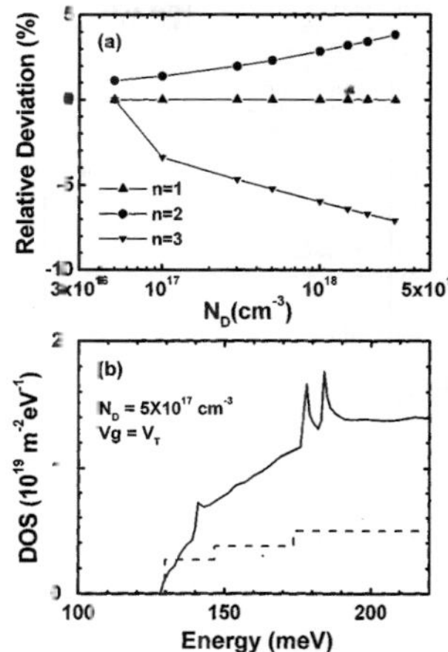

Fig. 1. (a) Relative deviation of subband energies of first three subbands at threshold voltage derived by the traditional one-band EMA from those of our six-band EMA, as a function of substrate doping N_D. (b) DOS for substrate doping 5×10^{17} cm^{-3} at threshold voltage. Solid and dashed lines are from our six-band EMA model and from traditional one-band EMA, respectively.

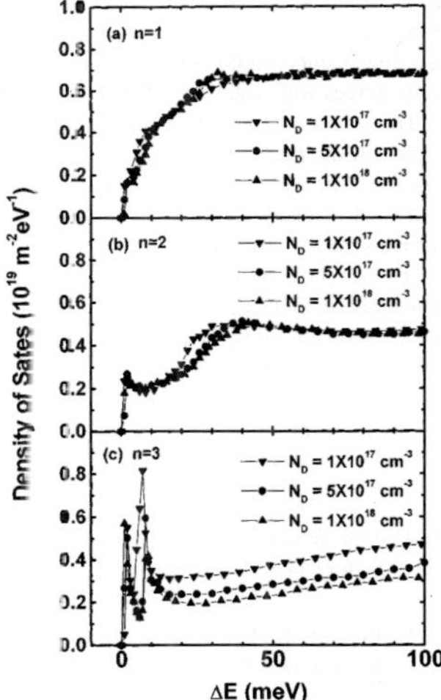

Fig. 2. DOS of first three lowest subbands at threshold voltage by our multiband EMA model for three typical substrate doping $N_D = 1 \times 10^{17}$ cm^{-3}, 5×10^{17} cm^{-3} and 1×10^{18} cm^{-3}. ΔE is the energy referenced from the subband edge.

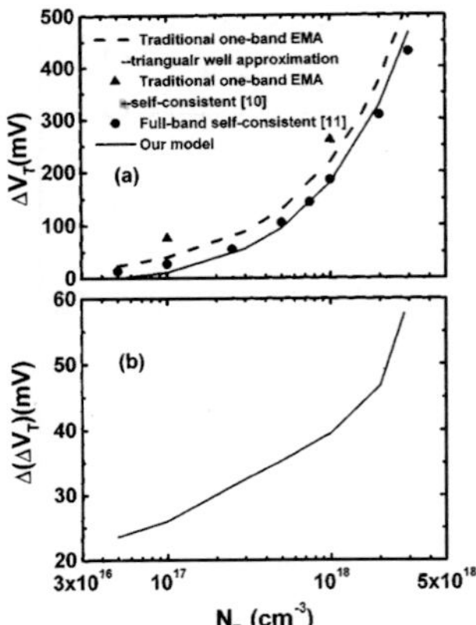

Fig. 3. (a) Threshold voltage shifts ΔV_T due to quantum effects at various substrate doping N_D with oxide thickness of 14 nm. Solid triangles and dashed line are from traditional one-band EMA, using self-consistent [10] and triangular well approximation, respectively. Solid circles are from the pseudopotential full-band self-consistent method [11], and the solid line is our result with six-band EMA model. (b) $\Delta \, (\Delta V_T)$ [defined by (8)] as a function of substrate doping N_D.

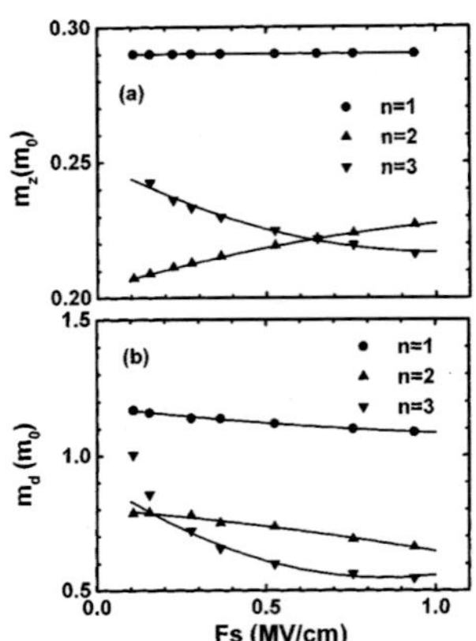

Fig. 4. Surface electric field F_s dependence of (a) empirical energy quantization effective mass m_z, and (b) empirical DOS effective mass m_d, determined from the numerical results of six-band EMA calculations (solid symbols). The solid lines are the fitting by polynomials up to second order of F_s with fitting parameters summarized in Table I.

is shown in Fig. 3(a). The same curve under the traditional one-band EMA is also shown as dashed lines. For comparison, we also presented the reported data from traditional one-band EMA calculation [10] and from pseudopotential full-band self-consistent formalism [11]. The result of our six-band model agrees well with the full-band pseudopotential calculation. This demonstrates that our simplified model remains accurate in the modeling of V_T shift due to hole QM effects. However, the traditional one-band EMA, either by self-consistent treatment or in triangular well approximation, both overestimates the V_T shift in pMOSFET, provided that the effective masses from bulk Si are used. It can be ascribed to its underestimation of the DOS as indicated in Fig. 1(b) and the inaccurate description of subband energy levels, as shown in Fig. 1(a). If we define $\Delta \, (\Delta V_T)$ as

$$\Delta \, (\Delta V_T) = (\Delta V_T)_{\text{traditional one-band EMA}}$$
$$- (\Delta V_T)_{\text{improved one-band EMA}}. \qquad (8)$$

Fig. 3(b) shows that $\Delta \, (\Delta V_T)$ is increased when the doping concentration N_D increases.

V. EMPIRICAL EFFECTIVE MASS— IMPROVED ONE-BAND EMA

As noted above, the DOS profiles of hole subbands are deviated from the simple step-like function of traditional one-band EMA. Its shape is not regular especially near the band minimum, where the carriers are mostly populated. Moreover, the detailed profile of DOS also depends on the

surface electric fields. Therefore, the hole subband structure cannot be adequately described by a simple analytical model (parabolic, etc.) and numerical method is necessary for accurate characterization. However, in some cases, we may not need to know the physical details of the hole subband structure but only of some macroscopic electrical parameters, such as V_T, are interested. In these cases, an improved one-band EMA assuming a parabolic dispersion is valuable. In such a treatment, (6) and (7) are still valid, however the m_z and m_{id} are not derived from the bulk Si, but from the numerical results of the six-band EMA calculation in an empirical way. The results are plotted in Fig. 4 as a function of the surface electric field, instead of substrate doping N_D, for general purpose. Attention is first paid to empirical energy quantization effective mass m_z for the first three subbands, as shown in Fig. 4(a). They are determined by (6) inversely when the numerical energies of the subband minima have been obtained by six-band effective mass equations. For $n = 1$, the empirical energy quantization effective mass is $0.29m_0$ and is independent of F_s because of its purity of heavy hole as discussed before. For $n = 2$ and $n = 3$ subbands, the empirical effective masses display an electric field dependent behavior due to the field dependence of band mixing. The empirical effective mass value of $n = 2$ subband increases as F_s increases. Near $k = 0$, this band is mainly light hole (bulk mass $0.20m_0$). As electric field increases, more and more split-off hole (bulk mass $0.29m_0$) is mixed into this band, manifesting itself as an increase of the empirical effective mass. On the other hand, the decrease of the empirical effective mass of $n = 3$ subband with increasing electric field corresponds to the increased mixing of light hole into this primarily split-off hole subband near $k = 0$.

1192   IEEE TRANSACTIONS ON ELECTRON DEVICES, VOL. 48, NO. 6, JUNE 2001

For subband DOS, we introduce the empirical DOS effective mass m_d which can be determined from (4) and (7). The inversion charge density occupying subband n is

$$N_s^n = \int e^{(E_f - E)/kT} D_n(E)\, dE$$
$$= kT \cdot e^{(E_f - E_{n0})/kT}$$
$$\times \int e^{-\Delta E/kT} D_n(\Delta E)\, d\left(\frac{\Delta E}{kT}\right). \quad (9)$$

Here, E_{n0} is the subband minimum and ΔE is the relative energy referenced from the subband edge. Combined with (7), the empirical DOS effective mass m_d is

$$m_d^n = \frac{\int e^{-\Delta E/kT} D_n(\Delta E)\, d\left(\frac{\Delta E}{kT}\right)}{(1/\pi\hbar^2)}. \quad (10)$$

The obtained empirical DOS effective masses are shown in Fig. 4(b). These effective mass values for the $n = 1$ and $n = 2$ subbands are weakly dependent of F_s in the range of our interests up to 1 MV/cm. This can also be clearly seen from Fig. 2, where the DOS profiles are nearly the same for different substrate doping concentration. For $n = 3$ subband, the N_D dependence (and therefore F_s dependence) of DOS is stronger as indicated in Fig. 2(c). We have fitted the electric field dependence of these empirical effective masses in polynomial form up to the second order of F_s. The obtained parameters are all summarized in Table I. In general, the hole quantization behavior can be described empirically in this improved one-band EMA using these electric field dependent effective masses. However, it must be noted that (7) is based on the Maxwell-Boltzmann statistical function and it is only applicable when the inversion is not very strong. Here the empirical DOS masse depends only on the electric field, and is independent of the Fermi level. In strong inversion case, Fermi-Dirac distribution should be used and the empirical DOS masses will also depend on the Fermi level. Moreover, in strong inversion case, the inversion charge also affects the surface electric field and surface potential.

Although some of these empirical effective mass values are electric field dependent, this dependence can be neglected in the first order approximation. The reason is that at room temperature most of inversion holes are occupied on the $n = 1$ subband (over 70%) and the contribution of $n = 2$ and $n = 3$ subbands is relatively small. From Fig. 4, both the energy quantization and DOS masses of $n = 1$ subband have a weak F_s dependence. This leads us to propose a set of constant empirical effective mass values to describe the V_T shift. They are found to be $0.29/1.14, 0.22/0.75, 0.24/0.66 m_0$ for the three subbands. In Fig. 5, we show the V_T shifts calculated from these constant empirical effective masses under one-band EMA. Comparing to the numerical results, it is verified that the V_T shifts can be obtained by this field independent empirical effective masses method with sufficient accuracy.

Finally, comparison with the experimentally observed V_T shift is given in Fig. 6. The experimental ΔV_T data for 23 nm and 15 nm oxides are from [6]. Our calculation of V_T shifts in pMOSFET by the improved one-band EMA using field independent empirical effective masses is in well agreement

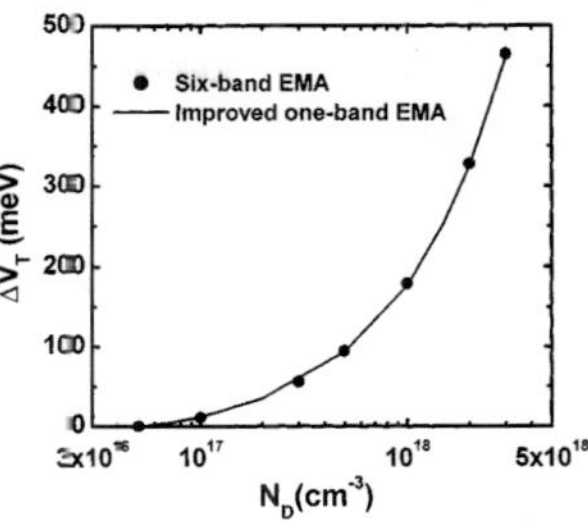

Fig. 5. Threshold voltage shifts ΔV_T at various substrate doping N_D. The numerical results by six-band EMA calculations are shown as solid circles while solid line is by improved one-band EMA using our new constant empirical effective masses listed in Table I.

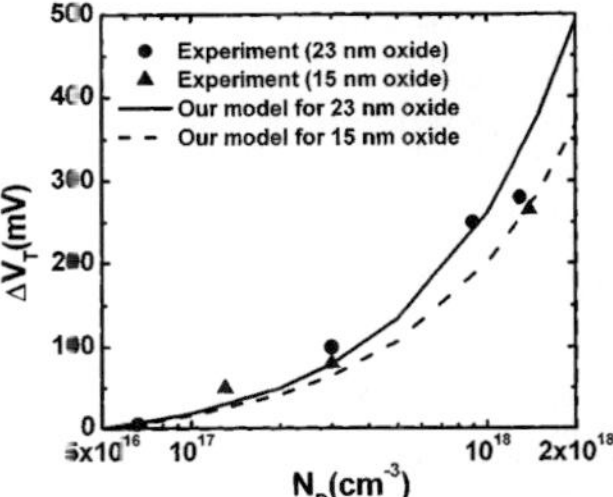

Fig. 6. Experimental and calculated threshold voltage shifts as a function of substrate doping. Solid circles and triangles are experimental data for 23- and 15-nm oxide [6]. Solid and dashed lines are their respective calculation using our improved one-band EMA with constant empirical effective masses listed in Table I.

with the experiments. This demonstrated that our improved one-band EMA method will be valuable for the modeling of hole quantization and V_T shifts in pMOSFETs. Our method has also been used to calculate the inversion layer capacitance and hole direct tunneling current giving rise to results that corresponds well with the experiments. The results will be published elsewhere.

VI. CONCLUSIONS

In conclusion, the hole quantization is studied by a six-band EMA, which takes into account the band mixing effect. The results demonstrate that the band mixing cannot be neglected. The traditional one-band EMA using bulk effective masses cannot describe the hole quantization accurately. It overall underestimates the 2-D hole DOS and overestimates V_T shift in pMOSFET. The V_T shift obtained by our six-band effective mass model in zigzag well approximation is found to remain sufficiently accurate compared to the reported pesudopotential full-band self-consistent calculation with more efficiency computationally. Based on this, an improved one-band EMA is developed in two hierarchies. In the first hierarchy, the empirical effective masses are derived from the numerical results of the six-band EMA and are electric field dependent. In the second hierarchy, further simplifications are made such that the electric field dependencies of the empirical effective masses are neglected. Hole quantization and V_T shift can be described adequately by the improved one-band EMA in either first or second hierarchy. The methods developed in this work with different approximations, due to their accuracy and computational efficiency, provide a firm ground in routine device simulation for deep submicron MOS devices.

REFERENCES

[1] F. Stern, "Self-consistent results for n-type Si inversion layer," *Phys. Rev. B, Condens. Matter*, vol. 5, no. 12, pp. 4891–4899, 1972.

[2] T. Ando, A. Fowler, and S. Stern, "Electronic properties of two-dimensional systems," *Rev. Mod. Phys.*, vol. 54, pp. 437–672, Apr. 1982.

[3] C. Moglestue, "Self-consistent calculation of electron and hole inversion charges at silicon-silicon dioxide interfaces," *J. Appl. Phys.*, vol. 59, no. 5, pp. 3175–3183, 1986.

[4] M. J. van Dort, P. H. Woerlee PH, and A. J. Walker, "Influence of high substrate doping levels on the threshold voltage and mobility of deep-submicrometer MOSFETs," *IEEE Trans. Electron Devices*, vol. 39, pp. 932–938, Apr. 1992.

[5] ——, "A simple model for quantization effects in heavily-doped silicon MOSFETs at inversion conditions," *Solid State Electron.*, vol. 37, no. 3, pp. 411–441, 1994.

[6] G. Chindalore, S. A. Hareland, S. Jallepalli, A. F. Tasch, Jr., C. M. Maziar, V. K. F. Chia, and S. Smith, "Experimental determination of threshold voltage shifts due to quantum mechanical effects in MOS electron and hole inversion layers," *IEEE Electron Device Lett.*, vol. 18, pp. 206–208, May 1997.

[7] Y. Ohkura, "Quantum effects in Si n-MOS inversion layer at high substrate concentrations," *Solid State Electron.*, vol. 33, pp. 1581–1585, Dec. 1990.

[8] T. Janik and B. Majkusiak, "Analysis of the MOS transistor based on self-consistent solutions to the Schrodinger and Poisson equations and on the local mobility model," *IEEE Trans. Electron Devices*, vol. 45, pp. 1263–1271, June 1998.

[9] Y. Ma, Z. Li, L. Liu, L. Tian, and Z. Yu, "Effective density-of-states to QM correction in MOS structures," *Solid State Electron.*, vol. 44, no. 3, pp. 401–407, 2000.

[10] C.-Y. Hu, S. Banerjee, K. Sadra, B. G. Streetman, and R. Sivan, "Quantization effects in inversion layers of pMOSFETs on Si (100) substrate," *IEEE Electron Device Lett.*, vol. 17, pp. 276–278, June 1996.

[11] S. Jallepalli, J. Bude, W.-K. Shih, M. R. Pinto, C. M. Maziar, and A. F. Tasch, "Electron and hole quantization and their impact on deep submicron silicon p- and n-MOSFET characteristics," *IEEE Trans. Electron Devices*, vol. 44, pp. 297–303, Feb. 1997.

[12] S. Rodriguez, J. A. Lopez-Villaneuva, I. Melchor, and J. E. Carceller, "Hole confinement and energy subbands in a silicon inversion layer using the effective mass theory," *J. Appl. Phys.*, vol. 86, no. 1, pp. 438–444, 1999.

[13] Y. Ma, L. Liu, Z. Yu, and Z. Li, "Characterization and modeling of threshold voltage shift due to quantum mechanical effects in pMOSFET," *Solid State Electron.*, vol. 44, no. 7, pp. 1335–1339, 2000.

[14] S. A. Hareland, S. Jallepalli, G. Chindalore, W.-K. Shih, H. Wang, A. F. Tasch, Jr., and C. M. Maziar, "A simple model for quantum mechanical effects in hole inversion layers in silicon pMOS devices," *IEEE Trans. Electron Devices*, vol. 44, pp. 1172–1173, July 1997.

[15] S. A. Hareland, S. Jallepalli, W.-K. Shih, H. Wang, G. L. Chindalore, A. F. Tasch, and C. M. Maziar, "A physically-based model for quantization effects in hole inversion layers," *IEEE Trans. Electron Devices*, vol. 45, pp. 179–185, Jan. 1998.

[16] Y. T. Hou and M. F. Li, "A novel simulation algorithm for Si valence hole quantization of inversion layer in metal–oxied–semiconductor devices," *Jpn. J. Appl. Phy*, vol. 40, pp. L144–L147, 2001.

[17] Y. Taur and T.H. Ning, *Fundamentals of Modern VLSI Devices*, Cambridge, U.K.: Cambridge Univ. Press, 1998.

[18] H. H. Mueller and M. J. Schulz, "Simplified method to calculate the band bending and the subband energies in MOS capacitors," *IEEE Trans. Electron Devices*, vol. 44, pp. 1539–1543, Sept. 1997.

[19] M.-F. Li, *Modern Semiconductor Quantum Physics*, Singapore: World Scientific, 1994, ch. 3, p. 243; ch. 5, p. 441.

[20] W. J. Fan, M. F. Li, T. C. Chong, and J. B. Xia, "Valence hole subbands and optical gain spectra of GaN/Ga$_{1-x}$Al$_x$N strained quantum wells," *J. Appl. Phys.*, vol. 80, no. 6, pp. 3471–3478, 1996.

[21] Landolt-Bornstein, *Numerical Data and Functional Relationships in Science and Technology. Group 3. Crystal and Solid State Physics*, O. Madelung, Ed. Berlin, Germany: Springer-Verlag, 1987, vol. 22a. p. 72 and p. 304.

Yong-Tian Hou received the B.S. and M.S. degrees in physics from Peking University, Beijing, China, in 1990 and 1993, respectively. Currently, he is pursuing the Ph.D. degree in electrical engineering at National University of Singapore.

In 1993, he joined Tianma Microelectronics Co. Ltd., Shenzhen, China. From 1998 to 1999, he was a Research Engineer in the Department of Electrical and Computer Engineering, National University of Singapore. His general research interests include QM effects, tunneling current, and reliability physics of ultrathin gate oxide in deep submicron MOSFETs.

Ming-Fu Li (M'91–SM'99) graduated from the Department of Physics, Fudan University, Shanghai, China, in 1960.

He joined the Department of Applied Physics, University of Science and Technology of China (USTC). In 1978, he joined the Graduate School faculty, Chinese Academy of Sciences, Beijing, where he became a Professor in 1986. He has also served as Adjunct Professor at the Institute of Semiconductors, Chinese Academy of Science, Fudan University, and USTC, Hefei.He was a Visiting Scholar at Case Western Reserve University, Cleveland, OH, in 1979, University of Illinois at Urbana-Champaign from 1979 to 1981, and was a Visiting Scientist at the University of California at Berkeley and Lawrence Berkeley National Laboratories from 1986 to 1987, 1990 to 1991, and 1993. He joined the Department of Electrical Engineering, National University of Singapore, in 1991, and became a Professor in 1996. His current research interests are in the areas of reliability physics in deep submicron CMOS devices, analog IC design, and wide energy gap group III nitride. He has published over 150 research papers and two books, including *Modern Semiconductor Quantum Physics* (Singapore: World Scientific, 1994). He has served on several international program committees and advisory committees in international semiconductor conferences in China, Japan, Canada, Germany, and Singapore.

IEEE TRANSACTIONS ON ELECTRON DEVICES, VOL. 48, NO. 12, DECEMBER 2001   2893

A Simple and Efficient Model for Quantization Effects of Hole Inversion Layers in MOS Devices

Yong-Tian Hou, *Student Member, IEEE*, and Ming-Fu Li, *Senior Member, IEEE*

Abstract—A simple and efficient model is introduced to study the hole quantization effects in the inversion layer of p-MOS devices. It is based on a six-band effective mass equation and a zigzag potential. The strong mixing between the heavy, light, and split-off hole bands is emphasized and quantitatively assessed. All subband dispersions are found to be anisotropic, far from parabolic, and electric field dependent. In addition, there are camel-back structures at some subband minima. The density of states (DOS) profiles show strong deviations from the step-like functions and one or two peaks may appear at the subband minima of the DOS, corresponding to the camel-back band structures. The traditional one-band effective mass approximation (EMA) using effective mass values extracted from bulk Si underestimates and oversimplifies the subband DOS. We justify this model by applying it to the capacitance of hole inversion layer and the threshold voltage shifts due to quantum mechanical (QM) effect. The model simulation shows good agreement with experimental results, demonstrating the accuracy of this model. The model and the characterization of the band structure of Si valence hole quantization lay the ground work in routine simulation of deep submicrometer MOS devices.

Index Terms—Inversion layers, MOS devices, quantization.

I. INTRODUCTION

WITH the scaling down of metal-oxide-semiconductor-field-effect transistors (MOSFETs), gate oxides are thinner and substrate doping concentrations are higher. This results in large electric fields at the Si/SiO$_2$ interface, even near the threshold of inversion. In this case, the electrons or holes occupying the quantized two-dimensional (2-D) subbands behave differently from the classical three-dimensional (3-D) case [1]–[4]. Because of the smaller density of states (DOS) in the 2-D system, it requires a larger gate voltage in order to populate the 2-D inversion layer with the same number of carriers as the corresponding 3-D system. This results in a higher threshold voltage (V_T) as compared to the classical prediction [5]–[8]. In addition, the peak of the 2-D carriers distribution is displaced away from the interface, which affects the inversion layer capacitance and thereby the performance of scaled MOSFETs [9]–[11].

Quantum mechanical (QM) effects of electron inversion layer have been well understood and extensively investigated by the one-band effective mass approximation (EMA) [1]–[5].

Manuscript received December 19, 2000; revised June 12, 2001. This work was supported by the Singapore NSTB/EMT/TP/00/001.2 and the National University of Singapore RP3982754 grants. The review of this paper was arranged by Editor M. Hirose.

The authors are with the Silicon Nano Device Lab, Department of Electrical and Computer Engineering, National University of Singapore, Singapore 119260 (e-mail: elelimf@nus.edu.sg).

Publisher Item Identifier S 0018-9383(01)10133-4.

According to the literature, such a treatment using effective mass values extracted from the bulk Si (or "traditional one-band EMA') is also applied to study hole quantization [3], [6]. However, such treatment is incorrect in physics due to the valence band mixing in the strong electric field. Few works have been published on the hole quantization using multiband Schrodinger-Poisson self-consistent equations (SPSC) [7], [12], [13]. While these methods are correct in physics, they are computationally prohibitive in routine device simulations. In this paper, we demonstrate a method which is correct in physics and simple and efficient in computation to investigate the hole QM effects. It is based on the six-band EMA with a periodic zigzag potential so the system can be ascribed to a superlattice system [14]. The simulation results are in good agreement with the available experimental data or the SPSC simulation.

II. DESCRIPTION OF THE MODEL

In the multiband $k \cdot p$ effective mass theory, the Hamiltonian of the system can be obtained by adding a diagonal electric potential energy term V to the unperturbed bulk band Hamiltonian [14]–[16]. For the inversion layer of a MOS device fabricated on (100) silicon substrate, V is only a function of z, where the z axis is perpendicular to the (100) face. The multiband effective mass equation can be written generally as

$$\sum_{j=1}^{N} (D_{ij}(-i\nabla) + V(z)\delta_{ij}) \varphi_j^n = E\varphi_j^n, \quad i = 1 \text{ to } N. \quad (1)$$

Here, N bulk bands are involved. $\{\varphi_j^n\}$ is the respective envelope function of nth subband. $D(-i\nabla)$ is a $N \times N$ matrix of the bulk Hamiltonian and it is obtained by $k \cdot p$ method [15], [16].

Regarding the potential energy term $V(z)$, the single triangular well potential approximation has been justified for the electron quantization in the electron inversion layer [1], [3], [17]–[20]. Using this approximation, the electron subband energies [17], the inversion layer capacitance [18], the V_T shift due to electron quantization [5], [19], and the electron direct tunneling current [20] in nMOSFETs have been simulated with great success. For the hole quantization in a hole inversion layer, in our approximation in (1), the single triangular well approximation is further replaced by a periodic triangular well approximation, i.e., a zigzag potential as schematically illustrated in Fig. 1. Then, the external electric potential energy $V(z)$ in (1) can be expressed as

$$V(z) = qF_s \cdot z', \quad 0 < z' < L,$$
$$z = z' + nL, \quad n = 0, 1, 2 \ldots \quad (2)$$

2894   IEEE TRANSACTIONS ON ELECTRON DEVICES, VOL. 48, NO. 12, DECEMBER 2001

Fig. 1. Schematically illustration of the multiple quantum wells with zigzag potential energy profile used in our model. $\Phi = qF_s \cdot L$. As the period L is large enough, it is equivalent to the single triangular well approximation of the inversion layer.

where

F_s surface electric field in the Si substrate;

q charge of the hole;

L periodic length of the zigzag potential.

The advantage of using a zigzag potential lies on the fact that the Hamiltonian reduces to a superlattice Hamiltonian with period L, so that the well-known technique and criterion already developed in the superlattice theory can be used [14], [16]. The wavefunction can be expanded in plane waves and the number of plane waves M, is greatly reduced, only for those with wave vector $k_z + 2\pi m/L$, m is integer [14], [16]. Following [14], we expand the nth subband envelope function $\varphi_{j,k}^n$ with wave vector $k = (k_x, k_y, k_z)$ into plane waves

$$\varphi_{j,k}^n = \exp\left[i(k_x x + k_y y)\right] \sum_{m=1}^{M} a_{j,k,m}^n \frac{1}{\sqrt{L}}$$
$$\times \exp\left[i\left(k_z + m\frac{2\pi}{L}\right)z\right]. \quad (3)$$

Combining (2) and (3), the matrix elements of $V(z)$ in the plane wave representation are

$$V_{m,m'} = \begin{cases} \frac{qF_s \cdot L}{2} & m = m' \\ \frac{qF_s \cdot L}{i \cdot 2\pi(m-m')} & m \neq m'. \end{cases} \quad (4)$$

The legitimacy of using a zigzag potential in replacing a single triangular well potential is supported by superlattice theory [16]. When L is large enough, the coupling of lowest few energy states between the neighboring wells disappears and each well in the zigzag potential can be considered as an isolated single triangular well. The coupling between neighboring well can be tested by k_z dispersion of the subband [16]. We have justified this point by using the one-band EMA and zigzag potential in the electron inversion layer problem. When L and M are large enough, the subband energies and DOS of our simulation converge to the well-known analytical solutions of a single triangular potential well [1], [17] with error less than 5%. The energy dispersion along k_z is less than 0.5 meV, which indicates that the coupling between neighboring wells is negligible. In the surface electric field range studied in this paper, L is typically 50–100 nm in order to achieve convergence of our zigzag well to the single triangular well in the calculation of the first six subbands.

To assess the band mixing effect quantitatively, the QM projection (or probability) functions are introduced

$$P_j^n(k) = \sum_{m=1}^{M} a_{j,k,m}^{n^*} a_{j,k,m}^n \quad (5)$$

and the following sum role holds:

$$\sum_{j=1}^{N} P_j^n(k) = 1. \quad (6)$$

$P_j^n(k)$ estimates the jth bulk band component of the nth superlattice subband wavefunction φ^n.

If the subband dispersion and DOS are known, the electrostatics of MOS system can be determined by the efficient parametric self-consistency method [17]. The carrier sheet density N_n of subband n, is related to the nth subband DOS $D_n(E)$ by

$$N_n = \int_{E_{n0}}^{\infty} D_n(E)f(E)dE \quad (7)$$

where $f(E)$ is the Fermi–Dirac distribution function. The carrier occupation factor of subband n is defined by

$$OF_n = \frac{N_n}{\sum_n N_n}. \quad (8)$$

III. Quantization Effects in the Hole Inversion Layer

A. General Formula

For valence bands of Si, there are three nearly degenerated bands: 1) the heavy hole (hh); 2) light hole (lh); and 3) spin-orbit split-off (so) bands. The spin-orbit split-off energy for Si is 44 meV, which is much smaller than the Si bandgap (1.12 eV) [16]. Therefore, the so band cannot be neglected. From the QM point of view, when there is a strong electric field perpendicular to the (100) interface, the symmetry of the point group of the Hamiltonian is reduced from O_h to C_{2v} and splitting and mixing between different bulk valence bands are expected [16]. We therefore use a 6×6 Hamiltonian in our analysis to include hh, lh, and so holes and their respective spin degenerate bands. The Hamiltonian of our model is given by [14] (9), shown at the bottom of the next page, where

$$P = \frac{\hbar^2}{2m_0}\left[(k_x^2 + k_y^2)(\gamma_1 + \gamma_2) + k_z^2(\gamma_1 - 2\gamma_2)\right]$$

$$L = \frac{\hbar^2}{2m_0}\left[(k_x^2 + k_y^2)(\gamma_1 - \gamma_2) + k_z^2(\gamma_1 + 2\gamma_2)\right]$$

$$\alpha = \frac{\hbar^2}{2m_0}2\sqrt{3}\left[k_z(ik_y - k_x)\gamma_3\right]$$

$$\beta = \frac{\hbar^2}{2m_0}\sqrt{3}\left[2ik_x k_y\gamma_3 - (k_x^2 - k_y^2)\gamma_2\right]$$

$$D = \frac{\hbar^2}{2m_0}\left[2(k_x^2 + k_y^2)\gamma_2 - 4k_z^2\gamma_2\right]$$

$$S = \frac{\hbar^2}{2m_0}\left[(k_x^2 + k_y^2 + k_z^2)\gamma_1\right] + \Delta_0 \quad (10)$$

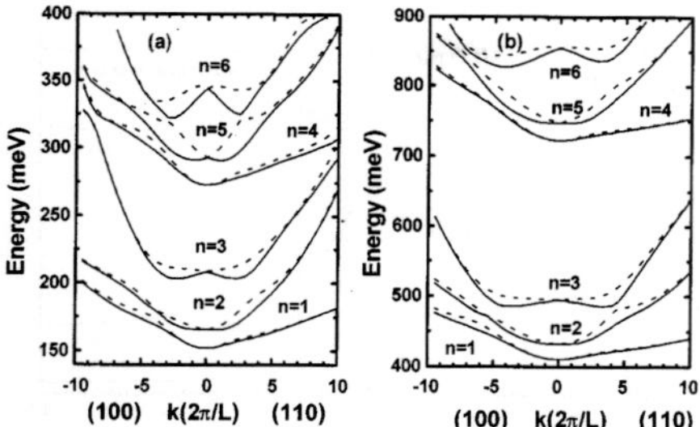

Fig. 2. In-plane dispersion of the first six subbands calculated by our model for hole inversion layer. The surface electric field is (a) $F_s = 0.5$ MV/cm and (b) $F_s = 2$ MV/cm. Both the dispersions along (100) and (110) in the $k_z = 0$ plane are shown (in the calculation, $L = 80$ nm). Our results in (a) are comparable with the SPSC simulation in [12, Fig. 6].

and $V(z)$ is expressed by (2) and (4). The Luttinger parameters [21] are: $\gamma_1 = 4.22$, $\gamma_2 = 0.39$, and $\gamma_3 = 1.44$ [22]. $\Delta_0 = 44$ meV is the spin-orbit splitting energy of bulk Si [22].

B. Valence Band Mixing

Fig. 2 shows the in-plane dispersions of the six lowest subbands at the surface electric field $F_s = 0.5$ MV/cm and 2.0 MV/cm. Convergence has been tested by using different L and M. The subband dispersions are found to be anisotropic, far from parabolic, and electric field dependent. The degeneracy of the hh and lh bands at Γ point is lifted by the electric field and their separation depends on the electric field. In addition, there are reversed camel-back structures (two reversed peaks) [23] with negative hole effective mass near $k = 0$ for $n = 2$, 3, 5, and 6 subbands.

Fig. 3 shows the projection functions P_{hh}^n, P_{lh}^n, and P_{so}^n of the first three subbands ($n = 1, 2, 3$). At $k = 0$, the $n = 1$ subband is purely hh and $n = 2$ and 3 subbands are mainly lh and so with some band mixing between lh and so bands. As the electric field is increased, the band mixing also increases. When $k \neq 0$, there are strong band mixing in all subbands.

C. Density of States (DOS)

Fig. 4 illustrates the simulation results of DOS of the three lowest subbands. The DOS profiles are deviated from the step-like function as predicted by the traditional one-band

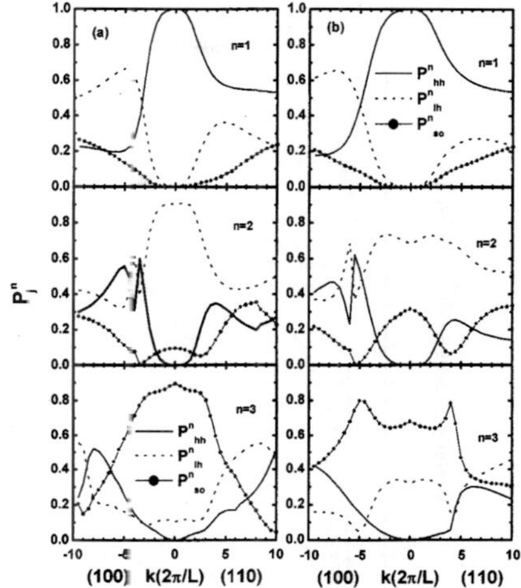

Fig. 3. Variation of the hh, lh, and so components in the three lowest subbands ($n = 1, 2, 3$) versus in-plane wave vector k. The surface electric field is (a) $F_s = 0.5$ MV/cm and (b) $F_s = 2$ MV/cm. P_j^n is the projection of the nth ($n = 1, 2, 3$) subband wavefunction to the j (hh, lh, or so) component defined by (5). (In the calculation, $L = 80$ nm).

EMA. Particularly, near the band minimum of the $n = 3$ subband, there are two peaks. They are caused by the camel-back structure [23] at the band minimum as observed in Fig. 2. The shapes of DOS are also electric field dependent. The DOS obtained from traditional one-band EMA (with the DOS masses of $0.65m_0$, $0.25m_0$, and $0.29m_0$ for hh, lh, and so bands, respectively, [3], [6]) are also shown in Fig. 4 which overall underestimates the DOS.

D. Subband Minimum and Carrier Occupations

From the above hole quantization model along with the parametric self-consistency method [17], the subband energy levels of the first six subbands and the occupation factors on the three lowest subbands are shown in Fig. 5 by solid lines. The corresponding data calculated by traditional one-band EMA is also shown (with dashed lines),

$$H = \begin{bmatrix} P+V(z) & \alpha & \beta & 0 & \frac{i\alpha}{\sqrt{2}} & -i\sqrt{2}\beta \\ \alpha^* & L+V(z) & 0 & \beta & \frac{-iD}{\sqrt{2}} & i\sqrt{\frac{3}{2}}\alpha \\ \beta^* & 0 & L+V(z) & -\alpha & -i\sqrt{\frac{3}{2}}\alpha^* & \frac{-iD}{\sqrt{2}} \\ 0 & \beta^* & -\alpha^* & P+V(z) & -i\sqrt{2}\beta^* & \frac{-i\alpha^*}{\sqrt{2}} \\ \frac{-i\alpha^*}{\sqrt{2}} & \frac{iD}{\sqrt{2}} & i\sqrt{\frac{3}{2}}\alpha & i\sqrt{2}\beta & S+V(z) & 0 \\ i\sqrt{2}\beta^* & -i\sqrt{\frac{3}{2}}\alpha^* & \frac{iD}{\sqrt{2}} & \frac{i\alpha}{\sqrt{2}} & 0 & S+V(z) \end{bmatrix} \qquad (9)$$

2896   IEEE TRANSACTIONS ON ELECTRON DEVICES, VOL. 48, NO. 12, DECEMBER 2001

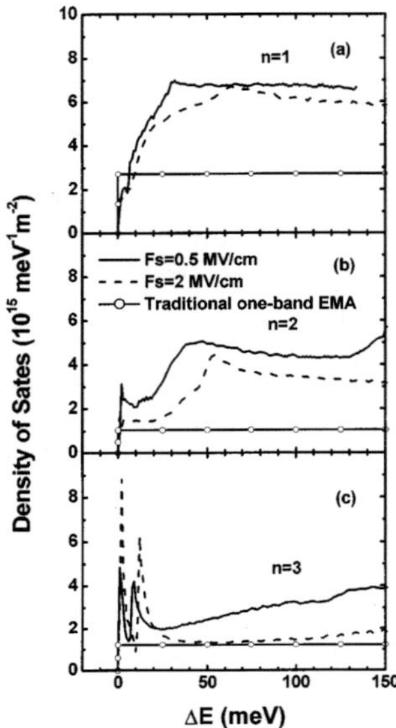

Fig. 4.   Obtained DOS of the three lowest subbands in hole inversion layer. The relative energy ΔE is the subband energy referenced from the subband edge. The solid and dashed curves are for surface electric field $F_s = 0.5$ and 2 MV/cm, respectively. The solid lines with open circles are the results from the traditional one-band EMA.

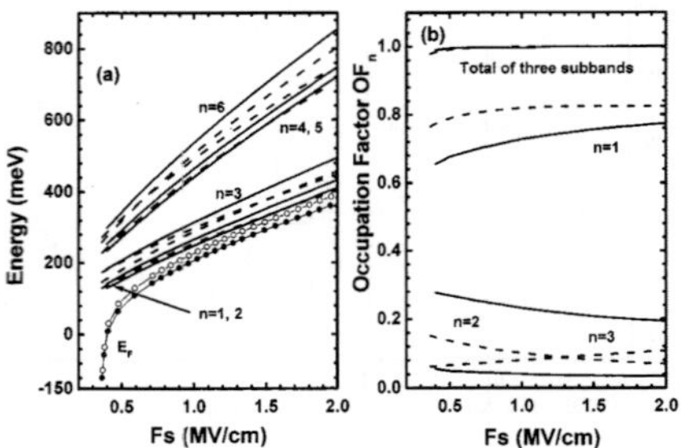

Fig. 5.   (a) Calculated subband energies of the first six subbands, and (b) occupation factors [OF_n defined by (8)] of the three lowest subbands for hole inversion layer in p-MOS device at various surface electric field. The substrate doping is 5×10^{17} cm^{-3}. The dashed curves are from the traditional one-band EMA. The results of our model are shown as solid lines. Fermi energy is also added in (a) for reference (solid circles for our model, open circles for traditional one-band EMA).

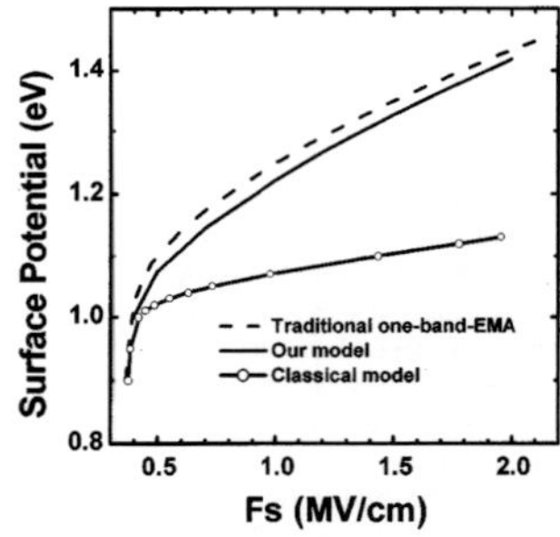

Fig. 6.   Calculated surface potential (band bending) of p-MOS structure at inversion. The substrate doping is 5×10^{17} cm^{-3}. The dashed curve is from the traditional one-band EMA. The results of our model are shown as solid line. The solid line with open circles is that from classical calculation with Fermi–Dirac statistics.

with the quantized $(m_\perp)$ and DOS $(m_{//})$ effective masses $m_\perp/m_{//} = 0.29/0.65, 0.20/0.25, 0.29/0.29m_0$ for hh, lh, and so holes, respectively [3], [6]. In Fig. 5(a), the energies of $n = 1$ subband obtained by two methods are in good agreement. This is expected because this subband is purely hh at the band minimum as shown in Fig. 3. However, the results are quite different for $n = 2$ or $n = 3$ subband due to band mixing. Furthermore, from traditional one-band EMA, there is a crossing between the lh and so subbands at about 1.5 MV/cm so the $n = 2$ subband will change from lh to so holes at electric field higher than 1.5 MV/cm. However, our calculation does not show such a crossing up to 3 MV/cm. For the occupation factors OF_n in Fig. 5(b), our calculation shows an overall lower occupation on the $n = 1$ subband and higher occupation on the $n = 2$ subband, compared to the traditional one-band EMA calculation, although both calculations predict that more than 98% of the carriers are distributed among the three lowest subbands. For subbands of higher order ($n = 4, 5, 6 \ldots$), the accuracy of the engergy levels determined in triangular potential approximation comparing to the self-consistent method will be reduced, because the actual potential profile is deviated from linear as the energy increases. Fortunately, from Fig. 5(b), 98% of holes occupy the lowest three subbands and the inaccuracy of higher energy levels will not lead to significant error for prediction of device parameters, especially for state-of-art

CMOS devices with high substrate doping concentrations. The surface potential (band bending) of the p-MOS structure with channel doping $N_D = 5 \times 10^{17}$ cm^{-3} is shown in Fig. 6. Due to the QM effects, the surface potential is not pinned as in the classical case. It is obvious that our calculation (solid lines) leads to lower band bending than the traditional one-band EMA (dashed lines). It is because the traditional one-band EMA underestimates the subband DOS as indicated in Fig. 4. In order to achieve the same inversion charge, the band needs to bend more in traditional one-band EMA and it leads to an overestimation of the band bending.

E. Hole Inversion Capacitance and Threshold Voltage Shift

We simulate the inversion capacitance and threshold voltage shifts due to QM effects. The results are shown in Figs. 7 and 8. Our model achieves a good agreement with the experiments. It has been found that the hole inversion capacitance is not

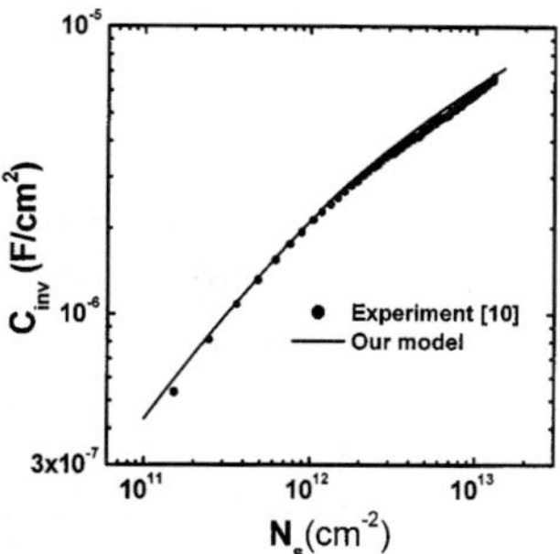

Fig. 7. Comparison of calculated and experimental capacitance of hole inversion layer. The solid line is our model simulation while the solid circles are experimental data from [10].

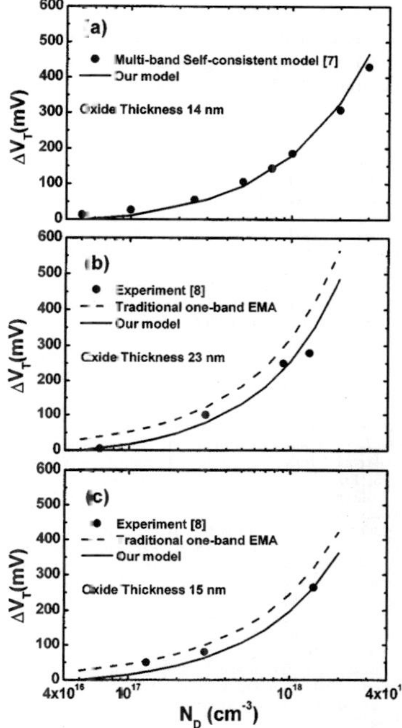

Fig. 8. Threshold voltage shifts due to hole QM effects ΔV_T at different channel doping concentrations N_D. (a) Comparison between our model simulation (solid line) and the SPSC simulation (solid circles) with oxide thickness 14 nm. (b) and (c) Comparisons between our model simulation (solid lines), the traditional one-band EMA simulation (dashed lines), and the experimental data (solid circles) for oxide thickness 23 and 15 nm, respectively.

very sensitive to the complicated valence band structure [10]. We actually found no considerable difference between our model and the traditional one-band EMA. However, the situation is different for the V_T shift. In our calculation, the classical V_T is defined as the gate voltage at surface potential $\varphi_s = 2kT\ln(N_D/n_i)$, where n_i is the intrinsic carrier concentration. The V_T shift due to QM effects is the additional gate voltage ΔV_T necessary to populate the quantized 2-D case to the same hole inversion charge as classical one [6]. In this traditionally subthreshold region, it is found that ΔV_T is not very sensitive to the exact definition of V_T [7]. ΔV_T is due to the lower DOS of 2-D carriers than the 3-D case and it is expected to be more sensitive to the subband DOS. Fig. 8(a) shows that the simulations conducted by our model and SPSC calculation [7] are in good agreement. Although the triangular well approximation used in our model is not accurate when applied to high-energy subbands, it does not affect the accuracy of the calculated V_T shift due to the QM effect [3], [19]. It is because most of the inverted carriers reside at the lowest three subbands as indicated in Fig. 5(b) and the results in Fig. 8 is not sensitive to the high-energy subband structure. Fig. 8(b) and (c) shows the comparison between the experimental data and the simulations. A significant discrepancy between the ΔV_T calculated by our model and the traditional one-band EMA is found. The traditional one-band EMA calculation gives larger ΔV_T than those by our model due to its underestimation of the subband DOS. Our model shows a better overall agreement with the experiments. In our calculation, the influence of tunneling current on V_T [24] has not been considered. The tunneling current has negligible disturbance to the carrier equilibrium in MOS system up to 1.5 nm gate oxide [4].

F. Improved One-Band Effective Mass Approximation (EMA)

In summary, the traditional one-band EMA underestimates the hole subband DOS and therefore overestimates the QM effects. It is possible to establish an improved one-band EMA equation with a set of effective masses extracted from the six-band EMA results to deduce the correct DOS and to assess the correct QM effects in many applications. Details will be published elsewhere.

IV. CONCLUSION

In conclusion, we present a simple and efficient model for hole quantization in inversion layer of the MOS devices, based on the six-band EMA and a zigzag electric potential. It is demonstrated that hole band mixing must be taken into account in order to describe the hole QM effect properly. Due to the band mixing, all subband dispersions are anisotropic, far from parabolic, and electric field dependent. In addition, there are camel-back structures at some band minima. Correspondingly, the DOS profiles are deviated substantially from the step-like functions. The traditional one-band EMA using bulk effective masses overall underestimates the subband DOS and therefore overestimates the hole QM effects. Using our model, the calculated hole inversion capacitance and threshold voltage shifts due to QM effects are both in good agreement with the experiments.

2898   IEEE TRANSACTIONS ON ELECTRON DEVICES, VOL. 48, NO. 12, DECEMBER 2001

ACKNOWLEDGMENT

The authors would like to thank Dr. S. Takagi of Toshiba Corporation of Japan for providing the experimental capacitance data of electron and hole inversion layers.

REFERENCES

[1] F. Stern, "Self-consistent results for n-type Si inversion layer," *Phys. Rev. B, Condens. Matter*, vol. 5, no. 12, pp. 4891–4899, 1972.

[2] T. Ando, A. Fowler, and F. Stern, "Electronic properties of two-dimensional systems," *Rev. Mod. Phys.*, vol. 54, pp. 437–672, Apr. 1982.

[3] C. Moglestue, "Self-consistent calculation of electron and hole inversion charges at silicon–silicon dioxide interfaces," *J. Appl. Phys.*, vol. 59, no. 5, pp. 3175–3183, 1986.

[4] S. -H. Lo, D. A. Buchanan, Y. Taur, and W. Wang, "Quantum mechanical modeling of electron tunneling current from the inversion layer of ultra-thin-oxide nMOSFETs," *IEEE Electron Device Lett.*, vol. 18, pp. 209–211, May 1997.

[5] M. J. van Dort, P. H. Woerlee, and A. J. Walker, "A simple model for quantization effects in heavily-doped silicon MOSFETs at inversion conditions," *Solid State Electron.*, vol. 37, no. 3, pp. 411–441, 1994.

[6] C.-Y. Hu, S. Banerjee, K. Sadra, B. G. Streetman, and R. Sivan, "Quantization effects in inversion layers of pMOSFETs on Si (100) substrate," *IEEE Electron Device Lett.*, vol. 17, pp. 276–278, June 1996.

[7] S. Jallepalli, J. Bude, W.-K. Shih, M. R. Pinto, C. M. Maziar, and A. F. Tasch, "Electron and hole quantization and their impact on deep submicron silicon p- and n-MOSFET characteristics," *IEEE Trans. Electron Devices*, vol. 44, pp. 297–303, Feb. 1997.

[8] G. Chindalore, S. A. Hareland, S. Jallepalli, A. F. Tasch, Jr., C. M. Maziar, V. K. F. Chia, and S. Smith, "Experimental determination of threshold voltage shifts due to quantum mechanical effects in MOS electron and hole inversion layers," *IEEE Electron Device Lett.*, vol. 18, pp. 206–208, May 1997.

[9] S. Takagi and A. Toriumi, "Quantitative understanding of inversion-layer capacitance in Si MOISFETs," *IEEE Trans. Electron Devices*, vol. 42, pp. 2125–2130, Dec. 1995.

[10] S. Takagi, M. T. Takagi, and A. Toriumi, "Accurate characterization of electron and hole inversion-layer capacitance and its impact on low voltage operation of scaled MOSFETs," in *IEDM Tech. Dig.*, 1998, pp. 619–622.

[11] Khairurrijal, S. Miyazaki, S. Takagi, and M. Hirose, "Analytical modeling of metal oxide semiconductor inversion-layer capacitance," *Jpn. J. Appl. Phys.*, vol. 38, pp. L30–32, 1999.

[12] S. Rodriguez, J. A. Lopez-Villaneuva, I. Melchor, and J. E. Carceller, "Hole confinement and energy subbands in a silicon inversion layer using the effective mass theory," *J. Appl. Phys.*, vol. 86, no. 1, pp. 438–444, 1999.

[13] A. Pirovano, A. L. Lacaita, G. Zandler, and R. Oberhuber, "Explaining the dependences of the hole and electron mobilities in Si inversion layers," *IEEE Trans. Electron Devices*, vol. 47, pp. 718–724, Apr. 2000.

[14] W. J. Fan, M. F. Li, T. C. Chong, and J. B. Xia, "Valence hole subbands and optical gain spectra of GaN/Ga$_{1-x}$Al$_x$N strained quantum wells," *J. Appl. Phys.*, vol. 80, pp. 3471–3478, 1996.

[15] W. Kohn, "Shallow impurity states in Si and Ge," in *Solid State Physics*. New York: Academic, 1957, vol. 5.

[16] M.-F. Li, *Modern Semiconductor Quantum Physics*, Singapore: World Scientific, 1994, ch. 3.

[17] H. H. Mueller and M. J. Schulz, "Simplified method to calculate the band bending and the subband energies in MOS capacitors," *IEEE Trans. Electron Devices*, vol. 44, pp. 1539–1543, Sept. 1997.

[18] S. A. Hareland, S. Krishnamurthy, S. Jallepalli, C.-F. Yeap, K. Hasnat, A. F. Tasch, Jr., and C. M. Maziar, "A computationally efficient model for inversion layer quantization effects in deep submicron n-channel MOSFETs," *IEEE Trans. Electron Devices*, vol. 43, pp. 90–96, Jan. 1996.

[19] Y. Ma, Z. Li, L. Liu, L. Tian, and Z. Yu, "Effective density-of-states to QM correction in MOS structures," *Solid State Electron.*, vol. 44, no. 3, pp. 401–407, 2000.

[20] N. Yang, W. K. Henson, J. R. Hauser, and J. J. Wortman, "Modeling study of ultrathin gate oxides using direct tunneling current and capacitance–voltage measurements in MOS devices," *IEEE Trans. Electron Devices*, vol. 46, pp. 1464–1471, July 1999.

[21] J. M. Luttinger, "Quantum theory of cyclotron resonance in semiconductors: General theory," *Phys. Rev.*, vol. 102, pp. 1030–1041, 1956.

[22] Landolt-Bornstein, *Numerical Data and Functional Relationships in Science and Technology*, O. Madelung, Ed. New York: Springer-Verlag, 1987, vol. 17a.

[23] ——, *Numerical Data and Functional Relationships in Science and Technology*. New York: Springer-Verlag, 1987, vol. 22a, p. 72.

[24] X. Liu, J. Kang, X. Guan, R. Han, and Y. Wang, "The influence of tunneling effect and inversion layer quantization effect on threshold voltage of deep submicron MOSFETs," *Solid State Electron.*, vol. 44, pp. 1435–1439, 2000.

Yong-Tian Hou (S'01) was born in Shandong, China, in 1969. He received the B.S. and M.S. degrees in physics from Peking University, Beijing, China, in 1990 and 1993, respectively. He is currently pursuing the Ph.D. degree in electrical engineering at National University of Singapore (NUS).

From 1993 to 1997, he was with Tianma Microelectronics Co. Ltd., Shenzhen, China, working on the development of STN LCDs. From 1998 to 1999, he was with the Department of Electrical and Computer Engineering, NUS, as a Research Engineer on optical characterization of III-nitride materials. His research interests are currently the quantum effect, tunneling current, and reliability physics in deep submicrometer CMOS devices.

Ming-Fu Li (M'91–SM'99) graduated from the Department of Physics, Fudan University, Shanghai, China, in 1960.

After graduation, he joined the University of Science and Technology of China (USTC) as a Teaching Assistant and then Lecturer. In 1978, he joined the Graduate School faculty, Chinese Academy of Sciences, Beijing, and became a Professor in 1986. He has also served as Adjunct Professor at the Institute of Semiconductors, Chinese Academy of Sciences, Fudan University, and USTC, Hefei. He was a Visiting Scholar at Case Western Reserve University, Cleveland, OH, in 1979, and at the University of Illinois at Urbana-Champaign from 1979 to 1981, and was a Visiting Scientist at the University of California at Berkeley and Lawrence Berkeley National Laboratories from 1986 to 1987, and 1990 to 1991, respectively. He joined the Department of Electrical Engineering, National University of Singapore in 1991, where he has been a Professor since 1996. His current research interests are in the areas of reliability and modeling of deep submicrometer CMOS devices, analog IC design, and wide energy gap group III nitride. He has published over 160 research articles and two books, including *Modern Semiconductor Quantum Physics* (Singapore: World Scientific, 1994). He has served on several international program committees and advisory committees in international semiconductor conferences in China, Japan, Canada, Germany, and Singapore.

Quantum Tunneling and Scalability of HfO$_2$ and HfAlO Gate Stacks

Y. T. Hou, M. F. Li, H. Y. Yu, Y. Jin[1], and D.-L. Kwong[2]

Silicon Nano Device Lab, Dept. of Electrical & Computer Engineering, National University of Singapore, Singapore 119260
Tel: 65 6874 2559, Fax: 65 6779 1103, E-mail: eleimf@nus.edu.sg
[1] Chartered Semiconductor Manufacturing Ltd, Singapore 738406
[2] Department of Electrical and Computer Engineering, The University of Texas, Austin, TX 78712, USA

ABSTRACT

We present a physical model for tunneling current through high-κ gate stack including the ultra thin interfacial layer between high-κ and Si substrate. The energy band offsets of high-κ are determined by XPS. The accurate carrier quantization in the substrate or gate is found to play a more significant role in tunneling through high-κ dielectric than in SiO$_2$. Excellent agreements between simulated and measured tunneling currents were achieved over several high-κ dielectrics with both poly-Si and metal gate electrodes. The model is also applied to analyze the scalability of HfO$_2$ and HfAlO gate stacks in future CMOS technology.

INTRODUCTION

Recently, HfO$_2$ has been considered as a promising high-κ dielectric. Although superior electrical characteristics have been demonstrated for HfO$_2$ [1-2], HfO$_2$ has the disadvantage of its low crystallization temperature. In order to improve the thermal stability, Al is incorporated into HfO$_2$ [3]. It is, therefore, technologically important to study tunneling currents through HfO$_2$ and HfAlO as well as their scalability in CMOS technology. However, physical modeling of tunneling current through HfO$_2$ gate stack is far from sufficient. Previous modeling efforts on tunneling through HfO$_2$ neglect the critically important interfacial layer (IL) between high-κ and Si substrate [4], lack of direct comparison with the experimental data over a wide range [5,6], or are empirical in nature [4]. In this paper, a physically based model is presented and demonstrated for tunneling current through high-κ gate stack and it is applied to analyze the scalability of HfO$_2$ and HfAlO gate stack.

PHYSICAL MODEL

In our model, electron and hole quantizations in both the substrate and gate electrode are accurately treated, including the band mixing effect in hole quantization [7-9]. The tunneling current from the discrete 2-D subbands in inversion or accumulation layers in MOS devices is obtained by a modified WKB approximation, which takes into account the wave function reflections at the dielectric-Si barrier interface [10]. In our model, the IL between high-κ and Si substrate, whose formation is formidable during high-κ deposition or post annealing, is readily included [9]. In addition, both tunneling through channel and source/drain extension region

(SDE) overlapped with the gate, as shown in Fig.1, can be calculated.

The model is first demonstrated by simulations of the tunneling current through ultra-thin SiO$_2$ in Figs.2-3, as well as oxynitride (SiON)/SiO$_2$ stack in Fig.4. The simulated results show good agreements with the carrier separation measurements in both n- and p-MOSFETs for different tunneling channels shown in Figs.2a,3a by conduction band electrons (CBEs), valence band electrons (VBEs) and valence band holes (VBHs) from either substrate or gate injections.

The valence band offsets ΔE_V between ALD grown (HfO$_2$)$_x$ (Al$_2$O$_3$)$_{1-x}$ and Si are determined from the valence band XPS and the energy gap Eg is determined by the O 1s energy loss spectra [11]. With the known Si energy gap 1.12 eV, the conduction band offsets ΔE_C can be readily obtained. Their dependences on the Hf composition are demonstrated to be linear as shown in Fig.5. Data in Fig.5 are used in tunneling simulations for HfO$_2$, Al$_2$O$_3$ and HfAlO.

SIMULATION RESULTS

For high-κ stacks, the simulations are compared with electrical data from ultra-thin CVD [1] and PVD HfO$_2$ [2] devices with and without a NH$_3$-based interface layer, and with either poly-Si or TaN as gate electrodes. ALD Al$_2$O$_3$ with poly-Si gate [12] was also used. The flat-band voltages for all samples were obtained from C-V measurements.

A. HfO$_2$ gate stacks

Fig.6 shows the comparison between simulation and experimental data of PVD HfO$_2$ on p-Si [2]. The physical thickness of HfO$_2$ (38Å) and IL (6Å) used in the simulation were determined by XTEM [2]. The IL is found likely to be SiO$_2$ by XPS [2]. The only fitting parameter in our simulation is the electron tunneling effective mass in HfO$_2$, m_{HfO}, and a value of 0.18 m_0 fits the experiments well in both the inversion and accumulation polarities. This m_{HfO} value is larger than that reported before (0.1 m_0) [3-4]. This is because in [3-4], the IL was neglected in their model, hence underestimating the m_{HfO}. Figs.7-8 are simulated and measured tunneling currents through CVD HfO$_2$ grown on NH$_3$-nitrided Si with poly-Si and TaN gates, respectively [1], using the same electron tunneling effective mass of HfO$_2$. The IL layers are oxynitride (SiON) with corresponding parameters from [9]. Good agreements between simulations and measurements are also obtained.

29.8.1

The importance of carrier quantization on the tunneling current is shown in Figs.9-10 by comparing simulations of classical [10] and quantum-mechanical (QM) models. In the QM model, the electron energy is lifted up due to the quantization, leading to larger tunneling current comparing to the classic model at the same substrate electrical field F_{sub} as indicated in Fig.9. This effect is enhanced for HfO_2 than SiO_2 dielectric by a factor of $\kappa_{HfO2}/\kappa_{SiO2} \sim 6$ approximately, because the electric field in the SiO_2 is higher than that in the HfO_2 by the same factor, assuming EOT and F_{sub} are the same. This demonstrates the importance of accurate carrier quantization in simulating tunneling through high-κ layers.

B. Al_2O_3 gate stacks

The simulation results are compared with experimental data of ALD Al_2O_3/SiO_2 stacks (Fig.11). A single value of m_{AlO} (0.28 m_0) is able to give the best fitting to the four experimental curves with different equivalent oxide thickness (EOT) and bias polarities. This value is comparable to that determined from independent band structure calculations of α–Al_2O_3 (0.35 m_0) [13].

C. $(HfO_2)_x(Al_2O_3)_{1-x}$ (with x=0 to 1)

The electron effective mass and the dielectric constant values of $(HfO_2)_x(Al_2O_3)_{1-x}$ are linearly interpolated from those of HfO_2 and Al_2O_3. As shown in Fig.12, Al incorporation increases the leakage current of HfAlO monotonically due to the reduction of the dielectric constant, hence Al concentration should be kept low for gate leakage. On the other hand, our XRD results indicate that a 30% mole fraction of Al incorporation into HfO_2 can raise the crystallization temperature to $\sim 900°C$. Therefore, we use HfAlO with [Al] = 30% as optimized HfAlO in the subsequent discussion of scalability of high-κ gate stack.

SCALABILITY

As shown in Figs.2-3, tunneling current in SDE, I_{SDE}, is the main contribution of transistor off-sate leakage I_{OFF} for devices with ultra thin gate dielectrics [14]. From Fig.13, our calculations show that tunneling current density in SDE is comparable with that from the channel. Since the SDE area is smaller than or comparable with the channel area and the criterions for I_{OFF} and gate leakage I_G are in same values from ITRS [15], the scalability of a dielectric can be obtained by analyzing the gate leakage in channel.

In the following, the gate leakage is calculated for each CMOS technology generation according to the ITRS 2001. For each generation, the gate leakage is estimated by the gate current value at $Vg=V_{DD}$ with corresponding EOT and operating voltage V_{DD} as listed in Table I. In Fig.14, we show the results for high performance CMOS technology. It is shown that SiO_2 or optimized SiON (Fig.12) can meet the ITRS target from leakage current viewpoint, indicating that the scaling of SiO_2 in high performance CMOS is probably not limited by its gate leakage and such conventional gate

dielectrics will be continually explored for high performance CMOS aggressively up to its physical limitation as a dielectric material. However, the situation is different for low power applications, where the gate leakage criterion is more stringent. Fig.15 is the results for low standby power (LSTP) application. It is shown that SiO_2 will approach the scaling limit in 2003 due to its high gate leakage, medium-κ dielectrics (SiON or Al_2O_3) can only extend the CMOS technology scaling by 1-2 generations. HfO_2 and HfAlO are demonstrated to be viable and potential candidates for long-term solutions as alternative high-κ gate dielectrics.

In practical applications, a molecular IL layer is generally needed in order to improve the interface quality and the channel mobility. We thus studied several interface layer materials including SiO_2, SiON and silicates, and their impacts on gate leakage of HfAlO gate stack. Considering the thinnest IL physical thickness of 3 Å (a single molecular layer), Fig. 16 shows that EOT of HfAlO stack with SiO_2 IL cannot be scaled down to below 1 nm. However, using SiON or silicates as IL, HfAlO stack is scalable to 25 nm node LSTP applications by 2016.

CONCLUSION

A physically based model for tunneling current through high-κ gate stack and its application for analyzing the scalability of HfO_2 and $(HfO_2)_x(Al_2O_3)_{1-x}$ gate stacks were presented. The interfacial layer (IL) between high-κ and Si substrate was considered, and energy band offsets of high-κ were determined by XPS. The carrier quantization plays a more significant role in tunneling through high-κ dielectrics than in SiO_2. Excellent agreements between simulations and experiments were achieved over HfO_2 and Al_2O_3. This model is used to predict when high-κ dielectrics are to replace SiO_2 and SiON according to 2001 ITRS for different applications. Recommendations for interface layer materials are also made for high-κ gate stacks that will meet ITRS roadmap.

Acknowledgement

This work was supported by the Singapore NSTB research grant EMT/TP/00/001,2.

References

(1) S.J. Lee, et.al VLSI Tech Symp.2001, p.133.
(2) L. Kang et.al *IEDM*-2000, p.35.
(3) W. Zhu, et.al IEDM 2001, p.463.
(4) Q. Lu et.al, *Semiconductor Research Symposium 2001*, p.377.
(5) Y.Y. Fan, et.al *58th Device Research Conference*, p. 63, 2000.
(6) E.M. Vogel, et al, *IEEE TED*. 45, p.1350, 1998.
(7) Y.T. Hou, M.F. Li, Y.Jin, and W.H. Lai, *J. Appl. Phys* 91, p.258, 2002.
(8) Y.T. Hou and M.F. Li, *IEEE TED* 48, p.1188, 2001.
(9) H.Y. Yu, Y.T. Hou, M.F. Li, D.L. Kwong, IEEE TED 49, p.1158, 2002.
(10) J. Cai and C.T. Sah, *J. Appl. Phys.* 89, p. 2272, 2001.
(11) H.Y. Yu et al, *Appl. Phys. Lett.*, 81, p.376, 2002
(12) A. Buchanan et.al, *IEDM*-2000, p.223.
(13) Y.N. Xu and W.Y. Ching, Phys. Rev. **B43**, p.4461, 1991.
(14) N. Yang et al., IEDM 1999, p.453.
(15) ITRS 2001, http://public.itrs.net/Files/2001ITRS/.

29.8.2

Table I: Scaling parameters for 2001 ITRS. The Node is shown as the MPU ½ Pitch in nm. The EOT (nm), operating voltage V_{DD} (V) and gate leakage Ig (nA/µm) are listed for both high performance (HP) and low stand-by power (LSTP) CMOS technology.

Year	2003	2004	2005	2006	2007	2010	2013	2016
Node (nm)	107	90	80	70	65	50	35	25
EOT (HP)	1.1-1.6	0.9-1.4	0.8-1.3	0.7-1.2	0.6-1.1	0.5-0.8	0.4-0.6	0.4-0.5
EOT (LSTP)	2.0-2.4	1.8-2.2	1.6-2.0	1.4-1.8	1.2-1.6	0.9-1.3	0.8-1.2	0.7-1.1
V_{DD} (HP)	1.0	1.0	0.9	0.9	0.7	0.8	0.8	0.6
V_{DD} (LSTP)	1.2	1.2	1.2	1.2	1.1	1.0	0.9	0.9
Ig (HP)	70	100	300	700	1000	3000	7000	10000
Ig (LSTP)	0.001	0.001	0.001	0.001	0.001	0.003	0.007	0.01

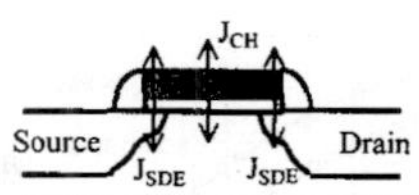

Fig.1: The gate current of a MOSFET is composed of tunneling currents from the channel (CH) and source/drain extension (SDE) region overlapped with the gate.

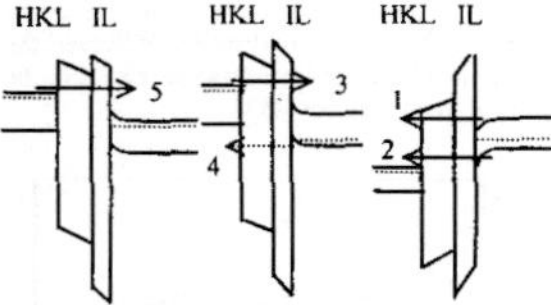

Fig.2a: Band diagrams of the n-MOSFET with gate stacks of high-κ layer (HKL) and interfacial layer (IL).

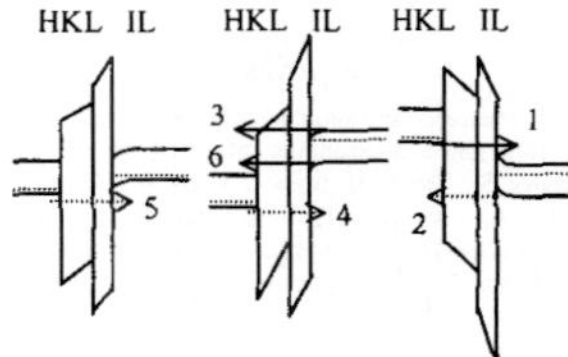

Fig.3a: Band diagrams of the p-MOSFET with gate stacks of high-κ layer (HKL) and interfacial layer (IL).

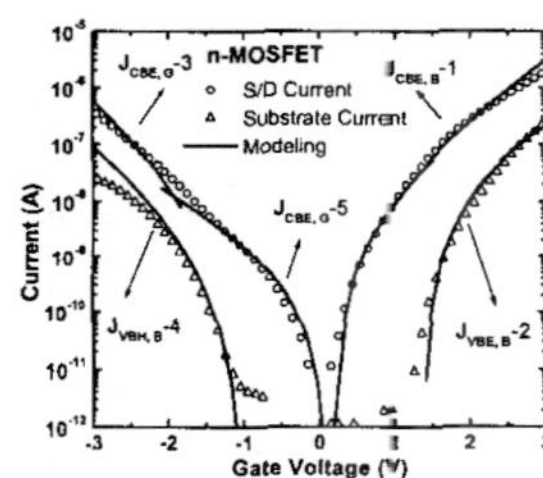

Fig.2b: Measured by carrier separation and simulated tunneling currents of n-MOSFET (L/W=10µm/0.18µm, SDE=5 nm with 1.65 nm SiO_2. CBE: conduction band electron; VBE: valence band electron; VBH: valence band hole. B is substrate injection and G gate injection. The simulated $J_{VBH,B}$-4 is lower than experiment at Vg ~ -3V, due to the compensation of impact ionization induced by $J_{CBE,G}$-3.

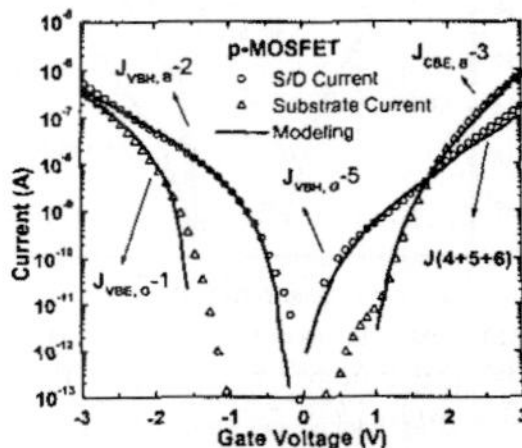

Fig.3b: Simulated and measured results for a p-MOSFET (L/W= 10µm/0.18 µm, SDE=6 nm) with 1.65 nm SiO_2. The s/d current at Vg>2V is the summation of tunneling channels 4, 5 and 6.

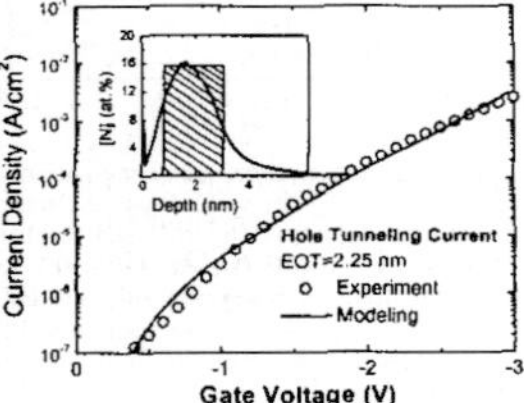

Fig.4: Tunneling hole current of p-MOS with $SiON/SiO_2$ stack. The stack structure was extracted from the SIMS profile of N in the stack (inset) and TEM measurement.

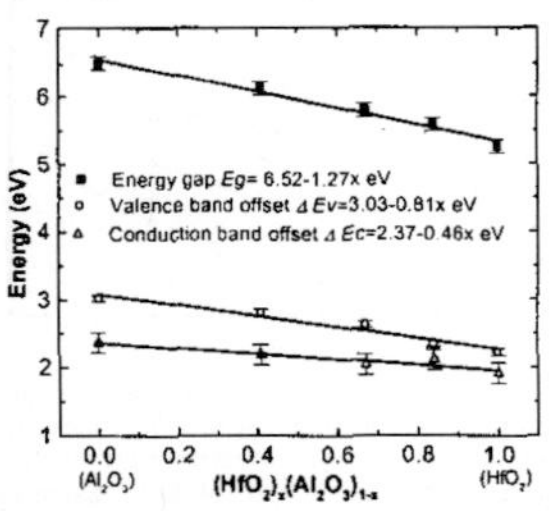

Fig.5: Dependence of energy gap Eg, valence and conduction band offset ΔEv and ΔEc for ALD HfAlO/Si determined by XPS on HfO_2 mole fraction x. Linear dependences are demonstrated.

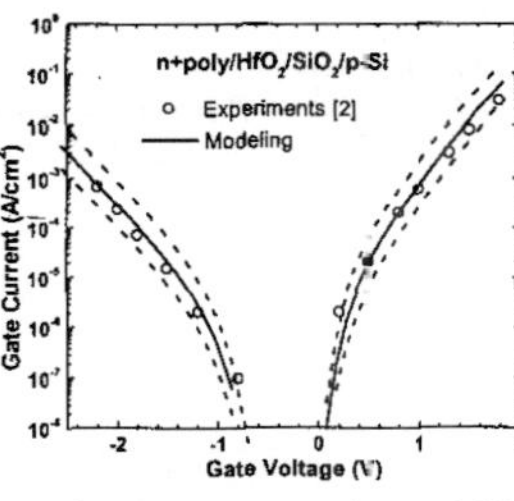

Fig.6: Simulated gate current of a n+ poly/HfO_2/SiO_2/p-Si device. The measured data are from [2], the physical thickness are HfO_2(38Å)/IL(6Å) from HRTEM and IL is likely SiO_2 from XPS [2]. The fitted tunneling mass m_{HfO} is 0.18 m_0. When the uncertainty of HRTEM is 1 Å, the resulted m_{HfO} error is ±0.02 m_0. The dashed lines are simulations with m_{HfO} = 0.20 and 0.16 m_0.

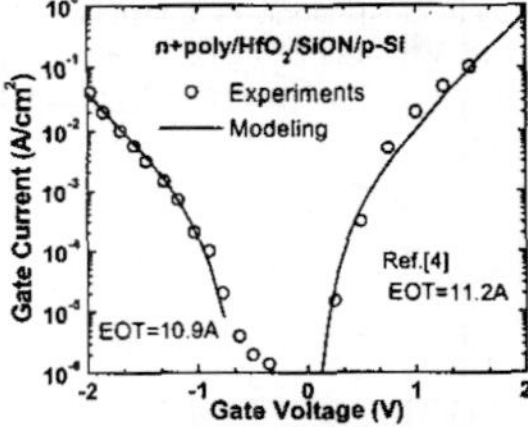

Fig.7: Simulated tunneling current of n+ poly /HfO_2 on NH_3 nitrided p-Si. The Vg < 0 data is from [4]. The IL is assumed as $(SiO_2)_{0.5}(Si_3N_4)_{0.5}$. Using m_{HfO}=0.18 m_0 and κ=22 for HfO_2, the effective IL physical thickness can be determined. The resulted 6.5 Å and 9 Å values for our capacitor sample and that of [4] are both in reasonable consistency with the HRTEM result, 6-7 Å [1] and 8.5 Å [4], respectively.

29.8.3

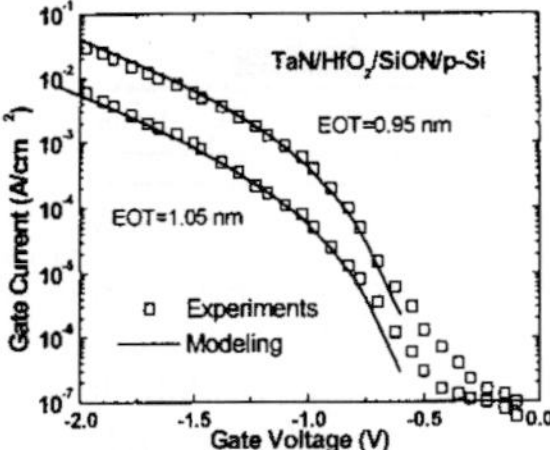

Fig.8: Measured and simulated tunneling currents of TaN gated n-capacitors of HfO_2 on NH_3 nitrided Si. The EOT from C-V are 9.5 Å and 10.5 Å, respectively. Assuming IL is $(SiO_2)_{0.5}(Si_3N_4)_{0.5}$, the physical thickness of IL are 6.5 Å and 7.5 Å, in consistent with the TEM measured 6-7 Å [1].

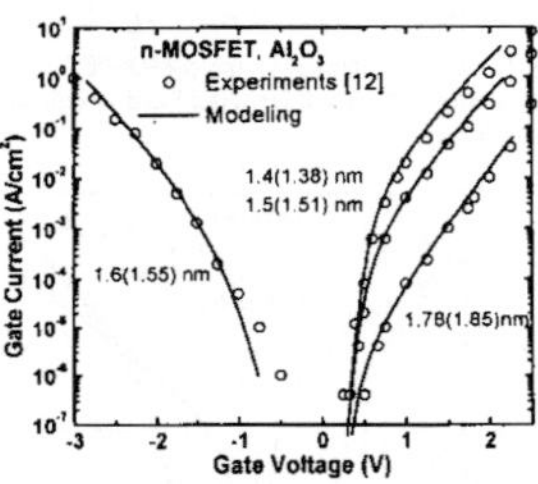

Fig.11: Simulated electron tunneling currents of n-MOS with Al_2O_3 gate dielectric. The data are from [12]. The tunneling effective mass is found to be 0.28 m_0 from overall fitting of all the data. The thickness values from best fitting to the measured data match well with those in [12] from C-V method (in parenthesis).

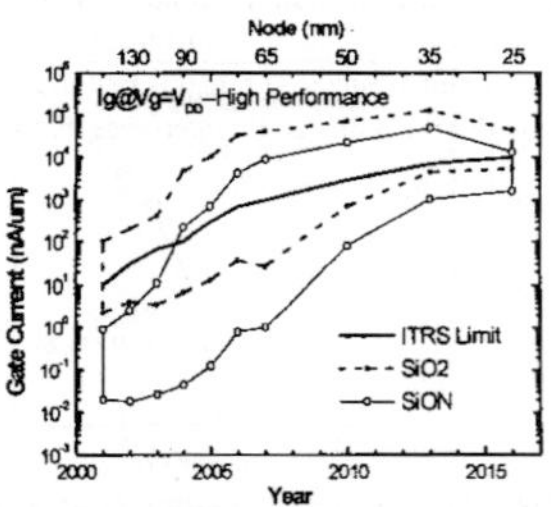

Fig.14: The calculated gate leakage of high performance CMOS. The V_{DD}, EOT and channel length are taken from Table 1 for each generation. The calculated high (low) gate leakage for each generation corresponds to the minimum (maximum) EOT proposed in Table 1. Si_3N_4 mole fraction is 40% for optimized SiON.

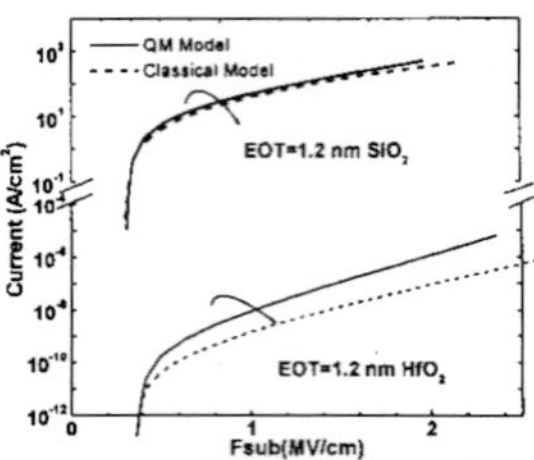

Fig.9: Calculated dependence of tunneling currents on substrate electric field for SiO_2 and HfO_2 by classical and quantum mechanical (QM) models. The quantization in inversion layer lifts the electron energy up, leading to larger tunneling current.

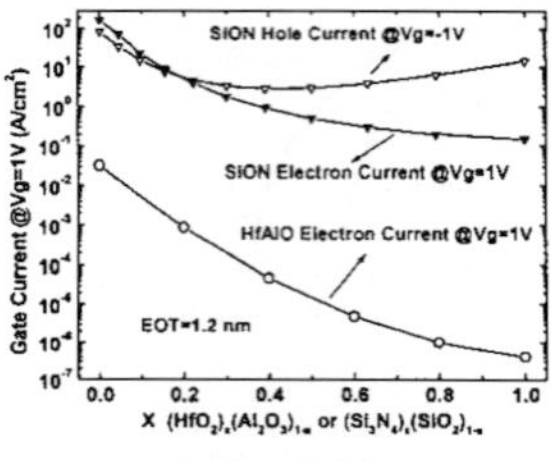

Fig.12: Calculated tunneling currents of SiON and HfAlO for various N or Hf compositions. For SiON, the optimized SiON (lowest tunneling leakage considering both n- and p-MOSFETs together) is at x~0.4 ([N]~34 at.%). For HfAlO, κ and tunneling mass m are linear interpolations of HfO_2 (κ=22 [4], m=0.18m_0) and Al_2O_3 (κ=10 [12], m=0.28m_0).

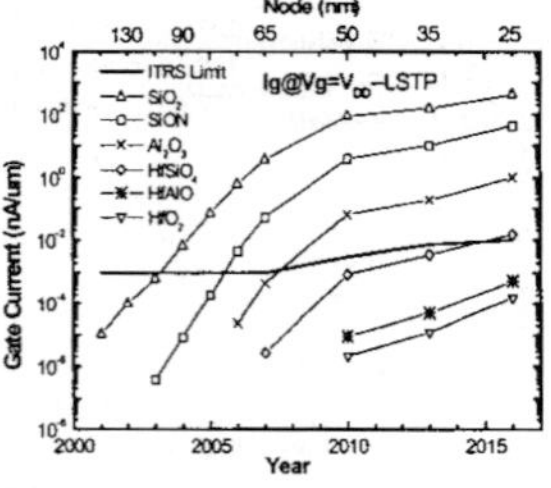

Fig.15: The calculated gate leakage for low standby power (LSTP) application. Here, an average value of the proposed maximum and minimum EOT from Table 1 is used for each generation. Al_2O_3 mole fraction is 30% for HfAlO and Si_3N_4 mole fraction 40% for optimized SiON. The $HfSiO_4$ parameters are same as in Fig.13.

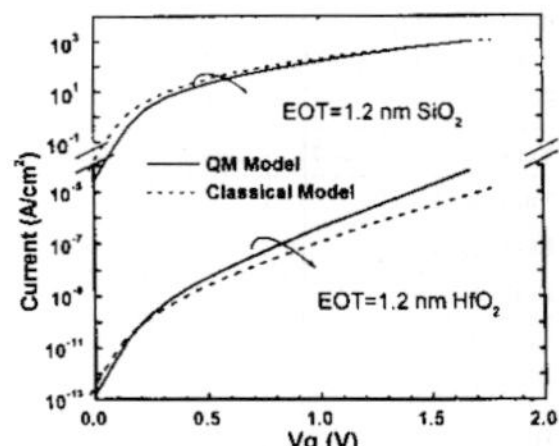

Fig.10: Calculated tunneling currents vs gate voltage from classical and QM models. The QM effect is compensated in some extent due to the larger band bending in QM model. However, the enhancement due to quantization can still be observed in I-V plot for HfO_2.

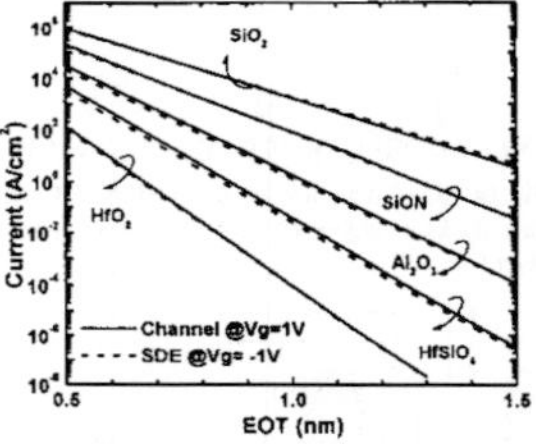

Fig.13: Simulated electron tunneling current of n-MOSFET vs EOT for various gate dielectrics. The substrate doping is 10^{18} cm^{-3} and flat band voltage is thus −1.0V. For $HfSiO_4$, κ=13 and m=0.34 m_0 from an average of SiO_2 and HfO_2 values are assumed. Al_2O_3 mole fraction is 30% for HfAlO and Si_3N_4 mole fraction 40% for optimized SiON.

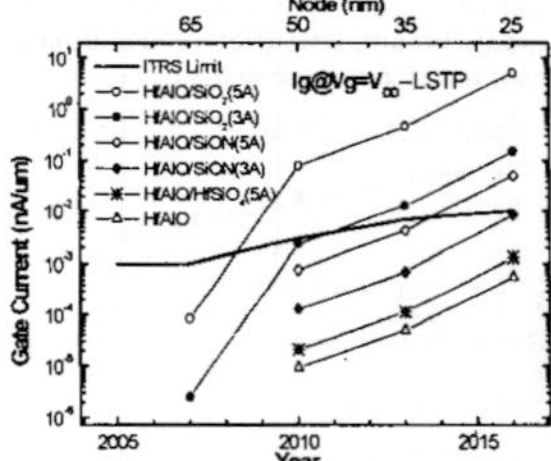

Fig.16: The calculated gate leakage for low standby power applications of HfAlO dielectric with different IL layers to improve the interface quality. Physical 5 Å IL of SiO_2, optimized SiON, $HfSiO_4$ were presented. A minimum 3 Å SiO_2 and SiON ILs are also shown to demonstrate the limit of SiO_2-based dielectrics as an IL layer.

29.8.4

JOURNAL OF APPLIED PHYSICS   VOLUME 91, NUMBER 1   1 JANUARY 2002

Direct tunneling hole currents through ultrathin gate oxides in metal-oxide-semiconductor devices

Y. T. Hou and M. F. Li[a)]
Department of Electrical & Computer Engineering, Silicon Nano Device Lab, National University of Singapore, 10 Kent Ridge Crescent, Singapore 119260, Singapore

Y. Jin and W. H. Lai
Chartered Semiconductor Manufacturing Ltd., Singapore 738406, Singapore

(Received 8 May 2001; accepted for publication 11 September 2001)

We present a physical model to calculate the direct tunneling hole current through ultrathin gate oxides from the inversion layer of metal–oxide–semiconductor field-effect transistors. A parametric self-consistency method utilizing the triangular well approximation is used for the electrostatics of the inversion layer. For hole quantization in the inversion layer, an improved one-band effective mass approximation, which is a good approximation to the rigorous six-band effective mass theory, is used to account for the band-mixing effect. The tunneling probability is calculated by a modified Wentzel–Kramers–Brilliouin (WKB) approximation, which takes the reflections near the Si/SiO_2 interfaces into account. It is found that the parabolic dispersion in the SiO_2 band gap used in the WKB approximation is only applicable for hole tunneling in oxides thinner than about 2 nm and for low gate voltage. A more reasonable Freeman–Dahlke hole dispersion form with significantly improved fitting to all experimental data for different oxide thickness and gate voltage range is adopted and discussed. © *2001 American Institute of Physics.* [DOI: 10.1063/1.1416861]

I. INTRODUCTION

As metal–oxide–semiconductor field-effect transistors (MOSFETs) quickly approach the deep submicron regime, aggressive scaling of gate dielectric thickness becomes necessary. An equivalent gate oxide thickness of less than 2 nm will be required for sub-0.1 μm generation devices.[1] For such an ultrathin gate oxide, direct tunneling current will dominate the gate leakage current and the off-state power dissipation of the transistor. The increased tunneling current through the thin gate oxide also hinders the extraction of transistor parameters by traditional techniques such as the capacitance–voltage ($C-V$) method.[2] Therefore, an accurate modeling of tunneling current through a trapezoidal barrier is important in transistor technology development.

Direct tunneling of the conduction band electron from inversion or accumulation layers has been extensively studied.[3–14] As for p^+ polysilicon gate p-MOSFETs, direct tunneling hole was recently found to dominate the gate current under channel inversion conditions.[15–17] However, little attention has been paid to the physical modeling of direct tunneling hole at present.[11–13,18] In a reliable physical model for direct tunneling hole current, two important characteristics should be present. First there should be an accurate treatment of the hole quantization effect in the Si substrate used in determining the hole densities at different energies and the respective voltage drops in the oxide layer and substrate. Second there should be a correct transmission formula for hole tunneling through the dielectric film. Among the existing models of hole tunneling, none of them satisfy these two

criteria.[11–13,18] Although these models can be fitted to experimental data, it may be the result of errors compensating each other.[3] In this article, we report a direct tunneling hole model that fulfills the two criteria mentioned above. In our model, we employ an improved one-band effective mass approximation (EMA) including valence band mixing effect,[19] which gives a more accurate description of the hole quantization in the inversion layer. A modified Wentzel–Kramers–Brilliouin (WKB) approximation with reflections at the interfaces being taken into account is used to compute the tunneling probability.[6–8,12,18] In the WKB approximation, it is crucial to use an appropriate hole dispersion in the SiO_2 band gap. All previous models assume a parabolic dispersion in the SiO_2 band gap during hole tunneling.[11–13,18] However their accuracy is questionable because the energy of valence hole tunneling aligns at the middle of the SiO_2 band gap. In this article, we shall show that a parabolic approximation is not always applicable to hole dispersion in the SiO_2 band gap. Instead, we use a more appropriate Freeman–Dahlke dispersion form in the calculation of direct hole tunneling.[20] It is physically more reasonable and it achieves a significant improvement in the matching of simulation results to the experimental data.

Other techniques developed in the direct electron tunneling model are readily transferred to the case of hole tunneling. The electrostatics of an inversion layer is calculated by a parametric self-consistency method based on the triangular well approximation.[7,21] This method is originally proposed for the calculation of the electrostatics in a n-MOSFET in inversion[21] and it has also been applied to electron tunneling successfully.[7]

[a)]Author to whom correspondence should be addressed; electronic mail: elelimf@nus.edu.sg

J. Appl. Phys., Vol. 91, No. 1, 1 January 2002   Hou *et al.*   259

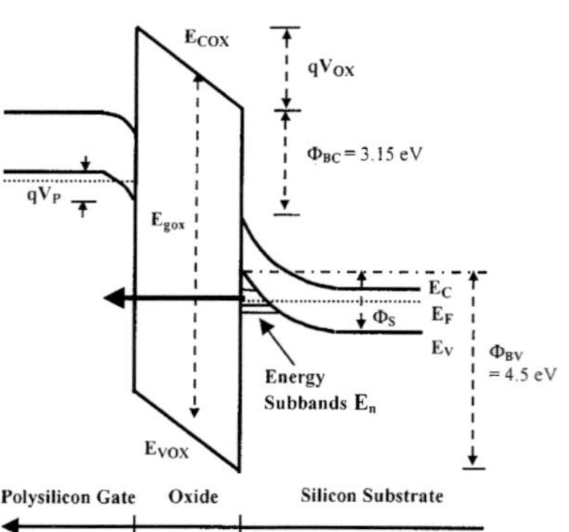

FIG. 1. A schematic of the band diagram of a p^+ polysilicon/SiO$_2$/n-Si MOS structure showing the hole quantization effect in the substrate and direct hole tunneling from the substrate inversion layer to the polysilicon gate.

II. PHYSICAL MODEL

In this section, the physical model used to calculate the direct tunneling hole current is described. In Fig. 1, a schematic band diagram is shown for a p-MOSFET in inversion. E_C and E_V are the Si conduction and valence band edges. E_{COX} and E_{VOX} are the SiO$_2$ conduction and valence band edges. E_F is the Fermi level in the Si substrate. Φ_{BC} and Φ_{BV} are the conduction and valence band offsets between Si and SiO$_2$. Φ_S is the total surface potential energy (band bending). V_{OX} is the oxide voltage drop.

A. Hole quantization in Si inversion layer

When the substrate is strongly inverted, carriers at the Si surface are confined near the surface. As shown in Fig. 1, the hole energy levels in the inversion layer form discrete subbands. This quantization effect is important for tunneling current calculation in deep submicron complementary metal–oxide–semiconductor (CMOS) devices with high substrate doping and ultrathin gate oxide.[3–7,9–11] In the existing direct tunneling hole models, the hole quantization effects are either neglected[12] or approximated with the method of electron quantization.[11,18] In Refs. 11 and 18, the authors have used the traditional one-band EMA[22,23] to calculate the hole subbands in the Si inversion layer. In this traditional one-band EMA, the heavy and light hole effective masses are extracted from the bulk Si valence bands. As demonstrated in previous studies,[24–26] when there is a strong electric field F perpendicular to the interface in the inversion layer along the [100] interface, the symmetry of the structure is reduced and mixing between different bulk valence bands is expected. From our calculation, the mixing of the heavy, light, and spin-orbit split-off hole bands of the bulk Si is so strong that the results derived by the traditional one-band EMA are no longer reliable.[26] We have developed a six-band effective mass approximation algorithm to calculate the hole quanti-

zation effect in the Si inversion layer.[26] It takes into account the band mixing effect of heavy, light, and spin-orbit split-off hole bands. According to our six-band calculation,[26] traditional one-band EMA underestimates the density of states and also results in large errors in the quantization energy levels. This in turn results in the computed threshold voltage shift due to hole quantization being higher than experimental results.[26] We have also simplified the six-band EMA into an improved one-band EMA, in which case the traditional formula of one-band EMA can still be used with a new set of effective mass values empirically determined from the numerical results of our six-band EMA calculation.[19] In our improved one-band EMA, the quantization effective mass of the nth subband $m_{\perp n}^*$ is determined from the hole subband minimum E_n by[19,21]

$$E_n = \left(\frac{\hbar^2}{2m_{\perp n}^*} \right)^{1/3} \left[\frac{3}{2} \pi q F \left(n - \frac{1}{4} \right) \right]^{2/3}, \tag{1}$$

where F is the electric field in the Si substrate. Furthermore the density of states effective mass m_{dn}^* can be obtained from the inversion hole charge density N_n of the nth subband by[19,21]

$$N_n = \left(\frac{kT}{\pi \hbar^2} \right) m_{dn}^* \ln \left(1 + \exp \left(\frac{E_F - E_n}{kT} \right) \right). \tag{2}$$

Although these effective masses are weakly electric field dependent, they can be treated as constant values independent of electric field in the first order approximation. The equivalent effective masses of the lowest three subbands obtained are: $m_{\perp}^*/m_d^* = (0.29/1.17)m_0$, $(0.24/0.71)m_0$, and $(0.22/0.57)m_0$ respectively, where m_0 is the free electron mass.

In Fig. 2, we show the subband energy levels, occupation factors of holes in the lowest three subbands, and the surface potential calculated by the improved one-band EMA at inversion. The solid circles are the corresponding results of the accurate six-band effective mass theory. Compared to the six-band calculation in Fig. 2, the accuracy of our improved one-band EMA is acceptable and the computation becomes more efficient. Furthermore by utilizing an improved one-band EMA, the hole quantization can be treated in the same frame as electron quantization and the band mixing effect is included in the empirical effective mass values.

B. Hole tunneling lifetime

For carriers confined in the quasibound states[27] in the inversion layer, the lifetime of an nth subband state is approximately given by:[3,5–7]

$$\frac{1}{\tau_n(E)} = \frac{T(E)}{\int_0^{z_n} \sqrt{2m_{\perp n}^*/[E_n - E_V(z)]} \, dz}, \tag{3}$$

where E_n is the subband energy for the nth quasibound state, $E_V(z)$ is the edge of the Si valence band, and z_n is the classical turning point for the nth bound state. $T(E)$ is the transmission probability of a particle. The exact solution of a trapezoidal barrier gives $T(E)$ in terms of Airy functions.[28] Justification of using WKB approximation in a direct tunnel-

260   J. Appl. Phys., Vol. 91, No. 1, 1 January 2002   Hou *et al.*

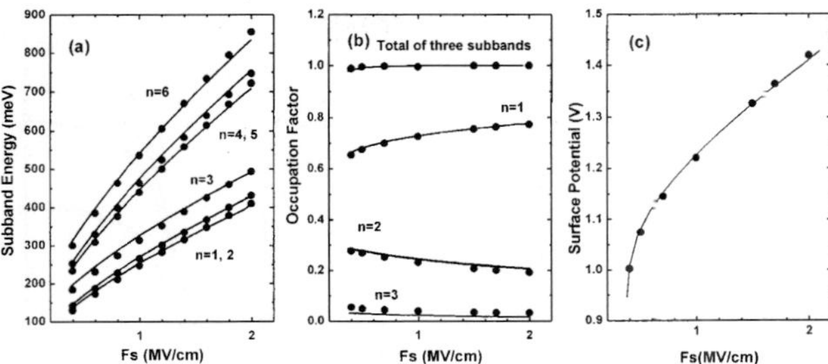

FIG. 2. Comparisons of: (a) subband energies, (b) occupation factors, and (c) surface potentials in the hole inversion layer calculated by the improved one-band effective mass approximation and the six-band effective mass theory. The substrate doping concentration in the calculation is $5 \times 10^{17}\,\mathrm{cm}^{-3}$.

ing regime has been provided.[5–7,12] In the calculation of direct tunneling hole current, we adopt a modified WKB approximation with a correction factor accounting for reflections at the boundary of the oxide layer, such that

$$T(E) = T_R(E) \cdot T_{\mathrm{WKB}}(E), \tag{4}$$

where T_{WKB} is the usual WKB approximation of the transmission probability and T_R the correction factor accounting for the reflections from boundaries of the oxide. T_{WKB} is given by

$$T_{\mathrm{WKB}}(E) = \exp\left(-2\int_0^{t_{\mathrm{ox}}} \kappa(E,z)dz\right), \tag{5}$$

where κ is the imaginary wave number within the oxide gap energy[29,30] and t_{ox} the oxide thickness, i.e., the tunneling distance. T_R can be expressed as a function of the group velocities of the tunneling carriers[6]

$$T_R = \frac{4\nu_{\mathrm{Si}}(E)\nu_{\mathrm{OX}}(E_{\mathrm{OX}i})}{\nu_{\mathrm{Si}}^2(E) + \nu_{\mathrm{OX}}^2(E_{\mathrm{OX}i})} \times \frac{4\nu_{\mathrm{Si}}(E+qV_{\mathrm{OX}})\nu_{\mathrm{OX}}(E_{\mathrm{OX}o})}{\nu_{\mathrm{Si}}^2(E+qV_{\mathrm{OX}}) + \nu_{\mathrm{OX}}^2(E_{\mathrm{OX}o})}, \tag{6}$$

where $\nu_{\mathrm{Si}}(E)$ and $\nu_{\mathrm{Si}}(E+qV_{\mathrm{OX}})$ are the group velocities of the carriers incident and leaving the oxide layer, respectively, V_{OX} the oxide voltage drop as well as $\nu_{\mathrm{ox}}(E_{\mathrm{OX}i})$ and $\nu_{\mathrm{ox}}(E_{\mathrm{OX}o})$ the magnitudes of the imaginary group velocities of carriers tunneling in and out of the oxide layer, respectively. T_R arises from the matching of wave functions and conservation of the carrier flux at the sharp boundaries. Such a correct factor in the same form was reported by other authors.[8,12]

Combining Eqs. (3)–(6) and the results from quantization calculation of the inversion layer, the tunneling current can be readily obtained as a function of the electric field F_{OX} in the oxide layer

$$J = \sum_n N_n / \tau_n(E_n), \tag{7}$$

where N_n is inversion charge density from the nth subband. The summation is taken over all of the subbands. The gate voltage V_g is determined from the voltage balance equation

$$V_g = V_{\mathrm{FB}} + V_{\mathrm{OX}} + V_P + \Phi_S, \tag{8}$$

where V_{FB} is the flatband voltage, oxide voltage drop $V_{\mathrm{OX}} = F_{\mathrm{OX}} \cdot t_{\mathrm{ox}}$, V_P the voltage drop in polysilicon gate due to poly depletion and Φ_S the substrate band bending.

The charge density N_n and subband energy E_n in Eq. (7), as well as the $V_g - F_{\mathrm{OX}}$ relation in Eq. (8) are all determined based on the improved one-band EMA as discussed in Sec. II A. In the determination of the electrostatics of the inversion layer by a parametric self-consistency method,[21] the electric field F used in Eq. (1) is replaced by an effective electric field F_{eff} in order to extend the triangular well approximation to the strong inversion condition. It is defined as

$$F_{\mathrm{eff}} = \frac{q(N_{\mathrm{depl}} + \eta N_{\mathrm{inv}})}{\epsilon_0 \epsilon_{\mathrm{Si}}}, \tag{9}$$

where N_{depl} and N_{inv} are depletion and inversion charge densities, respectively. The respective values of η for electron and hole inversion are 0.75[31] and 0.5.[25]

From Eq. (5), the dispersion $\kappa(E,z)$ relationship in the SiO_2 band gap is important for tunneling because it appears in the exponential factor. In principle, the imaginary $\kappa(E)$ value in the energy gap can be calculated.[29] Unfortunately, at present the exact dispersion relationship $\kappa(E)$ in the wide energy gap material SiO_2 is absent. For hole tunneling, we use the empirical $\kappa(E)$ dispersion introduced by Freeman and Dahlke[12,20]

$$\frac{1}{(\hbar k)^2} = \frac{1}{2m_{\mathrm{cox}}(E_{\mathrm{COX}} - E)} + \frac{1}{2m_{\mathrm{vox}}(E - E_{\mathrm{VOX}})}, \tag{10}$$

where m_{cox} and m_{vox} are the effective masses of conduction and valence band of SiO_2, respectively. Equation (10) gives equal weight to the SiO_2 conduction band and valence band on equal footing, but takes into account the difference of m_{cox} and m_{vox}. When E approaches E_{COX}, the first term on the right hand side of Eq. (10) becomes the major term and $E(k)$ reduces to a parabolic relationship with the conduction band effective mass. When E approaches E_{VOX}, the second term on the right hand side of Eq. (10) becomes dominant and $E(k)$ reduces to a parabolic relationship with the valence band effective mass. When $m_{\mathrm{cox}} = m_{\mathrm{vox}} = m_{\mathrm{ox}}$, Eq. (10) reduces to the Franz-type dispersion[32]

J. Appl. Phys., Vol. 91, No. 1, 1 January 2002

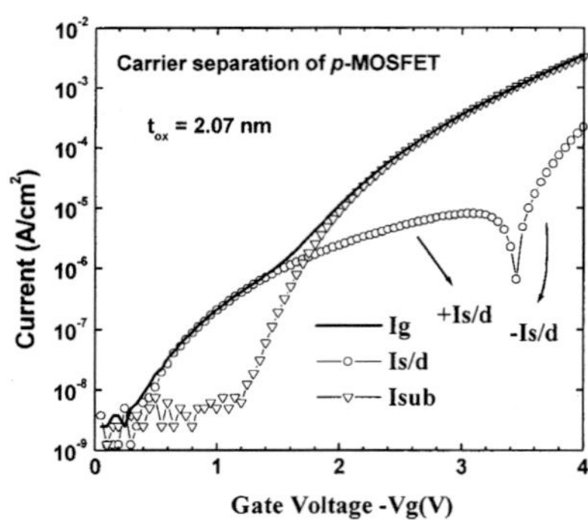

FIG. 3. Current–voltage characteristics obtained from carrier separation measurements for a *p*-MOSFET with t_{ox}=2.07 nm.

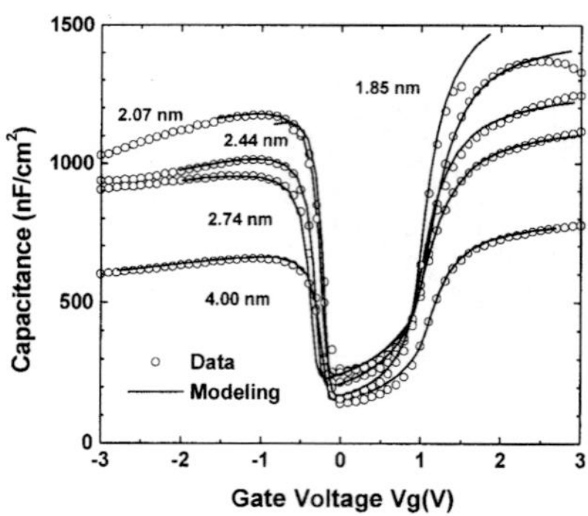

FIG. 4. The measured capacitance–voltage characteristics of the *p*-MOSFETs used. The solid lines are the fitting results using the QM-CV model of device group at UC Berkeley. The extracted oxide thickness is 1.85, 2.07, 2.44, 2.73, and 4.00 nm.

$$\kappa(E) = \left(\frac{2m_{\mathrm{OX}}}{\hbar^2}\right)^{1/2} E^{1/2} \left(1 - \frac{E}{E_{g\mathrm{ox}}}\right)^{1/2} \qquad (11)$$

where $E_{g\mathrm{ox}} = E_{\mathrm{COX}} - E_{\mathrm{VOX}}$ is the band gap of $\mathrm{SiO_2}$.

Since the $\mathrm{Si/SiO_2}$ conduction and valence band offsets are around 3.15 eV[4–7,33] and 4.5 eV,[13,16,17] respectively, the bottom of the Si conduction band is closer to the $\mathrm{SiO_2}$ conduction band E_{COX} than the $\mathrm{SiO_2}$ valence band E_{VOX}. It is therefore reasonable to use a simple parabolic dispersion or a Franz-type dispersion in the calculation of electron tunneling.[3–12] However the top of the Si valence band aligns at the middle of the $\mathrm{SiO_2}$ band gap. In this condition, the two items on the right hand side of Eq. (10) are comparable and neither of them can be neglected. This means that a parabolic approximation is not physically appropriate. The Freeman–Dahlke dispersion of Eq. (10) is expected to give more accurate results. This will be verified in Sec. IV when compared to the experiments.

III. EXPERIMENTS

In our experiments, the MOSFETs were fabricated by a standard dual-gate CMOS process. The oxide was grown by rapid thermal oxidation. The $C–V$ measurements were performed using a HP4284A *LCR* meter on large area (400×60 μm^2) MOS capacitors at a frequency of 100 kHz. The current–voltage ($I–V$) characteristics were measured using the HP4156A semiconductor parameter analyzer. For n-MOSFETs, the electron direct tunneling current was measured as the gate current at inversion (gate voltage $V_g > 0$). For p^+ polysilicon gate p-MOSFETs, the direct tunneling hole current was measured by a carrier separation method.[16] In the carrier separation measurement, the source and drain were tied together and grounded along with the substrate while a negative V_g was applied. Figure 3 shows a typical $I–V$ plot of a carrier separation measurement on a p-MOSFET with oxide thickness t_{ox}=2.07 nm. At low voltage ($|V_g| < 1.7\,\mathrm{V}$), the gate current is dominated by the source/drain current, which has been identified as direct tun-

neling hole current from the hole inversion layer.[16] At high voltages, the substrate current due to p^+ polysilicon gate valence band electron tunneling dominates the gate current. The change of sign of the source/drain current at about -3.5 V is due to the hole generation by impact ionization of the valence electron tunneling from the gate.[16] In our experiment, direct tunneling hole current is measured as source/drain current.

In our measurements, devices with areas of 20×0.5, 50×1, and 40×20 μm^2 were used and the current density exhibits no area dependence, which indicates a negligible edge effect.[34] The tunneling electron and hole currents were measured on the same wafer and oxide thickness nonuniformity within the wafer was not observed.

IV. RESULTS AND DISCUSSION

A. *C–V* characterization and extraction of device parameters

As discussed in Sec. II, our calculation of the tunneling currents is from the first principle. All the macroscopic device parameters, such as the substrate and gate doping, flatband voltage, and gate oxide thickness, are determined by $C–V$ data analysis. The electrostatics of the device is then calculated by the improved one-band EMA, based on these device parameters. Figure 4 shows the $C–V$ measurements of the p-MOS capacitors. The $C–V$ data were used to extract the substrate doping level, the polysilicon gate doping level, the flatband voltage, and the oxide thickness using a full quantum mechanical $C–V$ model developed by the device group at the University of California Berkeley. This widely used model takes the quantum mechanical and polysilicon depletion effects into account. Although this Berkeley model utilizes the traditional one-band EMA for hole quantization, it has been demonstrated previously that the capacitance values are not sensitive to the choice of the effective mass values.[35] From our six-band calculation, we also found

262   J. Appl. Phys., Vol. 91, No. 1, 1 January 2002

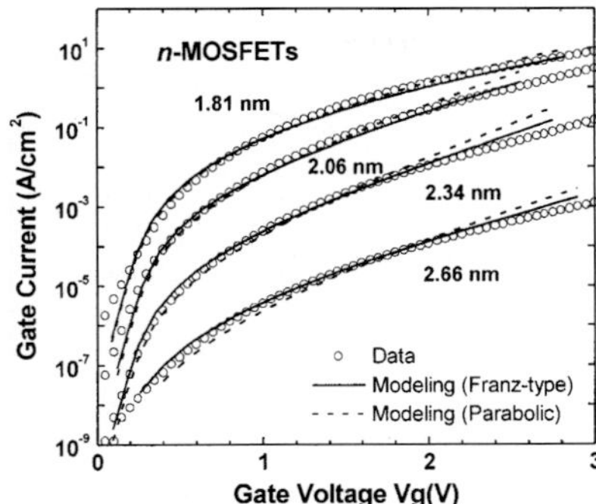

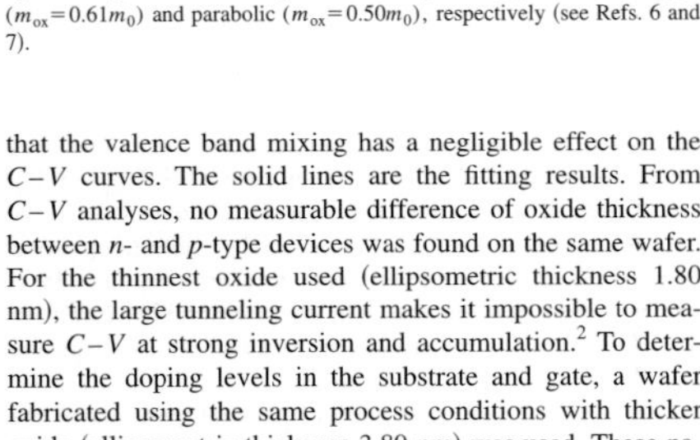

FIG. 5. The electron direct tunneling currents in *n*-MOSFETs. The open circles are the measurements. The solid and dashed lines are the calculations by assuming the electron dispersion in SiO_2 band gap to be Franz type ($m_{ox}=0.61m_0$) and parabolic ($m_{ox}=0.50m_0$), respectively (see Refs. 6 and 7).

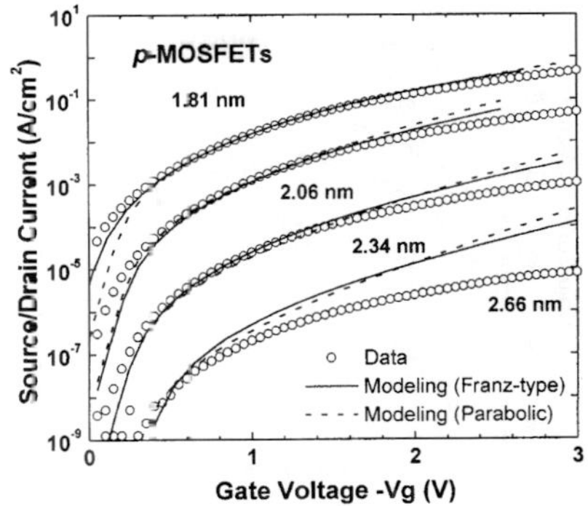

FIG. 6. The hole direct tunneling currents in *p*-MOSFETs. The open circles are the measured values. The solid and dashed lines denote the calculated values by assuming the hole dispersion in SiO_2 band gap to be Franz type ($m_{ox}=0.55m_0$) and parabolic ($m_{ox}=0.40m_0$), respectively.

that the valence band mixing has a negligible effect on the $C-V$ curves. The solid lines are the fitting results. From $C-V$ analyses, no measurable difference of oxide thickness between *n*- and *p*-type devices was found on the same wafer. For the thinnest oxide used (ellipsometric thickness 1.80 nm), the large tunneling current makes it impossible to measure $C-V$ at strong inversion and accumulation.[2] To determine the doping levels in the substrate and gate, a wafer fabricated using the same process conditions with thicker oxide (ellipsometric thickness 3.80 nm) was used. These parameters were used to calculate the direct tunneling current directly.

B. Extraction of oxide thickness from direct tunneling electron current from inversion layers of *n*-MOSFETs

Direct electron tunneling has been extensively studied.[3–12] The purpose of this section is to check our tunneling algorithm and extract the oxide thickness for verification. Figure 5 shows the results of direct tunneling electron current. The solid lines are the calculated results using an empirical Franz-type dispersion in the gap energy of SiO_2.[32] In the calculation, the conduction band offset between Si and SiO_2 is fixed at 3.15 eV and the $m_{ox}=0.61m_0$.[4,6,7] The best fit to the experimental results is obtained by adjusting oxide thickness t_{ox}. Similarly to previous studies,[7] the calculated $I-V$ characteristics are in good agreement with the experimental results at all voltages (0–3 V) and for all oxide thicknesses (1.8–2.7 nm). The calculated results using a simple parabolic dispersion ($m_{ox}=0.50m_0$) are also displayed in Fig. 5 as dashed lines. The fitting results using the simple parabolic dispersion are only slightly degraded for thick oxides. It indicates that the parabolic dispersion is a good approximation for direct electron tunneling. Furthermore the effective mass values in parabolic dispersion are also in

agreement with the band structure calculation of bulk SiO_2, in which case an effective mass of about $0.5m_0$ is demonstrated for the conduction band.[36,37]

The oxide thickness determined from the fitting of electron tunneling for our four samples are: 1.81, 2.06, 2.34, and 2.66 nm. They are close to the values determined from the $C-V$ method (1.85, 2.07, 2.44, and 2.74 nm, respectively).[18,38] The maximum deviation is about 0.1 nm and this is within the reported limits of different experimental methods, such as $C-V$, high-resolution transmission electron microscopy, and optical ellipsometry.[18,38] In the subsequent calculation of direct tunneling hole current, we used the values of t_{ox} extracted from electron tunneling rather than those from the $C-V$ method. It ensures consistency in the calculation of tunneling electron and hole currents.[18]

C. Direct tunneling hole current by parabolic dispersion

In this section, we shall first discuss the direct hole tunneling using a parabolic hole dispersion in the band gap of SiO_2. All previous works are based on such an approximation. The results obtained by our physical model calculations as discussed in Sec. II are displayed as dashed lines in Fig. 6. In the calculation, the valence band offset between Si and SiO_2 is fixed at 4.5 eV.[13,16] Although a Franz-type dispersion has never been used for hole tunneling, we also show the calculations using a Franz-type dispersion ($m_{ox}=0.55m_0$) as solid lines in Fig. 6 for comparison. Compared to the experimental data, the hole tunneling $I-V$ characteristics computed from either the parabolic or Franz-type dispersion are not as close to the experimental data as those for electron tunneling. The fit for thinner oxides (<2 nm) is better. However either the parabolic or Franz-type dispersion cannot fit the experimental data when the oxide thickness is larger than about 2 nm. The deviation is more obvious at a higher gate voltage. Assuming a parabolic dispersion, m_{ox} is found to be $0.40m_0$

J. Appl. Phys., Vol. 91, No. 1, 1 January 2002   Hou *et al.*   263

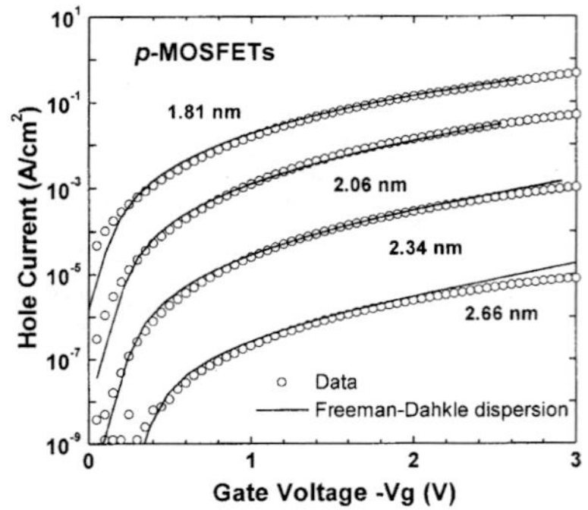

FIG. 7. The direct tunneling hole currents in p-MOSFETs. The open circles are the measured values. The solid lines denote the calculated values by assuming a Freeman–Dahlke form dispersion in SiO$_2$ band gap with $m_{\text{cox}}=0.50m_0$ and $m_{\text{vox}}=0.80m_0$.

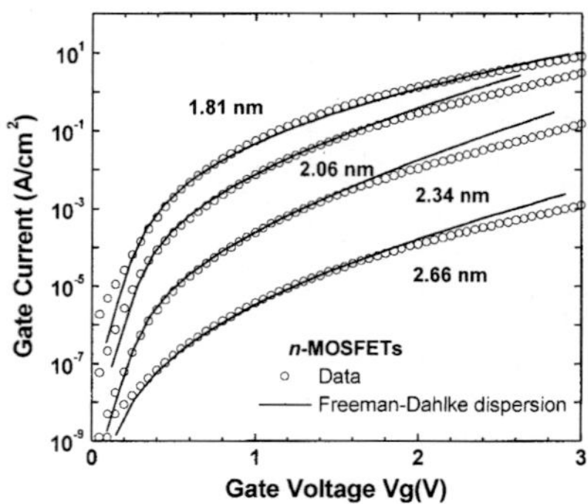

FIG. 8. The direct tunneling electron currents in n-MOSFETs. The open circles are the measured values. The solid lines denote the calculated values by assuming a Freeman–Dahlke form dispersion in SiO$_2$ band gap with $m_{\text{cox}}=0.50m_0$ and $m_{\text{vox}}=0.80m_0$.

in order to get the best fitting. This value is close to the previous results reported for valence band electron or hole tunneling $(0.35-0.50m_0)$.[11,16,18,34,39] It indicates a smaller m_{ox} for holes than for electrons. This is in conflict with the existing results of the band structure calculation of bulk SiO$_2$, in which case the effective mass of valence bands is found to be typically $3-10m_0$,[36,37] which is much heavier than that of the conduction band ($\sim 0.5m_0$).

D. Direct tunneling hole current by Freeman and Dahlke form dispersion

In this section we use the Freeman and Dahlke[20] form of Eq. (10) for hole dispersion in the SiO$_2$ band gap to calculate the direct tunneling hole current. Contrary to the parabolic or Franz-type dispersions, in which the WKB integration term Eq. (5) has a simple analytical formula, numerical calculation must be used for the Freeman–Dahlke form. In the calculation, there are two effective mass values. From the electron tunneling and SiO$_2$ band structure calculation, we have $m_{\text{co}}=0.5m_0$. m_{vox} is an adjustable parameter for best fit. A t_{ox} independent value of about $m_{\text{vox}}=0.8m_0$ can give the best results and the calculations are shown in Fig. 7 as solid lines. From Fig. 7, it is apparent that a much better fit of hole tunneling current to the experimental data can be achieved by using the Freeman–Dahlke form for the hole dispersion in SiO$_2$. Contrary to the parabolic assumption, the Freeman–Dahlke form gives, not only much better agreement with the experimental data, but also a reasonable correlation between electron and hole dispersion in the band gap of SiO$_2$. The m_{vox} in the oxide gap is heavier than m_{cox}, but it is still smaller than the available computed results of the effective mass at the top of the valence band of the bulk SiO$_2(3-10m_0)$.[36,37] This is not impossible, however further study is obviously needed.

At a fixed gate voltage V_g, a thicker oxide layer has a smaller oxide field, and hence a smaller surface potential and voltage drop in the polysilicon gate. According to Eq. (8), the total oxide voltage drop is larger. Correspondingly, there is a larger variation of the denominator term $(E-E_{\text{vox}})$ in Eq. (10) during hole tunneling from the Si substrate side to the polysilicon gate side. Therefore a simple parabolic approximation will induce a larger integration error in Eq. (5). A similar argument is applicable in the case of higher gate voltage. This explains the deviation between the experimental and simulation results in Fig. 6.

In Fig. 8, we also present calculations using the Freeman–Dahlke dispersion for direct tunneling electron current. Compared to the measured data, this dispersion also applies to electron tunneling with the same parameters. As discussed before, the contribution of the second term in Eq. (10) is negligible for conduction electron tunneling. As a result, calculation using the Freeman–Dahlke dispersion form is very similar to the traditional calculation using the parabolic dispersion form.

V. CONCLUSION

An efficient physical model is proposed to calculate the direct tunneling hole current from inversion layers of MOS-FETs. An improved one-band effective mass approximation, which includes valence band mixing, is used to compute the hole quantization effect in the carrier electrostatics in the inversion layer. A modified WKB approximation is used for the tunneling probability calculation. In this approximation, it is found that a Freeman–Dahlke dispersion form is more appropriate than the widely used parabolic dispersion in the oxide gap energy for the modeling of direct hole tunneling. After taking these factors into account, the agreement of the simulated tunneling hole currents with the experimental data is improved over a wide range of oxide thicknesses and gate voltage.

ACKNOWLEDGMENTS

This work was supported by the Singapore Grant No. NSTB/EMT/TP/00/001.2 and the National University of Singapore Grant No. RP3982754. The authors would like to thank the device group of UC Berkeley for the their online QM-CV software.

[1] Y. Taur *et al.*, Proc. IEEE **85**, 486 (1997).

[2] K. Ahmed, E. Ibok, G. C. F. Yeap, Q. Xiang, B. Ogle, J. J. Wortman, and J. R. Hauser, IEEE Trans. Electron Devices **46**, 1650 (1999).

[3] F. Rana, S. Tiwari, and D. A. Buchanan, Appl. Phys. Lett. **69**, 1104 (1996).

[4] S. H. Lo, D. A. Buchanan, Y. Tauer, and W. Wang, IEEE Electron Device Lett. **18**, 209 (1997).

[5] W. K. Shin, E. X. Wang, S. Jallepalli, F. Leon, C. M. Maziar, and A. F. Tasch, Solid-State Electron. **42**, 997 (1998).

[6] L. F. Register, E. Rosenbaum, and K. Yang, Appl. Phys. Lett. **74**, 457 (1999).

[7] N. Yang, W. K. Henson, J. R. Hauser, and J. J. Wortman, IEEE Trans. Electron Devices **46**, 1464 (1999).

[8] Khairurrijal, W. Mizubayashi, S. Miyazaki, and M. Hirose, J. Appl. Phys. **87**, 3000 (2000).

[9] E. Cassan, J. Appl. Phys. **87**, 7931 (2000).

[10] W. Magnus and W. Schoenmaker, J. Appl. Phys. **88**, 5833 (2000).

[11] A. Ghetti, C. T. Liu, M. Mastrapasqua, and E. Sangiorgi, Solid-State Electron. **44**, 1523 (2000).

[12] J. Cai and C. T. Sah, J. Appl. Phys. **89**, 2272 (2001).

[13] W. C. Lee and C. Hu, Symposium on Very Large Scale Integrated Technology, 2000, p. 198.

[14] M. Stadele, B. R. Tuttle, and K. Hess, J. Appl. Phys. **89**, 348 (2001).

[15] T. Matsuoka *et al.*, Tech. Dig. - Int. Electron Devices Meet. **34**, 851 (1995).

[16] Y. Shi, T. P. Ma, S. Prasad, and S. Dhanda, IEEE Trans. Electron Devices **45**, 2355 (1998).

[17] W. C. Lee, T. J. King, and C. Hu, IEEE Electron Device Lett. **20**, 268 (1999).

[18] K. N. Yang *et al.*, IEEE Trans. Electron Devices **47**, 2161 (2000).

[19] Y. T. Hou and M. F. Li, IEEE Trans. Electron Devices **48**, 1188 (2001).

[20] L. B. Freeman and W. E. Dahlke, Solid-State Electron. **13**, 1483 (1970).

[21] H. H. Muller and M. J. Schulz, IEEE Trans. Electron Devices **44**, 1539 (1997).

[22] C. Moglestue, J. Appl. Phys. **59**, 3175 (1986).

[23] C. Y. Hu, S. Banerjee, K. Sadra, B. G. Streetman, and R. Sivan, IEEE Electron Device Lett. **17**, 276 (1996).

[24] S. Jallepalli, J. Bude, W. K. Shin, M. R. Pinto, C. M. Maziar, and A. F. Tasch, IEEE Trans. Electron Devices **44**, 297 (1997).

[25] S. Rodriguez, J. A. Lopez-Villanueva, I. Melchor, and J. E. Carceller, J. Appl. Phys. **86**, 438 (1999).

[26] Y. T. Hou and M. F. Li, Jpn. J. Appl. Phys., Part 2 **38**, L331 (2001).

[27] L. D. Landau and E. M. Lifshitz, *Quantum Mechanics* (Pergamon, Oxford, 1958), Sec. 48.

[28] K. H. Gundlach, Solid-State Electron. **9**, 949 (1966).

[29] Y. C. Chang, Phys. Rev. B **25**, 605 (1982); or see M.-F. Li, *Modern Semiconductor Quantum Physics* (World Scientific, Singapore, 1994), Sec. 4.31.

[30] E. O. Kane and E. I. Blount, in *Tunneling Phenomena in Solids*, edited by E. Burstein and S. Lundqvist (Plenum, New York, 1969), p. 10.

[31] T. T. Ma, L. T. Liu, Z. P. Yu, and Z. J. Li, IEEE Trans. Electron Devices **47**, 764 (2000).

[32] J. Maserjian, J. Vac. Sci. Technol. **11**, 996 (1974).

[33] Z. A. Weinberg, J. Appl. Phys. **53**, 5052 (1982).

[34] E. N. Yang, H. T. Huang, M. J. Chen, Y. M. Lin, M. C. Yu, S. M. Ang, C. H. Yu, and M. S. Liang, Tech. Dig. - Int. Electron Devices Meet. **39**, 679 (2000).

[35] S. Takagi, M. T. Takagi, and A. Toriumi, Tech. Dig. - Int. Electron Devices Meet. **39**, 619 (1998).

[36] E. Gnani, S. Reggiani, R. Colle, and M. Rudan, IEEE Trans. Electron Devices **47**, 1795 (2000).

[37] E. M. Schneider and W. B. Fowler, Phys. Rev. Lett. **36**, 425 (1976).

[38] E. Ahmed, E. Ibok, G. Bains, D. Chi, B. Ogle, J. J. Wortman, and J. R. Hauser, IEEE Trans. Electron Devices **47**, 1349 (2000).

[39] E. K. Chananan, K. McDonald, D. D. Ventra, S. T. Pantelides, G. Y. Chung, C. C. Tn, J. R. Williams, and R. A. Weller, Appl. Phys. Lett. **77**, 2560 (2000).

1158   IEEE TRANSACTIONS ON ELECTRON DEVICES, VOL. 49, NO. 7, JULY 2002

Investigation of Hole-Tunneling Current Through Ultrathin Oxynitride/Oxide Stack Gate Dielectrics in p-MOSFETs

Hongyu Yu, *Student Member, IEEE*, Yong-Tian Hou, *Student Member, IEEE*, Ming-Fu Li, *Senior Member, IEEE*, and Dim-Lee Kwong, *Senior Member, IEEE*

Abstract—The systematic investigation of hole tunneling current through ultrathin oxide, oxynitride, oxynitride/oxide (N/O) and oxide/oxynitride/oxide (ONO) gate dielectrics in p-MOSFETs using a physical model is reported for the first time. The validity of the model is corroborated by the good agreement between the simulated and experimental results. Under typical inversion biases ($|VG| < 2$ V), hole tunneling current is lower through oxynitride and oxynitride/oxide with about 33 at.% N than through pure oxide and nitride gate dielectrics. This is attributed to the competitive effects of the increase in the dielectric constant, and hence dielectric thickness, and decrease in the hole barrier height at the dielectric/Si interface with increasing with N concentration for a given electrical oxide thickness (EOT). For a N/O stack film with the same N concentration in the oxynitride, the hole tunneling current decreases monotonically with oxynitride thickness under the typical inversion biases. For minimum gate leakage current and maintaining an acceptable dielectric/Si interfacial quality, an N/O stack structure consisting of an oxynitride layer with 33 at.% N and a 3 Å oxide layer is proposed. For a p-MOSFET at an operating voltage of -0.9 V, which is applicable to the 0.7 μm technology node, this structure could be scaled to EOT $= 12$ Å if the maximum allowed gate leakage current is 1 A/cm^2 and EOT $= 9$ Å if the maximum allowed gate leakage current is 100 A/cm^2.

Index Terms—Hole tunneling current, MOSFET, NO stack, scaling limits, silicon oxynitrides, ultrathin gate dielectrics.

I. INTRODUCTION

WITH THE continued scaling of the MOSFET gate dielectric, the resulting performance and reliability issues, such as excessive gate leakage current and boron penetration from the gate electrode, have become a pressing concern. Silicon oxynitride, silicon nitride, and high dielectric constant (high-k) materials have been proposed to replace silicon dioxide as the gate dielectric to address these issues [1], [2]. Due to the difficulties in integrating high-k material processing into the conventional CMOS process and reliability concerns associated with high-k materials, it is most likely for oxynitride/oxide and nitride/oxide stack films to replace pure oxide starting from the 0.13 μm technology node [1], [3]–[5].

Electron tunneling through ultrathin oxynitride/oxide and nitride/oxide stack films has been extensively studied [6]–[9]. To the best of our knowledge, work on hole tunneling through a dielectric was presented only in reference [8], in which hole tunneling in pure nitride is semi-empirically modeled. It has been demonstrated that hole tunneling would dominate the gate leakage current in a p-MOSFET if its gate dielectric is formed by an oxynitride/oxide or nitride/oxide stack under typical gate inversion biases ($|VG| < 1.7$ V) [8], [10]. It is also noteworthy that under such gate biases and if the gate dielectric made of nitride, gate leakage in a p-MOSFET (in which hole tunneling current dominates) becomes higher than that in a n-MOSFET (in which electron tunneling current dominates) [8]. Hole tunneling is more serious in nitride than oxide. This is expected since the hole barrier height (ΔE_V) at the nitride/Si interface is almost 2.6 eV lower than that at the oxide/Si interface, while the electron barrier height (ΔE_C) at the nitride/Si interface is only about 1 eV lower than that at the oxide/Si interface [6], [11]–[13]. As a result the scaling limit of nitride gate dielectric will probably be determined by the gate leakage in a p-MOSFET [10]. Hence there is a need for reliable prediction of the hole tunneling current through ultrathin oxide, oxynitride, nitride and the associated stack gate dielectrics in p-MOSFETs.

In this paper, we explain hole tunneling current through a stack gate dielectric film using a physical model. The results obtained by simulation using this model and results obtained experimentally for hole tunneling current through oxide, nitride, oxynitride/oxide and oxide/oxynitride/oxide stack films are reported and compared. The N concentration in oxynitride and oxynitride/oxide stack films that results in the lowest gate leakage current is obtained theoretically. Finally the theoretical scaling limit of the oxynitride/oxide stack gate dielectric in MOSFETs is obtained.

Manuscript received November 14, 2001; revised February 19, 2002. This work was supported by Singapore Grant NSTB/EMT/TP/00/001.2 and National University of Singapore Grant RP3982754. The review of this paper was arranged by Editor J. Vasi.

H. Y. Yu , Y.-T. Hou, and M.-F. Li are with the Silicon Nano Device Laboratory, Department of Electrical and Computer Engineering, National University of Singapore, Singapore 119260 (e-mail: elelimf@nus.edu.sg).

D.-L. Kwong is with the Department of Electrical and Computer Engineering, University of Texas, Austin, TX 78712 USA.

Publisher Item Identifier S 0018-9383(02)04885-2.

II. MODELING

In our model, gate stack dielectrics are assumed to be made up of several dielectric layers, each layer having a different dielectric constant, gap energy as well as electron and hole barrier height. For a p-MOSFET in inversion, the holes are confined in the inversion layer and form discrete two-dimensional (2-D)

0018-9383/02$17.00 © 2002 IEEE

subbands [14]. For such 2-D subbands, the hole tunneling current can be expressed as [15], [16]

$$J = \sum_n N_n/\tau_n(E_n) \qquad (1)$$

where N_n and τ_n are the hole density and hole lifetime of the nth subband, respectively. The lifetime of a quasi-bound state can be obtained semiclassically by

$$\frac{1}{\tau_n(e)} = \frac{T(E)}{\int_0^{z_n} \sqrt{2m^*_{\perp n}/[E_n - E_V(z)]}\, dz} \qquad (2)$$

where $T(E)$ is the hole transmission probability, E_n the subband energy for the nth quasi-bound state, $E_V(z)$ the edge of the Si valence band, and z_n the classical turning point for the nth bound state. It has been demonstrated that the lifetime evaluated by this formula is comparable to that obtained by rigorous quantum mechanical methods [17], [18].

The subband energies and hole densities in the inversion layer are determined by an improved one-band effective mass approximation (EMA) [15]. One feature of hole quantization that is distinct from electron quantization is the valence band mixing effect [14]. In the traditional one-band EMA [19], [20], the hole effective masses derived from the bulk Si are used. Band mixing effect is neglected in the traditional one-band EMA, thereby resulting in the underestimation of the density of states and overestimation of the quantum mechanical effect [14]. In our improved one-band EMA, the hole effective masses are extracted from the results of the rigorous six-band EMA with the valence band mixing effect included, thereby resulting in a more accurate modeling of the hole electrostatics in the inversion layer. The resulting sub-band energies, hole densities and surface potential for a p-MOSFET at inversion are consistent with those obtained from the more complex six-band EMA.

From (2), $T(E)$ is obtained by a modified Wentzel–Kramers–Brillouin (WKB) approximation. For gate stack dielectrics with N layers, it can be shown that

$$T(E) = T_R(E) \cdot T_1(E) \cdot T_2(E) \cdots T_N(E) \qquad (3)$$

where T_R is a correcting factor accounting for the wave-function reflections at the dielectric-Si barrier interface [21], [22], and $T_i(E)$ the tunneling probability through the ith layer such that

$$T_i(E) = \exp\left(-2\int_{t_i} \kappa_i(E, z)\, dz\right) \qquad (4)$$

where κ_i is the imaginary wave number within the ith layer and t_i the layer thickness.

III. EXPERIMENTS

P$^+$-poly/p-MOSFETs with three different types of gate dielectrics were fabricated using a standard dual-gate CMOS process. The gate dielectric in the first device type (Device 1) was formed by rapid thermal oxidation (RTO). The oxynitride/oxide (N/O) stack-gate dielectric in the second device type (Device 2) was obtained by forming a thin nitride layer

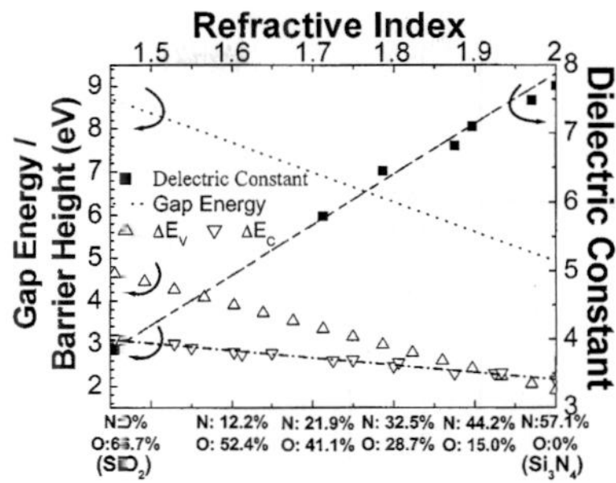

Fig. 1. Variation of the electron and hole barrier height at the oxynitride/Si interface (ΔE_C and ΔE_V), oxynitride gap energy, as well as dielectric constant with the oxynitride composition [6]–[13].

in NH$_3$ at 900 °C for 30 s, followed by oxidation in O$_2$ at 25 atm and 850 °C for 30 min in a vertical high pressure (VHP) furnace [23]. The oxide/oxynitride/oxide (ONO) stack gate dielectric in the third device type (Device 3) was obtained by forming a thin nitride layer in NH$_3$ at 900 °C for 10 s, followed by oxidation in N$_2$O at 950 °C for 30 s [24]. The capacitance–voltage (C–V) measurements were performed on large area (10 000 μm^2) MOS capacitors with an HP 4285A LCR meter at a high-frequency (100 kHz). The equivalent oxide thicknesses (EOT) of all of three types of gate dielectrics were determined in the strong accumulation region of the C–V curves in conjunction with quantum mechanical correction. Negligible flat band voltage shifts (<50 mV) from C–V simulations were revealed for all devices, indicating their negligible low density of oxide charges. The current–voltage (I–V) characteristics were measured using an HP 4156A semiconductor parameter analyzer. Hole tunneling current is obtained from the source/drain current in the carrier separation measurements [10]. In these experiments, the source, drain, and substrate were connected together and grounded.

IV. RESULTS AND DISCUSSION

Fig. 1 [6], [11]–[13] shows that the hole and electron barrier height at the oxynitride/Si interface, the oxynitride gap energy, as well as the dielectric constant vary linearly with the N concentration of the oxynitride film.

Figs. 2–4 show the proposed energy band diagrams of the p-MOSFETs with three different gate dielectric structures used in this study. The hole direct tunneling current formed by holes injected from the valence band of the Si substrate under inversion is also illustrated in each of these diagrams.

For Device 2, the oxynitride/oxide stack gate dielectric structure shown in Fig. 3 is determined by the following way: the inset of Fig. 3 shows the secondary ion mass spectrometry (SIMS) depth profile of N in the N/O stack gate dielectric of Device 2. The physical thickness and N concentration of the oxynitride layer can be determined from the shaded rectangular region and are respectively, 2.06 nm and 16 at.%. The EOT of the oxynitride layer is 2.06 × 3.9/5.2 = 1.55 nm, where

1160   IEEE TRANSACTIONS ON ELECTRON DEVICES, VOL. 49, NO. 7, JULY 2002

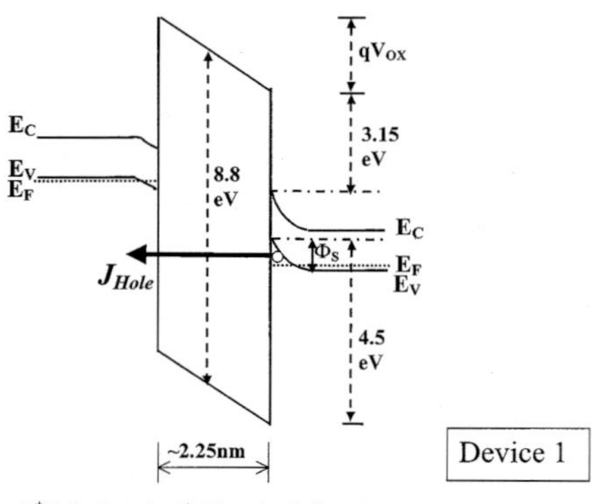

Fig. 2. Energy band diagram of a p-MOSFET with oxide gate dielectric (Device 1). Oxide thickness is determined as 2.25 nm from C–V measurements.

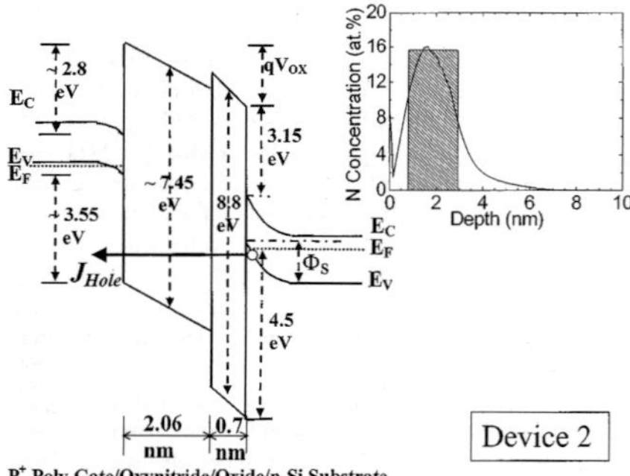

Fig. 3. Energy band diagram of a p-MOSFET with oxynitride/oxide stack gate dielectric (Device 2). Inset shows the SIMS depth profile of N in the oxynitride layer. The thickness and N concentration of the oxynitride layer can be determined from the shaded rectangular region. EOT of the oxynitride is $2.06 \times 3.9/5.2 = 1.55$ nm (3.9 and 5.2 are the dielectric constants obtained from Fig. 1 for the oxide and the oxynitride in device 2, respectively).

3.9 and 5.2 are the dielectric constant of oxide and oxynitride containing 16 at.% N, respectively. On the other hand the total EOT of the gate stack determined by the C–V measurement is 2.25 nm. Therefore, the oxide layer thickness is (2.25–1.55) nm = 0.7 nm, and the physical thickness of the gate stack is (2.06 + 0.7) nm = 2.76 nm, which is close to the physical thickness 2.5 ± 0.2 nm obtained by high resolution transmission electron microscopy (HRTEM) [23].

The energy band diagram for Device 3 as shown in Fig. 4 is obtained from the results of an angle resolved X-ray photoelectron spectroscopy (ARXPS) study of the ONO gate stack dielectric [24] and the C–V measurement, which determined the EOT of the stack to be 1.6 nm. The N concentration in the oxynitride

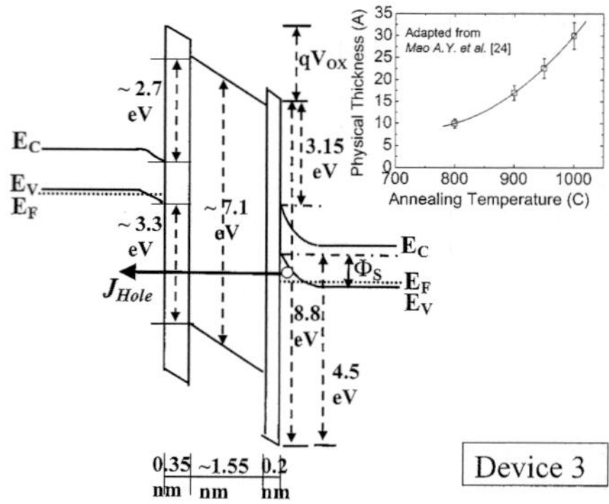

Fig. 4. Energy band diagram of a p-MOSFET with oxide/oxynitride/oxide stack gate dielectric (Device 3). Inset shows the variation of the physical thickness of the dielectric with annealing temperature. The 950 °C annealed gate stack in device 3 has a physical thickness of $\sim$2.25 nm [24]. The N concentration in the oxynitride is estimated to be around 21 at.% according to the XPS measurement. EOT of the oxynitride is $1.55 \times 3.9/5.7 = 1.05$ nm (3.9 and 5.7 are the dielectric constants obtained from Fig. 1 for the oxide and the oxynitride in device 3, respectively).

is estimated to be around 21 at.% based on the XPS measurement. Together with the results shown in Fig. 1, the EOT of the oxynitride layer is $1.55 \times 3.9/5.7 = 1.06$ nm, where 3.9 and 5.7 are the dielectric constant of oxide and oxynitride containing 21 at.% N, respectively. The physical thickness 2.1 nm of this stack structure is consistent with the thickness obtained by XPS measurement of the sample annealed at 950 °C, as indicated in the inset of Fig. 4 [24].

Fig. 5 shows the simulated and measured hole tunneling current through the oxide gate dielectric in Device 1 for an oxide thickness between 1.8 and 2.25 nm, as well as through the nitride gate dielectric in another device type, hereby denoted as Device 4, for an EOT of 1.42 nm [8]. In computing the tunneling current, an accurate determination of κ_i, which is obtained by the dispersion in the dielectric energy gap, is critical. When the oxide thickness is less than 2 nm or for tunneling current at a low gate voltage ($|VG| < 2$ V), a parabolic approximation for hole dispersion in dielectric bandgap is applicable [15], [25]. To obtain the best fit, a parabolic hole dispersion and effective mass of 0.41 m_0 in all the films are used [8]. As shown in this figure, the discrepancy of the slope calculated at higher gate voltages from the experiments might be due to the inaccuracy of the parabolic approximation used to describe the hole dispersion in the films [15].

Fig. 6(a) compares the simulated and measured hole tunneling current through the N/O and oxide gate dielectrics in Devices 1 and 2, respectively, for the same EOT of 2.25 nm. When the gate bias is low ($|VG| < 1.2$ V), the hole tunneling current through the N/O stack gate dielectric is almost one order of magnitude lower than the oxide gate dielectric. Fig. 6(b) shows the measured hole tunneling current through the ONO

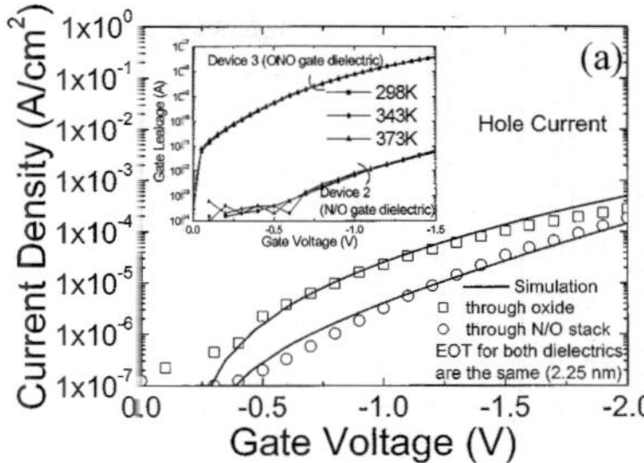

Fig. 5. Comparison of simulated (line) and measured (symbol) hole tunneling current through oxide and nitride gate dielectrics in p-MOSFETs at various dielectric thicknesses. In the simulation, $0.41 m_0$ is used for the hole effective mass in both nitride and oxide [8].

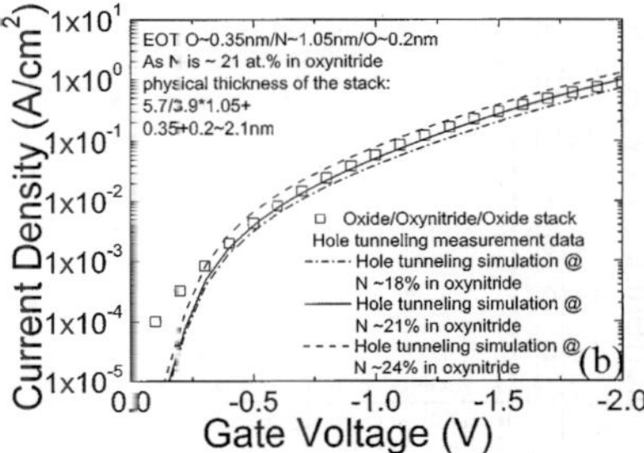

Fig. 6. Comparison of simulated (line) and measured (symbol) hole tunneling current through (a) oxide and oxynitride/oxide stack gate dielectrics in p-MOSFETs a ECT = 2.25 nm. Inset of Fig. 6(a): negligible change of gate leakage through N/O (device 2) and ONO (device 3) stacks with temperature up to 100 °C, indicating that the trap related leakage mechanism such as Frenkel–Poole hopping is not dominant in hole leakage current in our oxynitride samples. (b) Oxide/oxynitride/oxide stack gate dielectric in a p-MOSFET at ECT = 1.6 nm for various N concentrations in the oxynitride layer. In the simulation, $0.41 m_0$ is used as the hole effective mass in both oxide and oxynitride [8].

gate dielectric in Device 3 for an EOT of 1.6 nm. The simulated results are obtained by setting the N concentration to be 18 at.%, 21 at.%, and 24 at.%, respectively. The characteristic that gives the best fit to the measured data is associated with 21 at.% N, and hence the estimated N content (~21 at.%) in the ONO film is justified. The overall good agreement between the tunneling simulation and the experiments indicates that the trap related mechanism such as Frenkel–Poole hopping [6] is not dominant in hole leakage current in our oxynitride samples. In the inset of Fig. 6(a), the leakage currents were measured at different temperatures. The leakage currents are almost temperature independent which further support the tunneling nature of the leakage current and rule out the temperature dependent Frenkel–Poole hopping mechanism. As shown in Fig. 6(a) and (b), the measured gate current is slightly higher than the simulated current at very low gate voltage, probably due to the additional tunneling current to s/d [22], the tunneling current through interface states [26], and noise.

The good agreement between the simulated and measured results for all the four device types as shown in Figs. 5 and 6 therefore corroborates the validity of our physical model.

Using this model, the variation of hole tunneling current with N concentration in an oxynitride gate dielectric is investigated. Fig. 7(a) shows the simulated I–V characteristics at various N concentration for an oxynitride gate dielectric with EOT = 2.25 nm. Fig. 7(b) shows the simulated hole tunneling current at $V_G = -1$ V as a function of film composition for an oxynitride gate dielectric with various EOT. It can be readily seen that when $|VG| < 2$ V, the hole tunneling current through an oxynitride film with 33 at.% N is the smallest. This may be attributable to the competitive effects of the increase in actual thickness and decrease in hole barrier height with N concentration in an oxynitride film at a given EOT. In contrary, the electron tunneling current decreases monotonically with N concentration in an oxynitride film [6], which may be attributable to the smaller decrease in the electron barrier height with N concentration.

Fig. 8 shows the simulated hole tunneling current at $V_G = -1$ V as a function of film composition for an N/O stack film with EOT = 2.25 nm but formed by various oxynitride and oxide thicknesses. It can be readily seen that the hole tunneling current decreases monotonically with an increase in the oxynitride thickness when the N concentration in the oxynitride layer is kept constant. Furthermore the minimum hole tunneling current through an N/O stack film is again obtainable at 33 at.% N.

The interface state density at an oxynitride/Si interface is higher than that at an oxide/Si interface and the presence of N at a dielectric/Si interface degrades the peak channel mobility, which explains the rationale behind the N/O stack scheme for gate dielectrics [27], [28]. Together with our results, we therefore propose an optimized N/O stack gate dielectric being made up by an oxynitride layer with 33 at.%. N and an interfacial oxide buffer layer with a minimum thickness of 0.3 nm.

Fig. 9 shows the gate leakage current as a function of EOT at various operating voltages for the proposed N/O stack. Our calculation further indicates that under low gate biases ($|VG| < 2$ V), for the same total EOT, the electron tunneling current through this optimized N/O stack in n-MOSFET is lower than

1162   IEEE TRANSACTIONS ON ELECTRON DEVICES, VOL. 49, NO. 7, JULY 2002

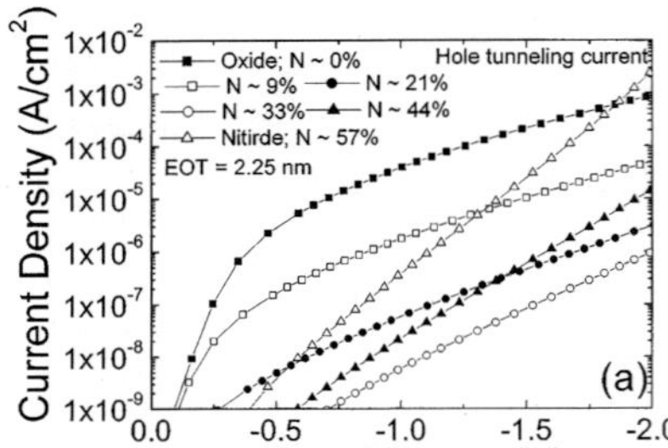

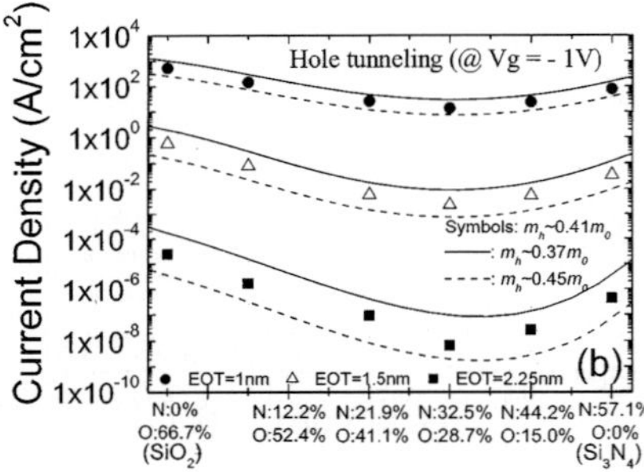

Fig. 7. (a) Simulated hole tunneling current through oxynitride gate dielectric in a p-MOSFET at EOT = 2.25 nm for various N concentrations in the dielectric. (b) Simulated hole tunneling current through oxynitride gate dielectric in a p-MOSFET at $V_G = -1$ V for various EOT and N concentrations in the dielectric. The effect of $\pm 10\%$ variation of oxynitride hole effective mass value ($0.41m_0$) [8] on hole tunneling current is demonstrated in this figure (solid and dashed lines). Hole tunneling current is lowest through the oxynitride with ~33 at.% of N for all of the cases.

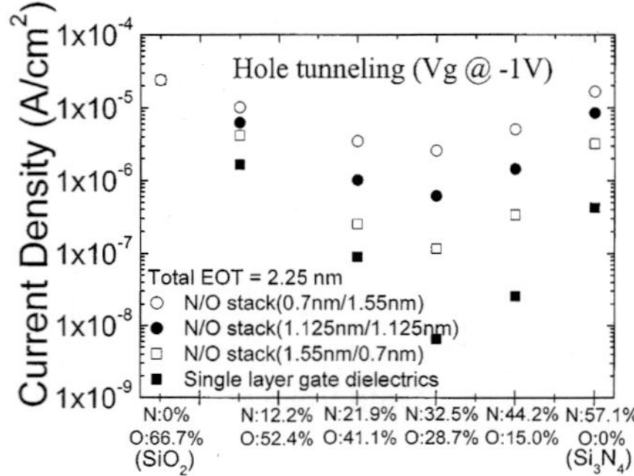

Fig. 8. Simulated hole tunneling current through oxynitride/oxide stack gate dielectric in a p-MOSFET at $V_G = 1.0$ V and EOT = 2.25 nm for various combinations of oxynitride and oxide thicknesses and N concentrations in the oxynitride layer (EOT of oxynitride /EOT of oxide data are given in the brackets).

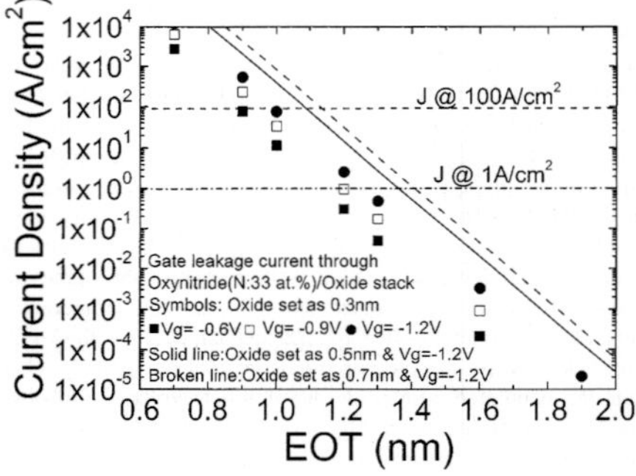

Fig. 9. Variation of the gate leakage current through an oxynitide/oxide stack gate dielectric formed by an oxynitride layer with 33 at.% N and a 0.3-nm-thick oxide buffer layer in a p-MOSFET with EOT at various operating voltages. The similar simulations with buffer oxide layer changed to 0.5 nm and 0.7 nm are also shown as references.

the hole tunneling current in p-MOSFET. As a result, we can use hole tunneling current shown in Fig. 9 to project the EOT scaling limit of the N/O gate stack dielectrics used in MOSFETs. It can be seen from Fig. 9 that at an operating voltage of -0.9 V applicable to the 0.7 μm technology node that is projected to be realizable by 2008 [33], the optimized stack structure could be scaled to EOT = 1.2 nm if the maximum tolerable gate leakage current is 1 A/cm² [29], [30] and EOT = 0.9 nm if the maximum tolerable gate leakage current is 100 A/cm² [31], [32].

V. CONCLUSIONS

Hole tunneling current through four different ultrathin gate dielectrics—oxide, nitride, N/O and ONO—in p-MOSFETs is explained using a physical model. The validity of this model is verified by the excellent agreement between the simulated and experimental results. Under typical inversion biases ($|VG| < 2$ V), an oxynitride film containing about 33 at.% N exhibits the lowest hole tunneling current, which predominates the gate

leakage current when $|VG| < 1.7$ V. The hole tunneling current through N/O stack films decreases with increasing oxynitride thickness for a given EOT and N concentration. To minimize gate leakage current and maintain an acceptable dielectric/Si interfacial quality, an optimized N/O stack structure consisting of an oxynitride layer with 33 at.% N and a 3 Å oxide layer is proposed. Under low gate biases ($|VG| < 2$ V) for the same total EOT, the electron tunneling current through this optimized N/O stack in n-MOSFET is lower than the hole tunneling current in p-MOSFET. Hole tunneling current is therefore used to project the EOT scaling limit of the N/O gate stack dielectrics used in MOSFETs. For a p-MOSFET at an operating voltage of -0.9 V, this stack structure could be scaled to EOT = 1.2 nm if the maximum allowed gate leakage current is 1 A/cm² and EOT = 0.9 nm if the maximum allowed gate leakage current is 100 A/cm².

ACKNOWLEDGMENT

The authors would like to thank W. H. Lai, Charted Semiconductor (CSM), for useful discussion and providing the CMOSFET wafer with 1.8 nm gate oxide.

REFERENCES

[1] G. D. Wilk, R. M. Wallace, and J. M. Anthony, "High-K gate dielectrics: Current status and materials properties considerations," *J. Appl. Phys.*, vol. 89, no. 10, pp. 5243–5275, 2001.

[2] A. I. Kingon, J. P. Maria, and S. K. Streiffer, "Alternative dielectrics to silicon dioxide for memory and logic devices," *Nature*, vol. 406, pp. 1032–1038, 2000.

[3] M. L. Green, E. P. Gusev, R. Degraeve, and E. L. Garfunkel, "Ultrathin ($<$4 nm) SiO_2 and Si–O–N gate dielectric layers for silicon microelectronics: Understanding the processing, structure, and physical and electrical limits," *J. Appl. Phys.*, vol. 90, no. 5, pp. 2057–2121, 2001.

[4] D. A. Buchanan, "Scaling the gate dielectric: Materials, integration, and reliability," *IBM J. Res. Develop.*, vol. 43, no. 3, pp. 245–264, 1999.

[5] H. Iwai, "Direction of silicon technology from past to future," in *8th Int. Symp. Physical and Failure Analysis of Integrated Circuits*, Singapore, 2001, pp. 1–35.

[6] X. Guo and T. P. Ma, "Tunneling leakage current in oxynitride: Dependence on oxygen/nitrogen content," *IEEE Electron Device Lett.*, vol. 19, p. 207, June 1998.

[7] Y. Shi, S. Wang, and T. P. Ma, "Tunneling leakage current in ultrathin ($<$4 nm) nitride/oxide stack dielectrics," *IEEE Electron Device Lett.*, vol. 19, p. 388, Oct. 1998.

[8] Y. C. Yeo, Q. Lu, W. C. Lee, T. J. King, C. Hu, X. Wang, X. Guo, and T. P. Ma, "Direct tunneling gate leakage current in transistors with ultrathin silicon nitride gate dielectric," *IEEE Electron Device Lett.*, vol. 21, p. 540, Nov. 2000.

[9] E. M. Vogel, K. Z. Ahmed, B. Hornung, W. K. Henson, P. K. McLarty, G. Lukovsky, J. R. Hauser, and J. J. Wortman, "Model tunnel currents for high dielectric constant dielectrics," *IEEE Trans. Electron Devices*, vol. 45, pp. 1350–1355, July 1998.

[10] Y. Shi, T. P. Ma, S. Prasad, and S. Dhanda, "Polarity dependent gate tunneling currents in dual-gate CMOSFETs," *IEEE Trans. Electron Devices*, vol. 45, pp. 2355–2360, Dec. 1998.

[11] D. M. Brown, P. V. Gary, F. K. Heumann, H. R. Philipp, and E. A. Taft, "Properties of $Si_xO_yN_z$ films on Si," *J. Electrochem. Soc.*, vol. 115, no. 3, p. 311, 1968.

[12] V. A. Gritenko, N. D. Dikovskaja, and K. P. Moginikov, "Band diagram and conductivity of silicon oxynitride films," *Thin Solid Films*, vol. 51, p. 353, 1978.

[13] Y. Y. Chen, M. Dardner, J. Fulford, D. Wristers, A. B. Joshi, L. Chung, and D. L. Kwong, "Enhanced hot hole degradation in p$^+$-poly p-MOSFETs with oxynitride gate dielectrics," in *Int. Symp. VLSI Technology, Systems, and Applications*, Taipei, Taiwan, R.O.C., 1999, pp. 86–89.

[14] Y. T. Hou and M. F. Li, "Hole quantization effects and threshold voltage shift in pMOSFET—Assessed by improved one-band effective mass approximation," *IEEE Trans. Electron Devices*, vol. 48, pp. 1188–1193, June 2001.

[15] Y. T. Hou, M. F. Li, W. H. Lai, and Y. Jin, "Modeling and characterization of direct tunneling hole current through ultrathin gate oxide in p-metal–oxide–semiconductor field-effect transistors," *Appl. Phys. Lett.*, vol. 78, pp. 4034–4036, 2001.

[16] Y. T. Hou, M. F. Li, Y. Jin, and W. H. Lai, "Direct tunneling hole currents through ultrathin gate oxides in metal–oxide–semiconductor devices," *J. Appl. Phys.*, vol. 91, pp. 258–264, 2002.

[17] W. K. Shih, E. X. Wang, S. Jallepalli, F. Leon, C. M. Maziar, and A. F. Tasch, "Modeling gate leakage current in nMOS structures due to tunneling through an ultra-thin oxide," *Solid-State Electron.*, vol. 42, pp. 997–1006, 1998.

[18] A. Dalla Serra, A. Abramo, P. Palestri, L. Selmi, and F. Widdershoven, "Closed- and open-boundary models for gate-current calculation in n-MOSFETs," *IEEE Trans. Electron Devices*, vol. 48, pp. 1811–1815, Aug. 2001.

[19] C. Moglestue, "Self-consistent calculation of electron and hole inversion charges at silicon–silicon dioxide interfaces," *J. Appl. Phys.*, vol. 59, no. 5, pp. 3175–3183, 1986.

[20] C.-Y. Wu, S. Banerjee, K. Sadra, B. G. Streetman, and R. Sivan, "Quantization effects in inversion layers of pMOSFETs on Si (100) substrate," *IEEE Electron Device Lett.*, vol. 17, pp. 276–278, June 1996.

[21] L. F. Register, E. Rosenbaum, and K. Yang, "Analytic model for direct tunneling current in polycrystalline silicon-gate metal–oxide–semiconductor devices," *Appl. Phys. Lett.*, vol. 74, pp. 457–459, 1999.

[22] J. Cai and C. T. Sah, "Gate tunneling currents in ultrathin oxide metal–oxide–silicon transistors," *J. Appl. Phys.*, vol. 89, pp. 2272–2285, 2001.

[23] T. Y. Luo, E. N. Al-Shareef, A. Karamcheti, V. H. C. Watt, G. A. Brown, M.-D. Jackson, H. R. Huff, B. Evans, and D. L. Kwong, "High performance PMOS devices using ultra-thin VHP oxynitride," *Ext. Abst. SSDM*, pp. 178–179, 2000.

[24] A. Y. Mao, J. Lozano, J. M. White, and D. L. Kwong, "N_2O oxidation kinetics of ultra thin thermally grown silicon nitride: An angle resolved X-ray photoelectron spectroscopy study," in *Proc. MRS Spring Meeting*, San Francisco, CA, 1999.

[25] K. N. Yang, H. T. Huang, M. C. Chang, C. M. Chu, Y. S. Chen, M. J. Chen, Y. M. Lin, M. C. Yu, S. M. Jang, D. C. H. Yu, and M. S. Liang, "A physical model for hole direct tunneling current in p$^+$ poly-gate pMOSFETs with ultrathin gate oxides," *IEEE Trans. Electron Devices*, vol. 47, pp. 2161–2166, Nov. 2000.

[26] A. Ghetti, L. Bude, and G. Weber, "T_{BD} prediction from measurements at low field and room temperature using a new estimator," in *VLSI Tech. Dig.*, 2000. pp. 218–219.

[27] N. Ikarashi and K. Watanabe, "Atomic structures at a Si–nitride/Si(001) interface," *J. Appl. Phys.*, vol. 90, no. 6, pp. 2683–2688, 2001.

[28] C. G. Parker, G. Lucovsky, and J. R. Hauser, "Ultrathin oxide–nitride gate dielectric MOSFETs," *IEEE Electron Device Lett.*, vol. 19, pp. 106–108, Apr. 1998.

[29] S. H. Lo, D. A. Buchanan, Y. Taur, and W. Wang, "Quantum-mechanical modeling of electron tunneling current from the inversion layer of ultra-thin-oxide nMOSFETs," *IEEE Electron Device Lett.*, vol. 18, pp. 209–211, may 1997.

[30] Y. Taur and E. F. Nowak, "CMOS devices below 0.1 μm: How high will performance go?," in *IEDM Tech. Dig.*, 1998, pp. 789–792.

[31] T. Ghani, E. Mistry, P. Packan, S. Thompson, M. Stettler, S. Tyagi, and M. Bohr, "Scaling challenges and device design requirements for high performance sub-50 nm gate length planar CMOS transistors," in *Proc. Symp. VLSI Technology*, Honolulu, HI, 2000, pp. 174–175.

[32] D. J. Frank, R. H. Dennard, E. Nowak, P. M. Solomon, Y. Taur, and H. P. Wong, "Device scaling limits of Si MOSFETs and their application dependences," *Proc. IEEE*, vol. 89, pp. 259–288, Mar. 2001.

[33] "The International Technology Roadmap for Semiconductors," Semicond. Ind. Assoc., Austin, TX, 1999.

Hongyu Yu (S'01) received the B.Eng. degree from Tsinghua University, Beijing, China, in 1999, and the M.Asc. degree from the University of Toronto, Toronto, ON, Canada, in 2001, respectively. He is currently pursuing the Ph.D. degree at the National University of Singapore.

His research interests include processing, characterization, and reliability of advanced gate dielectrics for future generation of devices.

Yong-Tian Hou (S'01) received the B.S. and M.S. degrees in physics from Peking University, Beijing, China, in 1990 and 1993, respectively. He is currently pursuing the Ph.D. degree in electrical engineering at National University of Singapore, Singapore.

From 1998 to 1999, he was with the Department of Electrical and Computer Engineering, National University of Singapore, as a Research Engineer on optical characterization of III-nitride materials. Currently, his research interests are the quantum mechanical effect, direct tunneling current, and reliability physics in deep submicron CMOS devices.

1164

IEEE TRANSACTIONS ON ELECTRON DEVICES, VOL. 49, NO. 7, JULY 2002

Ming-Fu Li (M'91–SM'99) graduated from the Department of Physics, Fudan University, Shanghai, China, in 1960.

After graduation, he joined the University of Science and Technology of China (USTC) as a Teaching Assistant and then Lecturer. In 1978, he joined the Graduate School Faculty, Chinese Academy of Sciences, Beijing, China, and became a Professor in 1986. He has also served as Adjunct Professor at the Institute of Semiconductors, Chinese Academy of Sciences, Fudan University, and USTC, Hefei. He was a Visiting Scholar at Case Western Research University, Cleveland, OH, in 1979, and at the University of Illinois, Urbana, from 1979 to 1981, and was a Visiting Scientist at the University of California, Berkeley, and Lawrence Berkeley National Laboratories from 1986 to 1987, 1990 to 1991, and 1993, respectively. He joined the Department of Electrical Engineering, National University of Singapore, Singapore, in 1991, and became a Professor in 1996. His current research interests are in the areas of reliability and quantum modeling of deep submicron CMOS devices and analog CMOS design. He has published over 180 research papers and two books, including *Modern Semiconductor Quantum Physics* (World Scientific: Singapore, 1994).

Dr. Li has served on several international program committees and advisory committees in international semiconductor conferences in China, Japan, Canada, Germany, and Singapore.

Dim-Lee Kwong (A'84–SM'90) received the B.S. degree in physics and the M.S. degree in nuclear engineering from the National Tsing Hua University, Taiwan, R.O.C., in 1977 and 1979, respectively, and the Ph.D. degree in electrical engineering from Rice University, Houston, TX, and receiving Best Dissertation Award, in 1982.

He was an Assistant Professor with the Electrical Engineering Department, University of Notre Dame, South Bend, IN, from 1982 to 1985. He was a Visiting Scientist at the IBM General Technology Division, Essex Junction, VT, during the summer of 1985, working on 4 Mb DRAM technology. He joined the Microelectronics Research Center and Department of Electrical and Computer Engineering, The University of Texas, Austin, in 1985 as an Assistant Professor. He was promoted to Associate Professor in 1985 and to Full Professor in 1990. He is the author of more than 470 referred archival publications and has been awarded more than 20 U.S. patents. His current areas of research interests include rapid thermal CVD technology for the growth and deposition of semiconductor materials compatible with ULSI processes, advanced dielectrics for logic, analog, and memory devices, metal gate electrode, shallow junctions, and high-K dielectrics. Thirty-five students received their Ph.D. degrees under his supervision.

Dr. Kwong received numerous awards, including the IBM Faculty Development Award in 1984 and the Engineering Foundation Teaching Award from the University of Texas, Austin, in 1994, and holds the Earl N. and Margaret Brasfield Endowed Fellowship.

1284 IEEE TRANSACTIONS ON ELECTRON DEVICES, VOL. 50, NO. 5, MAY 2003

Improved One-Band Self-Consistent Effective Mass Methods for Hole Quantization in p-MOSFET

Tony Low, Yong-Tian Hou, *Student Member, IEEE*, and Ming-Fu Li, *Senior Member, IEEE*

Abstract—An improved one-band self-consistent effective mass approximation (EMA) for hole quantization in p-MOSFET is presented. It is developed by extracting empirically a set of hole-effective masses based on the rigorous self-consistent six-band EMA. It is found that the self-consistent model using such improved one-band effective masses can provide accurate hole quantization characteristics. For further simplification, the triangular well approximation is also assessed. Fairly accurate MOS electrostatics is also obtained if introducing an effective field in the inversion layer in triangular well approximation. However, the triangular well approximation has its limitation in describing the hole centroid. In essence, the shorter computing time of the proposed improved one-band methods without sacrificing the accuracy of MOS electrostatics provides its potential in device modeling for hole quantization.

Index Terms—Hole quantization, improved one-band, MOSFET, self-consistent.

I. INTRODUCTION

WITH THE miniaturization of CMOS devices, quantum mechanical effects sets in and results in nonclassical behavior. The study of carrier quantization in the silicon surface region becomes essentially important for accurate modeling of CMOS device operation. A charge control problem based on the one-band effective mass approximation (EMA) self-consistent calculation for NMOS has been well developed and verified [1]–[3]. Such a traditional one-band EMA, which uses constant effective masses from bulk silicon, was also applied to hole quantization [2], [4], [5]. As we have pointed out in [6] and [7], such a treatment is incorrect in physics because a strong electric field will render the symmetry of the point group of the Si Hamiltonian to be reduced, which results in splitting as well as mixing between nearly degenerate valence bands [8]. Based on that, we have recently proposed an improved one-band EMA calculation for p-MOS, employing the use of empirically extracted effective masses, which are calculated by using a periodic zigzag potential approximation to the six-band Hamiltonian [6], [7]. In this paper, we discuss the use of improved one-band EMA with empirical masses extracted based on the rigorous six-band self-consistent calculation and an assessment is given to such an improved one-band EMA with potential profile in inversion layer treated by either self-consistent method

or triangular well approximation. The applicability, limitations, and improvements of the methods will be highlighted in this paper.

II. SIX-BAND SELF-CONSISTENT CALCULATIONS

A one-dimensional (1-D) Si–SiO$_2$ MOS system on a $\langle 100 \rangle$ silicon substrate is simulated. The potential profile of the p-MOS is solved self-consistently using the coupled Schrodinger and Poison equation [1].

Assuming z along the direction perpendicular to the Si/SiO$_2$ interface, the total charge density $\rho(z)$ in the 1-D Poisson equation for $z > 100$ nm, which is far away from the interface at $z = 0$, is obtained as follows:

$$\rho(z) = q(N_s(z) - n_{\text{free}}(z) + p_{\text{free}}(z)) \tag{1}$$

where $N_s(z)$ is the n-substrate doping concentration assumed to be fully ionized at room temperature. n_{free} and p_{free} are electron and hole densities calculated by classical method. For $0 < z < 100$ nm, holes are considered a quantized two-dimensional charge

$$\rho(z) = q(N_s(z) - n_{\text{free}}(z) + p_{qm}(z)). \tag{2}$$

$p_{qm}(z)$ is the quantized hole density and is the major topic of concern in this paper.

The hole energies and wave functions can be calculated by using the six-band Hamiltonian stated in [6], [9]. The Luttinger parameters [10] used in the six-band Hamiltonian for Si are: $\gamma_1 = 4.22$, $\gamma_2 = 0.39$, $\gamma_3 = 1.44$. $\Delta_0 = 44$ meV [8], [11] is the spin-orbit-splitting energy of bulk silicon. The boundary condition is defined to be infinite potential at the oxide-silicon interface and deep into the substrate (taken to be 2000 nm). The hole energy $E_i(k_x, k_y)$ and wave function $U_i(E_i, z)$ calculated are then used to compute $p_{qm}(z)$ as follows:

$$p_{qm}(z) = \frac{1}{(2\pi)^2} \int_{k_x, k_y} \sum_{Ei} |U_i(E_i(k_x, k_y), z)|^2$$
$$\cdot \left(1 + \exp\left(\frac{E_i(k_x, k_y) - E_f}{kT} \right) \right) dk_x \, dk_y. \tag{3}$$

With the carrier densities calculated by (1)–(3), we can calculate for the new voltage by Poisson equation, which is then fed back into the six-band Hamiltonian. An iterative process then brings the Poisson equation into convergence.

A. Valence Band Structure

The quantized subbands obtained by the six-band Hamiltonian demonstrate anisotropy as illustrated in Fig. 1. The

Manuscript received October 8, 2002; revised February 21, 2003. This work was supported by a Singapore A*STAR/EMT/TP/00/001.2 research grant and by the National University of Singapore under Grant R263-000-077-112. The review of this paper was arranged by Editor S. Datta.

The authors are with the Silicon Nano Device Lab, Department of Electrical and Computer Engineering, National University of Singapore, 119260 Singapore (e-mail: elelimf@nus.edu.sg).

Digital Object Identifier 10.1109/TED.2003.813469

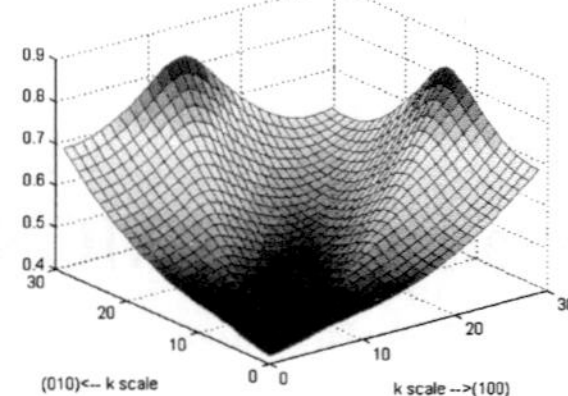 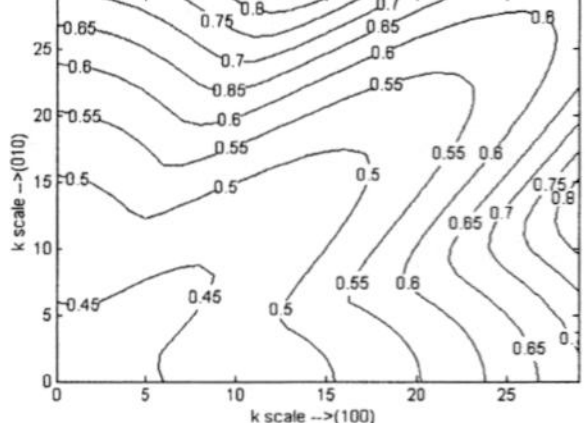

Fig. 1.   Contour plot for the lowest subbands $n = 1$, excluding its nearly degenerate pair.

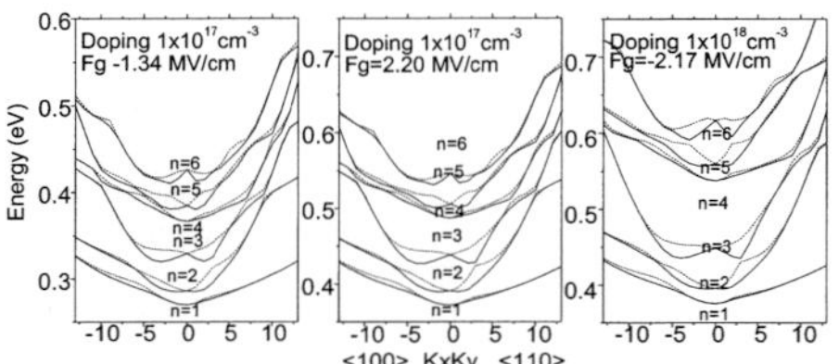

Fig. 2.   In-plane dispersion of the six lowest subbands, for different surface electric field F_s at the same doping for the left and center figures, and for a similar surface electric field F_s at different dopings for the center and right figures.

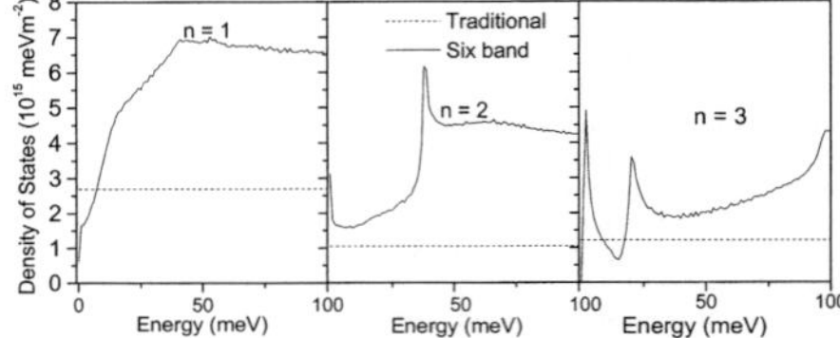

Fig. 3.   DOS profile of the six-band and traditional one-band (horizontal dashed lines) versus energy at silicon surface potential of 1.4 V.

in-plane dispersions of the six lowest subbands are also shown in Fig. 2. The Γ point ($\mathbf{k} = 0$) is characterized by reversed camel back structures (two reversed peaks) [12] with negative hole effective mass. Proper modeling of the dispersion near the Γ point is important as it provides the highest hole occupation. Such a negative hole mass has to be taken into account, especially in the simulation of hole transport, such as hole surface mobility in the channel of p-MOSFETs.

The dispersion for different doping and surface electric field is also illustrated in Fig. 2. They maintain similar dispersion characteristic for the first three subbands with field dependent amount of subband splitting. It is also apparent from the graph that the quantization energies for the first three subbands are almost independent of the doping level. However, higher energy levels $n = 4, 5, 6$ show larger energies due to the steepness of the potential well for higher doped substrates.

B. Hole-Carrier Distribution

The simulated results of density of states (DOSs) of the three lowest subbands are shown in Fig. 3. The DOS profiles deviates from the step-like function [1], [6] used in the traditional one-band modeling where the bulk Si effective masses used for the heavy, light, and split-off holes are 0.65, 0.25, and 0.29, respectively [6]. Its profile exemplifies the characteristic of its dispersion subjected to a particular potential well. The distinct peaks in DOS in Fig. 3 correspond to the camel-back structures in the band diagram in Fig. 2. The DOSs obtained from traditional one-band EMA are also shown in Fig. 3, which overall underestimates the DOSs.

The projection functions of n subband on $x = hh$, lh or bulk band P_x^n [6], [9] of the first three subbands are also shown in Fig. 4. Near the Γ point, the $n = 1$ subband is a purely heavy hole. The $n = 2$ subband is mainly light holes with some mixing with a split-off band. The $n = 3$ subband is mainly split-off having some mixing with a light hole. As the electric field is increased, the deeper potential wells will cause the holes to be strongly confined, resulting in stronger coupling between light hole and heavy hole bands as shown in the lower part of Fig. 4.

Comparing Figs. 2–4 with the similar results obtained in [6], we confirm that the simplified method using a zigzag potential proposed in [6], [7] can obtain overall important characteristics of the quantized hole band structures in a pMOS under inversion; however, it has some limitations as will be discussed in the following sections.

III. IMPROVED SELF-CONSISTENT ONE-BAND EMA

As indicated in our previous study [7], an improved one-band EMA can be used to obtain fairly accurate hole quantization data if we extract the hole effective masses empirically by comparing with the zigzag potential six-band EMA calculation results.

In this work, the DOS effective mass and quantization effective mass in improved one-band EMA (self-consistent calculation) are obtained empirically from the six-band self-consistent calculation results. This is achieved by adjusting the quantization mass in the one band Schrodinger equation to arrive at the correct subband energy (most importantly the first subband energies) obtained from six-band EMA results; whereas, the DOS effective mass is adjusted to arrive at the subband hole density result obtained from six-band EMA. The empirical masses obtained as such are plotted in Fig. 5. Our simulated

1286   IEEE TRANSACTIONS ON ELECTRON DEVICES, VOL. 50, NO. 5, MAY 2003

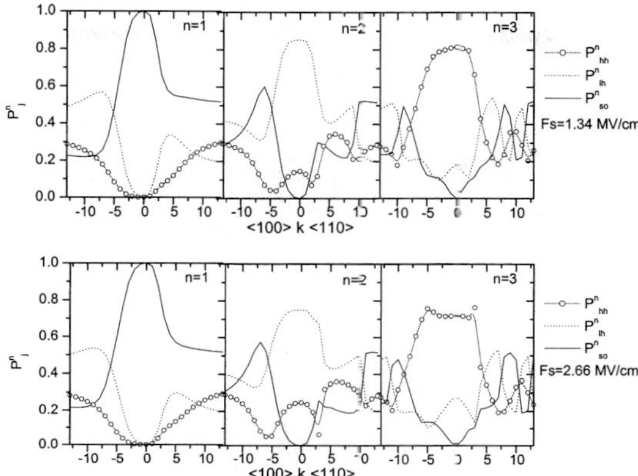

Fig. 4.   Projection function [6], [9] for the first three subbands subject to different surface electric field. $F_s = 1.34$ MV/cm for top graphs and $F_s = 2.66$ MV/cm for bottom graphs.

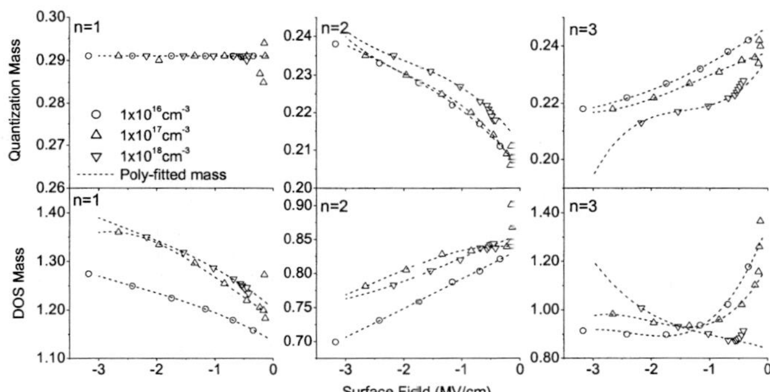

Fig. 5.   Empirical quantization effective masses and DOS effective masses in improved one-band EMA versus surface electric field at various doping levels. Dashed lines denote its respective poly-fitted effective masses.

result shows dependencies of quantization and DOS effective masses on doping and surface field. However, the quantization effective masses of the $n = 1$ hole band can be assumed to be independent of electric field and doping as shown. This is due to the purity of the $n = 1$ hole band. With this, a set of electric field and doping dependent effective masses is proposed, together with approximate constant effective masses in Table I.

As shown in Fig. 6, such an improved self-consistent one-band EMA can achieve fairly good results on the subband energy levels, carrier occupations and surface potentials, hole centroid, potential and carrier profiles into depth as compared with the numerical results of six-band simulation. The improved one-band results above are obtained from set of constant empirical effective mass. A more accurate result is obtainable with the fitted polynomials.

IV. TRIANGULAR POTENTIAL APPROXIMATION

Triangular potential well approximation is often used because it yields an analytic solution of the Schrodinger equation. The solution of Schrodinger equation in triangular potential well is documented in [1], where the jth energy level in the ith valley is given as

$$E_{ij} = \left(\frac{\hbar^2}{2m_i^*}\right)^{1/3}\left[\frac{3}{2}\pi q F_s\left(j + \frac{3}{4}\right)\right]^{2/3}.\qquad (4)$$

The triangular well approximation has been used to obtain the MOS electrostatics [13], but it has been shown that triangular well approximation is invalid for strong inversion if the surface electric field is used as the electric field in triangular well [2]. In order to extend the triangular well approximation into strong

TABLE I

EMPIRICAL EFFECTIVE MASSES. SURFACE ELECTRIC FIELD F_s IS EXPRESSED IN MV/cm IN ITS NEGATIVE FORM.
CONSTANT EMPIRICAL MASSES TAKEN AT SURFACE FIELD OF -2 MV/cm

Doping Density (cm^{-3})			Bulk Si	Field Dependent Empirical	Constant Empirical
1x10^{16} cm^{-3}	Quantization Mass	n=1	0.29	0.291	0.291
		n=2	0.20	$-0.0006F_s^3-0.0047F_s^2-0.0196F_s+0.2051$	0.230
		n=3	0.29	$0.0001F_s^3+0.0021F_s^2+0.0149F_s+0.2469$	0.225
	DOS Mass	n=1	0.65	$-0.0021F_s^3-0.0155F_s^2-0.0721F_s+1.1356$	1.235
		n=2	0.25	$-0.0002F_s^3-0.0006F_s^2+0.0429F_s+0.8348$	0.748
		n=3	0.29	$0.0409F_s^3+0.2905F_s^2+0.6500F_s+1.3577$	0.893
1x10^{17} cm^{-3}	Quantization Mass	n=1	0.29	0.291	0.291
		n=2	0.20	$-0.0018F_s^3-0.0103F_s^2-0.0268F_s+0.2036$	0.230
		n=3	0.29	$-0.0007F_s^3-0.0024F_s^2+0.0061F_s+0.2377$	0.222
	DOS Mass	n=1	0.65	$0.0065F_s^3+0.0145F_s^2-0.0725F_s+1.1865$	1.338
		n=2	0.25	$-0.0027F_s^3-0.0202F_s^2-0.0133F_s+0.8373$	0.805
		n=3	0.29	$0.0403F_s^3+0.2480F_s^2+0.4517F_s+1.1859$	0.952
1x10^{18} cm^{-3}	Quantization Mass	n=1	0.29	0.291	0.291
		n=2	0.20	$-0.0015F_s^3-0.0086F_s^2-0.0223F_s+0.2115$	0.234
		n=3	0.29	$0.0048F_s^3+0.0207F_s^2+0.0329F_s+0.2362$	0.215
	DOS Mass	n=1	0.65	$-0.0027F_s^3-0.0209F_s^2-0.1007F_s+1.2027$	1.342
		n=2	0.25	$-0.0015F_s^3-0.0045F_s^2+0.0314F_s+0.8569$	0.788
		n=3	0.29	$-0.0175F_s^3-0.0386F_s^2-0.0792F_s+0.8382$	0.982

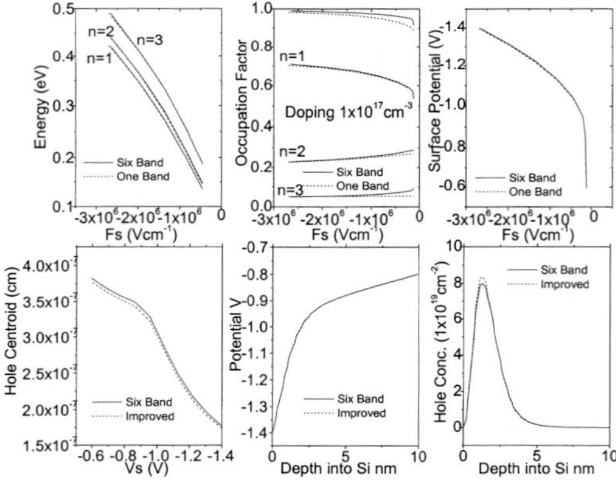

Fig. 6. (Top) The subband energies, occupation factor and surface potential versus surface electric field with self-consistent improved one-band EMA. (Bottom) The hole carrier centroid at different surface potential, the potential and carrier profile at surface potential 1.4 V with self-consistent improved one-band EMA.

inversion, an effective field is introduced, instead of the surface electric field. This electric field can be regarded as the mean field in the inversion layer. It can be expressed as

$$F_s = F_{\text{eff}} = q\left(\eta N_s + N_d\right)/\varepsilon_{si} \qquad (5)$$

where N_s and N_d are inversion carrier sheet density and depletion charge sheet density, respectively. η is a weighting coefficient to be properly chosen. For the modeling of the universality of the mobility of inversion layer carriers (for electrons), η is taken to be 0.5 [14], [15]. However, in the calculation of carrier sheet density, η is calculated to be 0.75 for inversion (electron) and 0.8 for accumulation layer (hole) in NMOS [16]. With the subband energy levels from the six-band result, we are able to obtain the value for η with (4) and (5). The weighting coefficient η for different doping over a range of electric fields is tabulated in Fig. 7.

Based on the data for the two lower concentrations, η is calculated to be 0.78 and 0.75 for $n = 1$ and 2 hole subbands, respectively, at a doping of 1×10^{17} cm^3. Apparently, η for $n = 3$ hole subband is an increasing function of surface field. Its strong dependency on the electric field is a characteristic of poor triangular well fitting for higher energy levels. Lower energy levels are able to achieve a relatively constant η.

Using the parameterized self-consistency [13] method for the triangular model, we are able to obtain the MOS electrostatics efficiently. Using the constant effective masses (improved one-band) as shown in Table I, a weighting coefficient of 0.77 (an

1288   IEEE TRANSACTIONS ON ELECTRON DEVICES, VOL. 50, NO. 5, MAY 2003

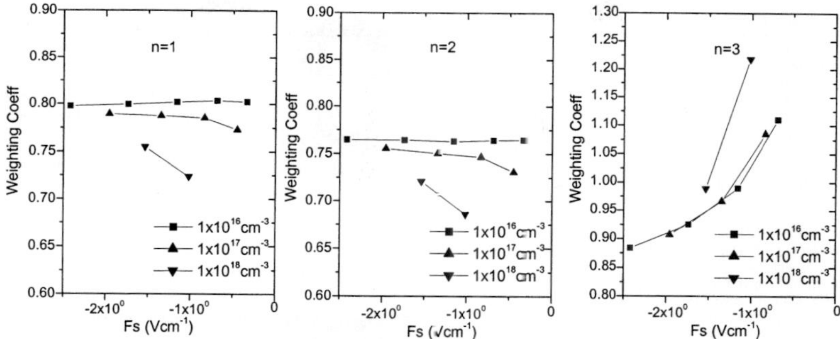

Fig. 7.   Weighting coefficient η for $n = 1, 2, 3$ hole subbands, respectively, for different doping concentration.

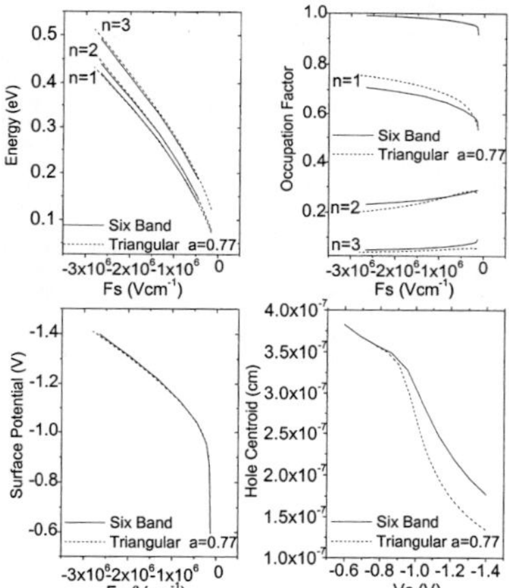

Fig. 8.   Results from improved one-band triangular approximation and six-band self-consistent EMA. Subband energies, occupation factor, and surface potential versus surface electric field. Hole carrier centroid at different surface potential V_s for improved one-band triangular approximation and self-consistent six-band.

average of $n = 1$, and $n = 2$), good fit for the subband energies, occupation factor, and surface potential can be obtained as shown in Fig. 8. Fig. 8 shows the improved one-band self-consistent calculation yields a good result for the hole carrier centroid. However, the for triangular model, the hole centroid falls short of the full band result at strong inversion although the results are consistent in depletion or weak inversion. This indicates that the triangular well approximation fails to describe the hole distribution in the inversion layer. Generally, the triangular gives a smaller hole centroid compared to the full-band self-consistent method. This point is similar to that found in electron quantization in nMOS [16].

V. CONCLUSION

In this paper, we have developed an improved self-consistent one-band EMA for accurate hole quantization by comparing the results with the rigorous six-band EMA self-consistent calculations. The empirical hole effective masses of quantization/DOS are found to be 0.29/1.33, 0.23/0.81, and 0.22/0.95 of m_0 (free electron mass) for $n = 1, 2$ and 3 hole subbands, respectively. The results of the improved self-consistent one-band EMA calculations have been demonstrated to satisfactorily fit the six-band self-consistent calculations. The triangular well approximation model is able to yield relatively good correspondence to our six-band self-consistent calculation in its subband energies, carrier occupations, and surface potentials with appropriate use of the weighting coefficient calculated to be 0.77. However, it fails to give good fitting to hole profile characteristics in the distance domain at strong inversion. This is an intrinsic disadvantage of the triangular potential approximation.

ACKNOWLEDGMENT

The authors will like to thank the device-modeling group of the University of California, Berkeley, for their one-band self-consistent CV program.

REFERENCES

[1] F. Stern, "Self-consistent results for n-type Si inversion layer," *Phys. Rev. B*, vol. 5, p. 4891, 1972.
[2] C. Moglestue, "Self-consistent calculation of electron and hole inversion charges at silicon–silicon dioxide interfaces," *J. Appl. Phys.*, vol. 59, no. 5, p. 3175, 1986.
[3] S.-H. Lo, D. A. Buchanan, and Y. Taur, "Modeling and characterization of quantization, polysilicon depletion, and direct tunneling currents in MOSFETs with ultrathin oxides," *IBM J. Res. Development*, vol. 43, p. 327, 1999.
[4] Y. Ma, L. Liu, Z. Yu, and Z. Li, "Characterization and modeling of threshold voltage shift due to quantum mechanical effects in pMOS-FETs," *Solid State Electron.*, vol. 44, p. 1335, 2000.
[5] C.-Y. Hu, S. Banerjee, K. Sadra, B. C. Streetman, and R. Sivan, "Quantization effects in inversion layer of pMOSFETs on Si (100) substrate," *IEEE Electron Device Lett.*, vol. 17, pp. 276–278, June 1996.
[6] Y. T. Hou and M. F. Li, "A simple and efficient model for quantization effects of hole inversion layers in MOS devices," *IEEE Trans. Electron Devices*, vol. 48, p. 2893, Dec. 2001.

[7] ——, "Hole quantization effects and threshold voltage shift in pMOSFET—Assessed by improved one-band effective mass approximation," *IEEE Trans. Electron Devices*, vol. 48, p. 1188, 2001.

[8] M.-F. Li, *Modern Semiconductor Quantum Physics.* Singapore: World Scientific, 1994, ch. 1, 3, 5, App. C.

[9] W. J. Fan, M. F. Li, T. C. Chong, and J. B. Xia, "Valence hole subbands and optical gain spectra of GaN/Gai.xAl,iN strained quantum wells," *J. Appl. Phys.*, vol. 80, p. 3471, 1996.

[10] J. M. Luttinger, "Quantum theory of cyclotron resonance in semiconductors: General theory," *Phys. Rev.*, vol. 102, p. 1030, 1956.

[11] Landolt-Bornstein, *Numerical Data and Functional Relationships in Science and Technology*, O. Madelung, Ed. Berlin: Springer-Verlag, 1987, vol. 17a.

[12] ——, *Numerical Data and Functional Relationships in Science and Technology*, O. Madelung, Ed. Berlin: Springer-Verlag, 1987, vol. 22a, p. 72 and 304.

[13] H. H. Mueller and M. J. Schulz, "Simplified method to calculate the band bending and the subband energies in MOS capacitors," *IEEE Trans. Electron Devices*, vol. 44, p. 1539, Sept. 1997.

[14] T. Janik and B. Majkusiak, "Analysis of the MOS transistor based on the self-consistent solution to the Schrodinger and Poisson equations and on the local mobility model," *IEEE Trans. Electron Devices*, vol. 45, p. 1263, June 1998.

[15] C.-K. Park *et al.*, "A unified current-voltage model for the long-channel nMOSFETs," *IEEE Trans. Electron Devices*, vol. 38, pp. 399–406, Feb. 1991.

[16] Y. Ma, L. Liu, Z. Yu, and Z. Li, "Validity and applicability of triangular potential well approximation in modeling of MOS structure inversion and accumulation layer," *IEEE Trans. Electron Devices*, vol. 47, Sept. 2000.

Tony Low received the B.S. degree (with first class honors) in electrical engineering from the National University of Singapore (NUS) in 2002.

Upon graduation, he was awarded sponsorships from Singapore Millennium Scholarship and the Chartered Semiconductor Manufacturing Company for his research at NUS. He is currently pursuing his research in Silicon Nano Device Laboratory (SNDL). His research interests include the study of quantum phenomena in nanoscale MOSFETs.

Yong-Tian Hou (S'02) received the B.S. and M.S. degrees in physics from Peking University, China, in 1990 and 1993, respectively. He is currently pursuing the Ph.D. degree in electrical engineering at National University of Singapore.

From 1998 to 1999, he was with the Department of Electrical and Computer Engineering, National University of Singapore, as a Research Engineer. At present, his research interests include the quantum mechanical effect in CMOS devices, tunneling phenomena in oxide, and high-K gate materials.

Ming-Fu Li (M'91–SM'99) graduated from the Department of Physics, Fudan University, Shanghai, in 1960.

After graduation, he joined the University of Science and Technology of China (USTC) as a Teaching Assistant and then Lecturer. In 1978, he joined the Graduate School Faculty, Chinese Academy of Sciences, Beijing, and became a Professor in 1986. He has also served as Adjunct Professor at the Institute of Semiconductors, Chinese Academy of Sciences, Fudan University, and USTC, Hefei. He was a Visiting Scholar at Case Western Reserve University, Cleveland, OH, in 1979, and at the University of Illinois, Urbana-Champaign, from 1979 to 1981, and was a Visiting Scientist at the University of California, Berkeley, and at the Lawrence Berkeley National Laboratories from 1986 to 1987 and 1990 to 1991, respectively. He joined the Department of Electrical Engineering, National University of Singapore, in 1991, and became a Professor in 1996. His current research interests include the areas of CMOS device technology, reliability, quantum modeling, and analog IC design. He has published over 200 research papers and two books, including *Modern Semiconductor Quantum Physics* (Singapore: World Scientific, 1994).

Dr. Li has served on several international program committees and advisory committees in international semiconductor conferences in China, Japan, Canada, Germany, and Singapore.

Investigation of Performance Limits of Germanium Double-Gated MOSFETs

Tony Low[1], Y. T. Hou[1], M. F. Li[1,2], Chunxiang Zhu[1], Albert Chin[3], G. Samudra[1], L. Chan[4] and D. -L. Kwong[5]

[1] Silicon Nano Device Lab (SNDL), ECE Department, National University of Singapore, Singapore 119260.
Tel: 65-68742559, Fax: 65-67791103, Email: elelimf@nus.edu.sg
[2] Institute of Microelectronics, Singapore 117684
[3] Dept. of Electronics Eng., National Chiao Tung Univ., Hsinchu, Taiwan
[4] Technology Development, Chartered Semiconductor Manufacturing, S738406, Singapore
[5] Dept. of Electrical and Computer Engineering, University of Texas, Austin, TX 78712, USA

Abstract

The performance limits and engineering issues of ultra-thin body (UTB) double gated (DG) Ge channel n-MOSFETs are examined in this paper. The non-equilibrium Green's Function (NEGF) approach, including both L and Δ conduction valleys, is employed for source to drain current, while the improved WKB tunneling is employed for substrate to drain (band-to-band BTB) and gate to channel current. All possible Ge surfaces and channel orientations are explored. *In terms of drive current I_{ON}*, highly anisotropic Ge<110> channel exhibits highest I_{ON} which increases with body thickness scaling; Ge<100> exhibits similar ballistic limit as Si<100> due to increasing Δ valley carrier dominance at UTB regime; Ge<111>exhibits higher ballistic limit but decrease at UTB regime due to the small density-of-states mass of L valley. Sub-threshold slope is worse for Ge<110> and Ge<111> as channel length is scaled down. *In terms of standby current I_{OFF} and gate leakage I_G* for low standby power (LSTP) devices, BTB tunneling is large due to the small energy gap of Ge. This imposes a limit on maximum tolerable supply voltage (of which Ge<111> is worst and Ge<100> is best) thus requiring low voltage operation. Body scaling is effective in suppressing BTB tunneling, since carrier quantization causes effective energy gap widening. The low voltage requirement demands small EOT for minimal oxide voltage drop. However, gate leakage will impose a limit for further EOT scaling, of which Ge <110> is worst and Ge <111> is best. Our results conclude that in addition to lower power supply voltage advantage, the engineered Ge<110> devices with suppressed BTB and gate leakages can achieve better intrinsic delay to OFF power ratio than Si<100> devices.

Introduction

Ge channel is a promising candidate for future advanced MOSFET devices due to its high mobility channel. Its low processing temperature also makes it compatible with advanced high-K [1,2] and metal gates [3] stack technology. These attributes propel recent research efforts into Ge MOSFETs. Hence, a study of Ge MOSFETs for future highly scaled advanced MOSFETs with UTB structure is pertinent for understanding its prospect as a viable long-term solution. Compared with Si, Ge has smaller carrier transport mass and gap energy, giving rise to higher drive current I_{ON} but also higher tunneling leakage I_{OFF}. Hence, a fair assessment requires both considerations. In order to accurately assess the prospect of Ge channel in highly scaled advanced MOSFETs, quantum simulations of I_{ON} and I_{OFF} of UTB (< 5nm) DG MOSFETs (Fig.1) under different surface and channel orientations are systematically investigated in this paper.

Modeling Methodology

In our work, the Ge effective masses m_i (i=x, y, z) for different valleys are derived according to [4] and listed in Table I. Carrier quantization for the various conduction valleys is analyzed by the

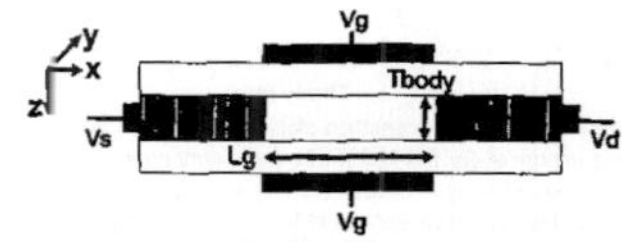

Figure 1. Straight DG MOSFET structure employed for simulation. This structure is the ultra-scaled version fabricated by Neudeck et al. [13]. The flared out S/D regions portion (replaced with metal contacts as shown) are treated as perfect absorber in the quantum simulations [7]. Channel doping of 1×10^{15} cm^{-3} and SD doping of 1×10^{20} cm^{-3} with abrupt junctions employed.

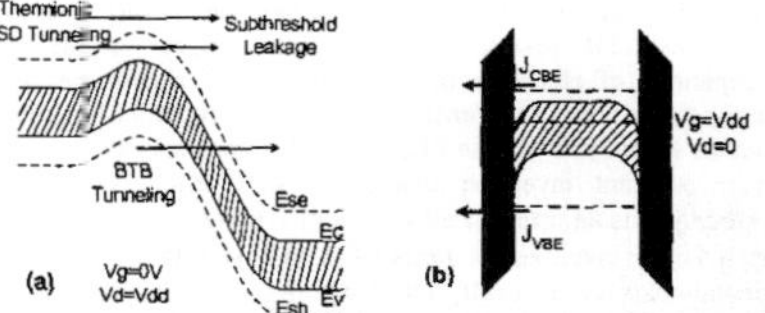

Figure 2. (a) Illustration of the leakage current components along the channel direction at Vd=Vdd and Vg=0. Ese and Esh are the electron and hole quantized energy levels due to the gate confinement. Subthreshold leakage is defined to be made up of thermionic and SD tunneling. BTB tunneling is due to the tunneling of valence sub-band electrons to conduction sub-band. (b) Illustration of gate tunneling components in channel at Vg=Vdd and Vd=0. CBE: conduction band electrons, VBE: Valence band electrons. J_{VBE} was found to be less important than J_{CBE} due to its larger barrier. Only J_{CBE} will be considered in this study.

self-consistent Schrodinger-Poisson solver [5]. The L and Δ valleys should both be included due to their relatively small valley splits [5]. Γ valley can be neglected due to its very small effective mass of $0.038 m_o$, which produces a rapid uplift of subband energy. The source to drain current (thermionic and tunneling, Fig.2a) is calculated using the well-developed NEGF method by Purdue's group [6-7]. NEGF formalism allows accurate calculation of the ballistic current and scattering can be efficiently treated using the Buttiker probes concept [8]. The substrate to drain (valence subband-to-conduction subband, BTB, Fig.2a) tunneling current, a major contribution to I_{OFF} in low power devices, is calculated, based on the profile along the channel, obtained from NEGF. BTB tunneling is calculated using the WKB approximation with Freeman and Dahlke dispersion relationship [9] to account for the difference in effective masses of conduction and valence band. The gate leakage (Fig.2b) is calculated using the method our group developed based on the improved WKB tunneling model [5,10]. The wave reflection at the abrupt silicon-oxide interface is accounted for in the model with good experimental corroboration with the model [10 and references therein].

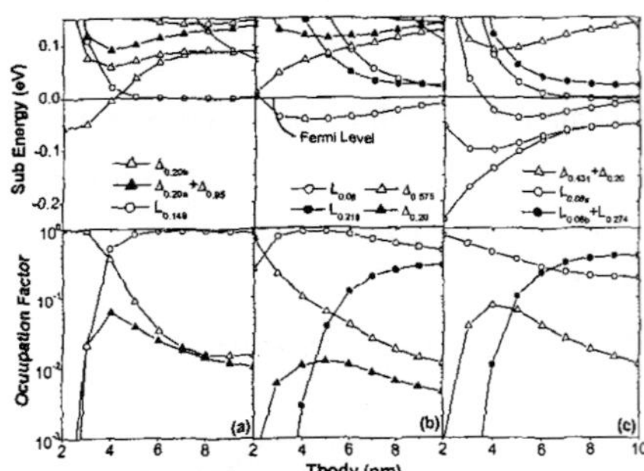

Figure 3. Different Ge surface orientation plotted in (a) <100> (b) <110> (c) <111>. (Top) Subband energy in Ge DG-MOSFETs (wrt Fermi energy) for various valleys minima as a function of body thickness (Tbody) at surface carrier concentration of $1\times10^{13}cm^{-2}$. The notation used for various valley minima is compiled in Table 1. (Bot) Occupation factor of electrons in Ge DG-MOSFETs for various valley minima, where sum of occupation factors from all valleys is 1.

Electron Quantization on L Valley Occupations

The high mobility of Ge channel is mainly attributed to the small transport mass from L valley. However, the strong quantization effects in UTB devices will have great impact on the relative occupations of electrons in L valley. Fig. 3 shows the subband energy (with respect to Fermi energy) and occupation factor each in relation to body thickness (T_{body}) for different surface orientations under constant inversion charge of $1\times10^{13}cm^{-2}$. The *Ge<100> surface* yields an uplift of all L valleys. This is due to the relatively much larger quantization mass of Δ valleys under <100> surface orientations. As a result, the electrons occupation in L valley reduced drastically as T_{body} scales beyond 5nm. The *Ge<110> surface* yields a small uplift in the lowest L valley such that the dominant carrier occupation still remains in the L valley. This is due to the relatively similar quantization mass of L and Δ valleys under <110> surface orientations. *Ge<111> surface* yields the most carrier occupation in the L valley due to its relatively large quantization mass. However, it is noted that its L valley ground state needs to stay much below the Fermi level in order to invert a given amount of inversion charge. This is due to the low density-of-states mass, m_d of L valley in Ge<111> channel (Table 1, $L_{0.08a}$ valley). Hence, a larger surface bending is required to achieve this. This is elucidated in Fig 4, which calculates the overdrive voltage characteristic for various orientations. Generally, body scaling enhances the overdrive characteristic of the devices. However, Ge<111> requires increasingly larger overdrive voltage for a given surface inversion charge as T_{body} scales down. This limitation will significantly compromise the current drivability of Ge<111> in highly scaled UTB MOSFETs.

Ballistic and Quasi-Ballistic Drive Current I_{ON}

I_{ON} for HP (LSTP) devices are analyzed with work function design based on the criterion of constant I_{OFF} of $1\mu A/\mu m$ ($10pA/\mu m$) at off-state condition [11]. Subthreshold slope (SS) for various orientations is calculated in Fig. 5. SS degrades with smaller channel length (Lg) especially for Ge<110> and Ge<111>. This is attributed to the larger direct source-drain tunneling current as a result of the small

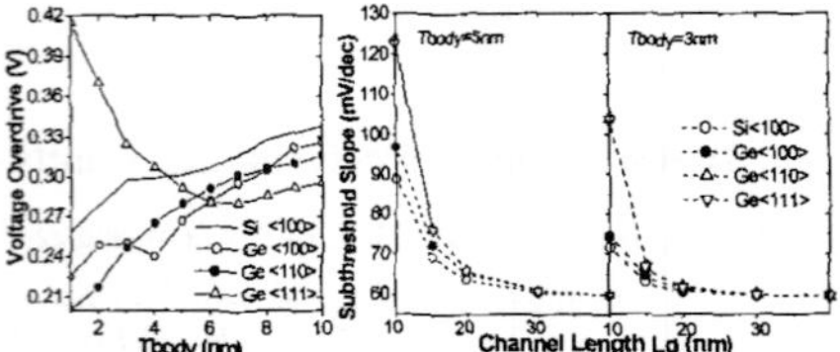

Figure 4. (Left) Comparing voltage overdrive (Vdd-Vt) for Ge and Si DG MOSFETs with surface carrier concentration of 1×10^{11} cm^{-2} and 1×10^{13} cm^{-2} at Vt and Vdd respectively. Figure 5. (Right) Subthreshold slope (SS) for Si and Ge DG-MOSFET at different Tbody (5nm and 3nm) as function of channel length (Lg). SS calculated for optimum channel directions as indicated in Table 1.

transport mass of L valley. Poor subthreshold slopes render the Lg scaling for Ge<110> and Ge<111> impractical beyond the 15nm regimes. Lg=20nm is used in our analysis as it provides relatively ideal subthreshold slope for both Si and Ge DG MOSFETs for T_{body}<5nm. In NEGF calculation, both thermionic and tunneling source to drain currents (Fig. 2a) depends on the conduction mass m_x, the density of state mass m_d and the valley occupations. The *ballistic current for all surface and channel orientations* are shown in Fig. 6. Drive current for Ge<100> and Ge<111> remains relatively isotropic at UTB regimes, unlike Ge<110>, which exhibits large degree of anisotropy. Its optimum current direction is [110] as explicitly outlined in Table 1, which will serve as an important guideline for integrated circuit layout designs. It is also reported that the optimum channel direction for hole transport in Si<110> is also [110] direction [12], which should also be the case for Ge<110> due to similar valence band structure. This allows optimum hole and electron transport in same channel direction under Ge<110> surface orientation, which is not plausible for Si<110> [12]. Fig. 7 depicts the impact of T_{body} scaling on the ballistic current for various orientations. Ballistic current of Ge<110> increases with T_{body} scaling whereas for Ge<100> and Ge<111> is not beneficial. For Ge<100>, since its Δ valleys dominate its ballistic current with thinning of T_{body} as shown in Fig. 8, its ballistic limit approaches that of Si at T_{body}=3nm (Fig. 7). Ge<111> exhibits slightly higher ballistic limit but decrease at UTB regime due to the degradation of current drivability (Fig. 4). The ballistic current characteristic for HP and LSTP devices are shown in Fig. 9 for various orientations. Ge<110> exhibits largest ballistic current, ~170% more than Si at Vg=Vd=0.5V for HP and ~150% for LSTP at Vg=Vd=0.7V. *Quasi-ballistic current* is evaluated by taking into account the carrier scattering in channel, source and drain using the efficient Buttiker concept [7,8]. Ge<110> yields an appreciably larger drive current (~ 360% of Si) which match the ballistic current of Si<100> as shown in Fig. 8 inset. Ge<110> is ~60% ballistic whereas Si<100> is only 40% at Lg=30nm. The higher ballistic component of Ge MOSFETs is mainly attributed to the less dissipative nature of the highly doped source drain as compared to Si counterpart. As a result, Ge<100> and Ge<111> DG MOSFETS may also outperform Si appreciably in quasi-ballistic regimes. As an intuitive appreciation of the relative performance advantage of Ge<110> over Si<100> DG MOSFETs, the intrinsic delay (using the CV/I metric at quasi-ballistic regime) [11] to a given OFF power is shown in Fig 9 insets for both HP and LSTP devices. The improvement in performance when using Ge<110> is an appreciable few fold enhancement in intrinsic delay. This is calculated for OFF power limited by the subthreshold current, which is generally true for HP devices only.

29.4.2

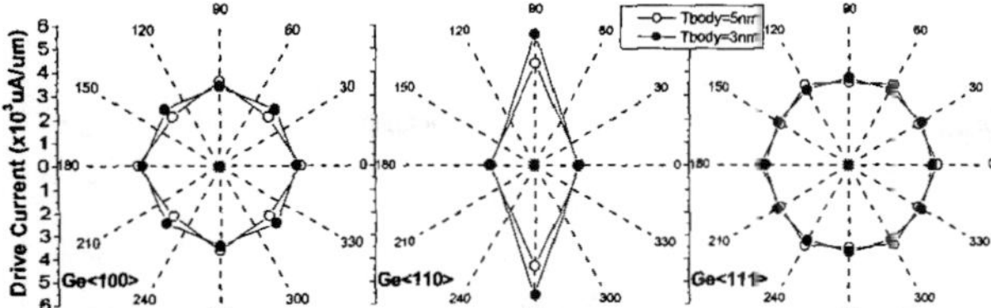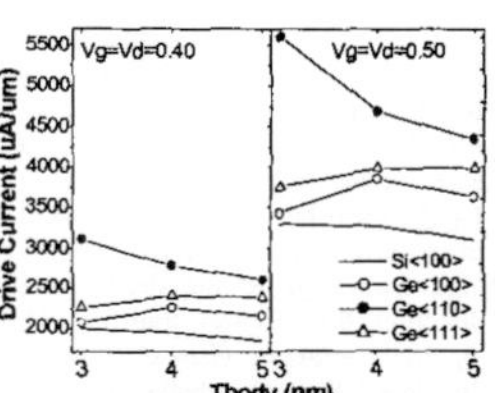

Figure 6. Exploring the impact of channel orientation on the ballistic current of Ge DG MOSFETs. Ballistic drive current is calculated using NEGF with effective masses listed in table 1. Various substrate orientations at Tbody=5nm and 3nm are considered. Lg=20nm, EOT=1nm used at Vg=Vd=0.5V condition. 0° denotes [100] channel direction for Ge<100> and Ge<110>, whereas for Ge<111>, it denotes [211] channel direction. The ballistic current is measured using the length of the line from the center to the point of interest, with scale indicated on the left axis.

Figure 7. Tbody scaling and its impact on ballistic limit of Si and Ge DG MOSFETs. Vg=Vd=0.4 and 0.5V investigated. Lg=20nm, EOT=1nm employed with optimized channel direction.

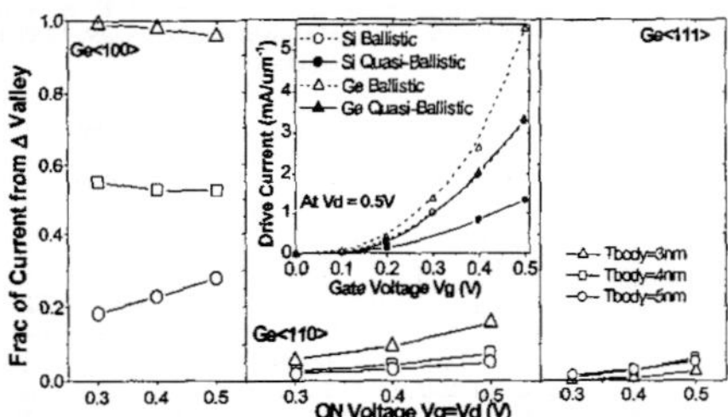

Figure 8. Fraction of ballistic drain current contribution from Δ valley vs. ON voltage (Vg=Vd) for various Ge surface orientations. Lg=20nm, EOT=1nm, with various Tbody and optimized channel direction. 1-(Fraction of current from Δ valley) will gives the contribution from L valley (Inset) Comparing drive current of Ge<110> and Si<100> HP DG MOSFET (Fig. 1) under ballistic and quasi-ballistic [7] regimes using NEGF. EOT=1nm, Lg=30nm (where it is less ballistic), Tbody=3nm, Vd=0.5V used. Channel low-field mobility for Si and Ge are 200 cm²/Vs [7] and 940 cm²/Vs [14] respectively. And SD doping dependent mobility is 40 cm²/Vs [15] and 400 cm²/Vs [15] respectively.

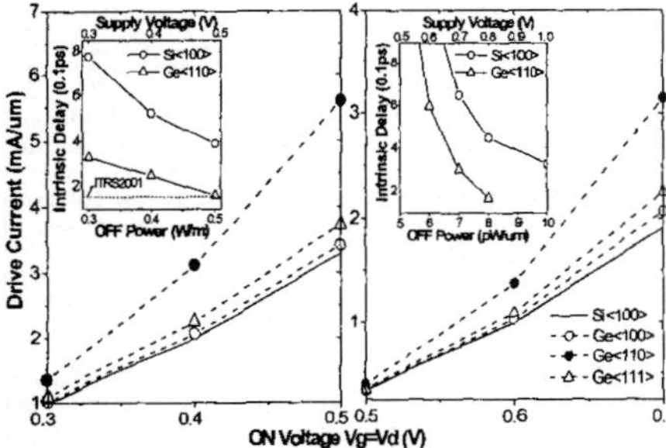

Figure 9. Comparing the ballistic drain current of Si and Ge HP (Left) and LSTP (Right) DG MOSFETs at ON voltage Vg=Vd. Lg=20nm, Tbody=3nm and EOT=1nm employed. Channel orientation optimized for various surface orientations as indicated in Table I. (Inset) HP and LSTP device performance of Ge<110> vs Si<100> based on intrinsic delay using CV/I metrix vs OFF power I_OFF x V_dd. EOT=1nm, Lg=30nm, Tbody=3nm used at quasi-ballistic regime with mobility values as indicated in Fig 8. The intrinsic delay specification is obtained from ITRS2001 at year 2016.

Leakage Current I_{OFF} and I_G in LSTP applications

For LSTP devices, since the I_{OFF} is very low, one has to consider leakage contributions not only from subthreshold current from source to drain, as calculated in the above section, but also the subband-to-subband (BTB) tunneling from substrate to drain (Fig.2a) and gate-to-channel tunneling (Fig.2b). Due to the smaller energy gap of Ge, BTB tunneling is an important consideration for LSTP devices. BTB tunneling and its dependency on supply voltage are calculated for various orientations as shown in Fig. 10. ***BTB leakage current*** has strong dependency on the supply voltage and hence imposes a limitation on the maximum allowable supply voltage for desired OFF leakage specification. As calculated, Ge<111> exhibits largest BTB leakage, and as a result, the supply voltage has to be kept below 0.4V, which severely limit its drive current. Fig 11 illustrates that when BTB is major component of I_{OFF}, the intrinsic delay of Ge<110> is comparable to Si under quasi-ballistic regimes. Hence, in order to harness the advantage of Ge<110>, BTB leakage must be suppressed by operating below the maximum allowable supply voltage (as defined as the voltage where BTB leakage surpass subthreshold leakage). The maximum supply voltage for the onset of BTB leakage can be increase by graded source-drain junction or widening of the band gap by body thinning. The energy gap widening effect due to electron and hole quantization is efficient in alleviating BTB tunneling. Fig. 12 shows that thinner T_{body} is a better architecture for reduced BTB due to body quantization effects. However, Ge<111> is unable to benefit much from the energy gap widening effect due to the high electron quantization mass of its ground state L valley. The requirement for lower supply voltage operation demands small oxide voltage drop with the use of High-K dielectrics. Fig. 13 illustrates the ***gate leakages current*** with 0.6nm EOT of HfO_2. Due to the larger valence band to oxide barrier for most general dielectrics, we only consider conduction band electron tunneling current in our study. Generally speaking, Ge exhibits larger gate leakage compared to Si due to its larger electron quantization energy [5]. However, Ge<111> exhibits the lowest gate leakage due to its lowest quantization energy shown in Fig.3. In addition, the gate leakage in Ge<110> increases with T_{body} scaling (Fig 14). Larger gate leakage of Ge<100> and Ge<110> impose a larger EOT requirement (~0.1nm less compared to Si) under constant gate leakage condition (Fig 15). However, in requirements for the very low voltage operation for Ge<110> as imposed by BTB leakage, an EOT of ~0.75nm (when HfO_2 is used) is required for small voltage drop across oxide while at the same time having sufficient suppression of gate leakages. This will thus entails the employment of quality high-K dielectric in Ge UTB MOSFETs.

29.4.3

Conclusions

In this paper, we have systematically assessed the performance limits and engineering issues of Ge UTB (<5nm) DG MOSFETs for future highly scaled device applications. Summarized results are as follows:

Ge<110>: It has high L valley occupation, yielding the largest drive current and advantageous with body scaling. Despite the high anisotropy of the drive current, its optimum channel direction for electron transport [110] is aligned with the reported optimum channel direction for hole transport in Si<110>. Several challenges are prominent. First, the fabrication of ultra-thin body (preferably<3nm) Ge channel with atomically smooth surface is critical for suppression of BTB leakage and achieving high current drivability. Secondly, the ultra-low voltage operation demands quality high-K dielectrics for small oxide voltage and efficient suppression of gate leakages.

Ge<111>: Despite the high L valley occupation for Ge<111>, its poor voltage overdrive characteristic significantly compromises its current drivability at T_{body}<5nm regime. In addition, Ge<111> suffers from too large BTB leakage making it inappropriate for LSTP applications.

Ge<100>: Due to the Δ valleys dominance at UTB regimes, further body scaling beyond 5nm is not advantageous for current drivability. At T_{body}=5nm, appreciable L valleys occupation is still obtainable and the use of Ge channel also appreciably enhances the ballistic components of the drive current due to its less dissipative source drain. Ge<100> also has relatively low BTB leakage and gate leakages, making it a viable candidate for LSTP applications.

References: [1] H. Shang et al., IEDM 2002, p.441 [2] C. O. Chui et al., IEDM 2002,p.437 [3] C. H. Huang et al., IEDM 2003 [4] T. Ando et al., Rev.Mod.Phys. 54, p.437, 1982 [5] T. Low et al., VLSI Tech. 2003, p. 117, 2003 [6] S. Datta, IEDM2002, p.703 and http://nanohub.purdue.edu [7] Z. Ren et al, IEDM2001, p.107 [8] R. Venugopal et al., JAP 93, p. 5613 [9] L. B. Freeman, W. E. Dahlke et al., SSE 13, p.1483, 1970 [10] Y. T. Hou et al., IEDM, p.29.8, 2002 [11] ITRS2001 roadmap 2001 [12] T. Mizuno et al., VLSI Tech. 2003, p.97 [13] G. Neudeck et al., IEDM 2000, p.169 [14] J. J. Rosenberg et al., EDL9, p. 639, 1998 [15] S. M. Sze, Phys. Semicon. Devices, p.29

		m_x	m_y	m_z	m_d	g
Surface <100> & Channel [001]	$\Delta_{0.20a}$	0.20	0.95	0.20	0.436	2
	$\Delta_{0.20b}$	0.20	0.20	0.95	0.20	2
	$\Delta_{0.95}$	0.95	0.20	0.20	0.436	2
	$L_{0.149}$ *	0.149	0.149	0.117	0.295	4
Surface <110> & Channel [110]	$\Delta_{0.20}$	0.20	0.95	0.20	0.436	2
	$\Delta_{0.575}$ *	0.575	0.20	0.33	0.34	4
	$L_{0.08}$	0.08	0.573	0.218	0.216	2
	$L_{0.218}$ *	0.218	0.117	0.08	0.357	2
Surface <111> & Channel [211]	$\Delta_{0.20}$	0.20	0.70	0.271	0.374	2
	$\Delta_{0.431}$ *	0.431	0.243	0.271	0.374	4
	$L_{0.08a}$	0.08	0.08	1.59	0.08	1
	$L_{0.08b}$	0.08	1.42	0.089	0.337	1
	$L_{0.274}$ *	0.274	0.105	0.089	0.337	2

Table 1. Electron effective masses of Ge calculated for different surface orientations (m_x: conduction mass, m_y: quantization mass, m_z: density of states mass) (direction specified in Fig 1) at both L and Δ valleys. g denotes the valleys degeneracy. Channel direction for optimal drive current is indicated in table. Effective masses derived according to [4] p. 460. The valleys denoted with * indicates the presence of off-diagonal components in its 2D effective mass tensors. This requires separate treatment and is not accounted for in this paper. However, its implication can be neglected in cases where theses valleys are not dominant.

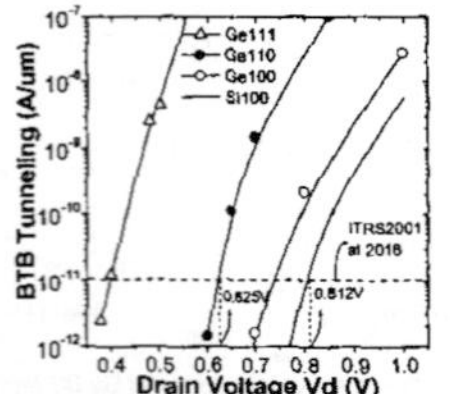

Figure 10. BTB tunneling current for LSTP Ge and Si DG MOSFETs and its dependence on Vd. Tbody=3nm, Lg=30nm and EOT=1nm used. ITRS 2001 OFF leakage spec for LSTP devices imposed a limit on supply voltage as indicated. Graded SD junctions will loosen this limit.

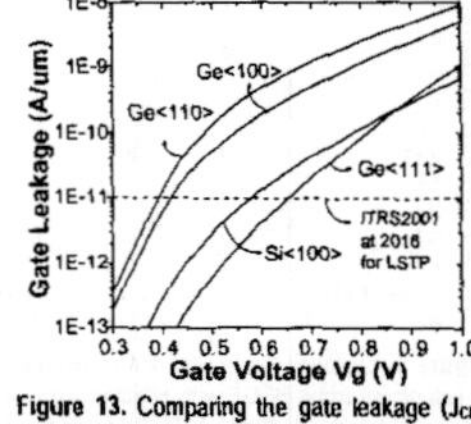

Figure 13. Comparing the gate leakage (J_{CBE} as depicted in Fig 2b) of Si and Ge DG MOSFET as function of gate voltage where device designed with work function for LSTP. Tbody=3nm, Lg=30nm, EOT=0.6nm with HfO₂ high-k dielectric constant of 22 [5]. Gate leakage is calculated using a modified WKB model depicted in [5,10].

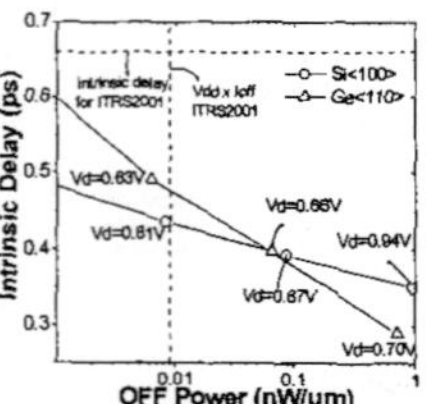

Figure 11. Intrinsic delay (CV/I) for given OFF power when BTB tunneling dominates I_{OFF}. Tbody=3nm, Lg=30nm, EOT=1nm and abrupt junction is used.

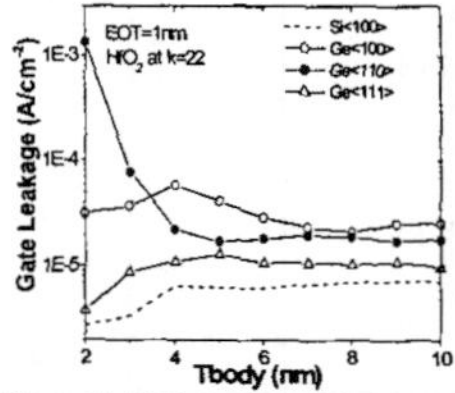

Figure 14. Tbody scaling and its impact on channel gate leakage for Si and Ge DG MOSFETs. Tunneling current calculated at constant inversion charge of 1×10^{13}cm⁻².

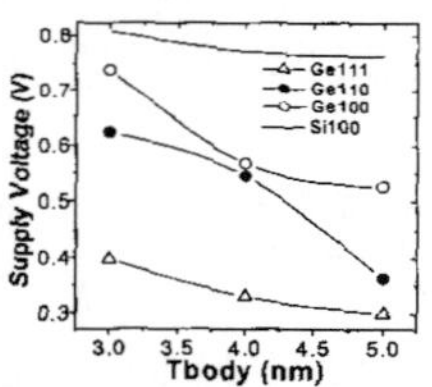

Figure 12. Maximum supply voltage of Ge and Si DG MOSFET for LSTP device at OFF leakage spec of 10pA/um [11] for BTB leakage as a function of Tbody. Lg=30nm, EOT=1nm and abrupt junction used.

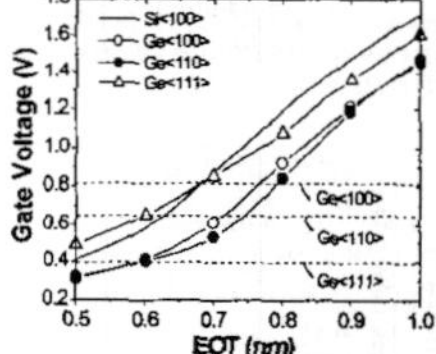

Figure 15. Gate voltage and EOT design requirements for gate leakage spec. of 10pA/um [11]. HfO₂ with k=22 [5], Tbody=3nm and Lg=30nm used. Maximum allowable supply voltage imposed by BTB leakage at Tbody=3nm is also shown as dashed line.

Acknowledgement: We gratefully acknowledge the program NanoMOS from Purdue University Computational Electronic Group and the help rendered by Prof Mark Lundstrom, Ramesh Venugopal and useful discussion with Rahman Anisur. This work is supported by Singapore A*STAR research grant R263000267305 and R263000266305. The author T. Low gratefully acknowledges the Scholarship from Singapore Millennium Foundation.

29.4.4

Germanium MOS: An Evaluation from Carrier Quantization and Tunneling Current

Tony Low, Y. T. Hou, M. F. Li, Chunxiang Zhu, D. -L. Kwong#, and Albert Chin*

Silicon Nano Device Lab, ECE Department, National University of Singapore, Singapore 119260.

Tel: 65-68742559, Fax: 65-67791103, Email: elelimf@nus.edu.sg

#Dept. Electrical and Computer Engineering, University of Texas at Austin, Austin, TX 78712, USA.

*Dept. of Electronics Eng., National Chiao Tung Univ., Hsinchu, Taiwan

Abstract

Ge is promising as an alternative channel material due to its high carrier mobility. In this work, we report an evaluation of Ge MOS from quantization and gate tunneling current simulations. Key findings are: *(1)* Electron quantization effect is stronger and thus more important in Ge than in Si, and results in smaller inversion capacitance in NMOS. *(2)* Using constant inversion charge for supply voltage V_{DD} scaling, moderate reduction in inversion charge is required to meet ITRS roadmap, which can be achieved with high mobility channel. *(3)* Due to its smaller electron mass and resulting higher electron quantization energy, Ge MOS shows considerably larger gate tunneling current than Si MOS for the same gate dielectric. *(4)* High-K gate dielectrics are required for low leakage; however, significant challenges exist in the formation of high quality interface layer between high-K and Ge.

Introduction

To improve the performance of scaled MOS devices, the improvement of hole mobility still remains a challenge [1]. Recently, MOS structures with Ge channel were fabricated and higher hole mobility was demonstrated [2,3], indicating the potential use of Ge as a channel material for future scaled CMOS. In addition, GeON and ZrO_2 have been studied as gate dielectrics for Ge MOS [2,3]. In this work, we investigate the carrier quantization and gate tunneling current in Ge MOSFETs for the first time and their impact on future device scalability.

What is New in Ge Quantization?

The Ge MOS electrostatics was obtained by solving self-consistently the Schrodinger and Poisson equations. For electrons in Ge, the conduction band minima are located at L. However, Δ valleys should also be considered due to their small energy splits from L and the larger quantization mass (Table I) [4]. Under carrier quantization, the L valleys will be uplifted and beyond Δ_2 under high field condition as depicted in Fig.3. Since the ground quantum state L in Ge is much higher than Δ_2 in Si, as indicated in Fig.3, the quantum effect in Ge is stronger than Si. For holes, the full band Kohn-Luttinger Hamiltonian is used, which takes into account the band mixing between the heavy and light holes [5,6]. Different from Si, the split-off band in Ge can be disregarded due to the large spin-orbit splitting energy Δ_{so}=0.28 eV. A typical hole subband dispersion of Ge is shown in Fig.1. Although the hole bands are mixed, an improved one band effective mass approximation still works well as depicted in Fig.2 by fitting to results from full band calculation [6].

Scalability on Supply Voltage

The capacitance density at EOT=1 nm are shown in Fig.4. Inversion capacitance of Ge NMOS exhibits more degradation compared to Si due to its smaller electron quantization mass. For a given inversion charge density of $10^{13}cm^{-2}$, quantization effect adds on 0.5Å EOT more to NMOS and 0.3Å EOT more to PMOS in Ge than its silicon counterpart. As a result, due to quantization, a significant portion of gate voltage is dropped across the inversion layer in Ge MOS, resulting in higher gate drive for a given inversion charge density. This will pose severe limitation on the scaling of supply voltage V_{DD}. Fig. 5 shows the V_{DD} for high performance devices calculated using constant inversion charge scaling [7] where metal gate is assumed. As can be seen, in order to meet the ITRS requirement on V_{DD} scaling at 22 nm technology node, the inversion charge density in Ge MOS must be reduced by a factor of 67% ~1/1.5. In terms of mobility, this translates to mobility enhancement by a factor of 1.5, to maintain the same drain current in the saturation region when backscattering r_C is taken to be 1 [8]. Even larger mobility enhancement is required if the back scattering factor is included [8].

Tunneling Current

The above unique band structure of Ge also has a great impact on gate tunneling current in Ge MOS. Intrinsic direct tunneling current is calculated using a tunneling model detailed in [9] with the tunneling parameters listed in table 2. Calculated electron tunneling current in both Si and Ge NMOS with HfO_2 gate dielectric is shown in Fig. 6. As can be seen, for 1 nm HfO_2, Ge MOS exhibits considerably higher leakage current than Si MOS, up to 1 order of magnitude. This is due to the higher energy uplift (smaller electron mass), as shown in Fig. 3a, which is independent of gate dielectric material. This is one of the most serious obstacles in scaling Ge MOS devices. Gate leakage of future generation devices with various dielectrics is compared in Fig. 7 and Figs. 8-9 for high performance and low standby power applications respectively. For Ge MOS, GeON and Ge_3N_4 gate dielectrics have been reported with large EOT (>7nm) [2-3, 5, 10]. However, from Fig. 8, GeON's potential use for future generation devices is unlikely due to high leakage. For low power CMOS applications (Fig.9), high-κ dielectrics must be employed and pure HfO_2 provides a viable solution. However, in practical applications, an interfacial layer (IL) between high-K and Ge substrate is inevitably formed during processing and its impact on gate leakage is shown in Figs. 8 and 9. Results show that the presence of an interfacial layer increases the gate leakage greatly. With Ge_3N_4 IL, the HfO_2 stack is applicable to high performance Ge MOS, but not for low power applications. There are significant process integration challenges and opportunities associated with Ge MOS gate stack technology.

Conclusion

An evaluation of Ge MOS is presented based on carrier quantization and gate tunneling current. Stronger electron quantization is found in Ge than in Si channel, resulting in more degradation of inversion capacitance and much higher tunneling current in Ge MOS. It is shown that mobility enhancement in Ge channel is an effective way to meet the ITRS requirement on V_{DD} scaling. Significant challenges exist in Ge MOS gate stack technology where a high quality high-K/Ge interface is required.

Acknowledgements: This work is supported by Singapore A*STAR EMT/TP/00/001,2 grant.

References
[1] K. Rim et al., VLSI, p. 2.1, 2002
[2] H. Shang et al., IEDM, p17.3, 2002
[3] C. O. Chui et al., IEDM, p17.4, 2002
[4] M. V. Fischetti et al., JAP**80**, p2234, 1996
[5] Y. Zhang et al., JAP**83**, p.4264, 1998.
[6] Y.T.Hou et al, TED**48**, p.1188, 2001.
[7] S. Takagi et al., IEDM, p22.2.1, 1998
[8] M. S. Lunstrom et al., EDL**22**, p293, 2001
[9] Y. T. Hou, M. F. Li, H. Y. Yu, Y. Jin, D.-L. Kwong, IEDM, p29.8, 2002
[10] G. A. Johnson and V.J. Kapoor, JAP**69**, p3616, 1991.
[11] W.Y. Ching and S.Y. Ren, PRB**24**, p5788, 1981.
[12] I. Chambouleyron and A.R. Zanatta, JAP**84**, p1, 1998.

Table I: Quantization mass (m_z), DOS mass (m_d) and relative valley split (Δ_E) for various valleys minima for conduction band electrons calculation [4] and the Luttinger parameters and effective masses (HH: Heavy hole, LH: Light hole) for full band calculation of valence band structure [5].

Electron	$m_z(m_0)$	$m_d(m_0)$	Δ_E(meV)
L Valley	0.12	0.30	0
Δ_2 Valley	0.95	0.2	150
Δ_4 Valley	0.2	0.44	150
Luttinger parameters			
γ_1=13.35	γ_2=4.25	γ_3=5.69	Δ_{so} =0.282eV
Empirical effective masses of Holes			
m_z=0.206(HH), 0.08(LH)		m_d=0.365(HH), 0.326(LH)	

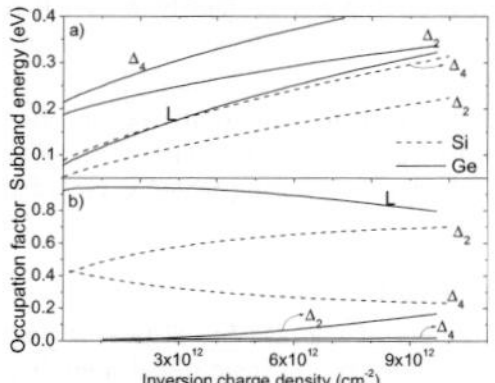

Fig 3: Subband energies (a) and occupation factor (b) of Si and Ge NMOS at inversion. The various conduction band valleys are indicated. The substrate doping is 10^{17}cm^{-3}.

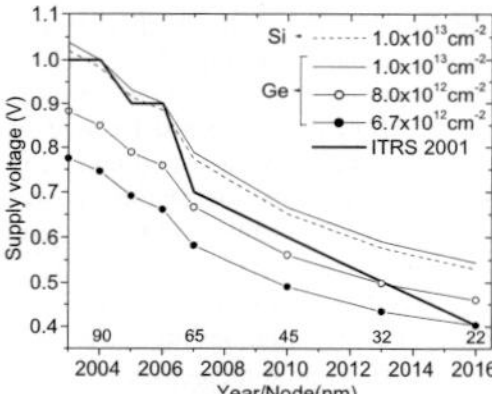

Fig 5b: Same as Fig.5a, however for PMOS.

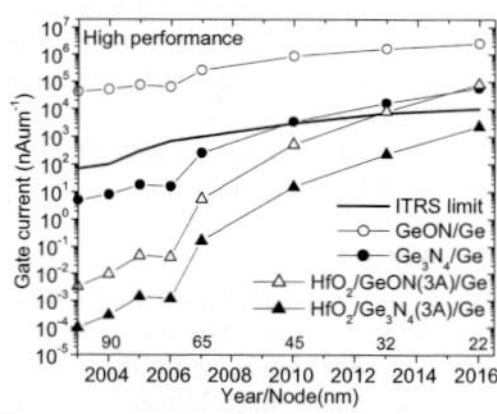

Fig8: Gate leakage of high performance CMOS for each generation for various gate dielectrics. The gate leakage is defined as the tunneling current at inversion charge 10^{13}cm^{-3}.

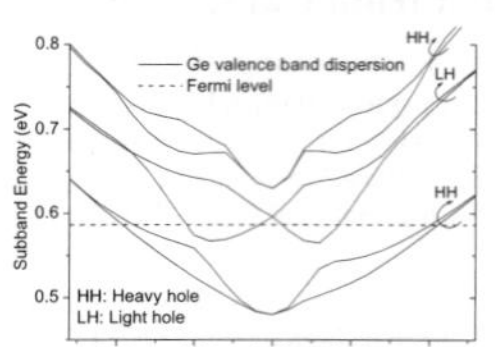

Fig 1: Valence band dispersion at surface potential 1.13V from full band effective mass method [5] for Ge with doping of 10^{17}cm^{-3}.

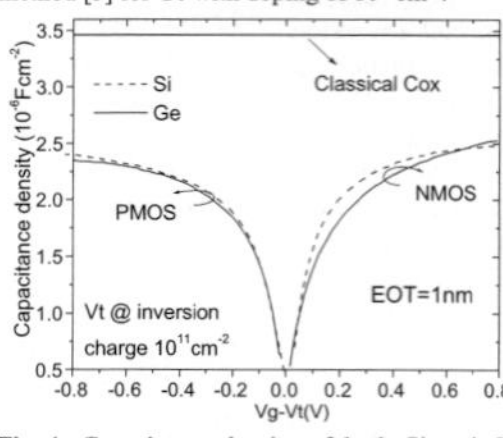

Fig 4: Capacitance density of both Si and Ge NMOS and PMOS under inversion. The substrate doping used for simulations is 10^{18}cm^{-3}.

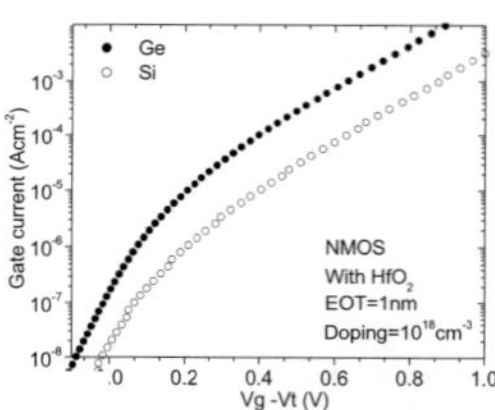

Fig 6: Calculated gate tunneling current of Si and Ge NMOS with substrate doping 10^{18}cm^{-3}. The inversion charge at Vt is defined to be 10^{11}cm^{-2}.

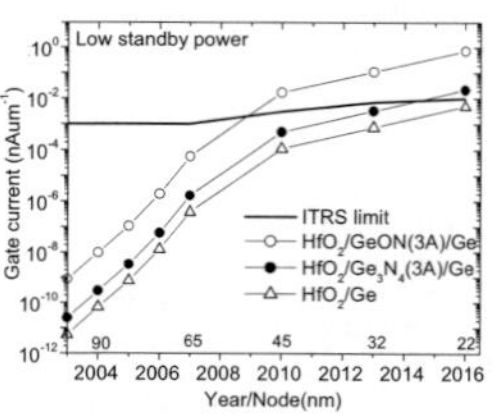

Fig 9: Gate leakage of low standby power CMOS for each generation for various gate dielectrics. The gate leakage is defined as the tunneling current at inversion charge 10^{13}cm^{-3}.

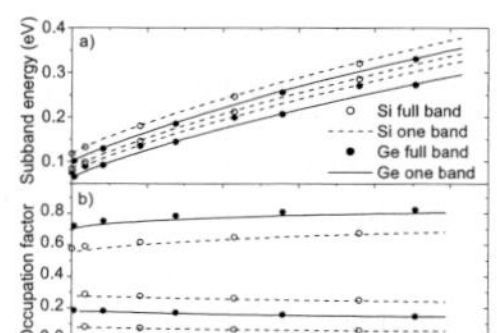

Fig 2: Hole subband energies (a) and occupation factor (b) of Si and Ge PMOS at inversion. Substrate doping is 10^{17}cm^{-3}.

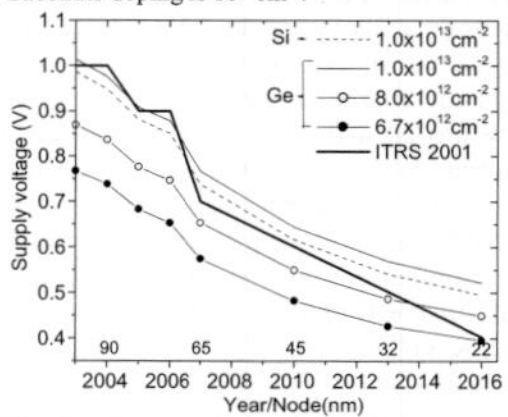

Fig 5a: Supply voltage for NMOS at doping 10^{18}cm^{-3} with constant inversion charge density (10^{13}cm^{-2}) scaling [7] at supply voltage Vdd. Reduced charge densities for Ge NMOS to fit the ITRS are also shown.

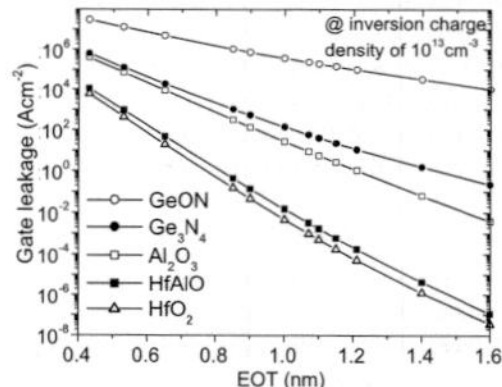

Fig 7: Calculated gate tunneling current at inversion charge density of 10^{13}cm^{-2} in NMOS vs EOT for various dielectrics for Ge.

Table II: Tunneling parameters for conduction band offset (eV), effective mass, dielectric constant for different dielectric materials HfO_2, Al_2O_3, HfAlO [9], GeON [11] and Ge_3N_4 [10][12].

High-K dielectrics	Conduction band offset	Effective mass m_0	Dielectric constant
HfO_2	1.9	0.18	22
Al_2O_3	2.3	0.28	10
HfAlO	2.0	0.21	18.4
GeON	1.5	0.50	4.8
Ge_3N_4	1.5	0.50	9.6

IEEE TRANSACTIONS ON ELECTRON DEVICES, VOL. 51, NO. 11, NOVEMBER 2004
1783

Metal Gate Work Function Engineering on Gate Leakage of MOSFETs

Yong-Tian Hou, *Member, IEEE*, Ming-Fu Li, *Senior Member, IEEE*, Tony Low, and Dim-Lee Kwong, *Senior Member, IEEE*

Abstract—We present a systematic study of tunneling leakage current in metal gate MOSFETs and how it is affected by the work function of the metal gate electrodes. Physical models used for simulations were corroborated by experimental results from SiO_2 and HfO_2 gate dielectrics with TaN electrodes. In bulk CMOS results show that, at the same capacitance equivalent oxide thickness (CET) at inversion, replacing a poly-Si gate by metal reduces the gate leakage appreciably by one to two orders of magnitude due to the elimination of polysilicon gate depletion. It is also found that the work function Φ_B of a metal gate affects tunneling characteristics in MOSFETs. It is particularly significant when the transistor is biased at accumulation. Specifically, the increase of Φ_B reduces the gate-to-channel tunneling in off-biased n-FET and the use of a metal gate with midgap Φ_B results in a significant reduction of gate to source/drain extension (SDE) tunneling in both n- and p-FETs. Compared to bulk FET, double gate (DG) FET has much lower off-state leakage due to the smaller gate to SDE tunneling. This reduction in off-state leakage can be as much as three orders of magnitude when high-κ gate dielectric is used. Finally, the benefits of employing metal gate DG structure in future CMOS scaling are discussed.

Index Terms—High-κ gate dielectric, metal gate, MOSFET, poly-Si depletion, scalability, tunneling current.

I. INTRODUCTION

AGGRESSIVE scaling of gate length and gate oxide thickness in CMOS transistors aggravates the problems of poly-Si gate depletion, high gate resistance and boron penetration from the p^+-doped poly-Si gate into the channel region. The poly depletion reduces the gate capacitance in the inversion regime and hence the inversion charge density, leading to a lower gate over drive and thus degrading the device performance. As a result, metal gate technology has recently been exploited [1]. On the other hand, novel device architectures, such as ultrathin body (UTB) or double gate (DG) structures fabricated on silicon-on-insulator (SOI) wafers may be utilized in future CMOS technology due to their excellent scaling capability [2]. In such SOI devices, in order to improve channel carrier mobility and eliminate threshold voltage (V_T) instability induced by dopants fluctuation, low doping in body is generally

Manuscript received January 19, 2004; revised May 12, 2004. This work was supported by the Singapore A * STAR under Grant R263-000-267-305. The review of this paper was arranged by Editor M.-C. Chang.

Y.-T. Hou and M.-F. Li are with the Silicon Nano Device Lab, Department of Electrical and Computer Engineering, National University of Singapore, Singapore 119260, and also with the Institute of Microelectronics, Singapore 177685.

T. Low is with the Silicon Nano Device Lab, Department of Electrical and Computer Engineering, National University of Singapore, Singapore 119260.

D.-L. Kwong is with the Department of Electrical and Computer Engineering, University of Texas, Austin, TX 78712 USA (e-mail: elelimf@nus.edu.sg).

Digital Object Identifier 10.1109/TED.2004.836544

desirable and metal gate with work function near midgap is thus required to obtain appropriate V_T [3], [4]. At the same time, the introduction of an alternative high-κ material is also underway in order to suppress gate leakage [5]. Therefore, an investigation of the impact of metal gate and its work function on the tunneling characteristics of MOSFETs, especially with high-K dielectric stack and DG device structure, is technically important and timely.

In this paper, we present a systematic investigation of the tunneling phenomena in metal gate MOSFETs with both conventional SiO_2 and high-κ gate dielectrics. First, the simulated tunneling currents will be compared to measurements on TaN metal gated devices with SiO_2 and HfO_2 dielectrics. Subsequently, the advantage of the metal electrode over poly-Si on the gate leakage characteristic is demonstrated in bulk CMOS. The advantage is mainly attributed to the elimination of poly-Si depletion. This is followed by a systematic study on how the change of metal gate work function affects the various tunneling components in CMOS transistors. Finally, the better scalability of SOI devices over bulk one in terms of tunneling leakage current is demonstrated for future CMOS technology.

II. PHYSICAL MODEL OF TUNNELING CURRENTS

The tunneling currents through SiO_2 and high-κ in poly-Si gated CMOS devices have been simulated using a physical model in our previous publications [6], [7]. Relevant material parameters with tunneling, including the band offset and effective mass values, were extracted from poly-Si gated devices with various gate dielectrics [7]. These parameters and the same simulation method will be used in the following calculations of tunneling currents in metal gate devices. In the simulations, free electron gas is assumed in the metal gate. The equivalent oxide thickness (EOT) is extracted from overall fitting of the capacitance–voltage (C–V) curves by using a C–V model including substrate quantization and poly-Si depletion [3]. The work function of metal electrode is determined from the flatband voltage (V_{FB}).

Fig. 1 shows I–V simulations on a metal-gated TaN/SiO_2 capacitor. From the calculations, for SiO_2 with an EOT of 2.15 nm [8], the gate current at accumulation is mainly comprised of the electron tunneling from the metal gate to the substrate ($J_{ME,G}$). However, the hole tunneling from the accumulated p-Si substrate to the gate ($J_{VBH,S}$) also contributes to the gate current appreciably and it even dominates over the electron tunneling current in low gate voltage regime. This is distinctively different from poly-Si gated SiO_2 devices. In poly-Si gate device, when biasing NMOS into accumulation,

1784   IEEE TRANSACTIONS ON ELECTRON DEVICES, VOL. 51, NO. 11, NOVEMBER 2004

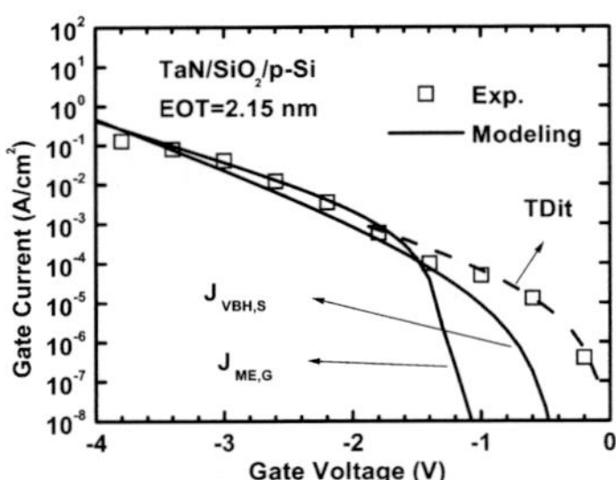

Fig. 1. Tunneling currents in a TaN–SiO$_2$/p-Si capacitor. The experimental data is from [8]. $J_{\mathrm{VBH},S}$ and $J_{\mathrm{ME},G}$ represents the valence band hole tunneling from substrate and metal electron tunneling from the gate, respectively. The dashed line is the simulated tunneling to interface state current ($\mathrm{T}D_{it}$) by the model of [9].

the electron and hole tunneling begin to occur from the same gate voltage (~ -1 V) and hole tunneling is always negligible compared to electron tunneling due to the larger tunneling barrier for holes. However, when the gate is metal with midgap work function, the electron barrier becomes higher. On the other hand, there is no hole tunneling energy gap at the gate side so hole tunneling occurs earlier than in the case of poly gate device. It results in the dominance of hole current in the electron tunneling forbidden region ($-1 < V_g < -0.5$ V in Fig. 1). Good agreements between simulations and experiments have been obtained at strong accumulation regime. In the regime of V_g between V_{FB} and 0 V, the much higher gate leakage experimentally observed is attributed to the electron tunneling through interface state (TDit) [9], which is verified by the simulation (dashed line) with the model of [9].

The simulations of gate currents in CMOS transistors with HfO$_2$ stacks are displayed in Fig. 2. Solid lines are simulations assuming parabolic dispersion in HfO$_2$ with effective mass values obtained from poly-Si gated devices [7]. Using one fitting parameter of 8 Å oxide interfacial layer (IL), overall agreements between simulations and experiments [10] were obtained for eight current–voltage (I–V) curves measured on both n- and p-FETs with different EOTs and gate voltage (V_g) polarities. Higher gate current values obtained from the simulations than the measurements are found at high V_g, which might arise partly from the nonparabolic effects, as discussed in [6]. The dashed lines are simulations using Freeman–Dahlke dispersion for HfO$_2$ [6], in which second adjusting parameter of hole mass (0.27 m$_0$) is introduced. Better agreement with experimental data at high V_g is obtained. Another reason for the discrepancy is that the modified WKB method used in the simulation is not accurate near the transition region from direct to F–N tunneling.

III. EFFECT OF POLY-SI GATE DEPLETION

For bulk CMOS, metal gates with work functions corresponding to poly-Si conduction and valence bandedges of Si are desirable for n- and p-MOSFET, respectively [1]. In this section, we first compared the gate leakage between metal and

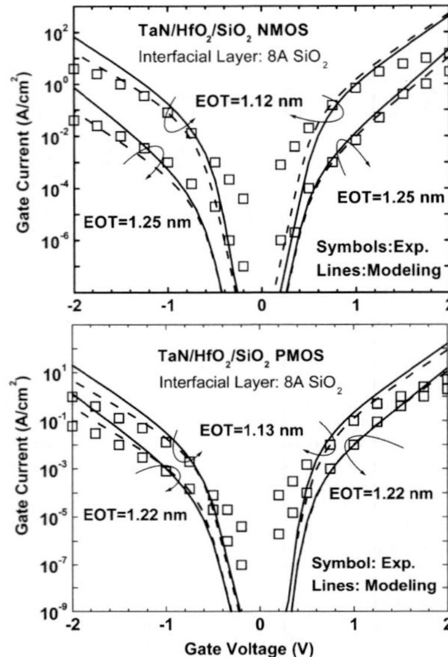

Fig. 2. Tunneling currents in TaN gated MOSFETs with HfO$_2$ stacks. Experimental data are from [10]. Solid lines are simulations assuming parabolic dispersion in HfO$_2$ with effective mass values of 0.18 m$_0$ [7]. Using one fitting parameter of 8 Å oxide interfacial layer, overall good agreements between simulations and experiments are obtained. Better agreement with experimental data at high V_g can be obtained by using Freeman–Dahlke dispersion for HfO$_2$ [6] (dashed lines).

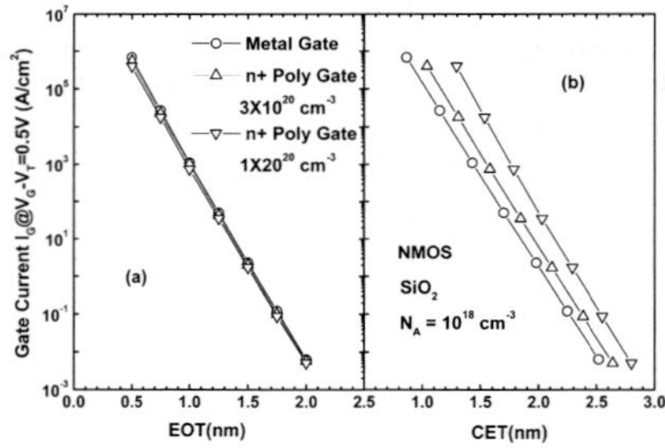

Fig. 3. Gate leakage of metal and poly-Si gate n-MOSFET with SiO$_2$ gate dielectric versus (a) EOT and (b) CET. The metal work function is assumed at Si conduction bandedge. In simulations, the substrate doping $N_A = 10^{18}$ cm^{-3}. EOT is extracted from overall fitting of the (C–V) curves by using a C–V model including substrate quantization and poly-Si depletion [3]. For SiO$_2$ gate dielectric, EOT is simply the physical thickness of SiO$_2$. CET is the capacitance equivalent thickness at inversion. It is defined by ε/C, ε is the permittivity of SiO$_2$ and C is the capacitance density at inversion (here, V_g–$V_T = 0.5$ V) [2].

poly-Si gated devices with the same gate work function. From Fig. 3(a), with the same EOT, poly-Si and metal gate devices show similar gate leakage in magnitude. However, at same

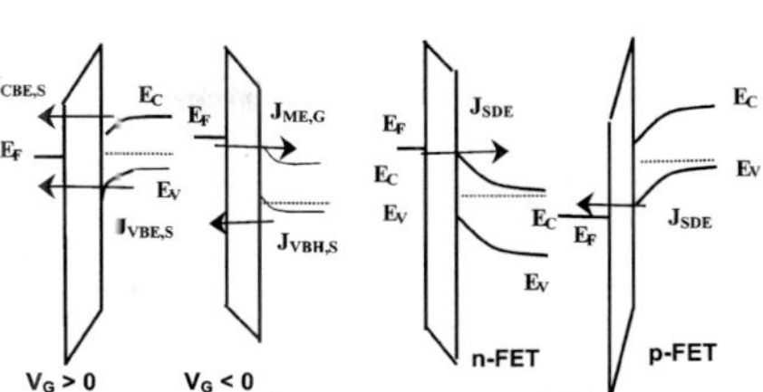

Fig. 5. (a) Band diagram schematics of tunneling in channel area of n-MOSFET. It is similar for p-FET except the substrate Fermi energy. (b) Schematics of band diagrams for gate to source-drain extension (SDE) tunneling, J_{SDE}, at accumulation bias of MOSFETs. In SDE region, the metal electrons tunnel to n^+ SDE in n-FET and valence band electrons tunnel from p^+ SDE to metal gate in p-FET.

Fig. 4. Gate leakage of n-MOSFET in future-generation CMOS for (a) high–performance application using SiO_2 and (b) low stand-by power application using $\text{HfO}_2/\text{SiO}_2$ stack with SiO_2 interfacial layer of 3 Å. In the calculations, the EOTs of gate dielectrics were selected to meet the required CET by ITRS 2001.

capacitance equivalent thickness at inversion (CET), metal gate device exhibits lower gate leakage than poly-Si gate, as shown in Fig. 3(b). CET is defined by ε/C, ε is the permittivity of SiO_2 and C is the capacitance density at inversion (here, $V_g - V_T = 0.5$ V) [2]. CET is thicker than EOT by adding the effective thickness of poly depletion at the gate side and the quantum effect at the substrate inversion layer. The lower gate leakage at the same CET in metal gate devices can be explained by the elimination of gate depletion by using metal gate. Similar results have been reported in [13, Fig. 13] using different approach. From Fig. 3, when poly-Si doping concentration is higher than 5×10^{20} cm^{-3}, the poly depletion effect is not significant. However doping in p^+ poly is difficult to reach such high concentration.

The advantage of metal gate over poly-Si in terms of gate leakage is also elucidated in Fig. 4, in which the gate leakage in future CMOS is predicted from the CET values required by ITRS 2001 [11]. It is projected that metal gate devices can lower the gate leakage by one to two orders of magnitude over poly-Si gated devices with typical poly-Si doping density in current processing technology ($\sim 10^{20}$ cm^{-3}). This reduction of gate leakage suggests the capability to scale gate dielectrics more aggressively (by additional $\sim$ 2–3 Å) by employing metal gate in bulk CMOS.

IV. Effect of Metal Work Function

For advanced CMOS with low doping body SOI structure, metal gate electrodes with work function near mid gap metal gates are required [3], [4]. In bulk CMOS, there also exists possibility to engineer the work function of metal gates in some extent around the poly-Si work function [13]. In this section, we focus on the impact of the metal gate work function on the tunneling currents in metal gated MOSFETs. For such studies, a double gate (DG) structure is selected as a typical example. The body thickness is assumed as 20 nm and body doping concentration 10^{16} cm^{-3}. Although some minor difference may exist, the following results obtained on DG structure are also applicable to other device architectures, such as bulk CMOS or ultra-thin body SOI. The impact of metal work function on gate leakage is studied by comparison of metal gate with work function at Si conduction/valence bandedge with that at Si midgap [14].

A. Gate to Channel Tunneling

The tunneling in channel area is first studied, with the tunneling mechanisms illustrated by the band diagrams in Fig. 5. The tunneling components are labeled by the type of the tunneling carrier (CBE: conduction band electron; VBE: valence band electron; VBH: valence band hole; ME: metal gate electron) and the electrode supplying the carriers (G: gate and S: substrate electrode). Fig. 6 compares the simulated tunneling currents between metal gate and channel in n- and p-MOSFETs with metal work function at Si conduction/valence bandedge (E_C/E_V metal), respectively, and at the midgap of Si. The tunneling currents are shown as a function of (V_g-V_{FB}), which is compared at approximately the same inversion level at operation. From the comparisons, it is clear that the gate current at inversion, which is dominated by $J_{\text{CBE},S}$ and $J_{\text{VBH},S}$ in n- and p-FETs, respectively, is independent of metal work function (Φ_B) after deducting the flatband voltage V_{FB} shift caused by Φ_B variation. However, the increase of Φ_B will reduce metal gate electron tunneling $J_{\text{ME},G}$ due to the increase of electron tunneling barrier. The most significant effect is observed in n-FET at $V_G - V_{\text{FB}} < 0$. As shown in Fig. 6, when metal Fermi energy changes from Si conduction band (E_C) to midgap, $J_{\text{ME},G}$ changes from larger to smaller than $J_{\text{VBH},S}$,

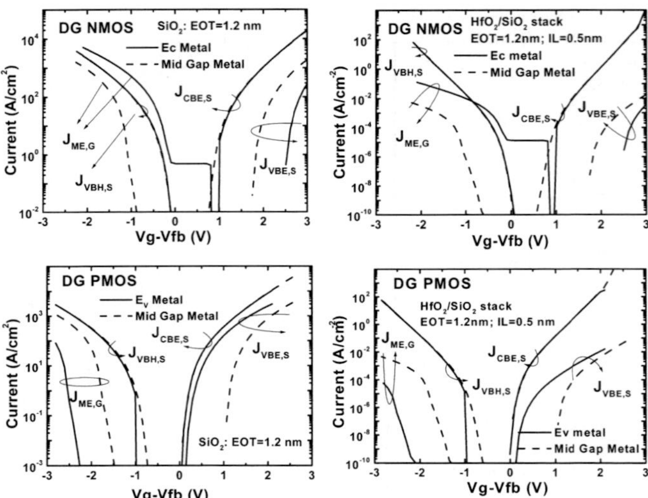

Fig. 6. Tunneling currents of metal double gate (DG) CMOSFETs with SiO_2 and HfO_2 stack as a function of $V_G - V_{FB}$. Solid lines are of metal work function Φ_B at Si conduction/valence bandedge (E_C/Ev metal) for n- and p-MOS, respectively, while dashed lines are those with midgap metal.

the hole tunneling current from substrate to the metal gate. As a result, the total gate current, as elucidated in Fig. 7, first decreases due to the reduced $J_{ME,G}$, then becomes independent of Φ_B after $J_{VBH,S}$ dominates over $J_{ME,G}$. The gate current in n-FET with high-κ dielectric has a less dependency on Φ_B than that with SiO_2. For p-FET, as also shown in Fig. 6, effects of Φ_B variation on various tunneling components are similar as that observed in n-FET. One noticeable characteristic is that the valence band electron tunneling $J_{VBE,S}$ at accumulation bias, which usually contributes negligibly to the gate current in p^+ poly-Si gate p-FET, becomes comparable in magnitude with the gate electron tunneling for SiO_2 gate dielectric when metal work function is at Si valence bandedge (E_V metal). It is worthy to note that $J_{ME,G}$ in NMOS remains flat in between 0–1 V of V_g–V_{FB} from tunneling simulation. In this bias region, the practical current is formed by the tunneling injection of electrons from the gate and the electron-hole recombination in the substrate depletion layer. For low doping substrate, the electron injection current is high and the observed current is determined by the hole generation rate in depletion layer [15].

B. Gate to Source/Drain Extension (SDE) Tunneling

In the following, we will discuss the tunneling between gate and the source/drain extension (SDE) region overlapped with the gate (J_{SDE}). In capacitors or long channel MOSFETs, tunneling in channel is forbidden in low V_g due to the band misalignment and tunneling through trap states may be manifested. However, in short channel MOSFETs with ultrathin gate dielectric, gate-SDE tunneling becomes the dominant current source [16], [17]. Since the SDE dimension is not so scalable as channel dimension, the contribution of gate-SDE tunneling

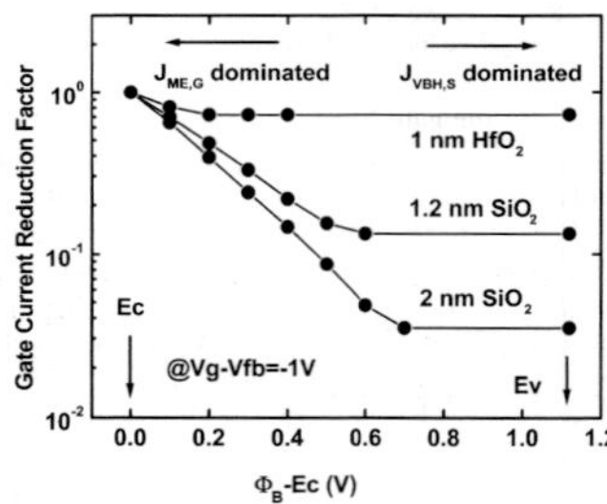

Fig. 7. Relative gate current of n-MOS at $V_G - V_{FB} = -1$ V as a function of metal work function Φ_B.

(J_{SDE}) to off-state leakage current becomes increasingly important as channel length is scaled down. J_{SDE} has been verified to be the dominant source of off-state leakage in FETs with ultrathin gate dielectrics [16]–[18].

J_{SDE} is important at the off-state of a MOSFET when $V_g = 0$ and V_d is biased at high voltage. The band diagrams for gate-SDE tunneling are illustrated in Fig. 5 for metal gate CMOSFETs at accumulation. Compared to that in poly-Si gate device, gate-SDE tunneling shows different mechanism for p-FET. In p-FET, tunneling in SDE comes from valence band electron tunneling from p^+ SDE to metal gate. The hole tunneling current, which dominates the SDE tunneling in p^+ poly p-FET, cannot occur at the presence of metal gate electrode. In n-FET, metal electron tunneling to n^+ SDE forms the gate-SDE tunneling.

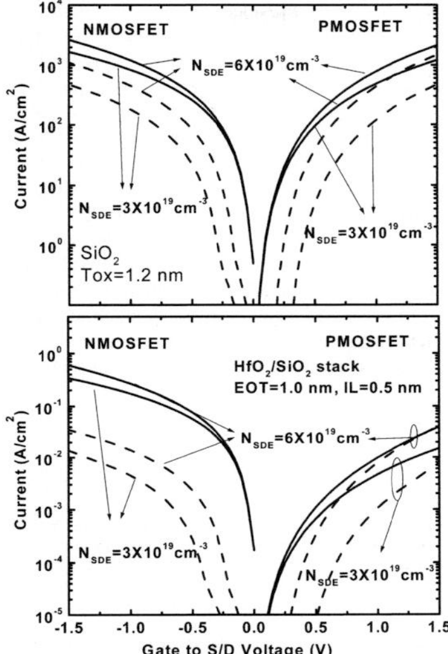

Fig. 8. Gate to SDE tunneling in metal double gate (DG) MOSFETs with SiO_2 and HfO_2 stack. Solid lines are those with E_C metal (NMOS) and E_v metal (PMOS) gates, while dashed lines are those using midgap metal gates.

Fig. 8 compares the gate-SDE tunneling with different metal gate work functions. It is shown that J_{SDE} is sensitive to Φ_B and the use of midgap metal gates reduces the J_{SDE} in both n- and p-FETs. J_{SDE} is also found to have a dependence of SDE doping concentration. Higher doping in SDE leads to higher magnitude of J_{SDE} because of the higher oxide field resulting from the less voltage drop in SDE at the same voltage. This indicates a trade-off between leakage and SDE resistance.

The dependence of J_{SDE} on Φ_B is summarized in Fig. 9. J_{SDE} in n-FET is always higher than that in p-FET in the whole range of Φ_B, indicating that the leakage limit is first reached in n-FET. It is also found that increasing Φ_B reduces J_{SDE} significantly and the reduction is further enhanced when using high-κ dielectric. This reduced J_{SDE} is expected to have an effect on reducing the FET off-sate leakage current.

V. ADVANTAGES OF DG STRUCTURE

Fig. 10 compares tunneling currents in DG and bulk FETs with the same V_T (0.2 V). The threshold voltage V_T is defined by the gate bias to induce inversion charge of 10^{11} cm^{-2}, which is obtained by a self-consistent model [19]. In DG, low body doping is assumed and V_T is achieved by metal gate Φ_B, whereas heavy uniform channel doping is used for V_T adjustment in bulk device. For FET on-state, DG shows slightly lower gate leakage, which is explained by the electric field lowering due to low body doping [20]. With respect to standard bulk MOSFET, DG FET exhibits significant advantage of reduced J_{SDE} due to the adjustment of metal Φ_B to near midgap. For n-FET, which is the limiting case as demonstrated in Fig. 9,

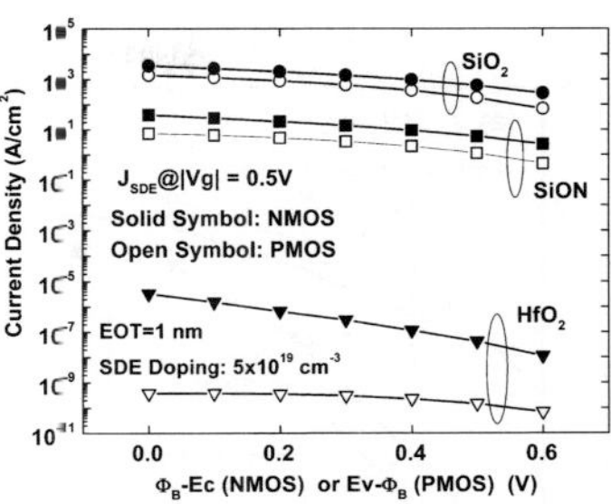

Fig. 9. Effect of metal gate work function on the gate to SDE tunneling J_{SDE} for various gate dielectrics.

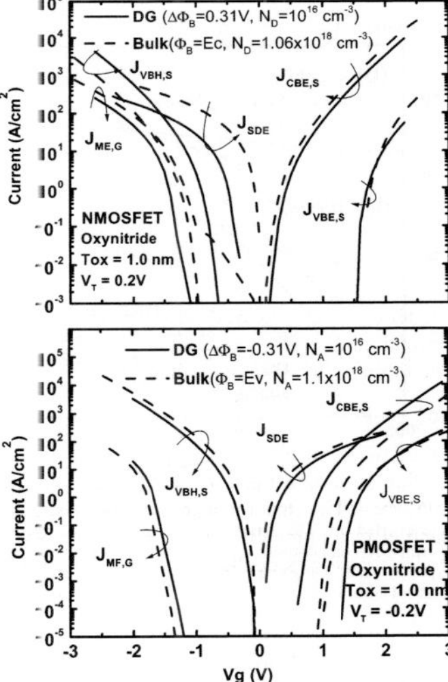

Fig. 10. Comparison of metal gate DG and bulk FET at same threshold voltage V_T (defined by inversion charge of 10^{11} cm^{-2}). In DG FET, Si body thickness is 20 nm, low body doping is assumed and V_T is adjusted by metal gate work function Φ_B while in bulk FET, V_T is tuned by channel doping. $\Delta\Phi_B$ is the metal gate work function shift from Si conduction/valence band for n/p MOSFET.

J_{SDE} is a ~ 1 order of magnitude lower for DG than for bulk FET, demonstrating the advantage of DG to suppress the off-state leakage induced by gate to SDE tunneling. The leakage current impacts adversely on circuit performance, such as the standby power reduction of an inverter in digital circuits and improvement of accuracy of sample/holder in analog circuits [21]. The reduced leakage current at both on- and off-states suggests the potential of using thin body SOI structure.

To study the impact of metal gate work function engineering on the device scalability, the off-state leakage (I_{OFF}) contributed by J_{SDE} in future CMOS technology is calculated

1788   IEEE TRANSACTIONS ON ELECTRON DEVICES, VOL. 51, NO. 11, NOVEMBER 2004

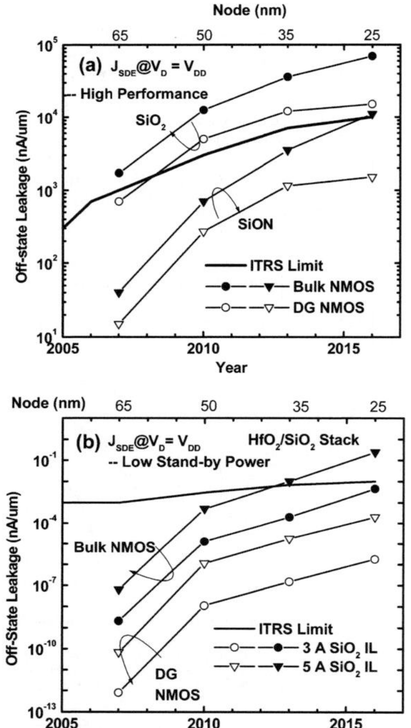

Fig. 11.   Off-state leakage contributed by gate-SDE tunneling in metal gate n-FET estimated by SDE dimension of 5 nm for (a) high-performance application using SiO_2 and optimized oxynitride [7], (b) low power application using HfO_2/SiO_2 stack. Values of EOT, V_T and V_{DD} from ITRS2001 were used in the calculation.

and presented in Fig. 11 for both bulk and DG SOI CMOS. In the calculations, values of EOT, V_T, and V_{DD} from International Technology Roadmap for Semiconductors (ITRS) 2001 were used for each generation. In DG, low body doping is assumed and metal gate work function is determined by V_T, while in bulk FET, metal work function is at Si conduction bandedge and uniform channel doping is used for V_T adjustment. From the simulated I_{OFF}, metal DG SOI demonstrates its potential in suppressing J_{SDE}. This reduction of J_{SDE} is expected to be as much as two to three orders of magnitude in low power application when high-κ is employed, suggesting the superior scaling capability from leakage perspective by utilizing metal gate DG SOI structure.

VI. CONCLUSION

Simulations of tunneling leakage current in metal gate MOSFETs have been carried out and the results show good agreements with measurements for both SiO_2 and HfO_2 gate

dielectrics with TaN electrodes. Further, the impact of metal gate on the gate leakage is studied. In bulk CMOS, the use of metal gate can reduce gate leakage appreciably by 1–2 orders of magnitude at the same CET (capacitance equivalent oxide thickness at inversion), due to the elimination of poly-silicon depletion. The effects of metal gate work function on various tunneling components in MOSFETs and the criterion for choosing suitable metal work function to reduce the gate current have been examined. In SOI CMOSFETs, the use of midgap metal gate results in significant reduction of gate to SDE tunneling in both n- and p-FETs. As a result, SOI FETs exhibit much lower off-state leakage than the bulk ones and this reduction of off-state leakage can be as much as three orders of magnitude when a high-κ gate dielectric is used, indicating their superior scalability in terms of leakage currents.

REFERENCES

[1] E. Josse and T. Skotnick, "Polysilicon gate with depletion or metallic gate with buried channel: What evil worse?," in *IEDM Tech. Dig.*, 1999, pp. 661–664.

[2] D. J. Frank, R. H. Dennard, E. Nowak, P. M. Solomon, Y. Taur, and H. P. Wong, "Device scaling limits of Si MOSFETs and their application dependencies," *Proc. IEEE*, vol. 89, pp. 259–259, Jan. 2001.

[3] L. Chang, S. Tang, T. J. King, J. Bokor, and C. Hu, "Gate length scaling and threshold voltage control of double-gate MOSFETs," in *IEDM Tech. Dig.*, 2000, pp. 719–722.

[4] Y. Taur, "An analytical solution to a double-gate MOSFET with undoped body," *IEEE Electron Device Lett.*, vol. 21, pp. 245–247, Mar. 2000.

[5] G. D. Wilk, R. M. Wallace, and J. M. Anthony, "High-κ gate dielectrics: Current status and materials properties considerations," *J. Appl. Phys.*, vol. 89, pp. 5243–5275, 2001.

[6] Y. T. Hou, M. F. Li, Y. Jin, and W. H. Lai, "Direct tunneling hole current through ultrathin gate oxides in metal–oxide–semiconductor devices," *J. Appl. Phys.*, vol. 91, pp. 258–264, 2002.

[7] Y. T. Hou, M. F. Li, H. Y. Yu, Y. Jin, and D. L. Kwong, "Quantum tunneling and scalability of HfO_2 and HfAlO gate stacks," in *IEDM Tech. Dig.*, 2002, pp. 731–734.

[8] Y. H. Kim, C. H. Lee, T. S. Jeon, W. P. Bai, C. H. Choi, S. J. Lee, L. Xinjian, R. Clarks, D. Roberts, and D. L. Kwong, "High quality CVD TaN gate electrode for sub-100-nm MOS devices," in *IEDM Tech. Dig.*, 2001, pp. 667–670.

[9] A. Ghetti, E. Sangiogi, J. Bude, T. W. Sorsch, and G. Weber, "Tunneling into interface states as reliability monitor for ultrathin oxides," *IEEE Trans. Electron Devices*, vol. 47, pp. 2358–2365, Dec. 2000.

[10] C. H. Lee, J. J. Lee, W. P. Bai, S. H. Bae, J. H. Sim, X. Lei, R. D. Clark, Y. Harada, M. Niwa, and D. L. Kwong, "Self-aligned ultra thin HfO2 CMOS transistors with high quality CVD TaN gate electrode," in *Symp. VLSI Tech. Dig.*, 2002, pp. 82–83.

[11] (2001) International Technology Roadmap for Semiconductors. [Online]. Available: http://public.itrs.net/Files/2001ITRS/

[12] Berkeley Device Group. [Online]. Available: www.device.berkeley.eecs.edu/qmcv

[13] W. P. Maszara, Z. Krivokapic, P. King, J. S. Goo, and M. R. Lin, "Transistor with dual work function metal gates by single fully silicidation (FUSI) of polysilicon gates," in *IEDM Tech. Dig.*, 2002, pp. 367–370.

[14] B. Cheng, B. Maiti, S. Samavedam, J. Grant, B. Taylor, P. Tonin, and J. Mogab, "Metal dates for advanced sub-80-nm SOI CMOS technology," in *IEEE Int. SOI Conf.*, 2001, pp. 91–92.

[15] A. Ghetti, C. T. Liu, M. Mastrapasqua, and E. Sangiorgi, "Characterization of tunneling current in ultrathin gate oxide," *Solid State Electron.*, vol. 44, pp. 1523–1531, 2000.

[16] N. Yang, W. K. Hension, and J. J. Wortman, "Analysis of tunneling currents and reliability of nMOSFETs with sub-2-nm gate oxides," in *IEDM Tech Dig.*, 1999, pp. 453–456.

[17] K. N. Yang, H. T. Huang, M. J. Chen, Y. M. Lin, M. C. Yu, D. C. Yu, S. M. Jang, and M. S. Liang, "Edge hole direct tunneling in off-states ultrathin gate oxide p-channel MOSFETs," in *IEDM Tech Dig.*, 2000, pp. 679–182.

[18] S. Song, H. J. Kim, J. Y. Yoo, J. H. Yi, W. S. Kim, N. I. Lee, K. Fujihara, H. K. Kang, and J. T. Moon, "On the gate oxide scaling of high-performance CMOS transistors," in *IEDM Tech Dig.*, 2001, pp. 55–58.

[19] S. Takagi, M. T. Takagi, and A. Toriumi, "Accurate characterization of electron and hole inversion-layer capacitance and its impact on low voltage operation of scaled MOSFETs," in *IEDM Tech. Dig.*, 1998, pp. 619–622.

[20] L. Chang, K. J. Yang, Y. C. Yeo, Y. K. Choi, T. J. King, and C. Ho, "Reduction of direct tunneling gate leakage current in double-gate and ultrathin body MOSFETs," in *IEDM Tech Dig.*, 2001, pp. 99–102.

[21] C. H. Choi, K. Y. Nam, Z. Yu, and R. Dutton, "Impact of gate direct tunneling current on circuit performance: A simulation study," *IEEE Trans. Electron Devices*, vol. 48, pp. 2823–2829, 2001.

Tony Low received the B.S. degree (with first class honors) in electrical engineering from the National University of Singapore (NUS) in 2002. He is currently pursuing the Ph.D. degree in electrical engineering and physics at the Silicon Nano Device Laboratory (SNDL), NUS.

Upon graduation, he was awarded sponsorships from Singapore Millennium Scholarship and the Chartered Semiconductor Manufacturing Company for his research at NUS. His research interests include the study of carrier transport in semiconductor and its band structure calculation.

Yong-Tian Hou (S'02–M'04) received the B.S. and M.S. degrees in physics from Peking University, Beijing, China, in 1990 and 1993, respectively. He is currently pursuing the Ph.D. degree in electrical engineering at the National University of Singapore (NUS).

From 1998 to 1999, he was with the Department of Electrical and Computer Engineering, NUS, as a Research Engineer. He is now with the Silicon Nano Device Laboratory (SNDL), NUS. His research interests include nano device integration and device modeling.

Ming-Fu Li (M'91–SM'99) received the degree from the Department of Physics, Fudan University, Shanghai, China, in 1960.

After graduation, he joined the University of Science and Technology of China (USTC), Hefei, as a Teaching assistant and then lecturer. In 1978, he joined the Graduate School faculty, Chinese Academy of Sciences, Beijing, and became a professor in 1986. He has also served as Adjunct Professor at the Institute of Semiconductors, Chinese Academy of Sciences, Fudan University, and USTC. He was a Visiting Scholar at Case Western Reserve University, Cleveland, OH in 1979, and at the University of Illinois at Urbana-Champaign from 1979 to 1981, and was a Visiting Scientist at the University of California at Berkeley and Lawrence Berkeley National Laboratories from 1986 to 1987, and 1990 to 1991, respectively. He joined the Department of Electrical Engineering, National University of Singapore in 1991, and became a Professor in 1996. His current research interests are in the areas of CMOS device technology, reliability, quantum modeling, and Analog IC design. He has published over 200 research papers and two books, including *Modern Semiconductor Quantum Physics* (Singapore: World Scientific, 1994).

Dr. Li has served on several international program committees and advisory committees in international semiconductor conferences in China, Japan, Canada, Germany, and Singapore.

Dim-Lee Kwong (A'84–SM'90) received the B.S. degree in physics and the M.S. degree in nuclear engineering from the National Tsing Hua University, Taiwan, R.O.C., in 1977 and 1979, respectively. He received the Ph.D. degree in electrical engineering from Rice University, Houston, TX, which won the Best Dissertation Award in 1982.

He was an Assistant Professor with the Electrical Engineering Department, University of Notre Dame, Notre Dame, IN, from 1982 to 1985. He was a Visiting Scientist with the IBM General Technology Division, Essex Junction, VT, during the summer of 1985, working on 4-Mb DRAM technology. He joined the Microelectronic Research Center and the Department of Electrical and Computer Engineering, The University of Texas, Austin, in 1985 as an Assistant Professor. He was promoted to Associate Professor in 1985 and to Full Professor in 1990. He is the author of more than 310 journal and 270 referred archival publications and has been awarded more than 22 U.S. patents. His current areas of research interests include rapid thermal CVD technology for the growth and deposition of semiconductor materials compatible with ULSI processes, advanced dielectrics for logic, analog, and memory devices, metal gate electrode, shallow junctions, and high dielectrics. Forty-three students received the Ph.D. degree under his supervision.

Dr. Kwong has received numerous awards, including the IBM Faculty Development Award in 1984 and the Engineering Foundation Teaching Award from the University of Texas, Austin, in 1994, and holds the Earl N. and Margaret Brasfield Endowed Fellowship.

Reprinted with permission from Tony Low, M.F. Li, Chen Shen, Yee-Chia Yeo, Y.T. Hou,
Chunxiang Zhu Albert Chin and D. L. Kwong, Appl. Phys. Letts,
Vol.85, pp.2402–2404 (2004). Copyright 2004, American Institute of Physics

APPLIED PHYSICS LETTERS　　　　　VOLUME 85, NUMBER 12　　　　　20 SEPTEMBER 2004

Electron mobility in Ge and strained-Si channel ultrathin-body metal-oxide semi conductor field-effect transistors

Tony Low, M. F. Li,[a] Chen Shen, Yee-Chia Yeo, Y. T. Hou, and Chunxiang Zhu
*Silicon Nano Device Laboratory, Department of Electrical and Computer Engineering National University
of Singapore and Institute of Microelectronics, Singapore 119260, Singapore*

Albert Chin
Department of Electronics Engineering, National Chiao Tung University, Taiwan

D. L. Kwong
Department of Electrical and Computer Engineering, University of Texas, Austin, Texas 78752

(Received 26 February 2004; accepted 2 July 2004)

Electron mobility in strained silicon and various surface oriented germanium ultrathin-body (UTB) metal-oxide semiconductor field-effect transistors (MOSFETs) with sub-10-nm-body thickness are systematically studied. For biaxial tensile strained-Si UTB MOSFETs, strain effects offer mobility enhancement down to a body thickness of 3 nm, below which strong quantum confinement effect renders further valley splitting via application of strain redundant. For Ge channel UTB MOSFETs, electron mobility is found to be highly dependent on surface orientation. Ge$\langle 100 \rangle$ and Ge$\langle 110 \rangle$ surfaces have low quantization mass that leads to a lower mobility than that of Si in aggressively scaled UTB MOSFETs. © *2004 American Institute of Physics.* [DOI: 10.1063/1.1788888]

Ultrathin-body (UTB) transistors with sub-10-nm-body thickness T_{body} is a promising candidate for device scaling into the sub-30-nm gate length L_G regime. However, degradation of electron mobility in UTB devices with sub-10-nm T_{body} was found experimentally.[1,2] Degration of mobility also leads to reduce current drivability in the linear regime[3] despite the improved gate inversion layer capacitive coupling with reduced body scaling.[3] Of particular concern is current drivability under high drain biases for decananometer channel length devices. With regard to this, Lundstrom[4] has pointed out, via a phenomenological approach, that the transport in decananometer metal-oxide semiconductor field-effect transistor (MOSFETs) is essentially source limited; hence the mobility at high vertical surface field, which embodies the effective scattering rate in the vicinity of the source, remains relevant. In addition, recent reports on aggressively scaled UTB devices have highlighted the importance of the interfacial perturbation attributed to the roughness of the Si/SiO$_2$ surface,[2,5] which is found to strongly limit the carrier mobility. For enhanced device performance, channel materials such as Ge and strained-Si (formed directly on insulator without a relaxed SiGe buffer layer) may be employed in UTB transistor.[6,7] Nevertheless, there is little work on their potential advantages. Little is also known about the carrier mobility in these UTB devices with advanced channel materials. In this letter, we perform a modeling study of the electron mobility in UTB transistor with sub-10-nm-body thicknesses employing strained-Si and various surface orientations of Ge as the channel material. A calibrated physical model that takes the effect of scattering due to optical phonons, acoustic phonons, surface roughness, and interface states into account is used.

Electronic structures for the two-dimensional electron gas are obtained by solving the coupled Schrödinger–Poisson equation self-consistently within the envelope function based

effective mass framework according to Stern *et al.*[8] Important bandstructure parameters such as the conduction valleys energy minima and their longitudinal and transverse masses used are obtained from Fischetti *et al.*[9] A unitary transformation is employed[8,10] to obtain the transport masses along the device coordinates for devices with various crystal orientations. The 2D density-of-states mass is preserved after the transformation in our context of low longitudinal field. The scattering matrix elements due to acoustic phonons (AP), optical phonons (OP), surface roughness (SR), and interface states (DIT) related scattering are then systematically formulated. The model for phonon spectrum in the bulk semiconductors are adapted from Jacoboni *et al.*[11,12] where the matrix elements of the electron-phonon interaction are considered in accordance with Price[11,13,14] Intravalley acoustic phonon (AP) with an effective isotropic deformation potential[11,15,16] intravalley optical phonon (OP) for L valleys[11,12] and intervalley phonons constraint within the selection rules for f and g type processes[11] are accounted for. Dynamic screening of phonons is disregarded.[17] Surface roughness (SR) scattering was conventionally treated by accounting for the localized perturbation potential due to variations of interface positions according to Ando's.[18–20] The perturbation Hamiltonian induced by energy level fluctuations has been obtained[21,22] for a rectangular quantum well potential, but this approach may not be accurate for the treatment of surface roughness in UTB devices. Issues also remain about the accurate treatment of perturbation potential due to change in charge density induced by SR. Consequently, we employed a phenomenological treatment as outlined by Gamiz.[23] The autocorrelation function of the asperities is assumed to be Gaussian. Intersubband transitions are left unscreened and the dielectric matrix is expressed according to Ref. 14 and in the quantum size limit when applicable, else it is left unscreened. Interface state (DIT) induced scattering potential according to Stern *et al.*[8] based on a perturbative approach is employed. By imposing appropriate boundary conditions, the scattering potential in all regions of interest can be obtained

[a] Electronic mail: elelimf@nus.edu.sg

Appl. Phys. Lett., Vol. 85, No. 12, 20 September 2004

Low *et al.*   2403

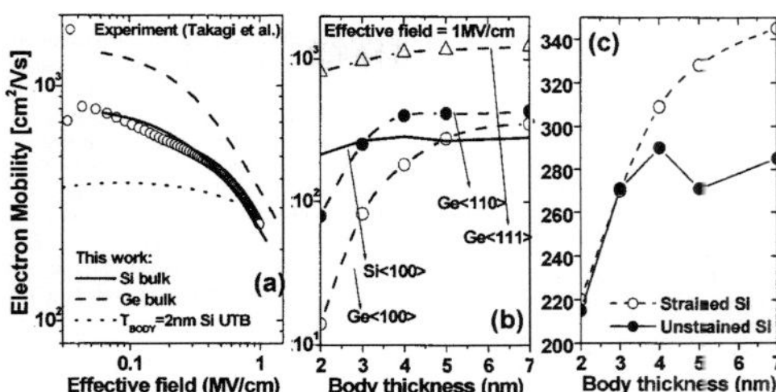

FIG. 1. (a) Calibration of our theoretical low-field mobility model with experimental results for Si (Ref. 24). Theoretical calculated mobility for a 2 nm T_{body} Si UTB MOSFET is also shown. Screening for SR scattering is accounted. (b) Electron mobility for various advanced channel UTB transistors as function of body thickness. Si channel is oriented in [010] direction. Ge$\langle 100 \rangle$ is oriented along [010] channel direction, Ge$\langle 110 \rangle$ at [1$\bar{1}$0] and Ge$\langle 111 \rangle$ is isotropic. We ignored the neighboring Δ valley in this work. (c) replots the mobility curve in linear scale for the Si$\langle 100 \rangle$ curve in the main figure. A peak is clearly shown, in good agreement with the experiment (See Refs. 1 and 2). A mobility curve for strained Si$\langle 100 \rangle$ is also plotted for comparison.

using the Nystrom method.[15] The scattering rate can be obtained by the Fermi Golden Rule. We then obtained the numerical solutions of the scattering time to the Boltzmann equation in the Ohmic regime by embracing the relaxation time approximation and imposing the appropriate scattering condition under detailed balance condition at equilibriums.[15,16]

Our physical model is calibrated using experimental Si mobility data,[24] showing good agreement [Fig. 1(a)]. An effective acoustic deformation potential of 15 eV[14,15] was used. As current processing technology is still unable to yield a reliable set of mobility data for Ge MOSFETs, a deformation potential of 15 eV for acoustic phonon intravalley process within valleys is assumed, yielding a reasonable two times mobility compared to Si counterpart[25,26] as shown in Fig. 1(a). A SR autocorrelation function with root mean square $\Delta = 4$ Å and correlation length $l = 10$ Å is assumed for Si and Ge[7] surfaces. These technologically dependent parameters are assumed to apply to UTB transistor technology. A conservative interface states density of 1×10^{11} cm^{-2} for each of the front and back interfaces is assumed. Our UTB device has a gate dielectric with an EOT of 1 nm, a metal gate electrode (which provides efficient charge screening), and back oxide thickness of 50 nm. The mobility for a 2 nm T_{body} Si UTB MOSFET is calculated as shown in Fig. 1(a). It is observed that its electron mobility at high surface field does not exhibit the same dependency on effective field as the bulk universal mobility. Perturbation Hamiltonian due to SR H_{SR} as obtained to first order approximation is:

$$H_{SR}(z, \mathbf{r}) \cong \frac{q_0 [V(z, \Delta_m) - V(z, 0)] \Delta(\mathbf{r})}{\Delta_m}, \qquad (1)$$

where the coordinates z (perpendicular to Si/SiO$_2$ interface, measured from back oxide interface) and $\mathbf{r}$ (vector in the plane of the Si/SiO$_2$ interface) are employed. $\Delta(\mathbf{r})$ is a function which effectively describes the sum of SR at the two interfaces and Δ_m is the statistical mean of the SR. $V(z, \Delta_m)$ is the electrostatic potential with a surface perturbation of Δ_m, which is also solved self-consistently accounting for the finite body thickness fluctuation Δ_m. However, at large body thickness, one obtains

$$T_{body \to \infty}^{lim} H_{SR} \cong q_0 \frac{V(z + \Delta_m, 0) - V(z, 0)}{\Delta_m} \Delta(\mathbf{r}) \cong q_0 \frac{\partial V}{\partial z} \Delta(\mathbf{r}), \qquad (2)$$

where H_{SR} is now proportional to the surface field $\partial V / dz$, accounting for the usual dependence of surface roughness limited mobility on effective field $\sim F_{eff}^2$ in the bulk

Si MOSFET. When T_{body} in the order of SR, deviation from usual electric field dependency is captured by H_{SR} in Eq. (1).

The electron mobility as a function of T_{body} is calculated for strained Si (with biaxial tensile strain of 2%, considerably larger based on current technology) and Ge with different surface orientations, as shown in Fig. 1(b). At high effective field of 1 MV/cm, SR induced scattering dominates the effective mobility. The limited mobilities due to AP, OP, SR, and DIT are also calculated as a function of T_{body} (Fig. 2), clearly indicating the dominance of SR limited mobility at an effective field of 1 MV/cm. The application of strain is found to lose its effectiveness at $\sim$3 nm of T_{body} as shown in the inset of Fig. 1(c). This is because, at small T_{body}, strong quantum confinement lifts the Δ_4 valleys much beyond that of the high mobility Δ_2 valleys, rendering further valley splitting via application of strain redundant. It has been assumed that the SR spectrum function in strained or un-

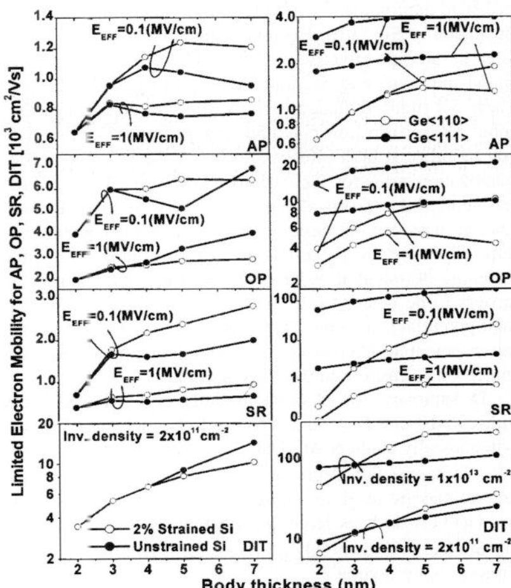

FIG. 2. Limited low field mobilities for strained Si and Ge UTB transistor, respectively. Acoustic phonons, optical phonons, SR, and interface charge limited mobilities are all systematically explored. All limited mobilities are plotted at constant effective field of 0.1 MV/cm (threshold condition) and 1 MV/cm (high inversion condition) except for interface charge limited mobility plotted at a constant electron density criterion.

2404　Appl. Phys. Lett., Vol. 85, No. 12, 20 September 2004　　Low *et al.*

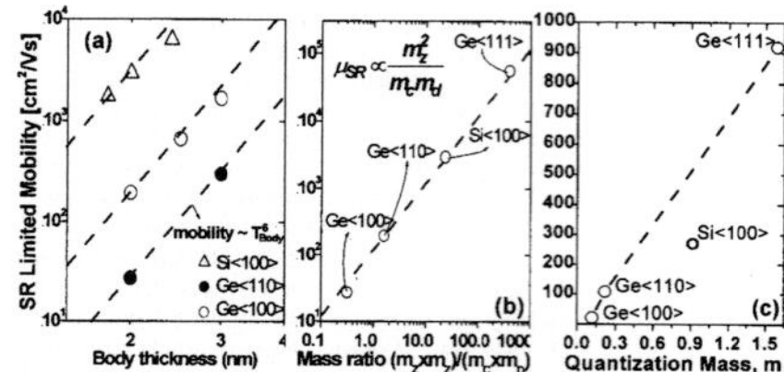

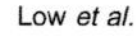

FIG. 3. (a) SR limited mobility plotted at effective surface field of 0.1 MV/cm, exhibiting approximately the T_{body}^6 dependency, (b) surface roughness limited mobility at effective surface field of 0.1 MV/cm as function of mass ratio as expressed in inset, with T_{body} =2 nm under same SR condition. (c) Electron mobility as function of quantization mass m_z. Simulated at T_{body}=2 nm and E_{eff} =1 MV/cm. High quantization mass m_z, is beneficial for aggressively scaled UTB device.

strained Si devices are the same. Of particular interest is the choice of channel surface orientations for optimum device performance. Figure 1(b) compares the total effective mobility of Ge and Si channel UTB devices with T_{body} of sub-10-nm. Interestingly, while the electron mobility in bulks Ge is higher than that in bulk Si, Ge⟨110⟩ and Ge⟨100⟩ UTB devices have lower mobility than Si⟨100⟩ at T_{body} below 3 nm. Ge⟨111⟩ shows better mobility than Si⟨100⟩ at all body thicknesses. Generally, it is observed that at sufficiently thin T_{body}, the electron mobility begins to degrade. The onset and amount of this degradation differs for the various orientations.

Sakaki *et al.*[22] has established the SR limited mobility with a T_{body} to a power of 6 dependency for the of quantum well.[22] Uchida *et al.* have also experimentally verified this dependency for Si UTB with small T_{body} at low inversion charge condition.[2] Our model predicts a similar dependency on T_{body} for the SR mobility at a low constant surface field of 0.1 MV/cm, as shown in Fig. 3(a). In addition, the SR limited mobility for UTB with small T_{body} at low constant surface field approximately follows a mass ratio relationship as elucidated in Fig. 3(b). However, since SR scattering dominates in the high effective surface field regime, it is of paramount importance to examine their mobility in this regime, where SR induced charge perturbation will contribute additional SR perturbation potential. Figure 3(c) plots the electron mobility at high effective surface field as a function of quantization mass for the various channel materials. A general trend of decreasing electron mobility with reduced quantization mass can be observed. This can be phenomenologically explained by the effect of quantization mass on the screening of potential. A larger quantization mass causes the carriers to be nearer to the interface resulting in more efficient potential screening. This reduces the overall SR perturbation potential. Conversely, a small quantization mass will render very sensitive to the SR condition.

In summary, the electron mobility of UTB MOSFETs with sub-10-nm T_{body} and advanced channel materials are systematically studied. At small T_{body}, Ge⟨100⟩ and Ge⟨110⟩ suffer large mobility degradation due to their very low quantization masses, resulting in high susceptibility to SR scattering. Ge⟨111⟩ with its large quantization mass and low density of state mass is highly desirable for high mobility in the ultrathin-body regime.

This work is supported by the Singapore A*STAR R263-000-267-305 and IME/03–450002 JML/SOI Grant. We gratefully acknowledge useful discussions with D. Esseni on Coulomb scattering. We appreciate useful discussions with M. V. Fischetti and D. K. Ferry pertaining to their published literatures. We also thank S. Takagi for providing the experimental data from his classic paper.

[1]K. Uchida, J. Koga, R. Ohba, T. Numata, and S. Takagi, Tech. Dig. - Int. Electron Devices Meet. **2001**, 633 (2001).

[2]K. Uchida, H. Watanabe, A. Kinoshita, J. Koga, T. Numata, and S. Takagi, Tech. Dig. - Int. Electron Devices Meet. **2002**, 47 (2002).

[3]See Ref. 1. Figure 10 shows the enhanced inversion layer capacitances with body scaling from 25 to 7 nm. Figure 13 shows the degradation of current drivability in linear regime, attributing mainly to degradation of low-field mobility.

[4]M. S. Lundstrom, IEEE Electron Device Lett. **22**, 293 (2001).

[5]Z. Ren, P. M. Solomon, T. Kanarsky, B. Doris, O. Dokumaci, P. Pldigies, R. A. Roy, E. C. Jones, M. Leong, R. J. Miller, W. Haensch, and H. S. Wong, Tech. Dig. - Int. Electron Devices Meet. **2002**, 51 (2002).

[6]K. Rim, K. Chan, L. Shi, D. Boyd, J. Ott, N. Klymko, F. Cardone, L. Tai, S. Koester, M. Cobb, D. Canaperi, B. To, E. Duch, I. Babich, R. Carruthers, P. Saunders, G. Walker, Y. Zhang, M. Steen, and M. Leong, Tech. Dig. - Int. Electron Devices Meet. **2003**, 49 (2003).

[7]S. Nakaharai, T. Tezuka, N. Sugiyama, Y. Moriyama, and S. Takagi, Appl. Phys. Lett. **83**, 3516 (2003).

[8]F. Stem and W. E. Howard, Phys. Rev. **163**, 816 (1967).

[9]M. V. Fischetti and S. E. Laux, J. Appl. Phys. **80**, 2234 (1996), see Figs. 1 and 4.

[10]A. Rahman, A. Ghosh, and M. Lundstrom, Tech. Dig. - Int. Electron Devices Meet. **2003**, 471 (2003).

[11]C. Jacoboni and L. Reggiani, Rev. Mod. Phys. **55**, 645 (1983).

[12]C. Jacoboni, F. Nava, C. Canali, and G. Ottaviani, Phys. Rev. B **24**, 1014 (1981).

[13]P. J. Price, Ann. Phys. (San Diego) **133**, 217 (1981).

[14]C. Jungemann, A. Emunds, and W. L. Engl, Solid-State Electron. **36**, 1529 (1993).

[15]D. Esseni and A. Abramo, IEEE Trans. Electron Devices **50**, 1665 (2003).

[16]D. Esseni, A. Abramo, L. Selmi and E. Sangiorgi, Tech. Dig. - Int. Electron Devices Meet. **2002**, p 719 (2002).

[17]M. V. Fischetti, and S. E. Laux, Phys. Rev. B **48**, 2244 (1993).

[18]T. Ando, J. Phys. Soc. Jpn. **43**, 1616 (1977).

[19]T. Ando, A. B. Fowler, and F. Stern, Rev. Mod. Phys. **54**, 437 (1982).

[20]D. Esseni, IEEE Trans. Electron Devices **51**, 394 (2004).

[21]C. Y. Mou and T. M. Hong, Phys. Rev. B **61**, 12612 (2000).

[22]H. Sakaki, T. Noda, K. Hirakawa, M. Tanaka, and T. Matsusue, Appl. Phys. Lett. **51**, 1934 (1987).

[23]F. Gamiz, J. B. Roldan, J. A. Lopez-Villanueva, P. Cartujo-Cassinello, and J. E. Carceller, J. Appl. Phys. **86**, 6854 (1999).

[24]S. Takagi, A. Toriumi, M. Iwase, and H. Tango, IEEE Trans. Electron Devices **41**, 2357 (1994).

[25]C. M. Ransom, T. N. Jackson, and J. F. DeGelormo, IEEE Trans. Electron Devices **38**, 2695 (1991).

[26]C. H. Huang, D. S. Yu, A. Chin, C. H. Wu, W. J. Chen, C. Zhu, M. F. Li, B. J. Cho, and D. L. Kwong, Tech. Dig. - Int. Electron Devices Meet. **2003**, 319 (2003).

Impact of Surface Roughness on Silicon and Germanium Ultra-Thin-Body MOSFETs

Tony Low[1,4], M. F. Li[1,2], W. J. Fan[3], S. T. Ng[3], Y.-C. Yeo[1], C. Zhu[1], Albert Chin[1*], L. Chan[4], D. L. Kwong[5]

(1) Silicon Nano Device Lab (SNDL), ECE Depart., National University of Singapore Email: elelimf@nus.edu.sg
(2) Institute of Microelectronics, Singapore (3) School of EEE, Naryang Technological University, Singapore
(4) Technology Development, Chartered Semiconductor Manufacturing, Singapore
(5) ECE Depart., University of Texas, Austin, USA *On leave from National Chiao Tung University

Ultra-thin body (UTB) SOI MOSFET is promising for sub-50 nm CMOS technologies [1]. However, recent experimental finding [2] suggests the need for serious reconsiderations of its long-term scaling capability into the sub-10 nm body thickness (T_{BODY}) regime. Two new phenomena attributed to surface roughness (SR) are identified [2]; they are enhanced threshold voltage (V_{TH}) shifts and drastic degradation of mobility with a T_{BODY}^3 dependence [2,3]. In this work, we detail a study of these two phenomena in UTB MOSFETs with sub 10 nm T_{BODY} Si and Ge channels. Firstly, the phenomena of enhanced V_{TH} shifts is modeled by accounting for the fluctuation of quantized energy levels due to SR up to second order approximation. Good corroboration with experimental results [2] is obtained. Our model is then applied to examine the impact of enhanced V_{TH} shifts on metal gate workfunction requirements. Secondly, we modeled the SR-limited electron and hole mobility and discuss their impact on the choice of surface orientations. Mobility anisotropy are also examined for the various surface orientations.

I. Physics and Modeling Methodology

NMOS electrostatics is solved self-consistently via the conventional one-band Schrödinger-Poisson equation. For PMOS, the multi-band Luttinger-Kohn Hamiltonian [4] is used and solved within the triangular well approximation [5]. As the valence band mixing with Γ_2 conduction band is well known in the study of Si-Ge quantum wells [7], we first need to ascertain the validity of the conventional six-band approach [4] in our study. Fig. 1 investigates this band mixing effect for a 3 nm quantum well structure, showing negligible deviations from the conventional six-band calculations except for the higher energy subbands in Ge. Therefore, a six-band approach will suffice. Important band structure parameters [6,7] for electrons and holes are listed in Table 1 and 2, respectively. Enhanced V_{TH} shift is modeled by accounting for the effect of SR induced energy levels fluctuation [10]. The energy level fluctuation has to be obtained up to the second order. In this non-linear approximation, a symmetric distribution of body thickness fluctuation due to SR can give rise to an effective increase of quantized energy level as shown in Fig 2a. This effective increase of quantized energy level is then taken into account during the computation of the inversion charge density according to Fermi-Dirac statistics. The resulting enhanced V_{TH} shift can thus be calculated. Fig. 2 shows the good agreement between the measured and calculated V_{TH} shifts of NMOS and PMOS devices with a root-mean-square roughness Δ_M of 0.85 nm, by taking the two-interfaces together. SR-limited hole mobility are calculated using a similar methodology as outlined by Fischetti et al [5]; by embracing the isotropic approximation for relaxation time while retaining of the full anisotropy of the electronic bandstructure. In the context of quantum well, the perturbation potential due to SR is usually modeled by expressing the perturbation Hamiltonian according to an energy level fluctuation term [3,9]. In particular, Meyerovich et al [19] obtained similar expressions via a canonical transformation of the system to that of smooth boundaries. This formulation is employed in this work. This model allows us to circumvent the difficulty of defining a value for the quantization mass as required in Ando et al.'s [8] model, which is made difficult due to mixing between the various hole subbands. SR-limited electron mobility is studied within a phenomenological model [18]. Fig. 3 shows the excellent fitting of electron mobility as function of temperature and of T_{BODY} at 25 K, with a root-mean-square roughness Δ_M of 0.6 nm and auto-correlation length L of 2.12 nm. Deviation at temperature larger than 100 K is due to the onset of

phonon scattering, which is unaccounted for in our model. The T_{BODY}^6 relationship with SR-limited electron mobility as reported in experiments [2,3] is also observed.

II. Enhanced V_{TH} for Various Bandstructure

Figs. 4 and 5 show the results of calculation of enhanced V_{TH} shifts for Si and Ge channel UTB MOSFETs with various surface orientations. T_{BODY} in the range of 2 to 5 nm is considered. The dependency of V_{TH} shift on T_{BODY} can be qualitatively analyzed by examining the case for that of a square quantum well where the quantized subband energy has a simple T_{BODY}^{-2} dependency. In the quantum limit, V_{TH} shifts have a T_{BODY}^{-2} dependency and the enhanced V_{TH} shift component will have a $T_{BODY}^{-4}\Delta_M^2$ dependency. Therefore, in the limit where the SR perturbation is of the same order as T_{BODY}, the enhanced V_{TH} shift component will contribute as significantly to the overall V_{TH} shifts. The amount of V_{TH} shift however, also depends directly on the carrier quantization mass m_Z since the quantization energy is approximately proportional to m_Z^{-1}. For NMOS devices, electron m_Z for both Si and Ge on various surface orientations are given in Table 1. V_{TH} shift is the smallest for the surface orientation with the largest m_Z, namely Si<100> for Si NMOS and Ge<111> for Ge NMOS as shown in Fig. 4a and 5a, respectively. With decreasing T_{BODY}, an anomalous retarded increase of V_{TH} shift is observed for Ge<100> at T_{BODY} of 3 nm. This is due to the transfer of carrier occupation from the L to the Δ valleys [12]. For PMOS devices, the result can be qualitatively interpreted in a similar fashion using the set of empirically fitted m_Z as shown in Fig. 15. The empirically fitted m_Z is calculated by fitting the subband energies at the zone center Γ to the analytical expression for energy dispersion of a quantum well. An interesting phenomenon of increasing effective quantization mass with decreasing T_{BODY} for the <110> surface is evident. The physical origin is identified to be the dependence of valence band mixing effect on the quantum well thickness. For Si and Ge PMOS devices, the <100> surface orientation shows the largest V_{TH} shift due to its lowest empirical m_Z. Due to the increasing m_Z, <110> surface advantageously exhibits a retarded increase of V_{TH} shift as T_{BODY} is scaled down, as shown in Figs. 4b and 5b. To assess the implications of enhanced V_{TH} shift on device performance, we shall evaluate its impact on metal gate workfunction requirement and V_{TH} variations σ_{VTH}.

III. Impact on Metal Gate Workfunction Requirement

Fig. 6 and 7 show the metal gate workfunction requirement for Si and Ge UTB MOSFETs, respectively, at all surface orientations calculated within a stipulated OFF-state charge density criterion as outlined in Fig. 6 caption, accounting for both the carrier quantization and SR effects. Carrier quantization and SR-induced workfunction shifts can amount to as large as 0.7 V for a T_{BODY} of 2 nm. This sets the gate workfunctions for NMOS and PMOS devices further apart, rendering metal gate workfunction engineering more challenging as current choice of metal gate materials only offers workfunction in the range of between ~4.1 eV to ~5.2 eV [13]. The smaller valence electron ionization energy of Ge advantageously sets its gate workfunction requirement for PMOS devices within the workfunction range that is currently achievable.

IV. Impact on Threshold Voltage Variation σ_{VTH}

Parametric mismatch and fluctuations have considerable impact on performance and yield [14]. UTB device with an undoped channel has the advantage of keeping σ_{VTH} due to random spatial fluctuation of dopants under control [15]. However, when T_{BODY} is scaled

                                                    IEDM 04-151

down, σ_{VTH} will be largely affected by T_{BODY} non-uniformity [16]. For a transistor with T_{BODY} = 3 nm, Fig. 8 shows the maximum allowable T_{BODY} variance in order to meet the industry target of σ_{VTH} of 20 mV [14]. It is interesting to note that the effect of SR reduces the maximum allowable T_{BODY} variance by as much as 50%. In Si or Ge UTB MOSFETs, the use of $<110>$ surface for NMOS or $<100>$ surface for PMOS would require a maximum T_{BODY} variation of ~1 atomic layer. This presents a significant challenge for substrate technologies.

V. SR-Limited Electron Mobility $\mu_{E,SR}$

A high mobility channel is essential for deca-nanometer devices to achieve ballistic operation [17]. Fig. 9 shows the calculated SR-limited electron mobility $\mu_{E,SR}$ calculated using a phenomenological model [18]. A simple mass ratio $(m_z \cdot m_z)/(m_D \cdot m_C)$ relationship for $\mu_{E,SR}$ is elucidated in Fig. 9b, where m_D is the density-of-states mass, and m_C is the conductivity mass. $Ge<111>$ yields the highest SR-limited electron mobility. In contrast, the small mass ratio for $Ge<100>$ and $Ge<110>$ leads to lower mobilities. This simple mass ratio provides a guideline for substrate selection (material and crystal orientation) for devices operating in quantum limit.

VI. SR-Limited Hole Mobility $\mu_{H,SR}$

Figs. 10 and 11 show the channel orientation dependence of the calculated SR-limited hole mobility $\mu_{H,SR}$ for Si and Ge devices with T_{BODY} = 3 nm. For $Si<100>$, we note that the anisotropy of the first two subbands neutralize each other, similar to the case for bulk $Si<100>$ [5]. Fig. 12 examines the dependence of the equi-energy lines on T_{BODY} for the $<100>$ surface. The ground state energy for $Si<100>$ is shown to maintain its strong anisotropy with the decrease of T_{BODY} down to 3 nm; presenting an optimum channel direction along $[100]$. For $Si<110>$, $\mu_{H,SR}$ exhibits high anisotropy with an optimum channel direction along $[011]$ (Fig. 12). Fig. 13 depicts an interesting phenomenon for $Si<110>$; where an increased radial carrier velocity along $[011]$ channel direction for the UTB device is observed as compared to its bulk counterpart. This is attributed to the dependence of valence band mixing effect on T_{BODY} for the $<110>$ surface, rendering its energy dispersion characteristics strongly dependent on T_{BODY}. These observations should also qualitatively apply to $Ge<110>$ since their energy dispersion characteristics exhibit similar behavior. For Ge devices, the surface and channel orientation dependence of $\mu_{H,SR}$ are generally very similar to that of Si devices except for the $<100>$ surface orientation, which exhibits an optimum channel direction along $[110]$. In general, $\mu_{H,SR}$ of Ge is little affected by higher energy subbands at T_{BODY} of 3 nm, due to the large energy quantization effects. The dependence of $\mu_{H,SR}$ on T_{BODY} and surface orientations is shown in Fig. 14. The impact of surface orientations on $\mu_{H,SR}$ can also be explained via its empirical masses. Fig. 15 plots the calculated empirical m_z. Fig. 16 shows the calculated empirical m_D. m_C is a tensor quantity and cannot be conveniently tabulated for the complicated hole bandstructure. However, m_D qualitatively depicts the average m_C for a given energy. $\mu_{H,SR}$ for the $<100>$ surface is the most limiting. Its small empirical m_z for the first subband; Si~$0.28m_0$ and Ge~$0.21m_0$, renders it very sensitive to SR scattering processes. Furthermore, the larger empirical m_D in $Si<100>$ compared to $Ge<100>$ results in a lower mobility. For the $<110>$ surface, its large m_z yields it the largest $\mu_{H,SR}$; hence it is most probable that SR-scattering will not be a very limiting mechanisms for UTB MOSFETs on $<110>$ surface. It also shows a retarded decrease of $\mu_{H,SR}$ with decreasing T_{BODY}, deviating sharply from the expected T_{BODY}^6 dependency. This can be explained by the increase of its empirical m_z with T_{BODY}. In addition, the empirical m_z of $Ge<110>$ is approximately half that of $Si<110>$, leading to a higher susceptibility to SR scattering. However, this is compensated by a smaller m_D for $Ge<110>$, resulting in comparable $\mu_{H,SR}$ for $Si<110>$ and $Ge<110>$. For the $<111>$ surface, comparable $\mu_{H,SR}$ for Si and Ge are also observed.

Summary: We outlined a simple model to account for the enhanced V_{TH} shifts in UTB MOSFETs, as recently observed experiments. The phenomena of enhanced V_{TH} shifts can be modeled by accounting for the fluctuation of quantized energy levels due to SR using a second order approximation. Our model is then used to examine the enhanced V_{TH} shifts phenomena in Si and Ge UTB MOSFETs. In particular, we examine its technological impact on metal gate workfunction requirements and the V_{TH} variations for UTB MOSFETs. SR-limited mobility for NMOS in the quantum limit can be qualitatively modeled by a simple relationship involving a mass ratio $(m_z \cdot m_z)/(m_D \cdot m_C)$. The crystal surfaces with the highest mobility for Si and Ge are $<100>$ and $<111>$, respectively. For PMOS, a similar relationship is followed, though the valence band structure is more complicated. In particular, the $<110>$ surface exhibits an increase of empirical m_z with reduced T_{BODY} due to the dependence of valence band mixing on T_{BODY}. This makes the $<110>$ surface least susceptible to SR scattering at ultra-thin T_{BODY}. The $<100>$ surface exhibits smallest SR-limited hole mobility. The $<110>$ surface also presents an increased mobility anisotropy as T_{BODY} is scaled down. Our study concludes that a channel material with a large m_z is an important selection criterion for scaling T_{BODY} into the sub-5 nm regime.

Acknowledgements: We greatly appreciate the insightful and useful discussions with M. V. Fischetti (T. J. Watson Research, IBM) and C. Y. Mou (National Tsing Hua University) on modeling of SR-limited mobility. This work is supported by Singapore A*STAR R263-000-267-305 grant and IME/NUS JML R263-000-221-112 grant.

References: [1] ITRS 2003 [2] K. Uchida, IEDM 2002 p. 47 [3] H. Sakaki, APL 51, p. 1934, 1987 [4] J. M. Luttinger, PR 97, p. 869, 1955 [5] M. V. Fischetti, JAP 94, p. 1079, 2003 [6] M. V. Fischetti, JAP 80, p. 2234, 1996 [7] S. Ridene, PRB 64, 085329, 2001 [8] T. Ando, RMP 54, p. 437, 1982 [9] C. Y. Mou, PRB 61, p. 12612, 2000 [10] T. Low, paper in preparation [11] D. Esseni, TED 51, p. 394, 2004 [12] T. Low, IEDM 2003, p. 691 [13] H.Y.Yu, EDL 25, p.337, 2004 [14] H. P. Tuinhout, Proc. ESSDERC 2002, p. 95 [15] C.T.Sah, ICSICT 2001, p.12 [16] G. Tsutsui, Si Nano. Elec. Workshop 2004 p. 4.2 [17] M. S. Lundstrom, EDL 22, p293, 2001 [18] F. Gamiz, JAP 86, p. 6854, 1999 [19] A. E. Meyerovich , PRB 51, p. 17116, 1995

	valley	$m_z\ (m_0)$	$m_D\ (m_0)$	g	$E_S\ (eV)$
$Si<100>$	Δ	0.916	0.190	2	0
	Δ	0.190	0.417	4	0
$Si<110>$	Δ	0.315	0.324	4	0
	Δ	0.190	0.417	2	0
$Si<111>$	Δ	0.258	0.358	6	0
$Ge<100>$	L	0.117	0.295	4	0
	Δ	0.200	0.436	2	0.15
	Δ	0.950	0.200	2	0.15
$Ge<110>$	L	0.218	0.216	2	0
	L	0.080	0.357	2	0
	Δ	0.200	0.436	2	0.15
	Δ	0.330	0.340	4	0.15
$Ge<111>$	L	1.590	0.080	1	0
	L	0.089	0.337	3	0
	Δ	0.271	0.374	6	0.15

Table1: Electron effective masses m_z (quantization) m_D (density-of-states). g: valley degeneracy. E_S is energy split reference conduction band minimum. Bandstructure parameters at conduction valley minima for Si and Ge; such as their ellipsoidal forms and valley energy splits, are obtained from [6]. Quantization mass and density-of-states masses for solving metal-oxide-semiconductor electrostatics on various surface orientations is calculated according to [8].

	γ_1	γ_2	γ_3	E_G	Δ	E_P	m_c
Si	4.285	0.339	1.446	4.185	0.044	21.60	0.528
Ge	13.38	4.24	5.69	0.898	0.297	26.30	0.038

Table2: Numerical values of bulk parameters for Si and Ge. The Kane energy E_P, energy gap E_G and spin-orbit splitting Δ are given in eV. γ_j (j=1,2,3) are Luttinger parameters and m_c is effective mass (in m_0) at the band edge of type Γ_2^-. Modified Luttinger parameters for eight-band model are calculated from [7].

6.5.2

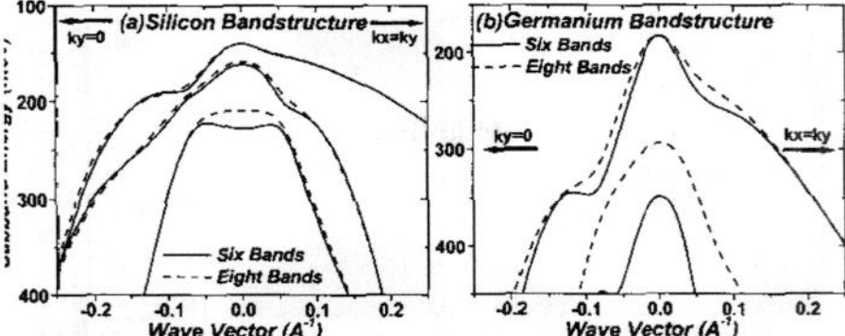

Fig1: *(a)* Electronic bandstructure for a 3nm *Si* quantum well with *<100>* surface. Calculations done with eight-band [7] and six-band [4] are compared. As the direct band-gap for *Si* is relatively large (See Table 2), the coupling of valence bands with conduction bands is negligible. *(b)* Same as Fig. 1, except for *Ge*. Deviation from results of six-band Hamiltonian is non-negligible at higher energies. However, for electrostatics calculation at threshold condition, we can ignore higher subbands for *Ge*, which are ~100meV larger than the first subband.

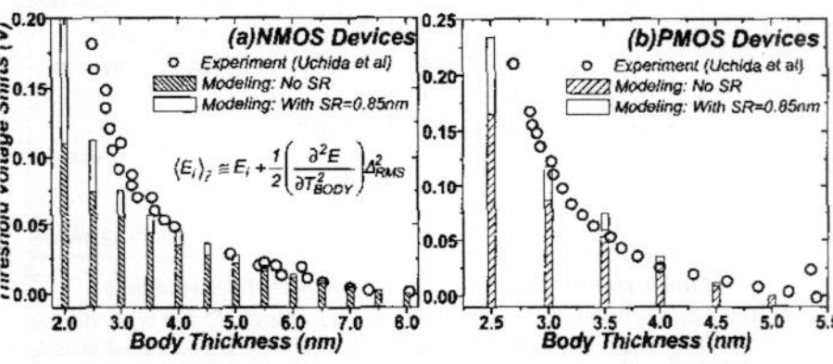

Fig2: *(a)* Enhanced V_{TH} shifts due to *SR* (with respect to the V_{TH} of device with T_{BODY}=8nm) for *NMOS*. Excellent corroboration with experimental [2] is obtained. An effective root-mean square *SR* Δ_{RMS} of 0.85nm for the two interfaces taken together is employed. *(b)* V_{TH} shifts (with respect to the V_{TH} of device with T_{BODY}=8nm for *PMOS* calculated using triangular model with same *SR* parameters.

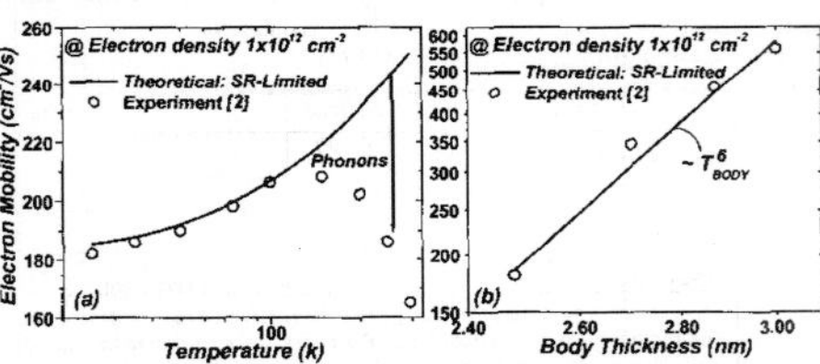

Fig3: *(a)* Measured *SR*-limited electron mobility with T_{BODY}=2.48nm at 25K [2] and simulated result with *SR* Δ_{RMS} of 0.60nm *and SR* auto-correlation length *L* of 2.12nm. Deviation at temperature larger than 100K is due to onset of phonon scattering. *(b)* Measured *SR*-limited electron mobility as function of T_{BODY} at 25K [2]. Same *SR* parameters used for calculation, with observed T_{BODY} to-power-of-six relationship as reported in experiment [3].

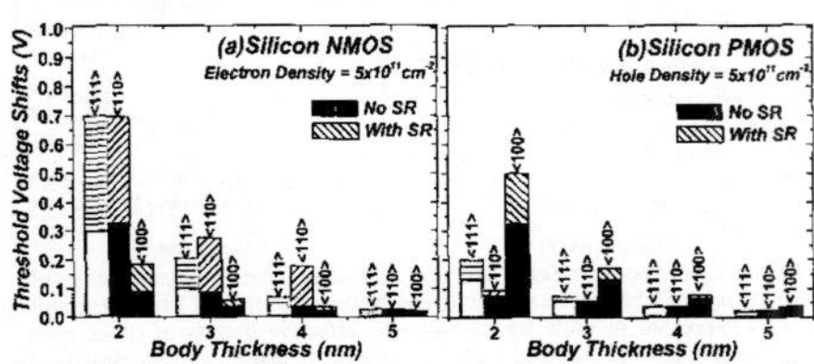

Fig4: Simulated V_{TH} shifts with respect to the V_{TH} of device with T_{BODY}=8nm. Calculated for *Si* *(a)* *NMOS* and *(b)* *PMOS* devices with 3 different surface orientations. Same *SR* parameters as in Fig. 2 are used. All simulations are performed at an electron inversion density of 5x10^{11}cm^{-2}.

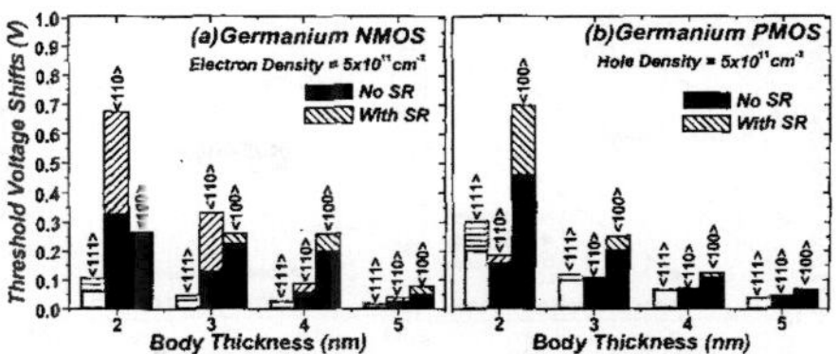

Fig5: Same as Fig. 4 except that these are for *Ge*; *(a)* *NMOS* and *(b)* *PMOS* devices.

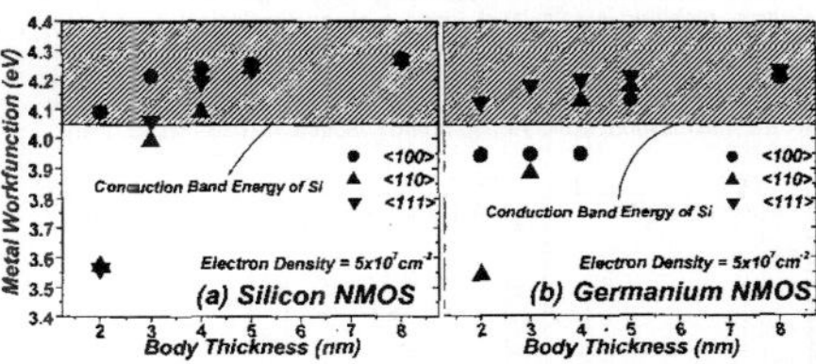

Fig6: Workfunction for *(a)* *NMOS* and *(b)* *PMOS* devices calculated for *Si*. Effect of enhanced V_{TH} shifts is taken into account with same *SR* parameters as Fig.2. *OFF*-state carrier density assumed to be 5x10^{7}cm^{-2} with ideal sub-threshold slope 60mV/decade. Shaded region is the energy values within the bandgap of *Si*.

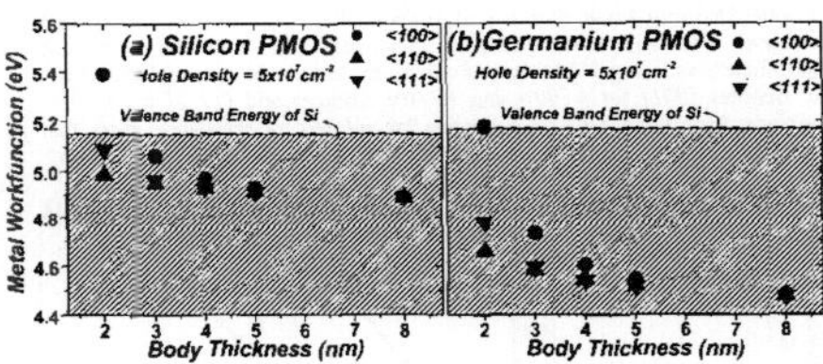

Fig7: Same as Fig. 6, except that these are for workfunction requirement of *(a)* *NMOS* and *(b)* *PMOS* *Ge* channel devices.

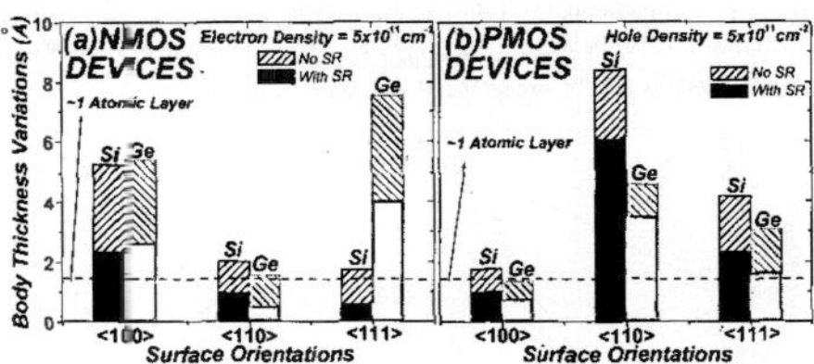

Fig8: Maximal tolerable amount of T_{BODY} variations in order to meet industry target of threshold variations σ_{VTH} of 20mV [14], calculated for *(a)* *NMOS* and *(b)* *PMOS* devices at T_{BODY}=3nm. Same *SR* parameters are used (Fig. 2). In undoped *UTB* devices, on-chip V_{TH} variations (σ_{VTH}) is dominated by T_{BODY} variations. V_{TH} defined at constant inversion carrier density of 5x10^{11}cm^{-2}. Dashed line shows T_{BODY} variations equivalent to 1 atomic layer for *Si<100>*. Maximal tolerable amount of T_{BODY} variations is reduced when *SR* is considered.

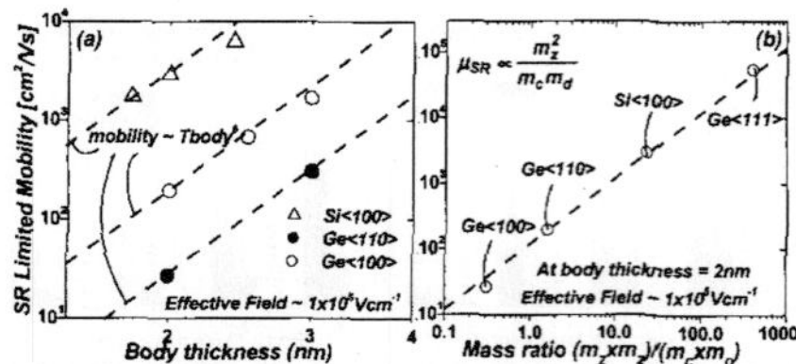

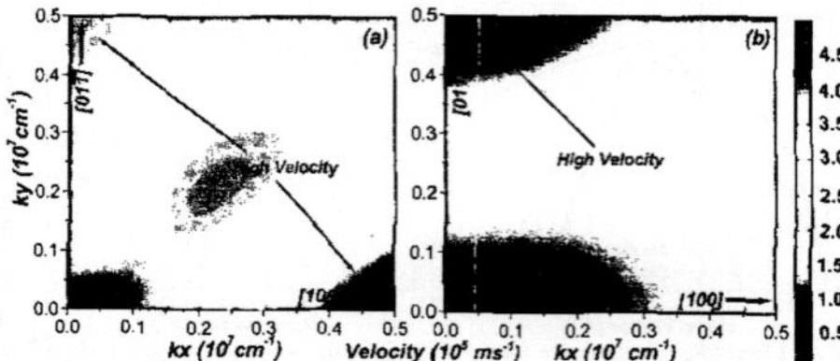

Fig9: *(a) SR*-limited electron mobility for selected channel materials at effective field of $1\times10^5 \text{Vcm}^{-1}$, exhibiting T_{BODY}^6 dependence. *SR*-limited electron mobility is calculated using a phenomenological model [18]. Mobility is shown for optimal channel direction. *(b) SR*-limited electron mobility at same effective field at T_{BODY}=2nm, showing a linear relationship with the mass ratio $(m_z \times m_z)/(m_D \times m_C)$. Where m_z, m_D and m_C are the quantization, density-of-state and conductivity mass respectively.

Fig 13: Intensity plot for hole carrier radial velocity of different in-plane wave vector. Plotted for the ground state energy of $Si<110>$ for *(a)* T_{BODY}=100nm and *(b)* T_{BODY}=3nm at Fs=1MV/cm. Radial velocity $v(k) = \hbar^{-1}\partial E/\partial k$, is obtained by taking the derivative of its energy dispersion in the radial direction.

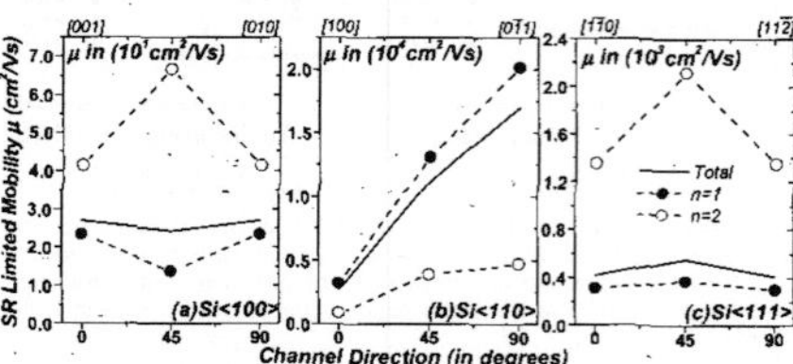

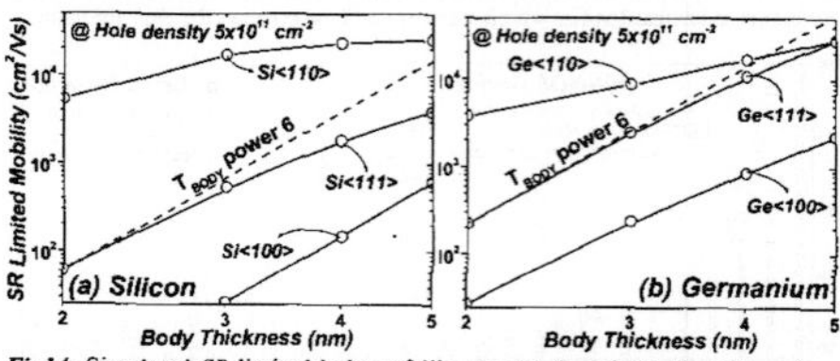

Fig10: *SR*-limited hole mobility of *Si* respectively, for various surface and channel orientations simulated at T_{BODY} of 3nm with the same *SR* parameters as in Fig.4. Mobility is calculated at hole density of $5\times10^{11}\text{cm}^{-2}$. 0^0 denotes *[001]* for *<100>* and *<110>* surfaces and *[11-2]* for *<111>* surface. Mobility for n=1,2 subbands are plotted for reference. Note that the mobilities are expressed in different scale for each surface orientation.

Fig14: Simulated *SR* limited hole mobility for *(a)* Si and *(b)* Ge with various orientations, with same *SR* parameters as Fig.3. Mobility is calculated at hole density of $5\times10^{11}\text{cm}^{-2}$ and result plotted for the optimum channel direction for T_{BODY}=3nm (see Fig. 10 and 11). Hole mobility on *<100>* surface found to be very limiting. The other orientations exhibit relatively high mobility. Mobility deviates from the T_{BODY}^6 dependence especially for the *<110>* surface.

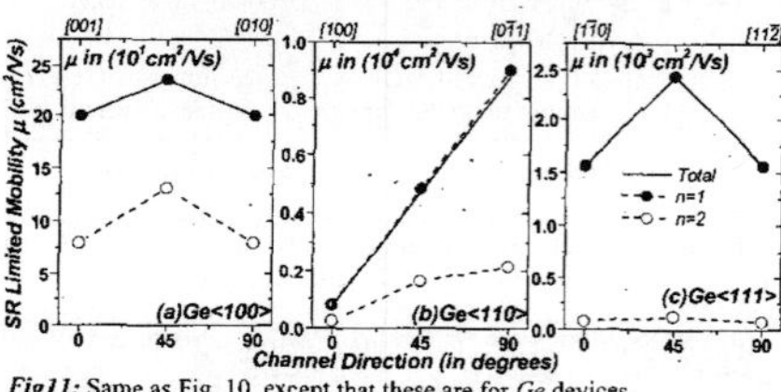

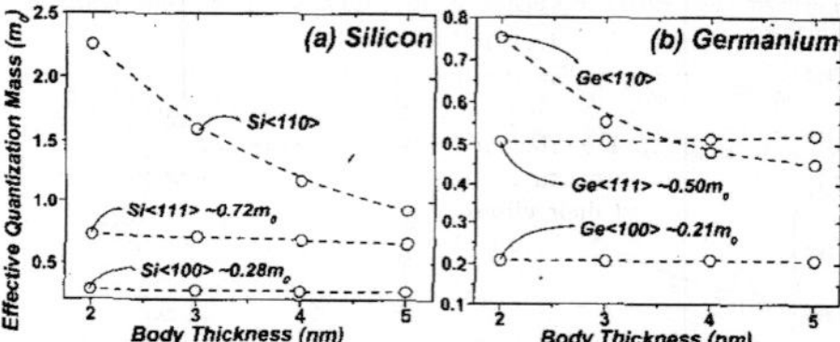

Fig11: Same as Fig. 10, except that these are for *Ge* devices

Fig15: Effective quantization mass m_z calculated for *(a)* Si and *(b)* Ge devices by fitting the subband energies at zone center to the analytical expression for that of a quantum well. Due to the mixing of the different hole bands, the calculated m_z is not constant especially for the *<110>* surface. Filled (unfilled) symbol is for the n=1 (n=2) subbands. Inversion carrier density is $5\times10^{11}\text{cm}^{-2}$.

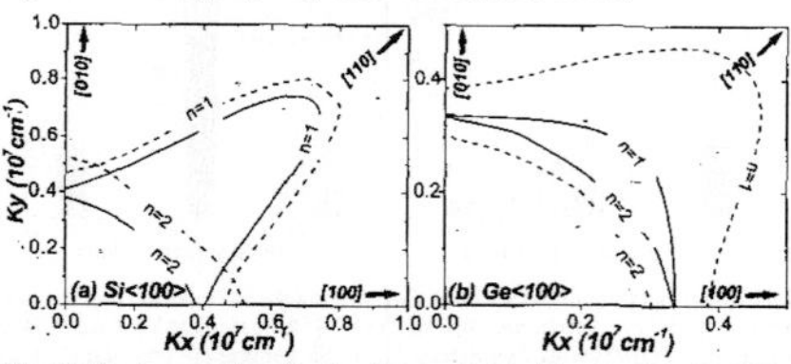

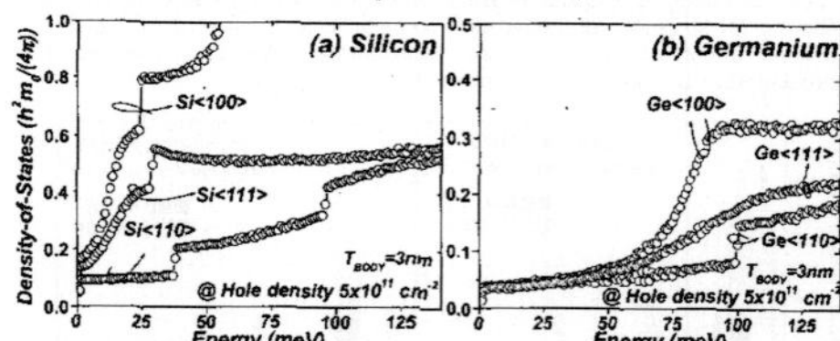

Fig 12: Equi-energy lines for the first two hole subbands of *(a)* $Si<100>$ and *(b)* $Ge<100>$ plotted at energy $(E-E_i, i=1,2)$ of 25meV and 50meV respectively, under (i) Bulk high field conditions: Fs=1MV/cm and T_{BODY}=100nm, represented by solid lines (ii) *UTB* high field conditions: Fs=1MV/cm and T_{BODY}=3nm, represented by dashed lines.

Fig16: Density-of-states for *(a)* Si and *(b)* Ge devices. The empirical density-of-states mass m_D for a particular spin-state can be directly read off from these plots since it is expressed in units for an isotropic effective density-of-states mass. Contributions from all subbands are summed. Inversion carrier density is $5\times10^{11}\text{cm}^{-2}$. The m_D for *Ge* are generally lower than that of *Si*.

6.5.4

Extended Abstracts of the 2004 International Conference on Solid State Devices and Materials, Tokyo, 2004, pp. 776-777

C-7-3

Study of Mobility in Strained Silicon and Germanium Ultra Thin Body MOSFETs

Tony Low [1,3], Chen Shen [1], M. F. Li [1,2], Yee-Chia Yeo [1], Y. T. Hou [1,2], Chunxiang Zhu [1], Albert Chin [4], L. Chan [3], D. L. Kwong [5]
(1) Silicon Nano Device Lab (SNDL), ECE Department, National University of Singapore, Singapore 119260. Email: elelimf@nus.edu.sg
(2)Institute of Microelectronis, Singapore, 117685;   (3) Technology Development, Chartered Semiconductor Manufacturing, Singapore
(4) Dept Electronics Eng., National Chiao Tung Univ., Taiwan; (5) Dept. Electrical and Computer Engineering, University of Texas, Austin, TX 78712, USA

1. Introduction

Electron mobility in strained Si and Ge ultrathin-body (UTB) MOSFETs with sub-10 nm body thickness T_{body} is studied in this paper. In the literatures, only little is known about the carrier mobility in such devices [1-3] despite the fact that low longitudinal field mobility is still important [4,5]. In this paper, an accurate and calibrated physical model that takes the effect of scattering due to optical phonons, acoustic phonons, surface roughness, and interface states into account is used. We found that biaxial tensile strained-Si offers mobility enhancement down to $T_{body} \cong 3\text{nm}$, below which quantum confinement effect gives the same benefit as strain effects, rendering the application of strain to Si redundant. In Ge channel UTB transistors, electron mobility is found to be highly dependent on surface orientation. Ge<100> and Ge<110> surface have low quantization mass that leads to high susceptibility to surface roughness, resulting in even lower mobility than Si in aggressively scaled UTB. Ge<111> with its higher quantization mass and low density of states mass is highly advantageous to channel mobility for aggressively scaled body.

2. Physical Model for Electron Mobility in UTB

Electronic structures for the two-dimensional electron gas are obtained by solving the coupled Schrödinger-Poisson equation self-consistently within the effective mass framework according to Stern [6]. Important bandstructure parameters such as conduction valleys energy minima and their ellipsoidal forms used are obtained from Fischetti [7] (Table 1). Transport masses in the device coordinate are derived by employing suitable unitary transformation [6, 8], such as to preserve the density-of-states mass in our context of low longitudinal field. The scattering matrix elements due to acoustic phonons (AP), optical phonons (OP), surface roughness (SR) and interface states (DIT) related scattering are formulated (Fig 10, 11 caption). The scattering rate is obtained by Fermi Golden Rule. We obtained the numerical solutions of the scattering time to the Boltzmann equation in Ohmic regime by embracing the relaxation time approximation under detailed balance condition [18, 19].

3. Experimental Results and UTB Mobility Modeling

Our physical model is calibrated using experimental Si mobility data [20], showing good agreement. An acoustic deformation potential of 15 eV [12, 18] was used. Using a deformation potential of 15 eV for intra-valley scattering within the L valleys of Ge, a mobility that is twice that of Si is obtained, in reasonable agreement with [21, 22]. A surface roughness autocorrelation function with root mean square $\Delta = 4\text{Å}$ and correlation length $l =10\text{Å}$ is assumed for Si [23] and Ge [3] surfaces. A conservative interface states density of $1\text{x}10^{11}$ cm^{-2} for front and back interfaces is assumed. Our UTB device has a gate dielectric with an EOT of 1 nm, a metal gate electrode, and back oxide thickness of 50 nm. Fig 2 shows the calculated mobility for unstrained Si UTB transistor. It has been reported that as the Si body thickness is reduced [1, 24], mobility is reduced and deviates from the universal relationship. This was due to a variation in body thickness, i.e. δT_{Si}-induced scattering. This phenomenon can be captured via a body perturbation Hamiltonian H_{SR} (Fig 2), where V_P is a perturbed potential by a δT_{Si} of Δ_m to be solved self-consistently, Δ_m the root mean square value, $\Delta(r)$ is the function describing the interface profile, z and r define distance perpendicular and parallel to the dielectric/silicon interface respectively. In the limit of large body thickness, universal relationship with effective field is thus obtained (Fig 2). This, in essence, explains the deviation from the universal relationship as body thickness is reduced to the order of the surface roughness, corroborating with experimental trend [1, 24] (Fig 3).

4. Strained Silicon for Mobility Enhancement

We then examine the impact of biaxial tensile strain of 2% on the band structure and electron mobility in UTB MOSFETs. The limited mobilities due to AP, OP, SR, and DIT are calculated as a function of T_{body} (Fig 10). Strain generally leads to enhancement for the dominant AP and SR limited mobilities for effective vertical field of 0.1MV/cm (threshold condition) and 1MV/cm (strong inversion condition) down to T_{body} of 3 nm. As T_{body} is reduced, the strong body confinement lifts the energy of the Δ_4 valleys, leading to more carrier occupation in the high mobility Δ_2 valleys and reduced carrier occupation at the Δ_4 valleys (Fig. 5). This effect is similar to the effect of biaxial tensile strain. When T_{body} is reduced below 3 nm, quantum confinement achieves the same effect as the strain considered, leading to comparable total mobilities for both strained and unstrained Si. This is illustrated for $T_{body} = 2$ nm in Fig. 4. In addition, the peak mobility observed in unstrained Si due to the subband level modulation [1] is smoothened out with applied strain. The mobility enhancement for SR limited mobility is appreciable for effective field at 0.5MV/cm (moderate inversion condition) and 1MV/cm, for T_{body} down to 4 nm. Subsequent drastic decrease of mobility at thinner body is attributed to the deviation of the perturbation Hamiltonian from direct effective field dependence.

5. Germanium UTB MOSFETs

We had systematically explored the performance limits of Ge UTB MOSFETs [25] relating to its electrostatics, ballistic current and leakage current. Here, we examine the impact of various crystal orientations on the mobility of Ge UTB transistors. Fig 6 compares the total effective mobility of UTB devices with Ge and Si channel at T_{body} of 2nm. While electron mobility in bulk Ge is higher than that in bulk Si, Ge<110> and Ge<100> UTB devices have lower mobility than Si<100>. Ge<111> shows better mobility than Si<100> at $T_{body} = 2$ nm. We then evaluated the limited motilities of Ge<111> and Ge<110> to gain an understanding of the underlying physics. Fig 11 plots the various limited mobilities for Ge<111> and Ge<110> as a function of T_{body}. As T_{body} is reduced, the gradual increase in form factor (the subband wave function overlap integral) for Ge<110> leads to a lower mobility [19]. However, the larger quantization mass of Ge<111> causes the carrier to reside nearer to the channel surface and hence its form factor is less sensitive to the body confinement effect. In particular, the SR limited mobility of Ge<110> is severely degraded as T_{body} is scaled down. Fig 7 highlights the general trend that at high effective field, the effective mobility decreases with reduced quantization mass. A larger quantization mass provides more efficient potential screening, hence reducing the overall SR perturbation potential (Fig 7 inset). Fig 8 illustrates the perturbing potential felt at the same surface roughness condition for Ge<110> and Ge<111>. The small quantization mass of Ge<110> renders it very sensitive to the surface roughness condition, resulting in exponential increase of perturbing potential with reduction in T_{body} and subsequent mobility degradation (Fig 11). Electron mobility of various UTB transistors are summarized in Fig 9.

References: [1] K. Uchida, IEDM, p.47, 2002 [2] K. Rim, IEDM, p.49, 2003 [3] S. Nakaharai., APL83, p.3516, 2003 [4] M. S. Lunstrom, EDL22, p293, 2001 [5] M. S. Lundstrom, IEDM, p.789, 2003 [6] F. Stern, PR163, p.816, 1967 [7] M. V. Fischetti, JAP80, p2234, 1996 [8] A. Rahman, IEDM, p.471, 2003 [9] C. Jacoboni, RMP 55, p.645, 1983 [10] C. Jacoboni, PRB 24, p.1014, 1981 [11] P. J. Price, Ann. Phys 133, p.217, 1981 [12] C. Jungemann, SSE36, p.1529, 1993 [13] M. V. Fischetti, PRB48, p.2244, 1993 [14] F. Gamiz, JAP86, p.6854, 1999 [15] Y. C. Cheng, Surf. Sci. 27, p.663, 1971 [16] T. Ando, J. Phys. Soc. Jpn43, p.1616, 1977 [17] J. Lee, JAP54, p.6989, 1983 [18] D. Esseni, TED50, p.1665, 2003 [19] D. Esseni, IEDM, p.719, 2002 [20] S. Takagi, TED41, p.2357, 1994 [21] C. M. Ransom., TED38, p.2695, 1991 [22] C. H. Huang, IEDM, p.319, 2003 [23] T. Sugano, Surf. Sci98, p.154, 1980 [24] K. Uchida, IEDM, p.805, 2003 [25] T. Low, IEDM, p.691, 2003

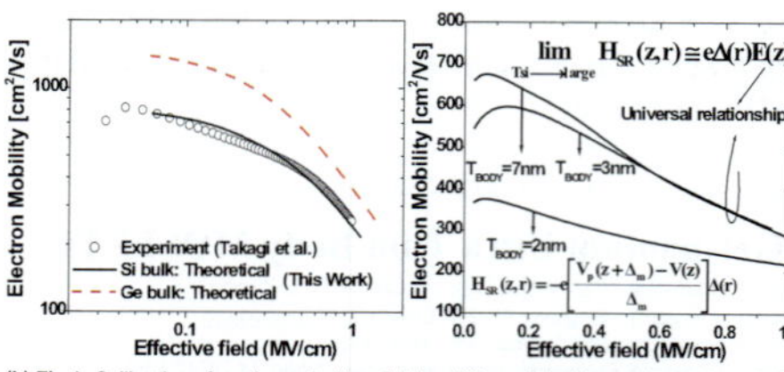

(L) Fig 1: Calibration of our theoretical low-field mobility model with experimental results for Si [15], showing excellent agreement. A two times mobility for Ge is obtained [21, 22] by fitting the technological dependent acoustic deformation potential for L valleys **(R) Fig 2:** Theoretical calculated total effective mobility curve for Si UTB at various body thicknesses demonstrating an explanation for the non-universality of mobility relationship with effective field (Fig 3). Effective field is the calculated mean electric field.

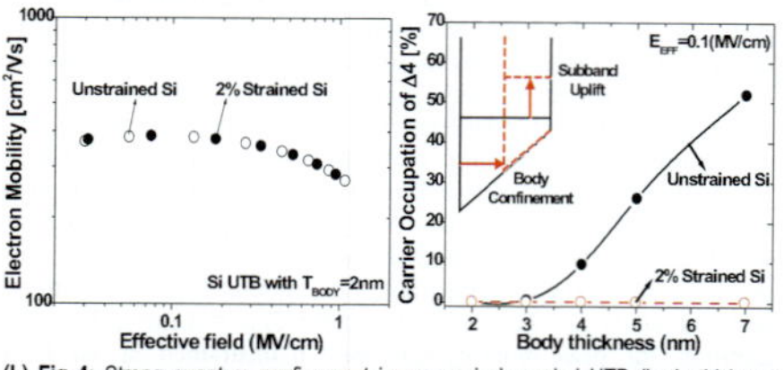

(L) Fig 4: Strong quantum confinement in aggressively scaled UTB (body thickness 2nm) renders the strained induced valley splitting using biaxial tensile strain (2%) redundant, leading to same low field mobility as unstrained device **(R) Fig 5:** Strong body confinement in unstrained Si results in subband energy uplift, reducing carrier occupation in $\Delta4$ valley (with lighter mz, Table 1). At body thickness 3nm, $\Delta4$ valley occupation is negligible, strain induced valley splitting will be redundant.

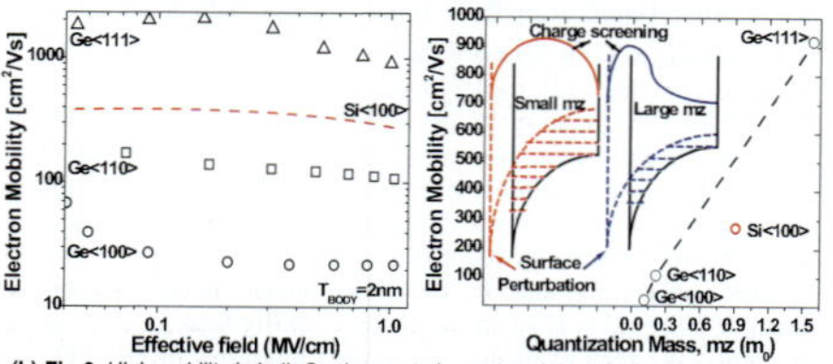

(L) Fig 6: High mobility in bulk Ge does not always translate to high mobility in Ge UTB transistor. Choice of surface orientation has a huge impact on device low field mobility. **(R) Fig 7:** High quantization mass mz, is critical for aggressively scaled UTB device. **Inset:** Energy band (along gate confinement) diagram illustrating effect of surface perturbation on small and large mz. A higher quantization mass propagates the electron nearer to the interface, providing more effective potential screening and reducing the overall perturbation potential. Simulated at body thickness 2nm and E_{EFF}=1MV/cm

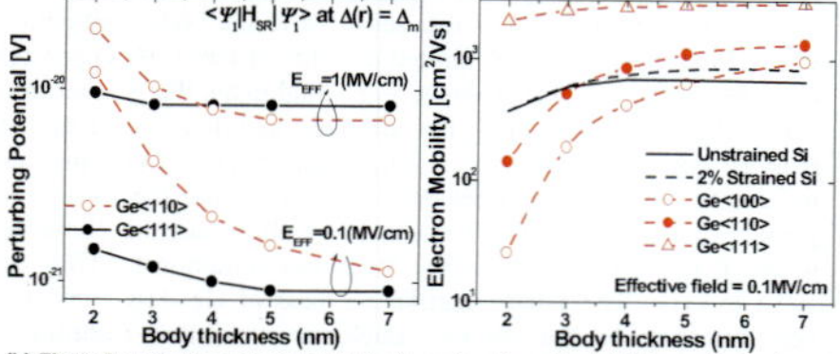

(L) Fig 8: Perturbation potential at $\Delta(r) = \Delta_m$ as function of body thickness for the lowest subband for a low mz (Ge<110>) and large mz (Ge<111>). Carriers experience larger perturbing potential as body is scaled down. Poorer charge screening for carriers with low mz render it very susceptible to surface roughness perturbation, aggravating at smaller body thickness. **(R) Fig 9:** Electron mobility for various advanced bandstructure UTB transistors as function of body thickness. Large mz and small md of Ge<111> (Table 1) provides the excellent high channel mobility.

Table 1: mz: quantization md: density of state mc: conductivity effective mass. Es: energy split (eV) ref. Ec of Si. g: degeneracy. (*): strain 2%

	valley	mz	md	mc	Es	g
Si<100>	$\Delta2$	0.916	0.190	0.190	0 (-0.17)*	2
	$\Delta4$	0.190	0.417	0.315	0 (+0.37)*	4
Ge<100>	L	0.117	0.295	0.149	0	4
	Δ	0.950	0.200	0.200	0.15	2
Ge<110>	L	0.218	0.216	0.140	0	2
	L	0.080	0.357	0.316	0	2
Ge<111>	L	1.590	0.080	0.080	0	1

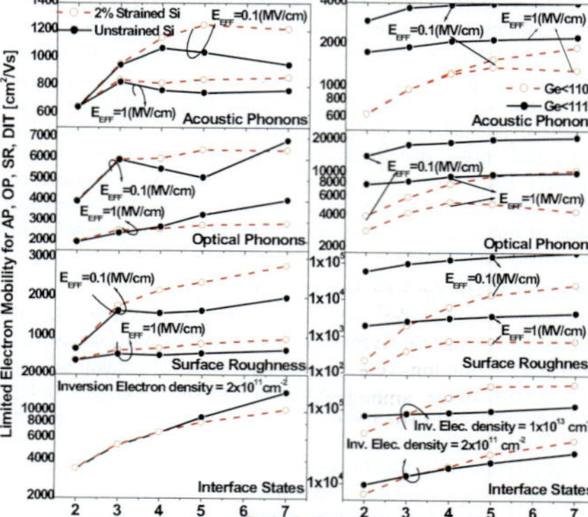

Fig 3: Experimental observation of non-universality of mobility for aggressively thin body [1] for both NMOS and PMOS. NMOS is more resistant to this phenomena, noticeable only at ~2.48nm whereas PMOS at ~3.57nm. This can also be consistently explained by our model as attributing to the smaller quantization mass of hole carriers (see Fig 7).

(L) Fig 10 and (R) Fig 11: Limited low field motilities for Strained Si and Ge UTB transistor respectively. Acoustic phonons, Optical phonons, Surface roughness and Interface charge limited mobilities are all systematically explored. All limited mobilities are plotted at constant effective field of 0.1MV/cm (threshold condition) and 1MV/cm (high inversion condition) except for interface charge limited mobility plotted at constant electron density criterion. **Formulation of scattering matrix elements:** Model for phonon spectrum in the bulk semiconductors are adapted from Jacobini et al [9, 10] where the matrix elements of the electron-phonon interaction are considered in a conventional way in accordance with Price [11, 12, 9]. Intra-valley acoustic phonon (AP) with an effective isotropic deformation potential [9, 12], intra-valley optical phonon (OP) for L valleys [9, 10] and inter-valley phonons constraint within the selection rules for f and g type processes [9] are accounted for. Dynamic screening of phonons is also disregarded [13]. Surface roughness (SR) scattering is treated in similar spirit as Gamiz et al [14] by extending Cheng's original treatment [15, 16] to account for the inefficacy of the linear expansion of the perturbed potential. The autocorrelation function of the asperities is assumed to be Gaussian. The static dielectric function for quasi-two-dimensional electron gas, based on a time-dependent perturbation approach [17] is employed for the treatment of screening [12]. Inter-subband transitions are left unscreened and the dielectric matrix is expressed according to [12] and in the quantum size limit when applicable. Interface states (DIT) induced scattering potential according to Stern et al [6] based on a perturbative approach is employed. By imposing appropriate boundary conditions, the scattering potential in all region of interest can be obtained using Nystrom Method [18].

Acknowledgements: This work is supported by Singapore A*STAR R263-000-267-305 grant. We gratefully acknowledge the useful discussion with D. Esseni (DIEGM) on coulomb scattering. We appreciate the useful discussion with M. V. Fischetti (IBM Corporation) and D. K. Ferry (Arizona State University) pertaining to their published literatures. We also thank S. Takagi (Tokyo University) for providing the experimental data to his classic paper [20]. T. Low gratefully acknowledge the scholarships from Singapore Millennium Foundation and Chartered Semiconductor Manufacturing.

Reprinted with permission from Tony Low, M.F. Li, Y.C. Yeo, W.J. Fan, S.T. Ng and D.L. Kwong, J. Appl. Phys., Vol.98, pp.024504-1-024504-8 (2005). Copyright 2005, American Institute of Physics.

JOURNAL OF APPLIED PHYSICS **98** 024504 (2005)

Valence band structure of ultrathin silicon and germanium channels in metal-oxide-semiconductor field-effect transistors

Tony Low, M. F. Li,[a] and Y. C. Yeo
Silicon Nano Device Laboratory (SNDL) Electrical and Computer Engineering (ECE) Department, National University of Singapore and Institute of Microelectronics, Singapore 119260, Singapore

W. J. Fan and S. T. Ng
Electrical and Computer Engineering (ECE) Department, Nanyang Technological University of Singapore, Singapore 63979 Singapore

D. L. Kwong
Electrical and Computer Engineering (ECE) Department, University of Texas, Austin, Texas 78712

(Received 8 September 2004; accepted 17 May 2005; published online 18 July 2005)

The ultrathin body (UTB) silicon-on-insulator metal-oxide-semiconductor field-effect transistor (MOSFET) is promising for sub-50-nm complementary metal-oxide semiconductor technologies. To explore a high-mobility channel for this technology, this paper presents an examination of Si and Ge hole sub-band structure in UTB MOSFETs under different surface orientations. The dependence of the hole subband structure on the film thickness (T_{Body}) was also studied in this work. We found that the valence-band mixing in the vicinity of the zone center Γ is strongly dependent on T_{Body} for both Si and Ge, particularly for the $\langle 110 \rangle$ surface orientation. This gives rise to the following two phenomena that crucially affect the electrical characteristics of p-MOSFETs: (1) an anomalous increase of quantization mass for $\langle 110 \rangle$ Si and Ge surfaces as T_{Body} is scaled below 5 nm. (2) The dependence of energy dispersion and anisotropy on T_{Body} especially for the $\langle 110 \rangle$ surface, which advantageously increases hole velocity along the [011] channel as T_{Body} is decreased. The density of states for different surface orientations are also calculated, and show that—for any given surface orientation—Ge has a smaller density of states than Si. The Ge $\langle 110 \rangle$ surface has the lowest density of states among the surface orientations considered. © *2005 American Institute of Physics.*
[DOI: 10.1063/1.1948528]

I. INTRODUCTION

The ultrathin body (UTB) silicon-on-insulator (SOI) metal-oxide-semiconductor field-effect transistor (MOSFET) is a promising candidate for the development of sub-50-nm complementary metal-oxide semiconductor (CMOS) technologies.[1] However, recent experimental work[2,3] has reported serious mobility degradation in Si UTB MOSFETs with T_{Body} less than 5 nm. This may present significant limitations to achieving ballistic transport in Si UTB MOSFET devices[4,5] with a sub-10-nm gate length.[6,7] To counteract this problem, several notable efforts are being made to implement a higher mobility channel. Ge UTB p-MOSFET devices with a T_{Body} of 4.5 nm have been fabricated using the local condensation techniques,[8] with large mobility improvement over Si MOSFETs. The technology for fabricating n-MOSFET and p-MOSFET devices on hybrid substrates[9] along their optimum surfaces and channel orientations has also been demonstrated.[9] To further this effort, it is of paramount importance to understand the band-structure characteristics of these thin-film channels, especially for the case of complicated valence bands. In the dissipative transport regime, it is advantageous to select and orient the band structure to yield a smaller two-dimensional (2D) density-of-

states mass (m_D) and conductivity mass (m_C).[10] However, from a current overdrive point of view,[11] a larger m_D is more desirable. A large quantization mass (m_Z) is also pertinent for the suppression of surface-roughness scattering for UTB MOSFETs with a sub-5-nm T_{Body}.[12]

The purpose of this article is to provide a detailed examination of the impact of T_{Body} on the hole subband structure of Si and Ge with sub-10-nm T_{Body} and discuss its implications for carrier transport attributes in its relation to their effective m_D, m_C and m_Z. The valence-band structure is calculated via the six-band Luttinger–Kohn Hamiltonian[13] for Si and Ge with common surface orientations—$\langle 100 \rangle$, $\langle 110 \rangle$, and $\langle 111 \rangle$—using a triangular-well approximation[14] and assuming an infinite potential barrier for the oxides at both interfaces of the semiconductors. In the following section, we will first consider the effect of coupling with the conduction band of type Γ_2^- and its implications on the use of a six-band Hamiltonian[13] description for band-structure calculation. Section III presents a brief outline of the numerical approach employed. In Sec. IV, we will provide an assessment of the energy quantization effect for the various surface orientations and its dependency on film thickness. Section V examines the energy anisotropy of the various Si and Ge thin films. Section VI concludes with a calculation of their 2D density of states.

[a]Electronic mail: elelimf@nus.edu.sg

024504-2   Low *et al.*   J. Appl. Phys. **98**, 024504 (2005)

TABLE I. Numerical values of bulk parameters used for the valence-band Hamiltonian for Si and Ge are obtained from Ref. 15. The Kane energy E_P, energy-gap E_G, and spin-orbit splitting Δ are given in units of eV. γ_j ($j=1,2,3$) are Luttinger parameters and m_c is the effective mass (in units of free electron mass, m_0) at the band edge of type Γ_2^-, modified Luttinger parameter (not listed in table) for eight-band Hamiltonian can be calculated from Ref. 15.

	γ_1	γ_2	γ_3	E_G	Δ	E_P	m_c
Si	4.285	0.339	1.446	4.185	0.044	21.60	0.528
Ge	13.38	4.24	5.69	0.898	0.297	26.30	0.038

II. COUPLING WITH CONDUCTION BAND OF TYPE Γ_2^-

Calculating valence-band structure is complicated by the strong interaction between the various hole subbands. When T_{Body} is continuously scaled down, the hole quantization energy becomes comparable with the energy gap; therefore, possible coupling with the conduction bands should also be considered. In this work, we began with an eight-band Hamiltonian[15] description, including the valence-band coupling with the conduction band of type Γ_2^-, to investigate the sufficiency of a six-band Hamiltonian[13] approach with the parameters used listed on Table I. Figure 1 shows the comparison of hole subband structure calculated with an eight- and six-band Hamiltonian for both Si and Ge quantum wells with a $\langle 100 \rangle$ surface and a T_{Body} of 30 Å. We observed that the hole subbands structure for Ge calculated with a six-band

Hamiltonian deviates noticeably from the eight-band description, particularly for the higher subbands. From an analysis of the wave-function components for Ge subbands (see Fig. 2), it is evident that there is notable coupling with conduction band of type Γ_2^- for energy subband $n=2$, contributing to about 10% of the probability function. From this, we may conclude that an eight-band Hamiltonian approach is necessary for an accurate description of Ge hole subband structure.

For our work, however, a six-band Hamiltonian approach will suffice if the biasing condition is such that the hole carriers predominantly populate only the first subband of Ge; this usually applies to electrostatics under threshold condition. Our argument is as follows:

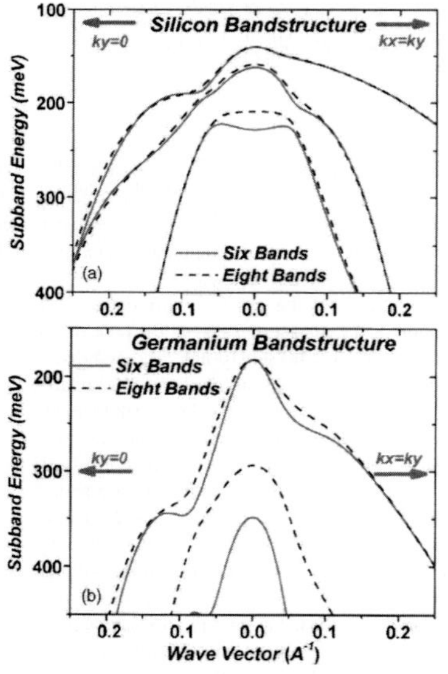

FIG. 1. Electronic hole subband structure for a 30-Å Si quantum well with $\langle 100 \rangle$ surface with an infinite energy barrier height. Calculation is done with an eight-band (see Ref. 15) and six-band (see Ref. 13) Hamiltonian for comparison. Energy plotted along wave-vector direction of [010] and [110]. Confinement direction is taken to be along z.

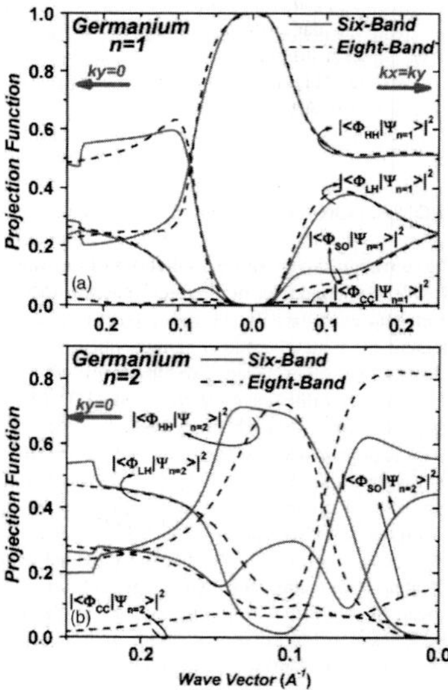

FIG. 2. Probability function projected on the various basis functions for 30-Å Ge quantum well with $\langle 100 \rangle$ surfaces, plotted for energy level (a) $n=1$ and (b) $n=2$. Calculation is done with an eight-band (see Ref. 15) and six-band (see Ref. 13) Hamiltonian for comparison. $|\psi\rangle$ is the wave function for energy subband n and $|\phi_i\rangle$ (i=HH: heavy hole, LH: light hole, SO: split-off hole, and CC: conduction band of type Γ_2^-) is the basis functions (Refs. 13 and 15) employed for our Hamiltonian.

024504-3 Low *et al.* J. Appl. Phys. **98**, 024504 (2005)

(a) At threshold condition, the Fermi energy is approximately a few kT ($\sim$0.025 eV; k: Boltzmann's constant and T: temperature) away from the lowest subband energy minimum, where threshold condition is usually defined to be the onset of carrier inversion. Hence, higher hole subbands can be disregarded if they are a few kT higher than the lowest subband energy minimum.

(b) Ge generally has relatively small quantization masses, resulting in larger energy separations for the various subbands. In particular, for a 30-Å Ge quantum well with $\langle 100 \rangle$ surfaces, we have an energy separation (between $n=1$ and $n=2$) of $\sim$0.1 eV.

Supported by these details, we shall adopt a six-band Hamiltonian approach in this work, exercising care when results pertaining to higher hole subbands of Ge are interpreted.

III. A NUMERICAL SOLUTION TO THE CHARGE CONTROL PROBLEM IN THE MOS SYSTEM

The numerical representation of the six-band Hamiltonian is obtained by following a discretization process outlined in Ref. 14. The six-band Hamiltonian is explicitly outlined in the Appendix where k_Z is represented by its differential form $-id/dz$ and coordinate z is taken to be the direction perpendicular to the semiconductor/oxide interface. When dealing with different crystal surface orientations of $\langle 100 \rangle$, $\langle 110 \rangle$, and $\langle 111 \rangle$, appropriate rotations of the **k** space must be performed. Although the charge control problem in the MOS system can be solved by obtaining the self-consistent solution to the Schrödinger and Poisson equations,[16] this approach is computationally intensive. In this work, we resort to the triangular-well approximation: $V(z)=eF_sz$, where F_S is the surface field and e the electronic charge. We must be aware of the limitations of the triangular-well approximation in describing MOS electrostatics under high inversion conditions,[16] where the charge-screening effect will alter the potential profile significantly. Despite this, the triangular approximation is computationally efficient and expected to be qualitatively correct, thus facilitating the study of a wider range of applications.

IV. IMPACT OF ACTIVE LAYER THICKNESS AND SURFACE ORIENTATION ON HOLE QUANTIZATION EFFECT

In this section, we will examine the effect of T_{Body} on the valence-band structure of Si and Ge with a sub-5-nm T_{Body}. In Fig. 3, hole subband energies (the lowest lying at $k=0$) are plotted as a function of T_{Body} for both Si and Ge channels with various surface orientations. At a high surface field of 1 MV/cm, the sub-band energies are generally less sensitive to the decrease of T_{Body} until a certain point ($T_{\text{Body}} \sim 4$ nm for $F_s=1$ MV/cm), where body quantization effects begin to be observed. We also noted that the energy quantization effect is strongest for $\langle 100 \rangle$ surface, followed by $\langle 111 \rangle$ and $\langle 110 \rangle$ surfaces.

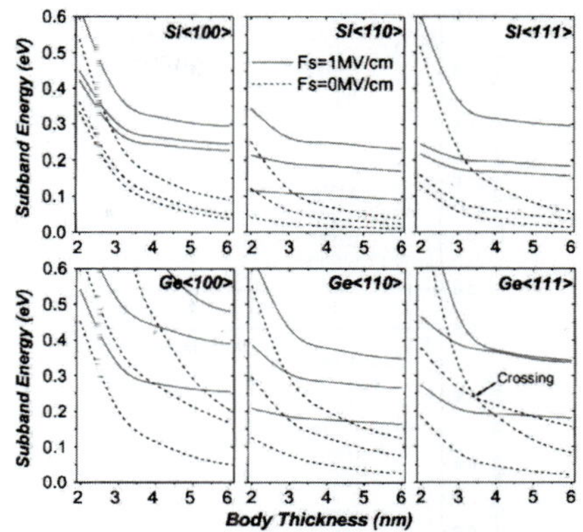

FIG. 3. Hole subband energies (at **k**=0) as function of body thickness T_{Body}, plotted for (a) Si and (b) Ge channels with various surface orientations. Surface field Fs of 1 MV/cm (bold lines) and 0 MV/cm (dashed lines) are compared. For each case, only the lowest three subbands are shown. There is apparently a crossing of the second and third subband energy for Ge$\langle 111 \rangle$ as T_{Body} is decreased.

It will be useful to provide a qualitative explanation for this observation. We begin by highlighting previous observations that the ground state of Si hole band structure are derived from the heavy-hole band.[17] Figure 4 shows the equienergy surface for the heavy-hole band of bulk Si, depicting the twelve distinct prongs. A particular crystal surface with prongs more aligned to the normal of the surface will effectively yield a larger "quantization mass." $\langle 110 \rangle$ surfaces have the prong perfectly aligned with the normal of the plane, which explains the small hole quantization effect on the $\langle 110 \rangle$ surface; this argument applies, with different results, for the $\langle 100 \rangle$ and $\langle 111 \rangle$ surfaces as well.

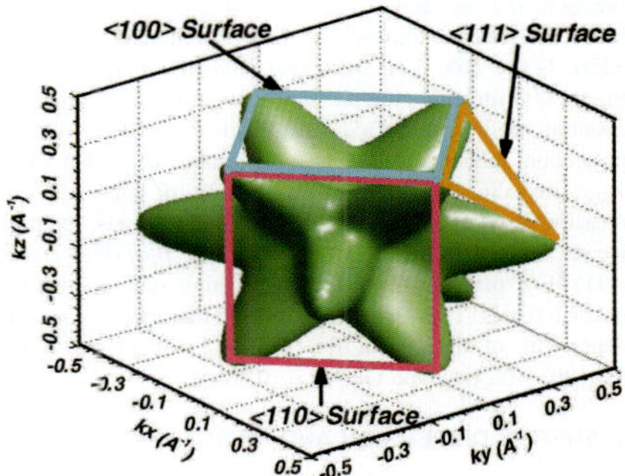

FIG. 4. Three-dimensional constant energy (at 0.1 eV reference from Γ point) surface plot for bulk Si band structure for the first valence energy band (or commonly known as heavy-hole band) depicting the twelve prominent *prongs*. The $\langle 100 \rangle$, $\langle 110 \rangle$, and $\langle 111 \rangle$ surface planes are illustrated and the axes are along [100] direction.

024504-4   Low *et al.*   J. Appl. Phys. **98**, 024504 (2005)

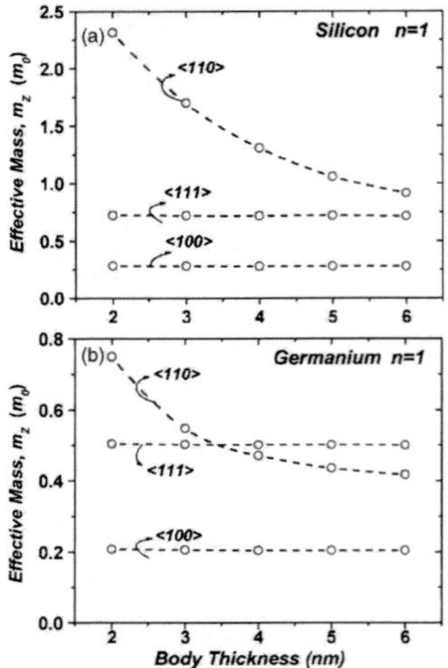

FIG. 5. Effective quantization mass m_2 calculated for (a) Si and (b) Ge quantum wells by fitting the subband energies at zone-center Γ to the analytical expression for that of a quantum well; $E_0 = (\hbar^2/2m^*)(\pi/T_{\text{BODY}})^2$.

Figure 5 computes the effective quantization mass (m_Z) for the ground-state subband in both the Si and Ge quantum wells under various surface orientations. This is calculated by fitting the subband energies at the zone-center Γ to the analytical expression for that of a quantum well with parabolic energy dispersion: $E_0 = (\hbar^2/2m^*)(\pi/T_{\text{BODY}})^2$. An interesting phenomenon—increasing effective quantization mass with decreasing T_{Body} for the $\langle 110 \rangle$ surface—is immediately revealed. The physical origin of this is the dependency of the valence-band mixing effect on the quantum well thickness. This effect can be analyzed by calculating the projection function, the probability function projected on the various basis functions employed for our Hamiltonian:[13] $|\Phi_i\rangle$ (i =HH: heavy hole, LH: light hole, and SO: split-off hole). Figure 6(a) illustrates the projection functions of the Si $\langle 110 \rangle$ quantum wells with a T_{Body} of 30 and 60 Å. Evidently, the projection function at Γ has a noticeable dependency on T_{Body}, revealing an increasing mix of split-off holes with decreasing T_{Body} around the vicinity of Γ. Conversely, the projection function in the vicinity of Γ for the $\langle 100 \rangle$ (Fig. 2) and $\langle 111 \rangle$ [Fig. 6(b)] surfaces does not exhibit such anomalous characteristics. Thus, we may conclude that their m_Z are independent of T_{Body} (Si$\langle 100 \rangle \sim 0.28m_0$, Si$\langle 111 \rangle \sim 0.72m_0$, Ge$\langle 100 \rangle \sim 0.21m_0$, and Ge$\langle 111 \rangle \sim 0.50m_0$).

V. ENERGY DISPERSION AND ANISOTROPY

Anisotropic energy dispersion and its effect on transport anisotropy in Si MOSFETs have been investigated theoretically.[14] There are also experimental studies of its transport characteristics in both the low-field[18,19] and high-field regime.[20] However, a study of hole energy dispersion

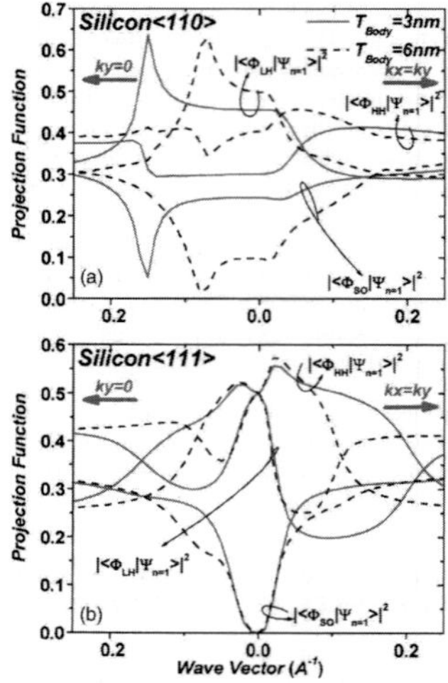

FIG. 6. Probability function projected on the various basis functions for 30 Å (bold line) and 60 Å (dashed line) Si quantum well, plotted for the first hole subband, with (a) $\langle 110 \rangle$ and (b) $\langle 111 \rangle$ surfaces. Calculation is done with a six-band (see Ref. 13) Hamiltonian. $|\psi_n\rangle$ is the wave function for energy subband n and $|\phi_i\rangle$ (i =HH: heavy hole, LH: light hole, and SO: split-off hole) is the basis functions (see Ref. 13) employed for our Hamiltonian.

anisotropy and its dependency on film thickness have not yet been conducted. Energy dispersion for the first two hole subbands of Si (Fig. 7) and Ge (Fig. 8) are plotted for two particular conditions as follows: (i) Bulk $F_s = 1$ MV/cm and $T_{\text{Body}} = 100$ nm, shown on the right and (ii) UTB $F_s = 1$ MV/cm and $T_{\text{Body}} = 3$ nm, shown on the left. Of particular interest are the following observations:

(1) $\langle 100 \rangle$ surface orientation: The energy dispersion characteristics of Si$\langle 100 \rangle$ for both bulk and UTB devices [as depicted in Fig. 7(a)] look similar except with a larger uplift of the $n = 2$ subband for the UTB device due to the body quantization effect. The ground-state energy ($n = 1$) for Si$\langle 100 \rangle$ maintains its strong anisotropy with the decrease of T_{Body} to 3 nm, presenting an optimum channel direction (smallest m_C along [100] as elucidated in Figs. 7(a) and 9(a). The $n = 2$ subband presents an apparent optimum channel direction along [110]. The anisotropy of the two subbands may neutralize each other, rendering the observed transport characteristics relatively isotropic. This is, in fact, the experimental observation for bulk Si MOSFETs.[19] However, the increased body quantization effect in UTB MOSFETs will result in carrier occupation dominance of the ground-state energy and it is very likely that the current anisotropy characteristics, due to the $n = 1$ subband may ultimately manifest. For Ge$\langle 100 \rangle$, it is noted in Fig. 9(b) that its energy anisotropy is less severe than Si$\langle 100 \rangle$, with a significant

024504-5   Low *et al.*   J. Appl. Phys. **98**, 024504 (2005)

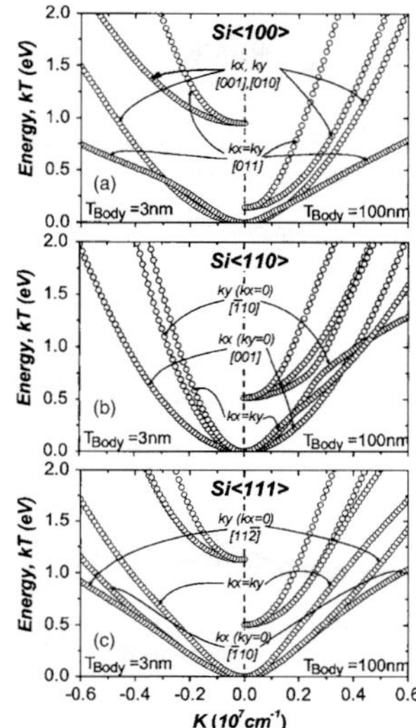

FIG. 7. Energy dispersion of Si (sub-band E_1 and subband E_2) plotted at surface field $F_S=1$ MV/cm for $T_{Body}=100$ nm (right) and $T_{Body}=3$ nm (left). For $\langle 100 \rangle$ and $\langle 110 \rangle$ surface, k_X is parallel to [001] channel direction. For $\langle 111 \rangle$ surface, k_X is parallel to [110] direction.

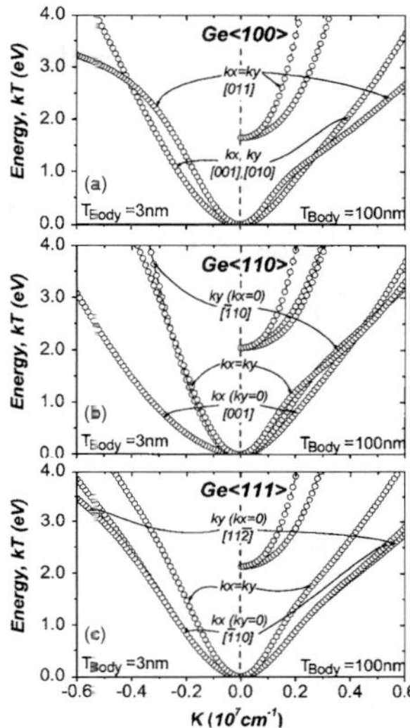

FIG. 8. Energy dispersion of Ge (subband E_1 and subband E_2) plotted at surface field $F_S=1$ MV/cm for $T_{Body}=100$ nm (right) and $T_{Body}=3$ nm (left). For $\langle 100 \rangle$ and $\langle 110 \rangle$ surface, k_X is parallel to [001] channel direction. For $\langle 111 \rangle$ surface, k_X is parallel to [110] direction. E_2 subband energy for $T_{Body}=3$ nm is too high to be shown in the figure.

relaxation of the energy anisotropy of ground-state energy as T_{Body} is decreased.

(2) $\langle 110 \rangle$ surface orientation: In Sec. IV, we noted the dependency of the valence-band mixing effect on the quantum well thickness for $\langle 110 \rangle$ surface thin film and its effect on m_Z. It can be anticipated that this dependency of valence-band mixing effect on the quantum well thickness will also have an important effect on the energy dispersion characteristics for the $\langle 110 \rangle$ surface. Figures 7(b) and 8(b) depict a notable change in the energy dispersion of the $n=1$ subband for Si$\langle 110 \rangle$ and Ge$\langle 110 \rangle$, respectively, as the T_{Body} is decreased to 3 nm. Specifically, this change is an increase of the energy gradient drawn along the k_y (at $k_x=0$, this is the [011] channel direction) and $k_x=k_y$ directions. The main effect of this will be in its carrier velocity, as elucidated in Fig. 10, which plots the radial carrier velocity [$\mathbf{v}(\mathbf{k}) = \hbar^{-1}\partial \mathbf{E}/\partial \mathbf{k}$ where the energy derivative is taken along the radial direction] of $n=1$ subband for both the (a) bulk and (b) UTB devices. A significantly higher intensity for carrier velocity is observed in the UTB device than in its bulk counterpart. Recently, it was experimentally observed that the optimum low-field mobility direction for Si UTB devices with a $\langle 110 \rangle$ surface is along the [011] channel direction (see Fig. 11 of Ref. 14). For Si$\langle 110 \rangle$ bulk MOSFETs, it was also experimentally found that the anisotropy characteristic of low-field mobility is not significant.[18] These experimental observa-

tions correlate with our theoretical calculations and should also qualitatively apply to Ge$\langle 110 \rangle$ since its energy dispersion characteristics are similar to Si$\langle 110 \rangle$.

VI. IMPACT OF SURFACE ORIENTATION ON DENSITY OF STATES

Figure 11 plots the 2D density of states for Si and Ge thin films with a T_{Body} of 3 nm. Density of states for Ge are several times smaller than for its Si counterpart, with the $\langle 110 \rangle$ surface having the smallest density of states, followed by the $\langle 111 \rangle$ and then $\langle 100 \rangle$. From a density-of-states point of view, the employment of Ge$\langle 110 \rangle$ crystal for the device's active layer will be helpful in achieving a high-mobility channel.

VII. CONCLUSIONS

In summary, our examination of the impact of film thickness on the hole subband structure of Si and Ge thin films within the framework of kp approximation via the Luttinger–Kohn Hamiltonian has shown an anomalous increase of "effective quantization mass" for the $\langle 110 \rangle$ surface as T_{Body} is decreased. This phenomenon is due to the body-thickness-dependent valence-band mixing effect in vicinity of the zone-center Γ. The other two orientations exhibit relatively constant effective quantization mass as T_{Body} is decreased, with Ge having a consistently smaller effective quantization

024504-6   Low *et al.*   J. Appl. Phys. **98**, 024504 (2005)

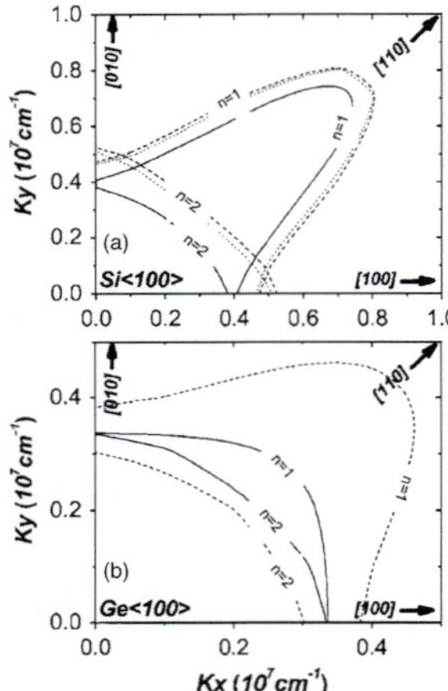

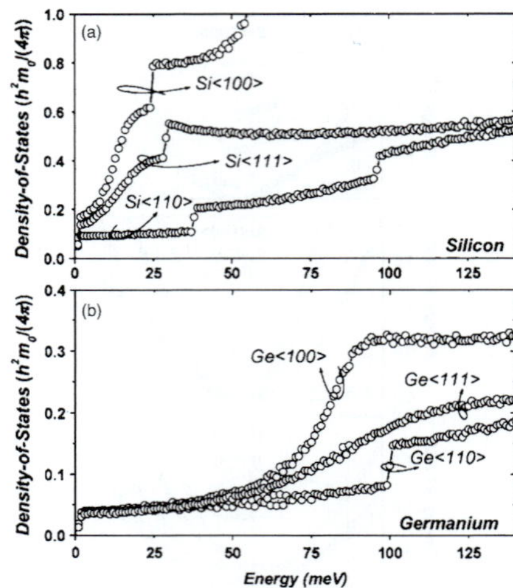

FIG. 9. Equienergy lines for the first two hole subbands of (a) Si⟨100⟩ and (b) Ge⟨100⟩ plotted at energy $(E-E_i, i=1,2)$ of 25, and 50 meV, respectively, under (i) bulk high-field conditions: $F_S=1$ MV/cm and $T_{Body}=100$ nm, represented by solid lines and (ii) UTB high-field conditions: $F_S=1$ MV/cm and $T_{Body}=3$ nm, represented by dashed lines.

FIG. 11. Density of states for (a) Si and (b) Ge quantum wells for a particular spin state at $F_S \sim 0.1$ MV/cm and $T_{Body}=3$ nm. Contributions from all the subbands are summed.

mass than its Si counterpart. It has also been observed that energy dispersion anisotropy exhibits a dependence on film thickness, especially for the ⟨110⟩ surface. In particular, its radial carrier velocity along the [011] channel increases as

the T_{Body} is decreased. Calculated density of states of Ge thin film are several times smaller than its Si counterpart, indicating that Ge should prove advantageous for the implementation of high-mobility channels. These findings should provide useful guidelines for the design of UTB *p*-MOSFETs.

ACKNOWLEDGMENTS

This work is supported by the Singapore A*STAR R263-000-267-305 and IME/03-450002 JML/SOI Grants. One of the authors (T.L.) gratefully acknowledges the scholarships from Singapore Millennium Foundation and Chartered Semiconductor Manufacturing.

APPENDIX

The basis functions employed for Kohn–Luttinger Hamiltonian[13] are depicted as follows:

$$\left| \frac{3}{2}, \frac{3}{2} \right\rangle = \frac{1}{\sqrt{2}} |x\rangle \uparrow + \frac{1}{\sqrt{2}} |y\rangle \uparrow,$$

$$\left| \frac{3}{2}, \frac{1}{2} \right\rangle = i\frac{1}{\sqrt{6}} |x\rangle \downarrow - \frac{1}{\sqrt{6}} |y\rangle \downarrow - i\frac{\sqrt{2}}{\sqrt{3}} |z\rangle \uparrow,$$

$$\left| \frac{3}{2} -\frac{1}{2} \right\rangle = \frac{1}{\sqrt{6}} |x\rangle \uparrow - i\frac{1}{\sqrt{6}} |y\rangle \uparrow + \frac{\sqrt{2}}{\sqrt{3}} |z\rangle \downarrow,$$

$$\left| \frac{3}{2} -\frac{3}{2} \right\rangle = i\frac{1}{\sqrt{2}} |x\rangle \downarrow + \frac{1}{\sqrt{2}} |y\rangle \downarrow,$$

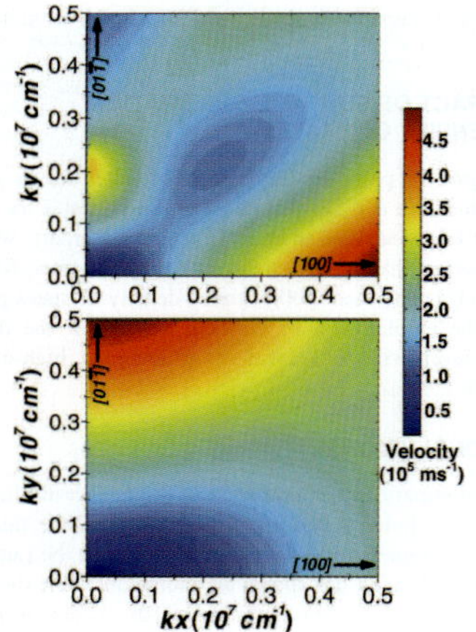

FIG. 10. Intensity plot for hole carrier radial velocity of different in-plane wave vector. Plotted for the ground-state energy of Si(110) for (a) $T_{Body}=100$ nm and (b) $T_{Body}=3$ nm at $F_S=1$ MV/cm. Radial velocity $v(k) = \hbar^{-1} \partial E / \partial k$, is obtained by taking the gradient of its energy dispersion in the radial direction.

024504-7   Low *et al.*

J. Appl. Phys. **98**, 024504 (2005)

$$\left|\frac{1}{2},\frac{1}{2}\right\rangle = \frac{1}{\sqrt{3}}|x\rangle\downarrow + i\frac{1}{\sqrt{3}}|y\rangle\downarrow + \frac{1}{\sqrt{3}}|z\rangle\uparrow,$$

$$\left|\frac{1}{2},-\frac{1}{2}\right\rangle = -i\frac{1}{\sqrt{3}}|x\rangle\uparrow - \frac{1}{\sqrt{3}}|y\rangle\uparrow + i\frac{1}{\sqrt{3}}|z\rangle\downarrow.$$

Where the functions $|X\rangle$, $|Y\rangle$, and $|Z\rangle$ are the basis functions of Γ_4 representation of the tetrahedral point group. The above choice of basis functions allows us to diagonalize the well-known spin-orbit coupling term, H_{SO} [see for, e.g., Ref. 13 eq. (V.1)]. Hence the Kohn–Luttinger Hamiltonian is

$$\begin{bmatrix} HH & \alpha & \beta & 0 & \dfrac{i\alpha}{\sqrt{2}} & -i\sqrt{2}\beta \\[2ex] \alpha^* & LH & 0 & \beta & \kappa & i\dfrac{\sqrt{3}}{\sqrt{2}}\alpha \\[2ex] \beta^* & 0 & LH & -\alpha & -i\dfrac{\sqrt{3}}{\sqrt{2}}\alpha^* & \kappa \\[2ex] 0 & \beta^* & -\alpha^* & HH & -i\sqrt{2}\beta^* & -\dfrac{i\alpha}{\sqrt{2}} \\[2ex] \dfrac{i\alpha^*}{\sqrt{2}} & \kappa^* & i\dfrac{\sqrt{3}}{\sqrt{2}}\alpha & i\sqrt{2}\beta & SO-\lambda & 0 \\[2ex] i\sqrt{2}\beta^* & -i\dfrac{\sqrt{3}}{\sqrt{2}}\alpha^* & \kappa^* & \dfrac{i\alpha}{\sqrt{2}} & 0 & SO-\lambda \end{bmatrix},$$

where:

$$HH \equiv \frac{h^2}{2m}[(\gamma_1 + \gamma_2)(k_x^2 + k_y^2) + (\gamma_1 - 2\gamma_2)(k_z^2)],$$

$$LH = \frac{h^2}{2m}[(\gamma_1 - \gamma_2)(k_x^2 + k_y^2) + (\gamma_1 + 2\gamma_2)k_z^2],$$

$$so = \frac{h^2}{2m}[\gamma_1(k_x^2 + k_y^2 + k_z^2)],$$

$$\alpha \equiv \frac{h^2}{2m}[-i2\sqrt{3}\,\gamma_3(k_x - ik_y)k_z],$$

$$\beta \equiv \frac{h^2}{2m}[\sqrt{3}\,\gamma_2(k_x^2 - k_y^2) - i2\sqrt{3}\,\gamma_3 k_x k_y],$$

$$\kappa \equiv \frac{h^2}{2m}i\sqrt{2}\,\gamma_2[-(k_x^2 + k_y^2) + 2k_z^2],$$

where λ is the spin-orbit splitting.

For $\langle 110\rangle$ surface, a rotation of the original **k** space is required:

$$k_x' = k_x,$$

$$k_y' = \frac{1}{\sqrt{2}}k_y + \frac{1}{\sqrt{2}}k_z,$$

$$k_z' = -\frac{1}{\sqrt{2}}k_y + \frac{1}{\sqrt{2}}k_z,$$

and similarly for $\langle 111\rangle$ surface:

$$k_x' = \sqrt{\frac{2}{3}}k_x - \frac{1}{\sqrt{6}}k_y + \frac{1}{\sqrt{6}}k_z,$$

$$k_y' = \frac{1}{\sqrt{2}}k_y + \frac{1}{\sqrt{2}}k_z,$$

$$k_z' = -\frac{1}{\sqrt{3}}k_x - \frac{1}{\sqrt{3}}k_y + \frac{1}{\sqrt{3}}k_z.$$

[1] D. Edenfeld, A. B. Kahng, M. Rodgers, and Y. Zorian, *2003 International Technology Roadmap for Semiconductors* (IEEE Computer Society, California, 2004).

[2] K. Uchida, H. Watanabe, A. Kinoshita, J. Koga, T. Numata, and S. Takagi, *International Electron Devices Meeting*, San Francisco, CA, 8–11 December 2002 (IEEE, New York, 2002), p. 47.

[3] K. Uchida and S. Takagi, Appl. Phys. Lett. **82**, 2916 (2003).

[4] This experimental observation is first reported in Ref. 2 for a Si n-MOSFETs, done at low vertical electric-field condition of 0.1 MV/cm at 25 K.

[5] H. Sakaki, T. Noda, K. Hirakawa, M. Tanaka, and T. Matsusue, Appl. Phys. Lett. **51**, 1934 (1987).

[6] M. S. Lundstrom, IEEE Electron Device Lett. **22**, 293 (2001).

[7] The physical mechanism for the drastic degradation of mobility is identified to be due to surface roughness (SR) related scattering mechanisms via a low-field mobility measurement at 25 K (Refs. 2 and 3). In addition, the electron mobility follows a T_{Body} to-power-of-six relationship (Ref. 4), which has also been observed in quantum well structures (see Ref. 5). Lundstrom has pointed out (Ref. 6), via a phenomenological approach, that the steady-state current in a decananometer MOSFETs is essentially source limited, although scattering processes at the drain side may affect the transport potential profile via the self-consistency loop. Hence the mobility at high vertical surface field, which embodies the effective scattering rate in vicinity of the source, remains relevant. Consider the electron at the source with carrier velocity of 1×10^5 m/s with an effective momentum relaxation time of 1×10^{-14} s, this yields a mean free path of merely 1 nm. SR scattering in aggressively scaled T_{Body} will further shorten the effective momentum relaxation time.

[8] T. Tezuka, S. Nakaharai, Y. Moriyama, N. Sugiyama, and S. Takagi, *Symposium on VLSI Technology*, Honolulu, HI, 15–17 June 2004 (IEEE, New York, 2004), p. 198.

[9] M. Yang *et al.*, *International Electron Devices Meeting*, Washington, DC, 8–10 December 2003 (IEEE, New York, 2003), p. 453.

[10] T. Low, M. F. Li, C. Shen, Y. C. Yeo, Y. T. Hou, C. Zhu, A. Chin, and D. L. Kwong, Appl. Phys. Lett. **85**, 2402 (2004).

[11] T. Low, Y. T. Hou, M. F. Li, C. Zhu, A. Chin, G. Samudra, L. Chan, and D. L. Kwong, *International Electron Devices Meeting*, Washington, DC, 8–10 December 2003 (IEEE, New York, 2003), p. 691.

[12] T. Low, Chen Shen, M. F. Li, Y. C. Yeo, Y. T. Hou, C. Zhu, A. Chin, L. Chan, and D. L. Kwong, *Study of Mobility in Strained Si and Ge Ultra-thin-Body MOSFET* (SSDM, Tokyo, 2004), p. 776.

[13] J. M. Luttinger and W. Kohn, Phys. Rev. **97**, 869 (1955).

[14] M. V. Fischetti, Z. Ren, P. M. Solomon, M. Yang, and K. Rim, J. Appl. Phys **94**, 1079 (2003).

[15] S. Ridene, K. Boujdaria, H. Bouchriha, and G. Fishman, Phys. Rev. B **64**, 085329 (2001).

[16] T. Low, Y. T. Hou, and M. F. Li, IEEE Trans. Electron Devices **50**, 1284 (2003).

[17] See Ref. 14. The nature of each sub-band was established by analyzing the shape of the equienergy lines in **k** space and by counting the nodes of the wave function. It is found that the ground state is always heavy-holelike

024504-8   Low *et al.*

J. Appl. Phys. **98**, 024504 (2005)

under a triangular potential well over a surface field range from 0 to 2.5 MV/cm.

[18]T. Sato, Y. Takeishi, H. Hara, and Y. Okamoto, Phys. Rev. B **4**, 1950 (1971).

[19]T. Sato, Y. Takeishi, and H. Hara, Jpn. J. Appl. Phys. **8**, 588 (1969).

[20]M. Kinugawa, M. Kakumu, T. Usami, and J. Matsunaga, *International Electron Devices Meeting*, Washington, DC, 5–8 December 1985 (IEEE, New York, 1985) p. 581.

2430

IEEE TRANSACTIONS ON ELECTRON DEVICES, VOL. 52, NO. 11, NOVEMBER 2005

Modeling Study of the Impact of Surface Roughness on Silicon and Germanium UTB MOSFETs

Tony Low, *Student Member, IEEE*, Ming-Fu Li, *Senior Member, IEEE*, Ganesh Samudra, *Member, IEEE*, Yee-Chia Yeo, *Member, IEEE*, Chunxiang Zhu, *Member, IEEE*, Albert Chin, *Senior Member, IEEE*, and Dim-Lee Kwong, *Senior Member, IEEE*

Abstract—We outlined a simple model to account for the surface roughness (SR)-induced enhanced threshold voltage (V_{TH}) shifts that were recently observed in ultrathin-body MOSFETs fabricated on $\langle 100 \rangle$ Si surface. The phenomena of enhanced V_{TH} shifts can be modeled by accounting for the fluctuation of quantization energy in the ultrathin body (UTB) MOSFETs due to SR up to a second-order approximation. Our model is then used to examine the enhanced V_{TH} shift phenomena in other novel surface orientations for Si and Ge and its impact on gate workfunction design. We also performed a calculation of the SR-limited hole mobility ($\mu_{H,SR}$) of p-MOSFETs with an ultrathin Si and Ge active layer thickness, $T_{Body} < 10$ nm. Calculation of the electronic band structures is done within the effective mass framework via the Luttinger Kohn Hamiltonian, and the mobility is calculated using an isotropic approximation for the relaxation time calculation, while retaining the full anisotropy of the valence subband structure. For both Si and Ge, the dependence of $\mu_{H,SR}$ on the surface orientation, channel orientation, and T_{Body} are explored. It was found that a $\langle 110 \rangle$ surface yields the highest $\mu_{H,SR}$. The increasing quantization mass m_z for $\langle 110 \rangle$ surface renders its $\mu_{H,SR}$ less susceptible with the decrease of T_{Body}. In contrast, $\langle 100 \rangle$ surface exhibits smallest $\mu_{H,SR}$ due to its smallest m_z. The SR parameters i.e., autocorrelation length (L) and root-mean-square (Δ_{rms}) used in this paper is obtained from the available experimental result of Si$\langle 100 \rangle$ UTB MOSFETs [1], by adjusting these SR parameters to obtain a theoretical fit with experimental data on SR-limited mobility and V_{TH} shifts. This set of SR parameters is then employed for all orientations of both Si and Ge devices.

Index Terms—Germanium, mobility, silicon, surface roughness, ultrathin-body (UTB) MOSFETs.

I. INTRODUCTION

ULTRATHIN-BODY (UTB) silicon-on-insulator (SOI) MOSFETs are promising for sub-50-nm MOSFET technologies [2]. However, recent experimental findings by Uchida *et al.* [1] suggests the need for serious reconsiderations of the scalability of the body thickness T_{Body} into the sub-10-nm regime. Two phenomena attributed to surface roughness (SR)

Manuscript received April 7, 2005; revised July 26, 2005. This work was supported by A*STAR under Grant R398-000-019-305 and Grant R263-000-267-305. The work of T. Low was supported by the scholarship from Singapore Millenium Foundation and Chartered Semiconductor. The review of this paper was arranged by Editor T. Skotnicki.

T. Low, G. Samudra, M.-F. Li, Y.-C. Yeo, C. Zhu, and A. Chin are with the Silicon Nano Device Laboratory, Department of Electrical and Computer Engineering, National University of Singapore, Singapore 119260, and also with the Institute of Microelectronics, Singapore 117685, (e-mail: elelimf@nus.edu.sg).

D.-L. Kwong is with the Department of Electrical and Computer Engineering, University of Texas, Austin, Austin, TX 73301 USA, and also with the Institute of Microelectronics, Singapore 117685, Singapore.

Digital Object Identifier 10.1109/TED.2005.857188

are experimentally identified in Si UTB MOSFETs [1], they are; an enhanced threshold voltage (V_{TH}) shift and a drastic degradation of mobility with a T_{Body}^6 dependency. The former is an additional V_{TH} shift unaccountable by the effect of energy quantization in the body alone. Although SR was suspected to be the main cause of enhanced threshold voltage shifts (ETVS) [1], no explicit quantitative model has yet been proposed to account for the experimental observations. The latter effect of drastic degradation of SR-limited mobility with a T_{Body}^6 dependency has already been experimentally observed in GaAs/AlAs quantum well structures [3] and well modeled in the context of quantum well structures [3]–[5].

On the technological front, there are several notable efforts on implementing a UTB transistor channel with enhanced carrier mobility. Ge UTB p-MOSFET devices with a T_{Body} of 7 nm have been fabricated using a local condensation technique [6], demonstrating large mobility enhancement over its Si counterpart. Novel technology for fabricating n-MOSFETs and p-MOSFETs devices on hybrid substrates with optimum crystal surface orientations has also been demonstrated [7]. Hence, it is of paramount importance to evaluate the impact of SR phenomena in these novel channel materials with aggressively scaled T_{Body}.

In this paper, we report a modeling study of SR-induced enhanced V_{TH} shifts and SR-limited mobility in Si and Ge UTB MOSFETs. This paper serves to communicate in detail the models employed in our published conference paper [8]. Section II discussed the modeling methodology used for the calculation of the subband structures of electron and hole inversion layers. Section III discusses the modeling of enhanced V_{TH} shifts, by taking into account the second order effect of SR on quantized energy level. Section IV provides a systematic study of this phenomena in all test devices and discuss its impact on gate workfunction design. Sections V and VI outlines the model used and the simulated results for SR-limited hole mobility respectively. Mobility anisotropy are also discussed. Last, this manuscript ends with a discussion and conclusion in Section VII.

II. MODELING OF SUBBAND STRUCTURES

The energy versus momentum vector k dispersion of Si and Ge in vicinity of the conduction band edge is of a parabolic nature, characterized by a transverse mass (m_t) and longitudinal mass(m_l) (values used are obtained from [9]). For a general surface orientation, the longitudinal and transverse axis of the ellipsoidal constant energy surface for a parabolic dispersion may

TABLE I

ELECTRON EFFECTIVE MASSES m_z (QUANTIZATION) AND m_d (DENSITY-OF-STATES), g IS THE VALLEY DEGENERACY, E_s IS THE ENERGY SPLIT REFERENCE FROM CONDUCTION BAND ENERGY MINIMUM. BANDSTRUCTURE PARAMETERS AT CONDUCTION VALLEY MINIMA FOR Si AND Ge; SUCH AS THEIR ENERGY ELLIPSOIDAL FORMS AND VALLEY ENERGY SPLITS, ARE OBTAINED FROM [9]. QUANTIZATION MASS AND DENSITY-OF-STATES MASSES FOR SOLVING MOS ELECTROSTATICS ON VARIOUS SURFACE ORIENTATIONS IS CALCULATED ACCORDING TO [10]

Device	Valley	$m_z(m_o)$	$m_d(m_o)$	g	$E_s(eV)$
Si $\langle 100 \rangle$	Δ	0.916	0.190	2	0
	Δ	0.190	0.417	4	0
Si $\langle 110 \rangle$	Δ	0.315	0.324	4	0
	Δ	0.190	0.417	2	0
Si $\langle 111 \rangle$	Δ	0.258	0.358	6	0
Ge $\langle 100 \rangle$	L	0.117	0.295	4	0
	Δ	0.200	0.436	2	0.15
	Δ	0.950	0.200	2	0.15
Ge $\langle 110 \rangle$	L	0.218	0.216	2	0
	L	0.080	0.357	2	0
	Δ	0.200	0.436	2	0.15
	Δ	0.330	0.340	4	0.15
Ge $\langle 111 \rangle$	L	1.590	0.080	1	0
	L	0.089	0.337	3	0
	Δ	0.271	0.374	6	0.15

TABLE II

NUMERICAL VALUES OF BULK PARAMETERS USED FOR THE VALENCE BAND HAMILTONIAN FOR Si AND Ge ARE OBTAINED FROM [14]. KANE ENERGY E_P, ENERGY GAP E_G AND SPIN-ORBIT SPLITTING Δ ARE GIVEN IN UNITS OF eV. γ_j (j = 1, 2, and 3) ARE LUTTINGER PARAMETERS AND m_c IS EFFECTIVE MASS (IN UNITS OF FREE ELECTRON MASS, m_o) AT THE BAND EDGE OF TYPE Γ_2^-, MODIFIED LUTTINGER PARAMETER (NOT LISTED IN TABLE) FOR EIGHT-BAND HAMILTONIAN CAN BE CALCULATED FROM [14]

Material	γ_1	γ_2	γ_3	E_G	Δ	E_p	m_c
Si	4.285	0.339	1.446	4.185	0.044	21.60	0.528
Ge	13.38	4.24	5.69	0.898	0.297	26.30	0.038

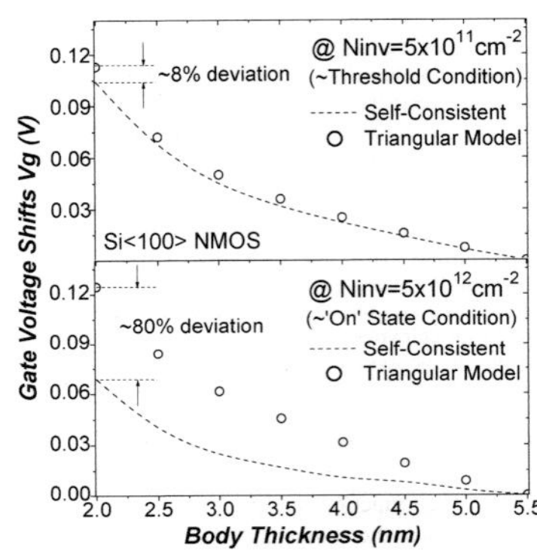

Fig. 1. Calculation of gate voltage shifts V_g (with respect to the V_g of device with $T_{\text{body}} = 5.5$ nm) for Si $\langle 100 \rangle$ nMOS devices with an EOT = 10 nm at a given electron inversion charge density of 5×10^{11} cm^{-2} and 5×10^{12} cm^{-2}. We compare the results calculated from both the triangular well approximation model and self-consistent model. The triangular well approximation model provides a good estimation of the actual gate voltage shifts V_g under electron inversion charge density of 5×10^{11} cm^{-2}, approximately the threshold condition.

not necessarily align with axes of the device coordinate system. Following the method by Stern *et al.* [10], the original Hamiltonian system can be unitarily transformed to decouple m_z (the effective mass in the quantization direction) from the three-dimensional effective mass tensor. The m_z and density-of-states mass $m_d = \left(\left(m_t^2 m_l \right) / m_z \right)^{0.5}$ obtained are shown in Table I. n-MOS electrostatics are then solved self-consistently via the conventional one-band Schrödinger–Poisson equation (see for e.g., [11] and [12]).

The strongly coupled valence subband structures with spin-orbit interaction are expressed by the Kohn–Luttinger Hamiltonian [13]. Numerical parameters used for the valence band Hamiltonian for Si and Ge are obtained from [14] and listed in Table II. The numerical representation of the six-band Hamiltonian follows the usual discretization process (see e.g., [15]). When dealing with different crystal surface orientations of $\langle 100 \rangle$, $\langle 110 \rangle$, and $\langle 111 \rangle$, appropriate rotations of the k space must be performed and we represent k_Z by its coordinate representation form of $-id/dz$, where direction perpendicular to the surface is taken to be z.

Although the charge control problem can also be solved by obtaining a self-consistent solution to the Schrödinger and Poisson equations [16], this approach is computationally intensive. In this paper, we resort to the triangular well approximation $V(z) = qF_s z$, where F_s is the surface field and q the electronic charge. Despite the limitations of the triangular well approximation at high inversion condition [16] where charge screening affects the potential profile significantly, it is computationally efficient and expected to be qualitatively correct in our context. Fig. 1 illustrates our argument. It plots the gate voltage shifts V_g (with respect to the V_g of a device with $T_{\text{body}} = 5.5$ nm) for Si$\langle 100 \rangle$ nMOS devices with an EOT = 10 nm (as per our experimental devices used in [1]) at a given electron inversion charge density of 5×10^{11} cm^{-2} and 5×10^{12} cm^{-2}. We compare the results calculated from both the triangular well approximation model and self-consistent model. The triangular well approximation model provides a good estimation of the actual gate voltage shifts V_g under electron inversion charge density of 5×10^{11} cm^{-2} (approximately the threshold condition) with an over-estimation of approximately 8%.

2432   IEEE TRANSACTIONS ON ELECTRON DEVICES, VOL. 52, NO. 11, NOVEMBER 2005

III. MODEL FOR ENHANCED V_{TH} SHIFTS

Experimental measurement of V_{TH} shifts in Si $\langle 100 \rangle$ n- and p-MOSFETs sub-10-nm T_{body} UTB transistors is obtained from [1]. It is apparent from [1] that the theoretical prediction of V_{TH} shifts without accounting for surface roughness effects underestimate the V_{TH} shifts experimentally observed for both n-MOS and p-MOS. In order to capture the physics of the enhanced V_{TH} shifts, SR-induced quantized energy level fluctuation in the Si body or quantum well has to be accounted for in the model. This fluctuation is usually expressed using a linear approximation [3], [5], which suffices in the study of low-field mobility in quantum well structures. However, a Taylor series expansion up to second order approximation is required in our context. As a result, a symmetric distribution of body thickness fluctuation due to SR can give rise to an overall energy shift, resulting in an enhanced V_{TH} shift. In similar fashion to [5], we shall also ignore the curvature effect due to roughness on two interfaces and conveniently set the back interface to $z = 0$ (where z is taken to be the gate confinement direction) in our UTB device with an average $T_{\text{body}} = L_o$. We are interested in the effect of energy level fluctuation due to the roughness. We expanded the subband energy E_i as function of body thickness about L_o up to the second-order terms as follows:

$$E_i\left(T_{\text{body}}\right) = E_i\left(L_o + \Delta\left(\vec{r}\right)\right)$$
$$\cong E_i\left(L_o\right) + \frac{\partial E_i}{\partial T_{\text{body}}}\Delta\left(\vec{r}\right) + \frac{1}{2}\frac{\partial^2 E_i}{\partial T_{\text{body}}^2}\Delta^2\left(\vec{r}\right) \quad (1)$$

where $\Delta\left(\vec{r}\right)$ is the function that describes the SR morphology. Taking average over the in-plane vector $\vec{r}$, one obtains

$$\left\langle E_i\left(L_o + \Delta\left(\vec{r}\right)\right)\right\rangle \cong E_i\left(L_o\right) + \frac{1}{2}\frac{\partial^2 E_i}{\partial T_{\text{body}}^2}\Delta_{\text{rms}}^2 \quad (2)$$

where Δ_{rms} is the rms value of SR and used as a fitting parameter. $\langle \ldots \rangle$ denotes the average taken over the in-plane vector $\vec{r}$, which is the plane normal to z. In the limit of only a single subband occupation, (2) effectively captures the enhanced V_{TH} shifts due to SR which one can easily calculate for the case of a square quantum well. It is interesting to note that for the case of a square quantum well where $E_i = \left(\hbar^2/2m^*\right)\left(\pi/T_{\text{body}}\right)^2$, V_{TH} shift has a T_{body}^{-2} dependency and the second order component contributing to enhanced V_{TH} shift will have a T_{body}^{-4} dependency. Therefore, in the case where the surface roughness perturbation is of the same order as T_{body}, the enhanced V_{TH} shift component $(1/2)(\partial^2 E_i/\partial T_{\text{body}}^2)\Delta_{\text{rms}}^2$ will have a $T_{\text{body}}^{-4}\Delta_{\text{rms}}^2$ dependency, contributing as significantly to the overall V_{TH} shift. However, in the general case of multisubbands with space charge effects, we have to compute it self-consistently for all occupied subbands. We shall illustrate the case for n-MOS, where its local inversion charge density can be expressed as follows:

$$N_{\text{inv}}(\vec{r}) = \sum_i D_i \times kT \ln\left(1 + \exp\left(\frac{E_F - E_i(L_o + \Delta(\vec{r}))}{kT}\right)\right) \quad (3)$$

where D_i is the density-of-states for subband i, E_F is Fermi energy, k is the Boltzmann constant, and T is the temperature.

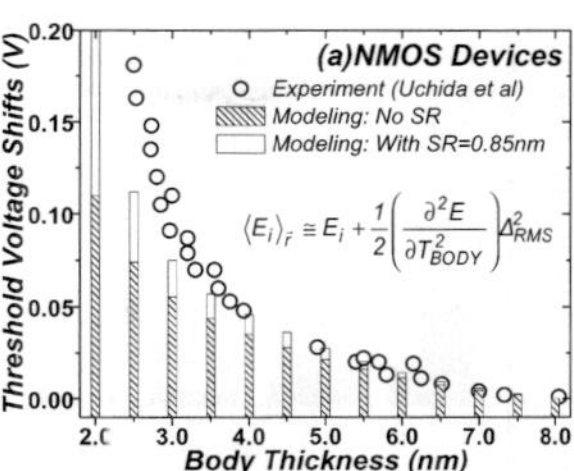

Fig. 2. Enhanced V_{TH} shifts due to SR (with respect to the V_{TH} of the device with $T_{\text{body}} = 8$ nm) for Si nMOS at $\langle 100 \rangle$. Excellent corroboration with experimental result [1] is obtained. An effective rms SR Δ_{rms} of 0.85 nm for the two interfaces taken together is employed.

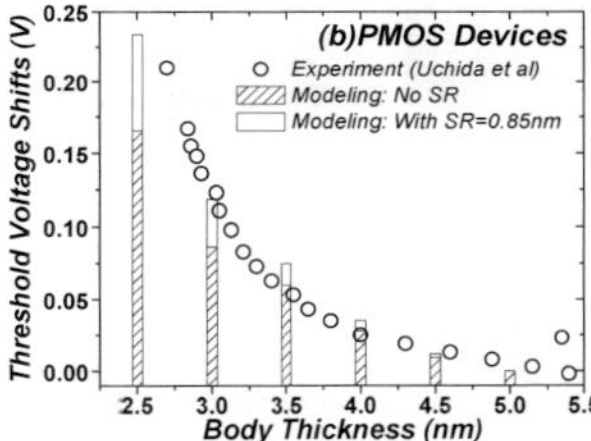

Fig. 3. V_{TH} shifts (with respect to the V_{TH} of device with $T_{\text{body}} = 5.5$ nm for pMOS calculated using triangular model with same SR parameters as those used in Fig. 2.

Averaging the inversion charge density $N_{\text{inv}}(\vec{r})$ in (3) and assuming that the exponent is much greater than 1, (3) becomes linear allowing us to just replace E_i with (2) as follows:

$$\left\langle N_{\text{inv}}(\vec{r})\right\rangle \approx \sum_i D_i \times \left(E_F - E_i(L_o) - \frac{1}{2}\frac{\partial^2 E_i}{\partial L^2}\Delta_{\text{rms}}^2\right). \quad (4)$$

This result obtained is applicable to device electrostatics is in the linear regime. Equation (4) can then be solved numerically.

IV. IMPACT OF ENHANCED V_{TH} SHIFTS

Figs. 2 and 3 show the good agreement between the measured and calculated V_{TH} shifts of nMOS and pMOS devices with a rms roughness $\Delta_{\text{rms}} = 0.85$ nm, by taking the two interfaces together. For the purposes of comparison with experiments, the V_{TH} is obtained by a linear extrapolation of the charge versus surface potential plot in the linear regime, in a similar spirit to a conventional experimental extraction of V_{TH}. Whereas the enhanced V_{TH} shifts attributed to SR is obtained by accounting for the shift of this curve in the linear regime, to be calculated with (4).

A. Enhanced V_{TH} Shifts for Various Channel Materials

Figs. 4 and 5 show the results of calculated enhanced V_{TH} shifts for Si and Ge channel UTB MOSFETs with various surface orientations T_{body} in the range of 2–5 nm is considered.

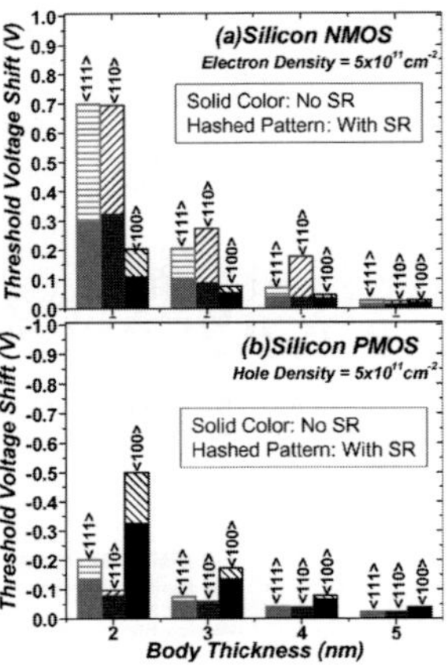

Fig. 4. Simulated V_{TH} shifts with respect to the V_{TH} of device with $T_{\mathrm{body}} = 8$ nm. Calculated for Si (a) nMOS and (b) pMOS with three different surface orientations. Same SR parameters as in Fig. 2 are used. All simulations are performed at an electron inversion density of 5×10^{11} cm^{-2}.

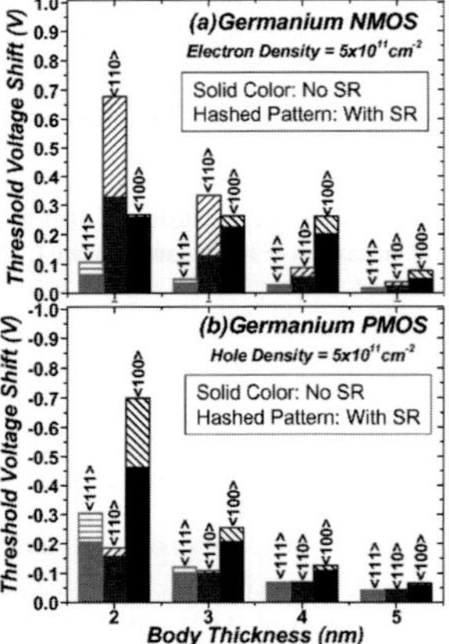

Fig. 5. Simulated V_{TH} shifts with respect to the V_{TH} of device with $T_{\mathrm{body}} = 8$ nm. Calculated for Ge (a) nMOS and (b) pMOS with three different surface orientations. Same SR parameters as in Fig. 2 are used. All simulations are performed at an electron inversion density of 5×10^{11} cm^{-2}.

The dependency of the V_{TH} shift on T_{body} can be qualitatively analyzed by examining the case for that of a square quantum well where the quantized subband energy has a simple T_{body}^{-2} dependency. In the limit of only a single subband occupation, V_{TH} shifts have a T_{body}^{-2} dependency and the enhanced V_{TH} shift component will have a $T_{\mathrm{body}}^{-4}\Delta_M^2$ dependency. Therefore, when the SR perturbation is of the same order as T_{body}, the enhanced V_{TH} shift component will contribute as significantly

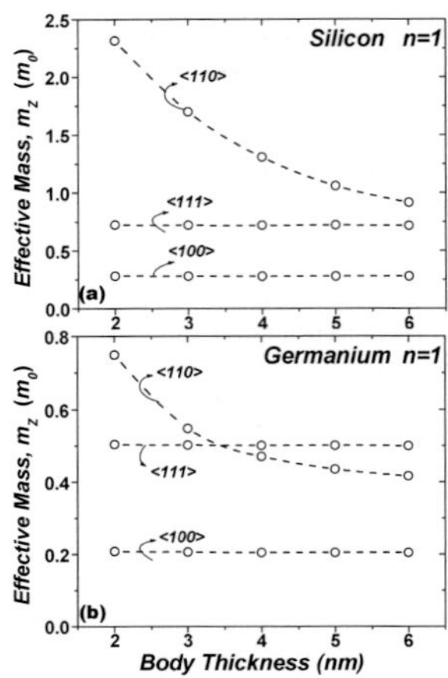

Fig. 6. Effective quantization mass m_Z calculated for (a) Si and (b) Ge quantum well by fitting the subband energies at zone center Γ to the analytical expression for that of a quantum well: $E_o = (\hbar^2/2m^*)(\pi/T_{\mathrm{body}})^2$.

to the total V_{TH} shift as the quantization energy $E_i(L_0)$. The amount of V_{TH} shift however, also depends indirectly on the carrier quantization mass m_Z since the quantization energy is approximately proportional to m_Z^{-1}. For nMOS devices, electron m_Z for both Si and Ge on various surface orientations are given in Table I. V_{TH} shift is the smallest for the surface orientation with the largest m_Z, namely Si $\langle 100 \rangle$ for Si nMOS and Ge$\langle 111 \rangle$ for Ge nMOS as shown in Figs. 4 and 5, respectively. With decreasing T_{body}, an anomalous retarded increase of V_{TH} shift is observed for Ge$\langle 100 \rangle$ nMOS at $T_{\mathrm{body}} = 3$ nm. This is due to the transfer of carrier occupation from the L to the Δ valleys [17]. For pMOS devices, the result can be qualitatively interpreted in a similar fashion using the set of empirically fitted m_Z as shown in Fig. 6. The empirically fitted m_Z is calculated by fitting the subband energies at the zone center Γ to the analytical expression for energy dispersion of a quantum well. An interesting phenomenon of increasing effective quantization mass with decreasing T_{body} for the $\langle 110 \rangle$ surface is evident. The physical origin has been identified to be due to the dependence of valence band mixing effect on the quantum well thickness [18]. For Si and Ge pMOS devices, the $\langle 100 \rangle$ surface orientation shows the largest V_{TH} shift due to its lowest empirical m_Z. Due to the large m_Z of the $\langle 110 \rangle$ surface of Si and Ge, V_{TH} shift does not increase rapidly as T_{body} is scaled down, as shown in Figs. 4 and 5.

B. Impact on Metal Gate Workfunction Requirement

Figs. 7 and 8 show the metal gate workfunction requirement for Si and Ge UTB MOSFETs, at all surface orientations calculated for a stipulated OFF-state carrier density of 5×10^7 cm^{-2} accounting for both the carrier quantization and SR effects. Simulation devices are assumed to have an ideal subthreshold slope

2434   IEEE TRANSACTIONS ON ELECTRON DEVICES, VOL. 52, NO. 11, NOVEMBER 2005

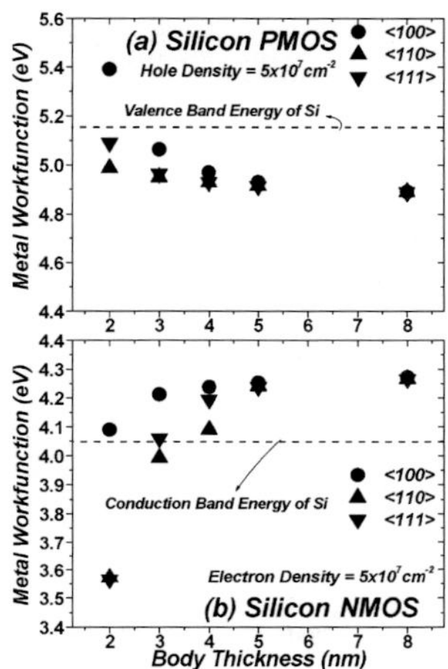

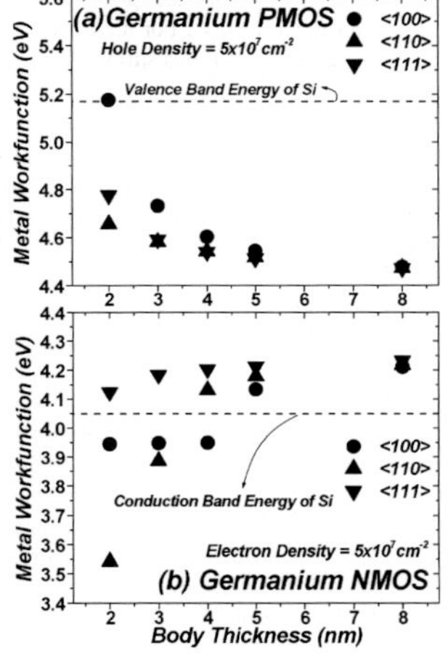

Fig. 7. Workfunction for (a) pMOS and (b) nMOS devices calculated for Si. Effect of enhanced $V_{\rm TH}$ shifts is taken into account with same SR parameters. OFF-state carrier density assumed to be 5×10^7 cm^{-2} with ideal subthreshold slope 60 mV/dec. Shaded region is the energy values within the bandgap of Si.

Fig. 8. Workfunction for (a) pMOS and (b) nMOS devices calculated for Ge. Effect of enhanced $V_{\rm TH}$ shifts is taken into account with same SR parameters. OFF-state carrier density assumed to be 5×10^7 cm^{-2} with ideal subthreshold slope 60 mV/dec. Shaded region is the energy values within the bandgap of Si.

of 60 mV/dec. Carrier quantization and SR-induced workfunction shifts can amount to a $V_{\rm TH}$ shift that is as large as 0.7 V for a $T_{\rm body} = 2$ nm. The use of such aggressively scaled body

thicknesses sets the required gate workfunctions for nMOS and pMOS devices further apart, rendering metal gate workfunction engineering more challenging as current choice of metal gate materials only offers workfunction in the range of between 4.1 to 5.2 eV [19]. The required workfunction for pMOS devices employing Ge channel is lower than that for Si due to the smaller valence electron ionization energy of Ge.

V. MODELING OF SR LIMITED HOLE MOBILITY

Simulation results for electron mobility are reported in [20], [21]. In this paper, we will discuss the hole SR-limited mobility in these devices with various $T_{\rm body}$. A methodology similar to that as outlined by Fischetti *et al.* [15] is used. We begin with the Boltzmann equation

$$\frac{\partial f_o}{\partial t} + \frac{dr_i}{dt}\frac{\partial f_o}{\partial r_i} + \frac{dk_i}{dt}\frac{\partial f_o}{\partial k_i} - \left.\frac{\partial f}{\partial t}\right|_c = 0 \qquad (5)$$

where $f \equiv f(\vec{r}, \vec{k}, t)$ is the particle distribution function, describing its probability density within a small volume of $\vec{r}$ and $\vec{k}$. Einstein summation over repeated indexes is implicitly implied. We denote $f_c \equiv f(\vec{r}, \vec{k}, 0)$. The last term is the collision integral. In our context, we have a homogeneous medium ($\partial f_o/\partial r_i = 0$) and a constant time-independent force ($\partial f_o/\partial t = 0$). Making use of the rules of semiclassical dynamics and "relaxation time approximation" [22], [23], we arrive at an important equation

$$f(\vec{k}) = f_o(\vec{k}) - T(\vec{k})\frac{\partial f_o}{\partial E_{\vec{k}}}\vec{F} \cdot \vec{v} \qquad (6)$$

where $T(\vec{k})$ is the relaxation time, $\vec{F}$ is the applied force and $\vec{v}$ is the particle velocity. Next, we write the collision integral as a detailed balance of in-scattering and out-scattering

$$\left.\frac{\partial f}{\partial t}\right|_c = \sum_{\vec{k'}}\left[S(\vec{k'}, \vec{k})f(\vec{k'})\left[1 - f(\vec{k})\right] - S(\vec{k}, \vec{k'})f(\vec{k})\left[1 - f(\vec{k'})\right]\right] \qquad (7)$$

where $S(\vec{k}, \vec{k'})$ represents the scattering rate from an initial state $\vec{k}$ to final state $\vec{k'}$. We assume the process is elastic for our particular case of SR scattering process ($E_{\vec{k'}} = E_{\vec{k}} = E$), allowing us to impose $S(\vec{k}, \vec{k'}) = S(\vec{k'}, \vec{k})$ and $f_o(\vec{k}) = f_o(\vec{k'})$ by embracing the isotropic approximation, $\left|\vec{k'}\right| = \left|\vec{k}\right|$ [15] (where its validity has been numerically investigated), arriving at

$$\frac{1}{T(\vec{k})} = \sum_{\vec{k'}} S(\vec{k'}, \vec{k})[1 - \cos\alpha] = D(E)\left\langle S(\vec{k'}, \vec{k})[1 - \cos\alpha]\right\rangle \qquad (8)$$

where α is the angle between wave vector $\vec{k}$ and $\vec{k'}$. The average is taken over all $\vec{k}$ at energy E. $D(E)$ is the density-of-states.

The scattering rate $S(\vec{k}, \vec{k'})$ is calculated using Fermi Golden rule, where the perturbation potential due to SR, as formulated by Ando *et al.* [24], accounts for the potential steps induced by the rough surface and shifts of inversion charge density. Recently, this approach has been applied to UTB transistor structures [15]. In the context of quantum well, this is often modeled by expressing the perturbation Hamiltonian according to an energy level fluctuation term [4], [3], [5]. In particular, Meyerovich

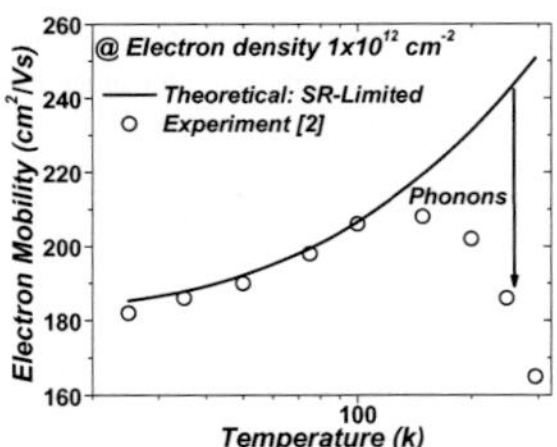

Fig. 9. Measured SR-limited electron mobility with $T_{\rm body}$ = 2.48 nm at 25 K [1] and simulated result with SR $\Delta_{\rm rms}$ = 0.60 nm and SR auto-correlation length L = 2.12 nm. Deviation at temperature larger than 100 K is due to onset of phonon scattering.

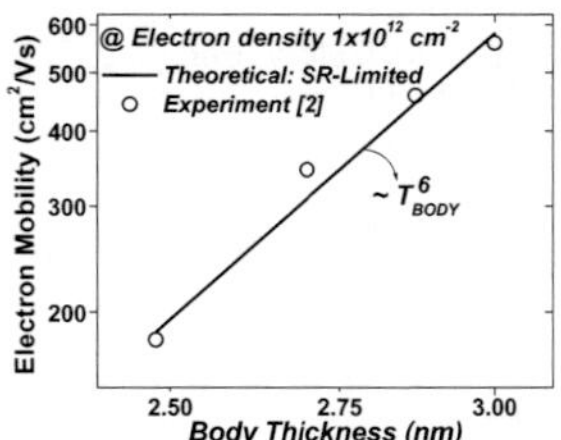

Fig. 10. Measured SR-limited electron mobility as function of $T_{\rm body}$ at 25 K [1]. Same SR parameters used for calculation, with observed $T_{\rm body}$ to the power-of-six relationship as reported in experiment [3].

et al. [4] obtained similar expressions via a canonical transformation of the system to that of smooth boundaries. In this paper, we employed the latter formulation

$$S(\vec{k}, \vec{k'}) = \frac{2\pi}{\hbar} \left(\frac{\partial E_s}{\partial T_{\rm body}} \right)^2 S_p(\vec{q}) \qquad (9)$$

where $S_p(\vec{q})$ is the roughness spectrum using an exponential model [25] with the energy level fluctuation numerically calculated. This approach allows us to circumvent the difficulty of defining a value for the hole quantization mass m_z (which is made difficult due to mixing between various subbands), in contrast to the electron inversion layer where $\partial E_s / \partial T_{\rm body} \propto m_z^{-1}$. In addition, we have also confined our study to the condition of low inversion charge density, allowing us to minimize the additional scattering due to change of charge density induced perturbation [5] and the two-dimensional dielectric screening altogether (both of which are unaccounted for in the model). The SR related parameters, such as the rms roughness (Δ_M) and auto-correlation length (L) are obtained from a theoretical fitting of the experimental result of electron mobility, to be discussed.

Hole mobility is then calculated using the following relation in [15], allowing us to express the xx component of the mobility tensor μ_{ij} for subband s as shown in (10) at the bottom of the page. where wavevector $k_s \equiv k_s(E_s, \theta)$, $n_{\rm inv}$ is the total carrier density in the hole inversion layer, E_S is the subband energy with subband minimum at $E_{S,\rm min}$, θ is the polar angle of the wavevector k_S, and $f_0(E)$ is the Fermi–Dirac distribution function. $T_s(k_s)$ is the anisotropic relaxation time for subband s along an applied electrical force $\vec{F}$ along x.

VI. Simulation Results of SR-Limited Hole Mobility

Simulation results on hole mobility is discussed in this section. Figs. 9 and 10 show the excellent fit of electron mobility as function of temperature in the 25 k to 100 k range and of $T_{\rm body}$

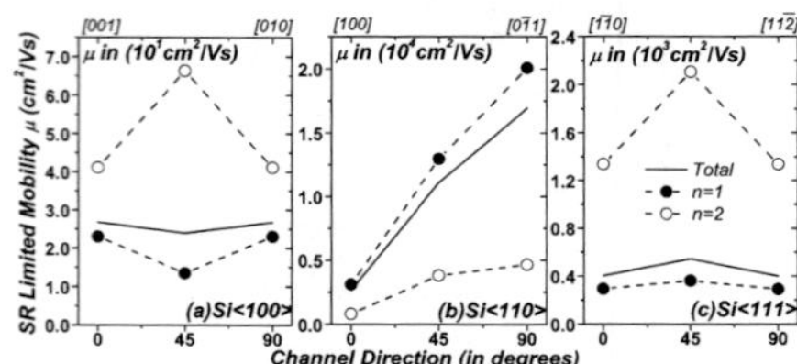

Fig. 11. SR-limited hole mobility of Si (solid line) simulated at $T_{\rm body}$ of 3 nm for (a) $\langle 100 \rangle$, (b) $\langle 110 \rangle$, and (c) $\langle 111 \rangle$ surface orientations as a function of channel orientation. SR parameters used are the same as those in Fig. 9. Mobility is calculated at a hole density of 5×10^{11} cm^{-2}. 0° denotes [001] for $\langle 100 \rangle$ and $\langle 110 \rangle$ surfaces and [11$\bar{2}$] for $\langle 111 \rangle$ surface. Mobility for $n = 1$, two subbands (dashed lines) are plotted for reference. Note that the mobilities are expressed in different scale for each surface orientation.

at 25 K, with a rms roughness Δ_M of 0.6 nm and auto-correlation length L = 2.12 nm. Deviation between the theoretical prediction and experimental data at temperature larger than 100 K is due to the dominance of phonon scattering, which is unaccounted for in our model. The $T_{\rm body}^6$ relationship with SR-limited electron mobility as reported in experiments [1]–[3], is also observed in Fig. 10.

A. Optimum Channel Orientation

Figs. 11 and 12 show the channel orientation dependence of the calculated SR-limited hole mobility $\mu_{\rm H,SR}$ for Si and Ge devices with $T_{\rm body}$ = 3 nm. For Si$\langle 100 \rangle$, we note that the anisotropy of the first two subbands neutralize each other, similar to the case for bulk Si$\langle 100 \rangle$ [15]. For Si$\langle 110 \rangle$, $\mu_{\rm H,SR}$ exhibits high anisotropy with an optimum channel direction along [01$\bar{1}$] (Fig. 11). Fig. 13 depicts an interesting phenomenon for Si$\langle 110 \rangle$, where an increased radial carrier velocity along a [011] channel direction for the UTB device is observed as compared to its bulk counterpart. This is attributed to the dependence of valence band mixing effect on $T_{\rm body}$ for the $\langle 110 \rangle$ surface, rendering its energy dispersion characteristics

$$\vec{\mu}_{xx}^s = \frac{q}{4\pi^2 \hbar^2 k T n_{\rm inv}} \int_0^{2\pi} \int_{E_{s,\rm min}}^\infty T_s(k_s) f_o(E_s)(1 - f_o(E_s)) k_s \left[\frac{\partial E_s}{\partial k_x} \right]_{k_s}^2 \left[\frac{\partial E_s}{\partial k} \right]_{k_s}^{-1} d\theta dE_s \qquad (10)$$

2436   IEEE TRANSACTIONS ON ELECTRON DEVICES, VOL. 52, NO. 11, NOVEMBER 2005

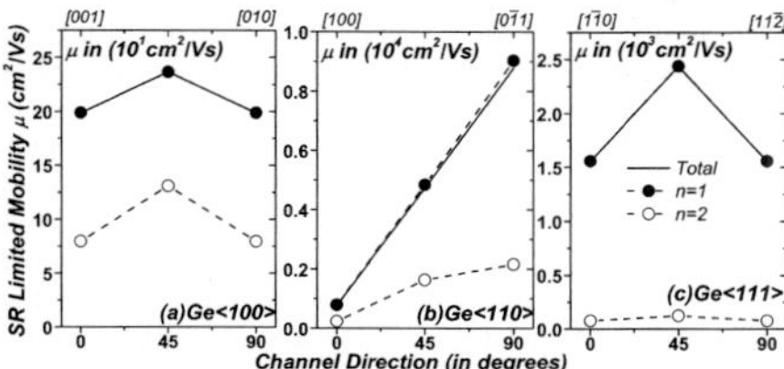

Fig. 12. SR-limited hole mobility of Ge (solid line) simulated at T_{body} of 3 nm for (a) $\langle 100 \rangle$, (b) $\langle 110 \rangle$, and (c) $\langle 111 \rangle$ surface orientations as a function of channel orientation. SR parameters used are the same as those in Fig. 9. Mobility is calculated at a hole density of 5×10^{11} cm^{-2}. $0°$ denotes [001] for $\langle 100 \rangle$ and $\langle 110 \rangle$ surfaces and [11$\bar{2}$] for $\langle 111 \rangle$ surface. Mobility for $n = 1$, two subbands (dashed lines) are plotted for reference. Note that the mobilities are expressed in different scale for each surface orientation.

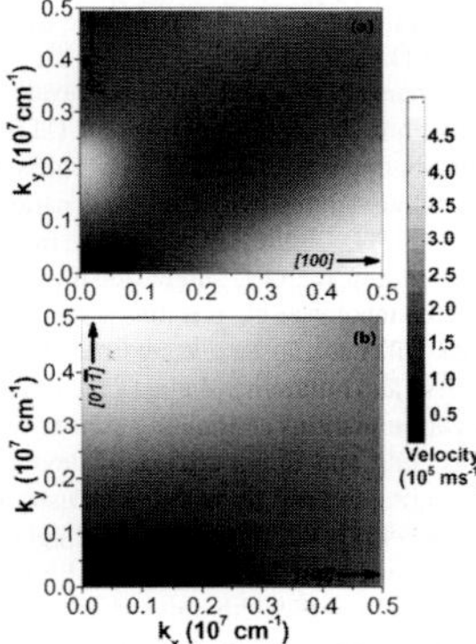

Fig. 13. Intensity plot for hole carrier radial velocity of different in-plane wave vectors. Plotted for the ground state energy of Si$\langle 110 \rangle$ for (a) $T_{\text{body}} = 100$ nm and (b) $T_{\text{body}} = 3$ nm at $F_s = 1$ MV/cm. Radial velocity $v_{\xi} = \hbar^{-1} \partial E / \partial k$, is obtained by taking its gradient of its energy dispersion in the radial direction.

strongly dependent on T_{body} [18]. These observations should also qualitatively apply to Ge$\langle 110 \rangle$ since their energy dispersion characteristics exhibit similar behavior. For Ge devices, the surface and channel orientation dependence of $\mu_{\text{H,SR}}$ are generally very similar to that of Si devices except for the $\langle 100 \rangle$ surface orientation, which exhibits an optimum channel direction along [110]. In general, $\mu_{\text{H,SR}}$ of Ge is little affected by higher energy subbands at $T_{\text{body}} = 3$ nm, due to the large difference in quantization energies of the first two subbands. The highly anisotropic hole transport characteristic for $\langle 110 \rangle$ may posed significant technology problems for engineers. For electron transport in Si$\langle 110 \rangle$, the optimum channel direction has been experimentally reported to be along [100] channel direction [26]. Fortunately, the optimum channel direction for electron transport in Ge$\langle 110 \rangle$ was predicted to be along [110] direction [17], [27].

B. Optimum Surface Orientation

The dependence of $\mu_{\text{H,SR}}$ on T_{body} and surface orientations is shown in Fig. 14. The impact of surface orientations on $\mu_{\text{H,SR}}$

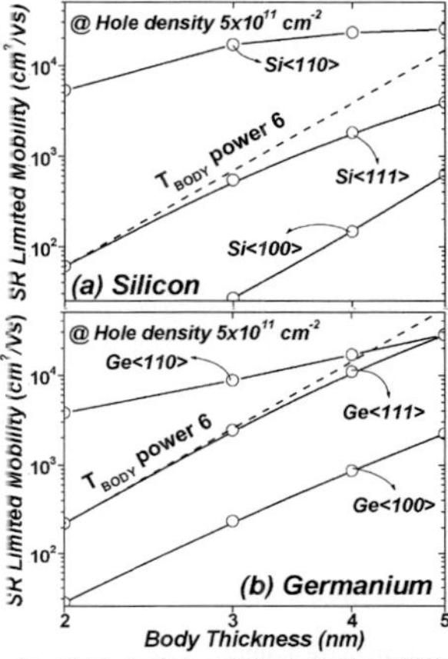

Fig. 14. Simulated SR limited hole mobility for (a) Si and (b) Ge with various orientations, with same SR parameters as Fig. 9. Mobility is calculated at hole density of 5×10^{11} cm^{-2} and result plotted for the optimum channel direction for $T_{\text{body}} = 3$ nm (see Figs. 11 and 12). Hole mobility on $\langle 100 \rangle$ surface found to be very limiting. The other orientations exhibit relatively high mobility. Mobility deviates from the T_{body}^6 dependence especially for the $\langle 110 \rangle$ surface.

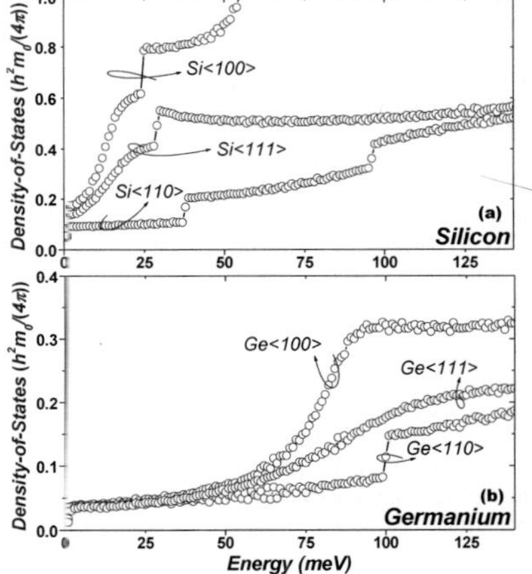

Fig. 15. Density-of-states for (a) Si and (b) Ge quantum well for a particular spin-state at $Fs = 0.1$ MV/cm and $T_{\text{Body}} = 3$ nm. Contributions from all the subbands are summed.

can also be explained via its empirical masses. Fig. 6 plots the calculated empirical m_Z. Fig. 15 shows the calculated density of states m_D in mass units of $h^2 m_o / 4\pi$. Conductivity mass m_C is a tensor quantity and cannot be conveniently tabulated for the complicated hole bandstructure. However, m_D qualitatively depicts the average m_C for a given energy. $\mu_{\text{H,SR}}$ for the $\langle 100 \rangle$ surface is the most limiting. Its small empirical m_Z for the first subband (Si($0.28m_o$) and Ge($0.21m_o$)), renders it very sensitive to SR scattering processes. Furthermore, the larger empirical m_D in Si$\langle 100 \rangle$ compared to Ge$\langle 100 \rangle$ results in a lower mobility. For the $\langle 110 \rangle$ surface, its large m_Z yields it the largest $\mu_{\mathrm{H,SR}}$. This is because the scattering rate $S(\vec{k}, \vec{k}')$ is inversely proportional to m_Z [see (9)]. Hence, it is most probable that SR-scattering will not be a very limiting mechanisms for UTB MOSFETs on $\langle 110 \rangle$ surface. It also shows a retarded decrease of $\mu_{\mathrm{H,SR}}$ with decreasing T_{body}, deviating sharply from the expected T_{body}^6 dependency. This can be explained by the increase of its empirical m_Z with T_{body}. In addition, the empirical m_Z of Ge$\langle 110 \rangle$ is approximately half that of Si$\langle 110 \rangle$, leading to a higher susceptibility to SR scattering. However, this is compensated by a smaller m_D for Ge$\langle 110 \rangle$, resulting in comparable $\mu_{\mathrm{H,SR}}$ for Si$\langle 110 \rangle$ and Ge$\langle 110 \rangle$. For the $\langle 111 \rangle$ surface, comparable $\mu_{\mathrm{H,SR}}$ for Si and Ge are also observed.

VII. DISCUSSION AND CONCLUSION

In this paper, we have assumed that the SR parameters to be the same for all devices. This approach, though unrealistic, it facilitates a simple means for comparing the relative merits of devices with different orientations, where only their intrinsic properties i.e., quantization mass, density-of-states mass, comes into play. In addition, the SR condition for each orientation may also be subjected to a specific processing condition to yield the optimum SR parameters, it is therefore beyond the scope of this modeling work to determine what these set of SR parameters are. Indeed, this has recently been studied [28] for the case of Ge, which reported Ge$\langle 111 \rangle$ with the best SR condition. Therefore, the impact of SR parameters on the threshold voltage V_{TH} shifts and SR-limited mobility warrants discussion. The enhanced V_{TH} shifts will have a Δ_{rms}^2 dependency as apparent from its average subband energy relation [see (2)]. SR-limited mobility is related to the SR parameters through the roughness spectrum $S_p(\vec{q}) = \pi \Delta^2 L^2 / [1 + (q^2 L^2 / 2)]^{3/2}$ [25]. Hence, the SR parameters will have significant impact on the SR characteristics studied in this paper, which to a certain degree may affect the conclusions presented. However, based on most recent studies on $\langle 100 \rangle$ and $\langle 111 \rangle$ Ge substrates [28], it was found that $\langle 111 \rangle$ has a better interface quality with high-κ dielectrics. Hence, this particular result would only reinforce our conclusion. More experimental studies in this aspect has to be conducted before a more quantitative evaluation accounting for the SR parameters dependence on surface orientation can be confidently carried out.

Our consideration of SR-limited hole mobility indicates that carrier transport under $\langle 100 \rangle$ surface orientation is more susceptible to SR scattering (our simulation assumes a condition of low inversion charge density of 5×10^{11} cm^{-2}) as compared to $\langle 110 \rangle$ and $\langle 111 \rangle$ orientations, suggesting that the latter two orientations are better candidates for aggressively scaled UTB p-MOSFETs from a mobility point of view, under the assumption that SR condition is independent of surface orientations. For $\langle 110 \rangle$ and $\langle 111 \rangle$ orientations, it is most probable that other scattering mechanisms may sets in to be more dominant. Especially for $\langle 110 \rangle$ orientation, which

is the most robust to SR scattering due to its increasing m_z with decreasing T_{Body}, we expect phonon scattering processes to be relatively more limiting. In dissipative transport where other scattering mechanisms i.e., phononscattering, coulomb scattering also have to be accounted for, Ge generally serves as better candidate for p-MOSFET as compared to its Si counterpart because of its lower density-of-states (see Fig. 15). For n-MOSFET, we had calculated the effective low-field mobility for Si and Ge UTB under all common surface orientations [20], [21]. Ge$\langle 111 \rangle$ offers the highest low-field mobility, whereas Ge$\langle 110 \rangle$ is predicted to have even lower low-field mobility than Si$\langle 100 \rangle$ for aggressively scaled T_{Body} down to 2 nm under the same SR condition [20], [21]. We should be reminded that these conclusions about the optimum channel orientation for dissipative transport does not applies to device in ballistic transport. In fact, it was predicted that Ge$\langle 110 \rangle$ would yield the optimum ballistic current with device structure of $L_G = 20$ nm at $T_{\mathrm{Body}} = 2$ nm [17].

SR-induced enhanced V_{TH} shifts will aggravates the impact of T_{Body} fluctuations in aggressively scaled UTB MOSFETs [8]. Therefore, SR-induced enhanced V_{TH} shifts phenomena should also be included in the lists of intrinsic parameter fluctuations [29], [30] in UTB MOSFETs structures; random discrete dopants in source/drain regions, single dopant or charged defect state in the channel region, gate line edge roughness etc. These variations will have an increasing crucial impact on the functionality, yield and reliability of electronic circuits at a time when the fluctuation margins continually shrinks due to supply voltage down-scaling and miniaturizing of device sizes. It is definitely challenging to keep up with the industry target of 20 mV V_{TH} variations [31]. We should also highlight that the experimental devices we used for this paper [1] had a relatively large active region of length/width dimension 500 μm/200 μm. Hence, the intrinsic parameter fluctuations due to random discrete dopants in source/drain regions, single dopant or charged defect state in the channel region and gate line edge roughness [30] are completely masked out. And it can be confidently deduced that the enhanced V_{TH} shifts captured by our modeling are dominantly SR-induced. More specifically, our model only accounts for the roughness of the T_{Body}. The devices used in the experiments [1] had a large EOT of 10 nm, henceforth, the impact of oxide thickness fluctuation can be ignored. However, our modeling assumes a SR roughness profile in a statistical manner, which is only applicable to devices with large active region. To accurately assess the impact of SR-induced enhanced V_{TH} shifts on devices with decananometer channel length would entail an actual construction of the SR and is beyond the scope of this work.

Impact of SR on Si and Ge UTB MOSFET devices have been investigated in this paper, where the phenomena of enhanced V_{TH} shifts and SR-limited mobility are systematically modeled. It is apparent from this simulation study that the choice of channel materials and surface orientations significantly impacts the manifestation of these SR-induced phenomena devices through their intrinsically different two-dimensional electronic bandstructures. Prudent selection of channel materials and surface orientations with a large quantization mass is an effective method for suppression of these effects.

IEEE TRANSACTIONS ON ELECTRON DEVICES, VOL. 52, NO. 11, NOVEMBER 2005

2438

ACKNOWLEDGMENT

The authors would to thank M. V. Fischetti and C. Y. Mou for attending to queries pertaining to their published papers.

REFERENCES

[1] K. Uchida, H. Watanabe, A. Kinoshita, J. Koga, T. Numata, and S. Takagi, "Experimental study on carrier transport mechanism in ultra-thin-body SOI n- and p-MOSFETs with SOI thickness less than 5 nm," in *IEDM Tech. Dig.*, 2002, p. 47.

[2] Process Integration, Devices and Structures [Online]. Available: http://www.itrs.net/Common/2004Update

[3] H. Sakaki, T. Noda, K. Hirakawa, M. Tanaka, and T. Matsusue, "Interface roughness scattering in GaAs/AlAs quantum wells," *Appl. Phys. Lett.*, vol. 51, p. 1934, 1987.

[4] A. E. Meyerovich and S. Stepaniants, "Transport in channels and films with rough surfaces," *Phys. Rev. B, Condens. Matter*, vol. 23, p. 17116, 1995.

[5] C. Y. Mou and T. M. Hong, "Transport in quantum wells in the presence of interface roughness," *Phys. Rev. B, Condens. Matter*, vol. 61, p. 12612, 2000.

[6] S. Nakaharai, T. Tezuka, N. Sugiyama, Y. Moriyama, and S. Takagi, "Characterization of 7-nm-thick strained Ge-on-insulator layer fabricated by Ge-condensation technique," *Appl. Phys. Lett.*, vol. 83, p. 3516, 2003.

[7] M. Yang, M. Leong, L. Shi, K. Chan, V. Chan, A. Chou, E. Gusev, K. Jenkins, D. Boyd, Y. Ninomiya, D. Pendleton, Y. Surpris, D. Heenan, J. Ott, K. Guarini, C. D. Emic, M. Cobb, P. Mooney, B. To, N. Rovedo, J. Benedict, R. Mo, and H. Ng, "High performance CMOS fabricated on hybrid substrate with different crystal orientations," in *IEDM Tech. Dig.*, 2003, p. 453.

[8] T. Low, M. F. Li, W. J. Fan, S. T. Ng, Y. C. Yeo, C. Zhu, A. Chin, L. Chan, and D. L. Kwong, "Impact of surface roughness on silicon and germanium ultra-thin-body MOSFETs," in *IEDM Tech. Digest*, 2004.

[9] M. V. Fischetti and S. E. Laux, "Band structure, deformation potentials, and carrier mobility in strained Si, Ge and SiGe alloys," *J. Appl. Phys.*, vol. 80, p. 2234, 1996.

[10] F. Stern and W. E. Howard, "Properties of semiconductor surface inversion layers in the electric quantum limit," *Phys. Rev.*, vol. 163, p. 816, 1967.

[11] F. Stern, "Self-consistent results for n-type Si inversion layer," *Phys. Rev. B, Condens. Matter*, vol. 5, p. 4891, 1972.

[12] C. Moglestue, "Self-consistent calculation of electron and hole inversion charges at silicon-silicon dioxide interfaces," *J. Appl. Phys.*, vol. 59, p. 3175, 1986.

[13] J. M. Luttinger and W. Kohn, "Motion of electrons and holes in perturbed periodic fields," *Phys. Rev.*, vol. 97, p. 869, 1955.

[14] S. Ridene, K. Boujdaria, H. Bouchriha, and G. Fishman, "Infrared absorption in $Si/Si_{1-x}Ge_x/Si$ quantum wells," *Phys. Rev. B, Condens. Matter*, vol. 64, pp. 085 329/1–9, 2001.

[15] M. V. Fischetti, Z. Ren, P. M. Solomon, M. Yang, and K. Rim, "Six-band k · p calculation of the hole mobility in silicon inversion layers: Dependence on surface orientation, strain, and silicon thickness," *J. Appl. Phys.*, vol. 94, p. 1079, 2003.

[16] T. Low, Y. T. Hou, and M. F. Li, "Improved one-band self-consistent effective mass methods for hole quantization in p-MOSFET," *IEEE Trans. Electron Devices*, vol. 50, no. 10, p. 1284, Oct. 2003.

[17] T. Low, Y. T. Hou, M. F. Li, C. Zhu, A. Chin, G. Samudra, L. Chan, and D. L. Kwong, "Investigation of performance limits of germanium double-gated MOSFETs," in *IEDM Tech. Digest*, 2003, p. 691.

[18] T. Low *et al.*, "Valence bandstructure study of metal-oxide-semiconductor-field-effect transistors with an ultrathin silicon and germanium channel," *J. Appl. Phys.*, vol. 98, p. 02 4504, 2005.

[19] H. Y. Yu, C. Ren, Y. C. Yeo, J. F. Kang, X. P. Wang, H. H. H. Ma, M. F. Li, D. Chan, and D. L. Kwong, "Fermi pinning-induced thermal instability of metal-gate work functions," *IEEE Electron Device Lett.*, no. 4, p. 337, Apr. 2004.

[20] T. Low, M. F. Li, C. Shen, Y. C. Yeo, Y. T. Hou, C. Zhu, A. Chin, and D. L. Kwong, "Electron mobility in Ge and strained-Si channel ultrathin-body metal-oxide semi conductor field-effect transistors," *Appl. Phys. Lett.*, vol. 85, p. 2402, 2004.

[21] T. Low, C. Shen, M. F. Li, Y. C. Yeo, Y. T. Hou, C. Zhu, A. Chin, L. Chan, and D. L. Kwong, "Study of mobility in strained silicon and germanium ultrathin body MOSFETs," in *Int. Conf. Solid-State Devices Materials*, 2004, p. 776.

[22] D. K. Ferry and S. M. Goodnick, *Transport in Nanostructures*. Cambridge, U.K.: Cambridge Univ. Press, 1997.

[23] D. Chattopadhyay and H. J. Queisser, "Electron scattering by ionized impurities in semiconductor," *Rev. Mod. Phys.*, vol. 53, p. 745, 1981.

[24] T. Ando, A. B. Fowler, and F. Stern, "Electronic properties of two-dimensional systems," *Rev. Mod. Phys.*, p. 437, 1982.

[25] S. M. Goodnick, D. K. Ferry, and C. W. Wilmsen, "Surface roughness at the $Si(100)$-SiO_2 interface," *Phys. Rev. B, Condens. Matter*, vol. 32, p. 8171, 1985.

[26] T. Mizuno, N. Sugiyama, T. Tezuka, Y. Moriyama, S. Nakaharai, and S. Takagi "(110)-surface strained-SOI CMOS devices with higher carrier mobility," in *Symp. VLSI Tech. Dig.*, 2003, p. 97.

[27] T. Low, Y. P. Feng, M. F. Li, G. Samudra, Y. C. Yeo, P. Bai, L. Chan, and D. L. Kwong, "First principle study of Si and Ge band structure for UTB MOSFETs applications," in *Proc. ISDRS*, 2005, to be published.

[28] M. Toyoma, K. Kita, K. Kyuno, and A. Toriumi, "Advantages of $Ge\langle 111 \rangle$ surface for high quality HfO2/Ge interface," in *Proc. SSDM*, Tokyo, 2004, pp. 226–227.

[29] A. Asenov, A. R. Brown, J. H. Davis, S. Kaya, and G. Slavcheva, "Simulation of intrinsic parameter fluctuations in decananometer and nanometer-scale MOSFETs," *IEEE Trans. Electron Devices*, vol. 50, no. 11, pp. 1837–1852, Nov. 2003.

[30] A. R. Brown, A. Asenov, and J. R. Watling, "Intrinsic fluctuations in sub 10 nm double-gate MOSFETs introduced by discreteness of charge and matter," *IEEE Trans. Nanotechnol.*, vol. 1, no. 1, pp. 195–200, Jan. 2002.

[31] H. P. Tuinhout, "Impact of parametric mismatch and fluctuations on performance and yield of deep-submicron CMOS technologies," in *Proc. 32nd Eur. Solid-State Device Research Conf.*, 2002, p. 95.

Tony Low (S'05) received the B.S. degree (with first class honors) in electrical engineering from the National University of Singapore (NUS) in 2002, where he is currently pursuing the Ph.D. degree in electrical engineering and physics.

Since 2002, he has been with the Silicon Nano Device Laboratory, NUS where his research mainly focuses on modeling and simulation of semiconductor devices. This involves the study of novel device structures and new channel materials for enhanced device performances, physics of transport in diffusive and ballistic regimes, bandstructure calculations, and *ab inito* simulations. His research was supported by a Chartered Semiconductor scholarship, Singapore Millennium Graduate scholarship, and the 2005 IEEE Electron Device Society fellowship.

Ming-Fu Li (M'91–SM'99) received the degree from the Department of Physics, Fudan University, Shanghai, China, in 1960.

After graduation, he joined the University of Science and Technology of China (USTC), Hefei, as a Teaching assistant and then Lecturer. In 1978, he joined the Graduate School faculty, Chinese Academy of Sciences, Beijing, and became a Professor in 1986. He has also served as Adjunct Professor at the Institute of Semiconductors, Chinese Academy of Sciences, Fudan University, and USTC.

He was a Visiting Scholar at Case Western Reserve University, Cleveland, OH in 1979, and at the University of Illinois at Urbana-Champaign from 1979 to 1981, and was a Visiting Scientist at the University of California at Berkeley and Lawrence Berkeley National Laboratories from 1986 to 1987, and 1990 to 1991, respectively. He joined the Department of Electrical Engineering, National University of Singapore, Singapore, in 1991, where he is currently a Professor in the Department of Electrical and Computer Engineering. Since 2003, he is also a Senior Member Technical Staff of the Institute of Microelectronics, Singapore. His current research interests are in the areas of CMOS device technology (high-κ/metal gate and Schottky S/D transistors), reliability, and quantum modeling. He has published over 300 research papers and two books, including *Modern Semiconductor Quantum Physics* (Singapore: World Scientific, 1994).

Dr. Li has served on several international program committees and advisory committees in international semiconductor conferences in Canada, China, Germany, Japan, Singapore, Taiwan, R.O.C., and the U.S.

Ganesh Samudra (M'86) received the Ph.D. degree from Purdue University, West Lafayette, IN in 1985.

He is currently an Associate Professor at the Department of Electrical and Computer Engineering, National University of Singapore (NUS), Singapore. He was a Visiting Professor at the Massachusetts Institute of Technology, Cambridge, in 2001, and worked for three years at Texas Instruments (TI) Incorporated before joining NUS. In TI, he worked on the development of TCAD tools linking device and circuit simulator and defining three-dimensional structures for simulation. At NUS, he is mostly involved in research projects in the area of device simulation. He has published about 70 technical papers in journals and conferences.

Yee-Chia Yeo (S'96–M'03) received the B.Eng. (first-class honors) and M.Eng. degrees from the National University of Singapore (NUS), Singapore, and the M.S. and Ph.D. degrees from the University of California, Berkeley (UCB), all in electrical engineering.

In 1998 to 2001 at UCB, his research was on sub-100 MOS transistor design and fabrication, strained-channel transistors, alternative gate dielectrics, and process integration of dual-metal gates for CMOS technology. In 2001 to 2003, he was with the Taiwan Semiconductor Manufacturing Company (TSMC), where he worked on exploratory transistor technologies. He is an Assistant Professor of Electrical and Computer Engineering, NUS, and a Research Program Manager at the Agency for Science, Technology, and Research, Singapore. He has authored or coauthored a book chapter and more than 90 journal/conference papers. He holds 21 U.S. patents, and has more than 50 U.S. patents pending.

Dr. Yeo was awarded the 1995 IBE Prize from the Institution of Electrical Engineers, U.K. In 1996, he received the Lee Kuan Yew Gold Medal and the Institution of Engineers, Singapore Gold Medal, He is also the recipient of the 1997–2001 NUS Overseas Graduate Scholarship Award, the 2001 IEEE Electron Device Society Graduate Student Fellowship Award, the 2002 IEEE Paul Rappaport Award, and a 2003 TSMC Best Invention Disclosures Award.

Chunxiang Zhu (S'97–M'02) received the B.S. and M.S. degrees in electrical engineering from Xidian University, Xi'an, China, in 1992 and 1995, respectively, and the Ph.D. degree in electrical engineering from Hong Kong University of Science and Technology, Kowloon, Hong Kong, in 2001.

He joined the Department of Electrical and Computer Engineering, National University of Singapore, as an Assistant Professor in 2001. His current areas of research interests include high-κ gate stacks for both Si and Ge MOSFETs, high-κ MIM technology for RF and mixed signal IC application, and nonvolatile memory with novel materials. He has authored or coauthored 70 publications in these areas.

Albert Chin (SM'03) received the Ph.D. degree from the Department of Electrical Engineering, University of Michigan, Ann Arbor, in 1998.

He joined AT&T-Bell Laboratories from 1989 to 1990, and General Electrics Electronic Laboratory from 1990 to 1992. He is a Professor at the National Chiao-Tung University, Hsinchu, Taiwan, R.O.C., and was Deputy Director of the Nanometer Center, University System of Taiwan, and Visiting Professor at the Silicon Nano Device Laboratory, National University of Singapore. He has published more than 200 technical papers and presentations. He invented the 3-D IC to solve the ac power consumption and to be able to extend the VLSI scaling, high mobility GOI technology for CMOS and optical devices, resonant cavity photodetector, and high mobility strain-compensated HEMTs. He is a pioneer in high-κ gate dielectrics research (Al_2O_3 and $LaAlO_3$) for low dc power consumption CMOS, NiGe, and IrO_2 metal gates. He also developed the high-performance RF passive devices on VLSI-standard Si substrate using ion implantation to convert into a semi-insulating, much-improved RF device performance close to GaAs that has been realized up to 100 GHz. The developed low-power consumed 3-D IC, metal-gate/high-κ/GOI MOSFETs and RF devices on process converted semi-insulating Si are followed by research laboratories and universities worldwide and in pilot runs at IC fabs. He is currently working on 3-D GOI/Si IC, metal-gate/high-κ nano-CMOS, 3-D III-V/VLSI IC, RF Si, and quantum-trap nano memory technologies.

Dr. Chin has given invited talks at the IEDM and other conferences in the U.S., Europe, Japan, and Korea (i.e., Samsung Electronics), etc.

Dim-Lee Kwong (A'84–SM'90) received the B.S. degree in physics and the M.S. degree in nuclear engineering from the National Tsing Hua University, Taiwan, R.O.C., in 1977 and 1979, respectively, and the the Ph.D. degree in electrical engineering from Rice University, Houston, TX.

He was an Assistant Professor with the Electrical Engineering Department, University of Notre Dame, Notre Dame, IN, from 1982 to 1985. He was a Visiting Scientist with the IBM General Technology Division, Essex Junction, VT, during the summer of 1985, working on 4-Mb DRAM technology. He joined the Microelectronic Research Center and the Department of Electrical and Computer Engineering, The University of Texas, Austin, in 1985 as an Assistant Professor. He was promoted to Associate Professor in 1985 and to Full Professor in 1990. He is the author of more than 310 journal and 270 referred archival publications and has been awarded more than 22 U.S. patents. His current areas of research interests include rapid thermal CVD technology for the growth and deposition of semiconductor materials compatible with ULSI processes, advanced dielectrics for logic, analog, and memory devices, metal gate electrode, shallow junctions, and high dielectrics. Forty-three students received the Ph.D. degree under his supervision.

Dr. Kwong has received numerous awards, including the Best Dissertation Award in 1982, the IBM Faculty Development Award in 1984, and the Engineering Foundation Teaching Award from the University of Texas, Austin, in 1994. He also holds the Earl N. and Margaret Brasfield Endowed Fellowship.

APPLIED PHYSICS LETTERS 87, 062105 (2005)

Negative-U property of oxygen vacancy in cubic HfO$_2$

Y. P. Feng[a] and A. T. L. Lim
Department of Physics, National University of Singapore, 2 Science Drive 3, Singapore 117542

M. F. Li
Silicon Nano Device Lab, Department of Electrical and Computer Engineering, National University of Singapore, Singapore 119260, and Institute of Microelectronics, Singapore 117685

(Received 8 February 2005; accepted 5 July 2005; published online 3 August 2005)

Oxygen vacancy in cubic HfO$_2$ was investigated using first-principles calculation based on density functional theory and generalized gradient approximation. Five different charge states (V^{++}, V^{+}, V^{0}, V^{-}, and V^{--}) were investigated. It was found that the oxygen vacancy in HfO$_2$ has negative-U behavior and it is energetically favorable for the vacancy to trap two electrons or two holes when the respective charges are injected into the oxide, due to large electron-lattice interaction. Therefore, oxygen vacancy is a main source of charge traps in both n- and p-type metal-oxide-semiconductor field-effect transistors based on HfO$_2$, and reducing such defects will be useful in limiting charge trapping and in improving the quality of the high-k dielectric in modern complementary metal-oxide semiconductor technology. © 2005 American Institute of Physics. [DOI: 10.1063/1.2009826]

With the continued downscaling of complementary metal-oxide semiconductor (CMOS) transistors, high-k dielectrics are urgently needed to replace SiO$_2$ as the gate material. HfO$_2$ has been demonstrated as one of the most promising high-k gate dielectrics.[1] However, a number of problems associated with HfO$_2$ must be solved before it can be used in devices. One of such problems is the threshold voltage shift of MOS transistors under stress at high temperature, or the so-called bias temperature instability (BTI) due to charge trapping in the gate dielectric.[2-4] Oxygen vacancy is probably one of the major traps contributing to charge trapping in HfO$_2$.[5]

In this letter, we report results of our first-principles study on different charge states of oxygen vacancy in HfO$_2$. It is known that HfO$_2$ gate dielectric has a monoclinic or cubic structure.[6] For convenience of computation, we used the cubic phase in our study. The results are likely applicable to the other phases of HfO$_2$ since the defect properties of oxygen vacancy are expected to be similar in those structures.

Our calculation is based on density functional theory (DFT) within the generalized gradient approximation (GGA) and the pseudopotential-plane-wave method.[7-9] The ultrasoft pseudopotentials[10] were employed in the self-consistent total-energy calculations. A plane-wave cutoff of 495 eV for the kinetic energy was used in the calculation. The Monkhorst-Pack scheme was used for k point sampling in the Brillouin zone, with a mesh of $9 \times 9 \times 9$ in the bulk calculation and a similar k point spacing in the defect calculation. Results of our calculation for bulk HfO$_2$ are in excellent agreement with those of previous pseudopotential calculations[11,12] as well as those obtained using the linearized augmented planewave method.[13] For cubic HfO$_2$, the calculated lattice constant in this study is 5.06 Å and the energy gap is 3.68 eV. That the gap energy is underestimated is a well known shortcoming of DFT calculations. Defect calculations were carried out using an 80-atom three-dimensionally periodic supercell, which consists of 3×3 $\times 3$ primitive cells of cubic HfO$_2$, with one oxygen atom removed to model the oxygen vacancy. The large supercell size separates the periodic images of the vacancy by over 10 Å, which is sufficient to eliminate the artificial Coulomb interaction between charged vacancies.[11] A neutralizing background was applied to the supercell for calculations of charged vacancy. All structures were thoroughly optimized, without symmetry constraints, through a minimization of the magnitude of Hellmann-Feynman forces acting on the atoms to within 0.05 eV/Å.

The calculated total energies of V^{++}, V^{+}, V^{-}, and V^{--} charge states of the oxygen vacancy relative to that of the neutral vacancy are listed in Table I. Based on the total energies, we calculated vacancy formation energies[14] for the various charge states. For the neutral oxygen vacancy, we obtained a formation energy of 9.33 eV, which is in good agreement with the results of Foster *et al.*[11] who obtained vacancy formation energy of 9.36 eV for the threefold-coordinated oxygen and 9.34 eV for the fourfold-coordinated oxygen in monoclinic HfO$_2$.

We consider the processes which take place within the gate dielectric of a metal-oxide semiconductor field-effect transistor (MOSFET) under applied gate bias. When a positive gate bias is applied to the n-MOSFET, electrons are injected into the dielectric from the Si substrate, as shown in Fig. 1(a).[15] The V^{0} trap captures an electron and becomes V^{-} center. From Table I, the reaction

TABLE I. Total energies of various charge states of the oxygen vacancy, relative to that of the neutral vacancy.

Charge state	Energy (eV)
V^{--}	13.73
V^{-}	7.02
V^{0}	0.00
V^{+}	−6.20
V^{++}	−13.35

[a]Electronic mail: phyfyp@nus.edu.sg

062105-2   Feng, Lim, and Li

Appl. Phys. Lett. **87**, 062105 (2005)

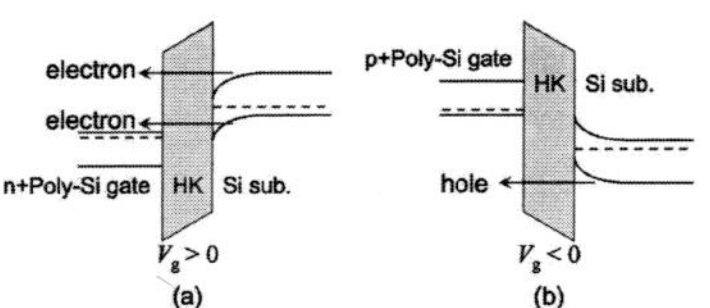

FIG. 1. Band diagrams of MOSFETs. (a) n-MOSFET with positive gate voltage. Electrons are injected from the Si substrate. (b) p-MOSFET with negative gate voltage. Holes are injected from the Si substrate (Ref. 15).

$$2V^- \rightarrow V^0 + V^{--} \tag{1}$$

in which an electron is transferred from one V^- center to another is exothermic, releasing about 0.32 eV in energy. This indicates that V^- centers are energetically less favorable as compared to the V^0 and V^{--} centers. On the other hand, if a negative gate bias is applied to the p-MOSFET, holes are injected into the dielectric from the Si substrate, as shown in Fig. 1(b).[15] This time, the V^0 trap captures a hole and becomes V^+ center. As in the case of V^- centers, the V^+ centers are also unstable with respect to the reaction

$$2V^+ \rightarrow V^0 + V^{++}. \tag{2}$$

From Table I, we can see that this reaction is highly exothermic, releasing about 0.94 eV in energy. Therefore, the charged states V^+ in the high-k oxide are also metastable and an electron will be transferred from one V^+ center to another to form a neutral and a V^{++} state that is more stable. In other words, oxygen vacancy traps are responsible for charge trapping in both n- and p-MOSFETs. Finally, when the gate bias is removed, no charge injection occurs. All charge in the oxygen traps will be de-trapped and the gate dielectric remains neutral. It is also noted that the reaction

$$V^{--} + V^{++} \rightarrow 2V^0 \tag{3}$$

releases 0.38 eV in energy, which indicates that in a neutral HfO_2 film, the oxygen vacancies prefer the neutral vacancy state V^0 rather than $V^{--}+V^{++}$.

Equations (1) and (2) and Table I indicate that oxygen vacancy is a negative-U trap in HfO_2. The concept of negative-U system was introduced by Anderson in 1975 to explain the failure of defect or impurity paramagnetism in semiconductor glasses with a high density of defects or impurities.[16] This idea was further developed by Baraff et $al.$ who used the self-consistent pseudopotential method to calculate the vacancy defect states V^{++}, V^+, and V^0 in p-type Si.[17] In a negative-U trap as shown in Eq. (1), capturing the first electron by a neutral defect, with a trap energy $E(-/0)$, leads to a metastable state. Capturing a second electron, with a trap energy $E(--/-)$ leads to a stable state with a negative-U energy, $U=E(--/-)-E(-/0)$.[17,18] Table I indicates a negative-U of -0.32 eV in the reaction of Eq. (1). Similar discussion can be applied to Eq. (2) with $U=-0.94$ eV. The negative-U is possible in the case of strong electron-lattice interaction which leads to large lattice relaxation. The released energy due to lattice relaxation over-compensates the electron-electron Coulomb interaction energy. Negative-U traps are commonly found in ionic materials with strong electron-lattice interaction. Recently, ab $initio$ calculations have shown that hydrogen and oxygen vacancy related traps in SiO_2 are negative-U centers.[19,20] HfO_2 is more ionic com-

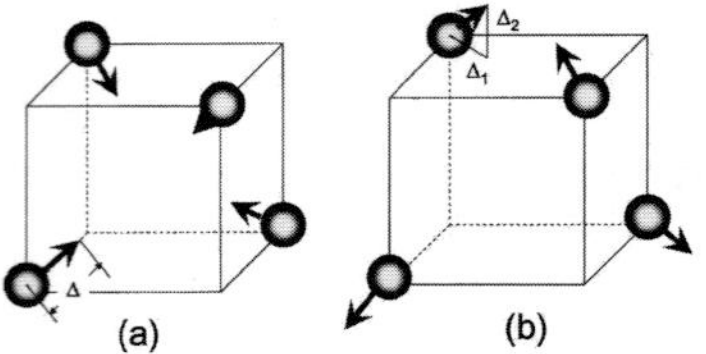

FIG. 2. (a) Breathing mode and (b) C_{2v} mode of relaxation of the nearest neighboring Hf atoms of the oxygen vacancy. The four nearest neighboring Hf atoms of the oxygen vacancy occupy opposite corners of a cube with the vacancy located at the center of the cube. All arrows in each figure are of the same length and represent the same amount of relaxation.

pared to SiO_2, and it is the strong ionic polarization that contributes to the high dielectric constant. Therefore, it is not surprising that certain traps in HfO_2 exhibit negative-U property. Recently, Kang et $al.$ investigated hydrogen related defect complexes in HfO_2 using first-principles method and found that hydrogen in monoclinic HfO_2 is a negative-U trap.[21]

Let us turn to consider the lattice relaxations around the oxygen vacancy in HfO_2 in different charge states. The calculated band structures show that the formation of a neutral vacancy introduces a new doubly occupied one-electron level in the middle of the band gap, similar to that found in monoclinic HfO_2.[11] This level is singly occupied in the charged state V^+. In the case of the neutral vacancy, all four nearest neighbor Hf atoms of the oxygen vacancy relax towards the vacancy site [breathing mode, Fig. 2(a)] by a small amount, 0.03 Å. Larger relaxations were found for oxygen atoms which are bonded to two of such Hf atoms. In the charged vacancy states, due to strong electron-lattice interaction, much larger atomic relaxations around the oxygen vacancy were found. In the positively charged states, V^+ and V^{++}, the nearest neighbor Hf atoms relax away from the vacancy by 0.08 Å and 0.16 Å, respectively. In the negatively charged states V^- and V^{--}, all nearest neighbor Hf atoms of the oxygen vacancy relax towards the vacancy site, but the relaxation is a mix of the breathing mode and a C_{2v} mode in which two Hf atoms move towards each other and the other pair move away from each other, as shown in Fig. 2(b). The breathing mode relaxation can be described by an amplitude, Δ, which is the amount of inward relaxation of each atom, while the C_{2v} mode can be described by two parameters, the amount of relaxation of each atom in a 2D plane, Δ_1, and the displacement, Δ_2, in the perpendicular direction, respectively, as shown in Fig. 2. Details of the atomic relaxation for the various charged states around the oxygen vacancy are given in Table II. It is noted that in the case of V^-, the distance between one pair of Hf atoms in the relaxed structure is 3.33 Å and that between the other pair is 3.58 Å, compared to 3.45 Å between any other two Hf atoms around the vacancy. The same distances in the case of V^{--} are 3.12 Å, 3.57 Å, and 3.35 Å, respectively. All these interatomic distances are smaller than that in the unrelaxed structure (3.58 Å) due to the large inward relaxation of the Hf atoms.

In conclusion, we carried out ab $initio$ total energy calculation to investigate the properties of oxygen vacancy in cubic HfO_2. Results of our calculation show that the oxygen

062105-3   Feng, Lim, and Li

Appl. Phys. Lett. **87**, 062105 (2005)

TABLE II. Details of relaxations of the nearest neighboring Hf atoms of the oxygen vacancy for various charge states. Here Δ is the amplitude of breathing mode relaxation (a negative amplitude indicates outward relaxation), Δ_1 and Δ_2 specify the in-plane and perpendicular relaxation in the C_{2v} mode. See Fig. 2.

Charged state	Breathing mode Δ (Å)	C_{2v} mode	
		Δ_1 (Å)	Δ_2 (Å)
V^{--}	0.14	0.11	−0.006
V^{-}	0.07	0.06	0.002
V^{0}	0.03	0.00	0.00
V^{+}	−0.08	0.00	0.00
V^{++}	−0.16	0.00	0.00

vacancy in cubic HfO_2 is a negative-U center, and energetically it favors trapping two electrons or two holes when electrons or holes are injected from the Si substrate, due to strong electron-lattice interaction and large lattice relaxation. The oxygen vacancy thus acts as charge traps in both n- and p-MOSFETs. It is based on the negative-U behavior of oxygen vacancy in HfO_2 that the mechanism for the charge trapping in MOSFETs can be well understood. Recently, frequency dependent dynamic charge trappings were reported for n- and p-MOS transistors based on HfO_2 gate dielectric.[22,23] This frequency dependence of dynamic charge trapping can also be fully understood based on the negative-U property of the oxygen vacancy traps in HfO_2, the details of which will be published elsewhere. Consequently, reducing such defects in HfO_2 will be useful in limiting charge trapping and in improving the quality of the high-k dielectric in modern CMOS technology.

[1]G. D. Wilk, R. M. Wallace, and J. M. Anthony, J. Appl. Phys. **89**, 5243 (2001).

[2]A. Kerber, E. Cartier, L. Pantisano, M. Rosmeulen, R. Degraeve, T. Kauerauf, G. Groeseneken, H. E. Maes, and U. Schwalke, International Reliability Physics Symposium Proceedings, 2003, p. 41.

[3]S. Zafar, A. Callegari, E. Gusev, and M. V. Fishetti, International Electron Device Meeting, Tech. Digest (2002), p. 517.

[4]K. Onishi, R. Choi, C. S. Kang, H.-J. Cho, Y. H. Kim, R. E. Nieh, J. Han, S. A. Krishnan, M. S. Akbar, and J. C. Lee, IEEE Trans. Electron Devices **50**, 1517 (2003).

[5]H. Takeuchi, H. Y. Wong, D. W. Ha, and T. J. King, International Electron Device Meeting, Tech. Digest (2004), p. 829.

[6]J. Aarik, A. Aidla, H. Mändar, T. Uustare, K. Kukli, and M. Schuisky, Appl. Surf. Sci. **173**, 15 (2001).

[7]M. C. Payne, M. P. Teter, D. C. Allen, T. A. Arias, and J. D. Joannopoulos, Rev. Mod. Phys. **64**, 1045 (1992).

[8]G. Kresse and J. Hafner, J. Phys.: Condens. Matter **6**, 8245 (1994); G. Kresse and J. Furthmüller, Comput. Mater. Sci. **6**, 15 (1996); G. Kresse and J. Furthmüller, Phys. Rev. B **54**, 11169 (1996).

[9]J. P. Perdew, K. Burke, and M. Ernzerhof, Phys. Rev. Lett. **77**, 3865 (1996).

[10]D. Vanderbilt, Phys. Rev. B **41**, 7892 (1990).

[11]A. S. Foster, F. L. Gejo, A. L. Shluger, and R. M. Nieminen, Phys. Rev. B **65**, 174117 (2002).

[12]R. H. French, S. J. Glass, F. S. Ohuchi, Y.-N. Xu, and W. Y. Ching, Phys. Rev. B **49**, 5133 (1994).

[13]A. T. L. Lim, Honors Project Report, National University of Singapore (2002).

[14]C. G. Van de Walle and J. Neugebauer, Nature (London) **423**, 626 (2003). In this paper, the formation energy of interstitial hydrogen is defined relative to a H_2 molecule. A similar definition is used in the present study but the formation energy of the oxygen vacancy is defined relative an isolated O atom, as in Ref. 11.

[15]Y. T. Hou, M. F. Li, H. Y. Yu, Y. Jin, and D.-L. Kwong, International Electron Device Meeting, Tech. Digest (2002), p. 731.

[16]P. W. Anderson, Phys. Rev. Lett. **34**, 953 (1975); International Reliability Physics Symposium Proceedings, 2004, p. 269.

[17]G. A. Baraff, E. O. Kane, and M. Schluter, Phys. Rev. B **21**, 5662 (1980).

[18]Ming-Fu Li, *Modern Semiconductor Quantum Physics* (World Scientific, Singapore 1994), Sec. 390, p. 316.

[19]D. J. Chadi, Appl. Phys. Lett. **83**, 437 (2003).

[20]P. E. Blochl and J. H. Stathis, Phys. Rev. Lett. **83**, 372 (1999).

[21]J. Kang, E.-C. Lee, and K. J. Chang, Appl. Phys. Lett. **84**, 3894 (2004).

[22]C. Shen, H. Y. Yu, X. P. Wang, M. F. Li, Y.-C. Yeo, D. S. H. Chan, K. L. Bera, and D. L. Kwong, International Reliability Physics Symposium Proceedings, 2004, p. 601.

[23]S. J. Rhee, Y. H. Kim, C. Y. Kang, H. J. Cho, R. Cho, C. H. Cho, M. S. Akbar, and J. C. Lee, International Reliability Physics Symposium Proceedings, 2004, p. 269.

Modeling Study of InSb Thin Film For Advanced III-V MOSFET Applications

Z. G. Zhu[1], Tony Low[1], M. F. Li[1,2], W. J. Fan[3], P. Bai[4], D. L. Kwong[2] and G. Samudra[1] (Email: eleshanr@nus.edu.sg)

(1) Silicon Nano Device Lab (SNDL), ECE Department, National University of Singapore (2) Institute of Microelectronics, Singapore
(3) School of EEE, Nanyang Technological University of Singapore (4) Institute of High Performance Computing, Singapore

Abstract: *Band structure of III-V material InSb thin films is calculated using empirical pseudopotential method (EPM). Contrary to the predictions by simple effective mass methods, our calculation predicts that the Γ valley (with the smallest isotropic bulk effective mass) in InSb remains the lowest lying conduction valley despite size quantization effects in the presence of competing L and Δ valleys which have larger quantization mass. Based on EPM, we computed the important electronic parameters (effective mass, valley minima) of InSb thin film as a function of film thicknesses. Our calculations reveal that the 'effective mass' of Γ valley electrons increases with the scaling down of film thickness. We then studied the transport of InSb thin film using Non-Equilibrium Green's Function. The calculation reveals that InSb is comparable but not superior to Si as channel material of ultra-thin body double gate n-MOSFET in the ballistic limit of these devices.*

Introduction: MOSFET structure based on thin-film semiconductor channel is a promising candidate for scaling devices into the nanometer regime [1] due to their superb short channel effect suppression. InSb is also being studied as a new channel material [2] to reap benefits of its high mobility (in fact, highest electron and hole mobility among common III-V compound semiconductors [3]). Theoretical investigation of InSb based on simple effective mass approximation (EMA) had been conducted to predict its performance limits [4]. However, EMA may not predict the physical result even qualitatively in thin film regime due to unaccounted effects like band coupling and non-parabolic dispersion. In this work, we will study InSb thin film using a more physically reliable model; the local empirical pseudopotential method (EPM), including spin-orbit coupling. We derive important parameters of its electronic structure, followed by an assessment of ballistic transport in MOSFET with InSb thin film channel.

Theoretical Background: The calculations in this work are based on the empirical pseudopotential method (EPM). This method is successfully applied in band structure calculations of metal, semiconductor or other material. Then it is extended to calculate band structures or other properties of quantum film and quantum dots for Si successfully [5, 6]. The key points of this method are (1) a supercell (Fig 1a) is constructed including the film and vacuum to form a translational invariance system in order to apply the conventional band structure calculation method; (2) Bulk empirical pseudopotential of elements are used for film calculation. In the following, we shall give a description of three important aspects.

(1) Empirical Model Potential (MP): Bulk empirical pseudopotentials (at special q^2 values, also known as Empirical Symmetry and Anti-symmetry Form Factors (ESAFF) shown in Table I) of the various atomic elements are used to derive a MP for InSb thin film calculation (Fig. 2). To obtain the MP of InSb, we need eight constraints, and seven constraints come from seven nonzero ESAFF

parameters at q^2 equal to 3, 4, 8, 11 and 12. The other one is provided by InSb workfunction which gives one nonzero ESAFF at q^2 equal to 0 [5]. It is reasonable because the workfunction describe the surface property which is important for quantum film [5]. The best fitting parameters for the MP are shown in inset of Fig 2. Table II further validates EPM based on this new MP as it reproduces the energy band minima derived from experiment. We shall point out that the MP of one element in semiconductor compounds is actually the atomic model potential, i.e. the volume of unit cell is divided by the number of atoms in this unit cell. So when it comes to the thin film calculations, there will be a renormalization of different unit cell volumes under the assumption that the overall pseudopotential of atoms is the superposition of the local pseudopotential of all the atoms in the thin film.

(2) Hydrogen Passivation: When we calculate band structure of thin film, we must ensure that surface dangling bonds are correctly passivated with Hydrogen (H) so that there will be no surface states within the band gap. By using MP of H in [6], we calculated the band structure of 8 atomic layers (atm) Si<100> thin film and compared the results with those derived from *ab initio* calculation [7] (Fig. 3), which yields consistent result establishing the validity and reliability of this H MP. In Si<100> thin film calculation, the renormalization of supercell volume is set as a ratio of the number of H atoms in the supercell and the number of Si atoms in the same unit cell. Atomic volume of H was determined as the volume which can reproduce known thin film Si band structure. For InSb thin film, the bond length of In-H is set as 1.75Å (from experiment [8]) and Sb-H as 1.711Å (from *ab initio* energy minimization calculation [7], which is very close to 1.67 Å in Ref. [9]). The angle between the bond of In-H and the top surface (θ) (Fig 1a), should be treated carefully. Fig 4a shows the calculated direct InSb bandgap (E_Γ) as function of θ for varying InSb atm. In principle, the stable θ is when E_Γ is stationary with respect to θ. Fig 4b shows E_Γ as function of film thickness at θ=0.77rad, and a fitting formula (similar to [6]) affirms that E_Γ approaches the bulk bandgap of 0.43eV as film thickness tends to bulk.

(3) Spin-Orbit Coupling (SOC): SOC treatment is considered by calculating the matrix element in plane wave representation and it is reduced to the spin orbit form factors (SOFFs) [10]. Firstly, SOFFs of bulk InSb is obtained. Then they are used in thin film calculation like pseudopotential form factors described above. So SOFFs have to be normalized with respect to supercell volume. Conventionally, bulk band structure of InSb is derived by considering a 2-atom unit cell in which the primitive vectors are non-orthogonal and the mutual angles between them are all $\pi/3$. In thin film calculation, a larger supercell is constructed in which the primitive vectors are mutually orthogonal. Hence this angular effect and volume renormalization must be considered in our treatment (Fig. 6 caption).

1-4244-0439-8/06/$20.00 ©2006 IEEE

Features of InSb Electronic Structures: Fig. 5 shows the calculated bandstructure of bulk InSb from our EPM. We derived the following effective masses; $m_L=2.45m_0$ (the longitudinal mass of L valley), $m_X=3.90m_0$ (the longitudinal mass of Δ valley) and $m_\Gamma=0.016m_0$ (the isotropic mass of Γ valley). Due to this much smaller mass of Γ valley, a simple EMA calculation would predict that this valley would be uplifted above the competing L and Δ valleys due to size quantization (see Fig 7). Fig. 5b shows the effective mass at Γ valley (m_Γ) calculated from the second derivative of energy dispersion. We note that the parabolic assumption for the energy dispersion at Γ is only valid for a very small k range. This will render the EMA approach highly unreliable. In addition, the small bandgap also entails a considerable amount of coupling between the conduction and valence bands, which render the uncoupled EMA approach unreliable. Fig. 6 shows the calculated bandstructure of thin film InSb from our EPM. These thin films still retain the direct bandgap properties in contrast to EMA calculation [4], which has L valley as the lowest lying conduction valley when film thickness is below 5nm. Fig. 7a shows variation of the energy minima at Γ, L and Δ valleys with film thickness. There was no crossing of the energy minima down to 6atm, in contrast to predictions by EMA methods. In addition, we note that the lowest lying Γ valley is separated from the other valleys with a gap >0.3eV, signifying that electrons are dominantly occupying Γ valley. Another stark difference with results of EMA is the bandgap. Although, the bandgap is enhanced by quantization effect, this decay is much slower than the well-known d^{-2} behavior predicted by EMA in an infinite deep well model [3] (For e.g. at Γ valley, see Fig. 7a). However, EPM also yields the result that the conduction band quantization effect is larger than that of valence band as predicted by EMA (Fig 7b). From a logic device point-of-view, one would desire to have a larger bandgap to curb the increasing band-to-band tunneling current with each new generation CMOS devices. The fact that InSb is a direct bandgap material may aggravate the problem. However, if one can achieve a sufficiently thin film ~8atm, which offers a bandgap of >1eV, band-to-band tunneling current should still be tolerable. Fig 8 shows the contour plot of the conduction and valence band energy dispersion for 8atm InSb. We confirm that the Γ valley in thin film still retains its isotropy unlike the case of Si [11]. Fig. 9 shows the Si and InSb thin film effective mass fitted from their 2D energy dispersion. We have attempted to fit a parabolic dispersion over the calculated energy dispersion to obtain the effective mass. In fact, we expect the effective mass also increase with wave-vector as depicted in Fig. 5b. But still, although effective mass is not a rigorously derived quantity, it serves as a very useful 'figure of merit'. Interestingly, the isotropic effective mass of InSb Γ valley increases with the decrease of film thickness. At ~1nm of InSb film, the effective mass is ~$0.1m_0$. Increase in m_Γ would retard the quantization effect in InSb thin film. The isotropy of the electron mass which translates to isotropy of the electron transport property should be advantageous as it affords engineer with more flexibility in orientating the n and p-MOSFET devices to yield the most optimum transport direction on the same substrate.

Ballistic Limit of InSb Nanoscale Devices: We calculated the InSb thin film devices' ballistic limit using the Non-Equilibrium Green Function (NEGF) method [12], under the framework of simple effective mass methods. Since the energy dispersion for <100>-surface InSb thin film is relatively isotropic (Fig 8), one can employ a decoupled 2D treatment to the problem and calculation is done in the framework of mode-space NEGF approach [12]. The effective masses used for the various InSb devices are as derived in Fig 9. For Si devices, due to the anisotropy, we simply assumed an effective mass of $0.20m_0$. A double-gated device structure is employed. Channel length of 20nm is used. An EOT of 1nm and metal gate is employed. No wave penetration into oxide is assumed. Source/drain and channel are doped at 1×10^{20}cm^{-3} and 1×10^{15}cm^{-3} respectively. The boundary conditions for the potential at the contact are set to floating, by imposing the condition that the potential derivatives are zero at these boundaries. Channel length of 20nm is used. The drive current vs. gate voltage is plotted in Fig 10. The Si devices yields a slightly larger ballistic current compared to its InSb counterpart. The main reason is due to the larger density-of-states mass in Si and the fact that it is doubly degenerate compared to non-degenerate InSb Γ valley. Currently, experimental study shows that sub 50nm Si devices only achieve ~40% ballistic performance [13]. A lower density-of-states mass should help damp the dissipative processes (which translate to higher mobility) to achieve fully ballistic performances in InSb devices and performance then could be better than Si.

Conclusion: Band structure of III-V semiconductor InSb thin films is calculated using empirical pseudopotential method (EPM). Γ valley in InSb remains the lowest lying conduction valley (a desirable trait if high mobility characteristic is required) despite size quantization effects but its isotropic effective mass increases with decrease in film thickness. A NEGF calculation of the InSb and Si double-gated devices reveals that they have comparable ballistic drive current.

References: [1] ITRS 2004 [2] T. Ashley et al., ICSICT 2004 [3] M. F. Li, 'Modern Semiconductor Quantum Physics', World Scientific, (1994); M. Levinshtein et al., "Handbook series on semiconductor parameters', World Scientific (1996) [4] A. Pethe et al., IEDM, p.605 (2005) [5] S. B. Zhang et al., PRB 48, p.11204 (1993) [6] L. W. Wang et al, J. Phys. Chem. 98, p. 2158 (1994); 100, p. 2394 (1994); C. Yu Yeh et al., PRB 50, p. 14405 (1994) [7] M. D. Segall et al., J. Phys.: Cond. Matt. 14, p. 2717 (2002) [8] K. Raghavachari et al. J. AM. Chem. SOC. 124, p. 15119 (2002) [9] Gábor Balázs et al., Organometallics 20, p. 2666 (2001) [10] J. R. Chelikowsky et al., PRB 14, p. 556 (1976); 30, p. 4828 (1984); S. Bloom et al., Phys. Stat. Sol. 42, p. 191 (1970) [11] T. Low et al, ISDRS, p. 384 (2004) [12] Z. Ren et al., TED 50, p. 1914 (2003) [13] A. Lochtefeld et al., EDL 22, p. 95 (2001).

Acknowledgement: This work is supported by Singapore A*STAR Nanoelectronics Program under research grants R-398000013305. We gratefully acknowledge the use of NanoMOS [12] from Purdue Computational Electronics Group for this work.

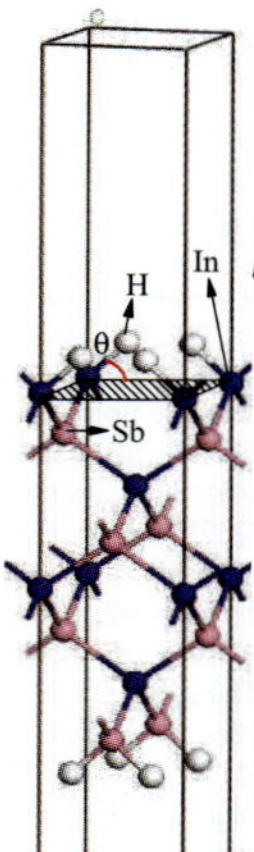

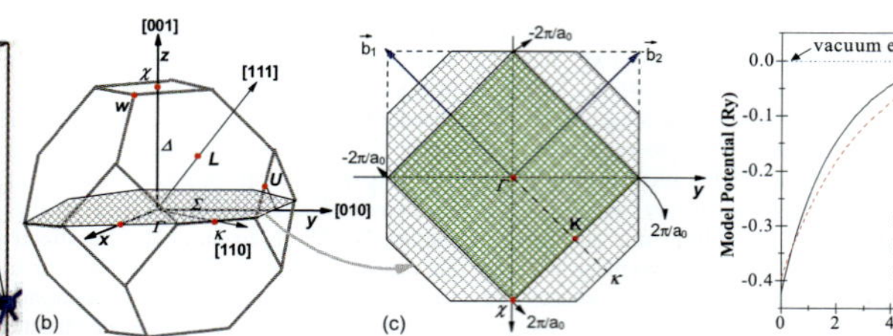

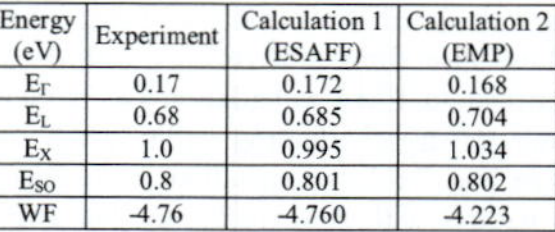

Fig.1. (a) Supercell of InSb used for our calculation. It consists of a thin film layer and a vacuum layer in the quantization direction. θ is the angle of the H-In bond as illustrated. (b) Three-dimensional Brillouin zone of InSb and illustration of the important symmetry lines and points (c) The two-dimensional Brillouin zone obtained by projecting onto the xy plane. The densest shaded region is the first Brillouin zone.

Fig.2. Atomic model potential of Sb and In. The functional form of the model potential is as shown in inset with their best fitting parameters a_1, a_2, a_3, and a_4 also given.

Form factor (Ry)		q^2=0	q^2=3	q^2=4	q^2=8	q^2=11	q^2=12
ESAFF	$V^S(q^2)$	-0.858	-0.20	0	0.018	0.034	0
	$V^A(q^2)$	0	0.035	0.032	0	0.011	0.013
EMP	$V^S(q^2)$	-0.816	-0.202	0	0.0179	0.03416	0
	$V^A(q^2)$	0	0.0353	0.0312	0	0.01221	0.0114

Table I. Symmetry $V^S(q^2)$ and anti-symmetry form factors $V^A(q^2)$ in our calculations. Unit of q^2 is $(2\pi/a)^2$, where a is crystal lattice constant. For InSb, we use a=6.47877 Å.

Energy (eV)	Experiment	Calculation 1 (ESAFF)	Calculation 2 (EMP)
E_Γ	0.17	0.172	0.168
E_L	0.68	0.685	0.704
E_X	1.0	0.995	1.034
E_{SO}	0.8	0.801	0.802
WF	-4.76	-4.760	-4.223

Table II. Experimental data [3] of various energy conduction valley minima (E_Γ, E_L, E_X), spin-orbit coupling (E_{SO}) and work function (WF) are compared with theoretical result from empirical pseudopotential method. Calculation 1 and 2 employs the Empirical Symmetry and Anti-symmetry Form Factors (ESAFF) and Empirical Model Potential (EMP) respectively.

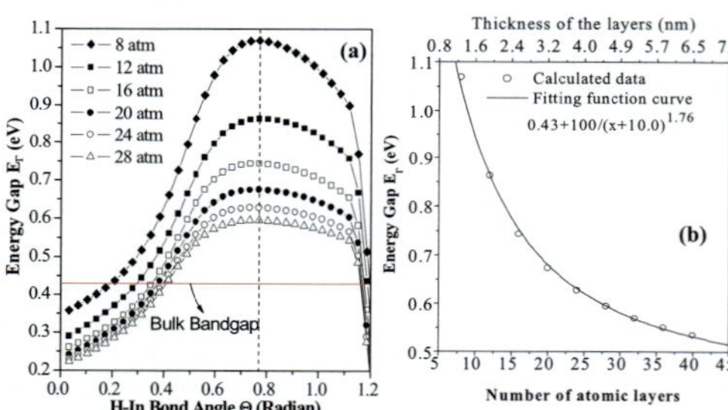

Fig.4. (a) The calculated InSb direct energy gap (E_Γ) as function of H-In bond angle θ, plotted for InSb thin film of varying atomic layers (atm). The graph illustrated that there is a common θ where the energy function is a stationary point. (b) The direct energy gap (E_Γ) at bond angle θ where the energy function is a stationary point, plotted as function of film thickness. A fitting formula which fits these data points concludes that the bandgap (E_Γ) gives correct asymptotic behavior when film thickness tends to bulk, yielding bulk bandgap value 0.43 eV.

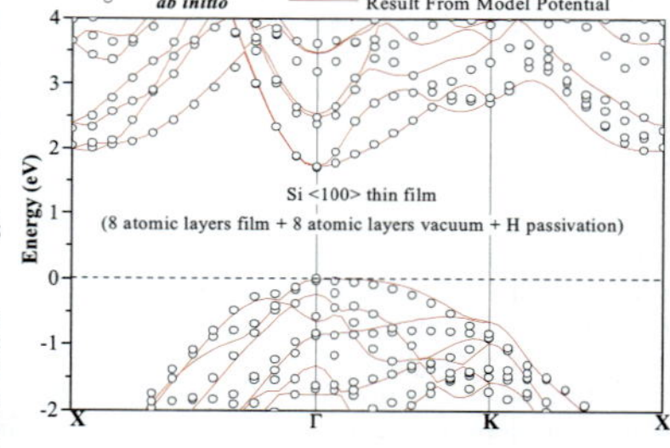

Fig.3. Energy bands of Si<100> thin film (8 atm Si film + 8 atm vacuum with H passivation) calculated with *ab initio* method (using CASTEP, Local Density Approximation) and via empirical pseudopotential method with model potential. Excellent agreement for the various conduction valleys minima is obtained.

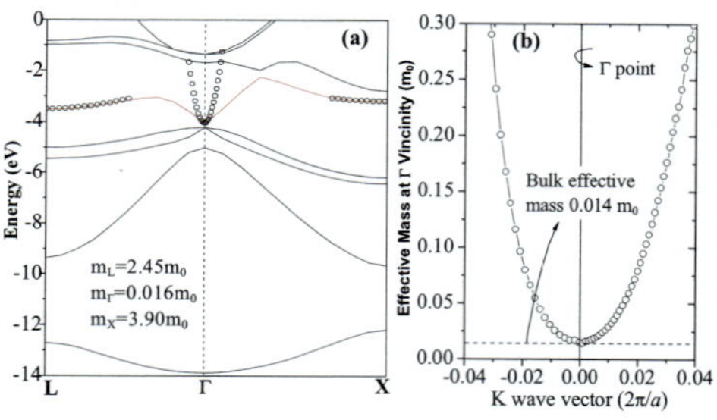

Fig.5. (a) Energy band structure of bulk InSb calculated using empirical pseudopotential method. We have calculated the various effective masses for the different conduction valleys. The longitudinal mass (derived from the empty circles in (a) at L, X and Γ) for L valley is found to be m_L=2.45m_0. The longitudinal mass for X valley is found to be m_X=3.90m_0. Lastly, the isotropic effective mass at Γ valley is found to be m_Γ=0.016m_0. (b) Effective mass Γ valley (m_Γ) calculated from the second derivative of energy dispersion. We note that the parabolic assumption for the energy dispersion at Γ is only valid for a very small k range. Consider a 1.3 nm InSb film, the wave vector spread according to uncertainty principle is ~0.04 ($2\pi/a$). From the plot, the parabolic assumption for EMA is only applicable up to a k range of ~0.01 ($2\pi/a$). Hence, in the ultra thin film regime, a parabolic EMA can not be used.

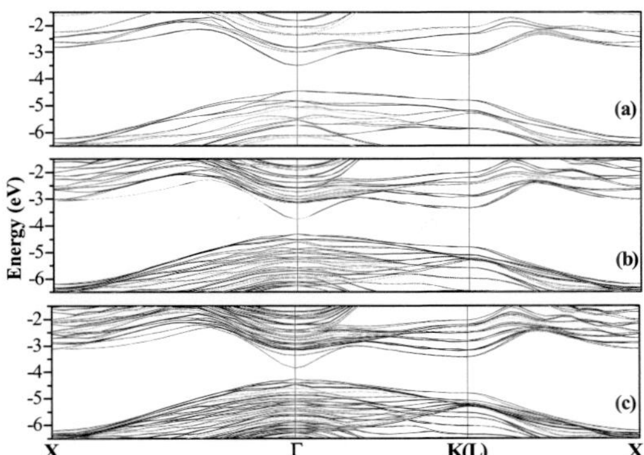

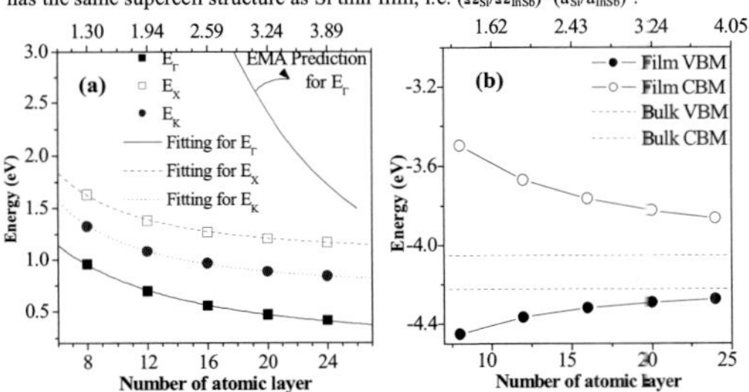

Fig. 6 Band structure of InSb thin film. There are 8 atm layer, 16 atm layer and 24 atm layer film in (a), (b) and (c), respectively. To arrive at the results, spin orbit Hamiltonian is divided into two parts, angular and magnitude part. When the supercell is changed from bulk 2-atom unit cell to <100> film supercell, the volume renormalization will appear in the second part like what has been done on form factors of pseudopotential of Sb and In. For the angular part, in our calculation, we treat it as a parameter. We determine this parameter by calculating the bulk band structure and spin orbit splitting. We use the supercell which is the film <100> supercell without the vacuum parts. Then we get this parameter for each film thickness. Then these parameters are used in film calculations in which a supercell is used by adding the corresponding vacuum parts. For Hydrogen, there will be a volume renormalization factor $V^{PS}_H(\Delta G)(InSb)=(\Omega_{Si}/\Omega_{InSb})$ $V^{PS}_H(\Delta G)(Si)$, where $\Omega_{Si(InSb)}$ is supercell volume of Si (InSb) and the InSb thin film has the same supercell structure as Si thin film, i.e. $(\Omega_{Si}/\Omega_{InSb})=(a_{Si}/a_{InSb})^3$.

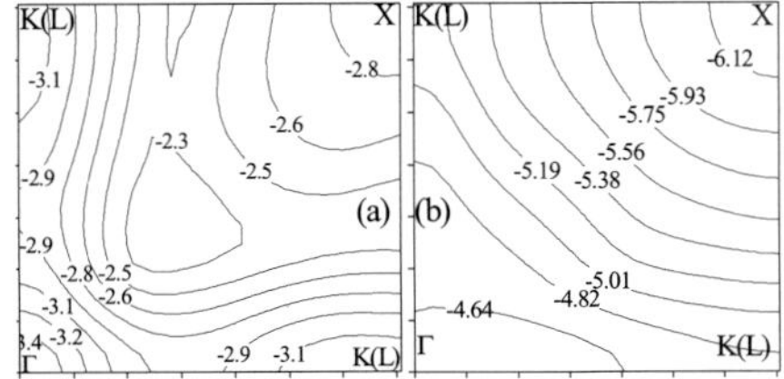

Fig. 7. (a) Variation of band gaps with the number of atomic layers. (Top scale characterizes the thickness of thin film in nm). We also plot the fitting curves in this figure. The fitting formulas are $0.16802+150/(x+9.7)^{1.83}$, $1.03395+92/(x+5.5)^{1.94}$, and $0.70439+94/(x+5.2)^{1.94}$. (b) The InSb thin film valence band maximum (VBM) and conduction band minimum (CBM) as function of the number of atomic layers. Bulk values are approached at thicker films.

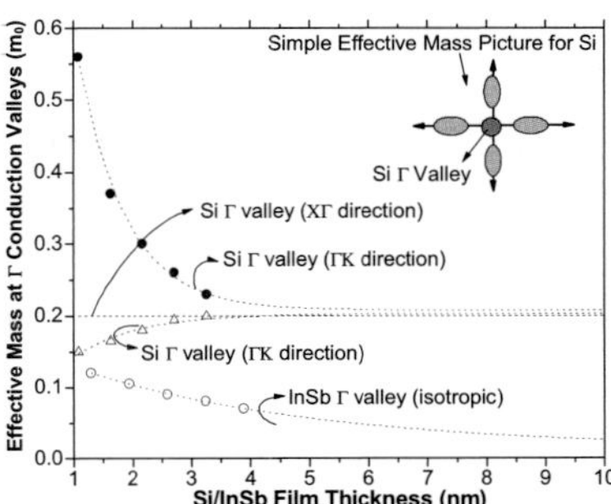

Fig. 9. Electron effective mass of Si and InSb Γ conduction valley. InSb Γ valley is isotropic, with increasing effective mass as film thickness is scaled down. Si Γ valley in thin film was originated from bulk Δ valley, projected onto the 2D k-space. We observe that the isotropy was reduced with decreasing of film thickness, with the ΓK direction effective mass diverging for each of the two degenerate bands. The larger ΓK effective mass is for the lower band of the two degenerate bands, and the smaller ΓK effective mass corresponds to the top band. The inset shows the constant energy ellipsoid for Si projected onto the 2D k-space, with the Γ valley being doubly degenerate.

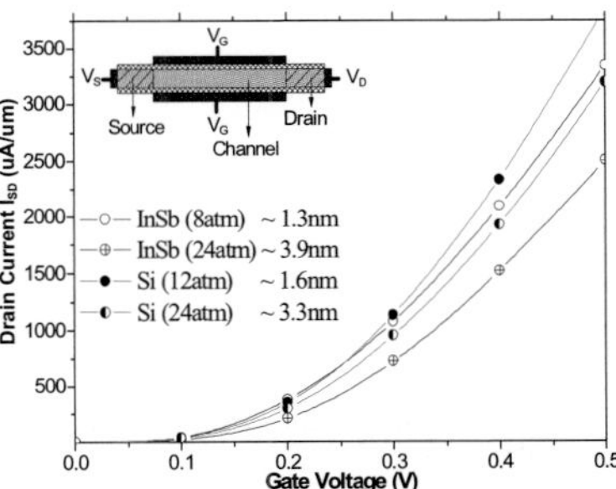

Fig. 10. Ballistic current calculated for InSb and Si double-gated devices using simple effective mass approximation. 8atm and 24atm InSb thin film devices are calculated. Effective mass used for InSb are derived from a parabolic fit of the bandstructure from EPM. For Si devices, an effective mass of $0.2m_0$ is assumed. An EOT of 1nm and metal gate is employed. No wave penetration into oxide is assumed. Source/drain and channel are doped at $1\times10^{20}cm^{-3}$ and $1\times10^{15}cm^{-3}$ respectively. The boundary conditions for the potential at the contact are set to floating, by imposing the condition that the potential derivatives are zero at these boundaries. Channel length of 20nm is used.

Fig. 8 (Left). Energy contour plot for 8atm InSb thin film for the first conduction and first valence band are plotted in (a) and (b) respectively. Values of the energy are indicated on each contour line. Important symmetry points are also indicated. It is apparent that the Γ conduction valley is relatively isotropic in nature even in the ultra-thin film regime.

Complete Publication List 1982–2008

Refereed Journal Papers

1. Li MF, Topics in modern IC operational amplifier design, *Bull Univ Sci Technol China* **8**:119, 1978 (in Chinese).
2. Li MF, KD-203 IC operational amplifier, *Bull Univ Sci Technol China* **9**:167, 1979 (in Chinese).
3. Li MF, Nonideal operational amplifiers, *Acta Electronica Sin* **4**:1, 1979 (in Chinese).
4. Li MF, Sah CT, A new method for the determination of dopant and trap concentration profiles in semiconductors, *IEEE Trans ED* **29**:309, 1982.
5. Li MF, Sah CT, New techniques of capacitance-voltage measurements of semiconductor junctions, *Solid State Electron* **25**:95, 1982.
6. Qin GG, Li MF, Sah CT, Deep level profiles in boron implanted n-Si, *J Appl Phys* **53**:4800, 1982.
7. Li MF, Mao DQ, Ren SY, Global predictions of T_2 symmetric deep level wavefunctions in semiconductors, *Solid State Commun* **48**:789, 1983.
8. Li MF, Ren SY and Mao DQ, Theory of T2 symmetric deep wavefunctions in Si, *Acta Phys Sin* **32**:1263, 1983 (in Chinese).
9. Li MF, Ren SY, Mao DQ, The behavior of deep level wavefunctions in shallow energy region, *Acta Phys Sin* **33**:738, 1984 (in Chinese).
10. Mao DQ, Li MF, Ren SY, A symmetric deep level wavefunction of substitutional defect pairs in GaP, *Acta Phys Sin* **33**:897, 1984 (in Chinese).
11. Li MF, Ren SY, Mao DQ, An investigation of deep level wavefunctions with energies near the band edges, *J Phys C: Solid State Phys* **17**:3415, 1984.

12. Li MF, Ren SY, Mao DQ, Theory of T_2 symmetric deep level wave functions in Si, *Chin Phys* **4**:294, 1984.

13. Li MF, Chen JX, Yao YS, Bai G, Au acceptor levels in Si under pressure, *J Appl Phys* **58**:2599, 1985.

14. Chen JX, Li MF, Li YJ, Defects and their thermal annealing behaviours in Si implanted with boron and phosphorus, *Chin Phys Lett* **2**:401, 1985.

15. Li MF, Chen JX, Yao YS, Bai G, The pressure coefficient measurement of gold deep levels in Si, *Acta Phys Sin* **34**:1068, 1985 (in Chinese).

16. Li MF, Pressure effects of deep levels in semiconductors, *Acta Phys Sin* **34**:1549, 1985 (in Chinese).

17. Wang JQ, Wu X, Li MF, Lin Y, Deep centers in VPE-GaAs grown under different atmospheres, *Chin J Semiconductors* **6**:556, 1985 (in Chinese).

18. Li MF, Ren SY, Mao DQ, Deep level wavefunction distributions in Bloch space, *Acta Phys Sin* **34**:547, 1985 (in Chinese).

19. Ren SY, Mao DQ, Li MF, The electronic structures of diavacancies in cubic semiconductors, Part I, *Acta Phys Sin* **34**:455, 1985 (in Chinese).

20. Li MF, Mao DQ, Ren SY, Binding energies of electrons by nitrogen pairs in GaP *Phys Rev B* **32**:6907, 1985.

21. Ren SY, Mao DQ, Li MF, Electronic structure of multivacancies in Si, *Superlattices Microstructures* **1**:375, 1985.

22. Chen JX, Li MF, Li YJ, Low dose B+ and P+ implantation associated defects in n-Si and their annealing behaviours, *Chin J Semiconductors* **7**:363, 1986.(in Chinese).

23. Bai G, Li MF, Wu X, A computer real time controlled system for experimental studies in solid state electronics, *Chin J Semiconductors* **7**:203, 1986 (in Chinese).

24. Mao DQ, Ren SY, Li MF, Electronic structures of diavacancies in cubic semiconductors, Part II, *Acta Phys Sin* **35**:808, 1986 (in Chinese).

25. Ren SY, Mao DQ, Li MF, Electronic structures of diavacancies in cubic semiconductors, Part III, *Acta Phys Sin* **35**:1457, 1986 (in Chinese).

26. Ren SY, Mao DQ, Li MF, Theory of the (100) uniaxial stress dependence of the deep levels in GaP and GaAs, *Chin Phys Lett* **3**:313, 1986.

27. Sun J, Chen J, Li MF, Electron traps induced by ion implantation in n-Si, *Chin J Semiconductors* **8**:545, 1987 (in Chinese).

28. Hu HM, Ren SY, Mao DQ, Li MF, Electronic structures of divacancies in cubic semiconductors, Part IV, *Acta Phys Sin* **36**:1321, 1987 (in Chinese).

29. Li MF, Yu PY, A new proposed method for determining inner or outer crossing lattice relaxation of DX center in AlGaAs based on pressure effects, *Solid State Comm.* **61**:13, 1987.

30. Li MF, Yu PY, Weber ER, Hansen W, Lattice relaxation of pressure induced deep centers in GaAs:Si, *Appl Phys Lett* **51**:349, 1987.

31. Li MF, Yu PY, Weber ER, Hansen W, Photocapacitance study of pressure induced deep donors in GaAs, *Phys Rev B* **36**:4531, 1987.

32. Sun JL, Chen JX, Li MF, Further investigation of electron traps induced by B+ and P+ ion implantation in n-Si, *Chin Phys* **8**:808, 1988.

33. Yu PY, Li MF, Comments on new evidence of small lattice relaxation for the DX center in AlGaAs, *Appl Phys Lett* **52**:1645, 1988.

34. Shan W, Li MF, Yu PY, Hansen W, Walukiewicz W, Pressure dependence of Schottky barrier height at Pt/GaAs interface, *Appl Phys Lett* **53**:974, 1988.

35. Li MF, Shan W, Yu PY, Hansen W, Weber ER, Bauser E, Pressure dependence of the DX centers in AlGaAs:Te in the vicinity of the Γ-X cross-over, *Appl Phys Lett* **53**:1195, 1988.

36. Martinez G, Li MF, Yu PY, Photoionization cross-sections of the DX centers in AlGaAs alloys and of the pressure induced deep donors in GaAs, *Defect Diffusion Forum* **62/63**:97, 1989.

37. Li MF, Yu PY, Shen W, Hansen W, Weber ER, Effect of boron on the pressure induced deep donors in GaAs:Si, *Appl Phys Lett* **54**:1344, 1989.

38. Li MF, Jia YB, Yu PY, Zhou J, Gao JL, Negative U property of DX center in AlGaAs, *Phys Rev B* **40**:1430, 1989.

39. Wang BS, Gu ZQ, Wang JQ, Li MF, *Ab initio* pseudopotential calculations of optical phonon deformation potentials in zincblende semiconductors, *Phys Rev B* **39**:12789, 1989.

40. Jia YB, Li MF, Zhou J, Gao JL, Yu PY, Kong MY, Chan KT, Discovery of new photoinduced electron trap state shallower than the DX state in Si doped AlGaAs, *J Appl Phys* **66**:5632, 1989.

41. Shan W, Yu PY, Li MF, Hansen WL, Bauser E, Pressure dependence of the DX center in GaAlAs:Te, *Phys Rev B* **40**:7831, 1989.

42. Jia YB, Li MF, Zhou J, Gao JL, Kong MY, Investigation of DX centers in AlGaAs by C-V method, *Chin J Semiconductors* **11**:801, 1990.

43. Gu ZG, Li MF, Wang JQ, Wang BS, Deformation potentials at the top of valence bands in semiconductors-*ab initio* pseudopotential calculations, *Phys Rev B* **41**:8333, 1990.

44. Li MF, Yu PY, Jia YB, Zhou J, Gao JL, Reply to comment on negative U property of the DX center in $Al_xGa_{1-x}As$:Si, *Phys Rev B* **42**:9711, 1990.

45. Li MF, Gu ZG, Wang JQ, Reinvestigation of shear-deformation-potential constant of the conduction band minima of Si: pseudopotential calculations, *Phy Rev B* **42**:5714, 1990.

46. Seguy P, Yu PY, Li MF, Leon R, Chan KT, Effect of light on the DX centers in Si and Te doped AlGaAs, *Appl Phys Lett* **57**:2469, 1990.

47. Li MF, Jia YB, Zhou J, Gao JL, Yu PY, A new point of DX center, *Chin J Semiconductors* **11**:81, 1990.

48. Wang J, Gu Z, Wang B, Li M-F, First principles calculations for quasi-particle energies of GaP and GaAs, *Phys Rev B* **44**:8707, 1991.

49. Wang J, Gu Z, Li M-F, Calculation of dielectric constants for semiconductors — an application of *ab initio* pseudopotential method, *Chin Phys Lett* **8**:21, 1991.

50. Wang J, Gu Z, Wang B, Li MF, First principle calculations of quasi-particle energies for Si and GaAs, *Chin Phys Lett* **8**:207, 1991.

51. Wang JQ, Gu ZQ, Li MF, First principle pseudopotential calculations of zone boundary phonon frequencies of III–V zincblende semiconductors, *Phys Lett A* **155**:506, 1991.

52. Li MF, Zhao XS, Gu ZG, Chan JX, Li YJ, Wang JQ, Shear deformation potential of the conduction band minima of Si: experimental determination by the deep level capacitance transient method, *Phys Rev B* **43**:14040, 1991.

53. Surh MP, Li MF, Louie SG, Spinorbit splitting of GaAs and InSb bands near Γ, *Phys Rev B* **43**:4286, 1991.

54. Li MF, Yu PY, Bauser E, Hansen WL, Haller EE, Deep level transient spectroscopy of DX centers in $Al_{0.38}Ga_{0.62}As$: Te under uniaxial stress, *Semiconductor Sci Technol* **6**:825, 1991.

55. Li MF, Yu PY, Weber ER, Simulation of effects of uniaxial stress on the deep level transient spectroscopy spectra of the DX center in AlGaAs alloys, *Appl Phys Lett* **59**:1197, 1991.

56. Wang J, Gu Z, Wang B, Li MF, First principle calculations of quasiparticle energies for semiconductors, *Commun Theor Phys* **15**:169, 1991.

57. Wang J, Gu Z, Li MF, Two delta function generalized plasma pole model for first principle calculations of quasiparticle energies in many electron systems, *Commun Theor Phys* **15**:395, 1991.

58. Li MF, Yu PY, Weber ER, Comment on: Occupancy of the DX center in $N-Al_{0.32}Ga_{0.68}As$ under uniaxial stress, *Appl Phys Lett* **60**:1404, 1992.

59. Wang J, Gu Z-Q, Li M-F, Lai W-Y, Intervalley gamma-X deformation potentials in III-V zinc-blende semiconductors by *ab initio* pseudopotential calculations, *Phys Rev B 46*, 1992.

60. Li MF, Chen X, Lim YC, Linearity improvement of CMOS transconductors for low supply application, *Electron Lett* **29**:1106, 1993.

61. Feng YP, Teo KL, Li MF, Poon HC, Ong CK, Xia JB, Empirical pseudopotential band-structure calculation for $Zn_{1-x}Cd_xSySe_{1-x}$ quaternary alloy, *J Appl Phys* **74**:3948, 1993.

62. Teo KL, Feng YP, Li MF, Chong TC, Band-structure of $Mg_{1-x}Zn_xSySe_{1-y}$, *Semiconductor Sci Technol* **9**:349, 1994.

63. Li MF, Du AY, Luo YY, Chua SJ, Comment on direct evidence for the negative-U nature of the DX center in AlGaAs, *Phys Rev B 49*:11479, 1994.

64. Li MF, Luo YY, Yu PY, Weber ER, Fujioka H, Du AY, Chua SJ, Lim YT, Two-electron state and negative-U property of sulfur DX centers in GaAsP, *Phys Rev B 50*:7996, 1994.

65. Poon HC, Feng ZC, Feng YP, Li MF, Relativistic band structure of ternary II-VI semiconductor alloys containing Cd, Zn, Se and Te, *J Phys Condens Matter* **7**:2783, 1995.

66. Du AY, Li MF, Chong TC, Chua SJ, Observation of carrier concentration saturation effect in n type AlGaAs, *Appl Phys Lett* **66**:1391, 1995.

67. Du AY, Li MF, Chong TC, Chua SJ, Carrier concentration saturation effect in n-AlGaAs, *Mater Sci Forum* **279**:196–201, 1995.

68. Fan WJ, Li MF, Chong TC, Xia JB, Electronic properties of zinc-blende GaN, AlN and their alloys GaAlN, *J Appl Phys* **79**:188, 1996.

69. Fan WJ, Li MF, Chong TC, Xia JB, Electronic structures of the zinc blende GaN/GaAlN compressive strained superlattices and quantum wells, *Superlattices Microstructures* **19**:251, 1996.

70. Fan WJ, Li MF, Chong TC and Xia JB, Band structure parameters of zinc-blende GaN, AlN and their alloys GaAlN, *Solid State Commun* **97**:381, 1996.

71. Zhao R, Lau WS, Chong TC, Li MF, A comparison of the selective etching characteristics of conventional and low-temperature-grown GaAs over AlAs by various etching solutions, *Jpn J Appl Phys* **35**:22, 1996.

72. Fan WJ, Li MF, Chong TC, Xia JB, Valence hole subbands and optical gain spectra of GaN/GaAlN strained quantum well, *J Appl Phys* **80**:3471, 1996.

73. Fan WJ, Li MF, Chong TC, Xia JB, Optical gain in zinc-blende GaN/GaAlN strained quantum well laser, *Solid State Commun* **98**:737, 1996.

74. Du AY, Li MF, Chong TC, Teo KL, Lau WS, Zhang Z, Dislocation related traps in p-InGaAs/GaAs lattice-mismatched heterostructures , *Appl Phys Lett* **69**:1, 1996.

75. Jia YB, Gremeiss H, Li MF, A transient capacitance voltage method for characterizing DX centers in Si doped AlGaAs, *Semiconductor Sci Technol* **11**:1787, 1996.

76. Przybtek J, Baj M, Slupinski T, Li MF, The distorted configuration of the EL2 defect stabilized under hydrostatic pressure in GaAsP, *Physica Status Solidi B* **198**:193, 1996.

77. Li MF, Yep SP, Lim YC, Analog Integrated Circuits and Signal Processing A novel integrated CMOS switch circuit for high precision sample-and-hold technique, *Analog Integr Circuits Signal Process* **12**:211, 1997.

78. Yeo YC, Li MF, Chong TC, Yu PY, Theoretical study of the energy band structure of partially CuPt-ordered $Ga_{0.5}In_{0.5}P$, *Phys Rev B* **55**:1641416419, 1997.

79. Teo KL, Li MF, Goo CH, Lau WS, Lim YT, New electron and hole traps in GaAsP alloy, *Int J Electron* **83**:29, 1997.

80. Du AY, Li MF, Chong TC, Xu SJ, Zhang Z, Yu DP, Investigation of dislocations and traps in MBE grown p-InGaAs/GaAs hetreostructures, *Thin Solid Films* **311**:7, 1997.

81. Jie BB, Li MF, Lou CL, Chim WK, Chan DSH, Lo KF, Investigation of interface traps in LDD pMOSTs by the DCIV method, *IEEE Electron Device Lett* **18**:583, 1997.

82. Liu W, Li MF, Chua SJ, Zhang YH, Uchida K, GaN exciton photovoltaic spectra at room temperature, *Appl Phys Lett* **71**: 1997.

83. Yeo YC, Chong TC, Li MF, Electronic band structures and effective-mass parameters of wurtzite GaN and InN, *J Appl Phys* **83**(3):1429, 1998.

84. Yeo YC, Chong TC, Li MF, Fan WJ, Analysis of optical gain and threshold current density of wurtzite InGaN/GaN/AlGaN quantum well lasers, *J Appl Phys* **84**(4):1813, 1998.

85. Yeo YC, Chong TC, Li MF, Fan WJ, Electronic band structures and optical gain spectra of strained wurtzite GaN-AlGaN quantum-well lasers, *IEEE J Quantum Electron* **34**:526, 1998.

86. Yeo YC, Chong TC, Li MF, Effect of the (1010) crystal orientation on the optical gain of wurtzite GaN-AlGaN quantum-well lasers, *IEEE J Quantum Electron* **34**:1270, 1998.

87. Yeo YC, Chong TC, Li MF, Uniaxial strain effect on the electronic and optical properties of wurtzite GaN-AlGaN quantum well lasers, *IEEE J Quantum Electron* **34**:2224, 1998.

88. Teo KL, Colton JS, Yu PY, Weber ER, Li MF, Liu W, Uchida K, Tokunaga H, Akutsu N, Matsumoto K, An analysis of temperature

dependent photoluminescemnce line shapes in InGaN, *Appl Phys Lett* **73**:1697, 1998.

89. Liu W, Teo KL, Li MF, Chua SJ, Uchida K, Tokunaga H, Akutsu N, Matsumoto K, The study of piezoelectric effect in wurtzite GaN/InGaN multilayer structures, *J Cryst Growth* **290**:648, 1998.

90. Liu W, Li MF, Xu SJ, Uchida K, Matsumoto K, Phonon-assisted photoluminescence in wurtzite GaN epilayer, *Semiconductor Sci Tech* **13**:769 1998.

91. Guan H, Zhang YH, Jie BB, He YD, Li MF, Dong Z, Xie J, Li WD, Non-destructive DCIV method to evaluate plasma charging damage in ultra-thin gate oxides, *IEEE Electron Device Lett* **20**:238, 1999.

92. Liu W, Li MF, Chua SJ, Akutzu N, Matsumoto K, Photo-reflectance study of Au-Schottky contacts on n-GaN, *J Electron Mater* **28**:260, 1999.

93. Liu W, Li MF, Chua SJ, Akutzu N, Matsumoto K, Photorelectance study on the surface states of n-GaN, *Semiconductor Sci Technol* **14**:399, 1999.

94. Liu W, Li MF, Teo KL, Akutzu N, Matsumoto K, Photovoltaic spectroscopic study of GaN epilayer and InGaN quantum well structures, *J Mater Res* **14**:2794, 1999.

95. Jie BB, Li MF, Chim WK, Chan DSH, Lo KF, DC voltage-voltage method to measure the interface traps in sub-micron MOSTs, *Semiconductor Sci Technol* **14**:621, 1999.

96. Hou YT, Feng ZC, Li MF, Chua SJ, Characterization of MBE grown GaAlAs alloy films by Raman scattering, *J Surf Interface Anal* **28**:163, 1999.

97. Liu W, Feng ZC, Li MF, Chua SJ, Material properties of GaN grown by MOCVD, *J Surf Interface Anal* **28**:150, 1999.

98. Feng ZC, Hou YT, Chua SJ, Li MF, Infrared reflectance studies of GaN epitaxial films on sapphire substrate, *J Surf Interface Anal* **28**:166, 1999.

99. Li MF, He YD, Ma SG, Cho BJ, Lo KF, Xi MZ, Role of hole fluence in gate oxide breakdown, *IEEE Electron Device Lett* **20**:586, 1999.

100. Hou YT, Feng ZC, Li K, Li MF, Chua SJ, Influence of Si-doping on the characteristics of GaN on sapphire by infrared relectance, *Appl Phys Lett* **75**:1, 1999.

101. Cho BJ, Xu Z, Guan H, Li MF, Effect cf substrate hot carrier injection on quasi-breakdown of ultra-thin gate oxide, *J Appl Phys* **56**:6590, 1999.

102. He YD, Guan H, Li MF, Cho BJ, Dong Z, Conduction mechanism under quasi-breakdown of ultra-thin gate oxide, *Appl Phys Lett*, **75**:2432, 1999.

103. Kwok SH, Yu PY, Tung CH, Zhang YH, Li MF, Peng CS, Zhou JM, Confinement and electronphonon interactions of the E1 exciton in self-organized Ge quantum dots, *Phys Rev B* **59**:4980, 1999.

104. Jie BB, Ng KH, Li MF, Lo KF, Correlation between charge pumping method and direct-current current voltage method in p-type MOSFETs, *Jpn J Appl Phys* **38**:67, 1999.

105. Lai WH, Li MF, Chan L, Chua TC, Growth characterization of rapid thermal oxides, *J Vac Sci Technol B*, **17**:2226, 1999.

106. Feng ZC, Hou YT, Li MF, Chua SJ, Wang W, Zhu L, Infrared reflectance investigation of un-doped and Si-doped GaN films on sapphire, *Physica Status Solidi B* **216**:577, 1999.

107. Hou YT, Teo KL, Li MF, Uchida K, Tokunaga H, Akutsu N, Matsumoto K, An analysis of temperature dependence piezoelectric Franz-Keldysh effect on AlGaN, *Appl Phys Lett* **76**:1033, 2000.

108. Guan H, Li MF, He YD, Cho BJ, Dong Z, A thorough study of quasi-breakdown phenomena of thin gate oxide in dual-gate CMOSFETs, *IEEE Trans ED* **47**:1608, 2000.

109. Xu Z, Cho BJ, Li MF, Annealing behaviour of gate oxide leakage current after quasi-breakdown, *Microelectron Reliab* **40**:1341, 2000.

110. Xu SJ, Liu W, Li MF, Effect of temperature on longitudinal optical phonon-assisted luminescence in hetero-epitaxial GaN layer, *Appl Phys Lett* **77**:3376, 2000.

111. Li MF, Dasgupta U, Zhang XW, Lim YC, A low-voltage CMOS OTA with rail-to-rail differential input range, *IEEE Trans Circuits Syst I* **47**:1, 2000.

112. Lai WH, Zhang XW, Li MF, Novel voltage-tunable linear CMOS transconductor, *Analog Integr Circuits Signal Process* **24**:123, 2000.

113. Lim YC, Lai WH, Zhang XW, Li MF, Improved cross-coupled quad transconductor cell, *Microelectron J* **31**:77, 2000.

114. Jie BB, Chim WK, Li MF, Lo KF, Analysis of the DCIV peaks in electrically-stressed pMOSFETs, *IEEE Trans ED* **48**:913, 2001.

115. Guan H, Cho BJ, Li MF, Xu Z, Ho YD, Dong Z, Experimental evidence of interface-controlled mechanism of quasi-breakdown in ultra-thin gate oxide, *IEEE Trans ED* **48**:1010, 2001.

116. Chen G, Li MF, Yu X, Interface traps at high doping drain extension region in sub-0.25 um MOSTs, *IEEE Electron Device Lett* **22**:233, 2001.

117. Ma SG, Zhang YH, Li MF, Li WD, Xie J, Sheng GTT, Yen AC, Wang JLF, Gate-induced drain leakage current enhanced by plasma charging damage, *IEEE Trans Electron Devices* **48**:1006, 2001.

118. Ha YJ, Li MF, Liu AQ, A new CMOS buffer amplifier design used in low voltage MEMS interface circuits. *Analog Integr Circuits Signal Process* **27**:5, 2001.

119. Hou YT, Li MF, A novel simulation algorithm for Si valence hole quantization of inversion layer in metal-oxide-semiconductor devices, *Jpn J Appl Phys Part 2* **40**:L144, 2001.

120. Hou YT, Li MF, Hole quantization effects and threshold voltage shift in pMOSFET-assessed by improved one-band effective mass approximation, *IEEE Trans ED* **48**:1188, 2001.

121. Hou YT, Li MF, Lai WH, Jin Y, Modeling and characterization of direct tunneling hole current through ultrathin gate oxide in p-metal-oxide-semiconductor field-effect transistors, *Appl Phys Lett* **78**:4034, 2001.

122. Hou YT, Li MF, A simple and efficient model for quantization effects of hole inversion layers in MOS devices, *IEEE Trans ED* **48**:2893, 2001.

123. Lim R, Koa KH, Wee ATS, Lai WH, Li MF, See A, Chan L, SIMS study of Silicon oxynitride rapid thermally grown nitride oxide, *Surf Rev Lett* **8**:569, 2001.

124. Chen G, Li MF, Jin Y, Electric passivation of interface traps at drain junction space charge region in p-MOS transistors, *Microelectron Reliab* **41**:1427, 2001.

125. Hou YT, Li MF, Jin Y, Lai WH. Direct tunneling hole current through ultrathin gate oxides in metal-oxide-semiconductor devices, *J Appl Phys* **91**:258, 2002.

126. Yu HY, Wu N, Li MF, Zhu CX, Cho BJ, Kwong DL, Tung CH, Pan JS, Chai JW, Wang WD, Chi DZ, Ramanathan S, Thermal stability of $(HfO_2)_x(Al_2O_3)_{1-x}$ on (100) Si, *Appl Phys Lett* **81**:3618, 2002.

127. Yu HY, Li MF, Cho BJ, Yeo CC, Joo MS, Kwong D-L, Pan JS, Ang CH, Zheng JZ, Ramanathan S, Energy gap and band alignment for $(HfO_2)_x(Al_2O_3)_{1-x}$ on (100) Si, *Appl Phys Lett* **81**:376, 2002.

128. Yu HY, Hou YT, Li MF, Kwong D-L, Investigation of hole tunneling current through ultrathin oxynitride/oxide stack gate dielectrics in p-MOSFETs, *IEEE Trans Electron Devices* **49**:1158, 2002.

129. Yu HY, Hou YT, Li MF, Kwong D-L, Hole tunneling current through oxynitride/oxide stack and the stack optimization for *p*-MOSFETs, *IEEE Electron Device Lett* **23**:285, 2002.

130. Hou YT, Li MF, Jin Y, Direct tunneling currents through gate dielectrics in deep submicron MOSFETs, *Chin J Semiconductors* **23**:449, 2002.

131. Loh WY, Cho BJ, Li MF, Evolution of quasi-breakdown in thin gate oxides, *J Appl Phys* **91**:5302, 2002.

132. Loh WY, Cho BJ, Li MF, Investigation of quasi-breakdown mechanism through post-QB thermal annealing, *Jpn J Appl Phys* **41**:2873, 2002.

133. Loh WY, Cho BJ, Li MF, Correlation between interface traps and gate oxide leakage current in the direct tunneling regime, *Appl Phys Lett* **81**:379, 2002.

134. Xu SJ, Liu W, Li MF, Direct determination of free exciton binding energy from phonon-assisted luminescence spectra in GaN epilayer, *Appl Phys Lett* **81**:2959, 2002.

135. Chen G, Li MF, Jin Y, Interaction of interface-traps located at various sites in MOSFETs under stress, *IEEE Trans Reliab* **51**:387, 2002.

136. Chen G, Li MF, Ang CH, Zhen JZ, Kwong DL, Dynamic NBTI of pMOS transistors and its impact on MOSFET scaling, *IEEE Electron Device Lett*, **23**:734, 2002.

137. Hu H, Zhu C, Lu YF, Li MF, Cho BJ, Choi WK, A high performance MIM capacitor using HfO_2 dielectrics, *Electron Device Lett*, **23**:514, 2002.

138. Hou YT, Li MF, Kwong DL, Modeling of tunneling currents through HfO_2 and HfAlO gate stacks and their scalability in CMOS technology, *IEEE Electron Device Lett* **24**:96, 2003.

139. Joo MS, Cho BJ, Wu N, Zhu C, Kwong DL, Yeo CC, Yu HY, Li MF, Balasubramanian N, Dependence of chemical composition ratio on electrical properties of HfO_2-Al2O3 gate dielectric, *Jpn J Appl Phys Pt 2* **42**:220, 2003.

140. Luo Z, Li MF, Lian Y, Rustagi SC, A new low voltage CMOS transconductor for VHF filtering application, *Analog Integr Circuits Signal Process* **37**:233, 2003.

141. Yu XF, Zhu C, Hu H, Chin A, Li MF, Cho BJ, Kwong D-L, Foo PD, Yu MB, A high density MIM capacitor (13 fF/μm^2) using ALD HfO_2 dielectrics, *IEEE Electron Device Lett* **24**:63, 2003.

142. Yu HY, Lim HF, Chen JH, Li MF, Zhu CX, Tung CH, Du AY, Wang WD, Chi DZ, Kwong DL, Physical and electrical characteristics of HfN gate electrode for advanced MOS devices, *IEEE Electron Device Lett* **24**:230, 2003.

143. Low T, Hou YT, Li MF, Kwong D-L, Improved one-band self-consistent effective mass methods for hole quantization in p-MOSFET, *IEEE Trans ED* **50**:1284, 2003.

144. Hu H, Zhu C, Yu XF, Chin A, Li MF, Cho BJ, Kwong D-L, Yu MB, Foo PD, MIM capacitors using atomic-layer-deposited high-K $(HfO_2)1$-$x(Al_2O_3)x$ dielectrics, *IEEE Electron Device Lett* **24**:60, 2003.

145. Fan WJ, Ng ST, Yoon SF, Li MF, Chong TC, Effects of tensile strain in barrier optical gain spectra of GaInNAs/GaAsN quantum wells, *J Appl Phys* **93**:5836, 2003.

146. Fan WJ, Yoon SF, Li MF, Chong TC, Investigation of optical gain of GaInNAs/GaAs compressive strain quantum wells, *Physica B* **328**:264, 2003.

147. Kim SJ, Cho BJ, Li MF, Yu XF, Zhu C, Chin A, Kwong DL, PVD HfO_2 for high-precision MIM capacitor applications, *Electron Device Lett* **24**:387, 2003.

148. Yang MY, Huang CH, Chin A, Zhu C, Li MF, Kwong D-L, High density MIM capacitors using $AlTaO_x$ dielectrics, *IEEE Electron Device Lett* **24**:306, 2003.

149. Lin CY, Ma MW, Chin A, Yeo YC, Zhu C, Li MF, Kwong D-L, Fully silicided NiSi gate on La_2O_3 MOSFETs, *IEEE Electron Device Lett* **24**:348, 2003.

150. Hu H, Zhu C, Lu YF, Wu YH, Liew YF, Li MF, Cho BJ, Choi WK, Yakovlev N, Physical and electrical characterization of HfO_2 MIM capacitors for Si analog circuit application, *J Appl Phys* **94**:551, 2003.

151. Loh WY, Cho BJ, Li MF, Chan DSH, Ang CH, Zhen JZ, Kwong DL, Localized oxide degradation in ultra-thin gate dielectric and its statistical analysis, *IEEE Trans ED* **50**:967, 2003.

152. Kim SJ, Cho BJ, Li MF, Zhu C, Chin A, Kwong DL, Lanthanide (Tb) doped HfO_2 for high density MIM capacitor, *Electron Device Lett* **24**:442, 2003.

153. Ding SJ, Hu H, Lim HF, Kim SJ, Yu XF, Zhu C, Li MF, Cho BJ, Chan DSH, Rustagi SC, Chin A, Kwong DL, High-performance MIM capacitor using ALD high-k HfO_2-Al_2O_3 laminate dielectrics, *Electron Device Lett* **24**:730, 2003.

154. Yu DS, Huang CH, Chin A, Chen WJ, Zhu C, Li MF, Kwong DL, Fully silicided NiSi and germanided NiGe dual gates on SiO_2 n- and p- MOSFETs, *Electron Device Lett* **24**:739, 2003.

155. Chan KT, Huang CH, Chin A, Li MF, Kwong DL, McAlister SP, Duh DS, Lin WJ, Large Q-factor improvement for spiral inductors on silicon using proton implantation, *IEEE Microw Wirel Compon Lett* **13**:460, 2003.

156. Chan KT, Chin A, Lin YD, Chang CY, Zhu C, Li MF, Kwong D-L, McAlister SP, Duh DS, Lin WJ, Integrated antenna on Si over 100 GHz

performance fabricated using optimized proton implantation process, *IEEE Microw Wirel Compon Lett* **13**:487, 2003.

157. Yang MY, Huang CH, Chin A, Zhu C, Cho BJ, Li MF, Kwong D-L, Very high density RF MIM capacitors (17 fF/um^2) using high-k Al$_2$O$_3$ doped Ta$_2$O$_5$ dielectrics, *IEEE Microw Wirel Compon Lett,* **13**:431, 2003.

158. Xu M, Tan C, Li MF, Extended Arrhenius law of time-to-breakdown of ultrathin gate oxides, *Appl Phys Lett* **82**:2482, 2003.

159. Yu HY, Ren C, Yeo YC, Kang JF, Wang XP, Ma HHH, Li MF, Chan DSH, Kwong DL, Fermi pinning induced thermal instability of metal gate work functions, *Electron Device Lett* **25**:337, 2004.

160. Yu HY, Li MF, Kwong DL, Thermally robust HfN metal as a promising gate electrode for advanced MOS devices applications, *IEEE Trans Electron Devices* **51**:609, 2004.

161. Yu HY, Kang JF, Ren C, Chen JD, Hou YT, Shen C, Li MF, Chan DSH, Bera KL, Tung CH, Kwong DL, Robust high-quality HfN-HfO$_2$ gate stack for advanced MOS device application, *IEEE Electron Device Lett* **25**:70, 2004.

162. Yu HY, Li MF, Kwong DL, ALD (HfO$_2$)$_x$(Al$_2$O$_3$)$_{1-x}$ high-k gate dielectrics for advanced MOS devices application, *Thin Solid Films* **462**:110, 2004.

163. Hou YT, Li MF, Low T, Kwong DL, Metal gate work function engineering on gate leakage of MOSFETs, *IEEE Trans ED* **51**:1783, 2004.

164. Zhu S, Chen J, Li MF, Lee SJ, Singh J, Zhu CX, Du A, Tung CH, Chin A, Kwong DL, N-type Schottky barrier source/drain MOSFET using ytterbium silicide, *Electron Device Lett* **25**:565, 2004.

165. Zhu SY, Yu HY, Whang SJ, Chen JH, Shen C, Zhu C, Lee SJ, Li MF, Chan DSH, Yoo WJ, Du A, Tung CH, Singh J, Chin A, Kwong DL, Schottky-barrier s/d MOSFETs with high-k gate dielectrics and metal-gate electrode, *IEEE Electron Device Lett* **25**:268, 2004.

166. Zhu SY, Yu HY, Chen JD, Whang SJ, Chen JH, Shen C, Zhu C, Lee SJ, Li MF, Chan DSH, Woo WJ, Du A, Tung CH, Singh J, Chin A, Kwong DL, Low temperature MOSFET technology with Schottky barrier source/drain, high-k gate dielectric and metal gate electrode, *Solid State Electron* **48**:1987, 2004.

167. Ren C, Yu HY, Kang JF, Hou YT, Li MF, Wang WD, Chan DSH, Kwong DL, Fermi-level pinning-induced thermal instability in the effective work function of TaN in TaN/SiO$_2$ gate stack, *Eelectron Device Lett* **25**:23, 2004.

168. Ren C, Yu HY, Kang JF, Wang XP, Ma HH, Yeo YC, Chan DSH, Li MF, Kwong DL, A dual-metal gate integration process for CMOS with sub-1nm EOT HfO_2 by using HfN replacement gate, *IEEE Electron Device Lett* **25**:580, 2004.

169. Kang JF, Yu HY, Ren C, Li MF, Chan DSH, Hu H, Lim HF, Wang WD, Gui D, Kwong DL, Thermal stability of nitrogen incorporation in HfNO gate dielectric prepared by reactive sputtering, *Appl Phys Lett* **84**:1588, 2004.

170. Kim SJ, Cho BJ, Li MF, Ding SJ. Zhu C, Yu MB, Narayanan B, Chin A, Kwong DL, Improvement of voltage linearity in high-k MIM capacitors using HfO_2-SiO_2 stacked dielectric, *Eelectron Device Lett* **25**:538, 2004.

171. Yu XF, Zhu C, Li MF, Chin A, Yu MB, Du AY, Kwong DL, Mobility enhancement in TaN metal-gate MOFETs using tantalum-incorporated HfO_2 gate dielectric, *Electron Device Lett* **25**:501, 2004.

172. Yu XF, Zhu C, Li MF, Chin A, Du AY, Wang WD, Kwong DL, Electrical characteristic and suppressed boron penetration behavior of thermally stable HfTaO gate dielectrics with polycrystalline-silicon gate, *Appl Phys Lett* **85**:2893, 2004.

173. Wu N, Zhang Q, Zhu C, Chan DSH, Li MF, Balasubramanian N, Chin A, Kwong DL, An alternative surface passivation on germanium for metal-oxide-semiconductor capacitors with high-K gate dielectric, *Appl Phys Lett* **85**:4127, 2004.

174. Wu N, Zhang Q, Zhu C, Yeo CC, Whoang SJ, Chan DSH, Li MF, Cho BJ, Chin A, Kwong D-L, Du AY, Tung CH, Balasubramanian N, Effect of surface NH_3 anneal on physical and electrical properties of HfO_2 films on Ge substrate, *Appl Phys Lett* **84**:3741, 2004.

175. Wu N, Zhang Q, Zhu C, Chan DSH, Du A, Balasubramanian N, Li MF, Chin A, Sin JKO, Kwong DL, A TaN-HfO_2-Ge pMOSFET with novel SiH4 surface passivation, *IEEE Electron Device Lett* **25**:631, 2004.

176. Ding SJ, Hu H, Zhu C, Li MF, Kim SJ, Cho BJ, Chen DSH, Yu MB, Du AY, Chin A, Kwong DL, Evidence and understanding of ALD HfO_2-Al_2O_3 laminate MIM capacitors outperforming sandwich counterparts, *IEEE Electron Device Lett* **25**:681, 2004.

177. Ding SJ, Hu H, Zhu C, Kim SJ, Yu X, Li MF, Cho BJ, Chen DSH, Yu MB, Rustagi SC, Chin A, Kwong DL, RF, DC, and reliability characteristics of ALD HfO_2-Al_2O_3 laminate MIM capacitors for Si RF IC applications, *IEEE Trans ED* **51**:886, 2004.

178. Yu DS, Huang CH, Chin A, Zhu C, Li MF, Cho BJ, Kwong DL, Al_2O_3-Ge-on-insulator n- and p- MOSFETs with fully NiSi and NiGe dual gates, *IEEE Electron Device Lett* **25**:138, 2004.

179. Yu DS, Chiang KC, Cheng CF, Chin A, Zhu C, Li MF, Kwong DL, Fully silicide NiSi: Hf-LaAlO$_3$/SG-GOI n-MOSFETs with high electron mobility, *IEEE Electron Device Lett* **25**:559, 2004.

180. Lin CY, Chin A, Hou YT, Li MF, McAllister SP, Kwong DL, Light emission near 1.3 um using ITO-Al_2O_3-$Si_{0.3}Ge_{0.7}$-Si tunnel diodes, *IEEE Photonics Technol Lett* **16**:36, 2004.

181. Low T, Li MF, Shen C, Yeo Y-C, Hou YT, Zhu C, Chin A, Kwong DL, Electron mobility in Ge and strained-Si channel ultra-thin-body MOSFETs, *Appl Phys Lett* **85**:2402, 2004.

182. Loh WY, Cho BJ, Joo MS, Li MF, Chan DSH, Mathew S, Kwong DL, Charge trapping and breakdown mechanism in HfAlO/TaN gate stack analysed using carrier separation, *IEEE Trans Device Mater Reliab* **4**:696, 2004.

183. Yu HY, Kang JF, Ren C, Li MF, Kwong DL, HfO$_2$ gate dielectrics for the future generation of CMOS device application, *Chin J Semiconductors* **25**:1193, 2004.

184. Shen C, Li MF, Yu HY, Wang XP, Yeo Y-C, Chan DSH, Kwong DL, A physical model for frequency-dependent dynamic charge trapping in metal-oxide-semiconductor field effect transistors with HfO$_2$ gate dielectric, *Appl Phys Lett* **86**:93510, 2005.

185. Ang K-W, Chui K-J, Vladimir B, Tung C-H, Du A, Balasubramanian N, Samudra G, Li M-F, Yeo Y-C, Lattice strain analysis of transistor structures with SiGe and SiC source/drain stressors, *J Appl Phys* **86**:93102, 2005.

186. Ren C, Yu HY, Wang XP, Ma HH, Chan DSH, Li MF, Yeo YC, Tung CH, Balasubramanian N, Huan ACH, Pang JS, Kwong DL, Thermally robust TaTb$_x$N metal gate electrode for n-MOSFETs applications, *IEEE Electron Device Lett* **26**:7577, 2005.

187. Kang JF, Yu HY, Ren C, Wang XP, Li M-F, Chan DSH, Yeo Y-C, Liu XY, Tung CH, Kwong D-L, Improved electrical and reliability characteristics of HfN/HfO$_2$ gated NMOS transistor with 0.95 nm EOT fabricated using a gate-first process, *IEEE Electron Device Lett* **26**:237, 2005.

188. Zhu SY, Li R, Lee SJ, Li MF, Du A, Singh J, Zhu C, Chin A, Kwong D-L, Germanian P-MOSFETs with Schottky barrier germanide source/drain, high-K gate dielectric and metal gate, *IEEE Electron Device Lett* **26**:81, 2005.

189. Feng YP, Lim ATL, Li MF, Negative-U property of oxygen vacancy in cubic HfO_2, *Appl Phys Lett* **87**:62105, 2005.

190. Low T, Li MF, Yeo YC, Fan WJ, Fan ST, Ng ST, Kwong DL, Valence band structure study of metal-oxide-semiconductor-field-effect transistors with an ultra-thin silicon and germanium Channel, *J Appl Phys* **98**:24504, 2005.

191. Ren C, Chan DSH, Wang XP, Faizhal BB, Li MF, Yeo YC, Trigg AD, Agarwal N, Narayanan B, Pan JS, Lim PC, Huan ACH, Kwong DL, Physical and electrical properties of lanthanide-incorporated tantalum nitride for n-channel metal oxide semiconductor field effect transistors, *Appl Phys Lett* **87**:73506, 2005.

192. Yang T, Shen C, Li M-F, Ang CH, Zhu CX, Yeo Y-C, Samudra G, Rustagi SC, Yu MB, Kwong D-L, Fast DNBTI component in p-MOSFET with SiON dielectric, *IEEE Electron Device Lett* **26**:826, 2005.

193. Yang T, Shen C, Li M-F, Ang CH, Zhu CX, Yeo Y-C, Samudra G, Rustagi SC, Yu MB, Kwong D-L, Interface trap passivation effect in NBTI measurement for p-MOSFET with SiON gate dielectric, *IEEE Electron Device Lett* **26**:758, 2005.

194. Low T, Li M-F, Samudra G, Yeo Y-C, Zhu C, Chin A, Kwong DL, Modeling study of the impact of surface roughness on silicon and germanium UTB MOSFETs, *IEEE Trans Electron Devices* **52**:430, 2005.

195. Yu DS, Chin A, Liao CC, Lee CF, Cheng CF, Li MF, Yoo WJ, McAlister SP, Three dimensional metal gate high-k GOI CMOSFETs on 1-poly-6-metal 0.18 um Si devices, *Electron Device Lett* **26**:118, 2005.

196. Lai CH, Chin A, Hung BF, Cheng CF, Yoo WJ, Li MF, Zhu C, McAlister SP, Kwong DL, A novel program-erasable high-k AIN-Si MIS capacitor, *IEEE Electron Device Lett* **26**:48, 2005.

197. Bai WP, Bae SH, Wen HC, Mathew S, Bera LK, Balasubramanian N, Yamada N, Li MF, Kwong DL, Three layer laminated metal gate electrodes with tunable work functions for CMOS applications, *IEEE Electron Device Lett* **26**:231, 2005.

198. Yu MB, Xiong YZ, Kim SJ, Balakumar S, Zhu C, Li MF, Cho BJ, Lo GQ, Balasubramanian N, Kwong DL, Integrated high-k MIM capacitor with Cu/Low-k interconnects for RF application, *IEEE Electron Device Lett* **26**:793, 2005.

199. Yu DS, Liao CC, Cheng CF, Chin A, Li MF, McAlister SP, The effect of IrO2-IrO2-Hf-LaAlO3 gate dielectric on the bias-temperature instability of 3D GOI CMOSFETs, *Electron Device Lett* **26**:407, 2005.

200. Kim SJ, Cho BJ, Yu MB, Li MF, Xiong YZ, Zhu C, Chin A, Kwong DL, Metal-insulator-metal RF bypass capacitor using niobium oxide (Nb2O5) with HfO_2/Al_2O_3 barriers, *IEEE Electron Device Lett* **26**:625, 2005.

201. Ding SJ, Zhu C, Li MF, Zhang DW, Atomic layer deposited AlHfO dielectric for metal-insulator-metal capacitor, *Appl Phys Lett* **87**:535011, 2005.

202. Chui K-J, Ang K-W, Chin H-C, Shen C, Wong L-Y, Tung C-H, Balasubramanian N, Li M-F, Samudra GS, Yeo YC, Strained SOI n-channel transistor with Silicon-carbon source/drain regions for carrier transport enhancement, *IEEE Electron Device Lett* **27**:778, 2006.

203. Shen C, Yang T, Li MF, Wang XP, Foo CE, Samudra G, Yeo YC, Kwong DL, Fast Vth instability in HfO_2 gate dielectric MOSFETs and its impact on digital circuits, *IEEE Trans ED* **53**:3001, 2006.

204. Xiong YZ, Yu MB, Lo GQ, Li MF, Kwong DL, Substrate effects on resonant frequency of Si-based RF on-chip MIM capacitor, *IEEE Trans ED* **53**:2839, 2006.

205. Wu CH, Chin A, Wang SJ, Li MF, Zhu C, Hung BF, Mc-Alister SP, High work function IrxSi gates on HfLaON p-MOSFETs, *IEEE Electron Device Lett* **27**:90, 2006.

206. Zhu SY, Singh J, Zhu C, Du A, Li MF, Fabrication of poly-Si TFT with silicided Schottky barrier source/drain, high-κ gate dielectric and metal gate, *Solid State Electron* **50**:232, 2006.

207. Shen C, Li M-F, Wang XP, Yeo Y-C, Kwong D-L, A fast measurement technique for MOSFET characteristics, *IEEE Electon Device Lett,* **27**:55, 2006.

208. Chen JD, Yu HY, Li MF, Kwong DL, van Dal MJH, Kittl JA, Lauwers A, Absil P, Jurczak M, Biesmans S, Yb-doped Ni FUSI for the n-MOSFETs gate electrode application, *IEEE Electron Device Lett,* **27**:160, 2006.

209. Ren C, Faizhal BB, Chan DSH, Li MF, Yeo YC, Trigg AD, Balasubramanian N, Kwong DL, Work function tuning of metal nitride electrodes for advanced CMOS devices, *Thin Solid Films* **504**:174, 2006.

210. Wang XP, Li M-F, Ren C, Yu XF, Shen C, Ma HH, Chin A, Zhu CX, Ning J, Yu MB, Kwong D-L, Tuning effective metal gate work function by a novel gate dielectric HfLaO for nMOSFETs, *IEEE Electron Device Lett* **27**:31, 2006.

211. Ren C, Chan DSH, Li MF, Loh WY, Balakumar S, Tung CH, Balasubramanian N, Kwong DL, Work function tuning and material characteristics of lanthanide-incorporated metal nitride gate electrodes for NMOS device applications, *IEEE Trans ED* **53**:1877, 2006.

212. Ding SJ, Zhang W, Li MF, High density and program-erasable metal-insulator-silicon capacitor with a dielectric structure of SiO_2/HfO_2-Al_2O_3 nanolaminate Al_2O_3, *Appl Phys Lett* **88**:42905, 2006.

213. Wang XP, Yu HY, Li MF, Zhu CX, Biesemans S, Chin A, Sun YY, Feng YP, Lim A, Yeo YC, Loh WY, Lo GQ, Kwong DL, Wide Vfb and Vth tunability for metal-gated MOS devices with HfLaO gate dielectrics, *IEEE Electron Device Lett* **28**:258, 2007.

214. Chen J, Wang XP, Li MF, Lee SJ, Yu MB, Shen C, Yeo YC, NMOS compatible work function of TaN metal gate with erbium-oxide-doped hafnium oxide gate dielectric, *IEEE Electron Device Lett* **28**:862, 2007.

215. Ang KW, Chui KJ, Madam A, Wong LY, Tung CH, Balasubramanian N, Li MF, Samudra GS, Yeo YC, Strained thin-body p-MOSFET with condensed silicon germanium source/drain for enhanced drive current performance, *IEEE Electron Device Lett* **28**:509, 2007.

216. Shen C, Pu J, Li M-F, Cho BJ, P-type floating gate for retention and P/E window improvement of flash memory devices, *IEEE Trans ED* **54**:1910, 2007.

217. Ang KW, Chui KJ, Tung CH, Subramanian N, Li MF, Samudra G, Yeo YC, Enhanced starin effects in 25-nm gate length thin-body nMOSFETs with Si-carbon spource/drain and tensile-stress liner, *IEEE Electron Device Lett* **28**:301, 2007.

218. Chui KJ, Ang KW, Balasubramanian N, Li MF, Samudra G, Yeo YC, n-MOSFET with Si-carbon source/drain for enhancement of carrier transport, *IEEE Trans ED* **54**:249, 2007.

219. Wang XP, Lim A, Yu HY, Li M-F, Ren C, Loh WY, Zhu CX, Chin A, Trigg AD, Yeo Y-C, Biesemans S, Lo GQ, Kwong DL, Work function tunability of refractory metal nitrides by lanthanum or aluminum doping for advanced CMOS devices, *IEEE Trans ED* **54**:2871, 2007.

220. Lousberg GP, Yu HY, Froment B, Augendre E, Keersgiester AD, Lauwers A, Li MF, Absil P, Jurczak M, Biesemans S, Shottky-barrier height lowering by an increase of the substrate doping in PtSi Schottky barrier source/drain FETs, *IEEE Electron Device Lett*, **28**:123, 2007.

221. Zhang G, Wang XP, Yoo WJ, Li MF, Spatial distribution of charge traps in a SONOPS-Typppe flash memory using a high-k trapping layer, *IEEE Trans ED* **54**:3317, 2007.

222. Wu CH, Hung BF, Chin A, Wang SJ, Wang XP, Li MF, Zhu C, Yen FY, Hou YT, Jin Y, Chen SC, Liang MS, Hight-temperature stable HfLaON p-MOSFETs with high-work-function Ir3Si Gate, *IEEE Electron Device Lett* **28**:292, 2007.

223. Zhu ZG, Low T, Li MF, Fan WJ, Bai P, Kwong DL, Samudra G, Pseudo-potential band structure claculation of InSb ultra-thin films and its applictaion to the n-metal-oxide-semiconductor transistor, *Semiconductor Sci Technol* **23**, 2008.

224. Wang XP, Li MF, Yu HY, Yang JJ, Chen JD, Zhu CX, Du AY, Loh WY, Biesmans S, Chin A, Lo GQ, Kwong DL, Widely tunable work function TaN/Ru stacking layer on HfLaO gate dielectric, *IEEE Electron Device Lett*, **29**:50, 2008.

225. Li M-F, Huang D, Shen C, Yang T, Liu WJ, Liu ZY, Understand NBTI mechanism by developing novel measurement techniques, *IEEE Trans Device Mater Reliab* **8**:62, 2008 (invited).

226. Ding S-J, Xu J, Huanbg Y, Sun QQ, Zhang DW, Li MF, Electrical characteristics and conduction mechanisms of metal-insulator-metal Al_2O_3-HfO_2 dielectrics, *Appl Phys Lett* **93**:92909, 2008.

227. Yang JJ, Wang XP, Zhu CX, Li MF, Yu HY, Loh WY, Kwong DL, Enhancement of the flatband modulation of Ni.-silicided gates on Hf-based dielectrics, *IEEE Trans ED* **55**:2238, 2008.

228. Huang D, Liu WJ, Liu ZY, Liao CC, Zhang LF, Gan Z, Wong W, Li MF, A modified charge-pumping method for the characterization of interface-trap generation in MOSFETs, *IEEE Trans ED* **56**:267, 2009.

Conference Papers

1. Li MF, Li YJ, Zhao XS, Chen JX, New results of stress induced effect on S+ centers in Si, *Proc 18th Int Conf Phys Semiconductors*, Stockholm, Sweden, p. 887, 1986.

2. Li MF, Chen JX, Zhao XS, Li YJ, Stress effects of deep centers in Si-New method to determine old parameter, *Mater Sci Forum* (Switzerland) **10**(12):469, 1986.

3. Li MF, Shan W, Yu PY, Hansen W, Weber ER, Bauser E, Lattice relaxation of the DX centers in AlGaAs and of the pressure induced deep donors in GaAs, in Stavola M, Pearton SJ, Davies G (Eds.), *Defects in Electronic Materials*, Proc. MRS Symposia, Pittsburgh, USA, Vol. 104, p. 573, 1987.

4. Li MF, Surh MP, Louie SG, Spin orbit interaction effects in zinc blende type semiconductors-the *ab initio* pseudopotential calculation, *Proc 19th Int Conf Phys Semiconductors*, Warsaw, Poland, p. 857, 1988.

5. Li MF, Yu PY, Shan W, Hansen WL, Weber ER, A comparison of the pressure induced deep donors in GaAs:Si and the DX centers in GaAlAs:Si alloys, *Proc 19th Int Conf Phys Semiconductors*, Warsaw, Poland, p. 1051, 1988.

6. Li MF, Yu PY, Shan W, Hansen WL, Weber ER, Reduction of capture barrier height of pressure-induced deep donors (DX center) in GaAs:Si, *Mat Sci Forum* (Switzerland) **38**(41):851, 1989.

7. Li MF, Shan W, Yu PY, Hansen WL, Weber ER, Bauser E, Pressure dependence of the DX center in AlGaAs:Te, *Mat Sci Forum* (Switzerland) **38**(41):1103, 1989.

8. Shan W, Li MF, Yu PY, Walukiewicz W, Hansen WL, Pressure dependence of Schottky barrier height at Pt/GaAs interface, *Mat Sci Forum* (Switzerland) **38**(41):1427, 1989.

9. Qin GG, Li MF, Some selected topics in high pressure semiconductor research in China, *Int Conf High Press Semiconductors*, Warsaw, Poland, 1988 (invited); *Semiconductor Sci Technol* **4**:225, 1989.

10. Li MF, Yu PY, Weber ER, Bauser E, Hansen WL, Haller EE, Studies of deep level transient spectroscopy of DX centers in AlGaAs:Te under uniaxial stress, *Mater Sci Forum* **83–87**:853, 1992.

11. Li MF, Yu PY, Probing the DX center in GaAs and related alloys by capacitance transient measurements under stress, *5th Int Conf High Pressure Semiconductors*, Kyoto, Japan, 1992 (invited); *Jpn J Appl Phys* **32** (Suppl.):200, 1993.

12. Feng YP, Teo KL, Li MF, Poon HC, Ong CK, Xia JB, Empirical pseudopotential band-structure calculation for $Zn_{1-x}Cd_xSpvypvSe_{1-y}$ quaternary alloy, *Eur Res Conf Electron Struct Solids*, Greece, September, 1993.

13. Teo KL, Feng YP, Li MF, Chong TC, Electronic band-structure of MgZnSSe semiconductor alloy, in Gumbs G, Luryi S, Weiss B, Wicks GW (Eds.), *Growth, Processing, and Characterization of Semiconductor Heterostructures*, *Proc Mat Res Soc Symp*, Boston, USA, Vol. 326, p. 139, 1993.

14. Li MF, Luo YY, Yu PY, Weber ER, Fujioka H, Du AY, Chua SJ, Lim YT, Sulfur DX centers in GaAsP: test of their negative U property at ambient pressure, Lockwood DJ (Ed.), *Proc 22nd Int Conf Phys Semiconductors*, Vancouver, Canada, p. 2303, 1994.

15. Du AY, Li MF, Chong TC, Chua SJ, Carrier concentration saturation effect in n-AlGaAs, *18th Int Conf Defects Semiconductors*, 2328 July 1995, Sendai, Japan; *Mater Sci Forum* **196201**:279, 1995.

16. Teo KL, Goo CH, Li MF, Lau WS, Lim YT, Defects in Te-doped GaAsP light emitting diode, *6th Int Symp IC Technol Syst Appl ISIC95*, Singapore, p. 27, 68 September, 1995.

17. Du AY, Li MF, Chong TC, Chua SJ, Carrier concentration saturation in n type AlGaAs, *6th ISIC95*, Singapore, p. 7, 68 September, 1995.

18. Li MF, Chong TC, Fang WJ, Teo KL, Feng YP, Energy bands of GaN based and II-VI alloy semiconductors for laser threshold analysis, *First Optoelectron Commun Conf OECC 96*, Chiba, Japan (invited), 1996.

19. Li MF, Chong TC, Fan WJ, Teo KL, Feng YP, Electronic structures of GaN based and II-VI alloy bulk and quantum well semiconductors for laser threshold analysis, *2nd Meet Sci Technol Crystal Growth Next-Generation Electro Opt Appl*, Japan Society for the Promotion of Science, Tokyo, Japan (invited) (1996).

20. Przybytek J, Baj M, Slupinski T, Li M-F, DLTS investigations of the distorted configuration of the EL2 defect stabilized under high hydrostatic pressure in GaAsP, *7th Int Conf High Press Semiconductor Phys*, Germany, July, 1996.

21. Liu W, Teo KL, Li MF, Chua SJ, Uchida K, Tokanaya H, Akutsu N, Matzumoto K, The study of piezoelectric effects in wurtzite GaN/InGaN quantum wells/AlGaN multilayer structures, *Proc Second Int Conf Nitride Semiconductors*, Tokushima, Japan, 2731 October, 1996.

22. Yeo YC, Li MF, Chong TC, YU PY, Theoretical study of the energy band structure of partially CuPt-ordered GaInP, *4th Int Conf Comput Phys*, Singapore, 24 June, 1997.

23. Jie B, Li MF, Lou CL, Lo KF, Chim WK, Chan DSH, Direct current measurements of interface traps and oxide charges in LDD pMOSFETs with an n-well structure, *IPFA97 Proc*, Singapore, p. 176, 2125 July, 1997.

24. Li MF, Du AY, Chong TC, Selected topics in defect studies in optoelectronic semiconductor materials, Sasaki T (Ed.), *Proc Int Symp Laser Nonlinear OptMater*, Singapore (invited), p. 245, 35 November, 1997.

25. Du AY, Li MF, Chong TC, Dislocations and traps in MBE grown lattice mismatched p-InGaAs layers on GaAs substrate, *MRS Fall Meet*, Boston, 15 December, 1997.

26. Liu W, Li MF, Chua SJ, Zhang YH, Uchida K, GaN room temperature exciton spectra by photovoltaic measurement, *MRS Fall Meet*, Boston, 15 December, 1997.

27. Chong TC, Yeo YC, Li MF, Fan WJ, Analysis of optical gain of strained wurtzite InGaN/GaN quantum well lasers, *MRS Fall Meet*, Boston, 15 December, 1997.

28. Yeo YC, Chong TC, Li MF, Valence band parameters for wurtzite GaN and InN, *MRS Fall Meet*, Boston, 15 December, 1997.

29. Du AY, Li MF, Chong TC, Teo TL, Lau WS, Deep level states caused by dislocations in MBE growth p-InGaAs/GaAs heterostructures, *Proc 7th Int Symp IC Technol Syst Appl*, Singapore, p. 239, September, 1997.

30. Zhang YH, Li MF, Liu W, Yen AC, Sheng TT, Zhao SP, Wong JLF, An anomalous temperature dependence of the electron-acceptor recombination in heavily Si-doped GaAs/AlAs quantum wells, *Proc 7th Int Symp IC Technol Syst Appl*, Singapore, p. 246, September, 1997.

31. Liu W, Li MF, Chua SJ, Zhang YH, Uchida K, Study on excitonic absorptions in wurztite GaN by photovoltaic spectra at room temperature, *Proc 7th Int Symp IC Technol Syst Appl*, Singapore, p. 243, September, 1997.

32. Yong YK, Li MF, Lim YC, Yep SP, Constant offset sample-and-hold CMOS switching circuit, *Proc 7th Int Symp IC Technol Syst Appl*, Singapore, p. 72, September, 1997.

33. Fang SJ, Tan KS, Li MF, An improved tuning circuit for continuous-time filters, *Proc 7th Int Symp IC Technol Syst Appl*, Singapore, p. 474, September, 1997.

34. Lai WH, Lim YC, Li MF, Improved cross-coupled quad transconductor cell, *Proc 7th Int Symp IC Technol Syst Appl*, Singapore, p. 273, September, 1997.

35. Liu W, Teo KL, Li MF, Chua SJ, The study of piezoelectric effect in wurtzite GaN/InGaN quantum wells/AlGaN multilayer structures, *Proc 2nd ICNS 97*, Tokushima, pp. 122, October, 1997.

36. Liu W, Li MF, Chua SJ, Mei T, Xu SJ, Lu D, Infrared reflection studies of MOCVD-grown GaN, *4th Nat Symp Prog Mater Res*, p. 157, March, 1998.

37. Chong TC, Yeo YC, Li MF, Comparison of the optical gain of wurtzite GaN/AlGaN quantum well lasers grown on the (0001)- and (1010)-oriented substrates, *SPIE Conf Optoelectron Mater Devices*, Taipei, Taiwan, Vol. 3419.0277.786x/98, p. 51, 1998.

38. Liu W, Li MF, Xu SJ, Uchida K, Matsumoto K, Investigation of phonon-assisted photoluminescence in wurtzite GaN eplilayer, *SPIE Conf Optoelectron Mater Devices*, Taipei, Taiwan, Vol. 3419.0277.786x/98, p. 27, 1998.

39. Hou YT, Feng ZC, Li MF, Chua SJ, Characterization of MBE grown GaAlAs alloy films by Raman scattering, *Asia Pac Surf Interface Anal Conf*, Singapore, December, 1998.

40. Liu W, Feng ZC, Li MF, Chua SJ, Material properties of GaN grown by MOCVD, *Asia Pac Surf Interface Anal Conf*, Singapore, December, 1998.

41. Liu W, Li MF, Chua SJ, Photoreflectance study on the surface states of n-type GaN, *Proc 24th Int Conf Phys Semiconductors*, Jerusalem, Israel, p. 173, August, 1998.

42. Ng KH, Jie BB, He YD, Chim WK, Li MF, Lo KF, A comparison of interface trap generation by Fowler-Nordheim electron injection and hot-hole injection using the DCIV method, *Proc IPFA99*, Singapore, p. 140, 1999.

43. Jie BB, Li MF, Lo KF, Energy dependence of interface trap density-investigated by the DCIV method, *Proc IPFA99*, Singapore, p. 206, 1999.

44. Guan H, Cho BJ, Li MF, He YD, Xu Z, Dong Z, A study of quasi-breakdown mechanism in ultra-thin gate oxide by using DCIV technique, *Proc IPFA99*, Singapore, p. 81, 1999.

45. Jie BB, Li MF, Chim WK, Chan DSH, A new DCVV method to measure the interface traps in deep sub-micron MOSFETs, *Proc IPFA99*, Singapore, p. 89, 1999.

46. Lai WH, Li MF, Chan L, Chua TC, Growth characterization of rapid thermal oxides, *1st Int Conf Adv Mater Process Microelectron*, San Jose, USA, 15 March, 1999.

47. Chollet F, Tang XH, Liu AQ, Ha YJ, Li MF, Micro-machined shutter and low-voltage electronics for optical displacement/acceleration sensing, *10th Int Conf Solid State Sensors Actuators (TRANSDUCERS99)*, Sendai, Japan, 710 June, 1999.

48. Ha YJ, Li MF, Liu AQ, Low-voltage high driving capability CMOS buffer used in MEMS interface circuits, *6th IEEE Int Conf Electron Circuits Syst*, Cypruss, 68 September, 1999.

49. Zhang XW, Li MF, Low voltage linear OTA with rail-to-rail differential mode input signal capability, *6th IEEE Int Conf Electron Circuits Syst*, Cypruss, 68 September, 1999.

50. Hou YT, Teo KL, Li MF, Pielectric Franz-Keldysh effect in a GaN/InGaN/AlGaN multilayer structure, *Int Symp Photonics Appl*, Singapore, November, 1999.

51. Feng ZC, Liu W, He YP, Chua SJ, Li MF, Photoluminescence characteristics of InGaN alloys with low in compositions, *Int Conf Silicon Carbide Relat Mater*, NC, USA, 1015 October, 1999.

52. Li K, Hou YT, Feng ZC, Chua SJ, Li MF, Comparative investigation of high resolution transmission electron microscopy and Fourier transform infrared spectroscopy for GaN films on sapphire substrate, *Int Symp Photonics Appl*, Singapore, November, 1999.

53. Guan H, Li MF, Zhang YH, Cho BJ, Jie BB, Xie J, Predicting plasma charging damage in ultra thin gate oxide by using non-destructive DCIV technique, *IEEE Int Integr Reliab Workshop Final Rep*, Lake Tahoe, CA, USA, October, 1999.

54. He YD, Guan H, Li MF, Cho BJ, Dong Z, Investigation of quasi-breakdown mechanism in ultrathin gate oxides, *MRS 1999 Fall Meet Symp Proc T*, Boston, December, 1999.

55. Lai WH, Li MF, Pan JS, Chan L, Chua TC, An XPS study of silicon dioxides rapid thermally grown in oxygen, nitrous oxide and nitric oxide, *MRS 1999 Fall Meet Symp Proc T*, Boston, December, 1999.

56. Li MF, He YD, Cho BJ, Lo KF, Xu Z, Roles of primary hot hole and FN electron fluences in gate oxide breakdown, *MRS 1999 Fall Meet Symp Proc T*, Boston, December, 1999.

57. Guan H, Xu Z, Cho BJ, Li MF, He YD, A study of quasi-breakdown mechanism in ultra-thin gate oxide under various types of stress, *MRS 1999 Fall Meet Symp Proc T*, Boston, December, 1999.

58. Ma S, Zhang YH, Li MF, Li W, Wang JLF, Yen AC, Sheng GTT, Gate induced drain leakage by plasma charging damage, *MRS 1999 Fall Meet Symp Proc T*, Boston, December, 1999.

59. Li MF, Selected topics in CMOS transistor reliability, *SRC-SEMATECK Top Res Conf Reliab*, Stanford University (invited), 1013 November, 2000.

60. Loh WY, Cho BJ, Li MF, Bipolar current stressing and electrical recovery of quasi-breakdown in thin gate oxides, *8th Int Symp Phys Fail Anal IC*, Singapore, p. 59, 913 July, 2001.

61. Teo WYJ, Lim HW, Jin Y, Huang JH, Chew WC, Leong CK, Gn FH, Li MF, Su G, Suppression of metal contamination by gettering, *8th Int Symp Phys Fail Anal IC*, Singapore, p. 216, 913 July, 2001.

62. Chen G, Li MF, Jin Y, Electrical passivation of interface traps at drain junction space charge region in p-MOS transistors, *Proc 12th ESREF*, Bordeaux, France, paper B12, 15 October, 2001.

63. Chen G, Li MF, Yu X, Monitoring degradation of source/drain extension in sub-quarter-micron MOSFETs, *Solid State Devices Mater*, Tokyo, p. 138, 2001.

64. Hou YT, Li MF, Lai WH, Jin Y, Physical model for hole direct tunnelling currents through ultrathin gate dielectrics in advanced CMOS devices, *SSDM 2001*, Tokyo, Japan, p. 144, 2001.

65. Chen G, Jie BB, Li MF, Investigation of interface traps located at different regions in p-MOS transistors using DCIV technique, *6th ICSICT*, Shanghai, China, p. 1047, 2001.

66. Hou YT, Li MF, Jin Y, Hole quantization and hole direct tunnelling in deep submicron p-MOSFETs, *6th ICSICT*, Shanghai, China (invite), p. 895, 2001.

67. Jin Y, Lim HF, Tong AF, Low AS, Teo WY, Hou YT, Li MF, Charge damage in dual gate oxide process, *6th ISCICT*, Shanghai, China, p. 970, 2001.

68. Xu AM, Li MF, A 1.2V rail-to-rail differential mode input linear CMOS transconductor, *Proc 2002 IEEE ISCAS*, Phoenix, Arizona, USA, May, 2002.

69. Zhang XM, Li J, Murukeshan VM, Choller F, Li MF, Liu AQ, MEMS MOVAble /rotational grating/mirror tunable lasers, MEMS 2002 Workshop Digest, *Pac Rim Workshop Transducers Micro/Nano Technol*, Xiamen, China, p. 99, July, 2002.

70. Liu AQ, Zhang XM, Jin X, Zheng JH, Lu C, Mei T, Tang DY, Li MF, Polysilicon micromachined fiber-optical attenuator for DWDM applications, MEMS 2002 Workshop Digest, *Pac Rim Workshop Transducers Micro/Nano Technol*, Xiamen, China, p. 577, July, 2002.

71. Hou YT, Li MF, Yu HY, Kwong DL, Quantum tunneling and scalability of HfO_2 and HfAlO gate stacks, *Int Electron Device Meet*, Technical Digests, p. 731, 2002.

72. Teo WY, Hou YT, Li MF, Chen P, Ko LH, Zeng X, Jin Y, Gn FH, Chan LH, Investigation on dual gate oxide charging damage in 0.13μm copper damascene technology, *Proc 7th Int Symp Plasma- and Process-Induced Damage (P2ID)*, Hawaii, USA, p. 14, 2002.

73. Jin Y, Teo WY, Hou YT, Gn FH, Lim HF, Han ZY, Li MF, Enhanced plasma charging damage due to AC charging effect, *Proc Int Reliab Phys Symp*, p. 359, 2002.

74. Yu HY, Li MF, Kwong D-L, Cho BJ, Pan JS, Ang CH, Zheng JZ, Investigation of $(HfO_2)x(Al_2O_3)1\text{-}x$ on (100) Si by XPS energy gap and band alignment, *The 2002 Int Conf Solid State Devices Mater*, Japan, p. 472.

75. Loh WY, Cho BJ, Li MF, Lek CM, Yong YF, Joo MS, Correlation between interface trap and gate leakage current in ultra-thin silicon dioxides, *Proc 9th IPFA*, Singapore, p. 246, 2002.

76. Yeo SB, Bordelon J, Chu S, Li MF, Tranchina BA, Harward M, Chan LH, See A, A robust and production worthy addressable array architecture for deep sub-micron MOSFETs matching characterization, *ICMTS*, Ireland, April, 2002.

77. Zhenying L, Li MF, Lian Y, Rustagi SC, CMOS transconductor design for Vhf filtering applications, *Proc 2003 IEEE ISCAS*, Bangkok, Thailand, May, 2003.

78. Chen G, Chuah KY, Li MF, Chan DSH, Ang CH, Zheng JZ, Jin Y, Kwong DL, Dynamic NBTI of pMOS transistors and its impact on MOSFET lifetime, *Proc IEEE Int Reliab Phys Symp Proc* , Dallas, TX, pp. 196202, 2003.

79. Li MF, Cho BJ, Chen G, Loh WY, Kwong DL, New reliability issues of CMOS transistors with 1.3nm thick gate oxide, *Seventh Int Symp Silicon Nitride Silicon Dioxide Thin Insulating Films*, 203rd Electrochemical Society Meeting, Paris, ECS Proceeding (invited), Vol. 200302, p. 228, April, 2003.

80. Yu HY, Lim HF, Chen JH, Li MF, Zhu CX, Kwong D-L, Tung CH, Bera KL, Leo CJ, Robust HfN metal gate electrode for advanced MOS devices application, *Symp VLSI Technol*, Kyoto, p. 151, 2003.

81. Low T, Hou YT, Li MF, Zhu C, Kwong D-L, Chin A, Germanium MOS: an evaluation from carrier quantization and tunneling current, *Symp VLSI Technol*, Kyoto, 9A.2, 2003.

82. Kim SJ, Cho BJ, Li M-F, Zhu C, Chin A, Kwong D-L, HfO_2 and lanthanide-doped HfO_2 MIM capacitors for RF/mixed IC applications, *Symp VLSI Technol*, Kyoto, Paper 6B3, p.77, 2003.

83. Huang CH, Yang MY, Chin A, Chen WJ, Zhu CX, Cho BJ, Li M-F, Kwong DL, Very low defects and high performance Ge-on-insulator p-MOSFETs with Al_2O_3 gate dielectrics, *Symp VLSI Technol*, Kyoto, Paper 9A3, 2003.

84. Yu X, Zhu C, Hu H, Chin A, Li MF, Cho BJ, Kwong D-L, Foo PD, Yu MB, MIM capacitors with HfO_2 and HfAlOx for Si RF and analog applications, *Mater Res Soc*, Spring Meeting, USA, 2003.

85. Chan KT, Chin A, Kuo JT, Chang CY, Duh DS, Lin WJ, Zhu C, Li MF, Kwong D-L, Microwave coplanar filters on Si substrates, *IEEE MTT-S Int Microw Symp*, p. 1909, 2003.

86. Huang CH, Yang MY, Chin A, Zhu C, Li MF, Kwong D-L, High density RF MIM capacitors using high-k AlTaOx dielectrics, *IEEE MTT-S Int Microw Symp*, p. 507, 2003.

87. Hu H, Zhu C, Lu YF, Zeng JN, Wu YH, Liew YF, Li MF, Cho BJ, Choi WK, Material and electrical characterization of HfO_2 films for MIM capacitors applications, MRS Spring, San Francisco, 2003.

88. Low T, Hou YT, Li MF, Zhu CX, Chin A, Samudra G, Kwong DL, Investigation of performance limits of Ge double-gated MOSFETs, *IEDM Tech Digest*, p. 691, 2003.

89. Yu HY, Kang JF, Chen JD, Ren C, Hou YT, Li MF, Chan DSH, Kwong DL, Bera KL, Tung CH, Thermally robust high quality HfN/HfO_2 gate stack for advanced CMOS devices, *IEDM Tech Digest*, p. 99, 2003.

90. Loh WY, Cho BJ, Joo MS, Li MF, Kwong DL, Analysis of charge trapping and breakdown mechanism in high-K dielectric with metal gate electrode using carrier separation,*IEDM Tech Digest*, p.927, 2003.

91. Hu H, Ding SJ, Lim HF, Zhu CX, Li MF, Kim SJ, Yu XF, Chen JH, Yong YF, Cho BJ, Chan DSH, Rustagi SC, Yu MB, Du A, My D, Fu PD, Chin A, Kwong DL, High performance ALD HfO_2-Al_2O_3 laminate MIM capacitors for RF and mixed signal IC applications,*IEDM Tech Digest*, p. 379, 2003.

92. Zhu CX, Hu H, Yu XF, Chin A, Li MF, Kwong DL, Voltage and temperature dependence of capacitance of high-K HfO_2 MIM capacitors: a unified understanding and prediction,*IEDM Tech Digest*, p. 879, 2003.

93. Huang CH, Yu DS, Chin A, Wu CH, Chen WJ, Zhu CX, Li MF, Cho BJ, Kwong DL, Fully silicide NiSi and Ge NiGe dual gates on SiO_2/Si and Al_2O_3/GOI MOSFETs,*IEDM Tech Digest*, p. 319, 2003.

94. Chin A, Chan KT, Huang CH, Chen C, Liang V, Chen JK, Chioen SC, Sun SW, Duh DS, Lin WJ, Zhu C, Li MF, McAlister SP, Kwong D-L, RF passive devices on Si with excellent performance close to ideal devices designed by electro-magnetic simulation, *IEDM Tech Digest*, p. 375, 2003.

95. Hou YT, Li MF, Low T, Kwong DL, Impact of metal work function on gate leakage of MOSFETs, *Int Semiconductor Device Res Symp Proc*, p. 154, 2003.

96. Wu N, Zhang Q, Zhu C, Li MF, Chanh DSH, Chin A, Kwong DL, Bera LK, Balasubramanian N, Du AY, Tung CH, Liu H, Sin JKO, Ge pMOSFETs with MOCVD HfO_2 gate dielectric, *Int Semiconductor Device Res Symp Proc*, p. 252, 2003.

97. Zhu S, Yu HY, Whang SJ, Chen JH, Shen C, Zhu C, Lee SJ, Li MF, Chan DSH, Woo WJ, Du A, Tung CH, Singh J, Chin A, Kwong DL, Low

temperature MOSFET technology with Schottky barrier source/drain, high-K gate dielectrics and metal gate electrode, *Int Semiconductor Device Res Symp Proc*, p. 254, 2003.

98. Zhang Q, Wu N, Zhu C, Li MF, Chan DSH, Chin A, Kwong DL, Bera LK, Balasubramanian N, Du AY, Tung CH, Haitao, Sin J, Germanium pMOSFET with HfON gate dielectric, *Int Semiconductor Device Res Symp Proc*, p. 256, 2003.

99. Hu H, Ding S-J, Zhu C, Rustagi SC, Lu YF, Li MF, Cho BJ, Chan DSH, Yu MB, Chin A, Kwong D-L, Investigation of PVD HfO$_2$ MIM capacitors for Si RF and mixed signal ICs application, *Int Semiconductor Device Res Symp Proc*, p. 328, 2003.

100. Yu X, Zhu C, Zhang Q, Hu H, Chin A, Li MF, Chan DSH, Wang WD, Kwong DL, Improved crystallization temperature and intergacial properties of HfO$_2$ gate dielectrics by adding Ta$_2$O$_5$ with TaN metal gate, *Int Semiconductor Device Res Symp Proc*, p. 62, 2003.

101. Loh WY, Cho BJ, Li MF, Chan DSH, Ang CM, Zhen ZJ, Kwong DL, Progressive breakdown statistics in ultra-thin silicon dioxides, *Proc 10th Int Symp Phys Fail Anal Integr Circuits*, Singapore, p. 157, 2003.

102. Yu HY, Wu N, Yeo C, Joo MS, Li MF, Zhu C, Cho BJ, Kwong DL, ALD (HfO$_2$)(Al$_2$O$_3$)$_{1-x}$ high-k gate dielectrics for advanced MOS devices application, *2nd Int Conf Mater Adv Technol IUMRS Int Conf Asia*, p. 562, December, 2003.

103. Lim HF, Kim SJ, Hu H, Yu XF, Yu HY, Li MF, Cho BJ, Zhu C, Chin A, Kwong DL, Study of PVD HfO$_2$ and HfO$_x$N$_y$ as dielectrics for MIM capacitor application, *2nd Int Conf Mater Adv Technol IUMRS Int Conf Asia*, p. 532, December, 2003.

104. Yu HY, Lim HF, Li MF, Zhu C, Kwong DL, The study of HfN as a gate electrode for advanced MOS devices, *2nd Int Conf Mater Adv Technol IUMRS Int Conf Asia*, December, 2003.

105. Kim SJ, Lim HF, Hu H, Yu XF, Yu HY, Cho BJ, Li MF, Zhu C, Chin A, Kwong DL, Properties of PVD HfO$_2$ films in MIM structure and the role of HfN barrier at dielectric/metal interface, *2nd Int Conf Mater Adv Technol IUMRS Int Conf Asia*, p. 513 December, 2003.

106. Zhang QC, Zhu C, Wu N, Chin A, Li MF, Cho BJ, Bera LK, Kwong DL, Gemanium MOS capacitors with ultra-thin hfO$_2$ gate dielectric, *2nd Int Conf Mater Adv Technol IUMRS Int Conf Asia*, p. 513, December, 2003.

107. Wu N, Zhu C, Balasubramanian N, Yeo CC, Joo MS, Yu HY, Zhang QC, Bera LK, Cho BJ, Li MF, Electrical and physical properties of

SiGe/HfO$_2$/Si MOS-capacitors, *Conf Mater Adv Technol IUMRS Int Conf Asia*, p. 535, December, 2003.

108. Luo ZY, Rustagi SCE, Li MF, Lian Y, A 1V 2.4GHz fully integrated LNA using 0.18um CMOS technology, *Proc 5th Int Conf ASIC*, Beijing, p. 1062, October, 2003.

109. Li MF, Kwong DL, Dynamic BTI in ultrathin SiO$_2$ and HfO$_2$ MOSFETs and its impact on device lifetime, *Int Workshop Dielectric Thin Films Future ULSI DevicesSci Technol*, Tokyo, May 2004 (invited); Published as: Li MF, Chen G, Shen C, Wang XP, Yu HY, Yeo YC, Kwong D-L, Dynamic bias-temperature instability in ultrathin SiO$_2$ and HfO$_2$ metal-oxide-semiconductor field effect transistors and its impact on device lifetime, *Jpn J Appl Phys*, 43(11B):7807, 2004.

110. Chen S, Yu HY, Wang XP, Li MF, Yeo YC, Chan DSH, Bera KL, Kwong DL, Frequency dependent dynamic charge trapping in HfO$_2$ and threshold voltage instability in MOSFETs, *IRPS*, p. 601, 2004.

111. Kang JF, Ren C, Yu HY, Wang XP, Li MF, Chan DSH, Yeo YC, Wang YY, Kwong DL, A novel dual-gate integration process for sub-1nm EOT HfO$_2$ CMOS devices, *Int Conf Solid State Devices Mater*, Tokyo, p. 198, 2004.

112. Yu HY, Ren C, Kang JK, Yeo YC, Chan DSH, Li MF, Kwong DL, Thermal stability of metal gate work functions, *SSDM*, Tokyo, p. 712, 2004.

113. Ang KW, Hou YT, Singh J, Li MF, Teo YC, Theoretical investigation of electrical performance and band structure of pMOSFETTs with SiGe source/drain stressors, *SSDM*, Tokyo, p. 722, 2004.

114. Yu DS, Cheng CF, Chin A, Zhu C, Li MF, Kwong DL, High performance fully silicided NiSi:Hf gate on LaAlO3/GOI nMOSFET with little Fermi-level pinning, *SSDM*, p. 746, 2004.

115. Low T, Shen C, Li MF, Yeo YC, Hou YT, Zhu C, Chin A, Kwong DL, Study of mobility in strained Si and Ge ultrathin body MOSFETs, *SSDM*, p. 776, 2004.

116. Ding SJ, Hu H, Zhu C, Kim SJ, Li MF, Cho BJ, Chin A, Kwong DL, A comparison study of high-density MIM capacitors with ALD HfO$_2$-Al$_2$O$_3$ laminated, sandwiched and stacked dielectrics, *7th Int Conf Solid-State Integr Circuits Technol Proc*, paper B1.9, 2004.

117. Kang JF, Yu HY, Ren C, Wang XP, Li MF, Chan DSH, Liu XY, Han RQ, Wang YY, Kwong DL, Characteristics of sub-1nm CVD HfO$_2$ gate dielectrics with HfN electrodes for advanced CMOS applications, *ICSICT*, paper B1.7, 2004.

118. Chin A, Kao HL, Yu DS, Liao CC, Zhu C, Li MF, Zhu S, Kwong DL, High performance metal-gate/high-k MOSFETs and GaAs compatible RF passive devices on Ge-on-insulator technology, *ICSICT* (invited), paper A6.1, 2004.

119. Hou YT, Low T, Xu B, Li MF, Samudra G, Kwong DL, Impact of metal gate work function on nano CMOS device performance, *ICSICT*, paper A1.8, 2004.

120. Liu XY, Du G, Xia ZL, Kang JF, Wang Y, Han RQ, Yu HY, Li MF, Kwong DL, Scaling properties of GOI MOSFETs in nano scale by full band Monte Carlo simulation, *ICICT*, paper D5.4, 2004.

121. Zhu S, Chen J, Yu HY, Whang SJ, Chen JH, Shen C, Li MF, Lee SJ, Zhu C, Chan DSH, Du A, Tung CH, Singh J, Chin A, Kwong DL, Schottky s/d MOSFETs with high-k gate dielectrics and metal gate electrodes, *ICSICT*, paper A1.7, 2004.

122. Li MF, Yu YH, Hou YT, Kang JF, Wang XP, Chen S, Ren C, Yeo YC, Zhu C, Chan DSH, Chin A, Kwong DL, Selected topics in HfO_2 gate dielectrics for future ULCI CMOS Devices, *ICSICT* (invited), paper B1.2, 2004.

123. Yang H, Sa N, Tang L, Liu XY, Kang JF, Han RQ, Yu HY, Ren C, Li MF, Chan DSH, Kowng DL, TDDB characteristics of ultra thin HfN/HfO_2 gate stack, *ICSICT*, paper C5.7, 2004.

124. 1Shen C, Li MF, Wang XP, Yu HY, Feng YP, Lim ATL, Yeo YC, Chan DSH, Kwong DL, Negative U traps in HfO_2 gate dielectrics and frequency dependence of dynamic BTI in MOSFETs, *IEDM Tech Digest*, p. 733, 2004.

125. Low T, Li MF, Fan WJ, Ng ST, Yeo YC, Zhu C, Chin A, Chan L, Kwong DL, Impact of surface roughness on Si and Ge ultra-thin-body MOSFETs, *IEDM Tech Digest*, p. 151, 2004.

126. Bae SH, Bai WP, Wen HC, Mathew S, Bera LK, Balasubramanian N, Yamada N, Li MF, Kwong DL, Laminated metal gate electrode with tunable work function for advanced CMOS, *Symp VLSI Tech*, p. 188, 2004.

127. Yu X, Zhu C, Wang XP, Li MF, Chin A, Du AY, Wang WD, Kwong DL, High mobility and excellent electrical stability of MOSFEs using a novel HfTaO gate dielectric, *Symp VLSI Tech*, p. 110, 2004.

128. Kim SJ, Cho BJ, Li MF, Ding SJ, Yu MB, Zhu C, Chin A, Kwong DL, Engineering of voltage nonlinearity in high-k MIM capacitor for analog/mixed signal ICs, *Symp VLSI Tech* p. 218, 2004.

129. Ang KW, Chui KJ, Bliznetsov V, Du A, Balsubramanian N, Li MF,

Samudra G, Yeo YC, Enhanced performance in 50nm N-MOSFETs with silicon-carbon source/drain regions, *IEDM Tech Digest*, p. 1069, 2004.

130. Yu DS, Chin A, Liao CC, Lee CF, Cheng CF, Chen WJ, Zhu C, Li MF, Yoo WJ, McAlister SP, Kwong DL, 3D GOI CMOSFETs with novel IrO2(Hf) dual gates and high-k dielectric on 1P6M-0.18 μm-CMOS, *IEDM Tech Digest*, p. 181, 2004.

131. Wu N, Zhang Q, Zhu C, Chan DSH, Li MF, Balasubramanian N, Bera LK, Du AY, Chin A, Sin JKO, Kwong D-L, A novel surface passivation process for HfO$_2$ Ge MOSFETs, *62nd Device Res Conf*, Notre Dame, Indiana, p. 19, June, 2004.

132. Cui JQ, Lian Y, Li MF, A low voltage dual gate integrated CMOS mixer for 2.4GHz band applications, *Proc 2004 IEEE Int Symp Circ Syst*, Vancouver, p. I-964, 2004.

133. Cho BJ, Kim SJ, Li MF, Yu M, Hafnium-oxide-based high-k metal-insulator-metal capacitors for RF/analog CMOS technologies, *Asia Pac Microw Conf*, New Delhi (invited), December, 2004.

134. Chin A, Lai CH, Lai ZM, Lee CF, Zhu C, Li MF, Cho BJ, Kwong DL, High performance RF MOSFETs and passive devices on Si, *Asia Pac Microw Conf*, New Delhi, India (invited), 2004.

135. Li M-F, Lee S, Zhu S, Li R, Chen J, Chin A, Kwong DL, New developments in Schottky source/drain high-k/metal gate CMOS transistors, *207th Electrochem Soc Meet Symp K*, *ECS Proc*, Quebec City, Canada (invited), Vol. 200505, p. 301, May 16, 2005.

136. Kim S-J, Cho B-J, Yu M-B, Li M-F, Kwong D-L, Xiong YZ, Chen JH, Zhu C, Chin A, High capacitance density (¿17 fF/m2) Nb2O5-based MIM capacitors for future RF IC applications, *Symp VLSI Tech*, p. 56, 2005.

137. Ren C, Chan DSH, Faizhal BB, Li M-F, Yeo Y-C, Trigg AD, Agarwal A, Balasubramanian N, Pan JS, Lim PC, Kwong D-L, Lanthanide-incorporated metal nitrides with tunable work function and good thermal stability for NMOS Devices, *Symp VLSI Tech*, Kyoto, Japan, 2005.

138. Li MF, Zhu CX, Yeo YC, Yu HY, Wang XP, Shen C, Kwong D-L, Investigation of high-k/metal gate for future nano scale CMOSFETs, *ICMAT*, Singapore (invited), p. 76, 2005.

139. Yu HY, Chen JD, Li MF, Lee SJ, Kwong DL, van Dal M, Kittl JA, Lauwers A, Augendre E, Kubicek S, Zhao C, Bender H, Brijs B, Geenen L, Benedetti A, Absil P, Jurczak M, Biesemans S, Modulation of the Ni FUSI work function by Yb doping: from midgap to n-type band-edge, *IEDM Tech Digest*, p. 645, 2005.

140. Yang T, Li MF, Shen C, Ang CH, Zhu C, Yeo Y-C, Samudra G, Rustagi SC, Yu MB, Kwong DL, Fast and slow dynamic NBTI components in p-MOSFET with SiON dielectric and their impact on device life-time and circuit application, *Symp VLSI Tech*, p. 92, 2005.

141. Wang XP, Li MF, Chin A, Zhu C, Chi R, Yu XF, Shen C, Du AY, Chan DSH, Kwong D-L, A new gate dielectric HfLaO with metal gate work function tuning capability and superior NMOSFETs performance, *ISDRS*, Washington DC, p. 242, 2005.

142. Low T, Feng YP, Li MF, Samudra G, Yeo YC, Bai P, Chan L, Kwong DL, First principle study of Si and Ge band structure for UTB MOSFETs applications, *ISDRS*, Washington DC, 2005.

143. Li MF, Zhu C, Shen C, Yu XF, Wang XP, Feng YP, Du AY, Yeo YC, Samudra G, Chin A, Kwong DL, New insights in Hf based high-K gate dielectrics in MOSFETs, 208th ECS Meeting, Los Angeles, *G3 Symposium Proceeding*, October 1621, 2005 (invited). Published in *ECS Transaction* **1**(5):717, 2006.

144. Cho BJ, Kim SJ, Li MF, Zhu C, Chin A, Yu MB, Xiong YZ, Kwong DL, Niobium oxide as a high-k dielectric for RF IC application, *3rd Int Conf Mater Adv Technol, Symp H: Silicon Microelectron: Process Packaging*, Singapore MRS, July, 2005.

145. Kao HL, Chin A, Hung BF, Lai JM, Lee JM, Li MF, Samudra GS, Zhu C, Xia ZL, Liu XY, Kang JF, Strain induced very low noise RF MOSFETs on flexible plastic substrate, *Symp VLSI Tech*, paper 9.4, p. 160, 2005.

146. Chin A, Chen C, Li MF, Yoo WJ, Zhu C, Samudra GS, 3D integrated metal gate/high-k CMOS for both DC and AC power consumption solution, *Proc 12th Symp Nano Device Technol*, Hinchu, Taiwan, May, 2005.

147. Yu XF, Zhu CX, Yu MB, Li MF, Chin A, Tung CH, Gui D, Kwong DL, Advanced MOSFETs using HfTaON/SiO$_2$ gate dielectric and TaN metal gate with excellent performance and low standby power application, *IEDM Tech Digest*, p. 31, 2005.

148. Chui KJ, Ang KW, Madan A, Wang H, Tung CH, Wong LY, Wang YH, Choy SF, Balsubramanian N, Li MF, Samudra G, Yeo YC, Source/drain germanium condensation for p-channel strained ultrathin body transistors, *IEDM Tech Digest*, p. 499, 2005.

149. Yu DS, Chin A, Wu CH, Li MF, Zhu C, Wang SJ, Yoo WJ, Hung BF, McAlister SP, Lanthanide and Ir-based dual metal gate/HfAlON CMOS with large work function difference, *IEDM Tech Digest*, p. 64, 2005.

150. Wu N, Zhang QC, Zhu CX, Shen C, Li MF, Chan DSH, Balasubramanian N, BTI and charge trapping in Ge p and n MOSFETs with CVD HfO_2 gate dielectric, *IEDM Tech Digest*, p. 563, 2005.

151. Ang K-W, Chui KJ, Bliznetsov V, Wang YH, Wong LY, Tung CH, Balasubramanian N, Li MF, Samudra G, Yeo YC, Thin body silicon-on-insulator N-MOSFET with silicon-carbon source/drain regions for performance enhancement, *IEDM Tech Digest*, p. 503, 2005.

152. Shen C, Yang T, Li M-F, Samudra G, Yeo Y-C, Zhu CX, Rustagi SC, Yu MB, Kwong D-L, Fast instability in high-k gate dielectric MOSFETs and its impact on digital circuits, *IEEE IRPS*, p. 653, 2006.

153. Wang XP, Shen C, Li M-F, Yu HY, Sun Y, Feng YP, Lim A, Sik HW, Zhu C, Chin A, Yeo YC, Lo P, Kwong DL, Dual metal gates with band-edge work functions on novel HfLaO high-κ gate dielectric, *Symp VLSI Tech*, p. 12, 2006.

154. Ang K-W, Chui K-J, Chin H-C, Li M-F, Samudra G, Balasubramanian N, Yeo Y-C, 50 nm silicon-on-insulator N-MOSFET featuring multiple stressors: silicon-carbon source/drain regions and tensile stress silicon nitride liner, *Symp VLSI Tech*, paper 8.4, 2006.

155. Yu HY, Kittl J, Lauwers A, Singanamalla R, Demeurisse C, Kubicek S, Augendre E, Veloso A, Brus S, Vrancken C, Hoffmann T, Mertens S, Onsia B, Verbeeck R, Demand M, Rothchild A, Froment B, van Dal M, Meyer KD, Li MF, Chen JD, Jurczak M, Absil PP, Biesemans S, Demonstration of a new approach towards 0.25V low Vt CMOS using Ni-based FUSI, *Symp VLSI Tech*, paper 12.4, 2006.

156. Shen C, Li MF, Foo CE, Yang T, Huang DM, Yap A, Samuidra G, Yeo YC, Characterization and physical origin of fast Vth transient in NBTI of pMOSFETs with SiON dielectric, *IEDM Tech Digest*, p. 333, 2006.

157. Zhu ZG, Low T, Li MF, Fan WJ, Bai P, Kwong DL, Samudra G, Modeling study of InSb thin film for advanced III-V MOSFET applications, *IEDM Tech Digest*, p. 807, 2006.

158. Wu CH, Hung BF, Chin A, Wang SJ, Chen WJ, Wang XP, Li MF, Zhu C, Jin Y, Tao HJ, Chen SC, Liang MS, High temperature stable In3SiTaN/HfLaON CMOS with large work-function difference, *IEDM Tech Digest*, p. 617, 2006.

159. Li M-F, Zhu C, Wang X, Yu X, Novel hafnium-based compound metal oxide gate dielectrics for advanced CMOS technology, *12th Workshop Gate Stack Technol Phys (Jpn Soc Appl Phys)* (keynote speech), p. 1, Mishima, Japan, 2007.

160. Li M-F, Shen C, Yang T, Gang C, Huang D, The physical origins of fast and slow components in NBTI degradation for p-MOS transistors with SiON gate dielectric, *9th Int Symp Silicon Nitride, Silicon Dioxide Thin Insulating Films and Emerging Dielectrics (Electrochem Soc USA)*, Chicago, USA (invited). Also published in *ECS Transactions*, **6**(3):163, ECS, Pennington, NJ, 2007.

161. Liu WJ, Liu ZY, Huang DM, Liao CC, Zhang LF, Gan ZH, Wang W, Shen C, Li MF, On-the-fly interface trap measurement and its impact on the understanding of NBTI mechanism for p-MOSFETs with SiON gate dielectric, *IEDM Tech Digest*, p. 813, 2007.

162. Wang XP, Li M-F, Yu HY, Yang JJ, Zhu CX, Hwang WS, Loh WY, Du AY, Trigg AD, Chen JD, Chin A, Biesemans S, Lo GQ, Kwong D-L, Highly manufacturable CMOSFETs with single high-k and dual metal gate integration process, *SSDM*, 2007.

163. Wang XP, Yang JJ, Yu HY, Li MF, Chen JD, Xie RL, Zhu CX, Du AY, Lim PC, Lim A, Mi YY, Lai DMY, Loh WH, Biesmans S, Lo GQ, Kwong DL, Practical solutions to enhance EWF tunability of Ni FUSI gates on HfO$_2$, *SSDM*, 2007.

164. Liu ZY, Huang D, Liu WJ, Liao CC, Zhang LF, Gan ZH, Wong W, Li M-F, Comprehensive studies of BTI degradations in SiON gate dielectric CMOS transistors by new measurement techniques, *IRPS Proc*, p. 733, 2008.

165. Wang XP, Yu HY, Yeo YC, Li MF, Chang SZ, Cho HJ, Kubicek S, Wouters D, Groeseneken G, Biesman S, Understanding and prediction of EWF modulation induced by various dopants in the gate stack for a gate-first integration scheme, *Symp VLSI Tech*, p. 162, 2008.

166. Li MF, Wang XP, Shen C, Yang JJ, Chen JD, Zhu CX, Huang DM, Some issues in advanced CMOS gate stack performance and reliability, *5th Int Symp Adv GateStack Technol (ISGAST)*, Austin, SEMATECH (invited), 2008.

Books and Book/Chapters

1. Ren SY, Li MF, Mao DQ, Hu WM, Theory of deep defect states in semiconductors, in Xide X (Ed.), *Statistical Physics and Condensed Matter Theory*, World Scientific, Singapore, p. 256, 1986.

2. Li MF, Lattice relaxation effects on deep defects in semiconductors, in Xia JB *et al.* (Eds.), *Lattice Dynamics and Semiconductor Physics*, Festschrift for Professor Kun Huang, World Scientific, Singapore, p. 224, 1989.

3. Lu Y, Li M-F, Jung A-L, EPM calculation of the electronic structure of the chalcogenide semiconductor Sb2Se3, in Huang L (Ed.), *Thin Films and Bean-Solid Interactions*, Elsevier, Netherlands, p. 129, 1989.

4. Li M-F, *Semiconductor Physics*, Chinese Science Publisher, Beijing, China, 1991.

5. Li M-F, *Modern Semiconductor Quantum Physics*, World Scientific, Singapore, 1994.

6. Li MF, Yu PY, High pressure study of DX centers using capacitance techniques, in Suski T, Paul W (Eds.), *High Pressure in Semiconductor Physics I, Semiconductors and Semimetals*, Academic Press, USA (invited review paper), Vol. 54, 1998.

7. Zhu C, Li MF, High-k MIM capacitors for silicon analog and RF IC applications, in Cai WZ (Ed.), *Si-Based Semiconductor Components for RF Integrated Circuits*, ISBN:81-7895-196-7. Transworld Research Network, Kerala, India, pp.79103, 2006.

Biography

Ming-Fu Li, received his degree from Department of Physics, Fudan University, Shanghai in 1960. After graduation, he was with the Department of Technical Physics, University of Science and Technology of China (USTC) as a Teaching Assistant, and then Lecturer. He joined the Graduate School, Chinese Academy of Sciences, Beijing in 1978 and became a Professor in 1986. He was also an Adjunct Professor with the Institute of Semiconductors, Chinese Academy of Science, USTC, and Fudan University.

He was a Visiting Scholar with the Engineering Design Center, Case Western Reserve University in 1979, and Electrical Engineering Department, University of Illinois, Urbana from 1980 to 1981, and a Visiting Scientist with the Department of Physics, University of California, Berkeley and Lawrence Berkeley Lab from 1986 to 1987, and 1990 to 1991 respectively. In 1991, he joined the National University of Singapore (NUS), where he became a Professor with the Electrical and Computer Engineering Department, and was the microelectronics division head in 1997–2001. He was a founding member of the Silicon Nano-Device Lab (SNDL) at NUS ECE Department. He also served as an Adjunct Senior Member of Technical Staff with the Institute of Microelectronics, Singapore. In Sept. 2006, he joined Fudan University as a Professor in the Microelectronics Department. He has worked in different areas of semiconductors by experimentation, theory, and electronic circuit design. Those areas include deep defects, band structure calculations, analog integrated circuit design, CMOS device technology, device reliability and nano-device quantum modeling.

He has published over 390 research papers, and two books, including *Modern Semiconductor Quantum Physics* (World Scientific, 1994). He has served on numerous international programs and advisory committees in semiconductor conferences in Canada, China, Germany, India, Japan, Singapore, Taiwan China and the USA, including the premier electron device conferences, International Electron Device Meeting (IEDM) and International Reliability Physics Symposium (IRPS).